# PLANT PHYSIOLOGY

## *Fundamentals and Applications*

**ARVIND KUMAR**
*Professor and Head,*
*Department of Plant Physiology*
*S. K. N. College of Agriculture*
*Rajasthan Agricultural University, Jobner, India*

**S. S. PUROHIT**
*Ex-Head, Department of Botany*
*Dungar College, Bikaner, India*

**SECOND ENLARGED EDITION**

AGROBIOS (INDIA)

***Published by:***
**AGROBIOS (INDIA)**
Agro House, Behind Nasrani Cinema
Chopasani Road, Jodhpur 342 003
Phone: 91-0291-2643993, 2642319
E. mail: agrobiosindia@gmail.com

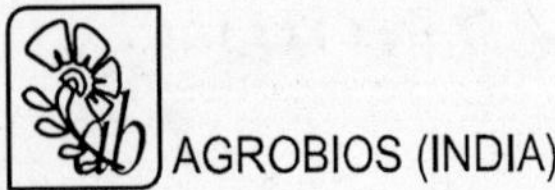


First Edition: 1998
Second Revised Edition: 2001
**Reprinted: 2025**



ISBN No. (10): 81-7754-094-7
ISBN No. (13): 978-81-7754-094-9

Published by: Dr. Updesh Purohit for Agrobios (India), Jodhpur
Laser typeset at: Yashee Computers, Jodhpur
Cover Design by: Reena
Printed at: AGROBIOS DIGITALS, Jodhpur

# Preface to Second Edition

We feel pleasure in presenting Second Edition of the book. New information keeps on expanding and enriching every sphere of knowledge. Plant Physiology is more like a rule of this dictum. The primary motivation for preparing this text was our own need for organising a document on the subject that presents the current scenario of advancements in fundamental concepts and the perspectives of application.

Amidst our endeavour, we came across a quotation of Albert Einstein. It conveyed the need of equal emphasis on fundamentals and applied aspects. Further, it projected a new dimension that possibly, concerns for *mankind* and *fate of mankind* are located at temporally segregated scales. A common set of priorities may not serve the man and his fate at the same time. Consequently, references to *green revolution* and *sustainability* have appeared in descriptions pertaining to role of plant physiology in agriculture (Chapter 1 and 35). These Chapters are likely to inculcate interest in students. The concepts pertaining to cell structure and function have been dealt with in Chapter 2.

Biochemistry often prevails over Bio-physics in books on plant physiology. Present book is slightly tilted in favour of bio-physics. In addition to an independent description, Chapters on water relations and photosynthesis also deal with biophysical aspects of water fluxes and photon induced electron transfer. The dynamic scenario in the areas of Inorganic transport, Organic transport and Transpiration have been discussed. The biochemical understanding of processes have been depicted in photosynthesis and photorespiration and Nitrogen metabolism. Physiology of Plant growth and development has evolved over a dichotomous path. Along with the conceptual advances in plant growth, plant growth regulators (Auxins, Gibberellins, Cytokinins, Abscisic acid and Ethylene) Seed physiology, Physiology and flowering, Photomorphogenesis, Fruit ripening and Senescence and their applied dimensions have been discussed in detail. A separate chapter have been included on Plant physiology in agriculture and industry. This Chapter also encompasses modern strides in biotechnology viz., cellular factories and molecular farming, and concerns for sustainable agriculture. Movements in plants, Circadian rhythms, and Physiology of abiotic stresses are the topics of current interest. The molecular aspects of these physiological activities have been added in this new edition.

We are optimistic that graduate and post graduate students shall find this book useful as an overview of the basic and applied aspects of plant physiology.

**Authors**

# Contents

## CHAPTER 1

# Plant Physiology: Fundamentals and Applications

*Concern for man himself and his fate must always form the chief interest of all technical endeavors. Never forget this in the midst of your diagrams and equations.*

***Albert Einstein***

## FUNDAMENTSALS

Every discipline has its characteristic hue. Plant physiology has been known to contribute more in analysis, understanding and enlightenment in comparison to applied, technological and utilitarian aspects. The linkage between fundamental science and applied technology is like the relationship between James Watt's steam power and Stephenson's steam engine. In agriculture it is noticed between Mendel's genetics and Borlaug's dwarf wheat varieties. Thus, the history of science is a testimony of a gradual and subtle transformation of science into technology or fundamentals into application.

Life is usually envisaged as a comparison between living beings and non living entities. As it is accomplished, a list of characteristics is obtained which are present in living organisms and absent in non living world. Analysis reveals that living beings are usually characterised by certain processes like respiration, enzymatic reactions, growth and reproduction. Modern definition of life also emphasized upon the functions as it states: *life is a peculiar series of functions associated with specially organised structures in which certain macromolecules are capable of reproduction, transfer and utilization of information and catalysis of metabolic reactions. All this is organized upto the level of a cell....*

At this juncture, it is important to realise that life is not merely a manifestation of series of functions, but is the demonstration of an exquisite coordination among myriad of processes and activities which are in operation at different levels of organisation. This situation is often referred as steady state. On a given point time, a plant is absorbing aqueous solutions from soil and transporting to aerial parts where as foliages are busy in the processes of energy transduction through a complex series of photophysical, photochemical and biochemical events. Simultaneously host of biosynthetic pathways are in operation at cellular level which are influencing the other processes and also getting affected by those activities. In nut shell, the operation of multifarious functions simultaneously in an exquisite state of coordination is the uniqueness of life which deserves the metaphors of orchestration and symphony.

Is life so unique that it defies the laws of physical sciences? The answer is an emphatic no. A discourse in plant physiology is essentially an exercise to understand the life processes in terms of laws of physics and chemistry. Photosynthesis and respiration, the inevitables of plant physiology are the two processes which bring about energy transformation and therefore follow and substantiate laws of thermodynamics. The area of plant water relation research derived much inference from Fick's law of diffusion, Raoult's law of decrease in thermodynamic energy and electric circuitry of wheatstone bridge.

Boyle's and Charle's laws of gases have contributed in understanding gaseous exchanges between plants and atmosphere where as laws of mass action and catalysis have been the underlying principles of large number of catabolic and anabolic reactions taking place inside cells. Thus, the three fundamental categories of cellular functions, namely osmotic, chemical and mechanical, are explicable in terms of established laws and theories of physical sciences. Among other diversions which are peculiar and original to biological silhouette- are structural economy, reproduction and evolution. In the context of plant physiology it is important to appreciate the structural economy or the efficiency of biological structures. The chloroplast and mitochondria are submicroscopic organelles which bring about energy conversions so efficiently that man made solar panels, hydroelectric and atomic power plants are far less efficient in terms of ratio of energy generated per unit space. Similarly, the most advanced models of aeroplane or helicopter are yet to achieve the proficiency of aerodynamic acrobatics a bird or a common house fly performs with extraordinary swiftness. The information contained in the form of DNA and RNA molecules shall require a library if it is to be preserved in the form of books. Besides, biological systems evolve spontaneously through processes of recombination and mutation. Thus, plant physiology offers an opportunity to view this intricate and finely interwoven fabric of interplay of functions- ***the gossamer of life***.

In cellular categories of life forms, namely, Monera (Prokaryotic bacteria, blue green algae and mycoplasma), Protista (Eukaryotic but mostly single celled plants and animals like algae and protozoans), Fungi (Heterotrophic and eukaryotic microorganisms), Plantae and Animalia (Eukaryotic multicellular animals), plants are distributed in Monera, Protista, Fungi and Plantae.

With an exception of heterotrophic forms, plants are primary transducers of solar energy into chemical energy. On account of this ability; plants, chlorophyll and photosynthesis have been designated as most important organisms, molecules and process on the planet earth. The entire heterotrophic form of life including *Homo Sapiens* is sustained by plants by the virtue of their autotrophism. The Sun is the ultimate source of energy which descends in abundance as radiation. But for the green mantle, this bounty of energy would have been dissipated wastefully. The electromagnetic form of energy, contained in the wavelengths of 400-700 nm spectrum of radiation, when transferred to chemical bonds of carbohydrate molecules, constitute the food of plants and rest of the non photosynthetic form of life. This food, is a form of energy which requires another conversion before its utilisation in the various forms of work a plant or living cell performs. Adenonine triphosphate (ATP) or such phosphorylated compounds are the ready source of energy. When the third phosphate group is broken apart about 10 K calorie energy is generated which can be utilised during osmotic, chemical and mechanical functions. In the following diagram the flow of energy in a plant has been shown along with energy conversion in photosynthesis and respiration. Sometimes such depictions mislead that all the functions are invariably mediated through ATP, where as there are several examples where functions are operated on account of gradients in energy level (e.g., diffusion) or through lowering of energy requirement (e.g., enzyme mediated processes). This notion has also been accommodated in the following model.

Plant physiology has remained very fascinating on account of its extra and sometimes sole emphasis on understanding the basic principles like structure-function relationship, energy-function relationship, physico-chemical explanations of functional manifestations and molecular basis of cellular regulation. It has always enriched the treasure of knowledge and elevated human minds with the joy of enlightenment.

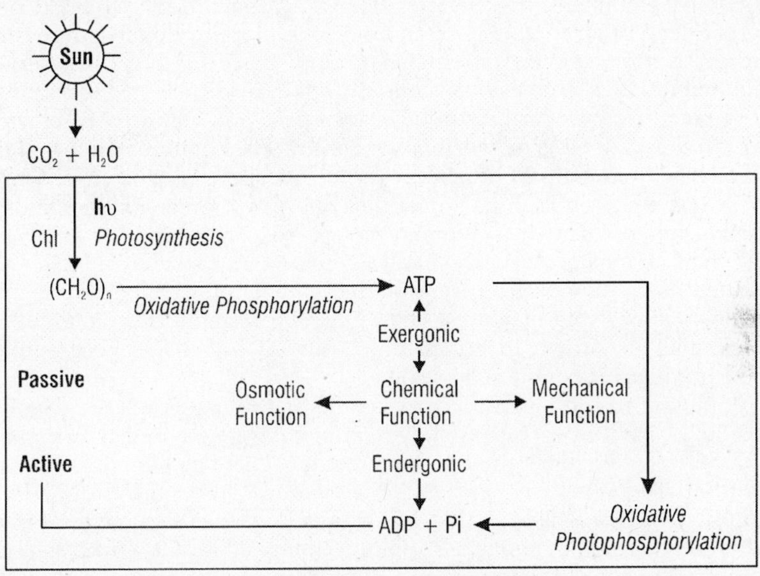


**FIG. 1.1:** Energy transduction and flow of energy in a plant cell at work.

Endeavours in the field of plant physiology made certain startly revelations about plant life that plants absorb carbon dioxide from air and release oxygen, in addition to normal dark respiration, photorespiration also occurs which is a destructive process, and there are many other types of respiration (like cyanide resistant pathway). Carbon dioxide is fixed at least through two alternative mechanisms in addition to universal Calvin cycle. Whole enzymatic gamut of symbiotic and non symbiotic nitrogen fixation, consequences of mineral deficiencies and intricacies of phloem transport were deciphered.

It was very rewarding to know that growth and development in plants are controlled by certain hormone like substances, which in infinitesimally minute concentrations altered growth proportions and were responsible for the ultimate expressions at whole plant level. In addition to the most important pigment the chlorophyll, a photomorphogenetic pigment phytochrome was discovered and was found responsible for myriad of light dependent morpho-physiological manifestations in plants.

History of science witnessed the rise and fall of Lysenko's doctrine and it was possibly the physiologists efforts which strived to strike a balance between the conflict of environmental and genetical control of plant life. Photoperiodism, vernalisation and thermosensitivity in plants are as important topics in plant physiology as that of induction and repression of enzyme synthesis through structural genes.

## APPLICATIONS

In this book, it is not intended to elaborate upon those relationships which naturally exist like functional explanations of agricultural practices e.g., mineral and nitrogen nutrition and fertilizer application, plant water relations and irrigation requirement, developmental physiology and crop phenotypes, photosynthesis and crop productivity, and such other related aspects. Although, this is also true that there is a great need of strengthening this kind of exploratory endeavours through which crop cultivation can be remodeled on the basis of their structural and functional characteristics. However, we initially tried to address this question *what is the area of main-stream agriculture where knowledge of plant physiology have contributed directly or have potentials to do so*? Premonitions of prophets of biotechnology constituted a part of answer to this question, because in recent past, biotechnology has emerged as a discipline that not only represents the coupling of basic biological functions with technology poised to change the life style, but to mark the union as an amalgamation so naturally imminent and unavoidable.

Interestingly, with the pace of research, the biotechnology is being reshaped in Biotics- the science of living machines and molecular farming in which hints from computronics shall be drawn to create new class of organisms programmed to biosynthesize products at commercial scale.

While reminding to the quotation of Einstein at the outset of this chapter we shall like to be aware that technical endeavours serving mankind may essentially not be a guaranty to safeguard the fate of mankind. This has happened in case of green revolution. The *seed-variety-fertilizer-chemoprotection* technology brought out a revolution which was important for meeting the challenge of feeding increasing human population and at one point it appeared that it would be possible to achieve the targets of green agriculture. Very soon the community of scientists started realizing very startling side effects of inorganic farming and chemoprotectants. The growing discipline of biotechnology offered a choice of chemical-free plant protection (Biological control) along with the options of rapidly providing genetic variability (transgenic plants). A speculation began that an era of '*gene-green revolution*' might come up as our saviour. This venture, too, lacked a holistic notion of sustainability which equally emphasized upon man and his fate. Sustainability dreams a green agriculture spread in time (future) and space (the entire globe-developed, developing and undeveloped countries).

Let us modify our question as: What is the areas in which plant physiology can be of use in realizing goals of a sustainable green agriculture? A modest answer can be: Plant growth regulators, breeding for drought and salt tolerance, hydroponics, tissue culture, transgenic plants and biotechnology are the areas in which plant physiological knowledge is extensively contributing in agriculture and industries. Eco-friendly agriculture, conservation of plant biodiversity and traditional wisdom, and promotion of peasants' and artisans' creativity are the areas contributory to sustainable development to which plant physiologists may participate both as scientists and inhabitants of planet earth.

CHAPTER 2

# The Plant Cell: A Symphony Epitomized

The term plant cell (Latin, *cellula* = small apartments) was first coined by Robert Hook in 1665 (he wrote a book *Micrographia*), when he observed cork cells under a microscope made by himself. After that a series of studies followed leading to the formation of the cell theory by Schleiden and Schwann (1838). There are three well recognized plans of cellular organizations:

1. Prokaryotic plan, 2. Eukaryotic plan and, 3. Mesokaryotic plan

## PROKARYOTIC PLAN

These cells are microscopic. The cells lack a nucleus. Most of the genomic DNA is in the central region of the cell; yet in many prokaryotes, such as *Bacillus subtilis*, the DNA is apparently attached to the plasma membrane at many points. The majority of the ribosomes are found near the boundary of the cell, possibly because they are excluded from the central regions, occupied extensively by DNA. The prokaryotic world includes bacteria, mycoplasmas and blue-green algae cyanobacteria. The structural organization of prokaryotic cells is simpler than that of the eukaryotic cells.

These are primitive type of cells and lack any kind of special apparatus of cell division. The cell is bound by selectively permeable membrane which produces numerous pockets called mesosomes. This membrane in addition to it, also performs the function of respiration. Photosynthetic thylakoids in the form of vesicles containing photosynthetic pigments are present near cellular boundary in photosynthetic prokaryotic cells like blue green algae. Outside the membrane, a cell wall is present which is rich in carbohydrate (polymucomers) and amino acids. A capsule may be present outside cell wall (e.g., bacteria) or absent (e.g., blue green algae).

## EUKARYOTIC PLAN

This type of cell is of advanced type and possesses a membrane-bound nucleus which contains genetic material in the nucleoplasm. They may be unicellular organisms, e.g., protozoans, and unicellular algae or the cells which make up the tissues and organs of multicellular organisms. The nucleus contains chromosomes which are made of DNA and various proteins. In addition to genetic material, the nucleus also possesses one or more nucleoli.

The cytoplasm is the ground substance which is amorphous and homogenous and includes various organelles in suspended position. The cytoplasm is bound by a plasma membrane both in plants and animals. In plants, the cell wall envelopes the plasma membrane. The cell organelles are membranous structures such as endoplasmic reticulum (ER), Golgi body, lysosomes, and mitochondria. The ER may harbour ribosomes, bound to the outer surface (rough ER), or it may be free of them (smooth ER). Among other organelles chloroplasts, mitochondria and Golgi bodies are present. Microbodies (peroxisomes and glyoxisomes and other partially membrane bounded bodies like spherosomes, oleosomes and protein bodies are also present. A cytoskeleton system, comprising of microfibrils, microtubules and intermediate filaments, is characteristically found in animal cells. In plant cells cytoskeleton like structures have been observed, specially in isolated protoplasts of several dicot and monocot species. The fibrillar cytoskeleton is made up of p-protein (phloem protein) instead of actin. The plant cytoskeleton fibrils may be involved in positioning the nucleus, but the exact role remains obscure.

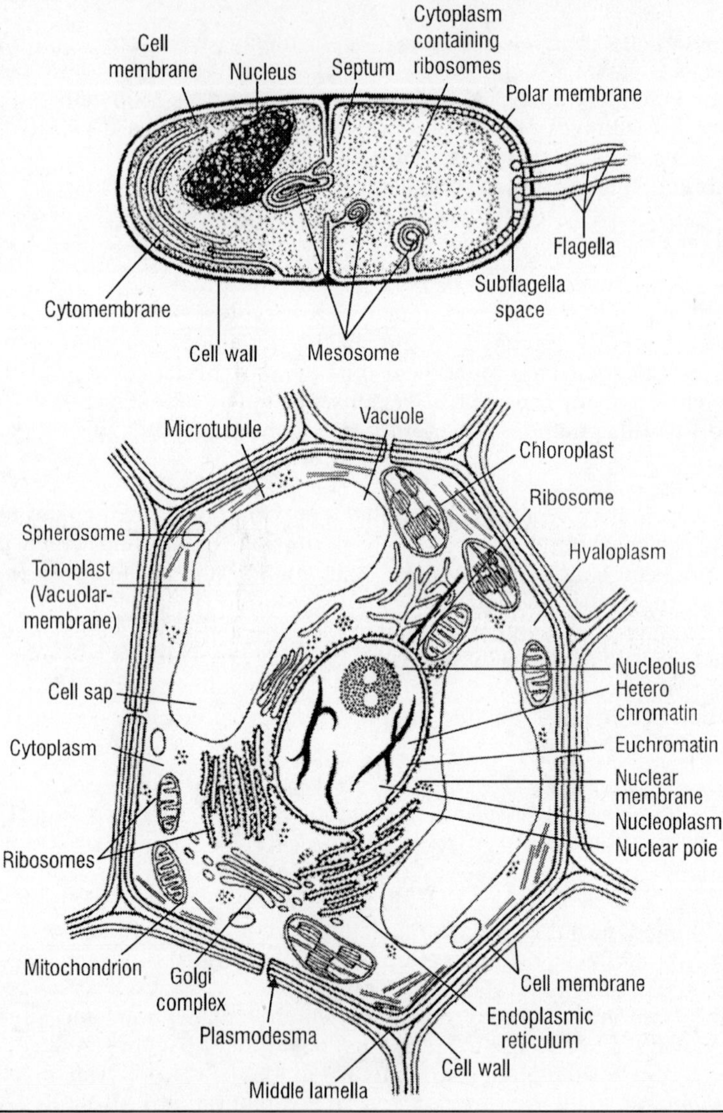


**FIG. 2.1:** Plant cell (a) Prokaryotic cell (b) Eukaryotic cell.

**TABLE. 2.1:** The Components of a Eukaryotic Plant Cell

I. CELL WALL[a]
  a) Primary Wall (about 1/4 cellulose); about 1 to 3 μm thick
  b) Secondary wall (about 1/2 cellulose + 1/4 lignin); may be 4 μm thick or more
  c) Middle lamella (cementing layer between cells, mostly pectin)
  d) Plasmodesmata (strands of plasma membrane penetrating wall); 30 to 100 nm in diameter
  e) Simple and bordered pits
II. PROTOPLAST (contents of the cell exclusive of the cell wall); 10 to 100 μm in diameter
  a) Cytoplasm (cytoplasm + nucleus + protoplasm)
    i) Plasma membrane (plasmalemma); 0.01 μm (10 nm) thick
    ii) Endomembrane system
      (1) Endoplasmic reticulum (ER); 7.5 nm thick (each membrane; cisternae with two membranes vary in thickness)
      (2) Golgi apparatus (consists of dictyosomes; 0.5 to 2.0 μm in diameter; membranes 7.5 nm thick)
      (3) Nuclear envelope (two unit membranes); 25 to 57 nm thick
      (4) Vacuolar membrane (tonoplast); 7.5 nm thick (see Vacuoles, below)
      (5) Microbodies; 0.3 to 1.5 μm in diameter
      (6) Spherosomes and protein bodies; 0.5 to 2.0 μm in diameter (surrounded by half of a unit membrane)
  b) Cytoskeleton
    i) Microtubules; 24 to 25 nm thick; core of 12 nm
    ii) Microfilaments; 5 to 7 nm thick
    iii) Other proteinaceous material
  c) Ribosomes; 15 to 25 nm in diameter (larger than in prokaryotes)
  d) Mitochondria (membrane-bound); 0.5 to 1.0 μm by 1 to 4 μm
  e) Plastids[b] (membrane-bount organelles)
    i) Proplastids (immature plastids)
    ii) Leucoplasts (colourless plastids); amyloplasts (contain starch grains, sometimes protein; proteinoplasts); elaioplasts (contain fats); etioplasts; and other food-storage plastids
    iii) Chloroplasts; 2 to 4 μm thick by 5 to 10 μm in diameter (may contain starch grains)
    iv) Chromoplasts (often red, orange, yellow, and other colours)
  f) Cytosol (fluid in which most of above structures are suspended)
III. Nucleus (cytoplasm + nucleus = protoplasm); 5 to 15 μm or more in diameter (see Nuclear envelope, above)
  a) Nucleoplasm (granular and fibrillar substance of the nucleus)
  b) Chromatin (chromosomes become apparent during cell division)
  c) Nucleolus; 3 to 5 μm in diameter
IV. Vacuoles (from nonexistent to 95 percent of cell volume; sometimes even more)
V. Ergastic substances (inclusions of relatively pure materials, often in plastids or vacuoles)[a]
  a) Crystals (such as calcium oxalate)
  b) Tannins
  c) Fats and oils (in elaioplasts or lipid globules)
  d) Starch grains (in amyloplasts and chloroplasts; see above)[b]
  e) Protein bodies
VI. Flagella and cilia; 0.2 μm thick, 2 to 150 μm long

[a]Occur in fungal, plant, and some protist cells, but seldom in animals
[b]Occur only in plant cells and some protistans.

**TABLE 2.2:** Differences between prokaryotic and eukaryotic cells

| Characters | Prokaryotes | Eukaryotes |
|---|---|---|
| Plasma membrane | Present | Present |
| Cell wall | Non-cellulosic; composed of amino-sugars and muramic acid | Absent in animal cells; present in plant cell - cellulosic |
| Capsule | When present, composed of mucopolysaccharides | Absent |
| Nuclear envelope | Absent | Present |
| DNA | Nacked | Combined with protein |
| Chromosomes | Single | Multiple |
| Nucleolus | Absent | Present |
| Nucleoplasm | Not differentiated from cytoplasm | Differentiated from cytoplasm by nuclear envelope |
| Division | Amitosis | Mitosis or meiosis |
| Ribosomes | Present; 70S (50S + 30S) | Present; 80S 60S + 40S) |
| Endomembranes (ER, Golgi body, mitochondria, chloroplast lysosomes) | Absent | Present |
| Respiratory enzyme | Enzymes in the plasma membrane | In mitochondria |
| Photosynthesis | In plasma membrane or thylakoids vesicles | In chloroplast |

## MESOKARYOTIC PLAN

Dodge (1964) used the term mesokaryotic cell to refer to nuclear status of dinoflagellates (a group of algae). It is characterized by the absence of deoxyribonucleo-histone (as in prokaryotic cells). Nucleus is, however, definite and shows peculiar type of division, called dinomitosis.

## THE PLANT CELL

### Cell Wall

Cell wall is comparatively thick and rigid envelope that surrounds all plant cells. It was first discovered by Hanstein. It is considered as secretion of protoplasm - the living portion of the cell. It is laid down in the telophase stage of the cell division, when the pectin rich vesicles of Golgi bodied (about 2.0 m) in diameter, migrate to the equatorial region of the cell. The membranes adjacent to vesicles fuse and form middle lamella-a cell plate. Cell wall is synthesized later to complete cytokinesis.

The plant cell wall is differentiated into three layers as follows:
1. Middle lamella, 2. Primary wall, 3. Secondary wall

1. **Middle lamella** is the cementing substance between two adjacent cells. It is made of pectic compounds, chiefly calcium and magnesium pectate. Pectic substances are the mixture of polygalacturon (D-galacturonic acid) and polysaccharides (neutral sugar). Sometimes lignin and hemicellulose also are found in the middle lamella. In mature ripe fruits, the softening is due to the dissolution of middle lamella by the action of pectolytic enzymes.
2. **Primary cell wall** is formed when cellulose fibrils produced by protoplasm, are accumulated in concentric rows on either side of middle lamella. It is thin and elastic in young enlarging cells and becomes thick and rigid during cell

maturity. It is composed of hemicellulose, cellulose, non-cellulosic polysaccharide, smaller amounts of pectic substances, fats and proteins.

3. **Secondary cell wall** is laid down between the primary cell wall and protoplast and is formed when the cell matures. It is formed by the deposition of cellulose on primary cell wall. It is inelastic and rigid. It is chiefly composed of cellulose and other non-cellulosic substances like lignin, fats and proteins. The secondary cell wall commonly has three layers e.g., outer layer, the middle layer and the inner layer.

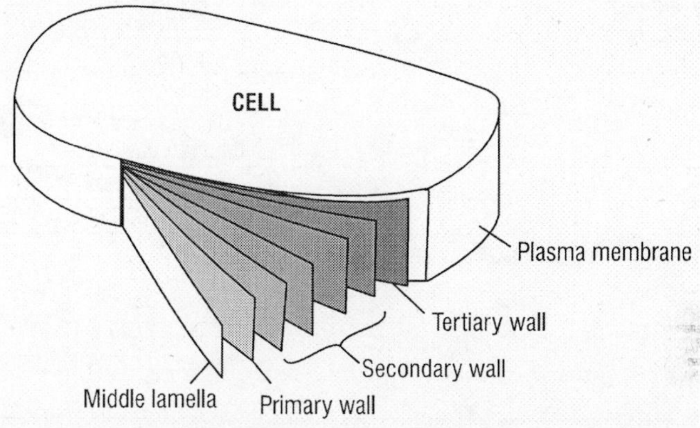


**FIG. 2.2:** Organisation of a plant cell boundaries. (*Note: Not all plant cells have all the layers.*).

### *Structural Organization of Cell Wall*

The cell wall is structurally thought as a fine interwoven network of cellulose strands. According to Siegel (1962) elementary fibrils or micelles are the smallest structural units of the cell wall. They are composed of approximately 100 individual cellulosic chains with a cross sectional area of about 3000 Å. The molecules of the cellulose, however, consist of long chains of about 3000 glucose units (hexose) and in the cell wall these units are associated parallel to one another forming microfibrils.

Microfibrils are the basic structural units. The cellulosic microfibrils are arranged differently within the primary and secondary cell wall. In the primary cell wall, the microfibrils are initially oriented in a transverse direction to the cell axis and during cell enlargement they become reoriented along the longitudinal axis of the growth. In the secondary wall, the microfibrils are arranged in three ways, in the three layers of the secondary wall. In the outer layer of the secondary wall, two spirals of microfibrils may be seen. In the middle layers, a step spiral in addition to concentric rings of spirals can be observed and in the inner layer, helical arrangement of microfibrils can be seen.

In cell wall, perforations are found in the form of simple bordered pits. Simple pits are thin walled spots which are formed at certain points where the deposition of the layers of the secondary wall does not take place. The pits of adjoining cell are usually opposite to one another and are separated from one another by the common middle lamella and the two primary cell walls.

**Bordered pits** are the pits, which are partly covered by the innermost layer of the secondary wall as found in dead tracheids and vessels.

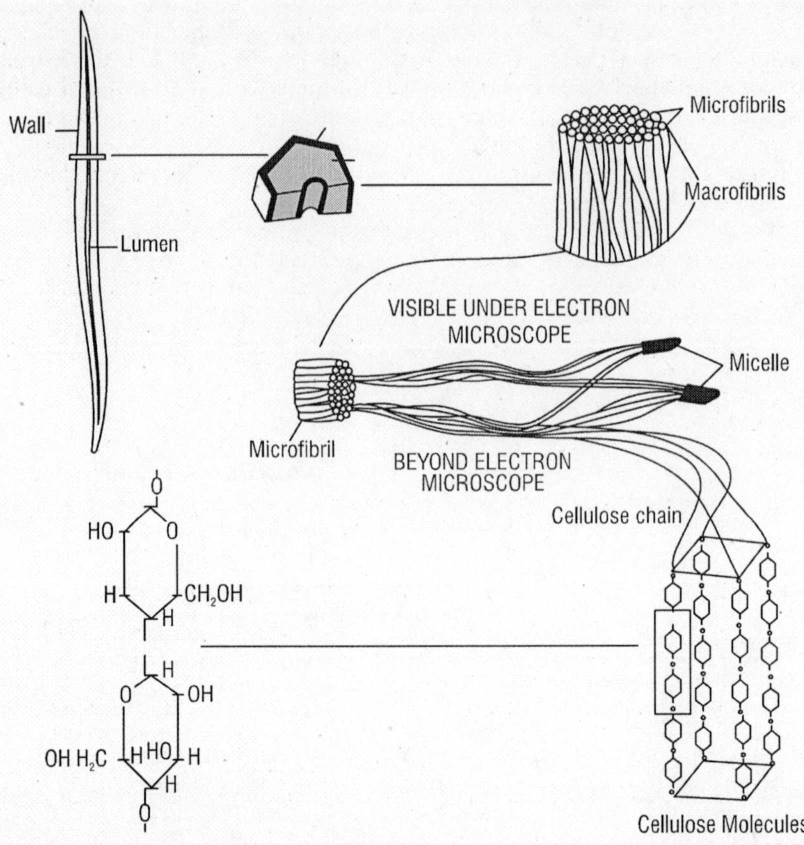


**FIG. 2.3:** Cell wall. (Note: microfibrils, macrofibrils, micelle, cellulose and gulcose molecules).

On maturation, certain fibrous structures develop called plasmodesmata (*sing.* plasmodesma), through which the adjacent cell are in communication with each other.

Plasmodesmata are also called tonofibrils. These are fibrous, unbranched, hollow, tubular structures, passing through pores of different sizes, present in cell wall. They are made of pectic substances. Since, they make connections between two adjacent protoplasm, they are also known as protoplasmic connections.

### *Chemical Composition of Cell Wall*

The cell wall comprises the following components:

| Component | Percentage |
|---|---|
| Cellulose | 10-15% |
| Hemicellulose | 5-15% |
| Water | 60% |
| Protein | 1-2% |
| Lipids | 0/5-3% |
| Pectic substances | 2-8% |

### Function of Cell Wall

1. It helps the aerial portion of the land plants to tolerate gravitational forces.
2. It being permeable, helps in the transport of materials in and out of the cell.
3. It expands and contracts physically with the change in turgor pressure.
4. It plays role in cell enlargement.
5. It provides a definite shape to the cell and protects internal organelles from external shocks.
6. It helps the aerial portion of the land plants to tolerate gravitational forces.
7. It being permeable, helps in the transport of materials in and out of the cell.
8. It expands and contracts physically with the change in turgor pressure.
9. It plays role in cell enlargement.
10. It provides a definite shape to the cell and protects internal organelles from external shocks.

## PROTOPLAST

All living material of a cell, except cell wall is the protoplast. It is composed of cytoplasm and nucleoplasm.

## CYTOPLASM

It is the part of the protoplast which lies outside the nucleus and inner to the cell wall. The term cytoplasm was given by Kolliker. It is composed of three parts; the *plasma membrane*, the *hyaloplasm*, and the *cytosomes*. Under an electron microscope the cytoplasmic matrix appears homogenous or finely granular. It possesses mechanical properties like contractivity, elasticity, cohesion and rigidity. About 14 elements have been actually detected in cytoplasm, and 12 of them are universally present. Water is the most abundant component of cytoplasm. Salts which dissociate into ions are important in the maintenance of the acid-base equilibrium of the cell and osmotic-potential. The common substances of cytoplasm are carbohydrate, lipids and nitrogen base derivatives such as DNA and RNA.

**Chemical Composition of Cytoplasm**

| Component | Percentage |
|---|---|
| Water | 75-85% |
| Protein | 10-20% |
| Lipid | 2.3% |
| Carbohydrates | 1% |
| DNA | 0.4% |
| RNA | 0.7% |
| Other organic matter | 0.4% |
| Inorganic matter | 1.5% |

### Plasma Membrane

This membrane has been variously named as: plasmalemma, cell membrane, and cytoplasmic membrane. It is the outermost layer of protoplasm lying just below the cell wall. It was discovered by Plowe. Its origin is considered from protoplasm.

The plasma membrane is a definite protoplasmic membrane, selectively permeable and its thickness is 75-100 Å. Under electron microscope, plasma membrane appears as a tripartite structure consisting of two parallel dark lines separated by central clear area (like rail road track). The two dark parallel lines are 20 Å thick each and made of proteins. The middle central clear area is 35 Å thick and remains sandwiched between protein layers.

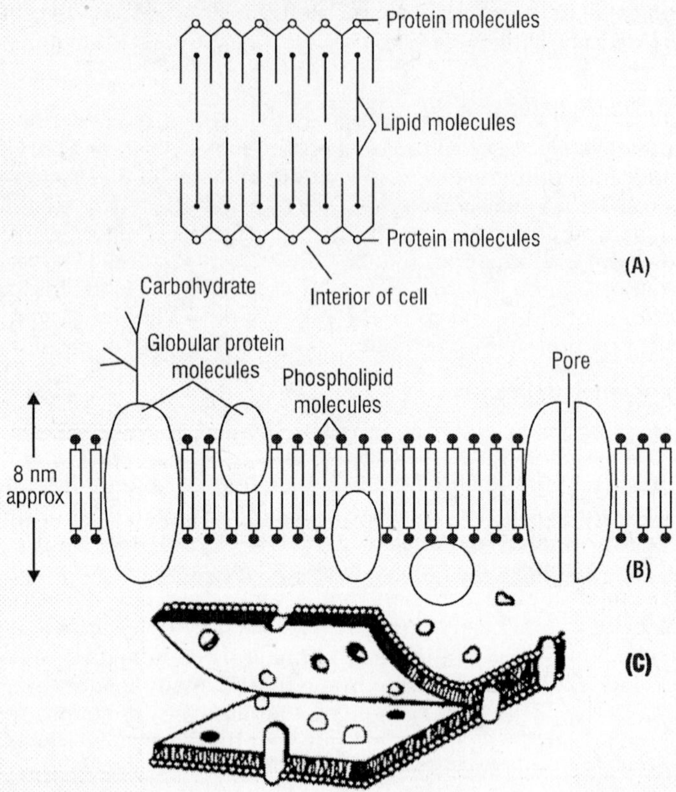


**FIG. 2.4:** Plasma membrane. (a) Danielli-Davson model (b) Fluid-mosaic model (c) Three dimensional configruation.

### *Structure of Plasma Membrane*

Overton (1895) studied the permeability of the plasma membrane and concluded that the plasma membrane is composed of a thin layer of lipid, which would account for the penetration of lipid soluble substances. Gorter and Grendel (1925) provided an additional support for Overton's ideas. They demonstrated that lipids form a single layer (mono layer) that occupies twice the surface area of the cell. They concluded that the surface lipids form a layer which is two molecules thick. Regarding the arrangement of proteins and phospholipids in plasma membrane, various molecular models have been proposed. These are as follows:

### *Lamellar Hypothesis (Sandwich Model)*

This hypothesis was proposed by Danielli (1935) which explains that the plasma membrane consists of a bimolecular layer of phospholipids lying between two layers of proteins. The phospholipids are arranged into two parallel chains. The protein layer contains both polar and non-polar parts. This model explains various properties of plasma membrane like permeability of lipid solvent and low surface tension. In 1954, Davson and Danielli modified lamellar model. The proteins were considered to be in the form of folded b-chains. Protein-linked polar pores of about 7 Å diameter were envisaged in the membrane. In still other variations the proteins are thought to be in the coiled form on both sides of the lipid layer or were thought to be asymmetrical with folded$_q$ -chains on one side and globular proteins on the other.

### Unit Membrane Hypothesis

The unit membrane hypothesis was promulgated by Robertson (1959) who had studied the electron microscopic appearance of membrane extensively. According to this model, the plasma membrane is trilaminar, with two layers of lipid molecules between two protein layers. The molecular chains of lipids remain parallel to each other thus forming a bimolecular structure. The two chains remain linked together by non-polar and hydrophobic inner ends of lipid molecules. Van der Waals forces bind together the two lipid layers. The lipid layer, in turn, forms bridges with the protein layers by their outer, polar and hydrophilic ends. In these connections, hydrogen-ionic-linkage or electrostatic forces bind the lipid layer with protein layer. Carbohydrate molecules also occur in association with protein molecules and provide stability to lipo-protein complex. The protein layer provides elasticity and mechanical resistance to plasma membrane. Robertson (1959), using a new fixative, ($KMnO_4$) provided evidence of this trilaminar composition, has shown such type of existence in other cell organelles, such as mitochondria, golgi complex, etc. It indicated its cellular universality. He believed that membrane have structural polarity, i.e., the inside and outside surfaces are different. Outer surface is covered with mucoprotein, while on the inner surface of the membrane unconjugated protein molecules are present. This shows the asymmetry of the protein surfaces on either side of the lipid bilayer. With improved preparation techniques and higher microscopic resolution the plasma membrane of most type of cells appears three layered. The two outer dense layers are about 20 Å thick and middle layer is about 35 Å (3.5 nm) thick. Out of 35 Å thickness of middle phospholipid layer, polar groups are 5 Å thick, and thus clear zone is only 25 Å thick.

The Robertson's unit membrane model was widely approved but remained unsatisfactory because it could not explain the dynamic natural and functional specificity of the membrane.

### Micellar Hypothesis (Model)

This model was proposed by Fernandez-Moran (1962) which explains that plasma membrane consists of globular units or elementary particles (micelles). The globules are 40 Å-70 Å in diameter. Each globule consists of a central core of lipid molecule whose hydrophillic polar ends are directed towards the periphery of the subunits and hydrophobic tails are facing towards centre. The protein forms a single molecular layer on either side of the lipid. The space between globular micelles are thought to represent water filled pores about 4 Å.

Electron microscopic studies have provided support for both the bimolecular layer structures and micellar structure. It can not be said with confirmation that which of the two cases represent the natural conditions. It is visualized that different membranes vary in their structures and their may be the transformation between the bimolecular and micellar states. It is also observed that micellar structure is mostly related to inter cytoplasmic membranes rather than plasma membrane.

### *Fluid Mosaic Model*

In the early years Benson, Lenard, Singer and Baums proposed mosaic membrane model. In 1972, Singer and Nicholson proposed the fluid mosaic model - the most widely accepted membrane model of today. They regarded biological membranes as fluid mosaic or quasi fluid structures. In such membranes the lipids and integral proteins were arranged in a mosaic manner. The membrane consists of a double layer or bilayer or molecular bilayer of phospholipids, with their hydrophobic hydrocarbon tails oriented inward and their hydrophilic polar heads constitute the top and bottom surfaces. The bimolecular layer is about 45 Å in thickness. The lipid bilayer is a fluid in which the globular proteins are dispersed to give a mosaic appearance to the whole membrane. It is much like variously weighted *ping-pong balls* in a puddle of viscous fluid. The protein components may be structural or functional (enzymatic). There are two kinds of proteins in the membrane, viz., the peripheral or extrinsic proteins and the integral or intrinsic proteins. The peripheral proteins are completely external and are loosely attached to the polar heads of the lipids on both the surfaces. The integral proteins penetrate the membrane surface and bimolecular lipid layer partially or wholly.

These either project partially from either surface of the membrane or span the entire thickness of the membrane and project from one or both surfaces.

The proteins may also be partially hydrophilic towards the surfaces of the membrane, exposed ends of protein are found primarily within the middle of the membrane, where the protein is clearly associated with the lipid layer, where the hydrophobic interconnections are found. That is the reason why the integral proteins can not be easily removed. According to Singer and Nicholson the membrane proteins can freely diffuse in a lateral direction within the lipid matrix. This is essential to enable them to interact with each other or with lipids. They can not, however, move from one side of the membrane to the other side (*flip-flop* diffusion). According to them three types of interactions, namely between protein and protein, protein and lipid and lipid and lipid are responsible for the structure and dynamism of the membrane.

Green and Capaldi (1974) have given a slightly modified version of the fluid mosaic model. According to them interactions between the membrane and the lipid bilayer lies in between the proteins in the form of patches.

It is now well established that proteins of the membranes act as carriers and enzymes. Carriers are concerned with the transport of ions etc., enzymes present in the membranes carry on some of the most important metabolic reactions e.g., those of photophosphorylation are present in thylakoids of the chloroplast and of oxidative phosphorylation are present in the inner membrane of the mitochondria.

### *Chemical Composition of Plasma Membrane*

The plasma membrane chiefly consists of protein and lipids. The percentage of these may vary in different cells, but usually percentage they are found as follows:

Lipids .................... 40%
Protein .................... 58-59%
Carbohydrate .................... 1%
Enzymes .................... 30 types

### *Function of Plasma Membranes*

1. The plasma membrane, being differentially permeable, and regulates the passage of diverse substances.
2. All cytosomes originate from plasma membranes.
3. It is responsible for aerobic respiration in prokaryotic cells.

**Hyaloplasm or Cytosol:** The matrix of the cell between nucleus and cell membrane is known as hyaloplasm. It was discovered by Kolliker. It is optically homogenous and exists as continuous phase with various inclusions embedded in it.

### *Chemical Composition of Cytoplasm*

**Water and Cytoplasmic inclusions:** The cytoplasmic inclusions includes, (i) **Organic compounds** which includes carbohydrates, lipids, proteins enzymes, hormones, vitamins, organic acids, glycogen, nucleic acids and amino acids. (ii) **Inorganic elements** includes - C, H, O, S, P, K, Mg, Ca, Fe, Na, Cl, Mn, Mo, B, Zn, Se, etc.

## CYTOSOMES

The cytoplasmic organelles are surrounded by their own membrane of a definite shape. These are living contents of cell and are specified to conduct specific physiological functions. They are: mitochondria, plastids (chloroplast), endoplasmic reticulum, and golgi complex. Organelles are enveloped by a double membrane. In addition, spherical bodies bounded by a single membrane (microbodies) or bounded by one half of a single membrane (spherosomes) are also present. Peroxisomes and glyoxysomes are two kinds of microbodies which range in diameter from 0.5 to 1.5 μm. Peroxisomes play role in photosynthesis and photorespiration. Glyoxysomes are involved in conversion of fats to carbohydrate during seed germination. Spherosomes range in diameter from 0,5 to 2,0 μm. They mostly contain fatty materials and are possibly are the sites of fat synthesis and storage. The other fat storing bodies are called oleosomes, or of protein, called protein-bodies.

## ENDOPLASMIC RETICULUM

The cytoplasm of the cell is interlaced by an elaborate membrane-bound vesicular system called the endoplasmic reticulum (ER). It was first observed by Porter, Claude and Fullam (1945) in the cytoplasm as a network of delicate strands and vesicle like bodies. The term endoplasmic reticulum was first used by Porter and Kalluman (1952). Watson (1955) described a continuity between the outer nuclear membrane and that of ER. The space of the ER opens into perinuclear space between the two nuclear membranes. Porter and Machado (1960) showed that ER is an extension of the nuclear membrane. The ER sometimes shows communications with the outside of the cell. It is also known as ergastoplasm. It is present only in eukaryotic cells. Like other cell organelles, it is originated from plasma membrane by the process of invagination and foldings. It is also believed that it originates from nuclear membrane by the process of evagination.

### Types of Endoplasmic Reticulum

It is of two types: (i) Rough ER and (ii) Smooth ER.

### *Rough ER*

The portion that has ribosomes on the outer surface is called rough or granular endoplasmic reticulum (RER). It is specially well developed in cells actively engaged in protein synthesis, such as the enzyme-producing cells.

### *Smooth ER*

The portion that lacks ribosomes is called smooth or a granular endoplasmic reticulum (SER). It plays a special and important role in the synthesis and assembly of glycolipids. It is mainly found inactive during protein synthesis.

ER is composed of following three kinds of structures, viz., (i) Cisternae, (ii) Vesicles, and (iii) Tubules.

I. **Cisternae:** These are broad, long, flattened, sac-like, unbranched tubules having diameter of 4 to 5 nm. They are arranged parallel in bundles or stacks to form lamellae. They are interconnected with each other.
II. **Vesicles:** These are oval, membrane bound vacuolar structures having the diameter of 2.5 to 50 nm. They often remain isolated in the cytoplasm.
III. **Tubules:** The tubules are irregular in shape and branched, smooth walled structures enclosing space. They form reticular system along with the cisternae and vesicles. They are 5 to 10 nm in diameter.

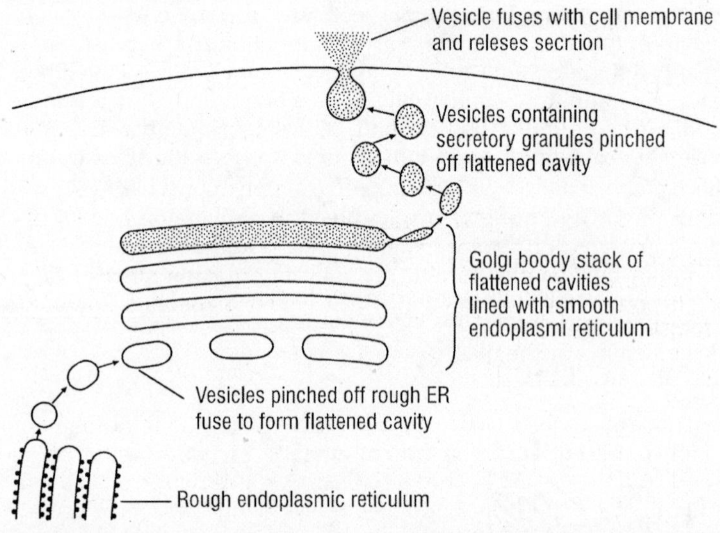


**FIG. 2.5:** Endoplasmic reticulum. (a) Endoplasmic reticulum and in dynamic equilibrium with Golgi body. (Note: Rouugh ER).

When cells are ultracentrifuged, the ER with its associated ribosomes, break up into small fragments of tubules, cisternae, which are called microsomes. This was first isolated by Claude (1941). They constitute 15 to 20 per cent of the total mass of the cells. They have a high percentage of lipids (phospholipids, inositol, acetyl phospholipids, and gangliosides), RNA and enzymes. The enzymes include peptidases, glycosyl transferases, hydrolases, and NADPH-cytochrome, C-reductase, etc. Microsomal enzymes are asymmetrically distributed across the membrane. They are involved in the biosynthesis of phospholipids, ascorbic acid and steroids and take part in a series of reactions which require $NADPH_2$ and $O_2$. They also play role at certain stages in the synthesis of glyceride, phospholipids and glycolipid.

**Chemical composition of ER**

| | |
|---|---|
| Protein | 70% |
| Lipids | 20% |
| Other | 10%. |

### *Functions of ER*

The ER helps in intracellular translocation. Therefore, it is also known as '*travelling agency*' of the cell.

1. It provides mechanical support to the colloidal structure of cytoplasm.
2. It helps in protein synthesis. It is evident from the fact that ribosomes attached to ER can actively synthesize protein than when found freely in the cytoplasm.
3. It makes continuity or relation between nucleus and cytoplasm.
4. It plays important role in the formation of a new nuclear membrane and cell plate during cell plate formation.
5. ER may also behave like the nervous system of the cell.
6. It helps in the secretion of waste products of the cell.
7. It forms certain cytosomes like Golgi bodies.

## GOLGI COMPLEX

Golgi complex or Golgi apparatus is a cluster of smooth membranes found associated with the endoplasmic reticulum of all cells of higher plants and animals except prokaryotic cells of mature mammalian erythrocytes.

Though the existence of Golgi apparatus was first reported by George (1867), but its morphological structure and details were presented by an Italian scientist, Camillo Golgi (1898) in the nerve cells of an owl. These structures are also known as dictyosomes, Golgisomes, idiosomes or lipochondria. They are present in eukaryotic cells. It originates from endoplasmic reticulum by the loss of ribosomes or the plasmalemma or the nuclear envelope.

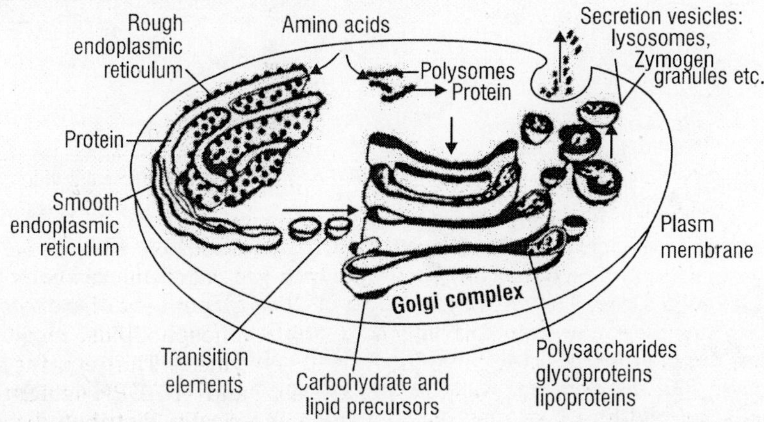


**FIG. 2.6:** Golgi complex. Arrangement of vesicles.

### Ultra structure

Dalton and Felix (1954) were the first to reveal the electron microscopic structure of the Golgi complex. It is composed of the following three distinct components:

I. **Cisternae or Saccules:** These are stacks of flattened or tubular elongated sac-like structures of smooth membranes. They are separated by the space of 20 to 30 nm which may contain rod-like elements or fibers. Often, the cisternae are arranged concentrically with a convex and concave faces.

II. **Tubules and Vesicles:** From the peripheral area of cisternae, a complex, anastomosing flat network of tubules 30 to 50 nm diameter, are present. The vesicles are small droplet-like sacs which remain attached to tubules at the periphery of the cisternae and may show continuity with the cisternae. They develop the cisternae by budding or constriction at their ends.

III. **Vacuoles:** They are large, spacious, rounded sac like structures occurring at the distal ends of cisternae.

### Chemical composition of Golgicomplex

Chemically it is characterized by the presence of following components: protein = 65-70%, phospholipids = 25-30%, enzymes like acid phosphatase, galactosyl transferase, thiaminepyro phosphatase, and adenine diphosphatase, etc.

### Functions of Golgicomplex

1. One of the major roles is secretion of enzymes and hormones.
2. It helps in the formation of hemicellulose and pectic groups of polysaccharides especially during cell division. These substances are laid down on the cell plate to constitute middle lamella.
3. The Golgi complex plays an important role in biogenesis of the plasma membrane by supplying it with glycosidated molecules.
4. Golgi complex also gives rise to the primary lysosome by budding.

## MICROBODIES

Microbodies are extremely small, dense, structures present in all cell substances (1-2 nm in diameter). The microbodies of a cell can be classified into different groups on the basis of their structure, location and functions. These are as follows:

(i) Lysosomes, (ii) Spherosomes, (iii) Peroxisomes, (vi) Glyoxysomes, (v) Lomasomes.

### (i) Lysosomes

The lysosomes (Gr. *Lysis* = dissolution, *soma* = body) are membrane bound vesicles, and are about 0.2 to 0.8 Å in diameter. They are single membranous cell organelle. Lysosomes were discovered by Christian deDuve (1965). Lysosomes contain enzymes of digestion. They are found in eukaryotic cells specially in yeast, fungi, and green algae. They are believed to originate from plasma membrane and ER, sometimes they may develop from Golgi complex. Currently it is believed that lysosomes are characteristically found in animal cells. In plant cells, some vacuoles may behave like lysosomes. Probably, the enzymes in the vacuole digest various materials absorbed into it or may digest cellular material after the death of a cell, like in wood cells where protoplast breaks down.

### *Ultra structure of Lysosomes*

The lysosomes appear as dense, granular structure under electron microscope. Their shape and size are not definite. They are more or less spherical in shape and are covered by a single lipoprotein membrane and contain enzymes. The membrane is impermeable, therefore, it insulates the enzymes from the rest of the cell. There are present about 50 hydrolytic enzymes in the lysosomes. Since they all require the presence of acid for their action. They are commonly known as acid-hydrolases. The enzymes are bound by a membrane which has a low permeability to the substrates of enclosed enzymes. The membrane thus protects the cell from autolysis (self-dissolution). Lysosomes have polymorphic nature (that is existence of a structure in more than one morphological forms).

### *Functions of Lysosomes*

Although they are called *atom bombs* or suicidal bags of the cell, they possess a built-in-mechanism that enables a cell to adapt metabolically. In plant cell, they play a major role in germination. They may be involved in intracellular and extra cellular digestion and also in the process of development. In seedlings, they are involved in the hydrolysis and removal of protein (aleurone bodies) and starch during germination. These changes are accomplished by the release of various lysosomal enzymes (e.g., α-amylase) that are able to hydrolyse the stored material to be used by the growing plant.

## (ii) Spherosomes

They are small, highly refractive, enzyme containing particles that are found in cytoplasm of plant cells. They were observed by Hanstein (1880) and were first discovered by Dangeard (1919). Later, on Perner (1952) gave the term spherosome. Spherosomes are now believed to be spherical bodies bound by one half of a single membrane. Their role is also believed to in the storage and synthesis of fats (Fig. 2.7)

### *Ultra structure*

They are more or less spherical structures surrounded by a single unit membrane and are 0.5 to 1.0 Å in diameter. In addition to the enzyme hydrolase, spherosomes contain other hydrolytic enzymes classified as protease, ribonuclease, phosphatase, and esterases.

### *Function*

The primary function of spherosomes in the cell appears to be the storage and transport of lipids. They are similar to lysosomes in that they possess a single membrane and some hydrolytic enzyme activity. However, although they do contain a number of similar enzymes, their overall enzyme complements are quite different, and the two particles should be thought as distinct from one another.

## (iii) Peroxisomes

These were first of all isolated by Tolbert and his coworkers in 1968 from a *broken chloroplast* fraction by density gradient centrifugation. They are found in yeast and in many cell types of higher plants. The number of peroxisomes varies from 70-100 in one cell (Fig. 2.8).

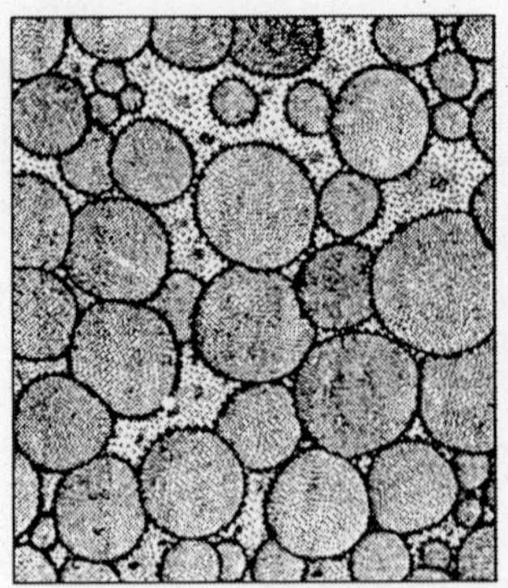

FIG. 2.7: Spherosomes

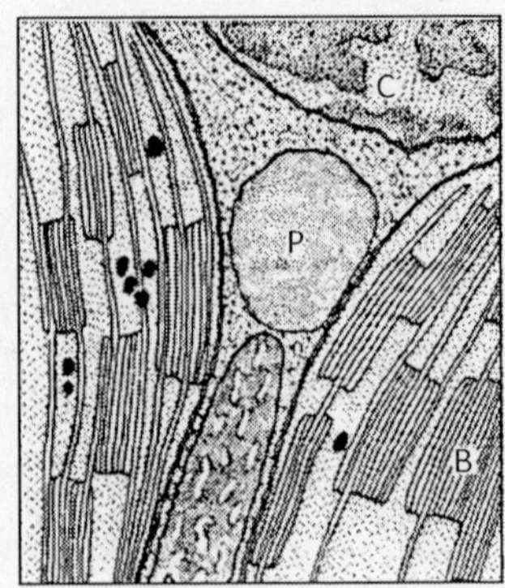


FIG.2.8: Peroxisome (P)

### Ultrastrucure

These are similar in appearance to glyoxysomes and are more or less spherical in shape. They have a single membrane cover which encloses a dense stroma.

Peroxisomes contain several enzymes like glycolateoxidase, glyoxylate reductase, catalase, glyoxylate-transaminase and malate dehydrogenase, etc.

### Functions

These are concerned with the metabolism of the glycolate produced by the chloroplasts during photosynthesis. They are also found associated with photorespiration - a process characteristic of $C_3$ plants. Localization of photorespiration in the plant cells corresponds to area of dense numbers of peroxisomes.

## (iv) Glyoxysomes

They were discovered by Breiclenback and Beevers (1967) from the extract of endosperm of germinating castor beans. They are primarily found in tissues of oil-bearing seeds and mainly contain the enzymes of glyoxylate cycle and hence they are termed as glyoxysomes. These are analogous particles to peroxisomes.

### Ultra structure

They have single layered membrane enclosing a finely granular stroma. They contain enzymes that are characteristic of the glyoxylate cycle, like isocitrate lyase, malate synthetase, aconitase, citrate synthetase, glycolate oxidase, malate dehydrogenase and catalase.

### Function

They convert fat into carbohydrate and thus carry out the glyoxylate cycle.

## (v) Lomasomes

They were discovered by Moore and McAear (1961) and as referred to **border bodies**. They are present between the cell wall and cell membrane of both lower and higher plants.

### Function

They play role in secretion, increasing the surface area for the diffusion of substances involved in cell wall formation or breakdown in membrane proliferation and in endocytosis. Very little is known about origin, significance and functions of lomasomes.

## VACUOLES

In young, immature cells several small vacuoles are present which appear under microscopic examination as clear droplets. As the cell enlarges or matures, the small vacuoles fuse together to form a large vacuole. It usually fills almost 90 per cent of the entire volume of the cell.

### *Structure*

The vacuole is bounded by a single membrane, the tonoplast, which is differentially permeable but encloses a solution in which numerous materials are suspended. The contents of the vacuoles are collectively referred to as the cell sap.

Regarding its origin, it is believed that vacuoles are produced because plants cannot produce enough protein to fill their enlarging cells with protoplasm. The vacuole is believed to be originated from cell membrane by invagination or from the endoplasmic reticulum.

### *Functions*

1. Storage of aqueous solutions and maintenance of the turgor pressure, which is important for structural support and control osmotic functions.
2. Storage of materials essential to the cell metabolism.
3. Accumulation of cellular metabolites, defensive substances, and toxic materials. Thus, the cell sap contains such substances as sugar, organic acids, mineral salts, gases, pigments (anthocyanins), alkaloids, fats, oils, tannins, and sometimes crystal (e.g., calcium oxalate).
4. The tonoplast of the vacuole undoubtedly plays an important role in biochemical activities of plant cells.
5. The anthocyanins present in the vacuoles are water soluble and are responsible for the colour of flowers, fruits and vegetables.

## RIBOSOMES

They are found associated with endoplasmic reticulum or are found *free* in the cytoplasm or in mitochondria and plastids. They are sub microscopic, spherical, amembranous particles- called ribosomes. They were first observed in plant cells by Robinson and Brown (1953). Palade (1953) also described such structures in animal cells. Porter gave the term ribosome to these structures, since they are rich in RNA. The number of ribosomes in a cell is related to RNA contents of the cell. They occur both in prokaryotic and eukaryotic cells. In prokaryotic cells ribosomes are formed inside the cytoplasm because of the absence of nucleolus whereas in eukaryotic cells, ribosomes originates in nucleolus. These are the smallest ultramicroscopic and negatively charged particles.

### Types of Ribosomes

Ribosomes are classified mainly into two types, depending upon the size and their subunits sedimentation coefficient (S) (which indicates how fast a cell organelle sediments in an ultracentrifuge). They are mainly of two types- (i) 70S ribosomes and (ii) 80S ribosomes. Each ribosome is made of two subunits, one is larger and other is smaller in size. They remain united at low concentration of $Mg^{++}$ ions (0.001M). If Mg concentration is decreased below 0.001M, the ribosomes dissociate into its subunits, whereas if it is increased above 0.001M, ribosomes can combine to form a dimer. During protein synthesis, the ribosomes are complexed with mRNA strand which are known as polyribosomes or polysomes.

### *70S Ribosomes*

These are found in prokaryotic cells. Their sedimentation rate is 70S (S is the Svedberg unit of sedimentation velocity). The 70S ribosome is 170 × 230× 250Å. The 70S ribosome consists of two subunits, a small 30S subunit and large 50S subunit.

The 30S and 50S subunits of the 70S ribosome can be dissociated into inactive core particles (23S and 40S respectively, and split proteins (SP30- 30 and SP50 respectively). The 23S core particle contains core proteins (23S) and rRNA of 16S. The 40S core particles contains core proteins (40S) and two RNAs of 23S and 5-S. The split proteins of the subunits called SP30 and SP50 which can be fractionated further into acidic (A) and basic (B) proteins. The total number of proteins isolated from 30S and 50S subunits are 21 and 35, respectively. The 70S ribosomes are inactivated by chloromycetin. In mitochondria and chloroplasts of eukaryotic cells, the ribosomes are similar to those of prokaryotic cells.

**Chloroplasts ribosomes (70S):** Ribosomes have been reported in chloroplasts of a wide variety of plant species. In general, the chloroplast ribosomes have a sedimentation coefficient of 70S. The 70S chloroplast ribosomes contain three classes of rRNA, 16S in the small subunit, and 23S and 5S in the large subunit. The ratio of chloroplast ribosomal 23S RNA to 16S RNA is 2:1.

### *Mitochondrial ribosomes*

Mitochondrial ribosomes are also known as mitoribosomes to distinguish them from cytoplamic ribosomes. Mitoribosomes occur in a wide variety of forms, the predominant form in multicellular organisms has a sedimentation coefficient of 55S. The 55S mitoribosomes consist of a small 30S subunit and a large 40S subunit. They are significantly smaller than 80S ribosomes. The small subunit of the mitoribosomes contains 12-13S rRNA and the large subunit contain 16-17S rRNA.

The nuclear ribosomes were first shown by Ts'O and Sato (1959) who on the basis of their chemical and physical characterisation stated that they are identical with cytoplasmic ribosomes.

### *80S ribosomes*

Eukaryotic ribosomes are called 80S ribosomes. Their actual S values may be lower or higher than 80S. The small subunit and the large subunit are 40S and 60S, respectively. The 40S and 60S subunits of the 80S ribosome are also dissociated into core protein particles and split protein. The 80S ribosomes are inactivated by cycloheximide compound.

Chemical composition of ribosomes: They are made of ribonucleoprotein. The major constituents of ribosomes are RNA and proteins, present in approximately equal amounts. Certain divalents metallic ions have been reported in ribosomes. The RNA/protein ratio is 2:1 in the 70S ribosomes, while in 80S ribosome it is 1:1. On the whole, the ribosomes contain 40-60% RNA and the rest is protein. RNA makes the core and protein forms its covering.

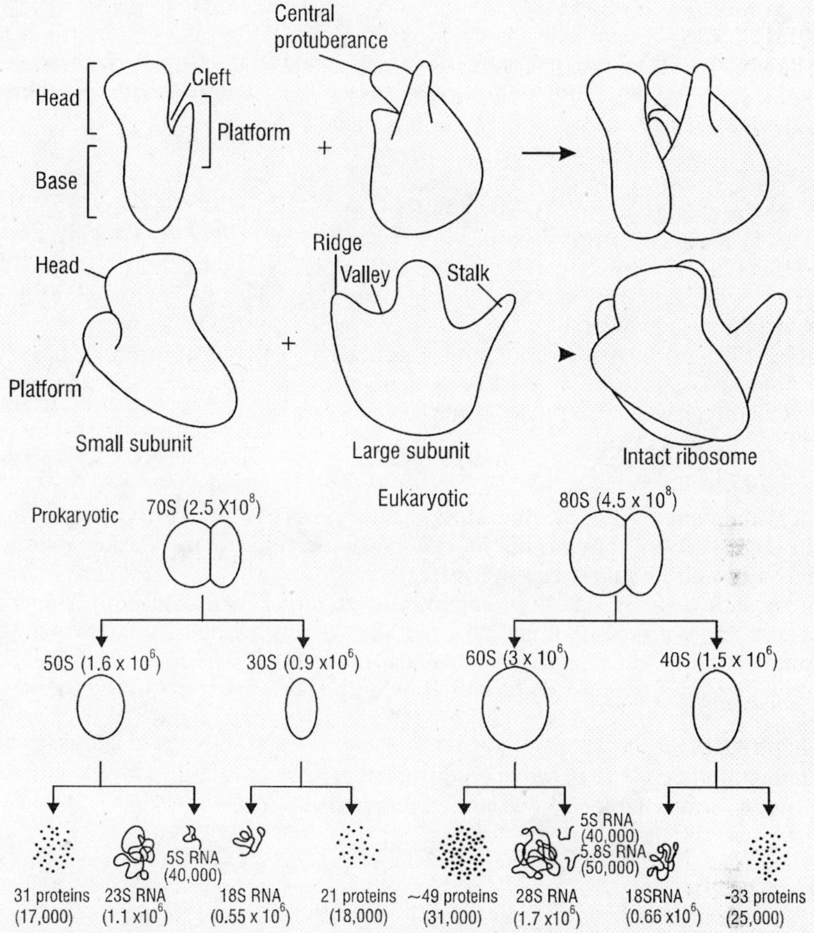

**FIG. 2.9:** Ribosomes. (a) Three dimensional structure of *E. coli ribosomes.* (b) Molecular constitution of 70s and 80s ribosomes.

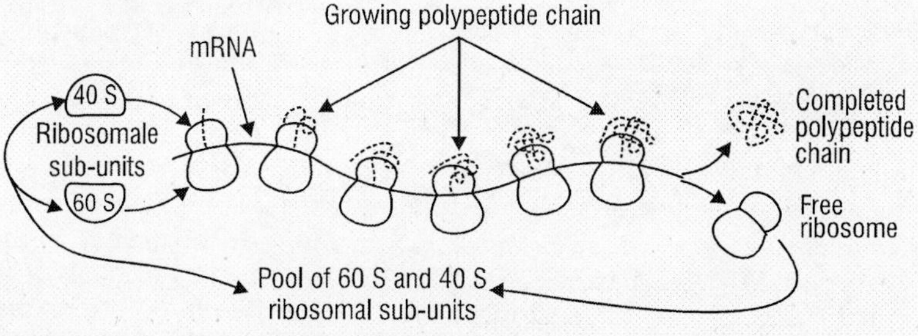


**FIG. 2.10:** Ribosomes. Polyribosomes and polypeptide chain.

### *Functions*

Ribosomes are known as the '**engines**' of a cell as they initiate protein synthesis. These are the only sights where amino acids are finally polymerised into proteins.

## MITOCHONDRIA

Mitochondria (Gr. *mito* = thread + *chondrion* - granule) are granular or filamentous organelle present in the cytoplasm of eukaryotic cells. Mitochondria were first observed by Kolliker (1850). The name was given by Benda (1898). Altmann (1894) described these as '**bioplasts**'. He considered them as symbionts like bacteria. The origin of mitochondria has remained controversial. They are believed to have arisen from plasma membrane, endoplasmic reticulum, nuclear membrane or pro mitochondria. The prokaryotic origin and symbiont hypothesis of origin of life have been put forward. The mitochondria are semi-autonomous bodies. They can be stained with Janus green. The mitochondria of whole cell are collectively known as chondriosome or chondriome.

### *Ultra structure*

Mitochondria appear as double membrane structures. They are ovoid, rounded, oblong or rod shaped. Mitochondria are pleomorphic (many formed) bodies, bounded by two unit membranes.

A mitochondrion consists of two membranes and two compartments. The outer membrane is smooth and uninterrupted. It shows similarity with smooth endoplasmic reticulum in structure as well as in chemical composition. It is freely permeable to various soluble metabolites, which enter the mitochondrial matrix from cytosol. It contains enzymes like cytochrome $b_5$ and enzyme system capable of oxidizing $NADPH_2$.

The inner membrane is extensively infolded towards the mitochondrial cavity to form many finger-like projections called cristae (*sing.* crista). The inner membrane shows on its inner surface, stalked particles attached on cristae. These particles are called elementary particles or Fernandex Moran particles or $F_1$ particles.

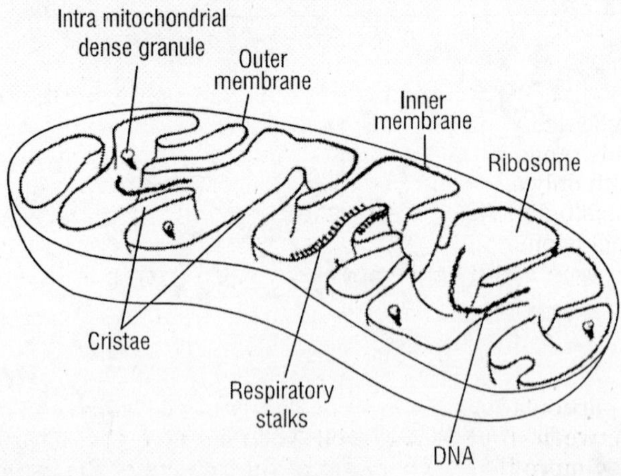


**FIG. 2.11:** A Three dimensional structure of mitochondrion.

Each particle consists of 3 parts, viz., a base, a stalk, and a head. The inner membrane divides the mitochondrion into two following chambers or compartments:

1. The **outer chamber** contained between the two membranes and in the core of the cristae.
2. The **inner chamber** filled with dense proteinaceous material called mitochondrial matrix. The matrix contains granules (consisting of insoluble inorganic salts), ribosomes and mitochondrial DNA. The presence of DNA in mitochondria (mDNA) was shown by Nass.

### *Chemical composition of mitochondria*

It contains, protein = 73%, lipids = 25-30%, RNA = 5-7%, DNA = in traces and enzymes of 60 types.

### *Functions*

They are commonly called **power houses** of the cell as the energy rich compound (ATP) are formed by them. These contain enzymes of the Krebs cycle, flavoproteins and cytochromes which are involved in respiration of the cell. Mitochondria are sites of oxidative phosphorylation.

## PLASTIDS

Plastids are characteristically present in plant cells. These are the largest cell organelles. These were discovered by A.F.W. Schimper (1883). Plastids (plastids = formed or moulded) were named by Kackel (1866). They are present in the cells of the leaf as well as in many other cells of the plant. The main types of plastids are: Chromoplasts (coloured), and Leucoplasts (colourless).

The chromoplasts are of various types:

1. **Chloroplast** (contains chlorophyll a and b, green colour; e.g., higher green plants).
2. **Phaeoplasts** (contain fucoxanthins, brown coloured; e.g., brown algae, diatoms, dinoflagellates).
3. **Rhodoplasts** (contain phycoerythrin).
4. **Blue green chromoplasts** (contain phycocyanin, phycoerythrin, chlorophyll a, carotenoids; e.g., cyanobacteria).

Certainly, no living cell of a higher plant has so far been described that completely lacks plastids which are described with the help of light- and electron microscopy. Plastids range in the size from tiny spherical proplastids less than 1nm in diameter and with only a few internal membranes, to pigmented chromoplasts 10 nm long having plastoglobuli (lipid droplets), complex membrane structure and often crystalline inclusions. All plastids are to some extent inter convertible, that is one morphological type can be shown to be derived from a at least one another morphological form.

## THE CHLOROPLASTS

In green leaf cells particularly in the chlorenchymatous cells, the chloroplasts lie in the cytoplasm between the central cell vacuole and the parenchyma. The chloroplasts occupy more than 90 per cent of the volume of the cytoplasm in leaf and are typically lens-shaped (in mosses, fern and seed plants), appearing biconvex, plano-convex or concavo-convex in profile. In algae they are heteromorphic. In higher plants they are usually 4-8 nm in diameter and 2-3 nm thick. Changes in shape and size can also be induced osmotically both in the leaf cells and chloroplast isolated from the cell. They often move in the cell under influence of illumination and as a result of cytoplasm streaming.

## The Functions of Chloroplast

An intact chloroplast of the leaf cell has one most important function - photosynthesis. This activity involves the photoreduction of $CO_2$ to phosphoglyceric acid, the photo oxidation of water, and photophosphorylation. Among other functions, limited protein synthesis, amino acid biosynthesis, chlorophyll biosynthesis, and carotenoid biosynthesis are important. In addition, the synthesis of complex lipids occurs on the outer envelope of the chloroplast. The well known endogenous inhibitor Abscisic acid (ABA) is also synthesized in chloroplast. Thus, in addition to fixing $CO_2$, the chloroplast carries out many other important functions.

## Number

Their number per cell varies greatly among plant species. The assimilatory parenchyma (palisade and spongy parenchyma) of leaves contains a specially large number of chloroplasts. In spinach leaves, for instance, 300-400 chloroplasts have been found per plastid parenchyma cell and 200-300 per spongy parenchyma. The dry weight of spinach chloroplasts is about $9 \times 10^{-12}$ g and that of corn chloroplasts is about $38 \times 10^{-12}$ g.

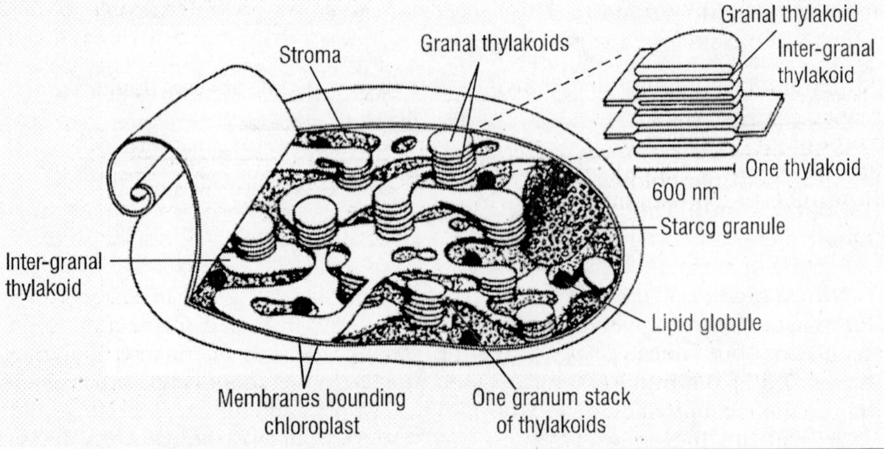


**FIG. 2.12:** A Three dimensional structure of chloroplast.

## Structure

The structure of higher plant chloroplast can be regarded as being bipartite- a membrane system and a matrix system. The membranes are of two different types:

1. An **external envelope** (characteristic of eukaryotic cell), and
2. An **internal pigmented membrane system** of grana and integranal connections.

The granal and the integranal membrane (*the frets*) form a continuous membrane known as the *fretwork.* Chlorophyll is not evenly distributed throughout the chloroplast, but remains contained in the inner membrane system of granum. This membrane system which gives rise to the stack-like configuration is called *grana*, and they are composed of individual flattened membrane sacs called *thylakoids.*

The surrounding material, which does not contain chlorophyll, is called *stroma*. It is a hydrophilic proteinaceous matrix. The $CO_2$-assimilating apparatus is present within the stroma (the non-green portion). The stroma along with thylakoids also contains particulate components such as *fraction protein I*, ribosomes and polysomes, and the fibrils of DNA, nucleotides (DNA), plastoglobuli (lipid reserve) and starch grains. Robert Huber, Johann Deisenhofer and Hartmut Michael were awarded the 1988 Nobel Prize for Chemistry for work in the photosynthetic process, for succeeding in making a structural analysis of a membrane linked to protein complex which opened up completely new possibilities in chloroplast structural research.

**Fraction protein I:** It accounts for 50 per cent of the soluble stroma and 50 per cent of total leaf soluble portion and contains photosynthetic carboxylating enzyme, ribulose diphosphate carboxylase (carboxydismutase, the enzyme responsible for catalysing the formation of phosphoglyceric acid from $CO_2$ now called RuBisco. Fraction protein I can be split into two distinct subunits: a large subunit of approximately 54,000 Dalton and a small subunit of 16,00 Dalton molecular weight.

**Chloroplast ribosomes:** The stroma also contains single ribosomes, clusters of ribosomes and helically arranged polysomes. The single chloroplast ribosomes are smaller than cytoplasmic ones measuring 17 nm in diameter (compare 25 nm for the cytoplasmic ribosomes). It has sedimentation coefficient of 70S and can be split into two subunits sedimenting at 55-60S and 40S, respectively. They resemble bacterial ribosomes more closely than eucaryotic cytoplasmic ribosomes.

**Chloroplast DNA:** It is stained with uranyl acetate and can be seen as 2.5 nm fibrils lying in clear zones of the stroma. The DNA is double stranded and the amount of DNA per chloroplast suggests that it corresponds to $2\text{-}3 \times 10^9$ Dalton.

**Plastoglobuli:** The plastoglobuli are large lipid droplets between 10 and 500 nm in diameter and are present in all profiles of chloroplasts viewed in the electron microscope. It is meant to store the lipid surplus to fulfil cellular requirements.

**Starch grains:** These are often single, largest inclusions found in the stroma of chloroplasts. They appear in profile as oval structures generally electron transluscent but occasionally they are electron opaque. Starch specifically accumulates in the bundle sheath chloroplasts of $C_4$ plants and only to a smaller extent in the mesophyll chloroplasts of $C_3$ and $C_4$ plants.

**Crystalline inclusions:** Crystalline inclusions or bodies are of 1 nm in size and lie in stroma. They are proteinaceous in nature and their function is unknown.

**Phytoferritin:** It is an iron protein complex which appears as an electron opaque aggregation of particles in the plastid stroma.

**Pyrenoids:** Algae (unicellular green algae like *Chalmydomonas sphagnicola*) possess pyrenoids which organize the deposition of reserve substances. Griffiths (1970) suggested that these are the regions of temporary storage for early product of photosynthesis which after overproduction converts into starch. They arise de novo or may be formed by the division of pre-existing ones. Electron microscopic observations reveal that they are traversed by thylakoids.

**Microfilaments:** They are 5-6 nm in diameter have occasionally been seen in electron micrographs and are thought to run lengthwise in the stroma near the plastid envelope. The functions of these microtubule-like structures is unknown.

## NUCLEUS

The nucleus (Latin *nux* - *nut*) is the most essential part of the cell which directs cellular activities and controls cell division. It was discovered by Robert Brown (1831). It is found in all eukaryotic cells whereas in the prokaryotic cells, the nucleus is not well defined (absent) and is known as *incipient nucleus* or *nucleoid*. In the immature cells, the nucleus is spherical body embedded in the cytoplasm of the cell. The nucleus is generally located to one side of the cell since it is pressed against the cell wall by the vacuole. Here, it appears to be slightly flattened. During cell division, the nucleus undergoes a series of changes. It is during interphase of mitotic (non-dividing) phase, the nucleus is known as interphase nucleus, shows significant structures.

### *Ultra structure of Nucleus*

Usually, the nucleus is 5 to 10μm in diameter. Its shape is variable, may be disc-shaped or elongated and C-shaped, etc. The following structures are generally recognized in the interphase nucleus:

1. Nuclear membrane or envelope, 2. Nucleoplasm

1. **Nuclear Membrane:** The nuclear membrane or envelope or karyotheca separates the nucleus from the cytoplasm. It is double layered and each layer has the same structure as that of plasma membrane. Each membrane is 70-80 Å thick and the perinuclear space between two membranes ranges from 100 to 700 Å in thickness. The nuclear envelope is perforated at intervals by pores of 600-1000 Å in diameter. Pores are thought to allow low communication between nucleus and cytoplasm. The pores are lined with annulus which fills the pore except for a narrow central channel. Sometimes these appear plugged with particulate material resembling subunits of ribosomes. The outer nuclear membrane is often continuous with ER, which connects with the plasma membrane.
2. **Nucleoplasm:** The nucleoplasm is composed of structure and a structureless phase and is composed of following parts: (i) nuclear sap, (ii) nucleolus, (iii) chromatin fibres.
   a) **Nuclear sap** is the structureless phase of the nucleoplasm which appears as a granular substance. It contains nucleic acid, nucleoproteins etc., and forms the matrix.
   b) **Nucleolus** (*pl.* - nucleoli) is found in nucleoplasm which may be single or more than one in each nucleus depending upon the species. The nucleoli are spherical bodies made of small granules containing proteins and RNA. It is primarily composed of ribosomal RNA (rRNA) subunits and proteins. Chromatin fibres form the structured phase of nucleoplasm. It does not form a true network as chromatin fibres are not interconnected. Sometimes, this phase of the nucleoplasm appears either as a net work or as a distinct chromosome, depending on whether or not the nucleus is in mitosis. During cell division, the chromatin network undergoes condensation and appears in the form of definite rod-like structures- the chromosomes.

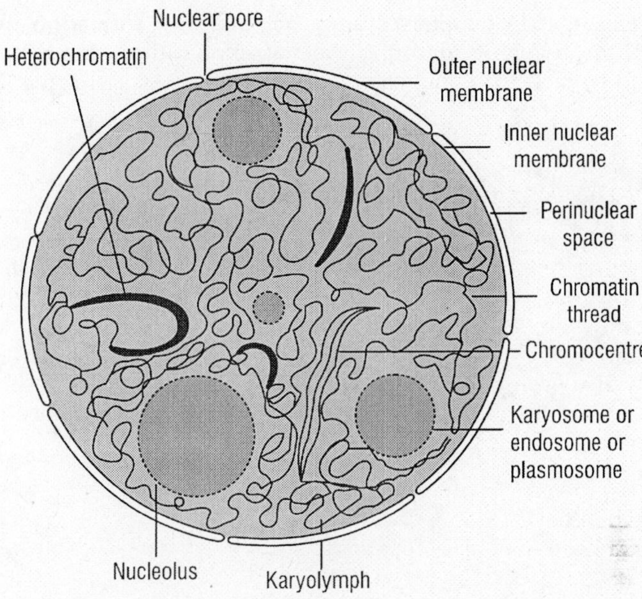


**FIG. 2.13:** Structure of nucleus.

**Chemical composition of nucleus**

| | |
|---|---|
| Protein | 65% |
| DNA | 9-12% |
| RNA | 5% |
| Lipids | 3%. |

### *Functions of Nucleus*

The nucleus issues directions for all cellular functions. Thus, it is also known as *Director of a cell.*

## MICROTUBULES OR MICROFILAMENTS

The term was given by Slautterback (1963). Their presence has been observed in many plant cells and are mostly found in peripherial cytoplasm immediately within the cell membrane. Microtubules are substructures of flagella and cilia in plant cells that are motile, such as gametes of lower land plants and algae. They are specially observed attached to the kinetochore (centromere) of chromosomes and in the spindle fibres during mitosis.

### *Ultra structure*

Microtubules are elongated, hollow, non-membranous structures, are approximately 10 to 20 nm in diameter. Their limiting boundary consists of distinct 12 or 13 longitudinal profilaments composed of globular protein subunits ($\alpha$ and $\beta$-tubulin).

### *Functions*

They are involved in the separation and migration of homologous chromosomes to opposite poles during telophase. They help in wall formation by directing the alignment of the cellulose microfibrils as these microfibrils are deposited. They are also believed to direct the movement of Golgi vesicles and bring them in contact with the cell membrane.

Should this whole gamut of functional multiplicity (musical notes) being played on diverse instruments, organelles and cellular sites in melodious orchestration (steady-state) be analogised with a symphony?

# CHAPTER 3

# Biophysics: Basic Concepts

## COVALENT BONDS

Covalent bonds involve the sharing of electrons between atoms. The more precise definition of a **molecule** is a combination of two or more atoms joined by covalent chemical bonds. A simple example of a covalent bond is the one joining two hydrogen atoms in a molecule of hydrogen gas, $H_2$. Each atoms of hydrogen has one electron, but two electrons are required to complete the first energy level. Each hydrogen atom has the same capacity to attract electrons, so one does not donate an electron to the other. Instead, the two hydrogen atoms share their single electrons so that each of the two electrons is attracted simultaneously to the two protons in the two hydrogen nuclei. The two electrons are thus under the influence of both atomic nuclei, and they join the two atoms together.

```
     H
     |
H—C—H
     |
     H
```

The carbon atoms has four electrons in its outer energy level. These four electrons are available for covalent bonding: -C-. When one carbon and four hydrogen atoms share electrons a molecule of methane ($CH_4$) is formed:

Each atom shares its outer-electrons with the others, thereby completing the first energy level of each hydrogen atom and the energy level of the carbon atom. When one electron pair is shared between two atoms, the covalent bond is referred to as a single bond.

```
     H
     |
H—N—H
```

The nitrogen atom has five electrons in its outer shell. When a nitrogen atom shares electrons with three hydrogen atoms, a molecule of ammonia, $NH_3$, is formed:

Two oxygen atoms may achieve stability by forming covalent bonds with one another. Each oxygen atom has six electrons in its outer shell. To become stable, the two atoms share to pairs of electrons, forming molecular oxygen. When two pairs of electrons are shared in this way, the covalent bond is referred to as a **double bond**. Some atoms form triple bonds with one another, sharing three pairs of electrons.

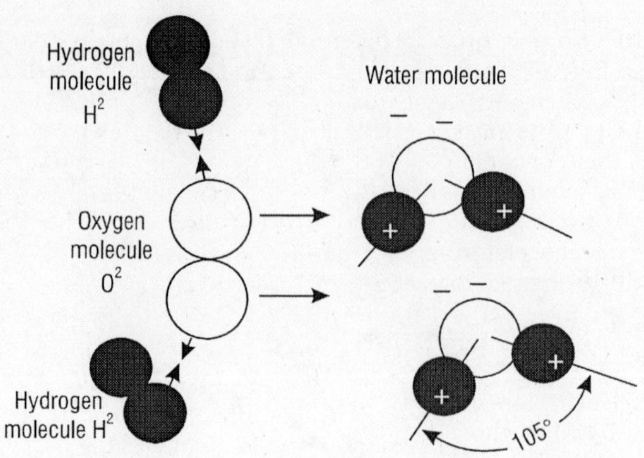


**FIG. 3.1:** The joining of diatomic hydrogen and oxygen molecules to produce water molecules of a polar nature.

**TABLE 3.1:** Some biologically important ions

| Name | Formula | Charge |
|---|---|---|
| Sodium | $Na^+$ | $1^+$ |
| Potassium | $K^+$ | $1^+$ |
| Hydrogen | $H^+$ | $1^+$ |
| Magnesium | $Mg^{2+}$ | $2^+$ |
| Calcium | $Ca^{2+}$ | $2^+$ |
| Iron | $Fe^{2+}$ or $Fe^{3+}$ | $2^+$ or $3^+$ |
| Ammonium | $NH_4^+$ | $1^+$ |
| Chloride | $Cl^-$ | $1^-$ |
| Iodide | $I^-$ | $1^-$ |
| Carbonate | $CO_3^{2-}$ | $2^-$ |
| Bicarbonate | $HCO_3^-$ | $1^-$ |
| Phosphate | $PO_4^{3-}$ | $3^-$ |
| Acetate | $CH_3COO^-$ | $1^-$ |
| Sulfate | $SO_4^{2-}$ | $2^-$ |
| Hydroxide | $OH^-$ | $1^-$ |
| Nitrate | $NO_3^-$ | $1^-$ |
| Nitrite | $NO_2^-$ | $1^-$ |

**Electronegativity** is a measure of an atom's attraction for the electrons in chemical bonds. A covalent bond between atoms of different electronegativities is known as a **polar covalent bond.** The polarity of compounds is very important in understanding the structure of biological membranes and their properties. Covalent bonds have many degrees of polarity, from **non polar**, in which the electrons are exactly shared (as in the hydrogen molecule), to those in which the electrons are much closer to one atom to the other.

## IONIC BONDS

One extreme of polarity is the ionic bond, in which electrons are pulled completely from one atom to the other. The number of protons in the nucleus remains unchanged, so the loss or gain of electrons produces an atom with a net positive or negative charge. Such electrically charged atoms are termed **ions.** Atoms with one, two, or three electrons in the outer shell generally donate electrons to other atoms. These atoms, then, become positively charged because of the excess of protons in the nucleus. For example, sodium has one valence electron. It tends to donate this electron, becoming a sodium ion ($Na^+$) with a one plus charge. Calcium tends to lose its two valence electrons to become a calcium ion ($Ca^{2+}$) with a two plus charge. Positively charged ions are termed **cations**.

Atoms with five, six or seven electrons in their outer shell tend to gain electrons from other atoms and become negatively charged anions (e.g., $Cl^-$, chloride ion). Charged particles, both anions and cations, play many important roles in biological systems.

An **ionic bond** is the force of electrical attraction between two oppositely charged ions. When held together by ionic bonds, oppositely charged ions form an ionic compound. Sodium chloride is a good example of an ionic compound. A sodium atom, with atomic number 11, has two electrons in its inner shell, eight in the second, and one in the third shell. A sodium atom cannot fill its third shell by obtaining seven electrons from other atoms because it would then have a vast excess of negative charge. Instead, it gives up the single electron in its third shell to some electron acceptor, leaving the second shell as the complete outer shell. A chlorine atom, atomic number 17, has two electrons in its inner shell, eight in the second, and seven in the third shell. The chlorine atom achieves a complete outer shell not by losing the seven electrons in its third shell (for it would then have a vast positive charge) but by accepting an electron from an electron donor, such as sodium, to complete its third shell.

When an electron donor, such as sodium, meets an electron acceptor, such as chlorine, the electron may be transferred completely from the donor to the acceptor. The sodium ion now has 11 protons in its nucleus and ten electrons around the nucleus, giving it a net charge of $1^+$. The chlorine ion has 17 protons in its nucleus and 18 electrons around the nucleus, so it has a net charge of $1^-$. These ions attract each other as a result of their opposite charges. They are held together by this electrical attraction to form sodium chloride (NaCl), common table salt. This transfer of one or more electrons from one atom to another and the binding together of two ions of opposite charge (the ionic bond) results in the formation of an ionic compound. Ionic bonds occur between electron donors and electron acceptors.

Whether an ionic compound is in solid form or is dissolved in water, its constituent particles (ions) do not share electrons. Because of this, the term molecule does not adequately explain the properties of ionic compounds such as NaCl. Chemists simply refer to them as compounds.

## HYDROGEN BONDS

Another type of bond that is important in biological systems is the **hydrogen bond**. This type of bond is formed when a hydrogen atom is attracted to two other atoms, one of which is usually oxygen or nitrogen. Hydrogen bonds tend to form between a hydrogen atom covalently bonded to oxygen or nitrogen and some other electronegative atom, usually oxygen or nitrogen. The atoms involved may be in two parts of the same molecule or in two molecules. When hydrogen is combined with an electronegative atom, such as oxygen, it has a partial positive charge because its electron is positioned closer to the oxygen atom. Hydrogen bonds are weak and are readily formed and broken. They have a specific length and orientation, which is very important in their role in helping to determine the three-dimensional structure of large molecules, such as nucleic acids and proteins. Though relatively weak individually, the large numbers of hydrogen bonds that occur in the double helix of DNA and in the a-helix of proteins are together very effective.

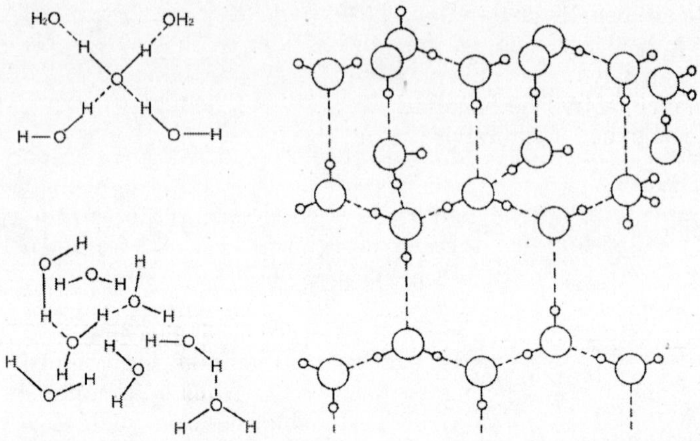

**FIG. 3.2:** Hydrogen bonding of water molecules, (a) Each water molecule forms 4H bonds, (b) H bonds in liquid water (c) H bonds in ice.

## OXIDATION-REDUCTION

Rusting- the combination of iron with oxygen- is a familiar example of oxidation and reduction:

$$4Fe + 3O_2 \longrightarrow 2Fe_2O_3$$

**Oxidation** is a chemical process in which a substance loses electrons. In rusting, iron is changing from its metallic state to its iron (III) ($Fe^{3+}$) state. We say it is being oxidized:

$$4Fe \longrightarrow 4Fe^{3+} + 12e^-$$

The $e^-$ stands for electron. At the same time, oxygen is changing from its molecular state to its charged state:

$$3O_2 + 12e^- \longrightarrow 6O_2$$

Oxygen accepts the electrons removed from the iron and is said to be reduced. **Reduction** is a chemical process in which a substance gains electrons. An oxidation cannot take place without a substance which gains electrons. An oxidation cannot take place without a reduction because electrons have to go somewhere; free electrons are not usually found in nature. Some substance must accept the electrons that are lost. Oxidation-reduction reactions are sometimes called **redox reactions.**

Electrons are not easy to remove from covalent compounds unless an entire atom is removed. In living cells oxidation almost always involves the removal of a hydrogen atom from a compound. Reduction often involves a gain in hydrogen atoms. As will be discussed in later chapters, redox reactions are essential part of photosynthesis, cellular respiration, and other aspects of metabolism.

## INORGANIC COMPOUNDS

Chemical compounds can be divided into two broad groups- inorganic and organic. **Inorganic compounds** are relatively small, simple substances. **Organic compounds** are usually large and complex and always contain carbon. Among the biologically important groups of inorganic compounds are water, simple acids and bases, and salts. Organisms depend upon appropriate amounts of these substances for fluid balance, acid-base balance, and many cell activities, such as transporting materials through cell membranes.

## BIOENERGETICS

Living system is a highly organised arrangement of matter. The synthesis and maintenance of the system is possible only by the availability of certain reactions. These reactions used by the living systems are the degradation of complex molecules of food materials to simple molecules and photosynthesis of materials from the simple compounds. Such spontaneous reactions produce useful energy for the performance of various vital activities of the organisms. The raw materials are used by the organism as forces of energy and as building blocks for its tissues. Here it is considered the way in which the organism gets its energy and utilizes it for the maintenance of life.

Energy is essential to every manifestation of life. Energy is required for differentiation and growth involving energy required to synthesize substances like nucleic acids, proteins lipids etc., mechanical work by muscles, generation of electrical impulses in nerves and transport of substances by active processes. The energy released by energy releasing reactions. Energy is defined as the capacity to do the work. The various types of energy are generally inter convertible except heat energy. The heat energy is partly convertible to other types of energy to some extent. On the other hand, all other kinds of energy are convertible to heat. In living organisms which maintain at a constant temperature, two types of energy are known- (a) heat energy which can serve to maintain the body temperature this is not useful for any other purpose and (b) free energy which is available for work. These two types of energy can be expressed in energy units (Calorie, Joules). It is described here about the free energy which are liberated in the chemical reactions.

### Nature of Energy and its Measurements

Energy is a fundamental physical concept. As mentioned above it is the capacity to do work. Many kinds of energy are known. The potential and kinetic energies, surface energy, chemical energy, electrical energy and radiant energy are examples. These may be converted from one form to another. When one form is created equal amount of another form disappears. This is the law of conservation of energy.

Energy can be understood clearly, if it is measured exactly. It is measured by various ways. The table (3.2) gives the different kinds of energies, units and their inter-conversion. The ultimate source of energy for life process is radiant energy is converted into chemical energy by biological mechanisms of transformations to vital kinds of work, mechanical, chemical biosynthesis and osmotic (transport). However, it is important to note that only a fraction of energy is released for vital processes. This is known as free energy.

## CHEMICAL ENERGY

Different organic compounds yield different amounts of energy when they are completely oxidized. This heat of combustion is determined by the elementary composition and structure of the substances and also its energy content. The total energy liberated is represented by the difference in energy contents of the substance oxidized and products of oxidation. The total amount of energy liberated is dependent upon the pathway of oxidation.

**TABLE 3.2:** Energy units and their inter conversion

| Unit | Definition | Equivalent in other unit |
|---|---|---|
| Erg | Work performed by a force of one dyne moving its point of application one cm. | |
| Joule | | $10^7$ ergs |
| Joule (electrical) | Energy developed when one coulomb of electrons (10.364 X $10^{-6}$ moles) passes through a potential of one volt. | 0.2390 calorie |
| Calorie | Energy (heat) required to warm one g of $H_2O$ from 25° to 26°C | 4.1840 joules |
| Electron volt | Energy gained by an electron passing through a potential of one volt | 1.6 X $10^{-12}$ ergs 1.6 X $10^{-19}$ joules 23.06 Kcal per mol |
| Quantum | Energy of an electromagnetic quantum calculated from its wave length expressed in millimicrons, m: eV= 1,239.8/m | 1,239.8 $m^{-1}$ ev |
| Kilo calories per Einstein | Energy of mol of photons (one Einstein) of wavelength in m: Kcal per mol = 28,589.7/m | 28,589.7 $m^{-1}$ Kcal per mol |

The different compounds yield different quantities of energy on combustion. When combustion takes place at constant pressure, the heat change is called the enthalpy change. The heat liberated in the combustion of ethane is higher than its derivative like alcohol, aldehyde and acetic acid. The later can be derived from ethane by one stage of oxidation. Heats of combustion represent the maximum energy from a substance by breaking and oxidizing its elements. The combustion of glucose in the living system is carried out by a series of oxidation process and ultimately liberates energy water and carbon dioxide. The energy is liberated step by step.

Chemical reactions are always associated with energy change. Some reactions proceed with heat liberation and these are called exothermic. Other reactions absorbed heat from the surrounding and these are called endothermic. Reactions that absorbed energy proceed only when energy (heat) is supplied. Many processes associated with break down of food stuffs are exothermic.

### Free Energy

Free energy is the energy of the system capable of doing work. The capacity of the chemical system to do useful work often depends on the downhill flow of electrons; $\Delta F$, the free energy available for such work can be measured in potential energy unit just as the ability to do work of falling water and just as the potential difference in the case of the aqueous system depends on the height of the water fall. So in the chemical system the free energy difference ΔF is the difference between the sum of the free energies of all the reactants ΣF and those for all the products, ΣFp

$$\Delta F = \Sigma Fp-$$

Free energy F is a property of a system just are as volume and mass; once the size and certain environmental conditions have been fixed, F is accurately defined. Neither the previous history of the system not the route where by a chemical system of energy $\Sigma F_R$ approaches another one of energy ΣFp, plays a part in determining the magnitude of ΔF. The change $\Sigma F_R \longrightarrow \Sigma Fp$ with consequent change ΔF in the free energy, is the thermodynamic quantity of much significance. Because its sign indicates whether a reaction theoretically able to proceed spontaneously. If the reaction is exergonic the magnitude of -ΔF indicates the maximum energy available for the work. In addition the driving force -ΔF of reversible reaction, declines as the mixture of reactants products move towards the equilibrium.

Although all chemical reactions are reversible in theory, the term irreversible is used in thermodynamics to designate those reactions which proceed spontaneously, the reactants being virtually converted to the products. As such reactions are usually characterized by large equilibrium constant ΔF becomes a large negative number. The chief biological significance of the determination of the free energy change lies in the fact that the negative ΔF of an exergonic reaction provides a measure of the maximum amount of work obtainable from this reaction. The positive ΔF of an endergonic reaction provides an estimate of the minimum amount of work to drive this type of reaction.

## REQUIREMENTS OF BIOENERGETIC SYSTEM

The fundamental requirement of a bioenergetic system is the sources of energy. This energy is a compound capable of being degraded to products of lower potential energy. As the most energy yielding reactions are oxidative, the materials are oxidizable foodstuffs. Secondly, the enzymatic mechanisms are needed to degrade the foodstuffs. There should be a facility for the collection and storage of the liberated free energy. The phosphorylated compounds store and distribute the energy. The final mechanism of utilization of stored energy is the conversion of free energy into such type of work as muscle contraction, secretion, synthesis etc.

## SCHEME OF BIOENERGETICS

The figure (3.3) gives the scheme of bioenergetics in the total metabolism of organism. Individual reactions are also shown in the figure. Degradation of foodstuffs of high potential energy to products of low energy is shown in coupled to a mechanism for collection and storage energy. This mechanism consists of the conversion of inorganic phosphate into high energy phosphate compound. This high energy phosphate compound can in turn be degraded back into inorganic phosphate after releasing energy. Another coupling mechanism ensures that this energy is made into useful energy.

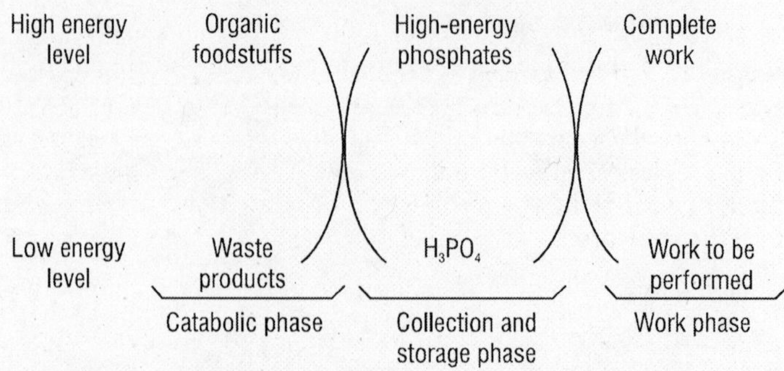


**FIG. 3.3:** Scheme of bioenergetics.

Catabolism of food stuffs may be divided into three phases. Firstly, food materials of high molecular weight (Polysaccharide, lipids) are digested to products more easily absorbed form into the alimentary canal. These reactions are hydrolytic and the small amount of energy liberated is not recoverable but is lost as heat. Products of this stage of catabolism include mainly common food sources like glycerol, fatty acids and many amino acids.

Second phase of catabolism converts this diverse collection of compounds into small number of intermediates for final oxidation in the next stage. Hexose and fatty acids are firstly activated by the formation of esters with phosphate and co-enzyme-A respectively. At this stage some free energy is liberated. In addition to the anaerobically producing lactic acid or alcohol depending upon the organism, lactic acid formation (glycolysis) makes available about 50000 calorie and alcohol formation (fermentation) about 62000 calories and the rest of the energy is available during later oxidation. Higher animals in general can oxidized glucose completely to carbon dioxide and water and liberate much free energy. Under aerobic conditions pyruvic acid rather than lactic acid is the main product and this is further oxidized to acetyl coenzyme A. Acetyl coenzyme-A is also produced from fatty acid. The glycerol of fat joins the carbohydrate path at the triose state.

Certain aminoacids are catabolized via special pathway, but most are initially deaminated to form the corresponding keto acid. The nitrogen of amino group follows its own pathway to eventually appear in the urine has urea and ammonia. Sulphur contained in the aminoacid (e.g. methionine) is oxidized to inorganic sulphate. The carbon skeletons of a few aminoacids are converted to ketone bodies or acetyl co-enzyme-A. But many aminoacids either yield pyruvic acid, $\alpha$-keto glutaric acid or oxaloacetic acid directly on deamination or form products readily convertible to these $\alpha$-ketoacid. During this second phase a small amount of recoverable free energy is liberated.

During third phase all of the products of the second phase are channeled into a single mechanism known as tricarboxylic acid or Kreb's cycle. This is a sequence of reaction which releases the carbon dioxide, while it feeds pairs of their hydrogen atom into oxidative chain to form water.

Free energy liberated or consumed in any reaction is the difference between the free energies of formation and reactants in the reaction. It has been possible to calculate the free energy change associated with many biochemically important reaction. Depending upon the type and number of coupling system, it is possible to estimate the amount of free energy recoverable by the cell from a given oxidation.

## OXIDATION-REDUCTION POTENTIAL

The importance of oxidations-reductions lies in the fact that they comprise all known energy yielding reactions which living organisms have at the disposal when compared to other reactions (hydrolysis, condensation etc.). Oxidations furnish the most energy per mole of compound. This is how the organisms of all levels have retained oxidations for reductions as source of energy during the long evolutionary period. The hydrogen atoms locked up in the food material, are oxidized from various linkages to carbon, oxygen and other elements in the organic compounds. In the oxidation-reduction process the hydrogen and oxygen are brought together in gradual step by step fashion; the energy is liberated in each step. The study of biological oxidation-reduction is largely a study of individual step by which hydrogen is brought together with oxygen to form water ultimately. Thus, biological oxidation is essentially a reverse process of photosynthesis as follows:

$$6CO_2 + 6H_2O \xrightarrow[\leftarrow \text{Biological oxidation}]{\text{Photosynthesis} \rightarrow} C_6H_{12}O_6 + 6O_2$$

Contrary to the earlier held view of the direct combustion of oxygen with the carbon, the intake oxygen is used primarily to form water with the hydrogen atom of the substrate. The oxygen contained in the $CO_2$ is derived from the substrate itself or from the body water by an indirect route.

**Oxidation-reduction potential and electron transfer**: Bio-chemically oxygen has the highest oxidation potential (electron affinities) of all systems in the living cell while the lowest values are assigned to the hydrogen atoms attached to the various substrates. Other substances have potential intermediate between these two extremes. The electrons are transferred from systems of lower potential to high potential. A coupled oxidation-reduction reaction can occur between two systems are follows:

System 2 has greater potential than system 1, hence can accept electrons from it. The reductant system 1 provides these electrons by changing to its oxidant form, while the oxidant system 2 accepts the electrons and in so doing changes to its reductant form. It should be noted that the effective oxidation potential of a given system may be higher or lower than its standard potential. This is due to variation in the oxidant-reductant ratio. Thus, if conditions within a cell maintain a high ratio NAD/NADH, then this system may well act as an electron acceptor (Oxidant). The amount of energy liberated in transferring electrons from one system to another is determined by the difference in oxidation potential between two systems. Electrons derived from metabolites such as malate or succinate move from one carrier to the next along the electron transport assembly to oxygen which is the final electron acceptor. The flow of electrons is always towards the oxygen, because electrons in the water molecule can be regarded at a lower energy level. As the electrons are transferred in steps from malate to oxygen, they are descending an energy gradient like a downhill task.

The tendency for substances to undergo oxidation reduction reactions differs considerably and depends on the relative with which they accept or give up electrons. This tendency can be measured and can be referred the ability of hydrogen gas is oxidised to hydrogen ions.

$$H2 \longleftrightarrow 2H^+ + 2e$$

**TABLE 3.3:** Oxidation-reduction potentials of important cellular reactions

| Oxidation-reduction reaction | E'o (25C; pH 7.0) (V) |
|---|---|
| Oxidised form ⟶ reduced form | |
| Succinate ⟶ α-ketoglutaric acid | -0.67 |
| $CO_2$ + α-ketoglutarate ⟶ isocitrate | -0.48 |
| Ferredoxin (Ox) ⟶ (red) | -0.43 |
| NAD ⟶ $NADH_2$ | -0.32 |
| Riboflavin ⟶ reduced riboflavin | -0.21 |
| Acetaldehyde ⟶ ethanol | -0.20 |
| Pyruvate ⟶ lactate | -0.18 |
| Oxaloacetate ⟶ malate | -0.17 |
| Fumarate ⟶ Succinate | +0.03 |
| Cytochrome b $Fe^{3+}$ ⟶ $Fe^{3+}$ | +0.04 |
| Ubiquinone (COQO ⟶ $COQH_2$) | +0.10 |
| Cytochrome C Fe3+ ⟶ Fe2+ | +0.22 |
| Cytochrome a $Fe^{3+}$ ⟶ $Fe^{2+}$ | +0.29 |
| $Cu^{2+}$ ⟶ $Cu^{+}$ | +0.35 |
| Oxygen, $y_2O_2$ ⟶ $H_2O$ | +0.82 |

The compounds in the upper part of the table (3.3) lose electrons more readily than the hydrogen gas, while the compounds in the lower part have lesser capacity to become oxidised. For each substance the tendency of its reduced form, to give up electrons and so become oxidised is called as the standard redox potential Eo and can be measured in Volts. The standard conditions are needed, because actual redox potential varies with concentration and pH, E'o refers to a molar solution containing equal quantities of the oxidised and reduced forms of the compound at pH7. When E'o is negative the substance tends to become oxidised more readily than hydrogen gas. A high negative E'o indicates a high electron donating or reducing capacity. On the contrary the substances with positive values of E'o tends to become reduced because they readily accept electrons and act as oxidizing agents.

The values of E'o for a number of donors and carriers of electrons, are presented in the table 3.3. The reduced forms of compounds at the upper side of the table loose electrons to the oxidized forms of those lower forms i.e., to those with more positive value of E'o. As electrons move down to low every levels of carriers with more and more redox potentials, a small amount of energy is dissipated as heat; but much energy is transferred to other systems.

Riboflavin as flavoproteins has wide range of oxidation-reduction potentials.

The energy change is associated with the movement of electrons down the redox gradient towards oxygen. This can be calculated in the following manner. When the electrons move from NADH to oxygen the distance in the redox scale is from -0.32 to +0.82 V, so the potential difference is E'o is 1.14V. The energy change can be said in calories. The equation:

$$\Delta G^\circ \text{ (in kcal)} = \frac{\text{n X } 96.500}{4.18 \text{ X } 1000} \text{ X } \Delta E^\circ$$

n is 2, two electrons are transferred therefore,

$$-\Delta G^\circ = \frac{2 \text{ X } 96.500 \text{ X } 1.14}{4.18 \text{ X } 1000} = 52.6 \text{ kcal.}$$

Since the reaction is exergonic, 52.6 kcal of energy is made available when one mole of NADH transfer its electrons.

Another important example is succinate oxidation and the oxygen is the final electron acceptor.

$\Delta E^\circ$ = 0.82-0.03 or 0.79V.

The number of electrons involved is again 2, and the amount of energy released:

$$\Delta G^\circ = \frac{2 \text{ X } 96{,}5008 \text{ X } 0.79}{4.11 \text{ X } 1000} = 36.5 \text{ kcal.}$$

The reaction is exergonic and small amount of energy is released compared to NADH is the electron donor. It is therefore expected that a fewer molecules of ATP are produced when succinate is oxidized.

The malate can reduce NAD, since its standard redox potential is lower than that of NAD on the redox scale. The malate, lactate and ethanol are able to reduce NAD in the presence of suitable dehydrogenases. The actual redox potential operative in vivo are slightly greater than the standard values. Thus when there is more reduced form (e.g. malate) than oxidized form (oxaloacetate), the redox potential is more negative than the standard of E'o. In the case of succinate/fumarate system, the redox potential is too low on the scale to reduce the limited amount of $NAD^+$. This carrier has a power potential than the system $NAD^-$/NADH.

## PLANT GROWTH: MODELING

Crop models are becoming important tool to asses the crop growth and development with respect to different factors that limit productivity. A model is any representation of a real system. On the other hand, a scientific model is the numerical quantification of the relationship in the form of a single equation or a set of equations. In other words model is an organised and refined expression of knowledge with regard to factors interacting with each other. They help (1) interpreting experimental results, (2) as an agronomic tool and (3) as a grower tool. Crop models are being formulated with the goal of their ultimate use by the grower.

**Types of Models**: The models can be broadly classified as-

(i). Relatively untested models, (ii). Partially tested models (Theories), (iii). Well tested models (Laws). The various types of models include-

1. **Empirical** (or correlative) model which describe relationship between variables without referring to any underlying biological or physical structure that may exist between the variables as has been discussed by Reynolds and Acock (1985).
2. **Mechanistic** (or explanatory) models attempt to explicitly represent causality between variables. In such types of models, mathematical functions represent the known or hypothesized mechanism that relates the variables and explain their observed behaviour. Many crop models infact contain a mixture of empiricism and mechanism although distinction between the two is very important. For example, a mechanistic crop model would be considered

empirical by a modeler working at a cellular level. Nevertheless, the amount of mechanism in a model is an important determinant of its utility.

In addition, the models may be deterministic or stochastic and dynamic or static. Deterministic models are those in which output is defined once the input is known. Most models in physiological ecology are deterministic mainly because of their greater simplicity and convenience but some stochastic models have been used to simulate random weather sequences, light penetration in canopies, spread of pathogens or ovules etc. Dynamic models include treatment of the time dependence of a process and are therefore particularly appropriate for simulating processes such as plant growth and yield production that integrate development and environmental changes over a long period. These include simple empirical and a complex physiological relationships.

Physiologically and physically based simulation models are those mechanistic models where plant or soil processes can be physiologically, physically, or chemically described.

Crop simulation models are gaining more importance because they behave like real crop in the field. An extensive work is going on all over the world to build the crop simulation models. Computer based simulation modeling of agriculture ecosystem is rapidly expanding scientific methodology for two reasons, on the one hand the volume of scientific knowledge on agricultural production systems in relevant scientific disciplines is expanding rapidly. Secondly, computer based simulation and modeling is an efficient tool to integrate this knowledge and to apply it to practical problems. System analysis and simulation modeling can be applied to a wide range of research questions related to improved crop production and to more efficient resource management. It can provide quantitative insight in to the interactions between the crop, its environment and its management. Among the crops which have been subjected to this type of modeling are citrus, peanuts, sugar beets, corn, sorghum, wheat, rice, barley, white potato, cotton, soybean and alfalfa.

## Models on Crop Physiology and Morphology

### *Primary Processes*

The maximum importance is given to the carbon fixation which may be due to the reason that it is the process, which is often researched and is therefore quite well understood but mainly because a mass balance is essential for yield estimates. Water uptake receives almost as much attention.

### *Light Interception*

Photosynthetically active radiation (PAR) supplies plant with the energy required for $CO_2$ fixation. It is fairly obvious that the amount of light intercepted depends on leaf area of the crop and this is usually expressed as leaf area of the crop. In 1953, Monsi and Saeki showed that the decrease in light flux density with depth in canopy approximately follows Beer's law. Beer's law strictly applies to the light interception by small particles that are uniformly distributed; as for example in a fog or muddy water, while leaves in a canopy do not meet these criteria, the equation works quite well as long as the canopy is closed i.e. the leaves of the plants in adjacent rows are touching. Sometimes part of the light reaching the field falls on the ground without ever passing through the crop canopy. There have been several attempts to model this situation by defining the outline of the crop canopy and applying Beer's law only to those rays of light which pass through it. Numerical integration to calculate overall crop light interception was used but this was expensive in computer time and probably unnecessary since there are approximate analytical solutions for some conceptual models. Most of these interpretations of the model contain equations to deal with both direct and diffuse radiation. Baker et al. (1978) found that proportion of light captured by the canopy each day was directly proportional to the heights of the plants divided by row spacing. This simple relationship holds during the period of canopy development but naturally fails during canopy defoliation at the end of growing season.

### *Carbon Dioxide Fixation*

It has been observed that crop dry weight gain is approximately proportional to total canopy light interception. This relationship is used in some models (e.g. SORG GOSSYM; GLYCIM; IRRIMOD; RICEMOD; CERES-WHEAT; POTATO; SOYGRO; CERES-MAIZE etc.) to predict dry weight gain but suffers from the problem that the parameter values need to be verified from one year to next. Because of this problem, most crop models deal mechanistically with the processes involved with the $CO_2$ fixation. In nearly all the cases, $CO_2$ fixation shows a hyperbolic dependence on light interception and regular hyperbola is often used to describe the relationship. It may not be necessary to deal separately with maintenance respiration for all the crops. Also the idea of fixed maintenance requirement for the crop gives rise to the possibility that maintenance respiration may exceed the amount of substrate available to support it. To avoid this situation, Thornley (1977), has proposed a model in which structural material that is degraded becomes part of supply of substrate available for growth. When this substrate is used for growth, the resynthesis is accompanied by the usual growth respiration. Growth respiration is associated with synthesis of protein, oil, and other plant biochemicals. It is closely related to amount of substrate available for growth and hence to net $CO_2$ fixation rate. When growth respiration is modeled separately, it is normally made a function of the compounds being synthesized by the plant at that time using the estimates of Penning de Vries (1975).

### *Partitioning*

Modelers Wilkerson et al. (1981) use empirical partitioning factors which according to them vary from stage to stage in the plants development. Other modelers partition carbon according to the potential of the various organs, on the plant for the growth (e.g. Baker et al., 1983). This approach is often linked to a system of priorities for various organs, so that, for example, seed may have first call on carbon available. Acock and Allen (1985) have suggested that plants have the following order of priorities for carbon use; (1) survival (2) reproduction (3) growth of existing organs (4) increase in number of organs (mostly by branching) and (5) storage of excess carbon for future use.

### *Growth and Development*

Horie (1991) developed a dynamic model (SIMRIW) for simulating rice growth and yield in relation to weather and climate on the basis of rational simplification of the individual physiological and physical processes of crop growth and development so that it may be universally applicable with a minimum set of crop parameters. The model does not predict the actual yield but gives potential yield that may be expected under a given climate from the physiology of a given cultivar grown under optimal conditions. Evaluating it, a close correlation was found between predicted and actual grain yield in Japan (1978-89). Kannegante (1991) parameterized the previously developed process based, warm season annual grass growth model (WANGRO) for maize and *Sorghum bicolor*. Dry matter production, crop emergence, canopy light absorption and shoot growth, above ground plant parts dry matter partitioning, evapotranspiration and growth were modeled. The model detail was limited to facilitate the models use in management. Only three crop specific parameters were modified from sowing to physiological maturity. Model yield predictions agreed well with field data from New York: predicted dry matter yield were within 15, 13 and 12% of observed yield for maize grain, maize silage and *S. bicolor* above ground biomass respectively. Yavtushenko et al. (1990), on the basis of mathematical modeling predicted the barley yield in relation to soil moisture and nitrate nitrogen and moisture contents in the top 50 cm layer than in top 30 cm layer. Optimum combination of soil moisture content and nitrate nitrogen content at which yield were highest were discussed. Lee (1991) developed a two dimensional non parameteric model using daily mean temperature and day length as predictor variable and estimated daily development rate between seedling emergence and heading for 26 rice cultivars using field data. Compared with 3 parametric models, the non parametric models gave better prediction of heading date.

Klepper and Rouse (1991) gave the procedure to reduce parameter uncertainty for complex models by comparison with real system output illustrated on potato growth model. For a simulation with increased irradiance, the model predicted the similar yields with similar uncertainty ranges as for calibration year; for a simulation with increased temperature predicted yield was much lower and relative uncertainty much greater.

Kleemola (1991) examined the effect of temperature on the phasic development of spring wheat in northern conditions using 3 cultivars grown in Finland. Using four mathematical models to simulate the length of development phases (sowing to heading, heading to yellow ripening and sowing to yellow ripening) accuracy of models was compared. Mathematical simulation of the sowing to yellow ripening period was more accurate than for the other periods; deviation with respect to the over assured length of the phase ($M_D$) was 2.7 days per year on average. Simulation of sowing to heading period ($M_D = 3.3$) was more accurate than for heading to yellow ripening period ($M_D = 5.3$). Mc Master et al. (1992) developed another simulation model named SPIKEGRO to simulate winter wheat shoot apices growth and development in a stand SPIKEGRO emphasizes the reproductive functioning of the shoot apex. The complete developmental sequence of the shoot apex is outlined and quantified. Spikelet and floret primordium initiation, growth and abortion, ovule fertilization and growth, and rachis and chaff growth were simulated on morphologically identified culms. The phyllochron interval rather than growing degree days, is used throughout the model to increase flexibility on predicting yearly and within stand variability in development. Validation results and sensitivity analysis suggested size modifications that should improve model relation and predictions. Most of the modifications were easy co-relations of simplified algorithms.

The CERES-Maize, CERES-Maize SAT and CORNF growth models were compared for accuracy and consistency in predicting various maize growth and yield parameters; to develop unified model capable of assessing the joint impact of shatter cane (*Sorghum bicolor*) and European corn borer (*Ostrinia nubilalis*) infestation on maize growth and yield (Retta et al., 1991). Simulated LAI, vegetative, grain, and total biomass and yield components were compared. CERES-Maize modified for leaf growth and phenology computation (VO/SAT) gave more accurate prediction of date of silking than CORNF or the original CERES-Maize. VO/SAT showed adequate sensitivity to reduction in light and water.

## *Conclusion*

Briefly, a model allows the synthesis and mobilization of existing knowledge about individual processes leading to yield and help pinpoint deficiencies of knowledge. A model that works can calculate crop response to environmental changes. Passioura (1973) has argued that models should be simple, be readily testable in final yield and in subsystem calculations, and amenable to improvement after experimental testing. Although, variation in yield to a greater extent is due to weather yet, the effects of management practices, soil and biotic factors must also be recognized. As a general strategy, the crop model should focus on factors responsible for most of the yield variations, and should address factors that can be manipulated by either farmer or breeder for increasing the productivity.

# CHAPTER 4

# Water Relations: The Dynamic Scenario

*Everything originated in the water, and everything is sustained by water.*
***Goethe***

## INTRODUCTION

All life-forms on earth depends on water. The origin of water on the earth is not clear so far. However, the current presumption is that the primordial earth had no oceans, and perhaps very little atmosphere. It is believed that the volatile constituents bound in the earth's crust, oozing to the surface through volcanoes, rock movements, and hot springs, condensed to form the ocean and the atmosphere. This way, perhaps this remarkable combination of hydrogen and oxygen - called water - came into being and eventually became an indispensable component of the earth's environment.

Water is not only essential to life but is the predominant inorganic constituent of living matter, forming in general nearly three quarters of the weight of the living cell. Organisms which contain relatively small amounts of water are generally in dormant state or show very slow development; seeds and certain invertebrates that live in arid environments are the examples. On the other hand, high rain fall over a land mass invariably means a large biomass per unit area.

Of the three states in which water occurs in the ecosphere gas, solid and liquid-only the last is an indispensable resource. The various forms of water are found in every section if ecosphere: the atmosphere, the lithosphere and the hydrosphere (a term designating the entire system of seas and inland waters).

In one way or the other plants use all available sources- inland water, ground water, and even ocean water. Water is the most abundant constituent of the higher plants. Living tissues of plants usually contain more than 70 per cent by weight of water; the actual percentage varying in different tissues, for example, being less in the xylem than in the phloem. Maintenance of satisfactory water contents is essential for the plants to remain alive. Even quite small variations in plant-tissue water-content can markedly influence processes of growth and metabolism.

The water can exist in three forms: as the solvent of a true solution in the vacuole, as the dispersal medium for the colloidal system in the cytoplasm, or as water held in the pores of the cell wall by capillary forces. The factors which affect the distribution of water between these phases differ. The most successful land plants of today can sustain continued water loss from aerial parts by maintaining an efficient uptake of water from the soil. Their tissue water balance is thus a dynamic one, a continual balancing of loss and gain. Therefore, in modern approach to plant water relations, much emphasis has been given on dynamics and movement of water.

## WATER: SOURCES

Water can be divided into four well marked classes:

### Atmospheric water

Rain water and water formed by snow are grouped under atmospheric waters. As the rain and snowfall, they wash the dusty-atmosphere and bring down dust, soot, and other suspended particles to the earth. Along with these non-living particles atmospheric water also carries air-borne bacteria to the earth's surface.

### Surface water

As soon as rain drops and snow flakes reach the surface of the earth, they quickly become contaminated with soil- micro flora of the area. Such waters are a few microorganisms but as it flows from higher regions to lower regions, it contains large amounts of organic matter and the population of microorganisms increases considerably. Such water facilitates the growth of saprophytic species and protozoa. River water shows their highest count during rainy season.

### Stored water

Water which is present in ponds, reservoirs, lakes or ocean for a considerable period is called stored water. The microbic population increases greatly in such waters.

### Ground water

When rain falls on the earth, two forces act on it. One is the earth's gravity, which pulls the water downward through any open path. The second force is the electrical attraction between water and the other materials to which water tends to stick. That is why water makes the things wet. The movement of water upward through small holes is called **capillary action.**

When rain drops fall on the dry soil they will wet the soil particles. As the rain continues after all the land surface will wet, gravity will pull the excess of water down through the spaces in the rock, sand, or gravel below. The larger the underground spaces, the faster the water will flow down. Finally, the downward flow of water will be stopped when the water meets rock that has no porosity. Since the water can go no further, it blocks up, filling all the pores in the rock above the barrier. This causes saturation of water in the soil and the zone is called **saturation zone**. The upper boundary of the zone of saturation is called the water table. Below the water table the rock is saturated; above it the rock may be moist but is not saturated.

Soil water is normally surrounded by a thin film of water. It is from these films that plant roots take up all the water they need. This water can be in the chemically combined or uncombined form. In the former, it exists as water of crystallization of mineral grains and water of hydration of clay mineral particles. Both of these kinds of combined water are not at all available to the living organisms. Uncombined water is not chemically combined with the soil particles and is held in **soil adhesion** (attraction between soil particles and water molecules) and **cohesion** (attraction between molecules of water).

## WATER IN SOIL

Consider a soil, originally fairly dry, that is recently wet from rain or irrigation. Immediately after wetting, the soil is nearly saturated in the wetted zone, and it is said soil is nearly saturated in the wetted zone, and it is said to be at **saturation**. Much of the water (called **gravitational water**) moves down through the pore spaces between the soil particles. If the concentration of ions is dilute (true of most soils), the water potential of this water is quite high, usually about -0.03 MPa. After several hours, or even a day or so, the only water left in the soil is that which can be held against the force of gravity by the adhesion between the water molecules and the soil particles. This remaining water exists in capillary pores that are often wedge-shaped and seldom cylindrical. Soil that contains all the **capillary water** it can retain against gravity is said to be **field capacity**.

As water is removed from the soil by drainage, evaporation, and plant roots, the remaining water is held in ever-thinner films, so its water potential becomes increasingly negative. When soil and root water potentials are the same, the roots no longer remove water from the soil, but water continues to transpire from the shoots, so the plant wilts. Even if transpiration is stopped by enclosing the shoot in an atmosphere with nearly 100 per cent RH, it is still not able to obtain enough water to overcome wilting; the amount of soil water is then said to be at the **permanent wilting percentage**. Soil scientists rather arbitrarily consider soil with a water potential of -1.5 MPa to be at the permanent wilting percentage, although most plants, especially those of the desert and those that can grow in salty soils, can remove water when the soil water potential is more negative than -1.5 MPa to be at the permanent wilting percentage, although most plants, especially those of the desert and those that can grow in salty soils, can remove water when the soil water potential is more negative than -1.5 MPa. The minimum plant water potential is apparently determined by the osmotic potential of plant cells and the designed leakage in the xylem. This minimum plant $\Psi$ determines the permanent wilting point.

## PLANT, WATER AND SOIL

Soil acts as a reservoir of water for the plants to use through the root zone system. When rainfall is inadequate or untimely, application of irrigation water becomes essential for successful growth of crops. And it enables satisfactory functioning of other crop production requirements, viz. solar radiation and plant nutrients.

The factors influencing plant growth and productivity may be grouped as follows:

1. **Soil factors:** soil moisture content, texture, structure, density, salinity, fertility, aeration, temperature and drainage.
2. **Plant factors:** type of crop, density and depth of rooting, rate of root growth, aerodynamic roughness of the crop, drought tolerance and varietal effects.
3. **Weather factors:** sunshine, temperature, humidity, wind and rainfall.
4. **Miscellaneous factors:** soil volume and plant spacing, crop and soil management.

Water is retained around the soil particles and in capillary pores by the forces arising out of surface tension and presence of salts. A plant has to overcome the retentive forces to absorb water from the soil moisture reservoir. The upper limit of water retention is limited to the **field capacity** of the soil, where ground water table is sufficiently low. When the soil texture is more of clay content, the field capacity value increases. Time required for draining off of excess soil water also depends on soil texture. When plants absorb moisture the amount is reduced to a **permanent wilting point**, at or below which plants no longer absorb water and survive, even though in physical terms, there may be some moisture present in the soil. The moisture content between field capacity and permanent wilting point is known as **available soil moisture regime** for plant growth.

Humid conditions enable plants to draw soil moisture almost up to the permanent wilting point, whereas in hot and dry weather conditions, plants may show wilting symptoms in spite of enough available water present in the root zone. Only 5% of total water absorbed by plant is utilized for physiological functioning. Major part of the absorbed water is lost in transpiration. Water absorption, conduction and translocation and transpiration are the main physiological processes in plant growth.

### Functions of Water in Plants

In living tissue, water is the medium for many biochemical reactions and extraction processes. Inorganic nutrients, photosynthate, bases and hormones are all transported in aqueous solution; evaporation of water can control the temperature of skin or leaf. Soil nutrients are available to plant roots only when dissolved in soil water. In short, water is essential for life and plays a unique role in virtually all biological processes. The significance of water to the life of the higher plants can be emphasized best by enumerating some of its functions.

1. It is a major constituent of protoplasm.
2. It is the solvent in which mineral nutrients enter a plant from the soil solution. Also, water is the solvent in which mineral nutrients are transported from one part of a cell to another and from cell to cell, tissue to tissue, and organ to organ.
3. It is the medium in which many metabolic activities occur.
4. It is a reactant and product in a number of metabolic reactions (e.g., certain reactions in the tricarboxylic acid cycle and photolysis of water).
5. In photosynthesis the hydrogen atom in the water molecule is incorporated into organic compounds and oxygen atoms are released as $O_2$.
6. It imparts turgidity to growing cells and thus maintains their form and structure. In fact, water can be regarded as a material that provides mechanical support and rigidity to nonlignified plant cells.
7. Gain or loss of water from cells and tissues is responsible for a variety of movements of plant parts. Included are the diurnal swelling and shrinking of stomatal guard cells, the nocturnal folding of leaflets of certain plants, the opening and closing of flowers of certain plants at various times of the day, and the sensitivity to touch of leaflets of plants, such as the sensitive plant (*Mimosa pudica*).
8. The elongation phase of cell growth depends on absorption of water.
9. It is a metabolic end product of respiration.
10. Considerable amount of water is absorbed by plants and greater amounts of water are lost (as water vapor) by plants than any other substance.

## CELLULAR WATER

Large fraction of Plant cell water is present in the form of vacuolar water that is probably quite similar in nature to a dilute salt solution in a beaker except that it is usually subjected to a hydrostatic pressure of several bars. In contrast, a small but sizable fraction of the water in plant cells is adsorbed to various plant cell constituents by dipolar and hydrogen bond forces. Especially important in this regard is the water held in microcapillaries in protoplasmic colloids and cell wall components. Moreover, the surfaces of all protoplasmic membranes can be visualized as encased in a one molecule thick sheet of water molecules; these water molecules can be considered integral components of protoplasmic membranes. Also, the surfaces of cellulose microfibrils and polysaccharide colloids in cell walls can also be considered coated with water molecules.

Water molecules that adhere to the surfaces of cellular constituents, such as proteins, protoplasmic membranes, and cellulose microfibrils, can exchange readily with water molecules in other parts of the plant cell.

## WATER: PHYSICOCHEMICAL PROPERTIES

Water acts as universal solvent because of unequal distribution of charge on a water molecule. It is a continuous medium in which most biochemical reactions take place. The unique biochemical properties of water are a reflection of its exceptional physicochemical properties.

### Polar Properties of Water

The physical and chemical properties of water permit life to exist on our planet. Water molecules are **polar;** that is, they bear a partial positive and a partial negative charge. The water molecules in liquid water and in ice are held together in part by hydrogen bonds. The hydrogen atom of one water molecule, with its partial positive charge, is attracted to the oxygen atoms of a neighboring water molecule, with its partial negative charge, forming a hydrogen bond. Each water molecule can form hydrogen bonds with a maximum of four neighboring water molecules.

### Properties of Water as a Solvent

Because its molecules are polar, water is an excellent **solvent**, a liquid capable of dissolving many polar substances. For example, the ions of a salt, such as sodium chloride, are held together by strong ionic bonds; in fact, they form a crystal lattice. Considerable energy is required to pull the positively and negatively charged ions apart. However, when the sodium chloride is placed in water, the strong electrical attractions between the polar water molecules and $Na^+$ and $Cl^-$ ions result in the formation of stable dissociated $Na^+$ and $Cl^-$ ions. Because of its solvent properties and its tendency to cause the ionization of compounds in solution, water is important in facilitating chemical reactions.

### Cohesive and Adhesive Forces

Water exhibits both cohesive and adhesive forces. Water molecules have a very strong tendency to stick to each other; that is, they are **cohesive.** This is due to the hydrogen bonds among the molecules. Water molecules also stick to many other kinds of substances (i.e., those substances that have charged groups of atoms or molecules on their surfaces). These **adhesive** forces explain how water makes things wet. Water has a high degree of surface tension because of the cohesiveness of its molecules; its molecules have a much greater attraction for other water molecules than for molecules in the air. Thus water molecules at the surface crowd together, producing a strong layer as they are pulled downward by the attraction of other water molecules beneath them.

Adhesive and cohesive forces account for the tendency, termed **capillary action**, of water to rise in very fine tubes. Capillarly action plays some part in the rise of water through the stems of plants to their leaves.

## Temperature Stabilization

Other properties are responsible for water's ability to minimize temperature changes. Water has a high **specific heat**. It is the amount of heat required to raise the temperature of 1 gram of water 1°C. The high specific heat of water is due to its hydrogen bonds. For the temperature of a substance to be raised, heat energy must be added to make its molecules move faster- to increase the kinetic energy of the molecules. Before water molecules can move freely, some of the hydrogen bonds holding the water molecules together must be broken. Thus a great deal of the heat energy added to water is used up in breaking the hydrogen bonds, and only a part of this energy is available to speed the motion of the water molecules (increase the temperature of the water).

Because so much heat input (or heat loss) is required to raise (or lower) the temperature of water, the oceans and other large bodies of water have relatively constant temperatures. Thus the aquatic environment provides the multitude of organisms that inhabit it with a relatively constant environmental temperature. The water within all organisms, even those that live on land, contributes to a relatively constant internal temperature. This is important because metabolic reactions can take place only within a relatively narrow temperature range.

Because its molecules are held together by hydrogen bonds, water also has a high heat of vaporization, another property that helps stabilize temperature. More than 500 calories are required to change 1 gram of liquid water into 1 gram of water vapor. A calorie is a unit of heat energy (defined as 4,184 joules). Because of the heat of vaporization, we can rid ourselves of excess heat by the evaporation of sweat. And plants can remain cool in the midday heat by evaporating water from their surface.

Although most substances become more and more dense as the temperature decreases, water reaches its maximum density at 4°C and then begins to expand again as the temperature decreases. Hydrogen bonds become more rigid and ordered, and ice floats upon the denser cold water. This important property of water explains why lakes and ponds freeze from the surface down rather than from the bottom up. The sheet of ice that forms at the pond surface insulates the water below from the wintry chill so that it is less likely to freeze. Further, organisms that inhabit northern lakes and ponds are able to carry on their life activities despite the frigid winter.

## Water and Surface Tension Property

Water has a high surface tension because its molecules are more closely bound together at the interspaces between water and other media. For this reason dissolved substances tend to be concentrated at the surface. Under such condition chemical reactions between various dissolved substances occur more rapidly. This property is of great biological significance because it enables the formation and maintenance of lipoprotein membranes. As the water has low viscosity, it serves as an efficient medium for the transport of substances both inside and outside the cell.

## WATER: CHEMISTRY

The chemical substances are usually classified into groups of compounds with similar properties such as salts, alcohol's and water. It would be solid (ice at room temperature, with a regular crystal structure. Clearly, liquid water is not solid at room temperature and, therefore, it is generally thought that it is made up of a large number of small clusters (or icebergs) or hydrogen-bonded water molecules separated by volumes containing free bonded molecules only. Each of these clusters, which have a rather open cage-like cluster structure is, therefore dynamic, in contrast to the static structure of crystalline solids or sugars. Water, however, is unique; there is nothing else like it. To understand the properties of water, it is best to start with its molecule, which consists of two hydrogen atoms bonded to an oxygen atom. The water molecule is planar and angled that, due to the high electronegativity of oxygen, there is a partial separation of charge leading to the formation of hydrogen bonds between molecules.

It is important to note that, although these hydrogen bonds are strong compared with other intermolecular forces (Van der Waals bonds), they are only about 1/24th as strong as O-H covalent bond.

One molecule of water is formed chemically by two atoms of hydrogen and one atom of oxygen, with two covalent bonds, holding two hydrogen atoms by one oxygen. Water is a dipole particle with positive and negative charges. The hydrogen atoms are negatively charged and the configuration of the bond angle makes electric charge on the molecule unevenly distributed. Therefore, it tends to orient itself as a dipole in an electric field. However, the bonds are weak (4.5 Kcal/mol) compared to covalent bonds (110 Kcal/mol). This property enables water to act as an excellent solvent. The charged character of water molecules permits it to form oriented layers around macromolecules such as protein that contain many charged groups. This action not only help in the separation and solution of protein but also forms as a protective field around them cutting down forces of the charged groups.

### Acid and Bases

An **acid** is a compound that ionizes in solution to yield hydrogen ions ($H^+$)[1] and an anion. An acid is a proton *donor.* Acid turn blue litmus paper red and have a sour taste. Hydrochloric acid (HCl) and sulfuric acid ($H_2SO_4$) are examples of inorganic acids. The strength of an acid depends upon the degree to which it ionizes in water. Thus HCl is a very strong acid because most of its molecules dissociate, producing hydrogen and chloride ions.

### Hydrogen Ion Concentration (pH or *Potentia hydrogenii*)

pH is a scale of intensity of acidity or alkalinity and measures the concentration of hydrogen ions in water. If free $H^+$ are more than $OH^-$ ions, the water will be acidic, or alkaline the other way round. The most chemically pure water is partly dissociated, and at a temperature of 22°C molecules of water per litre is equally dissociated into $H^+$ and $OH^-$ ions. The concentration of hydrogen ions thus being equal to hydroxyl ions. Instead of the inconvenient number $10^{-7}$ the pH value is expressed as the negative logarithm of the hydrogen ion concentration. Thus pH 7 indicates neutral water, pH. 7 to 14 alkaline, and pH below 7 is acidic.

pH of natural waters varies around, 7, generally over 7 (i.e., alkaline) due to presence of sufficient quantities of carbonates. It increases during day largely due to photosynthetic activity (consumption of carbon dioxide), whereas decreases at night due to respiratory activity. Factors like exposure to air, temperature, and disposal of industrial, wastes etc. also bring about changes in pH.

Reserve pH (RpH) is that value obtained by a second pH determination made after the sample has been thoroughly aerated. Free carbon dioxide in the sample is partly responsible for the initial pH reading; aeration causes the release of the free carbon dioxide and measurement after aeration shows a RpH value higher on the alkaline side of pH scale.

## Redox Potential

Redox potential is the measure of oxidating or reducing power of the water. In the process of oxidation a substance loses electrons while in reduction it gains electrons. As such, in a system processes of oxidation and reduction take place simultaneously. When a platinum electrode is immersed in such system, a potential, depending upon the ratio of oxidized and reduced states, is developed and called as redox potential.

The value of redox potential gives a picture of oxidation and reduction processes going on in water; a low value of redox potential indicates high reduction while a high value of it indicates low oxidation. For better aerobic treatment of waste water redox potential should be high (+200 to +600 mV) while for anaerobic treatment it should be low (-100 or -200 mV).

## Electrical Conductivity (EC)

Pure water is a poor conductor of electricity. Acids, bases, and salts in water makes it relatively good conductor of electricity. Such substances are called electrolytes; those with poor solubility in water are called weak electrolytes and those with high solubility the strong electrolytes. In other words, electrolytes in a solution dissociate into positive (cations) and negative (anions) ions and impart conductivity. Thus higher the concentration of electrolytes in water the more is its electrical conductance (i.e., lesser the resistance).

## Water Potential

As noted the term *water potential* is used by plant physiologists in place of **chemical potential of water**. The Greek letter psi $\psi$ designates the **water potential** in a system, whether a plant cell or tissue, or a sample in which plants are growing, or a solution in a beaker. Water potential is expressed either in bars (a pressure unit or megapascals (1 MPa = 10 bars). (Bars will be used in this book rather than megapascals because multiplicity of usage of these units in the literature (e.g, 760 mmHg = 1 atm = 1.013 bars).

Water potential helps the plant physiologist to assign a precise value to the water status in plant cells and tissues. The lower the water potential in a plant cell or tissue, the greater is its ability to absorb water. Conversely, the higher the water potential, the greater is the ability of the tissue to supply water to other more desiccated cells and tissues. Thus water potential is used to measure water status, water deficit and water stress in plant cells and tissues.

## HOW DO MATERIALS PASS THROUGH MEMBRANES?

### Properties of Cell Membrane

Whether a membrane will permit the molecules of any given substance to pass through it depends on two aspects: (i) the structure of the membrane and the size and (ii) charge of the molecules.

A membrane is said to be permeable to a given substance if it permits the substance to pass through and impermeable if it prevents passage of the substance. A selectively permeable membrane allows *some, but not other*, substances to pass through it. Depending on environmental conditions or cellular needs, the membrane may present a barrier to a particular substance at one time and then actively promote its passage at another.

When materials move passively through cell membranes (as they do through non living materials) by **diffusion and osmosis**. These are physical processes. However, in living cells materials can also be moved actively (require the expenditure of metabolic energy) by physiological processes, such as active transport, phagocytosis or pinocytosis.

## DIFFUSION

Diffusion may be defined as the movement of molecules from a region of their higher concentration to one of lower concentration brought about by the kinetic energy of the molecules. Some substances pass into or out of cells by simple physical diffusion. All molecules in liquids and gases tend to move (diffuse) in all directions until they are spread evenly throughout the available space. Molecules tend to move down a concentration gradient, that is, from where they are more concentrated to where they are less concentrated.

The phenomenon depends upon the random movement of individual molecules, propelled by collision with other vibrating molecules or with the side of the container. As it diffuses, each individual molecule moves in a straight line until it **bumps** into something- another molecule or the side of the container. Then it rebounds in another direction. The movement of the molecules is continued even when they have become uniformly distributed throughout a given space. The molecules may travel as fast in one direction, while others travel in the opposite direction.

### Diffusion: Rate

It is a function of the size and shape of the molecules, the temperature, and whether the medium is solid, liquid, or gas. Small particles move faster than larger ones at the similar temperature. However, heat energy causes molecules to move more rapidly, so as temperature rises, the rate of diffusion increases. The molecules of any number of substances will diffuse independently of each other within the same solution; eventually all become uniformly distributed.

Diffusion of gases may be demonstrated by opening a bottle of ammonia. After some time ammonia would have left the bottle and begun to diffuse through the air and odor would be evident throughout the room. In a closed class room (without air currents) the molecules would distribute themselves evenly throughout the room.

**TABLE 4.1:** Mechanisms for moving materials through cell membranes

| Process | How it works |
|---|---|
| **Physical Process** | |
| Diffusion | Net movement of molecules (or ions) from region of greater of greater concentration to region of lower concentration of concerned molecules |
| Facilitated diffusion | Carrier protein in cell membrane accelerates movement of relatively |
| | large molecules from region of their higher to region of their lower concentration |
| Osmosis | Water molecules diffuse from region of their higher to region of their lower concentration through differentially permeable membrane |
| Physiological Process | Molecules are transported across the membrane against concentration gradient at the expense of metabolic energy (ATP) |
| Active transport (pump) | Protein molecules in cell membrane transport ions or molecules through membrane; movement may be against concentration gradient (i.e., from region of lower to region of higher concentration of concerned molecules, and is supported by ATPase activity). |
| Endocytosis | Cell membrane encircles particle and brings it into cell by forming vacuole around it. |
| Pinocytosis | Cell membrane takes in fluid droplets by forming vesicles around them |
| Exocytosis | Cell membrane ejects materials; vesicle filled with material fuses with cell membrane |

## Diffusion in Cytoplasm

A large variety of substances are distributed throughout the cytoplasm by diffusion, and this process is also responsible for movement of many substances *in and out* of the cell across the cell membrane. Oxygen, carbon dioxide, water, and numerous other small ions and molecules can diffuse into or out of the cell readily. However, diffusion through the cell membrane is limited by both molecular size and the compatibility of substances with the lipid membrane. When carrier proteins in the cell membrane assist the passage of small solute molecules through the membrane the process is known as **facilitated diffusion**. The proteins serve as a passive conveyor belt that permits the substance to pass in either direction down a concentration gradient. The carrier proteins are through to combine temporarily with the solute molecule and to accelerate its movement through the membrane.

The uptake of substances by plant tissue can take place by physical diffusion. For a non-electrolyte the concentration inside the cell cannot exceed the concentration outside and diffusion will take place until the concentrations inside and outside are equal (i.e., the chemical potential gradient is zero). The amount of substance (ds) entering in time (dt) is given by Fick's law of diffusion:

$$ds/dt = P\,(C_o - C_i).\,A/x$$

Where A = area of the membrane, $C_o$ and $C_i$ = concentrations outside and inside the cell, and P = permeability coefficient, x = thickness of the membrane.

The quality of concentration will refer only to the form in which the substance is transported. If, after entry into the cell, the penetrating substance forms some chemical complex, or is absorbed at a surface, then the concentration of free substance inside the cell may be maintained lower than that outside, and transport into the cell by diffusion will continue, although the total concentration inside may exceed than outside. In the case of electrolytes, involving the movement of charged ions, the existence of electrical potential differences must also be taken into account. In the presence of a potential difference there will be an unequal concentration of ions on the two sides of the membrane, such that at equilibrium the electrochemical potential gradient is zero.

The study of the diffusion of un-ionized substances into the cell has shown that the rate of penetration is far less than that of water and is dependent on certain molecular properties. The correlation with molecular size indicates the existence of a porous membrane, and this is consistent with the high rate of water movement relative to that of larger molecules.

## TYPES OF MEMBRANES

1. **Permeable membrane:** This kind of membrane allow diffusion of both solvent and solute molecules or ions through them e.g., cellulose wall of cells, lignified cell walls are also quite freely permeable to water and when wet to solute molecules and ions as well.
2. **Impermeable membranes:** This kind of membrane prohibit the diffusion of both solvent and solute particles through them, e.g., heavily cutinised or suberised cell walls in plants.
3. **Semipermeable membrane:** This kind of membrane allows diffusion of solvent molecules but do not allow the passage of solute molecules e.g., membranes of collodion, parchment paper and copper ferrocyanide membrane.
4. **Differentially permeable membrane:** All membrane present in plants allow some solutes to pass through them along with the solvent molecules. This kind of membrane is known as differentially permeable, rather than semi-permeable, e.g., plasmalemma, tonoplast (vacuolar membrane) and membrane surrounding cell organelles.

## SELECTIVE PERMEABILITY AND DIFFUSION

If two solutions are separated by a porous membrane which has pores of such a size that water molecules can pass freely through it but sucrose molecules cannot, water will pass from the weaker to the more concentrated solution without the complication of sucrose moving in the opposite direction. Such a membrane would be termed a **semi-permeable membrane** (or better, a selectively permeable membrane). The selective permeability of plant membranes is only maintained under physiological conditions. Treatment with chloroform (which will extract lipid components from the membrane) or heating above 50°C, treatment with various salts, or extreme alkalinity or acidity, results in a loss of selective permeability. These treatments affect the denaturation of the protein component of the membrane. After such treatments, substances can freely diffuse in and out of the cell. Ions may also affect the permeability properties of a membrane. If two ions have opposite effects on the permeability properties of the membrane they are spoken of as **antagonistic**. A mixture of salts which has little net effect on the permeability of the membrane is spoken of as a **balanced solution**. There are two phases of penetration of either an electrolyte or non-electrolyte into cell and is found to be **biphasic**.

First phase is rapid movement into the cell and entry is to a large extent governed by the laws of diffusion and the rate of entry is not closely related to the metabolic activity of the cell. The addition of poisons or the removal of oxygen has relatively little effect. Most of the solute taken up can be removed by washing the tissue with water. It is believed that during this phase the ions or molecules are entering largely into the free space of the cell with an electrolyte, considerably more cation than anion equivalents enter, suggesting that the free space contains more immobile anions than cations. The free space has, therefore, been analysed into two parts - first, a water-free space into which cation and anion enter in equal concentration, and second, a phase containing an immobile anion which takes up excess cations. The equilibrium conditions resulting when one phase contains immobile ions was first considered in non-living systems by Donnan; hence the free space containing immobile anions is referred to as the Donnan free space. Much, if not all, of it is likely to be associated with the cell wall.

If the quantity, Q, taken up of a non-electrolyte (which can be shown neither to be accumulated nor metabolized by the cell) is measured, the volume of the water-free space can be calculated as Q/C, when C is the concentration in the external solution.

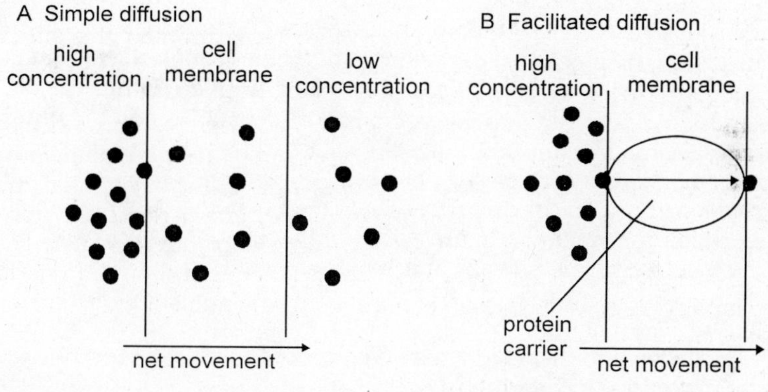


**FIG. 4.1: Diffusion** (a) Simple diffusion of $H_2O$ molecules, (b) Facilitated diffusion of solute molecules.

## OSMOSIS

Osmosis- a special kind of diffusion, is defined as the diffusion of water molecules across a differentially permeable membrane from a region where water molecules are more concentrated to one where they are less concentrated. Cell membranes selectively regulate the passage of most solutes, but water is able to move rather freely in and out of the cell.

When living cells are placed in a solution that has a solute concentration equal to that in the cells, the water molecule concentration is also equal, and therefore water molecules move in and out of cells at the same rate. The net movement of the water molecules is zero. Such a solution is known as **isotonic** to the cells (equal solute concentration).

Cells- especially of single-celled organisms- may find themselves in solutions that are of greater or lesser solute concentration relative to the solute concentration with the cytoplasm. If the solution has a greater solute concentration, it is said to be **hypertonic** (above strength) to that of the cell, whereas if it has a lesser concentration, it is **hypotonic** (under strength) compared with the cell. The terms hypertonic and hypotonic are relative to each other. A 5% solution is hypertonic to a 2% solution but hypotonic to a 10% solution.

**TABLE 4.2:** Osmotic terminology

| Solute concentration in solution A | Solute concentration in solution B | Tonicity | Solute diffusion | Solvent diffusion |
|---|---|---|---|---|
| Greater | Less | A hypertonic to B | A to B | B to A |
| | | B hypotonic to A | | |
| Less | Greater | B hypertonic to A | B to A | A to B |
| | | A hypotonic to B | | |
| Equal | Equal | Isotonic | Equal | Equal |

When we place some living cells in distilled water that contains 100% water molecules. If the total number of **solute** molecules in the cells amounted to 1% of the total molecules present, then water molecules would account for only 99% of the total. Since water molecules, like solute molecules, tend to move from a region where they are more concentrated to a region where they are less concentrated, they diffuse inward across the cell membrane. The solute molecules may diffuse in the opposite direction, then cell membrane prevents them from **leaking out** to any great extent. Instead, they may be thought of as being trapped within the cell and exerting an osmotic pressure upon the less-concentrated solution on the other side of the membrane. So much water may enter the cell that it swells and bursts.

On the other hand, when cells are placed in a solution that is hypertonic to them, water tends to flow out of them. The cells may become dehydrated, shrink, and die. When a solution contains more solute molecules, it has proportionately fewer water molecules, so the solvent and solute concentrations are reciprocally related.

Plant cells are also adapted to the hypotonic water that often bathes them. Their rigid cell walls enable them to withstand, without bursting, pressure exerted by the water that seeps in, filling their central vacuoles. As this internal pressure, called **turgor pressure**, increases, it forces water molecules back out of the cell. The turgor pressure levels off when the outward passage of water molecules equals the rate of inward movement of water molecules. When conditions become dry and the plant cell does not have enough water, the central vacuole decreases in size and the cell shrinks. Thus the plant wilts.

### Significance of Osmosis in Plants

1. It helps in water absorption by plants.
2. The turgidity of plant organs is dependent on water which is absorbed due to osmosis.
3. It helps in movement of water from one cell to another.
4. It also maintain the resistance of plants during drought and frost.
5. It helps in opening and closing of stomata
6. The expansion of cells is dependent upon turgidity

## OSMOTIC PRESSURE

The forces involved in the movement both of water and the solute are determined by the laws of diffusion. Solvent movement could be prevented by applying a pressure to the solution so that effectively the concentration (or activity) of water molecules was made equal to that in the solution. Then no movement would take place. The force required is defined as the osmotic pressure of the solution and is that force which must be applied to a solution to maintain it in equilibrium with pure solvent.

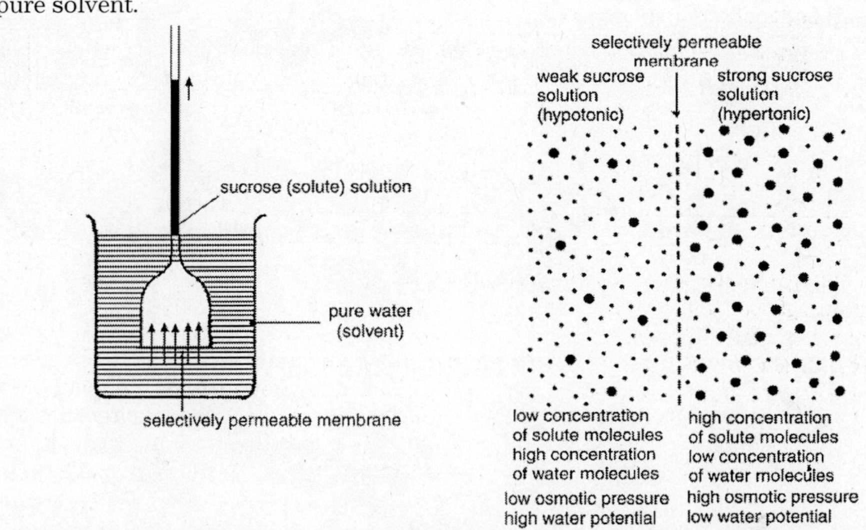


**FIG. 4.2:** Osmosis (a) A simple osmometer. (b) Molecular states across the membrane.

## Osmotic Pressure: Methods of Measurement

The osmotic pressure can be measured mechanically by determining the hydrostatic pressure required to maintain the solution in equilibrium with pure solvent. Measurements of the osmotic pressure at different temperatures and concentrations have shown that the fundamental relationships of a weak solution are of the same form as those for an ideal gas. For a gas the pressure P, Volume V and temperature T are related by PV = RT, where R is the gas constant; for a solution OP = RTC, where OP = osmotic pressure in atmospheres, and C = concentration in moles/litre and T = absolute temperature.

It follows that all molar solutions should have an equal osmotic pressure of 22.4 atoms at O°C. The osmotic properties of a solution are dependent on the number of dissolved particles and not on their chemical form. In the case of salt solutions which ionize, the number of particles is greater than the number of dissolved molecules. If the solute associates to form aggregates with itself or solvent molecules, there would be relatively fewer particles and the osmotic pressure would be less than expected. For this reason the equation is normally written OP = $i$RTC, where $i$ is the isotonic coefficient. For sodium chloride $i$ has a value near to 2 if that for sucrose is taken as 1.

## OSMOTIC RELATIONS OF A PLANT CELL

The osmotic pressure of the cell vacuole is determined by the amount of dissolved substances. On placing a cell in pure water there is an increase in the volume of the vacuole resulting from the diffusion of water from outside the cell into the salt solution of the vacuole. The cell wall, composed of microfibrils with innumerable sub microscopic interstices, is relatively permeable to water and most dissolved substances. The outer membrane of the protoplast, the plasmalemma, will in the first place be assumed to be relatively permeable to small molecules; then effectively an external solute may permeate up to the tonoplast. The *tonoplast acts as a semi-permeable membrane* and the changes of volume resulting when a cell is placed in a solution will, until the final stages, be confined to the vacuole. Ultimately, the vacuole increases in volume that the cytoplasm is pressed against the cell wall. The cell wall stretches within the limits of its elasticity but cannot expand indefinitely. Thus the final volume of the cell will depend on two opposing forces:

1. One resulting from the difference in osmotic pressure of the vacuole and the external solution.
2. The other being the wall pressure which prevents unlimited expansion of the cell.

## DIFFUSION PRESSURE DEFICIT OR SUCTION PRESSURE

The net resultant force which is the combined effect of the osmotic pressure of the cell sap and of the forces restraining the cell expansion is called the **suction pressure** (S.P.) or **diffusion pressure deficit** (DPD). It follows that S.P. = OP - WP, where WP is the force exerted by the wall. For a cell in a tissue WP will consist of the forces exerted not only by the wall of the cell under consideration but all the forces on the wall due to the surrounding cells. As wall pressure is equal to TP the equation can be written as DPD = OP-TP

If the cell is placed in a solution more concentrated than that in the vacuole, the volume of the vacuole will diminish; eventually the cytoplasm will no longer press against the cell wall. This stage of a cell is referred to **incipient plasmolysis**. If the vacuole continues to shrink the cytoplasm will eventually draw completely way from the cell wall and the cell will be **plasmolysed**. Under these conditions the wall of the cell and all surrounding cells can no longer exert any effect upon the suction pressure, and then S.P. = OP, because TP = 0.

The above discussion suggests that the vacuole alone has been considered to change its shape because the tonoplast has been regarded as the semipermeable membrane of the cell. Recent work shows that according to the nature of the solute, penetration may take place to varying depths into the cell. Then the limiting layer into which penetration does not take place represents the semi-permeable membrane. It is convenient to refer to the volume within the cell which changes with change in osmotic pressure as the **osmotic volume**. The space external to this limiting layer, which an external solution is regarded as freely penetrating, is called the **free space**. The osmotic volume of a plant tissue can be measured by using **Archimedes' principle**. If the weight of tissue ($W_1$ and $W_2$) is determined in two solutions of different strength but known densities ($D_1$, $D_2$) then, if W is the weight in air, $W_1 = W - VD_1$, and $W_2 = W - VD_2$. Therefore, $V = W_2 - W_1$. Where V is the volume not penetrated.

It is found experimentally that with discs cut from carrot root the osmotic volume is less for potassium nitrate than for sucrose. This indicates that the smaller molecule can penetrate further into the cell. The change in volume of a cell with time (dV/dt) during plasmolysis can be expressed in terms of the **osmotic pressure gradient**.

### Measurement of Suction Pressure or Diffusion Pressure Deficit

The suction pressure is the net force acting on the water in the plant cell and is dependent on the osmotic pressure of the contents, the extensibility of the cell wall, and any tissue tensions. These forces must remain unchanged during the measurement of suction pressure. Hence it is necessary to find the concentration of a solution which has no osmotic effect on the tissue; when this is done the net forces on the water in the tissue are balanced by the osmotic forces of the external solution, in which a tissue is unchanged, are equal to the average suction pressure of the tissue cells.

In practice, if sections are made from a root tissue such as carrot, it is convenient to determine the solution which results in no change of weight after immersion. If the section is thin, the error will be more. It is due to the change of tissue tensions upon cutting. However, if the section is thick, the rate of water penetration will be slow, the experiment will need more time, and more possibilities of error due to the uptake.

Slices of 1-2 mm thickness are most often used. Measurement of a single linear dimension is frequently used with tissue cut from such material as the hollow Pandelion peduncle. The cells on the interior side are relatively more permeable than those of the epidermal side. Therefore, uptake or loss of water results in an easily detectable change in curvature of the tissue.

## WATER RELATIONS OF AN INDIVIDUAL PLANT CELL (OLD AND NEW TERMINOLOGY)

Water relations of an individual plant cell can be described in old set of terminology (Meyer or Renner's terminology) which describes the cellular water status and its component forces a Diffusion pressure deficit (DPD), Osmotic pressure (OP) and Turgor pressure (TP), or in new set of terminology (Slatyer and Taylor's terminology). The water potential ($\psi$) concept is the new term for DPD. Both DPD and $\psi$ are numerically equal but are opposite in algebric sign.

$$\psi - (-)\ \text{DPD}$$

Similarly, osmotic potential ($\psi s$) is numerically equal to osmotic pressure

$$\psi s = (-)\ \text{OP}$$

Turgor presure is referred as turgor potential or pressure potential ($\psi p$) which is not only numerically equal to TP but similar in algebric sign also

$$\psi p = \text{TP}$$

Matric pressure (MP) is usually neglected but under new terminology ($\psi m$) it exists in following relationship-

$$\psi m = (-)\ \text{MP}$$

When a plasmolysed plant cell is placed in a hypotonic solution or pure water, it absorbs water and cell volume gradually increases as the vacuole swells in size and at a certain point plasmalemma exerts a pressure on cell wall, the turgor pressure. Thus, turgor pressure is a centrifugal pressure exerted by cytoplasm and cell membrane on cell wall. Using old terminology, the water status of the cell is described as DPD and its contributing pressures as follows

$$DPD = OP\text{-}TP$$

$$DPD = OP \text{ when } TP = 0$$

The same situation is explicable in new terminology and water potential equation

$$\psi = \psi s + \_p + \psi m$$

$$\psi = \psi s + \psi p \text{ when } \psi m \text{ is negligible}$$

$$\psi = \psi s \text{ when } \psi p \text{ is zero}$$

It is important to keep notice of algebric signs while attempting data handling and numerical problems.

$$(-)\ \psi = (-)\ \psi s + (+)\ \psi p + (-)\ \psi m$$

Hoefler's diagram can also be employed to understand water relations of individual plant cell and relationship of old and new terminology.

Figure shows that when a plant cell absorbs water essentially there are two phases when DPD ($\psi$) is equal to OP ($\psi s$) and when DPD ($\psi$) is higher than OP ($\psi s$) by a value of TP ($\psi p$). Thus, TP ($\psi p$) can be calculated as TP = OP - TP

$(+)\ \psi p = (-)\ \psi - (-)\ \psi s = \psi p = (-)\ \psi + \psi s$

## *Water Potential: The Components*

There are three following components of water potential:

1. **Osmotic potential** (or Solute potential): It is caused by the presence of solute particles. The symbol for osmotic or solute potential is $\Psi s$, however the symbol P can be used for pressure and s for solute potential. (The symbols $\pi$ or $\Psi\pi$ should not be used for osmotic potential; instead, $\pi$ should be reserved for osmotic pressure. Osmotic potential ($\Psi s$) is always negative (or zero in pure water). By adding solute particles, water potential is always decreases, below that of pure water.
2. **Pressure potential:** It is caused by the addition of pressure and is equal to the real pressure in the part of the system being considered. Pressure potential is the real pressure and is referred to pressure. The symbol for pressure potential is $\Psi P$, but P can also be used. In living a cell, value of the pressure is usually positive but is typically negative in dead xylem elements or in soil (positive below the water table).
3. **Matric potential:** This component describes water in cellular boundreis like cell wall and middle lamella or apoplastic water. On account of its low value it is often neglected but recently its importance has been recognised in indirect measurement of turgor potential.

Pressure may have any value. By convention P = 0 at atmospheric pressure. An increase of pressure results in a positive pressure, and **tension** (suction or pulling, the opposite of pressure) results in a negative pressure.

The values of water potential varies under different conditions:

1. Water potential ($\Psi$) can be negative, zero, or positive because pressure can be positive and very high, and osmotic potential can be zero or negative.
2. Water potential of pure water at atmospheric pressure is zero.
3. In a solution at atmospheric pressure, water potential is negative.

4. In pure water under some external pressure above atmospheric (for example, below the water table), water potential is positive.
5. In a solution under some pressure other than atmospheric, water potential may be negative (osmotic potential is more negative than pressure is positive), zero (pressure equals osmotic potential but is opposite in sign), or positive (pressure is more positive than osmotic potential is negative).

### Water Potentials in the Soil-Plant-Air System

When relative humidity is less than 100 per cent, water potential is highest in the soil and lowest in the atmosphere, with intermediate values in various parts of the plant; that is, there is a gradient from the soil, through the plant, to the atmosphere. But the components of water potential vary.

In a wet soil above the water table, P = 0 and $\Psi_s$ is only slightly negative because the soil solution is dilute, so water potential ($\Psi$) is also only slightly negative. Xylem sap is very dilute, so $\Psi_s$ is only slightly negative; but the water is virtually always under tension

In land plants $\Psi$ is virtually never positive. This is because $\Psi_s$ caused by solutes in cells always negative, and metric forces also lower $\Psi$ below zero. Pressure in cells can rise $\Psi$ close to zero, but these cells never becomes positive. Although it is nearly pure water and could have a positive $\Psi$ if it were under pressure, stays under tension (negative pressure) as it remains at least close to equilibrium with the living tissues and their negative water potentials.

### Water Potential: Units

The Gibbs free energy indicates the maximum energy available to do work. The free energy per mole is the chemical potential ($\mu$). Slatyer and Taylor (1960) defined - *the water potential is the chemical potential of a water solution* (free energy per mole) in a system minus the chemical potential of pure water at atmospheric pressure and at the same temperature. The components of water potential consist of solute and metric forces, which decrease the water potential, and pressure, which increases it.

A system's water potential expresses its ability to do work compared with the ability in a comparable quantity of pure water at atmospheric pressure and the same temperature. The osmotic potential of a solution is negative because the solvent water in the solution can do less work than pure water. As pressure on the solution increases, the solvent's ability to do work (and thus the water potential of the solution) also increases.

The energy and pressure units for water potential are easy to relate to each other, as was suggested by Taylor and Slatyer. When the energy units are divided by the partial molar volume (the volume of 1 mol of $H_2O$, which is 18,000 $mm^3$ $mol^{-1}$) or the specific volume of water ($10^6$ $mm^3$ $kg^{-1}$ = 1 cm3 $g^{-1}$), pressure units are obtained. Thus the osmotic potential of a 1.0 molal glucose solution at 28°C (or the water potential of the same solution at atmospheric pressure) can be expressed in energy terms as -2.5 kJ $kg^{-1}$ and in pressure terms as -2.5 MPa. With SI units, kilojoules per kilogram have the same numerical value as megapascals.

### Water Potential: Measurement of the Components

Water potential ($\psi$) is a relative expression. The measurements represent the difference between the chemical ($\mu$) potential in a system under investigation (e.g., a plant tissue) and that in a reference state. The reference state is the state of *pure liquid water at the same temperature and same atmospheric pressure,* as the system under investigation. Water potential of aqua puris is thus assumed a value of 0 bar.

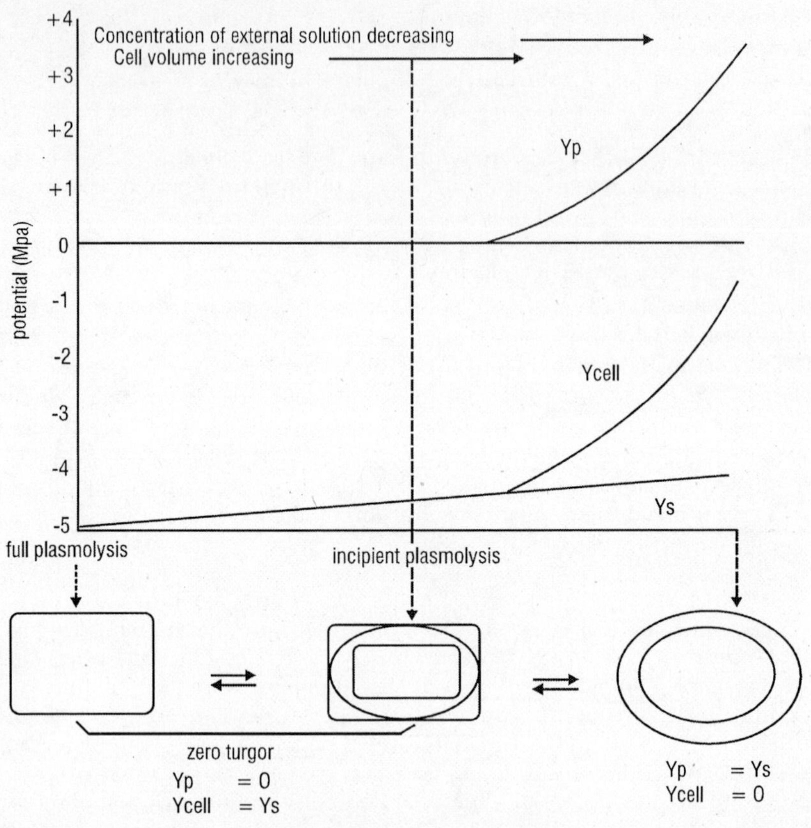


**FIG. 4.3:** Water relations of an individual plant cell and Hoefler's diagram.

Water potential in a plant tissue is always less than 0 bar and hence a negative number, as per following equations.

$$\mu = RT \text{ in } a$$

Where μ = chemical potential, R = gas constant (8.314 J mol-1 K-1), T= temperature (K), a = activity of molecule

$$\psi = (\mu_0 - \mu_\omega)/V_\omega$$

ψ = water potential; $\mu_0$ = chemical potential of water in the system under consideration; ω = chemical potential of pure water at minimum temperature and pressure; $V_\omega$ = partial molar volume of water (18 $cm^3$ $mol^{-1}$)

## *Water Potentials: Range in Representative Plant Tissues*

As a general rule, leaves of most plants rooted in well-watered soils are likely to have water potentials between about -2 and -8 bars. With decreasing soil moisture supply, leaf water potential will become more negative than -8 bars and leaf growth rates will decline. Most plant tissues will cease growth completely (i.e., will not enlarge) when water potential drops to about -15 bars, usually referred as permanent wilting point (PWP).

**Water Potential of herbaceous leaves:** Leaves of herbaceous plants may survive for relatively short periods of time-but will not grow- when water potential is less than -15 bars and they are not likely to recover if the water potential drops below about -20 to -30 bars.

**Water Potential of desert leaves:** Leaves of desert shrubs can survive for longer time under normal arid condition and also when subjected to low water potentials. In desert shrubs leaf water potentials may be very low under conditions of severe drought, perhaps in the range of -30 to -60 bars (sometimes as low as -100 bars).

**Water Potential of seeds:** Viable air-dried seeds are also likely to have very low water potentials, perhaps as low as -60 to -100 bars, or even much lower, depending on the extent of drying and the plant species.

## WATER RELATIONS OF THE WHOLE PLANT

The supply of water to the whole plant involves not only uptake by individual root cells but its subsequent movement to all the tissues of the plant. How water rises to the top of a tall tree has intrigued plant physiologists for many years. It is a biophysical problem and placed in perspective with other plant processes. The movement of water through the plant is dependent upon water potential gradient from root-shoots-air in soil-plant-atmosphere-continuum (SPAC).

Absorbed water moves through the root hairs is transported to the different parts of the plant and is subsequently lost from the leaves, petiole, and stem surface into the atmosphere.

### *Negative Turgor*

This question is often asked whether negative turgor is a myth or reality. On theoretical grounds the existence of negative turgor should be very rare that is when cell wall is sucked in on account of extremely plasmolysed state of cell under high degree of desiccation. But under actual measurements the negative turgor has been recorded more often than expected, even in mesophytes growing under optimal water supply. This situation has been ascribed to indirect measurement of cell sap during freezing. Direct measurements of turgor using cellular pressure probes are likely to reveal the true situation.

## MEASURING THE COMPONENTS OF WATER POTENTIAL

Soon after the water-potential concept was formulated by Renner in 1915, methods were developed for measuring water potential and its components. Newer methods have since been introduced, but the older methods can help us understand the water relations of plants. The newer methods are more useful. We will summarize both methods, nit so much as a cookbook of useful techniques as an illustration and application of the principles we have been discussing.

### Water Potential

Probably the most meaningful property we can measure in the soil-plant-air system is the water potential. It is not only the final determinant of diffusional water movement, but it is often the indirect determinant of bulk water movement, which occurs in response to the pressure gradients that are set up by such diffusional movement. Furthermore, in principle and in practice, water potential is probably the simplest component of an osmotic system to measure.

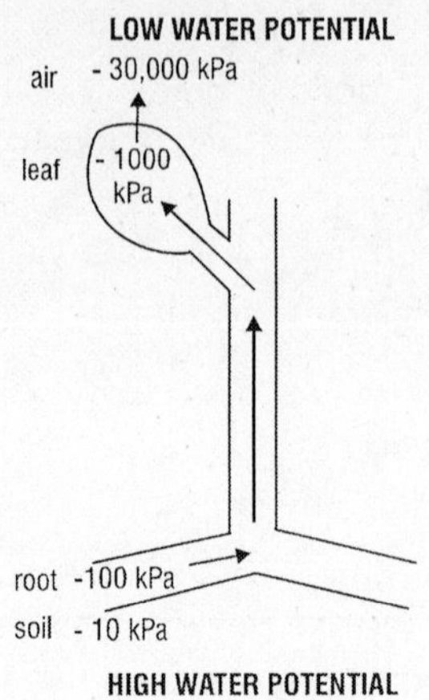


**FIG. 4.4:** Water potential gradiant from root-shoot-air in soil-plant-atmosphere continuum (SPAC).

Remember that at equilibrium $\Delta\ \Psi = 0$; that is, $\Psi$ is equal in all parts of the system. Thus, a plant part can be introduced into a closed system, and after equilibrium has been achieved, $\Psi$ may be known or determined for any other part of the system and therefore for the plant part. There are several possibilities for applying this principle.

In the tissue volume method a sample of the tissue in question is placed in each of a series of solutions of varying but known concentrations (usually sucrose; sorbitol; mannitol; or, best, polyethylene glycol, PEG). The best solute for such measurements is one that does not easily pass through membranes or harm the tissue. The object is to find that solution in which the tissue volume does not change, indicating no gain or loss in water, which means that the tissue and the solution were in equilibrium to begin with the water potential of the tissue was and is equal to that of the solution. At atmospheric pressure, when $P = 0$, $\Psi = \Psi_s$. $\Psi_s$ for the solution, with its known concentration, can then be calculated.

In practice, there are several ways to determine changes in volume. One is to measure the volume of tissue before placing it in the solution (usually standard volumes are cut), and then to measure the volume (or only the length) after sufficient time for exchange of water. Volume change can be plotted as a function of solution concentration, which indicates an increase in volume in relatively dilute solutions and a decrease in volume in relatively concentrated ones. On such a plot, the point at which the volume curve crosses the zero line indicates the solution that had the same water potential as that of the tissue at the start of the experiment.

In another approach (not illustrated), tissue samples in small closed containers are allowed to equilibrate with the vapor over solutions of known concentration, rather than with the solutions themselves. Thus, the solutions are not contaminated with solutes from the tissue. Mass rather than volumes are usually measured in this method.

Rather than measuring changes in the tissue, one might measure the concentration of the test solution. If it becomes less concentrated, the tissue will have lost water. This faster approach is better than measuring tissue volumes, which often gives Ψ values that are too negative because solutes are absorbed from the test solution by the tissue samples.

In 1948, a Russian scientist, V.S. Chardakov, devised a simple, efficient way to find the test solution in which no change in concentration occurs. More practical methods are now available for field work, but **Schardakov's method** nicely illustrates the principles we are discussing. Test tubes that contain solutions of graded concentrations are coloured slightly by dissolving in them a small crystal of a dye such as methylene blue. (Addition of the dye does not change osmotic potential significantly). Tissue samples are placed in test tubes that contain solutions of equivalent concentrations but no dye. Time is allowed for exchange of a certain amount of water, but it is not essential that the tissue reach equilibrium with the solution. Sufficient exchange may take place in as little as 5 to 15 min. Then the tissue is removed, and a small drop of the equivalent coloured solution is added to the test tube. If the coloured drop rises, the solution in which the tissue was incubated has become denser, indicating that the tissue has absorbed water; in this case, the tissue had a lower (more negative) water potential than the original solution. If the drop sinks, the solution has become less dense, having absorbed water from the tissue; the solution without rising or sinking, then no change in concentration has occurred, and the water potential of the solution equals that of the tissue. When several solutions with different concentrations are used, there is usually one in which the drop neither sinks nor floats. At $P = 0$, $\Psi_s$ is equal to Ψ and thus, the average Ψ of the tissue.

In the vapor-pressure method, tissue is placed in a small, closed volume of air. The water potential of the air comes into equilibrium with the water potential of the tissue, which changes only insignificantly in the process. The water potential of the air is then determined by measuring vapor density (humidity) at a known temperature.

## Relative Humidity (RH)

Relative humidity (RH) is the amount of water vapor in air at a particular temperature compared with the amount of water vapor that the air could hold at that temperature (the **saturation vapor density** or **pressure**). The saturation vapor density approximately doubles for each temperature increase of 10°C (or 20°F). If a volume of air that is saturated with water vapor (100 per cent RH) is warmed by 10°C, it can hold about twice as much water as before; therefore, its relative humidity is about 50 per cent. The scenario in plant water relations has been dominated by acceptance in new terminology which triggered so much interest in the subject that it lead to employment of concepts in several areas; especially the physiology of biotic and abiotic resistance and mathematical modeling of crop resources.

# CHAPTER 5

# Absorption of Water and Mineral Elements by Plants

## INTRODUCTION

It was formerly believed that mineral elements are taken into plants with the water that enter via the root hair. It is now known that part of the root most actively engaged in mineral absorption is not the root hair region but region of elongating cells just behind the root tip. Numerous experiments have also failed to show a correlation between rates of the two processes. Nevertheless mineral elements once absorbed by the root are distributed to the rest of the plant in the transpiration stream and hence the rate of water absorption can sometimes have an indirect influence on the rate of mineral uptake. It has been shown that if concentration of mineral elements in the external solution is low the rate of water absorption has little effect on the rate of mineral uptake. If, however concentration of elements in the external medium is high then rates of the two processes run approximately parallel to one another.

## WATER UPTAKE BY THE ROOT

Water moves from the outer surface of the root to the xylem. The differentiation of phloem takes place before the xylem and well-differentiated xylem, presumably functional in water transport which is present only a few mm from the tip. Absorbed water reaches the xylem through the epidermis, cortex, endodermis, and pericycle. At this stage each cell of the endodermis has **casparian strip**.

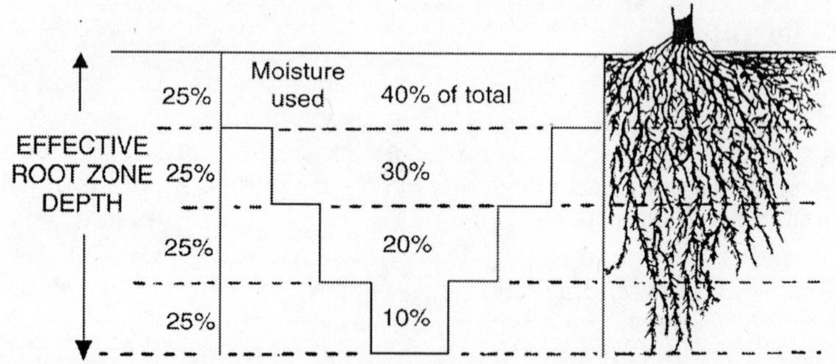


**FIG. 5.1:** Relative moisture use pattern as percent of total moisture absorbed by a plant from different unit depths of its total effective root depth (A representative, diagrammatic presentation).

**Casparian strip:** It is a region of the radial walls which is impermeable to water, due to impregnation with suberin or lignin or both. The plasmalemma of the endodermal cell is firmly attached to the wall at the Casparian strip. Hence at this point water cannot move radially through the wall, nor between the wall and the plasmalemma. Water can only move to the stele crossing the plasmalemma and protoplasm.

In dicotyledons and gymnosperms, secondary thickening occurs, therefore, the outer tissues of the root usually become dead throughout the life of the root. However, most monocotyledons are without secondary thickening, and the outer tissues of the root usually remain alive throughout the life of the root. In these species the endodermis often shows further development.

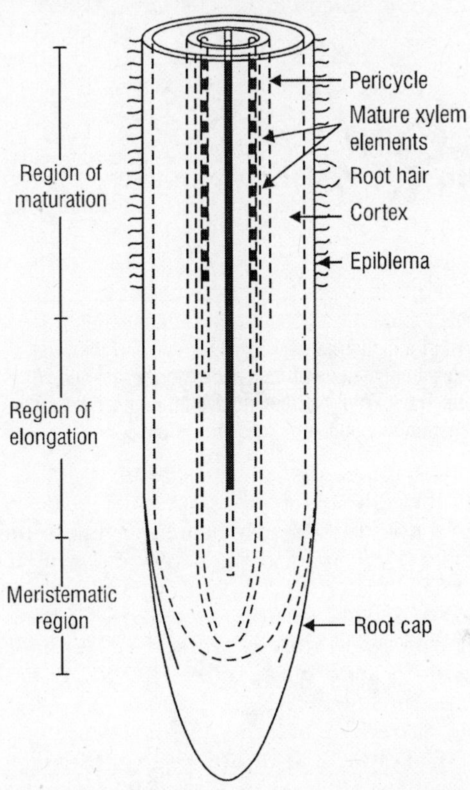


**FIG. 5.2:** Generalised anatomy of root tip with details of region of maturation. Most water is absorbed by roots from this region. (Diagrammatic).

## THE DRIVING FORCE FOR WATER MOVEMENT

The water movement across the root is due to the difference in water potential between the soil at the root surface and the xylem. This is the **driving force** for water movement. Entry of water in the root is dependent on the low root water potential as compared to the surrounding soil. Transpiration plays important role in developing negative pressure potential in xylem by withdrawal of water from the leaves during transpiration process. The osmotic potential of the xylem sap also make a minor contribution in this regard.

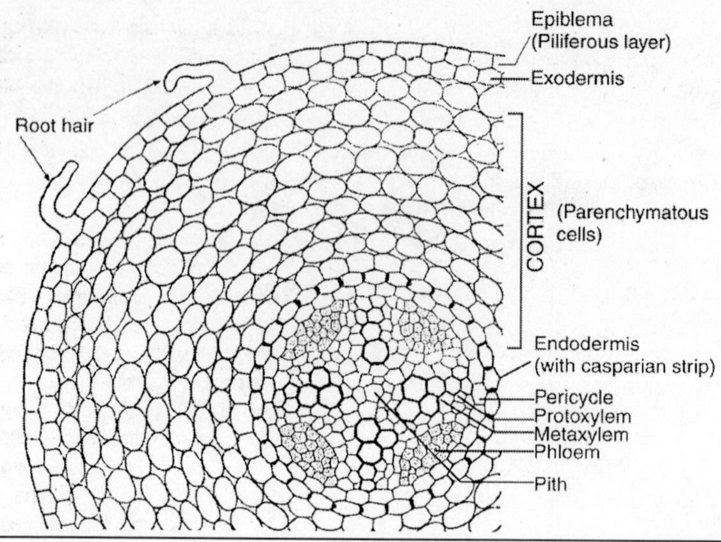


**FIG. 5.3:** A transverse section of tip of a young dicotyledonous root from its region of maturation, depicting (a) *Stele*, comprising endodermis, pericycle, proto- and meta-xylem (vessels), phloem and pith; (b) *Cortex*, comprising parenchymatous cells; (c) *Exodermis*; and (d) *Epiblema* with root hair. Water, absorbed by epiblema and its root hair, must pass through all these cell layers before arriving at the xylem vessels for its transportation to leaves through stems.

**Water potential** falls under dry soil condition which decreases the driving force for water uptake unless the plant water potential is also adjusted downwards. The reductions of water potential in plant causes wilting, stomatal closure, and reduction in water uptake. Plant wilts permanently under lower limit of soil water availability and will remain wilted in a humid atmosphere. Sometimes the plant may recover on re watering. This *permanent wilting point* is, about -15 bar soil metric potential for *Helianthus annuus* L. (a standard test plant).

The pathway of water movement across the root is still uncertain. The Casparian stip directs the movement of water across the root to the xylem via the cell walls, never passing through membranes or protoplasm.

**Metabolic rates of roots and their permeability:** There is also good evidence that water does have to cross membranes and protoplasm on its way to the xylem; because root permeability is often markedly influenced by root temperature and metabolic inhibitors which reduce root permeability (unless the concentration of inhibitor is high enough to kill the root). These effects are similar to those for single cells, strongly suggesting that membranes or protoplasm are the main resistance in both cases. Three other alternative routes have been suggested for the movement of water in roots:

1. Water may cross all the cells, passing from vacuole to vacuole (Symplastic movement).
2. The movement of water may be: -Entry in epidermal cells protoplasm (or pass inwards by the cell walls) → protoplasm of cortical cells → movement via the plasmodesmata from cell to cell → stele (Apoplastic movement).
3. It may remain in the walls except at the endodermis, where it enters the protoplasm.

Various attempts have also been made to obtain more direct experimental evidence on the pathway, but still confirmed results are awaited and difficult to interpret. According to Newman (1974) *until better evidence becomes available we cannot be sure which of these pathways is the main one, nor where in the root the main resistance to water movement lies.*

## WATER UPTAKE: ROOT AGE

Water uptake is largely confined to the young portions of the root, a short way behind the tip. Anatomical evidences suggest that older roots usually have either periderm or suberised endodermis and these tissues may be impermeable to water. The experiments conducted with reference to water uptake at various distances from the tip showed that in some species the water uptake rate remained high at the farthest point of measurement from the root tip.

R. Brouwer (1965) presented interesting data which suggest very complex situation of water uptake by broad bean root tips. A single root of broad bean, still attached to the plant, was inserted through a series of compartments each 2.5 cm long, so that water uptake by each 2.5 cm length of root could be measured separately. Under slow transpiration rate, uptake was highest near the root tip while under fast transpiration rate uptake was highest away from the tip.

At each transpiration rate the xylem water potential and root permeability were determined. The change in distribution of water uptake was due to a marked increase (up to eight-fold) in root permeability in the other parts of the root, whereas near the tip there was little change. This change is root permeability is unexpected; it means that the roots do no obey Darcy's law. On the basis of above evidences it may be suggested that *as the demand from the above-ground parts of the plant increases, the roots alter in some way (as yet not understood) to let water pass through more easily.*

It is now known that water can be taken up rapidly through roots with periderm or with suberised endodermis. In Loblolly pine young root tips and older root segments with periderm, when subjected to the same suction, showed similar rates of uptake per unit area.

**Water movement through the root periderm:** This occurs through tenticles and small cracks. In some cereal species water uptake can continue at quite a rapid rate through regions of the root whose endodermis is heavily suberised. In some species the suberised endodermis contains passage cells, which remain unsubmerised (apart from having a Casparian strip). But these are not found always. In barley, for example, passage cells are absent, yet water uptake by suberised roots occurs. An electron microscope study of the suberised endodermis of barley reveals that numerous plasmodesmata pass through the thickened wall. Perhaps these provide an *easy pathway* for the water.

## XYLEM

The xylem forms a continuous system running from near the tips of the roots, up the stem, and into the leaves, where it branches profusely. Xylem acts as the **long-distance transport system** of water in the plant, leaving only a minute distance, usually less than 1 mm, at each end, in the roots and the leaves, where the water has to pass through other tissues.

Xylem is made of four types of cell: fibres, tracheids, vessels and parencyma. Tracheids and vessels conduct water, and provide the mechanical strength to the xylem. Fibres are responsible for mechanical support only. Parenchyma cells function in storage and lateral transport.

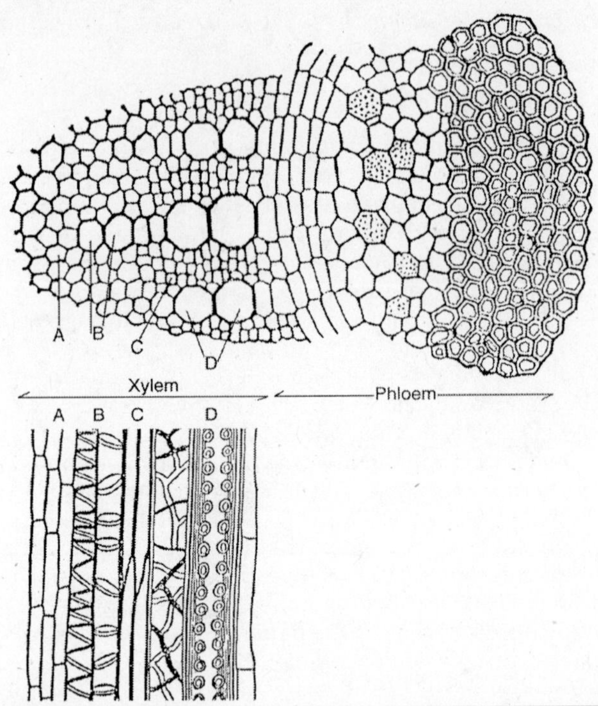


**FIG. 5.4:** A vascular bundle of sunflower (diet) stem in transverse (upper and longitudinal (lower) sections. The longitudinal section included here shars only the water transporting and some associated tissues, *viz.,* A, wood parenchyma; B and D, vessels; C, tracheids and wood fibres. In general, the stems are richer in vascular bundles than roots. Also, the monocot plants are in their conducting tissues than dicot plants, both in their roots and stems.

## Xylem Vessel

It is formed from numerous cells (vessel elements) that are arranged end to end. The end walls are perforated or completely lost, so that water can pass freely from one vessel to the next. The size of each vessel varies from several centimetres long to several metres. At the end of the vessel, however, the water must pass through a cell wall, which it does mainly through pits, the small areas where the wall is much thinner than elsewhere.

Water moving from tracheid to tracheid must always pass through pits, since the end walls never break down. In xylem the form and size of the pits varies considerably.

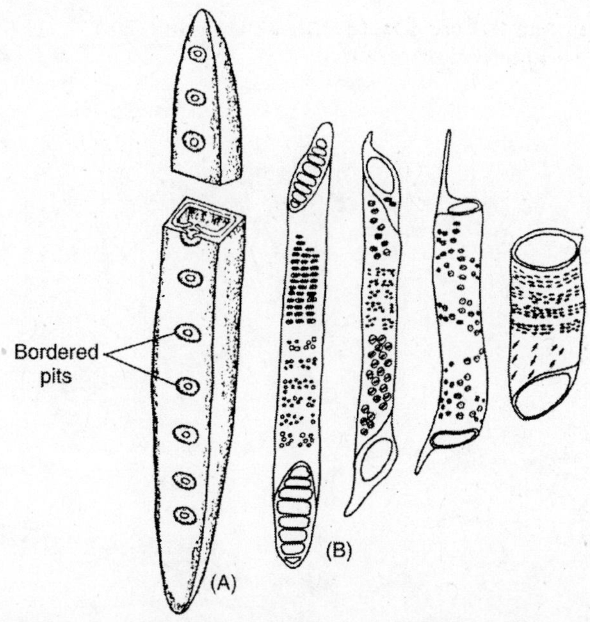


**FIG. 5.5:** The xylem conduits for conductance of water in plants comprise, (A) tracheids and (B) vessels or tracheae. Both are dead tissues, devoid of protoplasm and usually with lignified walls. The ends of trachides are walled while those of vessels are open. Both have wide central cavities, called lumen, for translocation of water (actually sap). In trachieds, water passes through pits present in their, both radial and end walls. Individually a tracheid may rarely exceed 1 mm in length. Vessels, with their open ends placed one above the other, may appear several centimeters long.

## COHESION THEORY

Xylem can conduct water even when it is dead because the driving force for water movement does not originate in the stem; and how transpiration pull could raise water up tall trees? This was explained on the basis of classical physics. Water is not sucked up in a tube by a vacuum at the top, but rather the air pressure at the bottom pushes the water up. If this is so then the maximum height to which water can be raised by suction is about 10 m, since this height of water column exerts a pressure of 1 atmosphere at its base. Yet, many trees are far taller than 10 m. The answer to this problem proposed by the *cohesion theory* is now generally accepted. This states that water in xylem can be pulled up from the top; the water possesses cohesion so that the column does not break, even when much more than 10 m long.

The cohesion of water in xylem is increased by two factors:

1. Water is effectively filtered when it enters the root, thus removing fine particles which might have acted as nuclei for bubble-formation.
2. Water remains in very fine tubes. Nevertheless, there is good evidence that bubble-formation (cavitation) does occur, and in many trees it is probably a normal occurrence. Once formed, a bubble will be isolated within a single vessel or tracheid. This is because a very large tension would be required to **pull** the air-water interface through the middle lamella which forms a barrier

across every pit-pair between one cell and the next.

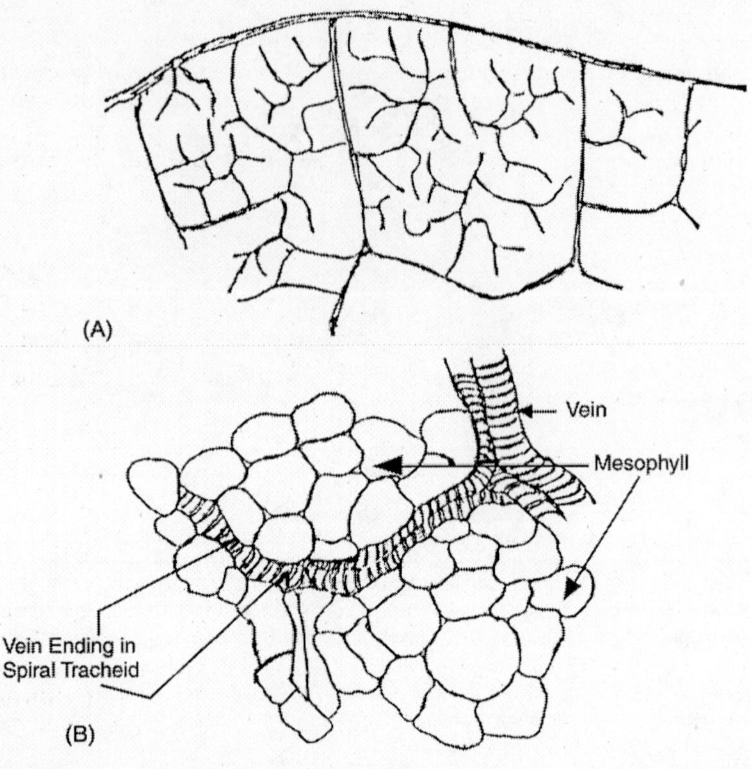


**FIG. 5.6 [A & B]:** The xylem conduits continue from stem into leaf petiols. When passing into the leaf, the vascular bundles may divide into a number of smaller bundles. Within lamina, the vascular bundles of veins run between the palisade and spongy mesophyll. The nework of midrib (vein) and veinlets of a leaf is called vein-inlets. Usually a short, spiral tracheid lirmenates the xylem into the mesophylls, as depicted figures A and B.

Pores in the middle lamella probably do not exceed 0.01 μm in radius, therefore, a tension of at least 150 bar would be required to pull the bubble through. Such high tensions are very unlikely in living plants. Whether bubbles once formed can be re dissolved is uncertain. It is quite possible that in narrow elements they can.

## PASSIVE AND ACTIVE TRANSPORT OF SOLUTES

Different solutes absorbed by cells are accumulated at higher concentrations within cells than outside them. For example, most of the plant cells (e.g., sieve tubes of the phloem) accumulate potassium, and accumulate sucrose before they translocate it. But a few solutes that are absorbed rapidly by cells never reach higher concentrations inside than out and never reach higher concentrations inside than out. A good example is gaseous $N_2$, but it probably moves in and out only by diffusion through the lipid bilayer.

When a plant tissue that has been immersed in a relatively concentrated salt solution is transferred to a dilute salt solution some of the ions already absorbed diffuse rapidly out into the external solution. If the tissue is then replaced in the concentrated salt solution, ions will diffuse rapidly back into the tissue. It is thus, evident that diffusion of ions can occur freely between the external solution and a certain part of the tissue, which is called the **free space** (free space of a tissue consists only of the volume not occupied by the protoplast i.e., the cell wall, the intercellular spaces, and any dead cells that may be present). Because the intercellular spaces are usually filled almost entirely with air they are not considered to form significant part of free space.

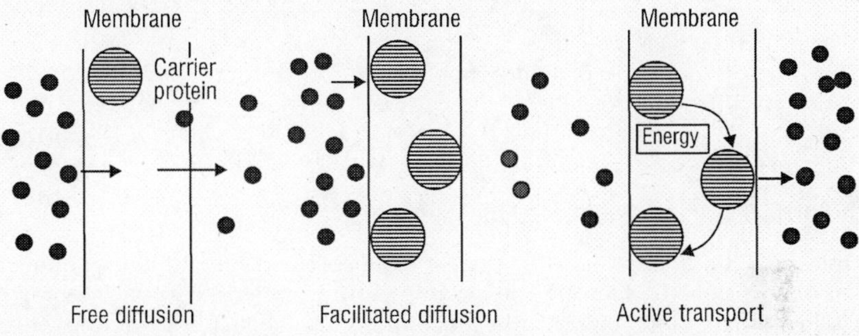


**FIG. 5.7:** Movement of molecules across the membrane. (*Note the molecular concentration as dots*).

Although rapid exchange of ions by passive absorption can account for salt accumulation within the plant tissues it is not adequate to explain the total accumulation that occurs. Many ions continue to enter the tissues at a considerably slow rate long after this initial phase of free diffusion is over. Such ions enter the cytoplasm and accumulate in the vacuole from which they cannot readily pass back again into the external solution. This slow steady accumulation of ions against a concentration gradient would seem to require metabolic energy because when the metabolic activity of the tissue is inhibited the rate of ion uptake is also inhibited. This phase of ion uptake which has a $Q_{10}$ of between 2 and 3 is called active ion uptake because the cells must do work to move the ions. There is extensive evidence that the necessary energy is derived ultimately from respiration (i.e., that ATP energy is used). Various experiments suggest that greater the rate of respiration the faster is the accumulation of inorganic salts. The dependence of salt uptake on the faster is the accumulation of inorganic salts. The dependence of salt uptake on respiration is supported by the fact that if respiration is inhibited by respiratory poisons such as cyanide, salt uptake is also inhibited.

Since ions once inside the vacuole, cannot readily pass back into the free space of the cell it is reasonable to assume that the cytoplasm form a permeability barrier across which ions must be pushed using energy supplied by respiration. The mechanism by which ions pass through the cytoplasm and accumulate in the vacuole is incompletely understood but most theories postulate the operation of so called is incompletely understood but most theories postulate the operation of so called **carriers**. These may be envisaged as enzyme like substances located in or on the two membranes plasmalemma and tonoplast. A carrier it is assumed combines on the outer surface of the membrane with an ion in the external solution, transports it across the membrane and then releases it on the inner surface. The ion once it has been transported across the membrane cannot pass back because the membrane is permeable only to the carrier-ion complex. The carrier however, returns to the outer surface of the membrane where it is available to repeat the cycle.

| Passive Absorption | Active Absorption |
|---|---|
| Movement of ions across membrane takes place according to concentration gradient | Movement of ions across membrane takes place against concentration gradient. |
| No energy is required | Expenditure of energy is required. |
| No carrier is necessary for this process. | A specific carrier is necessary in this process. |
| No enzymes are involved. | Enzymes may be involved. |

If it is further assumed that a given carrier combines only with certain ions, then the carrier concept provides a possible basis for explaining the selective absorption of different ions. To account for the dependence of active ion uptake on respiration it is postulated that the carrier is able to act only during active metabolism.

## Which Carriers are Involved in Solute Transport?

Principles led to the theory that transport proteins, specifically called carriers, control absorption. In fact all the principles mentioned are consistent with that theory. Proteinaceous channels are more likely to account for some of the results than are true carriers, but otherwise the theory that specific transport proteins in membranes control absorption of most essential solutes seems valid.

## Accumulation of Solutes inside Cells

A remarkable fact about all cells is that they can absorb certain essential solutes so fast and over such long periods of time that concentrations of these solutes become much higher within the cells than in the external solution. This is known as absorption **accumulation**. **Accumulation ratio** is the ratio between the internal and external concentration. Accumulation of certain solutes is a universal phenomenon of living cells.

## Solutes Absorption is Specific and Selective

Solutes are absorbed and accumulated by selective processes. If roots are provided with a solution containing dilute (about 0.2 mM) KCl and with about 0.2 mM $Ca^{2+}$ to maintain normal membrane functions, the rate of absorption of $K^+$ is unaffected by similar concentrations of $Na^+$ salts. This is true even though $Na^+$ is chemically similar to $K^+$. The process of $K^+$ accumulation is therefore selective and is not influenced by a related ion under these conditions. As expected, several others monovalent and divalent ions also have no influence on $K^+$ uptake. It has been observed that potassium absorption is inhibited competitively by Rubidium (both are monovalent) and penetration of membranes by these two ions is apparently governed by the same mechanisms.

Similar competitive results have also been obtained with $Cl^-$ and $Br^-$, and with divalent $Ca^{2+}$, $Sr^{2+}$. This selectivity of ion transport by roots also applies to organic compounds such as amino acids and sugars. It has been established that proteinaceous carries in membranes help to move solutes into cells. In this action enzymes (which are proteins) are involved and are known to recognize the ions or molecule selectively.

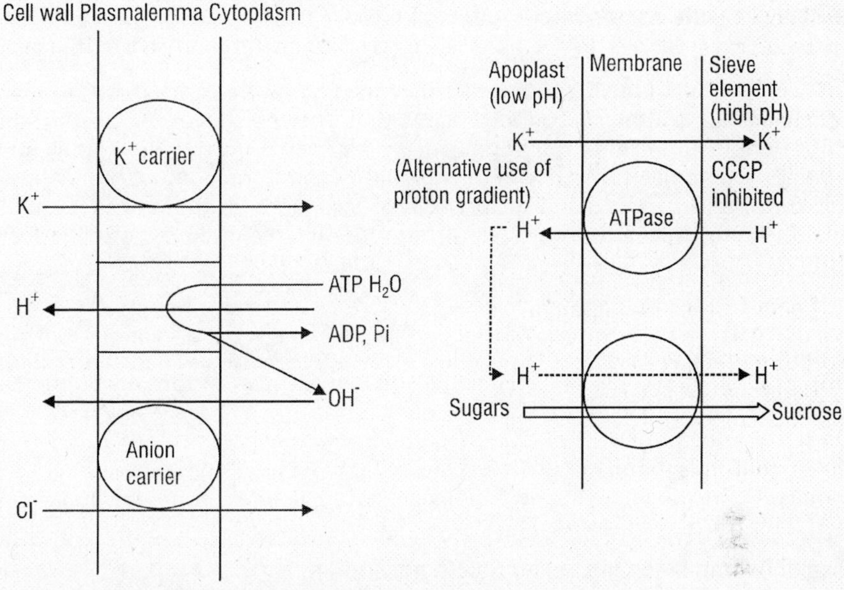


**FIG. 5.8:** Active transport accross plasmalemma. (a) Inorganic (b) Organic.

## Slow Leakage or Efflux of Absorbed Solutes

After arrival of absorbed ions or organic molecules into the cytoplasm or vacuoles of cells, they do not readily leak out (that is, efflux, or outward movement, is often slow). If the membrane is damaged by heat, poisons, or lack of $O_2$, then the rapid leakage can be induced and observed.

The following experiment helps to understand the leakage in plant tissue. If tissues are removed from the solution at a given time and are then placed in water at room temperature, and that the amount of ion retained inside is measured at various times. Only a small fraction of the ions present leak out rapidly, within a few minutes. Such ion losses occur mainly from cell walls, not from inside cells. Ions present in the cytoplasm and vacuole remain there longer, and they represent by far the majority of ions absorbed. If these experiments are continued for several minutes or hours the efflux from the cytoplasm and then from the vacuole be observed. Following are the established facts regarding the transport of ions:

1. Most of the ions are transported across membranes into cells.
2. The transportation is much faster than they move out.
3. Membrane carriers or unidirectional channels speed only their inward absorption.
4. The efflux of several ions becomes significant when their internal concentrations become high (non-low-salt roots).
5. For intact plants, efflux is especially active during darkness and in some cases may actually exceed influx.
6. The process is probably energy dependent Respiration and solute absorption are probably strongly related because respiration provides the ATP needed for soluble absorption.

## The Rate of Solute Absorption

The rate of solute absorption varies with solute concentration. Several experiments have been conducted using different concentration levels of fertilizers to learn the mechanisms of solute absorption. A relation between fertilizer absorption and fertilizer application rates was established which varies with solute concentration. In nature, plants use nutrients in abundance (nitrate, ammonium, phosphate, and potassium), and their **diffusion** to the root surface is the limiting factor. It also explains that absorption properties of the roots have limited importance for plant nutrition.

## How Solutes Cross Membranes

The mechanism of this aspect of mineral absorption was the problem for many years. E. Epstein and his associates investigated the absorption of specific ions by excised roots as a function of external concentration and time. They indicated that *proteins in membranes might act like enzymes (*transport proteins carriers). To a large extent this concept proved true, which means that membranes contain various proteins that specifically recognize certain solutes, combine with them, and speed their transport inward.

## Possible Overall Mechanisms for Solute Absorption

**Diffusion across the Membrane.** The diffusion of solute across the membrane is responsible for the rate of absorption and is directly proportional to the external solute concentration. Certain organic solutes (not available in nature, e.g., methanol, ethanol, urea, valeramide, and even to atmospheric nitrogen gas) have also been tested to strengthen the concept. This is an example of passive absorption and it applies to such solutes and is dependent on solute concentrations.

## How ATPase Pumps Transport Protons and Calcium?

Cells use energy stored in ATP to absorb solutes actively. The reaction of breakdown of ATP into ADP releasing energy (32 KJ or 7.6 kcal for each mole of ATP hydrolyzed) is catalyzed by an enzyme (**ATP phosphohydrolase** or **ATPase**) that is aparently present in every membrane of every living cell and is one of the transport proteins. Each molecule of ATP and ADP is chelated with one $Mg^{2+}$, which emphasizes one essential function of magnesium for all life. The reaction is strongly **exergonic** in the forward direction, with the release of approximately. In membranes most of the generated energy (through ATPase) is used to transport protons from one side of the membrane to the other against an electrochemical gradient. This transport of $H^+$ provides energy that is then used to transport essential mineral salts.

In cell, ATPase also pumps $Ca^{2+}$ out of the cytosol at the same time it pumps $H^+$ into the cytosol probably exists in plants (as it does in animals), with a slow speed. Still we need concrete evidences. Furthermore, there is evidence for still another calcium pump in the plasma membrane of some species. This is called a (**Ca + Mg**)-**ATPase** because it depends on a chelate of $Mg^{2+}$ and ATP to move $Ca^{2+}$ out of the cell. This pump does not move H+ in as $Ca^{2+}$ moves out. This process is probably depends on **calmodulin** for its activity.

The mechanism by which ATPases pump H+ and $Ca^{2+}$ from one side of a membrane still needs further confirmation. Although following evidences are available to give strength to the mechanism.

1. All ATPases are large enough to span any membrane (often more than once) and so they represent integral proteins.
2. A part of every ATPase molecule protrudes out of the membrane into the aqueous cytosol, where that part reacts with ATP and water.
3. ATPase contains a potential hole extending across the membrane, but that hole opens only when ATP is being hydrolyzed.
4. When ATP is being hydrolyzed, can the ATPase combine with a proton to move the proton across the membrane through the hole?.
5. The plasma membrane $H^+$ ATPase is thought to transport out of the cytosol and into the cell wall only one proton for each ATP hydrolyzed. This ATPase is the most energy-wasteful of the three major classes of ATPase known (much of the energy in ATP is lost as heat).

The plasma membrane is the first barrier for absorption and the solutes first pass through tonoplast membrane to enter into the vacuole. ATPase is present in the tonoplast which pumps protons into the vacuole. This makes the vacuole acidic (often near pH 5). The ATPase of tonoplast is different than that of plasma membrane in following ways:

1. It transports $H^+$ without combining with a phosphate from ATP.
2. It transports two more energy-efficient than the plasma membrane ATPase.
3. It is not $K^+$ dependent.

How $H^+$ is transported is not known. But it seems to be energy dependent process because when ATP is hydrolyzed energy is released. This probably changes the structure of tonoplast membrane by opening the holes through which protons then are forced to move. The membranes of endoplasmic reticulum and dictyosomes also contain an ATPase proton pump much like that of the tonoplast; this pump transports $H^+$ into the interior of organelles and makes them slightly acidic.

**ATPase Pyrophosphatase Pump:** In addition to the ATPase pump, the tonoplast has a **proton pumping** pyrophosphatase that uses energy in inorganic pyrophosphate. The pyrophosphate is present as chelate with $Mg^{2+}$. Probably there is still another kind of ATPase that transports protons across chloroplast and mitochondrial membranes.

## How Carriers and Channels Speed Passive Transport?

Passive transport is independent of energy but it is possible that carriers and channels could passively speed movement of solutes across membranes by taking advantage of the electro-potential gradient established by an ATPase or pyrophosphatase pump. Carriers have following features:

1. The term carrier implies that these proteins pick up solutes and move across the membrane with them, but this almost surely is not true. They may undergo a reversible structural change which help in solute transfer.
2. Probably, carriers can specifically bind one or a very few related solutes (much like an enzyme attaches to its substrate).
3. Possibly, carriers are also thought to be integral proteins that completely span the membrane.

4. The nature of these channels (open, closed, or partly open) is dependent on environment.

## Correlation's Between Root and Shoot Functions in Mineral Absorption

Except for certain desert species with unusually large root systems, most species invest most of their biomass in shoots. Therefore, it seems reasonable that absorption of mineral salts should be controlled in part by activities of the shoot. There are two following ways of looking at this control.

1. The shoot might increase root absorption of mineral salts by rapidly using them in growth products (for example, proteins, nucleic acids, and chlorophyll).
2. The shoot supplies carbohydrates via the phloem that the root must respire to produce ATP that drives mineral salt absorption. The shoot probably also supplies the roots certain hormones that affect absorption by roots.

## Physiological Limitations on Internal Transport and Use

At adequate external concentrations, nutrients can be taken up with minor energy cost. However, even then the nutrients are not assured of incorporation into plant tissue at maximal rates set by the ultimate limitation of photosynthesis. Some plants strategically restrict their maximal growth rates, and hence internal use rates of nutrients. Stress-tolerators do so dramatically. Even most crop plants, which are genetically derived from fast-growing opportunistic species, may have lesser but significant and hard-to-evaluate strategic limitations in nutrient use. Modern maize has been bred for increased yield, to which increased internal rates of nitrate reduction and nitrogen translocation have contributed. Other limitations may be hidden behind complex considerations of seasonality or the like. After all, crop cultivars are genetic melanges that are not finely adapted to any single environment of any contributing wild genotype.

Other nutrient-use limitations can be usefully classed as absolute, not to be overcome by any conceivable physiology without more than cancelling penalties in other functions. These represent the limit or asymptote to which improved crops may reach but not attain or exceed. The first absolute bottleneck is nutrient transport in xylem sap. The major mass fraction of metabolizing tissue is transpiring tissue, mostly leaves. New tissue is always concentrating the xylem sap about 50 to 200 fold. On the average, then, nutrients should be loaded into the xylem sap at about 1/50 of their final tissue concentrations. The loading ratio varies about this value, controlled by transpiration and nutrient demand. However, some nutrient compounds pose hazards of adverse mutual reaction at xylem concentrations near those required.

## Inter-organ Transport of Nutrients

Inter-organ transport of nutrients in phloem is necessary in several following situations.

- First, limited soil reserves can cause a decreasing nutrient uptake per unit growth late in the season. The decline must be compensated, so that no tissue drops to absolute metabolic insufficiency. Nutrients are remobilized from old tissues of declining utility, such as heavily shaded lower leaves, into new growth.
- Second, and perhaps infrequently, on a continuous deficient amounts of

nutrients- e.g., one portion may take up inadequate N, another inadequate P. To keep root and total plant growth high, the root portions must trade nutrients through the shoot via the phloem.

- Third, there is more or less massive scavanging of vegetative tissue nutrients for reproductive growth at the season's end. In the extreme case of herbaceous annuals, which includes most crops functionally if not botanically, the vegetative tissues and their nutrients lose all value when the plant dies; only the nutrients passed directly to the next generation in seeds or other reproductive structures have value.

# CHAPTER 6

# The Ascent of Sap: Realm of Hypotheses

The upward translocation of water from the roots to the tops of the plants is known as **ascent of sap**. Water is absorbed by the plants through roots from where it reaches the xylem via cortical cells and passage cells, through xylem it reaches the top of the plant where it is transpired by the leaves. The fact that the path of movement of water in xylem has been demonstrated by a number of experiments. The movement of water from the stem base to the tip of the plant is against the force of gravity.

## WATER MOVEMENT IN THE LEAF

Evaporation of water takes place mainly from the surfaces of mesophyll cells which are in contact with the air of the intercellular spaces. Water reaches to these surfaces after passing from the xylem across the bundle-sheath cells, and then from one mesophyll cell to another.

The distance between two veins is only 0.1-0.4 mm, when the finest veinlets are taken into account. Thus water has to move only about this distance from xylem to reach every cell. The pathway of water is similar to that in the root, but there is no Casparian strip in the leaves of seed plants (though there is one in *Pinus* leaf). Thus in angiosperms there is nothing to prevent water passing from cell to cell entirely via the cell wall.

Whether that is the pathway of least resistance is not certain. However, evidences suggest that at least some of the water does pass that way. Experiments conducted using dyes like gold sol and lead chelate to the xylem sap of transpiring plants indicates that dyes were found to accumulate in the cell walls of the leaf mesophyll and epidermis, where evaporation from the cells would be expected.

## ROOT PRESSURE

It is the pressure created in the roots of some plants to aid in transporting **sap** (it is a plant juice or fluid, generally moves through the vascular tissue or is contained within the cell vacuole) upward toward the stem and leaves. When stems are cut from certain plant, e.g. vine or marrow, liquid is forced out from the cut surface. If a manometer is placed on the cut surface, pressure of three to five atmosphere can be measured. This ability of a plant to force out liquid under pressure has been called **root pressure**.

Root pressure can also be measured by placing the roots in an external solution so that no liquid escapes from the cut surface. Then the root pressure is balanced by the osmotic pressure of the external solution, and this gives the most reliable method of measurement of root pressure. Root pressure shows both a diurnal and seasonal rhythm, in general, the pressure being greater at midday and in spring. Exudation of sap on account of root pressure is usually noticed in well watered plants of rejected species. In most of the plants the sap level retracts back into xylem which can be forced to surface as a droplet using external pressure. This situation is exploited to measure water potential with pressure chamber.

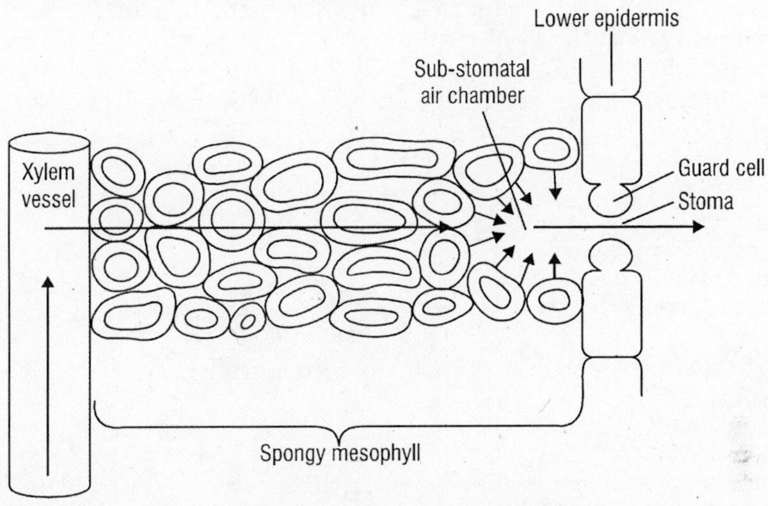


**FIG. 6.1:** Anatomy of dicot root. (a) Structure of dicot root showing different cells through which water (and mineral salts) may be drawn from the soil into the vascular tissues. (b) The three pathways through which water might flow from cell to cell: 1, vacuolar pathway; 2, symplast pathway; 3, apoplast pathway. (d) Casparian strip.

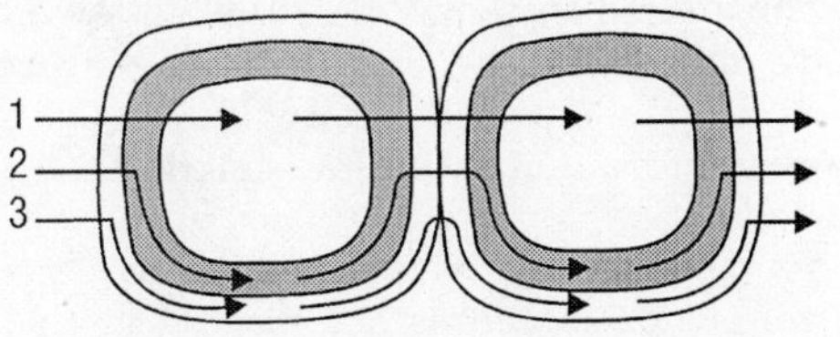


**FIG. 6.2:** The three pathways through which water might flow from cell to cell. 1, vacuolar pathway; 2, symplast pathway; 3, apoplast pathway.

## Root Pressure is the Metabolic Activity

Root pressure is related to the metabolic activity of the root because of the following reasons:

1. Roots placed in an atmosphere deficient in oxygen have a considerably lower root pressure.
2. When roots are treated with potassium cyanide, which suppresses respiration, root pressure is again considerably reduced.

Indoleacetic acid stimulates the root pressure of tomato roots at low concentrations but is inhibitory at high concentrations.

## Root Pressure as Subsidiary Force

The volume of the exudate from the cut stem is small, probably some 5 per cent of the total amount of water which is being transported through the plant. Root pressure is not therefore believed to be the primary mechanism by which water is moved through the root. It is a **subsidiary force**, of negligible magnitude at certain times of the year, producing only a small fraction of the total water movement required for maximal transpiration. Such force is absent in conifers. The primary movement upon which the plant depends for its water supply must be related to osmotic forces. Root pressure is an additional force dependent on the metabolic activity of the plant. The mechanism by which metabolism gives rise to root pressure is still unknown.

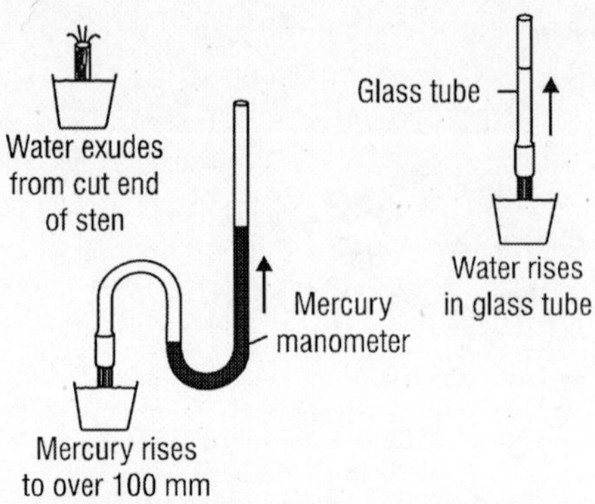


**FIG. 6.3:** Demonstration of root pressure.

## VITAL THEORIES

Those theories which considered the 'vital activities; in the stem to be responsible for the ascent of sap are celled vital theories. The belief arose due to the fact that the vessel and trachieds are closely associated with the living cells present in the xylem tissue (xylem parenchyma and xylem ray cells). However, there is no direct evidence in favour of such a view.

This view was first proposed by Westermeir (1883). He considered that the upward movement of water in the stem is affected by the wood parenchyma. The tracheids and vessels served as reservoirs only. Golglewski (1884) put forward the theory to explain the ascent of sap. He proposed that the living cells of xylem (wood parenchyma and medullary rays) caused the upward movement of water due to their pumping activity resulting by the periodic changes in their osmotic pressure. When osmotic potential of the living cell decreased, water from the tracheids was diffused into it. Thus, incerasing their osmotic potential, and then water would be pumped into the tracheid above. This caused a sort of staircase movement.

Janse (1887) supported the **relay pump theory** of Godlewski. He showed that when the lower portion of a branch was killed, the leaves above were affected.

But these theories were not supported experimentally. Thereafter. celebrated Indian scientist Sir J. C. Bose (1923) proposed **pulsation theory** and stressed strongly for a vital theory for explaining the ascent of sap,. According to him ascent of sap is due to the pulsatory activities of the innermost cortical cell lying just outside the endodermis.

For the experiment, he used a galvanometer, a thin copper wire, and a potted plant. He attached one end of the galvanometer to the plant and other end to the electric probe. When the fine needle of the electric probe was inserted into the stem, the needle of the galvanometer showed movement by oscillation. When the needle tip reached the innermost layer of the cortex, the oscillations became fast and sustained. This was not the case when the needle was inserted on either side of that particular layer. This led Bose to believe that the cells of this layer were puls a long and the pulsatory activity of them was responsible for the pumping of water into the next higher cells. Molisch (1928) supported Bose's theory.

### Objections to Vital Theories

Experiments by Strasburger (1891) and others, however have caused objections to the vital theories. He showed that stem in which the living cells have been killed by the exposure to high temperatures or by poisons, are still capable of water translocation. This experiment raised questions regarding the theory.

## ROOT PRESSURE THEORY

This theory is based on hydrostatic pressure which is developed due to the water accumulation by the root cells. The theory is based on the pressure created by the root cells, therefore, called root pressure theory. If a plant stem is cut near its base, xylem sap is seen flowing out through the cut end. This phenomenon is called exudation or bleeding.

Root pressure was defined by Stocking (1956) as a pressure developing in the trachiery elements of the xylem as result of metabolic activities to the roots. Thus, root pressure is referred as an active process. Root pressure develops largely due to an osmotic potential phenomenon that is created as a result of active absorption of the salts by the roots. It appears to be affected by the factors which also affect respiration (e.g., respiration inhibitors, supply of oxygen, poisons and auxin).

Several investigators have observed that root pressure causes autonomic diurnal fluctuation in exudations. There is close agreement between rate (the relative speed of a liquid released to the cut surface of the stem). Vaadia has suggested that diurnal fluctuation of exudation rates is caused by a periodicity of salt transfer into the xylem.

### Objection to Root Pressure Theory

1. The magnitude of root pressure is not sufficient to push water to the height reached by most trees e.g., Gymnosperms, which have the tallest trees of the world.
2. Root pressure may be significant factor in water movement, when the conditions for transpiration are poor. However, when the transpiration rate is high, root pressure is absent.

## PHYSICAL FORCE THEORIES

All these theories which consider the non living cells are responsible for the ascent of sap are called physical force theories.

### *Capillarity Theory*

Bochn (1809) proposed that the ascent of sap was due to the phenomenon of the capillarity of the tracheids and vessels. The capillarity theory faces the following objections: (1) Capillary forces can not account for the rise of water even through the trees of average height. (2) In the plants where vessels are absent e.g., Gymnosperms, the capillary theory is not applicable.

### Imbibitional Theory

The imbibitional theory was first proposed by Unger (1868) and later supported by Sachs (1878) and it considers the ascent of sap to be due to imbibitional forces in the thick walls of xylem cells.

The imbibitional forces ranging from 100 - 1000 atm were sufficient to carry water to any required distance. However, the movement of water through imbibition by colloids is extremely slow. The theory was discarded, as the ringing and phyging experiments demonstrated that the movement of water is through the lumen of the xylem vessels, rather than along the wall. Moreover, the imbibitional movement of water has been found to be too slow to account for the ascent of sap.

### Atmospheric Pressure Theory

This theory considers that the atmospheric pressure is responsible for the ascent of sap. Transpiration from the leaves causes fall in atmospheric pressure and water moves in order to maintain this pressure.

There are two objections to this theory: (1) The presence of free surface at the lower end of plant is essential for the creation of atmospheric pressure, which is not found in plants. (2) The atmospheric pressure alone is not adequate to rise water in trees 10-100 meters high.

This theory was proposed by Dixon and Jolly (1894) and was supported by Renner (1911, 1915), Curtis and Clark (1951) Bonner and Galston (1952) Kramer and Kozlowski (1960), Milburn and Johnson (1966), Hammel (1967) and Levitt (1969). These physiologists in the end of the 19th century formulated a model to account for the rise of sap in tall trees. One of its elements, the cohesion of water, was not familiar from everyday experience, so the model was quite controversial. There are three basic elements (forces) of the cohesion theory for the ascent of sap:

## COHESION - TENSION THEORY OR TRANSPIRATION PULL THEORY

(i) The driving force, (ii) hydration (adhesion), (iii) and the cohesion of water.

### Driving Force

The force in which the gradient in decreasing (more negative) water potentials from the soil through the plant to the atmosphere. Water moves in the pathway from the soil, through the epidermis, cortex, and endodermis, into the vascular tissues of the root, up through the xylem elements in the wood, into the leaves, and finally, by elements in the wood, into the leaves, and finally, by transpiration, through the stomates into the atmosphere. It is the special structure of this pathway (the relatively small diameters and thick walls that prevent collapse of the tubes), the low osmotic potentials of living leaf and stem cells.

### Hydration Force

The cell walls have hydration properties, especially in leaves, that make the system function. *The hydration force between water molecules and the cell walls is caused by hydrogen bonding* and is called **adhesion**, which is an attractive force between unlike molecules.

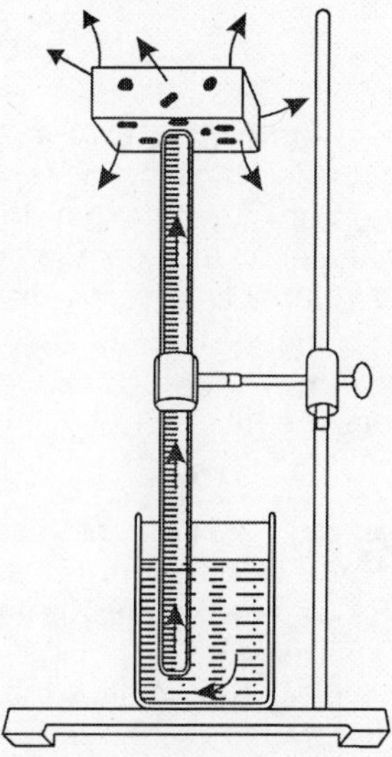

**FIG. 6.4:** A physical system demonstrating cohesion-tension theory.

## Cohesion Force

It is an attraction between like molecules is the key. This is the attractive *force between the water molecules in the pathway*. This is the tension in the xylem. It holds the water columns together. An Irish botanist, formulated the hypothesis that the tensions created by transpiration, osmotic uptake of water by living cells, and the hydration of the cell walls (all these drawing water from the pathway within the plant) were relieved by an upward movement of water from below, with the columns of water being held together by cohesion.

In that special environment these cohesive forces are so great (water has such high tensile strength) that when water is pulled, by osmosis and by evaporation from its holding points in the cell walls at the top of a tall tree, the pull extends all the way down through the trunk and the roots into the soil. Whereas a water column in a vertical pipe of macro dimensions would normally cavitate as just described, cavitation does not stop the flow of sap in the plant because of its highly specialized anatomy.

## Tensile Strength of Water

The problem is whether water can sustain tensions of up to -3.0 MPa without cavitating. It is very difficult for determination of the tensile strength of water. There are three possible approaches to understand the mechanism of tensil strength and the third one seems to be conclusive enough to provide a final answer.

### *First*

The knowledge about hydrogen bonding in water suggests that potential cohesive strength under ideal conditions is extremely high, enough to resist a tension of several hundred MPa. But our theories of liquids remain imperfect, therefore, some assumptions must be made.

### *Second*

High tensile strengths is experimentally observed e.g., by measuring the force required to separate steel plates held together by a water film and Fern annuli that are pulled apart by water under tension have been studied. Glass tubes were sealed while full of hot, expanded water, and then cavitation was observed as the water cooled and contracted. This produced values for the tensile strength of water, and even of tree sap with dissolved solutes and gases, on the order of -10 to -30 MPa (100-300 bars). Plant physiologists measured only lower values, on the order of -0.1 to -3.0 MPa. These results probably reflect interface conditions instead of tensile strengths of water (water tensions on the order of -1.5 to -3.0 MPa).

### *Third*

The experiments are based on the experimental studies of Briggs who used capillary glass tubes bent in the form of a Z. After centrifuging these tubes there is a development of a tension on the water column. He measured greatest tensions in the tubes of smallest diameter. With rather fine capillaries, values as negative as -26.4 MPa were measured, but even with capillary tubes of 0.5 mm diameter (the size of the largest vessels), air-saturated tap water did not cavitate under tensions of -2.0 MPa. Cavitation did occur when the center of the Z-tube was frozen with dry ice; because air is virtually insoluble in ice, freezing expelled the dissolved air. Briggs further indicated that cavitation is more likely to occur in large vessels than in small tracheids.

## Demonstration of Cohesion-tension Theory

In this experiment, a hollow, long glass tube is taken, with its one end submerged in a beaker of water the tube is filled with water so that there is an unbroken connection between the water in the beaker and in the tube. If the thoroughly soaked sponge is placed at the other end of the tube, a connection is made between the water held by the sponge and the water held by the sponge and the water in the tube; an unbroken column of water can be pulled up from the beaker. The process can be accelerated by using a fan and a heat lamp. The rat at which the water moves up the tube is directly related to its rate of evaporation from the sponge. Water evaporating from the sponge is replaced by water from glass tube, which in turn is replaced by water from the beaker.

This physical example is comparable with a plant growing in soil in a natural habitat. The glass tube is analogous to the tracheary tissue of the xylem, the evaporating surface of the sponge is analogous to the evaporating surface of the leaf mesophyll and the water in the beaker could be compared with the soil water.

Thus, in case of plant, as water transpires from the leaf mesophyll cells, it causes a decrease in the water potential of the mesophyll cells in contact with the air spaces of the leaf. The water moves from cell to cell within the mesophyll along water potential gradients. This replaces the water lost by the cell surfaces. Finally the water within the leaf will be transmitted to the water in the xylem elements in the veins, hence exerts a pull or brings the water in the xylem tissue in a state of tension. This state of tension continues through unbroken columns of water from the leaves to the root system. The unbroken or continuous column of water is due to tensile strength or cohesive property of water molecules. The water potential in the living cells of root system and along the cell walls become more negative in comparison to the water potential of the soil, which influences water movement greatly.

## FLOW VELOCITIES AND CONTINUOUS WATER COLUMNS REALLY

Two questions arise in relation to ascent of sap that - What are the flow velocities in xylem? And are the water columns really continuous in plants?

It has been experimentally demonstrated that the water columns in the xylem are continuous. The most elegant method to measure flow velocities is the **heat-pulse technique**, introduced by Rein (1928), for measurement of blood-flow velocities in animals. Many measurements were made of sap velocities with the method, which consists of heating the liquid briefly at a point along the stem and then measuring (with a sensitive thermocouple) the time of arrival of the heat pulse a few centimeters downstream.

It has been observed that velocities depend on distance from the wall of the capillary, for example, but heat-pulse methods can at least give comparative information. As expected, velocities in wide-vessel trees such as ash are much greater than those in narrow vessel trees such as birch. Measurement was made at different points along the trunk. It was found that velocities increase first at higher points along the trunk in the morning as transpiration begins, strongly suggesting that transpiration pulls the sap up the trunk. Reports are also available which suggests that peak velocities (at noon) range from 1 to 6 m $h^{-1}$ for narrow-vessel trees to 16 to 45 m $H^{-1}$ for trees with wide vessels. Slowest velocities occur in conifers.

## PRESSURE BOMB METHOD

Measurement of tension in stem can be performed by pressure bomb method. Scholander reasoned that *if the water in the xylem of a stem is under tension, then the pressure outside compresses the cell walls of the xylem.* Therefore, when the stem is cut, allowing pressure inside to equal pressure outside, the cell walls should expand, and the water columns in the xylem elements should recede from the cut surface. If the pressure difference is reestablished by increasing the pressure on the outside of the stem until the pressure difference is the same as it was before the stem was cut, the water should move back exactly to the cut. The cutting of twig from a tree or a stem with leaves from an herbaceous plant are placed it in a pressure bomb, and the gas pressure is increased on the branch until the water in the xylem can be observed with a hand lens to return to the cut surface. The pressure in the bomb should then be equivalent to the absolute value of the tension in the stem before the cut.

## CHAPTER 7

# Organic Transport in Plants

The solutes absorbed along with water, through roots, are transported to apical portions of the shoot through xylem ducts. These solutes are redistributed in many organs as well. Transport of the soluble products of photosynthesis is called **translocation**.

The present chapter describes how the products of photosynthesis (sugars, amino acids and the like) are transported from the leaves where they are formed to the parts of the plant where they are needed, because all the cells that are themselves unable to photosynthesize need a share of these materials, but the apices of the stem and side branches where cell division and growth are taking place are particularly needy.

**Growing points** of a plant need more food material in the spring and summer when growth is most prolific. The **perennating organs** like tubers, bulbs, corms also need food materials in the later part of the year. The food materials are transported to these organs for storage in soluble form to the growing points of the new plant.

Translocation of soluble products of photosynthesis occurs in the part of the vascular tissue known as the phloem. Phloem tissue consists of several types of cell, but only one, the sieve element, is modified for the massive longitudinal translocation of sucrose and similar solutes which is the primary function of the phloem.

In flowering plants phloem tissue consists mainly of elongated **sieve elements** placed end to end so as to form long **sieve tubes** running parallel with the long axis of the plant. Each sieve element is derived from a cell whose nucleus disintegrates during development. Its end walls, known as **sieve plates,** are perforated by numerous pores which allow the passage of materials from one sieve element to the next.

Closely applied to the side of each sieve element are one or **more companion cells,** which possess a nucleus, dense endoplasmic reticulum, ribosomes and numerous mitochondria. Plasmodesmata connect each sieve element with its adjacent companion cells. The latter are the site of intense metabolic activity, and one is metabolic needs. On purely structural grounds it wood seem that translocation of food materials takes place along the sieve tubes.

**Apoplastic *vs* Symplastic Transport:** A plant body consists of two anatomical parts: apoplast and symplast. Apoplast is in fact dead part of the plant and situated outside the plasmalemma barriers and include cell wall of all cells. Symplastic parts are delineated by the plasmalemma and essentially these are the living parts of the plant. These include cytoplasm of all cells in plants, although it would probably include vacuoles. The symplasts are interconnected through cytoplasmic connections called plasmodesmata. The translocation of solutes can take place either through apoplast or symplast.

## TRANSLOCATION OCCURS IN THE PHLOEM

Following are the evidences which suggest that translocation of solutes takes place in phloem.

### Ringing Experiment

Some of the earliest investigations involved **ringing experiments**. It so happens that in a mature tree trunk the phloem is confined to the bark, indeed the living component of bark is almost entirely phloem tissue. Now if a ring of bark is stripped off a tree trunk it can be shown that the sugar concentration increases immediately above the ring and decreases below it, indicating that the downward movement of sugars is blocked at that point.

### Radioactive Tracers Technique

More critical investigation have been carried out with tracer technique. If a plant is exposed to carbon dioxide labelled with radioactive $^{14}C$, the $^{14}C$ becomes incorporated into the end products of photosynthesis which are subsequently detected in the stem. That these substances are confined to the phloem can be shown by cutting sections of the stem, placing the sections in contact with photographic film and making autoradiographs. It is found that the sites of radioactivity correspond precisely to the positions of the phloem.

### The Aphid Technique for Analysis of Phloem Sap

In recent years number of aphids and coccids feeding on phloem tissue have been used to demonstrate the phloem translocation of carbohydrate. Several aphids are allowed to tap the phloem of an experimental plant. They are then anesthetized, and their bodies carefully cut away from the stylet. When this operation is successful, the stylet remains in a sieve tube element and continues to exude its contents, often for several days. It is now possible for the skilled, steady handed investigator to collect the exudate from the stylet stump and analyze its contents chemically. Phloem stream carries chiefly sucrose and also substantial amounts of other sugars containing three or four basic sugar units. The rate at which fluid exudes from an aphid proboscis may exceed 5 $mm^3$ per hour, which must delight the average aphid as it means that it does not even have to suck in order to get a plentiful supply of food; the pressures in the plant are sufficient to force out the fluid. For a botanist this provides a means of calculating the speed of translocation. An exudation rate of 5 $mm^3$ per hour means that an individual sieve element must be emptied the speed of translocation in an individual sieve tube is of the order of 1000 mm per hour, a figure which is about twice as great as those obtained from work with radioactive tracers.

**Speed of Nutrients Flow within the Sieve Tubes:** The speed at which the nutrients flow within the sieve tubes can also be estimated via the aphid stylet technique; however, radioactive tracer methods are probably more suitable for quantitative studies.

## CHARACTERISTIC FEATURES OF SIEVE ELEMENTS

Sieve element has the following characteristics:

1. A diameter equivalent to that of a transverse end wall sieve plate, with a plasmalemma closely apposed to the wall.

2. A sieve plate which, if free of callose, would have sieve pores pierced through the middle lamella of the average diameter for the species (0.1 to 15 μm each and usually occupying between 50 and 60% of the total plate area). Callose is present on the sieve pores, occluding them to a variable degree. The callose is bounded by a unit membrane continuous with the wall plasmalemma.

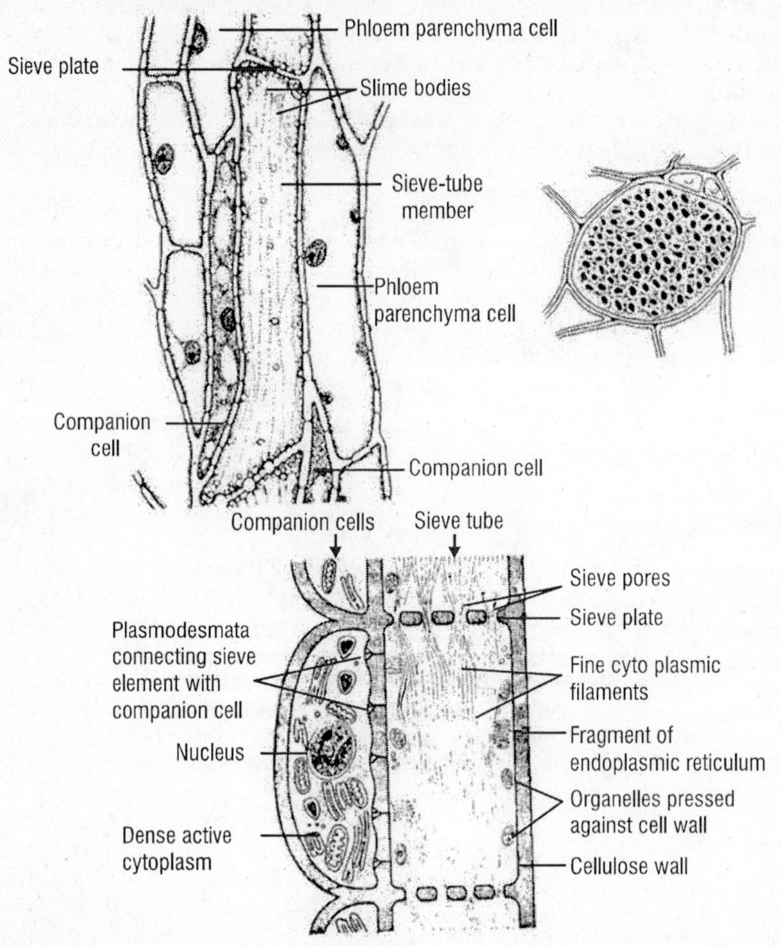


**FIG. 7.1:** (a) Longitudinal view of sieve element, (b) Superficial view of sieve plate, (c) Cellular inclusions of sieve tube and companion cell.

3. Plastids and mitochondria each with an intact double membrane boundary. The plastids usually contain starch grains and the mitochondrial inner membrane is folded into cristae in the normal manner.
4. Flattened cisternae arranged singly, or in parallel or patterned aggregates occur sporadically on the walls. The minimal width of the cisternae is within the normal range found for endoplasmic reticulum in adjacent parenchyma or cambial cells.
5. Phloem protein (P protein or slime) fibrils 8 to 10 μm in diameter and often

with 8-10 μm transverse banding occur in a meshwork in the lumen of the sieve tube and through the sieve pores. In the sieve pores the P protein fibrils may be packed much more tightly than in the sieve tube. Such fibrils have not been conclusively shown to exist in gymnosperms.

6. The only pores through the wall of the sieve tube connect to specialised cells developmentally and/or functionally related to the sieve tube which die when the sieve element dies, normally after one year. These pores characteristically have many fine branches on the specialised [albuminous (gymnosperms): companion (angiosperms)] cell side and a single wide pore, usually bounded by callose, on the sieve tube side.
7. There may be remnants of a nucleus consisting of dense clumps of material, but these are rare in angiosperms, more frequent in gymnosperms.
8. There are no golgi bodies, ribosomes, microtubules or tonoplast.

The conifers and ferns do not have sieve tube elements as such. Translocation in these plants occurs in cells known as sieve cells. Sieve cells, in contrast to sieve tube elements, are not associated with companion cells and do not have sieve plates located on their end walls. Sieve cells have many areas that contain groups of narrow pores, and these are distributed uniformly on all wall surfaces. Sieve elements are extremely sensitive to injury and plugging.

## PHLOEM TRANSPORT: BIOCHEMICAL ASPECTS

The conducting systems which are used by cormophytes for the long distance transport of assimilates are composed of living cells. They differ from normal parenchymatous cells in their cytological structure, but they still have a selectively permeable ectoplasm with a plasmalemma. The sieve elements are supposed to have a metabolism which differs from that of the other cells of the phloem. Since the transport substances in the sieve elements are quantitatively and qualitatively different from the compounds in the donor and acceptor tissues, the loading and unloading of the conducting cells must be accompanied by chemical transformation of the substances. Therefore, various biochemical and bioenergetic processes are involved in phloem transport.

### Biochemical Reactions Involved in Loading of the Sieve Elements

In most plant species sugars or sugar alcohols form the bulk of the transport substances in the phloem. About 400 species have been analysed for the sugar composition of the sieve tube sap, belonging to 43 orders of the Magnoliophyta (angiosperms). 145 species from 70 genera and 35 families were studied by Helga Kluge (1967). On the basis of the composition of the sugars in the sieve tube sap three main types of species can be distinguished.

1. Species with sucrose as the only or at least as the predominant sugar.
2. Species which have considerable amounts of oligosaccharides as well as sucrose.
3. Species which contain considerable quantities of sugar alcohol's in addition to the sugars.

### The Biochemical Capacity of the Sieve Tubes

Considering the special cytological structure of the sieve elements (lack of nucleus and of the tonoplast, scarcity and poor development of the mitochondria, special appearance of the cytoplasm) one would assume that these cells would not have very pronounced metabolic activities.

### *Enzymes*

The sieve tube sap shows activity of a considerable number of enzymes of the carbohydrate and the nitrogen metabolism. A metabolic utilisation of the sucrose (e.g. for the formation of starch in the sieve tube plastids and of callose) seems possible, however, by the action of the UDPG-fructose transglucosylase, which is weakly active in the sieve tube sap. The potential activity of the different enzymes indicate that sucrose can be transformed into starch ad vice versa, and that hexose-phosphates, once formed, would be metabolised in the sap at least down to the ethanol. It is possible that this effect is one of the functions of the high ATP content in the sieve tubes. This ATP pool, however, is not a static one, because the phosphate group(s) is substituted very fast by $^{32}PO^{3}_{4}{}^{-}$ when supplied to the phloem.

### *Respiratory inhibitors and phloem transport*

There are several reports that respiratory inhibitors block phloem transport. Surprisingly, exclusion of oxygen from the bundles does not inhibit this translocation. Most of the measured metabolic activity of the whole phloem should, therefore, be attributed to the companion cells (and to a lesser degree to the phloem parenchyma), which show an unusual density of mitochondria and of cytoplasm.

## PHLOEM LOADING AND UNLOADING

An excellent, quantitative investigation of the processes by which sugar is loaded into the sieve elements from the chloroplasts of the mesophyll cells has been described by Gunning, Pate and co-workers. There are two following possible parallel pathways from the mesophyll to the phloem:

1. The sugar may move through the symplasm, chiefly by diffusion. There are many plasmodesmata connecting the cells all the way from the mesophyll to the sieve element, so that symplasmic transport of sugar is an obvious possibility.
2. Sugar may move from the mesophyll to the phloem by diffusing through the apoplasm or extracellular free space.

Sugar is probably transported across the chloroplast membranes and the cell membranes by the action of carrier enzymes or permeases. Kursanov (1968) has speculated in some detail on possible biochemical reactions, he believes the membrane penetration requires a successive phosphorylation and dephosphorylation of the sugar with the involvement of ATP, the hydrolysis energy of which provides the metabolic energy input for transport. The only energy input to drive the symplasmic flux will be at the chloroplasts since it is generally considered to be unlikely that active transport can occur at plasmodesmata; therefore the high sugar concentrations in the phloem can only be achieved by symplasmic loading if there are even higher sugar concentrations just outside the chloroplasts. Hartt and Kortschak (1967) proposed apoplasmic loading of the phloem will be driven by additional energy transduction processes.

Water is a **second major component** of the phloem translocation stream. The one thing common to all present models of translocation mechanism is that all presume there to be a net flow of water along the sieve tubes from the leaves and obviously this water must be supplied to the sieve tubes in the leaves from the xylem. The driving force for water entry to the phloem is the osmotic gradient created by the accumulation of sugar, and we may therefore suppose that a second reason for the proliferation of membrane surface in the transfer cells is to ensure an adequate supply of water.

There are apparently two following parallel pathways to load sugar into the phloem:

1. Through the symplasm from the mesophyll
2. Through the apoplasm from the mesophyll.

They may be to some extent complementary; with strong illumination and active photosynthesis the symplasmic route may be the more important, because high sugar concentrations in the mesophyll will accelerate symplastic diffusion.

## PHLOEM TRANSLOCATION: MECHANISMS

Phloem is a morphologically distinct tissue type in the vascular system of the higher plants which conducts in its sieve tubes long distance translocation of various key substances from one region to another. By far the major constituent (other than water) of the phloem translocation stream is carbohydrate, sucrose in most plants, but raffinose, stachyose and verbascose have also been reported, and in a few cases the sugar alcohol's mannitol and sorbitol have been reported. The ionic contents of phloem sap are apparently under membrane regulation; potassium concentrations are much higher than sodium, magnesium levels are much greater than calcium, and the most common inorganic anions are chloride and phosphate.

### Translocation Mechanism: Models

It is generally agreed that the soluble products of photosynthesis, formed in the photosynthetic tissue, enter the sieve tubes by **active transport.** But what causes the movement of materials once they are in the sieve tubes? On this there is still no complete answer. Although sugars and amino acids tend to move along concentration gradients, the speed at which they travel is far too fast to be explained simply by diffusion.

Independent movement of different substances strongly suggests that translocation involves some sort of active mechanism. This is supported by other lines of evidence. For example, phloem tissue has a high rate of respiration, and there is a close correlation between the speed of translocation and the metabolic rate. Again, lowering the temperature and treatment with metabolic poisons both reduce the rate of translocation, suggesting that it is an active energy-requiring process which cannot be explained by physical forces alone.

There are presently four proposals of translocation mechanism, the Münch model, the electro-osmotic model, the peristaltic tubule model and the streaming model. All these models presume that translocation involves a bulk flor of sucrose solution along the sieve tubes.

#### *Münch Model*

The model is based on following aspects.

1. The sieve element-transfer cell (companion cell) complex is bounded by a semi-permeable membrane,
2. Sugar is accumulated in this complex, and

3. Xylem water tension is not too negative, it is certain that there will be osmotic water entry into the sieve tube at the leaf, along the other pathways.

In other regions of the plant where sugar is utilised or converted to osmotically inactive forms, there will be a reverse flow of water from phloem to xylem. Under these appropriate conditions, water must osmotically enter the phloem at the loading zone and leave at the unloading zone and indeed all the other proposals of translocation mechanism accept, at least tacitly, that water loading and unloading does occur osmotically much in the manner originally suggested by Münch (1930).

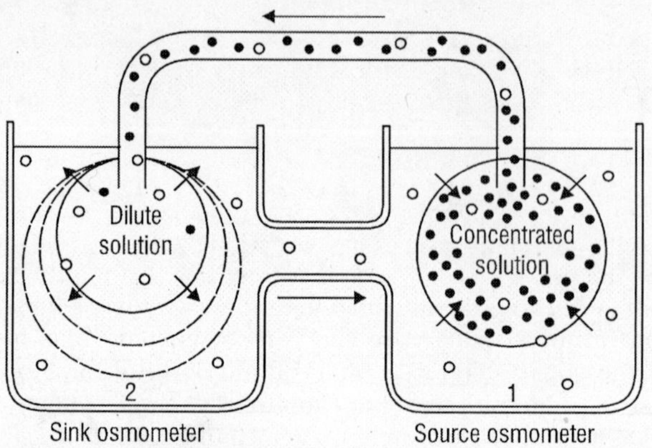


**FIG. 7.2:** A model, originally demonstrated by Münch, to illustrate the mass flow hypothesis.

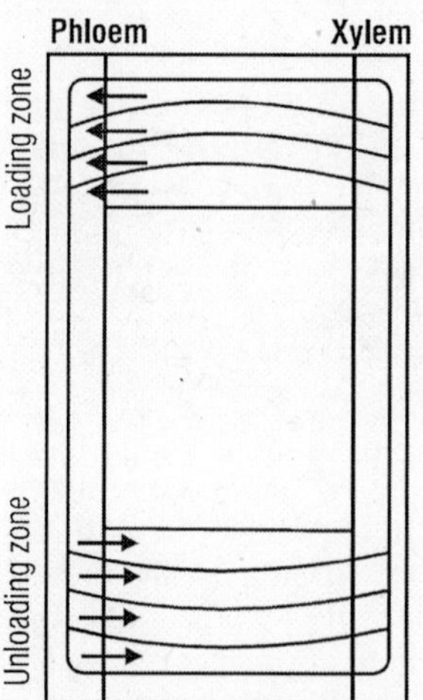


**FIG. 7.3:** A schematic representation of the circulation suggested on the Münch model. Solutes (represented by the heavy arrow) are accumulated into the phloem in the loding zone and are exported at the unloding zone, while water (represented by the light arrow) circulates around phloem and xylem in the manner indicated.

The only uncertainty about the Münch proposal is whether the hydraulic resistance of the sieve tube is sufficiently low that the turgor pressure in the phloem at the leaf, created by the osmotic entry of water there, can by itself drive translocation at the observed velocities. Water may be expected to exchange between the xylem and phloem of a vascular bundle all along its length, so that there will be at every point, local balancing between xylem and phloem water potentials. The continual water balancing between xylem and phloem, will always be taking place. In that sense the Münch model, like the poor, is always with us. The question is whether it is the only translocation mechanism or whether there are others operating.

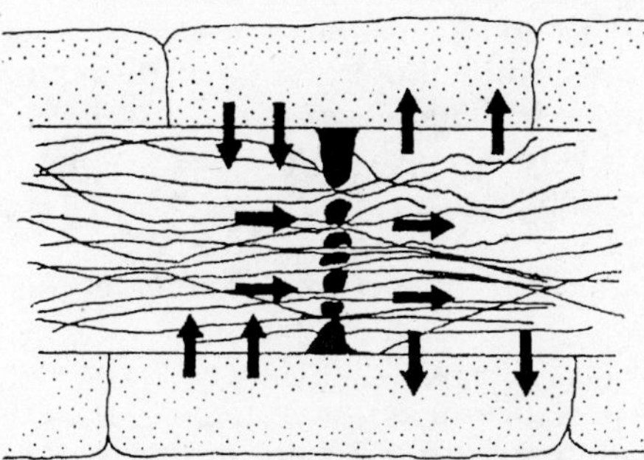

**FIG. 7.4:** The electro-osmotic model suggests that $K^+$ fluxes (here represented by the heavy arrows) circulate from the companion cell, through the sieve plate and back into the companion cells as indicated. The sieve pores are thought to contain many P-protein fibrils which cause the only available channels from sieve element to sieve element to be narrow and charged, thus maximising electro-osmotic efficiency. The translocation stream is dragged through the sieve pores as the $K^+$ flux is maintained.

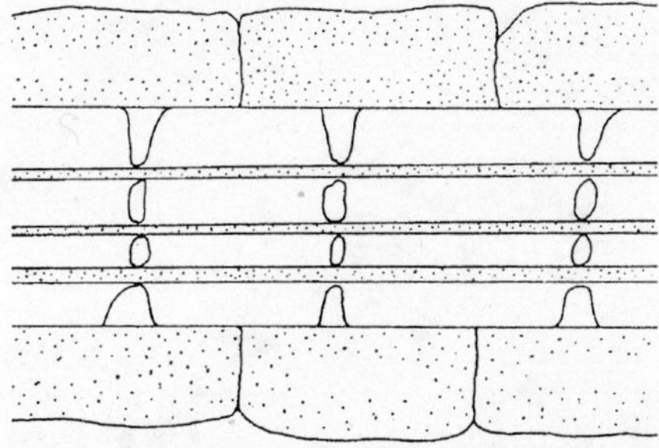

**FIG. 7.5:** A schematic representation of the structure of the sieve elements as proposed by Thaine (1964).

## *Electro-osmotic Model*

This model was originally suggested independently by Fensom (1957) and Spanner (1958). The basic idea is that sugar solution is driven through the sieve pores, and therefore along the sieve tubes, by the electro-osmotic drag of $K^+$ ions, accumulated and secreted in turn across every sieve plate by the action of the companion cells. The metabolic energy requirements, the magnitude of the $K^+$ fluxes across the companion cell membranes, and particularly the apparent impossibility of translocating anions in the phloem, all cause this proposal to seem highly unlikely.

### Peristaltic Tubule Model

Thaine (1964, 1969) suggested that the sieve tubes of the higher plants contain a much more highly organised system of transcellular tubules than the majority of anatomists would agree to. He further suggested that translocation mechanism based on this tubular system has been modified and quantified.

Aikman and Anderson (1971) showed that Thaine's proposal would require apparently acceptable rates of energy supply from phloem respiration to maintain observed translocation rates, even if all the flow were driven by peristalsis. The model basically supposes, following: (i) a series of tubules runs from sieve element to element, passing through the pores, and that these tubules are (ii) bounded by contractile microfibrils much like those in cilia or flagella. Rhythmic contraction of these microfibrils impels the translocation stream along the tubule. (iii) the turgor pressure gradient generated by the Münch mechanism can drive flow through the tubules.

There are several other attractive features to this model, which may be summarised as follows: (i) contractile systems are the most common and perhaps most efficient mechano-chemical transducers; (ii) sugar translocation is by connective solution flow driven by pressure waves within the tubules and is a hydrodynamically and thermodynamically efficient process; (iii) the metabolic energy input is continuously distributed along the sieve tube; (iv) simultaneous bi-directional transport is possible within a single sieve tube; (v) the number of tubules within a single sieve tube transporting in a particular direction may vary to give a very flexible translocation capability.

In summary, the model seems feasible and attractive but these features must be weighed against the overwhelming preponderance of anatomical evidence which denies the existence of transcellular tubules.

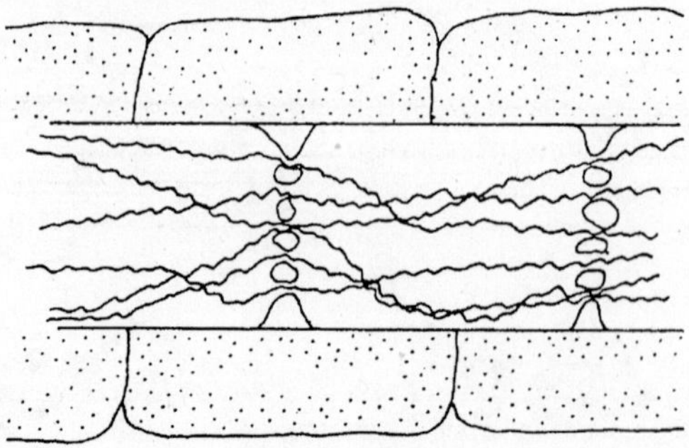

**FIG. 7.6:** Synchronised wave motions along the P-protein fibrils.

### Streaming Model

This model bears some resemblance to the PPM because there is thought to be continuous energy input along the sieve tube to drive translocation. However, it differs markedly from the former in that the sieve elements are not considered to contain organised tubules, but rather a network of P-protein which oscillate in a non-Brownian manner.

MacRobbie (1971) has gone into the possibilities of streaming in sieve tubes as a mechanism for phloem translocation. She has extensively reviewed streaming in slime moulds and algae, and suggests a very interesting comparison between P-protein in the sieve tubes and the contractile fibrils in other cells which produce shearing forces in cytoplasm and thus streaming. However, it must be realised that there is unlikely to be any uni-directional flow of sieve element cytoplasm, because there is no obvious mechanism to re-cycle it.

The streaming model (or the peristaltic P-protein model as it might be better named) is by far the most likely subsidiary translocation mechanism. It has all the advantages of the tubule model with the exception of allowing bi-directional flow of diffusion-segregated streams within a single sieve tube. It has been directly observed in operation in the living phloem of one species and the P-protein network is the structure most commonly observed in electron micrographs.

**Surface Spreading:** Another mechanism which has been put forward is surface spreading. The idea here is that solute molecules might spread over the interface between two different cytoplasmic materials, as oil spreads at a water-air interface. The molecular film so formed could be kept moving by molecules that the films would be so thin that a very large number of them would need to be formed to account for the known rates of translocation. However, sieve tubes do contain numerous membranes and filaments which collectively might provide the necessary surface.

## TRANSLOCATION BETWEEN SOURCES AND SINKS

The transport system in a plant is concerned with the movement of materials from sources to sinks. The terms source and sink can be ambiguous and although useful, can cause confusion. Moorby (1977) defined a source as a region of the plant which is supplying materials of any kind to the transport system, and a sink as a region where the material is being removed or lost from the system.

### Movement into Sinks

The onset of translocation into a region is accompanied by the initiation of growth and/or storage of materials and the direction of translocation can, in part, be controlled by the manipulation of the rates of growth of the sinks or the rates of carbohydrate storage. For example, mature leaves can be induced to impart carbohydrates by shading, a temporary inhibition in the supply of assimilates to *Solanum tuberosum* tubers can lead to the breakdown of previously stored starch and its transfer to regions of the tuber which are still growing and if tomato fruits are cooled they can start to export the carbohydrates already accumulated. In each of these three examples, the translocation pattern re-adjusts so that movement is towards a region of low carbohydrate concentration.

It is easy to imagine how this type of situation can control movement into a major sink such as a growing storage organ. In *Solanum* tubers, for example, the sucrose concentration is low because it is converted to starch. In contrast, in sugar cane, the sucrose is in a storage compartment, probably the vacuole. It is less easy to envisage how such a system operates in a whole plant with many sources and sinks. The simplest situation would be that suggested for tomato (*Lycopersicon*) by Khan and Sagar (1966) with all the leaves supplying assimilates into a general carbohydrate pool, to be drawn on by all the fruit. In most species, however, sinks tend to be supplied by specific leaves and there is usually no absolute dependence of one on the other. Once sinks are initiated, positive feedback relationships, which are presumably transmitted via the sieve tubes, appear to develop within the source-sink system.

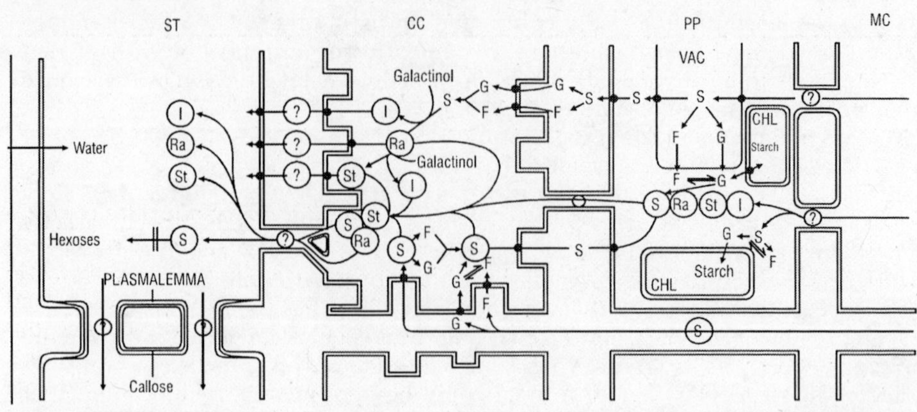


**FIG. 7.7:** Biochemical processes involved with the transfer of photosynthates to the phloem and with the entrance into the sieve tubes. MC, mesophyll cell. PP, phloem parenchyma cell (with amyloplasts and vacuole). CC, companion cell (developed as a transfer cell with wall protuberances); note the absence of larger vacuoles and in general of amyloplasts and the specialised plasmodesmos to the sieve tube member. ST, sieve tube member with plasmalemma. I Supposed carriers in the membranes. Sites, for which active transport is discussed, but not proved. CHL, chloroplast; VAC, vacuole.

## The Effect of Growth Substances

It has been recognized for a considerable time that there are high concentrations of growth substances in regions of active growth, i.e. the sinks, and there have been numerous demonstrations of transport into a region in response to the application of growth substances. Davies and Wareing (1965) suggested that in their experiments auxin (IAA) acted directly on the transport system in the sieve tubes because TIBA (tri-iodobenzoic acid), which inhibits the polar transport of IAA, also reduced IAA-stimulated transport. Mullins (1970), however, demonstrated that even within the short period of his experiments protein synthesis was stimulated. Yet a further possible interpretation of this type of experiment was made by Lepp and Peel (1970, 1971) when they suggested that IAA could stimulate the loading of sucrose into the sieve tubes.

# CHAPTER 8

# Transpiration: A Queer Navigation

The basic function of photosynthesis is served by almost every adaptation of the leaf, from its internal characteristics- such as transparency, shape, and air spaces- to the way the leaf is positioned on the plant. Yet each of these photosynthetic adaptations also promotes the greatest hazard facing the plant: dehydration. The leaves of the vascular land plants represent a design compromise between the conflicting requirements for exposure to atmosphere and light, and protection against water loss. For instance, in the hottest part of the day the leaves of *Eucalyptus* trees orient themselves parallel to the rays of the sun, limiting solar heating of their tissues and minimizing water loss by evaporation.

A forest interior can be as much as 10° to 15° F cooler than the surrounding countryside not just because of the shade but as a result of the head consumed by the evaporation of hundreds of gallons of water each day from large trees, each of which is estimated to have the cooling ability of five 10,000-watt air-conditioners. *This loss of water from the aerial parts of the plant is termed as transpiration and occurs almost entirely through the leaves.*

The loss of water from a leaf surface takes place mainly through the stomata, the water vapour originating from the wet surface of mesophyll cell walls which open onto the sub-stomatal cavity. If the air surrounding the leaf is not fully saturated, water will evaporate from the free water surface within the leaf, diffuse through the stomata or lenticels and out into the air.

Transpiration occurs from the leaves is known as **foliar transpiration**. However. Stomata are not only the sites of water loss but are also exclusive ports of entry for photosynthetic $CO_2$. Thus stomata are required to maintain a water-photosynthesis syndrome. Prof. Stocker described the situation as a queer navigation between the Cliffs of dehydration and starvation.

**Stomatal transpiration:** When water is lost through stomata is termed as stomatal transpiration. Water evaporates from their moist walls into the internal atmosphere of the intercellular spaces of the mesophyll, which becomes saturated with water vapour. The outer atmosphere is usually unsaturated. The result is the diffusion of water vapour from the intercellular spaces of the leaf into the outer air through the stomata. Consequent on the outward diffusion of water vapour, the concentration of water vapour in the intercellular spaces decreases with the result that more evaporation from the wall of the mesophyll cells takes place.

**Cuticular transpiration:** When a small amount of water vapour is lost from the leaves and herbaceous stem by direct evaporation from the epidermal cells through cuticle, the process is known as cuticular transpiration. The cuticle being almost impervious to water, the amount of water lost through it is comparatively small and varies inversely with the thickness of the cuticle. Cuticular transpiration represents only $1/10^{th}$ or $1/20^{th}$ of the total loss from a leaf, as shown by comparing the loss of weight of an untreated detached leaf with that of a similar leaf in which the stomata have been blocked. In herbaceous plants growing in humid and shady places the cuticle is poorly or not at all developed. Transpiration from stems, fruits and flower parts is mostly cuticular.

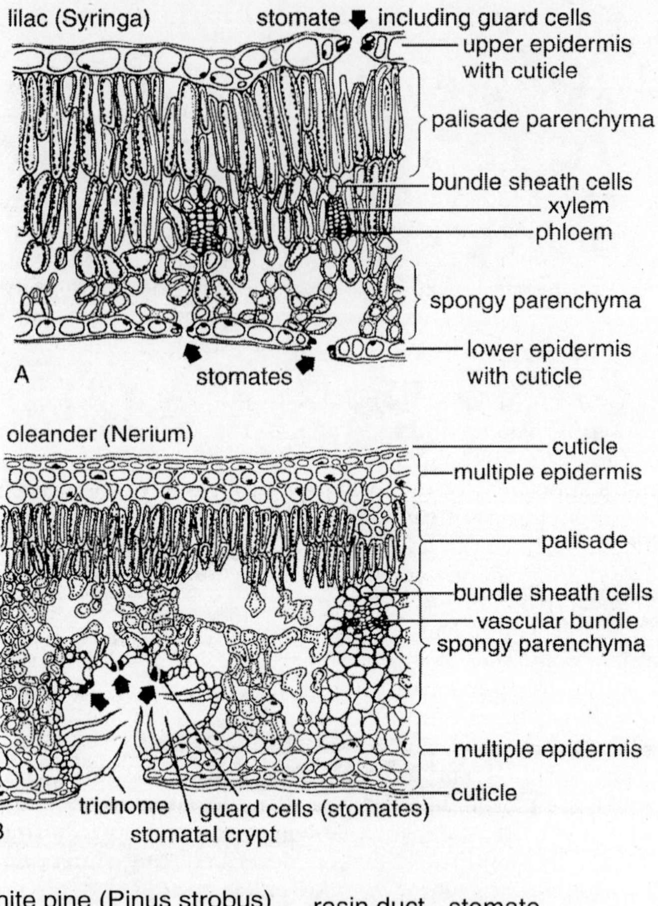


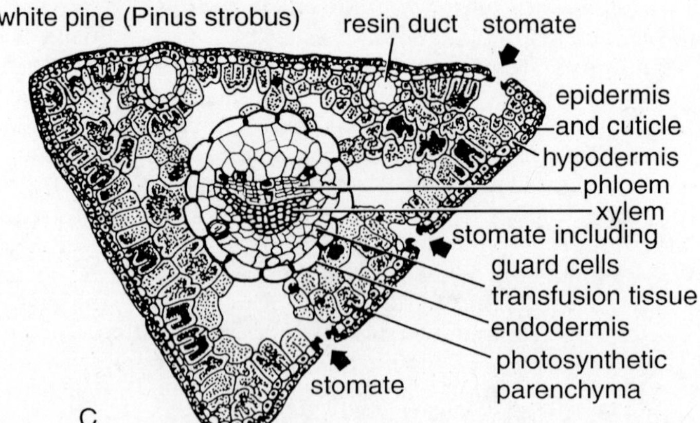

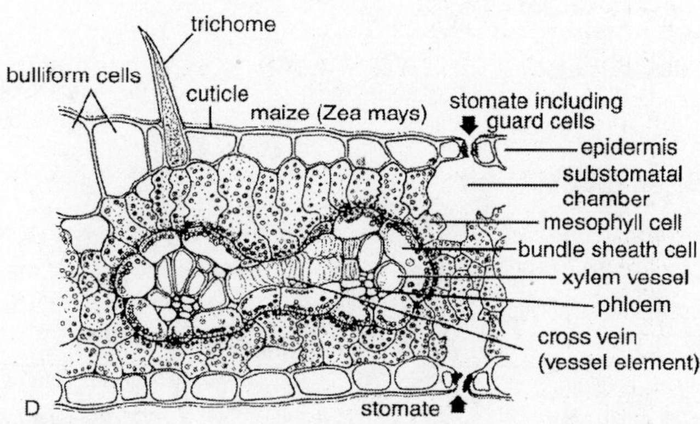


**FIG. 8.1:** Cross section through four representative leaves, one with "normal" stomates (A), one with stomates deeply sunken in a stomatal cavity (B), one a pine needle (leaf) with slightly sunken stomates (C), and one a grass leaf with about equal numbers of stomates on both surfaces (D). Arrows point to the stomatal pores, but the stomates include the guard cells. Spongy parenchyma and palisade parenchyma (as in A and B) collectively form the mesophyll. Note details of differing leaf anatomy; pine and grass leaves do not have a palisade layer, for example. Bulliform cells of maize (D) shrink in response to water stress, causing the leaf to roll into a cylindrical shape.

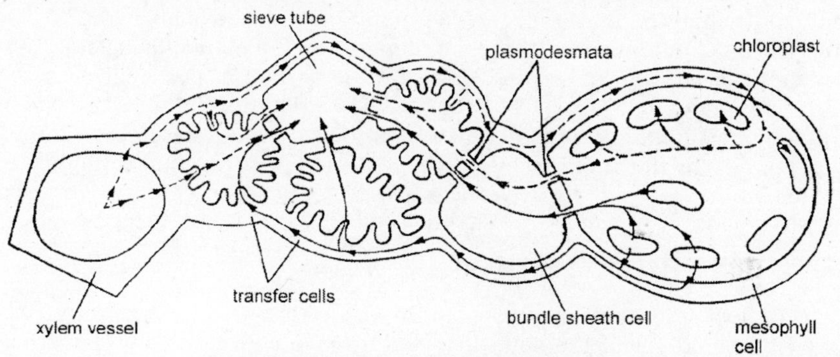


**FIG. 8.2:** Cross section through a typical leaf showing the movement of water during transpiration.

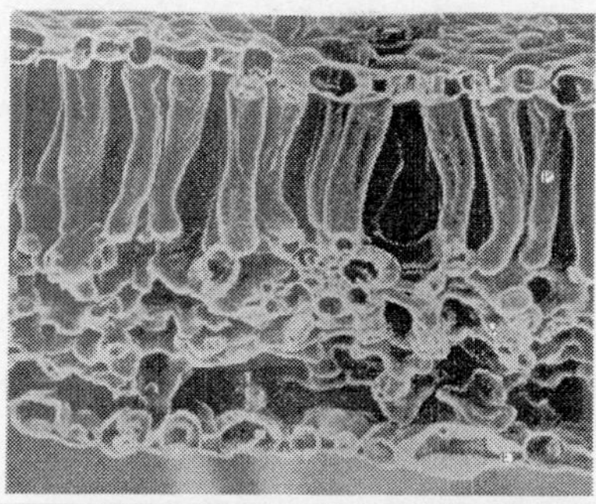

**FIG. 8.3:** Transverse view of a mature broadbean (*Vicia faba*) leaf. P = palisade cells, M = Mesophyll cells, E = Epidermis.

**Lenticular transpiration:** Water loss from lenticular spaces is known as lenticular transpiration. Loss of water from a woody stem with a covering of bark takes place largely through the lenticels, with only a small loss through the bark tissue.

## THE NATURE OF TRANSPIRATION

Evaporation differs from transpiration, however both concern with water loss. The main differences are as follows:

**TABLE 8.1:** Differences between evaporation and transpiration

| Evaporation | Transpiration |
|---|---|
| It is just a physical phenomenon. | It is a vital and physiological phenomenon. |
| Guard cells are not involved in it. | It is regulated by the activity of guard cells. |
| Any kind of pressures are not involved. | Transpiration involves different types of pressures. |
| No guard cells or stomata are involved | Guard cells regulate the surface area of transpiration. |
| It may take place from any surface. | It is a process of the living cells. |
| Evaporation causes dryness of the surface. | Transpiration keeps evaporation surface cool. |

**Evapotranspiration (ET):** It is the quantity of water transpired by plants during their growth, or retained in the plant tissue, plus the moisture evaporated from the surface of the soil and vegetation. That is, amount of water between field capacity and wilting point within the root zone becomes available for ET. This is nothing but the **Consumptive Use** (CU) of water by the plants, because water use for metabolic activities is very insignificant (less than 1 per cent of ET) and therefore it denotes use for losses due to ET. Under field conditions, incoming solar radiation supplies the energy for ET process. Wind promotes removal of water vapour from the field and the prevailing temperature and humidity result from the interaction of these processes and these affect the process of transpiration as well.

## TRANSPIRATION RATE: MEASUREMENT

The rate of loss of water by the plant in the process of transpiration can be measured by a number of methods. The water vapour given off by plants may be collected and condensed or determined in the vapour phase by physical or chemical methods. Alternatively, the change in total weight of pot plant can be measured, provided that loss from the soil is prevented. A suitable balance is required, since the relative change in weight may be quite small, if a large volume of soil is present. The loss in weight of a detached plant organ can be determined more easily and accurately. If the plant is in the light there is in addition a loss in weight due to respiration. In most conditions the changes in weight resulting from these two processes are negligible compared with the decrease in weight due to the loss of water vapour.

## TRANSPIRATION PATH

There are three major paths of transpiration: non-stomatal external diffusion path, stomatal path and non-stomatal internal diffusion path.

### The Non-stomatal External Diffusion Path

The diffusion path for transpiration originates at the mesophyll cell wall, continues across the sub-stomatal cavity, through the stomatal pore, and then through a layer of still air beyond the leaf surface. This layer of still air, is known as the **boundary layer.** It is the air path through which the diffusing water molecule must pass before it can be regarded as free and part of the external air. The thickness of shearing layer vary with wind velocity. The boundary layer resistance varies with wind velocity, vapour pressure deficit (VPD) and foliar geometry.

## Stomatal Path

It was thought at the end of the 19th century that whether the relatively small stomatal openings in the leaf could provide the major pathway for the gaseous exchange of plants. F.F. Blackman showed that when the stomata on the lower epidermis of leaves of *Nerium oleander* were blocked, the transpiration rate was reduced to 10 per cent of its previous value. This suggested that cuticular transpiration was only 10 per cent of the total.

## The Non-stomatal Internal Diffusion Path (Mesophyll Resistance)

The remaining part of the diffusion path is across the sub-stomatal cavity. Its relative magnitude is dependent on the anatomy of the leaf, mainly on both the total surface area of the cells abutting on to the sub-stomatal cavity and on the mean length of path across the sub-stomatal cavity. This can be studied by cutting series of sections and observing the three-dimensional structure of the sub-stomatal cavity.

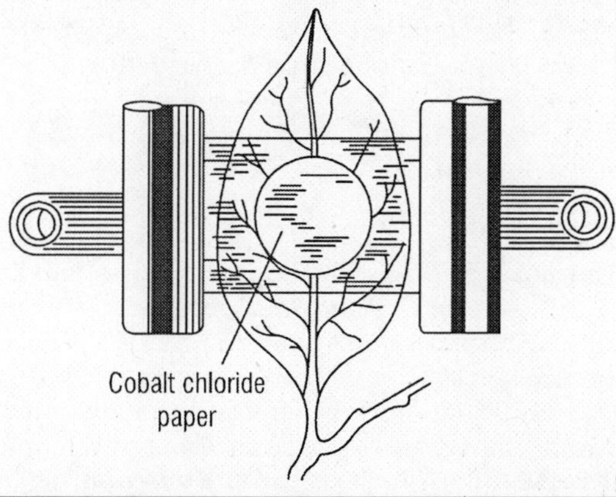


**FIG. 8.4:** Quantitative estimation of unequal transpiration using cobalt chloride method.

**Cobalt chloride method:** This method is used to study the sub-stomatal cavity and the stomata. In order to do this an indicator of the rate of loss of water vapour, such as a piece of paper impregnated with cobalt chloride (3%), is placed on the leaf surface. This will change colour from blue to pink as it takes up water vapour. The cobalt chloride is placed in immediate contact with the leaf surface and the time for a given colour change to take place is determined. Now the same piece of cobalt chloride paper is placed a known distance away from a free water surface. This is conveniently done by taking as a free water surface a piece of moistened blotting paper and placing the cobalt chloride impregnated paper at a fixed distance above it, using a plastic spacer. If the time for the change of colour in the cobalt chloride paper is the same for the leaf as for the physical model, the leaf may be regarded as having an equivalent air diffusion path equal to the thickness of the plastic spacer. Hence it is possible to determine the total sub-stomatal cavity combined. If the physical dimensions of the stomata are known the diffusion path equivalent to the stomata can be calculated and hence the relative contribution of the sub-stomatal cavity and stomata can be determined.

## Relative Magnitude of the Parts of the Diffusion Path

Thus it is possible to estimate approximately the relative contributions of the three parts of the diffusion path. For a sunflower leaf with widely open stomata these values are approximately one unit for the stomata, one unit for the boundary layer (but dependent on wind conditions), and three for the substomatal cavity i.e. the stomata represent only one-fifth of the total diffusion path. As the stomata close, their contribution to the total path will increase. It follows that since when fully open any change in the stomatal path is a change in only one-fifth of the total path, stomatal changes have little effect on the rate of transpiration. But when the stomata are nearly closed the stomata will exert a considerable degree of control over the rate of transpiration. Stalfelt experimentally demonstrated with leaves of *Betula pubescens* that the degree of control was less when widely open.

The same considerations determine the rate of diffusion of carbon dioxide into the leaf and hence the rate of photosynthesis. But the total diffusion path for carbon dioxide into the leaf must also include diffusion through the cell wall and entry into the chloroplast. Thus, in addition to the three parts of the diffusion path which were concerned in transpiration, namely stomata, shearing layer, and sub-stomatal cavity, there are two additional terms, the cell wall and the internal cell path, to be taken into account. By comparing observed rates of photosynthesis for a sunflower leaf with the calculated maximum rate of diffusion through the stomata alone, it appears that the total equivalent diffusion path for carbon dioxide is of the order of nine times that due to the stomatal path alone. Hence, with fully open stomata, changes in the stomata are changes affecting only one-tenth of the total diffusion path for photosynthesis and can have only a small effect on the rate. But as the stomata close they begin to have a greater effect; furthermore, closure will affect the rate of transpiration more than the rate of photosynthesis and will preferentially decrease transpiration relative to photosynthesis.

In the above discussion the important effect of the anatomy of the leaf on gas exchange has been emphasized. Plants with sunken stomata and/or hairy surfaces were regarded as potentially likely to have lower rates of transpiration and thus to be adapted to live in drier places. But measurements by Maximov showed that all such plants did not have significantly lower transpiration rates than plants without such characteristics when measured under the same conditions of adequate water supply. It is therefore important to distinguish between *xeromorphic* leaves which have specialised morphological characteristics that have sometimes been thought to limit water loss and leaves of those plants called *xerophytes* which actually grow under dry (xerophytic) conditions.

## Measurement of Stomatal Aperture

1. The simplest method of measuring stomatal aperture is by direct observation with a microscope at various depths. Difficulties arise in the case of hairy leaves, leaves with sunken stomata, and leaves in which the cuticle overhangs the guard cells.

2. An alternate method devised by Lloyd is to strip off the epidermis, plunge it rapidly into absolute alcohol, in an attempt to fix the stomata in the state in which they were on the leaf, and then examine it microscopically.
3. A simple qualitative method to determine the relative stomatal size is by the use of various liquids of different surface tension. A small drop is placed on the leaf surface and penetration or failure to penetrate is noted. Penetration into the sub-stomatal space is shown by a darkening of the leaf area surrounding the point of penetration owing to the air space becoming filled with liquid. Alcohol, benzene, and xylol form a progressive series, alcohol penetrating only widely open stomata but xylol entering those which are slightly open. An alternative method is to form an impression of the stomata using a silicon rubber liquid which hardens on the leaf and can then be stripped off.

**Porometer Method:** The classical instrument for studies of *relative* stomatal aperture is the porometer. In this method air is drawn bodily through the leaf and the rate of movement of air is determined under a fixed pressure gradient. A cup is sealed on to the leaf surface and the air pressure within it is lowered by suction. Air enters through the stomata in that part of the leaf outside the cup, moves through the intercellular spaces and enters the cup via the stomata within to restore normal atmospheric pressure in the cup. The time course of pressure within the cup is observed, using a manometer.

**Diffusion Porometer Method:** This method has been devised to avoid a mass flow of air through the leaf. The porometer cup is filled with a gas, such as hydrogen which passes from the cup through the stomata and time course of the subsequent fall of pressure in the cup is recorded. Successive measurements with either type of apparatus result in a flushing of the leaf with air or hydrogen and some drying out of the leaf takes place. Also, changes in the carbon dioxide content of the gas in the porometer cup during the experiment (particularly probable if the plant is illuminated) may affect the stomatal aperture. The time measured in the porometer depend on the combined resistance to gas flow offered by the stomata and the intercellular spaces which connect the sub-stomatal cavity within the cup with those outside. It is generally believed that the limiting effect is that of the stomata within the cup, but this may not always be the case.

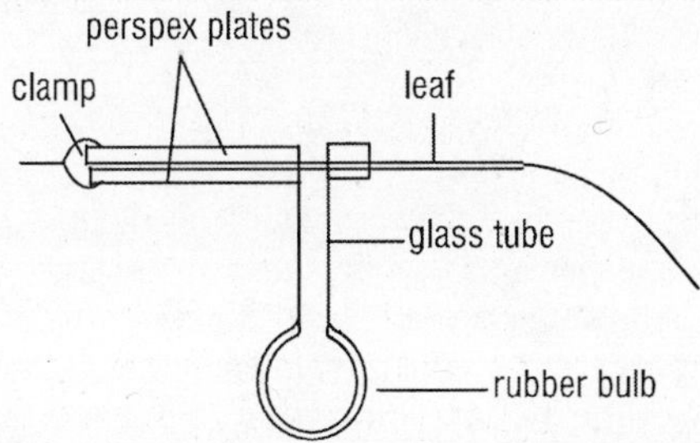


**FIG. 8.5:** A simple porometer.

An estimate of the stomatal contribution to the total diffusion path is most easily determined from the measurement of pore size and depth; these can be best obtained by direct observation combined with Lloyd's method.

## Estimation of Stomatal Index and Stomatal Frequency with Replica Method

Wolf and his associates (1979) developed this rapid, simple and suitable method and can be employed in most of the crop plants which are non waxy and non hairy. Blue/red correction fluid or transparent finger nail polish (collodion) is used as replicating material. Fluid-dipped brush is stroked across the area to be replicated with a single motion. After few seconds when fluid become firm, the replica can be removed. A small piece of celluloid tape is secured to a glass slide. The exposed sticky surface is then placed over the replica and with slight pressure the film is removed from the leaf surface. The replica is mounted for viewing as a imprint on reverse side of the image of the leaf surface. Using microscope with pre-calibrated grid, the number of stomata and epidermis cell can be counted.

The **stomatal index** (SI) can be calculated as follows:

1. Take a circular piece of paper of the size of eyepiece with a rectangular cut in the centre. Insert the paper in an eyepiece.
2. Count the number of epidermal cell (E) and stomata (S) within the rectangular area.
3. Plug in the values in the following equation and get the values of stomatal index.

$$SI\ (\%) = S/S+E \times 100$$

**Stomatal frequency** (*f*) can be calculated as follows:

1. Count the number of stomata within the area of rectangle/square. Half cut stomata may also be counted.
2. After knowing the scanned area and the stomatal number, *f* can be calculated in $mm^2$ per unit leaf area.

## STRUCTURE OF STOMATA

Leaf epidermis has several tiny pores called stomata which are microscopic and surrounded by two guard cells which control their opening and closing.

**TABLE 8.2:** Stomatal number, epidermal cell and SI of different field crops

| Crop | Stomatal number | Epidermal cell | SI (%) |
|---|---|---|---|
| Bean | 4 | 12 | 25.0 |
| Cajanus | 5 | 12 | 29.4 |
| Cotton | 6 | 32 | 15.8 |
| Groundnut | 9 | 36 | 20.0 |
| Sorghum | 8 | 20 | 28.6 |
| Tomato | 7 | 25 | 21.9 |

### Microscopic Anatomy of Guard Cells

Each guard cell is a crescent-shaped, semicylinder, and each pair is joined at the ends. One of their most striking features is the chloroplasts, which make them unique among epidermal cells, others of which are not photosynthetic at all. The cell wall of the guard cell is considerably thicker on the side nearer the stomatal opening. Additionally, fibers of cellulose in the cell wall are arranged in bundles, the radial micelles, which coverage at the thickened, stomatal side of the cell wall. This arrangement of the cell wall fibers causes the concave surfaces of the two guard cell to pull apart when the volume and pressure of the cells' contents increase, thus opening the stoma. Conversely, when this pair of cells becomes flaccid, it relaxes not uniformly but in such a way that the stoma closes.

### Stomata: Number and Density

The number and density of stomata are extremely variable even within a single species. Plants growing in the shade have fewer, but larger, stomata than those growing in exposed sunny conditions, whilst plants growing in permanently damp environments have more, but smaller, stomata than those growing in dry habitats. Despite these obvious differences between plants growing under contrasting environmental conditions, the total stomatal pore area is comparatively stable because there is an inverse relationship between stomatal size and density. Differences are found even within a single plant so that lower leaves tend to have larger but less numerous stomata than those of higher insertion.

### Internal Air of the Leaf

Air accounts for much of the volume of even the thinnest leaf. If a leaf is immersed in water and gradually warmed, tiny bubbles of gas can be seen to stream from the stomata as the trapped air expands. The total cellular surface exposed to this internal air may be more than 200 times greater than that of the leaf itself, but evaporation into it is relatively slow, since the internal air is at 100% relative humidity all of the time.

### Stomatal Complex: Development

Stomatal opening or closing is mediated by changes in the shape of the guard cells, which surround the stomatal pore. There are basically two morphologically different types of guard cells. The most common form are as follows:

#### Elliptic type

In this type, the guard cell pair are kidney-shaped with the aperture formed between the adpressed concave (ventral) sides of the two cells.

#### Graminaceous type

It is restricted to the Glumiflorae (grasses and sedges) and consists of a pair of dumbbell-shaped guard cells. Both types of guard cells are surrounded by epidermal cells which may or may not be of a distinctive shape. If the abutting epidermal cells are morphologically indistinguishable from other epidermal cells, they are called neighbouring cells, but if they are clearly distinct they are known as *subsidary cells*.

The guard cell pair and their associated subsidiary and/or neighbouring cells are collectively referred to as the stomatal complex and are typically subtended by a gap in the underlying mesophyll tissue known as the substomatal chamber.

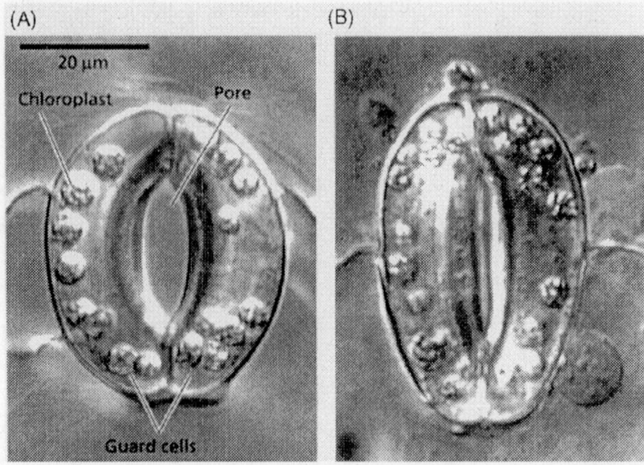


**FIG. 8.6:** Stomata in detached epidermis of *Vicia faba,* showing the open (A) and closed (B) states. Stomatal apertures are measured under a microscope as aperture widths. Measurements of the change in aperture as a function of experimental conditions allow the characterization of stomatal responses to different environmental stimuli.

## STOMATA: DEVELOPMENT

During leaf development, the leaf meristem cuts off, amongst, other things, protodermal cells which eventually mature into epidermal cells. Guard and subsidiary cells, as well as leaf hair and gland cells, do not arise directly from the meristem but are the products of protodermal cells. They become secondarily meristematic. Neighbouring cells, unlike subsidiary cells, are simply protodermal cells which become secondarily meristematic. Neighbouring cells, unlike subsidiary cells, are simply protodermal cells which mature in a position abutting onto guard cells.

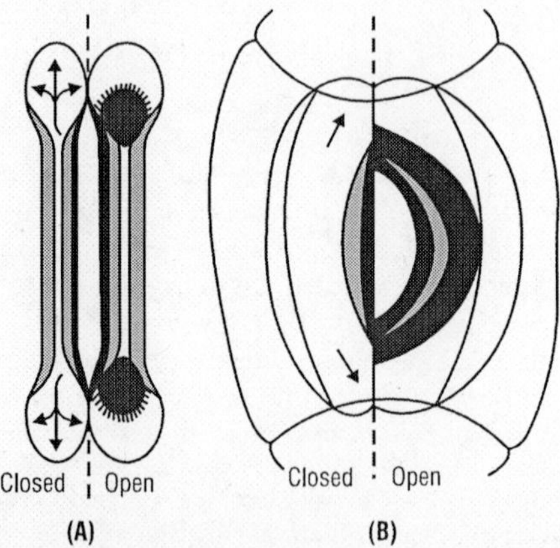


**FIG. 8.7:** Graminaceous (A) and elliptic (B) guard cell complexes. Stippled areas, cell walls: hatched areas, lateral subsidiary cells; solid black margin of stomata, cuticularized ledges. The arrows in the left half of each figure show the direction of cell expansion/extension which occurs during stomatal opening. The striations in the right half of each figure indicate the orientation of the cell wall micellae.

## Development of Stomatal Complexes

### *Agenous development*

The stomatal meristemoid divides symmetrically once to form a pair of guard cells. There are no subsidiary cells.

### *Mesogenous development*

The stomatal meristemoid divides more than once to produce a guard pair and at least one subsidiary cell. The mesogenous subsidiary cell(s) arise from asymmetric division(s) before the meristemoid finally divides symmetrically to form the guard cells.

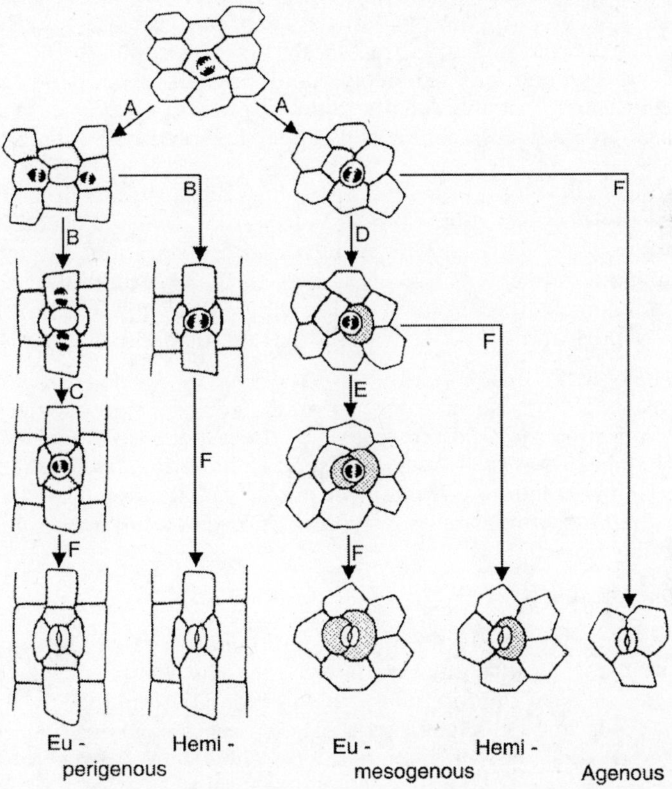


**FIG. 8.8:** Developmental pathways of stomatal complexes. Protodermal - epidermal/neighbouring cells are cross-hatched, stomatal meristemoids - guard cells are unshaded, perigene subsidiary cells are vertically-hatched and mesogene subsidiary cells are stippled. Illustrating development of: (A) stomatal meristemoid, (B) lateral perigene subsideary cells, (C) polar perigene subsidiary cells, (D) mesogene subsidiary cells, (E) second mesogene subsidiary cell and (F) guard cell pairs. If the guard cells are completely surrounded and contacted by subsidiary cells, their developmental pathway is prefixed eu,- if they are surrounded and contacted by both subsidiary and neighbouring cells, they are prefixed hemi- . Very occasionally guard cells may be found which are associated with both perigenous and mesogenous subsidiary cells and their developmental pathway is referred to as mesoperigenous.

### *Perigenous development*

As in the agenous pathway, the stomatal meristemoid undergoes a single symmetrical division to form the guard cell pair. However, protodermal cells abutting onto the stomatal meristemoid become secondarily meristematic to cut off daughter cell(s) against the stomatal meristemoid, before the latter, itself, divides, Perigenous subsidiary cells develop independently of the stomatal meristemoid.In all cases the guard cells develop from a stomatal meristemoid. In higher plants, it is smaller product of an asymmetrical protodermal cell division.

### *Number, shape and distribution of subsidiary cells*

These parameters are useful in classification of plants, although their functional role is poorly understood. Not all the ontogenetic pathways are found in all the major plant groups. Almost without exception, bryophytes are agenous, peteridophytes are mesogenous or rarely agenous, dicotyledons are either mesogenous or agenous and monocotyledons are either perigenous or agenous. These differences are probably related to the position of the leaf meristem which is typically basal in monocotyledons and apical and laternal in most other plant groups.

As guard cells develop in the epidermis, gaps or substomatal chambers appear in the mesophyll tissue immediately below the stomatal complexes. These chambers act as gaseous reservoirs which extend deeply into the mesophyll and thus maximize the diffusion of carbon dioxide to the photosynthetic tissues and at the same time increase the diffusion pathway of water vapour from the mesophyll to the stomatal pore and so reduce water loss from the photosynthetic tissues.

## ORGANELLES OF GUARD CELLS

Guard cells are rich in chloroplasts. The photosynthetic thylakoid membrane systems of the plastids are only weakly organized into granal stacks. The peripheral reticulum is probably involved in the transport of metabolites between the plastids and the surrounding cytoplasm; it is a typical feature of bundle sheath plastids in $C_4$ plants.

### Plastids of guards cells

Guard cells have about ten plastids per cell although there tend to be more in shade plants and as many as fifty plus in many ferns. The guard cell plastids in some species is a network of microtubules which extends around the periphery of the organelle. In epidermal cells, guard cell plastids are not inconspicuous and it has been observed that as guard cells open, the plastids become distributed around the periphery of the cells, although this migration could be attributed to displacement as guard cell vacolar volume increase. The plastids of guards cells are very rich in starch inclusion bodies. However, if a comparison is made between the starch inclusions of open guard cells in the light and closed guard cells in dark, it will be seen that that starch accumulates in the dark and decreases in the light; the reverse of what would be expected in normal photosynthetic tissue. Starch is an almost universal feature of guard cells although some *Allium* spp. lack any starch in their guard cells but their plastids are inconspicuous in any case. During stomatal opening, there are also reports that the volume of guard cell plastids becomes greatly reduced. This reduction in size could be connected to changes in the volume of starch inclusion bodies as they become hydrolysed.

### Nuclei of guard cells

The nuclei of guard cells are about the same size as those of other epidermal cells which, in effect, makes them relatively huge because of the reduced volume of guard cells. In *graminaceous guard* cells, the nucleus becomes wedged in the narrow central waist of the cell and protrudes from both ends, or may even become divided into two halves. In *elliptic guard* cells, the nuclei of may be in elongated form (closed state) to a round form (open state), but this may reflect a change in diameter of the guard cell during stomatal movements.

### Mitochondria of guard cell

The number of mitochondria in guard cells are between five and ten times and more numerous with guard cells on a per volume basis than in mesophyll tissues.

### Spherosomes in guard cells

Guard cells are rich in spherosomes which are membrane-bound inclusion which exhibit acid phosphatase activity.

### Other inclusions of guard cells

Guard cells also contrast cells may contain oil inclusions. They are often absent elsewhere in the epidermis (e.g. *Philodendron scandens*, sweetheart vine). The guard cells usually lack crystal inclusions. They may be present in considerable quantities in other epidermal cells (e.g. calcium oxalate crystals in *Commelina communis*, dayflower. Guard cells may lack anthocyanin pigments which pervale other epidermal cells (e.g. *Setcreasea purpurea*, purple heart).

## DEVELOPMENT OF GUARD CELLS

The early developmental phase of guard cells is identical in all plant groups with the immature guard cell pair fused together by their ventral walls. The principal changes which occur during the physical maturation of the guard cells are the differentiation of the cell walls and the development of the stomata.

### Thickening in Guard Cells

During early developmental process, microtubules aggregate in the cytoplasm adjacent to the guard cell plasmalemma. The microtublues are specifically orientated and decide the direction in which the cellulose micellae are laid down in the cell wall. The resulting micellae radiate out around the cell walls from the centre of the guard cell pair. At the same time, certain regions of the cell walls become especially thickened. The thickening in guard cells occurs mainly along the outer and inner margins of the ventral walls, with cell wall materials. This deposit is on the internal face of the walls which they eventually impinge on the contents of the cells.

### Stoma Formation

Stoma formation is initiated by an enzymatic breakdown of the pectinaceous middle lamella of the fused ventral walls of the guard cell pair which ultimately separate the pair. It is probably facilitated by an increase in turgor in the cells which forces the thickened regions of the two cells outwords so that a pore forms between the two. After the formation of the stoma, further differentiation of the cell walls in graminaceous and elliptic cells is very different.

## Elliptic guard cells

As elliptic guard cells continue to mature, there is a further increase in the cell wall deposits along the outer and inner margins of their ventral walls. Eventually these ridges become heavily cutinized and form thickened ledges which protect the throat of the stoma.

## Graminaceous guard cells

Just after the stoma of graminaceous guard cells are formed, the middle regions of the outer and inner guard cell walls become greatly thickened and rod-like to form the typical dumbell-shaped guard cells at maturity.

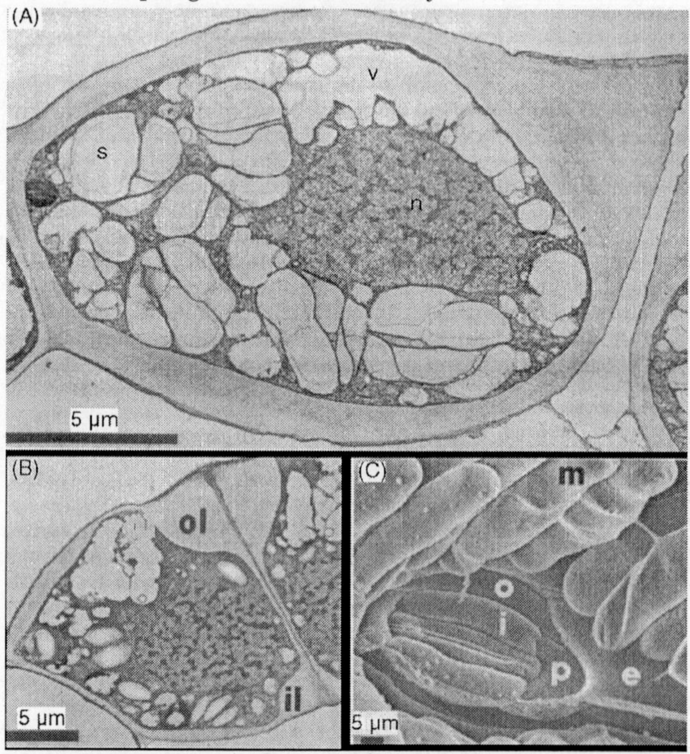


**FIG. 8.9:** (a) Transmission electron micrograph of a T.V.S. through a guard cell of *Polypodium vulgare* (common polypody fern) showing the massively-developed outer cuticularized ledge adjacent to the stoma, starch inclusion bodies (s), vacuoles (v), and nucleus (n). (b) Transmission electron micrography of T.V.S. through an immature guard cell of *Polypodium vulgare*, illustrating the deposition of cell wall material during the development of the outer and inner ledges (ol and il). The guard cell pair are just beginning to separate. (c) Scanning electron micrography of the stomatal complex of *Commelina communis* as viewed from within the leaf. Note how the palisade mesophyll cells (m) abut onto the epidermal cells (e) but not onto the polar (p), outer lateral (o) and inner lateral (i) subsidiary cells.

## STOMATAL MECHANICS

A knowledge of the structure of guard cells is an essential requirement to our understanding of stomatal behaviour and their physiology. The guard cell complexes as metabolic islands which have no direct connections (i.e. plasmodesmata) with their adjoining tissues. Therefore, any rapid stomatal response must arise from the guard cell's own metabolic machinery unless those responses result from environmentally-induced turgor changes within the guard cell complex itself.

The changes in osmotic volume in the guard cells activate the process of opening and closing the stoma.

In *elliptic guard* cells, the cellulose micellae radiate away from the stoma and encircle the cells so that as the volume of the cell increases radial expansion is constrained by the micellae and the bulk of the expansion in directed towards the poles. The resultant linear extension is restricted in the region adjacent to the stoma by the thickened ledges with the result that the cell bends away from the stoma to produce an opening movement. The expansion of the guard cell into their surrounding cells is facilitated.

The central regions of *graminaceous stomata* extremely rigid whilst the bulbous polar ends are comparatively thin-walled and have radiating micellae. As the volume of the guard cell pair increases, the polar ends balloon out and press against their opposite number. In this way, the rigid central regions bounding the stoma are forced apart.

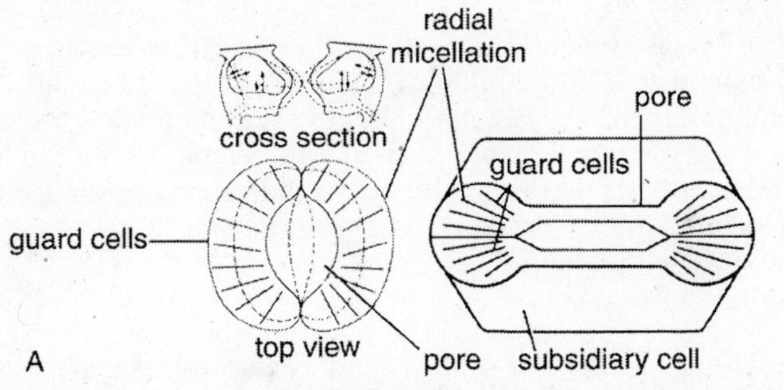


**FIG. 8.10:** Stomatal structures. (A) Schematic drawing of two stomas, showing radial micellation. Left, a dicot leaf (dashed lines show guard cells after loss of water, hence closing); right a monocot (grass) stomate.

### Plasmodesmata and Ectodesmata

Plasmodesmata are microscopic cytoplasmic strands which interconnect adjacent cells and function as transport pathways. These structures occur throughout the mesophyll and epidermal tissues. They are absent from mature guard and subsidiary cells, therefore, there can be no direct passage of metabolites to them from either the mesophyll or the epidermis, and vice versa. Any passage of metabolites to and from these cells must take place indirectly via the aqueous phase of the cell wall.

Ectodesmata are more problematic structures. They are of an extremely transient nature and can be best described as preferential transport channels through cell walls which lack any permanent structure. They are present on the outer guard cell walls, especially in the regions of their juncture with the dorsal walls. The number of ectodesmata varies with environmental factors and are most numerous in the dark and at low temperatures. Their function is uncertain. However, the suggestive functions is that they might be the sites for foliar absorption, cutine deposition or peristomatal transpiration.

Similar structures have also been found at the juncture of the dorsal and inner guard cell walls extending from just beneath the endocuticle and up the dorsal wall to just beneath the plasmalemma of the guard cell and probably act as channels for ion transport.

### Substomatal Ion-adsorbent Bodies

Discrete ion-adsorbent bodies occur beneath the poles of the guard cell pair in many ferns and the monocotyledon families Commelinaceae (spiderworts) and herbaceous Araceae (arums). These bodies are of a pectinaceous nature. They lie in an extracellular position between the inner guard cell walls and the endocuticle which, in the ferns, at least, is extended into an endocuticular sac. The itself interconnects with the endocuticle of the adjacent cells by pectin-filled folds or trabeculae. The function of the substomatal ion-adsorbent bodies is not understood although it has been established that they can absorb a wide variety of ions, including potassium, and may act as ionic reservoirs associated with stomatal movements.

## STOMATA: SIGNIFICANCE

Transpiration, stomatal movement, amount of water in leaves and osmotic value are important which reflect the ability of plant to take water under the prevailing atmospheric and edaphic conditions. The role of stomata is of fundamental importance for the growth and metabolism of land plants. They permit the entry of air, without which growth and eventually life would be impossible for the plant. Yet when the stomata are wide open, they allow water vapour to leave the leaf at a rate only 50% less than would occur if the parenchyma cells were directly exposed to the air. Thus, the open stomata represent a **glaring weakness** in the water-conserving adaptations of the leaf. Liverworts- whose leaflike structures are equipped only with pores for gas entrance and exit, are restricted to wet or very damp habitats. The key stomatal adaptation that allows more complex plants to live even in the desert is the guard cell. These cells enable the plant to close the stomata when the risks of water loss exceed the benefits of continuing photosynthesis.

## STOMATAL MOVEMENT: CONTROL

Stomatal movement is controlled by large number of external and internal factors. On the basis of controlling factors it can fall under five categories, viz., (i) photoactive (ii) scoto-active (iii) hydro-active (iv) passive and active, and (v) diurnal-autonomous movement.

Several hypotheses have been proposed from time to time to explain the mechanism of stomatal movement. The important are:

i) A correlation of stomatal aperture with sugar-starch content of the guard cell was proposed by Sayre (1926) and Scarth (1932). This classical hypothesis was then modified by Levitt (1967). But at present this hypothesis has been rejected because it could not explain the changes in pH and accumulation of $K^+$ inside the guard cells. Moreover, Scarth was unable to explain the replacement of starch by sugar.

ii) A new hypothesis was proposed by several workers like Das and Raghavendra (1974), Raschke (1975), Willmer and Mansfield (1970), Fujino (1967), etc. The hypothesis suggests that stomatal opening involves the participation of an active transport mechanism using ATP as energy source. Alkali cations, mainly $K^+$ ions are involved in such opening.

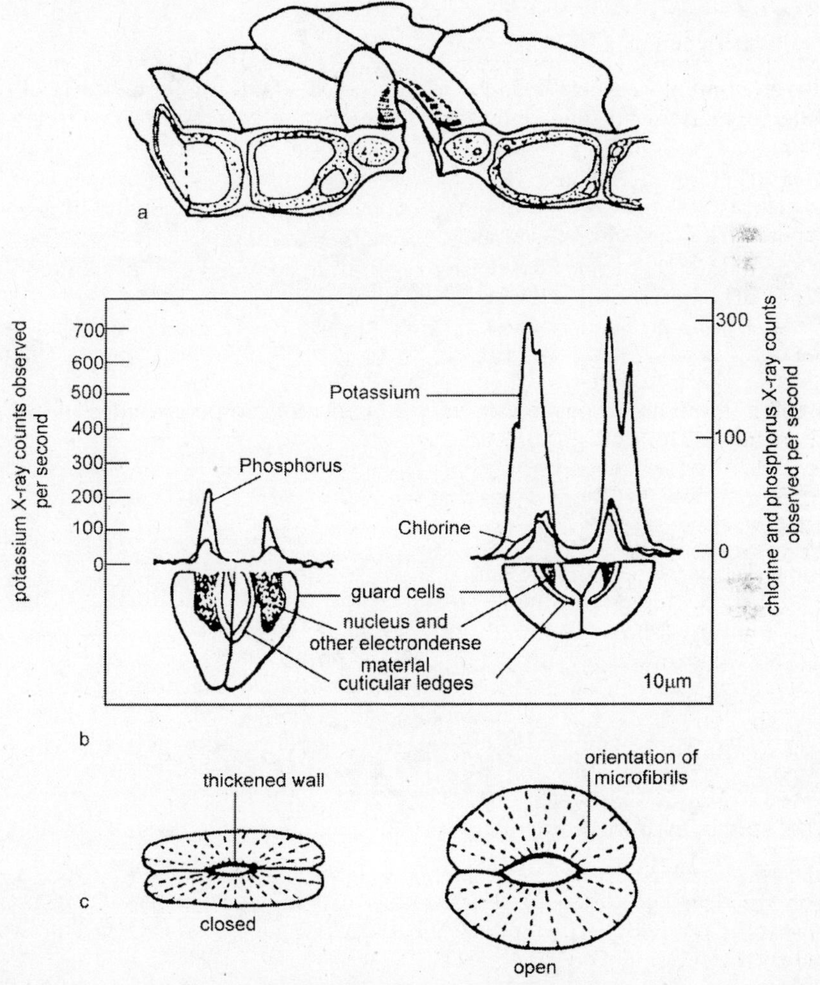


**FIG. 8.11:** Summary of the mechanism of opening and closing of stomata.

iii) Findings of Raschke and Humble (1973) demonstrated that no anions are needed to accompany the $K^+$ ions for entry of anions. It is maintained by the exchange of $K^+$ for $H^+$ ions which became available during the production of organic acids. The source of organic acids in the guard cells is starch.

iv) Stomatal movement is also closely coupled with leaf water potential and is found to be correlated with the synthesis of ABA. It is rapidly synthesized in wilty leaves (Raschke, 1975).

v) Zelitch (1969) indicated that endogenous hormonal levels play a key role in regulation of stomatal aperture, specially in controlling transpiration under condition of limited water supply. ABA plays principal part in controlling of guard cells activity. Wright (1969) demonstrated that ABA level increases when leaves are subjected to water stress.

## MECHANISM OF STOMATAL MOVEMENT

### Starch Sugar Hypothesis - Old Concept

The starch content of guard cells is high in the dark and low in light. However, in the other epidermal and mesophyll cells the case is reverse. Sayre (1926) made following observations:

1. High pH favoured opening of stomata while low pH closed stomata.
2. In light, $CO_2$ concentration of guard cells and neighbouring cells is lowered on being consumed in photosynthesis, hence the rise in pH.
3. In the dark, $CO_2$ of released during respiration accumulates in the guard cells, hence pH of guard cell is lowered.
4. A high pH favours conversion of starch into osmotically active reducing sugars, resulting in an increase of turgor in the guard cells and stomata open. The reverse reaction occurs when pH is lowered. The enzyme phosphorylase, present in the chloroplasts catalyzes this reaction in the presence of inorganic phosphate as follows:

$$\text{Starch} + P_i \xrightarrow[\text{pH5}]{\text{Phosphorylase (pH7)}} \text{Glucose-1-Phosphate}$$

Steward (1964) made comments on the Sayre's observations and pointed out that unless glucose-1-phosphate is further converted to glucose and inorganic phosphate, no appreciable change in osmotic potential can obtained. Inorganic phosphate on the left hand side of the above equation is just as osmotically active as glucose-1-phosphate on the right. Steward proposed the following scheme. At high pH starch is converted to glucose as follows to make the stomata open:

The stomatal closure needs metabolic energy in the form of ATP. In presence of ATP and oxygen, glucose is converted to glucose-1-phosphate. The enzyme hexokinase is responsible for this conversion. Glucose-1-phosphate is then converted to starch in the reaction.

The starch hydrolysis theory was not accepted well because of following drawbacks:

i) The decrease of the osmotic potential of the guard cells during stomatal opening cannot be explained by the formation of glucose from starch in its entirety.

ii) It does not account for the extra-effectiveness of blue light during the opening of stomata. In *Allium* sp. in which the guard cells are devoid of starch.

iii) The interconversion of starch sugar is very slow process and unable to explain the quick stomatal movement or responses to water levels.

## Metabolic Reactions Involved in Stomatal Movement

There are several metabolic reactions involved in the opening and closing of stomata. The difference arises under light and dark. They are summarized as under:

| Guard cell in light | Guard cells in dark |
|---|---|
| Respiratory $CO_2$ in the intercellular spaces is consumed in photosynthesis by the mesophyll | Respiratory $CO_2$ accumulates in the intercellular spaces as there is no photosynthesis |
| Guard cells show alkaline reaction: pH rises. | Guard cells show acid reaction: pH falls. |
| The alkaline reaction favours hydrolysis of starch and sugar | Acid reaction favours the formation of starch from soluble sugars in the guard cells. |
| The osmotic potential of sap of the guard cells increases | The osmotic potential of the sap of the guard cell rises. |
| Water enters the guard cells; its TP (yMV>p) and volume increases | Water leaves the guard cell; its T.P. ($y_{MV>p}$) and volume decreases. |
| Guard cells change their shape and stomata open. | Guard cells change their shape and become flaccid and stomata close |

## Modern Concept

The advancement of physiological knowledge, histological studies, development of instrumental technology and interdisciplinary approaches made it possible to know several aspects of mechanism of stomatal opening to certain extent. However, we still need more experimental evidences to understand it. Several anatomical, environmental and physiological factors are associated with opening and closing of stomata like role of subsidiary of cells, $CO_2$ concentration, ion transport (specially $K^+$ ions), malate switch, metabolic energy, rate of photosynthesis, endogenous hormone level, water stress, etc. All affect stomatal movement in one way or other.

## Ion Transport

It is a widely held concept that active uptake of cations like $Li^+$, $Na^+$, $Rb^+$ and $K^+$ by guard cells is responsible for stomatal movement (Iljin, 1957). The role of starch - sugar regulation in the guard cells has been considered to be an important cause of stomatal opening. Such relation is basically maintained by influx of $K^+$ ions *to and from* the guard cells was found to be responsible in the decrease of their water potential which in turn increases turgor required for stomatal opening. This arises a question about the nature of the anions required to balance $K^+$ ions. Allaway (1973) demonstrated that organic anions are generated from starch and indeed malate level in the guard cells do rise when stomata open. Although, presence of anions like $Cl^-$ has been detected in guard cells of open stomata.

## Malate Switch Hypothesis

Malate switch hypothesis was suggested by Bowling (1976). According to it:

1. Changes in pH of guard cells take place during opening or closure of stomata.
2. Balancing of $K^+$ by the amount of malate and $Cl^-$ present in epidermal tissue.
3. No change in the net amount of $K^+$, $Cl^-$ or malate in epidermal tissue.

According to Willmer et al. (1973) malate can be produced in the guard cells by carboxylation of PEP. Raschke (1975) suggested that incorporation of $CO_2$ into phospho enolpyruvate (PEP) in the cytoplasm could lead to malate accumulation and acidification. He considered that:

1. At low $CO_2$ concentration (less than 100 ppm) carboxylation of PEP contributes to malate requirements of the guard cells for stomatal opening.
2. At higher concentrations carboxylation leads to an excess malate production in cytoplasm which induces stomatal closure.

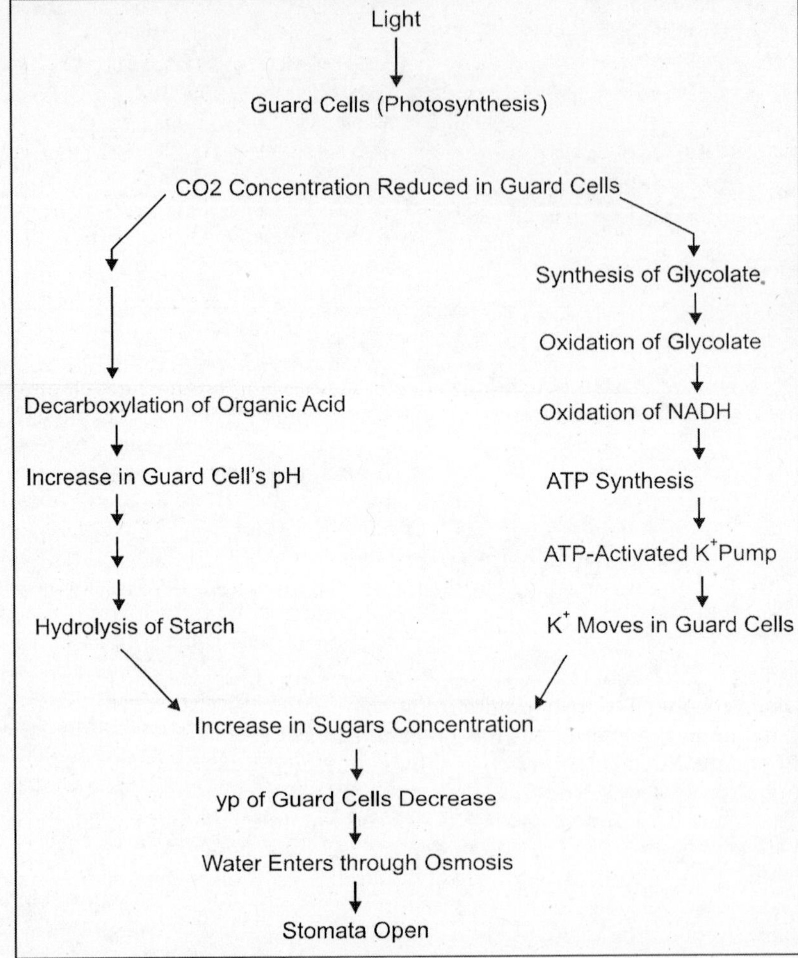


Raschke (1975) indicated that stomatal opening is operated by osmotic effect occurring due to accumulation of potassium salts and sometimes $Na^+$ in the guard cells, and at the same time chloride and malate are the important anions in this effect. Raschke and Schnable (1978) found a balance between KCl and potassium malate in the guard cells of *Vicia faba* which was determined by availability of $Cl^-$. Guard cells of *Vicia faba* contain chloroplast filled with starch while guard cells of *Allium cepa* possess poorly developed chloroplast and do not contain starch. Therefore, guard cells of *Allium cepa* appear to lack substrate for production of malate necessary for osmotic purpose. These cells possibly satisfy their anion requirement solely through import of $Cl^-$.

Das et al. (1977) indicated that in addition to malate, photosynthetic and/or respiratory electron transport-mediated $H^+/OH^-$ efflux could cause change in the pH and provide the necessary $H^+$ ions for exchange with $K^+$.

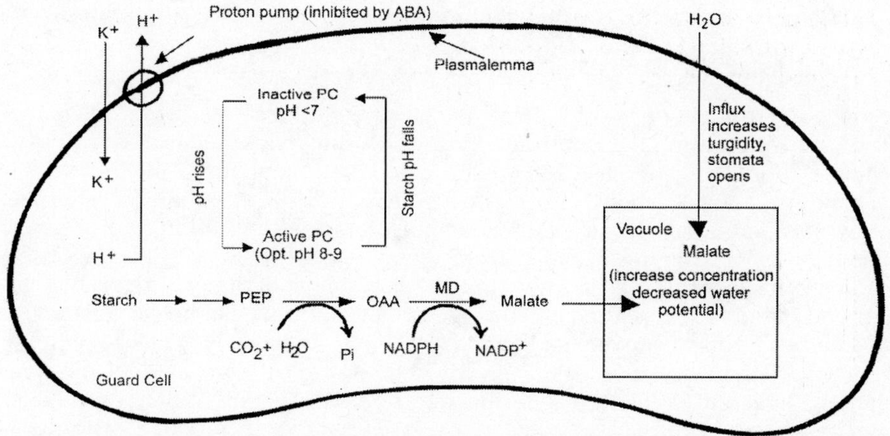


**FIG. 8.12:** Proposed mechanism for opening of stomata. The illustration presents a summary of ion movements and metabolism in guard cells during stomatal opening and closure.

## Fusicoccin and Stomatal Movement

Studies with Fusicoccin (FC) on ion transport, specially $H^+/K^+$ helped us to understand the role of cations in the stomatal opening. FC markedly increases secretion of proton and $K^+$ uptake (electrogenic $H^+/K^-$ exchange) in plant tissue. Therefore, it successfully mimics as natural hormones during physiological activities. FC stimulates uptake of monovalent cations ($Rb^+$, $Na^+$, $Cs^+$, $K^+$) in high salt concentration medium and at low salt concentration medium, uptake of $K^+$ is much greater than that of $Na^+$ and other cations. Therefore, it reveals that FC enhances the selectivity of the transport system for $K^+$. Stomatal opening by FC is mainly due to increase in cell turgor of the guard cells. This effect is dependent on the capacity of $K^+$ extrusion (Raschke and Humble, 1973). The effects of FC on $H^+$ secretion, $K^+$ uptake and potential difference (PD), all are depressed by respiration and inhibitors of phosphorylation. Thus it is evident from the above studies that opening of stomata is metabolism dependent mechanism.

**Ion movement and its inhibition:** ABA inhibits $K^+$ uptake in stomatal guard cells and interferes with energy metabolism in guard cells and consequently close stomata. In this respect $K^+$ uptake is highly coupled with $H^+$ extrusion and FC reverse such inhibitory effects of ABA. Auxin is known to influence energy dependent proton transport at the cell membrane level. Such effects of auxin (and other growth promoters) reveal that fluxes of $K^+$, other ions, and the transmembrane electropotential are closely associated with hormone-induced changes of $H^+$ transport. Little has been done to determine the direct or indirect effects of GA on ion transport.

**Phytochrome involvement in Ion transport:** Stomatal movement has been compared with nyctinastic movement. These systems being described as light-dependent turgor movement, mediated by phytochrome and involving changes in membrane permeability to $K^+$ ions. According to Raschke (1975) changes in stomatal pore were so small and can not be phytochrome-mediated change in the epidermis or mesophyll. The presence and action of phytochrome have not been conclusively shown to be essential for stomatal movement (Rustagi and Mohan Ram, 1977).

### Energy Relationship and Stomatal Movement

In mesophytes, stomata open during day and close with the set of night. This is the result of an **endogenous circadian rhythm**. During day (under light) a leaf experiences both an increase in temperature and reduction in inter $CO_2$ concentration brought about by photosynthesis. As with stomatal closure, reduction of radiation absorbed by plant canopies of the evaporation area of plant will also reduce photosynthesis and growth.

Photosynthesis provides energy through photophosporylation. According to Humble and Hsiao (1970) photosynthetic cyclic electron flow can be sufficient and possibly necessary for $K^+$ uptake and stomatal opening in light. Cyclic electron flow or cyclic photophosporylation (specially of photosystem II) is operated by blue region of the light and is effective in producing stomatal opening while far-red (730-720 mm) activates photosystem I and remains ineffective in producing stomatal opening (Hsiao *et al.*, 1973). At these high quantum (in blue light) there remains sufficient energy-input for active transport of $K^+$. How blue light stimulates stomatal opening is still a matter of research. However, Mauravieff (1973) suggested that in absence of blue light $CO_2$ causes hydrolysis of starch in guard cell chloroplast. This creates major dilemma in deciding the physiology of guard cells specially with reference to blue light.

## ROLE OF CYCLIC PHOTOPHOSPORYLATION

Das and Raghvendra (1974) have shown the control mechanism involved in the stomatal opening in light using isolated epidermal strips of *Commelina benghalensis* and *Petunia hybrida*. They demonstrated the key role of cyclic photophosporylation in causing the opening of stomata even under unfavourable conditions which normally induce stomatal closure (such as low pH, presence of mercuric chloride or PMA in medium). They further showed and confirmed that guard cell chloroplasts possess a highly functional machinery for the production of ATP through cyclic photophosporylation constantly with the requirement for the stomatal opening and maintenance.

**Histochemical investigation** of the guard cell chloroplasts suggests that they possess PS I and is highly active than that of PS II. The chloroplasts also show low rate of Hill reaction and non cyclic phosporylation but high rate of cyclic phosporylation. Investigation of Rustagi and Sankhla (1974) also reveals similar results. They studied the effects of cyclic adenosine monophosphate (C-AMP on stomatal opening in *Vicia faba* and suggested that in presence of KCl, C-AMP opening is greatly stimulated. ABA which induced stomatal closure, counteracted the action of C-AMP as well as $GA_3$.

**Stomatal opening occurs in darkness** in response to environmental stimuli also indicates that energy from non photosynthetic source can be utilized. Mansfield (1976) has forwarded a view while studying *Commelina communis* that stomata remain open in absence of $CO_2$ and chloroplast remains in close association with plasmalemma adjacent to the subsidiary cells, which is not normal feature. Therefore, it is evident that when chloroplasts have ready energy which is *surplus* to other requirements, is made available to plasmalemma to support active

$K^+$ transport for stomatal opening.

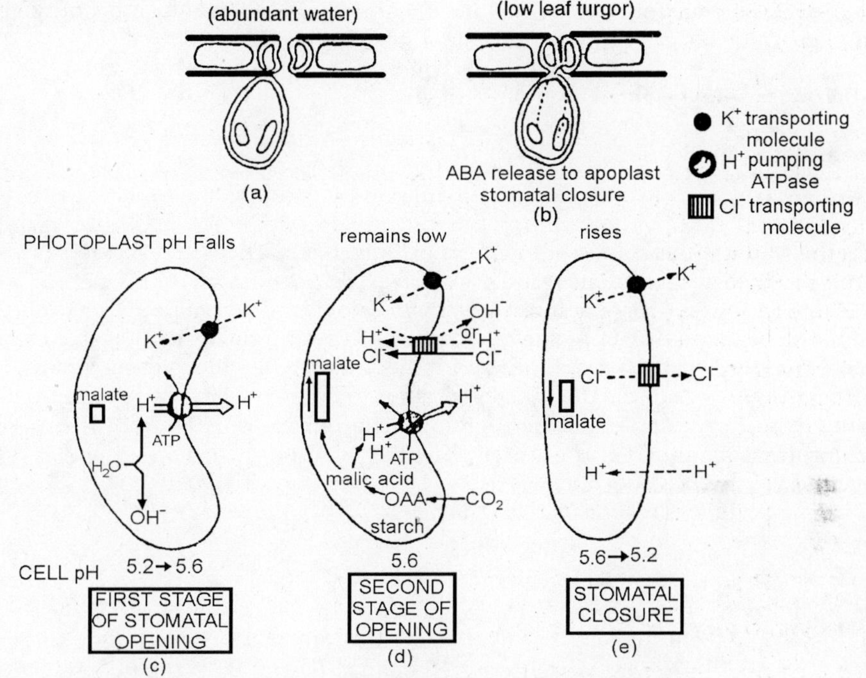


**FIG. 8.13:** Summary of physiological events in guard cells leading to stomatal movement. Note the major physiological steps (a to e) taking place at opening and closing of stomata.

## ROLE OF ABSCISIC ACID IN STOMATAL MOVEMENT

### ABA, Chloroplast Membrane and pH Gradient in Chloroplast

Degree of closure of the stomatal aperture could be controlled by exogenous ABA. In well watered plants most of the ABA is found in the mesophyll chloroplasts of leaves. The pH values of the different intracellular compartments of chloroplast determine the location of ABA. The pH of stroma chloroplast is higher than in other compartments and the protonated form of ABA, which moves across membranes, tends to accumulate there. Hartung et al (1990) have indicated that in leaf tissues the high membrane permeability of the unchanged protonated abscisic acid (species of ABA) and the impermeability of the anion species ABA results in an accumulation in **alkaline compartments**.

**The ABA concentration in the apoplast** is of particular importance because it is the only leaf compartment directly connected with the guard cells and because no plasmodesmata exist between fully differentiated guard cells and subsidiary cells.

Stomatal closure could also be due to initial water stress conditions which evokes the release of ABA in the chloroplast, lowers the pH of the stroma thus changing the activity of photosystems (Cowan et al., 1982). Thus, ABA would then move from the mesophyll to bring about a rapid response of the stomata which would be reinforced by a further synthesis of ABA if the water stress continues.

**ABA as an ideal antitranspirant:** ABA inhibits accumulation of solutes in the guard cells required for stomatal opening and attacks selectively on the essential processes of the guard cells which are responsible for stomatal behaviour. It also prevents accumulation of $K^+$ in guard cells which causes increase in starch. These functions suggest that ABA possesses the properties of an ideal antitranspirant.

## ABA, Water potential and Photosynthesis

As the water potential decreases there is noticed a marked increase in permeability of chloroplast membranes to ABA. The ABA synthesized by and stored in chloroplast then **diffuses** into mesophyll cytoplasm and moves from cell to cell through plasmodesmata to the guard cells where it induces stomatal closure. The drop in the level of stored ABA in the mesophyll chloroplasts induce **fresh biosynthesis** of more ABA which continues to be translocated to the guard cells as long as the water potential remains low. The release of ABA from the mesophyll chloroplast stops when the water potential is restored to normal and its rate of synthesis is markedly reduced.

The evidences for this are as follows:

i) The exogenous application of low concentrations (e.g., 1mM) of ABA to leaves causes stomatal closure within 3-9 minutes.
ii) The concentration of ABA in water stressed leaves rises apperecíably.
iii) Mesophyll chloroplasts can synthesize ABA but guard cell chloroplasts cannot
iv) The ABA of leaves of well-watered spinach was found to be contained in the chloroplasts but after 4 h of water stress the total ABA level was found to have risen 11 fold whilst that of the chloroplast had risen only 2 fold; this suggests that the *newly synthesized ABA is rapidly exported from the chloroplasts.*

**ABA and permeability of chloroplastic membrane:** Milborrow (1979) proposed that permeability of outer chloroplastic membranes to ABA increases, probably progressively, on wilting and so allows ABA to leak into cystosol. The ABA in cystosol would be subjected to turnover while the wilt continues. The plastids are assumed to become impermeable to ABA after re watering which consequently inhibits biosynthesis of ABA. This happens due to the restoration of the original high concentration of ABA.

According to Sharp (1990) primary roots of maize and other species show increase in endogenous ABA which act differentially to maintain root elongation and inhibit shoot elongation at low water potentials. It is now known that both roots and shoots can independently synthesize and accumulate ABA when subject to tissue water deficits (Mason et al., 1988).

## ABA and Photosynthetic Efficiency

ABA inhibits photosynthesis, however, the activity of several enzymes of $C_4$ pathway increases dramatically. Since ABA also brings about stomatal closure, it was speculated that ABA by increasing the activity of PEP-carboxylase possibly tends to compensate stomatal closure. ABA has direct effect on electron transport of photosystem II and on flow of $^{14}C$ into individual photosynthetic product which causes variations in stomatal behaviour (Bauer *et al*, 1977). ABA takes more time for closing stomata in $CO_2$-free air because in the absence of $CO_2$ fixation, energy is available from chloroplast as well as from mitochondria. Experiments conducted with photosynthetic and respiratory inhibitors support the present view.

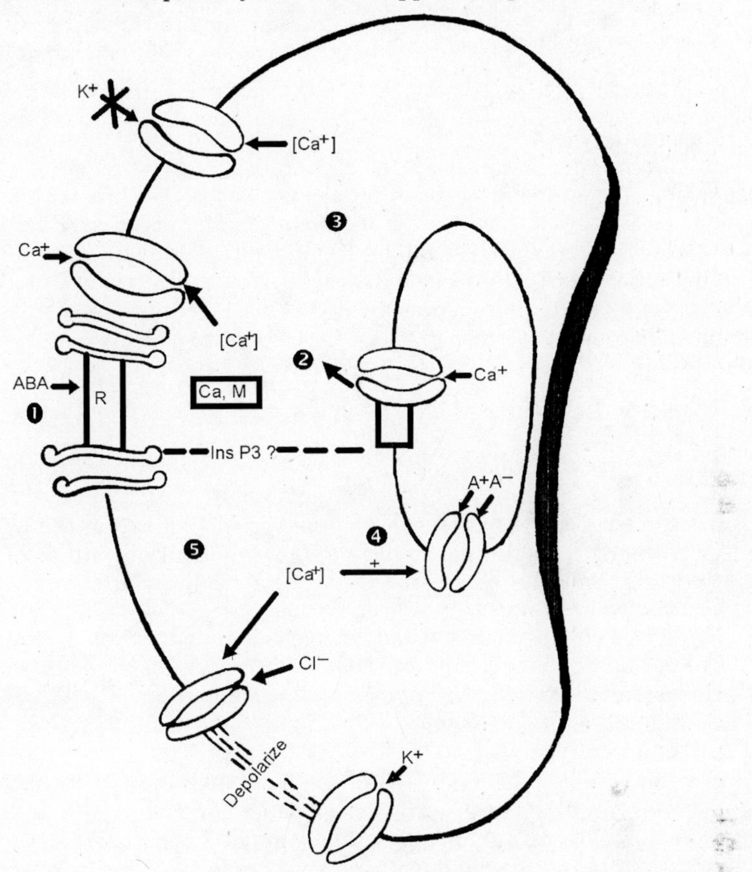


**FIG. 8.14:** $Ca^{+2}$ - ABA Second Messenger Model.

## Sensitivity to ABA and Control of Stomatal Movement

In considering the role of ABA in controlling stomatal movements, in is important to consider the possible changes in guard cell sensitivity to the compound. The notion that hormone sensitivity could be important in defining plant responses to hormones was brought to the fore by Trewavas (1981; 1982).

**Evidence for Changes in Stomatal Sensitivity During Water Deficit Stress:** There are many indications in the literature that stomatal sensitivity to ABA is affected by water deficit stress, although reports are mixed as to whether responses to ABA are increased or decreased following stress. Peng and Weyers (1994) made a detailed study of changes in stomatal sensitivity to ABA during three stress treatments contrasting in severity and duration. They found that the initial response to stress appeared to be sensitization to the hormone, while in the longer term, the stomata seemed to be desensitized. To some extent, this observation can explain some of the inconsistencies in past reports.

## CALCIUM - ABA SECOND MESSENGER MODEL

Calcium inhibits the stomatal opening if it is applied to isolated epidermis (De Silva et al, 1985). According to Mansfield and Atkinson (1990) calcium ions can strongly inhibit stomatal opening, and they can move rapidly from root to shoot in the xylem.

Mansfield (1990) suggested a hypothesis that an ABA-induced increase in cytosolic free Calcium ion ($Ca^{2+}$cyt) acts to trigger for the intercellular machinery responsible for stomatal closure. Mc Ainish et al. (1990) reported an immediate increase in $Ca^{2+}$cyt in guard cells of isolated epidermis of *Commelina communis in response to ABA.* $Ca^{2+}$ also inhibit $K^+$-induced swelling of guard cell protoplast.

The initial response to challenge with ABA is a stimulus-induced increase in $Ca^{2+}$cyt some form of receptor by which the cell can recognize the external stimulus is required. Action of ABA with respect to $Ca^{2+}$cyt can be summarized as follows:

1. The presence of high-affinity proteinaceous ABA-binding sites located on the **plasmalemma.**
2. ABA changes plasmamembrane permeability which is brought about by the interaction of ABA with patches of membrane phospholipid.
3. Increase in $Ca^{2+}$cyt may result from either the activation of inwardly diverting $Ca^{2+}$ channels in the plasma membrane allowing an influx of $Ca^{2+}$ into the cytosol and/or the release of $Ca^{2+}$ from the internal stores stimulated by a secondary messenger.
4. Evidences are available which suggest a tangible link between changes in $Ca^{2+}$cyt and the regulation of stomatal aperture. In order to achieve stomatal closure, an increase in $Ca^{2+}$cyt should have two following effects
   a) Inhibition of opening
   b) Inhibition of closure
5. For the response of inhibition of opening, inhibition of inwardly directed K+ μM concentration of $Ca^{2+}$ provides evidence for this.
6. Stomatal closure on the other hand might be initiated by a $Ca^{2+}$-mediated release of anions and cations from the vacuole (e.g., in sugar beet).
7. If this similar mechanism operates in stomatal guard cells, the parallel opening of outwardly directed Cl- channels in the plasma membrane which would accompany this event might cause sufficient depolarization of the plasma membrane to activate the outwardly directed K+-channel. This would provide the mechanism whereby a stimulus-induced increase in $Ca^{2+}$cyt could initiate stomatal closure

This hypothesis states that a stimulus-induced increase in guard cell $Ca^{2+}$cyt may trigger a loss of turgor in the guard cells of open stomata or inhibit the increase of turgor required to open stomata by the mechanism.

## CALMODULIN

There are evidences that Ca-binding protein - Calmodulin may also participate in the reaction mediated by $Ca^{2+}$. Immunoflorescence studies indicate that this protein is present in the guard cells and may participate in the regulation of stomatal aperture. Treatment of epidermal strips with calmodulin-inhibitors resulted in an increase in stomatal aperture. $Ca^{2+}$ is involved in signal transduction. In plants it is present in vacuole (store house) and plasmalemma but in animals stored in endoplasmic reticulum at low cone (less than 1 μm) in cytoplasm. At higher concentration (more than 1 μm) it react with $PO_4^-$ to form insoluble precipitate and would destroy the energy currency of cell (ATP). Plasma membrane is highly impermeable to $Ca^{2+}$ but regulation of $Ca^{2+}$ is through **gated channels**.

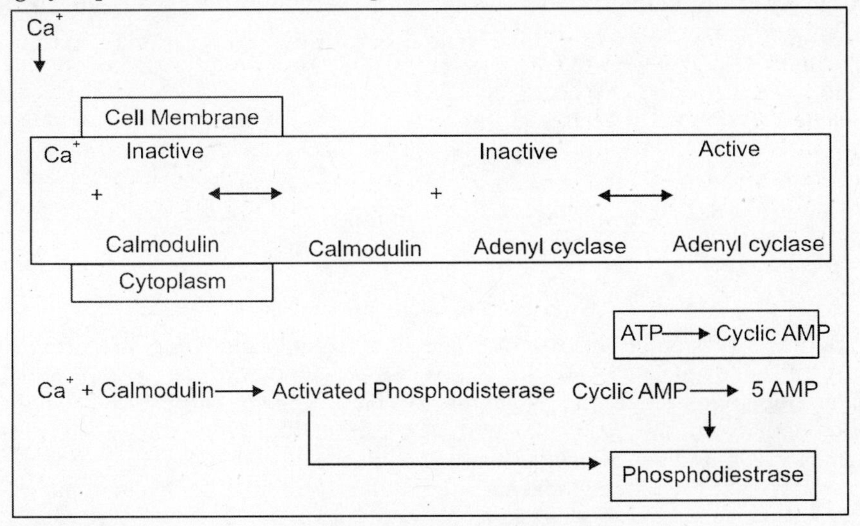


**FIG. 8.15:** Involvement of $Ca^+$ cAMP and calmodulin in stomatal movements

## Possible Involvement of cAMP in Stomatal Movement

Cyclic AMP is in fact a **secondary messenger** which after binding to corresponding receptor in the plasma membrane can illicit to do the synthesis of cAMP for ATP through a common pool of adenylate cyclase. The first report of cAMP in the tissues of higher plants was provided by Pollard, 1970, Salomon and Mascarennas (1972), Martin (1983). However, identification of this compound in higher plants did not come until 1981 (Brown and Neuton, 1981). A cAMP binding protein was also reported to occur in plants (Anderson and Pastan, 1973; Brown et al., 1979, 1980). Recently a soluble cytoplasmic fraction of adenylate cyclase was purified and characterised from roots of *Medicago sativa* (Carricarte et al., 1988).

cAMP together with adenyl cyclase, phosphodiesterase and the protein calmodulin may constitute the system of regulation of biochemical reaction by cAMP. Evidences are available that support the presence of adenylate cyclase in guard cells of *Vicia faba* and indicate the involvement of cAMP in physiological modulation of stomatal movement. Adenylcylase acts on ATP and form 3,5-cAMP and require pyruvate.

cAMP is now known to remove the inhibitory effect of $Ca^{2+}$ and ABA in stomatal aperture. cAMP was also able to enhance stomatal aperture of strips in darkness. This enhancement was paralleled with a stimulation of $K^+$ uptake by guard cells as demonstrated by 85% $K^+$ positive staining of guard cells in cAMP treated peels as compared to 15% in control peels.

It is conceivable that a hormone action for stomata opening/closing is regulated by stimulation/inhibition of the synthesis of adenyl cyclase by hormone. Degradation of cAMP or inhibition of its synthesis would leads to stomatal closure.

The depletion of concentration of cellular cAMP would also be results in the increased activity of cAMP phosphodiesterase via calmodulin activated by $Ca^{2+}$.

Endogenous cAMP enhanced stomatal opening in light and promoted in dark. It also removed the inhibitory effect of $Ca^{2+}$ probably by making negligible the action of endogenous phosphodiesterase presumably activated by $Ca^{2+}$.

Stomatal closure effect of ABA could occur through stimulation of enzyme

⇓

Hydrolyses cAMP phosphodiesterase

⇓

Possibly involve in participation of Ca2+ dependent calmodulin

or⇓

By inhibiting adenylate cyclase

By using the inhibitors of cAMP phosphodiesterase activity e.g., caffeine, the levels of endogenous cAMP increases which supports that cAMP enhances stomatal opening in light and promotes it in dark. The demonstration of adenylate cylase activity in the plasmalemma of guard cells and in the membranes of chloroplasts, mitochondria, ER, vacuoles further supports the argument in favour of cAMP role in the biochemical steps affecting the stomatal movement as well as in the other metabolic processes (e.g. cAMP can replace the role of $GA_3$ by stimulating $\alpha$-amylase synthesis.

## CONCEPT OF HORMONAL BALANCE

The periodic changes in water status may indicate periodic fluctuation of hormonal concentrations which influence stomatal and root resistance. Cytokinins (CK) decrease stomatal resistance by opening stomata and increase resistance of the root for water absorption. Kinetin controls stomatal aperture by altering epidermal potential and hence stomata open. Stomatal opening in barley is encouraged by kinetin which could partly be attributed to a reduction in $CO_2$ concentration by photosynthesis. The interaction studies between CK and ABA suggest that stomatal opening is encouraged by kinetin while ABA closes stomata. These findings reveal that there is an operation of feedback mechanism involving the opposing action of these two hormones. Rapid changes in the level of kinetin and ABA in guard cells have also been demonstrated. These studies indicate that stomatal movement depends on cytokinin: ABA balance in the guard cells.

## STOMATAL MOVEMENT AND ANTITRANSPIRANTS

The rate of transpiration in plants may be controlled by various methods e.g., the erection of windbreaks, by reducing the input of solar radiation to the leaf surface. Unfortunately, many of these techniques are only of limited applicability and therefore research has been intensified over the past twenty years to develop chemicals which can be applied to crops to reduce transpiration under a much wider range of conditions. Such compounds have been given the general name – *antitranspirants*. There are three major groups of these compounds:

1. Stomatal inhibitors
2. Film-forming compounds

3. Metabolic poisons

Stomatal inhibitors consist of a wide range of synthetic substances (herbicides, fungicides, plant growth regulators) which induce stomatal closure by a direct effect on the stomatal mechanism, or indirectly by some effect on a process not directly linked to the stomatal apparatus. They act by elevating the internal carbon dioxide concentration of the leaf as a result of inhibiting the photosynthetic activity of the mesophyll.

Plants have *self defence strategies.* Most crop plants close stomata in response to water stress and consequently control their own internal water status. Such defence mechanisms may not always be entirely adequate, particularly when the supply of water from the soil becomes limiting. Also from a theoretical point of view there are occasions when the artificial closure of stomata can lead to enhanced water use efficiency. Antitranspirants may not be the ideal method of solving the problem of improving water use efficiency, especially in view of the physiological complexity of the stomatal mechanism. Many of the chemicals tested so far have complicated side effects, which only came to light after detailed research studies on their mode of action.

## Antitranspirant Films

Film-forming compounds involve the application to the leaf surface of relatively inert films, made from diverse groups of materials like, wax emulsions, kaolin, polythene and higher alcohol's like hexadecanol dimethyl silicones. These thin film on the leaf surface constitute a physical barrier for evaporation. These films occlude the stomatal pore and reduce cuticular transpiration and they are one of the oldest recorded method of reducing transpiration. To be effective, such film forming substances should be non toxic, durable and stable.

## Use of Antitranspirant Films

The use of antitranspirant films has also been widely researched, although even here results have generally been inconsistent. Some successful applications have been made to vines to delay the harvesting of grapes and improve their yield. Films have also been used successfully to prevent desiccation and mortality during the transplantation of tress, and they also improve the survival and rooting of cuttings. Varied success has been achieved by their application to increase the shelf life of cut flowers and in preserving fruit after harvest It has also been suggested that their application could reduce smog damage and reports indicate that film antitranspirants could also lower the incidence of certain foliar pathogens and pests.

### *Limitations of Antitranspirant Films*

Antitranspirant films have selective permeability to water and carbon dioxide. From a practical point of view the permeability of the film to water vapour should be smaller or at least equal to its permeability to carbon dioxide, otherwise photosynthesis could be reduced far more than transpiration, with obvious effects on the production ratio.

To date antitranspirant films are available which are more permeable to water vapour than to carbon dioxide. We need selective permeable film structures. The films should be inexpensive, non-toxic, remain elastic for long periods and resist degradation by solar, UV radiation and microorganisms. Currently, there appear to be too many constraints for the widespread implementation of film antitranspirants in agriculture, although their successful application in specific situations will no doubt continue.

## Which Antitranspirant is Safe

The use as an antitranspirants is safe or nor? Therefore, it is necessary to know about its characteristics. The increasing antipathy towards the application of toxic elements such as mercury to food crops, the investigation of such compounds has now ceased. These studies have not, however, been entirely wasteful, because they have highlighted the criteria that are necessary for the development of commercially acceptable antitranspirants. Ideally antitranspirants should fulfil the following criteria:

1. They should be non-toxic to humans and other forms of life.
2. They should have low mobility within the plant, only exerting a specific action on the guard cells or the immediately surrounding them. Inhibitory effects on the mesophyll or elswhere in the plant should be minimal.
3. They should cause no permanent damage to the stomatal mechanism or to the structure of the guard cells.
4. It should be stable, long lasting and cheap. All these requirements are not met with, although a number of chemicals such as fluorides, azides, hydroxy sulphonates and phenylmercuric acetate are tested.
5. Their effect should perisit for several weeks.
6. They should be cost effective.

## Antitranspirants: Synthetic

### *Acetylsalicylic acid (a Salicylate)*

One of the more interesting chemicals which have been tested is acetylsalicylic acid (a salicylate), commonly known as aspirin. The presence of salicylates in plants is well known, and particularly high concentrations occur in willow and many other species. However, no physiological role has yet been ascribed to them. When low concentrations of aspirin were fed via the petiole to leaves, it caused stomatal closure and reduced transpiration. This discovery provides a scientific basis for the traditional hospital practice of adding aspirin to the water of cut flowers to preserve them longer.

### *Phenylmercuric acetate (PMA)*

Phenylmercuric acetate (fungicide) is a very powerful antitranspirants. It causes stomatal closure at very low doses and improving the water use efficiency of crops. Field trials indicated that it might also be of value in controlling transpiration in pine plantations. However, detailed studies on its mode of action revealed that in addition to closing the stomata it caused substantial physiological toxicity to both the guard cells and surrounding subsidiary cells and furthermore, it directly inhibited the photosynthetic activity of the mesophyll. It inhibits photophosphorylation in guard cells and causes stomatal closure but it also inhibits photophosphorylation in mesophyll cells and is phytotoxic (Mansfield, 1960).

## Antitranspirants: Naturally-occurring

Most of the inhibitors of stomatal transpiration which had been investigated up to

1970 were toxic. The sole exception was carbon dioxide.

### Carbon dioxide

Carbon dioxide is a good antitranspirant. If the plants are covered in enclosures and $CO_2$ concentration is increased, excess $CO_2$ causes partial closure of stomata. Secondly this also increases the rate of photosynthesis as it is a limiting factor. Use of $CO_2$ fulfills all the requirements of an ideal antitranspirant but the only disadvantage is that it cannot be used except perhaps in glass houses as plants have to be enclosed. In maize, for example, by raising the atmospheric carbon dioxide concentration by approximately 0.25% transpiration can be cut by 23% and net photosynthesis increased by 30%. However, the general application of this technology to field crops is questionable and its practical application many only be realized on crops grown in greenhouses and polythene tunnels, where the conservation of water may be of little significance.

### ABA

The naturally-occurring plant growth inhibitor, abscisic acid (ABA) accumulated in the leaves of wheat after they had been subjected to water stress. (see Chapter: *Stomatal Movement, Water Relations and Plant Growth Regulators*). One of the most significant physiological changes occurring in such stressed plants is the suppression of stomatal opening. Observation that low concentrations of externally applied ABA could simulate this response led to more detailed studies of the mode of action of ABA on the stomatal physiology of plants.

Preliminary results showed that single surface application of ABA (0.02 μg $cm^{-2}$ of leaf surface) could suppress stomatal opening for up to 9 days. At the same time it was found that young wilty mutants of tomato, which were unable to close their stomata, were unable to synthesize ABA. It therefore seemed probable that endogenous ABA might be playing an important role both in the normal behaviour of stomata, as well as regulating their behaviour after water stress.

The mode of action of ABA on the stomatal mechanism is still rather uncertain. The mode of action of ABA therefore closely approaches the ideal characteristics of an antitranspirants. Although ABA reduces the overall growth of the plant, this may not be too detrimental during conditions of extreme water shortage.

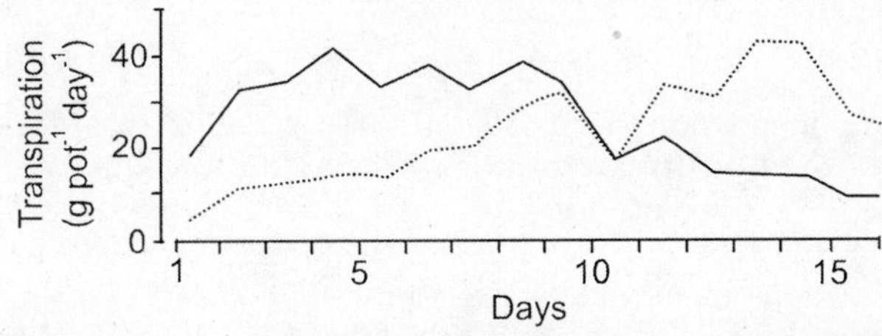


**FIG. 8.16:** The effect of the methyl ester of ABA on water loss of unwatered coffee plants. Untreated plants, (soiled line) treated plants (dotted line) (redrawn from Mansfield, Wellburn and Moreira (1978).

### Action of ABA as antitranspirant

1. ABA inhibits 70-80% of the $K^+$ uptake into the guard cells of some species, and recent evidence indicates that increased efflux of $K^+$ (estimated from 86 $Rb^+$ efflux experiments) may be involved in ABA-induced stomatal closure.
2. ABA treatments also inhibit starch breakdown within the guard cell

chloroplasts and enhance the incorporation of glucose into guard cell starch.

3. Measurements of the intercellular carbon dioxide concentration in ABA treated leaves indicate that although small increases occur they are initiated within 2.6 minutes; in other species closure it initiated within 8-9 minutes and complete closure can be observed in 30 minutes.

The observations on field trials on orchard trees and cereals, have not been too encouraging. The effects are too short-lived to be of commercial interest. In some plants the methyl ester of ABA can reduce transpiration for up to 9 days, possibly as a result of greater penetration into the plant or due to slower breakdown in the plant. Such results are an optimistic sign that analogues with an appropriate action may be synthesized in the near future.

## Farnesol

All-trans-farnesol (a sesquiterpenoid) is found in leaf cells. When All-trans-farnesol is applied externally to sorghum leaves it also induces stomatal closure.

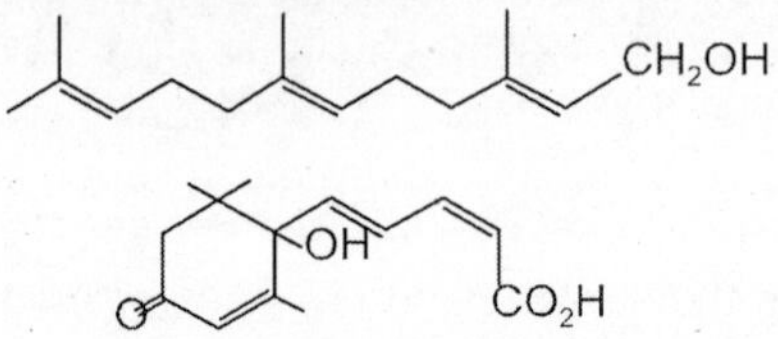

$CH_2OH$

OH

$CO_2H$

**FIG. 8.17:** Chemical structures of all-trans-farnesol (top) and abscisic acid (bottom).

## *Phenolic acids (Scopoletin and Chlorogenic)*

There is also evidence that phenolic acids such as scopoletin and chlorogenic acid which occur naturally as plant growth inhibitors and accumulate under various stress conditions, can also induce stomatal closure. They are believed to act as competitive inhibitors of the plant growth hormone indole-3-ylacetic acid (IAA) and recent reports that IAA may stimulate stomatal opening and antagoinze the closing response to ABA may indicate a very complex interaction between various growth regulators during the development of water stress. The mode of action of these compounds on the stomatal mechanism is till to be verified.

## Metabolic Poisons

The stomata remain open when the guard cells are turgid but close when the guard cells lose turgidity. The turgidity in turn depends upon the ability of guard cells to hold osmotica like $K^+$ and malate ions. It is the plasma membranes of guard cells which exercise this control. If the plasma membranes of guard cells which exercise this control. If the plasma membrane is denatured, the guard cells soon lose turgor and the stomata would close. This is the principle of using some of the metabolic poisons as antitranspirants. However, when such compounds are sprayed, the surrounding epidermal cells may also be killed along with guard cells. Often the epidermal cells are more affected than the guard cells as the latter are usually more resistant to metabolic poisons. When this happens instead of closing, the stomata may become wide open (Waisel et al., 1969).

The stomatal opening and closing are active processes requiring metabolic energy. This requirement is fulfilled by the ATP produced in the chloroplast of guard cells during light reaction. The guard cells also contain mitochondria and therefore mitochondrial ATP may also contributed to ATP requirement. Therefore, any chemical which inhibits electron transport chain in the chloroplast and the mitochondria may reduce ATP availability finally leading to stomatal closure. Application of DNP, CCCP and salicyladoxime can cause stomatal closure. As a result these uncouplers of electron flow which cause stomatal closure act as antitranspirants. Here again the action is not specific on guard cells but other mesophyll cells may also become affected.

It is frequently argued that if transpiration is reduced by antitranspirants, some of the beneficial effects of transpiration may be lost. But transpiration does not appear to have any beneficial effects and it is considered to be a necessary evil (Arnon, 1972). Therefore, reduction of transpiration should not seriously affect other physiological functions of plants.

Any reduction in transpiration should be beneficial by conserving moisture rather than being harmful. So far there is no method of cutting down transpiration without interfering with gaseous exchange. This might become possible when some film forming chemical which is more permeable to $CO_2$ but less permeable to water vapour becomes available.

## WHICH FACTOR(S) IS RESPONSIBLE FOR STOMATAL MOVEMENT?

Multiple factors are responsible to determine stomatal movement in plants. In general, a reasonably good correlation has been found between sugar-starch content of the guard cells, intercellular $CO_2$ content, glycolic acid content of the plant tissue and potassium and hormone content of guard cells in stomatal movement. The physiology of guard cell is controlled by massive movement of $K^+$ *to and from* the guard cells and role of ABA. Subsidiary cells also play an active role in ion transport. A malate switch hypothesis suggests that there is a movement of malate along with $K^+$ from the surrounding cells when stomata open. Recently, it has been investigated that $Cl^-$ ions are taken up by the guard cells from their environment and carboxylation of PEP (derived from starch) leads to synthesis of malate. The energy required for influx of ions becomes available from photosynthetic/respiratory source. Development of water stress in plant increases amount of ABA which inhibits accumulation of $K^+$ in guard cells and causes an increase in starch content and consequently closes stomata. Antitranspirants (Farnesol and PMA) cause damage of chloroplastic membrane and raise ABA level in it. This happens due to leakage of ABA in cytosol within chloroplast. For development of drought resistance in crop plants, uses of plant hormones and other chemicals in the fields are necessary. Our knowledge, pertaining to the physiological mechanism of stomatal opening and drought resistance for providing full protection to the crop plants in field is in its infancy. Introduction of new potential antitranspirants which could have direct action on hormonal balance of guard cells are still needed. It may be concluded that all these factors either independently or collectively influence the guard cell movement at primary or secondary level. The exact mechanism is still unknown.

### Guard and Subsidiary Cell

The opening and closing of stomata depends on changes in the suction pressure of the guard cells relative to that of the surrounding epidermal cells. The wall of the guard cells is thickened in such a way that changes in volume result in movement of the guard cells relative to one another; when they swell they open the aperture between them and when they shrink they close it. Early measurements of the osmotic pressure of the guard cells showed that the osmotic pressure of guard cell was greater when the stomata were open than when closed.

### *Role of Subsidiary Cells*

The subsidiary cells play an important role in stomatal opening (Penny and Bowling, 1974), because the $K^+$ ions across the complex of subsidiary cells into the unspecialized epidermal cells. An electrochemical gradient is maintained between guard cells and subsidiary cells, thus, there is an active transport of $K^+$ along a series of cells. In absence of subsidiary cells stomata behave differently. However, the role of subsidiary cells has not been properly evaluated and it is not known whether they play any essential role other than acting as pathways or reservoirs for osmotically active materials.

## $CO_2$ Concentration and CAM Plants

The literature is replete with examples which provide definite evidence for effects of $CO_2$ in the stomatal movement. Darwin found that a high concentration of $CO_2$ caused the stomata to close. Linsbauer showed that regardless of whether the leaf was in the light or dark, provided the concentration of $CO_2$ was sufficiently high, the stomata would close. An interaction between light intensity and $CO_2$ concentration showed that in darkness, 0.017% $CO_2$ was required to cause closure of the stomata; with an illumination of ninety foot-candles 0.049 per cent was required, and with eight hundred foot-candles 0.084% was required. Light and $CO_2$ factors influence the pH of the cells surrounding the sub-stomatal cavity. The effects of light and of $CO_2$ were partially dependent and partly independent.

Stomatal sensor for $CO_2$ is located in the guard cells. As the demand for $CO_2$ increases, the stomata open and this forms a close feedback system. Ortan and Mansfield (1976) indicated that 2 to 3 times increase in the atmospheric $CO_2$ level causes partial closure of stomata. The partially close stomata would offer protection from excessive water loss, and the only inhibition of photosynthesis would be due to reduced availability of $CO_2$ resulting from lower conductivity of the epidermis. The steeper concentration gradient for $CO_2$ diffusion causes increase in photosynthetic rate (Mansfield, 1976). In majority of crop plants stomatal opening is regulated by light-driven photosynthetic carbon assimilation. Thus they possess photoactive stomata.

Crassulacean acid metabolism (CAM) is characteristic in succulent plants. Stomatal movement in these plants is of scoto-active type. Recently it has been observed that CAM pathway also occurs in several non-succulent scrubland species which have the property of nocturnal stomatal opening and fixation of $CO_2$ in dark (Das, 1977). Guard cells of epidermal tissue of CAM plants are able to synthesize the malic enzyme, PEP-carboxylase, RuDP-carboxylase and NADP thus establishing their physiological similarities to CAM behaviour. Import of $K^+$ into the guard cells and production of organic acids (mainly malic acid) lead to stomatal opening. Organic acids produced in guard cells maintain electrochemical balance of surplus $K^+$ ions. In *Notonia grandiflora* (a CAM plant) stomatal closure is induced by ABA and is accompanied by $K^+$ ions accumulation in the guard cells. ABA can significantly suppress the KCl as well as acid-induced stomatal opening in CAM plants.

The suggestion that stress-induced, endogenous ABA may sensitize stomata to

$CO_2$ could help to resolve the controversy concerning the role of internal $CO_2$ concentration in stomatal response to environment. Stomatal closure by low concentrations of N-dimethyl aminosuccinamic acid (B-9) in normal air is largely reversed in $CO_2$-free air. The main effect of B-9 is due to an increase in internal $CO_2$ concentration. The mechanism of action of B-9 on stomata, therefore, has little in common with that of ABA which does not operate by increasing $CO_2$ concentration and does induce cellular damage (Orton and Mansfield, 1976).

## Light and Dark

Stomata open in the light and close in the dark. This was assumed to be the cause of the diurnal changes in stomatal aperture. Lloyd and Loftfield both attempted to show that the starch content of the guard cells showed a diurnal change, in a manner correlated with the change in aperture of the stomata. This has been most conclusively demonstrated recently by Williams, who devised an approximately quantitative method for measuring starch content.

## pH and Active Water Movement

The change in pH induced by some external factor such as light control stomatal movement. Williams postulated that active water movement is necessary to keep the stomata closed and that in light, processes must take place which partially overcome this active movement (refer to classical hypothesis).

## Water Balance of the Plant

The factors which influence the uptake of water by the root are not the same as those which influence the loss of water from the leaves and shoot by transpiration. During the day when the temperature rises, the transpiration rate increases, and there will be an increased tension on the water in the xylem which will tend to result in an increase in water uptake. However, the increase in water uptake may lag in time behind the increased transpiration, and there will be a period of net loss of water from the plant as a whole. Indeed, the rate of water uptake may reach a maximum value which is less than the maximum rate of transpiration, when there will be a continual loss in net water content of the plant and the plant will wilt.

## Silicon and Transpiration

One finding that can help to explain some of the symptoms associated with Si deficiency is that the rate of transpiration of Si deficient rice plants was increased by about 30% over the rate of Si supplied plants. These observations suggested a role for Si in the water economy of the plant. This could explain the wilting and the mineral toxicity's that occur in aerial parts of Si deficient plants. Such reduction in transpiration rate by Si was due to a well thickened layer of silica gel on epidermal cell walls, which does not allow water to escape at an accelerated rate. Both the distribution and content of Si accurately reflect the loss of water from various organs (Horst and Marschner, 1978). Application of Si at the rate of 0.75, 30 and 162 mg $SiO_2\ L^{-1}$ was found to reduce the transpiration coefficient in rice (by 15% at 30 mg $SiO_{\cdot 2}L^{-1}$, sugar-cane and wheat (by 21% and 10% respectively at 162 mg $SiO_2L^{-1}$).

## Drought and Related Stress Conditions and Stomatal Response

The onset of drought is determined by a relative deficiency of soil moisture under atmospheric conditions which favour high rates of transpiration. Its appearance will be delayed the larger the root absorbing surface and the smaller the leaf area of the plant. Continued water loss results in dehydration of the protoplasm. This degree of dehydration is irreversible and vary between species to species. These considerations determine the degree of drought resistance of a plant.

Stomata in higher plants begin to close as soon as they are subjected to mild water stress and completely close before there are any visible symptoms of wilting. The phenomenon is called **hydroactive** closure by Stalfelt; which may be defined as a **first line of defence** against the damaging effect of water shortage. Succulents represent the extreme examples of controlling water loss by stomatal closure. Generally, stomata close in response to stress, tending to conserve water and to prevent further stress buildup. The starch content of the guard cell is increased under internal deficit condition which leads to stomatal closure. Two aspects of such response can be identified viz., (i) an early reaction leading to closure long, before visible wilting is evident, (ii) prolonged after-effect of the duration which depends on severity of water stress (Mansfield, 1976).

## Hydroactive and Hydropassive Responses of Stomata

The distinction between hydroactive and hydropassive stomatal responses to stress receives very little attention and yet is fundamental to an understanding of ABA's role.

**Hydroactive** stomatal movements are defined here as involving a change in guard cell turgor caused primarily by change of protoplasmic solute potential. This will occur because of solute transport and metabolism processes triggered by a stimulus. While the actual movement of solutes down electrochemical gradients may be passive, the process is active in the sense that it requires metabolic energy at some stage, especially to create the proton motive force that drives the fluxes, opens ion channels and influences metabolic events (MacRobbie, 1991; Outlaw *et al.*, 1992). The hydroactive component to the stomatal response to water stress appears to operate due to the effect of ABA on specific ion transport processes (Raschke, 1987; MacRobbie, 1991; Gilroy *et al.*, 1992).

**Hydropassive** movements are defined here as arising from guard cell turgor changes which are due to alterations in the prevailing water potential. This will occur primarily because of a change in the balance between supply and demand for water in the epidermis. This process will be passive (it does not involve metabolic energy) although it has been observed that, in the longer term, hydroactive solute movements may follow hydropassive stomatal movements (Lösch and Schenk, 1978). The precise response is difficult to predict as it depends on the prevailing turgor relations of the cell types and the mechanical advantage of the functional subsidiary cells (Weyers and Meidner, 1990). These hydropassive responses will occur over the same timescale as responses to ABA, and it may be difficult to isolate the latter from the former in realistic treatments. The initial response to a loss of epidermal turgor can involve stomatal opening while the functional subsidiary cells maintain turgor, followed by stomatal closure in the longer term, after the functional subsidiary cells have lost turgor.

## Short Term Coordination of Stomatal Responses

Basal levels of ABA from essentially unstressed plants are sufficient to give rise to a closure response. Yet the stomata in such plants are capable of steady-state conductances at reasonably high levels. This demonstrates that the stomatal response is poised between both positive and negative stimuli, as envisaged by Raschke (1979). For instance, under normal unstressed conditions, the negative ABA signal may be counterbalanced by positive stimuli, such as high light level and low internal [$CO_2$]. It would appear that stomatal closure during stress occurs because the **ABA signal** becomes so strong that it overrides the influences of other stimuli. At present, there is little agreement on methods for comparing a given chemical's ability to affect a physiological process with that of other stimuli.

## Overriding Other Signals

The classic "Ivanoff effect", which follows leaf excision, exposes the complexity of stomatal responses to stress. It comprises a well-defined sequence of stomatal movements (Meidner, 1965) that are interpreted as a series of hydropassive and hydroactive responses. The initial rapid stomatal closure (1-2 min) is due to the release of xylem tension and consequent increase in epidermal turgor; the phase of stomatal opening which follows (2-20 min), results from turgor loss in the epidermis; and the final closure response (20 min onwards) is probably a mixed response, involving hydropassive stomatal closure induced by prolonged reduction of epidermal cell turgor, and hydroactive stomatal closure, possibly involving ABA. Such responses seem to be exaggerated in short-term stresses with detached leaves. Except the initial closure, they are also to be expected in longer-term stresses with intact plants, although over these time scales they can be expected to interact with the normal diel rhythm of stomatal movements.

To survive a period of drought, it is essential for higher plants to have an internal stimulus that can cause hydroactive stomatal closure. Without such a mechanism, stress-induced stomatal closure might never occur, or hydropassively closed stomata might reopen as the stress continues, because other opening stimuli may override the hydropassive closure.

The additional transpirational water loss this would involve could be very damaging to the plant. Examples of possible responses that would need to be counteracted include reactions to:

a) the decline in intracellular [$CO_2$] that would occur if photosynthesis were to continue at the same rate while the stomata close;
b) the irradiance, in the reasonably likely event of a stress episode being correlated with high light levels;
c) the raised leaf temperatures following a reduction in transpiration.

## ABA and Stomatal Heterogeneity

Heterogenous stomatal closure has been recognized in a wide variety of plants, particularly in heterobaric leaves (Terashima, 1992). These observations have focused attention on spatial heterogeneity in stomatal movements, and in particular on the dynamic nature of stomatal aperture/conductance patterns. While heterogenous patterns can certainly arise from variation in the micro-environment (e.g. boundary layer and sunflecks), they may also involve short-term rhythmic behaviour (Cardon *et al.*, 1994), as individual stomata, or groups of stomata, *hunt* their optimal condition in relation to the prevailing conditions (Weyers *et al.*, 1995). Such patterns of movements might perhaps be accentuated in the event of a traumatic stress, providing a need for an overriding chemical controller like ABA.

The main roles for ABA during periods of stress are:

1. As an overriding influence over other guard cell responses;
2. As a co-ordinator of the activity of individual stomata;
3. As a messenger communicating spatial signals to the guard cells;
4. As a messenger communicating a temporal signal to initiate and co-ordinate the plant's long-term responses.

### Roots send Signals to the Shoot and Control of Stomatal Movement

The phenomena of root pressure and stomatal movement are examples of physiological responses which are dependent on metabolism. There is evidence that ABA produced by roots in **drying soil** moves in the transpiration stream and accumulates at or near the guard cells, that is at the site of action. In non watered *Commelina*, the stomata closed without any change in turgor or ABA content. However, the ABA content of the roots increased several fold along with that of the lower epidermis of the leaves. Movement of ABA from roots to the shoot was observed, even under conditions of low transpirational flux (Davies et al., 1987). These results support the hypothesis that ABA produced in stressed roots may act as a **chemical signal to the shoot**, initiating closure of the stomata before any changes in the water status of the leaves occur, thereby optimizing the plant's water use under conditions of restricted availability.

Saab and Sharp (1990) have shown that the *roots sends signals to the shoot* and causes stomatal closure and inhibition of leaf growth in Maize when there is even slight reduction in soil water potential with no detectable change in leaf water relations. Moreover, no effect of soil drying was observed on stomata despite varying the rate and extent of soil drying, light, humidity or nutrition.

### Drying Leaf and Stomatal Movement

Although stomatal closure of plants in drying leaf usually coincides with leaf water deficit, experiments have indicated that a decrease in stomatal conductance can occur without a change in leaf water status. Recent work has indicated that stomatal conductance can depend directly on soil water status; stomata started to close as the soil water content decreased even though that leaf water status was maintained at a high level. This suggests that a **message from the roots** can control stomatal conductance independently of the leaf water status.(Gollan et al., 1986; Munns and King, 1988)

### Oxygen-deficient roots and Stomatal Movement

In bean plants with oxygen-deficient roots, stomatal closure is influenced by changes in leaf water relations and by increased fluxes of ABA (Neuman and Smit, 1990). The initial effects on stomata appear to be due to decrease in turgor. If turgor is maintained, events in plants with oxygen deficient roots appear to be metabolically regulated.

### Stomatal Movement under Salt Stress

Crop plants show a range of sensitivity of stomata to salt stress because cytokinin in leaves remains low due to its decreased translocation from roots to the leaves. This observation indicates that water balance in plants directly influence **cytokinin: ABA balance** and ABA content in the leaves depends upon turgor. The amount of ABA was found higher in the salinized tobacco leaves growing under low humidity than that of higher humidity. They also indicate that presence of water in leaves constitutes a **primary signal** for the modification of plant hormone balance. ABA level in salt-stressed plant remains higher and it immediately decreases when salinity stress is removed. There exists no direct information on the endogenous level of other hormones like GA or ethylene as influenced by soil salinity.

### Stomatal Response: Evolutionary and Ecological Perspective

Terrestrial mesophytes have evolved to occupy an environmental niche where water supply is variable. Set against the competitive advantages of being tall and having thin, broad leaves are the disadvantages of needing to pipe water to the moist gas exchange surfaces and the possibility of cell death when the supply is insufficient to meet evaporative demand. The stomata and cuticle represent the mesophyte plant's primary defence against the damaging effects of turgor loss and it is crucial for the plant's survival that the stomata respond rapidly to the first signs of wilting. Weyers and Peng (1995) outlined above why the complex system that controls stomatal function under normal, unstressed, conditions may be inadequate for occasions when water loss exceeds demands. This normal system needs to be countermanded by a strong signal, both to inhibit or reverse natural feedback processes causing stomatal opening and to co-ordinate the activity of all the pores so that they respond together. The evidence, while not complete, suggests that ABA can fulfill this role.

This requirement seems to be a specific need in mesophytes. Hydrophytes, by definition, are rarely stressed during their normal life cycle, and may lack the ABA defence system (see e.g. Dörffling *et al.*, 1977). Xerophytes have a more severe problem with water supply that is usually avoided by more radical changes in growth physiology and biochemistry. The fact that most crop plants are mesophytes lends importance to our understanding of ABA and its role (see Loveys, 1991; Quarrie, 1991).

## GUTTATION

The magnitude of root pressure seems to have little relationship to the rate of loss of water from the top of the plant. When a plant is placed under conditions where the air is saturated with water vapour so that the transpiration loss is reduced to zero, root pressure may continue to operate and force water out of the vein endings in the leaves. The phenomenon is called **guttation**. Water loss by guttation occurs through marginal leaf hydathodes, or as exudation directly through the epidermis in a wide variety of herbaceous plants and a number of woody species. Guttation also sometimes occurs from stems through leaf scars and lenticels. Water loss by guttation usually is small but under unusually favorable environmental conditions guttation water may actually drip from leaves for extended periods. Guttation is considered as expression of positive root pressure and occurs most commonly at night and may occur in the daytime in plants growing in humid air and moist, warm soils. Guttation does not occur in plants in cold or dry soils, or in soils with a high concentration of solutes, toxic substances, or with poor aeration.

**Guttated fluid:** The chemical composition of guttated fluid is varied. It contains minerals, sugar, glutamine, and various enzymes, etc. Hence, considerable injury to leaves may result when salts are concentrated on the leaf surface after the guttated liquid evaporates. Injury also is caused to mesophyll tissues as minerals left by guttation are redissolved and drawn back into the leaf (Curtis, 1943).

Guttation water, which does not appear to play an important physiological role in the life of plants, often is a medium for pathogenic microorganisms (Bald, 1952). Ivanoff (1961) stressed that, although it is not very conspicuous, laminal guttation through the epidermis occurs more frequently than guttation through hydathodes. For a good description of extensive laminal guttation injury on a variety of plants the reader is reffered to Ivanoff (1961).

Differences between transpiration and Guttation are as follows:

| Transpiration | Guttation |
|---|---|
| This takes place in day time. | This takes place in night. |
| Water is lost as vapours. | Water is lost as liquid. |
| Water is transpired through stomata. | Water lost in guttation is rich in minerals, etc. through hydathode. |
| It is a regulated phenomenon. Temperature of leaf and plant is decreased. | It is uncontrolled phenomenon. No such effect occurs. |

## BLEEDING

It is the slow exudation of watery solution from an cut or incision made in a plant tissue. It may be of four types:

1. Exudation of xylem sap from an incision made in the stem of well- watered plant.
2. It occurs due to root pressure.
3. In the stems of some plants, etc. when a cut is given phloem sap exudates out.
4. Exudation.

## WILTING

All plants are to a lesser or greater extent in a potential state of wilting during the day when conditions favour high transpiration. Negative water balance develops when conditions change during the evening and the rate of transpiration falls, the rate of water uptake may remain at its earlier value so that during this period there is an excess water uptake compared with water loss. This restores the plant to a state of water balance, allowing it to make a *net gain equal to the net loss* of water during the day.

Wilting is dropping and sagging of plant leaves due to water deficits in their cells. The cells then lose their turgidity and become flaccid. Wilting is not exhibited by plants having lignified tissues, though these do respond to water stresses. Sunflower is most sensitive to water deficits as it readily wilts. Therefore, it is used as an indicator plant for determining 'permanent wilting point' in soils.

When water loss exceeds uptake by the leaves the water menisci will retreat within the pores of the mesophyll cell wall. This will lower the vapour pressure and hence the rate of transpiration. At this stage the leaf may not show any visible signs of wilting and it is referred to as being in a state of **incipient wilting**. If the plant recovers when conditions are changed so as to reduce transpiration, the plant is said to be in a state of **temporary wilting**. If the plant can only be revived by supplying water to the roots, the state is one of **permanent wilting** and may lead ultimately to **irreversible wilting** and the death of the plant.

### Permanent Wilting Point (PWP)

Absorption of water by plants and its subsequent transpiration to atmosphere, combined with surface evaporation from soils, deplete the soil moisture pretty fast. Soils gradually become dry and there is simultaneous drop in their matric potential of water. Crop plants then find it increasingly difficult to absorb water from such soils and the water absorbed by them begins to fall short of their transpirational needs, causing water deficits in plant cells. At this stage many herbaceous plants may display *daytime wilting* of leaves though they recover at night when air is comparatively cool and moist. Moisture content of soil at this stage, on oven dry basis, is termed its *temporary wilting* or *transient wilting point*. It is the first landmark of soil moisture stress in plants. In semiarid conditions, a medium texture soil under "normal" cropping may reach this point from its 'field capacity' in about a week's time. When soils continue to lose their moisture further due to plant absorption and surface evaporation, the crop plants are further subjected to increasing soil moisture stresses (in the absence of rainfall or irrigation). Soon the plants reach a point when they remain wilted both during the day and night. The soil moisture content at this stage (on oven dry basis) is called permanent wilting point (PWP) of the soil. At PWP the plants absorb soil moisture extremely slowly which is just sufficient to keep them alive, though not growing. In other words, the plants are now in a state of suspended growth, awaiting renewal of soil moisture supplies. The matric potentials of soils at PWP are found to vary from –7 to –32 bars, depending upon chiefly the plant species. But for all practical purposes, in crop production a mean value of –15 bars is accepted as the matric potential of water in soils at PWP. Plants subjected to PWP soil moisture status will regain growth and development if moisture supply was resumed through irrigation or rainfall, although the crop damage already incurred due to water stress at PWP may not be made-up.

At PWP, moisture is present in soils only in their smallest of the micropores and as adsorbed on soil colloids in very thin films. The soil moisture movement at this time is 1000 times slower than at field capacity.

The average moisture contents at PWP in three representative texture soils are given in Table 2.2. These values may, however, vary with the prevailing soil condition and should be taken only as average values.

## Ultimate Wilting Point (UWP)

When moisture content of soil is reduced to below PWP, the crop plants finally reach a state of irreversible wilting when neither irrigation nor rainfall will revive their growth. Plants at this stage are dead. Moisture content of soil at this stage, on oven dry basis, is termed ultimate wilting point (UWP). Soil moisture is then present in soil in its vapour form, held tightly by force of adhesion to the soil colloidal surfaces. At this stage there is no moisture present even in the smallest of the soil capillaries. The matric potential of water in soils at this time is accepted by definition a low of about –31 bars. The moisture content of soils between PWP and UWP is termed **willing range**.

# CHAPTER 9

# Mineral Nutrition: The Inorganic Support

## INTRODUCTION

Mineral nutrition is a broad field which is concerned with the complex of biosynthetic events by which higher plants produce organic materials from inorganic materials absorbed from their environment. These events include the absorption process itself, the translocation of absorbed mineral nutrients from roots to stems, leaves, fruits and finally conversion of absorbed (inorganic) nutrients to organic form. The nutritional needs of plants were understood when the addition of farmyard manures, plant tissue debris, wood ashes, ground bones, composts, and other materials of plant and animal origin to soils were found to increase their productivity. In the early half of the nineteenth century it was realized that plant growth and development can proceed only when plants are supplied with certain chemical elements referred to as essential elements. These essential elements are mostly derived from mineral constituents of the soil and are absorbed principally by the roots as inorganic ions. Roots are the most important part of the plant as the nutrient uptake takes place mainly through them.

**TABLE 9.1:** Presents the average per cent of dry weight in a plant material

| Element | % of dry weight |
|---|---|
| Carbon | 45 |
| Oxygen | 45 |
| Hydrogen | 6 |
| Nitrogen | 1.5 |
| Potassium | 1.0 |
| Calcium | 0.5 |
| Magnesium | 0.2 |
| Phosphorus | 0.2 |
| Sulphur | 0.1 |
| Chlorine | 0.01 |
| Iron | 0.01 |
| Manganese | 0.005 |
| Boron | 0.002 |
| Zinc | 0.002 |
| Copper | 0.0001 |
| Molybdenum | 0.0001 |

## MINERAL ELEMENTS: GENERAL FUNCTIONS

The criteria as to which element should be considered as essential was given by Arnon as follows:

1. The plant is unable to grow normally and complete its life cycle, in the absence of the elements.
2. The element is specific and cannot be replaced by another element however close it may be in the periodic table.
3. The element plays a direct role in the metabolism of the plant.

The important functions of minerals are as follows:

1. **Protoplasmic elements:** Nitrogen, Phosphorus and Sulphur are considered as protoplasmic elements as they form part of protoplasm along with carbon, hydrogen and oxygen.
2. **Balancing elements:** Calcium, Magnesium and Potassium counteract the toxic effect of other minerals by causing ionic balance.
3. **Framework elements:** Carbon, Hydrogen and Oxygen are considered as framework elements because they constitute the carbohydrates which forms cell walls.
4. **Influence on the osmotic pressure of the cells:** Plant cells contain dissolved mineral elements in the cell sap which influences the osmotic pressure of the cell.
5. **Catalytic elements:** Some elements function as part of the enzyme i.e., without the presence of these elements some enzyme can not function.

According to McMurtney 1950 a comparative account of the deficiency symptoms of different mineral elements is as follows-

| | | | | |
|---|---|---|---|---|
| **I. Older or lower leaves of plant mostly effected** | | | | |
| | A. | Effect generalized over whole plant, drying or firing of lower leaves. Plant light or dark green. | | |
| | | (i) | Plant light green, lower leaves yellow, drying to light brown, stalks short and slender, little abscission of brown, stalks short and slender, little abscission of leaves, stunted growth. | Nitrogen |
| | | (ii) | Plant dark green, often developing red or purple, lower, leaves sometimes yellow, drying to greenish brown or black, stalks short and slender, Delayed growth, Root growth poor, Leaves abscise early. | Phosphorus |
| | B. | Effects localized, mottling or chlorosis, with or without spots of dead tissue on lower leaves. | | |
| | | (i) | Mottled or chlorotic leaves, may redden, sometimes dead spots. Tips and margins turned or cupped upwards, stalk slender. Blooming delayed. | Magnesium |
| | | (ii) | Mottled or chlorotic leaves, with small or large spot of dead tissue. | |
| | | (a) | Spots of dead tissue small, usually at tips and between veins, more marked at margins of leaves, stalks slender plants more susceptible to disease. | Potassium |
| | | (b) | Spots generalized rapidly enlarging involving areas between veins and eventually involving secondary and even primary veins. Leaves thick. | Zinc |
| **II.** | **Newer or bud leaves, affected, symptoms localized** | | | |
| | A. | Terminal bud dies, following appearance of distortions at tips or base of young leaves. | | |
| | | (i) | Young leaves of terminal buds first typically hooked, finally dying back at tips and margins so that later growth is characterized by a cut out appearance at these points. Stalk finally dies at terminal bud. | Calcium |

| | | | | | | |
|---|---|---|---|---|---|---|
| | | (ii) | Young leaves of terminal buds becoming light green, with final breakdown. Later leaves twisted, stalk finally dies at terminal bud. | | | Boron |
| | B. | Terminal bud commonly remains alive. Wilting or chlorosis of younger or bud leaves with or without spots of dead tissues. Veins of light green colour. | | | | |
| | | (i) | Young leaves permanently wilted (wither tip effect) without spotting or marked chlorosis. Stalk or twig just below tip unable to stand in later stages | | | Copper |
| | | (ii) | Young leaves not wilted. Chlorosis present with or without spot of dead tissue scattered over the leaf. | | | |
| | | | (a) | Spots of dead tissue scattered over the leaf smallest veins tend to remain green | | Manganese |
| | | | (b) | Dead spots not commonly present, chlorosis may or may not involve veins making them light or dark green in colour | | |
| | | | | (b1) | Young leaves with veins and tissue between veins light green in colour | Sulphur |
| | | | | (b2) | Young leaves chlorotic, principal veins typically green, stalks short and slender | Iron |

## NUTRIENT DEPLETION IN SOIL AND PLANT GROWTH

India has harvested 182 million tonnes of food grains during 1993-94. The increase in per hectare yields of crops is possible through scientific farming, including judicious use of fertilizers. Mineral nutrients play an important role in agricultural production, however, the problem of nutrient depletion with cropping is of paramount importance. Therefore, it is essential to develop those practices which can reduce nutrients depletion and increase crop production. Table 9.2. presents the amount of nutrients required (kg) to produce one quintal economic produce by different crops.

There are three basic processes through which nutrient is depleted in soil, they are (i) Crop removal, (ii) leaching losses and (iii) Loss in gaseous form.

**Crop Removal:** Roots absorb large amount of nutrient from soil. The NPK and micronutrient are required for crop growth. With the intensive cropping when more than one crop is grown during a year, the amount of nutrients removed correspondingly increases. Weed control, should therefore, be effective to save nutrient depletion.

**TABLE 9.2:** Nutrient required (kg) by different crops.

| Crop | N | $P_2O_5$ | $K_2O$ |
|---|---|---|---|
| Rice | 2.0 | 11.2 | 3 |
| Wheat | 2.45 | 0.06 | 3.28 |
| Barley | 2.06 | 0.78 | 2.52 |
| Maize | 2.63 | 1.39 | 3.58 |
| Sorghum | 2.24 | 1.33 | 3.4 |
| Pearl millet | 4.23 | 2.26 | 9.08 |
| Finger millet | 2.98 | 1.13 | 3.9 |
| Chickpea | 4.63 | 0.84 | 4.96 |
| Groundnut | 5.81 | 1.96 | 3.01 |
| Mustard | 3.28 | 1.64 | 4.18 |
| Linseed | 1.90 | 1.2 | 3.3 |

| Crop | N | $P_2O_5$ | $K_2O$ |
|---|---|---|---|
| Castor | 3 | 1.2 | 1 |
| Sugarcane | 0.16 | 0.04 | 0.2 |
| Cotton | 4.45 | 0.83 | 7.47 |
| Tobacco | 1.65 | 1.93 | 2.19 |

*Source:* Velayutham, M. and Reddy, K. C. K. (1987), Regional Agricultural Meeting, IFA, New Delhi, 39-45

**Leaching Losses**: Out of major essential nutrients, only nitrogen is subject to leaching, essentially as nitrate due to its readily soluble nature. This kind of leaching depends on texture of soil and intensity of irrigation. Nitrogen losses are higher in light-textured soils and increase with heavy irrigation. Fertilizer management is essential to avoid such depletion

**Losses of phosphorus** is also a serious problem. It is immobile in soils and is subject to loss through erosion and runoff only. Practices which increase organic residues in soil reduce erosion and runoff and thereby, save phosphorus losses.

**Losses in Gaseous Forms:** Nitrogen is subject to removal through gaseous forms from soil. Two processes are involved in this loss: (i) volatilization as ammonia and (ii) denitrification (under aerobic conditions, as in rice field).

## METHODS OF STUDYING PLANT NUTRITION: SOLUTION CULTURES

Since 1804, scientists began to appreciate that plants require calcium, potassium, sulfur, phosphorus, and iron. Then, about 1860, three German plant physiologists (W. Pfeffer, Julius Sachs, and W. Knop) indicated that how difficult it is to determine the kinds and amounts of elements essential to plants growing in a medium as complex as a soil. They, therefore, grew plants with their roots in a solution of mineral salts (a **nutrient solution**), the chemical composition of which was controlled and limited only by the purity of chemicals then available. Growing plants in this way is referred to as **hydroponics**, **soil-less culture**, or **solution culture**. Other investigators later showed that many plants grew much better if the roots were aerated.

These techniques were developed to purify salts and the water for hydroponic cultures, more exact control over the elements made available to plants became possible. The technique is helpful especially for several elements that are required in very small (trace) amounts. It is very difficult to demonstrate the requirements for molybdenum, nickel, copper, zinc, and boron for various plant species (seeds). In such cases, deficiency symptoms are more easily observed in the second generation grown from seeds taken from parents that have been grown without the added elements. For example to demonstrate importance of nickel in barley seeds, they were taken from plants grown for three generations without nickel.

In the past 20 years, advancement in various instrumental techniques has helped to understand the need of various elements by plants. **Atomic absorption spectrometers** are now used to measure metals and some nonmetals. Even more valuable (and expensive) are **optical emission spectrometers** in which elements are vaporized at temperatures above 5,000 K. Such high temperatures temporarily excite electrons from their ground-state orbits into higher orbits, and as these electrons move back to their original ground states electromagnetic energy is emitted at wavelengths that are different for each element. These wavelengths are measured and the energy quantified by the spectrometer. Concentrations of more than 20 elements in a single solution can be measured with great sensitivity in less than a minute.

### *Disadvantages Solution-Culture Technique*

1. The technique needs root aeration.
2. Need to replace or supplement the solution every few days if maximum growth is to be achieved, because the solution's composition changes continuously as certain ions are absorbed more rapidly than others. This selective uptake not only depletes certain ions but also causes undesirable pH changes.

**Use of washed white-quartz sand or perlite:** To avoid depletion of mineral nutrient in liquid cultures, many physiologists use washed white-quartz sand or a mineral called perlite (expanded pumice) as a medium for the roots. Nutrient solutions are simply poured or allowed to drip onto these media at suitable intervals in excess amounts to ensure leaching of the old solution through holes in each container.

The disadvantages of this technique is that the sand and perlite contribute unknown amounts of certain essential elements to plant. For some studies in which ion uptake measurements by roots are necessary, solution cultures provide the only adequate method. In addition to solution cultures applied to roots, fruit trees and some other horticultural crops are often fertilized with foliar sprays.

## NUTRIENT SOLUTIONS

Different types of useful formulations of nutrient solutions have been devised for studies of the composition of plants. Usually, all nutrient solutions are more concentrated than soil solutions. Plants can grow well in solutions having concentrations of essential elements as low as those dissolved in the soil solution in which they normally grow, provided that the solutions are replenished often enough to maintain those concentrations or that dilute solutions in large volumes flow over their roots from a recirculating tank. The high concentrations typically used in non flowing solutions avoid the need to change the solution more than once a day or once every several days, depending on growth rates. Of course, the concentration must be low enough to avoid plasmolysis of the root cells. Most solutions have osmotic potentials no more negative than -0.1 MPa, so this is not a problem.

**The Ideal Nutrient Solution**

| | Compound | Element Supplied | Milligrams of Compound per liter of solution |
|---|---|---|---|
| **Macronutrients** | $CaNO_3 . 4H_2O$ | Ca;N | 94.00 |
| | $KNO_3$ | K;N | 60.00 |
| | $MgSO_4 . 7H_2O$ | Mg;S | 24.00 |
| | $NH_4H_2PO_4$ | N;P | 23.00 |
| | Fe-EDTA | Fe | 0.69 |
| **Micronutrients** | KCl | Cl | 0.37 |
| | $H_2BO_3$ | B | 0.15 |
| | $ZnSO_4 . 7H_2O$ | Zn | 0.057 |
| | $MnSO_4 . H_2O$ | Mn | 0.033 |
| | $CuSO_4 . 5H_2O$ | Cu | 0.012 |
| | $H_2MoO_4$ (85% $MoO_3$) | Mo | 0.0081 |

## CHELATING AGENTS

The micronutrient cations iron, and to a lesser extent zinc, manganese, and copper, are relatively insoluble in nutrient solutions when provided as common inorganic salts, and they are also rather insoluble in most soils. This insolubility is especially marked if the pH is above 5, as it is in nearly all soils with low rainfall. Under these conditions, micronutrient cations react with hydroxyl ions, eventually precipitating insoluble hydrous metal oxides. An example in which the ferric form of iron yields the reddish-brown oxide (rust) is shown in the reaction below-

$$2Fe^{3+} + 6OH^- \rightarrow 2Fe(OH)_3 \rightarrow Fe_2O_3.3H_2O$$

Due to insolubility, these micronutrients must be held in solution by some other agents which is known as ligand or chelating agent or chelator. Reaction of a divalent or trivalent metal ion with a ligand forms a chelate (from the Greek, *claw like*). *A chelate is the soluble product formed when certain atoms in an organic ligand donate electrons to the cation.* Negatively charged carboxyl groups and nitrogen atoms possess electrons that can be shared in this way. In calcareous soils (those rich in $Ca2^+$ and with a pH generally of 7 or higher), more than 90 per cent of the copper and manganese and half or more of the zinc are probably chelated with microbially produced organic compounds.

## HYDROPONICS

Hydroponics or the soilless cultivation of plants is a term heard frequently. In fact, the technique of hydroponics is becoming commercially important. Its advantages are simply that it is possible to control carefully the diet of plants and to control more readily infections by disease-causing organisms that normally live in the soil and infect plants via roots. The disadvantage of this technique is high cost. At present it appears that massive agricultural farming using this method is economically unfeasible. Nevertheless, there is increasing interest in hydroponics in many sectors of the economy. In areas, where the soil is poor or no soil at all, or only rocky, this technique is useful. Thus, hydroponic gardens may also be developed. We might conclude that while soil is not essential for growing plants, it is at present the most economical medium in which to cultivate plants.

## PROPERTIES OF ROOT SURFACE

According to Clarkson (1985) most models that describe processes involved in nutrient uptake by roots in soil relate to values for root size (usually length) and its increase with time, nutrient inflow into the roots as related to nutrient concentration in the soil solution near the root surface (this incorporates both kinetic and plant demand factors), and nutrient transport in the soil by convection or diffusion.

Competition between roots depends on their density in the soil (usually expressed as length of root per unit volume of soil) and the pattern of their distribution. From a theoretical stand-point at least, both density and pattern influence prediction of nutrient uptake from soil. Roots compete with one another for mobile nutrients like sulphate or nitrate at much lower densities than they compete for P and Zn and perhaps K because the radius of the zone of soil from which roots extract ions is much greater in the former instances and hence depletion shells come to overlap one another.

Kinetic parameters for ion absorption by solution-grown plants were used to estimate nutrient inflows into roots growing in soil. Depletion of labelled nutrient from solutions follows a Michaelis-Menten relationship and can be used to derive values for $I_{max}$ (equivalent to $V_{max}$) and $K_m$. The net inflow ($I_n$) at a given nutrient concentration at the root surface can be calculated:

$$I_n = \frac{I_{max} \ (C_s - C_{min})}{K_m + (C_s - C_{min})}$$

When $I_{max}$ is the maximal inflow at high (saturating) levels of $C_s$

$C_s$ = concentration at root surface

$C_{min}$= concentration at which there is no net inflow, i.e., $I_n = 0$

$K_m = C_s - C_{min}$ where $I_n = 0.5 \ I_{max}$

This application of ion transport kinetics put forth by Clarkson (1985) attracted the attention of many scientists. When cultivars of barley differing in either $I_{max}$ or $K_m$ were grown in a soil of moderate P-deficiency, the most evident correlation found were between phosphate uptake from the soil and root length, but a model incorporating the kinetic parameters predicted uptake by different variety quite well.

## MYCORRHIZAE: SPECIAL METHOD FOR OBTAINING ESSENTIAL ELEMENTS

Some soils are deficit in mineral salts and plants living in them have special means of obtaining nitrogen and other essential elements, for example, the roots of many moorland and woodland plants, particularly coniferous trees, which live in rich humus but deficient in mineral salts possesses a mycorrhizae. Mycorrhizae (**fungus-roots**) are symbiotic associations between fungi and the roots of vascular plants. Two main groups of mycorrhizae are recognized: the **ectomycorrhizae** and **endomycorrhizae**, although the rarer group with intermediate properties, the **ectendotrophic mycorrhizae**.

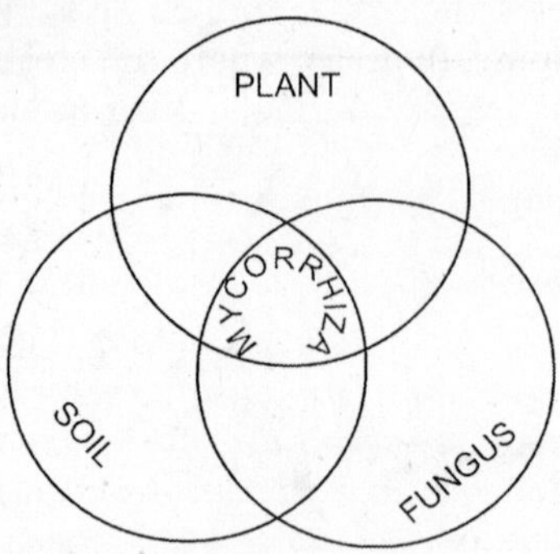


**FIG. 9.1:** Interrelationship betewwn mycorrhiza, soil and plant.

In the ectomycorrhizae, the fungal hyphae form a mantle both outside the root within the root in the intercellular spaces of the epidermis and cortex. No intracellular penetration into epidermal or cortical cells occurs, but an extensive network called the Harting net is formed between these cells. Ectomycorrhizae are common on trees, including members of the families pinaceae (pine, fir, spruce, larch, hemlock), Fagaceae (willow, poplar, chestnut), Betulaceae (birch, alder), Salicaceae (willow, poplar, and few others. Shows scanning electron micrographs of two Pinus contorta roots, one uninfected and having root hairs and the other infected with an ectomycorrhizae fungus.

Endomycorrhizae consist of three subgroups, but by far the most common are the vesicular arbuscular mycorrhizae (VAM). The fungi present in VAM are members of the Endogonacae, and they produce an internal network of hyphae between cortical cells that extends out into the soil, where the hyphae absorb mineral salts and water. Although VAM fungi seem to penetrate directly into the cytosol of cortical cells (where they form structures called vesicles and arbuscules, giving them their name), the hyphae are surrounded by an invaginated plasma membrane of the cortex cell. The VAM are present on most species of herbaceous angiosperms, whether monocot or dicot, annual or perennial crops, or native or introduced species; they also occur in the gymnosperm genera Cupressus, Thuja, Taxodium, Juniperus, and Sequoia and in ferns, lycopods, and bryophytes.

The importance of mycorrhizae was first recognized in connection with efforts to grow orchids in greenhouses. Orchids have microscopic seeds that germinate to form a tiny pad of tissue called a **protocorm**. Cultivators of orchids found that the plants seldom developed beyond the protocorm stage unless they were infected with a particular kind of fungus. Subsequently it was found that if seedlings of many forest trees are grown in nutrient solutions and then transplanted to prairie or other grassland soils, they fail to grow.

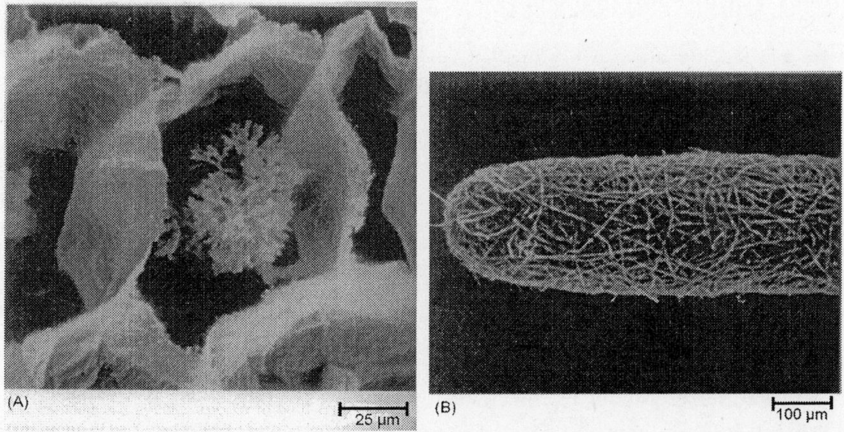


**FIG. 9.2:** Scanning electron micrographs of mycorrhize. (A) A fully developed arbuscule of an endomycorrhiza in a plant cell. (The term arbuscule means little bush). As the arbuscule decomposes, it releases accumulated nutrients from plant. (B) The mycelial mantel of a typical ectomycorrhizal fungus surrounding a eucalyptus tree root.

Eventually they may die from malnutrition even if a soil analysis shows that there are abundant nutrients in the soil. If a small amount (0.1 per cent by volume) of forest soil containing fungi is added to the soil around the roots of the seedlings, however, they will grow promptly and normally. Mycorrihizae are now though to occur in more than 90 per cent of all families of plants.

In some mycorrihizal associations known as endomycorrhizae, the fungal hyphae penetrate the cells of the root, forming coils, swellings, or branches. The hyphae also extend out into the surrounding soil. Endomycorrhizae occur in about 80 per cent of all vascular plants and the fungal component is usually a zygomycete.

In other associations known as ectomycorhizae, the hyphae form a sheath around the root but do not actually penetrate its cells. Ectomycorrhizae are characteristic of certain groups of trees and shrubs, including pines, beeches and willows the fungus is usually a basidiomycete but some associations involve ascomycetes including truffles. The exact relationship between roots and fungi is not known. Apparently the roots secrete sugars, amino acids, and possibly some other organic substances that are used by the fungi. Although the evidence is just now accumulating, it appears that the fungi convert minerals in the soil and decaying material into an available form and transport them into the root.

The majority of plants with mycorrhizal roots only obtain phosphorus and transfer phosphorus from the soil into roots, and perhaps nitrogen, compounds from the fungus. They can photosynthesize normally (e.g., bird's nest orchid - a siclkly looking inhabitant of dark woods) and can therefore, manufacture their own carbohydrate, some of which are absorbed by the fungus. There is evidence that water uptake is facilitated by the fungus.

A study of the fossils of early vascular plants has revealed that mycorrhizae occurred as frequently in them as they do in modern vascular plants. This has led to the interesting suggestion that the evolution of mycorrhizal associations may have been the critical step allowing plants to make the transition to the bare and relatively sterile soils of the then unoccupied land.

## ESSENTIAL AND DISPENSABLE ELEMENTS

Apart from the elements of water and carbon dioxide, the remaining elements are distinctively of mineral origin. Water is not usually considered to be a nutrient. However, it is by far the most abundant molecular species in cells and organisms. A nutrient is that which nourishes and out of which the substance of plants is built; in this sense water certainly performs an essentially nutritional role.

In rich variety of the plant kingdom, from the thallophyta to the angiosperms, the range and complexity of the physiological functions and biochemical reactions of plants are accomplished by utilizing the chemical properties of only a few elements of the periodic Table. Of all the elements nine macroelements C, H, O, N, P, S, K, Ca, Mg, and the seven well established micronutrient B, Mn, Cu, Zn, Fe, Mo and C and those which are beneficial like Cd, Na, Si, Co, V are easily available to some plants.

Though some elements are utilized by plants yet they are not most common. Carbon, by its small size and its ability to combine directly with itself, can form such repeating patterns alone. Carbon forms the essential skeleton and together with nitrogen and phosphorus forms the essential structure of proteins and nucleo-proteins, without which terrestrial life could not exist. The large molecules are built from silicon, with aluminium and oxygen which provide the repeating patterns in space, and are the basis of much inorganic form in minerals and in soils. But for all practical purposes the elements silicon and aluminium are dispensable to plants. Thus in general 16 elements including O, H, C are essential for all angiosperms and gymnosperms. With these elements and sunlight, most plants can synthesize all the compounds they require.

Sodium seems to be beneficial to desert species such as *Atriplex vesicaria*, a common weed of salty arid soils. It has been emphasized by Brownell and Crossland (1974) that those plants having the C-4 photosynthetic pathway probably require $Na^+$. Certain species that fix $CO_2$ in photosynthesis via CAM pathway also grow faster with sodium and for them sodium probably is essential.

Some plants probably require Ni because it is an essential part of the enzyme urease, which catalyzes hydrolysis (breakdown with $H_2O$) of urea to $CO_2$ and $NH_4^+$. Legumes of tropical origin, including soybean form ureides in root nodules during nitrogen fixation, and utilization of such ureides in leaves, stems, developing seeds and certain algae requires urease.

Selenium is absorbed and accumulated in fairly high concentration up to as least 0.5 per cent; i.e., 5 mg per g dry weight) by certain **accumulator species** of Astragalus. In bacteria and animals, both of which require selenium, a few essential proteins have been found that contain selenium. Most of these proteins are enzymes that catalyze oxidation reduction reactions and selenium is essential to their activity. Perhaps similar enzymes occur in selenium accumulators, but so far there is no evidence that it is true. Bollard (1983) observed that if selenium is essential for Astragalus sps, it will have to function at tissue concentrations at or less than 0.68 ppm of dry matter slightly lower than molybdenum levels.

## ESSENTIAL NUTRIENT ELEMENTS

The action of the nutrients in the growth processes of higher plants has been revealed sometimes by a study of the visible or anatomical effects that result from mineral deficiencies or by the effects of controlled levels of these elements on the chemical composition of the affected plants. Quite often the symptoms or chemical changes associated with deficiency or excess of the element in one plant may differ from those observed in another. It is therefore necessary to recognize the direct manifestations of the specific deficiency symptoms of mineral elements in different plants which may be masked or complicated by differences in the indirect or secondary effects of abnormal mineral concentrations or metabolism as a whole. Quantitative or qualitative differences in the activity and distribution of enzyme systems and possibly quantitative differences in the subsequent or prior metabolism of the constituents of cell that may be found to either accumulate or be depleted. As a result of such differences in enzyme activity such as occurs with amino acids would exert different effects on growth. Thus there may be a lot of variations in the overall behaviour which is revealed in visible, anatomical, or chemical characteristics.

The roles of mineral elements especially the micronutrients in specific enzyme activities is the main basis for their effect on growth, morphology, cell composition, and physiological activities. There are many aspects of nutrition such as the overall effects of mineral deficiencies and excesses on changes in morphology or composition, e.g., in carbohydrate or amino acid concentration, or on such activities as photosynthesis, respiration, and protein synthesis, which probably reflect the interactions of several factors which can be described as physiological responses.

## NUTRIENT DEFICIENCY SYMPTOMS AND FUNCTIONS OF ESSENTIAL NUTRIENT ELEMENTS

The deficiency symptoms for any element depend primarily on the function of the element and whether or not the element is readily translocated from old leaves to younger leaves.

Elements can be classified into **mobile** and **immobile** elements. The elements mobility in the phloem of vascular tissue is determined partly by the solubility of the chemical form of the element in the tissue and partly by how well it can enter the sieve tubes of the phloem. Some elements that readily move through the phloem from old leaves to younger leaves and then to storage organs include nitrogen, phosphorus, potassium, magnesium, and chlorine. Others such as boron, iron, and calcium are immobile, and the mobility of zinc, manganese, copper, sulphur and molybdenum is usually intermediate. The deficiency symptoms appear first and are more pronounced in older leaves when the element is mobile whereas in immobile element as in calcium and iron the symptoms first appear in younger leaves.

## NITROGEN

Two major ionic forms of nitrogen are absorbed from soils: nitrate ($NO_3^-$) and ammonium ($NH_4^+$) and because of its presence in so many essential compounds, it is not surprising that growth without nitrogen is slow. Soils are most often deficient in nitrogen than any other element. The most marked symptom is decrease in chlorophyll content and leaves are pale green. Older leaves become yellow green and often completely yellow, a condition associated with proteolysis without concurrent synthesis. The symptoms appear last in the younger leaves because of the high mobility of nitrogen in the plant.

One characteristic symptom which is found in many plants is the production of pigments other than chlorophyll. In tomato plants, a purple coloring of the petioles and veins caused by anthocyanin is observed. This response is also seen in the stems of many plants. High concentration of nitrogen has a tendency to increase the cell number and cell size with an overall increase in leaf production. We can thus assume that nitrogen is an essential constituent of protein and low nitrogen availability must cause a decrease in protein synthesis which then causes a decrease in cell size specially cell division.

Most recognized role of Nitrogen in the plant is perhaps its presence in the structure of protein molecule. It is found in purines and pyrimidines of nucleic acid RNA and DNA, essential for protein synthesis. The porphyrin structure is found in chlorophyll pigments and the cytochromes are essential in photosynthesis and respiration. Coenzymes are essential to the function of many enzymes.

A mixture of nitrate and ammonium has produced higher crop yields than with either source alone. An adequate supply of potassium enhances ammonium utilization and thus improves yields, when a mixed ammonium nitrate nitrogen is applied (Hagin et al., 1990). Presence of ammonium and nitrate in roots could stimulate formation of glutamic acid, which plays a central role in the synthesis of other amino acids. Enhanced rates of transport to leaves of these amino acids may permit the establishment of higher protein levels. Ammonium nitrate nutrition could maintain a higher leaf nitrogen level than nitrate and extend the photosynthesis period.

## PHOSPHORUS

It is absorbed as the monovalent phosphate anion ($H_2PO_4^-$) and less rapidly as the divalent ion ($HPO_4^{2-}$). It never undergoes reduction in plants and remains as phosphate, either free or bound to organic forms as esters. Many of the visible effects of phosphorus deficiency resemble those caused by lack of nitrogen. The similarities apply to such effects as acute leaf angles, prolonged dormancy of lateral buds, premature leaf fall, decreased size and number of flower primordia, delay or suppression of flowering, a few small fruits or seeds and purple or red anthocyanin pigmentation. Deficiency leads to dead necrotic areas on the leaves, petioles or fruits an overall stunted appearance and the leaves may have a characteristic dark to blue coloration. Because of the high mobility of the element, symptoms first appear on the older leaves. Eaton (1952) reports accumulation of carbohydrates in soybean and black mustard. Hewitt and Tatham (1960) found that acid phosphatase activity in leaves of tomato plants was increased tenfold by phosphorus deficiency. This may be due to decreased inorganic phosphate level, which inhibits phosphatase activity; if so, the effect must be on the synthesis of the enzyme during growth, but not by direct inhibition in vitro. Phosphorus deficiency would decrease DNA synthesis as well as that of RNA. This would restrict nuclear division or might cause defects in parts of chromosomes or gene suppression. There is conclusive evidence that Mo uptake by tomato plants is markedly enhanced by P deficiency.

It is an important constituent of nucleic acid, phospholipids, most important constituent of ATP and other high energy compounds, and the coenzymes NADP involved in respiration, photosynthesis, nitrogen metabolism, carbohydrate metabolism and fatty acid synthesis.

## CALCIUM

It is absorbed as divalent $Ca^{2+}$ easily available for plant growth, but acidic soils where high rainfall occurs are often fertilized with lime (a mixture of CaO and $CaCO_3$) to raise the pH.

The element is immobile in phloem and as a result deficiency symptoms are always pronounced in young tissues. Meristematic zones of roots, stems and leaves, where cell divisions are occurring, are most susceptible, perhaps because calcium is required to bind pectate polysaccharides that form a new middle lamella in the cell plate that arises between daughter cells, or because calcium is needed to form microtubules of the mitotic spindle apparatus. Twisted and deformed tissues result from calcium deficiency, and the meristematic areas die early. Roots may become short, stubby, and brown. Chlorosis occurs along the margins of younger leaves, which become necrotic. Malformation or distortion of the younger leaves is characteristic of calcium deficient plants, a hooking of the leaf tip being the most easily detected symptom.

Calcium is receiving renewed attention as it is now recognized that all organisms maintain unexpectedly low concentrations of free $Ca^{2+}$ in their cytosol, usually less then 1 mM. This is true even though calcium is as abundant in many plants, especially legumes, as phosphorus, sulfur, and magnesium. Most calcium is in central vacuoles and bound in cell walls to pectate polysaccharides. In vacuoles, it is frequently precipitated as insoluble crystals of oxalates, and in some species as carbonates, phosphates, or sulfates. Low cytosol concentrations of $Ca^{2+}$ must be maintained partly to prevent formation of insoluble calcium salts of ATP and other organic phosphates. Few enzymes are activated by $Ca^{2+}$, and many are inhibited, the inhibition providing further requirement for cells to maintain unusually low $Ca^{2+}$ concentrations in the cytosol where many enzymes exist.

Calcium ($Ca^{2+}$) can be used to stimulate $NH_4^+$ absorption by plants. The efficient use of N for crop production is important because N is normally the most expensive fertilizer input. As $Ca^{2+}$ concentration increased, $NH_4$ absorption and plant growth increases were greatest with young seedlings of radish and onion. In the mature growth stages, $Ca^{2+}$ -stimulated $^{15}NH_4^+$ absorption was less rapid than in the earlier growth stages but frequently exhibited a different response (i.e., altered metabolite translocation) to the added $Ca^{2+}$ concentration. However, at the intermediate and mature growth stages significantly increased N contents and plant growth also were noted in most cases. The $Ca^{2+}$ -increased N content in leaves and bulbs of the older plants had much less $^{15}NH_4^+$ suggesting that the newly absorbed $^{15}NH_4^+$ was being deposited in the roots replacing older N forms that were then translocated to the bulbs or leaves thus increasing N use efficiency. Plant growth stimulation in the presence of $Ca^+$ was also reported by Sung and Lo (1990) who found that the extra $Ca^+$ caused both a large increase in vegetative growth and also when applied at seed fill the kernels were up to 14% heavier.

The role of calcium has been implicated in guard cell physiology. Calcium salts added to the medium in which isolated epidermis is incubated can stimulate stomatal closure and inhibit opening in *Commelina*, tobacco, and *Vicia faba*. Calcium ions also inhibit $K^+$ induced swelling of guard cell protoplasts. Mansfield et al. (1990) have drawn interesting conclusions on the effects of both apoplastic and cytosolic calcium ions in the regulation of opening and closing of stomata.

The free $Ca^{2+}$ ion is now recognized as having an important role in animals and plants as a **second messenger** or **signal molecule** which couples various extracellular stimuli such as hormones, light, gravity, or stress with intracellular metabolic events. The coupling of stimulus to response by $Ca^{2+}$ implies that an external stimulus leads to a change in the rate at which $Ca^{2+}$ is transported into or out of a specific subcellular compartment. The resulting alteration of the intracellular free $Ca^{2+}$ concentration thus represents the signal which, through its ability to regulate the activity of specific target enzymes, leads to the appropriate metabolic response.

If $Ca^{2+}$ does act as a second messenger in plants it should be possible to demonstrate that its in vivo concentration changes in response to extracellular primary signals such as light or hormones. As yet, direct measurements have not been made. However, through manipulation of external or internal $Ca^{2+}$ concentrations it has been possible to indirectly demonstrate a role for $Ca^{2+}$ in stimulus-response coupling.

## CALMODULIN

**Occurrence:** Calmodulin has been isolated from lower and higher plants, the green algae *Chlamydomonas* and *Mougetia*; the fungi, *Neurospora cressa*, the yeasts *Candida elbucans* and *Saccharomyces cervisiae*, *Blastocladia emersonii*, two Basidiomycetes and different sources of higher plants including peas, mungbeans, wild carrots and peanut seeds, barley shoots and roots, Zucchni hypocotyls, spinach leaves, corn, etc.

**Concentration:** The concentration of calmodulin in most plant tissues varied between 2 to 4 μm Mol Kg. Measurements of the sub cellular calmodulin based on enzymatic and radioimmunoassays of wheat leaf cells indicated that 89 to 93% of the cellular calmodulin occurs in the cytoplasm, while more than 1% to 9% is present in the organelles like chloroplasts, mitochondria and microsomes. The following table indicates the association of calmodulin with oat tissue, cells and cell fractions as measured by radioimmunoassay.

The most important of all known $Ca^{2+}$ modulated proteins is calmodulin, a highly conserved, heat stable, small molecular weight protein found in all eukaryotic cells. Containing 4 $Ca^{2+}$ binding sites per molecule, calmodulin is composed of a single subunit and 148 amino acids. the binding of $Ca^{2+}$ induces a conformational change resulting in increased alpha helicity. This $Ca^{2+}$ induced conformational change allows the protein to interact with and alter the activity of calmodulin-dependent enzymes. The $Ca^{2+}$ calmodulin complex has a wide range of biochemical activities, including effects on both coarse and fine metabolic regulation. Because of these characteristics, calmodulin has become the standard against which all other Ca2+ modulated proteins have been compared. Plant enzymes which have been reported to be directly regulated $Ca^{2+}$ or $Ca^{2+}$ calmodulin include NAD-kinase, β-glucan synthase, $Ca^{2+}$ and $H^+$-ATPase, quinate; $NAD^+$-oxidoreductase and protein kinases.

## MAGNESIUM

Magnesium is absorbed as divalent $Mg^{2+}$. Since it is a constituent of the chlorophyll molecule, the most common symptom of magnesium deficiency in green plants is extensive interveinal chlorosis of older leaves and as the deficiency becomes acute, it eventually reaches the younger leaves. Chlorosis is often followed by anthocyanin pigments in the leaves leading to necrotic spotting.

$Mg^{2+}$ plays an essential role in photosynthesis and carbohydrate metabolism. Many of the enzymes involved in carbohydrate metabolism require magnesium as an activator in which ATP is involved. It is also an activator for those enzymes involved in the synthesis of nucleic acids, DNA, RNA from nucleotide polyphosphates. All these reactions involve phosphate transfer. Magnesium binds the subunits of ribosomes together and that EDTA causes dissociation by its removal of the magnesium ion from the ribosomal particle. Thus, it is assigned two roles in protein synthesis, as an activator in some of the enzyme systems involved in the synthesis of nucleic acids and as an important binding agent in ribosomal particles where protein synthesis takes place.

## POTASSIUM

After nitrogen and phosphorus, soils are usually deficient in potassium. $K^+$ is easily redistributed from mature to younger organs, so deficiency symptoms first appear on older leaves. A mottled chlorosis followed by the development of necrotic areas at the tip and margin of the leaf is observed. There is a tendency for the leaf tip to curve downward and, as in the case of the French bean and potato, marginal regions may roll inward towards the upper surface. Growth is stunted with a pronounced shortening of internodes. $K^+$ deficiency results in decrease of apical dominance since some plants like flax, cereals and barley develop a bushy habit due to the extension of numerous axillary shoots usually at the base of the plants.

The best known function of $K^+$ is its role in the opening of stomata. High concentrations of $K^+$ are found in meristematic regions of the plant. It is an activator for enzymes involved in the synthesis of peptide bonds. $K^+$ deficiency leads to accumulation of carbohydrates which may be due to impaired protein synthesis, that is, carbon skeletons that would normally go into protein synthesis are accumulated as carbohydrates. In addition to its role as an activator in protein metabolism $K^+$ can also act as an activator for several enzymes involved in carbohydrate metabolism.

## SULFUR

It is absorbed from soils as divalent sulfate ions ($SO_4^{2-}$). Sulphur deficiency is usually encountered in light-textured high-pH soils. Sulfur deficiency symptoms resemble somewhat those of nitrogen deficiency. There is general chlorosis followed by the production of anthocyanin pigments in some species. Unlike nitrogen deficient plants, deficiency of sulfur causes chlorosis in the young leaves first. Under severe conditions, however, all of the leaves may undergo some loss of green color.

Most of the sulfur in plants occurs in proteins, specially in the amino acids cysteine and methionine that are building blocks for proteins. Other essential compounds that contain sulfur are the vitamins thiamine and biotin and coenzyme A, a compound essential for respiration and for synthesis and breakdown of fatty acids.

## IRON

The most characteristic effect of iron deficiency is the failure to produce chlorophyll in young leaves. In most plants, chlorosis is initially interveinal and produces a fine reticulate pattern in partially expanded leaves. Fruits of tomato become silvery green and are orange rather than red when ripe. Iron deficiency affects cell division in the special meristem of excised pea roots but cell expansion and protein synthesis appeared to be continuing when cell division ceased.

Iron sprays on chlorotic leaves had no effect on size of chloroplast. Iron deficiency induces changes in the structure and functions of the whole photosynthetic apparatus of higher plants. It has an effect on thylakoid membrane, thylakoid composition, changes in photosystem II and photosystem I, in xanthophylls to chlorophyll ratio and their lipid composition. Chlorophyllase enzyme (chlorophyll-chlorophyllide hydrolase E.C.3.1.1.14) is involved in the metabolism of chlorophyll's increases notably with iron deficiency.

The occurrence of iron chlorosis is a serious problem in several crops grown in India and other countries. Iron deficiency chlorosis is common in groundnut growing areas in Chittoor district of Andhra Pradesh (India) and in lemon trees growing in the Segure basin (Murcia, Spain). Chlorosis is mainly observed on soils with high calcium carbonate immediately after irrigation or high rainfall. This induces iron stress, even though there is sufficient level of Fe in the soil, it remains mobilized forming insoluble precipitates in the soil and even inside the plant. At times the iron present in alkaline and calcareous soils in the ferric form and other insoluble forms are not readily available to many plants. Only some plant species or varieties with physiologically iron-efficient root metabolic systems can take up adequate iron from such soils as well as translocate it for utilization by the plant.

Iron deficiency resulted in significant effects on the structure and physiology of maize chloroplasts and changes in biomass accumulation and distribution between vegetative organs as well as changes in the content of chlorophyll a, chlorophyll b, and the carotenoids of young maize plants. The photosynthetic rate per leaf area was decreased. Part of $^{14}C$ incorporated in low molecular compounds was increased and the share of amino acids and organic acids in them was increased.

Onyezili and Ross (1993) used plant growth, root reductive capacity, rhizosphere pH, and phytosiderophore release capacity as indices to evaluate the responses of maize, millet, sorghum, barnyard grass, wheat, and white lupin to iron-deficiency stress. Significant iron-stress-induced phytosiderophore release was observed only in wheat and sorghum in which accompanying increases in rhizosphere pH were also evident. Such phytosiderophore release matched the severity of leaf chlorosis and iron uptake and depended on the form in which the element was supplied.

Scholz et al. (1992) report the accumulation of nicotinamine (NA) in iron deficient barley root tips. NA is a chelate for Fe (II) and other divalent transition metals. In tomato (strategy I), and probably also in strategy II plants, NA is essential for proper function of Fe(II) dependent processes, such as the regulation of iron deficiency response mechanism, iron enzymes, and phloem loading/unloading of iron.

## MANGANESE

The visible effects of a deficiency of manganese are more diverse than those of many other mineral elements. The symptoms may occur first in young or in old leaves and may exhibit a variety of chlorotic patterns and necrotic spotting in the interveinal areas of the leaf.

Manganese acts as enzyme activator in respiration and nitrogen metabolism. It can be substituted by other divalent cations such as $Mg^{2+}$, $Co^{2+}$, $Zn^{2+}$, and $Fe^{2+}$. It is the predominant metal ion of krebs cycle reactions. It acts as an activator for the enzymes nitrite reductase and hydroxylamine reductase. It is involved in the destruction or oxidation of indole-3-acetic acid (IAA), a natural auxin of plants.

Manganese plays a direct role in photosynthesis. The sensitivity or chlorophyll to light destruction increases under conditions of manganese deficiency and leads to chlorosis in *Chlorella pyrenoidosa*. It appears that the site of manganese activity is the oxygen producing step in photosynthesis. During the light reactions of photosynthesis it is involved in electron transfer from water to chlorophyll. However, manganese deficiency does not seem to impair photoreduction in photosynthesis per se.

## COPPER

The most pronounced symptoms of copper deficiency are those found in exanthema and reclamation. Exanthema is a disease of fruit trees that is characterized by gummosis (gummy exudates), accompanied by diebeck and glossy brownish blotches on leaves and fruits.

In the *White tip* or reclamation disease of cereals caused by copper deficiency in peat soils, the young leaves become limp and chlorotic and remain tightly rolled. Copper deficiency causes a necrosis of the tip of young leaves that proceeds along the margin of the leaf and gives it a withered appearance. Under severe conditions the leaves may be lost, and the whole plant may appear wilted.

It acts as a component of phenolases, laccase and ascorbic acid oxidase. Work by Neish (1939) has implied that copper may function in photosynthesis., It was found that chloroplasts of clover contain most of the copper of the plant. In addition, it was found that carbondioxide absorption is decreased in copper deficient tung trees. The chloroplast contains a copper containing protein called plastocyanin that is essential as an electron carrier in photosynthesis.

## ZINC

The most prominent symptom is an interveinal chlorosis of the older leaves starting at the tips and margins followed by white necrotic spotting, shortened internodes, small leaves and stunted growth are characteristic symptoms of zinc deficiency. Sometimes the leaves are clustered on short branches known as rosettes.

Zinc availability to plants generally increases as soil pH decreases. Ibrikci and Moraghan (1993), compared under greenhouse conditions, the growth and nutrient composition of soybean and navy bean grown in a calcareous soil low in available Zn and available N as affected by source of N ($KNO_3$ N + native soil N versus rhizobium-inoculated seed + native soil N), P and Zn fertilizers. Their data strongly indicate that the environment around $KNO_3$ treated navy bean roots was more acidic than that around soybean roots. Differential susceptibility of soybean to Zn deficiency apparently resulted from restricted translocation of Zn from roots to tops in the navy bean cultivar. Formation of Zn-protein complexes in roots, binding of Zn to root cell walls, and precipitation of Zn by P in roots reportedly suppress Zn translocation to plant tops. Any differential effect of N on the susceptibility of the two species to P and Zn fertilizers was obviously masked by N deficiency restricting the growth of inoculated plants.

Zinc may be involved in the biosynthesis of the plant auxin indole-3-acetic acid (IAA). Skoog (1940), observed a marked decrease in auxin content of zinc deficient tomato plants and a significant increase in the IAA content when zinc was added to the deficient plants. It appears that deficiency symptoms could be associated with the decrease in the plant auxin concentration. It is also reported that zinc reduces auxin content through its involvement in the synthesis of tryptophan, a precursor of auxin. Nason (1956), found that zinc deficient *Neurospora* showed low tryptophan synthetase activity. This enzyme catalyzes the reaction of the serine with indole to form tryptophan.

Zinc acts as an activator of several enzymes, alcoholic dehydrogenase, pyridine nucleotide dehydrogenase and carbonic anhydrase. Carbonic anhydrase is the first zinc containing enzyme found in some marine plants and animals (Keilin and Mann, 1940). It may also act as an activator for some phosphate transferring enzyme, such as hexosekinase or triose phosphate dehydrogenase. It has been found that zinc deficiency leads to accumulation of soluble nitrogen compounds such as amino acids and amides. It can thus, be assumed that zinc must play an important role in protein synthesis. Zinc deficient tomato plants, may have greatly increased polyphenol oxidase and peroxidase activities, and these changes might contribute to the accumulation of tannins that is observed in Zn deficient tissues.

## BORON

The most pronounced and visible symptom is the death of the shoot tip after boron withdrawal because of its requirement for DNA synthesis. This results in growth of lateral shoots the tip of which also die. The leaves may have a thick coppery texture and sometimes curl and become quite brittle. Boron deficiency produces in different plants a wide variety of symptoms such as heart rot of beet, stem crack of celery, brown heart of swede, water core of turnip, internal cork formation in apples and water core development in turnips. Severe necrosis occurs in root which become blackened and enlarged at the tips in beans, peas, beet and *Pinus*. Flowering is often totally suppressed and flower often fall without producing seed possibly due to the effect on pollen tube growth. Fruit formation is often abnormal and may be one of the most obvious effects in less severe conditions. Boron deficient apples become malformed with numerous irregular concave areas on the surface.

Though symptoms of boron are very striking, its role in plant metabolism as yet is to be ascertained. Gauch and Dugger showed its involvement in carbohydrate transport within the plant. They propose that sugar is transported more readily across cell membranes as a borate complex. As an alternative proposal they also suggest that the borate ion may be associated with the cell membrane where it could complex with a sugar molecule and facilitate its passage across the membrane. Thus it is generally accepted that it helps in the translocation of sugar and synthesis of DNA in meristems. It has also been implicated in cellular differentiation and development, nitrogen metabolism, fertilization, active salt absorption, hormone metabolism, water relations, fat metabolism, phosphorus metabolism and photosynthesis. Galloway and Dugger (1990), observed that when boron was deficient in peas, the activity of the enzymes catechol and polyphenol oxidases was found to be increased, but in the presence of boron, the activity of these enzymes was found to be inhibited. Phosphoglucose isomerase, 6-phosphogluconate dehydrogenase and KCl-stimulated ATPase activities have also been reported to be inhibited by boron as was alcohol dehydrogenase. a-amylase on the other hand, has been reported to be stimulated by boron. It has been found to inhibit the phosphoglucomutase reaction which was non-competitive with glucose 1,6- bisphosphate and the degree of inhibition was influenced by the pH of the reaction at 8.6, the inhibition in the presence of saturating glucose 1,6-bisphosphate was 20%, while at pH 7.3, the inhibition was greater than 40% (Galloway and Dugger, 1990).

Boron markedly affects nitrogen uptake and the metabolism of nitrogen compounds; the protein contents of young leaves decrease and soluble nitrogen compounds, particularly nitrate, accumulate under the condition of severe boron deficiency. The increased nitrate content may be the result of intensified nitrate uptake and/or decreased activity of nitrate reductase. Shen et al., (1993), have reported that the nitrate reductase activity in the leaves of rape plants increased with increasing nitrogen level in soil and the enhancement caused by nitrogen was more with boron than without boron added. Both with and without boron added, NR activity was obviously higher with $Ca(NO_3)_2$ than with $(NH_4)_2SO_4$. With boron added, NR activity in leaves increased from the basal to the top parts of the plant.

## MOLYBDENUM

Visible symptoms of molybdenum deficiency may start with chlorotic interveinal mottling of the lower leaves followed by marginal necrosis and in folding of the leaves. Mottled areas may become necrotic under severe conditions and the leaf may wilt. Flower formation is inhibited and if formed they abscise before setting fruit. A condition known as whiptail is demonstrated in cauliflower plants. The leaves at first show interveinal mottling and the leaf margins may become gray and flaccid and finally brown. The leaf tissues may wither and leave only the midrib and a few small pieces or leaf blade which gives the appearance of tail and hence the name.

Molybdenum has been implicated in gaseous nitrogen fixation and nitrate assimilation. Deficiency leads to a drop in ascorbic acid concentration in plants. Hewitt (1963), drew attention to the fact that chloroplast disorganization occurs with the appearance of whiptail symptoms a common disease of deficiency.

## SILICON

Silicon is one of the most abundant elements in the earth's crust and in the ash of plants. In soil solutions it exists (30-40 mg $SiO_2L^{-1}$) as monosilicic acid ($Si(OH)_4$) with a solubility of ~2 mm and its concentration tends to lower at pH> 7 and in the presence of large amounts of sesquioxides and dominant anion adsorption. It is deposited as silica gel in the epidermal cells and cell wall of many plant species of Equisetaceae, Cyperaceae, Gramineae and Urticaceae. In oat plant it is noncrystalline and isotropic. In rice 90-95% of total Si is in the form of silica gel, 0.5 - 8% as silicic acid and 0-3.3% as colloidal silicic acid. In addition, silicic acid forms complexes both with molecules of simple: Polyhydroxy organic compounds (Pectins, callose tannins and starch) and organic acids and complex: methemoglobin, albumin, collagen, gelatin, insulin, nylon, pepsin and laminarin. Varying amounts of silicon-pectin complexes are present in citrus fruits.

### Silicon and Plant Growth

The plants normally deposit considerable amounts of silica when grown under natural conditions are rice (5 to 20%) oat, barley, wheat (2 to 4%) and horsetails (16%). Silicon has been known as a beneficial element especially for rice in Japan. Si deficiency in rice plants developed necrosis, retarded growth and reduced fertility. In rice Si promotes the growth by increasing number of tillers; leaf area and photosynthetic activity of the lower leaves. The stimulation of growth may be due to beneficial soil conditions brought about by calcium and magnesium ingredients present in the carrier, prevention or depression of Mn and Fe toxicity, increased P availability in soil and increased P activity within the plant. Si was found to inhibit the action of the enzyme invertase, increased the phosphate into ATP, ADP and sugar phosphates and increased the rate of photosynthesis.

**Silicon and Lodging:** Si can modifying effects on some rice leaf characters particularly leaf openness. A sufficient supply of Si permits rice plant to keep the leaves erect, uncurved and thus improves the assimilatory system, counteracts the negative effects of a large nitrogen supply on light interception.

**Silicon and Pest Resistance:** Silica content in plants influence vulnerability to pests. As tissues (Particularly leaves) age lignification and/or the accumulation of Si in the epidermal cell layer may form an effective physical barrier to hyphal penetration. These processes provide the main structural resistance of plants to diseases especially in the leaves of grasses. Although there is no doubt the role of Si in the epidermal cells (as silica cellulose membrane or cuticle silica double layer) as a physical barrier to hyphal penetration, to the stylet and the mandibles of sucking and biting insects (Jones and Handreck, 1967; Sogawa, 1982), other mechanisms are also involved in the higher resistance of young leaves supplied with Si. Evidences also exist for specific interaction between Si and the content and metabolism of polyphenols in xylem cell walls and in leaves of rice. Si (100 mg $kg^{-1}$ soil) application to rice has been found to increase the leaf phenol content by 4 per cent.

**Silicon and Mineral Toxicity:** Silicon added to the growth medium alleviates Mn and Fe toxicity of plants. Si reduced the necrotic browning of barley plants although it did not appreciably alter the uptake of Mn by the plants. The effect of Si on the alleviation of the Fe and Mn toxicity of rice plants was due to a decrease in the uptake of Fe and Mn by the plants.

**Silicon and Yield Improvement:** Addition of Si in the form of soluble silicates increases the yield of many agronomic crops including sugar-cane, rice, sunflower, sugar beet and others. It has been reported that Si application to soil (100 mg $SiO_2$ $kg^{-1}$) was found to increase the grain yield by 25 per cent (Ashoka Rani and Saraswathi, 1992).

Cobalt is essential for many bacteria, including cyanobacteria. It is required for nitrogen fixation by bacteria in root nodules on legumes. Cobalt concentrations as low as 0.1 mg $l^{-1}$ were high for rapid growth, and neither vanadium, germanium, nickel nor aluminium could substitute for cobalt. Since cobalt is a component of vit $B_{12}$ it is found in many organisms including many animals. Higher plants and algae have been thought not to contain vit $B_{12}$ and hence show no cobalt requirement, but it is reported by Wilson and Nicholas (1967) that cobalt is needed by certain plants. An enzyme needed to form aromatic compounds is non-functional without cobalt.

## METAL TOXICITY IN PLANTS

High metal availability, arising from mining and industrial activities, disposal of sewage sludge or soil acidification, is an increasing problem in agriculture and forestry. Metal toxicity causes multiple direct and indirect effects in plants which concern practically all physiological functions.

The primary toxicity mechanisms of the different metal ions may be as different as their chemical properties, especially valence, ion radius and capacity to form complexes. Nevertheless, an excess of these metal ions or of soluble metal chelates may induce a series of biochemical and physiological alterations in plants which present some common characteristics. Membrane damage, alteration of enzyme activities and the inhibition of root growth are considered characteristic features of heavy metal stress. These may eventually lead to a large range of secondary effects such as disturbance of hormone balance, deficiency of essential nutrients, inhibition of photosynthesis, changes in photoassimilate translocation, alteration of water relations, etc., which further enhance the metal- induced growth reduction. In studies on metal toxicity in plants, the primary toxicity effects these secondary alterations which concern practically all physiological processes.

Plant cell membranes are generally considered primary sites of metal injury. In response to metal ion stress, the plasmalemma gets leaky to $K^+$. Toxic heavy metal concentrations have an effect on $K^+$ and $H^+$ extrusion and decreased $K^+$ content in root cells. Aluminium is found to affect the $K^+$ concentration mostly in the tip region of pea roots. The $K^+$ leakage is not believed to be a consequence of a general loss of membrane integrity, or of a general decrease of energy (ATP) available for ion pumping, but due to a direct interaction of the metal ion with the cell surface causing metal - sulphydryl bonds, which determine the effects on membrane leakage.

Analysis of the influences of excess heavy metals on plant water relations have to distinguish between effects on water availability in soils, effects on root growth limiting water uptake and other phytotoxic effects which may influence water relations.

Metal toxicity not only affects the length of the primary root, but changes the architecture of the entire root system. Toxic levels of Al, Pb, Cu, Zn, Mn, Cd or Cr have been found to enhance lateral root formation giving rise to a more dense and compact root system. Root hair density is generally decreased. These morphogenetic changes indicate that metal toxicity alters the root hormone balance, but the changes of endogenous hormone levels underlying these morphogenetic alterations are virtually unexplored.

## CARBON

The special role and function of carbon in plant life cause carbon to be discussed in the literature under the heading as being an essential element. The uptake of carbon by the plant is different from that of inorganic elements. Inorganic element uptake is mainly through the roots in the liquid phase, but carbon uptake is through the leaves and shoots in the gas phase. The toxicity symptoms of $CO_2$ in mature tomato and bean leaves, yellowing spots, bulkness, etc., resemble the signs of deficiency of P and toxicity of Mg and could be related to the accumulation of starch.

The strong pronounced zebra-like yellow stripes in corn and wheat are not comparable to any other known symptoms. There is no explanation for these symptoms. The full recovery of corn and wheat and no delay in flowering and grain development observed may indicate that no life processes has been interrupted by toxic amounts of $CO_2$. In all other species, a delay in plant development occurs. Hewitt (1966), has discussed some long term experiments with high root environment $CO_2$ concentration. No effects were observed in roots with regard to carbon toxicity. Improved growth under higher aerial $CO_2$ concentration has been observed in many culture $C_3$ and $C_4$ plants. A 1000 ppm concentration of $CO_2$ is recognized as above optimum. Increased $CO_2$ concentration (1000ppm) may compensate partly for water stress. The growth reduction, as a result of increased maintenance, respiration in some salt tolerant plants can be compensated for by a 2500 ppm $CO_2$ concentration.

## ALUMINIUM TOXICITY

Aluminium toxicity is one of the major growth limiting factors that affects plants in most of the acid soils, particularly in the developing countries of the tropics and subtropics, but also equally applicable to many cropland soils in developed countries. Aluminium is not present in metallic form in nature, but its compounds are present in almost all rocks and soils. Al is present as complex aluminosilicates in soils. It is released from the acid soils in the form of hexaqua-aluminium $Al(H_2O)_6$ called $Al^{3+}$ which is detrimental to plants, limiting growth and productivity.

Al tolerant cultivars and genotypes which perform well in Al toxic, acidic, and infertile soils could be the best strategy to circumvent Al toxicity. Al is one of the major constrains that limits plant growth and mineral uptake in the highly weathered acidic, infertile soils found in the tropical countries thereby decreasing yield. Moreover, Al is known to reduce the uptake and transport of many essential elements to a greater extent only in the susceptible genotypes. Though Al has been found to be detrimental for plants, it is found to be beneficial to some plants.

Some of the mechanism proposed to explain Al toxicity are:

i) binding of Al to the plasma membrane and calmodulin, thereby inhibiting their function.
ii) inhibition of cell division with a subsequent reduction in DNA replication.
iii) disturbed metabolism of mineral nutrition (Matsumoto and Yamamaya, 1986).

The foliar symptoms of some plants resemble those of P deficiency (Overall stunting; small, dark green leaves, and late maturity; purpling of stem, leaves and leaf veins; yellowing and death of leaf tip). In others, it shows symptoms like Ca deficiency (curling or rolling of young leaves and collapse of growing points or petioles). The roots are affected which become stubby and brittle and are inefficient in absorbing nutrient and water.

Plant species and varieties vary widely in tolerance to excess Al in the growth medium (Foy, 1974) which may be genetically controlled (Reid, 1971). However, the exact physiological mechanism that involves both Al toxicity or tolerance are still debated as these may be different in different species and varieties controlled by different genes through different biochemical pathways. If plants are tolerant to Al then it can prevent absorption of excess Al or detoxify the Al after it has been absorbed.

Wilkinson (1970, 1972) observed pH lowering in plant zones. These include H ion release, release and hydrolysis of $Co_2$, release of H ions from carboxyl groups of polygalacturonic acid residues of pectic acid and excretion of protons from micro-organism associated with root.

Many plants that are adapted to strongly acidic soils and hence tolerant to Al also tolerate $NH_4^+$ levels that are toxic to other plants and in some cases prefer $NH_4^+$-N to $NO_3^+$-N e.g. sugar cane, birch (*Betula verucora*). Both $NH_4^+$ and Al reduce the uptake of several essential elements.

Involvement of calcium nutrition in the toxicity of several plant species has been well documented. Al toxicity, a major growth limiting factor in many acid soils is often expressed as Ca deficiency. Increasing external Ca supply can counteract the antagonistic effect of Al on Ca uptake, thus eliminating Ca deficiency and the detrimental effect of Al on root elongation, but interactions with other nutrients especially Mg is ignored. In sorghum genotypes severity of Al induced root damage slightly decreased with increasing Ca level. Furthermore, Al toxicity with the $NO_3^-$ grown sorghum genotypes was mainly expressed as root impairment and $NH_4^-$ induced changes in growth response of sorghum plants to Al were due to a shift of growth control towards Mg deficiency.

$CaCO_3$ (0.5 g/1) to nutrient solutions has alleviated $NH_4^+$ toxicity by increasing the metabolism of tomato plants. Aluminium tolerance in certain cultivars of wheat, barley, soybean, snap bean and lupin has been associated with the ability to resist Al induced Ca deficiency or reduced Ca transport. Ca concentrations of nutrient solutions also decreased Al tolerance differences among corn inbred lines.

In many plants Al tolerance appears to be closely associated with P-use-efficiency. It has been observed in certain cultivars of wheat and inbred lines of corn, Al tolerance coincides with the ability to tolerate low P levels in nutrient solutions either in the presence or absence of Al, in certain pea cultivars it has been associated with higher P concentrations in plant roots. Naidoo et al (1976) suggested that Al was bound to esteric P in nucleic acid and to membrane lipids and that it reduced or inhibited cell division by alteration of nucleic acid molecules.

Aluminium concentrations (greater than I mM) was required to inhibit growth of emerging roots, shoots of germinating seeds and concentrations upto 10 mM did not affect mobilization of carbohydrate reserves in germinating seeds

Silicon has been suggested as a factor in aluminium tolerance of some species of the gramineae when grown on acid soils. Wallace (1993), has reported high concentrations of silicon (Si) in monocots than in dicots, and the phenomenon is related to the fact that mineral cation: mineral anion uptake ratio is much higher in dicots than in monocots. When large amounts of silicon, supposedly as silicate ($SiO_4^{46}$), participate in cation-anion balance to add to the excess of anion uptake, equivalent amounts of hydroxyl ions should be expelled from roots which can increase rhizosphere pH and decrease uptake of Al and iron (Fe). Silicon availability has been found generally to be increased with an increase of soil pH up to about pH 9.5. The simultaneous decreased uptake of Al with increase in soil pH to a limit could imply that there was no casual relationship other than soil pH between Al and Si concentrations in plants.

**Physiological and Biochemical Effects of Aluminium Toxicity:** Al toxicity has shown to interfere with cell division in plant root; fix less P, decrease root respiration, interfere with certain enzymes governing deposition of polysaccharides in cell walls, increased cell wall rigidity and interfere with the uptake, transport and use of several elements (Ca, Mg, P, K) and water by plants. Al is known to form strong complexes and precipitate nucleic acids; and are used to isolate polynucleotides from leaves. Al decreased hexosephosphorylation and increased concentrations of ATP in barley roots. Al inhibited activities of ATPase in plants, therefore, it directly or indirectly prevents the use of ATP in glucose phosphorylation.

In contrast, it was reported that Al increased the activities of ATPase and acid phosphatases in pea roots. It injured the cell walls by activating polygalacturonase, which hydrolyzes pectins and thereby promotes faster Al penetration. The increased activity of acid phosphatase was interpreted as a damage to membranes which leads to liberation of enzymes from lysosome, mitochondria and other phosphate containing cell structure.

**Beneficial Effects of Aluminium:** Though aluminium is not considered to be an essential element for plant growth but under some conditions, low concentrations can increase growth. This effect may be different for different plant genotypes and growth media. These responses may not result only from Al effects on plants per sc. Possible explanations include increasing iron solubility and availability in calcareous soils or in slightly acid, neutral or alkaline nutrient solutions by the hydrolysis of Al and a pH decrease; blocking negative charges on cell wall sites and thereby promoting P uptake (eucalyptus); correcting or preventing P toxicity; delaying a root deterioration in low Ca solutions by slowing growth and preventing depletion of Ca from the medium (sunflower) altering the distribution of growth regulators in roots (peach); preventing toxicity's of Cu (citrus) and Mn (atriplex); serving as a fungicide. Al accelerates root formation and might be transferred to leaves, and epidermal cells of the tea plant as waste material. Al markedly reduced or prevented Fe chlorosis caused by excess Zn or Mn influx.

## MANGANESE TOXICITY

Excess Mn unlike Al, affects plant tops more severely than roots. In addition, the symptoms produced by manganese are more definitive. Mn accumulates in the foliage somewhat in proportion to injury. Mn toxicity symptoms are quite diverse among plant species. The symptoms include marginal chlorosis and necrosis of leaves (barley, lettuce, alfalfa, rape, kale), leaf puckering (cotton, soybean, snapbean), and necrotic spots on leaves (barley, lettuce, soybeans). Specific plant physiological disorders associated with excess Mn include crinkle leaf of cotton, stem streak necrosis of potato 378, and internal bark necrosis (IBN) of apple trees. In severe cases of Mn toxicity plant roots turn brown, usually after the tops have been severely injured.

**Manganese Uptake and Transport:** Mn is reduced to the divalent (available) from under conditions of water logging of soil. In nutrient solution about 20 times as much Mn was required to produce toxicity in corn as in peanut. Peanut showed accumulation of Mn in its leaves but little or no accumulation in roots, stems or petioles except during the early growth stage. The superior Mn tolerance of corn (compared with peanut) was associated with reduced transport of Mn from roots and stems to leaves.

## IRON TOXICITY

Different plants show varied symptoms of iron toxicity. Excess Fe in flax produced dark green foliage and stunted top and root growth.

Roots were thick with accumulations of inorganically bound phosphate. The symptoms were intensified due to deficiency of phosphorus and excess of phosphorus reduced it. These symptoms are similar to Al toxicity. However, both the elements Fe and Al were different in their effects on roots of alpine and bean; root tops were flabby with excess Fe and brittle with excess Al. In rice Fe toxicity is characterized by brown spots starting from the tips of lower leaves. Finally the lower leaves turn grey or white. Excess Fe produce brittle, tender and dark brown to purple leaves in tobacco, which have poor burning qualities and flavour.

**General Occurrence and Importance of Iron Toxicity:** Fe toxicity is said to be involved in extremely complicated physiological diseases of rice under flooded conditions. The diseases include bronzing in Sri Lanka and elsewhere. High concentrations of reduced products, particularly $Fe^{2+}$ in the soil solution caused a physiological disorder of rice on three acid soils of pH 3.6 to 5.8 and addition of reduction retardants such as $MnO_2$ improved growth and corrected the disorder.

**Physiology of Iron Toxicity:** Fe toxicity is associated with increased Fe-uptake and translocation by plant tops. Tanaka and Yoshida (1970), associated bronzing disease of rice with accumulation of Fe, in necrotic spots on leaves. Freckle leaf disease in sugarcane has been associated with localized accumulation of Fe.

**Different Tolerance of Plants to Fe Toxicity:** Different plant species and genotypes differ in tolerance to Fe toxicity which may coincide with tolerance to waterlogging in soils. A number of factors such as age of plants, nutritional status and plant genotype can influence tolerance of plants to Fe toxicity. Young plants are more susceptible than older plants. If plants are subjected to high Fe levels for long periods, Fe accumulates gradually in shoots.

## OTHER METAL TOXICITIES

Though any one of the macro elements can be toxic to plants but only a few have been observed to cause phytotoxicity in soils. In soils most of the heavy metals occur as inorganic compounds or are bound to organic matter, clays, or hydrous oxides of Fe, Mn and Al. Mostly Zn, Cu, and Ni toxicity's have occurred frequently because of their precipitation and sorption of most metals by soils.

It has long been recognized that the different plant nutrients show complex interactions which influence their availability for plant uptake. Since agricultural soils are usually subject to anthropogenic interventions (amendments with fertilizers and sludges, application of pesticides, etc.) that are likely to increase the levels of some metals, and therefore changing the interactions among nutrients for plant uptake. The excess amount of copper increased root proton extrusion and membrane permeability in rice.

Zinc concentrations decreased with increasing Cu levels, while N, P, K, Na did not reveal any correlation. These variations according to them may be due to the interaction among Cu-uptake mechanism (s), plasma membrane-$H^+$ ATPase and root membrane permeability. It is also suggested that Cu effects the N, P, K, Na, Ca, Mg, B, Mo, and Zn concentrations in rice shoots mainly by changing their net uptake rate.

Wilson (1993), reported N x P x K interaction in macronutrient toxicity in the grass *Festuca ovina*, grown in a nutrient poor medium with 3 levels of nitrogen (N) fertilization, four of phosphorus (P), and three of potassium (K) in factorial combination. Toxicity effects were seen only when levels of the other nutrients were low. It was observed that N removed K toxicity, K removed N toxicity, and P removed K toxicity. NPK toxicity may occur in the field in conditions otherwise marginal for a species, such as at the margin of its pH tolerance.

## PHYTOXICITY

A number of soil factors affect metal toxicity and these in turn help to evaluate meaningful physiological and biochemical effects of toxic metals.

**Soil Factors:** The availability of a metal in the soil depends on soil adsorption strength as well as plant effectors such as roots, exudates for metal chelation or reduction. The activity of metals is dependent on pH. Metals which are sorbed by organic matter, clays and hydrous oxides of Fe, Mn and Al cause the metal activity to be lower than equilibrium values for known inorganic compounds. Normally only surface metals that are influenced to root pressure is available to plants. At acid pH, soil metals are more plant available but may not be in nutrient solutions.

**Absorption of Metals**: It is a well known fact that metal phytotoxicity can occur only if metals can move from soil to plant roots. Metals in the soil solution can move in the soil water as the plant absorbs water. Changes in soil pH near the root can strongly affect metal movement to the root. Another important aspect of phytotoxicity is the metals in the labile pool are in equilibrium with those absorbed by plants from the soil solution. Metals may be absorbed passively by diffusion due to ion activity gradients, or by an active mechanism by the help of which a macro nutrient is easily absorbed. Chelators reduce metal activity and uptake when added to nutrient solutions.

**Metal Translocation**: The study of translocation of metals from roots to top is not well characterized. Those metals having lower stability constants with natural xylem sap chelators, or with lower metal/Ca stability constant ratios e.g. Zn, Mn, Co generally do not form insoluble chelates. metal chelates are formed between organic and amino acids and the concentration of these two in flowing xylem sap should be high enough to hold metals soluble under normal conditions. However, amongst plant species there is lot of variation in levels of specific natural chelating compounds.

## MINERAL NUTRITION AND CROP PRODUCTIVITY

For good cultivation of plants especially, high priced field crops soil nutrients should be available at appropriate stages of plant development. Unless all the nutrients are present in available forms, crop yields will be limited. Since mostly these nutrients are limited in arable soils under natural conditions, soil fertility can be controlled by supplying as fertilizers, those ions that limit plant growth.

Crop yield can be increased by carrying out systematic investigations of fertilizer needs. Generally governmental agencies can provide data regarding schedule of fertilizing the soil, which nutrient has to be applied, when and how much. Thus deficiency of nutrients and as a consequence its negative impact on physiological processes can be avoided. The actual chemical fertilizer requirements of crop plants can be evaluated by exposing the plants to two tests- soil analysis and plant analysis. These two methods are complementary and each has a role in determining the nutrient situation for a given crop. While soil analysis reveals the low, or high status of nutrient supply, plant analysis on the other hand may be helpful in diagnosing a mineral nutrient deficiency. Thus in cases in which visual symptoms alone are not sufficient to pinpoint a deficiency plant analysis can be used to verify and specify deficiency.

## ROLE OF NUTRITION IN APICAL DOMINANCE

The growth habit of higher plants is determined to a large extent by the regulating influence exerted by the shoot apex on the growth and development of the lateral buds. This phenomenon, known as apical dominance, has been the subject of numerous investigations over the past 40 years yet there is still no general agreement as to the nature of the mechanism involved. Early investigators, believed that the shoot apex inhibited the buds by competing with them for a limited nutrient supply. Although this view, generally referred to as the nutritive hypothesis, was well supported experimentally Loeb (1918), the discovery of auxin and the demonstration that this substance could substitute for the shoot apex in maintaining lateral bud inhibition focused attention on possible mechanisms based on the influence and interaction of hormonal factors. Subsequent investigations, however, revealed serious objections to the concept of auxin as a direct inhibitor of lateral buds and, although various modifications of the direct inhibitor hypothesis have been proposed, all are open to certain criticisms and none can claim conclusive experimental support.

It is evident that plant food is organic in nature (which is the crux of photosynthesis) but inorganic nutrition, especially cationic in nature, greatly supports life processes.

## CHAPTER 10

# Enzymes: Bioaccelerators

## HISTORICAL BACKGROUND

The existence and power of enzymes were first recognized in the nineteenth century when reactions that has been considered to occur only in the presence of cells were found also to be mediated by cell-free extracts. It was the Buchner brothers who showed that yeast, a living organism, yielded a non-living cell-free extract capable of fermenting sugars. Subsequent investigators of this field showed that this was due to the presence in the extract of many powerful catalysts, each one is highly specific for one of the steps in the breakdown of glucose. Similar catalysts, or enzymes, are found in all living things. These are organic catalyst found within a living organism which alter the fate of a chemical change, but itself remains unchanged at the end of the reaction; and it is produced by the living organisms but is not itself alive.

**Enzyme** (Gr. *En* = **in** *zyma* = yeast): Any substance, protein in whole or in part, that regulates the rate of a specific biochemical reaction in living organisms. It is capable of catalyzing a reaction in which substrate(s) are converted to product(s) through the formation of an intermediate enzyme - substrate complex. As with other catalysts, enzymes are responsible for accelerating the rate of a chemical reaction.

**Endo Enzyme.** The enzyme which acts within the cell in which they are synthesized inner core of or substrate molecule are known as endoenzyme.

**Exoenzyme.** The enzyme which act outside the cell in which they are synthesized or at the outer ends of substrate molecule are termed exoenzyme.

The chemical nature of enzymes remained in dispute for a long time. Willstatter, working with peroxidase, had chosen an enzyme of such high catalytic efficiency that he believed that his active preparations were protein-free. By contrast, Sumner's crystalline urease was of such relatively modest activity that his critics attributed the catalysis to a highly active trace contaminant rather than the purified protein itself. Improvements in fractionation procedures have since allowed the purification of many hundreds of enzymes from diverse sources, leading to the realization that all enzymes are proteins. On the basis of chemical composition enzymes are of two types:

1. **Purely proteinaceous enzyme:** These are made up of only proteins e.g., proteases that split protein and amylase that split starch.
2. **Conjugate enzymes:** These are made up of protein molecule, to which a non-protein group (*prosthetic group* or cofactor) is also attached e.g., oxidising enzymes.

It is essential to understand the difference between the following important terms - substrates, prosthetic group, apoenzyme, coenzymes and cofactor.

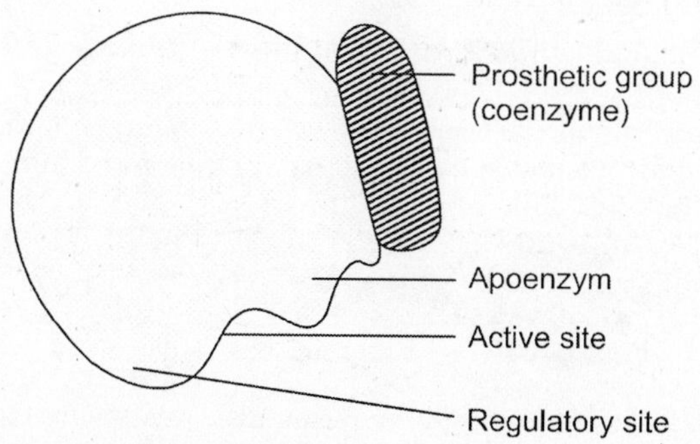


**FIG. 10.1:** Hypothetical morphology of an enzyme.

**Substrate:** The substance upon which an enzyme act is known as substrate. They are produced by the living protoplast of a cell.

**Prosthetic group:** It is the non-protein group of enzyme. The prosthetic group is firmly bound to protein component of the enzyme by chemical bonds and is not removed by dialysis. Haem, biotin and pyridoxal phosphate, like flavin, usually function as prosthetic groups.

**Apoenzyme:** If, however, the prosthetic group is removed, the remaining protein part of the enzyme is called as apoenzyme and it is inert.

**Coenzyme:** It is the co-worker of prosthetic group and apoenzyme because neither the apoenzyme nor the prosthetic group alone is enzymatically active. The prosthetic group, the coenzymes, as the name specifics, act as co-worker of an apoenzyme. They combine with the enzyme and leave it again in the course of a single catalytic cycle. The coenzymes are vitamins, DPN, NAD, tetrahydrofolate, Coenzyme A and adenosine phosphates, like $NAD(P)^+$.

**Cofactor:** When a prosthetic group consists of single atom of some metal like $Mg^{++}$, $Fe^{++}$, $Cu^{++}$, $Mo^{+++}$, $Zn^{++}$, then it is known as cofactor and can be easily separated from rest of the protein part. Many enzymes employ either metal ion cofactors or organic cofactors. Thus many oxidoreductases utilize haem, nicotinamide or flavin; the transferases use folate, coenzyme A, pyridoxal phosphate, thiamine and adenosine phosphates; some of the ligases use biotin etc. Among these cofactors, some may be regarded as integral parts of their enzymes; they are known as *prosthetic* groups.

## PROPERTIES OF ENZYMES

Enzymes are characherised by many bio-physical properties. Some common properties are given below:

1. **Proteinaceous nature:** Enzymes are made up of protein. They may be either purely made up of protein or associated with a non-protein part.
2. **Catalytic properties:** Enzymes act as catalysts and influence the speed of chemical reaction but themselves remain unchanged. A small quantity of enzyme can catalyse the transformation of a very large quantity of the substrate into end product e.g. the sucrase can hydrolyse 100,000 times of sucrose as compared with its own weight.
3. **Specificity of enzyme action:** The ability of an enzyme to catalyze one specific reaction and essentially no other is perhaps its most significant

property. Close examination reveals that most enzymes can catalyze the same type of reaction.

4. **Reversibility of enzyme action:** Most of the reactions catalysed by enzymes are reversible and the enzyme can catalyze the reactions in both directions e.g. the lipase can catalyse not only the hydrolysis of fats into fatty acids and glycerol but also can synthesize fats from fatty acids and glycerol as shown below.

$$\text{Fats} \xrightarrow{\text{Lipase}} \text{Fatty acids} + \text{Glycerol}$$

5. **Sensitivity of the enzymes:** The enzymes are sensitive to rise of temperature and are destroyed at higher temperature of 60-70° C. However, most of the enzymes act best between 20°C to 35° C. At 0° C or below, they are not destroyed but become inactivated.

## ENZYMES: NOMENCLATURE

A commission has been established for deciding the name of enzyme ans is known as **Enzyme nomenclature commission.** The terminology generally agreed for use in relation to enzymes. As in the example of urease, enzyme and classes of enzymes usually are named by appending the suffix *-ase* to the name or abbreviated name of a compound on which the enzyme acts. Enzymes that cause the cleavage of the polypeptide chain by reaction with water are referred to, as a class, as proteinase. Naming of enzymes is known as nomenclature. There are various ways of naming enzymes. The principles of nomenclature are as follows:

1. The enzymes are named by adding the *suffix-ase* to the name of the substrate on which they act e.g. Proteinases, sucrase, nucleases, which break up proteins, sucrose and nucleic acids respectively.
2. The enzymes caan be named according to the type of *function* they are performing e.g. dehydrogenases remove hydrogen, carboxylases help in adding $CO_2$, decarboxylases help in removal of $CO_2$, oxidases helping in oxidation and so on.
3. The enzymes are sometimes given *double name*, one after the nature of substrate upon which they act and second according to their function e.g. pyruvic decarboxylase catalyses the removal of $CO_2$ from pyruvic acid, alcoholic dehydrogenase catalyses the removal of hydrogen from alcohol.

## ENZYME CLASSIFICATION

A system of rules by which enzymes are classified on the basis of the substrate they react with and the type of reaction catalyzed. This systmm provides both a systematic name and a four-part number code. The first number of the code place the enzyme into one of six groups indicating the type of reaction involved. The next two numbers indicate the groups involved in the reaction and the fourth number provides the absolute identification of the enzyme. The enzymes have been classified into following types:

1. Hydrolysing enzymes, 2. Oxidation-reduction enzymes, 3. Ligases, 4. Group transfer enzymes, 5. Desmolases, 6. Isomerizing enzymes, 7. Carboxylation enzymes.

### Hydrolysing enzymes

They catalyse the hydrolysis of complex big molecules into simple, smaller molecules. Depending upon the nature of food substance upon which the enzyme act, they can be further classified into following types.

1. **Carbohydrases:** They hydrolyse the complex polysaccharides into simpler monosaccharides e.g. sucrases, maltases, lactase, cellulase etc.
2. **Esterases:** They hydrolyse the substances containing ester linkage e.g. lipase, phosphatase.
3. **Proteolytic enzymes:** They hydrolyse the proteins into peptones, and amino acids e.g. pepsin, peptidases.
4. **Amidases:** They hydrolyse amides into ammonia and acids e.g. urease.
5. **Oxidation-Reduction Enzyme:** They catalyse oxidation and reduction reactions e.g. Oxidases, reductases, catalases, dehydrogenases.
6. **Ligases:** They bring linkage between two molecules with the simultaneous breakdown of ATP molecule which supplies energy e.g. synthetases which join amino acids to RNA.
7. **Group transfer enzymes:** The catalyse the transformation of a group from one kind of molecule to another or from one position of molecule to another position in the same molecule and are called as transferases.
8. **Desmolases:** They catalyse the reactions in which long carbon chain is broken or lengthened e.g. aldolase.
9. **Isomerizing enzymes:** They catalyse reactions in which an organic molecule is transformed into its isomeric form and are known as isomerases.
10. **Carboxylation enzymes:** They catalyse the reactions in which $CO_2$ is added or removed and are known as carboxylases or decarboxylases.

## ENZYMES AS CHEMO-THERMO REGULATORS

The chemical reactions that take place within the living cell must be precisely controlled and coordinated. These reactions interlock somewhat as a multitude of assembly processes interlock on a modernday production line. For example glucose and other fuel molecules releases energy in a cells by burning. The cells could not toletate and survive in such high temperatures nor effectively utilize energy that is released in short sudden bursts. Therefore, cells need a slow, steady release of energy that they must be able to regulate to meet metabolic energy requirements. In cellular respiration, fuel molecules are slowly oxidized and energy is extracted in small amounts which includes sequences of 30 or more reactions (as many as 20 or 30 chemical transformations before it reaches some final state). Such chemical transformations require a system of flexible chemical control. The key elements of this control system are the remarkable enzymes.

## WHAT ENZYMES DO?

An enzyme increases the speed of a chemical reaction without being consumed itself. An enzyme affects the rate of a reaction by lowering the energy needed to activate the reactions. Even a strongly exergonic reaction that releases more than enough energy as it proceeds is prevented from beginning by an energy barrier. This is because before new chemical bonds can be formed, existing ones must be broken. The energy required to overcome this barrier and get the reaction going is called **activation energy.** An enzyme greatly reduces the activation energy necessary to initiate a chemical reaction. Enzyme promotes a chemical reaction. It does not influence the *direction of a chemical reaction* or the final concentrations of the molecules involved. They simply *speed up reactions*. For example carbonic anhydrase is an enzyme which promotes the interconversion of carbonic acid with carbon dioxide and water:

$$CO_2 + H_2O \xrightarrow[\text{Carbonic acid}]{\text{Carbonic anhydrase}} H_2CO_3$$

This reaction occurs in animal cells and tissues (kidneys and red blood cells). Carbonic acid ionizes, forming bicarbonate ions, the forms in which most of the carbon dioxide in the blood is transported. The reaction shown above could proceed without the enzyme but would be very slow. However, in presence of carbonic anhydrase, the reaction proceeds about 10 million times faster. This reveals that a single molecule of this enzyme can promote the conversion of an estimated 600,000 molecules of carbon dioxide into carbonic acid each second.

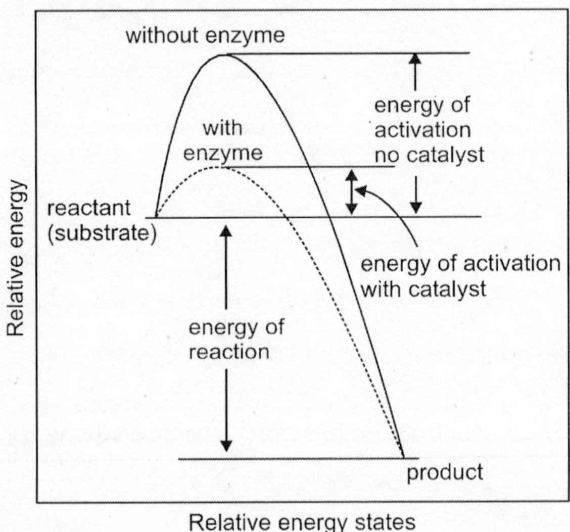


**FIG. 10.2:** An enzyme acclerates the rate of reaction through lowering activation energy.

## WHAT IS ENZYME KINETICS?

It is the study of enzymes in action. The extremely high rate of enzyme-catalysed reactions greatly facilitates this study. Consider, for example, the haem-containing proteins, haemoglobin and catalase. Haemoglobin binds oxygen. It may bind and release many oxygen molecules in the course of a minute but they remain oxygen molecules, and at any instant, only one is associated with each heam centre. Catalase, being an enzyme, has a cumulative effect. Again no more than one $H_2O_2$ molecule will be bound per haem, but while it is bound it may react, and one therefore observes a rapid evolution of oxygen - about a million molecules per minute per enzyme molecule.

## HOW ENZYMES WORK

The Chemicals upon which an enzyme operates are referred to as its **substrates**.

Enzymes are very **specific**. An enzyme will catalyze only a few closely related chemical reactions, or in many cases only one particular reaction. Enzymes form temporary chemical compounds with their substrates. Theses complexes then break up, releasing the product and regenerating the original enzyme molecules for reuse.

$$\text{Enzyme} + \text{Substrate 1} + \text{Substrate 2} \rightarrow \text{Enzyme Substrate Complex} \rightarrow \text{Enzyme} + \text{Product(s)}$$

*It is important to note that the enzyme itself is not permanently altered or consumed by the reaction.*

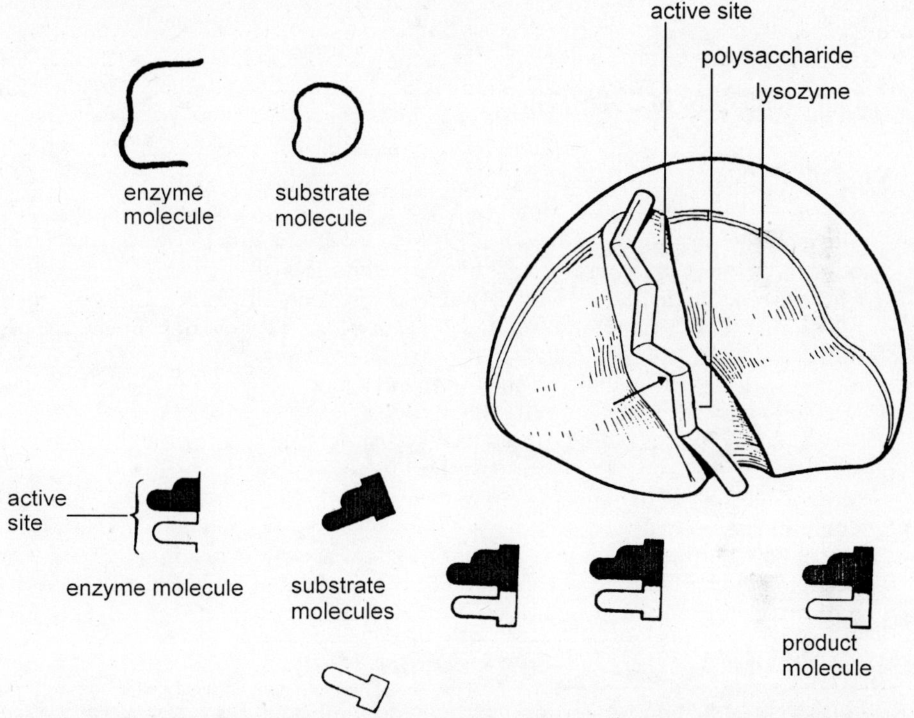


**FIG. 10.3:** (a) Enzyme-substrate (ES) complex formation, (b) Enzyme in an active site of substrate, (c) Induced-fit hypothesis (Lock a-s key mechanism).

Why does the enzyme-substrate complex break up into chemical products different from those that participated in its formation? As shown in Fig.10.1, each enzyme has one or more regions called **active sites,** which in the case of a few enzymes have been shown to be actual indentations in the enzyme molecule. These active sites are located close to one another on the enzyme's surface, so during the course of a reaction, substrate molecules occupying these sites are temporarily brought together and react with one another. It is thought that when the enzyme and substrate bind together, the shape of the enzyme molecule changes slightly. This produces strain in critical bonds in the substrate molecules so that these bonds break. The new chemical compound thus formed has little affinity for the enzyme and moves away from it. An enzyme can be thought of as a molecular lock into which only specifically shaped molecular keys- the substrates- can fit.

Similar to a lock and key, however, the enzyme and its substrate seem not to be exactly complementary shapes. A recent model of enzyme action, known as the **induced-fit model**, is based on data indicating that the active sites of an enzyme are not rigid. When the substrate binds to the enzyme, it may induce a change in shape in the enzyme molecule, resulting in an optimum fit for the substrate-enzyme interaction. The change in shape of the enzyme molecule can put strain on the substrate. This stress may help bonds to break, thus promoting the reaction.

An organic, nonpolypeptide compound that serves as a cofactor is called a **coenzyme.** Many coenzymes are synthesized from vitamins, particularly from the B vitamins. The coenzyme serves as an adaptor, permitting the enzyme to accept a substrate for which it would, by itself, have little affinity.

## ENZYMATIC ACTION: REGULATION

The chemical reactions are regulated by enzymes, but what controls the enzymes?

1. **First:** The mechanism of enzyme control depends upon the *amount of enzyme produced*. The synthesis of each type of enzyme is directed by a specific gene. The gene, in turn, may be switched on by a *signal* from a hormone or by some other type of cellular product. When the gene is switched on, the enzyme is synthesized. Then the amount of enzyme present influences the rate of the reaction. Up to a maximum value, the rate of an enzyme-dependent reaction increases as the concentration of the enzyme increases.
2. **Second:** Enzymatic control may depend upon the *activation* of enzyme molecules that are present in an inactive form in the cytoplasm. In the inactive form the active sites of the enzyme are inappropriately shaped, so the substrates do not fit. Among the factors that influence the shape (conformation) of the enzyme are acidity and alkalinity and the concentrations of certain salts. In a few enzymes activator site(s) (or an allosteric site). is also present. When a molecular activator, such as the substance cyclic AMP, occupies the activator site, the shape of the enzyme molecule changes, making the active sites better suited for binding with the substrate.

## ENZYME OPTIMA

Enzymes acts better under certain narrowly defined conditions known as *optima*. These conditions are optimum/appropriate temperature, pH, and salt concentration, etc. Enzyme pepsin (protein-digesting enzyme of the stomach) works best at the strongly acid pH of 2 whereas amylase (starch-digesting enzyme in saliva and pancreatic juice) works better at pH 8.5 (slightly alkaline). Strong acids or bases irreversibly in *activate* most enzymes. pH permanently changes the molecular conformation of enzymes. At low temperatures, enzymatic reactions occur very slowly or not at all, but their activity resumes when the temperature is raised to normal. The rates of most enzyme-regulated reactions increase with increasing temperature, within limits. Temperatures greater than 50°C or 60°C rapidly inactivate most enzymes by altering their secondary and tertiary level structures. The enzyme is said to be denatured at high temperature. This happens everyday while cooking an egg white; it changes in consistency as the protein is denatured.

## MICHAELIS-MENTEN CONSTANT (KM). VALUE: SIGNIFICANCE

The catalysis occurs on account of formation of transient enzyme-substrate (ES) complex. When substrate concentration is high, addition of enzyme can enhance the rate of reaction, till substrate concentration becomes limiting. Similarly, there is usually a direct proportionality between rate and substrate concentration until the enzyme concentration become limiting. The substrate concentration required to cause half the maximal reaction rate, a value named as Michaelis-Menten constant (Km). Km values are mostly constant and do not depend upon amount of enzymes. Usually the Km values for most of the enzymes studied vary from $10^{-3}$ to $10^{-7}$ molar Km value can be implicated as indicator of substrate concentration, affinity of enzyme with its substrate and it partly indicates enzymes substrate concentration in the cellular compartment, where reaction occurs. It is the affinity part, for which Km values are used most. Km values and inversely proporsional to the affinity of enzyme for its substrate. Therefore, higher Km values suffers lower stability of ES complex.

## INHIBITORS, ACTIVATORS AND INACTIVATORS OF ENZYMES

The rate of an enzyme-catalysed reaction may sometimes be altered in a specific manner by compounds other than the substrate(s). Activators increase the rate; inhibitors and inactivators decrease it. The study of such agents is of practical importance for several reasons:

1. Inhibition and activation of enzymes by key metabolites provides the normal means of metabolic *fine control* superimposed on the *coarse* control achieved by regulation of the synthesis and breakdown of active enzymes.
2. External interference with metabolism, whether by drugs, pesticides etc. or by undesirabale toxic agents, often depends on the inhibition of enzymes.

Inhibitors, and especially inactivators, provide a powerful tool for studying the chemical mechanisms of enzyme action.

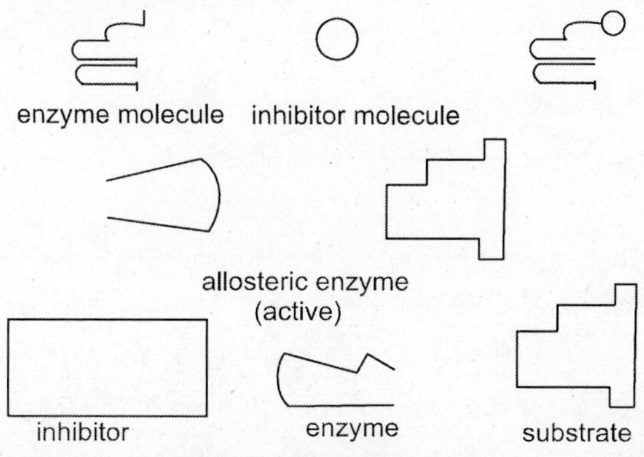


**FIG. 10.4:** Enzyme inhibition (a) non competitive, (b) Allosteric.

Enzyme inhibition may be reversible or irreversible. Reversible inhibitors can be competitive or noncompetitive. In **competitive inhibition** the inhibitor completes with the normal substrate for the active site of the enzyme. A competitive inhibitor usually is chemically similar to the normal substrate and so fits the active site and binds with the enzyme. However, it is not similar enough to the normal substrate to take its place effectively. The enzyme cannot act upon it to form reaction products. A competitive inhibitor occupies the active site only temporarily and does not damage the enzyme irreversibly.

In **noncompetitive inhibition** the inhibitor binds with the enzyme at a site other than the active site. Such an inhibitor renders the enzyme inactive by altering its shape. Many important noncompetitive inhibitors are metabolic substances that help regulate enzyme activity by combining reversibly with the enzyme.

Many poisons are **irreversible inhibitors** that permanently inactivate or even destroy the enzyme. Nerve gases are irreversible inhibitors that poison the enzyme cholinesterase, essential in the normal function of nerves and muscles. A number of insecticides and drugs are irreversible inhibitors. The antibiotic penicillin and its chemical relatives inhibit a bacterial enzyme necessary for bacterial cell wall construction. Unable to produce new cell walls, susceptible bacteria cannot multiply effectively. Since human body cells do not have cell walls (and so do not employ the susceptible enzyme), penicillin is harmless to humans, except for the occasional allergic patient.

## REVERSIBLE AND IRREVERSIBLE INHIBITION

Enzyme activity can be curtailed by various non-specific agents (acid or alkali, urea, detergents, proteases etc.) which disrupt protein structure. These agents interact with a protein at a small number of loci without markedly disrupting the three-dimensional structure. The two most important considerations in their classification are specificity and reversibility.

## MULTI-SUBSTRATE ENZYMES

The only true one-substrate enzymes are the isomerases, which catalyse reactions of the type A= B, and the lyases, which catalyse reactions of the type A - B + C, and are, therefore, one-substrate enzymes in one direction only. Hybrolases, catalysing reactions of the general type A-B + $H_2O$ = A-OH + BH may be regarded as honorary one-substrate enzymes, since one substrate, water, is normally present at constant concentration.

The remaining groups are:

$$\text{Oxidant} + \text{Reductant} \xrightarrow{\text{Oxidoreductases}} \text{Reduced Product} + \text{Oxidized Product}$$

$$A + BX \xrightarrow{\text{Transferases}} AX + B$$

This large group includes methyl transferases, transaldolase, transketolase, acyl transferases, glycosyl transferases, transaminases, kinases etc.

$$X + Y + ATP \xrightarrow{\text{Ligases}} XY + ADP + P_i \text{ (or } AMP + PP_i\text{)}$$

These three groups of enzymes by definition catalyse reactions involving more than one substrate.

## ENZYME SECRETION BY PLANT CELLS

Enzyme secretion is a common feature of most of the living cells. By this process, cells may act on their environment by modifying its chemical composition and ensure their defence againt external agressions. In pluricellulaar organisms, groups of cells (glands) become specialized in the secretion of enzymes which are utilized in a function useful for the whole organism. The control of such a release of enzyme is often mediated by chemical substances which migrate from one tissue to another one composed of secreting target cells. This coupling between stimulus and secretion is widely distributed in animal organisms.

## THE MECHANISM OF PROTEIN SECRETION

The synthesis, intracellular transport, and release of secretory proteins is a basic cellular function common to most eukaryotic cells (Chrispeels, 1976). In the pancreatic exocrine cell for example, the secretory proteins are the object of six steps or operations, which are: *synthesis*, *seggregation*, *intracellular transport*, *concentration*, *intracellular storage* and *discharge* (Palade, 1975). Proteins for export are generally synthesized on polysomes, attached to the membrane of the rough endoplasmic reticulum. In etiolated radish seedlings submitted to far red light, which is known to induce the synthesis of β-fructosidase, the enzyme activity is found associated to endoplasmic reticulum in a first time, then in Golgi apparatus and in cell wall. There is no evidence for the existence of a single intracellular route leading to enzyme secretion. Although the migration of proteins synthesized in endoplasmic reticulum, through Golgi towards the exterior of cell after exocytosis is likely to occur in some cases, direct transport from reticulum to plasmalemma and transfer of cytoplasmic enzymes accross plasmalemma cannot be excluded.

## THE CONTROL OF SECRETION

The secretory process is a complex mechanism. It is known that calcium is essential for the control of the secretory process (Rubin, 1982). Calcium plays a critical role in the exocytosis process, which is a **fusion-fission response** involving the interaction the of the plasmamembrane and secretory vesicle membrane. A second requirement for the occurrence of secretion is metabolic energy (Rubin, 1970). An adequate intracellular ionic equilibrium is also necessary. $Ca^{2+}$ is considered as a second messenger in animal cells, able to trigger the discharge of secretory products by regulated cells. In addition, non-regulated cells often requires $Ca^{2+}$ for their secretion. It would be, therefore, necessary to explore the knowledge concerning the regulation of $Ca^{2+}$ in plant cells. It is known that $Ca^{2+}$ in involved in the secretion of some plant enzymes, including peroxidase, α-amylase, and phosphatase. Recently, several articles have been published, which described modes of $Ca^{2+}$ transport across pant membranes. An ATP-dependent $Ca^{2+}$ uptake by isolated membranes vesicles was reported.

Some elements exist for substantiating the hypothesis that phytohormones regulate protein secretion through the mediation of second messenger such as c-AMP or $Ca^{2+}$/calmodulin. However, the existence of an exchange mechanism between $H^+$ and $Ca^{2+}$ suggests that plant growth regulators especially auxin could modify secretory process by a modification of the distribution of protons which indirectly affect $Ca^{2+}$ compartmentation.

## HORMONAL EFFECTS ON ENZYME SECRETION

There are a lot of works showing that treatment of whole plants, isolated plant organs or tissues with phytohormones of biological origin or with synthetic plant growth regulators results in changes in the activity of several enzymes (Barendse,

1983). A considerable number among these enzymes are of an exocellular nature but the process and control of their secretion has not been investigated in depth until now. On the other hand, there are much less papers establishing a correlation between endogenous hormonal status and enzyme levels or activities. It thus can be said that our knowledge of the hormonal control of enzyme secretion by plant cells is far from being well known.

## ALLOSTERY AND ITS ANTECEDENTS

Monod, Changeux and Jacob (1963) in their paper *Allosteric Proteins and Cellular Control Systems* proposed that in addition to active, substrate-specific sites, regulatory enzymes might possess separate **allosteric** sites specific for their regulator. Binding of a regulator molecule at such a site, they suggested, could influence events at the active site. The idea was not entirely new, but the paper drew together the information then available to provide a clear, persuasive argument for the widespread occurrence of regulation by this means. The name allosteric was intended to emphasize that the regulator need not bear any structural resemblance to the substrate.

## ISOENZYMES

Isoenzymes were discovered about 30 years ago and were at first regarded as interesting but of rare occurrences. Since then a wealth of information on enzyme heterogeneity has accrued and it now seems likely that at least half of all enzymes exist as isoenzymes. This is important in many areas of biological and medical science. Thus isoenzyme studies have provided the main experimental substance for the neutral drift controversy in genetics and evolution. Isoenzymes have greatly extended our understanding of metabolic regulation not only in animals but also in bacteria and plants; their existence has made available a multitude of highly sensitive markers for the study of differentiation and development, as well as providing indices of aberrant gene expression in carcinogenesis and other pathological processes. Isoenzymes are also being used increasingly in diagnostic clinical biochemistry. Isoenzymology crosses the traditional boundaries of the biological disciplines and it seems that advances and new applications in this field, are like to occur at a rapid pace.

## THE CAUSE OF ENZYME MULTIPLICITY

There are various causes of enzyme multiplicity and they may be divided into two categories. These are (a) **genetic** or **primary causes**, whereby the organism carries multiple genes each one encoding a different type of enzyme subunit; and (b) post-translational or secondary causes, whereby homogeneous enzyme subunits are modified differentially so as to produce a range of subunits from a single gene. There are, in turn, two types of genetic multiplicity; firstly, multiple alleles at a single genetic locus and secondly, multiple genetic loci.

## PRIMARY OR GENETIC ISOENZYMES

### Isoenzymes due to multiple alleles at a single genetic locus

In the diploid genome, each genetic locus is represented twice. For each locus, the individual will either be homozygous, possessing two identical alleles, or heterozygous, possessing two different alleles. Where the genetic locus encodes an enzyme subunit, the homozygous individual can only produce one type of subunit. However, the heterozygous individual with two different allelic variants will produce two different types of enzyme subunits. Within the individual the degree of enzyme multiplicity produced by multiple alleles is limited, as two different alleles per diploid locus is the maximum possible genetic variation of this type. However, from one individual to another, there may be considerable variation in the range of enzyme subunit types produced since there may be a variety of different alleles for the locus in the gene pool of the species.

The enzyme subunit types produced as a result of multiple alleles are likely to differ from each other only in minor ways, such as by individual amino acid substitutions caused by point mutations in the DNA nucleotide sequences. Where a genetic locus is active, both alleles will usually be expressed. Thus in the same individual, although the total activity of the enzyme may vary considerably between different kinds of cell, the isoenzyme profile will be constant throughout, homozygous individuals displaying only one subunit type, and heterozygotes possessing two different subunit types.

## ISOENZYMES DUE TO MULTIPLE GENETIC LOCI

Many enzymes are encoded at more than one genetic locus and, where this the case, each locus will produce a different type of enzyme subunit. As the expression of each genetic locus can be controlled independently, the organism may synthesize one type of enzyme subunit in a particular cell, and another enzyme subunit elsewhere. Furthermore, the expression of genetic loci may alter during the course of development, and therefore the type of enzyme subunit produced in a tissue may change. Thus multiple genetic loci permit **differences in isoenzyme profile** both from one tissue to another, and from one developmental stage to another even within the same tissue. Such variation in isoenzyme profile are not possible where the enzyme multiplicity is due to multiple alleles operating at only a single locus. However, as all members of the same species will possess the same genetic loci, multiple enzyme-encoding loci in the absence of multiple alleles cannot account for isoenzyme differences between one member of the species and the next. The isoenzyme subunits produced by multiple loci are likely to differ extensively as a result of numerous amino acid substitutions, and also deletions and additions of residues may have occurred resulting in small size differences between the subunits.

## SECONDARY OR POST-TRANSLATIONAL ISOENZYMES

Proteins may be modified in numerous ways following their synthesis. Such possible modifications include the addition of carbohydrate, limited proteolysis, and the covalent modification of amino acid side chains. Post-synthetic alterations affecting only part of the enzyme subunit population, so that modified and unmodified subunits are found in the same organism, will result in isoenzymes. Where the post-translational modification process is very active in some tissues but not in others, the result will be a tissue-specific distribution pattern of the secondary isoenzymes, mimicking the effect of multiple genetic loci. An example of this is one particular pyruvate kinase subunit type. There are three major types of pyruvate kinase subunit in mammals, each coded for by an independent genetic locus.

**TABLE 10.1:** Enzymes, coenzymes, and mineral activating substances

| Enzyme | Coenzyme or co-factor In processes of glycolysis | Activating mineral elements |
|---|---|---|
| Phosphatase | Phosphate | $Mg^{2+}$, |
| Phosphoglutamase | | $Mg^{2+}$ |
| Aldolase | | $Fe^{2+}$, $Zn^{2+}$, $Co^{2+}$ |
| Enolase | | $Mg^{2+}$, $Mn^{2+}$, $Zn^{2+}$ |
| Pyruvate kinase | ADP In the tricarboxylic cycle | $Mg^{2+}$, $K^{+}$, $NH_4^{+}$ |
| Decarboxylase and pyruvate dehydrase | Thiamine pyrophosphyte fatty acids, CoA | $Mg^{2+}$, $Mn^{2+}$ |
| Isocitrate dehydrogenase | NADP | $Mn^{2+}$, $Mg^{2+}$ |
| Succinate dehydrogenase | FAD | $Ca^{2+}$, $Cr^{3+}$, $Al^{3+}$ |
| Aconitase | | $Fe^{2+}$, -SH |
| Dehydrogenase | | $Mn^{2+}$, $K^{+}$, $Rb^{+}$, $NH_4^{+}$ |
| Pyruvic oxidase | | $Mg^{2+}$, $Mn^{2+}$ |

Among the three most important functions namely, the mechanical, osmotic and chemical, the later depends solely upon the bio acceleration of metabolic reactions caused by enzymes.

## APPARENT ENZYME MULTIPLICITY

It must be recognized that the term isoenzymes is often used loosely in an operational sense as it tends to be applied whenever enzyme multiplicity is observed. Multiple enzyme forms may therefore pass into the literature as isoenzymes when, for one reason or another, this term should not have been employed. The multiple forms may be artefacts resulting from laboratory manipulation of cells and cell extracts. Unphysiological ion concentrations may result in the non-specific binding of ligands with consequent alteration of the apparent properties of the enzyme, or liberation of proteolytic enzymes on disrupting the cell may result in the partial degradation of the enzyme. In this and other ways enzyme multiplicity which was not present in the intact cell may be created. As an example, heterogeneity of the glycolytic enzyme phosphoglucose isomerase was eventually shown to be due to artefactual oxidation of enzyme sulphydryl groups.

## ISOENZYMES AS GENETIC MARKERS

Isoenzyme analysis is now an important experimental technique in genetics. Each genetically determined isoenzyme subunit type, by definition, is the result of a different gene, whether we are considering an enzyme encoded by multiple alleles at a single locus, or multiple loci. Therefore, wherever the isoenzyme subunit type is found, the gene coding for it is not only present, but is being expressed. In this way the isoenzyme subunit is the marker for its own gene. Theoretically, this is not restricted to enzymes; any polypeptide is a marker for its encoding gene. However, in practice, since enzymes are catalytically active they can be assayed specifically so that enzyme variants can be detected more readily than the genetic variants of non-enzymic proteins. It must be emphasized that only isoenzymes due to genetic multiplicity are of interest here. Use of isoenzymes as markers in modern biochemical genetics is so widespread that the hypothesis **one gene-one isoenzyme subunit** have been proposed (Rider and Taylor, 1980). The usefulness of isoenzymes as genetic markers is illustrated in the following experiment devised to test the Lyon Hypothesis. Not only was this a particularly elegant application of isoenzyme markers, but also was in its own right an important advance in our understanding of genetic regulation

# CHAPTER 11

# Visible Radiations and Photosynthesis

## BASIC CONCEPTS

### Visible Radiations Have Characteristics of Both a Particle and a Wave

The radiation is the propagation of electromagnetic energy through space. A triumph of physics in the early twentieth century was the realization that light has properties of both particles and waves. A wave is characterized by a **wavelength**, denoted by the Greek letter lambda ($\lambda$), which is the distance between successive wave crests. The **frequency**, represented by the Greek letter *nu* ($\nu$), is the number of wave crests that pass an observer in a given time. A simple equation relates the wavelength, the frequency, and the speed of any wave:

$$c = \lambda v$$

Where $c$ is the speed of the wave-in the present case, the speed of light ($3.0 \times 10^8$ m $s^{-1}$). The light wave is a transverse (side-to-side) electromagnetic wave, in which both electric and magnetic fields oscillate perpendicular to the direction of propagation of the wave and at 90° with respect to each other.

Light is also a particle, which we call a **photon**. Each photon contains an amount of energy that is called a **quantum** (plural quanta). The energy content of light is not continuous but rather is delivered in these discrete packets, the quanta. The energy ($E$) of a photon depends on the frequency of the light according to a relation known as Planck's law:

$E = hv$

Where $h$ is Planck's constant ($1.58 \times 10^{-34}$cal/s or $6.63 \times 10^{34}$J/s).

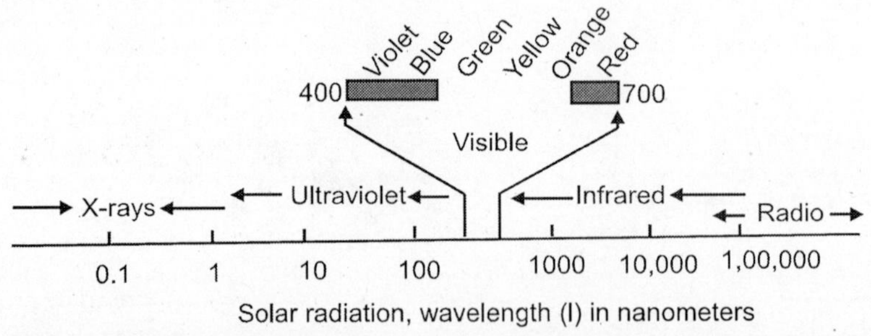


**FIG. 11.1:** Solar radiation spectrum ($\lambda$ in $\mu$m)

It is customary in biology to refer to the wavelength of the light wave $\lambda$, rather than the frequency $\kappa$; the two are related by the simple equation: $\kappa = c/\lambda$ Where $c$ *is* the velocity of light ($3 \times 10^{10}$ cm/s in a vacuum). Therefore, we can consider the energy in a mole of photons, E = $N\kappa\lambda$, where $N$ is Avogrado's number [$6.02 \times 10^{23}$

molecules (or photons) per mole]. Thus $E= Nh\nu= Nhc/\lambda$.

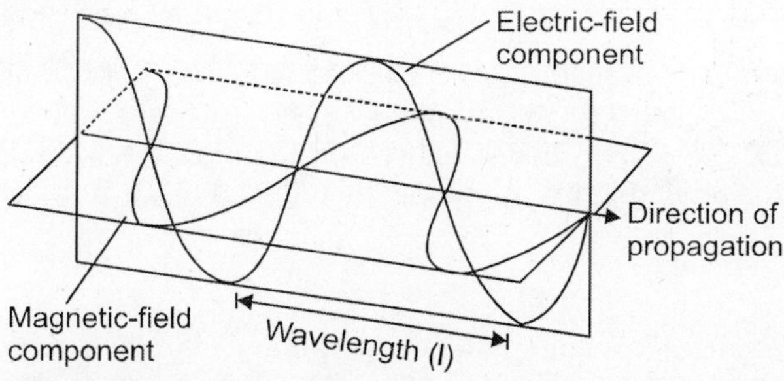


**FIG. 11.2:** Light is a transverse electromagnetic wave, consisting of oscillating electric and magnetic fields that are perpendicular to each other and perpendicular to the direction of propagation of the light. Light moves at a speed of 3 x $10^8$ m $s^{-1}$. The wavelength ($\lambda$) is the distance between successive crests of the wave.

**TABLE 11.1:** Energy of light quanta at different wavelengths

| Wavelength of light (nm) | Colour of light kcl/mol | Energy of quanta $E_o$'v |
|---|---|---|
| 350 Ultraviolet | 80 | 35 |
| 450 Blue | | |
| 550 Yellow | 602.6 | |
| 650 Red | | |
| 750 Infrared | 40 | 1.7 |

Sunlight is like a *rain of photons* of different frequencies. Our eyes are sensitive to only a small range of frequencies-the visible-light region of the electromagnetic spectrum. Light of slightly higher frequencies (or shorter wavelengths) is in the ultraviolet region of the spectrum, and light of slightly lower frequencies (or longer wavelengths) is in the infrared region. The output of the sun is shown in Figure 11.1, along with the energy density that strikes the surface of Earth. The absorption spectrum of chlorophyll *a* indicates approximately the portion of the solar output that is utilized by plants.

## Working with Light

Three light parameters are especially important when working with light: (1) the **amount**, (2) the **direction**, and (3) the **spectral quality**. The first two parameters are important with respect to the geometry of the part of the plant that intercepts the light. For a flat leaf, a flat or planar light sensor makes sense, and the amount of energy that falls on a flat sensor of known area per unit time is quantified as **irradiance**. Units can be expressed in terms of energy, such as watts per square meter (W $m^{-2}$). Time (seconds) is contained within the term watt: 1 W = 1 joule (J) $s^{-1}$. Light can also be thought of as a stream of particles, or **quanta**. In this case, units can be expressed in moles per square meter per second (mol $m^{-2}$ $s^{-1}$), where *moles* refers to the number of photons (1 mol of light = 6.02 × $10^{23}$ photons, Avogadro's number). This measure is called **photon irradiance**.

Quanta and energy units can be inter-converted relatively easily, provided the wavelength of the light is known. The energy of a photon is related to its wavelength as follows:

$$E = \frac{hc}{\lambda}$$

where $c$ is the speed of light ($3 \times 10^8$ m s$^{-1}$), $h$ is Planck's constant ($6.63 \times 10^{-34}$ J s) and $\lambda$ is the wavelength of light, usually expressed in nm (1 nm = $10^{-9}$ m). From this equation it can be shown that a photon at 400 nm has twice the energy of a photon at 800 nm. For simplicity, above can be reduced to the following:

$$E = \frac{1988 \times 10^{19}\ J}{\lambda}$$

where $\lambda$ is expressed in nanometers. From this calculation, we can determine that a photon of 400 nm light contains $4.97 \times 10^{-19}$ *J*.

Now, suppose we have 3 μmol of 400 nm light falling on 1 m$^2$ every second, or a photon irradiance of 3 μmol m$^{-2}$ s$^{-1}$. This quantity is approximately the amount of blue light at 400 nm that strikes the surface of Earth on a sunny day. If we want to convert photon irradiance to energy irradiance, we must first convert micromoles to moles (1 μmol = $10^{-6}$ mol): (3 μmol m$^{-2}$ s$^{-1}$) × ($10^{-6}$ mol μmol$^{-1}$) = $3 \times 10^{-6}$ mol m$^{-2}$ s$^{-1}$ of photons, or quanta. Calculating the number of quanta from Avogadro's number gives: ($3 \times 10^{-6}$ mol quanta m$^{-2}$ s$^{-1}$) × ($6.02 \times 10^{23}$ quanta mol$^{-1}$) = $1.8 \times 10^{18}$ quanta m$^{-2}$ s$^{-1}$. As we already calculated, each photon, or quantum, of 400 nm light contains $4.97 \times 10^{-19}$ *J*. Thus ($4.97 \times 10^{-19}$ J quantum$^{-1}$) x ($1.8 \times 10^{18}$ quanta m$^{-2}$ s$^{-1}$) = 0.9 J s$^{-1}$ m$^{-2}$. Since 1 J s$^{-1}$ = 1 W, therefore, the irradiance will be of 0.9 W m$^{-2}$.

Turning our attention to the direction of light, light can strike a flat surface directly from above or it can strike the surface obliquely. When light deviates from perpendicular, irradiance is proportional to the cosine of the angle at which the light rays hit the sensor. Thus, irradiance is maximal when light strikes a surface directly from above, and it decreases as light becomes more oblique- similar to the situation with a typical leaf. Sensors that correct for the angle of incidence of light are said to be *cosine corrected*. But there are many examples in nature in which the light-intercepting object is not flat (e.g., complex shoots, whole plants, chloroplasts). In addition, in some situations light can come from many directions simultaneously (e.g., direct light from the sun plus the light that is reflected upward from sand, soil, or snow). In these situations it makes more sense to measure light with a spherical sensor that measures light omnidirectionally (from all directions). The term for this omnidirectional measurement is **fluence rate**, and this quantity can be expressed in watts per square meter (W m$^{-2}$) or moles per square meter per second (mol m$^{-2}$ s$^{-1}$). It is obvious from the units whether light is being measured as energy (W) or as photons (mol). In contrast to a flat sensor, a spherical sensor is equally sensitive to light, independent of direction. Depending on whether the light is collimated (rays are parallel) or diffuse (rays travel in random directions), values for fluence rate versus irradiance can be quite different from one another. They are equivalent only under special conditions.

## Photosynthetically Active Radiation (PAR)

Photosynthetically active radiation (PAR, 400 to 700 nm) may also be expressed on the basis of energy (W m$^{-2}$) or quanta (mol m$^{-2}$ s$^{-1}$) (McCree 1981). It is important to note that PAR is an irradiance-type measurement. In research on photosynthesis, when PAR is expressed on a quantum basis, it is given the special term

**photosynthetic photon flux density (PPED)**. However, it has been suggested that the term *density* be discontinued (Holmes et al. 1985) because within the International System of Units (*Systeme Internationale d'Unites*, or SI units) *density* can mean area or volume. Moreover, area is contained within the term flux. PPED has in some cases been shortened to PPF, but it is not clear whether this abbreviation represents an irradiance-type or a spherical measurement.

When light moves through water such as lakes or oceans, it scatters in such a way that its direction of travel is randomized. Thus, aquatic photosynthetic organisms can receive light simultaneously from above, below, and the sides. In oceanography and limnology, the light that travels from the water surface toward the bottom is called *downwelling irradiance*, and the light that travels in the reverse direction is called *upwelling irradiance*.

Before the advent of specialized light meters suitable for plant physiology, light was measured in **lux** or **foot-candles**. These measurements are based on the perception of light by the human eye, which is maximally sensitive to light within the green at 555 nm. Sensitivity falls off on both sides of this wavelength and approaches zero within the blue and red, wavelength regions that are important for photosynthesis. Thus, lux and foot-candles are of little use in plant physiology. These units can be converted to the units of irradiance and fluence rate, but only if detailed knowledge about the spectral quality of the light is available, something that can be obtained only by use of an instrument called a *spectroradiometer*, which measures the amount of light at each wavelength.

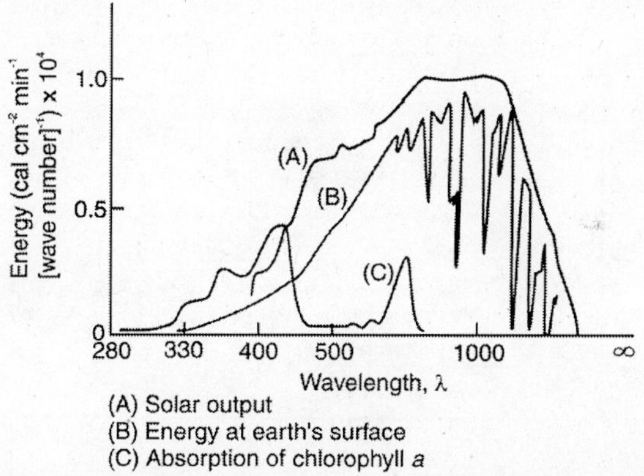


**FIG. 11.3:** The solar spectrum and its relation to the absorption spectrum of chlorophyll. Curve A is the energy output of the sun as a function of wavelength. Curve B is the energy that strikes the surface of Earth. The sharp valleys in the infrared region beyond 700 nm represent the absorption of solar energy by molecules in the atmosphere, chiefly water vapor. Curve C is the absorption spectrum of chlorophyll a, which absorbs strongly in the blue (about 430 nm) and the red (about 660 nm) portions of the spectrum. Since the green light in the middle of the visible region is absorbed only weakly, much of it is therefore reflected into our eyes and gives plants their characteristic green color. Note that curve C has absorbance rather than energy units.

Therefore, when choosing how to quantify light, it is important to match sensor geometry and spectral response with that of the plant. Flat, cosine-corrected sensors are ideally suited to measure the amount of light that strikes the surface of a leaf; spherical sensors are more appropriate in other situations, such as when studying a chloroplast suspension or a branch from a tree.

How much light is there on a sunny day and what is the relationship between PAR irradiance and PAR fluence rate? Under direct sunlight, PAR irradiance and fluence rate are both about 2000 μmol $m^{-2}$ $s^{-1}$, though higher values can be measured at high altitudes. The corresponding value in energy units is about 400 W $m^{-2}$. When light is completely diffuse, irradiance is only 0.25 times the fluence rate.

## When Molecules Absorb or Emit Light, they Change their Electronic State

Chlorophyll appears green to our eyes because it absorbs light in the red and blue parts of the spectrum, so only some of the light enriched in green wavelengths (about 550 nm) is reflected into our eyes. The absorption of light is represented by following, in which chlorophyll (Chl) in its lowest-energy, or ground, state absorbs a photon (represented by *hv*) and makes a transition to a higher-energy, or excited, state (Chl*):

$$\text{Chl} + h\nu \rightarrow \text{Chl}^*$$

The distribution of electrons in the excited molecule is somewhat different from the distribution in the ground-state molecule. Figure 1.1 describes the absorption and emission of light by chlorophyll molecules. Absorption of blue light excites the chlorophyll to a higher energy state than absorption of red light, because the energy of photons is higher when their wavelength is shorter.

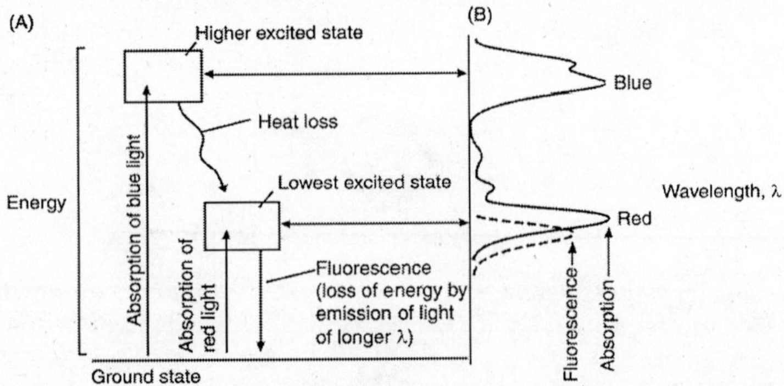


**FIG. 11.4:** Light absorption and emission by chlorophyll. (A) Energy level diagram. Absorption or emission of light is indicated by vertical lines that connect the ground state with excited electron states. The blue and red absorption bands of chlorophyll (which absorb blue and red photons, respectively) correspond to the upward vertical arrows, signifying that energy absorbed from light causes the molecule to change from the ground state to an excited state. The downward-pointing arrow indicates fluorescence, in which the molecule goes from the lowest excited state to the ground state while re-emitting energy as a photon. (B) The spectra of absorption and fluorescence. The long-wavelength absorption band of chlorophyll corresponds to light that has the energy required to cause the transition from the ground state to the first excited state. The short-wavelength absorption band corresponds to a transition to a higher excited state. (After Sauer 1975).

In the higher excited state, chlorophyll is extremely unstable, very rapidly gives up some of its energy to the surroundings as heat, and enters the lowest excited state, where it can be stable for a maximum of several nanoseconds ($10^{-9}$ s). Because of this inherent instability of the excited state, any process that captures its energy must be extremely rapid.

In the lowest excited state, the excited chlorophyll has several possible pathways for disposing of its available energy. It can re-emit a photon and thereby return to its ground state-a process known as **fluorescence**. When it does so, the wavelength of fluorescence is almost always slightly longer than the wavelength of absorption of the same electron state, because a portion of the excitation energy is converted into heat before the fluorescent photon is emitted. Conservation of energy therefore requires that the energy of the fluorescent photon be lower than that of the excitation photon-hence the shift to longer wavelength. Chlorophylls fluoresce in the red region of the spectrum.

Alternatively, the excited chlorophyll can return to its ground state by directly converting its excitation energy into heat, with no emission of a photon. A third process that deactivates excited chlorophyll is **energy transfer**, in which an excited chlorophyll transfers its energy to another molecule. A fourth process is **photochemistry**, in which the energy of the excited state causes chemical reactions to occur. The rate of the earliest steps in the photosynthetic energy storage process is among the fastest known chemical reactions. This extreme speed is necessary for photochemistry to compete with the other possible reactions of the excited state.

## The Quantum Yield Gives Information about the Fate of the Excited State

The process with the highest rate will be the one most likely to deactivate excited chlorophyll; slower processes only occasionally win out in the race to dispose of the energy of the excited state. This concept is expressed quantitatively by way of the quantum yield. The **quantum yield** ($\phi$) of a process in which molecules give up their excitation energy (or *decay*) is the fraction of excited molecules that decay via that pathway. Mathematically, the quantum yield of a process, such as photochemistry, is defined as follows:

$$\Phi = \frac{\text{Yield of photochemical products}}{\text{Total number of quanta absorbed (7.4)}}$$

The quantum yield of the other processes is defined analogously. The value of $\phi$ for a particular process can range from 0 (if that process is never involved in the decay of the excited state) to 1.0 (if that process always deactivates the excited state). The sum of the quantum yields of all possible processes is 1.0.

In functional chloroplasts kept in dim light, the quantum yield of photochemistry is approximately 0.95, the quantum yield of fluorescence is 0.05 or lower, and the quantum yields of other processes are negligible. The vast majority of excited chlorophyll molecules therefore lead to photochemistry.

The quantum yield of formation of the products of photosynthesis, such as $O_2$, can be measured quite accurately. In this case the quantum yield is substantially lower than the value for photochemistry, because several photochemical events must take place before any $O_2$ molecules form. For $O_2$ production the measured maximum quantum yield is approximately 0.1, meaning that 10 quanta are absorbed for each $O_2$ molecule released. The reciprocal of the quantum yield is called the **quantum requirement**. The minimum quantum requirement for $O_2$ evolution is therefore about 10. Quantitative measurements of the absorption of light and the fate of the energy contained in the light are essential to an understanding of photosynthesis.

## Photosynthetic Pigments Absorb the Light That Powers Photosynthesis

The pigments of the plant first absorb the energy of sunlight. All pigments active in photosynthesis are found in the chloroplast. The **chlorophylls** and **bacteriochlorophylls** (pigments found in certain bacteria) are the typical pigments of photosynthetic organisms, but all organisms contain a mixture of more than one kind of pigment, each serving a specific function. All chlorophylls have a complex ring structure that is chemically related to the porphyrin-like groups found in hemoglobin and cytochromes. In addition, a long hydrocarbon tail is almost always attached to the ring structure. The tail anchors the chlorophyll to the ring structure. The tail anchors the chlorophyll to the hydrophobic portion of its environment. The ring structure contains some loosely bound electrons and is the part of the molecule involved in electron transitions and redox reactions.

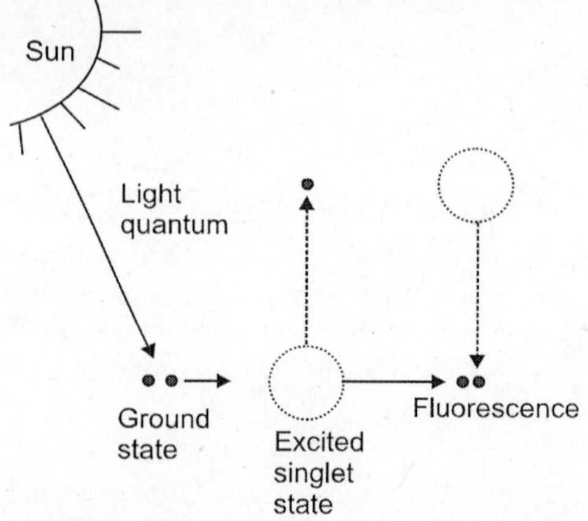


**FIG. 11.5:** Absorption of photon by chlorophyll atoms.

## Photosynthesis Takes Place in Complexes Containing Light-Gathering Antennas and Photochemical Reaction Centers

A portion of the light energy absorbed by chlorophylls and other pigments is eventually stored as chemical energy via the formation of chemical bonds. This conversion of energy from one form to another is a complex process that depends on cooperation between many pigment molecules and a group of electron transfers proteins. The majority of the pigments serve as an **antenna**, collecting light and transferring the energy to the **reaction center**, where the chemical reactions leading to long-term energy storage take place.

How does the plant benefit from this division of labor between antenna and reaction center pigments? Even in bright sunlight, a chlorophyll molecule absorbs only a few photons each second. If every chlorophyll had a complete reaction center associated with it, the enzymes that make up this system would be idle most of the time, only occasionally being activated by photon absorption. However, if many pigments can send energy into a common reaction center, the system is kept active a large fraction of the time.

In 1932, Robert Emerson and William Arnold performed a classic experiment that provided the first evidence for the cooperation of many chlorophyll molecules in energy conversion during photosynthesis (Emerson and Arnold 1932). They gave very brief ($10^{-5}$ s) flashes of light to a suspension of the green alga *Chlorella pyrenoidosa* and measured the amount of oxygen produced. The flashes were spaced about 0.1 s apart, a time that they had determined in earlier work was long enough for the enzymatic steps of the process to be completed before the arrival of the flashes and found that at high energies the oxygen production did not increase when a more intense flash was given: The photosynthetic system was saturated with light.

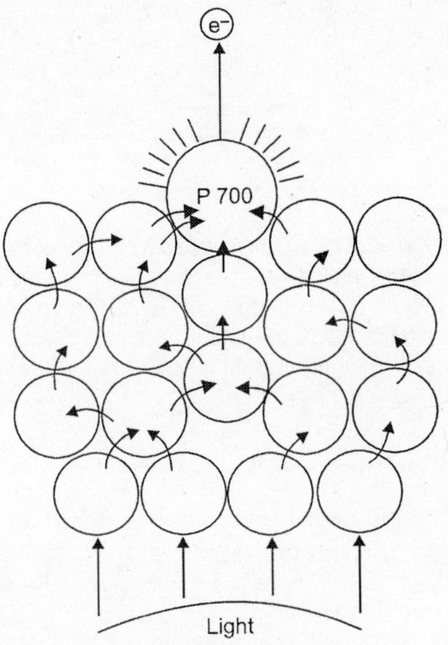


**FIG. 11.6:** Harvesting of solar light by chlorophyll *a* molecule.

The relationship of oxygen production to flash energy measured by Emerson and Arnold. They were surprised to find that under saturating conditions, only one molecule of oxygen was produced for each 2500 chlorophyll molecules in the sample. We know that several hundred pigments are associated with each reaction center and that each reaction center must operate four times to produce one molecule of oxygen; hence the value of 2500 chlorophylls per $O_2$.

The reaction centers and most of the antenna complexes are integral components of the photosynthetic membrane. In eukaryotic photosynthetic organisms, these membranes are found within the chloroplast; in photosynthetic prokaryotes, the site of photosynthesis is the plasma membrane or membranes derived from it.

*Many pigments together serve as an antenna*, collecting light and transferring its energy to the reaction center, where chemical reactions store some of the energy by transferring electrons from a chlorophyll pigment to an electron acceptor molecule. An electron donor then reduces the chlorophyll again. The transfer of excitation in the antenna is a purely physical phenomenon and involves no chemical changes. The first chemical reactions after photon absorption take place in the reaction center. Subsequent reactions stabilize the unstable products of the initial chemical reaction.

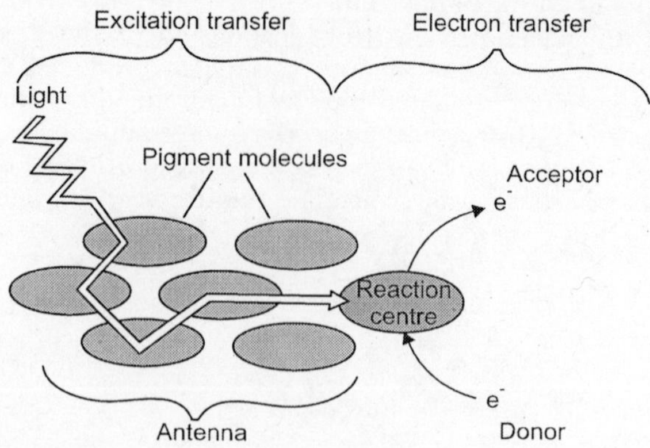


**FIG. 11.7:** The basic concept of energy transfer during photosynthesis. Many pigments together serve as an antenna, collecting light and transferring its energy to the reaction center, where chemical reactions store some of the energy by transferring electrons from a chlorophyll pigment to an electron acceptor molecule. An electron donor then reduces the chlorophyll again. The transfer of excitation in the antenna is a purely physical phenomenon and involves no chemical changes. The first chemical reaction after photon absorption take place in the reaction center. Subsequent reactions stabilize the unstable products of the initial chemical reaction.

## Action and Absorption Spectra

The visible radiation produced by the sun is a form of energy. In fact it is the ultimate source of virtually all the biological energy on the earth. However, there is a catch to understanding this because, in order for light energy to be useful in any biological process (photosynthesis most certainly included) it must first be absorbed and converted to a form of chemical energy useful to the cell.

Many molecules and atoms within cells have the ability to absorb units of light energy. But a given molecule or atom will not absorb light of just any wavelength of energy level. Rather, light is absorbed only when the photon is of just the right energy level to *move an electron* from one energy level to another, more energetic level. Different molecules vary drastically with respect to what is *just right energy level.* Some will absorb light of the visible spectrum, while others require different wavelengths. There are various wavelengths of sunlight, therefore, what portion of the sun's total spectrum is essential for photosynthesis?. This question is perfectly testable, assuming that we can separate various portions of the sun's spectrum- and such separation is indeed possible. That is, we can expose a plant to ray or visible rays or any other part of the solar radiation spectrum and note whether the plant responds by fixing $CO_2$ into food molecules. The result obtained would show that *visible radiation* was the effective form of the spectrum for inducing photosynthesis.

The visible radiation is really a mixture of light of various colours. What portions of the visible spectrum (e.g., red, orange, green, blue) are most efficient in supporting carbon fixation? Simply supply various colours of light to plants and measure the amount of photosynthesis that occurs in response to each type of light.

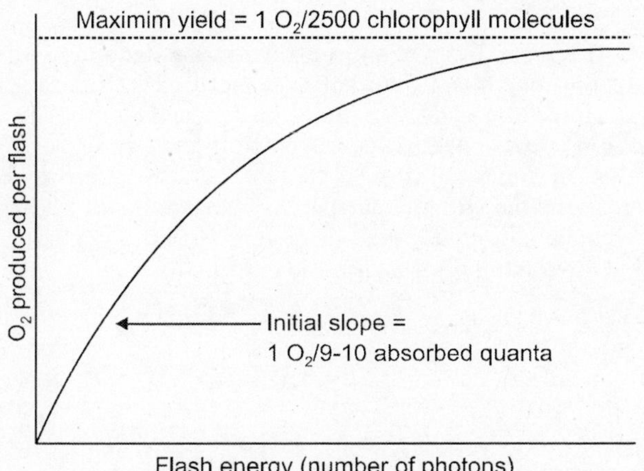


**FIG. 11.8:** The relationship of oxygen production to flash energy, the first evidence of the interaction between the antenna pigments and the reaction center. The light saturation curve, obtained in 1932 by Robert Emerson and William Arnold, showed that at saturating energies, the maximum amount of $O_2$ produced is 1 molecule per 2500 chlorophyll molecules. The slope of the low-intensity part of the curve gives the quantum requirement of oxygen production.

## Absorption Spectrum

It is the image recorder when electromagnetic radiation from a source emitting a continuous spectrum is passed through a substance. If the material is in the gaseous phase, bands will appear in the same position as the lines that occur in the characteristic emission spectrum of that substance in solids or liquids. These bands broadened and can be used to identify or quantify the material. The term **absorption** refers to complete retention of light, without reflection or transmission. **Absorbance** is the measure of the extent to which light is attenuated during passage through a coloured liquid or solid.

## Leaf Reflectance

Leaves are the primary scattering units in a vegetative canopy. Radiant energy incident on a leaf can be scattered from the leaf surface on to or away from the leaf, giving rise to secularly **reflected** and **diffusely reflected** and **transmitted radiation**. The leaf also will absorb radiant energy. Partitioning of radiation is a function of leaf cellular structure, leaf surface coating and roughness, and morphological parameters. Leaf anatomical structure and biochemical composition are critical to the amount of radiation absorbed, transmitted or reflected.

Relatively little incident visible radiation (350-700 nm) is reflected or transmitted in healthy leaves due to absorption of chlorophyll and other leaf pigments. Decreased visible reflectance with leaf development and maturation in cotton (*Gossypium hirsutum* L.), white oak (*Quercus alba* L.) and apple (*Malus domestica* Borkh) has been attributed to pigment accumulation. In contrast leaf absorbs relatively little incident near-infra-red (NIR) radiation (75-135 nm) because NIR scattering by the leaf mesophyll result in high reflectance and transmittance.

## Action Spectra

The use of action spectra has been central to the development of our current understanding of photosynthesis. An action spectrum is a graph of the magnitude of the biological effect observed as a function of wavelength. Examples of effects measured by action spectra are oxygen evolution and hormonal growth responses due to the action of phytochrome. Often an action spectrum can identify the chromophore responsible for a particular light-induced phenomenon. Action spectra were instrumental in the discovery of the existence of the two photosystems in $O_2$-evolving photosynthetic organisms.

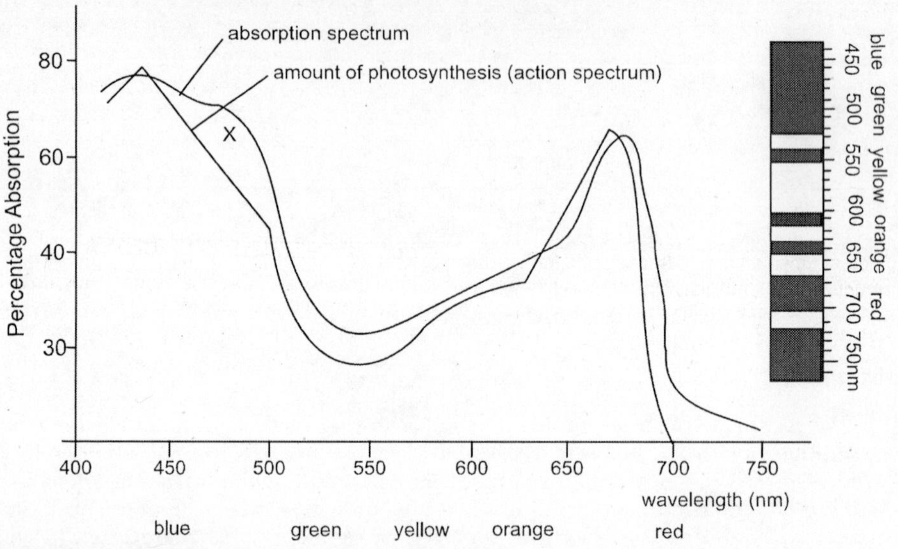


**FIG. 11.9:** (A) Absorption and action spectra of chlorophyll (B) Wavelenghts absorbed by chlorophyll shown by black areas.

Some of the first action spectra were measured by T.W. Engelmann in the late 1800s. Engelmann used a prism to disperse sunlight into a rainbow that was allowed to fall on an aquatic algal filament. A population of $O_2$-seeking bacteria was introduced into the system. The bacteria congregated in the regions of the filaments that evolved the most $O_2$. These were the regions illuminated by blue light and red light, which are strongly absorbed by chlorophyll. Today, action spectra can be measured in room-sized spectrographs in which the scientist enters a huge monochromator and places samples for irradiation in a large area of the room bathed by monochromatic light. But the principle of the experiment is the same as that of Engelmann's experiments.

## Difference Spectra

An important technique in studies of photosynthesis is light-induced **difference spectroscopy**, which measures changes in absorbance. In this technique, bright light, often called **actinic** light, is used to illuminate a sample, while a dim beam of light is used to measure the absorbance of the sample at wavelengths other than that of the actinic beam. In this way a difference spectrum is obtained, which represents the changes in the absorption spectrum of the sample induced by illumination with the actinic light. Absorption bands that disappear upon illumination appear as negative peaks; new bands that appear upon illumination appear as positive peaks. Difference spectra give important clues to the identity of molecular species participating in the photoreactions of photosynthesis. The difference spectrum of the photooxidation of P700 (a chlorophyll that absorbs light of wavelength 700 nm, as we'll discuss later in the chapter) is shown in Figure 6 (Ke 1973).

By the use of special flash techniques, it is possible to record the difference spectrum at a given time after flash excitation. Multiple difference spectra recorded at different times after flash excitation can be used to measure the kinetics of the chemical reactions that follow photon excitation of a reaction center. These techniques can have extraordinary time resolution, in some cases less than a picosecond ($10^{-12}$ s), and have provided great insights into the earliest events in the photosynthetic energy storage process.

## De-excitation

After absorption of photon, the chlorophyll molecule becomes excited. The process of absorption is extremely rapid and takes $10^{-15}$ sec to complete absorption. **What happens to the excited molecule?** The excited molecule remains in its excited state for a while and excess energy (absorbed by photon) comes back to its ground state. The phenomenon of coming back to the ground state from an excited state is called **de-excitation** and during this process energy is liberated in mainly four following possible pathways:

(i) Heat emission, (ii) Light emission (fluorescence, phosphorescence, delayed light emission), (iii) Energy transfer to another molecule, and (iv) photochemistry (or photoionization).

## Chlorophyll Fluorescence

Pigments that are involved in photosynthesis are capable of carrying on certain reactions even after they are extracted from the chloroplasts. They can absorb light and *fluorescen* that is, remit the light energy absorbed as light, but necessarily of longer wavelength (that is, of lower energy). The fluorescence photon has wavelength generally higher than that of the photon absorbed. In other words the energy of the emitted photon is lower than the energy of the absorbed photon. This can be explained by considering the substrates. The absorbed photon can lead the molecule to any substrate of $S_1$ depending on the energy of the absorbed photon. However, the molecule comes down quickly (10-12 sec) to the lowest substrate of $S_1$ in the stepwise fashion by heat emission (wavy vertical line). The molecule comes down to one of the substrates of $S_0$ by light emission (fluorescence). The molecule further goes down to one of the lower substrates of $S_0$ by heat emission. Thus, a part of the energy of the absorbed photon is emitted (de-excitation) as heat and the rest is emitted as fluorescence photon. Hence the energy of the photon emitted is less than the energy of the photon absorbed.

Chlorophyll solutions fluorescen dark red. If the reactions of photosynthesis are blocked by poisons, fluorescence will occur in vivo because the energy absorbed cannot be used. The fluorescence spectrum is characteristic of the pigment, so it is possible to tell which pigment is fluorescing, hence, which was activated. In addition, energy acceptors (that is, oxidized substrates capable of being reduced) can quench the fluorescence, demonstrating the ability of the pigment to transfer its energy to these acceptors instead of reemitting it as light.

Blue photons (blue light), carrying higher energy, raise electrons to the excited singlet state 2, red photons raise electrons to excited state 1. These, of course, are the photons, which are preferentially absorbed. Green photons are mostly not absorbed and it is the green light which is reflected (from and transmitted through) leaves which gives them their colour. Excited state 2 decays very rapidly to excited state 1 as electrons cascade down the energy gradient, losing energy by radiation less de-excitation. None of this energy is available for photosynthesis. It is the red, more long-lived, excited state created directly by a red photon (or indirectly by a blue photon) which is the starting point of photosynthesis and, therefore, of most of biological energy transducation. In addition the creation of the red excited state is accompanied by the generation of the massive oxidising potential, the positively charged *holes* within PSII, which are refilled by electrons drawn from water. Otto Warburg believed photosynthetic energy transduction to be perfect. It is not. Oxidation is inevitably associated with reduction. Chlorophyll fluorescence however, is only one minor channel of energy dissipation. Moreover, there are many factors, mostly still imperfectly understood, which can switch dissipation from one channel to another.

In functioning chloroplasts most of the incident energy is used photochemically. Therefore, in vivo we find only a small fluorescence yield (about 1%) in contrast to solutions in which photosynthesis does not take place. The fluorescence yield also increases in intact chloroplasts when photosynthesis is blocked by inhibitors such as diuron. In contrast to chlorophyll a solutions, intact chloroplasts show several fluorescence emission maxima which correspond to the various chlorophyll forms of photosystem I and II. At a temperature of - 196° C isolated chloroplasts show fluorescence emission maxima at 685 nm (F-685), 695 nm (F-695) and 735 nm (F-735). F-685 is associated with the antenna chlorophylls of LHCII, F-695 with the RC chlorophylls and F-735 with photosystem I. In vivo, virtually all the fluorescence is emitted by photosystem II. After previous dark adaptation intact leaves or chloroplasts show characteristic, temporal chlorophyll fluorescence changes upon irradiation, which were first, observed by Kautsky and are called variable fluorescence.

Measurement of the variable chlorophyll fluorescence is an important tool in photosynthesis research, which can be used to analyze partial reactions of the electron transport chain and the site of inhibitor activity. This technique has been used successfully to reveal the mechanisms of PSII herbicides, such as ureas and triazines. Fluorescence methods in general are highly advantageous due to their non-invasive nature, their high sensitivity and the fact that they can be applied easily to highly scattering samples like, e.g., intact leaves, needles, etc. Steady state fluorescence and/or quasi-steady state techniques like fluorescence induction have been used extensively in the past to probe a large variety of phenomena related to the energy absorption, energy distribution, etc.

Another pathway of de-excitation leads to **phosphorescence** which is also a light emission but having characteristics different from fluorescence. In this case the excited molecule first goes from $S_1$ state to the triplet state ($T_1$) by heat emission, and then comes down the ground state by light emission, called phosphorescence.

De-excitation can occur through another pathway leading to light emission known as **delayed light emission** or **delayed fluorescence.** Some of the molecules populated at the triplet state can come back to the $S_1$ state by absorbing the required energy from neighbouring molecules and then come down to the ground state emitting visible photon called delayed light.

## Energy Transfer

An excited molecule can, under suitable condition transfer its excess of energy to another molecule at the ground state. As a consequence the first molecule comes down to its ground state and the second molecule is raised to its excited state. This is called **energy transfer**. In this case no light emission takes place. As expected, the energy gap between $S_1$ and $S_0$ of one molecule should match with that of the other for energy transfer to occur. *The fluorescence wavelength of the donor molecule match with the absorption wavelength of the acceptor molecule.* Energy transfer takes place among *identical molecules* also, and molecules must be very close to each other (within several nanometers). During photosynthesis, chlorophyll molecule of the light-harvesting class absorbs photon, gets excited and the transfer its excited energy to another chlorophyll molecule of same type, and the event may be repeated for several times until excitation energy reaches the site where photochemistry (formation of chemical energy) takes place.

## PHOTOCHEMISTRY

In photosynthesis a molecule in its excited state can utilize its excess energy to drive a chemical reaction. The event involves other molecules apart from the one which absorbs a photon gets excited.

Through this reaction the energy of the photon is utilized to create stable reducing and oxidizing agents. This occurs through electron transfer and charge separation. The reaction is as follows:

$$XYZ \xrightarrow{\text{Photon}} XY^*Z \rightarrow XY^+Z \rightarrow X^+YZ$$

In above reaction there are three molecules (X, Y and Z) at ground states. For example Y absorb is a quantum of energy and is raised to an excited state. In an excited state the potential of the molecule becomes more negative (this happens because when an electron goes to a higher electronic level, the electron become strongly bound to the molecule, the molecule tends to lose it; in other words the affinity of the molecule for the electron is reduced). The affinity of the three molecules is in such a way that in the more negative state of potential, $Y^*$ transfers an electron spontaneously to Z. This makes C positively charged (its potential becomes more positive than in the ground state, and state drawing electron spontaneously from X) and Z becomes negatively charged. Thereby Y comes back to the ground state, but X is now positively charged. This causes a separation of charges in the system with the formation of reductant ($A^-$) and an oxidant ($X^+$) as a result of light absorption by Y. There is no net change in Y. This phenomenon is called **sensitizer of photochemical reaction**. At this stage, now $X^+$ has the property to oxidize and $Z^-$ has the property to reduce. This example explains the basis of conversion of light energy to chemical energy.

In chloroplast, Y may be considered equivalent to specialized chlorophyll molecule which is known as **trap** or **reaction centre**. The specialized chlorophyll is different from light-harvesting chlorophyll because in this chlorophyll there occurs no electron transfer (photochemistry) when it absorbs a photon; it only transfers its excitation energy to another molecule. In green plant photosynthesis, the electron from $Z^-$ is transferred through several redox couples eventually to NADP and $X^+$ becomes sufficiently positive to oxidize water producing protons, electron and oxygen. In this process water is the ultimate source of electrons needed for the reduction of NADP.

## Photochemical Quenching

Linear photosynthetic electron transport moves electrons from water through the two photosystems to ferredoxin. From ferredoxin these are passed to $CO_2$, via NADP, or to $O_2$ (the Mehler reaction or pseudocyclic electron transport.) If $CO_2$ is removed, NADP becomes rapidly reduced and the lack of NADP as a terminal electron acceptor affects other carriers in the chain. Once $Q_A$ (the primary electron acceptor in PSII) is reduced (i.e., once its open electron traps are closed) it no longer **quenches** fluorescence and the fluorescence signal rises. Although $Q_A$ is a quinone "Q" stands for **quencher** in this context. This was the name that Duysen and Swears (1963) first applied to a then unknown component in PSII which quenches fluorescence when it is oxidised but not when it is reduced. Such quenching is now referred to as $q_Q$ or *photochemical quenching.*

## CHAPTER 12

# Proteomics and Genetics of Photosynthetic Apparatus

## PLASTIDS

Plastids are self replicating organelles surrounded by an envelope composed of two membranes. They are found only in photosynthetic cells of eukaryotes but exceptionally found in generative and sperm cells of certain angiosperms too. Plastids are varied in size, shape and colour. Green coloured plastids are chloroplast, that contain the chlorophyll pigment and are site of photosynthesis. In green-leaf cells particularly in the parenchyma cells, the chloroplasts lie in the cytoplasm between the central cell vacuole and the parenchyma. The chloroplasts occupy more than 90 per cent of the volume of the cytoplasm in leaf and are typically lense-shaped (in mosses, fern and seed plants), appearing biconvex, plano-convex or concavo-convex in profile. In algae they are heteromorph. In higher plants they are usually 4-8 μm in diameter and 2-3 μm thick. Changes in shape and size can also be induced osmotically both in the leaf cells and chloroplast isolated from the cell. They often move in the cell under influence of illumination and as a result of cytoplasm streaming.

Other coloured plastids are found in flowers and fruits. On the basis of the storage products colorless plastids or leukoplasts are termed amyloplasts (Starch), proteinoplasts (protein) or elaioplasts (Lipids). Plastids have three compartments: (i) The intramembrane space between the outer and inner membrane, (ii) The stroma space and (iii) The thylakoid space.

### The Multifunctional Chloroplast

Chloroplasts in green algae and higher plants contain photosynthetic thylakoid membranes. Four multi-subunit protein complexes, Photosystem I (PSI), PSII, the ATP-synthase complex, and the cytochrome $b_6f$ complex, which together comprise 75 to 100 proteins, perform the photosynthetic reactions. It can be postulated that the thylakoid membranes contain a large number of other proteins that are involved in the biogenesis and regulation of the four multi-subunit complexes. These additional proteins would be involved in processes such as biosynthesis and ligation of cofactors, insertion of proteins in to membranes, and folding and degradation of proteins. Several such proteins have been identified in the thylakoid membrane. They include the proteases DegP and FtsH, protein translocation components, such as SecY, SecE, SecA, Alb3 and Hcf106, the PSII assembly factor Hcf B6 and a lumenal isomerase TLP40.

To regulate biogenesis and several thylakoid functions, kinases (Synders and Kohorn, 1999) phosphatases (Vener et al., 1998), and possibly other signal transducers are present in the thylakoid membrane. Several proteins without obvious function have been identified in the intra-thylakoid space, the thylakoid lumen.

Although best known for their role in Photosynthesis, chloroplasts also synthesize many essential compounds, including plant hormones (Lange, 1998) fatty acids and lipids (Essingmann et al., 1998), amino acids (Ho et al., 1999), Vitamins (B1, K1 and E, Belanger et al., 1995), Purine and pyramidine nucleotides (D Smith et al., 1998) and secondary metabolites such as alkaloids and isopernoids (Keller et al., 1998). In addition to it chloroplasts also are required for nitrate and sulfur assimilation (Heldt, 1997). Several of the enzymes in these biosynthetic pathways have been identified by analyzing mutants of *Arabiodopsis* and other plant species. However, more proteins and possible new pathways remain to be discovered, and it is not unlikely that several of these components are located in or at the surface of the thylakoid membrane.

## Number

Their number in a cell varies greatly among plant species. The assimilatory parenchyma (palisade and spongy parenchyma) of leaves contains a especially large number of chloroplasts. In spinach leaves, for instance, 300-400 chloroplasts have been found per plaisade parenchyma cell and 200-300 per spongy parenchyma. The dry weight of spinach chloroplasts is about $9 \times 10^{-12}$ g and that of corn chloroplasts about $38 \times 10^{-12}$ g.

## Other Inclusions

**Starch grains:** These are often single, largest inclusion found in the stroma of chloroplasts. They appear in profile as ovule structures generally electron translucent but occasionally they are electron opaque. Starch generally accumulates in the bundle sheath chloroplasts of C4 plants and only to a smaller extent in the mesophyll chloroplasts.

**Crystalline inclusions:** Crystalline inclusions or bodies are of 1 mm in size and lie in stroma. They are proteinaceous in nature and their function is unknown.

**Phytoferritin:** It is an iron protein complex which appears as an electron opaque aggregation of particles in the plastid stroma.

Algae possess **pyrenoids**, which organize the deposition of reserve substances. The microfilaments, 5-6 nm in diameter have been occasionally seen in electron micrographs are thought to run lengthwise in the stroma near the plastid envelope. The function of these microtuble-like structures is not well understood.

## Membrane System of Chloroplast

The structure of higher plant chloroplast can be regarded as a membrane system and a matrix system. The membranes are of two different types: (a) an external envelope (characteristic of eukaryotic cell), and (b) an internal pigmented membrane system of grana and integral connections.

The granal and the integral membrane (*the frets*) form a continuous membrane. Chlorophyll is not evenly distributed throughout the chloroplast, but rather contained in the inner membrane system. This membrane system gives rise to stack-like configuration of membranes called *grana*, and they are composed of individual flattened membrane sacs called *thylakoids*.

The surrounding material, which does not contain chlorophyll, is called *stroma*. It is hydrophilic protenaceous matrix. $CO_2$-incorporating apparatus is present within the stroma (the non-green portion). The stroma along with thylakoids also contains particulate components such as *fraction portion I*, ribosomes and polysomes, and the fibrils of DNA, nucleoids (DNA), plastoglobuli (lipid reserve) and starch grains. Robert Hüber, Johann Deisenhofer and Hartmut Michel were awarded the 1988 Nobel Prize for Chemistry for work in the photosynthetic process, for succeeding in making a structural analysis of a membrane linked to protein complex which opened up completely new possibilities in chloroplast structural research.

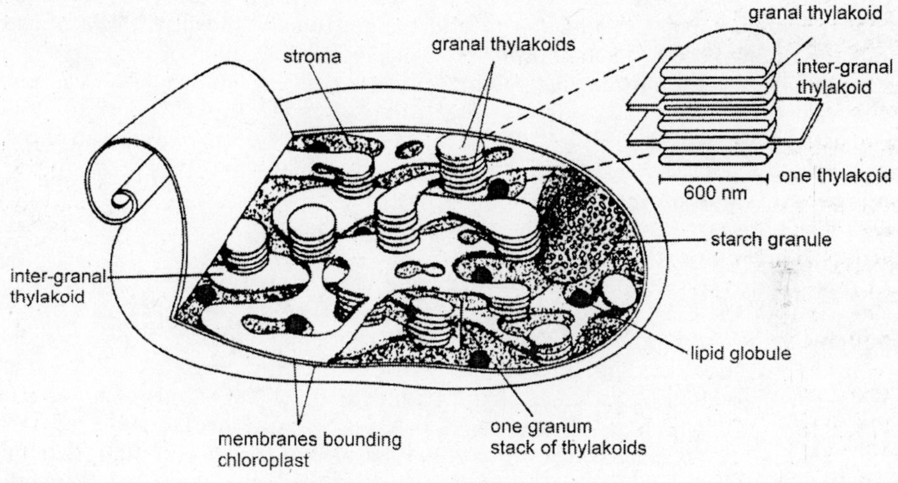


**FIG. 12.1:** A Three dimensional structure of chloroplast.

## TRANSPORT SYSTEM OF CHLOROPLAST MEMBRANE

The majority of the protein components of chloroplasts are encoded in the DNA. They are synthesized on cytoplasmic ribosomes outside the organelle and subsequently imported into the chloroplasts. *In-vitro* studies using the protein-synthesizing machinery of a wheat-germ extract coupled to polyadenyl messenger RNAs (poly $A^+$mRNA) isolated from algal or higher plants have demonstrated the transfer of a number of nuclearly encoded polypeptides into the chloroplast; examples are ferredoxin, plastocyanin, ferredoxin-NADP reductase, polypeptides associated with antenna pigment complexes, the small subunit of ribulose-bisphosphate carboxylase, and fructose 1,6-bisphosphatase. These proteins are assembled on the cytoplasmic ribosomes as a precursor of much higher molecular weight protein referred to as the transit peptide, with sequence extension, at the aminoterminal. The precursor of plastocyanin (molecular mass = 10.5 kDa), for example, is assembled with a transit peptide of molecular mas 15 kDa. The precursors are bound to the outer chloroplast membrane, translocated through the chloroplast envelope, processed to the mature polypeptide (mainly by proteolytic cleavage of the transit peptide), and then either assembled as membrane constituents or incorporated into the stroma. The import of precursors into the chloroplast is an energy-dependent process and consumes ATP synthesized by photophosphorylation. The requirements for binding of precursors to the envelope membrane, the nature of the receptors, and the mechanisms regulating the translocation and assembly of polypeptides are still being actively investigated.

There are two classes of precursor proteins based on their thylakoid assembly requirements. The integral membrane proteins (e.g. Chl *a/b* protein of LHC II) must cross the plastid envelope membranes, the aqueous envelope periplasm, the stromal compartment and then integrate into the thylakoid lipid bilayer. These precursor proteins contain a hydrophobic N-terminal extension called the *transit sequence* that contains topogenic information for targeting the precursor to the chloroplast and localizing the matured protein on the thylakoid. Lumenal proteins (e.g. plastocyanin) must cross all the plastid compartments and in addition must traverse the thylakoid bilayer membrane. The plastocyanin precursor contains two transit sequences as chloroplast-targeting components. A chloroplast-targeting domain (at the N-terminal) directs the pre-protein towards and into the chloroplasts; this domain is cleft by a stromal protease. A second, strongly hydrophobic, domain at the C-terminal directs the partially cleft precursor to the lumen where it is split off by a thylakoid protease.

Proteinaceous components exposed at the chloroplast surface may be involved in the import and binding of precursors.

At present there is no evidence for cytosolic factors being involved in protein translocation other than, may be, as chaperones. The molecular *chaperones*, incidentally, are a class of proteins found in prokaryotic and eukaryotic cells which play roles in transport, folding, and assembly of certain other proteins but are not themselves components of the final oligomeric structures. The 70-kDa *heat-shock proteins* (Hsp 70), which are synthesized in cells in response to exposure to elevated temperature (heat shock) and other 'stresses', are a typical group of chaperones. Hsp 70s have been detected in outer envelope membranes and stroma of chloroplasts.

## Exchange of ions and metabolites

Chloroplasts isolated from higher plants are bound by two distinct membranes which are together called the chloroplast envelope. These envelopes are the site of synthesis of a number of membrane constituents. These are involved in the transport of many metabolites.

The chloroplast envelope is the site of synthesis and assembly of membrane galactolipids, α-tocopherol and carotenoids. The polypeptide composition of the envelope is different from that of the thylakoids and of the stroma. The chloroplast envelope functions both as a barrier separating the stroma from the cytosol and as a bridge allowing the transport of specific metabolites between these two compartments. The chloroplast envelope is permeable to various molecules. The **outer envelope** membrane acts as a molecular sieve and is permeable to all molecules up to a molecular mass of 10 kDa. The **inner membrane** (osmotic barrier) is almost impermeable to charged species such as ferredoxin, NAD(P), $NAD(P)H_2$, acetyl CoA, pyrophosphate, many sugar phosphates, small cations ($Na^+$, $Mg^{2+}$), and to sucrose and sorbitol. $CO_2$, $O_2$, acetic acid, pyruvic acid, glycerol are uncharged molecules. They can diffuse through the inner membrane.

Pi and triose phosphates, ATP, and dicarboxylate anions (malate, aspartate, oxaloacetate, etc.) pass through the inner membrane in the presence of specific carriers present in the envelope called *translocators*. Of these translocators, the proteinaceous factor (phosphate or $C_3$ translocator), which allows the transport of triose phosphates from the stroma to the cytosol in exchange for the import of Pi into the stroma, has been studied in most detail.

The chloroplast envelope also mediates in the export of fatty acids synthesized in the chloroplast to other organelles (for lipid synthesis). The exact nature of the polar lipid interactions with plastid membrane proteins is not yet clear. It is also not clear that how the membrane components which are synthesized in the envelope are subsequently transferred and assembled in the thylakoids, how do the polyunsaturated fatty acids in the chloroplasts originate, what are the specific functions of various envelope proteins?

The **fraction protein I** accounts for 50 per cent of the soluble stroma and 50 per cent of total leaf soluble portion and contains photosynthetic carboxylating enzyme, ribulose diphosphate carboxylase (carboxydismutase, the enzyme responsible for catalysing the formation of phosphoglyceric acid from $CO_2$). Fraction protein I can be split into two distinct subunits: a large subunit of mol. weight approximately 54,000 and a small subunit of mol. weight 16,000.

## CHLOROPLAST DNA

The presence of DNA and ribosomes in chloroplasts was first demonstrated in 1962. Our present knowledge of the organization of genetic information in chloroplasts is based on the application of molecular cloning and nucleotide sequencing technologies. Restriction mapping and cloning confirmed the double-stranded circular nature of the chloroplast DNA (plastome). The size of the plastome varies between 120 and 160 kbp (kilobase pair) in land plants; the average molecular mass of the plastome is about $1 \times 10^8$ kDa. To date, the entire genome sequence has been determined for the cloned chloroplast DNAs of three plant species, viz. tobacco, liverwort and rice, and the sequences for defined segments of many other chloroplast genomes have been completed. These studies indicate that the chloroplasts probably code for all of their tRNA (transfer RNA), rRNA (ribosomal RNA) molecules, or many of the proteins in the photosynthetic electron transport chain and in the ATP synthase. The remainder of the electron transport chain proteins and some in the ATP synthase complex are encoded by the nuclear genome, synthesized in the cytoplasm, and imported into the chloroplasts.

### Chloroplast Ribosomes

Stroma show presence of single ribosomes, clusters of ribosomes and helically arranged show presence of polysomes after staining with uranyl acetate. The single chloroplast ribosomes are smaller than cytoplasmic ones measuring 17 nm in diameter (compare 25 nm for the cytoplasmic ribosomes). It has sedimentation coefficient of 70S and can be split into two subunits sedimentating at 55-60S and 40S respectively. They resemble bacterial ribosomes more closely than eukaryotic cytoplasmic ribosomes.

**Proplastids,** as the progenitor for all other types of plastids are also colourless. Plastids contain DNA, which appear as fibers in a matrix free region called nucleoid of the stroma which is back ground matrix material of the plastid. The plastid ribosomes may be arranged singularly or in polyribosome. Plastoglobuli, small lipid droplets are also present in stroma.

## PLASTID GENETICS

### Proteomics of Chloroplasts

**Defining Proteomics:** Proteomics is the large-scale study of proteins, usually by biochemical methods. The word proteomics has been associated traditionally with displaying a large number of proteins from a given cell line or organism on two-dimensional polyacrylamide gels. In 1990s, biological mass spectrometry, Matrix assisted laser desorption/ionization – Time of fright (MALDI-TOF.MS), and Electrospray ionization mass spectrometry (ESI-ms/ms), emerged as powerful analytical method that removed most of the limitations of protein analysis. Today, the term proteomics covers much of the functional analysis of gene products or *functional genomics*, including large-scale identification or localization studies of proteins and interaction studies. Therefore, Proteomics is the study of all the expressed proteins of an organism. The information that proteomics studies can provide includes expression levels, post-translational modifications, sub-cellular localizations, protein-protein interactions, and protein-nucleic acid interactions.

Proteomics already has become an important tool for drug discovery and the analysis of yeast and *E. coli* protein expression patterns; however, it has not been widely applied in plant biology. The total proteome of higher plants is estimated to consist of ~21,000 to 25,000 proteins and chloroplast contain 10 to 25% of this total, an illustration of the central importance of the chloroplast for the plant cell.

Recently, Peltier et al. (2000) studied proteomics of the chloroplast. The soluble andperipheral proteins in the thylakoids of pea have been systematically analyzed by usign two-dimensional electrophoresis, mass sectrometery, and N-terminal Edman seqnencing, followed by database searching. After correcting to eliminated possible isoforms and post-translational modifications, they estimated that there are atleast 200 to 230 different lumenal and peripheral proteins. Sixtyone proteins were identified, for 33 of these proteins, a clear function or functional domain could be identified, where as for 10 proteins, no function could be assigned. For 18 proteins, no expressed sequence tag or full-length gene could be identified in the databases, despite experimental determination of a significant amount of amino acid sequence. Nine previously unidentified proteins with lumenal transit peptides are presented along with their full-length genes; seven of these proteins possess the twin arginine motif that is characteristic for substrates of the TAT pathway. Logoplots were used to provide a detailed analysis of the lumenal targeting signals, and all nuclear-encoded proteins identified on the 2-D gels were used to test predictions for chloroplast localization and transit peptides made by the soft ware programs Chlorop, PSORT, and Signal P.

The 58 proteins (excluding and three stromal proteins) were classified in a pie diagram according to their function, following the categories used in theanalysis of a 1.9-mB contiguous sequence of chromosome IV of *Arabidopsis* (Bevan et al., 1998). This is a helpful tool to visualize the more general role of the lumenal and peripheral membrane proteins. For 17% of the proteins, no clear function could be predicted, whereas for 31% of the proteins, no expressed sequence tag or full-length gene could be assigned. The remaining 52% could be classified according to their function.

Naturally, a significant fraction of the proteins is involved in energy production (21%), either in photosynthetic electron transport or in ATP production. So far, only one transporter, Brittle-1, was found on the two-dimensional electrophoresis maps of both the lumenal and peripheral fractions. Brittle-1 has an adenlyate translocator function in the transfer of ADP glucose in amyloplasts. Brittle-1 was cloned using a transposon-tagged maize mutant (Sullivan et al., 1991) and was found to be localized in amyloplast membranes (Sullivan and Kaneko, 1995). Because starch also accumulates in chloroplasts, Brittle-1 is likely to have a similar function as in amyloplasts. Indeed, in vitro import assays showed that Brittle-1 could be targeted to the chloroplast inner envelope membranes, but the protein had not been found previously in chloroplasts in vivo (Li et al., 1992).

Transmembrane prediction programs indicate that Brittle-1 could be an integral membrane protein, but the predicted transmembrane domains are shorter and less hydrophobic than are observed usually. Any integral membrane protein was not detected on the two-dimensional electrophoresis maps, which is expected because the protein fractions were centrifuged at high gravity values to remove any membranes. Thus, it was surprising to find Brittle-1 on both the lumenal and peripheral maps. Brittle-1 is an integral membrane protein. Because the protein had not been detected previously in chloroplasts, further experimentation is needed to more precisely define the localization of Brittle-1.

The two-dimensional electrophoresis maps also revealed a lipid storage protein (PG1), which was classified in the cell growth and division category. PG1 recently was identified in plastoglobules extracted from pea chloroplast membranes, and in vitro import assays confirmed its localization in the chloroplast (Kessler et al., 1999). Plastoglobules have been observed in different types of plastids and are thought to serve as lipid reservoirs for thylakoid membranes. In addition to PG1, plastoglobules possess several other proteins associated with the outer surface, such as the 30-kD plastid lipid-associated protein (Pozueta-Romero et al., 1997). It is quite likely that several of those proteins are present on the two-dimensional electrophoresis maps, and they should be identified in the near future.

Other proteins involved in cell growth include a set of four histone H4-like proteins in the 10-to 14-kD size range. A large fragment of histone H4 is extremely well conserved among many eukaryotes, varying little in monocotyledons and dicotyledons, green algae, insects, and mammals. The identified peptides of the four proteins all corresponded to this conserved region. Two chloroplast proteins that are electrophoretically similar to histones H2A, H2B, and H3 of pea cell nuclei were part of the chloroplast nucleoids. The amino acid composition of these proteins demonstrates high similarity with teh HU proteins of *E. coli* (Yurina et al., 1995), but the identified peptides of the histone-like proteins in this study did not match those HU proteins. In total, ~ 30 polypeptides from 94 to 12 kD in size were detected in nucleoids. These nucleoids have been found on the thylakoid surface (Liu and Rose, 1992) and chloroplast inner envelopes (e.g., Sato et al., 1993). The histone-like proteins were found exclusively in two-dimensional electrophoresis maps of the peripheral proteins, which would be agreement with a fairly tight binding (i.e., sonication resistant) of plastid DNA to the thylakoid surface. An alternative explanation for these histone-like proteins is that a minor contamination with nuclei has occurred. To avoid such a contamination the intact chloroplasts purified on linear Percoll gradients, in which nuclei are known to sediment to the bottom, whereas intact chloroplasts can be collected between ~50 to 70% Percoll (Galagher and Ellis, 1982).

The finding of a known chloroplast-localized RNA binding protein points to the role of the thylakoid as a surface for transcription and translation, in agreement with observations of membrane-associated polysomes (Hattorie and Margulies,

1986; Jagendorf and Michaels, 1990) and plastid DNA.

One enzyme involved in fatty acid biosynthesis was identified and was the only protein classified under metabolism. It is indeed well known that fatty acid and lipid metabolism takes place both in the plastid as wella s in the endoplasmic reticulum (Miquel and Browse, 1992). Further studies are required to determine where in the chloroplast these enzymes are localized or not.

Approximately 12% of the proteins were classified under protein destioation and storage. They include the protease DegP, the assembly factor Hcf136, Ferritin, three chaperones, and a *cis-trans* isomerase that possesses a lumenal transit peptide. Ferritin concentrates and stores cellular iron to $\sim 10^{11}$ times the solubility of the free ion and earlier had been determined to be a plastid protein located at the stromal side of thylakoid membranes (e.g., Waldo et al., 1995). Ferritin accumulation is positively correlated to iron loading of the plant, is regulated at a transcriptional level (Gaymard et al., 1996), and is under developmental control (Lobreaux and Briat, 1991). The FKBP isomerase is only the second isomerase identified in the thylakoid; the other one was named TLP40 (Fulgosi et al., 1998).

The presence of fairly high amounts of the chaperones Hsp70, Cpn60, and Cpn21 on the lumen map but not on the peripheral map is intriguing. These ATP-dependent chaperones have been demonstrated to be functionally important for protein import across the envelope and for assembly of protein complexes (e.g., Lubeck et al., 1997; Keegstra and Cline, 19990. The MALDI-TOF or Edman sequencing data matched best to the genes for stromal chaperones.

The conclusion is that either (1) the genes for the lumneal chaperones have not been sequenced or (2) the gene products have a dual location due either to dual targeting to the stroma and lumen or to alternative splicing, as observed for stromal and thylakoid ascorbate peroxidases (Mano et al., 1997; Yoshimura et al., 1999). ChloroP predicts alternate stromal cleavage sites for the three chaperones if the program is provided with only the N-terminal sequences truncated before the experimentally determined (Hsp70) or postulated N terminus of the mature protein (Cpn60 and Cpn21; prediction for alternate cleavage site). This often is not the case for other substrates (data not shown). When these alternatively cleaved precursors then are analyzed by SignalP, lumenal cleavage sites are detected within the first 100 amino acid residues of the precursor proteins. Thus, if the actual stromal transit peptide is cleaved more toward the N terminus, it might be possible that these chaperones are targeted to the lumen. At this point, note that different stromal-processing peptidases may exist (Su and Boshchetti, 1994; Koussevitzky et al., 1998) and that a similar alternative cleavage behavior and dual targeting have been proposed for polyphenol oxidase (Koussevitzky et al., 1998).

A third possible explanation for the concentration of chaperones on thelumen map is that the proteins are bound to the stromal side of the thylakoid membrane. They might be released together with lumenal proteins upon sonication, just like $CF_1\alpha$ and $CF_1\beta$. It is likely that the stromal chaperones are involved in protein assembly at the thylakoid surface. Schlichter and Soll (1996) searched for lumenal chaperones and protease-treated thylakoid before release of the lumenal content; they found shorter isoforms of the stromal chaperones, based on immunodetection with antisera generated against the stromal chaperones. Their results might be explained by only partial digestion of the stromal chaperones when they are assembled in membrane-bound complexes at the thylakoid surface. Clearly, additional experiments are required to draw any final conclusions.

### Consensus Sequences and Prediction of Localization and Cleavage Sites

The neural network program ChloroP was most successful in localizing the newly discovered chloroplast proteins, whereas PSORT assigned 52% of the identified proteins to the chloroplast. What does this mean for the use of these programs to confirm localization of chloroplast proteins? If the proteins indeed are localized in the chloroplast, ChloroP recognizes ~94% of them, whereas PSORT probably identifies ~52%. ChloroP was trained with a positive test set of 75 known transit peptide containing chloroplast proteins (excluding lumenal proteins) and with a negative test set (75 proteins from other nonchloroplast localizations; Emanuelsson et al., 1999). When the authors tested 715 Arabidopsis entries in the SWISS-Prot database, they observed that 96% of those annotated in the database as being chloroplast localized were indeed predicted to be the chloroplast. By contrast, 11% of proteins annotated as nonchloroplast proteins were predicted to be in the chloroplast, which would result in ~ 2500 false positive for the full genome. In addition, the program is somewhat based toward the known chloroplast proteins, as exemplified by the 100% score for the 12 well-known photosynthetic proteins on our maps. The PSORT program is much more ambitious in that it predicts proteins to 17 different locations (three within the chloroplast), in the plant cell. Judging from information available on the PSORT the program has not been as rigorously tested as ChloroP has been (Nakai and Horton, 1999).

It is also possible that PSORT is more conservative in its prediction of proteins to be localized in the chloroplast. Testing newly identified proteins by both programs is probably the best option at the moment to get additional support/confirmation for chloroplast localization. It has been noted that in many cases, an additional amino acid residue needs to be removed to obtain the mature protein sequence (Richter and Lamppa, 1998; Emanuelsson et al., 1999).

## PLASTID INHERITANCE

In most angiosperms plastid inheritance is maternal because the sperm cells lack plastids. This condition may arise during the first pollen mitosis when all plastids go into vegetative cell resulting in an aplastidic generative cell.

In some species e.g. *Beta vulgaris*, *Solanum tuberosum* almost all plastids go into the vegetative cell during first pollen mitosis; a few plastids pass into the generative cell. And during the development of the sperm cells in these plants the plastids are lost and as a result plastid inheritance remains maternal. In a few angiosperm genera, e.g. *Oenothera* and *Pelargonium*, plastid inheritance is biparental because plastid distribution is equal during the first pollen mitosis and the plastids are not eliminated resulting in plastids being present in the sperm cells.

## PLASTID ULTRA STRUCTURE

### Proplastids

Proplastids, the progenitors of other plastids are found in root and shoot meristems, embryos, endosperms and in young developing leaves. The progenitor function of proplastids should be regarded as secondary to their simple structure which is a primary distinguishing feature. They are not structurally complex. They consist of an envelope, a nucleiod with DNA fibrils, a small number of ribosome, thylakoids, some time small starch molecule. Proplastids are varied in shape. They only slightly larger than mitrochondria. In light grown plants the proplastids in stems and leaves usually differentiate into chloroplasts. Proplastids may also have important biochemical functions such as the synthesis and storage of starch and other metabolites

## PLASTID DEVELOPMENT AND FUNCTIONS

### Chloroplast Development

In angiosperms (flowering plants), chloroplast biogenesis is usually studied by germinating seeds in total darkness and then exposing the seedlings to light. The changes occurring during illumination are followed by examination of the ultrastructure of the plastids by electron microscopy, the assay of photosynthetic electron transport and $CO_2$ fixation rates, and determination of the chromophore-containing components, e.g. cytochromes and ferredoxins, by spectroscopy. Such studies have shown that chloroplast develop from relatively undifferentiated structures known as *proplastids* via simple colourless organelles termed *etioplasts*.

The most important change occurring during illumination (greening) is the reduction of protochlorophyllide to chlorophyllide (chlorophyll without the phytyl side chain). The response of leaves to light involves changes in the expression of specific genes at the RNA level – RNAs complementary to many segments of the plastid genome are expressed more abundantly in light. Substantial increases in the content of carotenoids, cytochromes, ferredoxin, plastocyanin, ferredoxin-NADP reductase, etc. have been measured during the greening process. Photosystem I activity is expressed at an earlier stage than PSII activity. The enzymes of the $CO_2$ fixation pathway are present in the etioplasts, albeit at very low concentrations compared to mature chloroplasts. In higher plants the responses to light are thought to be mediated by three photoreceptors: protochlorophyllide, phytochrome (8.1) and a blue light receptor. The blue light receptor has not yet been isolated, although both flavoproteins and carotenoids have been proposed as likely molecules.

The early stages of chloroplast development can influence the expression of certain nuclear genes encoding chloroplast protein. At present it seems reasonable to suppose that the induction of some nuclear genes requires not only phytochrome, but one or more *plastid derived factors* as well.

At present, the molecular nature of plastid derived factor is unknown but the physiological characterisation of the factor by Oelmuller et.al. (1986) allows further insight in to its action. In experiments with *nor flurozon*, they found that in even brief periods of photo oxidative damage cause Rbcs and Cab mRNAs to decline rapidly and if damage occurs no recovery is found upon transfer in mustard seedlings to non photo oxidizing light conditions. So they concluded that the signal from the plastid is short lived and are required continuously and that even brief periods of damage prevent production of the signal.

Oelmuller et.al. (1986) pointed out that signal is a very like protein. In this regard, it has been suggested that plastid encoded RNA may regulate the cytoplasmic synthesis of chloroplast proteins. As we know chloroplast consists of a double outer envelope a stromal face where soluble proteins such as Rubisco are located and inner membrane, termed the thylakoid. Four major photosynthetic complexes are integrated into the thylakoid:

1. Photosystem I (PS I)

2. Photosystem II (PS II)
3. A cytochrome complex
4. ATP synthetase.

These complexes mediate light dependent vectorial electrone flow and the resultant generation of ATP and reducing power. The end products of photosynthetic electron transport are used in $CO_2$ fixation and other plastid functions. The plastid carries out numerous other functions e.g. steps of amino acid, fatty acid, pyrimidine, tarpene and tetrapyrole biosynthesis occurs in plastid. This organelle is also a site of plant growth regulator synthesis. The interdependence of cell and plastid function in photosynthetic as well as non photosynthetic tissue makes the regulation of genes encoding plastid involved in some problem of cell biogenesis and function. On the other hand, the genes encoding photosynthetic enzyme have special feature and these will be dealt with a border consideration of plastid origin and gene expression.

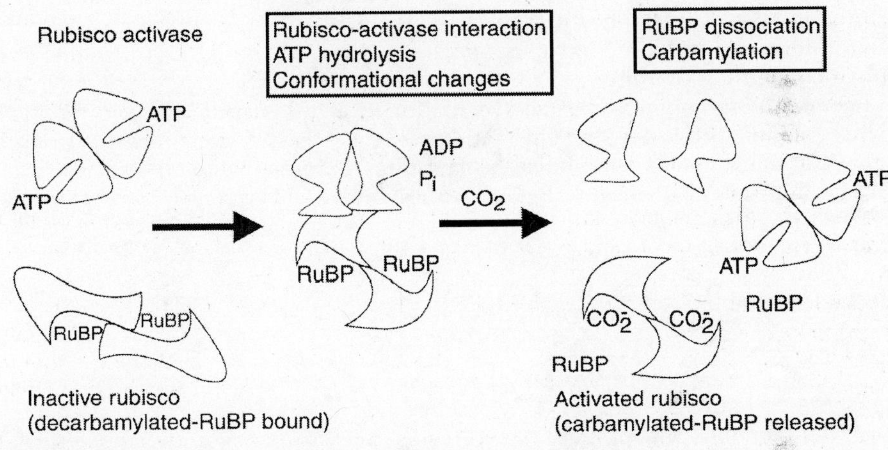


**FIG. 12.2:** Rubisco activase functions by removing bound RuBP from the decarbamylated inactive form of rubisco. The activase enzyme appears to couple the hydrolysis of ATP to the release of ribulose or other inhibitory pentose bisphosphates. Rubisco is then activated (carbamylated) by binding $CO_2$ and $Mg^{2+}$.

## Integration of Plastid and Cell Function

Plastid isolated from higher plants does not replicate and survive only a short period of time in vitro. This indicates that these organelles are not autonomous and are dependent on cytoplasmic functions and nuclear gene products. The inter-dependence of plastids and the cell in which they reside is relatively specific and the transfer of plastids into foreign plant cells results in impaired function and loss of the transferred organelles. It is cleared that extensive changes have occurred in the photosynthetic endosymbiont during the integration of process. For example, - Plastids of higher plant do not have peptidylglycan walls which are found in Bacteria. Presumably, the genes required for wall biosynthesis have been lost or inactivated. Interestingly, the cyanelles of *Cyanophora paradoxa* which resemble plastids and perform a photosynthetic function in this organism retain a partial peptidylglycan wall (Herdman and Denman, 1977). Some plastid functions are not found in prokaryotes and must have been acquired during the process of the host endosymbiont coevolution. The involvement of plastid in plant growth regulator biosynthesis is an example of this type of change. Biosynthetic pathway of host and endosymbiont have also become integrated like some steps of lipid biosynthesis in higher plants that occur on ER membrane while few other steps are localized on plastid envelope.

The integration of plastid and cell function must have involved in the transfer of many genes from the plastid to the nucleus. The fact that all known enzymes in chlorophyll and carotenoid biosynthesis are encoded by nuclear genes under scores this point. The expression of genes involved in photosynthesis in higher plants is regulated developmentally and shows organ, tissue, and cell specificity. Further more plastid populations in higher plants acquire specialized functions such as starch storage (amyloplast) carotenoid (chromoplasts) or lipid (leucoplasts) biosynthesis and storage.

Plastids with specialized functions are usually localized in specific organs or tissues such as tubers (amyloplasts) or epidermal layers of fruit and petals (chromoplasts). This implies that the host now regulates the expression of specific biosynthetic pathways, a control that was once localized in plastids. The discussion above concerning the origin of plastids provides a useful perspective when trying to understand the organization and regulation of genes encoding plastid function in higher plants. First, we should anticipate that plastid gene organization and decoding mechanisms will have prokaryotic features. Second, the transfer of genetic information from the plastid to the nucleus raises several questions. Why were some genes transferred but not others? What signals' are needed to direct nuclear gene products to the plastid compartment?

## Plastid Genome

The plastid genome in higher plants is circular and varies in size from 120 to 217 Kbp. Plastid gene content in higher plants is very constant and many polycistronic transcription units are conserved. A number of genes have been located on the circle and one of the principle features is the presence of two copies of the ribosomal DNA sequences. These sequences are often although not always present on a large inverted repeat. Other genes mapped include those for the large unit of RuBPcase, tRNAs, subunits of ATP synthase, and cytochrome f. The chloroplast genome does not have the capacity to code for all the proteins present in the chloroplast, and it is now known that there is a close co-operation between the nuclear and chloroplast genomes during chloroplast biogenesis.

### *Plastid Genome: Coding Capacity*

The complete sequence of the tobacco plastid DNA and the identification of gene by numerous investigators makes it possible to estimate that the higher plant plastid genome has the coding capacity for 123 genes (Shinozaki et al., 1986). Following table shows genes a table coded by plastid DNA. All these genes fall into 3 categories.

**TABLE 12.1:** Chloroplast genes

| **I. Genes required to decode plastid DNA transcription** | |
|---|---|
| (Genes for the genetic system) | (Gene product) |
| transcriptions rpo, A.B.C., | B,B[1] subunits of RNA |
| **Polymerase** | |

| | |
|---|---|
| Translation r DNA | 16S,23S,4.5S,5S Ribosomal RNA |
| trn (30 genes) | t RNAs |
| rps (12 genes) | 30 S ribosomal protein subunits |
| rpl (7 genes) | 50 S ribosomal protein subunits |
| infA | initiation factor 1, |
| **Putative secretory factor** | |
| tufA | elongation factor Tu (in algae) |
| **II Genes involved in photosynthesis** | |
| RbcL | large subunit of Rubisco |
| psaA, psaB | 80-85 kD PSI chlorophyll |
| **Apoproteins** | |
| psaC | 9kD PSI Fe S protein |
| psbA, psbD | 32-34 kD PSII reaction centre |
| chlorophyll apoproteins | |
| psbB, psbC | 47 kD, 43 kD PSII chlorophyll |
| **Apoproteins** | |
| psbE – psbF | Cytochrome 559 (PSII) |
| psbH | 10 kD PSII phosphoprotein |
| atpA, B,E,F,H,I | ATP synthetase subunits |
| ndh A,B,C,D,E,F,G,H,I,J,K | NADH dehydrogenase |
| 18>100 codons | unidentified |
| '18<100 codons | unidentified |

These genes fall into following three categories:

First category consists of genes that are involved in decoding the plastid genetic information. These genes encode 3 putative subunits of the chloroplast RNA polymerase, ribosomal RNAs, tRNAs, ribosomal proteins and proteins which may function in translation initiation (IFI) and protein targeting.

Genes of second category proteins involved in photosynthesis. The large subunit of Rubisco and several subunits of each proteins complex involved in photosynthetic electron transport (PSI, PSII, cytochrome complex, ATP synthase) are encoded by plastid DNA.

Third group of genes includes sequences for a putative NADH oxidoreductase complex and 18 open reading frames (> 100 codons) of unknown function.

**TABLE 12.2:** Genes for the photosynthetic apparatus

| Genes | Products | Tobacco | Pea | Chlamydomonas | Euglena |
|---|---|---|---|---|---|
| **Ribulose bisphosphate carboxylase (Rubisco)** | | | | | |
| rbcL | Rubisco, large subunit | + | + | + | + |
| rbcS | Rubisco, small subunit | - | - | - | - |
| Photosystem I complex (PS I) | | | | | |
| psaA | PS I, p 700apoprotein A1 | + | + | * | . |
| psaB | PS I, p700 apoprotein A2 | + | + | + | (+) |
| psaC | PS 1, 9 kDa polypeptide | + | + | | |
| **Photosystem II complex (PS II)** | | | | | |
| psbA | PS II, D 1 protein | + | + | ** | . |
| psbB | PS II, 47 kDa polypeptide | + | + | (+) | (+) |
| psbC | PS II, 43 kDa polypeptide | + | + | (+) | . |

| Genes | Products | Tobacco | Pea | Chlamydomonas | Euglena |
|---|---|---|---|---|---|
| psbD | PS II, D2 protein | + | + | + | |
| psbE | PS II, cytochrome $b_{559}$ (8 kDa) | + | + | | |
| psbF | PS II, cytochrome $b_{559}$ (4 kDA) | + | + | | |
| psbG | PS II, G protein | + | | | |
| psbH | PS II, 10 kDa phosphoprotein | + | + | | |
| **Cytochrome b/f complex** | | | | | |
| petA | b/f complex, cytochrome f | + | + | | |
| petB | b/f complex, cytochrome $b_6$ | * | * | | |
| petD | b/f complex, subunit IV | * | * | | |
| **ATP synthetase complex** | | | | | |
| atpA | $H^+$-ATPase, subunit $\alpha$ | + | + | +(+) | |
| atpB | $H^+$-ATPase, subunit $\beta$ | + | + | + | (+) |
| atpE | $H^+$-ATPase, subunit e | + | + | (+) | |
| atpF | $H^+$-ATPase, subunit I | . | . | (+) | |
| atpH | $H^+$-ATPase, subunit III | + | + | (+) | + |
| atpI | $H^+$-ATPase, subunit IV | + | + | | |
| **Putative NADH dehydrogenase components** | | | | | |
| ndhA | NADH dehydrogenase, ND1 | * | | | |
| ndhB | NADH dehydrogenase, ND2 | ** | | | |
| ndhC | NADH dehydrogenase, ND3 | + | | | |
| ndhD | NADH dehydrogenase, ND4 | + | + | | |
| ndhE | NADH dehydrogenase, ND4L | + | | | |
| ndhF | NADH dehydrogenase, ND5 | + | | | |

+, continuous genes present; split gene present; -, gene absent; (+), genes detected by heterologous hybridization. Numbers of + and symbols indicate copies per genome.

The plastid gene is identified for decoding of the plastid genome or -encode proteins involved in photosynthesis. Clearly, these two groups of genes are coupled, because to express genes involved in photosynthesis the plastids transcription and translation apparatus must be synthesized. This later process is no small undertaking which is shown by the fact that mesophyll leaf cells accumulate between $10^6$ and $10^7$ plastids ribosomes. Plastid ribosomes and mRNAs are present at much reduced levels in non photosynthetic plastids such as proplastids, chromoplasts or amyloplasts.

Further more, functions carried out by these non-photosynthetic plastids (plastid replication, carotenoid-biosynthesis, starch accumulation) are not known to require plastid gene expression and most likely involve primarily nuclear genes. The localization of these genes in the nucleus means that differentiation of specialized photosynthetic plastids does not require concomitant build up of plastid transcription and translation capacity. The plastid compartment in non-photosynthetic cells can therefore functions as other non-DNA containing organelles such as glyoxysomes or microbodies. Perhaps the advantages of this situations contributed to the present distribution of genes in the nucleus and plastid.

It has been suggested that some genes have been retained in the plastid because the proteins they encode would be difficult to transport through the chloroplast envelope. But this idea, although, is genious does not explain why some ribosomal proteins remain encoded by plastid DNA but not others of similar size and solubility. Further it has been shown that when psb A, a plastid gene, is moved to the nucleus of a transgenic plant and provided with appropriate expression signals, it will direct a functional psb A gene product into plastids (Cheung et.al., 1988). The data suggest that it is possible to the move that plastid genes are migrated to the nucleus without loss of function. Further support for this idea comes from the observation that although EF-Tu is encoded by a plastid gene in *Euglena*, in higher plants it is a nuclear gene. Similarly, rpoA is a nuclear gene in *Geranium* but a plastid gene in other higher plants. This suggests that gene transfer is still occurring, although slowly, possibly because gene transfer disrupts complex regulatory circuits which now coordinate plastid and nuclear gene expression.

## DNA Copy Number and Localization

Multiple copies of the plastid genome are found in each plastid. The amount of DNA per plastid varies with the stage of leaf and chloroplasts development. The polyploid nature of the plastid genome is even more striking when one considers leaf cells.

In pea, barley and spinach each leaf cell contains 9000-13000 copies of plastid DNA which are dispersed in 40 to 120 plastids. This means that rbcL, a single copy plastid gene which encodes the large subunit of Rubisco, is present in about 10,000 copies in a mesophyll cell.

The large copy number difference between some plastid genes and their counterparts located in the nucleus raise the question that why there are so many copies of plastid DNA per cell. Bendich (1987) proposes that the plastid genome is amplified to provide more copies of r-DNA there by stimulating r-RNA and ribosomes synthesis. Plastid DNA is associated with the chloroplast envelop or thylakoid membranes in aggregates of 5 to 20 DNA molecules. The mechanism of plastid DNA compaction is unknown.

## Plastid Genome: Decoding

**Transcription of Plastid Gene:** The genetic system in the chloroplast is essentially prokaryotic in nature, which has given rise to much discussion as to the possible prokaryotic origin of plastids. Control sequences at the 5' end of the gene which are involved in initiating transcription are very similar to the *pribnow* box and the '-35' region characteristic of bacterial genes. In addition, chloroplast genes do not usually contain intervening sequences characteristic of many eukaryotic genes, although some introns have been found, notably in tRNA genes of tobacco and maize. The mRNA produced from chloroplast genes is not usually polyadenylated, although short sequences of upto 20 residues have been reported, and, of course, no transport is required as the m RNA is produced in the same compartment as the ribosome on which it will be translated. There is no evidence that chloroplast m-RNA is transported out of the chloroplasts and translated on cytoplasmic ribosomes.

When *E. coli* is transformed with the plastid genes rbcL and psbA, transcription is initiated in *E.coli* at the site similar or identical to one that used by the chloroplast RNA polymerase. This shows that DNA sequences elements which direct transcription initiation of plastid genes (TTGaca and TAtaaT) similar to prokaryotic promoter elements. Furthermore, as in Bacteria, there promoter elements precede rRNA, tRNA and protein coding genes.

The only exceptions reported thus far are two tRNA genes ($trnS_1$, $trnR_1$) which lack there sequences and may use internal promoter elements similar to eukaryotic tRNA genes. Transcription of plastid genes by *E. coli* RNA polymerase, however is not identical to that found with the chloroplast RNA polymerase. The activity and initiation accuracy of the chloroplast enzyme are generally enhanced by supercoiled templates (Lam and Chua, 1987). Chloroplast RNA polymerase preparation contain from 5-14 subunits. At least 4 of these proteins are homologous to E. coli RNA polymerase subunits (B, $B^1$, ,). Sequence analysis indicates that plastid DNA encodes proteins homologous to the B-$B^1$ and subunits of *E coli* RNA polymerase (on the rpoA, B and C genes) (Shinozaki et al., 1986). It is likely that the chloroplast RNA polymerase contains at least 5 subunits (B, $B^1$, , , y) the presence of proteins homologous to *nus* A, nus B and *rho* and *tau* have not yet been reported.

Two lines of evidence indicate that the chloroplast RNA polymerase population may be heterogenous. First, it was reported that all subunits of an isolated RNA polymerase preparation were synthesized on cytoplasmic ribosomes. Second transcriptionally active complexes of DNA and RNA polymerase have been isolated which show preferential transcription of RNA.

Transcription termination of plastid genes has not been studied extensively. In *E. coli,* factor independent transcription terminators typically contain a G-C rich region of dyad symmetry followed by an AT rich stretch that often contains a run of thymidines. *In vitro* experiments indicate that DNA sequences with dyad symmetry located at the 3 end of rbcL do not cause transcription termination. Transcription was terminated with low efficiency by 3' regions of spinach psbA, rrnB and pet D. In contrast, trn S and trn H caused termination with high efficiency (80%) although these genes are not followed by regions of dyad symmetry. These results suggest that the stem loop structures found at RNA 3' ends may not play a significant role in factor independent transcription termination.

Chloroplast of higher plants contain gyrase and topoisomerase I activity which could alter the superhelicity of plastid DNA *in vivo.* Chloroplast ribosomes also resemble those from prokaryotes, being 70S particle consisting of 30S and 50S subunits, and ribosomes from *E. coli* and plant chloroplasts are immunologically similar. The tRNA chloroplast population is distinctly different from that in the cytoplasm, as are the aminoacylating enzymes. The prokaryotic nature of the protein synthetic machinery is also illustrated by its sensitivity to inhibitors of prokaryotic gene expression such as chloramphenicol. This antibiotic does not inhibit eukaryotic protein synthesis.

## Plastid Genome: RNA Processing

Transcripts containing unprocessed tRNA sequences are produced from the individual tRNA transpcription unit which encode proteins. Chloroplast genomes contain 20-40 tRNA genes. Unlike mitochondrial chloroplast genomes believed to encode all the tRNA species used in chloroplast protein synthesis. The tRNA gene (trn) maps were constructed by hybridizing purified individual tRNAs to chloroplast DNA fragments that hybridised to total chloroplast tRNAs. These have been sequenced, and thirty, different tRNAs genes were found. These thirty tRNA genes are probably all of the t RNA genes encoded in the chloroplast genome of tobacco and perhaps of other higher plants. The liverwort genome contains an additional tRNA Arg (CCG) sequence (Sugiura, 1992). The production of mature tRNAs involves 5' and 3' endonucleolytic cleavage of primary transcripts, addition of CCA to the 3' end of the processed RNA and base modification.

### Plastid Genome: Introns

Introns are DNA sequences within genes which disrupt protein coding regions termed exons. The presence of introns in chloroplast tRNA genes was first demonstrated in maize trn I and trn A located in the long spacer separating the 16S and 23S rDNAs. Six chloroplast tRNA genes from land plants are now known to harbor long single introns (0.5 to 2.5 kb). *Euglena* plastid protein genes contains numerous introns whereas tRNAs contain none. One plastid gene intron has been identified as a group I intron.

### Nuclear Genes Encoding Plastid Proteins

The majority of plastid localized proteins are encoded by nuclear genes. These genes are transcribed by RNA polymerase II and the resulting transcripts are spliced, capped and polyadenylated in the nucleus. The mRNAs then are translated by 80S ribosomes in the cytoplasm to produce proteins which can be transported into plastids post translationally. Following uptake into the chloroplast, the proteins are assembled with cofactors and other proteins to form functional complexes. Out of sixty different protein components of chloroplast ribosomes, one third of which are thought to be encoded by chloroplast DNA. An intron with in a potential chloroplast ribosomal protein gene was first found in *Nicotiana debneyi trpl* 2 but this intron is absent in spinach and some related dicots.

## PHOTOSYNTHETIC PIGMENTS

At global level, it has been estimated that each year the total chlorophyll production across the planet exceed $10^9$ tonnes with – 75 per cent from terrestrial plants, the reminder from aquatic organisms, largely marine phytoplankton. Production of carotenoids, the second most abundant plant pigment, is probably of the order of about $2 \times 10^8$ tonnes (Hendry, 1999). The distribution of the various chlorophylls and bacteriochlorophylls is shown in following table:

**TABLE 2.3:** The distribution of chlorophylls and bacteriochlorophylls

| Pigment | Organism | Light absorbed |
|---|---|---|
| Chlorophyll *a* | All plants, algae and the photosynthetic bacteria Cyanophyta and Prochlorophyta | Red and Blue-violet |
| Chlorophyll *b* | All plants, the algal Chlorophyta and Euglenophyta and bacterial Prochlorophyta | Red and Blue violet |
| Chlorophyll *c* | Algal Phaeophyta (brown algae), Pyrrophyta (dinoflagellates), Bacillariophyta (diatoms), Chrysophyta, Prasinophyta, Cryptophyta | Red and Blue violet |
| Chlorophyll *d* | Some Rhodophyta (red algae) and Chrysophyta | Red and Blue-violet |
| Chlorophyll *e* | Algal Xanthophyta | Near-red and Blue-violet |
| Bacteriochlorophyll *a* and *b* | Purple bacteria including Chromatiaceae and Rhodospirillaceae | Near-red and Blue-violet |
| Bacteriochlorophyll *c*, *d* and *e* | Green and brown sulphur bacteria – Chlorobiaceae and Chloropflexaceae | |
| Bacterioviridin | Green sulphur bacteria | Near-red and Blue-violet |
| Phycocyanin | Blue-green, red algae | Orange red |
| Phycoerythrin | Red, blue-green algae | Green |

| Pigment | Organism | Light absorbed |
|---|---|---|
| Carotenoids | Most plants, bacteria | Blue, Blue-green |

**TABLE 2.4:** Chemical composition of intact spinach chloroplasts

| Component | Percentage Chloroplast (Dry Weight) |
|---|---|
| **Total Protein** | 60 |
| Water insoluble protein | 27 |
| Water soluble protein | 33 |
| **Total lipids** | 20 |
| Chlorophylls | 4.3 |
| Carotenoids | 0.9 |
| Colourless lipids | 14.8 |
| RNA | 1.7-3.5 |
| DNA | ? |
| **Free Amino acids** | 2.5 |
| Orthophosphatase | 0.7 |
| Ions ($K^+$, $Na^+$, $Mg^{2+}$, $Ca^{2+}$, ($Cl^-$)* | 3.6 |
| Polysaccharides** | 0.6 |
| Hexosamine | 0.2 |
| Starch | variable |

*Data of *Beta vulgaris* chloroplast, ** Data of *Allium porrum* chloroplast.

**TABLE 2.5:** Lipid composition of spinach chloroplasts (based on 100 mol chlorophyll)

| Lipid | Mol. |
|---|---|
| **Chlorophylls** | 100 |
| Chlorophyll *a* | 70 |
| Chlorophyll *b* | 30 |
| **Carotenoids** | 39 |
| ß-carotene | 9 |
| Lutin | 14 |
| Viloxanthin | 10 |
| Neoxanthin | 4 |
| Antheraxanthin | 2 |
| **Quniones** | 11 |
| Plastoquinone | 4 |
| Lipid | Mol. |
| $\alpha$-tocopherol | 0.5 |
| Vitamin $K_1$ | 0.5 |
| Tocoquinone | 0.5 |
| Glycolipids | 447 |
| Monoglactosyldiglyceride | 270 |
| Digalactosydiglyceride | 154 |
| Sulfolipid | 23 |
| Phospholipids | 43 |
| Phosphatidylglycerine | 29 |
| Phosphatididylethanolamine | 2 |
| Phosphatidylinosite | 2 |

Chlorophyll a is found in all photosynthetic eukaryotes. The restricted distribution of the other chlorophyll (b, c, d, or e) may have evolutionary significance.

Most biological pigments are relatively large molecules with, typically, molecular masses ranging from about 200 (quinines), 300 (flavonoids), 400 (betalains) and 500 (carotenoid) to 800 (chlorophylls). Most pigments are composed, at least in part of unsaturated hydrocarbons. Many are ring (cyclic) structures. Of the 17 or so elements found in pigments biological materials, just four is: H, C, N and O predominate.

## PROPERTIES OF CHLOROPHYLL MOLECULE

The empirical formula of the chlorophyll molecule is $C_{55}H_{72}O_5N_4Mg$. The molecule is made up of five small rings, together forming a larger ring. In the centre of the large ring there is a Magnesium atom. Together they form a flat square measuring 1.5 × 12.5 nm. The square is equipped with a long *tail* of carbon atoms which can be thought to act as a handle or fixture which keeps the molecule properly positioned in the plant.

There are many double bonds between the atoms in the ring system. It is possible to circle the large ring in such a way that every second bond encountered is a double bond. Such a system of bonds is termed a *closed conjugated system.* It is this that provides the molecule with two important properties:

1. great light-absorbing power and
2. great stability power.

The tail, the magnesium atom and some peripheral atoms may break or loose if the molecule is denatured with heat, acids, etc., but the ring system is intact even in old fossils such as petroleum. A similar ring system as in chlorophyll is present in our red blood pigment, haemoglobin and in the cytochromes.

In addition to the pyrrole rings, *a-b* chlorophyll has a cyclopentanone ring with a methyl ester. A phytol ester is linked by a propionic acid with pyrrole ring D. Chlorophyll *a* is having formyl group at C atom 7 instead of a methyl group. Other chlorophylls, such as chlorophyll C*1* and $C_2$ are found in brown algae and chlorophyll *d* in red algae.

Photoautotrophic bacteria possess bacteriochlorophyll *a* which differs from chlorophyll in two features: it is hydrated in ring B and carries a vinyl group instead of the acetyl group on ring A. The green photobacterium, *Chlorobium* has been found to contain the bacteriochlorophylls *c*, *d* and *e* which have formerly been called *Chlorobium* chlorophylls.

### Chlorophyll Synthesis

The synthesis of chlorophylls and iron porphyrins are synthesized within the living cell has not, as yet been conclusively shown.

However, studies by many investigators on the metabolism of *heme and chlorophyll and on the biosynthesis of porphyrins have uncovered most of the steps involved in the synthesis of these compounds. There is a general agreement that succinyl CoA, a Krebs' cycle intermediate, and the amino acid glycine initiate the biosynthetic pathway that leads to chlorophyll formation. The condensation of these two compounds leads to the formation of the unstable a-amino-a-ketoadipic acid, which on decarboxylation yields c-aminolevulinic acid.

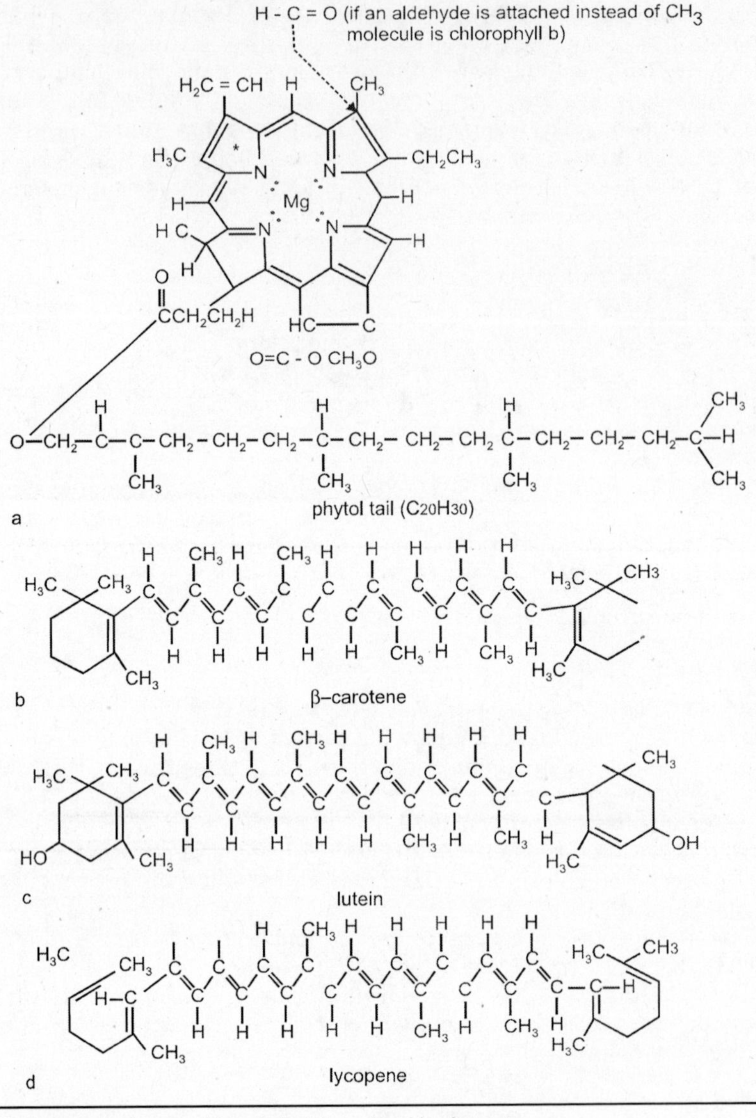
H - C = O (if an aldehyde is attached instead of CH3
molecule is chlorophyll b)
H2C = CH
CH3
H3C
CH2CH3
Mg
CCH2CH2H
HC—C
O=C - O CH3O
phytol tail (C20H30)
a
β-carotene
b
lutein
c
lycopene
d

**FIG. 12.3:** Structures of some chlorophyll and carotenoids pigments.

This reaction requires the presence of the cofactor pyridoxal phosphate and is catalyzed by the enzyme aminolevulinic acid synthetase. There is strong evidence that the synthesis of aminolevulinic acid is mediated by light.

## Physical Properties of Chlorophylls

The chlorophyll is a dark green waxy solid; but chlorophyll *a* is bluish black while chlorophyll *b* is dark-green solid. Both chlorophylls are soluble in organic solvents giving green solution; chlorophyll *a* melts between 117-129°C and chlorophyll *b* between 120-130°C.

The molecular formula of chlorophyll *b* is $C_{55}H_{77}ON_4O_6Mg$, that contains one oxygen atom more and two hydrogen atoms less than the chlorophyll *a* ($C_{55}H_{72}N_4O_5Mg$). The difference is due to the presence of -CHO group on position-3 in place of methyl group

Ordinary pigments preferentially absorb visible light of a certain spectral region and, upon de-excitation release the absorbed energy as heat and, possibly, fluorescence. In contrast, a photopigment can be defined as a chemical compound, capable of light absorption and subsequent conversion of excitation quanta into a form of energy to be used in some physical or chemical process. In the following description only photopigments of biological interest will be considered.

Apart from heat production and fluorescence emission, two phenomena may happen upon excitation: (1) converted energy is stored and used later on for driving energy requiring processes at a given moment, or (2) this energy is instantaneously used for chemical or physical processes. In case (2) there are three possibilities:

a) energy conversion is used for triggering a reaction,
b) this conversion results in the transformation of the photopigment into a catalyst and
c) a product suitable for *use* in chemical reactions is generated by the energy conversion.

Light quantum can be absorbed only if its energy content equals the energy difference between the ground state and one of the excitation levels of the pigment. If absorption occurs, an electron is raised to the level corresponding to its energy. This electron is one of the partners of an electron pair in the ground state. The lowest level is of special interest as regards excited states. Its lifetime is usually at least 103 times longer than that the higher ones. Thus the probability for release of the energy in a form other than heat is much increased as compared with that for higher excitation levels. This phenomenon can be easily demonstrated, e.g. by exciting chlorophyll with red, yellow or blue light. In all cases the fluorescence is the same: the emission spectrum is a mirror image of the red absorption band representing the lowest excited state. This shows that the lowest excitation level can be filled in two ways; either by raising an electron from the ground state to this level, or by electrons *tumbling down* from higher energy states to the lowest excitation one. The energy difference between these levels is released by internal conversion as heat. The lowest excited state is therefore of primary interest with regard to the function of a photo-pigment.

So far, it is assumed that the law of spin conservation holds for the transfer of an electron from a low-energy orbit to one of higher energy. If, however, for some reason the spin of the excited electron is reversed, the triplet state, a metastable state, is formed. This has a slightly lower energy than the above-mentioned lowest excited singlet state. Owing to spin reversal the direct transition from the triplet to the ground state is forbidden, i.e. the probability of its occurrence is small. Thermal quanta may reverse the spin again and raise the electron to the lowest excited singlet state, S(1), whence it can return to the ground state. Population of the triplet state means that the lifetime of the excited molecule is lengthened considerably, and thus the probability of converting the absorbed quanta to chemical energy is enhanced. Moreover, this triplet state may function as a new ground state and can be excited to higher-energy triplet levels.

## PHOTOPIGMENTS: FUNCTIONS

Upon de-excitation, conversion of the electronic energy can occur in various ways. Leaving aside internal conversion, which occurs concomitantly with the other possibilities and results in the production of heat, the following results of energy conversion can be envisaged:

**Fluorescence**, emitted by the return of an electron from the $S_1$ level to the ground state

a) **Delayed fluorescence:** This kind of the fluorescence is emitted if an electron is raised, with concomitant reversal of its spin momentum, from the triplet into the $S_1$ state by thermal energy, and returns from the latter level to the ground state. In this case the emission spectrum is identical with that of fluorescence.

b) **Phosphorescence:** This light emission occurs in the case of the forbidden transition from the triplet state to the ground state with reversal of the spin momentum. As the triplet state is of lower energy than the $S_1$ level, the emission consists of quanta of energy lower than those emitted when the $S_1$ state is emptied. Therefore, the phosphorescence spectrum is shifted to longer wavelengths as compared with that of fluorescence.

### Type of Biological Photoprocess

The major biological processes in which photopigments operate are:

**Photosynthesis:** Light energy is converted into chemical energy to be used for metabolism.

**Vision:** Light is used for providing information for the organism.

**Photomotive processes:** Light causes displacement in the following ways:

**Phototaxis:** Light induces motion or change of motion of a free organism or a cellular constituent. Two kinds of phototaxis occur:

**Topophototaxis:** The direction of motion is determined by the direction of the light beam. If the motion proceeds either towards or away from the light source the taxis is called positive and negative respectively. For animals, Kuhn (2) distinguished three kinds of topophototaxis which will not be further considered.

**Phobophototaxis:** The direction of motion is not determined by the direction of the light beam, but at the boundary of two light fields of different intensities-a shock reaction of the object occurs. This prevents it from crossing the boundary. If it occurs at a transition from strong to dim light, the taxis are called positive, whereas it is called negative at a transition in the reverse sense.

**Photokinesis:** The motility of an organism is changed by light. Photokinesis is indicated as positive or negative, if the motility is increased or decreased in the light respectively.

**Photodinesis:** The effect of light on protoplasmic streaming. Light may affect either the rate of streaming, or the amount of protoplasma taking part in it. The effect may be temporary or tonic. In case of acceleration or retardation, photodinesis is called positive or negative respectively.

**Phototropism:** The light-induced movement of a part of the organism resulting in bending towards the light source (positive), away from it (negative), or occupation of a certain angle to the light beam (diaphototropism), or in a bending not related to the light gradient, but to the structure of the organelle (photonasty). In most cases, the latter process is due to light-induced changes in turgidity, whereas the former ones are based on photic growth reactions.

**Photomorphogenesis:** Light also control certain formative proceses. It may do so either by itself, or in combination with endogenous rhythms, as occur in photoperiodicity. The reaction may be restricted to activation of secretary activities affecting the behavioural pattern of animals, or it may affect the entire organism as in seed germination.

Since it may not be one of the major photoprocesses, photorespiration, though it differs from dark respiration, is omitted from consideration. The same is true of photodynamic effects. Secondary photo-effects, such as light-induced changes of membrane properties are not considered either.

## PHOTOPIGMENTS: PROPERTIES

### Chlorophylls

Chlorophylls are esters of dibasic acids, the chlorophyllins.

These photopigments are active in (a) photosynthesis either by collecting light quanta or by the generation of free electrons leading to the production of reducing compounds and ATP, (b) phototaxis, (c) photokinesis.

#### *Chlorophyll a*

It occurs in all green plant's its formula is $C_{55}H_{72}O_5N_4Mg$. Chlorophyll a, like other chlorophylls, contains four pyrrole nuclei (I-IV), arranged in prophyrin ring structure. In addition to these nuclei an isocyclic, or cyclopentanone, ring (V) occurs adjoining to nucleus III. A Magnesium atoms is located in the centre of the porphyrin ring. In most cases, including chlorophyll a, the esterifying alcohols are methanol and phytol.

It is emphasized that the indication of double and single bonds in the Figure does not depict the actual situation. In reality, like other conjugated bond systems, chlorophyll is a resonance hybrid. According to there are major contributions of three resonating structures in this particular case.

A key position in determining various properties of this photopigment is occupied by the central Magnesium atom. Its presence is required for the oxidative process of allomerization. This reaction occurs upon storage of alcoholic solutions of chlorophyll under air. Allomerization consumes molecular oxygen in the molar ratio 1/1 with respect to chlorophyll. The presence of the central magnesium atom is indispensible for this process. The oxidation occurs in the cyclopentanone ring (V) at C-10 a follows: $H\text{-}C^{10} + O_2\text{-}HO\text{-}O\text{-}C^{10}$-, given that allomerization yields at least three products. This conclusion is based on the absence of the Molisch phase test with allomerized chlorophyll. It may be mentioned that this process is believed to be based upon a reversible shift of the hydrogen atom at C-10 to C-9. The sum total of these observation, namely (a) the absence of Mg, and of allomerization, (2) the fact that allomerization means fixation of the H atom at C-10, (3) that it also prevents the phase test, and (4) that this test implies a reversible shift of the H atom in question, suggests that the central magnesium atom is responsible for the mobility of the H atom at C-10.

Still, hydrogen exchange proceeds only very slowly. However, with the proton magnetic resonance technique, showed that hydrogen exchange occurs at C-8 exclusively under their experimental conditions.

### *Bacteriochlorophyll a*

This is the main photosynthetic pigment in both sulphur and non-sulphur purple bacteria. Its formula is $C_{55}H_{74}O_6N_4$ Mg. The structure differs from that of chlorophyll **a** in that C-3 and C-4 are hydrogenated, whereas the side group of C-2 consists of $H_3C$-C=O. Obtained evidence that the primary intermediate formed upon excitation is the triplet state of bacteriochlorophyll complexed with the reaction centre.

The following pigments will be mentioned only briefly.

### *Bacteriochlorophyll b*

This pigment was described by it occurs in a *Rhodopseudomonas* species as the main pigment.

### *Chlorophyll-770*

Chlorophyll-770 is found in green sulphur bacteria. It turned out to be identical with bacteriochlorophyll. It is suggested that this chlorophyll rather than the chlorobium chlorophylls functions as a main pigment.

### *Chlorobium chlorophyll-660*

Chlorobium chlorophyll-660 was previously called bacterioviridin. It is found in most of the green sulphur bacteria. It contains neither a methoxyl group at C-10, not phytol. Determined the magnesium content to be 2.92% on a dry weight basis.

### *Chlorobium chlorophyll-650*

Chlorobium chlorophyll-650 occurs less commonly than the previous pigment. The methoxyl group at C-10 and phytol are absent. Its magnesium content are 2.59% on a dry weight basis.

In green plants, including algae, chlorophyll a is usually accompanied by auxiliary chlorophylls. Such chlorophylls are:

### *Chlorophyll b*

Chlorophyll b, $C_{55}H_{70}O_6N_4Mg$, its structure is like that of chlorophyll **a**, except for a replacement of the methyl group at C-3 by a formyl group.

### *Chlorophyll c*

It is synonyms: chlorophyll g and chlorofucine and occurs in diatoms and other brown algae. It does not contain a phytol group two kinds, namely, chlorophyll $c_1$ and $c_2$ have been observed.

### *Chlorophyll d*

It is found in most of the red algae in rather small amount. It was suggested that this pigment is 2-desvinyl-2-formyl-chlorophyll a.

### *Chlorophyll e*

It was observed in yellow-green algae in very small amounts.

Precursors of chlorophylls might also be considered photopigments. Such may be the case with protochlorophyll, as this precursor needs light for conversion to chlorophyll by hydrogenation at C-7 and C-8.

In vivo chlorophylls are complexed with lipoprotein carriers. These complexes will be dealt with elsewhere.

Absorption spectra of chlorophylls are characterized by four independent electronic transitions responsible for bands between 350 and 900 nm.

## Carotenoids

These compounds act as photo pigments; in photosynthesis as auxiliary pigments; the absorbed energy is transferred to chlorophyll. It has been suggested that part of the carotenoids, the xanthophylls, act as oxygen transporters in photosynthesis. However, such function seems highly improbable. Their possible protective function in preventing photodestruction of chlorophyll will not be considered. Carotenoids function possibly as photoreceptors in phototaxis, photokinesis, and photomorphogenesis. The structure of xanthophyll lutein carotenoid provides an example of carotenoid structure. There are two classes of carotenoid: polyene hydrocarbons, called carotenes, and their oxygen derivatives, the xanthophylls. All are lipochromes; they are soluble in lipids and fats. The pigments are composed of isoprene units. At one or both ends of the isoprene chain carotenoids carry an ionone ring which may be closed b or open p Carotenoids usually occur in the all-trans configuration, whereas **is** isomers are encountered less frequently. Carotenoids participating in photosynthesis are as follows:

### *Carotenes*

α-carotene, $C_{40}H_{56}$, in green plants and algae, β-carotene, $C_{40}H_{56}$, in green plants and algae, γ-carotene, $C_{40}H_{56}$, in green plants and algae, and lycopene, $C_{40}$H56, in purple bacteria are present.

### *Xanthophylls*

Lutein, $C_{40}H_{56}O_2$, in green plants and algae, violaxanthin, $C_{40}H_{56}O_4$, in green plants and algae, neoxanthin, $C_{40}H_{56}O_4$, in green plants and algae, cryptoxanthin, $C_{40}H_{56}O$, in green plants and algae, zeaxanthin, $C_{40}H_{56}O_2$, in green plants and algae, fucoxanthin, in diatoms and other brown algae, (the exact structure is not yet established), spirilloxanthin, $C_{40}H_{54}$ $(OCH_3)_2$, in purple bacterial and rhodopin, $C_{40}H_{58}O$ ir C40$H_{56}O$, occurring in purple bacteria.

The absorption spectra of these pigments generally show two or three bands located at wavelengths in between 350 and 550 nm. Carotenoids occur in complexes with lipoproteins in the vivo state.

## Phycobiliproteins

These pigments act as photoreceptors in connection with (a) photosynthesis as auxiliary pigments by transfer of their excitation energy to chlorophyll, (b) phototaxis, (c) photokinesis (to a minor degree), and (d) photomorphogenesis as regards the special biliprotein phytochrome, capable of phototransformation. As their names indicate, these pigments are proteins with a chromophoric group. This consists of four pyrrole nuclei linked together by C atoms in a linear or open-cyclic arrangement. There are two major kinds of phycobiliproteins: the blue phycocyanins and the red phycoerythrins. Their chromophoric groups are phycocyanobilins and phycoerythrobilins, respectively. Both pigments coexist in blue-green algae, the Cyanophyceae, and red ones, the Rhodophyceae, but in considerably different proportions. Since phycocyanin as well as phycoerythrin from either blue-green or red algae show different absorption properties, the prefixes C or R are used to indicate their origin. Phycobilins are also observed in Cryptomonads.

Phycobiliproteins are complexes which leach readily out of damaged cells. A complex unit is known to be composed of two or more subunits. A detailed analysis of C-phycocyanin. This phycocyanin from *Oscillatoria agardhii* consists of two polypeptide chains, each of which carries at least one chromophoric group. The chains, of which the amino acid compositions are given below, are connected by a disulphide bridge. Assuming that each chain carries a single chromophoric group, the molecular weights for the chains were found to be 12,100 and 14,100 respectively.

The major phycobiliproteins are: C-phycocyanin, C-phycoerythrin, R-phycocyanin and R-phycoerythrin

B-phycoerythrin, occur in the red alga *Porphyra naiadum* allophycocyanin, and is also present in many red and blue-green algae, whereas phytochrome occurs in plants, including algae. It is active in photomorphogenesis and phototaxis.

## FLAVOPROTEINS

The primary reaction upon excitation by light consists of photoreduction. Probably flavoproteins are the photopigments of phototropism. According to riboflavin is most likely to act as the photoreceptor for the high-energy phototaxis of the *Vaucheria* chloroplast, although there is no conclusive evidence as yet for the photopigment function of flavins in this respect. Flavoproteins may also be responsible for the photosensitivity of fungi. As discussed it is less certain that these pigments function as photoreceptors for photomotion in microorganisms.

The difficulty in assigning the role of biological photopigment to flavoproteins unambiguously is due to the fact that their absorption spectra showing bands around 450 nm, overlap those of the carotenoids. Action spectra, therefore, are hard to interpret. It thus remains a challenge to try to obtain conclusive evidence about the proposed functions of flavoproteins in photobiology.

# CHAPTER 13

# Photosystems in Chloroplast

Significant advances have been made towards the understanding of the molecular structure of the thylakoid membrane, i.e. the proteins and lipids and their orientation, the polypeptides associated with light-harvesting pigment-protein complexes and their distribution in the core and peripheral parts of the antennae, etc. Membrane complexes can be isolated from higher plant chloroplasts, and their photochemical and biochemical properties studied *in vitro*. There is strong evidence of a lateral heterogeneity in the distribution of these complexes between stacked and unstacked regions of the thylakoid membrane. Photosystems I and II are now found to exist as two structurally (and even functionally) independent types designated as $PSI_a$, and $PSI_B$, and $PSII_a$ and $PSII_B$. $PSII_\alpha$ which are located in the grana membranes and $PSII_B$ is exclusively located in the stroma-exposed thylakoids. The light-harvesting capacity of $PSII_\alpha$ is greater (because of the presence of more antenna chlorophylls) than that of $PSII_B$. Little is understood of the mechanism of membrane stacking into grana and the biochemical and physiological significance of this phenomenon, although we know that the degree of stacking is influenced by the cation concentration and by the phosphorylation of light-harvesting proteins.

## EMERSON ENHANCEMENT EFFECT

In photosynthesis two photochemical systems are associated with two pigment system (i.e., pigment system I and Pigment system II). Chlorophyll *a* is found in two forms: (i) with an absorption maxima at short red wave length (673 nm - Chl *a* 673) and (ii) at long red wave length (683 nm - Chl *a* 683). Another form of chlorophyll *a* has also been reported which has an absorption maxima at 700 nm (called p 700), present in much smaller amounts than that of chl *a* 673 and chl *a* 683. PS I contains long wave lengths absorbing forms (chl a 683 + p 700 in addition to carotenoids) for collection of energy; while PS II contains chlorophyll b in addition to the short red wave length form of chl *a* 673 (in higher plants and green algae). In blue green and red algae phycobilins and other chl *a* are present as energy collecting centre (PS II). Emerson effect shows that when 653 nm and 700 nm wave lengths of radiations are given, the relative rate of photosynthesis is increased in comparison when 653 nm and 700 nm wave length of light is given successively, relative rate of photosynthesis is reduced.

## RED DROP EFFECT

Robert Emerson while working on *Chlorella* observed that wavelengths beyond 680 nm decreased rate of photosynthesis even though chlorophyll *a* could absorb them. This decrease in photosynthetic efficiency was said to be **red drop** effect. It is proposed that at this wavelength only one photosystem remains in operation.

The principal photosynthetic organ of higher plants is the foliage leaf. The leaf lamina is composed of three different tissues, epidermis, mesophyll cells and vascular cell. The mesophyll cells constitute the major photosynthetic tissue of the leaf since this tissue contains most of the chloroplast, the organelles in which photosynthetic processes proceed. The vascular cells and epidermis do not contain chloroplasts. Nevertheless, the cells of the epidermis are of importance for the process of photosynthesis because this outer layer of the leaf contains the stomatal complexes regulate the water balance and the photosynthetically required gas exchanges between the intercellular spaces of the mesophyll cells and the environment outside the leaf.

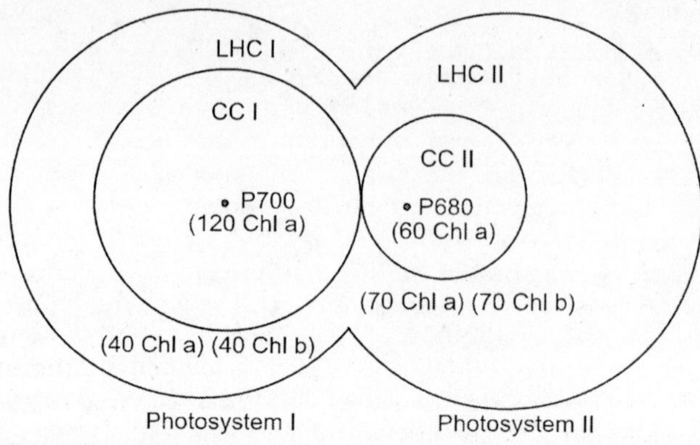


**FIG. 13.1:** Presentation of granum thylakoids, light harvesting center (LHC) with position of various components of the light reactions.

## PHOTOSYSTEMS

The light reactions in thylakoids are driven primarily by two photosystems namely Photosystem I (PS I) and Photosystem II (PS II) which are interdependent and are linked by membrane bound cyt b*f* complex and other mobile electron transport carriers. Both PS I and PS II consist of characteristic polypeptides of reaction center core complex, light harvesting antenna system and specific metal ions. PS II commissions the first step in the primary process of the photosynthesis. The light harvesting system of PS II absorbs photon, channels it to P680, the photon trap, where the photosystem converts the light energy to the chemical energy by photooxidation of water to oxygen and reduction of quinone pool to quinol, two major events of thylakoid membrane. Subsequently, the energy flows to produce ATP and NADPH via cyt bf complex and PS I. These substances are utilised in the Calvin cycle to fix carbon dioxide into carbohydrate.

Compared to PS I, PS II has been investigated in more details during last two decades. The structure of PS II is not only complex, but also extremely delicate. Secondly, it is one of the important players during electron transfer reactions of thylakoids because of its ability to split water molecule, the ultimate source of electrons. The photosystem is also capable of forming reaction centre oxidants with high oxidising potential (Barber, 1995). Because of the availability of oxygen at the vicinity, presence of pigments and reaction centers with ability of forming strong oxidants, PS II is extremely sensitive to all kinds of changes of its surrounding environment and is capable of creating oxidative stress in green leaves.

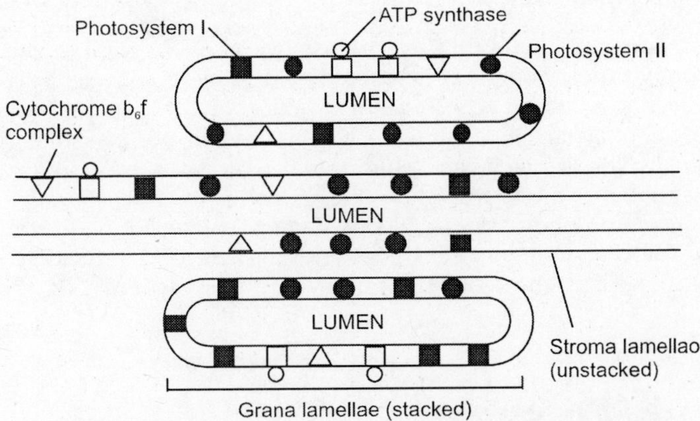


**FIG. 13.2:** Organization of the protein complexes of the thylakoid membrane. Photosystem II is located predominantly in the stacked regions of the thylakoid membrane; photosystem I and ATP synthase are found in the unstacked regions protruding in to the stroma. Cytochrome b₆f complexes are evenly distributed. This lateral separation of the two photosystems requires that electrons and protons produced by photosystem II be transported a considerable distance before they can be acted on by photosystem I and the ATP-coupling enzyme.

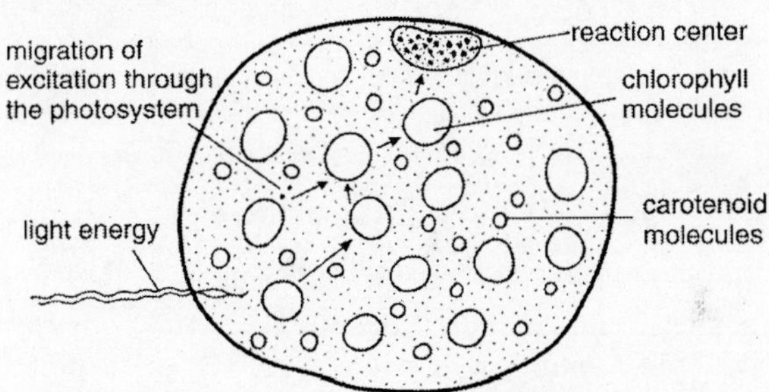


**FIG. 13.3:** Diagrammatic presentation of a photosystem showing transfer of energy from antenna molecules to reaction center.

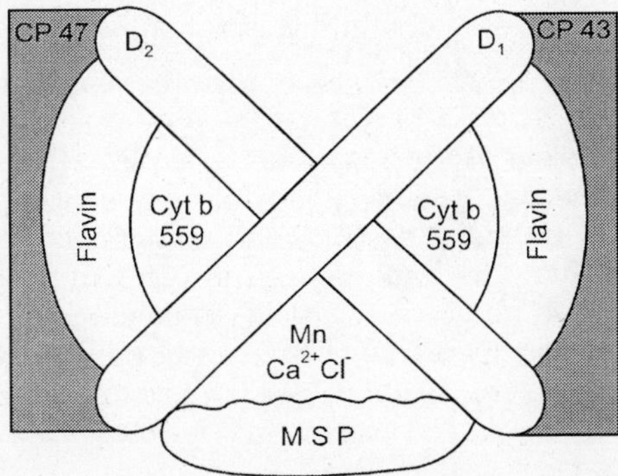


**FIG. 13.4:** A diagrammatic presentation of PS II showing its structural organization.

## Light Harvesting Complex (LHC)

Each photosystem (PS I and PS II) contains several chlorophyll molecules; yet only one chlorophyll per complex is capable of undergoing light-driven electron transfer. These specialized *reaction centre* chlorophyll *a* molecules are termed as $P_{680}$ (for pigment absorbing light at 680 nm) and in photosystem II and $P_{700}$ in Photosystem I. These specialized chlorophyll molecules are bound to an integral membrane protein which has molecular weight of 110,000 and contains the reaction centre for photosystem I. The molecular weight of protein-chlorophyll complex of the reaction centre for photosystem II is 47,000. Associated with each photosystem is a structure the **light harvesting complex** (LHC), or **antennae complex**.

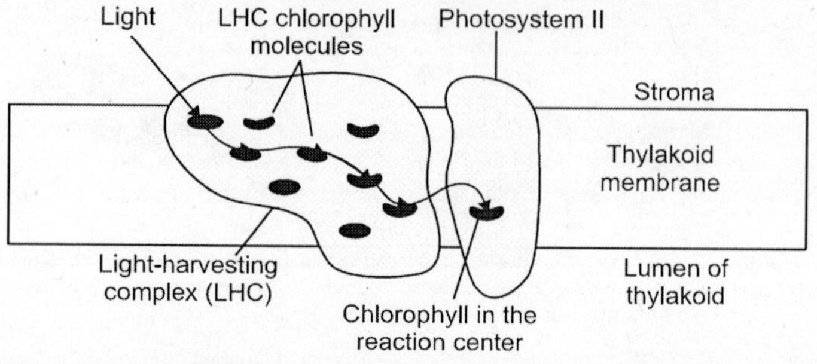


**FIG. 13.5:** Light harvesting complex (LHCs) are associated with each of the two photosystem complexes contain dozen of chlorophyll molecules. Any of these can absorb a photon. The energy is transferred to the reaction centre chlorophyll in the photosystem complex.

The primary function of LHC to capture light energy and to transfer that energy to PS I or PSII. It has no catalytic activity and contains at least two main polypeptides, each of molecular weight 32,000, which bind up half of the chlorophyll *a* and nearly all of the chlorophyll *b*. LHCs are situated on the inner surface of the thylakoid membrane. Dozens of chlorophyll molecules of each LHC absorb photons and transfer it rapidly (in less than $10^{-9}$ sec) to the specialized chlorophyll molecule in the reaction centres of the two photosystems. LHCs act to absorb light energy and to **funnel** it to a single reaction centre. The function of protein in membrane is to maintain an optimal orientation for energy transfer between chlorophylls. How it maintain this is not still clear.

## PHOTOSYSTEM II

Each photosystem consists of a central core complex (CC) and a light harvesting complex (LHC). By the use of detergents the CC can be further split into the pigmented reaction centre subunits (RC) and unpigmented proteins. Analogously, the LHC can be split into the light-harvesting pigment-proteins. In the CC (also called photosystems in some thylakoid models) the photochemical and biochemical responses occur which involve the chlorophylls and redox systems.

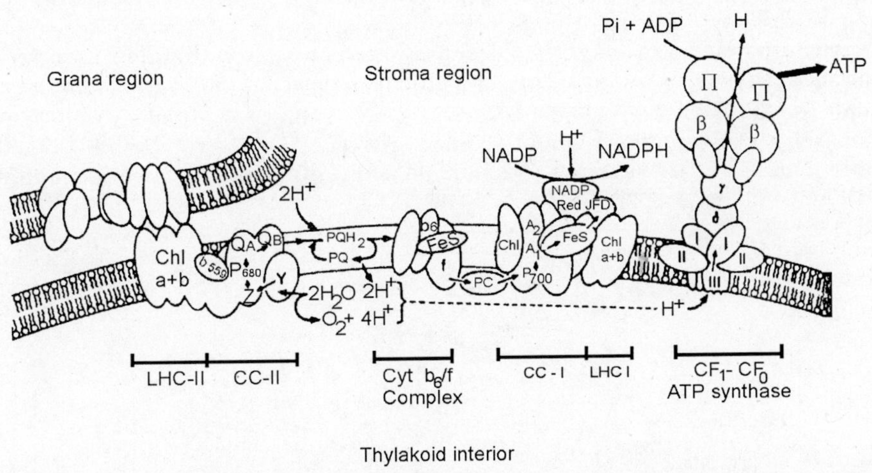


**FIG. 13.6:** Arrangement of chlorophyll protein complexes in photosynthetic unit.

An antenna consisting of pigment-protein complexes. The light-harvesting antenna or complex of PSII can be further subdivided into proximal and distal antennae. The proximal antenna, which surrounds the P680 core and is tightly coupled to it, consists of two pigment-protein complexes, CP47 and CP43. (CP = chlorophyll-protein complex.) Both complexes contain Chl *a* and β carotene bound to the apoproteins, but no Chl *b*. Excitation energy captured by the bulk antenna (proximal as well as distal) is transferred to P680 from the proximal antenna. The distal antenna is mainly composed of light-harvesting complexes (LHC II) and contains covalently bound Chl *a*, Chl *b* and xanthophylls.

The three-dimensional structure of LHC II at 6 Å resolution has recently been determined by electron crystallography. This low-resolution structure shows that LHC II forms a trimer, with each monomer containing 15 chlorophylls and several carotenoids. In addition to its function in light harvesting and energy transfer to P680, the LHC II also plays a role in energy distribution between the two photosystems by undergoing reversible phosphorylation-dephos-phorylation.

A **regulatory cap** made up of polypeptides extrinsically bound to the lumenal surface of the thylakoid membrane. It comprises of a set of hydrophilic extrinsic proteins (EP) that have apparent molecular masses of 33, 24 and 17 kDa and are designated as EP33, EP24 and EP17 respectively. The EP33 acts as a barrier between the water-oxidizing unit attached to the membrane and the aqueous lumenal phase. Spectroscopic and protein cross-linking studies suggest that EP33 is in contact with the reaction centre. It may also be required for the assembly and activation of the Mn complex in eukaryotic algae and plants *in vivo*.

## PSII Reaction Centre

A reaction center containing the primary reactants, plastoquinones, and the water-oxidizing complex. The PSII reaction centre can be functionally subdivided into the light-absorption and electron transfer segments and the water-oxidizing complex. The PSII *core* is the minimal unit able to perform light-induced charge separation and electron transport (but not to evolve $O_2$), and in this respect resembles the purple bacterial reaction centre. This core has been isolated now in many laboratories and comprises 4 to 6 Chl *a*, 2 Pheo, 2 β carotene, 1 $D_1$-$D_2$ heterodimer, 1 CP47, 1 CP43, 1 (or 2) cytochrome *b*559 and 1 10 kDa polypeptide (I) per P680. No quinone ($Q_A$) is present in the PSII core unit. The P680 and Pheo are believed to be bound to the $D_2$-$D_1$ (so called because they were first identified as two *diffuse* bands by gel electrophoresis and staining) proteins. In $O_2$-evolving PSII reaction centre complexes, the quinines ($Q_A$ and $Q_B$) and ferrous iron are also bound to the $D_2$-$D_1$ heterodimer.

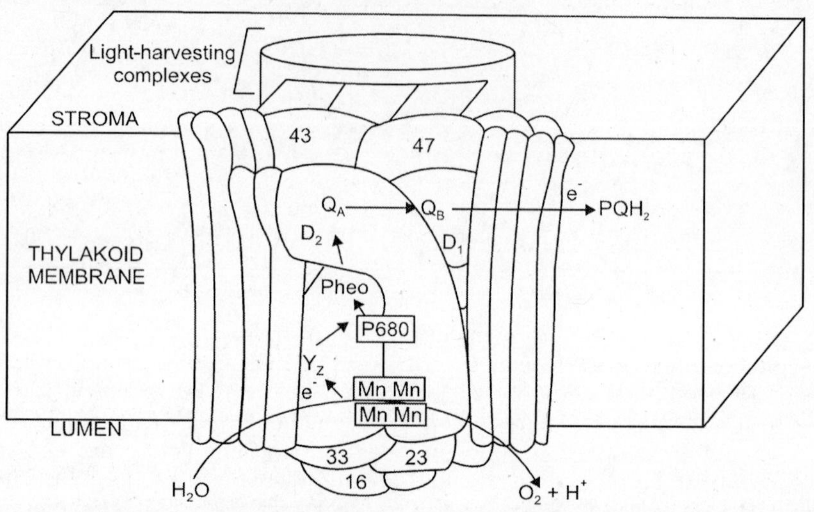


**FIG. 13.7:** Schematic model of the PS-II reaction center and the oxygen-evolving complex. P680 is associated with the D1 and D2 proteins, which are analogous to the L and M proteins of the bacterial reaction center. Electrons are transferred from P680 to pheophytin (Pheo) and then to $Q_A$ and $Q_B$ ($PQH_2$). P680+ (oxidized P680) is rereduced by $Y_Z$, a tyrosine residue in the D1 protein. The oxygen evolution complex involves Mn, $Ca^{2+}$, and $Cl^-$ as cofactors.

The primary charge transfer from excited P680 to Pheo takes place in 3 picoseconds. Recombination of the charged species is prevented by the rapid (250-500 μs) oxidation of $Pheo^+$ by a neighbouring $Q_A$ and the instant reduction of $P680^-$ by a redox active tyrosine residue $Y_Z$ of the $D_1$ polypeptide.

$$Pheo^- + Q_A \rightarrow Pheo + Q_A^-$$

$Q_A^-$ is oxidized within 100 to 200 μs by a second plastoquinone, $Q_B$, forming a $Q_B$ semiquinone. This semiquinone accepts another electron from a different $Q_A^-$ to form $Q_B^-$. (Note that $Q_A$ is a single-electron acceptor while $Q_B$ is a two-electron acceptor).

$$Q_A^- + Q_B \rightarrow Q_A + Q_B^-$$

$$Q_B + Q_A \rightarrow Q_A + Q_B^{-}$$

By analogy with the known sequences of the redox events occurring in purple bacterial reaction centers, it is believed that the fully reduced $Q_B^{-}$ accepts two protons from the stroma, forms $Q_BH_2$, and leaves the reaction centre as plastoquinol and then transfers the protons to the cyt *bf* complex. The $Q_B$ site in the

$D_1$ protein is then occupied by another $Q_B$ molecule from plastoquinone pool.

There is another redox active tyrosine, $Y_D$ in the $D_2$ polypeptide. Its function, if any, is unknown. Both $Y_Z$ and $Y_D$ were detected by EPR spectroscopy and their precise locations in the polypeptides were identified by site-specific mutagenesis.

Both $O_2$-evolving PSII preparations as well as the PSII core contain cytochrome *b*559, a heterodimer of 9-kDa ($\alpha$) and 4-kDa ($\beta$) polypeptides co-ordinated to a heam. The possible functions which have been assigned to cyt *b*559 are as a photoprotective agent and as a proton pump during cyclic electron flow around PSII.

## Location, Structure and Function

PS II is a transmembrane protein complex located in the thylakoid membrane. It consists of D1 and D2 intrinsic proteins, cyt b559, antenna proteins, Light harvesting proteins type I and type II and extrinsic proteins. Photosystem II is that part of the oxygenic photosynthetic apparatus which catalyzes the light-induced oxidation of water coupled to the reduction of plastoquinone. This function can be represented by the simple equation:

$$2H_2O + 2PQ \text{ light/PSII} \xrightarrow{\text{membranes}} O_2 + 2PQH_2$$

During this process, four photons are transferred across the thylakoid membrane from the stromal side to the lumenal side which are utilized in ATP synthesis. PSII is located within the thylakoid membrane in close association with lipids.

Major advances have been made over the last decade towards our understanding of the structure and function of PSII. These include following:

1. The isolation of almost pure PSII particles, free from other thylakoid components.
2. Separation and identification of the polypeptides in the purified complex by gel electrophoresis; determination of the primary structure of these polypeptides by isolating the genes, then cloning and sequencing them.
3. Most importantly, the determination of the structure of the reaction centers of photosynthetic purple bacterial reaction centers are structurally and functionally related.

The Central part of PS II consists of D1 (32 kD)/D2 (34 kD) heterodimer. The transmembrane helices, hydrophilic loops of D1 and D2 proteins. D1/D2 dimer embeds special pair of chlorophyll a, P680, two additional chlorophyll a, two pheophytins, quinone A(QA), quinone B(QB), non-heme iron and manganese cluster. P680 is the primary electron donor and pheophytin, the primary acceptor. The dimer is enfolded by CP47 (947 kD) and CP43 (43 kD) antenna polypeptides containing chlorophyll a molecules. The dimer is also closely associated with two units of cyt b559 polypeptides. A 33 kD extrinsic protein known as manganese stabilising protein (MSP) is associated with D1-D2-CP43-CP47 complex at donor side. D1-D2-CP43-CP47-cyt b559-33kD unit constitutes the core complex of PS II with 250-300 kD molecular mass, which is the smallest PS II unit that can split water.

Light harvesting chlorophyll a/b binding proteins (LHC II, CP 29, CP 26, CP 24) surround the PS II core complex. CP 29, CP 26 and CP 24 are closely associated with the antenna CP 47 and CP 43 9. The nonomeric light harvesting protein subunits trimerate to form functional unit associated with PS II. A theoretical model on the arrangement of protein subunits, distribution of pigments in individual protein subunits that best suit for quantum migration to the reaction center P680 in the photosystems are worked out in our laboratory (Raval et al, 1990). Electron crystallographic structure determination of LHC II reveals detail organization of seven chlorophyll a, five chlorophyll b and two lutein molecules in the protein.

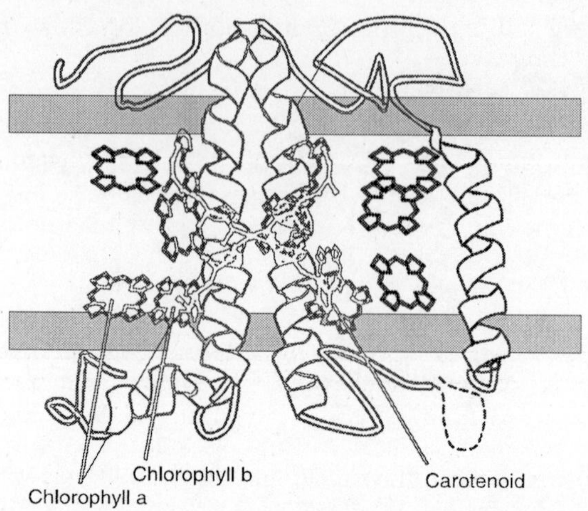


**FIG. 13.8:** Structure of the LHCII antenna complex from higher plants. The complex is a transmembrane pigment protein, with three helical regions that cross the nonpolar part of the membrane. Approximately 15 chlorophyll a and b molecules are associated with the complex, as well as several carotenoids. The positions of several of the chlorophylls are shown, and two of the carotenoids form and X in the middle of the complex. In the membrane, the complex is trimeric and aggregated around the periphery of the PS-II reaction center complex.

## Location and Geometry of Mn Cluster

A Mn cluster located in D1/D2 dimer oxidises water to oxygen. It has not been possible to discover the nature of this cluster, two dimers or a tetramer. Due to lack of direct experimental evidences, several models have been proposed for the Mn-binding site and oxidation of water. Gradually reports of site directed mutagenesis are coming in, identifying the amino acid residues binding to Mn. E69 in cyanobacteria (E 70 in higher plants) of D2 H 332, E 333, H 337 of D1 are reported to be the potential ligands to Mn cluster by mutant studies. D 59, D 61 may ligate to calcium and D59, D61, E65 and H 92 of D1 influence the properties of Mn cluster without significantly affecting its stability or ability to assemble. E 189, H 92 and D 342 influence the catalysis, stability and assembly of Mn. D 342 may bind to Mn or Ca or both. H 190 of D1 strongly influences the redox properties of secondary electron donor Z (y 161, D1) and either ligates to Mn or serves as a H-bond donor.

Studies with wild-type and mutant algae show that the EP33 polypeptide may be required for the assembly and activation of the Mn cluster and to optimize catalytic activity. Chemical modification of various amino acid residues in PSII polypeptides and subsequent measurement of the $O_2$ evolution rate and spectral properties of the modified PSII suggest that residues in the $D_1$-$D_2$ polypeptide (possibly histidine and carboxylate) may provide ligands for the Mn cluster. Similar conclusions were arrived at from site-directed mutations of amino acid residues in $D_1$ and $D_2$ polypeptides in the cyanobacterium *Synechocystis*.

## Charge Donor and Acceptor Components of PS II

The geometry of primary and accessory charge donor/acceptor components of reaction center of PS II is assigned in analogy with that of photosynthetic purple bacterial reaction center.

## Photochemistry

The primary photochemistry of PS II is initiated by charge separation between P680 (special chlorophyll a dimer) and pheophytin (pheo).

$$\text{P680, Pheo} \xrightarrow[3\text{ps}]{h\vartheta} \text{P680} + \text{Pheo} - 1.2\text{ eV}$$

Charge recombination is prevented by a series of downhill reductions during electron transfer from Pheo- to QA in D2 (200 ps), QB in D1 (0.3 ms) and free plastoquinone (PQ) pool. The orientation of the accessory chlorophyll a molecule determines the preference of path of electron transfer from P680 to pheophytin close to QA (D2 side) over the other route to QB (D1 side) (Heller et al. 1995).

There are two accessory donors YZ (Tyr 161, D1) and YD (Tyr 161, D2, 160 in cyanobacteria) to P680. Z reduces P680 + in 200 ns, half life time of YZ+ is 0.1 -1.0 ms. YZ+/YZ redox potential is between 810 mV and 1000 mv. YD is oxidised by water oxidising complex in S2 states. YD+/YD has redox potential 720-760 mV. Half life time of $YD^+$ is about 500 min (Vass and Styring, 1991). Site directed mutants suggest that Q 164, h 189 in synechocystis 6803, Q 165, H 190 in higher plant D2 and H 190 of D1 are in the vicinity of YD. Saturation-recovery electron spin resonance spectroscopy provides evidence for the symmetric location of YD and YZ at a distance of 37 ñ 5 Å from non-heme Fe (II) in PS II (Koulougliotis et al. 1995).

## ROLE OF CYT B 559

Cyt b 559 is closely associated with core complex of **reaction center** (RC II). Heber et al (1979) suggest participation of cyt b 559 in cyclic electron flow in PS II. Cyt b 559 exists in two states, low and high potential forms. Thompson et al (1987) propose a structural role of cyt b 559 in oxygen evolving comblex. Cyt b 559 is also suggested to have a protective role in photoinhibition. Cyt b 559 shifts between its high potential and low potential forms, triggered by unknown molecular switch, to act as an electron acceptor (for acceptor side protection) as well as electron donor (donor side protection) Barber and De Las Rivas, 1993). Cyt b 559 may be a sructural obligation during evolution of oxygen evolving RC II from bacterial reaction centers with cyclic electron flow. This vestigial polypeptide might have acquired a photoprotective role during evolution.

## Mechanism of Oxygen Evolution in PS II

Though there exists different models regarding organization of Mn-cluster, it is the catalytic site for oxidation of water.

**Energetics of Water Oxidation:** Free water at pH 7.0 is known to be oxidised in the following steps (see Radmer and Cheniae, 1977).

$$H_2O \longrightarrow OH + H^+ + e\text{- } 2.31V$$
$$H_2O + OH \longrightarrow H_2O_2 + H^+ + e \text{ ........................................ } 0.37\ V$$
$$H_2O_2 \longrightarrow O_2 + 2H^+ + e\ 1.00\ V$$
$$O_2 \longrightarrow O_2 + e\ \text{-o.45 V}$$
$$2H_2O \longrightarrow O_2 + 4H^+ + 4e^-\ 3.23\ V \text{ ............................ } (0.81\ V/e)$$

The energetics of bound water may be modified, but we can have some idea regarding mechanism of oxidation of $H_2O$ to $O_2$ from the above scheme. The photon absorbed by P680 (680 nm) has energy 1.8 ev. It is sufficient to carry primary charge separation by oxicising P680 to P680+ (Em, 7 = 1.2 V) and $P680^+/P680$ can carry out oxidation of water requiring potential 0.81 V.

## KOK'S OXYGEN CLOCK

Three photons will have sufficient energy to produce oxygen. It is observed in oxygen flash yield kinetics (Kok et al., 1970). This clock is based on independent chain model of oxygen evolution. The oxygen evolving system comprises five states labelled: $S_o$, $S_1$, $S_2$, $S_3$, $S_4$,. Number 0, 1, 2, 3, and 4 indicates the number of positive charges. After absorption of photon (*hv*) the S states advance to the next higher S state and the oxygen is evolved when $S_4$ reacts with $H_2O$., as presented below:

$O_2 + 4H^+$

*hv* *hv* *hv* *hv* *hv* ®$2H_2O$

$S_0$® $S_1$® $S_2$® $S_3$® $S_4$® $S_0$

It was previously thought that all four protons are released in the $S_4$ to $S_0$ reaction. However, recently it has been the established that charges produced on one centre are independent of all other centres. The scheme is as follows:

$S_0$ $S_1$ $S_2$ $S_3$ $S_4$ $S_0$

$IH^+$ $OH^+$ $IH^+$ $2H^+$

Another scheme is also proposed which explains that $H^+$ release pattern is in between 0, 1, 1, 2 and 1,1,1,1 which suggests that water may indeed have begun to chemistry before the last step contrary to the much mentioned belief. NMR analysis also provide evidence that protons could be released in several steps.

$O_2 + 4H^+$

*hv* *hv* *hv* *hv* *hv* ®$2H_2O$

$S_0$® $S_1$® $S_2$® $S_3$® $S_4$® $S_0$

It was previously thought that all four protons are released in the $S_4$ to $S_0$ reaction. However, recently it has been the established that charges produced on one centre are independent of all other centres. The scheme is as follows:

$S_0$ $S_1$ $S_2$ $S_3$ $S_4$ $S_0$

$IH^+$ $OH^+$ $IH^+$ $2H^+$

Another scheme is also proposed which explains that $H^+$ release pattern is in between 0, 1, 1, 2 and 1,1,1,1 which suggests that water may indeed have begun to chemistry before the last step contrary to the much mentioned belief. NMR analysis also provide evidence that protons could be released in several steps.

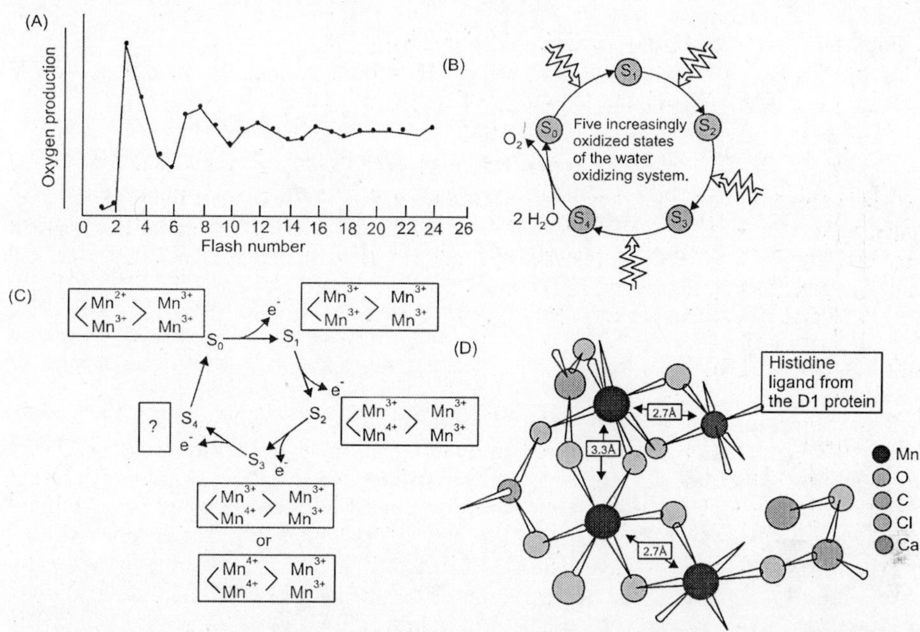


**FIG. 13.9:** The pattern of oxygen evolution in flashing light and the S state mechanism for oxygen evolution. (A. sample of dark-adapted chloroplasts is exposed to a series of short light flashes. The maximum production of oxygen occurs on the third flash, secondary maxima occur on even fourth flash, and the yield damps to a steady-state value after about 20 flashes. (B) The S state model developed b Kok and coworkers was advanced to explain the observations of part A. The oxygen-evolving system can exist in states: $S_0$ to $S_4$. Successive photons (Squiggly arrows) trapped by PS-II advance the system from $S_0$ to $S_1$, $S_1$ to $S_2$ and so on until state $S_4$ is reached. $S_4$ is unstable and read with two water molecules to produce $O_2$. (C) A structural model for the oxygen-evolving complex based on spectroscopic studies. The four Mn ions are linked to amino acids in the D1 protein, and to oxygen, chloride, and calcium. (One possible scheme for the redox states ($S_0$ to $S_4$) of the water-oxidizing Mn complex. Upon the release of oxygen $S_4$ converts into $S_0$. No data are yet available for $S_4$.

Evolution of oxygen per flash is a function of flash number and oscillates with the periodicity of 4. Therefore, any component that oscillates with that period and may be related to this system directly or indirectly. Some of these phenomenon are chlorophyll *a* -fluorescence which shows peaks at the 1st and 5th flashes; delayed light emission (DEL) due to the back reaction of PS II primary photochemistry and **thermoluminiscence.**

A detail stepwise mechanism for oxidation of water has been proposed by Raval and Biswal (1984, 1985) based on energetics and kinetics of oxygen evolution.

## THE PHOTOSYSTEM II OXYGEN-EVOLVING REACTION

This key reaction of plant, algal and cyanobacterial photosynthesis is the main source of all $O_2$ on earth but still continues to be one of the great unsolved mysteries of photosynthesis. However, 'light' has begun to get through the 'black box', and recent work has significantly helped to probe the mechanisms involved in the oxidation of water. The problem is being attacked from two main directions:

1. Isolation of PSII membranes with very good $O_2$ evolution rates and determination of the valence state and structure of the Mn atoms in the $S_n$ states in the membranes by a variety of spectroscopic techniques.
2. Synthesis of Mn clusters with spectroscopic characteristics that resemble the Mn spectra of PSII and which may even catalyze photo-oxidation of water.

Classical spectroscopy such as absorption, infra-red (IR), nuclear magnetic resonance (NMR) and EPR are being supplemented by more advanced techniques of electron nuclear double resonance (ENDOR), electron spin echo, X-ray absorption spectroscopy, etc. to identify the paramagnetic Mn species generated in PSII during laser flash photolysis. Considerable progress has been made in identifying the polypeptides surrounding the Mn cluster of the water-oxidizing complex. The Mn ions are located in a compartment towards the lumenal side of the membrane.

The requirement of manganese in the photoxidation of water has been known since 1970. On the basis of atomic absorption measurements and EPR quantification of Mn ions released by acid treatment of PSII, it is now almost accepted that the active $O_2$-evolving complex contains four atoms of Mn per P680 (although a few researchers believe there are six Mn per P680). Treatment with 0.8 M Tris buffer pH 8.5 releases most of the Mn from the complex, with the concomitant loss of the three polypeptides EP33, EP24 and EP17 of the regulatory cap. The Mn complex can be reconstituted in light; thekinetics of reassembly of the Mn tetramer appear to be a multistep process.

Joliot, and Kok, proposed the S state model of charge accumulation on the water oxidation catalyst to explain why the yield of $O_2$ from chloroplasts in response to a sequence of saturated light flashes oscillates with a period of 4. Most of the studies in recent years have tried to correlate the Mn signals from spectroscopic data with the S states of the water-oxidizing complex. Two EPR signals observable at very low temperatures ($<35$ K) have been attributed to the $S_2$ state. These are known as the multilane EPR signals with $g = 2.0$ and the $g = 4.1$ signal. The signals are perturbed by treating the samples with inhibitors of $O_2$ evolution, e.g. ammonia ions or $NH_2OH$. These EPR signals suggest that the Mn ions have a mixed valence in the $S_2$ state; however, definite assignment of the Mn oxidation state is difficult. The four Mn atoms may be arranged as a tetramer (or a pair of interacting dimers) or as a trimer which gives rise to the two EPR signals and an EPR = silent monomer. In dark-adapted PSII membranes, an EPR signals with $g = 4.8$ has been detected and has been proposed to arise from a precursor of the multilane signal and assigned to the $S_1$ state.

Valuable information regarding the ligands of Mn ions and Mn-Mn, Mn-O and Mn-N distances in the cluster has been obtained from X-ray absorption studies. These spectroscopic probes suggest that the atomic environment of Mn ions is heterogeneous. Flash-induced absorption changes in the UV that oscillate with a period of 4 have been assigned to the individual S state transitions. These absorption changes show correlations with the changes in the valence state of Mn from MnII to MnIII to MnIV. Spectroscopic data suggest that the Mn ions undergo oxidation during the $S_0 \rightarrow S_1$ and $S_1 \rightarrow S_2$ transitions, but it is still not clear whether Mn is oxidized on the $S_2 \rightarrow S_3$ transition.

The mechanism of water oxidation is still unknown. It is assumed that the Mn ions form the matrix to which two water molecules co-ordinate to form a dioxygen bridge. There are two possibilities for the chemistry of the oxidation.

1. A concerted four-electron oxidation occurs at the $S_3 \rightarrow S_4 \rightarrow S_0$ transitions. It is generally believed, but not unequivocally established, that water oxidation occurs at the $S_4 \rightarrow S_0$ transition.
2. A two-step reaction with the formation of a peroxide intermediate prior to the $S_3 \rightarrow S_4 \rightarrow S_0$ transition. Sequential one-electron oxidation of water with the formation of $OH^-$ intermediates is considered to be energetically unfavourable.

## Role of Ions Other than Manganese during Evolution of Oxygen

Other ions associated with evolution of oxygen in photosynthesis are $Ca^{2+}$, Cl and $HCO_3$. $Ca^{2+}$ binds to D1/D2 heterodimer and influences the stability and assembly of Mn-cluster as suggested by site directed mutants (Chu et al 1995 a, b) $Ca^{2+}$ may play a role in reduction of $P680^+$ by YZ (Satoh and Katoh, 1985). S3_So transition is also known to be modulated by $Ca^{2+}$. It may be speculated that $Ca^{2+}$ binding introduces structural modification suitable for Mn binding. Moreover, $Ca^{2+}$ may provide an electrostatic microenvironment to the redox species on the donor side of PS II.

The roles of $Ca^{2+}$ and Cl in the water oxidation are not definitely established. Atomic absorption measurements have shown that PSII membranes contain two to three bound Ca ions per P680. Bound Ca can be released with a loss of $O_2$ evolution by washing the membranes with 1-2 M NaCl supplemented with 1-5 mM EDTA, in light. Measurement of $O_2$ evolution of Ca-depleted PSII membranes as a function of added Ca has shown that PSII contains one high-affinity and two low-affinity Ca-binding sites per P680. X-ray absorption spectroscopy has indicated that Ca ions are located very close to the Mn complex. Whether the role of Ca is only to protect the Mn complex or whether it directly participates in the $O_2$ evolution reaction is still a moot question. The influence of Ca ions in electron transfer and $O_2$ evolution in cyanobacteria may be different from that in plants.

The molecular mechanism of Cl effect on PS II electron transport is understood to some extent by 35 Cl-NMR studies, which suggests a stereo selective binding of Cl to oxygen evolving site itself or very close to it. Coleman and Govindjee (1987) proposed that Cl has two binding sites, intrinsic and extrinsic. Binding of Cl to intrinsic sites (H 332, H 337 of D1 and H 337, D2) enhances the abstraction of $H^+$ from water by increasing pKa of imidazole (Histidine side chain) rings. Chloride may not act simply a passive counter ion of charged species formed at the active site rather participate actively by affecting pKa and conformation of the polypeptides.

The binding of Cl ions to PSII requires the prior binding of Ca. Binding values were attained by NMR spectroscopy using $^{35}Cl$ and isotopic labelling with $^{36}Cl$. Reported values for the number of $Cl^-$ per P680 required for maximum $O_2$ evolution have varied from 1 to 5. Some EPR measurements indicate that $Cl^-$ is essential for the $S_2 \rightarrow S_3$ transition. Requirement for $Cl^-$ is not specific; it can be replaced by $Br^-$.

From the foregoing discussion it is obvious that we still need to learn a lot more about the location, structure, chemical reactions, and the role of cofactors of the water oxidation complex in order to elucidate the mechanism of O2 evolution.

Non-heme iron of QA - Fe -QB complex, situated on the acceptor side of the reaction center II is identified as HCO3 binding site by spectroscopic studies (Diner and Petrouleas, 1990; Semin et al, 1990). HCO3 regulates the electron transfer between QA and QB in PS II and HCO3 binding is stabilized by positively charged aminoacid residues namely arginine and lysine (Govindjee and Van Rensen, 1993). Regulation of electron transport by HCO3 may be a strategy of the system to sense the concentration of carbon dioxide which is converted to glucose as the end product of the process initiated by PS II.

### Extrinsic 33 kD and Oxygen Evolving Polypeptide

Importance of 33 kD extrinsic polypeptide in oxygen-evolution has entered into controversy by recent results on mutants. Earlier reports suggest necessity of 33 kD extrinsic polypeptide for oxidation of water (Kuwabara et al., 1985; Kuwabara and Murata, 1982; Miyao and Murata, 1984). Isolation of 33 kD protein with 1 M $CaCl_2$ or $MgCl_2$ MgC retains Mn cluster in the RC II, however, requires Cl for oxygen evolution (One and Inoue, 1983, Bricker, 1992). Studies on mutants of Synechocystis Sp PCC 6803 in which psbo gene encoding 33 KD protein is detected, indicates that 33 kD protein is not essential for oxygen evolution, however, it is required for stability of Mn cluster (Burnap and Sherman, 1991, Philbrick et al., 1991). Thermoluminescence study on IC2 psb O less mutant of Synechocystis pcc 6803 shows damping of S2QB- and S3QB- charge recombination observed by flash induced oscillation of the B-band indicating inhibition of higher oxidation states of Mn-cluster (S3 and S4) in the absence of 33 kD polypeptide (Vass et al., 1992). On treatment with B-mercapto ethanol, inactivation of oxygen evolution is observed (Raval et al., 1994). The 33 kD polypeptide has two cysteine residues (C28, C51) forming a disulfide bridge which is supposed to maintain the conformation of the polypeptide and ability to regind to the PS II preparation (Tanaka and Wada, 1988; Raval et al., 1994). However, site directed mutant (C51A) and double mutant (C28A, C51A) show that both are capable of binding 33 kD to PS II (Betts et al., 1996). Mutant (C51A) is unstable in light where as double mutant (C28A) has properties similar to wild type PS II. This indicates that disulfide bridge does not affect the binding of 33 kD to PS II reaction center. However, sensitivity of mutant (C51A) to light suggests a light dependant interaction that occurs between disulfide bridge and PS II during the redox-cycling of Mn cluster. This may be explained with the model proposed by Raval et al. (1990b) where disulfide bridge interacts with the Mn (II) -Mn (III) dimer of manganese electron trap not directly involved in binding and photooxidation of water. Absence of disulfide bridge does not affect oxygen evolution However, it may participate in the redox reaction of Mn (II) - Mn(III) dimer in S3, S4 state.

In spite of the established fact that Mn cluster catalytically oxidises water to oxygen in PS II, geometry of Mn-cluster, its binding site, oxidation state and precise mechanism of catalysis are yet to be deciphered. Molecular modelling and site directed mutagenesis have thrown diffused light to dig out the facts embedded in the black box of oxygen evolving complex of PS II.

## PHOTOSYSTEM I

Photossystem I is a multisubunit membrane-protein complex. It catalyzes electron transfer from the reduced plastocyanin in the thylakoid lumen to theoxidized ferredoxin in the chloroplast stroma or cyanobacterial cytoplasm. In cyanobacteria, plastocyanin and ferredoxin are replaced, depending on the nutritional conditions. Purified cyanobacterial PS I complex contains at least eleven different proteins: the core proteins (PsaA and PsaB), three peripheral proteins (PsaC, PsaD, and PsaE) on the stromal side and six integral membrane proteins (PsaF, PsaI, PsaJ, PsaK, PsaL, and PsaM).

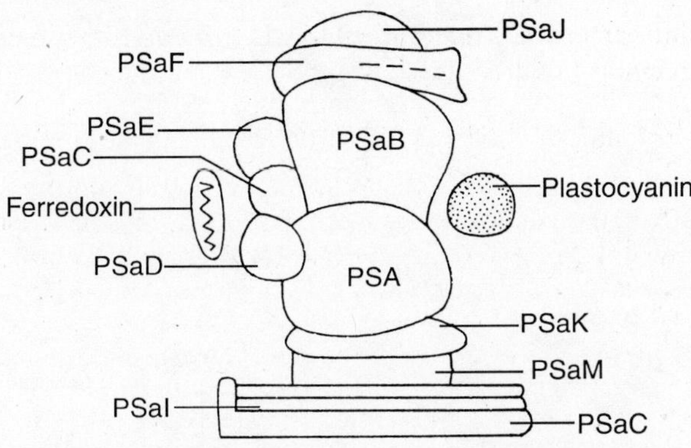


**FIG. 13.10:** Diagrammatic presentation of PS I, showing its structural organization.

## Additional Proteins PS I

1. Stromal peripheral protein (PsaH).
2. Lumenal peripheral protein (PsaN).
3. Integral membrane protein (PsaG).

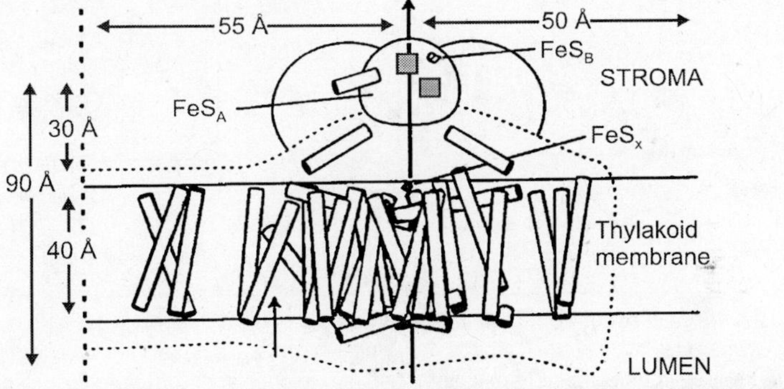


**FIG. 13.11:** Structure of Photosystem I from cyanobacteria at 4.0 Å resolution. The figure shows an arrangement of $\alpha$ helices (shown as tubes) viewed parallel to the plane of the membrane. The stromal side of the membrane is at the top, and the lumenal side is at the bottom of the figure. The three FeS centers that are electron acceptors in Photosystem I, $FeS_A$, $FeS_B$, and $FeS_X$ are visible on the stromal side as cubes. Pigments are removed for clarity, molecular dimension in Å are shown.

These PS I proteins, their characteristics and their functions are listed in Table 2.1. The PsaA and PsaB proteins form the heterodimeric core around which other proteins are organized to form the complex. The PS I complex also contains approximately 100 chlorophyll a molecules, 12-16 –carotenes, two phylloquinones, and three [4Fe-4S] clusters. These cofactors are bound to the PsaA, PsaB, and PsaC proteins. During the photosynthetic electron transfer across the thylakoid membrane, electrons are transported from the electron donor plastocyanin to the P700 reaction center, which is a dimer of chlorophyll *a* molecules, then through a chain of electron acceptors: a chlorophyll *a* monomer $A_o$, a phylloquinone $A_1$, [4Fe-4S] cluster $F_X$, $F_A$, and $F_B$, finally to the electron acceptor ferredoxin. The energy for the excitation of the P700 reaction center is provided by the photons absorbed by the antenna chlorophyll a molecules PS I as antenna and by the associated membrane-embedded light-harvesting complexes (LHCI). PS I converts the energy of photons into chemical energy with a remarkable quantum efficiency of 39 to 44%.

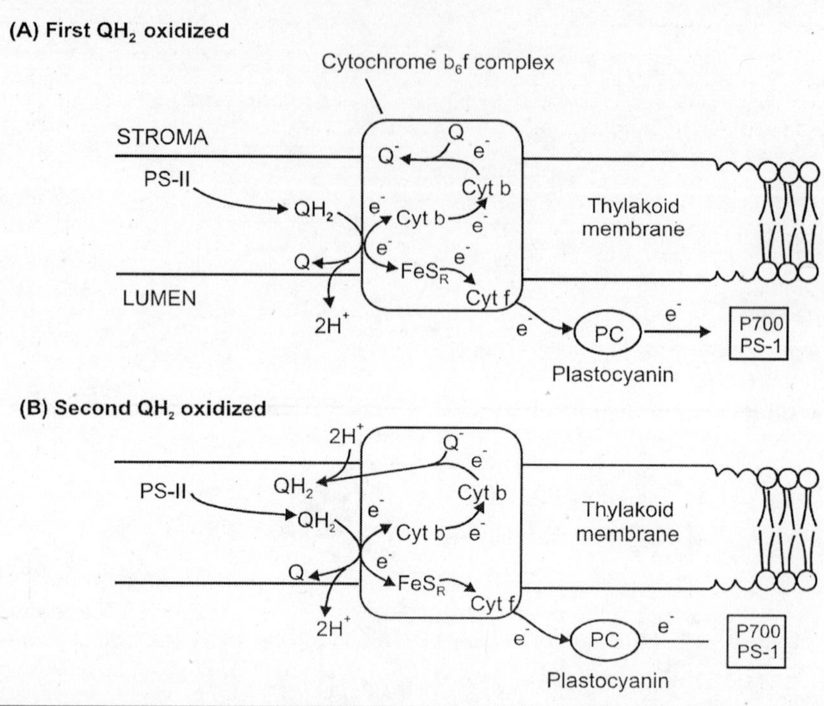


**FIG. 13.12:** Mechanism of electron and proton transfer in the cytochrome $b_6f$ complex. This complex contains two *b*-type cytochromes (Cyt *b*), a c-type cytochrome (Cyt *c*, historically called cytochrome *f*) a Rieske Fe-protein ($FeS_R$), and two quinone oxidation-reduction sites. (A) A plastohydroquinone ($QH_2$) molecule produced by the action of P II is oxidized near the lumenal side of the complex, transferring its two electrons to the Rieske Fe-S protein and one of the *b*-type cytochromes and simultaneously expelling two protons to the lumen. The electron transferred to $FeS_R$ is passed to cytochrome *f* (Cyt *f*) and then to plastocyanin (PC), which reduces P700 of PS-I. The reduced *b*-type cytochrome transfers an electron to the other *b*-type cytochrome, which reduces a quinone (Q) to the semiquinone ($Q^-$) state. (B) A second $QH_2$ is oxidized, with one electron going from $FeS_R$ to PC and finally to P700. The second electron goes through the two *b*-type cytochromes and reduces the semiquinone to the plastohydroquinoen, at the same time picking up two protons from the stroma. Over a four protons are transported across the membrane for every two electrons delivered to P700.

**TABLE 13.1:** Some subunit proteins of PSI complex (for details refer to Ikeuchi, 1992)

| Component | Molecular mass (kDa) | Gene | Location | Function |
|---|---|---|---|---|
| PSI-A | 81-84 | psa A© | Transmembrane | Binding site for Chl a, P700, $A_0$, $A_1$ and $F_X$ |
| PSI-B | 81-83 | psa B© | | |
| PSI-C | 9 | psa C© | Peripheral | Binding site for $F_A$ and $F_B$ |

| Component | Molecular mass (kDa) | Gene | Location | Function |
|---|---|---|---|---|
| PSI-D | 15-18 | psa D | | Ferredoxin-docking |
| PSI-E | 8-11 | psa E | | Ferredoxin- docking |
| PSI-F | 16-19 | psa F | Transmembrane and peripheral | Plastocyanin-docking |
| PSI-I | 4-5 | psa I© | Transmembrane | ? |
| PSI-J | 4-5 | psa J© | Transmembrane | ? |

*Notes:* The protein protein subunits are usually separated by SDS-PAGE of the PSI complex and identified by protein sequencing and/or gene cloning techniques. Genes denoted by © are coded by the chloroplast, and others by the nuclear genome.

In the past 10 years, multidisciplinary research efforts have led to characterization of the components, the electron transfer reactions, and the structure of PS I complex. In this chapter, we will discuss recent progress in the understanding of the function and structure of PS I complex. Detailed discussion of LHCs proteins, the electron transfer and redox cofactors in PS I genes can be found in the recent reviews ADDIN ENRfu (Brettel, 1997; Chitnis; 1996; Chitnis, 1997; Fromme, 1996; Green and Durnford, 1996; Ort and Yocum, 1996).

It also participates in the PS II-independent cyclic electron transport pathway which results in proton pumping and ATP synthesis.

## PHOTOSYSTEM I: FUNCTIONS

1. PS I is a unique light-driven generator of the reducing power.
2. It functions with PS II in a linear electron transport pathway to generate strong reductants, the reduced ferredoxin and NADPH.
3. The products of PS I are among the most electronegative reductants in a biological system. These reductants are used in diverse biochemical processes, including Calvin cycle, nitrate reduction to ammonia, fatty acid desaturation, pyruvate decarboxylation, glutamate synthesis, sulfite reduction, and enzyme regulation through reduced thioredoxin.
4. It plays a vital role in cyanobacteria and chloroplast metabolism.

### Linear Electron Transfer in PS I

In the linear electron transfer, PS I and PS II function according to the Z-scheme. The PS II and PS I reaction centers form the light-driven uphill sections where the energy of a photon is used to increase the reducing ability. The electron transport between the two photosystems involves cytochrome *b6f* complex. Among these three integral membrane complexes, the electron transfer is linked by mobile electron carriers: plastoquinone transfers electrons from PS II to the cytochrome bef complex which, in turn, donates electrons to PS I via plastocyanin. In the linear electron transfer, PS I is involved in three steps of electron transfer:

1. From plastocyanin to the oxidizing side of PS I.
2. Within the PS I complex.
3. From the reducing side of PS I to ferredoxin.

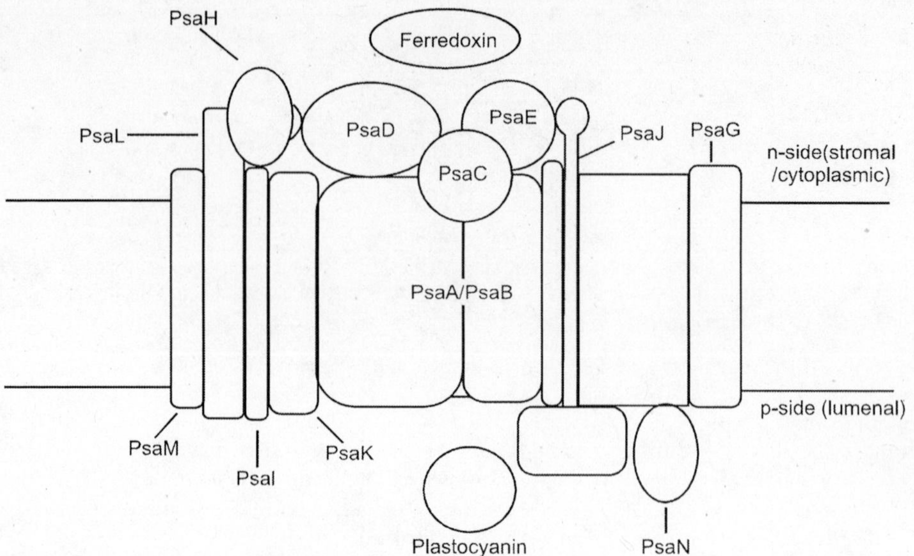


**FIG. 13.13:** A model for organization of PS I proteins. A side view of PS I proteins.

## Interactions on the oxidizing side

On the lumenal side of thylakoid membranes, PS I accepts electrons from plastocyanin. In cyanobacteria and algae, cytochrome $c_6$ can serve as an alternative electron donor to the P700 reaction center.

## Electron carriers of the PS I complex

The electron transfer within the PS I complex involves a series of electron carriers with increasing electronegative redox potentials $E_m$ from –1290 to –530 mV. The redox centers in the PS I complex are as follows:

**P700:** It is a dimer of chlorophyll *a* molecules oriented perpendicular to the thylakoid membranes. Upon transfer of excitation energy from the core antenna system, the primary electron donor P700 undergoes charge separation, producing $P700^+$. The electron lost by P700 is gained by oxidizing plastocyanin.

**$A_o$:** The charge separation of P700 is stabilized by spatial displacement of electrons through a series of low potential redox centers. The primary acceptor $A_o$ is reduced to the anion form upon transfer of an electron from the excited P700. The spectroscopy experiments have supported identity of $A_o$ as a monomeric form of chlorophyll *a*. There are four chlorophyll molecules arranged in two-fold symmetrical positions between P700 chlorophyll pair and the Ex Fe-S cluster. A pair of chlorophyll *a* molecules positioned 16 Å from P700 towards the stroma is approximately half-way between P700 and $F_X$.

Another pairs of chlorophyll *a* monomers are located approximately half-way between P700 and the $A_o$ equivalent chlorophyll molecules. They are assigned as accessory chlorophylls. Thus there are two possible branches of electron transfer pathway between P700 and $F_X$. It is not known if only one or both sides of possible electron transfer pathway are used during the PS I function. It is also not known what is the role of the accessory chlorophylls in the electron transfer.

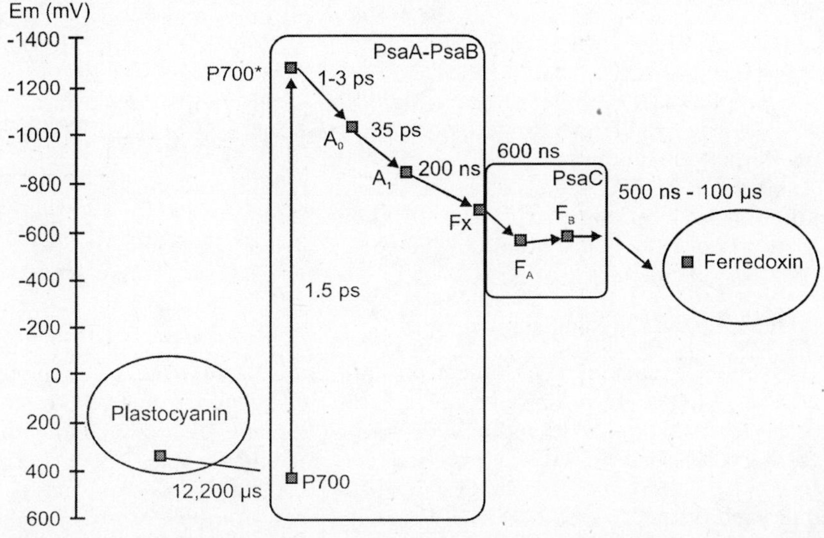


**FIG. 13.14:** Electron transfer in PS I. The electron carriers are arranged according to the electron flow on the scale of midpoint redox potential (Em). Typical values for the half-time for the forward reactions are indicated.

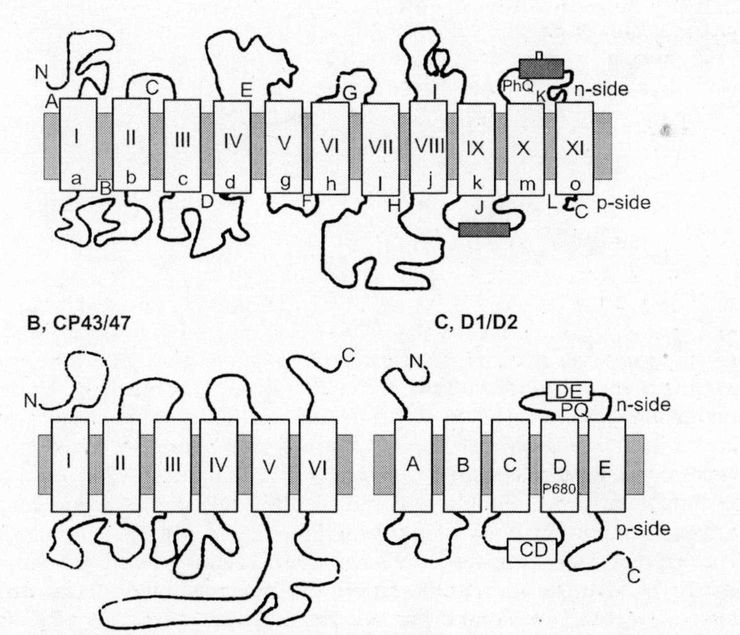


**FIG. 13.15:** The topographic models of PS I and PSII core proteins.

**$A_1$:** During electron transfer in PS I, $A_1$ serves as an intermediate electron transfer center that oxidizes $A_o$ and reduces $F_X$.

**$F_X$:** The next intermediate acceptor in the electron transfer pathway within PS I is the [4Fe-4S] cluster $F_X$. It is clearly identified in the crystal structure due to its high electron density. It is located near the stromal surface at the edge of membrane plane. $F_X$ is found at the center of the PS I complex from the top view. $F_X$ is 32 Å away from P700 along the membrane normal and 14 Å from the center of the phylloquinone binding pocket.

**$F_A$ and $F_B$:** The $F_A$ and $F_B$ clusters were the first to bound electron acceptors, identified in the PS I complexes. They are bound to the PsaC subunit, unlike other electron carriers which are bound to the core proteins. Like the $F_X$ center, they are also [4Fe-4S] in cluster.

## Interactions on the reducing side

On the reducing side of PS I., electrons are transferred from $F_A/F_B$ clusters to ferredoxin. The PS I-ferredoxin complex formation precedes electron transfer and the rate constants for complex formation depend on ionic concentration, suggesting electrostatic interaction between ferredoxin and PS I. Ferredoxin accepts electrons from $F_A/F_B$ clusters of PsaC, implying that these proteins should contact each other. Therefore, PsaD, PsaE and PsaC are the essential components of the reducing side of PS I. Structure of the reducing side of PS I is not understood at a fine resolution.

## Cyclic Electron Transfer

In addition to the PS II-dependent linear electron transfer, PS I also participates in the cyclic electron flow that results in proton pumping and ATP production without the participation of PS II. Cyclic electron flow is essential in the bundle sheath cells of C4 plants that predominantly contain PS I, but its contribution to ATP generation in C3 plants and in cyanobacteria is rather limited.

Cyclic electron transfer has also been proposed be involved in photoprotection and in fine-tuning of the balance between ATP and NADPH production (Bendall and Manasse, 1995).

Overall, the PS I complex utilizes energy of the photons trapped by either its core antenna system or the associated LHCs to function in linear or cyclic pathway producing strong reductants and ATP, and plays an important role in nourishing the whole world.

## The Catalytic Subunits

As a light-driven oxidoreductase, PS I catalyses the electron transfer from the reduced plastocyanin to the oxidized ferredoxin. The electron carriers participating the electron transportation are bound to the PsaA, PsaB, and PsaC proteins.

**PsaA and PsaB:** These proteins form the hydrophobic core of PS I. The hydropathy analysis of PsaA and PsaB indicated that each protein contains 11 potential transmembrane domains. In higher plants, the genes encoding PsaA and PsaB are present in the chloroplast. The PsaA and PsaB proteins might have evolved be gene duplication and are thought to have similar structural features. The deduced molecular masses of PsaA and PsaB are approximately 83 kDa and 82 kDa, respectively.

**PsaC:** It is an acidic peripheral protein with 80 amino acids. It coordinates the terminal electron transfer centers $F_A$ and $F_B$ of PS I. In higher plants, the *psa*C gene is located in the chloroplast genome. PsaC is an extraordinarily conserved protein. PsaC obviously interacts with the core proteins. Chemical cross-linking studies have demonstrated that PsaC can be cross-linked to PsaD and PsaE proteins. So far, many studies have been performed to describe the structure-function relationships in PsaC. However, two questions are still ambiguous:

1. Are both [4Fe-4S] clusters both essential for electron transfer?
2. Which cluster, $F_A$ or $F_B$, is proximal to $F_X$?

## The Docking Subunits

PS I complex accepts electrons from plastocyanin in the thylakoid lumen and donates electrons to ferredoxin in the chloroplast stroma or cyanobacterial cytoplasm. PsaD and PsaE are involved in the docking of ferredoxin, whereas PsaF may interact with plant plastocyanin to form complex for fast electron transfer.

**PsaD:** The PsaD subunit of PS I is conserved peripheral protein on the reducing side of PS I. In higher plants, the PsaD protein is encoded by the nuclear gene *psa*D. In general, the cyanobacterial PsaD protein contains 139-144 amino acids while the plant PsaD has about 23 additional residues forming an extension at the N-terminus. This extension of eukaryotic PsaD is accessible to proteases. The insertional inactivation of the *psa*D gene of Synechocystis sp. PCC 6803 indicated that PsaD is essential for efficient function of the cyanobacterial PS I (Chitnis et al., 1989).

Involvement of PsaD in the function and organization of PS I:

1. It interacts with at least three proteins of PS I and stabilizes their organization within the complex.
2. PsaD is required for the stable binding of PsaC to the PS I core proteins (Chitnis *et al.*, 1996).

Crystallographic studies at limited resolution have shown that PsaD contains a single, short –helix located near the common interface with PsaC and the integral membrane core subunits which may be important for the stable association of both PsaC and PsaD with the PS I core (Krauß *et al.*, 1996).

**PsaE:** PsaE is a moderately conserved peripheral subunit of PS I. Together with curved ridge on the stromal surface of PS I (Schubert *et al.*, 1997). In higher plants, *psa*E gene is located in the nucleus. Cyanobacterial PsaE contains 70-75 amino acid residues whereas the plant and algal PsaE has an N-terminal extension. Similar to the plant accessible to exogenous proteases (Lagoutte and Vallon, 1992). PsaE facilitates the interaction between PS I and ferredoxin. PsaE may also be required for the cyclic electron transport around PS I (Chitnis *et al.*, 1995a).

**PsaF:** PsaF is a membrane-integral protein with substantial N-terminal domain exposed on the lumenal side of PS I. The crosslinking product of PsaE and PsaF demonstrated the transmembrane nature of PsaF. Based on biochemical depletion and chemical cross-linking experiments, PsaF has been proposed to function as the plastocyanin or cytochrome $c_6$ docking site on the oxidizing side of PS I.

## Other Proteins of Photosystem I

The other subunits of PS I do not bind electron carriers. They are not involved in docking of soluble electron carriers. However, some of these proteins provide essential structural roles in the organization of PS I complex. Also, these subunits contribute to a roughly continuous ring of transmembrane helices around the core antenna system of PS I. Thus, they may provide a partially enclosed framework to the antenna system, thereby providing protection and stability.

PsaI, PsaJ and PsaM are small hydrophobic proteins in PS I complex, which usually contain one putative transmembrane helix flanked by short hydrophilic, charged domains. These proteins may function as nuts and bolt in the organization of PS I complex. The may assist in the correct organization of other subunits by anchoring peripheral proteins and stabilizing transmembrane helices in lipid bilayer (Xu *et al.*, 1995b). The PsaJ-PsaF interaction may involve stabilization of the transmembrane helix of PsaF (Xu *et al.*, 1994d). PsaI stabilize the association of PsaL with the PS I stabilize the association of PsaL with the PS I core (Xu *et al.*, 1995b). Whether PsaM has the similar role in organization of other subunits has not been identified.

**PsaI and PsaL:** PsaL proteins contains a greater diversity than seen in other subunits. The protein contains several conserved regions of homology, most notably in the hydrophobic regions. Function of PsaL was revealed by the inactivation of the *psa*L gene.

PsaI is an integral membrane protein with one transmembrane helix. This presequence of the *Anabaena variabilis* PsaI is predicted to assist the thylakoid translocation and consequently the N-terminus of PsaI may be located in the lumen. PsaI in intact thylakoids is not accesible to external proteases.

**PsaJ:** PsaJ is a 4.4-kDa hydrophobic protein. It has been identified in PS I preparations from cyanobacteria and higher plants. Analysis of the deduced amino acid sequence reveals a highly conserved sequence of 22 amino acids which may span the membrane. The hydrophilic N-terminal domain of PsaJ is only 4-5 amino acids long while the C-terminal domain has overall negative charge and more than 15 amino acids.

**PsaK:** It is an integral membrane protein that is tightly associated with the PsaA-PsaB complex.

**PsaM:** PsaM is a very small protein of about 29 amino acids and has been detected only in cyanobacterial PS I. It helps in stabilizing the trimer in co-operation with PsaK, a similar 'nuts and bolt' function as PsaI and PsaJ.

**PsaH, PsaG, and PsaN:** PsaH, PsaG, and PsaN are the three eukaryote-specific subunits of PS I. The conding genes for these three polypeptide *psa*H, *psa*G, and *psa*N are all present in the nucleus.

- **PsaG** is an integral-membrane protein with about 100 amino acids, which has been only found in eukaryotic PS I complex.
- **PsaH** is another PS I subunit that has only been detected in algae and higher plants. The mature PsaH protein contains an N-terminal acidic domain, a central hydrophobic domain and a C-terminal basic region. PsaH is an extrinsic subunit on the stromal side of PS I.
- **PsaN:** The mature PsaN is 85 amino acids in length, with no putative transmembrane domain. It is a 9 kDa extrinsic polypeptide of plant PS I located on the *p*-side of thylakoids. It does not bind any cofactors and is not required for $NADP^+$ photoreduction. The cDNA clones encoding the PsaN subunits have been isolated from barley and maize.

## THE CYTOCHROME $B_6F$ COMPLEX: THE Q CYCLE

The cyt $b_6f$ complex transfers electrons from PSII to PSI during linear electron flow from water to NADP. It can be considered as a plastoquinol-plastocyanin oxidoreductase. Recent work by J. Anderson (in Canberra) and others suggests that the cyt $b_6f$ complexes are distributed in three domains of the thylakoid membrane: the appressed granal stacks (along with PSII), the non-appressed stromal thylakoids (along with PSI), and the non-appressed grana margins.

The $b_6f$ complex can be isolated from chloroplasts by detergent fractionation and then further resolved into individual protein components by treatment with denaturing agents.

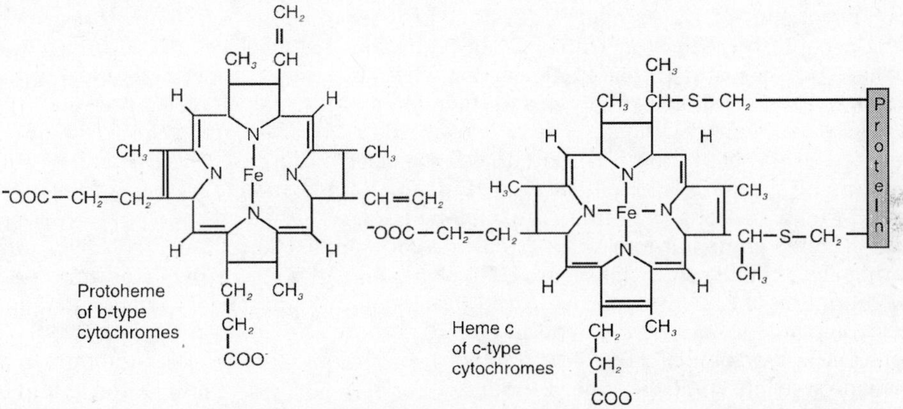

**FIG. 13.16:** Structure of prosthetic groups of *b*- and *c*- type cytochromes. The protoheme group also called proto-porphyrin IX) is found in *b*-type cytochromes, the heme *c* group in *c*-type cytochromes. The heme *c* group is covalently attached to the protein by the other linkages with two cysteine residues in the protein; the protoheme group is not covalently attached to the protein. The Fe ion is in the $2^+$ oxidation state in reduced cytochromes and in the $3^+$ oxidation state in oxidized cytochromes.

The isolated complex contains four major polypeptides: cyt $b_6$ (molecular mass 23 kDa, *pet* B gene product), cyt *f* (31 kDa, *pet* A), the Rieske (discoverer) iron-sulphur protein $FeS_R$ (20 kDa, *pet* C), and subunit IV (17 kDa, function unknown). The electron transfer sequence within the complex is: $PQH_2 \rightarrow FeS_R \rightarrow cytf \rightarrow PC \rightarrow PSI$. Cyt*f* is held in the thylakoid by a single transmembrane span located near its hydrophobic C-terminus, with a globular hydrophilic segment protruding into the lumen. Concomitant with electron transfer, the complex also translocates protons from the stromal to the lumenal side of the membrane.

Other functions of the $b_6f$ complex are electron transfer from reduced Fd to PC via $b_6$ during cyclic photophosphorylation, and possibly in the regulation of light energy distribution between the two photosystems (in response to ATP demand) via LHC II phosphorylation-dephosphorylation reactions. It has been postulated that the lateral movement of phosphorylated LHC II from the grana ($PSII_a$) membranes is accompanied by a simultaneous movement of the cyt $b_6f$ from the grana to the stromal thylakoids ($PSII_B$ and PSI).

The mechanism of proton translocation across the thylakoid membrane mediated by the $b_6f$ complex is still not understood. There are many similarities (and differences) in the structure, composition and function of the chloroplast $b_6f$ complex and the *bc* complex in the mitochondrial respiratory chain of mammals and bacteria. Cytochrome *f*, so named because it was detected by its unique absorption at 555 nm in leaves (Latin-*frons*), is truly a *c*-type cytochrome. Mitchell had proposed a *Q cycle* to explain the mechanism of proton translocation via the mitochondrial ubiquinone-$bc_1$ complex, the key step being an electrogenic electron transfer through the site of quinol oxidation. Does a similar Q cycle operate in the chloroplast cyt $b_6f$ complex? The electron and protein transfer across the membrane can be measured from light-induced changes in the absorption at 320 nm due to proton uptake. The participation of the b-type haems in the reaction sequences can be detected from the electrochromic absorption changes at 515/524 nm that occur when thylakoids are illuminated by single turnover flashes.

Proponents of the Q cycle operation propose that during electron transport $FeS_R$ accepts one electron from $PQH_2$ ($Q_B$ doubly reduced by PSII picks up two stromal protons and forms $PQH_2$) and donates it to cyt*f*. Simultaneously, $PQH_2$ transfers its second electron to one of the b haems (low potential) and releases the two protons to the lumen. The electron from the low potential b haem is transferred to PQ via the high potential b haem of the cyt $b_6$. The PQ gets doubly reduced after two cycles of electron transfer; doubly reduced PQ picks up two more stromal protons and enters the quinone pool.

Although some authors believe that the $b_6f$ complex functions electrogenically during electron transport, there are others who consider the evidence as equivocal. Furthermore, if the Q cycle is truly operative, $2H^+$ should be translocated per electron transferred to cyt *f*; experimentally determined values of $H^+/e^-$ range from 2 to 3. The mechanism of proton translocation across the thylakoid membrane is still an open question; may be a modified catalytic Q cycle or a proton pump is involved in the translocation.

# CHAPTER 14

# Bacterial Photosynthesis

Microorganisms are classified on the basis of light/organic/inorganic molecules requirements as phototrophs (use light as their energy source), chemotrophs (obtain from the oxidation of chemical compounds, either organic or inorganic). Microorganisms also have only two sources fork hydrogen atoms or electrons.

**Lithotrophs (*rock-eaters*):** They use reduced inorganic substances as their electron source.

**Organotrophs:** They extract electrons or hydrogen from organic compounds.

Photolithotrophic autotrophs (often called photoautotrophs): They use light energy and $CO_2$ as carbon source. Eucaryotic algae and blue-green bacteria (cyanobacteria) employ water as the electron donor and release oxygen. Purple and green sulfur bacteria cannot oxidize water but extract electrons from inorganic donors like hydrogen, hydrogen sulfide, and elemental sulfur.

**Chemoorganotrophic heterotrophs** (often called chemohetertophs, or even heterotrophs): They use organic compounds as source of energy, hydrogen, electron, and carbon for biosynthesis. Essentially all pathogenic microorganisms are chemoheterotrophs. The other two classed have fewer microorganisms but often are very important ecologically.

**Photoorganotrophic heterotrophs:** Some purple and green bacteria are photosynthetic and use organic matter their electron donor and carbon source. These are common inhabitants of polluted lakes and streams. Some of these bacteria also can grow as photoautotrophs with molecular hydrogen as electron donor.

**Chemolithotrophic autotrophs:** They oxidizes reduced inorganic compounds such as iron, nitrogen, or sulfur molecules to derive both energy and electrons for biosynthesis. Carbon dioxide is the carbon from organic energy sources. The carbon sources may be varies but sometimes $CO_2$. These organisms contribute greatly to the chemical transformations of elements (e.g., the conversion of ammonia to nitrate or sulfur to sulfate) that continually occur in the ecosystem.

## NUTRIENT REQUIREMENTS FOR MICROBIAL GROWTH

Ninety five per cent of cell dry weight of an organism is made of a few major elements; carbon, oxygen, hydrogen, nitrogen, sulphur, phosphorus, potassium, calcium, magnesium, and iron. These are called **macroelements** or macronutrients because they are required by microorganisms in relatively large amounts. The first six (C, O, H, N, S, and P) are component of carbohydrates, lipids, proteins, and nucleic acids. The remaining four macroelements exists in the cell as cations and play a variety of roles.

The microorganisms require severa **trace elements** or microelments or micronutrients besides macroelments. The trace elements- manganese, zinc, cobalt, molybdenum, nickle, and copper- are needed by most cells. Trace elements are normally a part of enzymes and cofactors, and they aid in the catalysis of reactions and maintenance of protein structure. Vitamins are needed for the promotion of physical growth (as growth factor).

The major elements (C, O, H, N, S, P) are needed in gram quantities in a liter of culture medium. The minor elements (K, Ca, Mg, Fe) often are required in milligram quantities. Trace elements (Mn, Zn, Co, Mo, Ni, Cu) must be available in microgram amounts. The important elements are:

**Carbon:** It is required for the skeleton or backbone of all organic molecules, and molecules serving as carbon sources usually also contribute both oxygen and hydrogen atoms. One carbon source for which this is not true is carbon dioxide ($CO_2$) because it is oxidized and lack hydrogen. Probably all macroorganisms can fix $CO_2$- that is, reduce it and incorporate it into organic molecules. However, by definition, only **autotrophs** can use $CO_2$ as their sole or principal source of carbon. Many microorganisms are autotrophic; many of these are photosynthetic, but some autotrophs, oxidize inorganic molecules to obtain energy.

The reduction of $CO_2$ is a very energy expensive process. Thus, many microorganisms cannot use $CO_2$ as their sole carbon source but must rely on the presence of more reduced, complex molecules for a supply of carbon. Organism that reduce preformed organic molecules as carbon sources are called **heterotrophs.** These preformed molecules normally come from other organisms. Most heterotrophs use organic nutrients as a source of both carbon and energy. For example, the glycolytic pathway traps energy as ATP and NADH and also produce carbon skeleton for use in biosynthesis.

**Tabel 14.1:** Source of Carbon, Energy, and Hydrogen/Electrons

| | | |
|---|---|---|
| Carbon Source | Autotrophs | $CO_2$ sole or prinicple biosynthetic carbon source |
| | Hetertophs | Reduced, performed, organic molecules from other organisms |
| Energy sources | Phototrophs | Light |
| | Chemotrophs | Oxidation of organic or inorganic compounds |
| Hydrogen or Electron Sources | Lithotrophs | Reduced inorganic molecules |
| | Organotrophs | Organic molecules |

Nutritional characteristic of microorganisms is extraordinary flexible, specially with respect to carbon sources. They utilise different carbon sources

| | |
|---|---|
| Actinomycetes | It can degrade amyl alcohol, papraffin, and even rubber. |
| *Pseudomonas cepacia* | It can use over 100 different carbon compounds. |
| Methylotrophic bacteria | They can metabolize only methane, methanol, carbon monoxide formic acid, and a few related one-carbon molecules. |
| *Leptospira* (parasitic members) | They use only long-chain fatty acids as their major source of carbon and energy. |

## AUTOTROPHIC OR AUTOPHYTIC BACTERIA

Those bacteria which are able to synthesize carbohydrate from carbon dioxide and water like the green plants are referred as autotrophic bacteria. They are of two types:

### Photosynthetic or Photoautotrophic Bacteria

There are few species of bacteria which contain two photosynthetic pigments for manufacturing organic food from inorganic sources. They are principally of two kinds viz., *purple sulphur* or red bacteria and *sulphur bacteria*. These are often found in sulphur springs where hydrogen sulphide is often present. Purple or red bacteria contain only one type of chlorophyll pigment called bacteriochlorophyll along with carotenoids. Green bacteria contain two types of chlorophyll:

i) chlorobium and
ii) chlorophylls (not chlorophyll a)

Therefore, bacteria are only plants which manufacture their food without chlorophyll a, in presence of light. Moreover, the process of photosynthesis in photosynthetic bacteria is for example of *Rhodosprillum* spp. and *Rhodoseudomonas* spp. is different to those of photosynthetic plants. Free oxygen is not liberated during photosynthesis. In them hydrogen sulphide serves as a source of hydrogen e.g., *chlorobium limicola*. The involvement of chemical changes in bacterial photosynthesis may be represented as follows:

$$6CO_2 + 12H_2S \xrightarrow[\text{Bacterial Chlorophyll}]{\text{Light}} C_6H_{12}O_6 + 6H_2O + 12S$$

## PHOTOSYNTHESIS IN BACTERIA AND CYANOBACTERIA

While the green plants have been the prime exploiters of the energy of light, photosynthetic bacteria are capable of using light have also evolved. These bacteria possess a chlorophyll system that is certainly analogous to that of the green plant, and chlorophyll that differs from plant chlorophyll only in minor chemical details. Bacterial chlorophyll is green. In some photosynthetic bacteria which are purple, the colour is caused by an abundance of red or brown pigments which mask the green colour of the bacterial chlorophyll. But this system seems to incapable of producing oxygen. Only green plant produce oxygen, not the photosynthetic bacterium. Photosynthesis in purple bacteria and green bacteria is driven only by **one photoreaction** during which no oxygen is evolved. The electron donors are sulphur compounds like $H_2S$ or organic substances taken up from the environment, rather than $H_2O$.

Cyanobacteria or Blue green algae which are true in prokaryotic organization have an oxygenic photosynthesis with two light reactions and are closely resemble that of higher plants. Similar to that of higher plants, electrochemical gradient is generated for ATP synthesis in photosynthetic bacteria also. Plant photosynthesis is part of more general reaction, as follows:

Plant........................................................ $CO_2 + 2H_2O \longrightarrow (CH_2O) + O_2 + H_2O$
Bacteria.................................................... $CO_2 + 2H_2A \longrightarrow (CH_2O) + 2A + H2O$

Here it is apparent that oxygen in the plant system is equivalent to the "2A" of the more general system, which is, indeed, the **oxidized fragment** of $H_2A$.

The photosynthetic bacteria may be separated on the basis of nature of $H_2A$ which they are capable of using.

**First**: Those which can utilize organic materials as $H_2A$. A simple example is a reaction carried out by *Rhodopseudomonas* spp. which in light can dehydrogenate isopropyl alcohol but can not use the resulting acetone.

**Second**: Those capable of using an inorganic material as $H_2A$. This material is characteristically $H_2S$, and in light the bacteria generally form S (compare $2H_2S \longrightarrow S_2$ with $2H_2O \longrightarrow O_2$). There are two kinds of such bacteria: (i) Purple bacteria and (ii) Green bacteria.

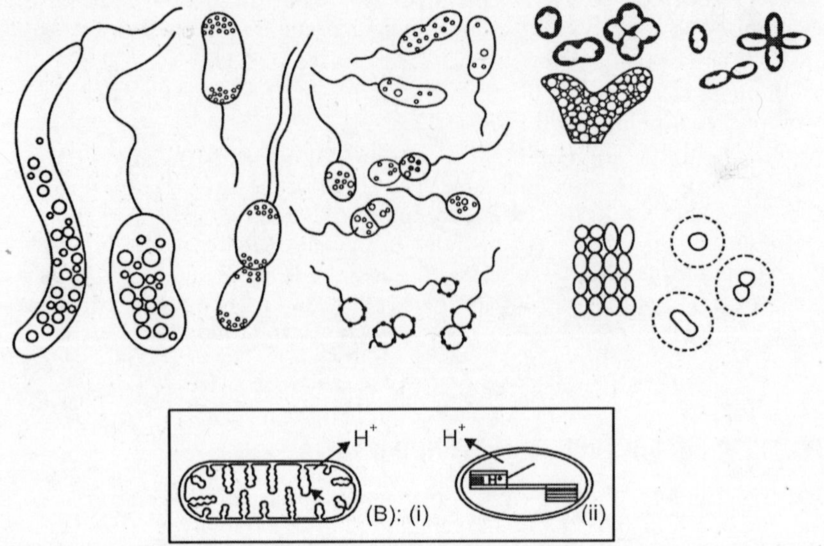


**FIG. 14.1:** (A) Some sulphur purple bacteria (Chromatiaceae) and non-sulphur becteria (Rhodosprillaceae). (B) Proton transport in illuminated cell and organelles (i) *Rhodopseudomonas sphaeroides* (ii) chloroplast of higher plant.

## PURPLE NON-SULPHUR BACTERIA

They use organic material as a source of $H_2A$. Their phorosynthetic metabolism involves the reduction of $CO_2$ at the ultimate expense of organic compounds. They cannot use $H_2S$ as a hydrogen donor. These bacteria are classified into sulphur-containing Chromatiaceae (use alcohol and carbonic acids as electron donors) and the sulphur-free Rhodospirillaceae (usually use $H_2S$ as electron donor).

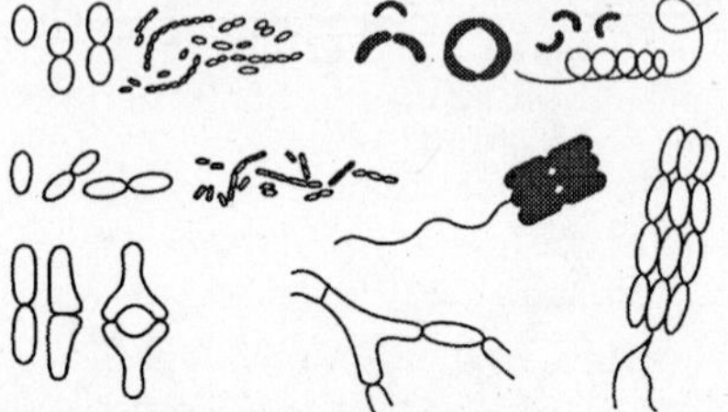

**FIG. 14.2:** Phototrophic green sulphur bacteria (Chloroblaceae).

### Photosynthetic Apparatus of Purple Bacteria

The photosynthetic apparatus of these bacteria is situated in chromatophores which resemble thylakoids. The origin of vesicles is from invaginations cytoplasmic membrane and are interconnected. It is proposed that light-harvesting protein complex consists of the bacteriochlorophyll protein B-800-850 and $B_{870}$ (maximum absorption at $P_{870}$). Such orientation of antenna pigments allows a directed excitation transport (similar process is found in higher plants). There are several cytochromes available in the reaction centre complex.

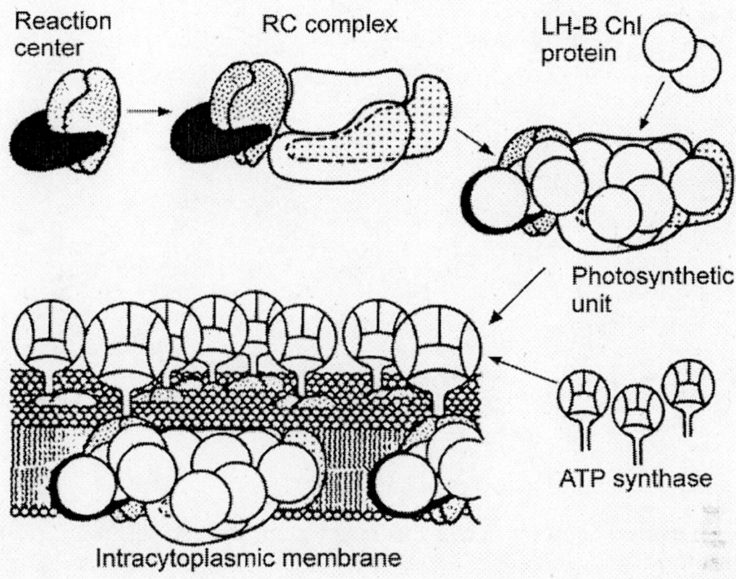


**FIG. 14.3:** Intracytoplasmic membrane (photosynthetic part) of photosynthetic purple bacteria and the individual particles shown separately (from left to right). LH-B-Chl = Light harvesting bacteriochlorophyll protein.

## Purple Bacteria: Electron Transport Chain

After excitation the electrons are transported from $P_{870}$, which is assumed to operate as a dimer, to acceptor X, presumably a bacteriophaeophytin (in analogy to phaeophytin in PS II of higher plants). The electrons follow a cyclic via the secondary acceptors $Q_a$ and $Q_B$, the quinone pool and several cytochromes and back to $P_{870}$. *Chromatium* probably use napthoquinone as $Q_A$ while other purple bacteria use ubiquinone. Similar to the higher plants, the protons are transported through the membrane into the chromatophore interior. In purple bacteria the pH difference between cytoplasm and the inner surface of the chromatophore is only 2 pH units (in thylakoids of higher plants the difference of pH remains 4 pH units in). The electrochemical gradient of the membrane (membrane potential) remains 200 mV which help in ATP synthesis. The reduction of NAD takes place by an inverted electron transport driven by ATP.

## GREEN BACTERIA

The cells appear green because the colour of the chlorophyll is not masked by other pigments. Their photosynthetic metabolism involves the reduction of $CO_2$ at the ultimate expense of $H_2S$ and other reduced sulphur compounds. No growth factors are required, growth occurring in mineral medium. They are obligately photosynthetic and strictly anaerobic.

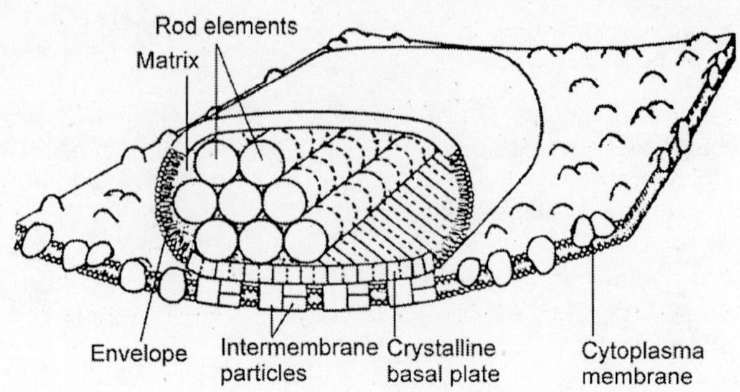


**FIG. 14.4:** Model of a chromosome with the associated plasma membrane of the green bacterium *Chlorobium limicola*.

## Photosynthetic apparatus of Green bacteria

These bacteria contain chlorosomes (closely structurally resemble the cyanobacterial phycobilins) as photosynthetic organelles. These are vesicles, closely connected with the cytoplasmic membrane by a crystalline nasal plate. The cytoplasm membrane thought to contain the non-crystalline light harvesting complex and the crystalline basal plate- the bacteriochlorophyll *a*-protein complex. In side the chlorosomes, which is surrounded by a single layer of galactolipids, rodlike elements are present and thought to contain bacteriochlorophyll *c*-protein complexes.

The reaction centre pigment is $P_{840}$ which passes electrons to a primary acceptor with a very negative potential. In green bacteria the acceptor can pass electrons to NAD via ferredoxin. Electrons are received from $H_2S$ and molecular sulphur is produced and deposited outside the cell. Usually non-cyclic electron transport occurs, however, cyclic one is also been reported.

# BIOCHEMICAL REACTIONS OF PHOTOSYNTHESIS (DARK REACTION)

Dark reaction is the pathway by which $CO_2$ is reduced to sugar. Its components are found in the stroma of the chloroplast. This reaction does not actually require darkness; it is merely light-dependent. Since $CO_2$ is an energy poor compound, its covnersion to en energy-rich carbohydrate involves a sizable jump up the energy ladder. This is accomplished through a series of complex steps involving small bits of energy. The $CO_2$ assimilation takes place both in light and darkness when the substrates NADPH and ATP are available. Because of the need for NADPH as a reductant and ATP as energy equivalent $CO_2$ fixation is closely linked to the light reactions.

The end products of photosynthesis, namely sugars or starchs are synthesized from PGA by reversed sequence of EMP pathway of glycolysis. Some PGA is used to regenerate RuDP for accepting some more $CO_2$ molecules. This is a regular cyclic-reactions that ultimately produce the end-products of photosynthesis. and this cyclic metabolic pathway has been termed Calvin cycle, after the work of Calvin who eventually received Nobel Prize for his work in 1961.

## The Reductive Pentose Phosphate (RPP) or Calvin Cycle: Carboxylation Phase

This scheme has been grouped into three phases.

### *First Phase*

The first phase is the carboxylation phase, involves the single reaction catalyzed by ribulose-1,5-bisphosphate carboxylase-oxygenase (also called rubisco). The reaction involves the carboxylation of five-carbon compound, ribulose-1,5-bisphosphate, and yield two molecules of 3-phosphoglyceric acid. In the presence of the enzyme rubisco, $CO_2$ adds to the enediol form of ribulose-1,5-bisphosphate to form an unstable β-keto acid that undergoes hydrolytic cleavage to form two molecules of 3-phosphologlyceric acid. The equilibrium of the reaction is far to the right.

### *Second Phase*

A second phase of the RPP cycle is known as **reduction phase**. It consists of two reactions previously encountered in glycolysis. ATP and a reduced nicontinamide nucleotide are utilized. The first reaction involves the phosphorylation of 3-phosphoglycerate by ATP to form 1,3-bisphosphoglycerate. while in the second reaction 1,3-bisphosphoglyceric acid is reduced by NADPH in the presence of a NADP-specific glyceraldehyde-3-phosphate dehydrogenase. ATP and NADPH activate the chloroplast enzyme. In these two reactions all of the NADPH and two-thirds of the ATP are required to drive the RPP cycle.

### *Third or Regeneration Phase*

The third reaction of the RPP cycle is the regeneration phase of ribulose-1,5-bisphosphate which is necessary to keep the cycle in normal operation. Three molecules of $CO_2$ being assimilated. At the end of the reduction phase, a total of 18 carbon atoms will be present in six molecules of glyceraldehyde-3-phosphate.

One of these can be considered as the photosynthetic products after incorporating 3 $CO_2$ molecules through the RPP cycle. The 15 carbon atoms in the other five molecules of glyceraldehyde-3-phosphate are, in turn, converted by the reactions of the regeneration phase into three moles of ribulose-5-phosphate. The phosphorylation of these pentose phosphates into ribulose- 1,5-bisphosphate completes the RPP cycle and produces the substrates for assimilation of another three moles of $CO_2$. At the last ATP is consumed in the RPP cycle. Examination of steps 2,3, and 13 show that a total of 2 NADPH and 3 ATP are required for each $CO_2$ assimilated by this process.

# CHAPTER 15

# Path of Carbon

## SUPPLY AND DEMAND OF $CO_2$ IN PHOTOSYNTHETIC PROCESS

The rate of photosynthetic carbon assimilation is determined by both supply and demand of $CO_2$. The supply of $CO_2$ to the chloroplast is governed by diffusion in the gas and liquid phases. It can be limited at several points in the pathway from the surrounding the leaf to the site of carboxylation in side. The demand for $CO_2$, is determined by the rate of processing the $CO_2$ in the chloroplast which is governed by the structure and biochemistry of the chloroplast, by environmental factors such as irradiance and plant demand for carbohydrates. Limitations imposed by either supply or demand can regulate overall rate of carbon assimilation.

Dark reaction is the pathway by which $CO_2$ is reduced to sugar. Its components are found in the stroma of the chloroplast. This reaction does not actually require darkness; it is merely light-independent. Since $CO_2$ is an energy poor compound, its conversion to an energy-rich carbohydrate involves a sizable jump up the energy ladder. This is accomplished through a series of complex steps involving small bits of energy. The $CO_2$ assimilation takes place both in light and darkness when the substrates NADPH and ATP are available. Because of the need for NADPH as a reductant and ATP as energy equivalent, $CO_2$ fixation is closely linked to the light reactions. During evolution three different ecological varients have evolved with different $CO_2$ incorporation mechanism: $C_3$, $C_4$ and CAM plants. In $C_3$ plants the first detectable $CO_2$ fixation product is a $C_3$ compound - phosphoglyceric acid, in $C_4$ plants we find $C_4$ compounds such as malate or oxalate. in CAM plants, most of which are of crassulaceae, COx is incorporated into $C_4$ acids during the night: these are used as $CO_2$ reservoir for $CO_2$ fixation during day-time, which follows the same pattern in $C_3$ plants. Plants such as crown, sugar cane, crabgrass, and other that can grow in a hot, dry environment must keep their stomata (the gas exchange pores in the leaves) closed much of the time to prevent excessive loss of moisture.

This causes the $CO_2$ level within the leaf to fall below the $K_m$ of ribulose 1,5-bisphosphate carboxylase; under these conditions, photorespiration is greatly favoured over photosynthesis. To avoid this problem, $C_4$ plants have evolved two steps pathway of $CO_2$ fixation that involves two types of cells: mesophyll cells - adjacent to surface of the leaf, and interior bundle sheath cells.

### $CO_2$ Response Curve

The response of photosynthetic rate to $CO_2$ concentration is the basic tool to analyze the demand for $CO_2$. The graph giving $CO_2$ assimilation (A) as a function of intercellular $CO_2$ partial pressure ($p_i$) is generally referred to as the A-pi curve, or A-$c_i$ when $CO_2$ concentration is given as mole fraction in air. There is no net $CO_2$ assimilation until the production of $CO_2$ in respiration (mainly photorespiration, but also some dark respiration that occurs in the light) is fully compensated by the fixation of $CO_2$ in photosynthesis. The $CO_2$ pressure at which this is reached is the $CO_2$-compensation point ($\tau$). In $C_3$ plants this is largely determined by kinetic properties of Rubsico, with values for $\tau$ in the range of 4 to 5 Pa.

## $CO_2$ – Stomatal and Boundary Layer Conductances

The rate of supply of $CO_2$ to the chloroplast can be described as a diffusion process (**supply function**). To analyze diffusion limitations it is convenient to use the term resistance because resistance for can be summed to arrive at the total resistance for the pathway. When considering fluxes, however, it is more convenient to use conductance, which is the reciprocal of resistance, because the flux varies in proportion to the conductance. Leaf conductance for $CO_2$ transport can be derived from measurements on leaf transpiration, which can also be described by Fick's law. The total leaf resistance for water vapor transfer, is largely composed of two components that are in series: the **boundary layer resistance** and the **stomatal resistance**. For the transport of $CO_2$ from the substomatal cavity to the chloroplast, an internal conductance also play a role. As a consequence, the internal resistance and partial $CO_2$ pressure in the chloroplast is less than $p_i$. The evidence for this aspect largely comes from carbon isotope ratio of the products of photosynthesis.

## CARBON: PRINCIPLES OF METHODOLOGY

Several techniques have been largely responsible for the successful analysis of the thermochemical dark reaction in photosynthesis. One of these is the use of radioactive carbon making it possible to tag the $CO_2$ ($C^{14}O_2$) which is incorporated into organic compounds during photosynthesis. The basic experimental approach was as follows:

In a plant that is carrying out photosynthesis at a steady rate, $CO_2$ is being converted to carbohydrates and other plant compounds through a series of intermediates.

$$CO_2 \rightarrow \text{Compound A} \rightarrow \text{Compound B} \xrightarrow[\text{Acetyl CoA}]{\text{Amino acid}}$$

$$\text{Compound C} \xrightarrow[\text{Lipids}]{\text{Proteins}} \text{Carbohydrate}$$

If, at time zero, radioactive $CO_2$ ($^{14}CO_2$) is introduced into the system, some of the labelled carbon atoms are be converted to these different products, and during the time it takes for this to occur all the intermediates are become labelled. After a relatively shorter period of time, the photosynthesizing plant is plunged into hot alcohol to inactivate its enzymes and stop all reactions. If the reaction is stopped in a short period (just after the formation of first stable intermediate), only compound A will be labelled and will be known as **first product** of $CO_2$ fixation.

Paper chromatography technique also helped in separating small amount of substances which appear even after second illumination can be identified.

### Method

Using these tools Calvin's group was able to identify the early stable intermediates in the path of carbon from $CO_2$ to plant metabolites. They used suspension of *Chlorella pyrenoidosa* or *Scenesdesmus obliques* (because they could be handled with ease) grown in culture medium, at a constant rate in presence of light and $CO_2$. Radioactive $CO_2$ was introduced into the reaction mixture at zero time, and a period of time was allowed to elapse. The cells were killed and extracted in boiling alcohol. The soluble constituents of alcohol solution were analysed by 2-dimentional paper chromatography and autoradiography. The compounds were identified in the chromatographs both their radioactive emissions and by chemical reactions. For confirmation, the spots were cut, extracted and rechromatographed with authentic samples.

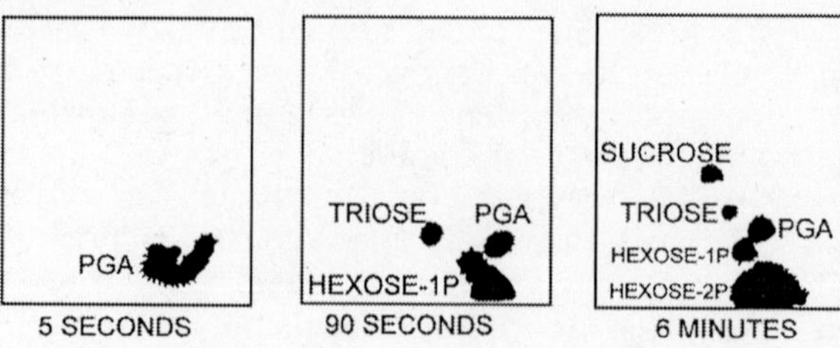


**FIG. 15.1:** $^{14}CO_2$ supplied to the algal suspension. after an interval of 5, 90 seconds and 5 minutes different chemicals were synthesized. Black spots indicate $C^{14}$ incorporation in different compounds.

When algae were exposed to $^{14}CO_2$ for 30 seconds, hexose phosphate, triose phosphate and phosphoglyceric acid were labelled. With long periods, these compounds as well as amino acids, organic acids, lipids, and proteins were labelled. With 5 second exposure, however, most of the radioactive carbon was located in 3-phosphoglyceric acid and within this compound, the carboxyl group contained the majority of the radioactivity. The results reveal that phosphoglyceric acid was formed by the carboxylation of some unknown compounds containing 2-carbon atoms. Further closer analysis of photosynthetic products revealed that sedodoheptulose-7-phosphate and several phosphorylated pentose are also produced during the assimilation of $^{14}CO_2$. Since PGA is a 3-carbon compound, it was thought that $CO_2$ was accepted by a 2-molecule to form the 3-carbon PGA. But no such compound was detectable in plant.

Exposure time (second) or $^{14}CO_2$ product in which $^{14}C$ appeared as metabolite of Calvin cycle

| Product | Time of exposure (Second) |
|---|---|
| 3-phosphoglyceric acid (3-PGA) | 5 |
| Triose phosphates | 30 |
| 8-hexose phosphate | 60 |

Later on, A. A. Benson working in the same laboratory found that a 5-carbon sugar ribulose diphosphate (RuDP) was the acceptor molecule of $CO_2$. Benson, further conducted many experiments in this regard. In one experiment in which light was switched off after a short period of $^{14}CO_2$ feeding. PGA accumulated and RuDP disappeared completely. In another complementary experiment in which the light was kept on but $^{14}CO_2$ was withdrawn RuDP accumulated and PGA disappeared. After accepting $CO_2$, plant synthesize many unstable carbon intermediates which cleave into 2 molecules of PGA. In vitro studies further suggested that unstable 6-carbon compound is formed when RuDP accepted $CO_2$.

The end products of photosynthesis, namely sugars or starchs are synthesized from PGA by reversed sequence of EMP pathway of glycolysis. Some PGA is used to regenerate RuDP for accepting some more $CO_2$ molecules. This is a regular cyclic-reaction that ultimately produce the end-products of photosynthesis and this cyclic metabolic pathway has been termed **Calvin cycle.** Calvin received Nobel

Prize for his work in 1961.

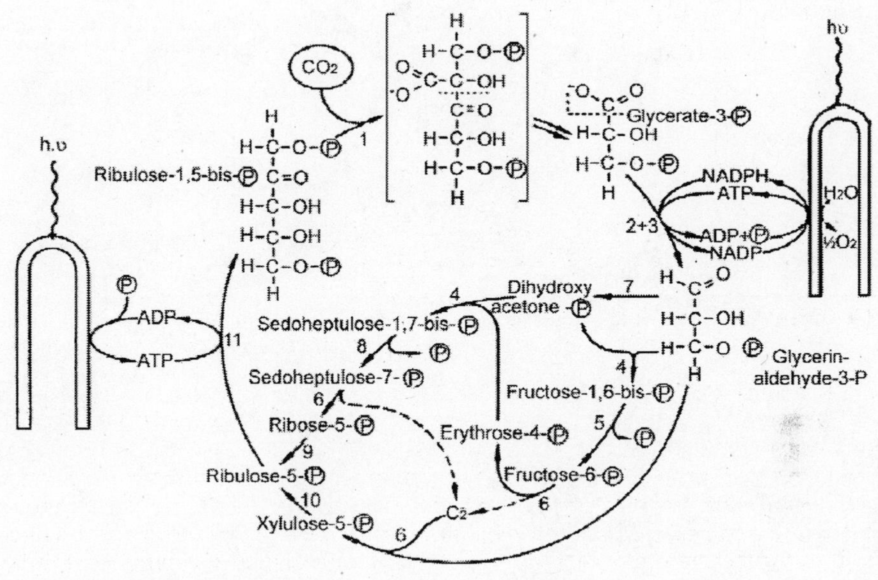


**FIG. 15.2:** Calvin cycle.

## THE REDUCTIVE PENTOSE PHOSPHATE (RPP) OR CALVIN CYCLE

Horecker, Racker, and others established the relationships between trioses, tetroses, pentoses, hexoses, and heptoses. Studies on $^{14}C$ labelling of the sugars produced during short periods of photosynthesis permitted Calvin's laboratory to postulate the operation of a reductive pentose phosphate (RPP) cycle. This cycle involves essentially only one new reaction, the **carboxylation** of ribulose- 1,5-bisphosphate. The remainder of the reactions are identical or similar to reactions encountered previously in glycolysis and pentose phosphate metabolism. The enzymes required to catalyze the reactions in the RPP cycle occur in the stromal phase of chloroplasts.

### Carboxylation Phase

This scheme has been grouped into three phases. The **first** phase is the carboxylation phase, involves the single reaction catalyzed by ribulose-1,5-bisphosphate carboxylase-oxygenase (also called **rubisco**). The reaction involves the carboxylation of five-carbon compound, ribulose-1,5-bisphosphate, and yield two molecules of 3-phosphoglyceric acid. In the presence of the enzyme rubisco, $CO_2$ adds to the enediol form of ribulose-1,5-bisphosphate to form an unstable beta-keto acid that undergoes hydrolytic cleavage to form two molecules of 3-phosphologlyceric acid. The equilibrium of the reaction is far to the right.

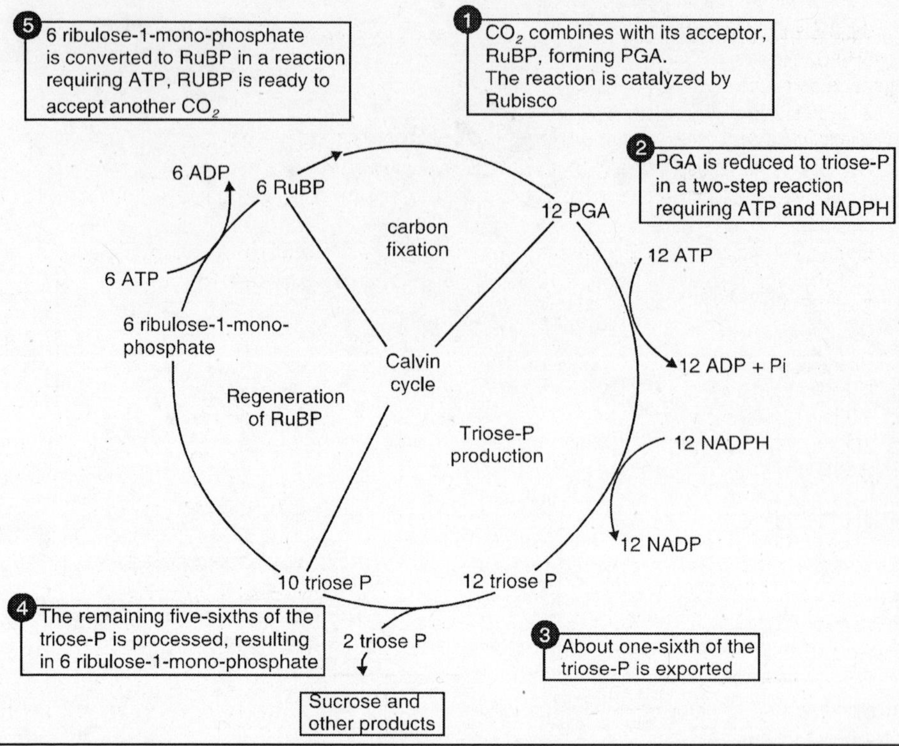


**FIG. 15.3:** Schematic representation of the photosynthetic carbon reduction cycle (Calvin cycle) that shows major steps: carbon fixation, triose-P production, and regeneration of RuBP. 1: $CO_2$ combines with its substrate, ribulose-1,5-bisphosphate (RuBP), catalyzed by ribulose bisphosphate carboxylase/oxygenase (Rubisco), producing phosphoglyceric acid (PGA). 2: PGA is reduced to triose-phosphate (triose-P), in a two-step reaction; the reaction for which ATP is required is the conversion of PGA to 1,3-bisphosphoglycerate, catalyzed by phosphoglycerate kinase. 3 and 4: part of the triose-P is exported to the cytosol, in exchange for $P_i$; the remainder is used to regenerate ribulose-1-monophosphate. 5: ribulose-1-monophosphate is phosphorylated, catalyzed by ribulose-5-phosphate kinase, producing RuBP.

## Reduction Phase

A **second** phase of the RPP cycle is known as **reduction phase**. It consists of two reactions previously encountered in glycolysis. ATP and a reduced nicontinamide nucleotide are utilized. The first reaction involves the phosphorylation of 3-phosphoglycerate by ATP to form 1,3-bisphosphoglycerate. While in the second reaction 1,3-bisphosphoglyceric acid is reduced by NADPH in the presence of a NADP-specific glyceraldehyde-3-phosphate dehydrogenase. ATP and NADPH activate the chloroplast enzyme. In these two reactions all of the NADPH and two-thirds of the ATP are required to drive the RPP cycle.

**Carboxylation phase**

1. 3 Ribulose- 1,5-bisphosphate + 3 $CO_2$ + 3 $H_2O$ $\rightarrow$ 6 3-Phosphoglycerate

**Reduction phase**

2. 6 3-Phosphoglycerate + 6ATP$\rightarrow$ 6 1,3-Bisphosphoglycerate + 6 ADP

3. 6 a,3-Bisphosphoglycerate + 6 NADPH + 6 $H^+$ → 6 Glyceraldehyde-3-phpsphate+ 6 $NADP^+$ + $6H_3PO_4$

**Regeneration phase**

4. 2 Glyceraldehyde-3-phosphate → 2 Dihydroxy-acetone phosphate
5. Glyceraldehyde-3-phosphate + dihydroxy-acetone phosphate → Fructose-1,6-diphosphate
6. Fructose-1,6-bisphosphate + $H_2O$ → Fructose-6-phosphate + $H_3PO_4$
7. Fructose-6-phpsphate + glyceraldehyde-3-phosphate → Xylulose-5-phosphate + Erythrose-4-phosphate
8. Erythrose-4-phosphate + dihydroxy-acetone phosphate → Sedoheptulose-1,7 bisphosphate
9. Sedoheptulose-1,7-bisphosphate + $H_2O$ → Sedoheptulose-7-phosphate + $H_3PO_4$
10. Sedoheptulose-7-phosphate + Glyceraldehyde-3-phosphate → Ribose-5-phosphate + xylulose-5-phosphate
11. Ribose-5-phosphate → Ribulose-5-phosphate
12. 2 Xylulose-5-phosphate → 2 Ribulose-5-phosphate
13. 3 Ribulose-5-phosphate + 3 ATP -→- 3 Ribulose-1,5-bisphosphate + 3 ADP
14. $3CO_2$ + 9 ATP + 6 NADPH = 6 $H^+$ + 5 $H_2O$ → Glyceraldehyde-3-phosphate + 9 ADP + 6 $NADP^+$ + 8 $H_3PO_4$

## Regeneration Phase

The **third** reaction of the RPP cycle is the **regeneration phase** of ribulose-1,5-bisphosphate which is necessary to keep the cycle in normal operation. At the end of the reduction phase, a total of 18 carbon atoms will be present in six molecules of glyceraldehyde-3-phosphate.

One of these can be considered as the photosynthetic product after incorporating 3 $CO_2$ molecules through the RPP cycle. The 15 carbon atoms in the other five molecules of glyceraldehyde-3-phosphate are, in turn, converted by the reactions of the regeneration phase into three molecules of ribulose-5-phosphate. The phosphorylation of these pentose phosphates into ribulose-1,5-bisphosphate completes the RPP cycle and produces the substrates for assimilation of another three moles of $CO_2$.

At the last ATP is consumed in the RPP cycle. Examination of steps 2,3, and 13 show that a total of 2 NADPH and 3 ATP are required for each $CO_2$ assimilated by this process.

## RUBISCO: STRUCTURE AND FUNCTION

Ribulose bisphosphate carboxylase-oxygenase (RuBisCO) is probably the most abundant protein on earth; it constitutes about half of the leaf proteins. RuBisCO catalyzes the biological fixation of $10^{11}$ tonnes of $CO_2$ annually from the atmosphere. In the last two decades considerable research effort has been expended towards the elucidation of the mechanism of carboxylation by the enzyme and in the determination of its three-dimensional structure. Besides fundamental understanding, a main objective behind this research is to design, if at all possible, a RuBisCO molecule with an increased carboxylation to oxygenation activity ratio than is found in the naturally occurring enzyme.

RuBisCO isolated from oxygenic photosynthetic organisms is an octamer consisting of eight large subunits (LSU, molecular mass 50-55 kDa) and eight small subunits (SSU, molecular mass-12-15 kDa).

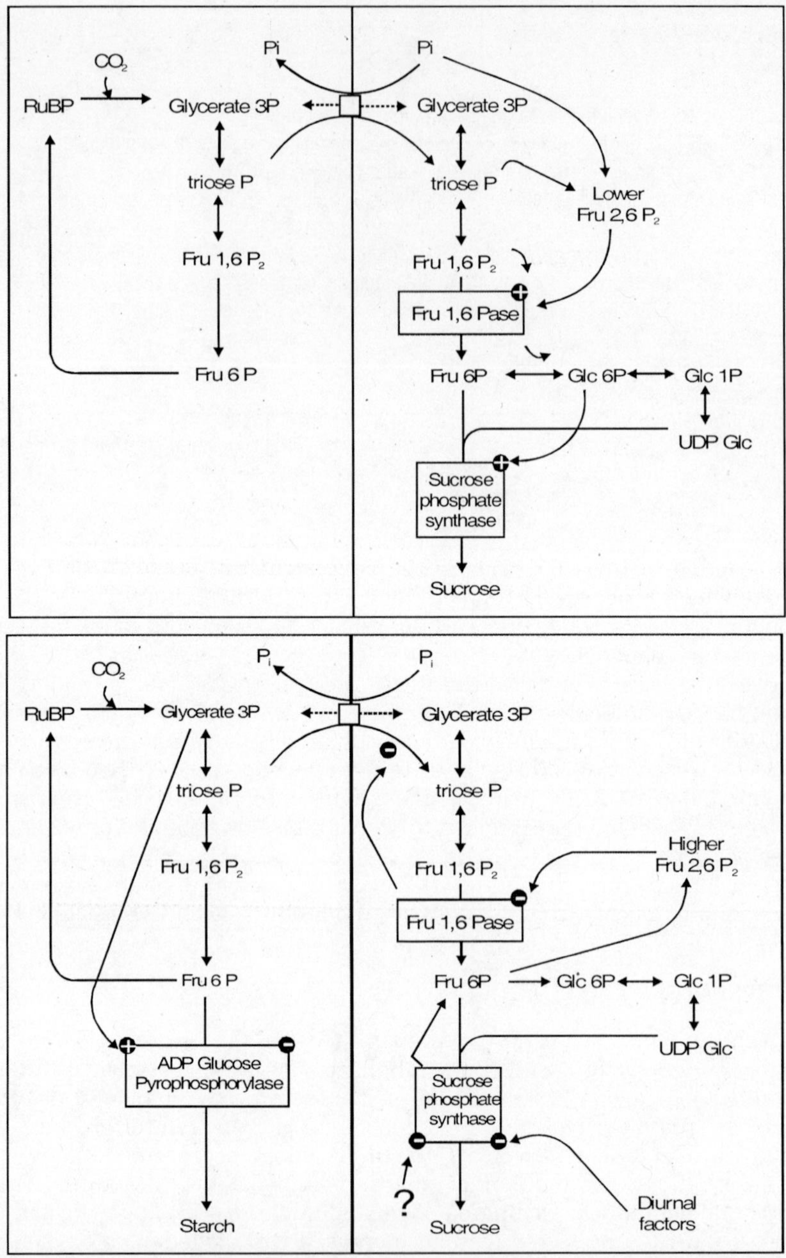


**FIG. 15.4:** The formation of triose-phosphate in the Calvin cycle. Triose-P is exported to the cytosol, in exchange for Pi (upper part), or used as a substrate for the synthesis of starch in the chloroplast (lower graph), when export of triose phosphate stagnates. Fructose 2,6 bisphate is a regulatory molecule, and not an intermediate glycolysis. The – and + signs refer to inhibition and slation of enzyme activities, respectively.

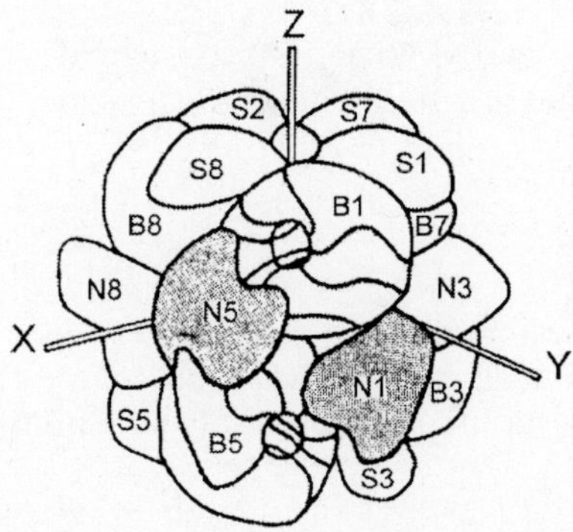


**FIG. 15.5:** The three dimensional relationship between two L-subunits and associated S subunits of Ribisco (Molecular configuration).

The LSU, which is coded by the chloroplast DNA, is synthesized within the chloroplasts; the catalytically active site(s) is located in the LSU. The SSU is coded by the nuclear genome and is assembled in the cytoplasm as a pre-protein that contains a domain of 50 extra amino acids at the aminoterminal end; this domain is cleaved off when the preprotein enters the chloroplast through the envelope. Assembly of the matured SSU and LSU within the chloroplast is mediated by a *chaperone*, the RuBisCO assembling protein.

In contrast to the RuBisCO enzymes of chloroplast stroma, the functionally active RuBisCO of *Rhodospirillum rubrum* is a dimmer of 50 kDa subunits which lack any SSU. The X-ray structure of R. *rubrum* enzyme has also been determined, and many residues in the enzyme molecule have been altered by site-directed mutagenesis. These studies have helped to identify the activator and substrate-binding sites on the RuBisCO molecule.

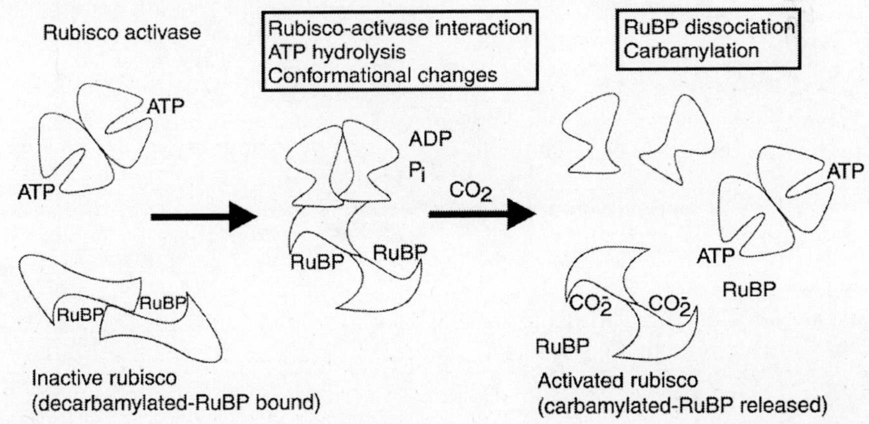


**FIG. 15.6:** Rubisco activase functions by removing bound RuBP from the decarbamylated inactive form of rubisco. The activase enzyme appears to couple the hydrolysis of ATP to the release of ribulose or other inhibitory pentose bisphosphates. Rubisco is then activated (carbamylated) by binding $CO_2$ and $Mg^{2+}$.

## Activation of RuBisCO

Prior to catalysis (either carboxylation or oxygenation), RuBisCO has to be converted from an inactive state to an active carbamylated form by condensation with a molecule of $CO_2$ and by subsequent addition of a divalent metal ion:

$$E + CO_2 \rightarrow E.CO_2 \xrightarrow[\text{Enzyme}]{-H_2O\ +\ Mg^{2+}} E.CO_2.Mg^{2+}$$

The carbamylation (condensation between $CO_2$ and enzyme – $NH_2$ group) occurs at a specific Lys residue at the activator site of the enzyme which is distinct from the catalytic site. Activation of RuBisCO also requires light. Recent work with a mutant of *Arabidopsis* (a plant amenable to mutations and genetic engineering) has led to the discovery of a protein named *rubisco activase* which is involved in the light-regulated activation of RuBisCO.

The action of rubisco activase is dependent on the availability of ATP and a membrane pH gradient, which are generated during light-induced electron transport. Many plant species produce 2-carboxy arabinitol-phosphate, an analogue of a six-carbon intermediate formed during RuBisCO catalysis, which inhibits the enzyme by occupying the catalytic site. It has been proposed that rubisco activase functions by removing this inhibitor from the catalytic domain of RuBisCO during transition from darkness to light. RuBisCO is one of the light-regulated chloroplast enzymes not controlled by the Fd-thioredoxin system.

## Catalysis by RuBisCO

The catalysis takes place in following three steps:

Translocation of a proton from the $C_3$ position of the substrate RuBP to the C2 position and formation of an enediol.

The enediol reacts with a molecule of CO2, forming the intermediate adduct 3-keto-2-carboxy arabinitol bisphosphate (CABP). This intermediate can be isolated in a stable form. Crystal structure analysis of the quaternary complex of spinach RuBisCO co-ordinated to CABP has confirmed the nature of the substrate-binding sites on the enzyme. These had been previously deduced from comparative sequence analysis, chemical modifications, and site-directed mutagenesis. The residues involved are thought to be a specific Lys and Glu.

The intermemdiate is hydrolytically cleaved into one molecule of PGA and a $C_2$ carbanion; the carbanion is immediately protonated to yield the second molecule of PGA.

Athough many derivatives of naturally occurring RuBisCO have been prepared by chemical or genetic modification of the molecule, none of these derivatives has manifested an enhanced ratio of carboxylation to oxygenation activity compared to the native enzyme.

As mentioned in 8.3, RuBisCO is one of the proteins subjected to extensive gene sequence determination in recent years. The genetic dissection of the enzyme in the green alga *Chlamydomonas reinhardtii* had identified three large-subunit (L) regions that should provide loci for future investigations via genetic engineering.

## Fallover of Rubisco

In vitro, the activity of purified rubisco from higher plants decreases with a half-time of several minutes after contact with substrate, ribulose-$P_2$. This loss in activity is termed as fallover. Three following reasons have been given for this fallover:

1. Rubisco becomes catalytically competent only after carbamylation by $CO_2$ at the catalytic site.
2. Readily reversible binding of ribulose-$P_2$ at a regulatory site.
3. Slow, tight-binding inhibition by isomerization or degradation products which accumulate in ribulose-$P_2$ preparations during storage.

## Carboxyarabinitol 1-Phosphate

Carboxyarabinitol 1-phosphate (**CA1-P**) a low molecular weight carbohydrate accumulates in the chloroplasts of numerous plant species when plants are placed in darkness or allow irradiance. Because of structural similarities to the carboxylation reaction intermediate, 2-carboxy-3-ketoarabinitol-1,5-bisphosphate, CA1-P binds tightly to the active site of Rubisco. CA1-P accumulation is readily detected in leaf extracts by measuring the inhibition of $Mg^{2+}/CO_{2-}$ activated Rubisco activity, since the decrease in maximal catalytic rate is a function of the proportion of active sites occupied by the inhibitor. Among species in which CA1-P accumulates, maximal synthesis of the inhibitor in the dark can cause a decrease in carboxylation activity ranging from 25 to 95 per cent. When leaves are subject to high irradiance CA1-P is metabolized, causing inhibition of Rubisco activity to be fully reversed within 10 min.

The metabolic reactions that synthesize CA1-P in the chloroplast of some plant species have not been elucidated. Metabolism of CA1-P by the stromal protein exhibited an apparent requirement for NADPH.

## Carbonic Anhydrase

Carbonic anhydrase (EC 4.2.1.1) catalyzes the freely reversible hydration of $CO_2$ to bicarbonate and it is present in both animals and plants. Since bicarbonate is a substrate for phosphoenolpyruvate carboxylase (the primary carboxylation enzyme in $C_4$ species), carbonic anhydrase is regarded as the first enzyme of the $C_4$ photosynthetic pathway. In $C_4$ plants this enzyme is localized in cytosol of mesophyll cells while it is essentially absent in the bundle sheath cells. In $C_3$ species it is found to be localized in chloroplasts. Recent immunological studies of this enzyme from $C_3$ and $C_4$ plants have indicated that there are isozymic forms of the enzyme in leaf tissue in both $C_3$ and $C_4$ plants which are similar in size but differ antigenically. Further studies have revealed that both $C_3$ and $C_4$ species show considerably high activity of this enzyme. Although the high activity of this enzyme in $C_4$ species is quite understandable, further attempts are needed to explore the role of this enzyme in $C_3$ plants.

## Light Activation of Photosynthetic Enzymes

Light was considered as an energy source for the primary photochemical processes for the production of NADPH and ATP after irradiation. In contrast to the primary reactions, the biochemical secondary reactions in the Calvin cycle reach their optimal activity only after a **lag phase** of a few minutes. The delay in $CO_2$ fixation is due to light activation of enzymes. Two mechanisms have been observed:

1. During initiation of photoreaction enzyme activation occurs. One of the

examples for this process is the activation of triose phosphate dehydrogenase by reduced thioredoxin. Reduced ferredoxin, generated by photosynthetic electron transport, reduces thioredoxin by means of thioredoxin reductase. Thioredoxin in turn reduces the disulfide groups of an enzyme, such as fructose diphosphatase or NADP glyceraldehyde dehydrogenase, causing activation by a conformational change. In darkness these light-regulated enzymes are deactivated by oxidation by, for instance, oxidized gluthathione, which is found in the chloroplast.

2. Change in pH or ion concentration in the stroma activates enzymes. Examples for this mechanism are RubP and PEP carboxylase which operate optimally at a pH of 8 and an increased $Mg^+$ concentration. Such conditions are generated in light by vectorial proton transport into the thylakoid interior and antiparallel $Mg^+$ transport into the stroma. This activity is also affected by a change in $CO_2/O_2$ partial pressures. Furthermore, RubP carboxylase is thought to be induced by visible light and UV-A radiation according to the above-described activation by mediators.

## REGULATION OF THE RPP CYCLE

The assimilation of $CO_2$ by the enzymes of the Reductive Pentose Phosphate (RPP) cycle is regulated primarily by the activity of the carboxylation enzyme, Rubisco which exists in an active from, that is a complex of $CO_2$ and $Mg^{2+}$ formed sequentially as follows.

i) The preliminary binding of $CO_2$ is a relatively slow process, but the reaction with $Mg^{2+}$ is more rapid and is accelerated at higher pH (8.5).
ii) Light strongly affects the above process, since the stromal compartment becomes more alkaline in the light as protons move into the thylakoids.
iii) The intermembrane space of the thylakoid becomes acidic by movement of $Mg^{2+}$ ions out into the stroma where they can complex to form the active form of rubisco.
iv) Since formation of the active Rubisco is reversible, the enzyme is converted to its inactive form in the dark when protons move out of the thylakoids and $Mg^{2+}$ leaves the stroma.

$$\underset{\text{Active}}{\text{Enzyme} + CO_2} \xrightarrow{\text{Slow}} \underset{\text{Inactive } Mg^{2+}}{\text{Enz-}CO_2} \xrightarrow{Mg^{2+}} \underset{\text{Active}}{\text{Enz-}CO_2\text{-}Mg^{2+}}$$

### Thioredoxin

Buchanan discovered thioredoxin in plants which contains a cysteine residue and has the property to oxidize and reduce. Carbon flow through the RPP cycle is also controlled at the steps catalyzed by fructose-1,6-bisphosphate phosphatase, sedoheptulose-1,7-biphosphate phosphatase, and phosphoribulose kinase. Such control involves the conversion of inactive forms of these enzymes to active enzymes by thioredoxin- a small molecular weight protein.

The reduced form of thioredoxin can activate the three enzymes listed above, presumably by reducing essential thiol groups in these proteins. Thioredoxin in turn is reduced in the light in the presence of thioredoxin reductase, an enzyme that catalyzes the reduction of oxidized thioredoxin by reduced ferredoxin produced by photosystem I. The overall process whereby these three enzymes are activated in the light may be represented as

The operation of the RPP cycle, of course, requires NADPH and ATP, both of which are produced in the light. Thus, the concentration of these co-substrates required in the reduction phase of the cycle increases in the light and is available for the reactions described.

## $C_4$ CYCLE (HATCH AND SLACK CYCLE)

In addition to Calvin cycle, an alternative (supportive to RPP) pathway for $CO_2$ fixation in photosynthesis was discovered by Kortschak et al. (1954). They reported the formation of $C_4$ dicarboxylic acids as primary products of photosynthesis in sugarcane. When leaves of sugarcane were treated with $^{14}CO_2$ for a short period, 3 types of 4-carbon dicarboxylic acids (oxaloacetate, malate and aspartate) were produced at initial stage. When the period of exposure was extended, label appeared in 3-phosphoglycerate. M.D. Hatch and C.R. Slack (1966) further investigated the incorporation of $^{14}C_4$ incorporation in several plants and then proposed an alternative pathway, which way later found supportive to Calvin cycle, because $CO_2$ was converted to carbohydrates through Calvin cycle in bundle sheath cells of C4 plants.

The cycle begins with the fixation of $CO_2$ by phosphoenol pyruvate carboxylase (PEP-carboxylase). $CO_2$ is first accepted by PEP molecule and is carboxylated by the addition of $CO_2$, hence form oxaloacetate (4-carbon product). Oxaloacetate is an unstable product and breaks down readily to malate (MDH enzyme) and aspartate through transamination (GOT enzyme). These $C_4$ acids are enzymatically degraded to yield free $CO_2$ and pyruvate. The latter 3-carbon intermediate compound is phosphorylated to produce PEP. The operation of $C_4$ cycle can continue in this way as the new $CO_2$ enters the system.

It is further proposed that $CO_2$ released from $C_4$ cycle is incorporated in RuDP (or RuBP) carboxylase system in the $C_3$ cycle where COx is reduced to carbohydrates. This reveals that $C_4$ plants have an additional $CO_2$-fixing sequence which is coordinated with the $C_3$-cycle and RuDP-carboxylase. Thus $CO_2$ is fixed by PEP carboxylase and is finally handed over to RuDP carboxylase within the leaf. Basically $C_4$ plants are also $C_3$ types but primary $CO_2$ fixation in mesophyll cell is of $C_4$ type and classification of plants in to $C_3$ and $C_4$ depends upon primary $CO_2$ fixation.

## $C_4$ PLANTS

Over 117 genera of plants belonging to 13 families possessing $C_4$ mechanism of $CO_2$ fixation are known and out of 13 families, 11 were dicotyledons while the remaining two were monocotyledons (Downton, 1975). At least 11 genera had both $C_3$ and $C_4$ species. All these plants have some features in common: similar anatomical structure (Kranz type of leaf anatomy) and capability to grow successfully in environments with low levels of $CO_2$ in the air. $C_4$ plants possess several adaptive features such as low rates of photorespiration, higher water use efficiency, higher nitrogen use efficiency, temperature independent quantum yield and higher light saturation point which make them markedly superior to $C_3$ species with respect to harvest of solar energy.

In $C_4$ plants where modifications of the primary $CO_2$ trapping systems have been noticed. The secondary mechanisms are associated with distinctive anatomical characteristics and seem to have arisen independently several times in different angiosperm families during evolution.

**TABLE 15.1:** Some important crop plants and their characteristics*

| Common name | Scientific name | Photorespirat-ion type* | Photoperiodic reaction* |
|---|---|---|---|
| Alfalfa | *Medicago sativa* L. | $C_3$ | L |
| Barley | *Hordeum vulgare* L. | $C_3$ | L;N |
| Bean, field | *Phaseolus vulgaris* L. | $C_3$ | S;N |
| Bean, lima | *Phaseolus lunatus* Macf. | $C_3$ | S |
| Bean, mung | *Phaseolus aureus* Roxb. | $C_3$ | S |
| Beet, sugar | *Beta vulgaris saccharifera* L. | $C_3$ | L |
| Bentgrass, creeping | *Agrostic palustris* Huds. | $C_4$ | L |
| Bermudagrass | *Cynodon dactylon* (L.) Pers. | $C_4$ | - |
| Birdsfoot trefoil | *Lotus corniculatus* L. | $C_3$ | L(?) |
| Bluegrass, Kentucky | *Poa pratensis* L. | $C_3$ | N |
| Bromegrass, smooth | *Bromus inermis* Leyss. | $C_3$ | L |
| Broomcorn | *Sorghum vulgare technicum* Pers. | $C_4$ | S |
| Buckwheat, common | *Fagopyrum esculentum* Gaertn. | $C_3$ | L;N |
| Cassava | *Manihot utilissima esculenta* Pohl. | - | N |
| Castorbean | *Ricinus communis* L. | $C_3$ | L |
| Clover alsike | *Trifolium hybridum* L. | $C_3$ | L |
| crimson | *Trifolium incarnatum* L. | $C_3$ | L |
| ladino | *Trifolium repens* L. | $C_3$ | L |
| red | *Trifolium pratense* L. | $C_3$ | L |
| sweet, white | *Melilotus alba* Med. | $C_3$ | L |
| sweet, yellow | *Melilotus officinalis* Lam. | $C_3$ | L |
| white | *Trifolium repens* L. | $C_3$ | L |
| Corn or maize, dent | *Zea mays* L. | $C_4$ | S |
| Corn or maize, sweet | *Zea mays saccharum* L. | $C_4$ | L |
| Cotton, upland | *Gossypium hirsutum* L. | $C_3$ | N |
| Cotton, Egyptian | *Gossypium barbadense* L. | $C_3$ | N |
| Cowpea | *Vigna sinensis* Endl. | $C_3$ | S |
| Crotalaria, showy | *Crotalaria spectabilis* Roth. | $C_3$ | S |
| Crownvetch | *Coronilla varia* | $C_3$ | S |
| Flax | *Linum usitatissimum* L. | $C_3$ | L |
| Guar | *Cyamopsis psoralides* DC. | - | - |
| Guayule | *Parthenium argentatum* Gray | - | W |
| Hemp | *Cannabis sativa* L. | $C_3$ | S |
| Hop | *Humulus lupulus* L. | $C_3$ | L |
| Kenaf | *Hibiscus cannabinua* L. | $C_3$ | S |
| Kudzu | *Pueraria thunbergiana* Benth. | - | - |
| Lentil | *Lentilla lens* | $C_3$ | L |
| Lespedeza, common | *Lespedeza striata* Hook and Arn | $C_3$ | S |
| Lespedeza, Korean | *Lespedeza stipulacea* Maxim | $C_3$ | S |
| Lespedeza, sericea | *Lespedeza cuneata* (Dum. de Cours) | $C_3$ | S |
| Lupine, blue | *Lupinus angustifolius* L. | $C_3$ | N;S |

| Common name | Scientific name | Photorespirat-ion type* | Photoperiodic reaction* |
|---|---|---|---|
| Lupine, white | *Lupinus albus* L. | $C_3$ | N |
| Lupine, yellow | *Lupinus luteus* L. | $C_3$ | N;L |
| Millet, foxtail | *Setaria italica* (L.) Beauv. | $C_4$ | S |
| Millet, pearl | *Pennisetum glaucum* L. | $C_4$ | S |
| Mint (peppermint) | *Mentha piperita* L. | $C_3$ | - |
| Mint (spearmint) | *Mentha spicata* L. | $C_3$ | - |
| Oats, common | *Avena sativa* L. | $C_3$. | L;N |
| Oats, red | *Avena byzantina* C. Koch | $C_3$ | L;N |
| Orchardgrass | *Dactylis glomerata* L. | $C_3$ | N;L |
| Pea, field | *Pisum sativum* L. | $C_3$ | L |
| Peanut | *Arachis hypogaea* L. | $C_3$ | S |
| Potato | *Solanum tuberosum* L. | $C_3$ | L;N |
| Ramie | *Boehmeria nivea* Gaud. | $C_3$ | - |
| Rape, oilseed | *Brassica napus annua* Koch | $C_3$. | L |
| Rape, winter | *Brassica napus biennis* (Schubl.and Mart.) | $C_3$ | - |
| Redtop | *Agrostis alba* L. | $C_3$ | - |
| Reed canarygrass | *Phalaris arundinacea* L. | $C_3$ | L |
| Rice | *Oryza sativa* L. | $C_3$ | S |
| Rye | *Secale cereale* L. | $C_3$ | L;N |
| Ryegrass, Italian | *Lolium multiflorum* Lam. | $C_3$ | L |
| Ryegrass, perennial | *Lolium perenne* L. | $C_3$ | L |
| Safflower | *Carthamus tinctorius* L. | $C_3$ | L;N |
| Sesame | *Sesamum indicum* L. | $C_3$ | S |
| Sorghum (feterita) | *Sorghum bicolor* (L.) Moench | $C_4$ | S |
| Sorghum (hegari) | *Sorghum bicolor* (L.) Moench | $C_4$ | S |
| Sorghum (kafir) | *Sorghum bicolor* (L.) Moench | $C_4$ | S |
| Sorghum (milo) | *Sorghum bicolor* (L.) Moench | $C_4$ | S |
| Sorghum (sorgo) | *Sorghum bicolor* (L.) Moench | $C_4$ | S |
| Sorghum × sudangrass | | $C_4$ | S |
| Soybean | *Glycine max* Merrill | $C_3$. | S;N |
| Sudangrass | *Sorghum bicolor drummondi* | $C_4$ | S |
| Sugar beet | | | |
| Sugarcane | *Saccharum officinarum* L. | $C_4$ | S |
| Sunflower | *Helianthus annuus* L. | $C_3$ | N |
| Teosinte | *Euchlaena mexicana* Schrad. | $C_4$ | - |
| Timothy | *Phelum pratense* L | $C_3$ | L |
| Tobacco | *Nicotiana tabacum* L. | $C_3$ | N |
| Vetch, common | *Vicia sativa* L. | $C_3$ | L |
| Vetch, hairy | *Vicia villosa* Roth | $C_3$ | L |
| Wheat, common | *Triticum aestivum* L. | $C_3$ | L;N |
| Wheat, durum | *Triticum turgidum* L. | $C_3$ | L;N |
| Wheatgrass, crested | *Agropyron cristatum* (L.) Gaertn. | $C_3$ | - |
| Wheatgrass, slender | *Agropyron trachycaulum* (Link) | $C_3$ | - |

| Common name | Scientific name | Photorespirat-ion type* | Photoperiodic reaction* |
|---|---|---|---|
| | Malte | | |
| Wheatgrass, western | *Agropyron smithii* Rybd. | $C_3$ | L |

*Photorespiration type: $C_3$ = Low net assimilation rate (high compensation point); $C_4$ = High net assimilation rate (low compensation point)
*Photoperiodic reaction: L = Long day; S = Short day; N = Day neutral or indeterminate.

It can be observed that crop plants represent both by $C_3$ and $C_4$ plants. The entire group of poaceous grasses is exclusively $C_4$. Several dicot families have $C_4$ representatives but in leguminosae and compositae, no $C_4$ plant has been identified till date.

On the basis of photosynthetic carbon fixation pathway some workers recognize three groups within $C_4$ plants. These are as follows:

1. Maize, sugar-cane
2. *Panicum maximum* and *Chloris gayana*
3. *Atriplex spongiosa*

The important aspects of $C_4$ cycle are:

1. Mesophyll cells of $C_4$ plants contain few chloroplast.
2. The vascular bundles are surrounded by concentric cylinders called vascular bundle sheath, whose cells are packed with chloroplast and starch and the sheath is surrounded by mesophyll cells which are rich in chloroplast as well.
3. Mesophyll and vascular bundle sheath (VBS) cells take part in photosynthesis.
4. In $C_4$ plants the $C_4$ acids and their precursors have been identified in the mesophyll cells chloroplast whereas bundle sheath cells abound in Calvin cycle compounds.
5. Chloroplasts of VBS cells contain ribulose-1,5-diphosphate carboxylase (causes addition of $CO_2$ to $C_5$ sugars) while mesophyll chloroplast has PEP-carboxylase ($CO_2$ + PEP).

Electron microscopic studies of $C_4$ and $C_3$ plants conducted by Professors W.M. Leatch and C.C. Black revealed that mesophyll chloroplast of $C_4$ plants has well developed grana.

The significance of two photosynthetic carbon cycles operating in the same individual appears to assist the mechanism by which $CO_2$ in the environment is utilized most efficiently. The two types of cells (mesophyll cells and VBS cells) are associated with each other by a complicated transport process. Consequently malate is exported from the mesophyll and pyruvate is imported in turn. At present not much is known about the process precisely.

PEP-carboxylase has strong affinity to combine with $CO_2$ than RuDP carboxylase. The $CO_2$ concentration for maximum photosynthesis in vivo is lower for PEP carboxylase. Therefore, the rate of photosynthesis is higher and $CO_2$ compensation point is lower in $C_4$ plants compared with $C_3$ plants. Thus $C_4$ plants can utilize $CO_2$ from weak concentration with the help of PEP- carboxylase. Obviously $C_4$ plants are more efficient for increased photosynthesis under low $CO_2$ concentration. The role of $C_4$ plants appears to concentrate $CO_2$ for RuDP-carboxylase of $C_3$.cycle.

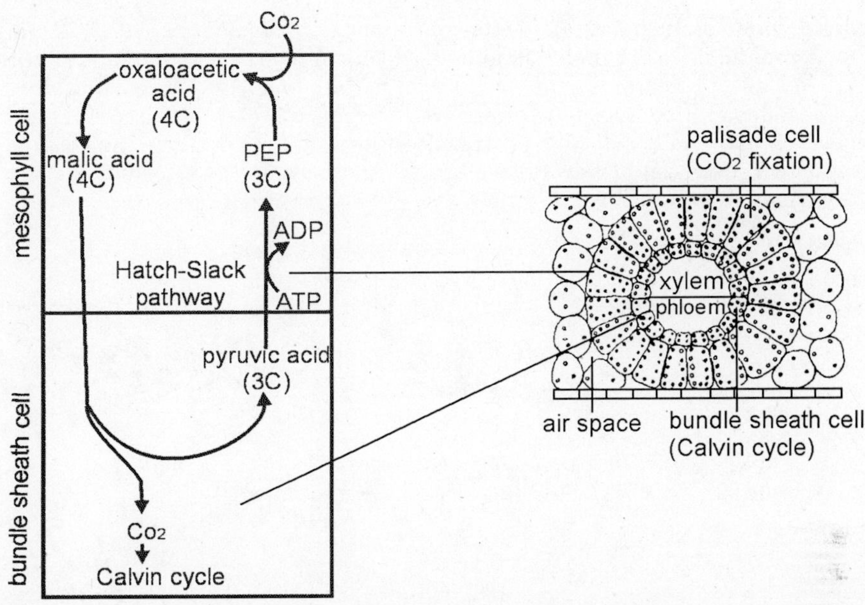


**FIG. 15.7:** (A). A brief summary of $C_4$ dicarboxylic acid pathway and Kranz anatomy, (B) Diagrammatic transverse section of part of the leaf of a $C_4$ plant showing the close relationship between the palisade cells, which here form a circle round the vascular bundle, and the bundle sheeth cells.

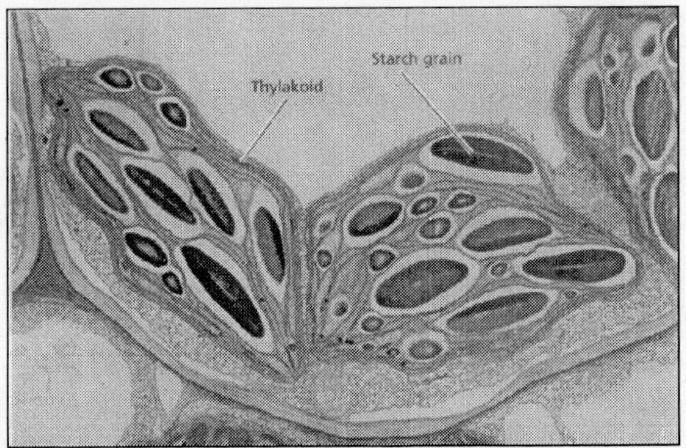


**FIG. 15.8:** Electron micrograph of a bundle sheath cell from maize, showing the starch graisn in the chloroplasts. (15,800x).

## SUCROSE SYNTHESIS IN CYTOSOL

Sucrose synthesis in cytosol is the main result of photosynthesis. The sucrose so produced has a number of possible fates. There are three major fates: translocation, storage in the vacuole or conversion into fructan. **Fructan** synthesis is particularly important in sucrose stpring species especially at low temperatures. The rates of synthesis of sucrose, starch and fructan and the vacuolar storage and translocation of sucrose and their diurnal changes will thus determine the carbohydrate status of a leaf and its contribution to the plant.

A pre-requisite for understanding the control of translocation and partitioning is then a description of the major fluxes of carbon in a leaf during light-dark cycles. Such a description has to distinguish between at least two types of pools of sucrose: that readily available for translocation and thus partly in the cytosol of the mesophyll cells and that not available for translocation which is generally considered to be vascular.

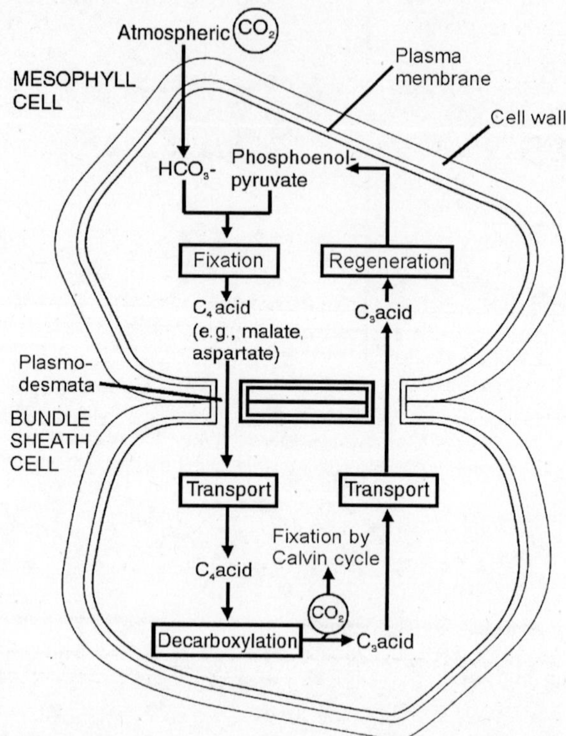


**FIG. 15.9:** The basic $C_4$ photosynthetic carbon cycle involves two cell types and proceeds in four stages: (1) *fixation* of $CO_2$ into a four-carbon acid in the mesophyll cells; (2) *transport* of the four-carbon acid from the mesophyll cells to the bundle sheath cells; (3) *decarboxylation* of the four-carbon acid, generating a high concentration of $CO_2$ in the bundle sheath cells; and (4) *transport* of the residual three-carbon acid back o the mesophyll cells, where the original $CO_2$ acceptor, phosphoenolpyruvate, is regenerated. The $C_4$ cycle concentrates $CO_2$ in the bundle sheath cells, to which rubisco and the other enzymes of the Calvin cycle are restricted. This high concentration of $CO_2$ in the bundle sheath cells suppresses the oxygenation of the ribulose-1,5-bisphosphate by rubisco.

## INTRACELLULAR TRANSPORT IN $C_4$ PLANTS DURING PHOTOSYNTHESIS

In leaves, the rate of photosynthesis is about 10-fold higher than the rate of dark respiration, which illustrates the potential for high rates of metabolite flux between cellular compartments during photosynthesis. For example, in C3 plants there is a high rate of movement of $CO_2$ into the chloroplast, and of $O_2$ out of the chloroplast by free diffusion, equivalent to the rate of photosynthesis. The next most rapid exchange across the C3 chloroplast envelope is $P_i$/triose-P on the phosphate translocator, which needs to function at rates near one-third of the rate of photosynthesis when sucrose is the major photosynthetic product.

In C4 photosynthesis, movement of metabolites across organelles occurs in association with the C4 cycle, the reductive phase of the C3 pathway and photorespiration. As much as half of the PGA, the product of RuBP carboxylase, may be shuttled from bundle sheath to mesophyll chloroplasts for reduction, in which case this shuttle would e operating at the same rate as $CO_2$ fixation.

Evidence that this transport must be occurring is based on studies of enzyme compartmentation and metabolism by isolated organelles. However, direct studies on metabolite transport, in order to identify carrier-mediated mechanisms and their kinetic properties, has been limited mainly to the mesophyll chloroplasts. The phosphate translocator is the only carrier protein identified in C4 photosynthesis.

### Phosphate Translocator

A single phosphate translocator in the mesophyll chloroplast is considered to mediate PEP/$P_i$ exchange associated with the C4 cycle and PGA/triose-P exchange associated with the C3 pathway. Thus the specificity of this translocator is different from that in C3 chloroplasts such as those of spinach, which can effectively exchange $P_i$/truise-P but not PEP/$P_i$. DIDS (4,4-diisothiocyanatostilbene- 2,2'-disulphonic acid), a disulphonic acid stilbene derivative, is a potent, irreversible inhibitor of the phosphate translocator. Incubation of chloroplasts with tritiated DIDS results in specific labelling of a polypeptide of c 29 kDa in the envelopes of C3 and C4 mesophyll chloroplasts, which is considered to be the phosphate translocator protein.

### Pyruvate Translocator

The translocator facilitates the uptake of pyruvate into the mesophyll chloroplast. It shows Michaelis-Menten type kinetics and is inhibited by analogues of pyruvate and by low temperature. There is light-induced uptake and accumulation of pyruvate by C4 mesophyll chloroplasts, which is driven by a proton or sodium gradient, depending on the species. In mesophyll chloroplats from monocots of NAD-ME and PEP-CK species there is pyruvate uptake in the light, addition of $Na^+$ and substitute for the light requirement. This light-dependent accumulation of pyruvate may be associated with an active $Na^+/H^+$ exchange, or active $Na^+$ efflux from the chloroplast which drives co-transport of $Na^+$ and pyruvate into the chloroplast. In NADP-ME type monocots there is enhanced uptake of pyruvate by mesophyll chloroplasts following a pH drop, suggesting a $H^+$/pyruvate import. Sodium does not stimulate uptake of pyruvate by chloroplasts of these species. In this case, the light-enhanced pyruvate uptake may be linked to light-dependent alkalisation of the stromal compartment.

## OXALOACETATE TRANSLOCATOR

There is a specific carrier in C4 mesophyll chloroplasts for uptake of oxaloacetate. It has a high affinity for oxaloacetate ($K_m$ = 45 mm for maize chloroplasts), which is considered important since oxaloacetate is present at low levels in the cell.

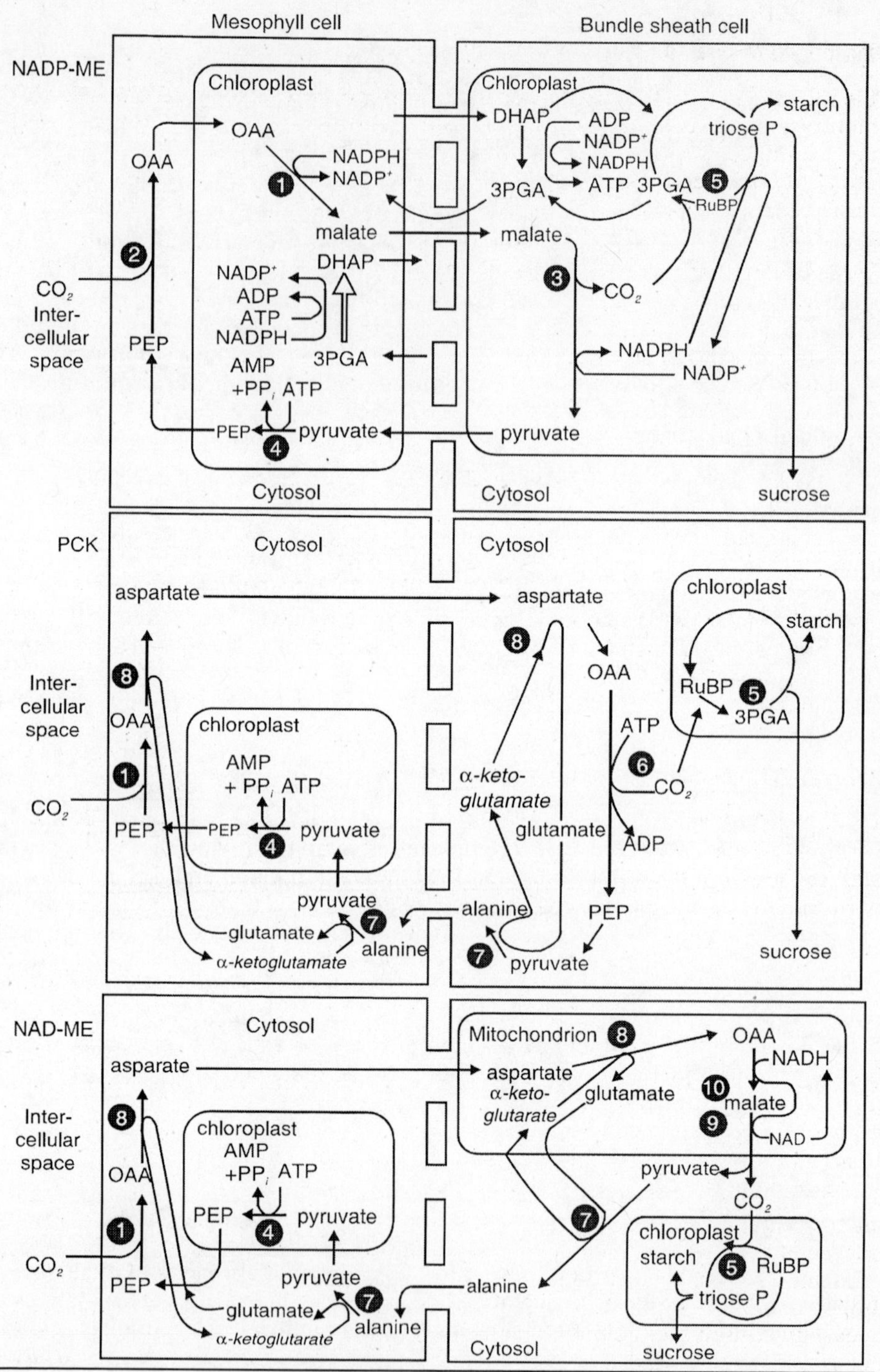


**FIG. 15.10:** A relation of mesophyll cells and bundle sheath cells during C4 cycle.

## Dicarboxylate Translocator

This translocator catalyses the exchange of dicarboxylic acids by the C4 mesophyll chloroplasts, as seen with *Digitaria sanguinalis* and maize. In vivo, its function may be linked in some way to the oxaloacetate translocator and involves the uptake of oxaloacetate and export of malate by the C4 mesophyll chloroplast. The dicarboxylate translocator(s) may also have other functions in association with movement of other dicarboxylic acids across the chloroplast envelope (for example, aspartate, 2-oxoglutarate and glutamate).

## Glycerate Translocator

There are evidences for a glycerate/glycolate translocator in C4 mesophyll and bundle sheath chloroplasts which may function in the export of glycolate from the bundle sheath chloroplasts and import of glycerate into the mesophyll chloroplasts. Some stimulation of uptake of glycerate by light may be attributed to an alkalisation of the stroma and a $H^+$/glycerate$^-$ import.

## INTERMEDIATE BETWEEN $C_3$ AND $C_4$ PLANTS

More than 20 plant species exhibit photosynthetic traits that are intermediate between $C_3$ and $C_4$ plants (e.g., species in the genera *Alternanthera*, *Flaveria*, *Neurachne*, *Moricandia*, *Panicum* and *Parthenium*). These show reduced rates of photorespiration and $CO_2$-compensation points in the range of 0.8 to 3.5Pa, compared with to 4 to 5 in $C_3$ and 0 to 0.5 in $C_4$ plants. They have a weakly developed Kranz anatomy, compared with the true $C_4$ species, but Rubisco is located both in the mesophyll and the bundle sheath cells.

Two types of intermediates have been distinguished. In the first type (e.g., *Alternanthera ficoides*, *Alternanthera enella*, *Moricandia arvensis*, *Panicum milioides*) the activity of key enzymes of the $C_4$ pathway (i.e., pyruvate-Pi-dikinase, PEP carboxylase, NAD-malic enzyme, and PEP carboxykinase) is very low, and they do not have a functional $C_4$-acid cycle. Their low $CO_2$-compensation point is due to the light-dependent recapture by mesophyll cell released in photorespiration in the bundle sheath cells. The bundle sheath cells contain a large fraction of the organelles involved in photorespiration, compared with that in $C_3$ species. The $CO_2$-compensation point is negatively correlated with the percentage of photorespiratory organelles that occur in bundle sheath cells. These $C_3$-$C_4$ intermediates have evolved a system to utilize $CO_2$ that escapes from the bundle sheath cells but do not have the $CO_2$-concentrating mechanism of true $C_4$ species. In the leaves of these intermediate species, glycine decarboxylase, which is a key enzyme in photorespiration that releases the photorespiratory $CO_2$, occurs exclusively in the cells that surround the vascular bundle sheath. The mitochondria of the mesophyll cells, like those of true $C_4$ plants, have a low activity of glycine decarboxylase. Immunogold labeling shows a four times greater amount of one of the subunits of glycine decarboxylase cell. It is likely that products of the oxygenation reaction, including glycine, move to the bundle sheath cells. The products are presumably metabolized in the bundle sheath; therefore, serine can move back to the mesophyll. Due to the exclusive location of glycine decarboxylase in the bundle sheath cells, the release of $CO_2$ in photorespiration occurs close to the vascular tissue.

## $C_4$ Species: Evolution and Distribution

$C_4$ species represent approximately 5 per cent of all plant species, $C_3$ species accounting for about 85 per cent and CAM species for 10 per cent. The $C_4$ pathway evolved nearly synchronously on several continents in the late Miocene 5 to 7 million years ago, probably in response to a reduction in atmosheric $CO_2$ concentration. This change decrease in atmospheric $CO_2$ concentration resulted from both the photosynthetic activity of plants and tectonic and subsequent geochemical events. To put it briefly, the collision of the Indian subcontinent caused the uplift of the Tibetan Plateau, causing new earthy crust that consumed $CO_2$ to become exposed over a vast area. The reaction $CsSiO_3 + CO_2$ matic decline in atmospheric $CO_2$ concentration.

Low atmospheric $CO_2$ concentration would increase photorespiration and favor the $CO_2$-concentrating mechanisms and lack of photorespiration that characterize $C_4$ species. Considering the three subtypes of $C_4$ species and their occurrence in plants of at least 18 different families of diverse taxonomic groups, $C_4$ plants must have evolved from $C_3$ ancestors independently a number of times. The physiology of $C_3$-$C_4$ intermediates suggests the mechanism to remechanism. Morphological and ecogeographical information combined with molecular synthesis has evolved twice in different lineages within the genus *Flaveria*. The phylogeny of *Flaveria* species deduced from an analysis of the nucleotide sequences that encode a subunit of glycine decarboxylase, suggests that $C_4$ species are derived from $C_3$-$C_4$ intermediates.

Apart from the decreasing atmospheric $CO_2$ concentration, high temperatures and water shortage may also have been the driving forces for the evolution of $C_4$ species. This makes it likely that $C_4$ species evolved primarily in tropical areas, which continue to be their centers of distribution e.g., tropical and temperate grasslands, with abundant warm-season precipitation, are dominated by $C_4$ species.

The high concentration of $CO_2$ at the site of Rubisco allows net $CO_2$ assimilation at relatively high temperatures, when photorespiration results in low net photosynthesis of $C_3$ species due to the increased oxygenating activity of Rubisco. This explains why $C_4$ species naturally occur in warm, open ecosystems, where C3 species are less successful. There is no proper reason, however, why $C_4$ photosynthesis could not function in cooler climates. The lower quantum yield of $C_4$ species at low temperature may be important in dense canopies where light limits photosynthe-sis (and where quantum yield is therefore important). Quantum yield, however, is less important at higher levels of irradiance, and there is quite a wide temperature range where the quantum yield is still high compared with that of $C_3$ plants. The high sensitivity to low temperature of pyruvate-Pi-dikinase, a key enzyme in the $C_4$ pathway, may be the main reason why $C_4$ species have rarely expanded to cooler places. Compatible solutes can decrease the low-temperature sensitivity of this enzyme; this could allow the expansion of $C_4$ species into more temperate regions in the future. On the other hand, rising atmospheric $CO_2$ concentration may offset the advantages of the $CO_2$-concentration mechanism of $C_4$ photosynthesis.

## BREEDING OF $C_3$ AND $C_4$ PLANTS

Because of the fundamental relationship between photosynthesis and yield, there has been considerable interest in enhancing photosynthetic capacity through breeding.Active research is presently going on in Plant breeding laboratories where incorporation of characters of $C_4$ plants to the $C_3$ plants is done and thereby augmenting the amount of metabolites they form. Probably limited number of genes are involved which need to be brought together. Incidentally, crosses between $C_3$ and $C_4$- *Atriplex* species have failed to yield progeny with integrated $C_4$ characters. In view of the non-availability of plants with $C_4$ characteristics closely related to cereals, attempt is being made to screen different seed plant for $C_4$ types. The possibility of breeding entirely new plants from the genetic pool of $C_4$ types can also not ruled out.

## $C_2$ PLANTS?

RubP splits under $O_2$ consumption into the $C_2$ compound phosphoglycolate and a $C_3$ compound phosphoglycerate by the activity of an oxygenase. After dephosphorylation, glycolate can be oxidized to glyoxylate either directly in the chloroplast or in the peroxisomes. This reaction also produces the aggressive $H_2O_2$ which is converted into water and $O_2$ by catalase, the key enzyme of the peroxisomes. The glyoxylate glycine transaminase produces the amino acid glycine. Two molecules of glycine can be condensed to generate $CO_2$ to form serine, either in the mitochondria or in the chloroplasts. Since these reactions are dominated by $C_2$ compounds, photorespiration is also called the $C_2$ cycle. The significance of the $C_2$ cycle is not yet completely clear, since some plants have a photorespiration cycle while others do not. In addition to the production of amino acids it may have a role in photoinhibition. According to this concept the excessive oxygen produced at extremely high fluence rates would be deactivated by the $C_2$ cycle and the pigments and protein, especially in PSII, would be protected from oxidation. The disadvantage of photorespiration is the loss of fixed carbon, which can amount to 25 per cent of the primarily fixed $CO_2$. Experiments to genetically eliminate or reduce photorespiration have not yet been successful.

# CHAPTER 16

# CAM Plants

In addition to $C_3$ and $C_4$ species, there are many succulent plants with another photosynthetic pathway: Crassulacean Acid Metaboism (CAM). Primary $CO_2$ fixation is followed by a diurnal acid metabolism. This pathway is named after the Crassulaceae, which is a family in which any species show this type of metabolism. CAM, however also commonly occurs in other families, such as the Cactaceae, Euphorbiaceae, Orchidaceae, and Bromeliaceae (e.g., *Ananas comosus*, pineapple). There are about 10,000 CAM species from 25 to 30 families, all Angiospermae, with the exception of a few fern species, which also have CAM characteristics.

The unusual capacity of CAM plants to fix $CO_2$ into organic acids in the dark, which causes nocturnal acidification, with deacidification during the day, has been known for almost two centuries. A full aperciation of CAM as a photosynthetic process was greatly stimulated by analogies with $C_4$ species.

**TABLE 16.1:** Taxonomic survey of flowering plant families known to have species showing crassulacean acid metabolism

| Agavaceae | Geraniaceae |
|---|---|
| Aizoaceae | Gesneriaceae |
| Asclepidiaceae | Labiatae |
| Asteranceae | Liliaceae |
| Bromeliaceae | Oxalidaceae |
| Cactaceae | Orchidaceae |
| Clusiaceae | Piperaceae |
| Crassulaceae | Polypodiaceae |
| Cucurbitaceae | Portulacaceae |
| Didieraceae | Rubiaceae |
| Euphorbiaceae | Vitacege |

The productivity of most CAM plants is fairly low; however, this is not an inherent trait of CAM species because *Agave mapisaga* and *Agaave salmiana*, for example, may achieve an average productivity of 40 Mg $ha^{-1}$ $year^{-1}$. An even higher productivity has been observed for both irrigated and fertilized and carefully pruned *Opuntia amyclea* and *Opuntia ficus-indica* (46 Mg $ha^{-1}$ $year^{-1}$). These are among the highest productivities reported for any species. In a comparison of two succulent species with similar growth forms, *Cotyledon orbiculata* (CAM) and *Othonna optima* ($C_3$), during the transition from the rainy season to subsequent drought, the daily net rate of $CO_2$ assimilation is similar for the two species. This shows that rates of photosynthesis of CAM plants may be as high as those of $C_3$, plants, if morphologically similar plants adapted to the same habitat are compared.

As with $C_4$ plants, none of the reactions of CAM is really unique to these species; however, they proceed at different times of the day from those in $C_3$ and $C_4$ species. Two subtypes of CAM species air discerned based on differences in the major decarboxylating enzyme. In addition, there are intermediate forms between $C_3$ only and CAM, as well as facultative CAM plants.

## DISTRIBUTION AND EVOLUTION OF CAM SPECIES

CAM is undoubtedly an adaptation to drought. They close their stomates during most of the day. There are two major ecological groupings of CAM plants:

1. Succulents from arid and semiarid regions
2. Epiphytes from tropical and subtropical regions

Although CAM plants are uncommon in cold environments, this may reflect their evolutionary origin in warm climates rather than a temperature sensitivity of the CAM pathway. Roots of some Orchidaceae, which lack stomata, also show CAM. CAM species are well adapted to dry environments and naturally occur in semiarid and tropical environments. Some submerged aquatics also show CAM photosynthesis.

## CARBON ISOTOPE COMPOSITION OF CAM SPECIES

The enzyme from CAM plants discriminates against $^{13}CO_2$ (similar to Rubisco from $C_3$ and $C_4$ plants). The fractionation is considerably less than that of $C_3$ plants and similar to that of $C_4$ species. This is expected, as the stomata are closed during malate decarboxylation and fixation of $CO_2$ by Rubisco. Hence, only a small amount of inorganic carbon diffuses back from the leaves to the atmosphere. Upon a shift from $C_3$ to CAM photosynthesis in the stomates are closed during most of the day and open at night. The carbon isotope discrimination consequently decreases, in accordance with the smaller extent of discrimination in CAM species.

## ANATOMICAL, PHYSIOLOGICAL AND BIOCHEMICAL ADAPTATIONS

CAM plants are characterized by their succulence (*succulence is defined as the volume of water in the leaf at a relative water content of 100 per cent per unit leaf area*), but this is not pronounced in epiphytic CAM plants. The capacity to fix $CO_2$ at night via PEP-carboxylase, the accumulation of malic acid in the vacuole, and the subsequent deacidification during the day, when $CO_2$ is released from malic acid and fixed in the Calvin cycle, using Rubisco.

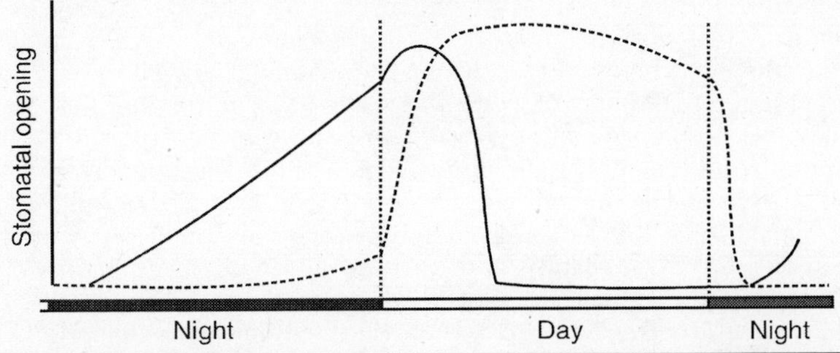


**FIG. 16.1:** A diagram representing the daily changes in stomatal aperture in a succulent (continuous line), aand mesophyte (broken line).

CAM plants show a strong fluctuation in pH of the cell sap because of the synthesis and breakdown of malic acid. The concentration of this acid may increase to 100mM. By isolating vacuoles of the CAM plant *Kalanchoe daigremontiana*, it has been demonstrated that at least 90 per cent of all the acid in the cells is in the vacuole. In addition, the kinetics of malic acid efflux from the leaves of *K. daigremontiana* provide evidence for the predominant location of malic acid in the vacuole.

At night, $CO_2$ is fixed in the cytosol which is catalyzed by PEP-carboxylase and produces oxaloacetate. PEP originates from the breakdown of glucose in glycolysis is formed from starch. Oxaloacetate is immediately reduced to malate which is catalyzed by malate dehydrogenase. Malate is transported to the large vacuoles in an energy-dependent manner, which involves a proton pump. An $H^+$-ATPase and a pyrophosphatase pump protons into the vacuole so that malate can follow passively. It becomes malic acid in the vacuole.

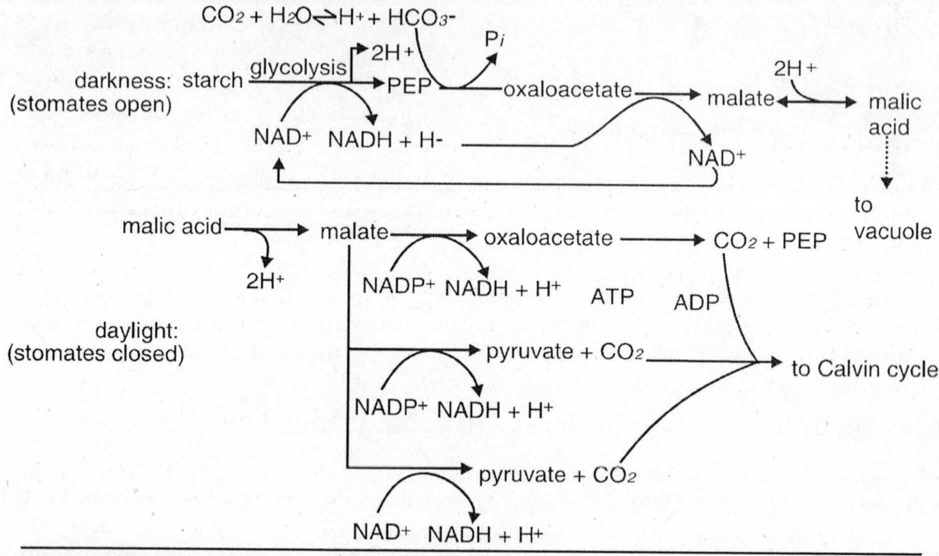


**FIG. 16.2:** Crassulacean acid metabolism in light and dark.

The release of malic acid from the vacuole during the day is supposedly passive. Malate is then decarboxylated by malic enzyme (NAD- or NADP-dependent) or by PEP-carboxykinase. CAM species are subdivided like $C_4$ species, depending on the decarboxylating enzyme. The malic enzyme subtypes (ME-CAM) have a cytosolic NADP-malic enzyme, as well as a mitochondrial NAD-malic enzyme; they use a chloroplastic pyruvate-Pi-dikinase to convert the $C_3$ fragment that originates from the decarboxylation reaction into carbohydrate. Unlike $C_4$ species, ME-CAM species are not further subdivided into NADP- and NAD-malic enzyme subtypes. PEPCK-type CAM plants have very low malic enzyme activities (as opposed to PEPCK-$C_4$ plants) and no pyruvate-Pi-dikinase activity, but high activities of PEP-carboxykinase.

The $C_3$ fragment (pyruvate or PEP) that is formed during the decarboxylation is converted into starch, and the $CO_2$ that is released is fixed by Rubisco, much the same as it is in $C_3$ plants. The stomata are closed during the decarboxylation of malic acid and the fixation of $CO_2$ by Rubisco in the Calvin cycle. They are open during the nocturnal fixation of $CO_2$.

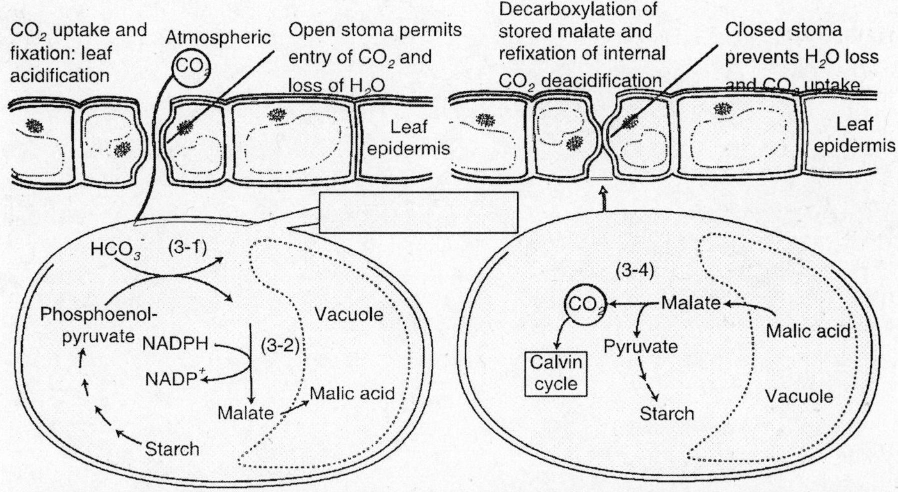


**FIG. 16.3:** In crassulacean acid metabolism (CAM), the reactions of photosynthesis and $CO_2$ uptake are temporally separated: $CO_2$ uptake and fixation take place at night, and decarboxylation and refixation of the internally released $CO_2$ occur during the day. CAM is an adaptation primarily to minimize the quantity of water that is lost when stomata are opened to permit the entry of $CO_2$. In CAM plants, the stomata are opened in the cool of the night. $CO_2$ is fixed as malic acid, which is stored in the vacuole. As malic acid accumulates, the leaf vacuoles acidify in the dark. Upon illumination, the stomata close, and the leaf deacidifies. The malic acid is recovered from the vacuole and undergoes decarboxylation. The $CO_2$ that is released is prevented from escaping by stomatal closure and is assimilated via the Calvin cycle using photochemically generated ATP and NADPH.

The CAM traits can be summarized as follows:

1. Flucturation of organic acids, mainly of malic acid, during a diurnal cycle;
2. Fluctuation of the concentration of sugars and starch, opposite to the fluctuation of malic acid;
3. A high activity of PEP-carboxylase (at night) and of a decarboxylase (during the day);
4. Large vacuoles in cells that contain chloroplasts;
5. Some degree of succulence;
6. The gas exchange of the leaves occurs predominantly at night.

## PHASE I

Phase I, the carboxylation phase, starts at the beginning of the night. The rate of carboxylation declines towards the end of the night, and the malic acid concentration reaches its maximum. The stomatal conductance and the $CO_2$ fixation change more or less in parallel; carbohydrates are broken down.

## PHASE II

Phase II, at the beginning of the day, is characterized by a high rate of $CO_2$ fixation, which generally coincides with an increased stomatal conductance. $CO_2$ fixation by PEP carboxylase, along with malic acid formation, coincide with the fixation of $CO_2$ by Rubisco. Fixation by PEP carboxylase is gradually taken over by Rubisco. $C_3$ photosynthesis predominates in the last part of phase II, using exogenous $CO_2$ as the substrate.

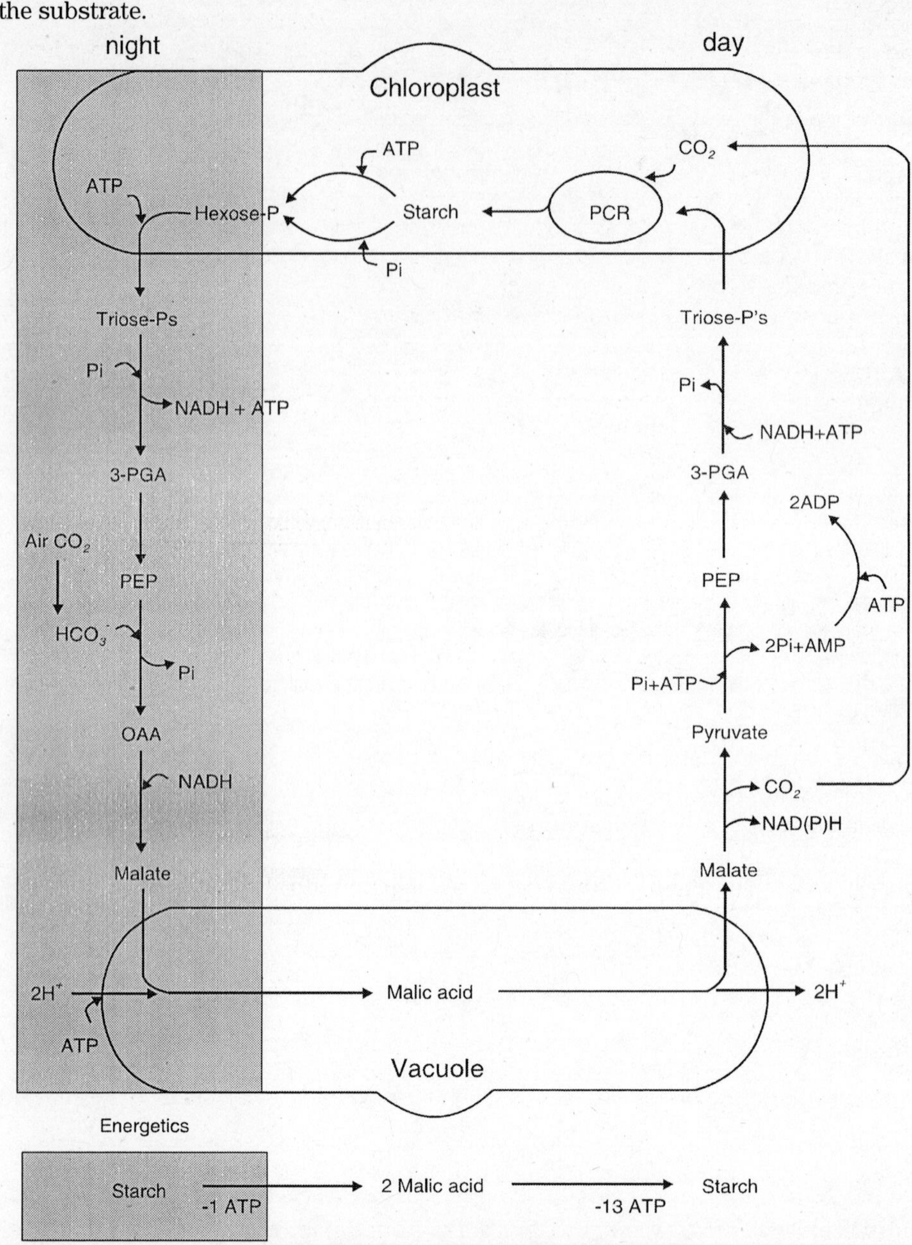


**FIG. 16.4:** Metabolic pathway and cellular compartmentation of Crassulacean Acid Metabolism, showing the separation in night and day of carboxylation and decarboxylation.

## PHASE III

In Phase III the stomata are fully closed and malic acid is decarboxylated. The pi may then increase to values above 1000 Pa, which occurs when the normal $C_3$ photosynthesis takes place and when sugars and starch accumulated. When malic acid is depleted, the stomata open again, possibly because pi drops to a low level; this is the beginning of Phase IV.

## PHASE IV

Gradually more exogenous and less endogenous $CO_2$ is fixed by Rubisco. In this last phase, $CO_2$ may be fixed by PEP-carboxylase again, as indicated by the photosynthetic quotient (PQ); that is, the ratio of $O_2$ release and of $CO_2$ uptake. The PQ is 1.0 over an entire day, but deviations from this value occur, depending on the carboxylation process.

In phase III, when the stomata are fully closed, malic acid is decarboxylated, and the pi is very high, photorespiration is suppressed, as indicated by the relatively low rate of oxygen uptake (as measured using $^{18}O_2$;). In phase IV, when malic acid is depleted and the stomata open again, photorespiration does occur, as demonstrated by increased uptake of $^{18}C_2$.

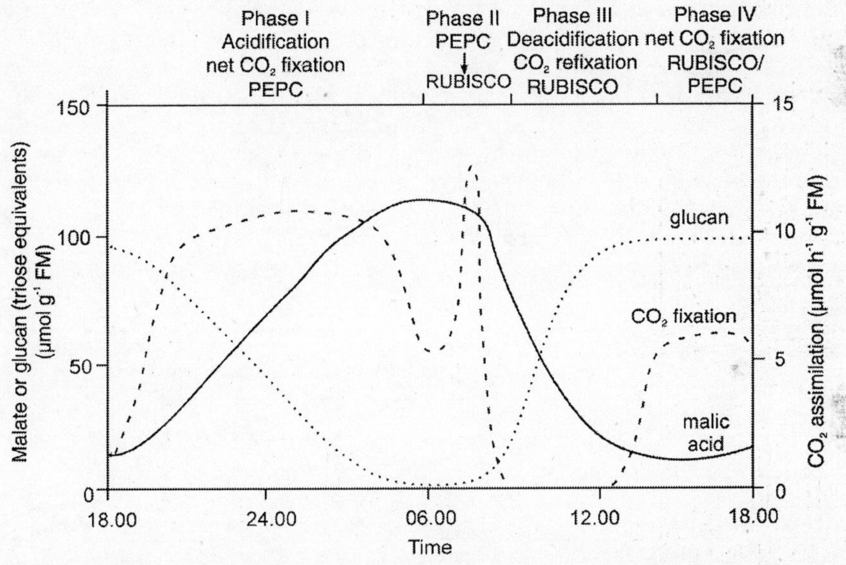


**FIG. 16.5:** $CO_2$ fixation in CAM plants, showing diurnal patterns for stomatal conductance, net $CO_2$ assimilation, malic acid concentration, and carbohydrate concentrations; PEPC is PEP carboxylase.

How do CAM plants regulated the activity of the two caroboxylating enzymes and decarboxylating enzymes in coordinated way to avoid futile cycles? Rubisco is inactive at night for the same reason as in $C_3$ plant: This enzyme is lightactivated and therefore cannot function in the dark. In addition, the kinetic properties of PEP-carboxylase are modulated. In Mesembryanthemum crystallinum and in Crassula argentea, PEP-carboxylase occurs in two configurations: a *day-configuration* and a "night-configuration." The night-configuration is relatively insensitive to malate (the Ki for malate is 0.06 to 0.9 mM, depending on pH) and has a high affinity for PEP (the Km for PEP is 0.1 to 0.3 mM). The day-configuration is strongly inhibited by malate (the Ki for malate is 0.004 to 0.07 mM, again depending on the pH) and has a low affinity for PEP (the Km for PEP has increased to 0.7 to 1.25 mM). When malate is rapidly exported to the vacuole at night in phase I, therefore, the carboxylation of PEP readily takes place, whereas it is suppressed during the day in phase III.

**TABLE 16.2:** Accumulative $CO_2$ and $O_2$ exchange and Photosynthetic Quotient (PQ) in the dark and in light [in mmol plant$^{-1}$ (12 h)$^{-1}$] of one shoot of *Ananas comosus* (pineapple), measured over two consecutive days*

| | Dark | | Light | | |
|---|---|---|---|---|---|
| | $CO_2$ assimilation | $O_2$ consumption | $CO_2$ assimilation | $O_2$ release | PQ |
| Day 1 | 10.6 | 6.4 | 10.4 | 27.1 | 0.99 |
| Day 2 | 11.1 | 6.3 | 10.7 | 27.5 | 0.98 |

*PQ is the ratio of the total net amount of $O_2$ evolved in the light to total amount of $CO_2$ fixed in the light plus dark period.

The fluctuations in the kinetic properties of PEP-caroxylase, do not require protein synthesis. In *Crassula argentea* the change in configuration involves a change from a dimmer, with subunits of 100 kDa, during the day, to a tetramer, with subunits of the same molecular mass, at night. When the enzyme is isolated from leaves collected during the day (dimmer, sensitive to tetramer (insensitive to malate inhibition) takes place during storage during 49 days at-70° C. If malate is added to the enzyme, after 11 days at low temperature, then a large fraction of the enzyme is transformed into the dimmer, which increases the sensitivity for malate inhibition. Addition of PEP vitro has the opposite effect; formation of the tetramer.

**TABLE 16.3:** Effects of malate and glucose-6-phosphte (G-6-P) on the kinetic parameters of PEP-carboxylase

| | Vmax mmolmg$^{-1}$ (Chl) min$^{-1}$ | Ratio | Km mM | Rate |
|---|---|---|---|---|
| Control | 0.42 | 1.0 | 0.13 | 1.0 |
| + 1 mM G-6-P | 0.45 | 1.07 | 0.08 | 0.61 |
| + 2 mM G-6-P | 0.47 | 0.12 | 0.05 | 0.39 |
| + 5 mM malate | 0.31 | 0.74 | 0.21 | 1.60 |
| + 5 mM malate and 2 mM G-6-P | 0.34 | 0.81 | 0.05 | 0.39 |

*Source:* Kluge & Ting 1978. *The ratio gives the ratio of the values in the presence and absence of effectors.

The sensitivity for malate inhibition tends to be lost during storage PEP-carboxylase from the CAM plant *Kalanchoe daigremontianic*. This effect is stronger in the presence of inorganic phosphate, PEP, 3-phosphoglycerate (PGA), and glucose-6-phosphate. The presence of malate slows down the rate of loss of sensitivity to malate inhibition. In this case, there is no change from dimmer to tetramer mass of 200 kDa. There is some evidence that the modification of the kinetic properties involves the phosphorylation and dephosphorylation of PEP-carboxylase.

Though the modification of its kinetic properties, the inhibition of PEP-carboxylase is likely to prevent a futile cycle of carboxylation and concomitant decarboxylation reactions. Further evidence that decarboxylation reactions. Further evidence that such a futile cycle does not occur comes from studies on a labeling with 13C of the first or fourth carbon atom in malate. This pattern of labeling is constant during de-acidification. If a futile cycle were to occur, then doubly labeled malate should appear because fumarase in the mitochondria would randomize the label in the malate molecule. Such randomization does occur during the acidification phase, which indicates rapid exchange of the malate pools of the cytosol and the mitochondria, before malate enters the vacuole.

Next to malate, glucose 6-phosphate is also an effector of PEP-carboxylase: It enhances the Vmax and decreases the Km (PEP). The physiological significance of this effect is that glucose 6-phosphate, which is produced from glucose, during its conversion into PEP stimulates the carboxylation of PEP.

Temperature has exactly the opposite effect on the kinetic properties of PEP carbxylase from a CAM plant and that from a $C_4$ plant. Such temperature effects help to explain why a low temperature at night enhances acidification.

## SIMILARITIES BETWEEN CAM AND $C_4$ PLANTS

Recent studies have shown sufficient similarity between the CAM and $C_4$ plants in regard to carbon fixation and photosynthesis. In plants of categories, during photosynthesis $CO_2$ is fixed in to $C_4$ dicarboxylic acid. However, they differ in respect to the place and time of carboxylation and decarboxylation. In CAM plants, carboxylation occurs during the night while decarboxylation takes place during the day time (temporal seggregation), however, in $C_4$ plants carboxylation reaction occurs in mesophyll cells while the carbohydrate synthesis takes place in the bundle sheath cells (spatial seggregation). Moreover, both these reactions occur during the day or night conditions. The organic acids formation in CAM plants is temporary during the night/dark.

## INCOMPLETE AND FACULTATIVE CAM PLANTS

### CAM idling

A few CAM plants may not open their stomata during the night when they are exposed to severe water stress. Under such conditions, they may continue to show a diurnal fluctuation in malic acid concentration, as first found in *Opuntia basilaris*. The $CO_2$ they use to produce malic acid at night does not come from the air. It is entirely derived from respiration which is released again during the day, which allows some Rubisco activity. This metabolism is termed CAM idling. It occurs typically is ordinary CAM plants that are exposed to severe water stress and have a very low stomatal conductance throughout the day and night.

Fluorescence measurements have indicated that the photosystems remain intact during severe drought. This has led to the suggestion that CAM idling prevents photodamage, in addition to recapturing respiratory $CO_2$. In the absence of an electron acceptor, damage of PSII might occur at the high irradiance to which these plants are exposed. CAM idling can be considered as a modification of normal CAM. The plants remain *ready to move* as soon as the environmental conditions improve, but they keep their stomates closed during severe drought.

### CAM Cycling

Some plants show a diurnal fluctuation in the concentration of malic acid without a net $CO_2$ uptake at night, but with normal rates of $CO_2$ assimilation during the day. These plants are capable of recapturing most of the $CO_2$ that is derived from respiration and to use this as a substrate for PEP-carboxylase. This is termed CAM cycling. It conversely, occurs in plants that have a high stomatal conductance and normal $C_3$ produced in dark respiration at night, which ordinary $C_3$ plants lose to the atmosphere.

## INCOMPLETE AND FACULTATIVE CAM PLANTS

In a limited number of species, CAM only occurs upon exposure to drought stress. They are known as facultative CAM plants (e.g., *Agave desertil Clusia uvitana, Mesembryanthemum crystallinum, and Portulacaria afra*), Irrigation with saline water or drought can change from a virtually normal $C_3$ photosynthesis to the CAM mode. There is at least one genus that contains $C_4$ species that can shift from a normal $C_4$ mode under irrigated conditions to a CAM mode under water stress: *Portulaca grandiflora, Portulaca mundula, and Portulaca oleracea.*

ABA can also induce CAM photosynthesis at a comparable rate. The transition from the $C_3$ or $C_4$ to the CAM mode coincides with an enhanced PEP-carboxylase activity and of the mRNA coding for this enzyme.

Upon removal of NaCl from the root environment of *Mesembryanthemum crystallinum*, the level of mRNA that encodes PEP-carboxylase declines by 77 per cent in 2 to 3 hours. The amount of the PEP-carboxylase enzyme itself declines more slowly: The activity declines after 2 to 3 days to half its original level.

## ENVIRONMENTAL FACTORS AND CAM PLANTS

### Temperature Sensitivity

Photosynthetic electron transport reactions of succulent plants from hot desert are able to tolerate extremely high temperatures and to acclimate to seasonal increase in temperature. Tissue temperatures of desert succulents in their natural habitats can be 10 to 22° C higher than ambient air temperatures. The highest non-lethal temperatures recorded over 60° C for various species of *Agave* and *Opuntia*, although most physiological processes are impaired when tissue temperatures of higher plants exceed 50 °C. It was reported that high temperatures tolerated by agaves and cacti increase in response to increasing growth temperatures

### Plants Growing in Shade or Below Forest Canopy

Plants growing in a shade or below the forest canopy are characterised by thin leaves with large surface area, more chlorophyll and few stomata per unit area. They have higher leaf area ratio and the ability to use small amounts of light with efficiency. They are also characterised by the following:

1. Low light compensation point.
2. Greater amount of photosynthetic tissue.
3. Ability to maintain a net uptake of $CO_2$.
4. More efficient to use incident light and,
5. lower level of ribulose 1,5-diphosphate carboxylase enzyme.

The plants that grow in direct or sharp sunlight have opposite characters than those of (i) to (v)

### Environmental Factors and Structural Changes

Environmental factors greatly influence the structural and physiological adaptations of leaves in plants. Some of the factors like light, temperature and water greatly affect the structural and biochemical change in the leaf. Thus plants growing under forest canopy and deserts have different structural adaptabilities.

Xerophytic plants have adaptation which help the plant to minimize water loss and at the same time maintain sufficient net uptake of $CO_2$ from the surroundings. In these plants, either the cuticle is extremely thick, abundant sclerenchymatous tissue is present and their leaves are small or well dissected or their leaves are swollen, and succulent. In both the categories, the stomata are few, and sunken. Thus transpiration is reduced but the diffusion resistance for $CO_2$ intake is increased. As a result xerophytes are less efficient photosynthetically as far as per unit area is concerned.

## Water-Use Efficiency

CAM plants have following characteristics in relation to water-use efficiency:

1. They keep their stomates closed during the day when the VPD between the leaves and the surrounding air is highest, and open at night when VPD is lowest.
2. They have a very high water-use efficiency.
3. As long as they are not severely stressed, which leads to complete closure of their stomata, the WUE of CAM plants tends to be considerably higher than that of both $C_3$ and $C_4$ plants.

## ECOLOGICAL SIGNIFICANCE

These plants are ecologically significant, because they reduced transpiration during day-time, and are well adapted to dry and hot habitats. The important features are as follows:

1. The stomata remain closed during the day and open at night when water loss is little due to prevailing low temperature.
2. CAM plants have parenchyma cells which are large and vacuolated. These vacuoles are used for storing malic and other acids in large amounts.
3. CAM plants increase water-use efficiency of the plant and secondly through its enzyme PEP carboxylase, they are adapted to extreme hot climates.
4. CAM plants can also obtain a $CO_2$ compensation point of zero at night and in this way accomplish a steeper gradient for $CO_2$ uptake compared with $C_3$ plants.
5. They lack of a *real* photosynthesis during day-time the growth rate is far lower than in all other plants (with the exception of pineapples).

CAM plants are not agriculturally important. The CAM mechanism is the characteristic fearure of many families of angiosperms of **Monocotyledons** (Agavaceae, Bromeliaceae, Liliaceae, Orchidaceae) and **Dicotyledons** (Aizoaceae, Asclepiadaceae, Cactaceae, Caryophyllaceae, Chenopodiaceae, Compositae, Convolvulaceae, Crassulaceae, Euphorbiaceae, Portulacaceae, Vitaceae). Pineapple (*Ananas comosus*) of Btomaliaceae family is an economically important CAM plant.

This $CO_2$ is fixed via PEP-carboxylase initially into oxaloacetic acid and then to other 4-carbon acids e.g., malic acid, etc. This is called dark carbon dioxide fixation and is most efficient at 10-15°C. CAM plants have parenchyma cells which are large and vacuolated. These vacuoles are used for storing malic and other acids in large amounts.

$$RuDP + CO_2 \rightarrow 2\ PGA \text{ -- } 2\ PEP$$

$$2\ PEP + CO_2 \rightarrow 2 \text{ oxalocetic acid} \rightarrow 2 \text{ malic acid}$$

These organic acids temporarily keep $CO_2$ in bound form. During the day, when the stomata are closed, the $CO_2$ is released from these organic acids and photosynthesized through the Calvin cycle. These acids are decarboxylated and the $CO_2$ released is fixed to produced PGA.

# CHAPTER 17

# The Photorespiratory: Carbon Oxidation Cycle

All land plants have an additional metabolic activity and lose a considerable amount of their photosynthate as $CO_2$ within a few seconds of its being fixed. It occurs when plants are illuminated. This process of $CO_2$ release is light dependent and, because of its otherwise superficial resemblance to respiration, it has been given the name *photorespiration*. During the process $O_2$ is consumed and $CO_2$ generated in light. It is independent of mitochondrial respiration but occurs in chloroplasts and peroxisomes and only a few reactions occur in mitochondria. However, unlike true respiration (frequently called dark respiration) it performs no useful function in making energy available to the plant. Instead, its occurrence leads to a *loss* of energy and this is responsible for a considerable reduction in the yield of many of our crops.

Photorespiration is an additional metabolic activity that is an important feature of all plants and occurs only when the plants are illuminated. Because photorespiration, like dark respiration, results in $CO_2$ evolution and $O_2$ consumption, it is the **opposite of photosynthesis**. It has the net adverse effect on decreasing plant growth and crop yield.

Photorespiration results from the ability of rubisco to catalyze the oxygenation of ribulose-1,5-bisphosphate as well as its carboxylation. This occurs because $O_2$ and $CO_2$ occupy the **same catalytic site** and **compete** in reacting with the sugar phosphate. The functioning of the RPP cycle is prevented by the production of one molecule of phosphoglycolic acid and only one molecule of phosphoglyceric acid. During photorespiration recycling of 75 per cent of the carbon in phosphoglycolate (25 per cent is lost as $CO_2$) and back to phosphoglyceric acid to permit the continued operation of the RPP cycle occurs. The recycling occurs by the enzymes located in three separate organelles.

Photorespiration probably arose in evolution due to the gradual loss of $CO_2$ from the Earth's atmosphere to a level at which most land plants cannot take it up without losing some of what has already been fixed. However, there is evidence that some of the newer arrivals on the evolutionary scene, the so-called $C_4$ plants, have evolved a $CO_2$ concentrating mechanism which generates a high $CO_2$ concentration for their photosynthetic enzymes, and have in this way almost eliminated photorespiration.

**TABLE 17.1:** Differences between photorespiration and true respiration

| Photorespiration | True Respiration |
|---|---|
| **The Effect of Oxygen Concentration** | |
| It increases steadily with oxygen concentration all the way from zero to one hundred per cent. | It is saturated at oxygen concentrations as low as two or three per cent and virtually no increase in rate is observed above these concentrations. |

| Photorespiration | True Respiration |
|---|---|
| **Requirement for the Photosynthetic Apparatus** | |
| It is intimately linked to photosynthesis and occurs only in cells which are capable of photosynthesis. | It occurs in any living cell |
| **The nature of the substrate** | |
| The nature of the substrate is different. | The nature of the substrate is different. |
| **Energy relationship** | |
| Biochemistry of photorespiration differs fundamentally from that of the true respiration. Photorespiration, however, does not lead to a net production of ATP or reduced nucleotides. Glycolet arises as an unwanted byproduct of photosynthesis. The bulk of the carbon of Glycolet, after oxidation, is converted to carbohydrate, but the remainder is converted to $CO_2$ which is released to $CO_2$ of photorespiration. | Biochemistry of true respiration differs fundamentally from that of the photorespiration. True respiration brings about the oxidation of its substrates to yield $CO_2$ and energy, mainly in the form of ATP and reduced pyridine nucleotides. |

## PHOTORESPIRATION: MEASUREMENT

During photorespiration green plants take up oxygen and release $CO_2$ in the light. Therefore, it can be measured by measuring either of these two activities. Unfortunately, most photorespiring tissues also show an oxygen uptake and $CO_2$ release in the light due to true respiration. This leads to an error in measurements of photorespiration, but this error is small, the rate of photorespiration for a tissue being normally several times greater than its rate of true respiration as measured in the dark. The rate of true respiration is reduced to an even lower level in the light. The main problem is that, under normal precise circumstances, photosynthesis and photorespiration occur simultaneously in the same tissue and carry out opposite overall reactions.

$$CO_2 + H_2O \xrightarrow{\text{Photosynthesis}} \xrightarrow[\text{Photorespiration}]{} \text{Photosynthate} + O_2$$

Some of the simpler ways of measuring photorespiration are to inhibit photosynthesis and to measure $O_2$ uptake or $CO_2$ output. One means of doing this is to put a plant or leaf in the light into a $CO_2$-free air stream, so that photosynthesis is inhibited by lack of $CO_2$ and photorespiration can be measured as the rate of $CO_2$ output. This is not accurate measurement leaves some of the $CO_2$ released by photorespiration is refixed by photosynthesis so that the rate of photorespiration is underestimated. Although the degree of this underestimate can be minimized by using fast air flow rates to sweep as much as possible of the $CO_2$ away from the leaf before it is refixed, the error cannot be eliminated altogether by this means.

Photorespiration is also measured to turn off the light and to means the $CO_2$ released immediately afterwards. This is because photorespiration goes on for a minute or so after photosynthesis has stopped and takes the form of a sudden outburst of $CO_2$ at the onset of darkness which rapidly declines to the much lower level of $CO_2$ release due to true respiration. The peak rate of $CO_2$ release is taken as a measure of photorespiration. However, this too could lead to an underestimate if the cessation of photosynthesis is not as abrupt as assumed and some of the outburst is refixed.

A more sophisticated methods of measuring photorespiration is also known. It works without difficulties. Because of the criticism that the above two methods measure photorespiration when photosynthesis is absent, or at least greatly reduced, and that its rate may not normal under these circumstances, attempts have been made to measure photorespiration actually during active photosynthesis by the use of isotopic tracers. For example, photorespiration has been measured as the rate of $^{18}CO_2$ uptake from an atmosphere containing this isotope, but again this leads to an underestimate of photorespiration because some of the $^{16}O_2$ released by released by photosynthesis is taken up by photorespiration in addition to the supplied $^{18}CO_2$.

All the above methods for measuring photorespiration are somewhat inaccurate and most lead to an underestimate of the true rate. Despite this, it has been generally found that measured rates of photorespiration are very high, commonly three or four times the rate of true respiration measured for the same tissue in the dark.

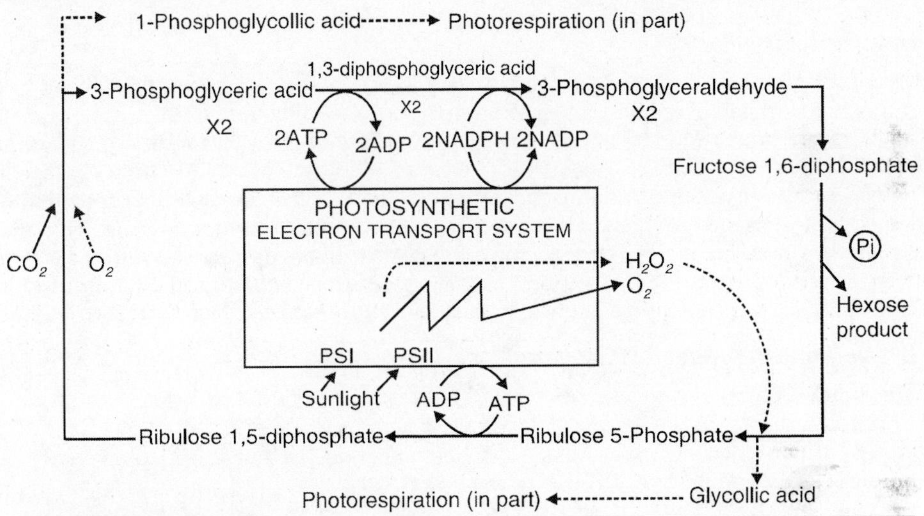


**FIG. 17.1:** A simplified scheme of photorespiration.

## PHOTORESPIRATION BIOLOGICAL FUNCTION IS UNKNOWN

Although the PCO cycle recovers 75 pert cent of the carbon originally lost from the Calvin cycle as 2-phosphoglycolate, why does 2-phosphoglycolate form at all? It may be due to the formation of 2-phosphoglycolate which is a consequence of the chemistry of the carboxylation reaction. It requires an intermediate that can react with both $CO_2$ and $O_2$. Such a reaction would have had litter consequence in early evolutionary times if the ratio of $CO_2$ to $O_2$ in air were higher than it is today. However, the low $CO_2$-to-$O_2$ ratios prevalent in modern times are conducive to photorespiration, with no other function than the recovery of some of the carbon present in 2-phosphoglycolae.

It is also possible that photorespiration is important under conditions of high light intensity and low intercellular $CO_2$ concentration (for example, when stomata are closed because of water stress), to dissipate excess ATP and reducing power from the light reactions, thus preventing damage to the photosynthetic apparatus. *Arabidopsis* mutants that are unable to photorespire grow normally under 2 per cent $CO_2$, but they die rapidly if transferred to normal air. Furthermore, recent evidence with transgenic plants suggests that photorespiration protects $C_3$ plants from photooxidation and photoinhibition.

## PHOTORESPIRATION: METABOLISM

### Evidence for Glycolet as the Substrate for Photorespiration

Glycolet is the source of much of the $CO_2$. It is given off in photorespiration comes from a number sources. The green leaves produced large quantities of Glycolet in the light but not in the dark. The leaves could oxidize artificially supplied Glycolet to yield, amongst other things, $CO_2$. The inhibition of an enzyme which oxidizes Glycolet appears to inhibit photorespiration and causes a temporary increase in the apparent rate of photosynthesis in leaves which photorespire, but not in leaves which do not.

#### *Glycolet: Biosynthesis*

Glycolet synthesis appears to be widespread occurrence in photosynthetic organisms and has been demonstrated in many of the algae and in higher plants.

It is an unwanted metabolic excrete of many algae it into the surrounding water, however, higher plants cannot excrete it in this way, but they oxidize it instead by the process of photorespiration. In algae and higher plants the origin of the Glycolet seems to be similar under similar environmental factors. Glycolet production in both the groups occurs only in the light and is stimulated by high oxygen and low $CO_2$ concentrations and the pattern of labeling in Glycolet, when $^{14}CO_2$ is being photosynthesized is also similar.

#### *Glycolet: Site of production in chloroplast*

The isolated chloroplasts are capable of synthesizing Glycolet. It is an early product f photosynthesis and becomes labeled within seconds when $^{14}CO_2$ is being photosynthesized. The mechanism has been proposed by a group of workers in Illinois (Ogren and Bowes, 1971; Bowes *et al.*, 1971).

They obtained that a derivative of Glycolet, namely, 2-phosphoGlycolet, was produced as a result of the apparent malfunctioning of the enzyme RuDP carboxylase. The proper function of this enzyme is to $CO_2$ for the Calvin cycle. It does this by adding a molecule of $CO_2$ to the 5-carbon compound ribulose-1, 5-diphosphate (RuDP) to form the two 3-carbn molecules of phosphoglyceric acid (PGA) which are further metabolized into sugars etc. by the Calvin cycle. However, these workers showed that oxygen was able to function as an alternative substrate for this enzyme in place of $CO_2$. Using purified preparations of RuDP carboxylase they showed that when oxygen underwent the reaction in place of $CO_2$, *phosphoGlycolet* was produced. The exact mechanism of the reaction was not investigated but it is probably an oxidative splitting of the 5-carbon RuDP into a 2-carbon and a 3-carbon unit:

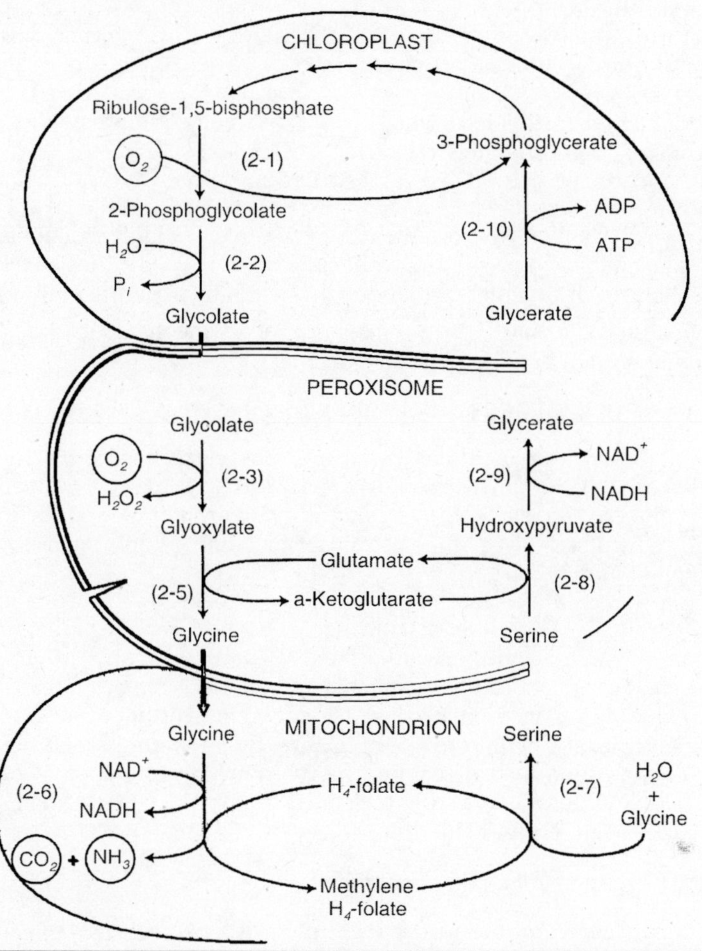


**FIG. 17.2:** Operation of the photorespiration carbon oxidation cycle involves cooperative interaction among three organelles: chloroplasts, mitochondria, and peroxisomes. Oxygen is consumed in each organelle. Shutting of metabolites between organelles is evident from the participating enzymes.

$$\text{RuDP (5-carbon)} + CO_2 \xrightarrow{\text{Normal}} \text{PGA (3-carbon)} + \text{PGA (3-carbon)}$$

$$\text{RuDP (5-carbon)} + O2 \xrightarrow{\text{Malfunction}} \text{Phospho Glycolet (2-carbon)} + ?\ \text{(3-carbon)}$$

PhosphoGlycolet, produced by this means is converted within the chloroplasts to free Glycolet by phosphoGlycolet phosphatase. This enzyme is shown to be located in the chloroplasts and is irreversible, highly active and specific for phosphoGlycolet. The enzyme produces Glycolet from phosphoGlycolet. This indicates Glycolet arises in nature from phosphoGlycolet. The above proposed mechanism of Glycolet biosynthesis explains the following:

1. High oxygen and low $CO_2$ concentrations stimulate Glycolet production. If Glycolet arises due to oxygen competing with $CO_2$ for an enzyme, then anything which increases the $O_2/CO_2$ ratio would be expected to increase

Glycolet synthesis.

2. Glycolet is produced only in the light. The enzyme RuDP carboxylase, when in its natural state in the chloroplasts, is only active in the light (Jensen and Bassham, 1968).
3. Glycolet becomes labeled when $^{14}CO_2$ is being photosynthesized after approximately the same time interval as RuDP.
4. The capacity to produce Glycolet has been retained during evolution, even though it appears to be detrimental to the plant. Naturally, the loss of the enzyme which leads to the production of Glycolet, namely RuDP carboxylase, would also lead to the loss of the plant's ability to photosynthesize.
5. The mechanism of Glycolet production is backed by experimental evidence.

It remains possible that there may also be other mechanisms, but these have yet to be demonstrated in quite such unequivocal terms.

## METABOLISM OF GLYCOLET BY THE PEROXISOMES

Green leaves contain enzymes mainly, Glycolet oxidase. It oxidizes the hydroxyl group of Glycolet to the corresponding aldehyde using oxygen derived from the atmosphere and, it the same time, generating hydrogen peroxide for Glycolet metabolism. Glycolet leaves the chloroplasts and enters small cytoplasmic organelles called peroxisomes.

Peroxisomes can be regarded as packets of enzymes, at least one which generates hydrogen peroxide as a byproduct. Catalase is also present to destroy this peroxide, which might otherwise be toxic. There are many different kinds of peroxisomes and they occur in both animals and plants.

The product of this reaction, glyoxylate, is then converted to the corresponding amino acid, glycine, still within the peroxisome. By an amino-transferase reaction. The donor of the group in this reaction may be one of a number of other amino acids, but glutamate seems to be the most active in this respect, becoming converted to $\alpha$-oxoglutarate in the process.

### Metabolism of Glycine by the Mitochondria

The peroxisomes carry out no further reactions with glycine. Instead the glycine is released and enters the mitochondria for the next stage in its metabolism. This stage is complex, probably involving many enzymes, but the overall reaction is the conversion of two molecules of glycine to one of the amino acid serine, with the loss of one molecule of $CO_2$ and one of ammonia. This process, which involves an oxidation, appears to be coupled to the oxidative phosphorylation system of the mitochondria so that two molecules of ATP are produced for each molecule of serine formed. It is the $CO_2$ given off at this stage which is evolved in the process of photorespiration.

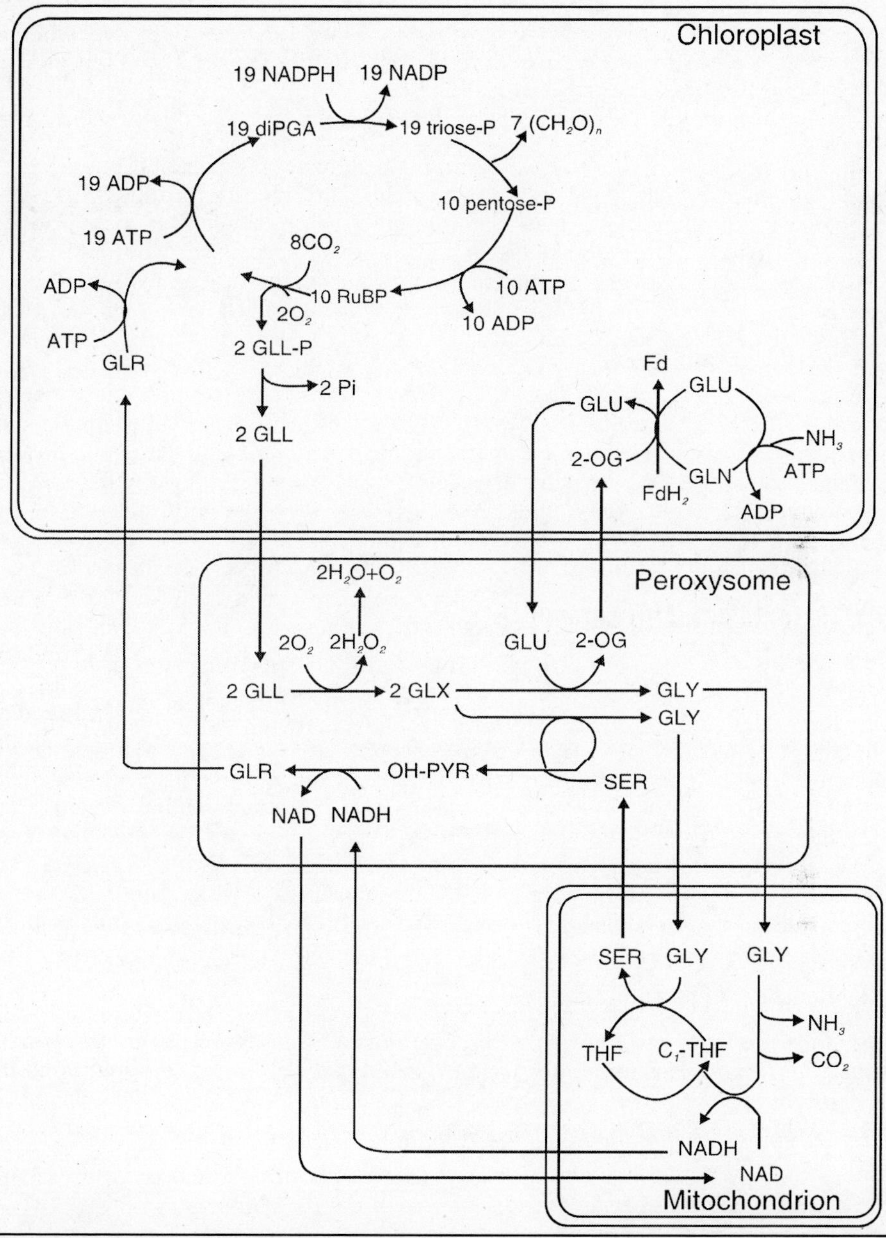


**FIG. 17.3:** Reactions and organelles involved in photorespiration. In $C_3$ plants, at 20°C, 20,000 Pa $O_2$ and 25 Pa CO2 in the chloroplast, 2 out of the 10 RuBP molecules are oxygenated, rather than carboxylated. The oxygenation reaction produces phosphoglycolate (GLL-P), which is dephosphorylated to glycolate (GLL). Glycolate is subsequently metabolized in peroxisomes and mitochondria, in which glyoxylate (GLX) and the amino acids glycine and serine (SER) play a role. Serine is exported from mitochondria and converted to hydroxypyruvate (PYR) and then glycerate (GLR) in the peroxisomes which it returns to the chloroplast.

### Conversion of Serine to Carbohydrate

In tobacco leaves, some of the label from photosynthesized $^{14}CO_2$ first entered glycine and serine and then became incorporated into carbohydrate. Similarly, artificially added serine-3-$^{14}C$ also became incorporated into carbohydrate, but only in the light.

The conversion of serine to carbohydrate, is an energy requiring process which can use the energy of light. However, since the carbohydrate products of serine metabolism accumulated both inside and outside the chloroplasts, it could not be certain whether the formation of carbohydrate from serine was confined to the chloroplasts. There is evidence that at least some of the stages in this reaction sequence can take place in peroxisomes since these particles have been shown to contain some of the necessary enzymes.

Peroxisomes derived from spinach leaves contain an aminotransferse which converts serine to hydroxypyruvate and a NAD-linked dehydrogenase which converts hydroxypyruvate to D-glycerate. The exact location of the remainder of the pathway is not clear, but a D-glycerate kinase enzyme has been demonstrated in higher plant leaves which uses ATP to convert D-glycerate into 3-phosphoglycerate. This compound (PGA) is, of course, an intermediate of the Calvin cycle and would, if it entered the chloroplasts, become converted to carbohydrate, using ATP and NADPH, by the normal processes of photosynthesis.

## PHOTORESPIRATION AND CROP PLANTS

Photorespiration is common in those crops (all C3 plants) that are an important part of the world's food supply such as wheat and rice, many legumes, and sugar beets. In some of these, photorespiration reduces as much 50 per cent of the net assimilation of $CO_2$ by photosynthesis. Therefore, it is essential to find out the means to inhibit photorespiration to increase the crop yield. Other equally important food crops - corn, sorghum, sugarcane exhibit much less photorespiration. This difference can be explained, in part, by the fact that these species are $C_4$ plants carrying out $C_4$ photosynthesis. In this process, the high concentration of $CO_2$ in the bundle-sheath chloroplasts will selectively favour **carboxylation** rather than **oxygenation** and little phosphoglycolate will be produced.

### Partial Recovery of Fixed Carbon

In photorespiratory carbon oxidation cycle-the reactions result in the partial recovery of fixed carbon, and the intracellular location of the participating reactions.

#### *Photorespiratory Oxygenation and Photosynthetic $CO_2$ Fixation*

The oxygenation of ribulose-1,5-bisphosphate results in the incorporation of one atom of molecular $O_2$ into one of the products of the reactions, 2-phosphoglycolate. The ability to catalyze the oxygenation of ribulose-1,5-bisphosphate is a property of all rubiscos regardless of taxonomic origin. Even the rubisco from anaerobic, autotrophic bacteria catalyzes the oxygenase reaction when exposed to oxygen. As alternative substrates for rubisco, $CO_2$ and $O_2$ compete for reaction with ribulose-1,5-bisphosphate, since carboxylation and oxygenation occur within the same active site of the enzyme. Offered equal concentrations of $CO_2$ and $O_2$ in a test tube, angiosperm rubiscos fix $CO_2$ about 80 times faster than they oxygenate.

The **photorespiratory carbon oxidation** (PCO) cycle acts as a scavenger operation to recover the fixed $CO_2$ lost by the oxygenase reaction of rubisco. The 2-phosphoglycolate formed in the chloroplast by oxygenation of ribulose-1,5-bisphosphate is rapidly hydrolyzed to glycolate by a specific chloroplast phosphatase. Subsequent metabolism of the glycolate involves the cooperation of two other organelles: peroxisomes and mitochondria. Glycolate leaves the chloroplast via a specific transporter protein in the envelope membrane and diffuses to a peroxisome. There it is oxidized to glyoxylate and hydrogen peroxide by an oxidase. The peroxide is destroyed by the action of catalase, and the glyoxylate undergoes transamination. The amino donor for this transamination is probably glutamate, and the product is the amino acid glycine.

The glycine leaves the peroxisome and enters a mitochondrion. There two molecules of glycine are converted to a serine and $CO_2$. The amino acid glycine thus serves as the immediate source of the photorespiratory $CO_2$. The newly formed serine leaves the mitochondrion and enters a peroxisome, where it is converted first by transamination to hydroxypyruvate and then by reduction to glycerate. Finally, glycerate reenters the chloroplast, where it is phosphorylated to yield 3-phosphoglycerate.

Therefore, overall, two molecules of phosphoglycolate (four carbon atoms), lost from the Calvin cycle by the oxygenation of RuBP, are converted into one molecule of 3-phosphoglycerate (three carbon atoms) and one $CO_2$. In other words, 75 pert cent of the carbon lost by the oxygenation of ribulose-1,5-bisphosphate is recovered by the PCO cycle and returned to the Calvin cycle.

Evidence in support of the PCO cycle stems from several types of experiments. Using mutant *Arabidopsis* plants lacking phosphoglycolate phosphatase. Glycolate synthesis in vivo is dependent on $O_2$ and is inhibited by $CO_2$ in a competitive manner, as expected from the primary role of rubisco in the cycle. Several studies have shown that, as with the Calvin cycle, the enzymes of the PCO cycle catalyze all the proposed reactions in vitro at rates higher than the in vivo rate of photorespiration. In addition, the participating enzymes are localized in the choroplast (rubisco, phosphoglycolate phosphatase, glycerate kinase), the peroxisome (glycolate oxidase, catalase, hydroxypyruvate reductase, and the two aminotransferases), and the mitochondrion (glycine decarboxylase and serine hydroxymethyltransferase).

## Competition between Carboxylation and Oxygenation

Since photorespiration is concurrent with photosynthesis, it is difficult to measure the rate of photorespiration in intact cells. Two molecules of 2-phosphoglycolate (4 carbon atoms) are needed to make one molecule of 3-phosphoglycerate, with the release of one molecule of $CO_2$; so theoretically one-fourth of carbon entering the PCO cycle is released as $CO_2$.

The rate of *true* photosynthesis is approximately 120 to 125 per cent of the rate of net photosynthesis. The ratio of carboxylation to oxygenation in air at 25°C is computed to be between 2.5 and 3.

## Interlocking of Carboxylation and Oxygenation

Photosynthetic carbon metabolism in the intact leaf reflects the integrated balance between two mutually opposing and interlocking cycles. The Calvin cycle is capable of independent operation, but the PCO cycle depends on the Calvin cycle for a supply or ribulose-1,5-bisphosphate. The balance between the two cycles is determined by three factors: the kinetic properties of rubisco, the concentrations of the substrates $CO_2$ and $O_2$ temperature.

As the temperature increases, the concentration of $CO_2$ in a solution in equilibrium with air decreases more than the concentration of $O_2$ does. Consequently, the concentration ratio of $CO_2$ to $O_2$ decreases as the temperature rises. As a result of this property, photorespiration (oxygenation) increases relative to photosynthesis (carboxylation) as the temperature rises. This effect is enhanced by the kinetic properties of rubisco, which also result in a relative increase in oxygenation at higher temperatures. Overall, then, increasing temperatures progressively tilt the balance away from the Calvin cycle and toward the photorespiratory carbon oxidation cycle.

### Further Metabolism of Glycolet

Phosphoglycolet is produced by accident, from an essential intermediate of photosynthesis. As far as can be judged at present, the remainder of the pathway leading from phosphoglycolet is concerned with recapturing as much of this byproduct as possible and converting it to more useful materials. Experiments in which labelled Glycolet was fed to leaves and leaf preparations indicate that it is oxidized and converted and converted via a number of intermediates to carbohydrate. During the process, some of its carbon is lost as $CO_2$ and it is this which gives rise to $CO_2$ of photorespiration. This metabolism can be conveniently divided into three stages.

## PHOTORESPIRATION: ENERGETICS

The total amount of energy lost by the plant will depend on the amount of glycolet produced. Zelitch (1959) measured the rate of Glycolet synthesis by following the rate of its accumulation when its further metabolism was blocked by an enzyme inhibitor. He found that, in tobacco about half of the carbon fixed by photosynthesis was passing into Glycolet. Other measurements, based on the rate of $CO_2$ release by photorespiration suggest that this proportion might be even higher. It can be seen, therefore, that photorespiration must represent a considerable loss of the plant's available energy.

Similar to true respiration (which makes energy available in the form of ATP and reduced pyridine nucleotides) photorespiration uses these substances and entails a net entails a net energy loss. The convert two molecules of Glycolet to carbohydrate generates two molecules of ATP in the conversion of glycine to serine and $CO_2$, but the subsequent conversion of serine to carbohydrate *uses* two molecules of ATP and two of reduced pyridine nucleotide (NADH and NADPH). If all the carbon of the original two Glycolet molecules is to be recovered, then the $CO_2$ released during serine formation must be refixed by normal photosynthesis and this requires two further molecules of NADPH and three of ATP. Thus, to restore to the level of carbohydrate the four carbon atoms lost in the form of a pair of glycolet molecules, the plant must use a net total of four molecules of reduced pyridine nucleotides and three ATP.

## CHAPTER 18

# Respiration: The Second Energy Transduction

## INTRODUCTION

Plant usually immobilized by their own root systems, exhibits surprising multiplicity of functions which can be deciphered at different levels of organisation. At cellular level also the versatility and synchrony among various functions is marvellously akin to most complex organisms. To study plants at work is the domain of plant physiology and if we enlist and conceptualise the *works* which are performed by plant cells, it would naturally convene into a comparison of living and nonliving systems depicting plants unique by the virtue of growth, development, reproduction, synthetic and degradative reactions, ability to absorb, transport and exchange water, minerals and gases. Primarily this chapter will be concerned with the controlled release and utilization of stored energy by the processes of respiration and fermentation for the support and maintenance of the living system.

During the process of photosynthesis, light energy is converted to chemical energy and stored in the bonds of complex organic molecules. The major portion of stored energy in plants is found in the form of carbohydrates, such as starch and glucose. The weakening or breaking of the carbon bonds of such compounds releases a considerable amount of energy for utilization by the plant. However, the total amount of energy contained within a compound such as, for example, glucose, is not released all at once, but slowly, in a stepwise series of reactions controlled by enzymes. Generally, a series of reactions in the cell leading to the synthesis or breakdown of organic compounds is referred to as *metabolic pathway*. In this chapter, then, we shall discuss the metabolic pathways of respiration and fermentation and their connection with the controlled release of stored energy.

## ADENOSINE TRIPHOSPHATE: THE ENERGY CURRENCY

Both energy yielding and energy consuming reactions occur with the living cell. The potential or stored energy of one compound (e.g. glucose) is released and utilized, in a most efficient manner, to drive the synthesis of other compounds (e.g. proteins). This energy, now stored in the newly synthesized compound, can, in turn, be made available for other synthetic reactions. What is described here is the coupling of energy yielding and energy consuming reactions. However, energy yielding reactions of the cell occur in many instances in the absence of energy consuming reactions. The energy released in such a situation would be in the form of heat and loss to the organism. However, nature has provided the cell with a means of temporary energy storage in the form of adenosine triphosphate (ATP). Thus, the energy released in the oxidation of compounds such as carbohydrates, lipids, proteins, etc., is immediately utilized in the synthesis of ATP from adenosine diphosphate (ADP) and inorganic phosphate (iP). The chemical energy transferred to ATP can be used to drive synthetic reactions, ADP and iP being released in the process. The bond joining the last phosphate group to ATP is referred to as a *high energy bond* (~). This last phosphate group of ATP is readily transferred from one compound to another. In this way energy is transferred from one compound to another compound through ATP. This is of obvious since ATP can be formed in the oxidation of a variety of compounds and can be used to drive the synthesis of a variety of compounds. In this process loss of energy in the form of heat occurs very less and this is because of the cell has very efficient energy transfer.

Energy locked in a biological compound may be transferred repeatedly. Thus in a dynamic system, such as the living cell, the stored energy of glucose may be found at one time in ATP and at another time locked in the bonds of a protein or lipid molecule. A schematic representation of the cyclic manner in which ATP is synthesized and broken down as an intermediate between energy releasing and energy consuming reaction is shown in Fig. 18.1.

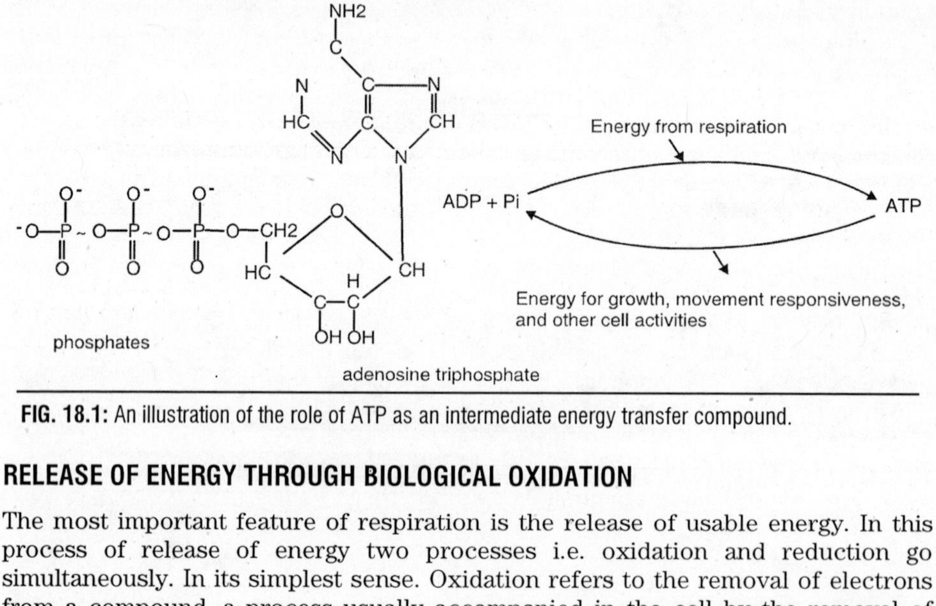

**FIG. 18.1:** An illustration of the role of ATP as an intermediate energy transfer compound.

## RELEASE OF ENERGY THROUGH BIOLOGICAL OXIDATION

The most important feature of respiration is the release of usable energy. In this process of release of energy two processes i.e. oxidation and reduction go simultaneously. In its simplest sense. Oxidation refers to the removal of electrons from a compound, a process usually accompanied in the cell by the removal of hydrogen. Conversely, the reduction of a compound refers to the addition of electrons to that compound, usually accompanied in the cell by the addition of hydrogen.

Biological oxidation proceeds in three stages if we introduce a carbohydrate, such as ubiquitous glucose, into the process.

## A METABOLIC DEGRADATION

A. **Glycolysis: Degradation of glucose to pyruvate Oxidative Decarboxylation of Pyruvate:** A relatively short step during which a 2-C body arises from the 3-C body pyruvate with the elimination of $CO_2$.

B. **The Citric Acid Cycle.** Degradation of the 2-C body furnished by stage B to $Co_2$. It is also known as the krebs cycle after krebs who, together with other scientists, elucidated this pathway; a third name is tri carboxylic acid cycle

(TCA cycle) which it owes to the participation of acids with three carboxyl functions.

C. **Terminal Oxidation in the Respiratory Chain.** The hydrogen which is extracted from the substrates in stages A to C is finally combined with oxygen to form water. For this to occur a hydrogen transport takes place along a chain of redox systems, i.e., via an electron transport chain. Energy set free by this transport is utilized for the formation of ATP.

## GLYCOLYSIS

Glycolysis is a term used to describe the sequential series of reactions present in a wide variety of tissues that starts with a hexese number cusually glucose and ends with pyruvic acid. The equation of the overall reaction may be written as follows. The reaction takes place in cytoplasm.

$$\underset{\text{Hexose}}{C_6H_{12}O_6} \;?\; \underset{\text{Pyruvate}}{2C_3H_4O_3 \text{ x } H_4}$$

### *Glycolysis I*

Glucose is converted to gloucose-6-phosphate which is converted to fructose-6-phosphate by isomerisation. Phosphofructokinase attaches an additional phosphate unit so that fructose 1, 6 diphosphate is formed. Aldolase converts fructose 1, 6-diphosphate to triose phosphate which is a mixture of 3-phosphoglyceraldehyde and dihydroxyacetone phosphate.

### *Glycolysis II*

The equilibrium of the aldolase reaction lies on the side of the hexose and that for the triose phosphates favours dihydroxyacetone phosphate. However, the equilibrium is displaced in the direction of glycolytic degradation by removal of 3-phosphoglyceraldehyde from the system. This removal is brought about by phosphotriose dehydrgenase, a HS-enzyme, i.e., an enzyme with a functionally important HS group. The enzyme adds to the carboxyl function of the 3-phosphoglyceraldehyde and $NAD^+$ takes up 2H from the addition product, thus giving rise to an energy rich thiol ester bond. The enzyme is none displaced by phosphate at this thiol ester bond.

In this way phosphotriose dehydrogenase catalyzes the formation of 1, 3-diphosphate glyceric acid in which an **energy-rich** phosphate has become bound to Cl. This is an important step because in the next reaction this energy-rich phosphate is cleared to from ATP. The second product of this reaction is 3-phosphoglyceric acid. Phosphoglyceromutase then catalyzes the conversion of 3-phosphate glyceric acid into 2-phosphoglyceric acid. Removal of the elements of water from the latter leads to 2-phospheonol pyruvate, a compound with an energy rich phosphate bond (enzyme enolase). In the next step and with the help of pyruvate kinase, the energy of this bond is utilized to form ATP. Concomitantly the enole form of pyruvate, which exists in equilibrium with its keto form, is set free.

## Energy Balance Sheet of Glycolysis

Two ATP are consumed in the phosphxylation of glucose, two ATP are furnished by pyruvate kinase for each glucose introduced. This phosphate is derived from the ATP utilized in the initial phosphoxylation + 2 ATP.

Up to this point the ATP account is balanced. However, 2 additional ATP are furnished per glucose molecule by the transition from 1, 3-disphospho glyceric acid to 3-phosphoglyceric acid + 2 ATP

Since this ATP formation occurs directly on the substrate it is referred to as *substrate level phosphoxylatin.* It supplies 2 ATP per molecule of glucose introduced.

One more point: She phosphotriose dehydrogenase supplies two NADH+H$^+$ per molecule of glucose. The fate of this reduced NAD$^+$ is varied. One possibility is for it to be utilized in the respiratory chain for the formation of more ATP. The NADH + H$^+$ furnished by glycolysis can, however, also be used in fermentation processes.

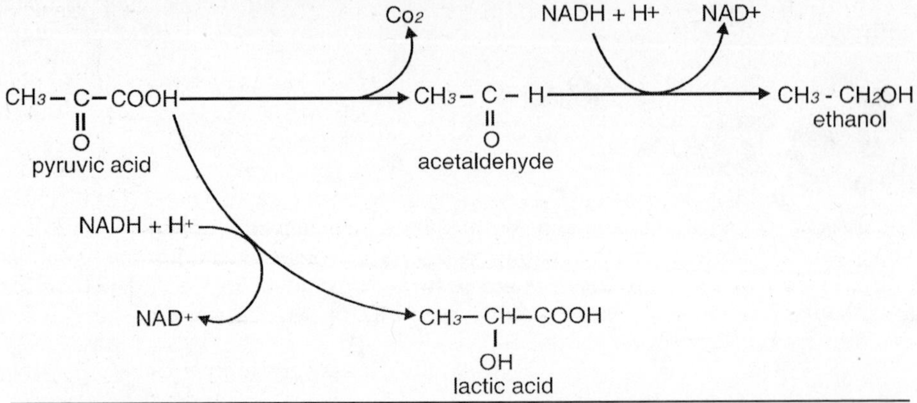


**FIG. 18.2:** Fermentation of pyruvate to form ethanol or lactic acid. Enzymes involved are (1) carboxylase (2) alcohodehydrogenase, and (3) lactic acid dehydroganase.

Fermentation, like glycolysis, is a requential series of reactions that occurs *in the absence of oxygen.*

NADH+H$^+$ is consumed in both kinds of fermentation. Thereby the possibility of a further gain of energy as would have been the case had the reduced coenzyme fed into the respiratory chain is last. Thus, the energy gain in case of both kinds of fermentation is limited to the two ATP molecules, which are derived from the substrate level phosphoxylation.

It should be mentioned here that the products of fermentation, ethanol and lactic acid, still contain a considerable amount of energy locked within their structures. The plant does not benefit from this unreleased energy, which undicated that anaerobic respiration, as fermentation is sometimes called, is a relatively inefficient process.

Fermentation is the major energy yielding process of a variety of micro organisms. The microorganisms in such a case are called *anaerabes*, since they are capable of existing and breaking down organic compounds in the absence of oxygen. Indeed, some of there organisms will die if exposed to any appreciable amount of oxygen. In this case they are called *obligate anaerabes*. Some organisms which ferment glucose to alcohol can exist either in the presence or absence of oxygen. These microorganisms are called *facultative anaerobes* e.g., yeast.

## *β-Oxidative Decarboxylation of Pyruvate, Formation of Active Acetate*

If sufficient oxygen is present, oxidative decarboxylation of pyruvic acid to form *acetyl coenzyme A* takes place.

This reaction is very complex, requiring the presence of atleast five essential cofactors and a complex of enzymes (Gunsalus, 1954; Korkes et al., 1951; Korkes et al., 1952). The five cofactors necessary for the successful formation of acetyl coenzyme A are as follows:

1. Thiamine Pyrophosphate (TPP)
2. $Mg^{2+}$
3. Coenzyme A (CoA)
4. $NAD^+$ and
5. Lipoic acid amide (LAA)

Beside these, flavoprotein (Fb) also involves in the process.

Gunsalus (1954) has suggested four steps in the formation of acetyl coenzyme A (HS-CoA) from pyruvic acid.

i) **I$^{st}$ Step:** Involve the decarboxylation of pyruvate to acetaldehyde takes place on TPP. The acetaldehyde bound to TPP is called "active aldehyde". From its function TPP derives the name carboxylase.
ii) **II$^{nd}$ Step:** In this step acetaldehyde is dehydrogenated to the acetyl group on LAA and formation of acetyl lipoic acid complex takes place. It is brought about by the reductive cleavage of the disulfide bridge. The newly formed bond is "energy rich" (newly formed acid forms a thioester with lipoic acid). Thus, here lipoic acid is reduced and aldehyde is oxidised to an acid and TPP is liberated.
iii) **III$^{rd}$ Step:** It involves the release of the acetyl group from lipoic acid to CoA. Here HS-CoA accepts the acetyl residue with its sulphydryl group (HS) group. The acetyl residue linked to CoA through an energy rich thiol ester bond is known as acetyl CoA or active acetate. This active acetate is a key substance in metabolism.
iv) **IV$^{th}$ Step**: The final step involves the regeneration of oxidised lipoic acid by the transfer of electrons from reduced lipoic acid to $NAD^+$.

The $NADH+H^+$ so formed can then enter the respiratory chain where it is utilized for ATP formation.

So summarise the above four steps, we have the reaction

$$\text{Pyruvate} + \text{CoA} + NAD^+ \rightarrow \text{Acetyl CoA} + Co_2 + NADH+H^+$$

Since TPP and LAA are returned to their original state during the reaction sequence, they have been excluded from this summarising equation.

## CITRIC ACID CYCLE

It is also know as the krebs cycle or Tricarboxylic acidic cycle. This cycle (so named because of the cyclic manner in which the starting compound, *oxaloacetate*, is regenerated) is named after the English biochemist, H.A. Krebs, who played a major role in its discovery.

By means of the Krebs cycle and the **electron transport system**, pyruvate is oxidised to $Co_2$ and $H_2O$. Thus, the complete oxidation of glucose to $Co_2$ and $H_2O$ may occur through the mediation of glycolysis, the krebs cycle, and the electron transport system. Through its association with the electron transport system. The oxidations of krebs cycle can account for the formation of 24 ATP molecules. Thus, the krebs cycle is far more efficient in the release of energy than either glycolysis or fermentation. The reactions of the krebs cycle and the electron transport system require the presence of oxygen and are confined to the mitochondria.

### Formation of Citric Acid

The first reaction of the krebs cycle is the condensation of acetyl CoA with oxaloacetate to form citric acid and release HS CoA the. The result of this reaction, catalysed by condensing enzyme, is that a four carbon dicarboxylic acid is converted to a six carbon tri carboxylic acid (so named Tricorboxylic Acid cycle).

### Regeneration of Oxaloacetic Acid

Through a series of reactions involving four oxidation steps and three molecules of $H_2O$ cone utilized in the condensation reaction), oxaloacetic acid is regenerated from citric acid. In process two molecules of $Co_2$ and eight molecules of H atoms are produced.

### Four Oxidation Steps in the Krebs Cycle

1. Isocitrate dehydregenares transfer hydrogen from the secondary hydroxyl group of isocitrate to either $NAD^+$ or $NADP^{+\cdot}$ $NADH+H^+$ then enters the respiratory chain and $NADPH+H^+$ can be utilized for synthesis. Product is oxalosuccinate.
2. α-ketogluterate dehydrogenase (a multienzyme complex) transfer hydrogen from _-ketogluterate to $NAD^+$ and $NADH+H^+$ formed. The mechanisms of action of this multienzyme complex is similar to the one involved in the oxidative decarboxylation of pyruvate to active acetate. The coenzymes are TPP, LAA, FAD, $NAD^+$, and finally, HS-CoA. The products are $Co_2$, $NADH+H^+$, and succinyl CoA.

Succinate dehydrogenase transfer hydrogen from succinate to $FAD^+$ and $FADH+H^+$ is formed. Product is molonate.

Succinate dehydrogenase is inhibited by malonate a structural analog of succinate. This is an outstanding example of competitive inhibition.

Fumerate hydratase (Fumerase) catalyzes the addition of water to the double bond of fumerate to give malate which is then dehydrogenated to acyl acetate by malate dehydrogenase. $NAD^+$ serves as the hydrogen acceptor.

### Formation of ATP

Succinyl CoA synthetase breaks down succinyl CoA into succinate and HS CoA. In so doing, the energy of the thiol ester bond is used for the formation of ATP. In animals GTP is formed at this point, which then transfer its terminal phosphate residue to ADP, thus leading indirectly to ATP.

### Decarboxylation

1. Oxalosuccinate is decarboxylated to α-ketoglutarate by isocitrate dehydrogenase. With that, one of the two e atoms introduced into the cycle as active acetate has been lost as $CO_2$.
2. α-keteglutarate is decarboxylated by multienzyme complex α-ketoglutarate dehydrogenase and with this second carbon atom of active acetate has been eliminated.

With the formation of oxaloacetate the cycle is complete.

## Balance Sheet

| Two C atoms introduced in TCA cycle are eliminated as $CO_2$ | | |
|---|---|---|
| | ATP | |
| Per molecule of active acetate ATP formed = | 1 | 1 |
| $NADH+H^+$ formed= | 3 | 9 |
| $FADH+H^+$ formed= | 1 | 2 |
| | | 12 |
| 12 ATP produced per active acetate molecule | | |
| 24 ATP produced per glucose molecule (2 active acetate mol.) in Krebs cycle. | 4 | 24 |

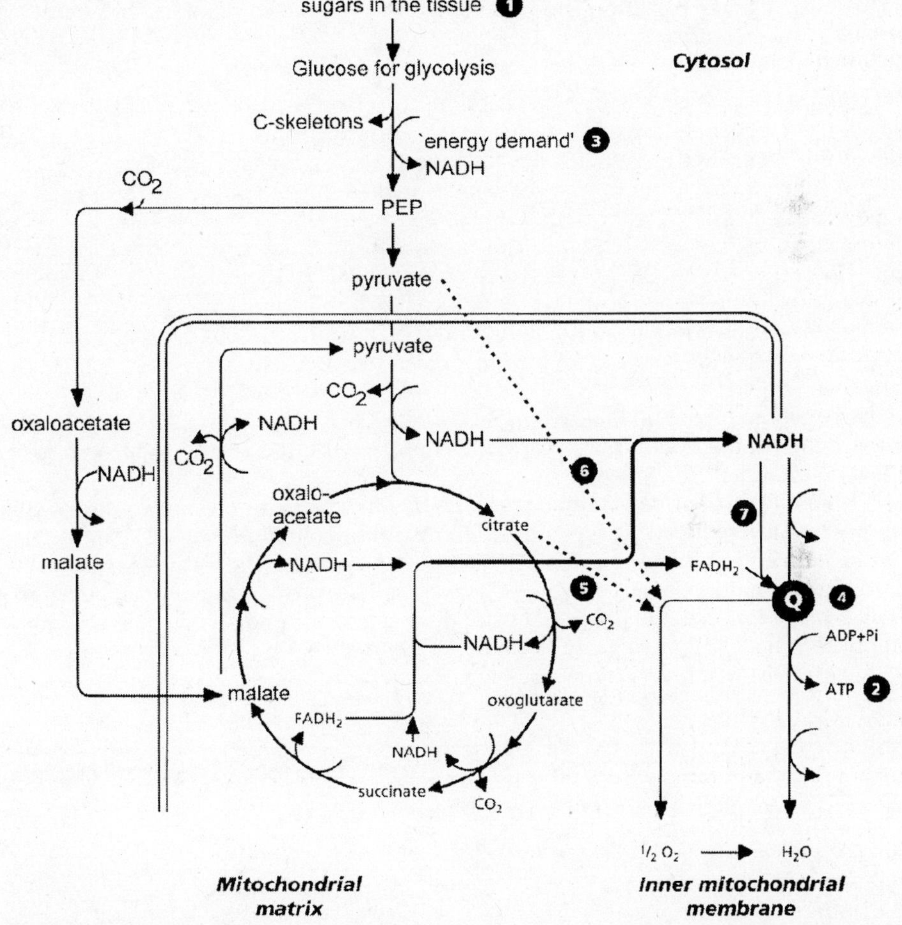


**FIG. 18.3:** Diagram the Krebs or tricarboxylic cycle in mitochondria.

## THE RESPIRATORY CHAIN

The electron transport chain into which the hydrogen of the reduced coenzymes of the citric acid cycle is channelled is called the *respiratory chain.*

In this chain hydrogen or electrons are conducted downhill from redox systems of high to redox systems of low electron pressure. The energy set free in the process is utilized for the formation of ATP. The end of the chain hydrogen is oxidized to water.

Following redox systems are components of the respiratory chain of higher plants

1. $NAD^+$
2. Flavoproteins
3. Ubiquinone
4. Cytochromes to (cytochrome to complex, typical of higher plants
5. Cytochromes c ($^c549$ and $^c547$. The naming of individual cytochromes of the a-, b-, c- groups is bared on the characteristic absorption maxima that appear in the reduced state.)
6. cytochromes a (a and $a_3$ = cytochrome oxidase complex)

$NADH+H^+$ is supplied by the reactions of citric acid cycle and hydrogen or, rather, the electrons derived from it are transported from left to right. At the right end of the chain is the cytochrome oxidase complex, consisting of cytochrome a and cytochrome $a_3$. Cytochrome oxidase $a_3$ makes direct contact with oxygen and is therefore known, together with a few other enzymes (Peroxidases, catalases, and phenolases), as a *direct* oxidase. The name cytochrome oxidase comes from the fact that the complex accepts electrons from cytochrome, its immediate predecessor in the series, and, thus, oxidises it.

The electrons come from the hydrogen which was introduced initially ($2H= 2H^+ + 2e^-$). By means of them cytochrome oxidase reduces oxygen to $0^-$ which then combines with $2H^+$ to form $H_2O$.

Reduced flavoprotein is also supplied by the TCA cycle it transfers its electrons to the cytochrome B complex. They are then transported to the right as outlined above.

Three ATP are formed for each $NADH+H^+$ introduced into the respiratory chain and 2 ATPs for reduced flavoprotein. That is understandable because the reduced flavoprotein enters the respiratory chain somewhat later than $NADH+H^+$.

ATP formation in the respiratory chain is known as *oxidative phosphoxylation*. Whole process has a dual function, namely, the production of energy and the supplying of intermediates for synthesis.

**TABLE 18.2:** Overall balance sheet for biological oxidation with regard to ATP production when glucose as starling material

| Stage | | in the terminal oxidation |
|---|---|---|
| **A Glycolysis** | | |
| 3 Phosphoglyceraldehyde | | |
| 1, 3-diphosphoglyceric acid | 2 NADH+H+ | 6 ATP |
| 1, 3-diphosphoglyceric acid | | |
| 3-phosphglyceric acid | | 2 ATP (Substrate chain phosphoxylation) |
| **B. Formation of Active Acetate** | 2 NADH+H+ | 6 ATP |
| **C. Citric Acid cycle** | | |
| Isocitrate $\alpha$-ketoglutarate | 2 NADH+H+ | 6 ATP |
| $\alpha$-Ketoglutarate Succinyl CoA | 2 NADH+H+ | 6 ATP |

| Stage | | in the terminal oxidation |
|---|---|---|
| Succinyl CoA Succinate | | 2 ATP (Substratechain phosphoxylation) |
| Succinate Fumerate | 2 FAD-H2 | 4 ATP |
| Malate Oxaloaetate | 2 NADH+H+ | 6 ATP |
| 1 Glucose molecule (Biological oxidation) | Sum | **38 ATP** |

## MITOCHONDRIA AS POWER PLANTS

Site of TCA cycle and ETS. The enzyme of glycolysis are located in the cytoplasm. On the other hand, the citric acid cycle and the respiratory chain including the systems of oxidative phosphoxylation, are found in the mitochondria. The enzymes for the ß-oxidation of fatty acids are also stationed there. Individual components are also known to occur in the cytoplasm but the complete systems are, nonetheless, limited to the mitochondria.

Often they are elongated and ellipsoids. Their number also varies from about 10 to 200, 000. Whereas they are present in aerobic cells, they are absent from anaerobic cells. The introduction of the electron microscope has helped to reveal the detailed structure of the organelle. The diagrammatic representation and an electron micrograph of a typical mitochondrion reveal that the mitochondrion consists of a smooth outer membrane surrounding an involuted inner membrane. The infolding of the inner membrane gives rise to structures known as cristae. Mitochondria from different types of cells show different arrangements of the cristae. Most animals and higher plants have cristae arranged like baffles. Two membranes form the inner and outer surfaces of the mitochondria. Both membranes are made up of unit membrane. Protein layer lipid layer (double) Protein layer. The outer membrane covers the surface of the mitochondrium tightly but the inner membrane is deeply folded and can form, for example, **mitochondrial cristae**. The space enclosed within the inner membrane is filled with a cytoplasmic ground substance, the matrix.

### Location of Respiratory Chain

There are various hypotheses as to where the systems described above are localised. Even though the details are unsettled it has been established that the components of the components of the respiratory chain lie in close spatial association with those of oxidative phosphoxylation and that both systems are localised in or on the inner mitochondrial membrane. Thus, e.g., spherical bodies, which show ATPase activity have been detected electron microscopically on that surface of the inner membrane that is oriented towards the matrix. Probably they are involved in ATP synthesis *in vivo* and thus are components of the citrate cycle and those of the ß-oxidation pathway of the fatty acids are found in the matrix. Just as with the chloroplasts two construction principles are apparent in the mitochondria

1. Surface enlargement by investigations of the inner membrane.
2. Compact mentalization by localization of particular systems in the membranes or in the matrix.

#### *Cristae and Surface Area*

Mitochondria from algae such as Chrysophyceae possibly have cristae in the form of microvilli. Whatever their form the cristae appear, it represents a device for increasing the surface area of the internal membrane of the mitochondrion. There is a correlations between the density of the cristae and the respiratory rate in plant and animal tissues. The respiratory activity of mitochondria isolated from the sterile spadix of *Arum maculatum* rises as the flower matures, this dramatic increase in respiratory ability is accompanied by an increase in the surface area of the inner membrane of the mitochondria.

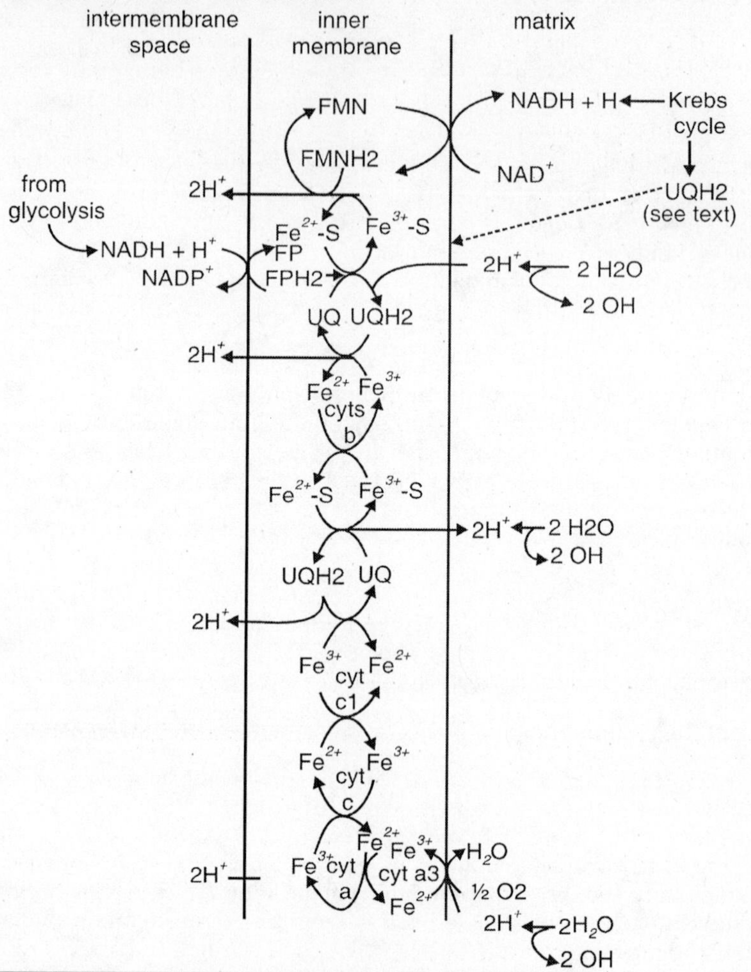


**FIG. 18.4:** Reactions of the mitochondrial electron-transport system.

## *Morphology of Inner and Outer Membranes*

The morphology of both the inner and outer membranes of the mitochondria have been studied in great detail, using the electron microscope. High-resolution electron micrographs of sections of the mitochondrion show that when positively stained both membranes appear to be similar and show a **trilamellar structure** similar to the classical three-layer unit membrane described by Robertson (1967). This structure is almost ubiquitous in biological membranes suggests an underlying unity of structure in these membranes which is consistent with the original model described by Danielli and Davson in 1935. The model consists of a **continuously bilayer**. This layer of hydrophilic groups on the surface of the lipid bilayer can interact with a layer of hydrophilic protein molecules or as layers of fully extended protein molecules.

## THE INTRAMITOCHONDRIAL LOCATION OF ENZYMES

Advancement in the separation of the inner and outer mitochondrial membranes has made it possible to study the localisation of the individual enzymes responsible for the metabolic activity of the mitochondrion. Reports on the localisation of enzymes came from studies of the permeability of the membrane to substrates, co-enzymes and antibodies. Table 2 presents a survey of the distribution of various key enzymes within the organelle. The enzymes located on the outer membrane form a rather heterogeneous group.

### *Monoamine oxidase and kyneurine hydroxylase*

These enzymes are key enzymes used as marker enzymes in mammalian mitochondria. These enzymes are also found in mitochondria *Neurospora* (fungi), however, they are not present in mitochondria higher plants.

### *NADH-cytochrome c reductase*

This enzyme is found on the outer membrane. This enzyme is closely related to the enzyme situated in the endoplasmic reticulum. The presence of NADH-cytochrome c reductase enzyme on the outer membrane of plant mitochondria further suggests the similarities between the outer membrane of the mitochondrion and the endoplasmic reticulum.

### *Adenylate Kinase*

It is the most interesting enzyme located in the inter-membrane space. Possibly it is connected with the relationship between organelle and cytoplasm, converting the AMP produced in the cytoplasm into ADP by reacting AMP with ATP. The ADP can then be translocated into the matrix where it is converted to ATP by oxidative phosphorylation. There is no known translocator that can bring about the direct entry of AMP into the matrix of the mitochondrion.

### *Inner Membrane Bound Enzymes*

The inner membrane contains most of the enzymes responsible for electron transport and oxidative phosphorylation, however, not all the enzymes are equally firmly associated with the membrane system. **Succinic dehydrogenase** and **NADH dehydrogenase** are firmly bound to the inner membrane, while other flavoproteins such as **lipoyl dehydrogenase** and those responsible for the oxidation of fatty acids can easily be dissociated from the membrane. The enzymes bound to the inner membrane have a very precise position with respect to the structure of the membrane. Thus **succinic dehydrogenase** appears to be bound to the inner surface of the outer membrane while **cytochrome c** is loosely bond to the outer surface of the inner membrane.

In the plant mitochondria there are two enzyme systems bound tightly to the inner membrane. One appears to be associated with the inner face of the inner membrane, is sensitive to inhibitors such as **rotenone** or **piericidin A** and is coupled to oxidative phosphorylation at the first site of ATP synthesis. The other system appears to be bound to the outer surface of the inner membrane, is insensitive to inhibition by piericidin A and by-passes the first site of ATP synthesis in oxidative phosphorylation.

### *Mitochondrial Inner Membrane and Ion Permeability*

The inner membrane is believed to be the main permeability barrier to ions contains specific translocators responsible for the uptake of charged ions of physiologically important compounds which would otherwise diffuse readily through the membrane. The matrix contains most of the enzymes connected with the operation of the tricarboxylic acid cycle and many of the enzymes responsible for the oxidation of fatty acids. There is increasing evidence that although their is no apparent subdivision within the matrix not all the enzymes act as if they were simply floating freely around in the matrix space and some of the soluble enzymes may associate with specific areas of the inner membrane.

## The Inner and Outer Mitochondrial Membranes are structurally and functionally different

The inner and outer mitochondrial membranes define the two submitochondrial spaces; the intermembrane space between the two membranes, and the central compartment, called the matrix. Because the different membranes and compartments of the mitochondrion can be purified and studied separately, it has been possible to localize each reaction to a specific membrane or space.

The inner membrane and the matrix are the sites of most of the reactions involving the oxidation of pyruvate and fatty acids to $CO_2$ and also the coupled synthesis of ATP from ADP and $P_i$. These very complex processes involve many discrete steps, but they can conveniently be subdivided into three groups of reactions, each of which occurs in a discrete membrane or space within the mitochondrion.

1. Oxidation of pyruvate or fatty acids to $CO_2$, coupled to the reduction of the electron carriers $NAD^+$ and FAD to NADH and $FADH_2$. These reactions occur in the matrix, or on inner-membrane proteins that face the matrix.
2. Transfer of electrons from NADH and $FADH_2$ to $O_2$. These reactions occur in the inner membrane and are coupled to the generation of an electrochemical proton gradient across the inner membrane.
3. Utilization of the energy stored in the transmembrane proton gradient for synthesis of ATP. This step is analogous to the step in photosynthetic generation of ATP and utilizes extremely similar protein complexes in the inner mitochondrial membrane called the $F_0F_1$- ATPase complex.

Each of the latter two groups of interrelated operations involves sets of multisubunit proteins that are in a highly asymmetrical orientation within the inner mitochondrial membrane.

The outer mitochondrial membrane defines the smooth outer perimeter of the organelle. It is an unusual membrane in that it is freely permeable to most small molecules (m.w. 10,000); in particular, it is freely permeable to protons. It does not act as a barrier to diffusion of metabolites.

### *Outer Membrane Protein or Porin*

The permeability is due to an outer membrane protein, **porin**, that forms a transmembrane channel. Porin is similar in structure and function to the protein lining the pores in the outer membrane of gram-negative bacteria which also confers permeability to most small molecules. Thus, the inner membrane is the only effective permeability barrier between the cytoplasm and the matrix of the mitochondrion.

The inner mitochondrial membrane contains a higher fraction of protein -76 per cent of the total weight of the membrane- than that in any other cellular membrane. Cardiolipin (diphosphatidyl glycerol), the lipid concentrated in this inner mitochondrial membrane, is believed to reduce the permeability of the lipid bilayer to protons, thus making it less *leaky*. Freezefracture studies on the inner mitochondrial membrane indicates that it contains many protein-rich intramembrane particles that are laterally mobile in the plane of membrane.

## Enzymes and Small Molecules Involved in the Citric Acid Cycle

Most of the enzymes and small molecules involved in the citric acid cycle are soluble in aqueous solution and are localized to the matrix of the mitochondrion. The exceptions are coenzyme A (CoA), acetyl CoA, and succinyl CoA, in which the hydrophobic CoA segment is bound to the inner membrane, with the acetyl or succinyl part facing the matrix. Succinate dehydrogenase is also localized to the inner membrane.

## Use of Molecular Oxygen

The reactions detailed thus far result in the conversion of one molecule of glucose into six molecules of $CO_2$. Concomitantly, 10 $NAD^+$ molecules are reduced to 10 of NADH, and 2 of FAD to 2 of $FADH_2$. Two NADH are formed in conversion of glucose to two pyruvate molecules. Oxidation of each pyruvate to acetyl CoA generates one NADH (or a total of two per glucose). Oxidation of each acetyl CoA to $CO_2$ generates three molecules of NADH and one molecule of $FADH_2$, or six of NADH and two of $FADH_2$ per glucose. Molecular oxygen is used to reoxidize these reduced coenzymes, but this does not occur in a single step. Rather, NADH and $FADH_2$ transfer their pairs of electrons to acceptor molecules in the inner mitochondrial membrane. The loss of electrons regenerates the oxidized forms of $NAD^+$ and FAD as well as the reduced form of the acceptor. The electrons thus released are transferred along a chain of electron carriers, all of which are integral components of the inner mitochondrial membrane. Eventually they are transferred to oxygen, the ultimate electron acceptor, thus forming water.

Bacteria lack mitochondria or any other internal membrane system; yet aerobic bacteria carry out the same processes of oxidative phosphorylation that occur in eukaryotic mitochondria. Enzymes catalyzing the reactions of both the Embden-Meyerhof Pathway and the citric acid cycle are all localized to the bacterial cytosol. The enzymes that oxidize NADH to $NAD^+$ and transfer the electrons to the ultimate acceptor, $O_2$, are localized to the bacterial plasma membrane. Movement of electrons through these membrane carriers is coupled to the pumping of protons out of the cell. Movement of protons back into the cell down their concentration gradient is coupled to the synthesis of ATP; the membrane proteins involved in ATP synthesis have a structure and function very similar to those of the mitochondrial $F_0F_1$ complex.

In both photosynthesis and oxidative phosphorylation, the movement of electrons through the membrane is coupled to the generation of an **electrochemical proton gradient**. The processes of electron transport, proton pumping, and generation of ATP are interdependent. The generation of the electrochemical proton gradient- also called the *proton-motive force* -and its subsequent disssipation in the formation of ATP normally occur simultaneously, and the processes are closely coupled.

## INNER-MEMBRANE PROTEINS

Inner-membrane proteins allow uptake of electrons from cytosolic NADH. During conversion of glucose to pyruvate in the cytoplasm, two molecules of $NAD^+$ are reduced to NADH. These electrons are ultimately transferred to oxygen, concomitant with generation of ATP. However, the inner mitochondrial membrane is impermeable to NADH. Several **electron shuttles** are employed.

Cytosolic NADH reduces dihydroxyacetone phosphate to glycerol-3-phosphate. The latter diffuses to the inner mitochondrial membrane; there it is reoxidized to dihydroxyacetone phosphate by an FAD-linked glycerol-3-phosphate dehydrogenase that is an integral protein of the inner mitochondrial membrane. The $FADH_2$ thus formed transfers two electrons directly into the electron transport system. The electrons from cytosolic NADH enter the electron transport system at the level of $FADH_2$, and only two ATPs are generated per pair of electrons transferred to $O_2$, rather than three, as in oxidation of mitochondrial NADH. The energy not used to form a third ATP, required for transport of the electrons into the mitochondrion. Thus not all of the energy generated by oxidation of pyruvate or NADH is available for synthesis of ATP; as much as 25 to 30 per cent is used for import of metabolites into the mitochondrion, against concentration gradients, and for export of ATP.

## METABOLIC REGULATION

All enzyme-catalyzed reactions and all metabolic pathways are tightly regulated. Mitochondria synthesize ATP only as needed by the cell. The import of metabolites into the mitochondrion and the export of ATP are also coordinated with ATP synthesis and, the generation of heat. Similarly, the conversion of glucose, fatty acids, and other metabolites to acetyl CoA is tightly regulated so that only the needed amount of this substrate for the citric acid cycle is produced. **Gluconeogenesis** is the process of the conversion of pyruvate and other small molecules to glucose- which is the apparent reverse of the Embden-Meyerhof pathway.

## UNCOUPLING OF RESPIRATORY CONTROL

Certain poisons such as 2,4-dinitrophenol uncouple oxidation of NADH from synthesis of ATP from ADP and $P_i$. Uncouplers allow oxidation of NADH and reduction of oxygen to continue at high levels, but there is no ATP synthesis. Dinitrophenol (DNP) acts as a membrane transporter for $H^+$, bypassing the ATP synthesis system normally associated with the $H^+$ movement. Both the neutral and negatively charged forms of DNP are soluble in phospholipid membranes and in aqueous solution, so DNP can act as a shuttle for protons. By making the membrane leaky to protons, DNP dissipates any transmembrane proton concentration or potential gradient. Uncouplers such as DNP abolish synthesis of ATP- and dispense with any requirement for ADP in oxidation of NADH or in transport of electrons. The energy released by the oxidation of NADH in the presence of DNP is converted to heat.

## ENERGY METABOLISM IN BACTERIA

In certain species of bacteria, energy metabolism is similar to the processes just described for eukaryotic cells. Anaerobic bacteria such as *Lactobacillus* species (which cause spoilage of milk) ferment glucose to lactic acid by a glycolytic pathway identical to that used in muscle cells. Many aerobic bacteria, such as the spore-forming soil bacterium *Bacillus subtilis*, oxidize glucose to $CO_2$ by a combination of the Embden-Meyerhof pathway and the citric acid cycle. But in two key respects, bacterial energy metabolism is different from that in eukaryotes:

1. bacteria lack mitochondria and possess only a plasma membrane; and
2. many bacteria are able to grow in unusual environments and do not use glucose or other sugars as a source of energy.

Although the details of energy metabolism in many bacteria are not well understood, it appears than most, if not all, bacteria utilize a proton-motive force across their plasma membranes. Electron transport chains are often used to generate this proton-motive force, even in certain anaerobic organisms.

## RESPIRATION IN ROOTS

Soils are composed of various mixtures of clay, sand, silt, and the partially decayed organic matter called humus. Sandy soils are **well oxygenated** but hold little water or exchangeable nutrients. Silt and clay soils are **poorly oxygenated** but retain water and nutrients well. Loam soils are mixtures of these components and often combine their desirable characteristics. Almost any soil can be improved by the addition of humus, which must, however, be continuously renewed because it continuously decays. A large proportion of the soil is, therefore, not solid at all but composed of either water or air. Without both, the plant could not exist.

Roots respire like any other part of the plant and depend mostly upon diffusion to carry oxygen to their cells. This oxygen must be immediately available to the roots or they will suffocate, as can occur in very wet soils.

Some plants, however, are especially adapted to live in aquatic habitats or those with boggy soil. The water lily is an extreme example. This plant is equipped with floating leaves that have stomata on their upper sides, as might be expected. Its tissues are also more spongy than usual, which airs the flotation of the leaves and also serves to carry air via the stems to the roots in the mud at the bottom. There are even parasitic insects that live on the submerged roots and stems, dependent on the plant not only for food but for air as well.

### Pneumatophores

In certain plants, such as some mangroves, air can reach the roots by special protruding organs called **pneumatophores** that reach above the water surface into the air and can form a small forest around the base of the plant. The protruding **knees** of cypress trees may have a similar function. The roots of some tropical trees also serve to prop up the plant. Logically named prop roots, some of these may also serve as respiratory organs, and there is some evidence that respiration may actually be their major function.

## CHAPTER 19

# Nitrogen Metabolism: From Bricks to Building

## INTRODUCTION

Nitrogen is the most important element for plant life. The framework elements like C, H and O which are constituents of carbohydrates though find their places prior to nitrogen yet it is the most prevalent element in the living organisms. It is found in essential compounds like proteins, nucleic acids, some of the plant growth regulators, vitamins etc. It constitutes about 80% of the earth's atmosphere. Ironically, nitrogen is not utilized by all plants and is unavailable in this form. Only certain microorganisms are capable of using nitrogen in this form. Nevertheless, nitrogen is an inert element because of N = N structure; excessive energy is required to break triple bond. It reacts with other elements only under high temperature and pressure conditions.

Nitrogen is one of the chief and important constituents of protein and nucleic acid molecules that play the basic role in cell metabolism, growth, reproduction and transmission of heritable characters. Therefore, without a constant supply of this unavoidable element, life can not go on.

Gaseous nitrogen is the most abundant element of atmosphere (78.08% v/v), and seems to have a highly complex nutrient cycle in the terrestrial and aquatic ecosystems. There is an interrelationship between different elements required for plant growth in environment. These all are affected by microbes. Atmospheric nitrogen is chemically inert. Therefore, it can be used, as such by most of the living organisms.

The common source of nitrogen in the nature is the nitrate ions ($NO_3$). The nitrate ion is often absorbed by the plants as a mineral raw material from the environment. The absorbed nitrate ions are then converted into amino groups (-$NH_2$). The conversion of -$NO_3$ to $NH_2$ group is only possible by plants and not by animals. Minerals accumulation in soil, dissolution of rock by water, addition of fertilizers in soil are also the sources of $NO_3$ in the soil.

Apart from these sources, atmospheric nitrogen and oxygen combine together in presence of lightening which come down with the help of rain-water to the ground, and serve as $NO_3$ source on the earth. Absorbed nitrogen (in the form of $NO_3$) incorporates into organic structures of plants which remain intact until death. Animals also obtain nitrogen by eating plants. Finally, both plants and animals die or decay. Through gradual decay of all nitrogen of dead plants and animals, it gets concentrated into ammonia ($NH_3$). The so forming $NH_3$ is then utilized as a source of nutrition for nitrifying bacteria. Usually, these bacteria are of two kinds:

1. Those bacteria which absorb ammonia and convert it into nitrite ions i.e., $NH_3 \rightarrow NO_2$.
2. Those bacteria which absorb nitrite ions and convert it into nitrate i.e., $NO_2 \rightarrow$

$NO_3$.

Thus, the environment again gets $NO_3$ by the activities of these nitrifying bacteria. The available $NO_3$ in the environment acts upon denitrifying bacteria which converts it into molecular, atmospheric nitrogen. The free molecular nitrogen is then, utilized by nitrogen fixing organisms in which few species of bacteria and blue green algae have been included. They are often present in soil and water. They absorb nitrogen and incorporate it into ammonia and protein. The process is called **Nitrogen fixation**.

Nitrogen fixation has always been a subject of great importance in biology, but since two decades the recent biotechnological development has lent its practical urgency. Nitrogenous fertilizers have become more and more expensive in terms of fossil energy, and hard cash; the population pressure have increased the need for high protein plant food. Biotechnologists are actively engaged in improving the efficiency of microorganisms to fix atmospheric nitrogen in fields. This will improve the functioning of the nitrogen cycle in biosphere which will ultimately increase plant productivity. The programme of improvement also includes the creation of nitrogen-fixing symbiotic association other than those existing between leguminous plants and bacteria of the genus *Rhizobium.* This will help to improve agricultural production.

## NITROGEN: SOURCE

Inspite of the large quantities of nitrogen present in the atmosphere soil is source of nitrogen to most of the plants. Soil contains 0.1 to 1% of nitrogen. This is evident that the soil nitrogen comes from atmospheric nitrogen. Soil nitrogen can be divided as organic or inorganic nitrogen.

### Inorganic nitrogen

Nitrates, nitrites and ammonia are the chief forms of inorganic nitrogen. Nitrogen is absorbed in the form of nitrates by the roots of most higher plants. It is converted into ammonia first and then incorporated into the nitrogenous compounds of the plant.

### Organic nitrogen

Amino acids and urea are the two important soluble forms of organic nitrogen. Many soil microorganisms and higher plants take up amino acids chiefly.

### Molecular nitrogen

Nitrogen constitutes about 80% of the atmosphere but only few bacteria, blue green algae and some leguminous plants with root nodules can utilize atmospheric or molecular nitrogen which is converted into nitrate, ammonia or ammonium before it is used by higher plants.

The utilization of atmospheric nitrogen by higher plants in the form of organic nitrogen compounds is called nitrogen fixation. It is also defined as the conversion of free nitrogen into nitrogenous absorption by plants. Nitrogen fixation is of two types on the basis of the agent carrying out the process. It can be either physical nitrogen fixation or biological nitrogen fixation.

**TABLE 19.1:** Forms of nitrogen

| | |
|---|---|
| Organic Nitrogen | All nitrogen present in organic compounds is considered to be organic nitrogen. The nitrogen-containing organic compounds are derivations of ammonia, the oxidation of which forms ammonia nitrogen. |
| Ammonia Nitrogen | The ammonia nitrogen is a result of bacterial decomposition of organic matter. Fresh sewage is generally high in organic nitrogen and low in ammonia nitrogen. The sum of organic and ammonia nitrogen should remain constant for the same liquid wastes, unless ammonia is allowed to escape to the atmosphere because of septic action. The total concentration of the two serves as a valuable index for evaluating the strength of liquid waste and for determining the type of treatment process to select. |
| Nitrite Nitrogen | Nitrite nitrogen is formed by bacterial oxidation of ammonia nitrogen. It is not present in fresh wastes but appears after bacterial activity has taken place. The presence of nitrite nitrogen indicate that the waste has undergone partial decomposition and is unstable. Nitrites can either be reduced back to ammonia or oxidized to nitrates. |
| Nitrate Nitrogen | Nitrate nitrogen is formed by the oxidation of nitrites and represents the most stable form of nitrogen. It is an indication of stability and is a determination of the completeness of the biological decomposition process. |

## NITROGEN FIXATION: PHYSICAL

It does not involve the participation of any biological agent and is merely a physical phenomenon. It is carried out in many steps and involves the combination of atmospheric nitrogen with $O_2$ to from nitric oxide;

1. $N_2 + O_2$ nitric oxide 2NO electric discharge
   Nitric oxide is then oxidised by $O_2$ to form nitrogen peroxide.
2. $2NO_2 + O_2 \rightarrow 2NO_2$ or $N_2O_2$
   This nitrogen peroxide then combines with water to form nitric acid or nitrous acid.
3. $2NO_2$ + Rain water $\rightarrow HNO_3$ or $HNO_2$
   This nitric acid, or nitrous acid then combines with alkali radicals, which come along with rain water to the ground and then they are absorbed by the roots of the plants.

Apart from *Azatobacter*, other non-symbiotic bacteria like *Rhodospirillum pneumoiniae*, *Rhodopseudomonas*, *Chlorobium*, *Diplococcus pneumoiniae*, *Azatobacter aerogenes*, *Micrococcus sulfurens* have also been known to fix atmospheric nitrogen asymbiotically.

The ability to fix nitrogen by microorganisms was confirmed by means of the technique known as *acetylene reduction* to *ethylene* by diazotrophic microorganisms. The mechanism of this conversion is controlled by an enzymatic complex (or nitrogenous enzyme) which can reduce gaseous nitrogen to ammonia.

The enzyme nitrogenase is sensitive towards oxygen. There are reports which reveal that several groups of microorganisms fix nitrogen in presence of minute quantities of oxygen. Such microorganisms are known as **microaerobic fixers**. They oxidize methane. The examples are bacteria, Spirilla (*Aquaspirillum* and *Azosprillium)*, *Xanthobacter autorophicus* and *S. flavus*, *Thiobacillus ferrooxidans* and non-heterocystous microaerobic cyanobacteria (*Oscillatoria, Plectonema*). Certain coliform bacteria also fix nitrogen only in aerobic conditions (for example, *Citrobactor*, *Enterobacter*, *Erwinia)*. A comprehensive lists of microorganisms have been presented in the book, but in near future they may need revision in light of findings of biochemical and physiological researches. It is established that oxygen not only inhibits the activity of nitrogenase enzyme but also regulates its biosynthesis, so that in conditions that are physiologically hardly favourable to nitrogen fixation, the enzyme is not synthesized.

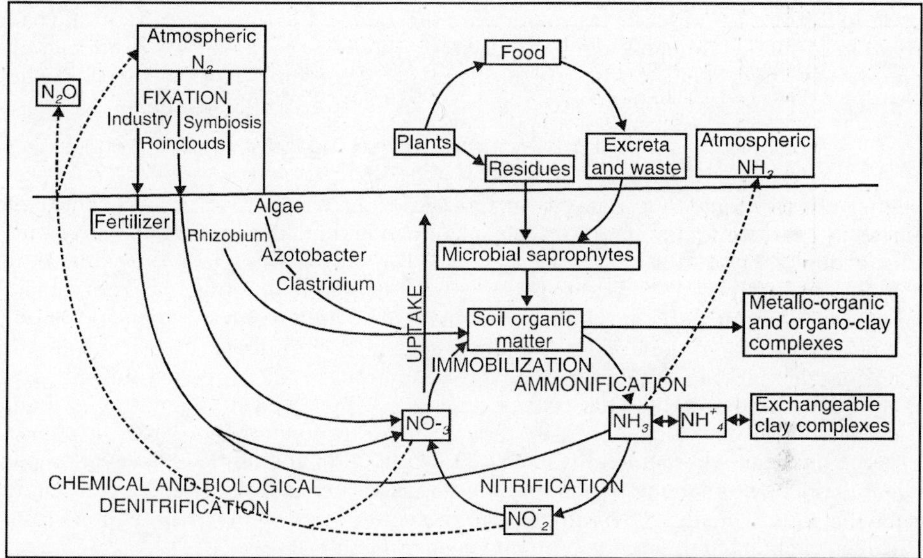


**FIG. 19.1:** The chemical conversion in the nitrogen cycle. The main activities are indicated together with some of the organisms concerned.

**TABLE 19.2:** Comparative idea of the capabilities of some microbial species with approximate quantities of nitrogen fixed by different symbiotic systems.

| Host plant | Nitrogen fixed (kg/ha/yr) |
|---|---|
| *Leucaena leucocephala* | 500 |
| *Gliricidia sepium* | 13 |
| Red clover | 200 |
| Soybean | 100 |
| Pea | 120 |
| Cow pea | 100 |
| Alfalfa | 600 |
| *Pueraria phaseoloides* | 650 |
| **Casuarina equisetifolia* | 100 |
| **Casuarina littoralis* | 250 |
| **Alnus glutinosa* | 50 |
| **Myrica gale* | 10 |

*In respect of non-leguminous trees the nodulating organism is mostly *Frankia* as against *Rhizobium* in legumes. *Frankia* is an actinomycete.

## REQUIREMENTS OF NITROGEN FIXATION

1. Oxygen and carboxyl group of the catalysts play an important role in nitrogen fixation.
2. A very small amount of chemical energy (ATP) is used up during nitrogen

fixation. About 1 kilo calories fix 2 mg of atmospheric nitrogen. It shows the relationship between respiration and nitrogen fixation.

Some strict anaerobic bacteria are more sensitive to oxygen. *Clostridium pasteurianum* is an obligate anaerobe, therefore, the mechanism of nitrogen fixation of such organism is similar to that *Azatobacter*, because they produce large amount of hydrogen during their metabolic reaction. The various possible pathways of nitrogen fixation have been shown in different illustrations.

## NITROGEN CONVERTERS IN THE SOIL NITROGEN CYCLE

Higher plants do not use nitrogen in free form but use nitrogen only in the fixed forms i.e., nitrate and ammonium ions. The elementary nitrogen is converted into nitrogenous compounds by a process called $N_2$ fixation which may be symbiotic or asymbiotic. Nitrogen fixation takes place through various microorganisms (nitrogen fixing bacteria and blue green algae etc.) As a result of nitrogen fixation, the ultimate product obtained is ammonia irrespective of the pathway taken by the microorganism. This ammonia is coverted into many different organic nitrogen compounds of the plant. This organic nitrogen is ingested by animal as essential components of their diet. Since the animal can not convert inorganic nitrogen to organic nitrogen when animal wastes and dead plants and animals decay, complex organic nitrogenous compounds are decomposed into various simpler compounds such as amino acids. Thereafter, ammonification takes place in which amino nitrogen is changed to ammonia by different soil bacteria.

In this way the organic nitrogen is returned to the soil in the form of ammonia which then combines with $CO_2$ and water of soil and makes ammonium salts (ammonium carbonate). The ammonia is also converted to nitrate by the process of nitrification. Hence the ammonification is followed by nitrification taking place by two kinds of bacteria. (i) bacteria of *Nitrosomonas* group (*Nitrosomonas* spp.) oxidizing ammonia of the ammonium salts to nitrite and (ii) bacteria of *Nitrobacter* group (*Nitrobacter* spp.) oxidizing the nitrite to nitrate. The oxidation of ammonia to nitrate and then to nitrate is called nitrification.

$$NH_4 \xrightarrow{\textit{Nitromonas}} NO_2 \xrightarrow{\textit{Nitrobactor}} NO_3^-$$

Both groups of bacteria bringing about nitrification were isolated in 1891 by Winogradsky.

After nitrification, denitrification may take place. Denitrification is the conversion of nitrogen through the mediation of various soil organisms. Thus, nitrogen gas is returned to the atmosphere by the denitrifying bacteria and this all make the nature's complex Nitrogen cycle. Although smaller amounts of combined or fixed nitrogen are contributed to the soil without the participation of living organism.

## NITROGEN FIXATION: SYMBIOTIC

Agriculturists have noted that a relatively large group of plants- the legumes, are capable of fixing atmospheric nitrogen through a symbiotic association with soil bacteria called *Rhizobium*. These bacteria or nodule bacteria, usually dwell in the roots of various beans, peas, lupines, cloves and other leguminous plants.

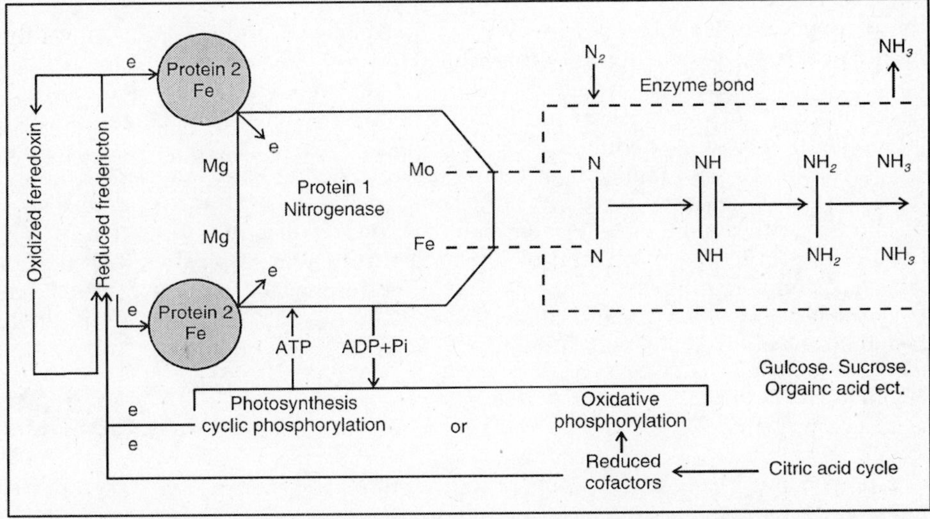


**FIG. 19.2:** A generalised scheme for the action of nitrogenase (incorporating results from studies on various organisms).

Hellriegal and Wilfarth (1888) first of all demonstrated that an increase in nitrogen content in soil is due to the presence of small tumour-like outgrowth on the roots of leguminous plants. It was found that in absence of root nodules, growth of succeeding plant-crop was retarded. The results of various experiments reveal that root nodule forming plants can feed upon atmospheric nitrogen and this capacity of plants can be observed only in those plants which are growing in non-sterilized or bacteria containing soils. These symbiotic bacteria e.g., *Rhizobium* utilize the free atmospheric nitrogen and synthesize it into new nitrogenous compounds, which are utilized by plants for their growth; and bacteria get their food from these plants. Such a mutual beneficial association of bacteria and plants is referred to symbiosis.

Symbiosis was originally defined by Bary in 1879. Symbiosis is the living together of any two organisms. Mutulism is a relationship where both partners benefit, though it grades into those where the advantage or disadvantage to one of the partner may not always be obvious. Commensal relationships are usually rather loose association of the benefit of another without any effect on itself.

## FORMATION OF NODULE IN LEGUMINOUS PLANTS

Roots of most of the leguminous plants possess circular out-growth or swelling called nodules or tubercules, which are formed after infection of certain types of nitrogen fixing bacteria particularly the different species of *Rhizobium*. These bacteria have the capacity to fix free atmospheric nitrogen of soil in nodule. They infect or get their entrance in root through the soft-hair or other epidermal cells by damaging them. In the first stage of the infection, bacteria grow very profusely at the tip of the root hair and form a long filament in the root hair- called *infection thread*. This thread reaches the endodermis and pericycle area through cortex tissue. Cells of this area (cortex) go on dividing and form a young nodule. The newly formed young nodule pushes the overlying parenchyma and the epidermal tissue towards outside and produces a small swelling on the surface of the root. Anatomically, a nodule is made of thin walled parenchymatous cells which is filled up with the nitrogen fixing organisms. The size and shape of the root nodule varies according to plants in which occur. It is not necessary that all the bacteria, which infect the root, produce nodule.

According to Wipf and Cooper (1939) the root nodule always contains double number of chromosomes as against of normal somatic tissue. If the root lacks such cells (cells with double chromosome number), there will be no formation of root nodule, even after the formation of infection thread. Those plants which do not bear root-nodules will never be able to fix atmospheric nitrogen in the plants. It has been observed by Allen and Allen (1947) that out of 429 genera of family Leguminosae about 179 genera containing 949 species form nodules.

## Nitrogen Fixing Organisms found in Nodules

### *Rhizobium*

The presence of these bacteria in the root nodule can be seen under high power microscope by crushing a washed nodule between two glass slider. Frank (1877) and Beijernick (1888) discovered *Rhizobium* a free living or symbiotic bacteria of leguminous as well as non-leguminous plants. This kind of bacteria, preferentially, infect legumes as compared to non-legumes. Structurally, *Rhizobium* are rod shaped but great variation can be observed during their life cycle. These are coccoid, very small, highly motile and ellipsoidal forms. The bacteroids are usually irregular with X, Y, star and club-shaped forms. They are 0.5 to 0.9 micron by 1.2 to 3.0 micron in size. All the species are non-sporing and non-acid fast. These symbiotic bacteria (*Rhizobium*) are difficult to cultivate in ordinary culture media but they grow on manitol agar.

*Rhizobia* are susceptible to antibiotics (produced by other microorganisms of soil) and also to the action of bacteriophages. Fungicides, herbicides and other plant protectants are also toxic to *Rhizobia*. Occurrence of different species of *Rhizobium* in leguminous plants can be enumerated in Table 19.3.

### *Agrobacterium (A. radiobacter, A. tumefacions)*

It has also been included in the species: *Rhizobium radiobacter*, but this splitting of *Rhizobium* at the generic level into *Agrobacterium* is not accepted by most of the microbiologists.

One more kind of slow growing asymbiotic bacteria of lupin, soybean, cowpea have been identified and placed under a separate genus *Phytomyxa* (*P. japonicum*). In short, the root nodule genera can be delimited into two genera: *Rhizobium* and *Phytomyxa*: the former comprises of three species: *R. leguminosarum*, *R. meliloti* and *R. radiobacter*, while the latter bears only one species *P. japonicum*.

**TABLE 19.3:** List of some plants in which *Rhizobium* is found.

| Name of the plants | *Rhizobium* species |
|---|---|
| Clover | *R. trifolii* |
| Bean | *R. phaseoli* |
| Lupinus | *R. lupini* |
| Ornithopus | *R. lupini* |
| Soybean | *R. japonicum* |

| Name of the plants | *Rhizobium* species |
|---|---|
| Melilotus, Trigonella | *R. meliloti* |
| Pea | *R. leguminosarum* |

According to Elken (1969) *Rhizobium* can also be classified on the basis of DNA composition, for example guanine and cytosine composition of *Rhizobium leguminosarum* and *Rhizobium meliloti* is 56.6 to 63.1 per cent while G+C composition of *P. japonicum* is 62.8 to 65.5 per cent (Delay and Rassel, 1965).

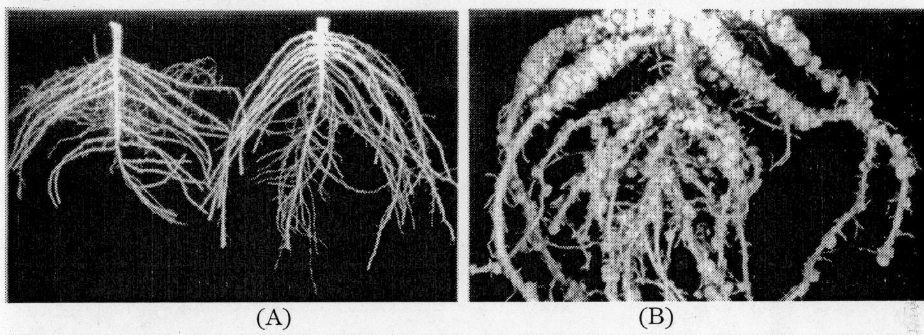


**FIG. 19.3:** Nodulated roots of soybean (B) inoculated with *Rhizobium.*

## STRUCTURE AND FUNCTION OF NODULE

The outermost layer of the nodule constitutes the bacteroid zone which is enclosed by several layers of cortical cells. The rate of nitrogen fixation of nodule is directly proportional to the volume of the effective nodule. Sometimes ineffective strains produce ineffective nodules. These nodules are relatively small and contain poorly developed tissue associated with morphological abnormalities.

### Leghaemoglobin

The effective nodules are usually larger and pink in colour. This is due to the presence of red coloured leghaemoglobin (leg indicating its presence in legume root nodule). This pigment is similar to haemoglobin of blood and is found in nodules between bacteroids and membrane envelops, enclosing them. Chemically, it is haem-protein and it contains a haem meoiety attached to a peptide chain and represents the globin part of the molecule. The molecular weight of leghaemoglobin is 16,000 to 17,000 daltons.

The amount of leghaemoglobin in nodules has the direct relationship between amount of atmospheric nitrogen fixed by legumes. The functions of leghaemoglobin are as follows:

1. It represents an active site of nitrogen absorption and reduction.
2. It acts as a specific electron carrier in nitrogen fixation.
3. It regulates the oxygen-supply in the nodule.
4. It acts as an oxygen-carrier.

According to Scholander (1960), leghaemoglobin acts as a biological-valve in regulating the supply of oxygen to bacteroids in nodules.

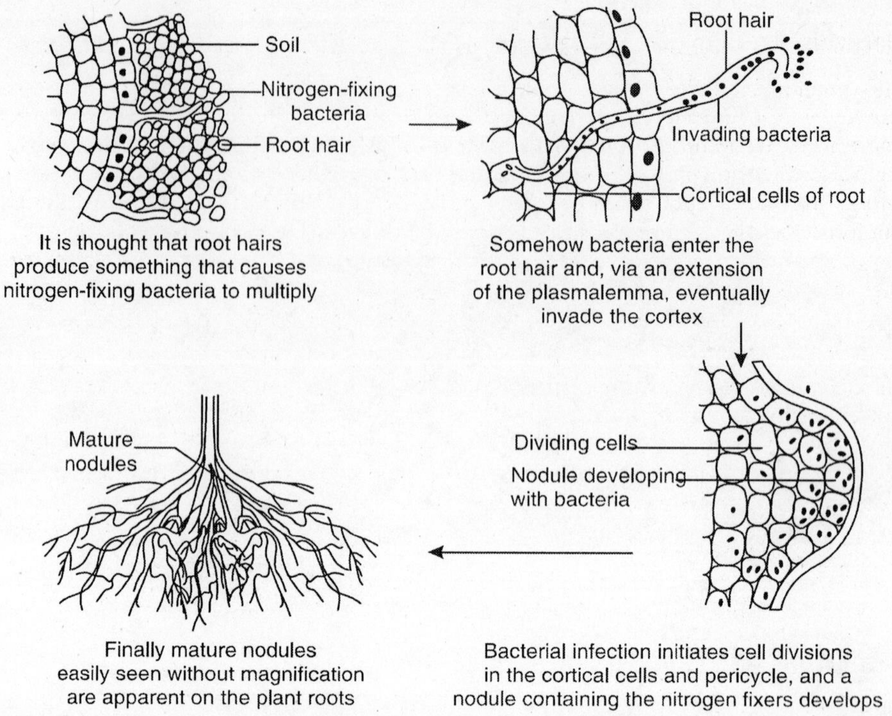


**FIG. 19.4:** Nodule formation by *Rhizobium* on legume plants. The root hairs are the site of infection. Note the change in the root hair. A characteristic deformation is the curling of the root hair tip to produce a Shepherd's crook. The nodule can be seen on the root.

## Site of Nitrogen Fixation in Nodules

According to Bergersen (1969) the membrane envelops surrounding the groups of bacteroids are the site of nitrogen fixation, but in recent years it has been proved that bacteroids are the seat of nitrogen fixation.

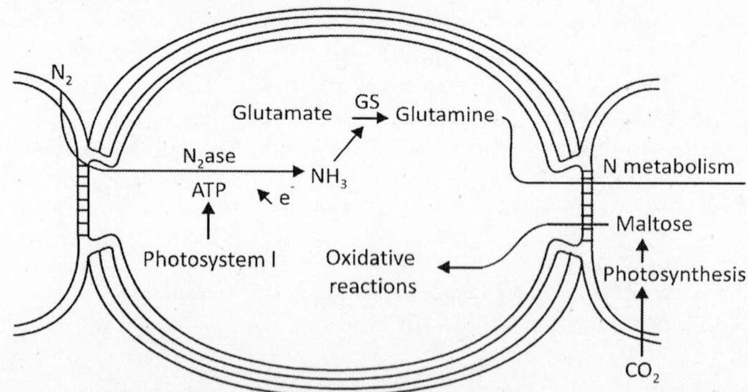


**FIG. 19.5:** Nitrogen and carbon flow associated with nitrogen fixation in the heterocyst of blue green alga. $N_2$ase = the nitrogenase enzyme as GS = glutamine synthetase, oxidative reactions including the oxidative pentose phosphate pathway. Polar nodule are not shown.

## Mechanism of Nitrogen Fixation in Legumes

The mechanism of nitrogen fixation in legumes has not yet been understood. However, the basic prerequisite for nitrogen fixation in extracts of nodule-bacteriods are similar to those that have been described for asymbiotic bacteria such as *Azatobacter*. It has not been well established that how in both asymbiotic and symbiotic system $^{15}N$ is converted into $NH_3$ and then to amino acids, because the intermediate products between $^{15}N$ to $NH_3$ have not been determined.

A good number of theories have been proposed by various workers. Two of them are important, which are as follows:

## Theory of Virtanen

Nitrogen fixation in root appears immediately after nodule formation. Virtanen (1948) proposed the following scheme of nitrogen fixation in root nodules:

According to him young plants fix more amount of nitrogen than the old plants. A great part of nitrogen is converted into L-aspartic acid and L-glutamic acid. Apart from these, β-alanine is also present in the nodule which is produced from L-aspartic acid by decarboxylation. In addition to this, small amount of Oxime-N and Nitrite-N are also present.

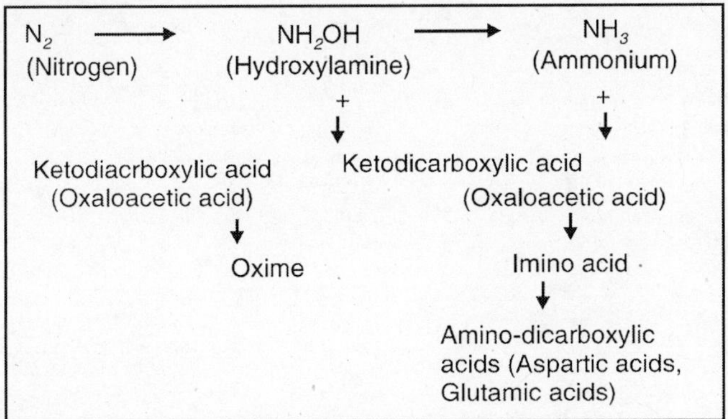


**FIG. 19.6:** Mechanism of nitrogen fixation (Virtanen, 1948).

## THEORY OF BURRIS AND WILSON

According to this theory hydroxylamine is the central compound of nitrogen fixation, from which ammonia is formed through reduction or it can also directly react to the keto acid. The proposed scheme of Burris and Wilson (1945) is as follows:

Both these theories reveal the formation of amino dicarboxylicacid as the primary products in biological nitrogen fixation

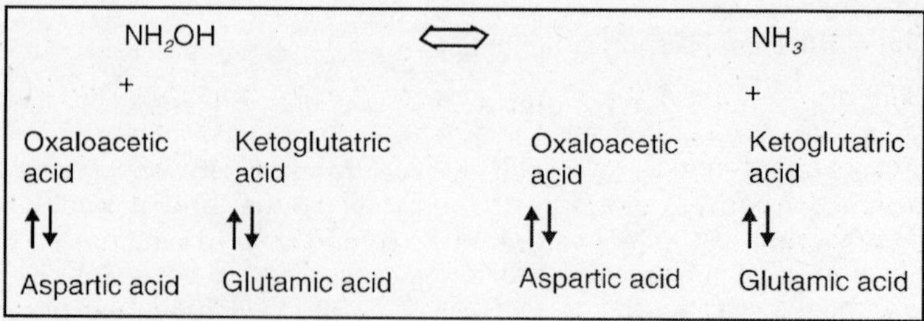


**FIG. 19.7:** Theory of Burris and Wilson

## NITROGEN CONVERTERS IN THE SOIL

The oxidation of ammonia to nitrate in the soil is one of the important phases in nitrogen cycle. This is brought about by two groups of bacteria- *Nitrosomonas* and *Nitrobacter*. The required energy for the growth of these bacteria is obtained through the oxidation of ammonia or nitrate. Winogradsky (1891) first isolated these autotrophic bacteria and described the conversion of ammonia to nitrite, and nitrite to nitrate. This is called **nitrification**. The conversion of nitrate to nitrous oxide and nitrogen gas by various kinds of soil organisms is called **denitrification**. The process releases nitrogen gas to the atmosphere and completes the complex nitrogen cycle in nature.

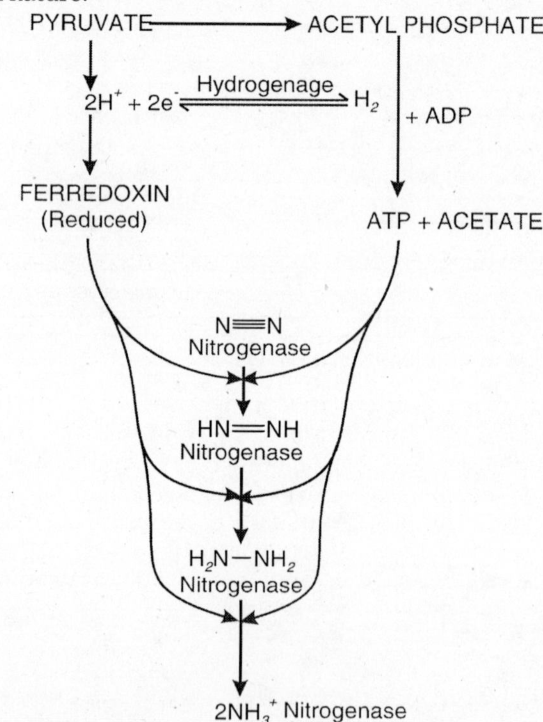


**FIG. 19.8:** Scheme proposed for mechanism of nitrogen fixation.

## Cyanobacteria and Nitrogen fixation

Frank (1889) first reported the ability of nitrogen fixation by blue green algae. According to Drewes (1928) *Nostoc punctiforme* and *Anabaena variabilis* have the ability to fix nitrogen in the soil. The principal genera of this group known to fix nitrogen are *Nostoc* spp. *(N. cummune, N. punctiforme, N. muscorum); Anabaena* spp. (*A. ambigua, A. cylindrica, A. fertilissima. A. circularis,* etc.), *Cylindrospermum* spp. (*C. gorakhporense, C. licheniforme)*- all of them belong to the family Nostocaceae. A few of members of the family Microchaetaceae and Rivulariaceae like *Aulosira fertilissima* and *Calothrix brevissima* respectively are known to fix atmospheric nitrogen in rice fields. Significantly most of the free living nitrogen fixing members of this group produce **heterocyst** which is considered to be the site of nitrogen fixation (Kulasooriya, 1972).

The free living diazotrophic cyanobacteria are the largest contributors of the process of biological nitrogen fixation, the second most important biological process on this planet. Their nitrogen fixing ability was first related to the presence of specialized non-oxygen evolving cells called heterocysts in which nitrogenase; a highly oxygen-sensitive enzyme is produced from oxygen.

## Nitrogen Fixing Fungi

The fungi also independently play an important role in fixing free nitrogen in soil. The fungi that are responsible for this function are of species of *Mactosporium commune, Cladosporium herbarum, Phoma, Alternaria tenuiis, Rhodotorula* spp. etc. The occurrence of these fungi has been noted in forest-land.

## BIOCHEMICAL ASPECTS OF DIAZOTROPHY

An American company (DuPont) in 1960 started the fundamental research on the biochemistry of nitrogen fixation specially on nitrogenase activity in cell free extracts of *Clostridium pasteurianum.* The experiments were conducted in absence of air because of the very great oxygen sensitivity of nitrogenase. The techniques like chromatography, electrophoresis and fractionation under anaerobic conditions, helped the researchers to isolate nitrogenase from *Clostridium pasteurianum.* Thereafter, nitrogenase was easily isolated from other anaerobic nitrogen fixers while this extraction was more difficult with *Azotobacter vinelandii.* The so isolated enzyme has the property to accept electorn from sodium dithionite and was oxygen-tolerant and evolved hydrogen in the absence of nitrogen. In 1981, nitrogenase in the form of crude preparations was isolated from about 30 microbes and out of which about 6 of them were purified.

Biochemically, nitrogenase is a binary enzyme and consists of two brown metalloproteins, of which the joint activity is essential to reduce nitrogen to ammonia.

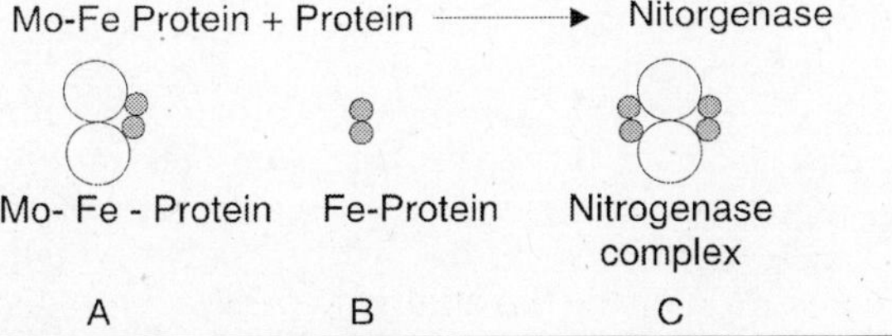


**FIG. 19.9:** Nitrogenase enzyme.

During nitrogen conversion, nitrogenase binds to molybdoferroprotein (MoFe-protein) which was named **disnitrogenase**. When nitrogen binds at the molybdenum level in a separable fragment of protein, called **FeMoco**. The two component proteins of the enzyme nitrogenase are quickly and irreversibly destroyed by exposure to oxygen.

Nitrogenase is also able to reduce other substrates in the place of nitrogen, including hydrogen ions and, in this case it evolves hydrogen.

## Nitrogenase Producing *E. coli* Cells

Because of the close relationship between *E. coli* and *Klebsiella pneumoniae*, *E. coli* is specially qualified for the investigation of the nitrogenase system. With *E. coli* C-M74 cells, high nitrogenase activity can be attained in culture medium, supplemented with 50mg/l ammonium sulfate, 1% aspartic acid, and 0.1% yeast extract.

During reduction of nitrogen to ammonia, electrons are required which are transferred to nitrogenase by a specific transport proteins; the *Ferrodoxins* or the *Flavodoxins*. ATP serves as energy source. The consumption of ATP molecules is 12 to 15 per nitrogen molecule. The reason for this high consumption of ATP has not been entirely elucidated.

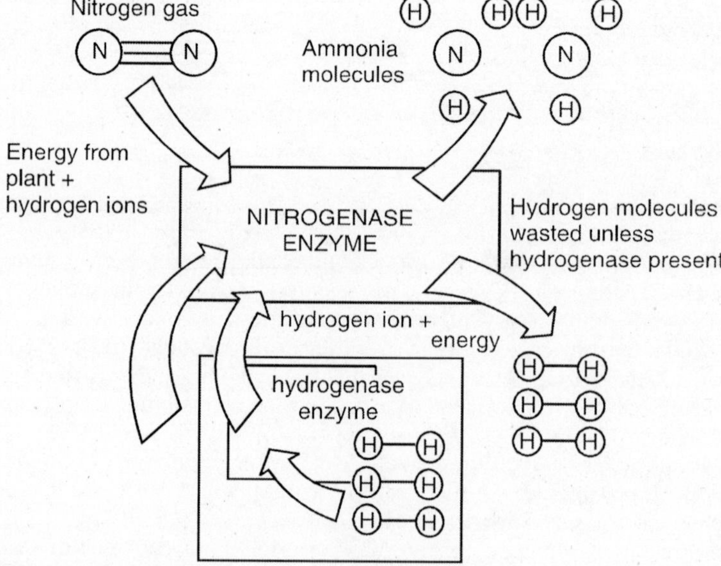


**FIG. 19.10:** Representation of the mechanism of nitrogen fixation and the action of hydrogenase.

As evident from the reactions that hydrogen is produced during nitrogen fixation. Certain nitrogen-fixing aerobic bacteria possess hydrogenase enzyme which has the ability to recycle hydrogen produced by nitrogenase in vitro. Therefore, the re-utilization of this hydrogen generates more ATP (by oxidative phosphorylation) and consequently improves the efficiency of nitrogen fixation.

*Rhizobium* is also associated with soybean plants which contain hydrogen uptake (or *hup*) genes which have the ability to recycle hydrogen and back into the nitrogenase system and fix nitrogen. Through this mechanism the energy is harvested by plants which is otherwise lost.

The nitrogenase enzyme of bacteria fix nitrogen by converting nitrogen gas into ammonia using hydrogen ions and energy by host plant. During the process there is a production of energy rich hydrogen gas. Some bacteria possess hydrogenase enzyme which can capture hydrogen gas and converts it into hydrogen ions and release energy which can feed back into nitrogenase enzyme.

## GENETICS OF FREE-LIVING AND SYMBIOTIC DIAZOTROPHS

### Organization of Nitrogen-fixation Genes

In molecular terms, very little is known of the means by which the prokaryote specifically recognizes and invades its respective host-plant. However, there has been a growing interest in approaching this problem by identifying and analysing the relevant symbiotic genes and their gene products. Because of the agricultural importance of the *Rhizobium*-legume interaction, the most detailed studies have been conducted on this system. Also, because of bacteria are simpler than plants. Our knowledge of the symbiotic genes of *Rhizobium* is better than that of the plant genes. There have been several recent advances in the analysis of the *Rhizobium* genes that are involved in nodulation, host-range specificity and nitrogen fixation. By dissecting the structure, organization and regulation of these genes, a better understanding of their functions may be forthcoming.

In the fast-growing *Rhizobium* species, which nodulate temperate legumes such as clover, alfalfa, peas and field beans, the genes for nodulation and nitrogen fixation are clustered on large symbiotic plasmids. One such plasmid, pRL1JI, was isolated from a strain of *R. leguminosarum* (which nodulates peas) and in this case the genes that determined the ability to nodulate peas were located between two groups of genes required for nitrogen fixation. The study of the nitrogen-fixation (*nif*) genes in *Rhizobium* has been greatly facilitated by the fact that at least some of them are very similar to the corresponding *nif* genes in the free-living nitrogen-fixing bacterium - *Klebsiella pneumoniae*.

In *Klebsiella*, 17 *nif* genes have been identified which are variously involved in *nif* gene regulation, the synthesis of the iron-molybdenum cofactor (FeMoCo). Using DNA sequence comparisons, *Rhizobium* species have been shown to contain the *nifHD* and *K* genes (nitrogenase proteins) *nifA* (regulation) and *nifB* (synthesis of FeMoCo). Further studies may well demonstrate similarities between other *Rhizobium* and *Klebsiella nif* genes.

The study of nitrogen fixation genes was first carried out in *Klebsiella pneumoniae* (strain M5al); and spectaular progress achieved in 1970s. Although the first report came from the mutants of nitrogen fixing bacteria which was isolated and biochemically characterized as early in 1959. The genetic research by Streicher et al. (1971) of U.S.A. and Dixon of Agric. Research Council, Unit of Nitrogen fixation (U.K.) added more knowledge in this field. Dixon and Postgate (1971) successfully transferred the nitrogen fixation genes, called *nif*, to *Klebsiella pneumoniae.* Later on they could able to transfer these genes to *Escherichia coli.* This work encouraged the researchers of this field specially for *Rhizobium* and the presence of *nif* genes in this microorganism. More than a dozen genes, termed the *nif* genes are involved in the assembly of nitrogen-fixation apparatus. The *nif* genes are not found in a free state but are clustered in one region (they are not scattered throughout the bacterial DNA). Therefore, it is much easier to stretch of DNA in a *Rhizobium* chromosome and insert the whole batch into another organism. This has also been done in *Agrobacterium tumefaciens* (Ti plasmids). In 1981, the number of *nif* genes was 17 and they were organized in eight transcription units (or operons), one of which was monocistronic. Brill (1980) also included two other genes: *nif*-R between L and F, and *nif*-W between F and M.

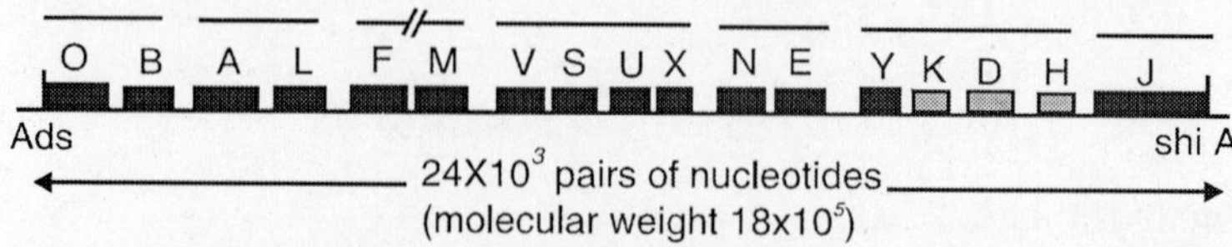


**FIG. 19.11:** Map of the *nif* gene cluster of the *Klebsiella pneumoniae* chromosome. The capital letters denote the genes (cistrons), the *nif* cluster being located between the genes for biosynthesis of histidine (his) and those for shikimic acid uptake (shi A). The arrow indicate the separate operations and the direction of their transcription. In 1981 the status of F and M as one or two distinct operons was not clear. The size of most of the genes, however, were known.

## *nod*-Genes for Nodulation

Concerning the genes (called *nod* genes) for nodulation and host-range specificity, an important observation is that, despite the morphological complexity of the infection process, only a few genes on the symbiotic plasmid are required for nodulation. In the *R. leguminosarum* symbiotic plasmid pRL1JI, less than 10kb appears to be required for nodulation and for the determination of host-range specificity for peas. The reason has been analysed by DNA sequencing (and thus identifying genes as long open reading frames), isolation and characterization of mutants and by studying the regulation of the *nod* genes.

The organiztion of *nif*-genes of other free-living diazotrophs from which *nif*-mutants were isolated (*Azatobacter vinelandii, Azospirillum lipoferum, Clostridium pasteurianum, Rhodopseudomonas capsulata* and a few cyanobacteria) was as like to that of *Klebsiella pneumoniae.*

## REGULATION OF NITROGEN-FIXATION GENES

The regulation of *nif* genes is rather complex. Ammonia and nitrogenous compounds (like nitrates and amino acids) inhibit the expression of nitrogenase activity. Excess of aeration in the culture medium also inhibits the expression of *nif*-genes. For induction of *nif*-genes, the presence of nitrogen is not necessary, because these genes can express themselves in cells cultivated in an argon atmosphere. In fact, the *nif*-genes expression is the result of a de-repression rather than that of an induction

Eady et al. (1978), Roberts et al. (1978) reported that all the products of *nif*-genes are not present in the cultures carried out in the presence of ammonia. When ammonia is added to a completely derepressed cultures, a rapid repression of the *nif*-phenotype occurs and nitrogenase activity as well as the enzymic proteins disappear within 45 minutes.

The presence of ammonia in the medium plays very important role in the synthesis of various biomolecules, e.g., in presence of ammonia, glutamine synthetase reacts with ATP and seems to lose the enzymic and regulatory functions of *nif*-genes.

The presence of oxygen in the culture medium represses the biosyntheis of nitrogenase enzyme in *Klebsiella pneumonia*. Similar results have also been observed in aerobic and anaerobic bacteria. However, the mechanism of oxygen repression seems to be independent from that of ammonia in such microorganisms.

Molybdate does not have any direct effect on the regulation of b-*nif*-genes, however, molybdoprotein is probably involved in the regulation of at least one of the *Klebsiella nif* operons (*nif* YKDH). The presence of molybdenum transport and storage proteins is known in *Clostridium pasteurianum* and other diazotrophs.

It is now established, after number of current reports that the regulation of the *nif*-genes of *Klebsiella pneumoniae* is carried out by both *nif* A and *nif* L genes as well as by genes that are remote from the *nif* cluster (*gln* genes). However, there are other genes which are involved in the expression of *nif*-genes, they are as follows:

*nar D* = which is involved in molybdenum processing.
*unc* and *a* gene in his operon = which influence ATP supply.
*nim* gene = gene of the uncertain function near tip.

## GENETICS OF SYMBIOTIC DIAZOTROPHS

Much works has been done on the fast growing nitrogen-fixing bacteria-*Rhizobium*. The genetic map of *Rhizobium* is now established. The studies reveal that *R.leguniosarum*, *R. phaseoli* and *R. trifolii* have very similar gene organization.

The genes for the component protein subunits of nitrogenase (molybdoferroprotein and ferroprotein) are formed by a single operon in *Rhizobium leguminisarum*. Similar situation has been reported in *Azatobactor vinelandii* and in *Klebsiella pneumoniae*.

## TRANSFER OF *nif*-GENES TO MICROORGANISMS

As mentioned earlier that *nif*-genes are not found in a free state but are clustered in one region (they are scattered throughout the bacterial DNA). Therefore, it is much easy to cut out the relavent stretch of DNA in a *Rhizobium* chromosome and insert the whole batch into another organism.

The construction of self-transmissible plasmids (Ti-plasmids) bearing *nif*-genes made it possible to shift the nitrogen-fixing capacity to non-fixing species or to non-fixing mutant of diazotrophic species like *E. coli (E. coli* C-M74), *Salmonella typhimurium*, *Serratia marcescens*, *Pseudomonas flurescence*, and *nif*-mutant of *Azatobacter vinelandii*.

It is quite easy to transfer *nif*-genes from one species to another, or form one microbial genus to another under laboratory conditions. This may be done by three methods of gene-transfer: conjugation, transduction and transformation. Therefore, the transfer of *nif*-genes between the plasmid and bacterial chromosomes can take place under laboratory as well as under natural conditions.

### Transfer of *nif*-genes and Development of New Nitrogen-fixing plants

It is easy to transfer grouped genes and the *nif*-gene cluster. Plant viruses could be used as vectors for this purpose. The cauliflower mosaic virus is a deoxyribovirus to which it is possible to join certain *nif*-genes which could become integrated into the plant genome or which could be transcribed into the plant cell at the same time as those of the virus when the latter's replication takes place.

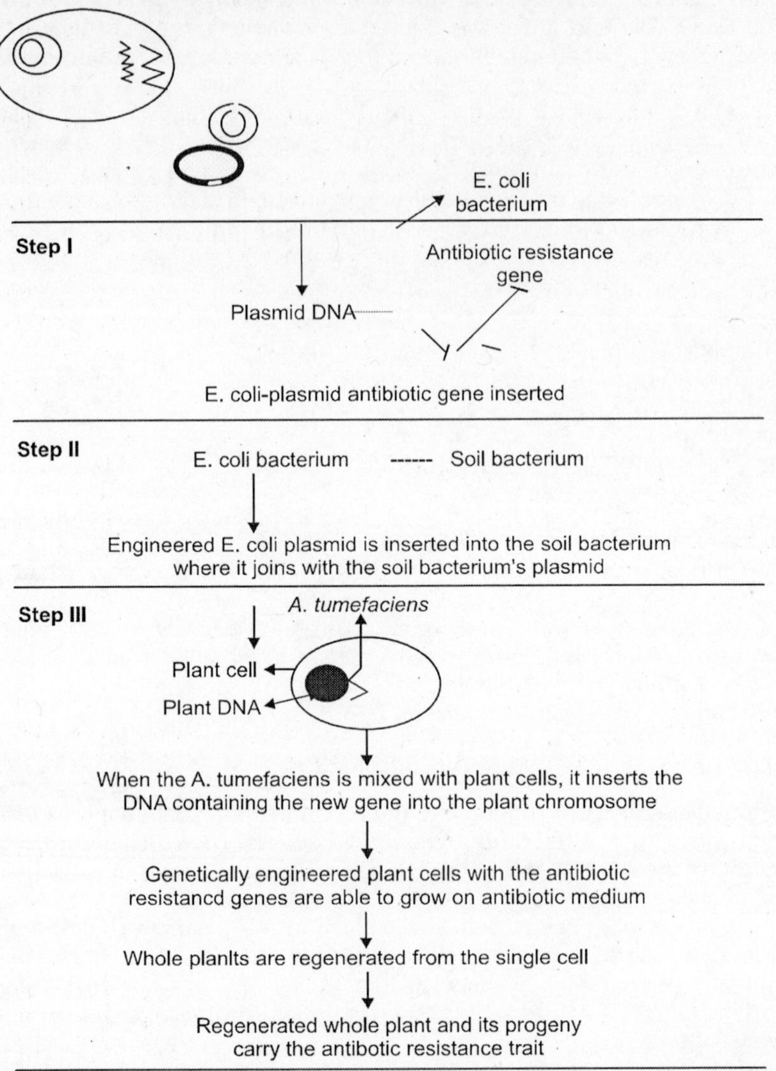


**FIG. 19.12:** Schematic presentation of plant genetic engineering with Ti plasmids from *Agrobacterium tumefaciens.*

*Agrobacterium tumefaciens* is a pathogenic bacterium, which contains a large plasmid, Ti and is responsible for the induction of tumours in plant tissues on the site of the infection by the bacterium. These plasmids are of 150 to 230 kilobases and can able to develop tumours, therefore, they are called Ti or **tumour inducing plasmids**. Ti -plasmids are introduced by either transformation or by conjugation. The DNA of Ti-plasmids have two following regions:

1. The larger region contains information that makes it possible for the bacterium to catabolize derivatives of the basic amino acids, or opines, synthesized by the

tumour cells.

2. It corresponds to the T-DNA and is incorporated into the genone of the plant cell. This T-DNA includes the ability of utilization of the latter as well the transformation of its metabolism in order to produce opines. The molecular weight of T-DNA is 15 x $10^6$ daltons.

The development of tumours in plant tissue by Ti -plasmid of *Agrobacterium tumefaciens* is brougnt about by the transfer of a segment of bacterial DNA (from a prokaryotic) and its integration into the genome of eukaryotic cells. This introduction of Ti -plasmid was successfully made by Schell and his coworkers. Out of the seven genes of T-DNA of Ti -plasmid, they found that five of them are responsible for tumorization, that is in blocking cell differentiation. It is of interest to mention that *Agrobacterium tumefaciens* infects only dicotyledonous plants, however, it is not easy to introduce Ti plasmids into the cells of monocotyledonous plants like cereals.

Kemp (1981) used nucleic acid recombination and cloning techniques to transfer genes to plant cells. He has transferred genes for the bean protein-**phaseolin** to sunflower plants. This transfer was followed by the synthesis of mRNA of the storage protein in the sunflower cells, made tumoral by the T-DNA of *Agrobacterium tumefaciens* which was used as a vector for the gene.

The development of plants which can able to fix atmospheric nitrogen independently of symbiotic microbes remains a difficult and complex task. However, in the present situation, it is possible to develop and gain better knowledge of the factors that induce nitrogen-fixing symbiosis, with a view of improving their efficiency.

## NITROGEN FIXATION RESEARCH IN INDIA

The word on biochemistry of nitrogen fixation has mainly been done at Haryana Agriculture University, Indian Agriculture Research Institute, Punjab Agriculture University and BARC. At HAU, a pathway for ureide biogenesis in nodules of pigeonpea has been proposed. The further metabolism of ureides in aerial parts particularly leaves and developing pods has also been worked out. Detailed biochemical and physiological studies explaining nodule senescence have also been carried out at IARI. Attempts have been made to understand the regulation of uptake hydrogenase so as to reduce the energy cost of biological nitrogen fixation. In addition, studies are concentrated on increasing nitrogen fixing efficiency by insertion of uptake hydrogenase genes or by increasing nitrogenase activity in nodules. At BARC, early interactions in legume- *Rhizobium* symbiosis have been investigated, according to which, host-induced alterations of capsular polysaccharides and de novo synthesis of specific proteins by both the partners in response to each other, appear to be among the first steps in the series of signal response interactions. At PAU, the role of extracellular invertase of *Rhizobium* in free sugar metabolism in developing root nodules of legumes has been ascertained (Adopted from Dr. Randhir Singh, In: Souvenir, Int. Cong. Plant Physiology, Feb. 15-20, 1988, New Delhi).

The activity of nodule is also controlled by nitrogenase enzyme. This enzyme is about 75% efficient. This is because, when soil aeration is poor and nitrogen availability is limited, electrons instead of being used to reduce nitrogen, now combine with protons forming hydrogen. As a result, nodules evolve hydrogen instead of fixing nitrogen. This is a wasteful process. Sufficient progress is being made to identify strains of *Rhizobia* which evolve less hydrogen and fix more nitrogen.

The most ideal situation would be the one in which the higher plant is able to fix nitrogen without the help of the bacteria. Towards this end, attempts have been made to transfer nitrogen fixing genes (*nif* -genes) to higher plants from *Rhizobium*. However, it appears to be a long way before it is achieved.

In India, intensive forest cultivation of nitrogen fixing trees, with consequent short rotation is needed. This will require replacement of nitrogen reserves at a greater rate than the present secondary succession leading to old-growth forests. With the increased use of trees, genetically selected, it will be necessary to maximize the potential for biological nitrogen fixation in future forest management, so as to maintain the long term productivity of forest lands without significantly accelerating the rate of nutrient depletion. India is in urgent need of additional professional training and better transfer of technology in this context. Further, improved strains and new species of nitrogen-fixing microorganisms should be investigated/developed in order to reduce the use of chemical fertilizers in the fields.

## BIOCHEMISTRY OF $N_2$ FIXATION

Nitrogen fixing organisms possess a special enzyme system called nitrogenase. The enzyme takes nitrogen gas, using plant energy (obtained from photosynthesis, oxidative. Phorphorylation, converts the gas into ammonia.

### Transport of Fixed Nitrogen to the Host Plant

The fix nitrogen transport is assured by the radial differentiation between vascular tissues connecting the nodule with the main vascular strands of the host plants. However, the mode of transfer of symbiotically fixed nitrogen from the nodule to the host plant is not precisely known. It is generally thought to be due to excretion of soluble nitrogenous compounds by the bacterial cells.

### Assimilation of Ammonia

Most of the inorganic nitrogen available to a plant is converted into the amide amino group of glutamine with the help of enzyme glutamine synthetase. Then, this amino amide group can be transferred to the 2-position of 2-oxoglutarate give rise to glutamate via the enzyme glutamate synthetase (Glutamine: 2-oxoglutarate amino transferase). This process or organic nitrogen interconversion takes place especially in developing and germinating seeds during synthesis and mobilization of nitrogenous reserve compounds. The reductive amination of 2-oxoglutarate catalysed by enzyme glutamate dehydrogenase has also been considered as the main reaction of ammonia assimilation in plants. The enzyme glutamate dehydrogenase is reported from mitochondria and in some heterotrophic tissues. It is also reported from plastids.

The ammonia rapidly becomes incorporated into organic combination in amino acids and amides. The carbon skeleton required for the formation of amino acids is derived from the respiratory metabolism of carbohydrates.

### Amino acids and Amides

Amino acids are essential components of all living cells as they are essential components of all living cells as they are the building blocks of proteins which in turn are the building blocks of protoplasm. Amino acids are composed of an amino group and carboxylic acid as evident from their name. The total number of amino acids known so far is 175. Out of them about 20 are protein amino acid which form the structural components of protein. The non protein amino acids form the structural components of protein. The non protein amino acids have more of nitrogen than the protein amino acids and possibly they act as storage nitrogen in the seeds. The non protein amino acids also act as an intermediate in the synthesis of protein amino acids. They constitute a group of secondary metabolities. Still, much remains to be understood in their metabolism.

Here the amino group is attached to the α-carbon atom (Carbon atom adjacent to the carboxyl group), hence they are also called α-amino acids. The length and configuration of -R in the general formula is variable. Amino acids are optically active and leavorotatory (L-form with the exception of glycine which is the simplest amino acid.

Amino acids may be neutral, acidic or basic in reaction depending upon the number of amino group and carboxyl group. Since amino acids contain both acid and base in the form of carboxyl and amino group, they can react with acids and bases to form salts. Such compounds are said to be amphoteric or ampholytes. Due to the amphoteric nature of amino acids, the proteins are the primary buffers in living systems. Depending on the pH, the alpha -amino and alpha- carboxyl functional group of an amino acid may exist in one of the following formula.

In the first case, in solution with low pH (acidic), an amino acid is a cation because of the dominating -$NH_3^+$/-COOH functional groups. In the second case, an amino acid is consisting of both a positive and negative charge or it is a Zwitterion (German, ion of both kinds) or dipolar ion. Here it can function both as an acid or base (amphoteric). The pH at which the Zwitterion form exists is called isoelectric pH (pl). The Zwitterion solution of amino acids exhibit zerocharge and do not migrate to any pole in an electric field. In the case, in solution with high pH (basic), an amino acid is an anion because of the domination -NH2/-COO- groups. The chain of the two amino acids is called dipeptide while a chain of many amino acids is called polypeptide.

## Biosynthesis of Aminoacids

The amino acids have a carbon skeleton and ammonia. They are generally considered as, initial products of nitrogen assimilation using radiotracers, it has been shown that in most cases, the recipients of the nitrogen of these compounds are the free α-keto acids in the cytoplasm. There are two ways by which nitrogen can be incorporated into α-ketoacids

1. Reductive amination, 2. Transamination.

### *Reductive amination*

It is the direct amination of keto acids which occurs by the combination of ammonia with ketoacids in the presence of $NADH_2$ and enzyme dehydrogenase. The reaction is reversible and proceeds as follows in glutamate synthesis where ammonia is incorporated into α ketoglutarate (an intermediate of Kreb's cycle).

This reaction is of great importance to the nitrogen metabolism of the plants. Since the glutamate or glutamic acid is important in the synthesis of other amino acids and high proportion of glutamate is formed in this manner by the plant.

Reductive amination is of little importance in the synthesis of amino acids other than glutamate. But, some workers like Smith et al. (1961) and kretovech (1965) suggested that photosynthetic tissue contain enzymes for the reductive amination of oxaloacetate and pyruvate to form aspartate and alanine, respectively. In general, there are four ways by which ammonia nitrogen is incorporated into organic compounds to form amino acids:

$$\text{Pyruvate} + NH_3 + NADPH_2 \rightarrow \text{Alanine} + NADP + H_2O$$
$$\text{Oxaloacetate} + NH_3 + NADPH_2 \rightarrow \text{Aspartate} + NADP + H_2O$$
$$\alpha\text{-Ketoglutarate} + NH_3 + NADH_2 \rightarrow \text{Glutamate} + NAD + H_2O$$
$$\text{Fumarate} + NH_3 \rightarrow \text{Aspartate.}$$

### *Transamination*

It is the process of amination of ketoacids which occurs by the transfer of amino group of an aminoacid to the Keto group of a Ketoacid according to the general formula below:

Here, amination of retoacid takes place at the cost of deamination of an amino acid. The aminogroup CH ($NH_2$) and Retonic group - C = O of two acids exchange their places and the reaction is catalysed by the enzyme aminotransferase or transaminase. It has been established that transamination reactions involve the participation of pyridoxal phosphate or pyridoxamine phosphate as a coenzyme. It has been observed that transamination reactions with glutamate give rise to 17 different amino acids. The reaction proceeds as follows in transmination between glutamic acid and other amino acids.

The transamination reactions involving the glutamic acid are more prevalent in the plant. These involve glutamic acid or α-Ketoglutaric acid as an essential compound. However, other transamination reaction have also been found in higher plant which involve aspartic acid and aloaine.

The appropriate keto acid precursor formation is necessary for the formation of an amino acid, provided the capacity of aminogroup transfer. The keto acid analogues of many aminoacids have been identified in plants and the synthesis of amino acids proceed through the activities of amino transferases. Using radiotracers it has been shown that most of the amino acids have similar carbon skeleton. A definite metabolic relation exists between the precursors and products of different amino acids.

## DEGRADATION OF AMINOACIDS

Aminoacids are the primary source of carbon and nitrogen which is utilized by plants and microorganisms. Besides, they provide energy, their degradation or catabolism may occur either by oxidative deamination or by non-oxidative deamination.

$$R - CH - NH_2\ COOH \rightarrow R.\ Co.\ COOH + NH_3 \rightarrow \text{Urea}$$

In the first stage of their degradation -$NH_3$ group is removed and corresponding Reto acid is formed. The removal of ammonia from the amino acid is called deamination. The oxidative deamination is the removal of amino group by oxidation. It occurs in alanine, glutamic acid, glycine or histidine. For example:

$$\text{Glutamate} + NAD^+ + H_2O \rightarrow NH_3\ \alpha\ \text{Ketoglutarate} + NADH + H^+$$

This reaction is reversible and it is catalysed by flavin containing enzymes called aminoacid oxidase. Non oxidative deamination is catalysed by aminoacid ammonia lyase. In this process, $NH_3$ is removed with the formation of a double bond in the remaining carbon skeleton. For example;

Aspartate → fumarate + $NH_3$ Aspartate → ammonia lyase.

## AMIDES

Amides are the derivatives of amino acids in which the hydroxyl group (-OH) of one of the carboxyl groups (-COOH) is replaced by another amino group ($-NH_2$). The general formula is-

R. CH. ($NH_2$). CO $NH_2$

The process of formation of amides is known as amidation. Amides like asparagine and glutamine are found in large quantities in many plants. They play an important role in the transport and storage of nitrogen.

### Biosynthesis of Amides

The synthesis of amides may involve direct oxidation or transoxidations.

**Direct amidation:** It is the direct synthesis of amides by combination of ammonia with an amino acid. This occurs only in the synthesis of glutamine where the enzyme glutamine synthetase catalyse the reaction with $Mg^{+2}$ as metal cofactor. Moreover, A.T.P. is also required. According to Webster and Warner (1955). A.T.P. activates the enzyme which then forms a complex with glutamic acid. In the complex, the aminoacid is in the activated form and can combine with ammonia to form amide.

**Transmidation:** In transmidation, the amides are formed by the transfer of amino group.

## PROTEIN

Proteins are essential constituents of the living cells. The term protein (GK. Protein meaning primary or holding first place) was first suggested by Berzelius (1836) in view of their primary importance in the living system. They are the building blocks of protoplasm and comprise the structure of cell. They are also concerned with metabolic activities of the cell like respiration, catalysis of enzyme reaction, transport of materials control of movement and defence reactions. They are composed of repeating units, the aminoacids and have high molecular weight ranging from few thousands to many millions. They contain C, H, O, N and sometimes S and other elements like Fe in hemoglobin, Mg, Zn in muscle protein and I in thyroxine. Fischer and coworkers found that proteins on refluxing in concentrated HCl for many hours, give rise to free amino acids. The aminoacids are linked together by bonds connecting the carboxyl group of one aminoacid with the amino group of another.

The (NH.CO) bond is called peptide bond which is repeated many times in protein. Evidence for peptide linkage in protein compound is obtained from biuret test. Substances containing two or more peptide linkages produce blue colour with dilute copper solution in a strong alkali (color due to a comp.) All proteins exist in a colloidal state with large molecule and they are unable to diffuse through membrane such as cellophane and exhibit the properties in association with the colloidal state of matter. It is difficult to isolate and purify them. They behave both as acidic and basic in properties forming salts with the acid as well as alkali. Most of the protein coagulate on heating. All protein are soluble in strong acid.

## Types of Proteins

Proteins are classified on the basis of solubility properties and on known chemical and physical properties. Proteins may be simple, conjugated and derived.

Simple proteins are compounds that, on hydrolysis yield only amino acids. These proteins may be structural or may be in the form of reserve proteins. The structural or protoplasmic proteins are connected with functioning of the living cells. They can be classified into enzymatic and non-enzymatic proteins. The enzymatic proteins are commonly globular and while the non-enzymatic proteins are fibrous both can have the same chemical nature.

The reserve or storage proteins are present as food reserve in seeds and other storage organs. They occur as crystalloid grains. Some are rearlly sensitive and get coayulated. All are soluble in aqueouo solution. These are of six types: Albumins, Glubulins, Prolamines, and Glutelins, Histones, Proteminase.

1. **Allumins:** The albuminous are soluble in water and in dilute salt solutions. They are coagulated by exposure to heat e.g., $\alpha$-amylase of barley, legumelin in leguminosae, leucosin in cereal grains and recin.
2. **Globulins:** The globulins are insoluble or sparingly soluble in water and are soluble in dilute salt solutions. They are coagulated by exposure to heat. e.g., legumin ad tuberin.
3. **Prolamines:** The prolamines are insoluble in water but are soluble in 70 to 80 per cent ethanol. They are called prolamine because they yield relatively large quantities of proline and ammonia on hydrolysis e.g., zein of maize, gliadin of wheat and rye and hordein of barley.
4. **Glutalins:** The glutalins are insoluble in water, alcohol and salts solutions (neutral solution) but are soluble in weak acid of basic solutions, e.g. Glutamin in wheat grains, oryzenin in rice.
5. **Histones:** The histones are soluble in water but are insoluble in dilute ammonia. They are rich in basic aminoacids, such as a ginine and lysine; e.g., Globin, hemoglobin and protein components of most nucleoproteins.
6. **Conjugated proteins:** Conjugated proteins are proteins which contain non-amino acid components in addition to the aminoacids. The non-aminoacid components or non-proteinaceous groups are called prosthetic groups. The conjugated are named in accordance with their associated prosthetic group. The conjugated proteins are divided into seven major types on the basis of nature of prosthetic groups nucleoproteins, glycoproteins, lipoproteins, chromoprotein and metalloproteins.
7. **Nucleoproteins:** They are found in the nucleus conjugated with nucleic acids. They are called nucleoproteins as they give rise to simple proteins and nucleic acids on hydrolysis. These are weakly acidic and soluble in water.
8. **Glycoproteins:** They are found in cell membrane. They are proteins containing small amounts of carbohydrates as prosthetic groups and hence called glycoproteins.
9. **Lipoproteins:** They are found in membranes of nucleus, mitochondria, chloroplast and cell. They are conjugates of lipids and proteins and the prosthetic group of them are lipids. They are generally insoluble in water.
10. **Chromoproteins:** They are found in flavo-proteins, carotenoids, chlorophylls

and hemoglobins. They possess pigments as their prosthetic groups and are coloured. The colour is because of metals such as Cu and Va or metals with organic groups such as in Fe and Mg porphyrins.

11. **Metalloproteins:** They are found in respiratory enzymes which require metals as activators.

## Derived Proteins

They are derived from simple proteins and are actually products of their partial hydrolysis with seeds, alkalies and enzymes. They are intermediated products and they are not accumulated. They are two types.

1. Primary derived proteins, 2. Secondary derived proteins

1. Primary derived proteins: These are of two types:
   a) **Metaproteins:** Metaproteins are insoluble in water and in dilute salt solution but soluble in acids and alteralies. These are produced by hydrolysis of simple proteins by alkalies or acids.
   b) **Coagulated proteins:** Coagulated proteins are insoluble in water and are produced by the prolonged treatment with alcoholor heat. These are coagulated on heating.
2. **Secondary derived proteins:** These are of three types:
   a) **Proteoses:** Proteoses are soluble in water but are not coagulated by heat. These are produced by hydrolysis of metaproteins.
   b) **Peptones:** Peptones are soluble in water and are not coagulated by heat. Peptones are produced by hydrolysis of proteoses by action of HCl, $H_2SO_4$ or certain enzyme.
   c) **Peptides:** Peptides are soluble in water and are not coagulated by heat. These are produced by prolonged hydrolysis of simple proteins with HCl or $H_2SO_4$. They do not give biuret test.

## Protein: Structure

The biological properties of a protein molecule are dependant on its structure. The structure of protein in turn is related to the order in which aminoacids are joined together in a peptide chain, called the aminoacid sequence of that chain. Amino acid sequence of the smallest protein known, is that of insulin which was studied by Friendrick Sanger in 1954. There are two chains which were held together by disulphide bonds. Aminoacid sequence have also been reported for ribonucleus (124 resudue), T.M.V. protein (150 resudue) and theman hemoglobin. The number of each kind of amino acid and their sequence in the peptide chain is specific for each protein and protein may differ in the number of polypeptide chain in the lodging of their sequence lies the mysteries of gene.

The protein has following structures: (i) Primary structure (ii) Secondary structure (iii) Tertiary structure (iv) Quarternary structure.

### *Primary structure*

The amino acid forming peptide bond are present in straight line. Besides peptide bond, other necessary feature of the protein molecure is the disulfide (-S-S) bond between two cysteine molecules.

It is evident that peptide and disulfide bonds are not the only links involved in protein structure. It is found that dissociation of many proteins may occur without disturbing the peptide or disulfide bond.

### *Secondary structure*

The peptide chains are twisted in a helix i.e., spirally twisted around a central axis. The coiling or twisting of polypeptide chain into spiral helix is called secondary structure of protein. The polypeptide chains also exhibit pleated sheet and random arrangement besides the helical type. But the most common type is alpha helix or alpha structure which is maintained by some secondary forces such as hydrogen bonds.

Hydrogen bonds are formed by electrostatic interaction between the molecules. They are non-covalent links that occur as a result of the hydrogen atom sharing electrons with two oxygen atoms. In this case hydrogen bonds are formed between the carbonyl oxygen and imide hydrogen along the polypeptide backbone (N - N...O = C). It is commonly found in proteins of wool, hair and other fibrous structures.

The pleated sheet arrangement or Beta structure is formed between polypeptide chains are arranged side by side either parallel or antiparallel with respect to one another. β-structure are common in nature. Example of β-structure is found in silk and other synthetic fibres such as nylon and orlon.

In the random coil arrangements, there is no consultant geametrical order resulting due to the folding of aminoacid side chains rather then the folding of polypeptide backbone. Salt bond, vander waals fores and hydrogen bonds are responsible for helical structure.

**Tertiary structure:** The further folding of the barian secondary structures and random segments in number of ways in called tertiary structure of proteins. The existence of such tertiary structure was recognized quite early from studies of enzyme active site. The tertiary structure seems to be emergetically most stable by vander waals forces, salt like and hydrogen bonding. It differ from secondary structure in that it involve interactions of the R-groups. Tertiary structure is exhibited by most of the biologically active proteins and it is intimately reboled with molecules biological function. When the secondary and tertiary structures are disrupted, certain specific functions like enzyme activity are lost irreversibly. The disruption of structure takes place when proteins are exposed to relatively high temperatures, changes in pH ultraviolet radiation etc. These conditions cause denaturation followed by loss of many of the properties of the protein molecule viz., solubility specific activity and crystallizability. Myoglobin is an example of tertiary structure.

**Quaternary structure:** Association of two or more polypeptide chains give rise to the quaternary structure of protein molecule. The compined polypeptide chains forming a three dimensional geometry constitutes quaternary structure. Hydrogen bonds are responsible for the maintenance of this structure. The association of hydropholic groups which combine and remove water is also involved in holding the quaternary structure together.

# CHAPTER 20

# Growth of the Green Plants

## INTRODUCTION

Growth is the self multiplication of living material, the protoplasm it self. Growth is an increase in size (volume or length) due to cell divisions and subsequent enlargement. It is an increase in dry weight of bulk of an organism associated with development. Development is defined as *an ordered change* or *progress, often towards a higher, more ordered or more complex state.* Development may take place with growth and growth may take place with development. However, these are often quite integrated processes.

In most organisms, growth involves cell division, expansion, differentiation and morphogenesis. Increase in height and dry weight are the most obvious manifestation of growth, e.g., in forage and pasture crops. In grain crops, dry weight increases in nongrain components are important only in producing a plant capable of high grain yield. Growth is also expressed as the advance of a plant from one development stage to another. A seed germinates, the seedling develops roots, leaves, and stem, and the plant grows to maturity. The plant flowers and produces seed to complete its life cycle.

Morphological and chemical differentiation takes place during the development. These processes of cells are complicated and involve synthesis of many organic compounds like protein, cellulose, and nucleic acids and requiring physical forces that cause cell enlargement. The process becomes more complicated because it involves division and expansion of many different cell types. Growth includes assimilation and the formation of new protoplasm, permanent change in size, and increase in weight, either of the plant as a whole or of some organ or tissue.

The growth pattern and division of the cells in root types, leaf buds, and developing seeds occur in basically in the same way, but they serve vastly different functions for the plant. Several abiotic and biotic factors influence growth of an organ (leaf, fruit, root, etc.) or whole plant at the cellular level.

## WHY GROWTH OCCURS?

Growth is expressed as the division of a cell to form two cells and the enlargement of the newly divided cells. When we say that cells double in all their constituents and then divide in half, we are obviously describing the average case. When we observe individual cells, however, we find deviations from the average. In some instances the growth rate is constant rather than accelerating. In others, the growth rate is high at first and then decreases.

A mass of cells doubles and halves only in an average way and not in an exact way. Cells divide when they are **ready**, and they are ready only when they have completed certain preparations for division.

**Citrical Mass Theory:** In 1908, the German biologist Richard von Hertwig proposed the theory of the citrical mass. In Hertwig's view a growing cell eventually reaches a size at which the ratio between the masses represented by cytoplasm on the one hand and the cell nucleus on the other becomes limiting. When that limiting ratio is reached, he proposed, some instability sets which triggers cell division. The nucleus-cytoplasm relation theory, as Hertwig called it, does not hold up very well when mass alone is considered to be the factor that triggers cell division. Many exception to this hypothesis have accumulated that, although the division of cells usually parallels an increase in cytoplasmic mass, this increase alone is not generally accepted as the fundamental mechanism controlling the cell cycle. Another idea advanced for a time was that transition from G1 to S and the reminder of the cell cycle is moderated by the appearance of the enzymes and other factors required for DNA replication. Cells in G1 lack these enzymes and factors. Activation of the genes coding for them constitutes the trigger of cell division.

Speculation and experimentation have now focused on other cellular factors as possible cell cycle controls. These include changes in the concentration of cyclic nucleotides, cAMP and cGMP.

## KINETICS OF GROWTH OR MEASUREMENT AND PATTERN OF GROWTH

Growth of a plant is usually measured in terms of increase in dry weight, length or height and such measurement indicates that growth-rate varies with the age of the organism. The growth of most plants follows a similar pattern, generally S-shaped or sigmoid shape curve. The size versus time plot is the most direct way of handling growth data. One can also plot the logrithum of the size as a function of time. In the size plot the units are length, volume or weight, but in the rate curve the units are length, volume or weight per unit time. A relatively slow growth rate characterizes early or seedling growth. The rate of growth increases as the plant becomes larger, is greatest just before or in early stages of flowering, and then decreases as the plant matures.

The plotted growth curves have two following phases (Fig. 20.1).

### *First rapid phase*

During this phase the rate curve as well as size curve increases. It is also known as logarithmatic or exponential phase.

### *Maximum growth rate phase*

Growth curves of many plants also show another evident phase called maximum growth rate or the linear phase. Sachs called it the grand period of growth, and it lies between the logarithmatic phase and the senescence phase.

### *Last phase*

During this phase, the size continues to increase but more slowly so that rate decreases. This phase is known as senescence or decreasing phase.

According to Blackman (1919) growth of plants can be represented by equation:

$$W_1 = W_0 e^{rt}$$

($W_1$ = final size [weight, height, etc. after time t], $W_0$ = initial size at the beginning of the time period, r = rate at which plant substance is laid down during time t, and e = base of natural logarithms). It should be noted that r is the relative growth rate. The above equation also indicates that the size of an organism ($W_1$)

depends on the initial size ($W_0$).

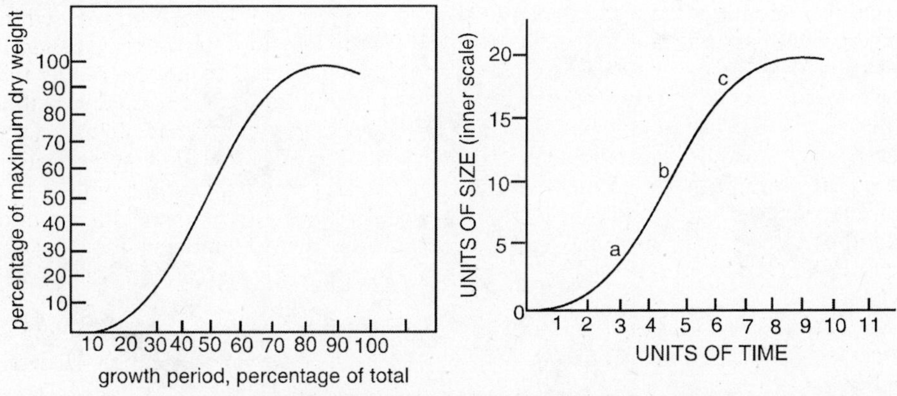


**FIG. 20.1:** (A) Sigmoid shaped growth curve, (B) Phases a = exponential phase, b = linear growth phase and c = senescence phase when the rate of growth decreases.

From this equation it is seen that plant size also depends on the magnitude of r (relative growth rate). r is the measure of the ability of a plant to produce new plant material. Blackman called r as the **efficiency index**. The value of r differs among plant species and is not constant during the life of a plant. The efficiency index is used as a predictive factor in evaluating performance of a plant.

## WHAT HAPPENS WHEN GROWTH OCCURS?

Growth of a plant is initiated in the meristems and the cellular changes in these tissues. The three main changes in the complete development of a cell are cell division, cell elongation and cell differentiation. After cell division, daughter cell enlarges to the size of the mother cell before they again divide. During cell growth uptake of water occurs which produce turgor pressure sufficient to overcome the force of attraction between the cell particles in cell wall. Consequently, wall stretches and become thin and the osmotic potential increases because of water uptake which dilutes cell contents. The consequent increase in the cell's water potential brings water uptake (and therefore growth) to a stop. Besides water and solute absorption, new cell wall material is also added and wall become thick due to opposition (accumulation of new material on the inner surface of the wall). This happens during cell maturation.

**Biochemical Changes:** The important biochemical changes that occur during growth are: synthesis of protein, nucleic acid, phospholipids, multiplication of organelles and utilization of energy in the form of ATP, etc.

## DRY MATTER ACCUMULATION

The dry weight accumulation pattern of plants in the early stages is like the accumulation of compound interest in bank account. The weight of the plant at any given time is likened to the size of the bank account and the increase in weight equated with the interest. In the early stages when plants are very small, the actual dry weight increase per day (interest) in small (first 4 weeks), but at the plant becomes larger, its weight gain per day increases (first 10 weeks).

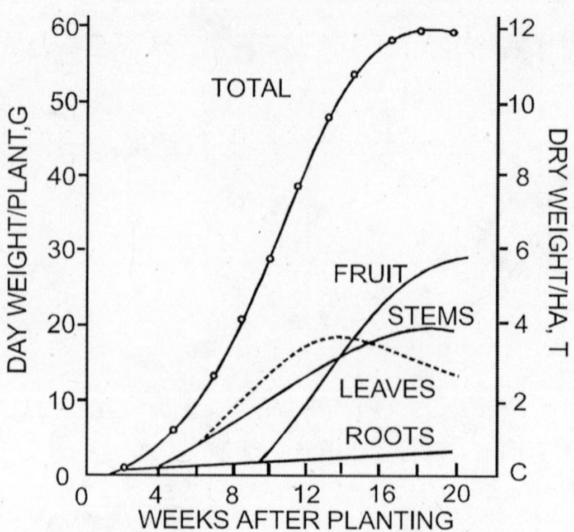


**FIG. 20.2:** Dry weight accumulation patterns for a hypothetical crop plant and its parts. The dry weights per unit of land area are for the same plants as the individual plant weights, but plant population is assumed to be 200 000 plants/has. Plants mature at 18 to 20 weeks after planting.

This logarithmic increase in growth cannot continue indefinitely, of course. All plants mature and growth ceases (approximately 18 weeks) because of completion of the life cycle or the onset of dormancy.

The slow growth rate of seedlings relative to large plants may be attributed to the relatively small number of cells that can divide, the small leaf area available for light interception and photosynthesis, and perhaps to the relatively large percentage of photosynthate (photosynthetic products) going to roots. As plants increase in size, larger meristematic regions (areas of cell division) per plant exist, and more leaves are present to act as photosynthetic energy source. Thus each plant is capable of capturing and utilizing more energy as it increases in size.

The decline in growth rate near maturity may be attributed to several factors. Many species have determinate growth characteristics; that is, vegetative growth ceases, prior to or concurrent with reproductive growth. In these species, such as corn[1] or small grains, the formation of fruit signals an end to growth of other plant parts. Even in species such as cotton where vegetative and reproductive growth occur simultaneously, the onset of reproduction reduces vegetative growth rates.

## AGE AND PLANT GROWTH

As plants become older, a large portion of the plant structure becomes inactive. Lower leaves may become heavily shaded or nonphotosynthetic due to senescence. Leaves may fall from the plant, representing loss of dry weight. Large portions of the plant may be comprised of stem or other tissue relatively low in metabolic activity and therefore not contributing to growth. In addition, competition from neighboring plants for water, nutrients and light may cause a reduction in growth rates when plants in crop stands become large.

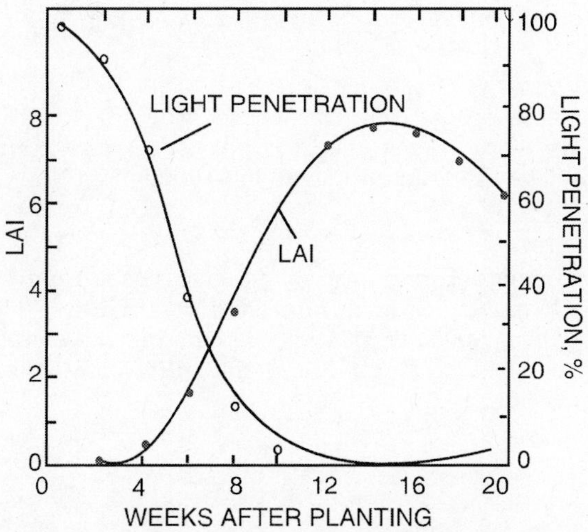


**FIG. 20.3:** Increase in leaf area index (LAI) and decrease in light penetration to the soil surface during growth of the crop.

## Leaf Area Expansion

Leaf growth is a part of the total dry matter accumulation, and leaf itself is the part of the dry weight of a plant. Leaf provides a plateform for photosynthesis. Photosynthesis and dry matter production of a plant are proportional to the amount of leaf area on the plant as long as some leaves are not heavily shaded by others. Leaf weight and leaf area follow patterns similar to that of total dry weight during the first half of the growing season.

## Leaf Area Index (LAI)

It is the ratio of leaf area (one side of leaf only) to soil area it occupies. The rate of increase of leaf area determines the rate of increase in the photosynthetic capacity of the plant. The growth rate of plants during the seedling and up to midseason stages may depend on the expansion of new leaves.

## Plant Height

Increase in height is the most obvious change in growth of most plants which helps a plant to compete with other plants in a community. The pattern of increase in plant height with age may be very similarly to the dry weight increase in some species, but not in others. Corn, for example, reaches its maximum height before the fruit starts to develop, but in other plants, height continues to increase during fruiting.

## Root Growth

Roots are important component of growth in all plants. The function of roots, is more important than their weight. Roots anchor the plant, support its vertical stand. It absorbs water and mineral nutrients used by plants. Therefore, the extent of the root system is important, if not necessarily its weight. The youngest growing portions of the roots are most active in mineral nutrient uptake is reason that the root system must continually grow to provide adequate nutrition for the plant.

## Growth Analysis

In field, growth of a crop has been characterized by a system of growth analysis based mostly on dry matter accumulation rates. The most meaningful analysis of crop growth is on a land area rather than an individual plant basis.

## Crop Growth Rate

The dry matter accumulation rate per unit of land area is referred as crop growth rate (CGR), normally expressed as g (m of land area)$^{-2}$ day$^{-1}$. CGR is measured by harvesting plants at frequent intervals and calculating the increase in dry weight from one harvest to the next. Roots are excluded in many studies of CGR. for a given time interval.

$$CGR = \frac{(W_2 - W_1)}{SA\ (t_2 - t_1)}$$

($W_1$ and $W_2$ = crop dry weight at beginning and end of the interval, $t_1$ and $t_2$ = corresponding days, SA = soil area occupied by the plants at each sampling)

CGR can be measured throughout a growing season by sampling at frequent intervals.

The maximum CGR occurs when plants are large enough to exploit all the environmental factors to the greatest degree. In favorable environments, this maximum CGR occurs when leaf cover is complete and may represent the maximum potential dry matter production and solar energy conversion rate for a given species. Tropical or warm-season grass crops such as corn, sorghum, millet, and sugarcane have higher CGR than the other crop species. Such difference is due to more efficient photosynthesis in most tropical crop with the $C_4$ photosynthetic pathway. Rice is a tropical grass but lacks $C_4$ photosynthesis.

CGR analysis is to interprerate the yield differences among crop varieties and cultural practices. Since crop yields may be affected by many factors during the growing season. Analysis of growth throughout the growth cycle may be helpful in explaining yield differences.

**TABLE 20.1:** Maximum crop growth rates reported for some crop plants*.

| Crop | Maximum CGR+gm$^{-2}$day$^{-1}$ |
|---|---|
| Corn | 51 |
| Sorghum | 51 |
| Millet | 54 |
| Sugarcane | 42 |
| Rice | 27 |
| Soybean | 23 |
| Peanut | 28 |
| Alfalfa | 20 |
| Sugar beet | 32 |

*These values are not valid for exact comparisons among species, since climatic conditions were quite different for the crops listed.

## Relative Growth Rate (RGR)

Relative growth rate (RGR) is the increase in plant weight per unit of weight already present and may be expressed as g (g of dry weight)$^{-1}$ day$^{-1}$. At any given instant during growth,

$$RGR = \frac{1\, \delta\, w}{w\, \delta\, t}$$

(w = plant dry weight and $\delta w/\delta t$ = change in dry weight per unit time)

RGR decreases with plant age due to the fact that an increasing part of the plant is structural rather than metabolically active tissue and as such does not contribute to growth and impart to shading and increased age of lower leaves.

## Net Assimilation Rate

Net assimilation rate (NAR) is a measure of the average efficiency of leaves on a plant or in a crop stand or dry matter accumulation rate per unit of leaf area. Since leaf surfaces intercept sunlight and absorb $CO_2$ in photosynthesis, it is desirable in some cases to express growth on the basis of leaf area. It may be expressed as follows:

$$NAR = \frac{1\, \delta\, w}{A\, \delta\, t}$$

(A = leaf area and $\delta w/\delta t$ = change in plant dry weight per unit time). The NAR is expressed as g (m of leaf area)$^{-2}$ day$^{-1}$.

When all leaves are exposed to full sunlight NAR remains highest. It also remains highest when plants are small and leaves are few enough that none are shaded by others. As plants grow, more and more leaves are fully or partially shaded; thus, NAR decreases during the growing season and with plant's age due to low photosynthetic efficiency of older average leaf. Older leaves at the bottom of the plant or crop stand may be contributing nothing to growth, while young upper leaves may have an efficiency much greater than indicated by the NAR. The leaf petioles (usually not included in leaf area), stems, and inflorescence also carry on photosynthesis and contribute substantially to fruit formation. In cereals such as barley the inflorescence may contribute up to 50% of the dry weight of the grain.

# CHAPTER 21

# Plant Growth Regulators

## STATUS AND DEFINITION

The conventional definition of plant hormone originally proposed by Fitting (1909) has met with several difficulties of explaining the different responses and reactions of several compounds, both synthetic and natural. Pincus and Thimann (1948) proposed definitions for these two categories of substances. Plant hormone has been defined as *organic substance produced naturally in the higher plants, controlling growth or other physiological functions at a site remote from its place of production, and active in minute amounts.* To distinguish it from animal hormone, it was termed as phytohormone. However, Trewavas (1982) has started a discussion about abolishing the name of plant growth substances as hormone. This idea was hotly debated at the XI International Conference of Plant Growth Substances (1982). The crisis of the phytohormone theory is caused mainly by the following problem: a poor correlation was found between growth rate and endogenous hormone levels in developing plants. Further, unexpected results were frequently found with exogenously applied hormone, i.e., in some cases, promoters cause an inhibition and inhibitors a promotion of growth. Plant growth regulators are defined as *organic compounds other than nutrients which in small amount promotes/inhibit or otherwise modify any physiological response in plants.* The definitions adopted by the American Society of Plant Physiology in 1954 are as follows:

### Plant Hormones

Plant hormones are regulators produced by plants which in low concentration regulate a physiological plant processes.

*Hormones usually move within plant from a site of production to site of action.* There are five classes of phytohormones: auxins, gibberellins, cytokinins, abscisic acid and ethylene.

#### *Auxins*

These are organic substances which at low concentration (less than 0.001 M) promote growth along the longitudinal axis, when applied to shoot of plants freed as far as placing from their own inherent growth promoting substances.

#### *Gibberellins*

These substances are having gibbane ring skelaton capable of producing the same physiological responses as gibberellic acid that it must be active in specific gibberellin bioassay. The gibberellins are phytohormones which are active in regulating dormancy, flowering, fruit setting, and stimulating germination of seeds and extending growth of shoots.

#### *Cytokinins*

These are substances composed of hydrophilic group of higher specificity (adenine) and one lipophilic group without specificity. The CKs form a group of plant hormones having similar effects as those of GA in breaking the dormancy of a wide range of seeds and in increased fruit set. These hormones mainly stimulate cell divisions and prevent chlorophyll degradation.

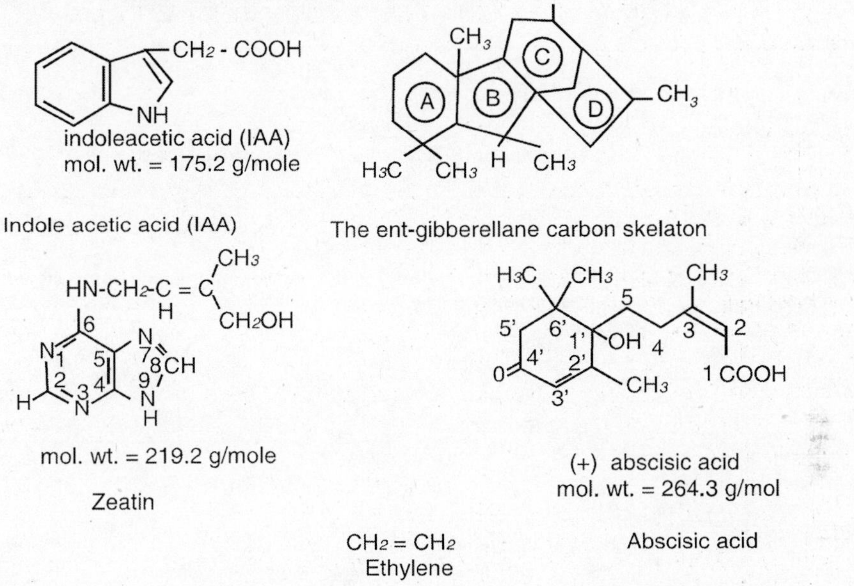

indoleacetic acid (IAA)
mol. wt. = 175.2 g/mole

Indole acetic acid (IAA)

The ent-gibberellane carbon skelaton

mol. wt. = 219.2 g/mole

Zeatin

(+) abscisic acid
mol. wt. = 264.3 g/mol

Abscisic acid

$CH_2 = CH_2$
Ethylene

**FIG. 21.1:** Structures of plant hormones.

Various synthetic and natural plant growth regulators are listed in Table 21.1.

**TABLE 21.1:** Names and structures of plant growth regulators. (Modified from Halmann, M., 1990).

| Abbreviation or Common Name | Chemical Name or Structure | Major uses |
|---|---|---|
| ABA | Abscisic acid | Defoliant |
| ACC | 1-aminocyclopropane-1-carboxylic acid | |
| Acifluorfen | 5-2-chloro-4-(trifluoromethyl) phenoxyl-3-nitrobenzoic acid | GR Herb. |
| AVG | aminoethoxy vinyl glycine | |
| Alachlor | 2-chloro-2', 6'-diethyl-N-(methoxymethyl) acetanilide | Herb. |
| Ancymidol | a-cyclopropyl-4-methoxy- a-(pyrimidin-5-yl) benzyl alcohol | GR |
| Atrazine | 2-chloro-4(ethylamino)-6-(isopropylamino)-1,3,5-triazine | |
| BSAA | (Benzo-b-selenienyl)-3-acetic acid | DB |
| Barban | 4-chloro-2-ynyl-3-chlorocarbanilate | Herb. |
| Brassinolide | | |
| Chlormequat, CCC | 2-chloroethyl-trimethyl ammonium chloride | LP, SR |
| Cycloheximide | 3-[2-(3,5-dimethyl-2-oxocyclohexyl)-2-hydroxyethyl]-glutarimide | AS |
| 3-CPA | 2-(3-chlorophenoxy) propionic acid | FT |
| 4-CPA | 4-chlorophenoxy-acetic acid | FT |
| CPPU | N-(2-chloro-4-pyridyl)-N'-phenylurea | Herb. |
| 2,4-D | (2,4-dichlorophenoxy) acetic acid | Herbs. |
| Daminozide | Succinic acid-2,2-dimethyl hydrazine | GR |

| Abbreviation or Common Name | Chemical Name or Structure | Major uses |
|---|---|---|
| SADH | | |
| Dichlopentezol | (E)-1-(2,4-dichlorophenyl)-4,4-dimethyl-2-(1,2,4-triazol-1-yl)-1-penten-3-ol | GR Fung. |
| Dikegulac sodium | Sodium-2,3-4,6-di-O-isopropylidene-$\alpha$-L-xylo 2-hexulofurasonate | GR |
| Diquat | 6,7-dihydrodipyrido pyrazinediium dibromide | FS |
| Diuron (DCMU) | 3-(3,4-dichlorophenyl) 1,1-dimethylurea | |
| DPX-1840 | 2-(4-methoxyphenyl)-33a-hydro-8H-pyraxolo[5,1-a] isoindol-8-one | GR |
| Etacelasil | 2-chloroethyl-tris-(2-methoxyethoxy) silan | FL |
| Ethephon | 2-chloroethylphosphonic acid | SR |
| Glyphosate | N-(phosphonomethyl) glycine | SR Herb. |
| Glyphosine | N,N-bis(phosphonomethyl) glycine | SR |
| $GA_3$ | Gibberellic acid | GE |
| Kinetin | 6-furfurylaminopurine | DB |
| IAA | Indole-3-acetic acid | Enlarg. |
| IBA | Indole-3-butyric acid | Enlarg. |
| Maleic hydrazide | 1,2-dihydro-3,6-pyridazinedione | Herb. GR |
| Mefluidide | N-2,4-dimethyl-5-(trifluoro methyl) sulfonyl-amino-phenyl acetamide | GR Herb. |
| Mepiquat chloride | N,N-dimethyl piperidinium chloride | GR |
| Metolachlor | 2-chloro-N-(2-ethyl-6-methylphenyl)-N-(2-methoxyl-1-methylethyl) acetamide | Herb. |
| Metribuzin | 4-amino-6-tert-butyl-4,5-dihydro-3-(methylthio)-1,2,4-triazine-5-one | Herb. |
| Monuron | 3-(p-chloro-phenyl)-1,1-dimethylurea | Herbs. |
| NAA | 1-naphthaleneacetic acid | FT GR |
| NOXA BNOA | 2-naphthalenyloxyacetic acid | GS |
| Paclobutrazol | (2RS, 3RS)-1-(4-chlorophenyl)-4,4-dimethyl-2-(1,2,4-triazol-1-yl)-pentan-3-ol | GR |
| Paraquat | 1,1-dimethyl-4,4-bipyridinium dichloride | Herb. |
| Pretilachlor | 2-chloro-2',6'-diethyl-N-(2-propoxyethyl) acetanilide | Herb. |
| Propachlor | 2-chloro-N-isopropylacetanilide | |
| Propazine | 2-chloro-4,6-bis(isopropyl-amino)-1,3,5-triazine | |
| Release | 5-chloro-3-methyl-4-nitro-1-H-pyrazole | As |
| Terbutryn | 2-(tert-butylamino)-4-(ethylamino)-6-(methylthio)-5-triazine p-hydroxyphenyl-aceticv acid | |
| 2,4,5-T | (2,4,5-trichlorophenoxy) acetic acid | Herb. |
| Tetcyclacis | 5-(4-chlorophenyl)-3,4,5,9,10-pentaaza-tetra-cyclo-(5,4,$10^{2,6}0^{8,11}$) dodeca-3,9-diene | GR |
| Thidiazuron | N-phenyl-N'-1,2,3-thiadiazol-5-ylurea | CD GR |
| TIBA | 2,3,5-triiodo-benzoic acid | GR |
| TRIA | 1-hydroxy-triacontane triacontanol | GS |
| Triadimefon | 1-(4-chlorophenoxy)-3,3-dimethyl-1-(1,2,4-triazol-1-yl)-2-butanone | GR |
| Triadimenol | ß-(4-chlorophenoxy)-_-(1,1-dimethylethyl)-1H-1,2,4-triazole-1-ethanol | GR Fung. |

| Abbreviation or Common Name | Chemical Name or Structure | Major uses |
|---|---|---|
| Triapenthenol | (E)-(RS) 1-cyclohexyl 4,4-dimethyl 2-(1H 1,2,4-triazol-1-yl) pent-1-en-3-ol | |
| Uniconazol | (E)-(4-chlorophenyl)-4,4-dimethyl-2-(1,2,4-triazol-1-yl)-1-penten-3-ol | GR Fung. |

**Abbreviations:** AD= Apical-dominance reducer; AS= Abscission stimulant; CD= Cotton defoliant; DB= Dormancy breaker; Enlarg.= Plant cell enlarger; FL= Fruit loosener; FS= Flowering suppressant, in sugarcane; FT= Fruit thinner; Fung.= Fungicide; Gamet= Gametocide; GE= Grape enlarger; GR= Growth retardant; GS= Growth stimulant; Herb= Herbicide; LP= Lodging preventor; PF= Pineapple flowering agent; SR= Sugarcane ripener.

### *Abscisic acid (ABA)*

ABA is a naturally occurring sesquiterpene which regulate plant growth and metabolism in various ways and have been detected in nearly all plants. It is involved in the abscission of plant organs, induction and vegetative buds, in regulation of fruits repining and generally in reduction of growth.

### *Ethylene*

It is the only gaseous hydrocarbon hormone which plays an important role in the ripening of fruits, inhibition of root growth, abscission and other growth processes.

Unlike the other hormones, ABA and ethylene are not discovered through any interaction with fungi.

### *Brassinosteroids*

In 1979 a novel plant growth-promoting steroidal lactone, termed brassinolide was found in the pollen of *Brassica napus*. Since then chemists and biologists have isolated brassinolide and identified related compounds in various higher and lower plants (Mandava, 1988). Brassinolide and its analogues have been synthesized and used in laboratory and field trials. Further structure-activity evaluation of these compounds has also been undertaken. Nothing precisely is known regarding the biosynthetic pathway for the brassinosteroids. Workers all the world over are using these substances to increase crop yield and biomass.

In the 1930s and 1940s, USDA workers recognized that pollen extracts promoted growth. Mitchell and his associates used bean first internode bioassay and demonstrated that extracts from immature bean seeds elicited growth. These workers demonstrated that pollen from rape plant and *Alnus glutinosa* caused an unusual response that combined elongation with swelling and curvature. The group of hormones isolated from rape plant pollen was lipoidal and was termed, brassins. Subsequently, Mandava, Mitchell and coworkers reported an active fraction predominantly containing the glucosyl esters of fatty acids.

In 1972 Mitchell and Gregory demonstrated that brassins could enhance crop yield, crop efficiency, and seed vigour and in 1975 several location trials on cereals, oilseeds and potatoes were conducted in USA. Subsequently brassinolide was isolated. In pollen brassinolide is present at 200 parts-per-billion (ppb). The discovery of brassinolide demanded more than 10 years and cost about a $1 million, mainly supported by USDA.

According to Mandava (1988) Brassins, comprise a complex mixture of lipids purified from *Brassica napus* pollen. Brassin activity refers to biological activity comprising elongation, curvature, and splitting of internode in the bean second-internode bioassay following application of brassin. Brassinolide and or its related compounds are known collectively as brassinosteroids. Briefly BR are a group of naturally occurring polyhydroxysteroids. As many as 16 natural BRs we are brown as a group of naturally occurring polyhydroxysteroids. As many as 16 natural BRs have been identified and of these 15 have been isolated from higher plants. We may state that $BR_1$, $BR_2$ ........ BRn are designated to be naturally occurring brassinosteroids. However, brassinosteroids analogues are included in this classification. $BR_1$ to $BR_9$ are all $C_{29}$ steroids whereas $BR_{14}$ and $BR_{15}$ are $C_{27}$ steroids. Brassinosteroids are known to increase ATPase activity and increases the level of soluble proteins and reduce those of sugars. BR application has several advantages in increasing crop productivity and also overcoming stress. Mandava (1988) has proposed as many as five distinct functions in plant tissues and fruits.

## Other plant compounds which affect growth

There are several other plant products which have been shown to modify plant growth in different bioassays. Most of these compounds are inhibitors though their natural role remains to be established *in vivo*. In some cases their interaction is well characterized as growth hormones has also been postulated. In this category phenolic compounds are also included. We shall discuss the significance of these compounds in interacting with IAA oxidase and regulating the IAA levels in later discussion. Of the 550 secondary plant constituents nearly more than 200 are shown to be inhibitory in nature. For details the reader may consult a treatise by MacMillan (1980). We shall briefly mention some of these constituents:

### *Aromatic compounds*

These compounds include simple phenols and benzoic acid to the polymeric tannins. It also includes cinnamic acids, the coumarins and the flavonoids. In fact next to carbohydrates these compounds are met with in abundance in plants. In plants these compounds occur as glycosides. The role of phenolics as potent plant growth regulators has been discussed (Malik *et al.*, 1986). Some of the phenolics are shown to affect flowering and series of other physiological characteristics. Recent studies in salicylic acid are of considerable interest (Raskin, 1992). The role of phenolic compounds on plant growth is generally attributed to their interaction with IAA oxidase and hence regulating IAA levels *in vivo*. The prevalent postulate is that monophenols stimulate IAA decarboxylation and hence inhibit growth where as *o*-dihydroxy phenolics inhibit IAA oxidation and are thus stimulatory for growth (see Basra and Malik, 1984). Irrespective of hydroxylation pattern phenolics at high concentrations are inhibitory. The role of flavonoids on photomorphogenesis is well documented. Several of the coumarins are strong inhibitors of germination. It has been demonstrated that coumarins react with enzyme sulphahydryl groups though some workers do not support this contention. Several of the unsaturated lactones are shown to inhibit root growth and block mitosis. Potato tubers contain 1,4- and 1,6-dimethyl naphthalene which inhibit sprouting.

### *Nitrogen containing compounds*

Hydrogen cyanide is liberated enzymatically from cyanogenic glucosides and inhibits growth. Amygdalin is also highly inhibitory and is believed to be responsible for the ‘peach re-plant’ problem. Several of the alkaloids are reported to affect plant growth processes e.g. colchicine, delcosine, lycorine, cocaine, cinchonin, quinone, caffeine, codeine, berberine,.etc.

Several of the non protein amino acids are also shown to have variable effects on plant growth. Recently role of polyamines as potent growth regulators has been widely discussed and for details a reference be made to a review by Evans and Malmberg (1989). These nitrogenous compounds are of wide occurrence in plants and are shown to stimulate several growth processes in plants.

### *Terpenoid compounds*

Several naturally occurring terpenoids from plants are shown to affect plant growth. These include monoterpene-essential oils (growth inhibitors), α-methylene lactones (both growth inhibition and root stimulation), unsaturated lactones. Of the several dehydrocostus lactones from *Saussurea lappa* (compositae) have been widely shown by Kalsi and his associates to be potent plant growth regulators with divergent functions. This group has also demonstrated that α, ß-unsaturated ketones also promote rootings. Besides sesquiterpenoids, diterpenoids, non-lactone diterpenoids with variable plant regulating activities are also reported. Steroids and triterpenoids are also shown to affect plant growth.

### *Aliphatic alcohols*

Several of the naturally occurring aliphatic alcohols, aldehydes, ketones, acids and esters are reported to affect plant growth. Besides triacontanol several other aliphatic alcohols produce variable responses in plant systems and for some recent studies the reader is advised to look up the review by Malik *et al.* (1987), Malik *et al.* (1991, 1993). Different aliphatic alcohols are mixed in variable proportions to use them commercially. Some of these mixtures are being used to enhance yields in various crop species.

Besides the compounds mentioned thus far, extracts of leaves of *Eucalyptus grandis* have yielded rooting inhibitors; ascorbic acid is shown to have variable effects on plant growth and has been suggested as plant hormone.

### *Triazoles*

New types of plant growth retardants developed (as fungicides) in recent years include triazoles, a group of compounds known as ergosterol biosynthesis inhibitors. These compounds include paclobutrazol, Lab-117-682, uniconazol, triapentenol, Bas-111, flurupimidol, etc. These compounds have greater broad spectrum retarding ability than earlier retardants-CCC, Mepiquat chloride and diaminozide, etc. They inhibit endogenous production of gibberellic acid; control excessive shoot growth. Their application alters several physiological and morphological traits. In many fruits like peaches, tomato, mango, cherry, fruit bearing increased, size became bigger than control, and bearing was hastened. In rice and *Brassica* stem growth was considerably dwarfened. Some of the triazoles protect the plants against stresses. For details refer to a recent review by Malik and Thind (1994).

Over the years it has been widely accepted that plant growth and development is controlled by chemicals produced by the plant itself i.e. naturally occurring or endogenous plant hormones. However, synthetic plant growth regulators are also increasingly used for improving the productivity of plants. The plant growth regulators (PGR's), either naturally occurring or synthetic, are applied directly to the target plant to alter its life processes or its morphology to improve quality, increase yield or facilitate harvesting. The chief purpose of using plant growth regulators in agriculture and horticulture is to optimize plant production by modifying growth, development and stress behaviour and the qualitative and quantitative yield of crop plants. It is generally suggested that the synthetic PGRs cause their effects through changing the endogenous level of naturally occurring hormones and hence modifying growth and development of plants in the intended direction and to a desired extent. The experience has also shown that the synthetic PGRs were likely to prove increasingly important in both agriculture, horticulture and plant propagation through tissue culture. Their role in hardening of seedlings raised through biotechnology before transferring to the field deserves special mention.

### *Polyamines*

There is controversy whether these compounds should be classified as plant hormones. It was during the International Conference on Plant Growth Substances held in 1982 that they were first tentatively accepted by their inclusion as hormones. According to Galston (see in Davies, 1987) there is full justification in classifying them as hormones on the following basis: They are widespread in all cells and exert regulatory control over growth and development at micromolar concentrations. In plants where polyamine content is genetically changed including mutation, development is severely affected. In tissue culture of *Vigna* or carrot, when polyamines level is low, only callus exists and embryoids are formed only when polyamine content is high. In tobacco plants, which are overproducers of spermidine, anthers are produced in place of ovaries. Such developmental control is affected by hormones than nutrients such as amino acids or vitamins.

Polyamines have a wide range of effects in plants and seem to be essential for cell division, normal morphology and hence plant development. It appears that polyamines are ubiquitous and universally present in all cells rather than having a specific site of synthesis. For details the reader is referred to the recent review by Sankhla *et al.* (1988).

With increasing cost of energy, PGRs were sure to play an increasingly important role in energy conservation as a consequence of enhanced yield following their use. The dire need to double world's food supply by the end of the 20th century is also another significant incentive to improve agricultural production through novel approaches e.g. PGRs technology.

Plant growth regulators are known to bring about diversity of effects which include effects on plant functions e.g. induction of rooting, control of flowering, regulation of sex and control of ageing. Additionally PGRs are also shown to control the shape, size or growth rate. Several of the biochemical processes and suitable metabolic changes are affected by PGRs leading to alterations in both quality and quantity of economic products.

Striking economic successes have already been achieved through the judicial use of PGRs. Some of the instances are: enhancement in latex flow in rubber trees, ripening of sugarcane, control of premature sprouting in potatoes and onions, control of buds in tobacco, dwarfing of wheat and *Brassica* to prevent lodging, preventing premature ripening and deterioration of fruits, improving fruit quality, control of deterioration of fruits, improving fruit quality, control of timing of maximum utilization of crops, exudation of gums, etc.

The coming years will witness increasing efforts in the control of adverse effects of various environmental stresses e.g. drought and low temperature, through the selective use of PGRs.

In the near future it is very likely that novel PRGs will be developed and rationally used for crop improvement. In fact these chemicals open new opportunities to improve agriculture and horticulture by circumventing several barriers imposed by genetics and environment.

## Other growth regulators

New other growth regulators have been identified in different plant species and these are briefly described below:

### *Lunaric acid*

In liverworts, lunaric acid is present in gemmae formed in gemma cups and prevents germination of these gemmae unless the acid is leached out. Lunaric acid is reported to be present in many species of lower plants excluding algae.

### *Batasins*

These chemicals cause dormancy in bulbils of *Dioscorea batatas* and arise from swelling of the aerial lateral buds. The precise mechanism of its action and transport is not well understood.

### *Jasmonic acid*

Jasmonic acid and its methyl ester are widely distributed in plant species and are isolated from oil of jasmine. They are biosynthesized from free linoeic acid through the action of lipoxygenase. These compounds promote senescence and inhibit growth of some plant species.

### *Turgorins*

Some German chemists have isolated and identified compounds that activate pulvini in *Mimosa* and *Acacia* leaves (see Salisbury and Ross, 1992). Compounds like ß-glucosides of gallic acid with the glucoside bonding at its para-hydroxyl group, ß-D-glucoside-6-sulphate, ß-D-glucoside-3, 6-disulphite of gallic acid, etc., are included in this category. These compounds are called turgorins since they act on the turgor of pulvinus cells and are active at very low concentrations (10-5 to 10-7 M). Hermann Schildknecht and his associates (1990) have demonstrated the presence of a receptor - a protein in the outer side of the plasma membrane of *Mimosa*.

### *Tuberonic acid*

Tuberonic acid is closely related to jasmonic acid and has strong potato tuber-inducing activity. It has been isolated from the leaves of the potato plant and identified as 3-oxo-2-(5-ß-D-glucopyranosyloxy-2-in-pentenyl) cyclopentane-1-acetic acid. The substance isolated is a glucoside and aglycone of the substance is called 'tuberonic acid'.

### *Cucurbic acid*

This is an inhibitor isolated from cucumber seedlings.

## HOW HORMONE ACTS?

When a hormone acts upon a responsive plant system, it of course, enters into some direct and molecular interactions, which results, eventually in manifestation of measurable effects (biochemical or physiological responses). However, there are two aspects of hormone action:

1. **Mechanism of action:** The direct and specific molecular interaction, and

2. **Mode of action:** The succeeding series of steps (events and processes) which results in the measurable biochemical or physiological responses.

According to Paleg (1960) the two terms *Mechanisms* and *Mode* of action should be taken as different terms rather than synonyms. The mechanism by which minute amounts of plant hormones, relatively simple organic compounds, can do dramatically controlled growth is one of the pressing and challenging problems in plant physiology. The progress in this field has been quite spectacular but still it needs a proper synthesis and analysis of various concepts given in this regards.

In the last years, numerous physiological and biochemical responses to plant hormones have been observed which have a bearing on the mechanism of hormone action. Some of the responses have been enumerated by Gordon (1952), they are as follows:

1. Increased plasticity of the shoot and elasticity of the root cell wall.
2. Increased permeability to water.
3. The enhanced capacity to retain water taken up.
4. The *active* uptake of water and solute apparently independent of osmotic forces and occurring even against an osmotic gradient.
5. A decrease in protoplasmic viscosity.
6. An accelarated rate of respiration and cyclosis.
7. More rapid synthesis of protein, with lowered levels of free amino acids.
8. The increase of monosaccharide at the expense of reserve polysaccharide.
9. An increment of (intermicellar) wall pectin and of cellulose and a stimulation of alcohol and malic dehydrogenase, catalase, phosphatase and ascorbic acid oxidase.
10. These large varieties of effects which hormone evokes depict the multiplicity of sites.

**TABLE 21.2:** List of hormones and their physiological responses*

| Effects | Auxin | GA | CK | ABA | Ethylene |
|---|---|---|---|---|---|
| Cell elongation | + | + | + | - | - |
| Cell division | + | | + | | - |
| Apical dominance | + | | - | | |
| Leaf growth | | | | | |
| In length | | + | | | |
| In breadth | | | + | | |
| Chlorophyll formation | | | + | | |
| Vein elongation | + | | | | |
| Leaflet movement | + | | | | |
| Root formation | + | | | | |
| Fibre growth in cotton seed | + | + | - | | |
| Abscission | - | - | | + | + |
| Sex expression | | | | | |
| Maleness | - | + | | - | |
| Femaleness | + | | + | | |
| Fruit growth and division | + | + | + | + | + |

| Effects | Auxin | GA | CK | ABA | Ethylene |
|---|---|---|---|---|---|
| Xylem formation | + | + | | | |
| Flower | + | + | + | - | - |
| Dormancy break | | | | | |
| Seed dormancy | | + | + | + | - |
| Bud dormancy | | + | + | + | - |
| Senescence | | - | - | + | + |
| Tropism | + | + | | + | |

+ promote, - inhibit, * After Sen, S.P. (1984).

An understanding of hormone action is dependent on location of sites of action (Sen, 1984). According to Overbeck (1966) *We are beginning to learn that the site of action of hormones is close to gene.* Usually two sites of plant hormone action at the subcellular/molecular level are discussed: an early (*primary*) effect at the cell membrane and a subsequent (*secondary*) effect at genetic level (Parthier, 1985).

## Influence of Hormones on Nucleic Acid and Protein Metabolism

The fundamental action of hormones concerns with nucleic acid metabolism. Auxin stimulation of nuclear Gibberellin stimulation of nuclear RNA synthesis was shown by Varner and Chandra (1964). Kinetin stimulates nuclear RNA synthesis. ABA inhibits RNA synthesis. ABA, according to some workers blocks transcription by inhibiting RNA synthesis.

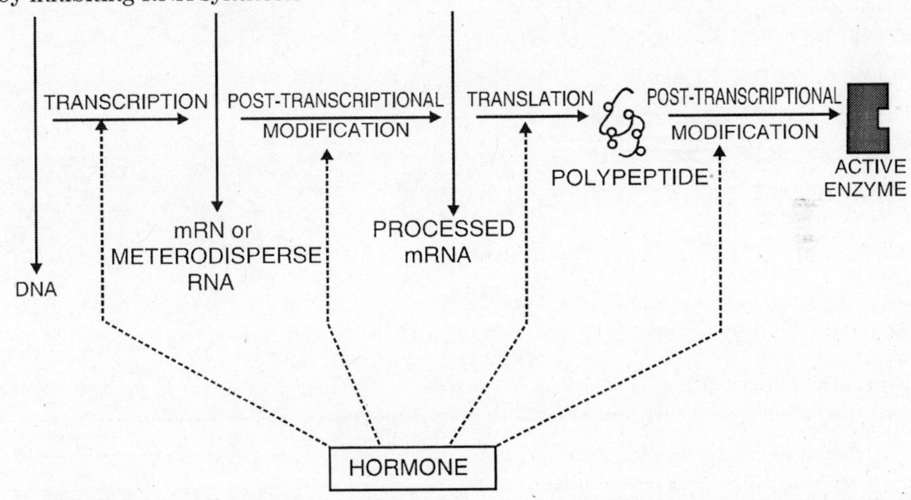


**FIG. 21.2:** Basic action of plant hormone.

Hormone effects on protein synthesis are generally achieved through enhnced RNA synthesis, formation of polyribosomes or through an interaction with the t-RNA and ribosomes. Hormonal stimulation of nuclear protein synthesis was first shown by Datta and Sen (1965). The dependence of hormone action on the central dogma involving DNA, RNA and protein is necessary for both qualitative and quantitative changes in plant function due to hormone action. Since all enzymes are destroyed after a certain period of time and their m-RNA usually decays with relatively short half lives, the synthesis of RNA and protein have to continue all the time; that RNA synthesis is an important link in the chain of events originating from hormone application and culminating in the expression of physiological response, is evident from the fact that dual effects of hormone concentrations on growth are also exhibited by RNA synthesis, protein synthesis, respiration, $CO_2$ fixation and other metabolic processes.

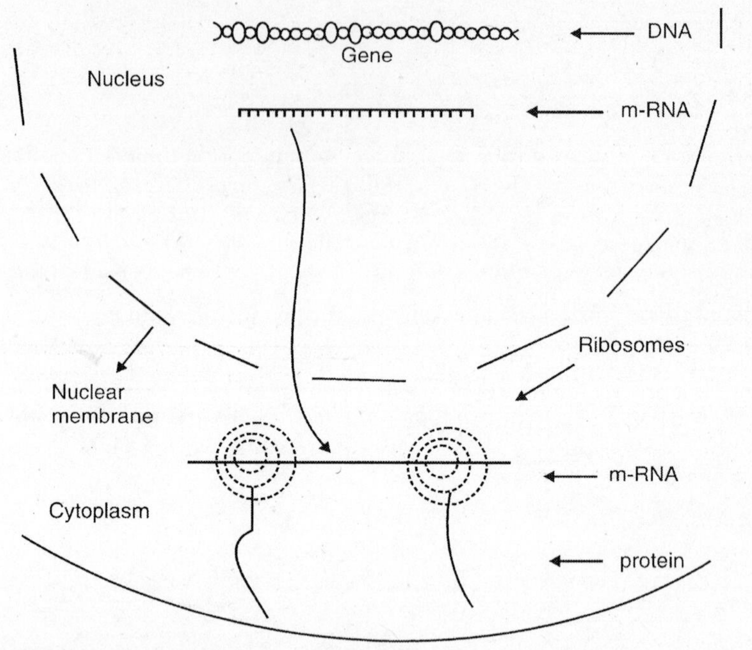


**FIG. 21.3:** Scheme of genetic direction of protein synthesis.

## Influence of Plant Hormone on Gene Expression

Plant hormones play a pivotal role in the control of gene expression at the transcriptional and post-transcriptional levels. In the transcriptional control, the phytohormone enhances the preferential synthesis of gene transcripts which eventually results in the induction of enzymes. There are various steps at which plant hormones could influence gene expression. These are as follows:

1. Synthesis of pre-mRNA transcripts
2. Post-translational processing of RNA
3. Turnover of functionally active mRNA
4. Post-translational modification of the preformed protein

Amount (biosynthesis) or function (activity) of enzymes or other polypeptides involved in gene activation, transcription, gene product modification, translation or post-translational steps are reported to change upon hormone addition.

These diversity of effects suggest multiple sites and modes of phytohormone actions, rather then a unique event that regulates gene expression. It may, however, reflect the limited sensitivity of available methods. This field is now rapidly improving due to the use of recombinant DNA techniques and subtle detection methods for gene products, which enables to analyse hormone actions in gene expression with greater sensitivity and in more detail.Following examples of phytohormone responses are considered under the aspects of their action in gene expression.

**TABLE 21.2:** Plant growth substances interacting in physiological responses and in gene expression (transcription)

| Physiological Process | Plant Growth Substance |
|---|---|
| Cell Enlargement | Auxin |
| $\alpha$-amylase synthesis in aleurone cells | Gibberellin |
| Plastid Development | Cytokinin |
| Embryogenesis and seed development | ABA |
| Fruit ripeninbg | Ethylene |

## Secondary Messenger Hypothesis

According to this hypothesis, auxin does not directly act on the cell wall. Addition of auxin to isolated cell walls or to cells with non-functional protoplasts has no effect on the wall extensibility (Cleland 1967, 1977). This has led to the conclusion that the initial action of the hormone is at a site on or within the protoplast, an action which leads indirectly to wall loosening. Thus, there must be a signal or *Secondary messenger* of some type which travels from the protoplast to cell wall, initiating wall loosening.

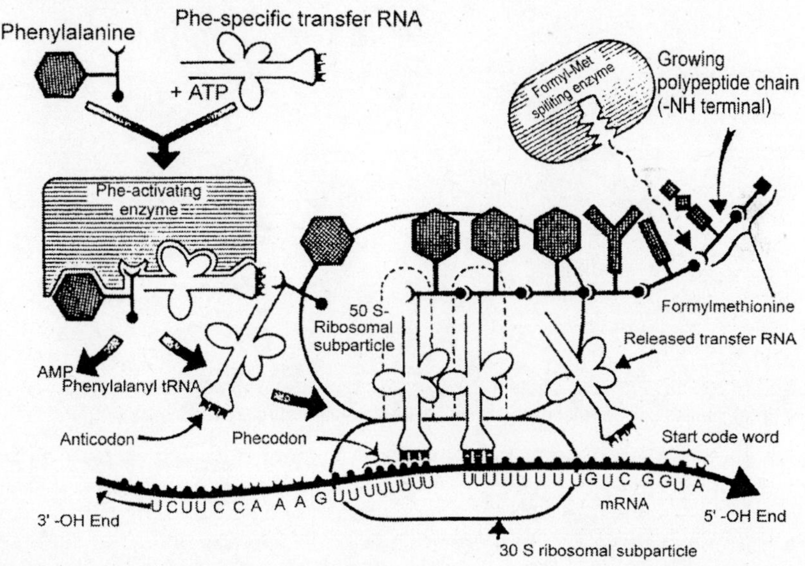


**FIG. 21.3 (A):** Schematic mechanism of protein synthesis.

## HORMONE RECEPTOR CONCEPT

This concept finds its origin in work on animal hormones which can not elicit physiological responses unless they are recognised by their specific receptor sites. Each animal hormone has a specific site (gland) of synthesis and a target organ having affinity receptor (s) located in the plasma membrane and cytosol of the cell. These receptor molecules have been shown to be proteinaceous in nature.

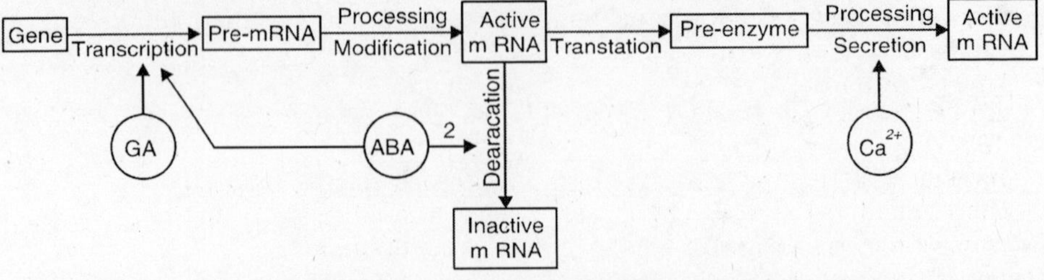


**FIG. 21.4:** Possible site of action of GA and ABA in α-amylase gene expression.

Based on analogy with animal systems, it is currently thought that most of the physiological and biochemical responses elicited by plant hormones are also inhibited by binding of hormones to their respective receptors. A hormone must be recognized by the cell in order to evoke its specific physiological and biological response. The binding sites, often referred to as receptors have been characterised as protein moeities, with a dissociation constant ranging from $10^{-8}$ M to $10^{-5}$ M.

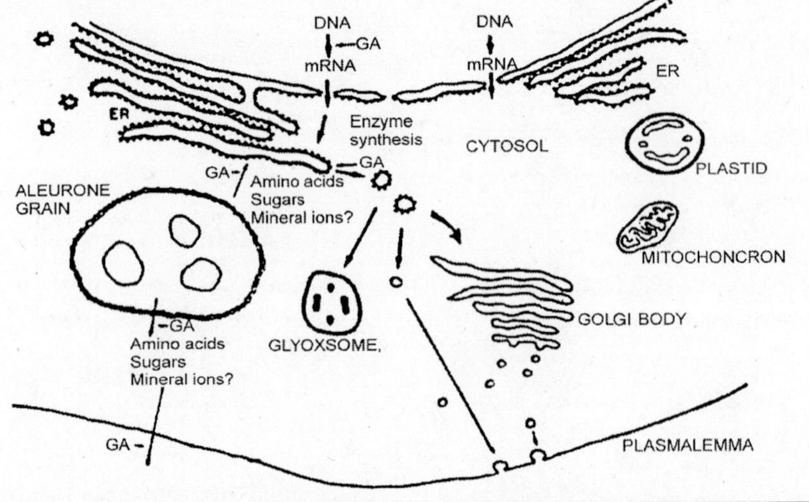


**FIG. 21.5:** Suggested sites of response to gibberellin in cereal aleurone cell.

Thus, according to the basic scheme, the mechanism by which plant hormones exert their effect(s) is by first binding to specific proteins or receptors, commonly envisaged to be an allosteric protein. As a consequence of binding to the hormone, forms a hormone-receptor complex which initiates the biochemical events leading to a response. Binding to such a receptor would, therefore, represent the primary level of interaction of a hormone within a cell. The hormone-receptor complex may bind to the template DNA or to RNA polymerase or both and results in improved transcriptional activity. Such a participation of hormone receptor complex in the transcriptional process has not been fully established. Thus, in summary following could be various alternatives based on receptor concept in explaining the hormone regulation of gene expression of enzyme proteins.

1. Hormone receptor complex interacts with chromatin at specific intiation sites and thereby enhances the transcriptional activity.
2. Hormone receptor complex may bind to RNA polymerase and thereby result in enzyme activation; the modified RNA polymerase could then accelerate transcription activity.
3. An authentic receptor system should satisfy certain minimum criteria (Venis, 1980), namely:
4. binding should be reversible, of high affinity and of finite capacity;
5. the saturation range of binding should be consistent with the concentration range over which the physiological response saturates;
6. binding specificity for different analogues should reflect their relative biological activity, and binding should lead to a hormone specific response.

Until recently, it is difficult to obtain a plant system that approximated to even remotely accepted concept of hormone receptors. However, based on research during the two decades and specially because of the availability of labelled hormones and high affinity chromatography, receptor molecules all of proteinaceous nature have been isolated, and identified for their molecular weights and dissociation constants of the several hormone-receptor complexes.

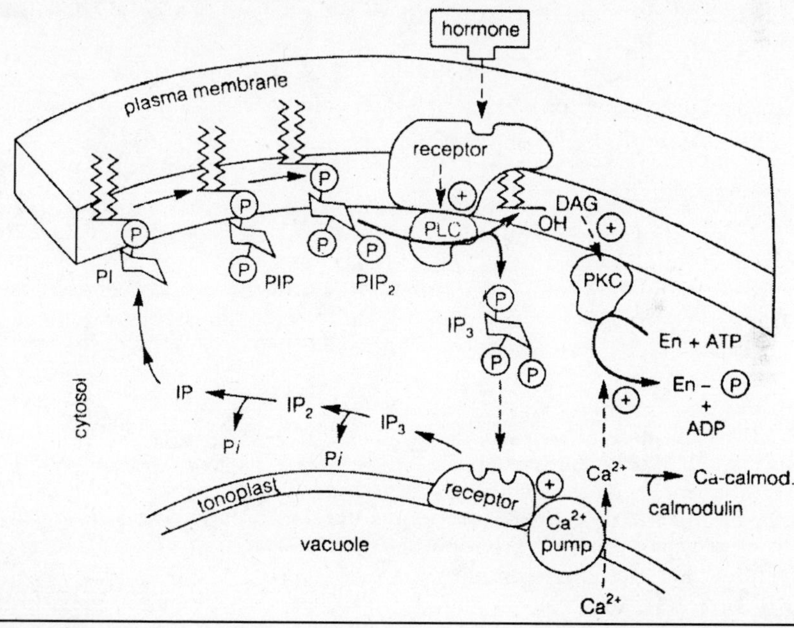


**FIG. 21.6:** Model for initial hormone transduction at the plasma membrane. Binding of a hormone to its receptor cause activation (+) of nearby phospholipase c (PLC). PLC hydrolyzes a membrane lipid, phosphatidylinositol-4,5-bisphosphate ($PIP_2$) to release inositol-1,4,5-trisphosphate ($IP_3$) and diacylglycerol (DAG). $IP_3$ moves to the tonoplast in plant cells, where it combines with a receptor that activates (+) a $Ca^{2+}$ pump or tranporter that moves $Ca^{2+}$ from the vacuole to the cytosol. DAG, which remains membrane-bound, activates protein kinase c (PKC). PKC is also activated by $Ca^{2+}$ released from the vacuole, so various enzymes, when free or bound with calmodulin. $IP_3$ loses phosphates by hydrolysis to form $IP_2$. and IP, which is then converted back to phosphatidylinositol (PI) and other phosphoinositide lipids (PIP and $PIP_2$.) in the plasma membrane.

## Auxin Receptors

By analogy with animal hormones it has been presumed that auxin-induced responses are triggered by binding of auxin to a receptor molecule of high specificity and high affinity. Auxin triggers various biochemical responses through their interaction with receptor molecule. This interaction may bring about a change in receptor conformation, micro-aggregation, redistribution, internalization or chemical modification. This is followed by specific responses such as regulation of enzyme activity through release of secondary chemical messenger (cAMP or cGMP), opening or closing of ion channels, phosphorylation, or increased RNA synthesis. Receptors are proteinaceous in nature and differ from cell to cell. These can be subclassified on the basis of signal transduction system they employ. On the basis of their localization, these can either be soluble (intracellular) or bound (of plasma membrane localization). Several auxin-binding proteins have been reported both from the particular as well as soluble fractions of cells of various plant tissues.

### Membrane-Bound Auxin Receptors

Investigations of membrane-bound auxin-binding sites have relied on the postulate that the auxin receptor should:

i) be able to act with full strength within 15 min as by the early kinetics of auxin stimulated growth, and
ii) activate proton extrusion as predicted by acid growth hypothesis.

Search for auxin receptor has mainly been concentrated in maize coleoptile by studying binding of labelled auxins.

### Soluble Auxin Receptors

Interest in soluble of chromatin bound auxin receptors has been evoked by the hypothesis that auxin must primarily regulate gene activity. Two IAA-binding proteins from the nucleoplasmic fraction (n-IRP) and the non-histone chromatin, protein fraction (c-IRP) of nuclei isolated from the immature coconut encosperm, have been reported. It has been found that n-IRP factor stimulates in-vitro RNA synthesis.

Dextran-bound charcoal-binding assays have been used extensively in the analysis of auxin binding in soluble fractions in soybean nuclei and in wheat coleoptiles. Such high affinity soluble binding sites have also been detected in number of other plant materials like dwarf bean seedlings, etiolated mungbean and pea seedlings and maize cytosol, but details vary so greatly that it would appear that no two of these proteins are identical. Several attempts have been made to isolate soluble auxin-binding protein by affinity chromatography on 2,4-D-linked sepharose. Most of these proteins have been reported to alter the transcription of either chromatin or DNA in a homologous or heterologous in-vitro systems.

A study on auxin binding to cellular receptors in 4-day-old seedling of *Cicer arietinum* revealed the presence of both membrane-bound and cytoplasmic auxin receptors. Both showed almost similar specificity, kinetics and chemical characteristics.

## Gibberellin Receptors

Binding site for gibberellin ($GA_1$) was detected in subcellular fraction enriched in aleurone isolated from wheat endosperm. This site showed a specific binding for [$^3$H]-$GA_1$. The dissociation constant ($K_d$) was within the physiological dose response of $GA_3$ induced a-amylase production. A soluble fraction isolated dwarf pea epicotyls showed a specific binding for [$^3$H]-$GA_1$ binding of [$^3$H]-$GA_1$ with 2 protein fractions was observed. The hormone-protein interaction was non-covalent in nature. The results were further confirmed following some what different procedures. Dwarf pea epicotyls were incubated in radioactive $GA_1$ for 3 days at 0°C. [$^3$H]-$GA_1$ bound to 2 soluble proteins of molecular weight 600 and 40-7- Kd. The binding site possessed high affinity for $GA_1$. [$^3$H]-$GA_4$ binding site in the soluble protein fraction from cucumber hypocotyls was detected and was found to be specific, saturable and could be reversed unlabelled $GA_4$.

The studies so far made have helped to some extent in understanding the basic concept operative during enzyme regulation by GA. GA could control gene expression at transcriptional, post-transcriptional, translational or post-translational level. But it is still far from clear as to how exactly GA is regulating the gene expression.

## Cytokinin Receptors

CK binding to the ribosomes from higher plants, namely of chinese cabbage, was first reported by Berridge et al. (1970), who observed reversible but non-specific and unsaturable binding of low-affinity. Fox and Erion (1975) later reported a CK-binding protein from ribosomes of wheat germs and tobacco callus, and noted the existence of two types of binding sites on higher plant ribosomes. One of these, a high-affinity site, bound CKs at low concentrations, was saturated at one CK molecule per ribosome, was shown to be specific for substances with CK activity (although only a limited number of CKs were tested), and that it could be washed from the ribosomes with 0.5 M KCl. Subsequently they showed that the protein existed as both loosely attached to ribosomes and as a soluble entity (Fox and Erion 1977). Furthermore, they were able to resolve a 93 kD high-affinity CK-binding protein (CBF-1) from the ribsomal fraction from wheat germ. In addition, a very similar protein (probably identical to CBF-1), and approximately 30 kd high-affinity CK-binding protein (CBF-2), and high MW (250kD), low-affinity CK-binding protein (CBF-3) were resolved from the post-ribosomal supernatant fraction from wheat germ.

While CK-binding proteins have been detected in a number of higher plants system, CBF-1 is undoubtedly the best characterized CK-binding system. This has led Brinegar et al. (1985) to suggest a dual function for CBF-1; first as a storage protein, and second as a regulator of CK availability (a sequestering protein).

## ABA Receptors

The evidence supporting the existence of ABA receptors in guard cell is strong. A high-affinity binding site of ABA on plasmalemma of *Vicia faba* guard cell was found. There were $2 \times 10^6$ binding sites per protoplast. Three proteins that were covalently linked to ABA were separated by gel-electrophoresis. One protein had highest affinity for dissociated ABA, while the other 2 bound preferentially to protonated forms. One of these proteins could be modulator of $H^+$-ATPase which is inhibited.

### Ethylene Receptors

A calculation of the amount of ethylene receptors in plant tissue by the Abeles (1973) revealed 10 mg. per metric ton or approximately 500 attachment sites per cell, assuming the receptor has a molecular weight of 100,000. Based on kinetic studies of ethylene binding and displacement in tobacco leaves, Sisler (1979) estimated 4000 binding sites per cell.

Although direct proof is lacking, some indirect evidence suggests that ethylene binds reversibly to a metal containing receptor site (Burg and Burg, 1967). Cell free extracts of unripe bean cotyledons exhibit ethylene binding activity with a high specificity and activity (Bengochea et al., 1980 a, b). The binding site seems to be membrane bound. It was possible to isolate the binding site without loss of activity by means of Triton x-100. It proved to be an assymetrical protein with a molecular weight between 50,000 and 60,000. Binding sites for ethylene have been shown to exist in a variety of plant species. Whether some or all of them are requisite for the action of ethylene can not be stated with certainty at present. One of the best characterised, that in the cotyledons of *Phaseolus vulgaris* has been shown to be a hydrophobic integral membrane glycoprotein. Characterisation of sedimentation properties and identification of marker enzymes in the preparations suggest three different binding sites, possibly on protein body membranes, on smooth and rough endoplasmic reticulum. These results were confirmed with high resolution autoradiography.

## PLANT HORMONE CONJUGATES

Plant hormone conjugates or conjugated plant hormones are defined as *those plant hormones which are metabolically coupled to other low molecular weight compounds by covalent binding* (Sembdner, 1974; Sembdner et al., 1986). This definition excludes non covalent associations between plant hormones and high molecular weight particles like protein, cell particles and membranes. These complexes should be called *bound* plant hormones (Schneider, 1983). Non-identified polar fractions which possibly belong to the conjugates are also often termed *bound hormones*. The plant hormone conjugates are widely distributed in plants and occassionally present in high concentration. Investigations on the isolation and identification of plant hormones have led to the detection of these conjugates from which free hormones can be liberated by hydrolysis.

Conjugated plant hormones are those hormones which are metabolically coupled to other low molecular weight compounds by covalent binding. Plant hormone conjugates are widely distributed in plants and generally occur in high concentrations. Plant hormones or free hormones are released from conjugates through hydrolysis. In fact conjugation is a part of metabolism of plant hormone but unlike biosynthesis, they occur in irreversible state. Conjugates are different from the corresponding free hormone physically, chemically and in their biological properties. Thus, conjugates may be concerned with the regulation of the level of free and active hormone. Plant hormones could form conjugates with lipids, carbohydrates, myoinositol, amino acids, etc. Membrane associated steroids may form hydrogen bond with the carboxyl or a negative charge center of the hormone molecule with the phosphate hydroxyl: the lecithin in the other side.

IAA is reported to conjugate with glucose, arabinose, rhamnose; the carboxyl group of the IAA is involved in the esterification with sugars, however, it does not bind to cellulose or pectin, the cell wall components. Glucose is generally conjugated to the OH or carboxyl group of GA. The cytokinins are usually ribosylated in the N-9 position whereas ABA is reported to occur as its B-D-glucopyranosyl ester. In several plant systems IAA glucosyl ester is the precursor of IAA myoinositol and is very active in the tissue culture medium. Auxin conjugates amino acids to form indoleacetyl aspartate. IAA also conjugates with glutamic acid, lysine, glycine, alanine and valine. GA conjugates with proline, glycine whereas cytokinins also conjugate with some amino acids. In the past few years a concept dealing with the plant hormone sensitivity has been proposed by Trewavas (1987) who has suggested that a crucial factor for a chemical to be a hormone is its being sensitive controller of a given process over it is physiological concentration range. Here sensitivity pertains to a measure of the influence of changes of hormone concentration on physiological processes. This author has proposed that plant tissue are ordinarily responsive to alteration in endogenous hormone level under extreme conditions which cause enhancement in sensitivity to hormone(s).

The transduction of signals probably occupies the activities of the majority of the plant cell protein. Sequence of signal transduction begins with the signal receptor. The cellular sensitivity changes during development and is dependent on receptor number and distribution. Hence, there may be several receptors for the same signal, which permits for more complex responses. Receptors to auxin, red light and calcium are well characterized and those for blue light, ethylene are nearly identified. Anderson *et al.* (1994) have attempted to characterize the site of perception of ABA and GA in guard cells. It is demonstrated that both the hormones are perceived by outward-facing plasma membrane receptors. These pass microinjected ABA directly in the closed guard cell cytoplasm. Microinjection up to 200 μM ABA failed to prevent subsequent opening of the guard cell suggesting lack of a cytoplasmic receptor for ABA action. Gilony and Jons (1994) microinjected GA in aleurone cells of barley and monitored α-amylase expression/secretion with specific *β*-glucouronidase constructs on starch digestion. Exogenous GA application stimulated amylase formation, whereas microinjected GA failed to do so. GA-stimulated amylase could be antagonized by exogenous ABA; whereas microinjected ABA had no effect. The conclusion drawn favoured outward-facing GA and ABA receptors. Allan and Trewavas (1994) believed that GA and ABA are amphipathic molecules which could easily penetrate membranes. The rate of penetration is regulated by the concentration of the permeant species. Further, inside the cell the distribution of ABA/GA partly reflected the pH of the different compartments.

In guard cell ABA accumulates in plastids whereas GA accumulates in vacuoles. Hence it is difficult to discuss cytoplasmic perception. The data on subcellular distribution of phytohormones are urgently needed, and emerging technologies could fill up this lacuna. These worders suggest separate receptors for ABA promotion of closure and ABA inhibition of opening. Thus, outward facing plasma membrane receptors are concerned with raped response of cell to transient signals. Cytoplasmic perceptions may be slow with extended minimal occupancy times and concerns more extended but qualitative different facets of the response.

As stated earlier the initial step of signal transduction is perception or recognition of the signal by a specific receptor on the membrane. We have briefly discussed the hormone binding proteins which have been variously characterized and sequenced in some instances as putative receptor(s) for that specific hormone. In some instances the perception of elicitor by its receptor couples to intracellular $Ca^{2+}$ signalling. Similarly, phytochrome is the extensively investigated photoreceptor of divergent red light triggered physiologic and developmental processes in plants. Some of these include rotation of *Mougeotia* chloroplasts, rotational streaming of *Vallisneria* cytoplasm, spore germination in *Onoclea* and pollen germination in *Arachis hypogaea* and in some of the processes $Ca^{2+}$ participation has been clearly characterized. In some instances red light perception of phytochrome usually modulates intracellular $Ca^{2+}$ content and thus initiates $Ca^{2+}$ signalling pathway through the activation of $Ca^{2+}$ channel or suppression of $Ca^{2+}$ pump in the plasma membrane. However, the mode of phytochrome modulation of $Ca^{2+}$ transport remains to be deciphered. Recent studies in *Avena sativa* indicate involvement of G protein in phytochrome action (Muto, 1992). In fact GTP binding proteins are pivotal to the transduction of several receptor signals in animal cells and are highly conserved membrane associated proteins comprising α, *ß* and *α* subunits (Gα, Gβ, Gα). The G protein cycles between inactive GDP-bound and active GTP-bound forms. Gα binds GDP to GTP. Gβ and Gα complex of G protein is needed for efficient receptor-catalysed guanine nucleotide exchange on Gα. Using different techniques subunits of G proteins have been shown in plant membranes though a heterotrimeric G protein as such has not been isolated from plants. Recent report from the plasma membrane of maize seedling roots has shown G protein with a molecular weight of 61 kDa and comprised two subunits of 34 and 27 kDa. However, a peptide corresponding to the γ subunit was not detected.

Through ADP-ribosylation of Gα catalysed by pertussis and cholera toxins, G protein was identified in etiolated *Arabidopsis sativa* seedlings. Further, a gene *(GPAI)* for a Gα from *A. thaliana* was isolated using a sequence from yeast Gα. *GPAI* gene product has 383 amino acid residues. The *GPAI*-encoded mRNA is most abundent in vegetative plant tissues. A gene encoding a Gβ-like protein was isolated from *Chlamydomonas reinhardii* and the predicted amino acid sequence is similar to that of yeast $G_β$.

Irrespective of the fact whether G proteins were detected immunologically or by ADP-ribosylation or deduced from the gene sequence consist of heterotrimeric G proteins.However, molecular masses of such proteins were smaller than those of animal Gα. It is essential to carry out purification and characterization of native heterotrimeric G proteins to resolve whether the small molecular masses of plants Gα are native polypeptides or are artificially produced during extraction from cells.

In the microsomal fraction of *C. pepo* GTP-binding proteins with molecular masses of 23.4-28.5 kDa were detected. These proteins have been found in the outer envelope membrane of pea chloroplasts. Proteins with molecular masses of 28 to 24 kDa were detected in soybean plasma membrane. All these proteins resemble animal proteins with molecular masses of 29-30 kDa which possess conserved GTP-binding and hydrolysis domains. These may have significant role in signal transduction and in mediating various intracellulr interactions including aspects of organelles traffic in cells. The small GTP-binding proteins have also been detected as substrates for ADP-ribosylation by certain botulinum toxins.

In *Zea mays* DNA sequences homologous to the oncogenes *myb*, *ras*, *src* have been detected.

The basic role of G proteins in $Ca^{2+}$ channel and phospholipase C on the plasma membrane, which leads to an enhanced cytoplasmic $Ca^{2+}$ concentration. GTP stimulation of $Ca^{2+}$ release from plant microsomes has been detected. Some workers have demonstrated that GDPBS enhance inward $K^+$ channel in the plasma membrane of *V. faba* guard cells in comparison to GTPvS which has opposite effect. Reduction in inward current by GTPvS was eliminated by a calcium chelator. On the contrary, inward $k^+$ current could be reduced with the application of cholera toxins. From these data it is inferred that G protein is involved in the inhibition of inward $K^+$ current which possibly results from an associated increase in $Ca^{2+}$ concentration. The enhancement in $Ca^{2+}$ cytoplasmic concentration may be caused by $Ca^{2+}$ influx across the plasma membrane and/or by inositol 1,4,5,-triphosphate-induced release from intracellular stores via the $Ca^{2+}$ channel. We have been mentioning time and again that $Ca^{2+}$ plays a vital role as a second messenger in intracellular signalling. The $Ca^{2+}{}_{cyt}$ concentration must be kept as low as $10^{-7}$ M since even a small quantitative change could cause a 10-100-fold enhancement in the concentration.

The cytoplasmic $Ca^{2+}$ concentration is kept low by active extracellular extrusion and sequestration to intracellular organelles. On stimulation of cells influx of extracellular $Ca^{2+}$ and or efflux of $Ca^{2+}$ sequestered in organelles occurs causing an increase in $Ca^{2+}{}_{cyt}$ concentration. Ion channels are suggestively mediated $Ca^{2+}$ fluxes.

## CALCIUM PUMP AND $H^+/CA^{2+}$ ANTIPORTER

### Plasma membrane

ATP-dependent $Ca^{2+}$ transport system has been shown in microsomal fractions or partially purified plasma membrane fractions from several higher plants. $Ca^{2+}$ transport was $Mg^{2+}$ and ATP-dependent. In addition to ATP, GTP could also drive transport at 50% of the total level reserved for ATP. Evidences also point out that secondary $Ca^{2+}$ transport system driven by a proton gradient was absent. $Ca^{2+}$ accumulated in the membrane was released by an ionophore. In some very recent studies the presence of both $Ca^{2+}$-ATPase and a tonoplast type $H^+/Ca^{2+}$ antiport was reported in highly purified maize plasma membrane vesicles, and these observations were supported by several evidences. It is generally suggested that the plasma membrane $Ca^{2+}$-ATPase was identified as a $Ca^{2+}$ pump based on the similarities between kinetics of the $Ca^{2+}$-ATPase and $Ca^{2+}$ uptake into the membrane vesicles. However, this fact has not been clearly supported. Through a series of experiments it has been concluded that $Ca^{2+}$-ATPase functions as a $Ca^{2+}$ pump on the plasma membrane. It may be stated that calmodulin activation of $Ca^{2+}$ uptake into the plasma membrane vesicles has been reported. However, the activation effect is contradictory. Further, calmodulin activation of $Ca^{2+}$-ATPase activity in the plasma membrane has also been reported. Further, calmodulin-dependent $Ca^{2+}$-ATPase has been purified from corn coleoptiles. Most studies have demonstrated the presence of both calmodulin-dependent and independent $Ca^{2+}$-ATPase in corn plasma membrane, and that distribution of these enzymes is different among tissues.

### Endoplasmic reticulum

In ER ATP-dependent $Ca^{2+}$ uptake has also been demonstrated from roots or *Lepidium sativum*. $Ca^{2+}$ transport was directly coupled to ATP hydrolysis. Further, $Ca^{2+}$ transport of the plasma membrane and ER has fairly high affinity for $Ca^{2+}$ and possibly are concerned with keeping the basal cytoplasmic $Ca^{2+}$.

## VACUOLE

$Ca^{2+}$ transport across vacuolar membrane is clearly established. It is reported that two non-mitochondrial systems involved in ATP-dependent $Ca^{2+}$ accumulation in membrane vesicles from pea exist. Indication for the $H^+/Ca^{2+}$ antitransport system in plants also exists. Further data are available which suggest that a pH gradient generated by the $H^+$-ATPase drives $Ca^{2+}$ accumulation into the tonoplast vesicles via a $H^+/Ca^{2+}$ antiport. The antiporter is thermodynamically competent to account for $Ca^{2+}$ accumulation in cacuoles. The vacuole occupies nearly 90% of the mature plant cell volume and has a high capacity for $Ca^{2+}$ uptake and indicates its significance in intracellular $Ca^{2+}$ sequestration.

### Mitochondrion

Respiratory substrates play a vital role for $Ca^{2+}$ uptake. It requires inorganic phosphate. Mitochondrial $Ca^{2+}$ transport varies in different systems. Based on many observations *in vivo* contribution of mitochondrial maintenance of cytoplasmic $Ca^{2+}$ concentration is doubted.

### Chloroplasts

Light-dependent $Ca^{2+}$ infolux into chloroplast is believed to be electrogenic and is possibly mediated by a uniport-type carrier. The precise significance of this $Ca^{2+}$ transport in the regulation of $Ca^{2+}$*cyt* concentration is doubtful.

### $Ca^{2+}$ channels

It is through activation of $Ca^{2+}$ channels located on the plasma membrane and the membranes of $Ca^{2+}$ sequestering organelles, that cytosolic $Ca^{2+}$ concentration is enhanced. First report on the occurrence of $Ca^{2+}$ channels was provided from algae. In animal systems $InsP_3$ releases $Ca^{2+}$ from intracellular stores by binding to a specific receptor in the ER and may also contribute to the activation of plasma membrane $Ca^{2+}$ channels. An $InsP_3$-binding protein having properties consistent with the receptor has been isolated from various animal systems. In plants $InsP_3$-induced $Ca^{2+}$ release from microsomes and vacuoles has been reported from various systems. Some workers have reported that $InsP_3$ did not release sequestered $Ca^{2+}$ in ER of some systems and possibly acts on the vacuole to release $Ca^{2+}$ from vacuoles. Even though $InsP_3$ in animals and plants have different locations they share a common $Ca^{2+}$ release mechanism. In tobacco vacuole $K^+$ and $C1^-$ channels have also been reported. A voltage-gated $InsP_3$-insensitive outward rectified $Ca^{2+}$ channel with a 12-pS conductance has been found in some systems. In tobacco an inward rectified having a 19-pS conductance has also been reported. This channel was inhibited by cadmium but was insensitive to nickel, nifedipine, etc,

Binding of $Ca^{2+}$-channel blocker to plant membrane fraction has been reported. In several plants ion channels that are activated by stretching of the plasma membrane have been demonstrated. Recently, three stretch-activated channels permeable to $C1^-$, $K^+$ and $Ca^{2+}$ have been found on plasma membrane of *V. faba* guard cells. Such channels are differentiated from spontaneous channels and may play a role in osmoregulation and sensing mechanical pertubation of plant cells.

### Phosphatidylinositol turnover

Through stimulation by some agonists, in animal cells, PI turnover occurs since agonists bind to specific receptors. A specific phospholipase C coupled to a G protein cleaves phosphatidylinositol 4,5-bisphosphate, and hence produces a second messenger $InsP_3$ and diacylglycerol (DG). DG stimulates protein kinase C. The $InsP_3$ mobilizes $Ca^{2+}$ from intracellular stores and causes an increase in cytoplasmic free $Ca^{2+}$ concentration. Hence PI turnover is tightly linked to $Ca^{2+}$ signalling.

Added evidences are being accumulating in plant systems suggesting the importance of PI turnover in signal transduction. Following points must be satisfied if the PI turnover functions during transduction and these are: localization of metabolism in the plasma membrane; transient changes in levels of polyphosphoinositides and their hydrolysis products by stimuli and enhancement of free $Ca^{2+}$ in cytosol. Further activation of protein kinase C by $Ca^{2+}$ and DG must also occur. Activation of other $Ca^{2+}$-dependent enzymes should also occur.

### Inositol phospholipids and inositol phosphates

The presence of both the substances has been estalished in plant tissues, and phosphatidylinositol is the most common phosphoinositide in plants but their quantities in plant cells are low. $InsP_3$ extraction requires great care since its cellular level is very low.

### Phosphatidylinositol and phosphatidylinositol 4-phosphate kinase

Phosphoinositide kinases have been established in plant membranes. The microsomes from wheat phosphorylated PI to form PIP and PIP to $PIP_2$ when supplied with ($V$-$^{32}P$) ATP. In plants DG kinase, PI and PIP kinases have been reported.

### Phospholipase C

In plants both soluble and membrane-bound forms of phospholipase C have been found. In some systems phospholipase C activity was markedly activated at a submicromolar $Ca^{2+}$ range. This indicates the regulation of PI turnover in the plasma membrane by $Ca^{2+}$. Coupled to a signal, phospholipase C hydrolyses $PIP_2$ and produces $InsP_3$ which induces $Ca^{2+}$ mobilization from the intracellular pool raising the $Ca^{2+}$ concentration in the cytoplasm. $Ca^{2+}$ in turn activates $Ca^{2+}$-dependent enzymes, especially phospholipase C.As a result more $InsP_3$ is synthesized and cytoplasmic $Ca^{2+}$ further increases. However, Pi and PIP kinases are suppressed by $Ca^{2+}$ and consequently less of $PIP_2$ is synthesized. Thus, activated phospholipase C is limited in its substrate and hence $InsP_3$ content is reduced. Clearly $Ca^{2+}$ acts as a feedback inhibitor for the regulation of cytoplasmic $Ca^{2+}$ concentration via PI turnover. The involvement of G protein in activation of phospholipase C has been suggested.

### Hydrolysis of inositol 1,4,5-triphosphate

In mammalian systems the $InsP_3$ response decreases by the rapid metabolism of $InsP_3$.The prevalent inositol phosphate isomers found immediately after antagonist stimulation are $Ins(1,4)P_2$, $Ins(1,3,4)P_3$ and $Ins(1,3,4,5)P_4$. The enzymes that rapidly convert $(1,4,5)P_3$ are also present in plants though pathway of inositol phosphate metabolism may be different in plants from that reported in mammals.

### Inositol,4,5-triphosphate-induced $Ca^{2+}$ released from intracellular stores

In plants it has been demonstrated that vacuoles are the target for $InsP_3$. In several animal systems, $InsP_3$ is the link between membrane receptors and the release of $Ca^{2+}$ from intracellular stores especially ER. However, in plants $InsP_3$ receptors has not yet been investigated though in plants and animals it is of different location.

### Physiological responses involving phosphatidylinositol turnover

During cell cycle rapid alteration in PI metabolism is reported. Cytokinin is also reported to induce PI synthesis though $PIP_2$ was not detected in the cells. 2,4-D also stimulated rapid changes in $InsP_3$ and $InsP_2$ in the cells arrested at G phase. GA stimulation of Pi metabolism has been reported in barley aleurone layers. The involvement in stomatal movement has also been demonstrated. Further enhancement of light-induced stomatal opening and induction of stomatal opening under darkness by DG has also been reported. DG is also reported to activate an ATP-dependent voltage independent current of *V. faba* guard cell protoplast.

From several studies available todate it is inferred that an alternate mechanism exists that involves inositol phospholipids as direct membrane effector rather than as a source of $InsP_3$ or DG.

In plant cells, $Ca^{2+}$ appears to be central in signal transduction whereas cAMP messenger system seem to be absent. This is remarkable difference between plant and animal signal transduction pathways. However, some reports on the occurrence of cAMP in plants are available.In algae cAMP appears to work as second messenger. Thus, cAMP may be involved in some growth and developmental processes in plants The presence of other intracellular systems has been suggested e.g. elicitor induced rapid changes in membrane potential and proton transport, $H_2O_2$/superoxide production, lipid peroxidation, regulation of enzymes pertinent to photosynthesis and auxin induced phospholipid breakdown of phospholipase A.

In a very recent communication (Jones, 1994) new signal perception mechanisms in plants have been discussed. It is brought out how plants perceive blue light, salicylic acid (SA), ethylene and auxin. For SA (catalase), ethylene (ETR, kinase cascade), blue light (phospholipase) and auxin (AXRI) different receptors have been identified. Added efforts are being made to identify new kinds of receptors and signal transduction pathways for eukaryotic cells. Some of these mechanisms may be unique to plants and have evolved to sense the changing environmental conditions.

## PLANT HORMONE SENSITIVITY CONCEPT

Although the possible role of changes in plant hormone sensitivity in the physiological responses of plants has been appreciated from the early years of phytohormone studies, yet there is little doubt that such debate as currently surrounds the notion of hormone sensitivity is mainly due to the efforts of A.J. Trewavas. He has advanced three subtly different hypotheses involving the concept.

- **First,** it has been suggested that control over physiological events affected by plant hormones might be exerted by changes in plant's ability to respond to endogenous compounds rather than by changes in their levels within plant tissues (Trewavas, 1981; Trewavas and Jones, 1981; Trewavas, 1983).

- **Second**, Trewavas (1986, 1987) maintained that a crucial criterion for a substance's being a hormone is that it be a sensitive controller of a given process over its physiological concentration range. In this case *sensitivity* is intended as a measure of the influence of realistic changes of hormone concentration on physiological events.
- **Third**, Trewavas (1987) proposed a more radical thesis to the effect that plant tissues are not significantly responsive to changes in endogenous hormone levels except under extreme conditions; these conditions were found to cause increases in sensitivity to hormones. He argued that changes in sensitivity could represent a functional response to adverse conditions.

It is clear that the notion of *sensitivity* must be very clearly defined if such hypothesis as Trewavas are to be tested adequately and if his criteria for hormonality is to be applied.

## Past Definitions of Sensitivity

From the early literature it is found that there is no consensus as to the meaning of sensitivity and remarkably few attempts have been made to treat the concept in a quantitative manner, and it stems from the fact that different workers have used a range of different notions of sensitivity. *Sensitivity* appears a number of times in Went and Thimann's *phytohormones* (1937) but it is not defined anywhere yet it is evident from the contexts in which it is used that the authors are speaking of greater or lesser curvature in the *Avena* coleoptile auxin bioassay at given auxin concentration.

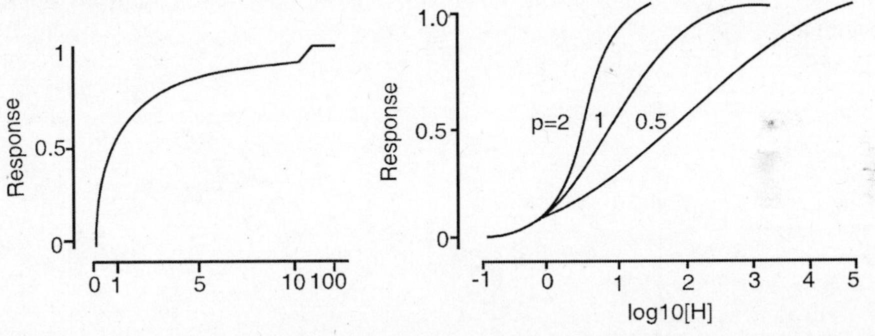


**FIG. 21.7:** (A) Idealised DR curve, (B) The effect of a change in the parameter, p on the D-R curve in semi-log plot.

Thus, sensitivity is defined in terms of the absolute value of the response evoked by a specific hormone concentrations, and the greater the response, the more sensitive the tissue. Such a usage of the word seems entirely legitimate. In 1981, Trewavas and Jones used the notion of sensitivity without explicit definition but seemingly in the sense of *capacity to respond*. It will be already clear that this type of usage is insufficiently precise but these authors also provide a further pointer to defining sensitivity; they observed that if the molecular event underlying a response is the binding of a hormone molecule to a protein receptor, the magnitude of the response will be dependent on the concentration of both entities. Trewavas and Jones suggest that changes in the levels of receptor could underlie changes in sensitivity to their ligands. From their ideas it is evident that alterations in sensitivity will be evident in differences in dose-response (D-R) curves which are obtained under appropriate conditions.

Another, aspect of sensitivity is seen in researches like that of Weyers and Hillman (1979) while studying the closure response of *Commelina* stomata to ABA, these authors were interested to study the minimum amount required to initiate stomatal closure i.e. the threshold value along with the degree of response elicited by various amounts of the substance.

It is quite natural to speak of one tissue being more or less sensitive than another by virtue of possessing a lower or higher threshold value. Further, the threshold values could be compared with the endogenous hormone levels. A threshold value can be obtained only by studying the responses at different hormone concentrations.

Snaith and Mansfield (1982) while working on the theme of stomatal responses to ABA asked the question- *stomatal sensitivity* to abscisic acid: can it be defined?" Instead of defining sensitivity, they presented plots of stomatal aperture against log ABA concentration which indicate that D-R relationship can change substantially depending on the conditions of the assay. The semi- logarithmic D-R curves obtained are all sigmoidal in shape, with the range of response taking place over roughly four decades of ABA concentration, but they differ from each other in various ways: the threshold concentration for response can vary; the absolute value of the aperture at a given concentration of ABA can vary; and finally it can be said that the *shape* of the curve can vary. It is apparent that two of the kinds of sensitivity change have already been discussed (extent of response at a given hormone level and threshold value for minimal response) and, further, and element of complexity (i.e. concentration- dependence of the sensitivity variation) has been observed.

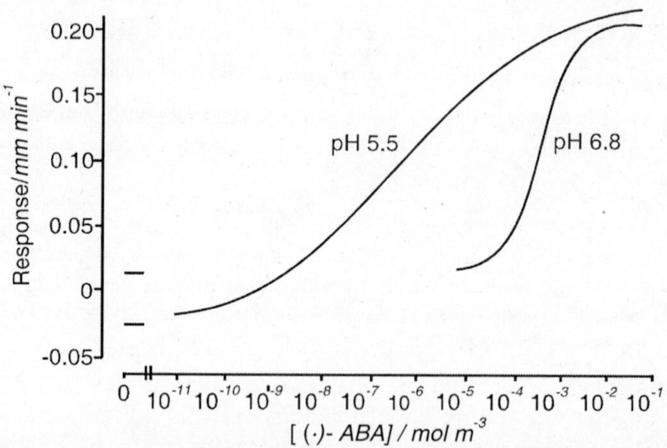


**FIG. 21.8:** (A) D-R curves for ABA-induced stomatal closure of *Commelina communis* at pH 5.5 and pH 6.8. (B) Response coefficient for ABA-induced stomatal closure in *Commelina communis* at pH 5.5.

Trewavas (1987) has presented another possible approach so as to define sensitivity. According to him, *sensitivity to control* or *control strength* has been considered as the individual contribution to overall control of the process or event of interest by anyone process or chemical substance; it is to be measured by making tiny changes in the contributing component and by measuring any resultant effects upon the response interest. Such work would again result for a given substance, in a D-R curve. To obtain a value for "*sensitivity*" in this sense one must presumably measure the slope of the D-R curve and this slope would normally vary with concentration.

Plant hormones, though present in very minute concentrations in the plants, have a very complex mechanism of action and they evoke diverse physiological effects. The large variety of effects explain the multiplicity of sites of action of hormones. Generally, two sites of action are considered. One is at the cell membrane and another is at genetic level. The biological activity of hormone is greatly affected by their molecular structures. Each of the *classic* plant growth substances seem to be involved in one or more steps of gene expression. But, mostly the transcription process is considered solely, because many new data convincingly support the old idea that hormones induce RNA synthesis, as systems and *Northern blot analysis* of mRNA hybridised to cDNA probes. Several criteria favour this view, especially in the cell enlargement response to auxins that site and mode of plant hormone action in gene expression resemble that of animal hormones: Rapid response specificity, and blockade by transcription inhibitors.

The hormones do not act directly but indirectly after forming a hormone-receptor complex with an allosteric protein. Plant hormones form conjugates which represent a part of their metabolism. They may be regulating the level of active hormones. Every hormone has its specific action at cellular and sub-cellular levels owing to its different site of action, specific binding and absorption on cell. Molecular specificity of hormonal responses imply the existence of specific receptors. Much is yet to be revealed in this field. With the advancement of recombinant DNA technology coupled with availability of clones of hormone regulated genes and plant derived systems (*transcriptional factors and hormone-receptor complex*), it should be possible to construct the models elucidating hormone regulation of molecular event in plants.

Here, an analytical approach combined with appropriate statistical methods has been outlined, which can be used to approach the problem of plant hormone sensitivity in a quantitative way. Further, it is believed that, via the notion of sensitivity, the tools presented here can be of assistance in investigating the central questions of hormonal control of plant processes, including the problems of hormone-hormone interactions. Application of this approach to structure-activity studies may also prove fruitful. It may be thought that the use of these techniques can be extended to a wider variety of experimental systems than the two already under study and thereby refine the techniques.

## GROWTH RETARDANTS

These are defined as synthetic organic chemicals that cause a retardation of cell division or cell elongation by specifically blocking some steps in pathways of hormone biosysnthesis without evoking substantial growth distortions. For the past four decades several growth retardants have been discovered. These chemicals have great significance in shortening stem internodes and reduce plant height and hence prevent lodging. Plant growth retardants have significant uses in agriculture and horticulture. These retardants evoke several biochemical and physiological alterations, including suppression of key steps in gibberellin synthetic pathway and also regulation of sterol biosynthesis in plants. We shall briefly describe the action of retardants on gibberellin and sterol biosynthesis as well as physiological effects on overall plant growth and development.

Several growth retardants are available which are active in a broader range of plant species than most retardants known hitherto. These include norbornanodiazetines viz. tetcyclasis (BAS 106); triazoles such as uniconazole (S-3307; XE -1019), paclobutrazol (pp 333, Cultar), triapenthenol (RSW 0411), and BAS 111; pyrimidines e.g. flurprimidol (EL 500) and 4-pyridines such as inabenfide (CGR-811).

There are other 'old time' retardants also and these include AMO-1618, CCC, Phosphon-D, C-111, B-Nine and Maleic hydrazide.

The new compounds of the norbornanodiazetine, triazoles-, pyrimidines-, 4-pyridine types as well as imidazole (e.g. n-decyl-imidazole) growth retardants are described as inhibitors of plant monooxygenase containing cytochrome P-450. As a common structural feature, these compounds have a lone pair of electrons on the $sp^z$-hybridized nitrogen of their heterocyclic ring. In each case this electron pair is located at the periphery of the molecule and thus might easily interact with the protoheme iron of cytochrome P-450 and hence displacing oxygen. These compounds effect oxidative reactions that are component of terpenoid metabolites.

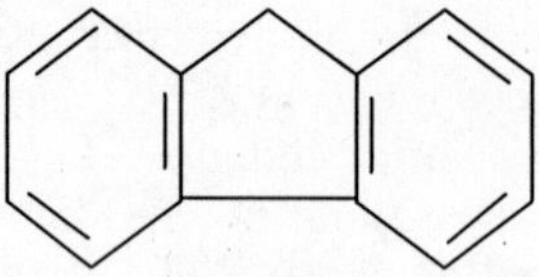

COOH

Fluorene

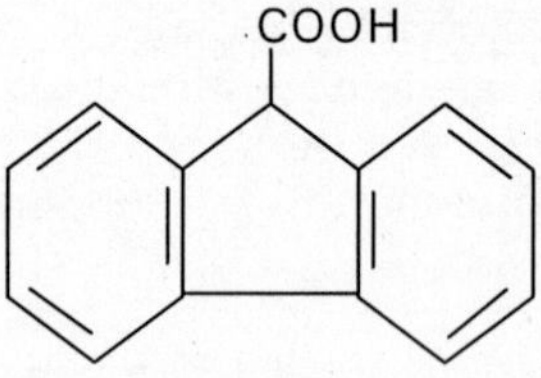

Fluorene-9-carboxylic acid

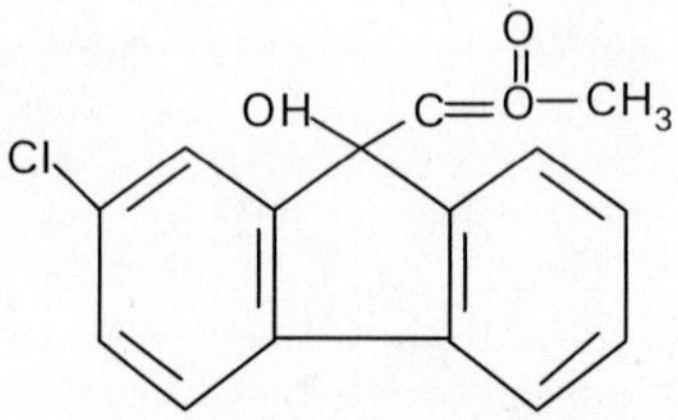

OH C=O–$CH_3$ Cl

Chlorflurenol methyl

**FIG. 21.9:** Various types of growth retardants.

## Physiological Effects

### *Enzymatic reactions*

The effect of ancymidol, tetcyclasis and some triazole-type growth retrdants on several cytochrome P-450 containing enzymes has been described, recently. The reactions from *ent*-kaurene to *ent*-kaurenoic acid were most drastically affected by these compounds. In most instances other enzymes (cinnamate 4-monooxygenase and oxygenases related to phenylpropane metabolism) needed very high doses for inhibition. Some of the retardants appear to have some effect on the oxidative detoxification of xenobiotics; when the metabolite of a herbicidal agent was blocked, it increased. Recently, it has been demonstrated that the oxidation of the herbicide chlorotoburon was inhibited in cell suspensions of cotton by relatively low concentrations of tetcyclasis. In wheat seedlings the bentazone inactivation was delayed in the presence of this compound. Malik and Thind (1994) have discussed the effect of some triazoles on the activity of some enzymes.

### *Control of lodging*

Compounds such as triazoles, paclobutrazol, uniconazole, triapenthenol, BAS 110, BAS 111 and 4-pyridine inabenfide have shown excellent results in some cereals and oil seed plants. Most of the new and old type of retardants cause shortening of internodes and reduction in height in treated plants; they suppress mitotic activity in these plants (see Thind and Malik, 1994; Kaufman and Song, 1988).

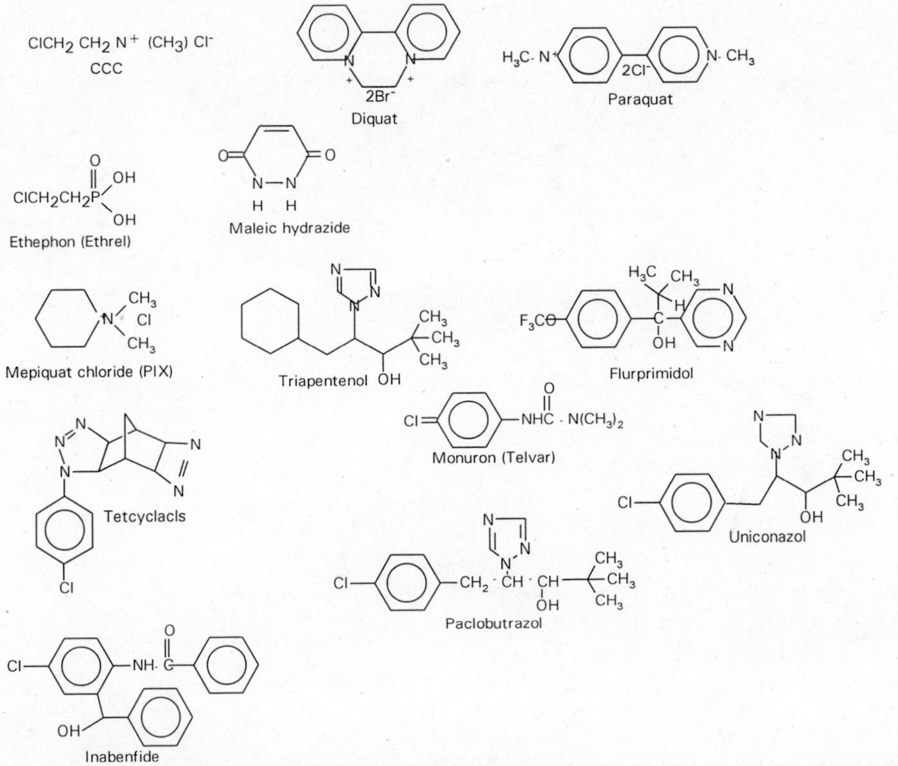

**FIG. 21.10:** Various types of growth retardants.

Even though retardants have divergent chemical structures, they have similar effects on plants. The most common response is inhibition of stem elongation and also suppression of flower-initiation. Most workers suggest that many of the retardants are suppressors of gibberellin biosynthetic pathway and hence are useful compounds to investigate the role (s) of GAs in metabolism, and plant growth and development. Paclobutrazol is successfully used to lessen the height of several fruit trees by inhibiting cell elongation and induction of dark green colour in the leaves (see Davis *et al.* 1986; Malik and Thind, 1994).S-3307 retards the growth of leaves in rice plants. Tetcyclasis inhibits cell division in cell suspension cultures of various plant species. However, application of GA reverses the inhibitory effect of paclobutrazol.

## Reduction in vegetative growth

Paclobutrazol, uniconazole, triapenthenol and flurprimidol reduce the vegetative growth of various fruit trees and orchard trees. The requirement for prunning decreases and the flowering pattern is altered favourably and fruiting is profuse. Their use is especially beneficial in fruit trees where dwarfing rootstocks or compact scion cultivars are not available. Ancymidol inhibitis stem elongtation and GA-induced growth in cereal internodes.

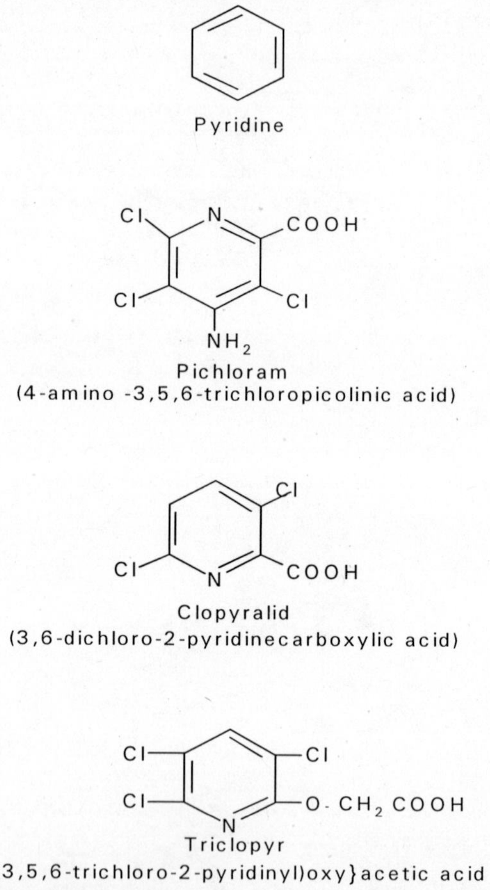

Pyridine

Cl N COOH Cl Cl $NH_2$

Pichloram
(4-amino -3,5,6-trichloropicolinic acid)

Cl Cl N COOH

Clopyralid
(3,6-dichloro-2-pyridinecarboxylic acid)

Cl Cl Cl O. $CH_2$ COOH N

Triclopyr
{(3,5,6-trichloro-2-pyridinyl)oxy}acetic acid

**FIG. 21.11:** Various types of growth retardants.

It enhances stem diameter, increases chlorophyll formation in leaves and stimulates root initiation in shoot cuttings. This retardant causes a substantial reduction in the linear growth of oat stem segments that are stimulated to grow with gibberellin and had very little effect on linear growth of stem segments treated with 0.1 M sucrose. Ancymidol decreased $GA_3$-promoted extension growth and caused enormous expansion of pulvinus tissues that surround the internode.

Kaufman and Song (1988) have demonstrated that the lateral expansion of the leaf-sheath pulvinus tissue is caused by radial enlargement of cells of the inner andouter epidermis. The swelling of internodal and pulvinus tissues induced by ancymidol is morphologically similar to the effects of ethylene on stem growth in other plants. In fact these authors demonstrated doubling of ethylene production in segments treated with sucrose ancymidol compared with segments treated with sucrose alone. When GA was added along with ancymidol and sucrose, the ethylene content increased three-fold. From these data Kaufman and Song (1988) inferred that swelling in oat stem segments tissue enhanced as a consequence of ethylene production induced by ancymidol in the stem tissues. The general conclusion is that $GA_3$ and ancymidol possibly operate at different target sites and this parallels similar instances in which hormones like cytokinins, IAA and ABA were also interacting in a non-competitive fashion with $GA_3$. Further, ancymidol is also shown to reduce $GA_3$ promoted cell elongation by partially increasing ethylene levels, which in turn promotes lateral cell expansion.

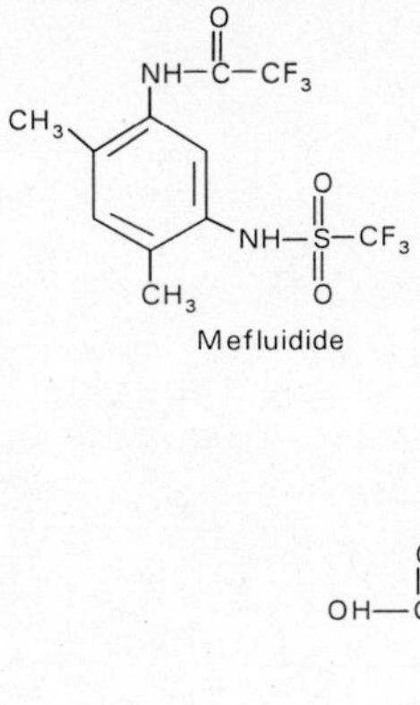

Mefluidide

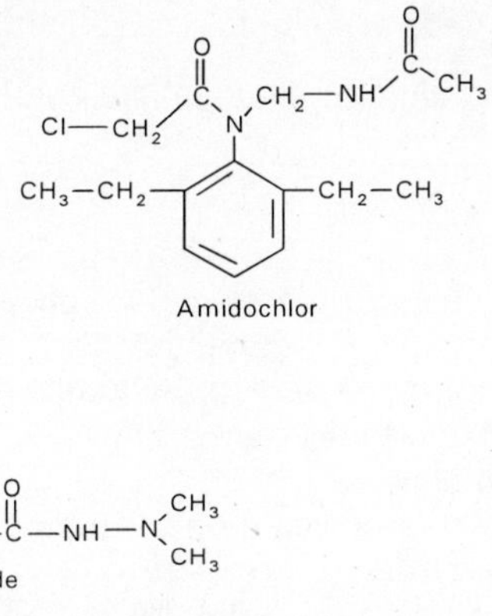

Amidochlor

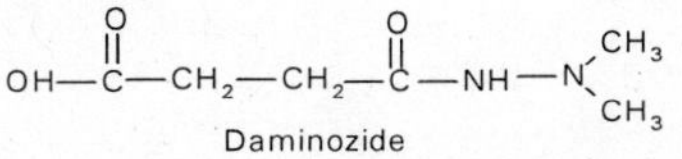

Daminozide

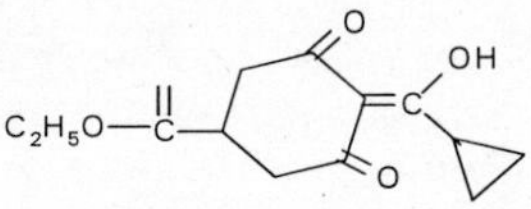

Cimectacarb

**FIG. 21.12:** Various types of growth retardants.

## Growth reduction in ornamental species

Trees which grow under telecommunication or powerlines need periodic trimming. The application of growth retardants e.g. paclobutrazol, uniconazole etc. could reduce their height and save tremendous expenses. Active ingredients have also been injected in the plants. Several of the ornamental trees have been shortened in regard to their height.

### Regulation of turf growth

In several of the developed and advanced countries where manual labour is very expensive mowing costs are extremely high while maintaining an aesthetically pleasinglawn, thus suitable turf growth regulators are sprayed. Here again paclobutrazol and flurprimidol have shown excellent results for shortening internodal length. When combined with some herbicides it could prove increasingly beneficial.

### Chlorophyll biosynthesis and photosynthesis

Leaves of plants treated with retardants are deep-green in colour due to enhanced synthesis of chlorophyll or possibly due to the shrinking of leaf area. Not much work has been undertaken to investigate their effects on enzymes concerned with chlorophyll biosynthesis. There are mixed reports on the effects of retardants on photosynthesis though most reports point towards their reduction role. Nevertheless most retardants delay senescence and hence photosynthetic activity of a given leaf continues for a longer period. Further, leaves are also retained on the treated plant for a longer duration. In oilseed rapes, paclobutrazol and CCC alter the canopy structure and hence influence light penetration and absorption. This aspect deserves detailed study under field conditions.

### Carbohydrate content, metabolism and translocation

In general paclobutrazol alters photosynthate partitioning from leaves to roots and elaborated sinks.The content of starch enhances possibly due to decreased hydrolysis.

Not much information exists on respiration though total root respiration of treated plants was unaffected and most part was through cyanide resistant pathway. Shoots from treated plants exhibited reduced respiration. However, roots treated with PCB showed enhanced activity of malate dehydrogenase though the precise significance of these observations is still indistinct.

In some treated plants have increased levels of minerals uptake though more data are needed for clear picture to emerge. In any case PCB may be altering the uptake of some minerals whereas others may be unaltered. Retardants delay iron chlorosis in some legumes.

### Water relations

Retardants treated plants use less of water than untreated plants possibly due to the reduced leaf area of treated plants or altered conductance to water vapour flux. Treated plants are shown to withstand drought in a better way than controls. Compared with controls PCB treated plants evolved less of ethylene suggesting that treated plants were less stressed.

### Stress tolerance

Treated plants withstand different stresses and are more tolerant to stresses (Davis *et al.* 1986; Malik and Thind, 1994). The retardant treated plants withstand low and high temperature stresses. The exact mechanism is still unclear.

## SENESCENCE

Retardants in general delay leaf senescence and prolong the metabolic activity of a given leaf. In fact PCB delayed the onset of various hydrolases concerned with senescence e.g. proteases, RNases, etc. also is shown to reduce lipid peroxidation in both light and dark and hence senescence is delayed (see Malik and Thind, 1994).

### Improvement of canopy structure and yield in oilseeds

Paclobutrazol, BAS 111, triapenthenol treated plants of *Brassica* have reduced plant height and number of primary and secondary branches are affected leading to altered canopy structure that permits better penetration of light to lower branches and leaves borne by them. Reduction in height causes resistance of the plants to several diseases and lessens physiological constraints against yields. Reduced lodging is also correlated with fewer fungal infections (see Malik *et al*,. 1993).

### Improving the quality of seedlings used for transplanting

Tetcyclasis when applied as a seed soak improves the quality of rice seedlings used for mechanical transplanting. The seedlings become stout and retain optimal size suitable for transplanting for an optimal time; are resistant to transplantation stress and have superior performance under field conditions. In maize also, similar results have been obtained where mechanical transplanting is done.

### Resistance against climatic stress

As stated earlier treated plants have improved resistance to low temperatures especially in oats, rice etc. Tetcyclasis is especially effective in this regard. Oilseed rape treated with PCB or BAS 111 performed better under drought conditions (see Malik and Thind, 1994).

### Resistance against fungal infections

In oilseed rape, plants treated with some of the retardants like triapenthenol, the rate of fungal infections was considerably reduced. In water melon infection to *Fusarium* wilt also decreased. Some of the effects could be attributed to antifungal activity of some of the retardants especially those belonging to triazole group.

## GROWTH RETARDANTS AND BIOSYNTHETIC PATHWAYS

We shall describe four different pathways:

### Sterol formation

Pyrimidines, triazoles and imidazoles are structurally related to certain growth retardants and are used as fungicides and often exhibit growth regulatory side activity.

Sterol biosynthesis in plants is highly complex and is still incompletely understood. The biosynthesis involves cyclization and oxidation steps which also occur during GA biosynthesis. Sterols regulate membrane permeability and maintain their integrity. They are also concerned with the uptake of some compounds and also protein synthesis. The available data indicate that growth retardants have the potentiality to inhibit key steps in sterol biosynthetic pathways. The N-containing heterocyclic growth retardats e.g. triazole compounds, ancymidol and NDA inhibit sterol C-14 oxidative demethylation during ergosterol biosynthesis. These compounds like triazoles inhibit the oxidative 14-demethylation in the course of fungal ergosterol synthesis. Several trizoles fungicides and PCB also reduce the formation of 14-demthylated sterols in higher plants. Tetcyclsis at high doses reduces cell division in the cell suspension cultures of maize. There was also reduction in sterol qualitatively and quantitatively.When cholesterol or other phytosterols were added, cell division was restored. It is inferred that higher doses of growth retardants interfere with the formation of sterols which leads to reduced cell division especially in meristematic tissues. It also affects GAbiosynthesis

## GIBBERELLIC ACID BIOSYNTHESIS

### Inhibitors in mevalonic acid to ent-kaurene

Early steps in the GA biosynthetic scheme are from mevalonic acid (MVA) to *ent*-kaurene. The possible sites for regulation in this scheme are enzymes which control these steps. Biosynthesis of *ent*-kaurene from MVA is demonstrated in cell-free systems of several plant species and also in *Gibberella fujikuroi* (Malik and Thind,1994). *ent*-kaurene is an intermediate in GA biosynthesis. Two stages in the biosynthesis of *ent*-kaurene from geranylgeranyl pyrophosphate (GGPP) are shown to proceed via a two step cyclization that is catalyzed by kaurene synthetase:

$$\text{GGPP} \xrightarrow[\text{Synthetase A}]{\textit{ent}\text{-kaurene}} \text{CPP} \xrightarrow[\text{Synthetase B}]{\textit{ent}\text{-kaurene}} \textit{ent}\text{-kaurene}$$

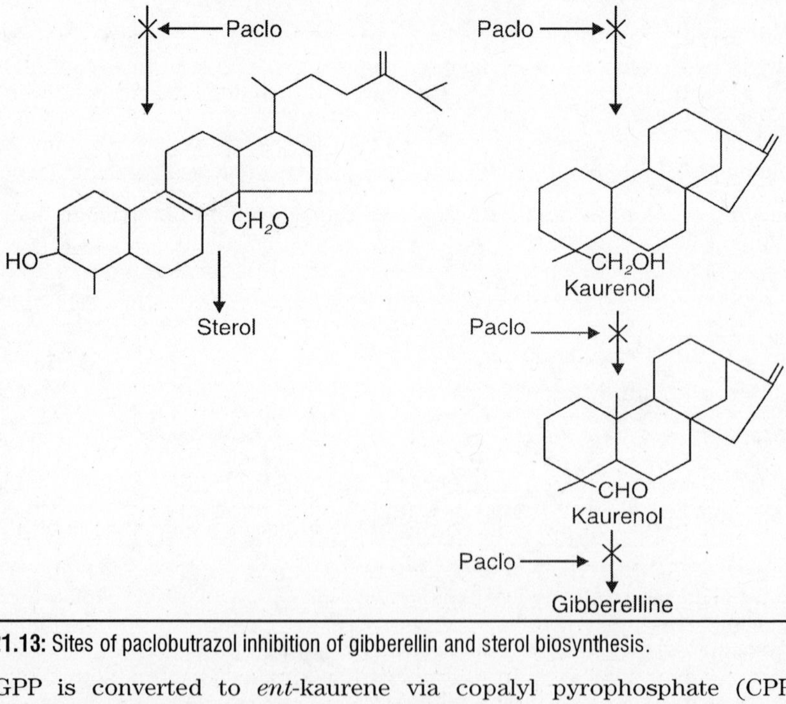

**FIG. 21.13:** Sites of paclobutrazol inhibition of gibberellin and sterol biosynthesis.

GGPP is converted to *ent*-kaurene via copalyl pyrophosphate (CPP) and kaurene synthetase is regarded as a primary site for the regulation of GA biosynthesis. In higher plants metabolite GGPP is a branch point for several biosynthetic pathways, GA pathway is a minor one. A and B activity operates in *G.*

*fujikori* though in higher plants cell cultures only B activity is present. The reactions catalyzed by kaurene synthetase are sites of inhibition by growth retardants and structurally related compounds. Phosphon-D, phosphon-S strongly inhibit the A activity of kaurene synthetase whereas Amchem, SKF-525A inhibit B activity while Deoxy-cholate inhibits both A and B activities.

## Inhibitors in kaurene to kaurenoic acid

The sequence of steps from MVA to GAs is still evasive in intact plants though it is suggested that these pathways do occur in higher plants in nature. Gibberellin biosynthesis has been investigated in cell free systems of higher plants and the sites of action of growth retardants investigated. The three steps are given below:

*ent*-kaurene ⟶ *ent*-kaurenol ⟶ *ent*-kaurenal ⟶ *ent*-kaurenoic acid

These steps are all oxidations catalysed cytochrome P-450 mixed function oxygenase. Some of the retardants interact with the P-450 component and these interactions possibly involve the binding of one of the pyrimidines nitrogen atoms to the heme iron of P-450 and hence preventing the binding of oxygen to it.

Compounds like PCB, tetcyclasis, ancymidol suppress GA action in many plants and are hence used as growth retardants. These compounds are reported to inhibit the three oxidative steps from *ent*-kaurene to *ent*-kaurenoic acid. PCB is shown to bind to both ent-kaurene and cytochrome p-450, and hence preventing p-450 from releasing oxygen from *ent*-kaurene. Thus, oxidation to *ent*-kaurenol is blocked. Possibly, ancymidol and tetcyclasis have the same enzymatic sites of action in these oxidation steps as PCB.

Reduction in stalk or shoot length is a most striking visible parameter for testing the activity of a retardant. This in fact led to the idea that these retardants interfere with the GA status of the plant since GAs are known to cause stem elongation. Further retardation of shoot caused by retardants can be reversed with the application of biologically active GA. The general conclusion is that norbornanodiazetines, trizoles and 4-pyridines inhibit GA biosynthesis at the same sites as pyrimidine-and imidazole-type retardants. Retardants reduce the GA content and the shoot length.

## Metabolism of ABA

Tetcyclasis application inhibits the tranformation of ABA into phaseic acid and therefore, when tetcyclasis is applied to rice plants, there is accumulation of ABA.Similarly when trizoles like triadimefon and triapthenol are applied to intact plants there is accumulation of ABA (Malik and Thind, 1994). ABA thus, reduces water use in barley and oilseed rape independent of its effect on leaf area. Tetcymidol and several triazoles and imidazoles however, had little or no activity as far as transpiration is concerned.With tetcyclasis, ABA level increased in intact plants. In general increase in ABA was likely to improve the resistance of plants to various stresses.

## Interference with other phytohormones

Tetcyclasis and trizoles like uniconazole influence several of the phytohormones in rice and sunflower seedlings. In rice uniconazole raised the evolution of ethylene whereas in cell suspension cultures tetcyclasis at high doses inhibited ethylene release. IAA content was not appreciably altered with any of the growth retardant though cytokinin levels were enhanced with retardants. The latter situation may account for the delay in senescence caused by retardants. We are still far from clear to understand how retardants affect ethylene and cytokinins levels.

**TABLE 21.1:** Effects of several of PGRs on different yield parameters are tabulated

| Crop | Chemical | Method of Application | Dose/Remarks |
|---|---|---|---|
| **Callus formation in tissue culture** | | | |
| *Crotalaria medicagenia* | 2,4-D; kn; NAA | MS medium +2,4-D, kn subcultured in MS +NAA | Root formation |
| | | High kn, BA, Adenine with or without NAA | Embryoids/ shoot |
| *Cyamopsis tetragonoloba* | 2,4-D, NAA; BA | Treated cotyledons yielded protoplast which divided and formed cell wall | |
| | 2,4-D; NAA + kn | Modified kao and Michayluk | 6.2 pH significant |
| | PVP | Prevented browning and necrosis of protoplast | |
| *Cuminum cyminum* | 2,4-D; NAA; kn; BA | Gamborg's B5 medium | Hypocotyl explants grew rapidly; leaf explants were slow |
| Cauliflower | kn; BA; NAA | Callus formed in 7-10 days | 5 mg BA $1^{-1}$ was best |
| *Echinochloa colonum* | 2,4-D; CM | Segments of inflorescence subcultured developed into plantlets in MS medium | 5 mg $1^{-1}$ + CM (10%) 0.2 mg $1^{-1}$ |
| *Arachis hypogaea* | kn + IAA; BA and IAA | Callus from embryos on MS medium, plantlets developed | |
| *Vigna radiata/aconitifolia* | Vit.; 2,4-D; kn; amino acids | MS medium; callus formed green patches in zeatin; plantlets developed in MS medium devoid of PGRs | |
| *Zea mays* | 2,4-D | Glumes at uninucleate stage of pollen development showed high callus formation on subculturing, plantlets developed | |
| *Cucumis melo* | IAA + Kn | MS medium; hypocotyl segments; shoot buds formed when ABA, $GA_3$ + BA added to the medium | |
| *Pinus strobus* | BA + NAA | 2-4 week old shoots of 15-18 years old plant. Callus sub-cult on MS medium for 15 months | |
| *Zingiber officinale* | PGRs | stem cutting from rhizome, shoot bud primordia on MS medium developed into callus | |
| *Carum coplicum* | 2,4-D; IAA; NAA; kn | IAA +MS medium, callus formed; NAA +kn organogenesis | |
| *Triticum hordeum* | PGRs | PGRs essential for regeneration from older calluses | |
| *Solanum tuberosum* | NAA; kn | Infected plants subcultured on MS medium + NAA + kn produced virus-free callus | |
| | kn; NAA; BAP; | Potato leaf callus in kn + NAA + kn | |

| Crop | Chemical | Method of Application | Dose/Remarks |
|---|---|---|---|
| | Zeatin | | |
| | BA | PM medium | |
| *Chrysanthemum morifolium* | | Shoot tips in March/April proliferated without callus | |
| *Azadirachta indica* | 2,4-D; IAA | Cotyledonary leaves and stem segments cultured on MS medium formed callus | |
| Sesame | NAA; BA | Protoplast cultured on MS medium + NAA + kn produced multicellular colonies | |
| *Albizia lebbek* *Cassia fistula* *C. siame* | IAA; BA; NAA; BA | Stem, petiole explants formed callus with PGRs | |
| *Piper nigrum* | 2,4-D; kn | Suspension culture of callus cells in MS medium; piperine formed | |
| Sugar cane | 2,4-D | Bud apices (healthy/diseased) on MS medium | |
| *Dianthus caryophyllus* | kn; NAA,2,4-D; BA | MS medium for callus formation | |
| *Dioscorea floribunda* | 2,4-D; NAA; BA; kn | MS medium; White's medium Tuber nodes - explants, plantlets | |
| *Cinnamomum tamala* | kn; NAA | Apical meristem, callus differentiation | |
| **Embryoid formation in Tissue culture** | | | |
| *Momordica charantia* | NAA; Adenine | 23-month colt; with NAA + Ad embryoids; | NAA 1 mg $1^{-1}$ Ad 33.75 mg $1^{-1}$ |
| *Cucumis melo* | BA +IBA | 26-month cult; BA +IBA plantlets | |
| *Phoenix dactylifera* | 2,4-D; BA; thiamine | Explants cultured on MS medium, somatic embryoids leading to viable plants | |
| *Brassica juncea* | 2,4-D; NAA; BA, CM | Mesophyll cells protoplast; embryoids, shoot buds | |
| **Rooting** | | | |
| *Mentha citrus* | IAA; IBA | Rooting enhanced | |
| *Pelargonium sp.* | IAA; IBA | IAA produced better rooting from terminal cuttings | |
| *Averihoa carambola* | IBA; IAA | Rooting by layering technique | 10,000 ppm |
| *Gardenia florida* | IBA; NAA | 100% rooting | 50 ppm |
| *Vigna sp.* | IBA; GA, acryflavin | Hypocotyl cuttings rooted with IBA; GA, acryflavin inhibitory | |
| *Jasminum sp* | IBA | 100% rooting | 4000 ppm |
| *Pyrus aurifolia* | IBA | 3-yr old cutting (61.2%),25-yr old cuttings (22.5%) | |
| *Phaseolus mungo* | IBA; NAA | Hypocotyl rooting; IAA + rifampcin synergistic | |
| *Peach* | IBA | Hard wood gave 100% rooting | 2000-2500 ppm |
| Apple | IBA + Cinnamic acid | Synergistic | 2500 ppm |

| Crop | Chemical | Method of Application | Dose/Remarks |
|---|---|---|---|
| Tea | UPASI-RF 7A | 13.2% rooting | |
| *Phaeseolus mungo* | Ch; Act-D, FU; IAA, DNP; sodium azide, ammonium sulphate | Low conc of inhibitor promoted rooting, Inhibition of oxidative phosphorylation stimulated rooting | |
| *Hibiscus rosa-sinensis* | Nicotinic acid, ascorbic acid | 90% rooting | 100 ppm |
| *Mangifera indica* | IBA | rooting in air layers | 50 ppm |
| | IBA | | $3X10^{-5}$ M |
| | IBA | | 100 mg $1^{-1}$ |
| Guava | ethephon IBA | 100% rooting | 50-100 ppm |
| *Piper nigrum* | IBA | Stem cuttings after dipping for 45 hour | |
| *Callistemon laceolatus* | IBA | Best in Feb. | 4000 ppm |
| *Poinciana regia* | IBA | IBA applied in lanolin paste in girdled mother shoots | |
| *Camphore* | Phenolics Gallic acid | 75% rooting | 1000 ppm |
| Peach and Almond hybrids | IBA | Washing of cuttings for 72 h followed by IBA, gave 20.3% rooting | 2000 ppm |
| *Artocarpus heterophyllous* | NAA +IBA | Air layering followed by PGR; juvenile trees gave best results | 2500 + 5000 ppm |
| Coffee | | Maximum in June; next best in August | |
| *Cassia fistula* | IBA/NAA | NAA gave 60% rooting | 2500 ppm |
| *Bougainvillea* | NAA/IBA | Stem cuttings gave 100% rooting | |
| | | | |
| Tea | IBA + NAA; NAA + Thiram | Dipping improved rooting | 6000 +40 + 80 ppm |
| Cherry | IBA | Root stock the best | 2500 ppm |
| *Impatiens balsamina* | Phenols-IAA $\alpha$-naphthol; Salicylic acid | | 1 mg $1^{-1}$ |
| *Ipomoea fistulosa* | AMP + IBA | Rooting profuse | |
| Lemon | NAA | | 4000 PPM |
| *Bamboosa arundinacea* | IBA or NAA | Best results in March | 100 ppm |
| *Citrus lemon* | Ferulic acid, NAA | | 200 & 5000 PPM |
| *Cucumis sativus* | Folic acid | Inhibited rooting | |
| *Pisum sativum* | 4-C1-IAA 4,6-dichloroindole acetic acid | Induced tooting | |
| *Rauvolfia serpentina* | IAA, IBA, NAA, 2,4-D, TIBA, MH,2,4,5-T | Promotory Inhibirory | |
| Casurina | IBA | Cladodes rooted | 4000 ppm |
| *Prunus* | IBA | Best rooting | 100,200 ppm |
| *Achras zapota* | IBA | Best rooting in air layers | 10,000 ppm |
| *Dombeya malalensis* | IAA | 90% rooting | 4000 ppm |

| Crop | Chemical | Method of Application | Dose/Remarks |
|---|---|---|---|
| Rose apple | CCC + Fth + IBA | 90% rooting; PGRs synergistic action | 5000 (CCC) + 5000 (IBA) + 200 (Eth) ppm |
| *Melia sp*. | IBA + PCB | Synergistic in Feb. | 2500 ppm |
| *Bignonia pyrostegia; Beauomontia sp.* | NAA | Improved rooting | |
| **Disease Related Nitrogen fixation** | | | |
| | IAA | High activity indicated resistance | |
| | | Parthenin inhibited the growth of *Rhizobium phaseoli* and *Azatobacter* vinelandii | |
| Tomato | GA, NAA | Leaf and infected plants showed high yield 15.3 fruits per plant against 7.9 in control | 200 ppm |
| Rice | $GA_3$ | RTV infected plants recovered with $GA_3$ | 2 $\mu$g ml$^{-1}$ |
| | Cytokinins | Application caused green island to *Helminthosporium oryzae* infected leaves | |
| | Gibberellins | Virus infection (tingo) enhanced $GA_3$ | |
| Cuscuta | Cytokinins | Haustoria had maximum cytokinins and linked with high RNase activity | |
| Cowpea | IAA, 2- thiourea, $GA_3$ | Cowpea plants showed delayed symptomology following virus infection; when sprayed with IAA, $GA_3$ or thiourea | |
| *Cucumis sativa* | IAA | Watery galls formed by *Syncytrium lagenaria* due to high IAA. Auxins and phenolics involved in hypertrophy and hyperplasia | |
| Coffee | Oxalic acid | *Mycena citricolor* causing American leaf spot, triggered by oxalic acid, stimulated IAA oxidase and macerating enzymes | |
| *Azospirillum* | IAA | Involvement of carbofuran in high IAA accumulation suggested | |
| Soybean | IAA | Plants infected with *Bacillus firmus* produced more IAA and high P content | |
| Rice | IAA, IBA, GA, Kn, 2,3-4-T | With PGRs polygalactouronates transaminase and pectin methyl transaminase checked; aureofungin decreased hydrolases phenols inhibited pathogen growth | 10,000 ppm |
| Tomato | IAA, CYT | PGRs increased in roots following inoculation with root knot nematode | |

### *Biochemical Components*

| Crop | Chemicals | Enzyme | Remarks |
|---|---|---|---|
| *Phaseolus mungo* | IAA, MH | Po | Two new isoenzymes observed in hypocotyl cuttings with IAA; but not with MH |
| Cucumber | BA | IAAO; Po | Ba inhibited activities of both oxidases |
| *Vigna sinensis* | Malathion (1 ppm) | Phosphatases; RNase; pyro-phosphatase | Low doses stimulated enzyme activities and seedling growth |
| Guava | $Ca^{2+}$, $K^{+}$, Alar (1000 ppm) | PME, cellulase | Activity of two hydrolases retarded during storage with Alar |
| Rice | Kn, ABA | Diamine oxidase; polyamine oxidase | Enzyme activities increased with loss of seed viability. DNA, RNA content decreased. kn enhanced polyamine content, ABA reduced it |
| | Morphactin | IAAO; amylase, catalase | malformation is attributed to decreased levels of auxins |
| Mango | Mangiferin | Po; catalase; IAAO; amylase chlorophyll; nucleic acids | Chemical inhibited enzymes; chlorophyll increased |
| Banana | IAA, IBA; GA, Kn | $\alpha$-amylase, starch phosphorylase acid phosphatase, catalase, Po | GA or Kn retarded enzymes activities whereas IAA, IBA promoted |
| *Cucumis sativus* | Polyamines (agmatine, put.) | Arginine decarboxylase (ADC) | Agm inhibited ADC by 50% and Put by 65% |
| Apple | Phenols | IAAO; Po | Phenols decreased during onset of ripening; IAAO, Po showed two peaks |
| Pigeonpea | NAA; SA; CA | NRA | SA, CA decreased NRA whereas NAA enhanced leaf, podwall, seed NRA |
| Cowpea | Kn, CCC, colchicine | Po | Chemicals increased Po in nodules |
| *Vigna mungo* | $KNO_3$; $NH_4SO_4$; glutamine, Kn ($10^{-4}$ M) | polyamine | 2.5 mM $NH_4SO_4$ enhanced polyamines and Kn did same thing in germinated seeds |
| *Cucurbita pepo* | Kn | Enzymes of lipid and carbohydrate metabolism | ICL; Inv enhanced in cotyledons by Kn |
| Cotton | GA | Amylase, invert, AP, aldolase G-6-PDH, MDH, hexokinase | Activities of enzymes increased in callus with GA |
| *Phaseolus mungo* | IAA, GA | IAAO; Po; NRA | Enzymes increased in nodules by IAA/GA |
| *P.vulgaris* | IAA | AP;ß-gluco; amylases | All enzymes enhanced with IAA |
| *Helianthus annuus* | BA | Chl content; NRA | 2-3 application stimulated NRA |
| *Commelina communis* | ABA; phytochrome | Mg-ATPase | ABA inhibited APTase |
| *Mungbean* | PCB | IAAO; Poly O; Po; AP; protease; amylase | PCB reduced amylase; enhanced IAAO; Po; AP |
| *Cicer arietinum* | IAA | RNA polymerase | IAA stimulated RNA polymerase in shoots |

| Crop | Chemicals | Enzyme | Remarks |
|---|---|---|---|
| Wheat | $GA_3$ | Po | Radioprotective effect |
| *Cowpea* | BA; $CaC1_2$ | lipoxygenase | delayed senescence |

### *Effect of various PGRs on physiological functions*

| Crop | Study | Results |
|---|---|---|
| *Vigna unguiculata* | IAA content | IAA content increased up to 50 days old nodules and then declined |
| *Saccharum* sp (Co 419; Co 6806) | ABA (1x10.5M) | ABA induced stomatal resistance and maintained better tissue hydration through minimal water loss. |
| *Cicer arietinum; Glycine max* | Endogenis levels of cytokinins | cytokinins levels imparted resistance to salinity |
| *Saccharum officinarum* | Cytokinins induction of plantlet differentiation | cytokinin like compounds isolated from differentiating calli of var. CoS 8118 could induce plantlet differentiation in calcitrant var. Co 1158 |
| *Mangifera indica* cv. *alphonso* | Estimation of endogenous growth substances (HPLC) | IAA attained two peaks (15/30 days preceding period of rapid fruit growth); ABA/ethylene levels were high initially and ethylene was high during fruit maturity |
| *Quercus serrata* | Effect of $GA_3$ on N-metabolium and leaf biomass | $GA_3$ (5 ppm) enhanced N-metabolites and biomass. A positive correlation between free amino acids and protein was noted |
| *Vigna unguiculata* | BA effect on nodules, nitrogenase activity leghemoglobin | BA stimulated nitrogenase activity, leghemoblobin and nodules number increased |
| *Cucurbita maxima* | Cytokinin activity in male and female flowers | Cytokinin activity (zeatin, ribosylzeatin, cytokinin glucoside) higher in female flowers. This may be associated with determination. |
| *Arachis hypogaea* | Tonoplast ATPase: interaction with phytohormones and phenolics | except phenylacetic acid, all other auxin failed to stimulate ATPase, $GA_3$, ABA and Kn were inhibitory; cinnamic acid stimulated; chlorogenic acid completely, hydroxybenzoic acid hydroxyquinaline partly inhibitory |
| *Vigna unguiculata* | Leaf growth, nodulation and yield parameters influenced by PGRs | IAA, GA and Kn applied foliarly stimulated nodulation, delayed their senescence; seed yield also enhanced |
| *Oryza sativa* "Ratna & CSR-4 | Endogenous ABA levels in salt-sensitive and salt-resistant was by NaCl salinity | In salt-sensitive free ABA enhanced in germinating seeds component salt-resistant, the increase in ABA correlated with precursor. High ABA could be associated with mitigation of salt-stress. Low levels of *ent*-kaurene indicated low levels of GA under stress. |
| *Oryza sativa* | Endogenous levels of GA and ABA in salt-sensitive rice seedlings | ABA increased in germinating rice seed subjected to salt-stress. |
| *Rosa hybrida* | CCC, SADH, ethrel preharvest spray affect life of cut flowers | CCC reduced soluble sugars in corolla whereas total free amino acids decreased |
| *Glycine max* cv. Bragg | PGR mitigating adverse effects of salinity | Soybean seeds pre soaked in $GA_3$, IAA and then sown in parts with variable salinity showed: salinity |

| Crop | Study | Results |
|---|---|---|
| | | increased $Na^+$ contents, decreased $K^+$ and $Ca^{2+}$. PGRs mitigated the adverse affects of hypersalinity-IAA (200ppm) was most effective. |
| *Cicer arietinum* | seed phenalies and *in situ* germination | Total phenolic contents in seed coat, cotyledon of six genotypes, of connection with germination. In cotyledon *p*-hydrobenzoic acid was predominant while seed coat had galic acid. |
| *Dianthus chinensis* | Effect of ABA on senescene in cut leafy flowering shoot of *Dianthus* under water stress | PEG was used to induce stress; water content decreased in petas, ABA decreased water lon. $\alpha$-amino nitrogen and proline increased. |
| *Cajanus cajan* | Improved translocation of 1-$^{14}$C-arginine towards with monophenols and arnostemon | Translocation of 1-$^{14}$C-arginine was stimulated with the two PGR's from leaf to peduncle, pod wall and the seed. Possibly membranes were made none permeable. |
| *Cartanun tinctorisus* | Stomatal movement as influence by PGRs and electrolytes | Methyl and butyl stress of morphactin reduced stomata size and opening; ABA closed stomata completely. Response varied in draught-semitance and drought tolerant species. |
| *Vigna radiata* | IAA effect on various photosynthetic characterstics under salt stress | Folian spray with IAA, GA & kinetin on seedlings subjected to NaCl stress metabolic alteration caused by salt stress. Kinetin was most effective in combating metabolic disorders caused by salinity. IAA also released stress caused by NaCl. |
| *Eugenia jambiolina; Terminalia bellerica Dolicharotron atrovireses* | GA effect on biomass of some trees | 6-month old saplinging foliar sprayed with $GA_3$ increased sheath length, leaf area, chlorophyll content- RuBpC, PAA; biomass production |
| *Psophocarpus tetragonolobus* | Root nodules and phytohormons | Infected roots had high levels of IAA, GA, CK and ABA- more than uninfected roots nodules had high tryptophan and hence more IAA metabolium |
| *Arachis hypogaea* | Water-logging and $GA_3$ affecting yield of peanut | Deleterious effects of waterlogging on various growth and yield parameters were partially alleviate by exogenous |
| *Vigna unguiculata* | B-9 and corbohydrate metabolim under saline condition | GA application, starch, soluble sugars and amalyse activity was determined in cotyledon and embryo axis of seeds B9 increased soluble corbohydrates by salinity |
| *Cicer arietinum* | Foliar application of NAA and BAP on *in vitro* pollen germination collected from salt-stressed plants. | NAA, BAP enhanced pollen germination and the growth in salinized plants. |
| *Triticum aetivum* | PGRs effects on grain development in wheat sown on different dates | Kn and ethrel stimulated DM accumulation in wheat grain growth period in early sown but decreased in late sown. PGRs stimulatesd $\alpha$-amalyse and starch synthatase in grains, response of starch and protein accumulation to PGRs varied. Eathrel and Kn decreased TSS and TFAA content. |

| Crop | Study | Results |
|---|---|---|
| *Glycyrrhiza glabra* | PGR initiation of rooting of cuttings | IAA, IBA, NAA significantly promoted rooting; response varied; IBA highly promotery. |
| *Vigna radiata* | PGR effect on biochemical parameters under salinity. | Kn decreased inhibitory effect of salinity on chlorophyll content, proline accumulation. |
| *Cicer arietinum* | NAA, Kn on biochemical constituents | Kn improved chlorophyll content but not yield contributing NAA features enhanced sugar content and yield. |
| *Cajanus cajan* | Cycocel treatment, gas exchange and water relation in response to water stress | CCC caused recovery on re-irrigation; improved exchange, stomatal conductance. |
| *Phaseolus vulgaris* | Phenoli effects on some hydrolases and oxidases in senescing beans | Three phenolic e.g. p-hyodroxybenzoic acid, protocatechinic acid and gentisic acid decreased chlorophylls, RNA, total protein, enhanced activities of RNase, protease, perpolyphonal oxidase, decreased catalase. |
| *Phaseolus mungo* | Eth and BA effect on lipid peroxidation in beans | Lipid peroxidation analysed in chl, mit and microsomal fractions in membrane lipid, Eth increased lipid peroxidation while BA reduced in all fraction. |
| *Phaseolus mungo* | Eth induced senescence and lipid peroxidation | Eth induced senescence PO, $H_2O_2$, phenols, added in system delayed senescence. Phenolic dependent lipid peroxidation caused senescene. in plants and into could be the primacy site which caused loss of membrane integrity of chl. |
| *Gibberella fujikuroi* | UV induced mutants (2) were isolated *G. fujikuai* | Fungal spores exposed to UV gave 2 strain which produced high content of gibberella MH and PCB, inhibited gibberellin systems. When transferred to inhibitor free media-overproduced gibberellin. |
| *Cicer arietinum* | Triadimefon protects membrane components injured by UV-B. | Triademofen protects seedling against UV-5 damage and maintained levels of glycoprotein hijajraten. |
| *Allium cepa* | GA, NAA, Ethaphon, affects yield characters of onion seed. | $GA_3$, NAA increased plant ht., no and size of seeds, umbles preplant. Eth decreased plant ht. NAA treated plants produced highest seed wt. and yield of seeds. |
| *Lens culinaris* | Naphthyl acetic acid (50 and 100 $\mu$g ml$^{-1}$). | Stem ht reduced, no of branches increased, pods per plant enhanced and seed yield increased. TDM, HI, partioning coefficient also enhanced. Seed had high total amino acids, reduced sugars, starch and decreased seed coat thickness, NAA stimulated phloem elements. |
| *Pennisetum americanum* | CCC, phosphon -D; MH; ß-nine affects water relation and yield. | Increased, yield under temperature stress, treated plants remained turngid; CCC was most effective in improving plant water status. |
| *Gossypium hirsutum* | Triacontnol, Vipuil, Hicco IIOR; Agrostemin, cytozyme, Atonic, cycocel, Miriculin on yield and quality of fibere. | CCC adversely affected growth, Agrostenin, seasal enhanced yield. Quality parameters remained unaffected. |
| *Arachis hypogaea* | GA, fusicoccin, Kn, Zeatin, methyl jasmonate, ABA | Methyl jasmonate inhibited seed germination, Kn the inhibitory effect. |

| Crop | Study | Results |
|---|---|---|
| | effects on seed germination. | |
| *Abelmoschus esculentus* L. | Aliphatic alcohols affecting photosynthesis, translocation, yield and metabolite levels in seeds. | Aliphatic alcohols enhanced number of seeds pre fruit, sugars, amino acids and starch decreased but amino acids and protein increased. |
| *Arachis hypogaea* L. | Aliphatic alcohols stimulated $^{14}CO_2$ incorporation and stimulated assimilates translocation to sinks. | AA delayed senescence of leaves, enhanced $^{14}CO_2$ fixation and stimulated mobilization of photosynthates to the seeds. |
| *Arachis hypogaea* L. | ABA affecting uptakre and partitioning of sucrose in developing seeds. | ABA (concentrations 1,10 $\mu$mol $m^{-3}$) stimulated $^{14}C$ sucrose uptate and radioactivity was partitioned towards carbohydrates or proteins whereas ABA directed radioactivity towards lipid and storage proteins. |
| *Lens culinaris* | NAA alteration of yield components | NAA (50, 100 $\mu$g $ml^{-1}$) reduced stem ht but increased pods and seed yield, TDM, partitioning coefficient and HI. Amino acids increased with NAA which improved phloem elements in pedicel. |

# CHAPTER 22

# Auxins

## HISTORY OF AUXIN RESEARCH

Phytohormones have been classified into 5 major classes. Auxins are the oldest and best known among plant growth regulators. Indole auxins are only identifiable natural auxins although non-indole auxins cannot be excluded on chemical grounds. All indole auxins can be changed to IAA in plant cell. Similarly, all the side of active auxins can be degraded to acetic acid. One of the important distinctions between natural auxins like IAA and some of the synthetic herbicidal one like -2,4-dichlorophenoxyacetic acid (2,4-D), or 2,4,5-trichlorobenzoic acid (2,4,5-T) is that the synthetic compounds are more stable. Auxins occur in two forms:

### Free auxins

These are diffusible auxin, which move out of the tissue quite readily (e.g., auxin diffusing out of coleoptile tips into agar). They are readily extractable in various solvents like diethyl ether at 0° to 5°C.

### Bound auxins

These auxins are released from the plant tissues only after they are subjected to hydrolysis, autolysis or enzymolysis. Bound auxins include *reserve*, or *storage*, forms and *detoxification* forms. Detoxification products are not active *per se*. They are often formed from excess IAA or from high levels of synthetic auxins that may be introduced into plant tissues.

The definitive discovery of auxin was made by a young Dutch Botany student of Utrecht named F.W. Went in 1926. He designed very simple experiment with *Avena* coleoptile and named as *Avena* test. According to Went, if intact tips of *Avena* coleoptile were allowed to stand for an hour or two on small blocks of agar, then growth hormone activity could readily be detected afterwards in the Agar.

He adapted and improved Stark's procedure of applying the test material to one side of decaptitated coleoptile, in darkness (source of irradiation was weak red light, which had not been found to be phototropically active) then the resulting curvature could be easily measured and thus the quantitative concept was demonstrated experimentally.

This experiment enabled the Cholodny theory of tropisms (in 1927, Growth hormone and tropism in plants) to be directly verified:

1. Different tropisms in plant can be traced to growth regulation by so called growth hormones.
2. It is most probable that unequal (assymetric) growth observed in various plant organs under the influence of definite stimuli (photoinduction, geoinduction etc.) and that results in the tropistic curvatures is caused by unequal distribution of growth hormones which are produced by definite cells in the organ continuously and independently of the above mentioned stimuli.

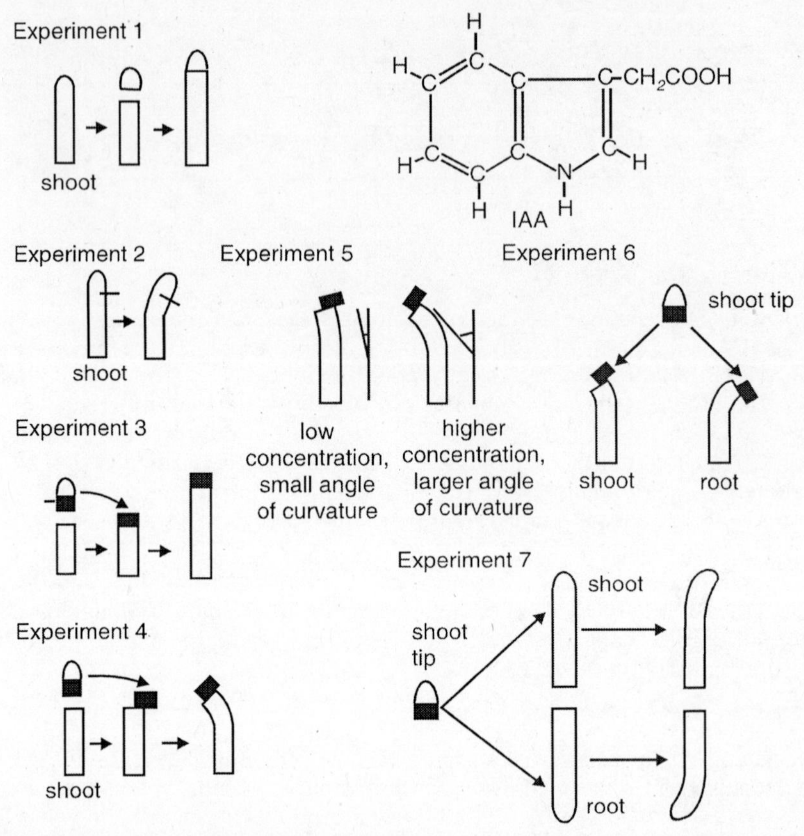


**FIG. 22.1:** Early experiments with *Avena* coleoptile.

Went succeeded in isolating auxin by diffusion technique but he was unable to isolate it in pure form. Thereafter, Kögl and Haagensmit (1931) obtained an active substance called auxin 'A' (auxinotriolic acid) from human urine. They concentrated about 150 litres of urine to 5.7 kg dark-brown syrupy liquid and crystalized it in the form of auxin A. During the course of investigation with urine Kögl et al. (1934) isolated auxin A and auxin B + heteroauxin. On subsequent analysis. heteroauxin was found similar with Indole-3-acetic acid (IAA). Later, it was isolated from yeast by Kogland Kostermans (1935). Isolation of IAA at large scale was made by K.V. Thimann from *Rhizopus suinus* culture in 1935. Thereafter, many purely synthetic and physiologically active auxins were synthesized but they were not named as hormones.

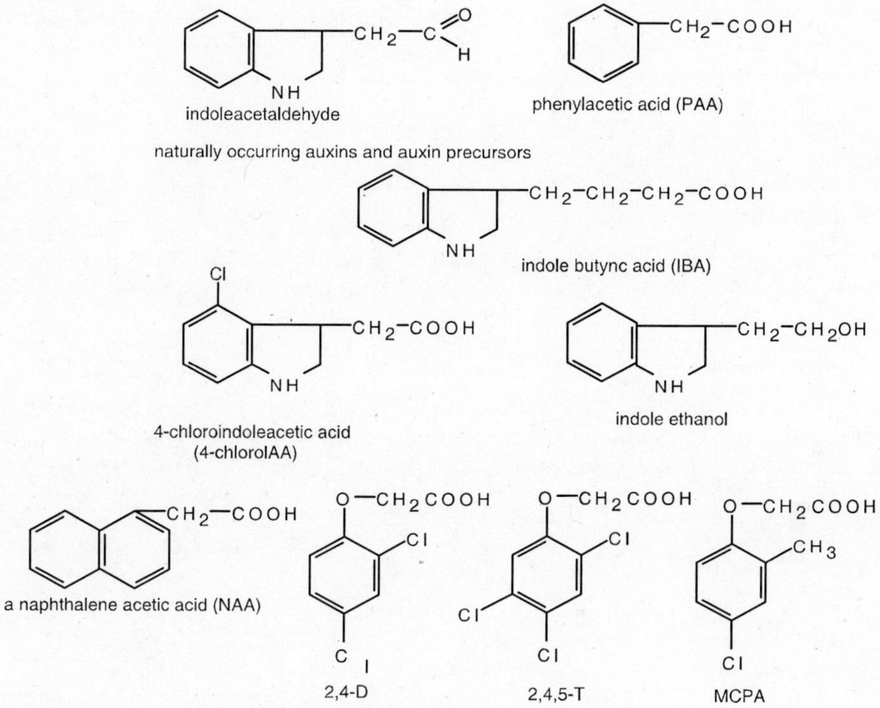

indoleacetaldehyde

phenylacetic acid (PAA)

naturally occurring auxins and auxin precursors

indole butync acid (IBA)

4-chloroindoleacetic acid (4-chloroIAA)

indole ethanol

a naphthalene acetic acid (NAA)

2,4-D

2,4,5-T

MCPA

**FIG. 22.2:** Various types of auxins.

## CHEMICAL NATURE OF AUXIN AND BIOLOGICAL ACTIVITY

If a chemical has a function similar to auxin, it should possess the following characteristics:

1. An unsaturated ring.
2. A carboxylic (-COOH) group.
3. Atleast one carbon atom between the ring and carboxylic group.

The molecule should possess a negative and positive charge and the distance between the charges should be 5.5 Å. If an auxin molecule has to attach to a reactive site through positive and negative charges, then the region of the reactive site which would be involved in this interaction, should have a complementary pattern having opposite charges, the same distance apart. Although, sites which would satisfy these requirements have not been isolated or characterized; and charge distribution data of such auxin molecules as *ortho*-napthalene acetic acid (NAA) and several other auxins are not available, these studies provide a base to formulate the nature of the reactive sites. Depending on pH, the major cellular polymers- proteins and nucleic acids- would be differentially protonated providing a point of attachment for the negative charge on the carboxyl group of the auxin molecule; the small negative charge density due to the lone pair of electrons on the *O*-atom of the carboxyl or hydroxyl group or the phosphate residues may attract the functional positive charge centers of the auxin molecule. This would provide several regions of the macromolecules for selective associates with the auxins. If the auxin molecule interacts with a receptor protein, which then undergoes allosteric modification, a larger region of the chromatin will be involved in the interaction.

**Antiauxins:** Those compounds when applied to the plant and nullify the effects of auxin are known as antiauxins, e.g., 2,4-dichlororanizole (DCA). These compounds compete with auxin for the same site of action in the cell. Therefore, the inhibition of auxin activity by antiauxins is competitive in nature.

### Site of Synthesis

Auxin is characteristically synthesized in the stem tip and in young tissues and moves mainly down stem. It thus, seems to form a gradient from shoot tip to the root.

## TECHNIQUES OF AUXIN BIOASSAY

A bioassay is a method of determining the presence or (more often) the amount of substance by measuring the substance's effect on a designated biological process. Hormone and vitamins are particularly likely to be measured by bioassay because they occur at very low concentration to be detected by the usual physical or chemical method.

### *Avena* Curvature Test

*Avena* seedlings are grown in dark. After a height of 14-30 mm, the plants are decapitated to remove the auxin source in the tip, leave them for 3 hours and redecapitated to remove the regenerated tissue which now produces auxin.

The primary leaf is pulled loose but leave partly in. Applied auxin is dissolved in 1.5% of agar. It is applied as a 10 $mm^3$ block on one side of the cut surface. The side of the coleoptile below the agar block now elongates more than to opposite side, and the resulting curvature reaches an optimum value in about 110 minutes. Curvature is measured by recording the angle made by a vertical line and drawn parallel to the curved position of the stem.

There is a linear relationship between the concentration and the amount of curvature obtained.

### Avena Section Test

It is based on the effect of auxin to stimulate cell elongation. Coleoptile, 20-30 mm length, are used. The apical 4 mm portions are removed, and a section of 3-5 mm in length is cut from each coleoptile. All the sections are first soaked in distiled water for one hour and then are placed in petri plates containing auxin solution. After an interval of 24 or 48 hours, the length of each section is measured. The measurements are made with an ocular micrometer and a desecting microscope. The capacity of the solution to increase the growth of section is directly proportional to the logarithm of the concentration of auxin used. The disadvantage of this test is that quite a depressingly long list of substances that can give auxin-like responses in the section bioassay. Another disadvantage of this bioassay is the likelihood of bacterial contamination and resulting complications from incubating nonsterlized tissues in a 2% sucrose solution for 24 hours.

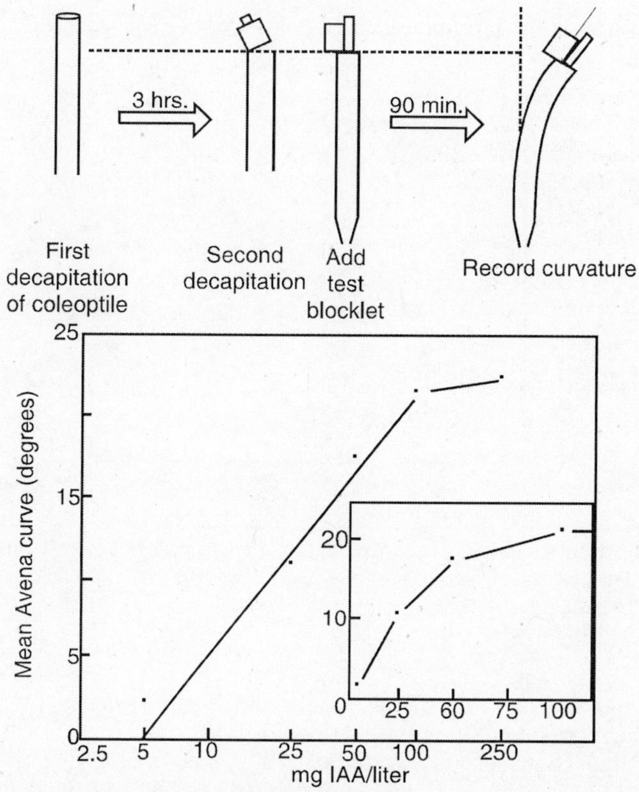


**FIG. 22.3:** (A) The procedure and timing of the *Avena* curvature bioassay. The shadowgraphs are made 90 minutes after test blocklet was applied (after Went and Thimann, 1937). (B) Dose-response curve of the *Avena* curvature bioassay, with the X-axis representing the concentration of IAA added in calibration blocklets and the Y-azis showing the average coleoptile curvature resulting.

## Cross Root Inhibition Test

Roots are more sensitive to auxin than stems, and their growth is inhibited by those concentrations of auxin that normally stimulate stem growth. Low concentrations of auxin induce root growth. The value, then of cross inhibition test is that the effect of extremely low concentrations of auxin, such as found in plant extracts, may be measured. It is conducted as follows:

1. The sterilized seeds are germinated on moist filter paper.
2. When the seedling have reached a desired length, they are placed in perti dishes containing 15 ml of test solution.
3. Root growth is measured after 48 hours.

## TRANSPORT

The polar transport of auxin was noted as early as 1928 by Went using *Avena* coleoptile and agar blocks. The polar transport of IAA shoot is in basipetal direction after following tropic stimulation. There is evidence that lateral transport is possible. The transport is inhibited by 2,3,5- triidobenzoic acid and inhibited by polar transport.

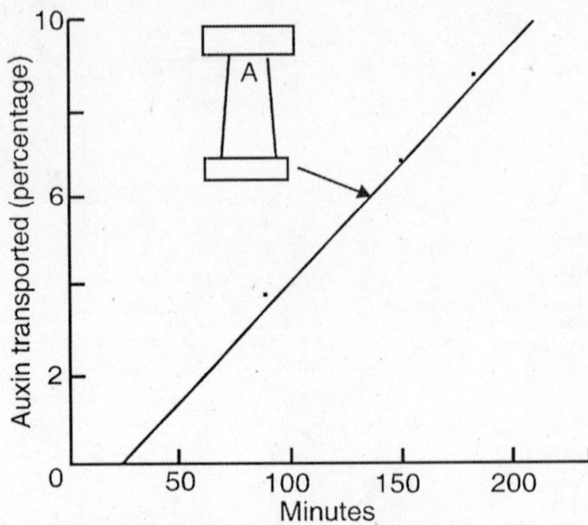


**FIG. 22.4:** Graph illustrating the principle of auxin-transport experiments. Auxin added in an agar block at the apical (A) end of an excise coleoptile section moved basipetally through the section and accumulated in the basal receiver block.

The specific transport takes place in intact plants. Auxin moves quite slowly, not faster than would expected for diffusion from cell to cell across the cell walls. Not all auxin movement is basipetal; numerous recent experiments have shown that apical radioactive IAA will move acropetally with considerable velocity. This is rather very difficult to measure how much auxin moves in this way, or what portion of the total mobile pool of auxins is involved. It is possible to propose that auxin might move simultaneously in opposite, but in different or in different columns of cells.

Auxin transport is an active process therefore, it needs metabolic energy and is inhibited by metabolic inhibitors like cyanide and 2,4-dinitrophenol (DNP).

## ACCUMULATION OF AUXIN

The concentration of auxin is highest in stem tip and gradually decreases downwards. The basal portion of the stem contains lowest concentration of auxin. The production/biosynthesis of auxin by some tissue of plant is a continuous process.

However, they do not accumulate it in large amount. This means that some process or processes of inactivation or destruction must take place. Inactivation of auxin is one of the important processes of the plant system by which growth is controlled. Auxin concentration at a given site is proportional not only to its rate delivery but also to its rate of destruction.

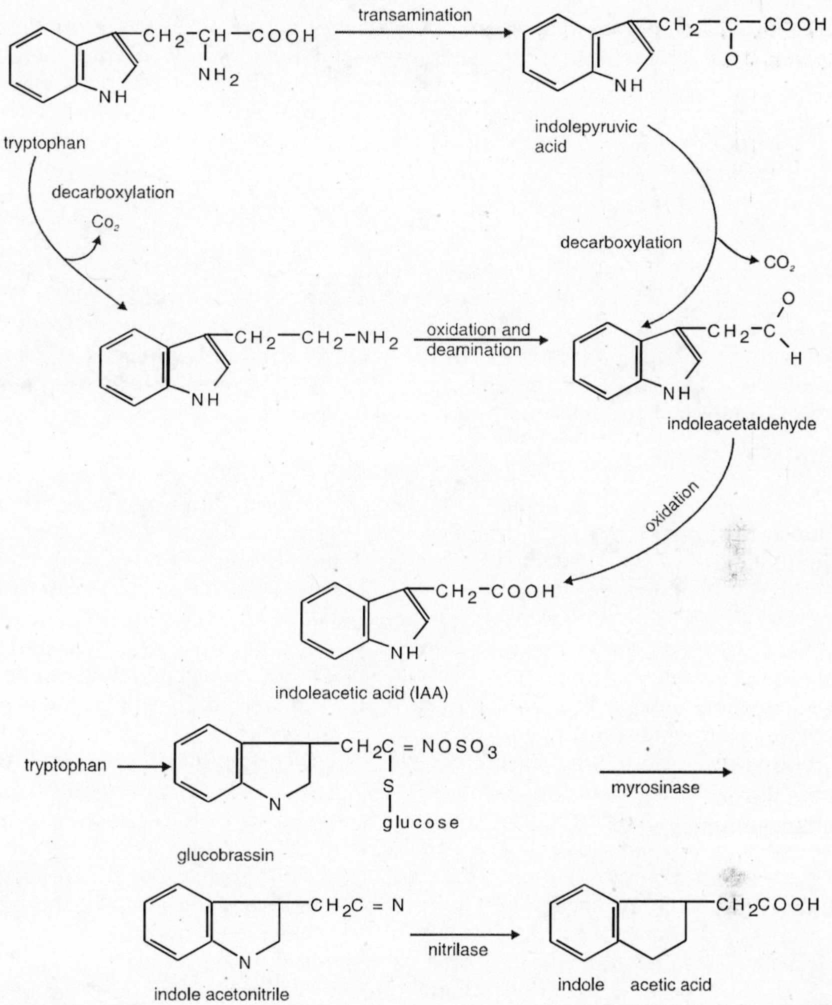

tryptophan → (transamination) → indolepyruvic acid → (decarboxylation, $CO_2$) → indoleacetaldehyde → (oxidation) → indoleacetic acid (IAA)

tryptophan → (decarboxylation, $CO_2$) → tryptamine → (oxidation and deamination) → indoleacetaldehyde

tryptophan → glucobrassin → (myrosinase) → indole acetonitrile → (nitrilase) → indole acetic acid

**FIG. 22.5:** (A) Pathway of auxin biosynthesis from tryptophan. (B) An alternate pathway of tryptophan biosynthesis in some plants.

## BIOSYNTHESIS

Naturally auxin is produced by *Rhizopus suinus* when grown in peptone-containing medium. It synthesizes auxin by oxidation of peptone. It has also been demonstrated that this mould could convert tryptophan into auxin. Ferri and Bonner indicated that enzymes which are involved in conversion of tryptophan to IAA in *Avena* coleoptile are similar in distribution of IAA.

There are several pathways for synthesis of IAA for example, it is possible to begin with tryptophan pathway for derivation of metabolic indole ring. Tryptophan is the precursor of IAA biosynthesis which is converted into IAA by the removal of amino acid and carboxylie acid. Tryptophan is converted into tryptomena, that is -

COOH group of tryptophan is replaced by $NH_2$. However, tryptophan may also be converted into indole 3 pyruvic acid. Both the compounds are independently converted into Indole 3- aldehyde which may be further converted into Indol-3-ethanol. The indole 3-aldehyde and indole-3- ethanol are interconvertable. In the last, the indole 3- aldehyde is converted into Indole-3-acetic acid. The biosynthesis of IAA is well established.

## METABOLISM OF IAA

In plants, IAA is metabolized by conjugation and oxidation via different pathways which vary with species and probably also with the organ and stage of growth. The synthesis, conjugation and hydrolysis, oxidation and utilization during growth are determinants of IAA concentrations in plants. Limited evidences show that metabolism of IAA can be effectively regulated by environmental factors, such as light, or by chemical applications, such as auxin, cytokinin, glyphosate or phenol.

## CONJUGATES OF IAA

Plant hormones are sometimes available as bound, watersoluble, or polar hormones from which free hormones can be liberated by hydrolysis. These compounds are widely distributed in plants and occasionally present in high concentrations. Some of them have been characterized as *conjugated plant hormones*. This term includes plant hormones which are metabolically coupled to other low molecular weight compounds by covalent binding. This definition excludes noncovalent association between plant hormones and high molecular weight particles like protein, cell particles and membrane. These complexes are known as *bound* plant hormones.

Another important way in which IAA can be rapidly inactivated is by its conjugation with a variety of chemical normally available in the cell. IAA is metabolized in plants by oxidation and by conjugation with amino acids, sugars and the sugar alcohol *myo*-inositol. The existence of *bound IAA* was recognized long before chemical characterization. The discovery of the formation of IAA-aspartate from applied IAA by Andreae and Good (1955) open new direction in the field of IAA-conjugates. Since then, IAA-aspartate has been reported in several plant species, and conjugates of IAA with other amino acids have been found. Conjugation of IAA by ester linkage with glucose was first reported by Zenk (1961). Subsequently, Bandurski et al made extensive studies of ester conjugates of IAA with *myo*-inositol. Good et al. (1956) demonstrated the formation of indole-3-acetamide in tissues of several plants after incubation with IAA. In all endogenous auxin conjugates isolated from plant material up to now, the low molecular weight groups are covalently bound to the carboxylic group of the plant hormone, linkage to N-atom of the indole ring, e.g., by formation of N-glucosides has been observed. The endogenous auxin conjugates can be classified as either peptidyl conjugates, or sugar and *myo*-inositol derivatives, or alkyl esters.

## DESTRUCTION OF AUXIN

There are two major means of incativation or destruction of auxin in plants. (i) Enzymatic oxidation and (ii) Photooxidation.

### Enzymatic Oxidation of Auxin

The enzyme involved in destruction of IAA in vivo is known as *IAA-oxidase*. It is ubiquitously isolated from plants. The oxidation of IAA in pea epicotyls appears to be catalyzed by a peroxidase in which one molecule of oxygen is consumed (hence the termed as *oxidase*) and $CO_2$ is released for one molecule of IAA oxidized. A flavin protein is also found closely associated with the peroxidase which possibly generate $H_2O_2$ needed for reaction. There is a basic difference between peroxidae and oxidase activity. During peroxidase reaction there is *no* need of oxygen, however, in the destruction of IAA, the enzyme possessing peroxidase activity also operates as an oxidase, and oxygen is consumed in the reaction. The reaction is as follows:

The end products of destruction of IAA are 3-methyleneoxindole or indolealdehyde. $Mn^{++}$ ions and phenol (2,4-dinitrophenol) are required during the reaction.

The chlorogenic acid, caffeic acid, scopoletin, and ferulic acid are the inhibitor of IAA-oxidase activity. The activity of first two compound on IAA oxidase can be reversed by the addition of $H_2O_2$.

### Photooxidation of Auxin

IAA is inactivated by ionizing radiation. Pure IAA is inactivated by X- and gamma-radiation due to oxidation by peroxidases formed during irradiation. The radiation directly affects the enzyme system responsible for the conversion of tyrptophan to IAA. Ultraviolet radiations also inactivates IAA.

## MECHANISM OF ACTION

The initial event of IAA action is the absorption of IAA to the hormone specific binding site. The IAA binding site complex then initiates the reaction. This is evidenced as follows:

1. Membrane phenomena leading to media acidification.
2. Nucleic acid related phenomena leading to longer changes involved in growth and morphogenesis - one of the results of (a) and (b) would be changes in the plasticity of the wall so that growth could occur. This increases in plasticity which requires following changes:
   a) In protein matrix of the wall
   b) In cellulose matrix
   c) Hemicellulose matrix

   The working hypotheses is given below:

### Nature of Wall Loosening Signals

Davis (1973) and Evans (1974) presented reviews to describe the nature of the wall loosening signal moving from protoplast to cell wall. They presented the following suggestions:

1. Auxin accumulates synthesis of pectic fraction of the cell wall or enhance methylation of carboxyl group.
2. Auxin acts to stimulate the de novo synthesis of wall loosening enzyme by the induction of transcription (the *gene expression hypothesis*) or translation.
3. Auxin stimulates total wall synthesis perhaps by stimulation of glucan

synthetase activity.

4. Auxin stimulates wall turnover, especially the xyloglucan component.
5. Auxin stimulates the outward pumping of hydrogen ions leading to wall acidification (the *acid growth hypothesis*).

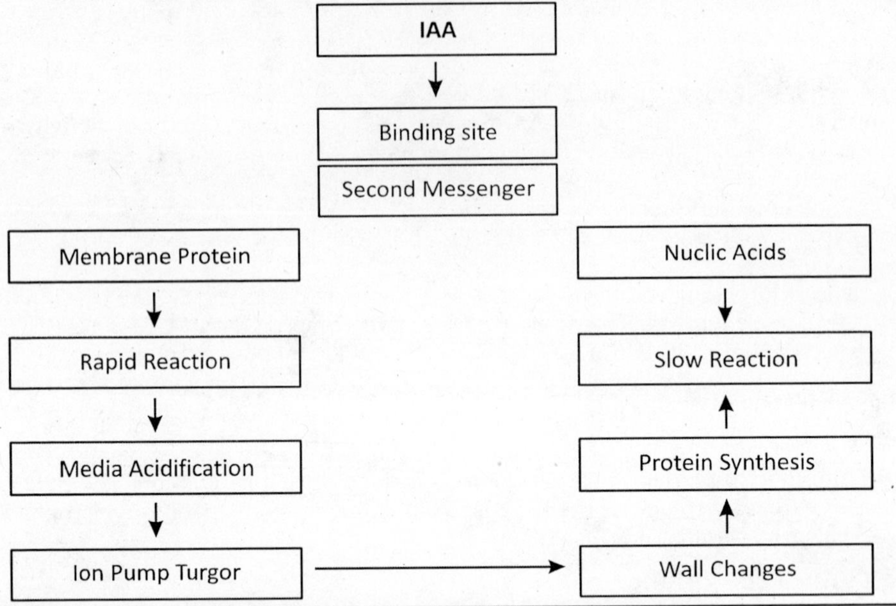


**FIG. 22.6:** Events in cellular enlargement and possible relationships of auxin action

In order for an auxin induced effect to be a candidate for the cause of growth promotion it must be shown, of course, that the effect occurs before the growth promotion begans. Among the most rapidly occurring effects of auxin are:

1. Stimulation of protoplasmic streaming (almost immediate),
2. Apparent increase, in membrane permeability to water (about 6 minutes),
3. Stimulation of beta-glucan synthesis activity (about 15 minutes),
4. Stimulation of wall xyloglucan turnover (15 to 30 minutes), and
5. Enhancement of $H^+$efflux (0 to 20 minutes depending on material and conditions).

The binding of IAA with cell wall/cell components is however reported by several workers, yet no plant hormone receptor has been isolated with certainly that the binding proline is involved in hormone action.

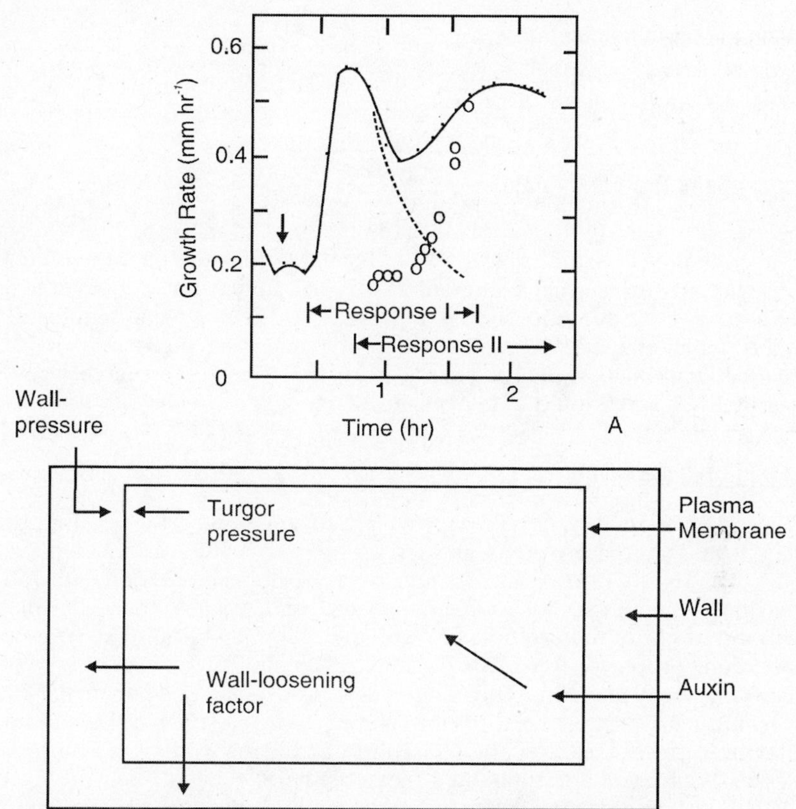


**FIG. 22.7:** (A) Rate kinetics show two separate elongation responses by soybean hypocotyl segments. (B) Auxin initiates wall loosening.

## MEMBRANE AND ION MOVEMENT LEADING TO MEDIA ACIDIFICATION

When auxin sensitive tissue is incubated in the solution of low pH, it promotes growth. The pH optimum for acid-induced growth is 4.5 to 5.0 for those tissues whose cuticle is removed. Hager, Melzel and Krauss (1971) proposed acid growth hypothesis for this low pH action. According to them IAA activates plasma membrane-ATP which stimulates active proton efflux from the cell. The low pH activates enzymes capable of hydrolysing wall polysaccharides to soften the wall and allow cell extension. The uncoupler like carboxyl cyanamide m-chlorophenyl hydrozone (CCCP) prevents both cell extension and media acidification. Similarly KCN, 2,4-dinitrophenol, valnomycine, abscisic acid and cyclohexamide also act similar to CCCP.

The acid growth hypothesis is supported by the finding that neutral buffers prevents IAA induced growth. The above discussion reveals that media acidification is some how related to IAA induced growth.

## Mechanism Media Acidification

The exact mechanism of media acidification is still obscure. However calcium ions are required for this process. The ATP:ADP ratio decreased while ADP:ATP ratio is increased. Respiration rate also induced acidification.

## Nucleic acid and Protein Changes

The growth of a plant by IAA may be °stimulated by an increase in RNA synthesis which may depress certain genes, thus develops change the rate of RNA and protein synthesis. It has been demonstrated that inhibitors of RNA and protein synthesis also inhibit auxin induced growth. It now seems within reach to determine, what enzymes are encoded in sequences of messenger RNA which becomes active after the plant is treated with IAA. There are several aspects like changes in RNA and protein, changes in cell wall plasticity which are triggered after IAA treatment.

## Changes in Wall Plasticity

Plant cell is enveloped by a cell wall which stretches or grow as the protoplast enlarges. It may increase in size, about 50-fold, with a little or no increase in dry weight. It was Hyen (1930) who first observed the effect of IAA on cell wall which resulted into stretching of cell wall. When the wall softens, the cell will expand owing to the turgor of the protoplast. Went and Thimman (1937) also discussed cell wall stretching property after IAA treatment. The cellulosic microfibrils of cell wall are almost at right angles to the long axis of the plant prior to cell expansion, however, after expansion these fibrils become more nearly parallel to the plants axis. How this action occurs is still a matter of active research. It is also established that the wall rigidity is determined by four polymeric substances: wall protein, the cellulosic beta-1-4 glucane, beta-1-3 cellulose type glucane and lichemine type glucane. They also determine the wall plasticity.

It may, therefore, concluded that there is no single determinant of the structural rigidity of the wall when IAA-induced growth occurs; the solution bathing the cell wall becomes more acidic and thus wall is easily deformed. This change in the plasticity is not only due to low pH but also by enzyme induced changes in the wall, however, the exact mechanism of cell wall plasticity needs further research.

## Secondary Messenger Hypothesis

According to this hypothesis, auxin does not directly act on the cell wall. The addition of auxin to isolated cell walls or to the cells with non-functional protoplasts has no effect on wall extensibility. This has led to the conclusion that the initial action of the hormone is at a site on or within the protoplast, an action which leads indirectly to wall loosening. Thus, there must be a signal or *secondary messenger*of some type to enrich travels from the protoplast to cell wall, initiating wall loosening.

## Composite Model?

There are certain difficulties in proposing a composite model for auxin action, because of a lot of variabilities in different systems have been studied so far. Biswas and Roy (1978) proposed a working hypothesis based on their studies with

IAA using coconut endosperm cells. According to them auxin may first form complex with receptor protein in the membrane and may activate membrane bound ATPase initiating a proton pump, thus loosening of cell wall occurs. Auxin is then detached from this complex and transferred to cytoplasmic receptor. Lipids might play a role in this transfer. This complex than translocates to the nucleus and then to chromatin. It is proposed that this nucleoplasmic receptor-hormone complex is associated with the chromatin. This binding may be observed by non-histone proteins in the chromatin. After transfer to the auxin to the chromosomal acceptor site, the genome would be switched on to be transcribed. However, this model needs further experimental evidences.

## Theologis Model

In an attempt to elucidate the relationship between early auxin-induced mRNA accumulation, $H^+$ secretion, and cell elongation, a different model was proposed by Theologis (1986). According to it, auxin action involves a complex metabolic pathway, with auxin regulating at head of the pathway (*gene transcription*) and the secretory processes at the tail. The model involves $H^+$ extrusion systems. One system, activated indirectly by auxin, and is linked to secretory system. The second system, activated directly by fusicoccin, is isolated on cell membranes and is independent of secretory processes.

**TABLE 22.1:** The physiological responses of auxins in cereals

| Phenomenon | Effects |
|---|---|
| **Effects at cellular level** | |
| (a) Cell division | Stimulates cambial cell division |
| (b) Cell elongation | Stimulates shoot growth, inhibits root growth. |
| (c) Cell differentiation | Stimulates xylem differentiation, stimulates phloem differentiation, promotes root initiation, regulates callus tissue morphogenesis |
| **Effects at organ and whole plant level** | |
| (a) Seedling morphology | Reverses red light of initiation of mesophyll elongation |
| (b) Geotropism | IAA is transported to lower side of the shoot |
| (c) Apical dominance | Replaces apical bud |
| Leaf senescence | Delay |
| Leaf abscission | Auxin inhibits abscission, auxin applied proximal to abscission layer promotes abscission. |
| Flowering | Promotes |
| Fruit setting | Allows parthenocarpic fruits |
| Fruit ripening | Delays |

## Coleoptile and Stem Growth

Auxin causes growth of the coleoptile of *Avena*, wheat, maize and other cereals. Such elongation is only possible when the tip of the coleoptile is removed before its treatment. This is because of when the tip is intact on the coleoptile, the tip itself produces the auxin, hence the intact coleoptile shows rapid growth. After the removal of tip, the growth of the coleoptile stump ceases in the absence of endogenous supply of auxin from the tip. Externally applied auxin can promote the growth of detopped coleoptile. When auxin causes growth of a coleoptile, it is exclusively due to the elongation of pre-existing cells. Stem growth is also promoted by auxin-treatment through cell elongation.

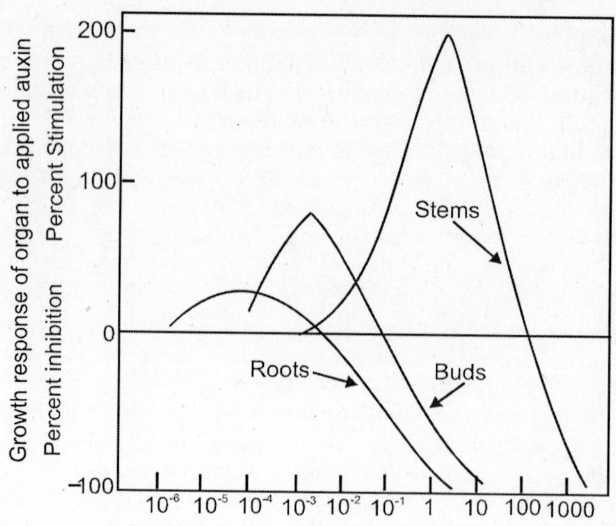


**FIG. 22.8:** Organ dose response curves effect of different concentration of growth of three plant.

## Cell Elongation

Auxin causes cell elongation by loosening of cell wall. The rigid cell wall becomes softened and its plasticity (irreversible capacity to stretch) increases. The loosening of cell wall begins due to following reasons:

i) dissolution of cell wall material,
ii) breaking of chemical bonds between cellulose and other wall materials, and
iii) fresh synthesis of new cell wall material and its incorporation into existing cell wall.

The loosening of cell wall decreases the wall pressure (WP) which is equal to turgor pressure (TP). This action causes greater uptake of water and increase in the size of vacuole; and cell stretches.

## Secondary Growth

When the tip of the sunflower seedling is removed (source of auxin) it unables to show any symptoms of secondary growth. However, when it is treated with auxin, the formation of secondary vascular cambium resumes. Promotion in cambium activity by auxin continues secondary growth. It also controls total thickness of the stem.

## Rooting of Cuttings

When the shoot-tip is removed it reduces the shoot's growth. In contrast, removal of the root-tip does not result in decreased root growth. The removal of less than 1 mm of the root tip results in a very small but a significant stimulation of growth rate. When the tip is replaced, the root growth is also retarded.

The action of auxin in root is similar to that of stem, but the difference is dependent on the concentration of auxin which is stimulatory to stem and inhibitory to root growth. This also reflects that roots are more sensitive to auxins than that of stems.

When high concentration of auxin is applied to root, it not only retards root elongation but causes a remarkable increase in the number of branch roots. Application of IAA + linolin paste to the severed end of a young stem initiates the rate of root formation and develops large number of roots. This aspect has its commercial value.

## Apical Dominance

The suppression of the emergence of lateral buds by the presence of shoot apex is called apical dominance. It occurs due to auxin produced in the shoot tip, which is transported downwards basipetally and reaches to the axillary buds. The concentration of auxin in the lateral bud becomes *supraoptimal* which retards the growth of lateral buds into branches.

Such phenomenon occurs in different types of organs including stems and tubers. The root apex may also inhibit branching in roots. The influence of apical bud on the growth of lateral buds may be shown by removing it from the plant. In the absence of apical bud, lateral buds grow vigourously. Exogenous application of auxin could replace the apical bud in inhibiting the growth of lateral buds. The auxin diffuses from the apical bud causes correlative bud inhibition. The removal of leaves and stems can stimulate the growth of other parts of the plant. This is an example of *compensatory growth.* It was proposed by Thimann (1937) that apically synthesized auxin enters axillary buds and directly inhibits their growth due to greater sensitivity of lateral buds to auxin. This concept is known as *direct theory of auxin inhibition.* According to another concept, apically secreted auxin has an indirect action in bringing about correlative inhibition of buds and shoots. This idea is now known as *indirect theory of auxin inhibition.* It has further suggested that auxin concentration in correlative inhibited buds may be *sub-optimal* rather than *supra-optimal.* This concept does not support the direct theory.

Gregory and Veale (1957) given a new theory with reference to Thimann's theory. They investigated the nutritional aspects of apical dominance. They indicated that effect of auxin on lateral bud growth which is controlled by the nutritional status of the plant. If nitrogen requirements of a *flax* plant are completely supplied during its growth, than at the optimal period of growth, lateral bud inhibition by exogenous supply of auxins can not be demonstrated. Under limited/inadequate supply of nitrogen, the effects of exogenously treated auxin on lateral bud can easily be observed.

## Parthenocarpy

Fruit setting starts after pollination and subsequent fertilization of the ovule of a flower. Growth of the ovary wall and, in some cases, the tissue associated with receptacle is enhanced. The acceleration in growth is due to the cell enlargement induced by endogenous auxin. The pollination and fertilization events are associated with the proper development of fruit, probably these events release some stimulus. Fruit development does not take place without *pollination-action.* Such fruit will be than without seed. The development of fruit in this manner is called *parthenocarpic development* and the fruit is known as parthenocarpic fruit.

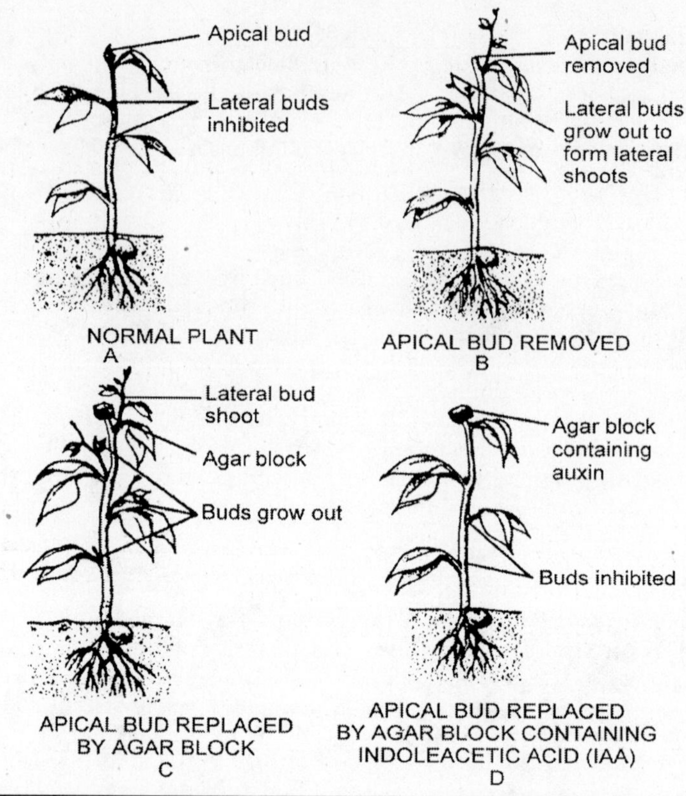


**FIG. 22.9:** Auxins and apical dominance.

Fertilization is the natural phenomenon of all seed plants. Fruit development does not occur without fertilization. How does fertilization of the ovule trigger off responses leading to fruit set? Various possible explanation have been presented from time to time, they are as follows:

1. Swelling of ovary wall of orchids could be stimulated by dead pollen grains (Massart, 1902).
2. Water extracts of pollen are capable of inhibiting floral abscission and stimulating ovary wall swelling in orchid (Fitting, 1909).
3. Parthenocarpic fruits developed after application of pollen extracts of cucumber flowers, which contained auxin (Yashuda, 1934).
   Application of IAA in linolin paste to stigma of flower could able to induce parthenocarpic fruits (Gustafuson, 1939).
4. A sharp rise in auxin content of tobacco ovaries has been observed just after pollination, however, in absence of pollination such increase was not observed (Muir, 1942). He further demonstrated that pollen tube increases the amount of extractable auxins in the style of tobacco plants. It shows that pollen tube contains certain enzymes which may catalyzes the production of auxin.

Pollen tube secretes an enzyme which can convert tryptophan to auxin (Lund, 1956).

The above experiments reveal that auxin is an important factor in the fruit development. Pollination, growth of pollen tube and fertilization all contribute to the cause that auxins are responsible for fruit development.

## Fruit Size

Exogenous application of auxin can alter fruit size and the period of maturity in plums, pineapples and fig. Auxin can increase the size of these fruits and rate of maturity. Auxin-palm application stimulated growth of strawberry fruits also.

## Abscission

Abscission is the shading, detachment, separation or dehiscence (a specialized type of abscission) of plant parts, that is, organs, other structures, or tissues.

It has been known for nearly 55 years now that auxin retards abscission. The level of auxin in leaf blades (*Gossypium*) decreases sharply with age and immediately prior to their shading. Jacobs (1962) showed that it is possible to substitute the abscission-retarding effect of the lamina by an appropriate amount of exogenously applied IAA. Applying IAA to explant systems usually retards abscission and a number of workers have either demonstrated or confirmed that when applied distal to the abscission zone, there is an increasing retardation of abscission with increasing levels of IAA. What terms a monophasic response that sustains the inhibition elicited by endogenous auxin?.

However, in certain systems and under certain conditions, IAA when applied distally results in a diphasic response: acceleration of abscission at low, and retardation at high levels. Such effect has been shown with the synthetic auxin-NAA.

Various theories have been proposed by different physiologists regarding the involvement of auxin in abscission process are as follows:

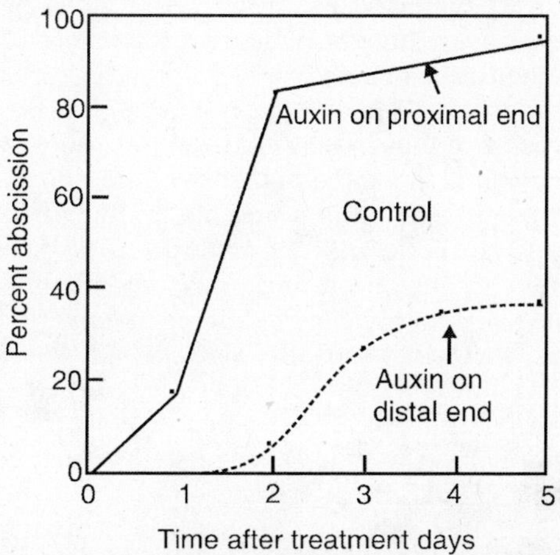


**FIG. 22.10:** Effect of proximal and distal applications of auxin on rate of abscission.

### *Auxin gradient hypothesis (Addicott et al., 1955)*

The proximal-distal gradient of auxin across the abscission zone is the principal factor in the regulation of the onset and rate of abscission. In the young leaf of a healthy, vigourously-growing plant the level of auxin in the tissues distal to the abscission zone is characteristically higher than that in positions proximal to it and abscission does not occur. When through age, shading, deblading, treatment with accelerants, etc. there is a decrease in the ratio of distal to proximal auxin, abscission occurs. Lower to this ratio, the greater will be acceleration of the process. The ability of IAA when applied proximally to the abscission zone to elicit this monophasic acceleration of abscission has been shown often.

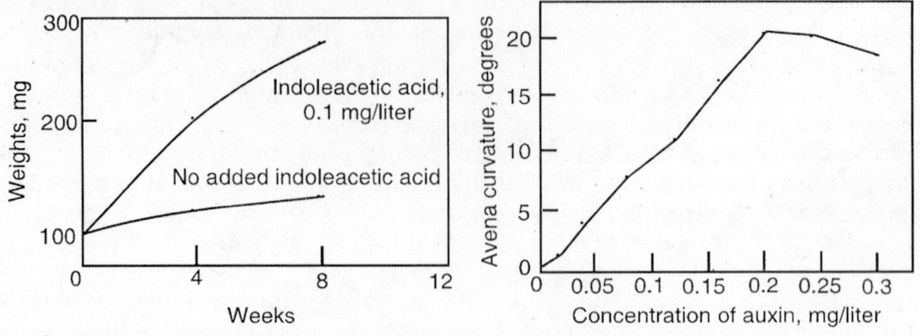


**FIG. 22.11:** Relation between the auxin gradient across the abscission zone and abscission.

### *Auxin concentration hypothesis (Gaur and Leopold, 1956)*

In contrast to the concepts involving gradient and balance, low concentrations of auxin are thought to promote and high concentrations to retard abscission.

### *Auxin-senescence factor hypothesis (Osborne, 1955)*

Leaf abscission is controlled not only by endogenous auxins but also by some substances derived from senescing petioles (leaves of *Phaseolus*). This substance, called senescence factor, which accelerates abscission, has been detected in *Gossypium* and *Coleus*. It was thought by Osborne (1968) not to be ABA, although this was the working hypothesis of Jacob's laboratory.

The understanding of the role of auxin in abscission on the one hand is complicated by its diverse effects such as those of stimulating respiration and ethylene evolution, retarding the development of the separation layer but than, once formed, enhancing cell division in it and maintaining cell membrane integrity. There are two stages of abscission (see Addicott, 1970). During **Stage I** applied auxin prevent abscission, an indication that one of the many physiological changes taking place in explants presumably is a decline or inactivation of endogenous IAA. These changes appear to lie in tissue distal to the abscission zone *per se* and characterize a stage spanning that from full maturity of the organ (leaf, petal, fruit, etc.) to just before the onset of final senescence. During **Stage II** the tissue distal to the abscission zone undergo the final stage of senescence while in the zone itself there is intense biochemical activity as witnessed by the synthesis of RNA, protein and various cell wall degrading enzymes. Application of auxins and other hormones during this stage, accelerate the completion of abscission.

There is as yet no clear explanation of auxin's dual role, but Addicott (1970) suggested that applied auxin maintains the ongoing biochemical pattern in plant.

Auxin is also an inhibitor of the synthesis activity of the cell wall degrading enzyme- cellulase, it the principal retardant of abscission.

## Respiration and Auxin

Auxin increases respiratory rate of the plant cell. Its activity only takes place in the presence of oxidative metabolism. A correlation between increased growth and auxin treatment has been established. How auxin induce the rate of respiration is a matter of further research? French and Beevers (1953) suggested that respiration may be increased by substances that have no effect or an inhibitory effect on growth. Dinitrophenol (DNP) inhibits oxidative phosphorylation but rate of respiration is increased. This also inhibits growth rate. The rate of respiration is normally limited by the supply of ADP, and treatment of plant tissue by DNP should increase the availability of ADP, and thus stimulates respiration. Auxin may also increase the supply of ADP by causing ATP to be used up in the expanding cell, thus increasing the supply of ADP. These evidences indicate that auxin has indirect role in stimulation of respiratory rate in plant cell.

## Organogenesis and Auxin

Auxin causes swelling in the callus tissue (rapidly dividing parenchyma cell) in vitro. Auxin is added in the culture medium for procurement of continued callus growth. Skoog and Miller (1957) established that in interaction between kinetin and the auxin- IAA, IAA alone, or high ratio of IAA to kinetin lead to adventitious root formation. The application of NAA to tobacco explant and on many other plants promotes root formation. Exogenous auxin can have two major effects, therefore, in the initiation and/or enhancement of root organogenesis and as an inhibitor of cytokinin-induced shoot initiation. Another application of auxin is in the rooting and the shoot obtained through micropropagation. Auxin ($5.10^{-6}$ to $10^{-5}$ concentration) initiates rhizogenesis. IAA, NAA or IBA (indole butyric acid) are generally more effective than the phenoxyacetic acids like 2,4-D to 2,4,5-T which induce callusing even at very low concentrations ($5.10^{-8}$ to $10^{-7}$ M). 2,4-D is more effective inhibitor of shoot initiation than IAA, IBA or NAA.

Removal of auxin may induce organogenesis. In carrot cell cultures and certain monocotyledonous callus cultures root formation only occurs after the auxin, removal (essential for growth). 2,4-D is the most effective auxin for the formation of somatic embryo and helps in major advancement of plant tissue culture, namely plant regeneration from cell cultures of the Gramineae via somatic embryogenesis. In alfa-alfa, growing the callus on 2,4-D is the pre-requisite to allow shoot organogenesis after transfer to auxin-free or hormone-free medium. Exogenous supply of auxin in culture medium is necessary for the induction of adventitious shoot formation.

## Crown Galls

Callus like tissue, known as crown galls, are formed in sunflower, marigold, tomatoes and beans, etc. These are produced by an external stimulus, namely by parasitic bacteria. Mature crown gall tissues produce large amount of auxins to continue the growth.

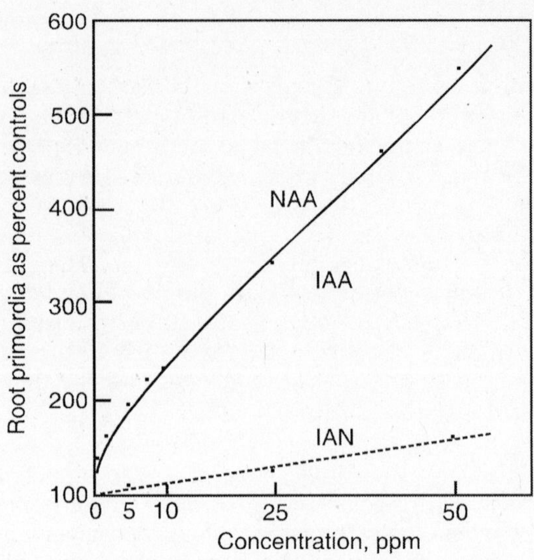


**FIG. 22.12:** The effects of auxin on the differentiation of ground meristem and root formation.

## AGRICULTURAL USES OF AUXINS

### Auxins as Herbicides, Ripeners, and Rooting Inducers

The most widespread commercial application of auxins as herbicide is well known. Auxins like (2,4-dichlorophenoxy) acetic acid (2,4-D) and (2,4,5-trichlorophenoxy) acetic acid (2,4,5-T), which act as herbicides when used in large doses.

Dicotyledonous weeds are particularly sensitive to very high auxin level. Some of the most valuable and widely used selective herbicides in the weed control are auxins, particularly the phenoxyacetic acid analogous (e.g., 2,4-D, 2,4,5-T, and McPA). 2,4-D is one of the most widely used herbicides. It is highly selectively, non-corrosive, effective at low concentrations, safe to handle, relatively easy to formulate, and economical to use. Several benzoic acid analogs (e.g., dicamba, chloramben, and substituted picolinic acid, picloram [Tandon]) are also important herbicides.

R. J. Weaver (1972) proposed another use of auxin in agriculture by using the principle of inhibition of abscission layer formation, auxin (e.g., NAA, 2,4-D) are effective to control fruit drop in apple and pear. Auxin including 2,4-D, induced ethylene formation and fruit set in pineapple.

Biennial bearing (light and heavy crops in alternate year) is common to many tree crops. This is a great problem for agriculturists. By corrected and timely thinning in the heavy years and timely application of NAA or other auxins can solve the problem.

2,4-D enhances fruit set and fruit ripening particularly in pineapple cultivation. Induction, or forcing of flowering is the major problem in pineapple. This was achieved initially in the field by 1-naphthyl-acetic acid (NAA), which is also used as a rooting agent. Ethephon is also preferred for this purpose.

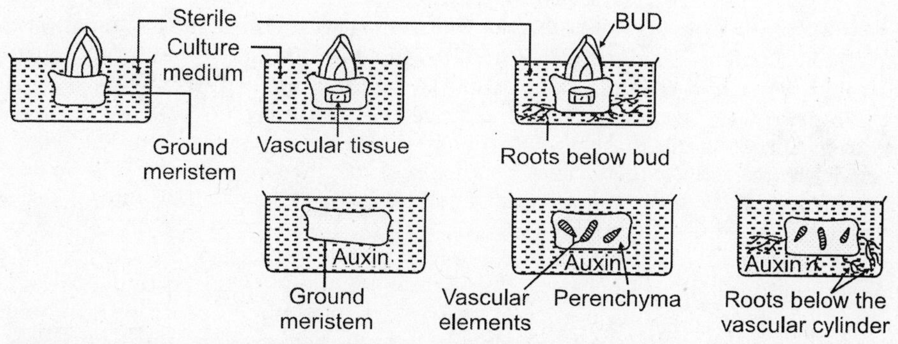


**FIG. 22.13:** Auxin in rooting (A) root formation (B) rooting in curring.

Commercial use of rooting auxin compounds is very common. They promote callus and root formation, which can improve establishment from cuttings. Species and cultivars difficult to root are enhanced by dipping the cut surface into rooting compounds. Auxins are also used in the prevention of sprouting of stored potatoes by dipping them in NAA solution or by sprinkling with telc or fuller's earth containing auxin, or stored with strip of paper impregnated with an auxin solutions. Indolyl butyric acid (IBA) is also used in inducing the rooting of cuttings. In vitro rooting of shoots of Antonovka 313 apples was achieved by using a medium containing 0.25 mM indolyl butyric acid and 1.5% sucrose. Promotion of rooting by auxins is sometimes enhanced by the presence of some phenolic compounds, which have a synergistic effect. More roots were produced in cuttings of *Phaseolus vulgaris* L. in the presence of both indole butyric acid and 4-chlororesorcinol, an effect explained by the inhibition of polyphenol oxidase (PPO) activity.

**Glyphosine** [N, N-bis(phosphonomethyl) glycine], an auxin-like substance is also used as sugarcane ripeners. Its action, primarily, is as herbicides, to prevent late growth of the cane so as to cause a major part of the assimilate to be stored as sugar.

$$HOOC\text{-}CH_2\text{-}N[CH_2\text{-}PO(OH)_2]_2$$

The compound has enabled a marked increase in the yield of sugar. Other important sugarcane ripeners are ethephon, 2-chloroethyl-phosphonic acid, and glyphosate N-(phosphonomethyl) glycine.

$$HOOC\text{-}CH_2\text{-}NH\text{-}CH_2\text{-}PO_3H_2$$

The use of ethephon has enabled an increase in the yield of rubber from *Hevea brasiliensis* by 200% to 300%.

Synthetic auxins or auxin-synergists like 2,4-D, 4-CPA (4-chlorophenoxyacetic acid), and N-metatolylphathalamic acid are used widely to increase fruit set in vegetables, for example, with tomatoes grown in glasshouses or under plastic tunnels, or with aubergines. Highly effective synthetic auxins, which are derivatives of benzotriazoles, were recently synthesized, using a design approach based on QSARs.

### *Menadione Sodium Bisulfite (MSB)*

Since the presence of auxin is necessary for plant growth, an inhibition of the degradation of this hormone may enhance growth. Decreased activities of the enzymes involved in IAA oxidation were observed after the application of Menadione Sodium Bisulfite (MSB). Following the addition of MSB to tomato, cucumber, corn, and capsicum plants, the concentrations of IAA in the plant cells were increased by up to four fold relative to the untreated plants. Enzymatic studies in tomato plants sprayed with $10^{-5}$ M MSB indicated significant decreases in the activities of indole acetic acid oxidase and of peroxidase, but had almost no effect on ascorbic acid oxidase and of peroxidase, but had almost no effect on ascorbic acid oxidase and polyphenol oxidase.

The application of MSB also resulted in remarkable increases in the growth of tomato plants, alfalfa callus, and the rooting of mung bean cuttings. Spraying tomato plants twice during their growth with a $10^{-5}$ M solution of this compound caused an increase of 40-80% in fruit yield per plot, relative to the untreated plants. The water-soluble part of MSB is the sodium sulfite adduct of menadione (vitamin $K_3$, 2-methyl-1,4-naphthaquinone). Soil application of a powder containing a mixture of menadione bisulfite. Tween 20, and diatomaceous earth enhanced the growth of roots and aerial parts of rice by 32% and 22%, relative to the control.

# CHAPTER 23

# Gibberellins

## HISTORY OF GIBBERELLINS RESEARCH

Study of Gibberellins (GAs) is mainly based on experimental work of Japanese plant physiologist E. Kurosawa working in Formosa, discovered GA in 1928. After conducting several trials, he was able to reproduce symptoms of the *bakanae disease* with the culture medium in which *Gibberella fujikuroi* (the perfect form of a variant of *Fusarium moniliforme*) had been grown. There was no fungal infection involved, yet the treated plants showed typical excessive elongation, pale-green colour and inhibition of root growth-characteristic of *bakanae*. Infect, the knowledge of GA was made available by a semiliterate Japanese farmer Konishi in the last decade of the $19^{th}$ century (1898) described in an agricultural book. He had described the *bakanae* (foolish seedling) disease of rice (*Oryza sativa* L.) characterized by unusual elongation of seedlings. The description of disease and its causal fungus (*Gibberella fujikuroi*) was produced by Hori (1898-1910). Kurosawa (1926) grew *G. fujikuroi* in cell-free culture medium. Wollenweber (1931) corrected the taxonomy of *G. fujikuroi* and established that Bakanae is caused by *Fusarium moniliforme* (later the imperfect stage of fungus was identified as *F. moniliforme* Sheld. and the perfect stage as *Gibberella fujikuroi*. Kurosawa (1926) further suggested that growth of rice seedling is stimulated by an active principle toxin which was isolated by Yabuta, Hayashi and Kannbe. The active principle was assigned the name *gibberella* obtained from *G. fujikuroi* culture by Yabuta in 1935. Thereafter, Yabuta and Sumiki (1938) isolated and crystallized two biologically active substances- Gibberelins A and B. These efforts could not attract attention of researchers engaged in this field till 1950.

Studies on physiology and biochemistry of GA were carried out independently in United States and England. F.H. Stodala et al. (1955) of U.S. Department of Agriculture (USDA) purified GAs from fungal culture and simultaneously scientist of Imperial Chemical Industries (ICI), England (including P.W. Brain, A. Borrow, B.E. Cross, G.W. Elson and others) started large scale production of GAs from fungus cultures.

Scientists of USDA (1955) named the culture filterate as giberellin X while ICI. group (1954) named it as gibberellic acid. Later on both the names were universally accepted as gibberellic acid and presently known as $GA_3$. C.A. West and B.O. Phinney (1956) discovered GAs as natural products of higher plants. Infect, the first isolation and chemical identification of natural $GA_1$ from higher plants (from immature seeds of *Phaseolus coccineus* L.) was made by MacMillan and Suter (1958). It is now well established that GAs occur naturally in large number of species and their presence has been ubiquitously established in all angiosperms and gymnosperms and some other plants. At present, the number of GAs known from all sources is 94.

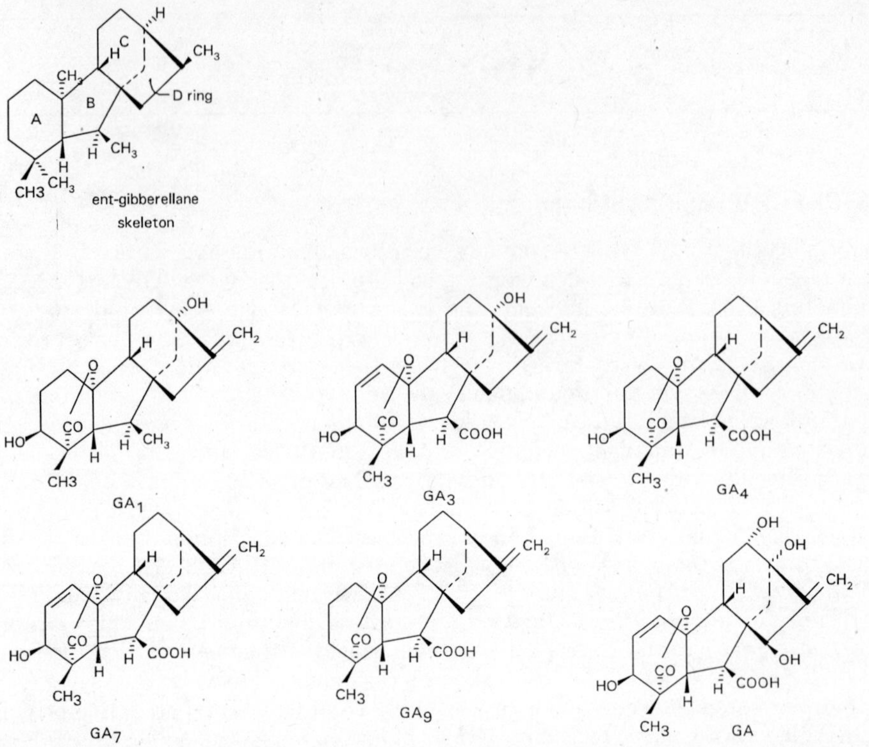

**FIG. 23.1:** Structure of the *ent*-gibberellins skeleton and of six active gibberellins. The gibberellins are numbered as for *ent*-gibberellane, except that of the carbon 20 methyl group of *ent*-gibberellane has been oxidized and then released as $CO_2$ to form the 19-carbon gibberellins shown here. Furthermore, in these gibberellins (and some others), the carbon 19 methyl group of *ent*-gibberellane has been oxidized to a carboxyl and then used to form a lactone ring.

It may be summarised that in last few years we were knowing about 4 types of GA's.

1. Gibberellin $A_1$ (similar to gibberellan described by Stodola) $C_{19}H_{24}O_6$
2. Gibberellin $A_2$ - $C_{19}$ $H_{26}$ $O_6$
3. Gibberellic acid - (Gibberellin $A_3$ similar to the gibberelline x) - $C_{19}H_{26}O_6$
4. Gibberellin $A_4$ - ($C_{19}H_{24}O_5$)

The nomenclature of $A_1$, $A_2$, $A_3$, and $A_4$ was given by Japanese workers and they have added such members like $A_5$-$A_{17}$. They further estimated that the number may go up to unlimited digit. Therefore, they described them as $A_1$-$A_n$.

The nomenclature of gibberellin is dependent on gibbane nucleus but recently it has been evaluated that the nomenclature should also be considered on the basis of one or more compounds known as gibbalane. Tani (1957) isolated three gibberelline like substance from a fungus *Helminthosporium sativm* and gave three different names- helmenthospori, helmenthosporic acid and hydrohelmenthosporic acid.

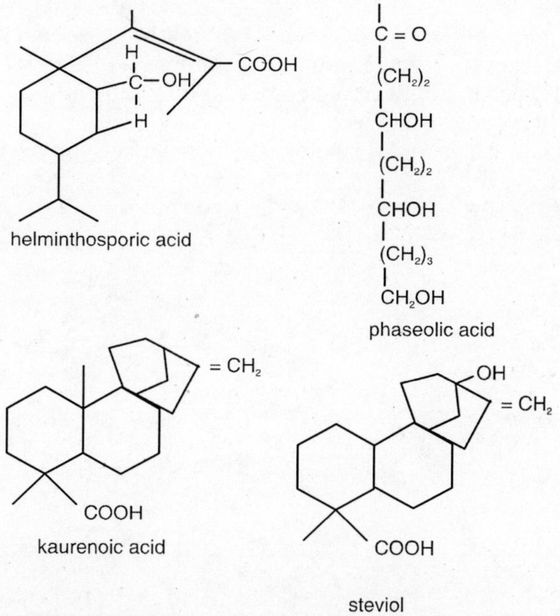

**FIG. 23.2:** Structure of some compounds which show gibberellin like activity.

## CHEMISTRY OF GIBBERELLINS

At present about 74 gibberellins are known which have been isolated from plant sources, e.g., $GA_{17}$, $GA_{38}$, $GA_{44}$, $GA_9$, $GA_{29}$, and $GA_{51}$, have been isolated from *Pisum sativum.*

As of 1990, 84 gibberellins have been discovered in various fungi and plants (Takahashi *et al.*, 1990). Of these 73 occur in higher plants, 25 in the *Gibberella* fungus, and 14 in both (Salisbury and Ross, 1992). Seeds of the cucurbit *Sechium edule* has at least 20 gibberellins and seeds of *Phaseoles vulgaris* contain as many as 16, but most other species have very few gibberellins.

All gibberellins have derivation from the *ent*-gibberellane skeleton. They are acidic and are named GA with a different subscript to distinguish them. Almost all GAs are chemically related to a large group of naturally occurring compounds-*terpenoids* (sterol and carotenoids are occur in plant tissues). The terpenoids are built from 5-carbon *isoprene units*, two unit from *monoterpene* (C-10), three units from a *sesquiterpene* (C-15) and four units from *diterpene* (C-20). *Kaurene* is the precursor of gibberellin. Gibberellins are made of basic skeleton of *ent*-gibberellane (20-carbons) or *ent*-20 norgibberellane (C-19). They are differentiated from one another by the presence or absence of the lacton configuration (*internal ester*) in the ring A and the substituents, mainly -OH (hydroxyl) groups, about the whole ring structure.

1. Gibberellins are synthesized from mevalonic acid in young tissues of the shoot. Its synthesis in the roots is still uncertain.
2. GAs are transported in the phloem and xylem tissues.
3. Gibberellins have several effects in plants and some of these include: hyper-elongation of stems by stimulating cell division and cell elongation; bolting in

long day plants; induction of seed germination especially in seeds which require stratification or light; stimulate production of several enzymes notably hydrolases in germinating cereal seeds; fruit setting and growth; and induction of maleness in dioecious flowers.

4. Gibberellins differ from indoleacetic acid in structure and properties.

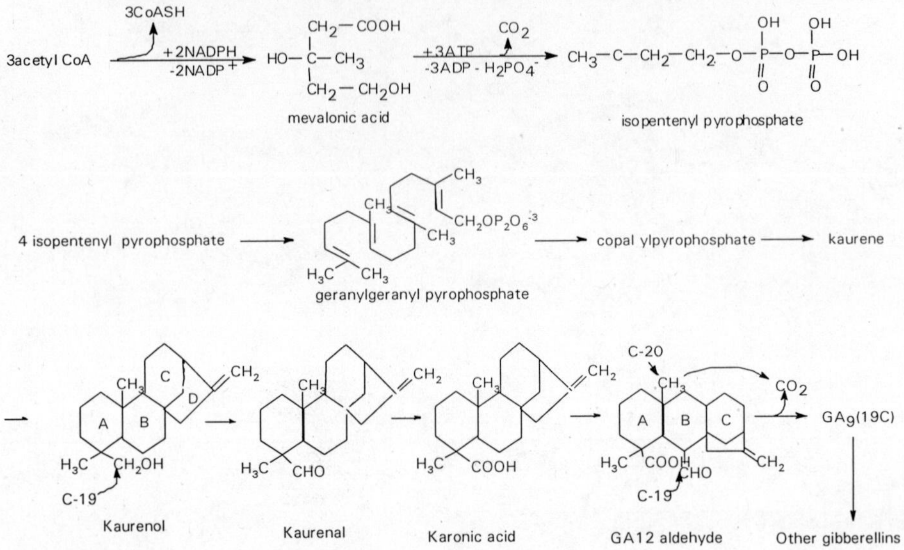

**FIG. 23.3:** Some reaction of gibberellin biosynthesis. Many steps indicated as single arrows actually involve more than one enzyme-catalyzed reaction, especially those before kaurene.

## BIOSYNTHESIS OF GIBBERELLINS

Although less is known about GA biosynthesis in higher plants than fungus *Gibberella fujikuroi*. A large amount of information is now accumulating. The initial work of GA biosynthesis was carried with liquid endosperm of immature seeds. Use of radioactive isotopes has solved the problem of GA-biosynthesis pathway.

Acetate is the precursor of GA-biosynthesis which is further transferred into active acetate groups involves the acetyl transfer coenzyme, coenzyme (CoA). The first few steps of GA-biosynthesis involve in the formation of three acetyl CoA molecules. This compound is transferred into mevalonic acid after condensation process in presence of 2 ATP molecules and an enzyme kinase. Mevalonic acid is phosphorylated in a two steps reaction to form mevalonic acid pyrophosphate. $CO_2$ is removed (decarboxylation) from in presence of decarboxylating enzyme and ATP, yields isopentyl pyrophosphate (IPP; it is a 5-carbon isoprenoid unit). IPP is the key compound from which carotenoids, gibberellinns, ABA and portion of cytokinin are derived.

IPP forms dimethyl pyrophosphate after isomerization. This is the *first* step of reaction for the formation of terpenoids in presence of IPP-isomerase.

## METABOLISM OF GA

Immature seeds are the major source of GA, therefore, our most of the information related to its biosynthesis are based on the studies conducted with immature seeds.

Metabolic studies carried out in different laboratories shown in figure from mevalonic acid to $GA_{12}$ - GA aldehyde have been established in enzyme preparation from seeds of *Cucurbita maxima, Marah macrocarpus.*

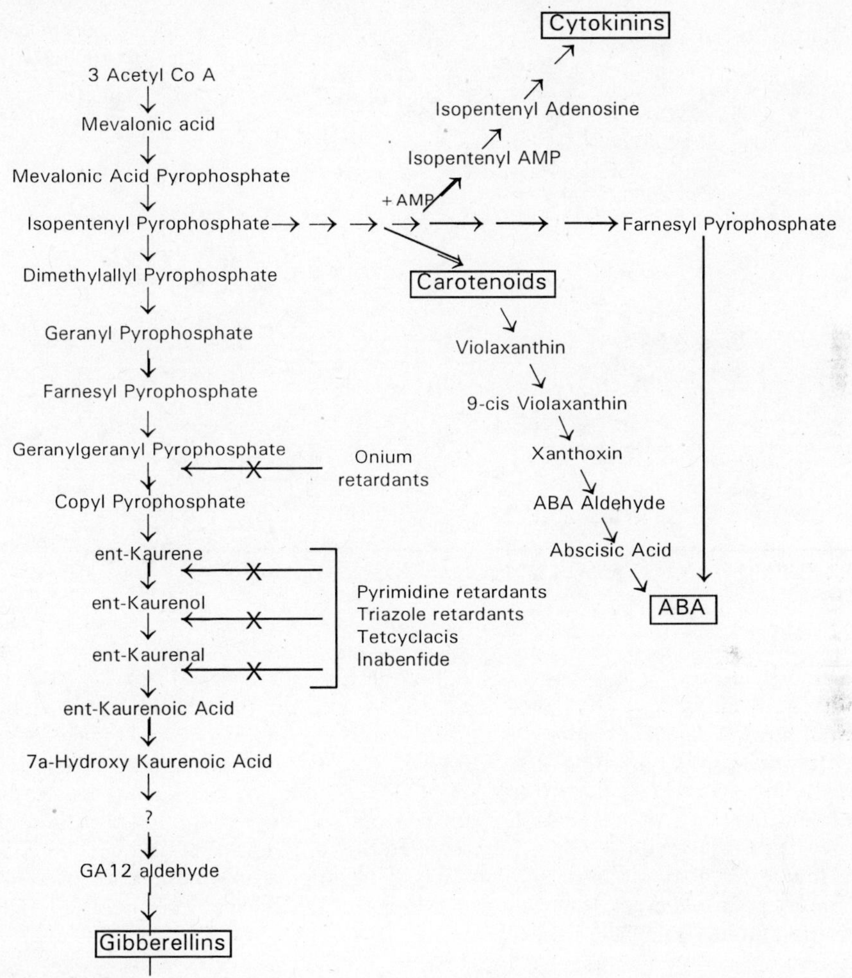


**FIG. 23.4:** Mevalonic acid pathway for the biosynthesis of gibberellin, cytokinins and ABA.

The branch from end 7-a-hydroxy kaurenaic acid to 6ᵦ -12 hydroxy phenolic acid have been observed from cucurbita maxima and culture of *G. fijikuroi.* Several pathways have been suggested which mainly differ in the position and sequence of hydroxylation.

Geranyl geranyl pyrophosphate is the precursor of all types of terpenes. The enzyme which catalyses its formation from mevalonic acid has not been studied in detail yet.

There are some indications that anti kaurene synthetase may be the site of regulation. The enzymes that catalyse the step from mevalonoic acid to ant-kaurene are soluble and require $O_2$ and reduce pyridine nicotine (preferably NADN) and appear to be dependent upon cytochrome $b_{250}$ with electron transport system.

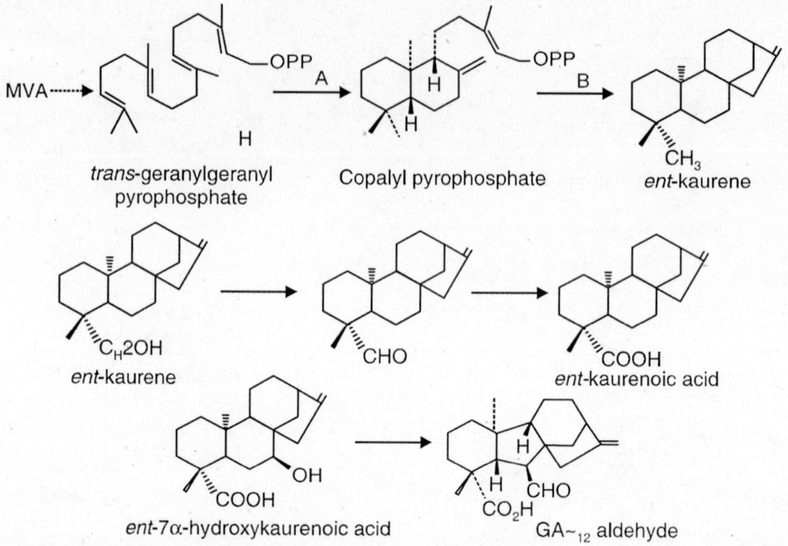

MVA → *trans*-geranylgeranyl pyrophosphate —A→ Copalyl pyrophosphate —B→ *ent*-kaurene

*ent*-kaurene ($C_H2OH$) → (CHO) → *ent*-kaurenoic acid (COOH)

*ent*-7α-hydroxykaurenoic acid → $GA_{\sim 12}$ aldehyde

**FIG. 23.5:** The biosynthetic pathway of gibberellin $A_{12}$-aldehyde.

## BIOASSAY

The most important bio-assay of gibberellins is increased growth of dwarf seedlings. Some strains of maize and pea are shorter (dwarf) in length than the normal strains. These strains are mutants and are believed to be dwarf because they have negligible amount of endogenous gibberellins. Application of exogenous gibberellins overcomes the dwarfness of the seedling by increasing the length of the stem. In dwarf maize seedlings, gibberellin solution is applied to the ligule of the first leaf. This causes marked increase in the length of next node or leaf sheath, in a few days. This bioassay is sensitive to as little as 1.0 nanogram of GA. The hormone is applied to the dwarf seedling and increase in epicotyl length is measured after 4-5 days.

Another widely used bioassay of gibbrellins is the barley endosperm test. By this method, as little as 0.2 nanogram of GA can be detected within 24 h. Barley aleurone (de-embryonated half seeds) layers are incubated in a solution of GA for 24 h. The gibberellins stimulate amylase activity which causes hydrolysis of starch to glucose. The glucose is secreted out in the incubation medium. This accelerated secretion of glucose (reducing sugar) solution.

Bioassays are primarily designed to detect the presence of hormone in plant tissue extract. However, now a days more sophisticated biochemical techniques are available to detect and quantitate the plant hormones. For example, gibbrellins can be determined by using gas chromatography, high performance liquid chromatography (HPLC) and other such analytical techniques.

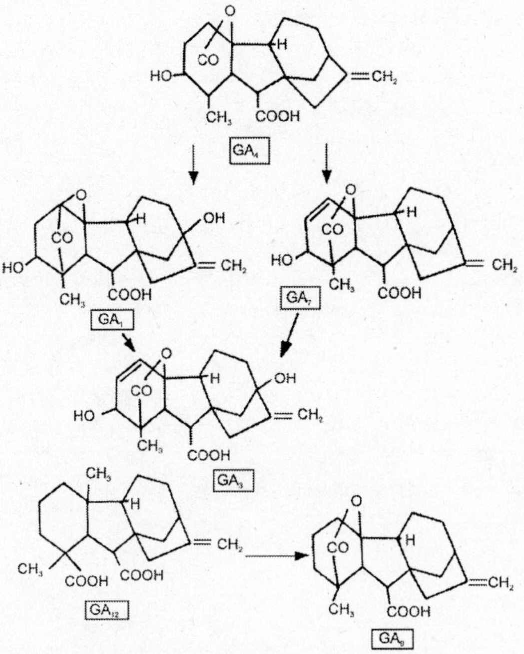

**FIG. 23.6:** Some known interconversions of gibberellins.

## MECHANISM OF ACTION OF GA

### GA and Growth: Role in Cell Division

It has been argued that GA can promote growth of plants by affecting either cell expansion or cell division or both. It must be emphasised that cell division alone cannot result in growth of organisms. Cell division contributes to growth by producing more cells which can undergo expansion. The effect of $GA_3$ on cell division can be readily accounted by an effect on cell cycle.

Jacqmard (1966) has proposed that one of the effects of $GA_3$ is to promote the onset of DNA synthesis in cells which are arrested a $G_1$ phase of cell cycle. It has been established that capacity of cell to respond to $GA_3$ by undergoing division is the function of their state of differentiation. According to Cosgrov (1961) the cell expansion after $GA_3$ treatment is based on biophysical phenomenon. The knowledge is further expanded on the basis osmotic potential. The mechanism for GA3 activity in growth involving the conversion of starch to sugar was inferred by analogue with the known effects of $GA_3$ by increasing in length and allowing auxin treatment, there was increase in cell wall extensibility but GA-treatment did not result in a change in cell wall yielding properties. Since $GA_3$ induced elongation was not correlated with cell wall elongating properties.

## Extensibility Changes in $GA_3$ Treated Tissues

Although the early experiments of Cleaveland (1968) suggested that $GA_3$ did not affect extensibility property of cucumber hypocotyl; evidence from several laboratories using many plant species indicate that $GA_3$ have marked influence on extensibility, using the osmotic method to determine the extensibility of tissues examined that effect of light, $GA_3$, KCl on wall yielding property of hypocotyl sections. Light reduced the extensibility of hypocotyle tissues which darkness and $GA_3$ overcome this inhibition.

Although cell expansion can be explained in biophysical terms, the events leading to cell wall extensibility are dependent on metabolism. It is well established example that cell expansion rapidly inhibited by respiratory poisons. This shows a direct relation between cell wall extensibility and readily available metabolites of plant system.

The plant cell wall contains proteins and inorganic ions in addition to complete polysaccharides. There have been several proposals that both proteins and ions specially $Ca2^+$; play a role in wall extensibility. However, $Ca2^+$ has long been known to cause decrease in wall extensibility in lettuce hypocotyl.

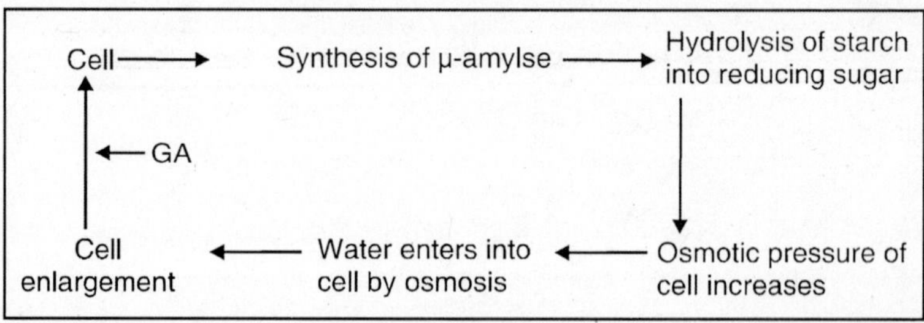


**FIG. 23.7:** Mechanism of action of GA on osmotic potential of a cell.

Moll and John (1981) indicated that $GA_3$ promotes uptake of $Ca2^+$ ion upto cytoplasm. It was proposed that in this tissue $Ca2^+$ activity in cell wall governs tissue extensibility and that extensibility could be regulated by uptake of ion into cytoplasm. Fary (1980) proposed that cell wall extensibility can be influenced by formation of cross bridge between adjacent ferulic acid (phenolic acid) moieties in the wall. The formation of diferuloyl bridges could be catalysed by peroxidase resulting in gelatinization of cell wall. Fery further demonstrated that $GA_3$ promotes cell expansion and inhibits release of peroxidase of cell. However this hypothesis may not explain how extensibility is controlled in dicot tissues since the primary cell walls of dicot contain negligible amounts of phenols.

Atleast two cell wall loosening factors have been identified and has shown to involve in hormone-induced cell wall expansion in plants. The role of proton ($H^+$) is described by acid growth hypothesis which is based on the observation that auxin-treatment results in rapid proton pumping from elongating tissue and that H+ alone can promote growth of auxin treated tissues.

Hydrolytic enzymes have also shown to be involved in modification of higher plants cell walls and their activity is thought to be modified by hormones.

## $GA_3$ and Synthesis of RNA Proteins

It is well established that there is an increase in level of RNA and protein in GA-treated plant tissue which causes an increase in total metabolites in plant cell.

## The Cereal Aleurone Layer

Alurone layer of cereal seeds is responsible for degradation of starchy endosperm. Paleg in Australia and Yamo in Japan showed simultaneously that endosperm breakdown in de-emleryonated Barley caryopsis could be stimulated by addition of $GA_3$ to the embryo less endosperm. The function of $GA_3$ is to control production of hydrolyting enzymes by aleurone layer. In barley, the aleurone layer is 2-4 cells layer thick white oats, rice and wheat, it is one cell layer thick. The aleurone layer respond to $GA_3$ by producing a wide spectrum of hydrolytic enzymes when isolated aleurone layers are incubated in optimal $GA_3$ concentration (10-5 -10-4m) containing 10-20 minimol $Ca2^+$ numerous hydrolytic enzymes are produced. Among newly synthesized enzymes is a-amylase which makes up to 60-70% of total protein produced by these cells. After 24 hours of incubation in $GA_3$ there occurs ultra-structural changes in aleurone cells which accompany hydroloses synthesis and secretion. There is no evidence for an increase in total amount of lipid in $GA_3$ treated barley aleurone.

The ER is site of α-amylase synthesis and $GA_3$ treated tissues in ER contain 4 iso-enzymes of α-amylase while in untreated tissues it contains only 2. The ER is also involved in transport of α-amylase from barley aleurone cell. There is compelling evidence that $GA_3$ effects the synthesis of hydrolytic enzymes in cereal aleurone and that it does by controlling the transcription and possibly translation of new messenger RNA. The synthesis of RNA is a pre required for the production of GA-induced enzymes. The α-amylase of barley produced in vivo has a molecular weight of 44000 while that produced in vitro has a mol weight of 46000.

The events leading to synthesis and secretion of GA-induced in cereal aleurone can be summarised as follows:

Transcription of a specific group of genes is promoted by $GA_3$ and newly transcribed equals are transported to the cytoplasm and translated on ER. Following translational cleavage of single sequence by the ER membrane, the completed -amylase molecule is transported from ER to the exterior of the cell by an as yet unidentified route. However, there is no evidence that $GA_3$ control post translational processing of proteins or their transport despite of the fact that on barley $GA_3$ effects the organisation of ER system.

## Dicotyledonous Fruits and Seeds

The effect of $GA_3$ on germination of dormant and non dormant dicotyledonous seeds and on growth of seedlings are well known. The studies are based on fat storing seeds, in particular to those of almond, castor bean.

In the endosperm of *Ricinus communis* seedlings; $GA_3$ stimulates the activity of about 16 enzymes involved in gluconeogenesis but not the activity of a other enzymes of intermediately metabolites that are not involved in gluconeogenesis.

Martin and North Coat (1982) observed the effect of GA on protein synthesis. They purified the m-RNA for isocitrate ligase from control to $GA_3$ treated castor bean endosperm showed that relative amount of mRNA of isocitratelyase is lower in $GA_3$-treated tissue than untreated control.

$GA_3$ influences the activity of enzymes of intermediatory metabolism in castor bean seeds. The action of $GA_3$ in castor bean aleurone. of the synthesis of some protein and inhibitor of others seems to be conflict with the non-specific effect of

$GA_3$ on RNA and protein synthesis in castor bean endosperm. Additional experiments are need before it can be concluded that action of $GA_3$ in these seeds is indeed different.

It can be concluded that effects of $GA_3$ on growth, development and metabolism appear to involve action of GA at several biochemical levels, it is possible that molecular mechanism whereby this hormone acts as the same. There is now convincing evidence that auxins are involved in control of synthesis of specific mRNA in elongation tissue.

Since GA and auxin promote growth in similar fashion $GA_3$ may also promote changes in RNA synthesis during elongation. GA could quantitatively effect and accurate the synthesis of all kinds of RNA's.

1. migrates into the aleurone layer,
2. where hydrolytic enzyme production is stimulated these enzymes are released into the starchy endosperm,
3. where they serve to hydrolyse endosperm reserves,
4. producing solutes that can inhibit further enzyme production, and
5. function to nourish the growing embryo.

## PHYSIOLOGICAL EFFECTS AND PRACTICAL APPLICATIONS OF GIBBERELLINS

Since the discovery of GAs in the 1920s a considerable body of information has emerged regarding the roles these hormones play in plant growth and development. As these compounds and analogs became available, either by fermentation or chemical synthesis, a number of potential practical applications were identified.

Gibberellins have been primarily used for manipulating production practices and insuring the quality of high value specialty crops such as grapes, citrus, cherries, and apples. Increasing market demands for a greater variety of quality fruit may provide new opportunities for the expanded use of Gas (i.e. tropical fruits, thinning and sizing wine and new table grape varieties, and as components of integrated pest management programs for postharvest quality of fruits). Despite considerable research, GAs have not found utility in major agronomic crops. However, increased understanding of the interrelationship between GAs and other PGRs, concurrent with the development of new GAs or GA-mimics, may offer the unique activities necessary to penetrate agronomic markets.

However, despite extensive research with GAs the number of commercial uses has fallen short of their predicted impact and growth in agriculture. Unique opportunities may still result from new research and development with GAs to meet the future demands for increased food quality and production.

Although over 97 different GAs have been identified in plants, only three ($GA_3$ and a mixture of $GA_{4+7}$) are available via fermentation by the fungus *Fusarium monilifome* (= *Gibberella fujikuroi*) in reasonable quantity for extensive applied research. The complexity of GA chemistry has precluded the entirely synthetic routes of production. With the diversity of physiological processes in which GAs are involved, it is not surprising that their practical application spans the entire range of plant growth and development. The majority of these uses however, represent specially niche markets in high cash value horticultural commodities. Despite the extensive research with $GA_3$ over the past 30 years, no defined GA product has been registered for use on a major agronomic crop. This may be due in part to a lack of specificity of response where multiple effects, often detrimental (i.e. lodging), are manifested from exogenous application of GA.

**TABLE 23.1:** Physiological response and current commercial uses of gibberellin an cytokinin products registered in the USA and Europe

| Physiological response | Crop | PGR Used[a] |
|---|---|---|
| Breakage of dormancy | Seed potatoes | $GA_3$ |
| | Rhubard | $GA_3$ |
| Acceleration of seed germination/hydrolytic enzyme production | Barley malting | $GA_3$ |
| Stimulation of bud break and | Tart cherries | $GA_3$ |
| increased lateral branching | Apples | $GA_{4+7}$/BA |
| | Conifers | BA |
| | Carnations/Roses | PBA |
| Stimulation of fruit set | Tangerine hybrids | $GA_3$ |
| | Blueberries | $GA_3$ |
| Reduction of fruit set | Grapes | $GA_3$ |
| | Tart and sweet cherries | $GA_3$ |
| Increased fruit size and quality | Grapes | $GA_3$ |
| | Apples | $GA_{4+7}$/BA |
| | Sweet cherries | $GA_3$ |
| | Apples | $GA_{4+7}$ |
| Increased vegetative | Sugar cane | $GA_3$ |
| growth and yield | Spinach | $GA_3$ |
| | Hops | $GA_3$ |
| Delayed senescence/ fruit ripening | Banana | $GA_{4+7}$ |
| | Sweet cherries | $GA_3$ |
| | Citrus | $GA_3$ |
| | Bermuda grass | $GA_3$ |
| Stimulation of abscission/ senescence | Cotton | TDZ |

[a]Abrreviations: $GA_3$: gibberellic acid; $GA_{4+7}$: gibberellin $A_4$ and $A_7$; BA: benzyladenine; PBA: tetrahydropyranylbenzyladenine; TDZ: thidiazuron.

## Floriculture

$GA_3$ is also used to thin approximately 60% of the flowers because almost every flower tends to set, resulting in a tightly packed cluster. The physiological basis for grape flower thinning by GA is unknown, but could be due to pollenicidal activity.

The result of adequate thinning almost appears excessive, but with two subsequent applications of GA, berry size is increased by 60%. The final product is a loose cluster that does not require hand thinning and is less susceptible to *Botrytis* bunch rot.

$GA_3$ also inhibits flowering on several stone fruits. In young sweet and tart cherries, $GA_3$ is applied to prevent excessive flowering to minimize the competitive effect of early fruiting on vegetation growth. Since cherry has true flower buds, the node following flowering will not support vegetative growth. If flowering is excessive, fewer spurs develop, thereby resulting in lower yield potential during the production life of the tree. The manipulation of flowering by $GA_3$ has also been integrated into a management program to suppress a pollen transmitted ring spot virus infection as well as the maintenance of the desired ratio of vegetative to flower bud formation in mature tart cherries. The consequence of continuous $GA_3$ applications over a five year period has been a 22% increase in yield.

A mixture of $GA_{4+7}$ is commercially available. It affects on varying structure in the flowering of *Lolium temulentum*, a LD-requiring plant and *Pharbitis nil*, a SD-requiring plant, as well as for flowering of Cupressaceae and Pinaceae conifers. While it is too early to predict commercial implications of such specificities in flowering response for GAs.

## Fruit

An additional application of $GA_3$ to sweet cherry when fruit is light green to straw coloured will delay harvest, increase fruit size and firmness. While this application is not new, the recent development of an export market to Asia for cherries has significantly increased this use.

### *Fruit Reset*

The other commercially used GAs produced by *Fusarium moniliforme* cultures are a mixture of $GA_4$ and $GA_7$. An application of $GA_{4+7}$ of increasing importance is the prevention of fruit russet, a superficial disorder in which the exocarp is interrupted by raised corky outgrowths over the surface of green and yellow apple cultivars. Although somewhat obscure, a number of factors have been implicated for russet, such as high humidity, presence of free water, light frosts, pesticides and powdery mildew. The multiple applications on `Golden Delicious' effectively increase the fruit finish quality and this correspond with higher packout gradens. Recent work in the US and Canada has suggested that the $GA_4$ component of $GA_{4+7}$ may be the primary active ingredient in the suppression of russet, while $GA_7$ appears to inhibit flowering in apple and reduce return bloom. In these studies, $GA_4$ appeared to have no effect on return bloom.

### *Grape Industry*

$GA_3$ has been used in the seedless table grape industry since 1960 to manipulate three physiological events; rachis cell elongation, flower thinning, and berry enlargement. The first is a pre-bloom spray to promote elongation of the rachis to provide a larger and looser framework for the grape cluster. In wine grapes, there is a need for cluster loosening by either rachis elongation and/or flower thinning to prevent *Botrytis* bunch rot. To date, there is no chemical thinning agent for wine grapes. Results with $GA_3$ have been variable with some varieties exhibiting a dramatic reduction of return bloom the year following application; others are relatively insensitive to increasing rates of $GA_3$. The response of wine grapes to other GAs certainly needs to be examined.

### *Citrus*

The effect of delayed fruit senescence by $GA_3$ has also been exploited in the citrus industry. $GA_3$ delays maturity of citrus rind without altering the maturation process of the internal fruit, thus preventing several rind disorders associated with fruit ripening. In lemons, this facilities scheduling of harvesting to synchronize with market demands. As harvest date is delayed in grapefruit, there is added benefit in combining 2,4-D to retain the fruit on the tree. Recent findings suggest that the delay in citrus rind maturity is associated with the quantity and quality of epicuticular wax deposition and its influence on respiratory gas exchange.

### *Apple*

More recent results suggest that $GA_4$ may promote apple flowering. The differential effect of $GA_4$ and $GA_7$ in apple serves to illustrate the specificity associated with a specific GA. However, separation of these compounds on a large scale is presently cost prohibitive.

### *Seed Production*

There are also minor practical uses of GAs in seed production of commercially important conifers. $GA_3$ can be used to promote flowering, and thereby significantly increase female cones and seed per tree in seed production orchards of Cupressaceae and Taxodiaceae species, and is used commercially in Japan and western British Columbia, at least, for this purpose. For Pinaceae species, which include most of the commercially important conifers, only the $GA_{4+7}$ mixture can significantly promote flowering.

## Protection against Plant Pathogens

The use of GA to modify growth for protection against plant pathogens was described previously in cherries and grapes and is becoming an increasingly recognized potential benefit of PGRs in a number of crops.

### *Insect Control*

Recent laboratory studies have shown that $GA_3$ application significantly reduces the number of probes and egg fecundity of the fruit fly due to increased exposure to toxic peel oils contained in the juvenile rind. If this effect is manifested in the field, a high level of resistance may preclude the need for postharvest treatment with an insecticide/larvicide.

## Applications on Agronomic Crops

The successful introduction of a GA product in an agronomic crop is dependent upon the specificity of a physiological response over a wide range of environmental/cultural conditions, and is complicated by the inherent genetic variability amongst cultivars. For example, the genetic reduction of plant height has led to dramatic increases in cereal grain yield, due principally to increased reduction of lodging. Consequently, these varieties have poor emergence rates and poor seedling vigor. In rice, GA application promotes seedling establishment in the principle semi-dwarf commercial varieties utilized. However, in wheat, the incorporation of the Rht1 and Rht2 dwarfing genes have rendered many commercial wheat varieties to be insensitive to applied GA. The high productivity of hybrid maize and GA-insensitive wheat varieties may be associated with enhanced endogenous levels of GAs. However, the differences in the response of maize to exogenous $GA_3$ may be a function not only of the endogenous concentrations of GAs, but also of varying rates of GA metabolism. In the GA insensitive wheat varieties compartmentalization may influence response to exogenous GA.

## CHAPTER 24

# Cytokinins: Juvenile Hormones

## HISTORICAL BACKGROUND

Cytokinins (CKs) were initially reecognised by their ability to induce cell division in certain plant tissues cultured in defined media containing an optimal concentration of auxin. The idea that specific chemical substances are involved in cell division was put forward by J. Wiesner (1892) and later, it received some experimental support by the work of Haberlandt (1913, 1921) who observed that wounding induced cell division in many plant tissues, e.g., potato parenchyma. The stimulus was lost if the tissue slices were washed with water, and the application of the crushed tissue or phloem diffusates restored cell division activity. Studies by Skoog and coworkers on the growth requirements of the plant tissue cultures finally led to the identification of CKs (initially called kinins). They observed that cell division in tobacco stem pith tissue grown on medium supplemented with auxin occurred only if a piece of vascular tissue was in contact with the pith, and that the effect of vascular strands could be replaced by coconut milk, malt extract and extracts of vascular tissues.

Later, a compound with high cell division promoting activity was isolated from autoclaved herring sperm DNA and identified as 6-(furfuryl amino) purine, commonly known as kinetin - $C_{10}H_9N_5O$ (Miller et al., 1956). Although kinetin (K) induced cell division in tobacco pith assay at concentrations as low as 1mg/l, it does not occur per se in living tissues and was an artefact of the autoclaving process due to dehydration and migration of a deoxyribose moiety from the 9-position of adenine to the $N^6$-position.

Although extracts from a wide variety of plant sources were found to contain kinetin-like activity, it was only in the early 1960s that the first naturally occurring CK, Zeatin (Z) was isolated from immature kernels of *Zea mays*, and later identified as 6-(4-hydroxy-3-methyl but-*trans*-2-ethylamino) purine (Letham et al., 1964). This compound turned out to be identical with the substances isolated but not fully characterized by Miller in 1961 (Letham and Miller, 1965). Letham (1966) also identified Zeatin riboside ([9R]Z) and its 5'-monophosphate ([9R-5'P]Z), and a number of related compounds with weak CK activity (Letham, 1973) from corn kernels. Dihydrozeatin (DiH)Z, a Zeatin analogue with saturated side chain, was isolated from lupin seeds soon after the identification of 6-(3-methyl but 2-ethylamino) purine (iP), first as the ribonucleoside from two serine tRNAs of yeast and then as a free base from culture filtrates of *Cornebacterium fascians* (Klambt et al., 1966). iP and its riboside, had previously been synthesized and shown to be about ten times more active than K in the tobacco callus test.

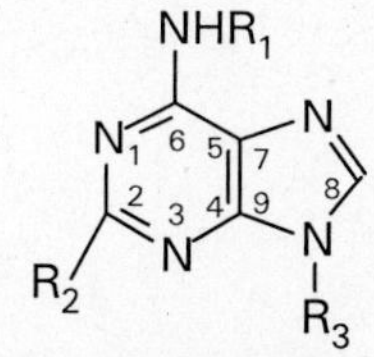

NHR$_1$ (purine ring, positions 1–9; $R_2$ at C-2, $R_3$ at N-9)

| Substituents | | | Trival name | Abbr. |
|---|---|---|---|---|
| $R_1$ | $R_2$ | $R_3$ | | |
| 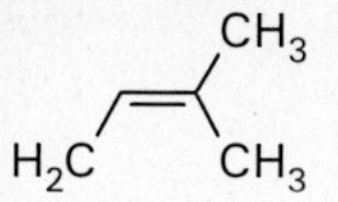 $H_2C$–CH=C($CH_3$)$CH_3$ | H<br>H<br>$CH_3S$ | H<br>ribofuranosyl<br>ribofuranosyl | $N^6$($\Delta^2$-Isopentyl) adenine<br>$N^6$($\Delta^2$-Isopentyl) adenosine<br>2 methylthio $N^6$($\Delta^2$-Isopentyl) adenosine | $i^6$ Ade<br>$i^6$ A<br>$ms^2$ $i^6$ A |
| 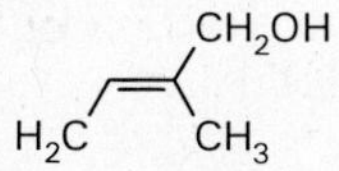 $H_2C$–CH=C($CH_2OH$)$CH_3$ | H<br>H | H<br>ribofuranosyl | Zeatin<br>Zeatin riboside | $io^6$ A |
| 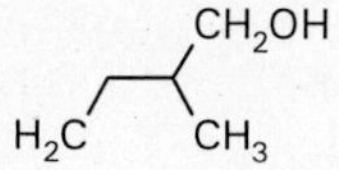 $H_2C$–$CH_2$–CH($CH_2OH$)$CH_3$ | H | H | dihydtozeatin | $H_2i^6$ Ade |
| 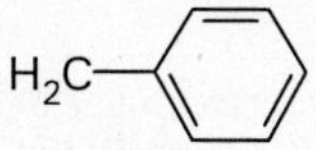 $H_2C$–(phenyl) | H | H | $N^6$ (benzyl) adenine | $bzi^6$ Ade |
| 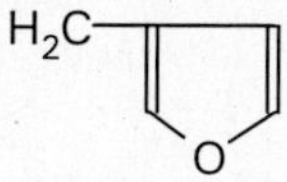 $H_2C$–(furyl) | H | H | kinetin | |
| H | H | 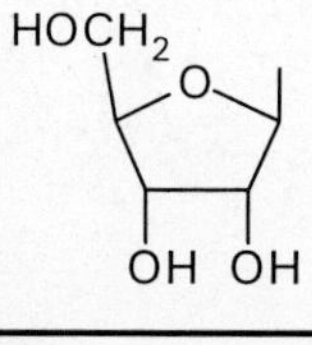 $HOCH_2$–(ribofuranose, OH OH) | β-D-ribofurranosyl adenosine | |

**FIG. 24.1:** Chemical structures, trivial names and IUPAC-IUB abbreviations for some naturally occurring and synthetic cytokinins.

Todate the number of naturally-occurring free CKs identified in plants or in microorganisms is over 30.

## CHEMICAL NOMENCLATURE

All naturally occurring cytokinins (CK) like zeatin L (Z) and isopentyl adenine(iP) and many synthetic CKs, like kinetin (K) and 6-benzyl-amino purine (BAP), are 6 substituted amino purines. However, there are reports that BAP and its derivatives exist in plants (Nandi *et al.*, 1989). A number of synthetic non-purines including substituted ureas, amides and pyrimidines (see Matsubara, 1980) are known to exhibit CK-like activity, e.g. N-phenyl-N-1,2,3-thio-dizol-5-yl urea (thydiazurone) shows CK activity comparable to that of Z (Mok *et al.*, 1982). Even though dihydro-coniferyl alcohol (Lee *et al.*, 1981) and dehydro-diconiferyl glucosides (Lynn *et al.*, 1987) are naturally occurring non-purines which promote cell division, they are not classified as CKs.

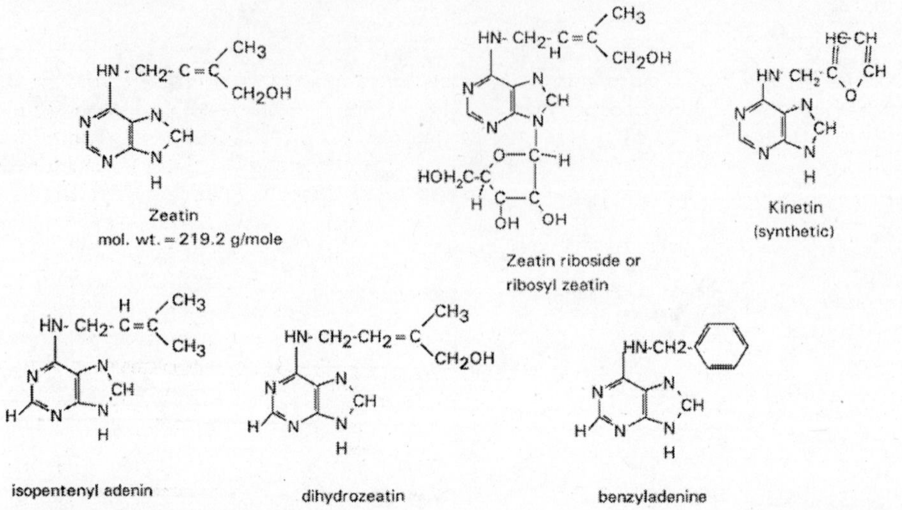

Zeatin
mol. wt. = 219.2 g/mole

Zeatin riboside or ribosyl zeatin

Kinetin (synthetic)

isopentenyl adenin

dihydrozeatin

benzyladenine

**FIG. 24.2:** Structures of common natural and synthetic (kinetin) cytokins. All these are adenine derivatives in which the purine ring is numbered as shown for zeatin (upper left). Zeatin and zeatin riboside can exist with groups arranged about the side-chain double bond either in the *trans* (as shown) or *cis* (with $CH_3$ and $CH_2OH$ groups interchanged) configuration. The *cis* form predominates in tRNA-bond cytokinins, but the *trans* form exits in free zeztin and zeatin riboside.

All the known naturally occurring CKs are virtually substituted purines. The system used for increasing the purine nucleus in also used in CK nomenclature to define the position of substitutents. Many systems of abbreviations for CKs, including those based on established system for nucleic acid derivatives IUPACIUB commission on Biochemical Nomenclature (1970) have been used. The nomenclature proposed by Letham (1978) is now widely adopted. Zeatin (Z), isopentenyl adenine (P) kinetin (K) and 6-benzyl aminopurine (BAP) are regarded as parent compounds, and substitutions on the purine nucleus are shown in square brackets, e.g. [9R]Z for zeatin riboside, and substitutions on the side chain are denoted in round brackets e.f. (diH)Z for dihydrozeatin and (OG) [9R]Z for O-Glucosyl derivative of zeatin riboside.

## Cytokinin levels in leaves and xylem sap

Analysis of CK levels in xylem exudate and leaves of plants during development revealed that leaf senescence might be under the regulation of CKs in sites. The induction of leaf senescence is possibly due to the environment or soil bacteria that reduce CK production in total leaves in xylem (Skene, 1975). The endogenous levels of CKs in the xylem sap seem to be an important factor in regulating leaf senescence. in sunflower CK has been reported to decline markedly at the commencement of flowering which precedes induction of leaf senescence (Sitton *et al.*, 1967; Kende and Sitton, 1967). Furthermore, reduced CK levels in the leaves of some plants were reported during the reproductive phase (Hewett and Wareing, 1973a; Oritani and Yoshida, 1973; Lindoo and Nooden, 1978; Ray *et al.*, 1983).

The above data provide strong support for the control of leaf senescence in many species by endogenous CK, a particularly root derived CK moving in the xylem.

## PHYSIOLOGICAL EFFECTS OF APPLIED CYTOKININS

Externally applied CK's are known to exert spectacular effects on the growth and development of plants and excised plant tissue. The regulation of all organisation in cultured tissue is one of the most dramatic effect of CK's and an example of delicate interaction of auxins and CK in mosses beed formation is induced markedly by CK's. Exogenous CK's have also a pronounced effect on stimulation of cell enlargement in cotyledons of diverse species and on leaf disc. They have been found to substitute for light in some cases e.g. CK's can induce germination of light sensitive seeds. Pigment [Beta-cyanin] synthesis in cotyledons of *Amaranthus* seedling, chlorophyll synthesis in detached yellow leaves and etiolated cotyledons, chloroplast development and opening of stomata, reduction of leaf senescence and release of lateral buds by apical dominance by CK application are some instances which found commercial use. CK can also induce nutrient mobilization and recently have been found to influence unloading of phloem derived photosynthate into cytoplasm of bean seed coats (Palni, 1990).

## BIOSYNTHESIS OF FREE CYTOKININS

Two mechanisms have been proposed for free CK biosynthesis, one resulting from the turnover of CK containing t-RNA molecules and the other-de novo biosynthetic pathway.

### Biosynthesis via tRNA turnover

Transfer RNA is a potential source of free CK's. It has been suggested that t-RNA hydrolysis may result in release of kinins in dying, autolyzing cells of differentiating vascular tissue. The free CK of bacteria may arise from t-RNA turn over. CK can also be produced from the culture of *Agrobacterium tumefaciens* and from *Lactobacillus acidophilus* and primary roots of *Zea mays*.

### De novo Biosynthesis

Free CK biosynthesis may also occur independently of tRNA breakdown obtained a cell free preparation from saline mold *Dictyostium discoideium* and showed adenosine 5'aminophosphate, to be the acceptor molecule of isopentyl group resulting in formation of [R-5'P] is the first step of CK biosynthesis. Enzymes 3-isopentyl pyrophosphate-3-isopentyl transferase community referred as CK synthase or isopentyl transferase, catalysing the above reaction, is considered to be key reaction of biosynthetic pathway to free CK. The existence of an alternative pathway of CK biosynthesis in plant tissue is also supported from in vivo labelling studies where radio labelled precursors (commonly adenine) were shown to be incorporated into CK's.

### Site of CK Biosynthesis

CK activity has been detected in extracts of almost all plant hormones and in many organisms. Although root apices have been reecognised as a major site of CK biosynthesis, there are circumstances to indicate that developing seeds, growing buds and shoot apices and leaves are additional sites of CK biosynthesis for the more cambium and embryonic axis have also been suggested as possible sites of biosynthesis. Developing leaves have been shown to contain high CK activity. **(Refer to Fig. 23.4).**

## CYTOKININ METABOLISM

When CKs are exogenously applied to plants they rapidly metabolise to a variety of different compounds. Most of the metabolites have been identified by co-chromatography with authentic compounds. Only few have been identified by mass spectrometry. CK-metabolism has been extensively studied in the past 15 years and a generalized scheme of metabolic interconversion has emerged. (Fig. 24.4).

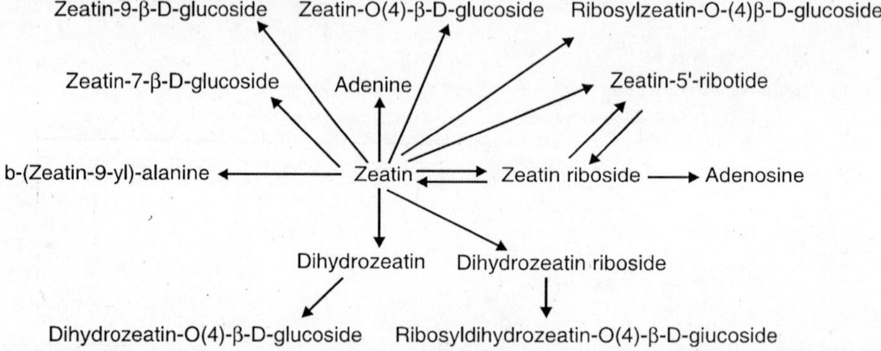


**FIG. 24.4:** Some of the known routs by which plants can metabolize zeatin. The pathways illustrates do not all occur in any single plant tissue but represent the results of metabolic experiments with several species of plant.

## ROLE OF CYTOKININS IN PLANT DEVELOPMENT

Externally applied CK's markedly influence a diversity of physiological processes and thus affect plant growth and development. Selected areas of plant development which has been studied in some detail and appear to be influenced by CK's are as follows:

### Seed Germination and Early Seedling Growth

Phytohormones play a regulatory role in the imposition of seed dormancy, in the release of seeds from dormancy, in the reserve mobilization during seed germination and in the subsequent development of the seedling. (Van Staden 1983) CK's are important in seed germination. This is indicated by the following evidences:

a) In cultured embryos of zea mays, the radical growth was retarded by the removal of the endosperm. This could however be partially restored by the addition of a plant extract reported to contain CK-O- fkucoside. (Smith and Van Staden, 1979).
b) Pinfield and Stobart (1972) observed that the elongation of the radicle of sycamore seeds could be induced by the external application of CK but not GA.
c) The endogenous CK levels increased markedly in the embryo sporophyte prior

to onset of radicle elongation in germinating *Podocarpus henkelii* seeds.

Thus the endogenous CK's would appear to be one of the key factors in the initiation of radicle growth. Endogenous CK appears to be important in reserve mobilization in germinating seeds, particularly in dicots. Seeds store large amount of food reserves which are localized in the endosperms or in the cotyledons. These reserves are degraded by a series of hydrolytic enzymes following imbibition. The embryonic axis is known to play an important role in the formation of these enzymes for the mobilization of reserve. The axial effect can be replaced, at least in part, by CK's - an ability not usually shown by other phytohormones with possible exception of GA in few cases.

There is evidence to indicate that CK's are produced in the embryonic axis from where they are transported to the cotyledons and develop varied physiological responses. The involvement of CK in monocot seeds, particularly in cereals, has also been studied and discussed comprehensively by Bewtey and Black (1978) and Halmea (1985).

## Cotyledon Expansion

The most dramatic and well-studied response to CK's in immature seedlings is cotyledon expansion in dicots. Some cotyledons respond to both GA and CK (Xanthium) and some markedly to CK but only weakly to GA (e.g., Radish) and some only to CK (e.g., Lupin) and Trigonella and cucumber in darkness.

## Chlorophyll Synthesis

Another important effect of CK's is their ability to stimulate chlorophyll synthesis and accelerate chloroplast differentiation in the detached cotyledons in light. Recent work by Nandi et al (1989) indicated that the stimulus required for chlorophyll formation and cotyledon expansion in the germinating seeds would appear to be a CK emanating from the axis.

## Cell Division and Organ Differentiation

CK's are associated with the induction of cell division in cultured tissues and in combination with auxin, and exhibit a striking quantitative relationship in the regulation of morphogenesis.

The organ formation in tobacco pith cultures can be controlled by the concentration and ratio of auxin and CK in the medium (Skoog and Millar (1957). It was found that low CK auxin ratios promoted bud (shoot) formation and the intermediate ratios resulted in callus formation. A good correlation between hormone level and morphogenesis can be provided by the following evidences:

1. A large number of buds are produced in several moss species *Physcomitrella pateus*, when they are exposed to high CK levels. The enhanced bud formation seen in a mutant of this moss is indeed associated with an increased level of endogenous CK.
2. The levels of endogenous auxins and CK's appear to control the altered morphology of shoot/root bearing crown gall tumour in agreement with the concept of Skoog and Miller, 1957).
3. In Begonia plants and cuttings it has been shown that a reduction in temperature or day length results in elevated CK and lowered auxin levels and also induced bud formation. Under non - inductive conditions (long days and

high temperatures) externally applied CK's can induce bud formation.

## Control of Leaf Senescence

All the phytohormones are probably involved at some stage of leaf senescence and may interact in its control, but much attention has been given to the role of CK's because of their dramatic effect when applied exogenously.

Chibnall (1954) was the first to postulate that some hormone like factors produced in roots may be involved in control of leaf senescence. Today Chibnall's **factor** is generally equated with CK. The overall evidence for the involvement of CK's in the control of leaf senescence is outlined below under four headings:

(1) Application of CK to excised leaves and leaf discs. (2) Application of CK to intact and detopped plants. (3) Studies involving root development or root excision. (4) CK levels in leaves and xylem sap.

1. **Application of CK to excised leaves and leaf discs:** CK's are very effective in retarding senescence of excised leaves of various genera. They induce regreening of detached yellow tobacco leaves. Bioassays for CK's have been based on their senescence retarding effect with excised oat (Thimann and Sachs (1966), Varga and Bruinsma (1973), and barley leaves which were kept in darkness (Carr and Burrow, 1966). This research house of CK's is known as Richmond and Large effect (1957).
2. **Application of CK to intact and detopped plants:** Frequently CK's have no effect on attached leaves that have a normal connection to a root system. In the classical study of Kulaeva (1962), kinetin when applied directly to the blades of excised tobacco leaves, retarded senescence. However, it had no effect on the leaves of intact plants or detopped plants. Such lack of response is attributable to supply of CK by the root system. CK's retard senescence of attached leaves of some plants for e. g., soybean, french beans, rice, wheat and oat. These results show that in the intact plant, leaves possess sub- optimal levels of CK before the onset of leaf senescence.
   Detopping of tobacco and bean plants induced regreening of lower leaves, an effect also evoked by application of exogenously applied CK. This indicates that senescence of lower leaves is not due to insensitivity to CK, but is probably due to CK defficiency possibly caused by preferential transport of root produced CK to the upon green leaves.
3. **Studies involving root development or root excision:** Root produced CK's appear to play a significant role in retardation of leaf senescences. This is proved by the following results.
   a) Rooting of tobacco leaf cuttings and CK application to the blades of excised leaves have similar retarding effects on leaf senescence.
   b) Secondly, the development of roots on the petiole of excised *Phaseolus vulgaris* leaves caused a very marked increase in CK level in the leaves.
   c) Thirdly, excision of lateral roots from soybean and pea seedlings induces leaf senescence which can be prevented by exogenous CK applied to the leaves.
   d) Fourthly, xylem supplied CK's, known to occur in soybean xylem exudate, retard leaf senescence in soybean stem explants and with nutrients essentially replace a root system in a physiological sense.
   e) Finally partial root excision causes flag leaf senescence in wheat and can be reversed by CK application.

4. **Cytokinin levels in leaves and xylem sap:** The endogenous levels of CK's in xylem sap seems to be an important factor in regulating leaf senescence. The CK activity in the xylem exudate of sunflower plants has been reported to decline at the commencement of flowering which precedes induction of leaf senescence. The environment or soil factors reduce CK production by roots levels in xylem sap. Usually causes induction of leaf senescence.

When considered collectively, the above evidence of diverse types provides strong support for the control of leaf senescence in many species by endogenous CK, particularly root derived CK moving in the xylem

### Root-Shoot Interactions

In plants many aspects of shoot development are affected markedly by roots and these may be mediated by CKs produced in the roots and present in significant amounts in the xylem sap. CKs often substitute for the effect of roots on the shoot development and physiology, convincingly in some cases GAs appear to act in this way.

Since root derived CKs play an important role in the control of leaf senescence, it is suggested tht root produced CK may also regulate other aspects of shoot development including release of lateral buds from apical dominance, development of inflorescence, DNA synthesis in shoot apex, expression in dioecious plants and photosynthetic activity (Goodwin, 1978).

CKs also promote and control the cell division in seed and the high concentrations reported in seed and fruit may be necessary for the creation of a strong physiological sink for nutrients, exogenously applied CKs have been shown to increase fruit set (Clifford, 1981), seed size (Michael and Beringer, 1980), total seed Weight (Clifford, 1981), promote phloem unloading of bean seed coats (Clifford *et al.*, 1986) and to affect seed storage proteins (Schroeder, 1948).

## MODE OF ACTION OF CYTOKININ

It comes as no surprise to find that cytokinin have been evaluated extensively for their role in nucleic acid metabolism beginning with the study of Guttman (1957) who found that the kinetin treatment of onion roots caused a quick increase in the amount of RNA in the nuclei. Kinetin caused a doubling of the RNA levels within 30 min in onion root tip cells although it reduced DNA.

### Cytokinin-dependent plastid development

Cytokinins are shown to stimulate plastid development. The general inference is that cytokinins do not control plastogenesis in cotyledons though hormones control cotyledon development superimposed by cytokinin promoted plastid differentiation through activated nuclear genes that code for organelle polypeptide. Thus CK regulate etiolated tissues whereas in light gene expression is regulated by phytochrome.

The general inference is that cytokinins and ABA do not regulate plastogenesis in cotyledons via selective transcription of plastid genes. However, this may take place in the CK-dependent cell cultures.

By the contrast, kinetin alone induces a rapid increase in the DNA of tobacco pith cells although the effect can be duplicated by auxin.

In leaves of tobacco, kinetin causes marked increase not only in RNA but DNA and protein synthesis as well, which implies that in the deferral of senescence, kinetin acts on the over all synthetic activities of the leaf. That cytokinin plays a role in the protein synthesis was, of course pointed out in the original paper by Richmond and Lang (1957) and strongly suggested by the work of Mothes and his colleagues. It has also been observed that kinetin stimulated increase in protein synthesis in the leaf discs and felt that a primary action of kinetin maight be to regulate DNA synthesis with a resulting regulation of one of the steps leading to protein synthesis. The cytokinin action is prevented effectively by inhibitors of RNA and protein synthesis.

## PRACTICAL USES OF CYTOKININS

### Mixture of BA and $GA_{4+7}$

The synergistic combination of BA with other GAs (especially $GA_5$) has also been demonstrated since 1977. A mixture of BA and $GA_{4+7}$ is used for commercial purpose of a cytokinin in agriculture. Shortly thereafter, the tetrahydropyranyl derivative of BA (PBA) was registered on ornamentals for induction of lateral branching. The mixture of $GA_{4+7}$ and BA is used to induce the extension of apple calyx lobes for the increased size and length/diameter ratio associated with a high calyx lobes for the increased size and length/diameter ratio associated with a high quality `Red Delicious' apple fruit. Additional research with this mixture demonstrated that applications to non-bearing apple trees increased lateral bud break and improved branch angles to provide a better framework and bearing surface for early cropping. Bud break and branch angles were primarily due to the BA component, but the combination with $GA_{4+7}$ promoted shoot growth to a much greater extent than either of the two compounds alone. The mixture is used in the promotion of flowering of *Chrysanthemum* under non-inductive conditions.

### Fascicular Bud Break in Conifers

BA is registered for use in promoting fascicular bud break in conifers to reduce the required mechanical shearing and altering the rotation period required to grow marketable Christmas trees. While this use has not been commercially significant, the compound's availability has encouraged research in a number of areas.

### Extension of Vegetable Shelf-life

Potential applications such as the extension of vegetable shelf-life, prevention of crop senescence, and increased soybean pod set have not been pursued due to cultivar differences or lack of field efficacy.

### Commercial Uses of Cytokinin-like Activity Compounds

In contrast to GAs, a number of compounds have been made synthetically which exhibit cytokinin-like activity including $N^6$-substituted adenine analogues, like, pyridylureas and thiadiazolylureas and pyrimidines. Of these the highly active thiadiazolyl urea, thidiazuron (TDZ) was found to have cotton defoliant properties and was subsequently the first cytokinin-like product to be successfully developed for widespread major crop use. TDZ is unique as a harvest aid in cotton because it promotes green leaf abscission and prevents regrowth, thus allowing growers more flexible harvesting schedules. Its mode of action in the cotton leaf abscission process appears to be associated with a sustained increase in ethylene production coupled with the disruption of auxin transport.

## Thinning of Apples

Recently, promising results were reported for thinning of apples with BA alone or in combination with carbaryl. Another cytokinin-like compound currently under development which also thins apples is **2-chloropyridylphenylurea** (CPPU). A number of potential applications have been examined with CPPU which overlaps with several GA effects including grape berry enlargement, and increased fruit size and weight in apple. The future of this compound as a commercial product is yet to be determined and may require an identified use in a major agronomic crop to justify further development.

## Agronomic Crops

The successful introduction of a CK product in an agronomic crop is dependent upon the specificity of a physiological response over a wide range of environmental/cultural conditions, and is complicated by the inherent genetic variability amongst cultivars. For example, the genetic reduction of plant height has led to dramatic increases in cereal grain yield, due principally to increased reduction of lodging. The current characterization of endogenous CKs in correlation with developmental events in major crop plants will certainly aid in the identification of application timings for candidate compounds.

A key factor in the successful development of thidiazuron (TDZ) in cotton was the availability of a large number of synthetic cytokinin-active compounds. With different classes of cytokinin chemistry, a systematic approach toward structure activity relationships could be employed.

The legacy of CK applications to major crops has always been the expectation of increased yield. More realistic opportunities may reside in identifying developmental modifications that are associated with improved production efficiency, thus reducing cost inputs.

# CHAPTER 25

# Abscisic Acid

## INTRODUCTION

Abscisic acid (ABA) was originally described as dormancy inducing and abscision-accelerating substance. To what extent endogenous ABA functions as an inhibitor in intact plants is still not clear. Several fungi produce ABA. In the biosynthetic studies of fungi ionylidene derivatives are precursors of ABA but there is no evidence for this pathway in higher plants. In the indirect pathway ABA is a breakdown product of xanthophyll with xanthoxin as an intermediate. A reduction in the level of ABA, a presumed growth inhibitor, leads to a reduced growth rate, in particular of leaves, which can be restored to normal by exogenous ABA. The effect of water-stress on changes in ABA concentrations in different plant organs particularly the leaves and roots is well known.

## DISCOVERY AND CHARACTERIZATION

The systematic search for an endogenous plant growth inhibitor was initiated by the work of Bennett-Clarke and his colleagues in the early 1950's. They discovered that the acid fraction of an alcoholic extract from a variety of plant tissues displayed growth inhibitory properties which they termed inhibitor B. In the early 60's, ABA was discovered as a result of two independent investigations, one by Addicott and coworkers who found naturally occurring compounds that accelerate leaf abscission and the other by Wareing and his coworkers into compounds that promote bud dormancy in woody plants.

In 1963 Addicott reported the isolation of a substance from young cotton (*Gossypium hirsutum*) fruit that accelerated the abscission of excised debladed cotton petioles.

The substance had a molecular formula of $C_{15}$ $H_{20}$ $O_4$ and was named **abscisin II**, was in deference to **abscisin I**, another abscission accelerating substance that had been isolated from mature cotton burs in 1963 by Liu and Carns. Abscisin II was also shown to inhibit the IAA-induced straight growth of *Avena* coleoptiles but had no GA- activity in the dwarf maize bioassay.

In 1963 Wareing and Eagles also isolated a substance from birch (*Betula pubescens*) leaves which caused the buds of growing seedlings to go dormant when applied to them and named it **dormin**. Wareing and Cornforth and his colleagues found the molecular structure of abscisin II and dormin to be identical. In 1967, Addicott's, Wareing 's and Cornforth's groups together decided to name the putative growth inhibitor as abscisic acid (ABA).

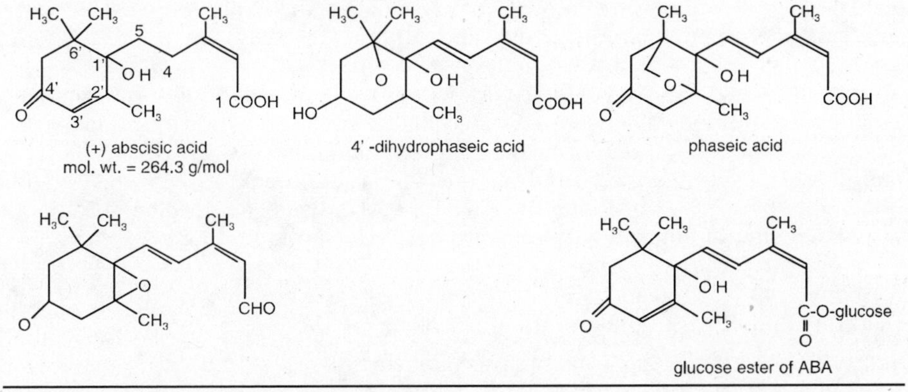

**FIG. 25.1:** Structure of ABA and related compounds.

## DISTRIBUTION OF ABA IN NATURE

Abscisic acids (ABA) is ubiquitous in higher plants, and is produced by algae and fungi (Tietz et al. 1987). However, the source of ABA in animals may be dietary in origin.

ABA has been identified in number of *angiosperms* both monocotyledonous and dicotyledonous plants. It is also identified in many gymnosperms (e.g. Yew, Silver fir, Pine, Sitka spruce) and a few *pteridophytes* (e. g., *Anemia phyllitides, Equisetum arvense*). Among *bryophytes* it has been found in a moss, but not in any liver worts. It has not been found in any *algae*. The place of ABA as a growth inhibitor in liverworts and algae appears to have been taken by lunularic acid.

ABA has been found in many higher plants tissues like roots, stems, buds, leaves, fruits and seeds. It has also been found in phloem and xylem sap and in nectar. Its concentration varies from tissue to tissue and also within a given tissue during the course of its development. Most tissues contain 20-100 mg per g wet weight but concentration of 10 mg and 20 mg per g fresh weight have been found in avocado fruit pulp and dormant cocklebur (*Xanthium*) buds respectively.

## CHEMISTRY OF ABA

Abscisic acid is a sesquiterpenoid composed of 3-isoprene residues (or fifteen carbon atoms and characterized by a six membered ring; with a chiral centre and an unsaturated six-carbon substituent). In addition the carbon skeleton of molecule bears a close similarity to the terminal rings of carotenoid such as violaxanthin

The chiral centre is responsible for the occurrence of two enantiomorphic forms - (R) - Abscisic acid and (S) - Abscisic acid - Among these two (S) - Abscisic acid is the naturally occurring active form. It has (+) and (-) enantiomers. Both these enantiomers are optically active. Both the (R) - and (S) Abscisic acid may be extracted from the plant tissue although only the latter enantiomorph exhibits biological activity.

### Techniques

### *Quantification of ABA*

For purification of ABA in crude extracts, HPLC has largely been replaced TLC.

Actual quantification is usually accomplished by Gas Chromatography with an electron capture detector, (Loveys and Milborrow, 1984) as well as by radio-immuno assays and enzyme-immunoassays. Antibodies of high specificity and affinity against ABA have been produced by number of laboratories and a monoclonal antibody against S-ABA is available commercially. Several laboratories have developed an indirect enzyme-linked immunoassay in which much less antibody is used than required in the Idetek protocol. An enzyme amplified immunoassay with a sensitivity of 0.05-2.5 pg ABA has been developed to measure the ABA content of mesophyll and guard cells (Harris et. al., 1988).

**Separation of S- and R- Abscisic Acid:** Commercial sources of ABA, including both $^{14}C$ - and $^{3}H$ labeled analogs, are racemic mixtures of S-ABA (natural) and R-ABA (unnatural). Catabolism of these two enantiomers is quite different. Only the natural S-enantiomer is used for metabolic studies.

Two types of methods have been described for the resolution of the enantiomers of ABA:

### High performance liquid chromatograph (HPLC) with a chiral stationary phase

In this method, methyl-RS-1, 4'-*cis*- ABA diol was resolved with two different chiral Pirkle columns in series - ABA was regenerated by oxidation with $MnO_2$ and basic hydrolysis (Vaughan and Millborrow, 1984). Railton (1987) separated the enantiomers of ABA as their methyl esters using a stationary phase of cellulose tris (3, 5- dimethyl phenyl carbamate coated on silica gel.

### Immunoaffinity chromatograph

In this method, antibodies selective against S-ABA were bound to sepharose. The S-enantiomer was retained on the column and subsequently eluted with organic solvent (Mertens et al., 1982). The column is reusable, but because of the low capacity, this technique is more for the resolution of radioactive RS - ABA than for the unlabelled racemic mixture.

## BIOASSAY OF ABA

ABA has been estimated by many bioassays and physical methods. Bioassays based on following phenomenon have been developed.

1. The acceleration of abscission of cotyledonary petioles in explants of 14-day old cotton seedlings.
2. The inhibition of IAA-induced straight growth of Oat (*Avena*) coleoptiles.
3. The inhibition of germination of isolated wheat (*Triticum*) embryos.
4. The inhibition of the growth of a duckweed species (*Lemna minor*)
5. The inhibition of the GA-induced synthesis of $\alpha$-amylase by barley (*Hordeum*) aleurone tissue.
6. The stimulation of stomatal closure.

Physical methods include ultraviolet absorption, optical rotatory dispersion and circular dichroism, gas-liquid chromatography of methyl ester or less frequently the trimethyl silyl ester of ABA on various silicon liquid phases and combined gas liquid chromatography- mass spectrometry.

## TRANSLOCATION OF ABA IN THE PLANT

The translocation of endogenous ABA in higher plants at various stages of growth is not absolutely clear. Numerous approaches have been used to investigate this problem. These include:

1. Experiment involving the local application of ($^{14}C$) ABA to the exterior surface of one part of the plant followed by determination of the distribution of radioactivity some time later.
2. Experiments involving the application of ($^{14}C$) ABA to the cut ends of stem, petiole and root segments followed by the determination of how much radioactivity reached the other end.
3. The examination of phloem and xylem sap from various vascular organs for the presence of ABA.
4. From these experiments it appears that externally applied ABA is able to get into tissues rapidly and is distributed freely in all directions within the plant. Cell to cell transport of ABA is slow and non-polar. ABA is present in the phloem and xylem sap and is probably transported throughout the plant in this way.
5. ABA synthesized in the root cap, moves basipetally in the tissues of the steles.
6. There is also some evidence for lateral ABA movement in roots.

## BIOSYNTHESIS OF ABA

### In Fungi

Several fungi produce ABA as secondary metabolite. Biosynthesis of ABA is studied on *Cercospora rosicola*, *C.cruenta* and *Botrytis cinerea.*

According to Zeevart and Creelman (1988), when ($^{3}H$) Mevalonic acid was fed to cultures of *C. rosicola* the main products were ABA and 1'-deoxy-ABA, radioactive 1'-deoxy-ABA was converted to ABA with 11% yield. NMR analysis of ABA, obtained from culture filtrates of *C. rosicola* fed (1,2-$^{13}C$) acetate, demonstrated that biosynthesis occurred via a isopyrenoid pathway. Incorporation of radioactivity from Farnesyl phosphate (FP) and Farnesyl pyrophosphate (FPP) into ABA was observed. FPP was dephosphorylated to FP by the fungal cultures but the further dephosphorylation of FP to farnesol was not reported. It cannot be unmistakably concluded that exogenous FPP and FP are precursors of ABA in fungal cultures. However, $C_{15}$ compound with carbon skeleton similar to that of ABA are found to be the intermediates in ABA biosynthetic pathway Fungi. Evidence from labelling studies indicates that ABA is formed from-a- ionylidene derivatives. α-Ionylidene ethanol, (I) α-ionylidene acetic acid (II) and 4'-hydroxy-α-ionylidene acetic acid (III) were detected in *C. rosicola*, and converted to 1'- deoxy -ABA (IV) and ABA. The 2- transisomers of these compounds, as well as β-ionylidene ethanol and b-ionylidene acetic acid were not converted (Neill and Horgan, 1983). Both epimers of 4'-hydroxy-α-ionylidene acetic acid were metabolised to 1'-deoxy-ABA and ABA. 1'-deoxy-ABA was stereospecifically hydroxylated at the 1' position to give ABA (Neill et al., 1987).

1'-hydroxy α-ionylidene acetic acid (V) is also metabolized in ABA (Norman et al., 1985). In cultures of *C. cruenta* a-ionylidene acetic acid was also converted to 4'-hydroxy-α-ionylidene acetic acid, 1'-deoxy-ABA and ABA, but 4'-hydroxylation was mainly trans in relation to the 1' side chain. In cultures of *B. cinerea* the 1',4'-t-diol of ABA (VI) was shown to be a precursor of ABA. In *C. pini- densiflorae*, 1', 4'-t-diol of ABA was more easily converted to ABA than 1'-deoxy ABA. This led Okamoto et al (1987) to propose that conversion of α-onylidene ethanol to ABA via 4'-hydroxy-α-ionylidene acetic acid and 1', 4'-t-diole of ABA as the major pathway in fungus.

Recently, another series of precursors based on the α-ionylidene structure has been described. It was found that cultures of *C. cruenta* produced a-ionylidene ethanol, (VII) 4'-hydroxyα-ionylidene acetic acid (VIII) and 1',4'-dihydroxy-α-ionylidene acetic acid (Oritani and Yamashita, 1987; Oritani et al., 1985). (IX) Radio-active Mevalonic acid was incorporated into 4'- hydroxy-α-ionylidene acetic acid, which is converted to 1'-deoxy-ABA and 1',4'-dehydroxy-α-ionylidene acetic acid. It was suggested that in *C. cruenta* the main biosynthetic pathway is via 1', 4'-dihydroxy-α-ionylidene acetic acid rather than via 1'-deoxy -ABA as in *C. rosicola*. (Oritani and Yamashita, 1985) because former pathway is more efficient than the latter.

## Biosynthesis in Higher Plants

ABA (15 carbon compound) is considered to be a sesquiterpenoid. Sesquiterpenoid are a class of terpenes derived from the condensation of three 5- carbon isoprene units. Mevalonic acid (MVA) is the first compound of the terpenoid pathway and is converted to isopentyl pyrophosphate (IPP), which isomerizes to form 3, 3-dimethyl allyl pyrophosphate. These two isomeric isopentenyl pyrophosphates (each 5 carbons) then undergo condensation to form 10 carbon compound Geranyl, pyrophosphate (GPP). A further condensation of IPP with GPP forms Gernayl geranyl pyrophosphate (GGPP), a 20 carbon compound. Carotenoids are formed by the condensation of two molecules of GGPP together to form phytoene; the first non-coloured carotenoid. These phytoene is further catabolized to carotenes and *Xanthophylls* (oxygenated carotenes).

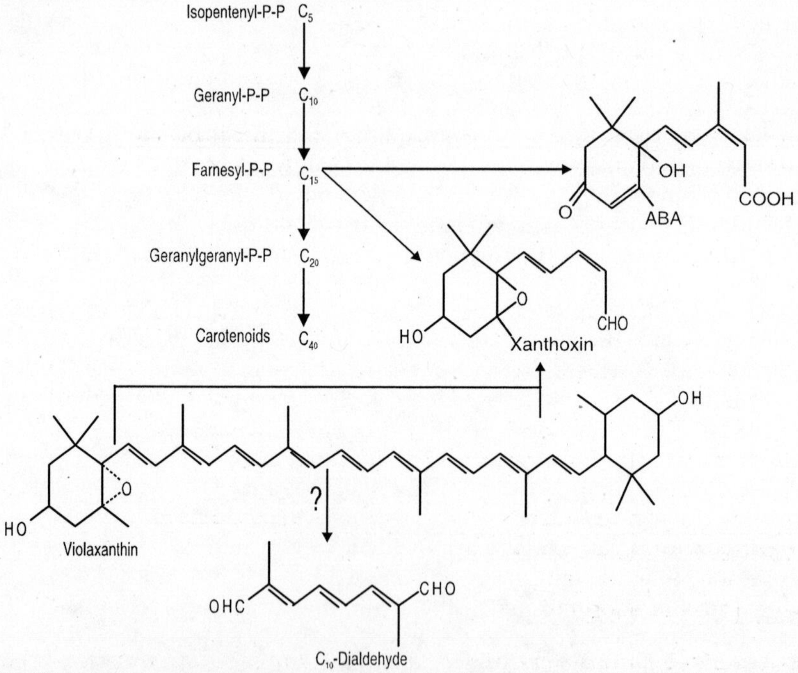

Isopentenyl-P-P $C_5$
Geranyl-P-P $C_{10}$
Farnesyl-P-P $C_{15}$
Geranylgeranyl-P-P $C_{20}$
Carotenoids $C_{40}$
ABA
Xanthoxin
Violaxanthin
?
$C_{10}$-Dialdehyde

**FIG. 25.2:** Theoretical biosynthetic pathway for ABA in higher plants.

There are two pathways of ABA biosynthesis:

1. Direct pathway involving a $C_{15}$ precursor derived from FPP.
2. Indirect pathway involving a precursor derived from a carotenoid. The relative importance of each pathway is not known but there is possibility that both these pathways exist in the higher plants at the same time. In both the cases the Mevalonic acid is the ultimate precursor.

It now seems that there exists a universal, indirect pathway for ABA biosynthesis in higher plants, and that the pre-cleavage precursor is a xanthophyll such as violaxanthin. The immediate precursor to ABA has recently been identified as ABA-aldehyde (ABA-ald) and most work to date is consistent with xanthoxin (Xan) being an intermediate between the isomer xanthophyll and ABA-ald.

Most leaf xanthophylls exist in the thermodynamically most stable *trans*-configuration but one exception may be neoxanthin, whose stereochemistry has been the subject of much debate (Parry et al., 1990).

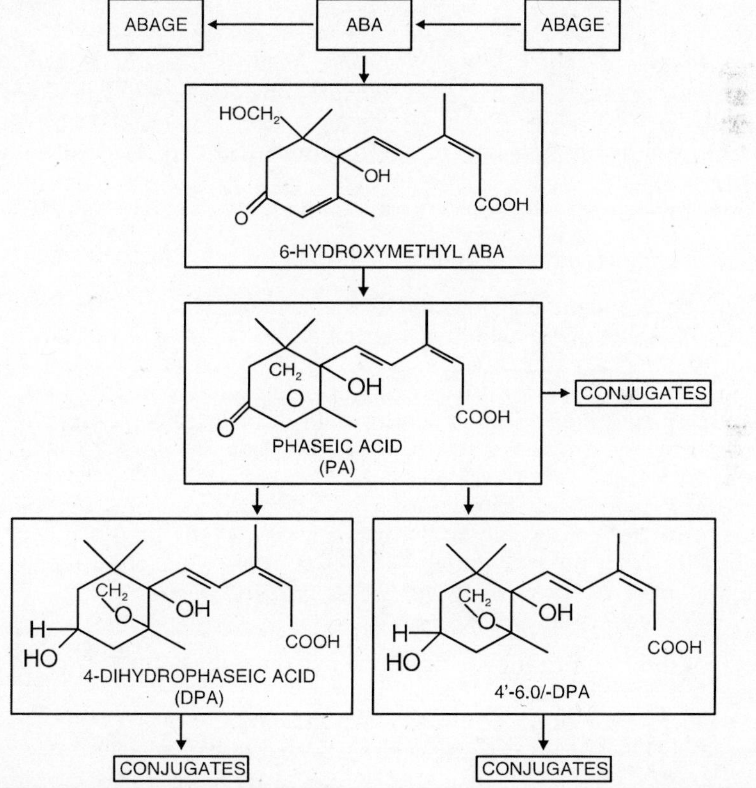


**FIG. 25.3:** Proposed pathway of ABA metabolism.

## Carotenoid Biosynthetic Inhibitor

If ABA is derived from a carotenoid, then blocking the carotenoid biosynthetic pathway and reducing the carotenoid pool size should prevent ABA accumulation. Inhibitors of carotenoid biosynthesis such as fluridone and norflurazon which block the conversion of phytoene to phytofluence also prevent accumulation of ABA. Fluridone did not inhibit the accumulation of ABA in water-stressed green leaves, but the specific activity was greatly reduced.

### Wilty or Droopy Mutants Deficient in ABA

ABA deficient mutants are green because they contain normal carotenoid and chlorophyll levels. Since the decreased ABA contents are not due to increased catabolism, the lesions are thought to be in enzymes involved in ABA biosynthesis between carotenoids and ABA. Normal water-stress induced accumulation of ABA does not occur in *droopy* (potato), sitiens and *flacca* (tomato) and *wilty* (pea). It is possible that these mutants are already stressed and have reached their maximum accumulation of ABA, or may not be able to accumulate stress-induced ABA. If the latter case is true, it implies that in higher plants two pathways exist, one operating in turgid and the other operating in water stressed leaves.

## PHYSIOLOGICAL RESPONSES OF ABA

ABA was discovered as a result of its ability to promote abscission in explants of cotton seedlings, induced bud dormancy, seed dormancy and germination, tuberization, leaf senescence and fruit ripening, the best authenticated examples of its hormones activity are:

1. Its role as growth inhibitor in the downward growth response of roots to gravity,
2. Its role in causing stomatal closure in response to water stress.

### Stomatal Closure and Water Relations

The drop in water potential is detected by the chloroplasts of the mesophyll cells of leaf, whose photosynthetic activities are thought to be the most sensitive of all the plant's metabolic processes to water stress. As the water potential decreases there is marked increase in permeability of chloroplast membranes to ABA. The ABA synthesized by and stored in chloroplast then diffuses into mesophyll cytoplasm and moves from cell to cell through plasmodesmata to the guard cells where it induces stomatal closure. The drop in the level of stored ABA in the mesophyll chloroplasts induce fresh biosynthesis of more ABA which continues to be translocated to the guard cells as long as the water potential remains low. The release of ABA from the mesophyll chloroplast stops when the water potential is restored to normal and its rate of synthesis is markedly reduced.

The evidences for this are as follows:

1. The exogenous application of low concentrations (eg. 1mM) of ABA to leaves causes stomatal closure within 3-9 minutes.
2. The concentration of ABA in water stressed leaves rises markedly.
3. Mesophyll chloroplasts can synthesize ABA but guard cell chloroplasts cannot
4. The ABA of leaves of well-watered spinach was found to be contained in the chloroplasts but after 4 h of water stress the total ABA level was found to have risen 11 fold whilst that of the chloroplast had risen only 2 fold; this suggests that the newly synthesized ABA is rapidly exported from the chloroplasts.

#### *Guard Cells*

Guard cells showed equal closure when exposed to ABA solutions over the pH range 5.0-8.0, while no significant uptake of the hormone took place at pH 8.0. This indicates that the sites of ABA action are on the outer surface of the plasmalemma. ABA prevents stomatal opening by rapidly blocking $H^+$ extrusion and $K^-$ influx and it initiates closure by the rapid release of osmotica, in particular $K^+$, $Cl^-$ and malate (Raschke, 1987). The latter phenomenon can also be observed in the shrinking of guard cell protoplasts (Fitzsimons and Weyers, 1987).

### *Chemical Signals for Stomatal Closure*

Saab and Sharp (1990) have shown that the roots send signals to the shoot and causes stomatal closure and inhibition of leaf growth in Maize when there is even slight reduction in soil water potential with no detectable change in leaf water relations. They observed no effect of soil drying on stomata. In bean plants with oxygen-deficient roots, stomatal closure is influenced by changes in leaf water relations and by increased fluxes of ABA (Neuman and Smit, 1990)

### *ABA and Potassium Uptake*

The primary action of ABA on guard cells seems to be its inhibition of potassium uptake by the guard cells. Certain substances, such as a fungal toxin fusicoccin, are known to overcome the effect of ABA by stimulating potassium uptake and proton release. The flow of other solutes also seem to be influenced by ABA during the stimulated closing of stomata. For example, Malic acid is released with potassium from the guard cells more quickly in the presence of ABA. ABA inhibits the $H^+/K^+$ exchange and promotes the leakage of Malic acid and accounts for loss of turgidity in the guard cells. The reduction of osmotically active solutes would of course render the guard cells flaccid and keep the stomata closed.

ABA produced by roots in drying soil moves in the transpiration stream and accumulates at or near the guard cells, ie. at the site of action (Davies et al., 1987). In non watered *Commelina*, the stomata closed without any change in turgor or ABA content. However, the ABA content of the roots increased several fold along with that of the lower epidermis of the leaves (Zhang et al., 1987). ABA produced in stressed roots may act as a **chemical signal** to the shoot, initiating closure of the stomata before any changes in the water status of the leaves occur, thereby optimizing the plant's water use under conditions of restricted availability.

ABA and other hormones can alter the response of stomata to $CO_2$. The increase in stomatal sensitivity to $CO_2$, occurring as an early reaction to the arrival of a chemical signal from the roots, could be important in curtailing transpiration in wind.

## Photosynthesis

There are several reports that ABA applied via the transpiration stream to leaves affected photosynthesis both via stomatal closure and via a direct effect on carbon fixation. As a possible mechanism, inhibition of the carboxylation of ribulose -1,5-bisphosphate has been suggested, but extractable ribulose-1,5-bisphosphate carboxylose activity from ABA-treated leaves was as high as that of control leaves. Raschke and Patzke (1988) found that the effects of ABA on photosynthesis in *Xanthium* resulted from non-uniform distribution of stomatal pore sizes and that ABA had no direct effect on the photosynthetic machinery in this species.

## Growth

### *Root Growth*

Reports on the effects of ABA on root growth are contradictory. Since exogenous ABA can inhibit root growth as well as promote growth of this organ. The effect of high concentrations of ABA (0.01-1mM) on root growth is triphasic. A **period of promotion**, which lasted approximately 12 h was followed by a **period of inhibition** (12 h) and finally a **gradual recovery** to about 80% of normal growth rate occurred after 24 h.

With low concentrations of ABA (0.1 mm), only a transient period of stimulation occurred. Pilet and Saugy (1987) found that the effect of exogenous ABA on root growth in maize depended on the initial elongation rate. Roots with a low growth rate (0.4 mm $h^{-1}$) were inhibited at a low ABA concentration (5 nM), whereas fast growing roots (0.8 mm $h^{-1}$) were stimulated. By quantifying ABA in the elongation zones of single maize roots by Gas Chromatography- Negative Chemical ionization- Mass spectrometry (GC-NCL-MS), a negative correlation between ABA content and growth rate was demonstrated (Reymond et al., 1987). However, a role of endogenous ABA in root growth is questionable, since ABA content and growth rate, as modified by fluridone and temperature, varied independently.

### *Elongation of Maize Coleoptile*

ABA decreased the elongation rate in maize coleoptile and also inhibited growth-induced by auxin, fusicoccin or acid. In all cases, this effect was due to an inhibition of cell wall loosening. A similar effect of ABA has been demonstrated in light-stimulated expansion of leaves.

The opposite effects of ABA on the growth of roots and shoots may be advantageous for survival of plants under stress conditions. In shoots, ABA induced growth inhibition provides the possibility of maintaining turgor pressure. On the other hand, a stimulation of root growth by ABA would enlarge the root system and this combined with increased hydraulic conductance would increase water uptake.

### *Root Gravitropism*

The root cap responds to light and gravity by synthesizing or accumulating growth inhibitors. It is believed that the inhibitors are produced in the lower portion of root cap in response to gravity and are translocated into the zone of cell enlargement. They may inhibit the cell enlargement of cells on the lower side of the root. Differential growth results in **geotropism**. The build up of ABA in the cap appears to require light and gravity. ABA produced in maize root caps seems to translocate basipetally and to stimulate a positive geotropic response. The roots of a horizontally held plant grow downward under the influence of gravity. It has long been held that this is result of an asymmetric redistribution of growth inhibiting substance produced in the root cap. Both Indole-3-acetic acid (IAA) and ABA have been proposed as the agent causing gravicurvature (Moore and Evans, 1986). It is thought that ABA synthesized in the root cap, is normally translocated in the vascular tissue in basipetal direction but that when the root is held horizontally it is pulled by gravity to the lower side where it inhibits growth. Since growth of the upper side of the root is inhibited, the result is down-ward curvature.

The evidences for this are:

1. Removal of the root cap destroys the geotropic response,
2. Removal of half of the root cap results in curvature towards the side with remaining half,
3. ABA has been extracted from the root caps and is known to be synthesized in them.
4. Application of ABA to the root surface at concentrations of $10^{-8}$-$10^{-4}$ M causes inhibition of growth and
5. Asymmetric application of ABA to the root surface causes curvature towards the side of application.

Recent evidence, however, argues against a role for ABA in gravitropism. Treatment with fluridone and norflurazon reduced ABA levels significantly, while allowing curvature to occur (Feldmand and Sun, 1986). The levels of xanthoxin were unaffected by treatment with norflurazon which is unexpected if xanthoxin is derived from carotenoid. ABA cannot be the inhibitor that causes gravitropic curvature (Mulkey et al., 1983).

## Bud Dormancy

*Innate* or *true dormancy* (i.e. temporarily suspended growth due to internal conditions) is exhibited by the buds of temperate-zone trees during winter, stem tubers (eg. potato), root tubers (e.g., Dahlia), Corms, bulbs and rhizomes. Hemberg (1949) found that the dormancy of buds of ash (*Fraxinus*) and potato tubers was positively correlated with the concentration in them of growth inhibitors, identified as ABA. The external application of ABA to buds of growing seedlings caused them to become dormant. According to these findings ABA was synthesized in the plant, in the response of some **environmental trigger** and this ABA travelled to the buds and induced them to go dormant.

In the buds of temperate-zone trees (e.g., birch - *Betula pubescens*; sycamore - *Acer pseudoplatenus*) the environmental trigger was the shorter days of autumn. Short days (photoperiods) were shown to cause increased levels of ABA in leaves and buds. The leaves appeared to play a crucial role in this response and it seemed that short days induced leaves to synthesize more ABA and export it to the buds. When ABA was measured by physical means, it was found that ABA in short days exposed leaves is not higher than those that had remained under long days. On the other hand physical assay methods have confirmed the presence of high levels of ABA in the dormant buds of several species and have shown that these levels progressively decline throughout the winters.

A role for ABA on the induction and maintenance of bud and seed dormancy has been neither unequivocally demonstrated nor disproven (Walton, 1980). This is still true with respect to bud dormancy in woody species. Studies with seedlings of *Salix* spp. have shown that the day length does not regulate cessation of growth by affecting the ABA levels in the plants. It is possible, however, that the sensitivity to the hormone is altered by short-day conditions, but factors other than ABA are probably also involved in the control of dormancy (Barros, Neill, 1986).

Tucker (1977) suggested that ABA inhibits lateral bud growth in tomato plants. High ABA levels also have high levels of auxin (IAA) activity Tucker (1978). On these basis Tucker suggested that IAA is responsible for maintaining the levels of ABA sufficiently high for strong apical dominance. The reverse is true for other tomato mutations. However, the correlation between high IAA and ABA levels in tomatoes is not a general phenomenon in that ABA levels do not always change when IAA levels drop.

## Seed Dormancy

ABA builds up in embryos of developing seeds. It is believed that increased ABA level comes primarily from *de novo* synthesis. However, there may be translocation of ABA or its precursors from the leaves since the radioactivity from exogenously applied $^{14}$C-ABA to leaves can be recovered in developing seeds. ABA in developing ovules inhibits the formation of germination enzymes in the embryo (Dure, 1975). Thus, ABA plays an important role in developing seeds by inhibiting vivipary. In the case of immature seeds, several evidence indicate that ABA prevents precocious germination (vivipary) of the developing embryo. Following are the evidences:

1. When immature soybean embryos were excised and cultured *in vitro*, a high percentage germination was observed only when the ABA content was less than 4 mg $g^{-1}$ fresh weight (Ackerson, 1984). The ABA contents of embryo in situ did not decline to that level until the onset of dehydration. The high ABA content of embryos may result partially from in situ ABA synthesis, but an important factor appears to be the pH differential between seed coats and embryos (Barthe et al., 1986). In cotton the seed coat was upto 1.4 pH units more acidic than the developing embryos which resulted in diffusion of ABA from the seed coat to the embryo.
2. Exogenous ABA prevented precocious germination of immature embryos of several species when cultured *in vitro*, e.g., in cotton, maize, rape-seed, soybean and wheat. ABA not only prevented embryo germination, but it often caused embryo growth and accumulation of storage proteins as well. However, as the embryo matured the endogenous ABA declined.
3. ABA-deficient mutants show vivipary under suitable conditions (Groot, 1987). In reciprocal crosses between wild-type and ABA-deficient mutants of *Arabidopsis*, it was demonstrated that the ABA produced in the embryo controls seed dormancy; maternal ABA had only a minor role in this process (Karssen et al., 1983). Taken together, these data clearly point to a role of seed-produced ABA in preventing precocious germination during the early stages of embryogenesis.

ABA only appears to play a role in the dormancy of seeds that require stratification (i.e. seeds that will not germinate unless they are exposed to periods of low temperature) and especially those that are *innately* dormant (e.g., *Taxus baccata* - in contrast to those which exhibit *seed coat-imposed dormancy*.

It has been shown that dormant and immature embryo of yew (*Taxus baccata*) can be induced to germinate without stratification if they are placed in a nutrient solution which causes ABA-like compound to be leached from them. Dormancy can be reimposed by immersing the leached embryos in an ABA solution. Similar results have been obtained with apple seed embryos.

However, attempts to correlate endogenous ABA levels with the state of dormancy of the seed have given contradictory results. In some species the ABA level is high in the dormant seed and declines during stratification whereas in no other such decline occurs during stratification. It is probably that ABA is important in the maintenance of dormancy in seeds. Recent work with stratification requiring seeds suggests that *dormancy is regulated by an antagonistic interaction between ABA and GAs and cytokinins*. The role of GAs and cytokinins in breaking seed dormancy has been indicated by the observations that:

i) Externally applied GAs and/or cytokinins stimulate germination in many types of dormant seeds and
ii) Treatments, such as stratification, that overcome dormancy result in increased levels of GAs and cytokinins in the seed.

Exogenous ABA inhibits seed germination even in the presence of gibberllins and cytokinins. Some scientists have suggested that exogenously applied ABA to mature, non dormant seeds may inhibit germination by suppressing nucleic acid-directed synthesis of specific germination enzymes. In fact, Dure (1975) suggested that ABA inhibits the translation of specific mRNA and in this way blocks protein synthesis. This hypothesis requires considerable experimentations, even though similar evidence implicates the action of most other phytohormones through their nucleic acid interaction.

## Abscission

Abscission term is synonymous with shading. Two functions are achieved by the abscission. One is, it facilitates vegetative propagation in the case of *Riella* and *Kalanchoe*. Second is the shading of mature plant parts. ABA was first isolated and characterized by abscission-accelerating bioassay using debladed petiole stumps of the cotyledons of cotton seedlings as a convenient explant system. The ABA containing solutions under test were applied dropwise to the cut surfaces of the stumps, and the time taken for separation of petiole from stem was recorded. In all these tests on explants systems, ABA reduced the time to zone cell separation compared with controls.

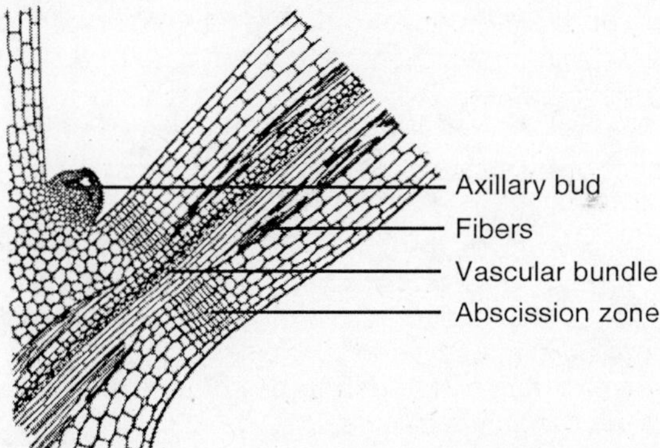


**FIG. 25.4:** The abscission layer.

ABA is regarded by some as one of the principal accelerators of abscission and by others as having an unclear hormonal role in abscission and bud dormancy (Milborrow, 1974). However, the suggestion that ABA's role in abscission is questionable According to Dorffling (1978) *there is no evidence available that ABA plays a role in leaf abscission.* Evidence appears to be conclusive that in the promotion of abscission in young fruit of Lupinus, *Gossypium* and in mature fruits of *Citrus*, ABA plays a hormonal role. Addicott and Wiatr (1977) point out that the relatively few cases where applications of ABA to plants have failed to accelerate abscission were possibly due to the low dosages applied. They draw an analogy to the early work on auxin, where IAA, when applied to intact plants, often failed to promote growth. Dorffling's view (1978) was probably based on the fact that he found no correlation between ABA contents in *Coleus* petioles and the latter's ability to abscise. It may be that tylosis in the xylem vessels induces water stress which in turn causes an increase in the level of ABA. However, the opposite sequence of events is possible since it has not been resolved with certainty whether tylose formation is casual to or a consequence of abscission.

Correlative studies of the content of extractable ABA from organs and tissues of intact plants prior to abscission (compared with controls) have shown rises in ABA levels prior to the initiation of cell separation. These include cotton fruits, abscinding lupin flowers, leaf petioles of deciduous trees (*Acer pseudoplatanus*), herbaceous leaves (soybean), and abscinding petals and stamens of citrus fruits.

## STRESS, ABA AND ABSCISSION

The rise in ABA can equally well be correlated with a stress condition of the leaves, flowers or fruit under test. Usually the stress is associated with senescence and a common feature is lowering in water status of the tissue and a decline in water potential. Membrane becomes leaky and the cell lose turgor. The latter are precisely the conditions that lead to a rise in ABA levels in plant tissues. These rises can be rapid and manifold as recorded by Wright and Hiron for a 9 per cent water loss in wheat leaves.

Analysis of the contents of free and conjugated ABA in the abscission zones of young cotton fruits during periods of water stress and subsequent irrigation in the field do show a convincing correlation with the number of shading bolls. ABA level rose as the xylem water potential fell, and all but 3 per cent of tagged fruit were shed, while 3-days after watering, when the water potential and free ABA had recovered to the pre-stress values, most (78 per cent) of young bolls were then retained.

It is not believed that ABA is not concerned in the abscission of leaves but that it may have a regulatory role in the abscission of flowers and fruits.

The evidence for this are as follows:

1. ABA, applied externally to the intact plant accelerates the abscission of the mature fruits of grape, olive, citrus and apple and the flowers of grape but only promotes leaf abscission at very high concentrations (2-4 mM) and,
2. There are two peaks in the endogenous level of ABA in developing cotton fruits, the first coinciding with the dropping of immature fruits, the second with the final dehiscence of mature fruits.

### Heterophylly

Aquatic plants often produce two distinct types of leaves on the same plant. Submerged leaves are highly dissected or linear and they have undifferentiated mesophyll and an epidermis with few stomata, if present. In contrast, aerial leaves are entire and broad and they have differentiated mesophyll and a stomatous epidermis. It is interesting that the ABA treatment can induce the formation of aerial leaves on submerged shoots. Such changes in leaf morphology due to ABA have been reported for a fern, a monocot, and several dicots.

Leaves produced under ABA treatment were similar to aerial leaves not only in shape but also in stomatal density, and anatomy. A high external solute concentration also caused aerial-type formation in submerged *Callitriche heterophylla* and *Hippuris vulgaris*, suggesting that the effect of low water potential was mediated by stress-induced ABA. Likewise, it is possible that in emerging leaves a water deficiency causes an increase in the ABA content which in turn results in aerial leaf characteristics. In *Proserpinaca palustris* both ABA and GA play a role in the control of leaf morphology. To explain the different leaf forms produced in response to these hormones, it was suggested that GA is produced under long days, while ABA would be produced in aerial leaves that are under short-day conditions.

### Response Mutants

Several mutants have been isolated that exhibit a decreased response to exogenous ABA. A viviparous mutant of maize, vp 1, has green seedlings and normal ABA and carotenoid levels. The embryos of this mutant are insensitive to ABA. Therefore, it has been classified as a response mutant. The vp-1 mutant shows a pleiotropic phenotype and, in addition to being viviparous, has decreased levels of activities of flavonoids enzymes. However, if vp 1 is a response mutant, then expression of this mutation response to ABA are capable of producing anthocyanins.

### Tuberization

ABA appears to be involved in the process of tuberization. Wareing has shown that ABA formed by the leaf of a one-node cutting taken from a short day-induced plant of *Solanum andigena* (a wild species of potato which requires about 20 short day cycles to fully induce tuberization) is exported to the axillary bud and causes it to develop rapidly into a small tuber. Externally applied ABA has also been shown to promote the formation of these tubers in Dahlia and Jerusalem artichoke.

### Senescence

ABA may also be involved in senescence and ripening processes. Externally applied ABA accelerates the senescence of detached leaves; however, its effects on the senescence of leaves still attached to a healthy plant is minimal. Externally applied ABA accelerates the ripening of young fruit; moreover, there is a rise in the level of endogenous ABA during the ripening of strawberries and grapes.

### Resistance to Frost Damage

ABA may also be involved in increasing the resistance of temperate-zone plants to frost damage; the external application of ABA has been shown to increase the frost hardiness of seedlings of box elder (*Acer nigundo*), apple and alfalfa (*Medicago sativa*) and to ameliorate chilling injury in cucumber seedlings.

## ABA RECEPTOR SITE(S)?

ABA shows two types of responses in plants. The closure of stomata is an example of a rapid (5 min) response that can be induced by S-ABA, while R-enantiomer is almost without effect. In this case, the site of action is the plasmalemma. On the other hand, in the slow responses (30 min) which appear to involve RNA and protein synthesis, S-ABA are equally effective. Since the fast and slow responses have different molecular requirements, (Milborrow, 1980) concluded that the receptor sites must be different and the fast response to ABA cannot be a prerequisite for the slow response. Evidence for ABA binding proteins has so far only been obtained with protoplasts from guard cells Hornberg and Weiler, 1984). However, in an extensive study with a variety of ABA sensitive tissues, Smart et al

(1987) failed to obtain any evidence for ABA-receptor proteins by a number of reversible binding assays. Not all genetic data are compatible with the hypothesis that, depending on the type of response shown, there are at least *two different ABA receptor sites*. Although observations with the mutants vp 1 in maize and abi-3 in *Arabidopsis* support this view, whereas results with the abi-1 and abi-2 mutants of *Arabidopsis* shows variation with this hypothesis (Koornneef et al., 1984).

## Barley Aleurone Cells

Among the most studied effects of hormones on gene expression is the GA induced production of hydrolytic enzymes and its inhibition by ABA in barley aleurone layers. When ABA is applied 25-fold excess of GA the synthesis of a-amylase, protease b-glucanase and ribonuclease is suppressed. The best characterized enzyme produced by barley aleurone layers is α -amylase. It consists of two isozyme families (low and high pI groups) which have many different characteristics and are encoded by two structural genes on different chromosomes. Therefore, using total enzyme activity as a marker for α -amylase synthesis may obscure changes in one isozyme group over the other ABA-inhibited the synthesis of both groups of enzymes when added at the same time as GA, but when added 12 h after GA the high-pI group was more inhibited than the low pI group. Evidence for a role of ABA in regulating the expression of α -amylase activity at the transcriptional and possibly at the translational level are known.

ABA causes the production of a large number of ABA inducible polypeptides in aleurone layers Lin and Ho, 1986). The accumulation of all ABA induced polypeptides was inhibited by treatment with GA.

Exogenous phaseic acid (PA) is as active as ABA in preventing the accumulation of α -amylase, which raises the question of whether ABA is active by the virtue of its conversion to phaseic acid. However, phaseic acid is probably not the sole agent preventing α -amylase synthesis; R-ABA, which is not converted to phaseic acid in leaves of several species, is just as active as S-ABA (Mertens et al., 1982). There are conflicting reports on induction of ABA specific proteins by phaseic acid.

## Developing Embryo

ABA not only prevents premature germination during embryogenesis, but it has also been implicated in the regulation of storage protein synthesis. In embryos of rapeseed, soybean and wheat cultured in vitro, exogenous ABA promoted the synthesis of embryo-specific proteins and mRNAs. In rapeseed, a high concentration of sorbitol produced an effect similar to that of ABA, an effect that did not result from increased ABA synthesis in the embryos. Treatment with floridone inhibited accumulation of q -conglycinin in cultured soybean cotyledons and of several storage proteins in embryos of rapeseed. In rapeseed, the inhibition could be overcome by applied ABA and also by a low osmotic potential. Thus, it would appear form these results that there may be other factors besides ABA that regulate the synthesis of embryo-specific proteins.

# STRESS, PROTEIN AND ABA

ABA has been called as stress hormone since it enhances adaptation to various stresses. Freezing tolerance could be increased by ABA both in plants and in cell cultures. Increased cold tolerance was not induced in the presence of cycloheximide which indicates that synthesis of new proteins is involved with stress adaptation. Furthermore, during cold hardening, plants showed a transient increase in ABA content, suggesting that ABA may be an endogenous regulator of adaptation to cold stress. ABA also accelerated adaptation of cultured cells to salt at least in part by osmotic adjustment.

Associated with ABA mediated salt adaptation was the synthesis of a 26-KD protein. It has been suggested that this protein plays a role in salt adaptation. ABA accumulation in stressed leaves was not required for accumulation of another stress metabolite.

Water stress and ABA induce incorporation of $^{35}$S-methionine into a 26-KD proteins. Water stress inhibits the incorporation of $^{35}$S-methionine into a 14.4-KD protein. It is assumed to be the small sub unit of Rubisc-proline accumulation caused by water stress appears to parallel the accumulation of the 26-KD protein. Exogenous ABA does not induce proline accumulation despite the increase incorporation of $^{35}$S-methionine into 26-KD protein. Quarrie (1990), reports that ABA induces the formation of water stress proteins. Several proteins which are induced by ABA in developing seeds have now been found in other parts in response to water stress. Other proteins not found in seed tissue have also been identified in shoots and roots during drought stress.

Rapid changes in the synthesis of proteins due to water stress of rice leaves are being studied to identify proteins responsible for drought induced ABA accumulation.

The physiological role for Abscisic acid in plants under stress was first suggested by Wright (1969). He found that *inhibitor*-B (later shown to be ABA) accumulated rapidly in the leaves of wheat when they were allowed to wilt. The amount present was dependent on the duration and severity of water stress and temperature. Later Wright and Hiron (1972) reported a survey of 70 sps. of higher plants, all of which showed an increase in ABA after a period of wilting. At the same time it was shown that ABA is a powerful inhibitor of stomatal opening.

ABA plays an important role in plants during water stress and during drought conditions. Exogenously applied ABA promotes stomatal closing. ABA accumulates in water stressed mesophytic plants and that the ABA level decreases when the plant is no longer under stress. A fact that indicates the importance of ABA in water-stressed plants is the existence of the wilty mutant of tomato. These mutant plants seem to contain very low levels of ABA; and these levels do not change when water is withheld. The tendency of these plants to wilt is reduced by exogenous application of ABA. ABA acts directly on the stomatal apparatus.

## Water Stress and Chloroplast

ABA is produced in chloroplasts and remains there in unstressed leaves. When the plant is under water stress, the permeability of the chloroplast membranes to ABA increases and allows movement of the hormone to the epidermal cells, including the guard cells. ABA then acts on the guard cells to promote additional hormone biosynthesis, and this promotion continues until the water stress is relieved. When the water stress is relieved, the chloroplast membranes become impermeable to ABA and ABA becomes restricted to the chloroplasts. Presumably the build up of ABA is eventually turned off by the end product inhibition.

Under stress, in the leaf tissues the high membrane permeability of the uncharged protonated abscisic acid species ABAH and the impermeability of the anionic species occurs. ABA results in an accumulation in alkaline compartments. In the experiment of Jovanovic and Quarrie (1990) the pot grown maize plants were given cycle of drought stress and recovery to study the distribution of ABA in different plant tissue. Older leaves contained the most ABA. Young leaves, apical tissues and roots had less ABA and degraded it most rapidly after rehydration.

According to Walton, the establishment of plant lines that produce high levels of ABA may be highly significant to the growing of crops in arid parts of the world.

## MODE OF ACTION OF ABA

The molecular mechanism of ABA is unknown. However, there are two major biochemical effects of the hormone which are well known. One is an alteration of plasma membrane which is manifested in a change in biometrical potential across the membranes and a leakness and efflux of $H^+$ which is involved in some action of ABA for e.g., regulation of stomatal opening and closing. This closure of stomata cannot be accounted for by the RNA synthesis and RNA directed protein synthesis as the kinetics show that the response is quite rapid and appears within a minute. ABA may act on plant system as follows:

i) ABA does not exert a general inhibition of DNA transcription but may inhibit the synthesis of a limited number of specific mRNA species, eg, the a-amylase mRNA of the cells of the aleurone layer in barley seeds, and
ii) ABA exerts its main effect by blocking protein synthesis at a post transcriptional level such as mRNA processing or translation rather than by preventing DNA transcription, e.g., ABA has been shown to inhibit the synthesis of protease and isocitrate lyase in the embryos of cotton seeds even though appropriate mRNA's are present.

The other major effect is inhibition of RNA and protein synthesis. This in turn inhibits the synthesis of specific enzymes for e.g. inhibition of a number of enzymes like $_{\mathrm{ß}}$ -amylase, protease $_{\mathrm{ß}}$ -glucanase and ribonuclease has been reported to be suppressed by ABA. In most of the studies, ABA has been shown to inhibit the accumulation of labelled RNA precursors into all fractions of RNA as also in specific RNA fractions. This could be due to an increased activity of nucleases, decreased availability of template or inhibition of polymerase activity. In addition, inhibition of polysome formation, possibly through the reduction of mRNA supply, has also been attributed to ABA.

### DNA, RNA and Protein

ABA binds and it interacts with DNA, increasing its stability to alkali and heat denaturation. ABA has also been found to bind with membranous preparations of bean leaves, the binding increasing with increasing concentration. Two classes of binding sites were known for higher affinity site which had a dissociation constant of $3.5 \times 10^{-10}$ M. ABA regulates nucleic acid synthesis in a variety of plant tissues. In several studies, ABA has been shown to inhibit the accumulation of labelled RNA precursors into all fractions. This could be due to an increased activity of nucleases, decreased availability of template or inhibition of polymerase activity. In addition, inhibition of polysome formation, possibly through the reduction of mRNA supply, has also been attributed to ABA.

## Synthesis of Enzymes

Synthesis of a number of enzymes like α-amylase, protease, β-glucanase, and ribonuclease has been reported to be suppressed by ABA. Some recent reports involving the use of cDNA clones from hormone-treated tissues have indicated that ABA is regulating the expression of α-amylase activity at the transcriptional level, while some other reports implicate ABA's role possibly at the translational level. ABA itself can induce the formation of new proteins in the barley aleurones and some of these proteins may be involved in the metabolic conversion of ABA to phaseic acid (PA), which in turn may inhibit $GA_3$-triggered production of α-amylase. This hypothesis was based upon the findings that **cordycepin** -an inhibitor of RNA synthesis, prevents the conversion of ABA to phaseic acid (PA) as well as the formation of ABA-induced proteins. Effects of ABA and PA on the ABA-inducible polypeptides are identical at the mRNA and protein level. However, the role of phaseic acid as ABA mediator is still a matter of debate.

## Gene Expression

Most of the work on the effects of ABA on gene expression has been done on the GA-induced production of hydrolytic enzymes and its inhibition by ABA in barley aleurone layers. To study the pattern of controls that $GA_3$ and ABA exert on gene expression in barley aleurone layers, the effects of hormones on the levels of 8 different mRNAs have been determined. Northern hybridization using cDNA specific for enzymes indicated that 3 of these mRNA species encoding alcohol-dehydrogenase, a probable α-amylase or protease inhibitor, were $GA_3$-suppressible whereas 3 mRNAs encoding high pI α-amylase, low pI α-amylase and aleurain were $GA_3$-inducible. Two other transcripts were constitutive. ABA counteracted the $GA_3$ effects for both the inducible and suppressible mRNA species.

cDNA clones of early methionine labelled polypeptides (Em) and other ABA-regulated wheat gene (encoding 7S globulin) were used to measure levels and synthesis of these specific mRNAs in mature wheat embryos treated with ABA either alone or in combination with α-amanitin. Results from this study indicated that the regulation of both ABA-induced wheat genes is controlled at transcriptional level. The regulation of Em expression by ABA does not seem to be tissue-specific as a rice suspension culture also accumulates Em-like mRNA when incubated in ABA for 16 hr. Another recently characterized rice-gene (RAB21) also appears to be resulted by ABA in a tissue-independent manner.

Regulatory sequences involved in hormonal activation of genes have not been described from any plant. However, normal regulation of phytohormone ABA-inducible promoter from wheat has been studied in a transient expression assay using rice protoplast. The Em-promoter (650 bp), without its normal coding and 3' regions, is sufficient for ABA induction of the reporter gene. This response in rice protoplasts is similar to ABA regulation of the maternal gene in wheat embryos. The results provide direct evidence for phytohormone to regulate the initiation of gene transcription.

Most of the work on the effects of ABA on gene expression has been done on GA-induced production of hydrolytic enzymes and its inhibition by ABA in barley aleurone layers. ABA thus, inhibits germination as these enzymes hydrolyse the reserve food materials into soluble from which is an important event in germination. Some recent reports involving the use of cDNA clones from hormone treated tissues have indicated that ABA is regulating the expression of α-amylase activity at the transcriptional level, while some other reports implicate ABAs role possibly at the translational level.

ABA might alter levels of other endogenous GA, since GA and ABA have been found antagonistic in several biological terms as stem growth, synthesis of amylase in barley aleurone cells, dormancy of seedling. The antagonism may be because of the following:

i) The competitive interaction of both GA and ABA for the same site.
ii) Inhibition of GA biosynthesis by ABA.
iii) Promotion of GA inactivation by binding.

In vitro experiments with developing wheat embryos suggests that high levels of ABA present in the developing grain enhance the expression of genes encoding the lectin wheat germ agglutinin and other apparently globulin-relation storage protein and an albumin-like protein found in mature embryos. These effects of ABA are thought to involve transcriptional and post transcriptional events. In developing soybeans cotyledons, ABA influences the level of transcript for the β-subunit of B-conglycin, although apparently it does not effect the levels of transcripts for the other two subunits of this storage proteins.

## ABA RECEPTORS

The evidence supporting the existence of ABA receptors in guard cell is strong. A high-affinity binding site of ABA on plasmalemma of *Vicia faba* guard cell was found. There were $2 \times 10^6$ binding sites per protoplast. Three proteins that were covalently linked to ABA were separated by gel-electrophoresis. One protein had highest affinity for dissociated ABA, while the other 2 bound preferentially to protonated forms. One of these proteins could be modulator of $H^+$-ATPase which is inhibited.

## ABA, PROTOPLAST AND GUARD CELL

Current opinion is that the protoplasts of guard cells does not synthesize their own ABA. The ABA content of isolated protoplasts is much higher for guard cells than for mesophyll cells, and it increases still further when leaves are subjected to water stress. This implies that guard cells accumulate ABA that is manufactured elsewhere in the plant Stomata on pieces of isolated epidermis respond sensitively to the ABA, and the inhibition of stomatal opening was accompanied by reduced accumulation of $K^+$ ions in the guard cells. Later on it was found that isolated stomata (functional guard cell pairs surrounded by disabled subsidiary and epidermal cells) also responded to ABA, but their sensitivity was highly dependent on the concentration of $K^+$ in the surrounding medium. In the intact leaf it must be assumed that the subsidiary and/or epidermal cells have an important role in the shuttling of $K^+$ associated with stomatal movements.

Raschke (1979) suggested that ABA might operate by inhibiting the proton pump in the plasma membrane. On the assumption the $K^+$ intake depends on the efflux of $H^+$ this would mean that the action of ABA would be on the influx of K+ rather than on its efflux from the guard cells. However, Mac Robbie (1981) found that in the first 20 mts after application of ABA there were high efflux rates of rubidium and bromide (analogous to $K^+$ and $Cl^-$) from isolated guard cells. Little change in the influx rates of ions into guard cells could be detected after treatment with ABA.

Stomata on isolated epidermis open more readily if $K^+$ ions are present in the medium, and $Na^+$ ions (Jarvis and Mansfield, 1980). This suggests that there are mechanisms for ion expulsion which show a high specificity for $K^+$. Weyers et al. (1982) reported that ABA induced stomatal closure can be blocked by metabolic inhibitors such as azide and cyanide suggesting that the expulsion of solutes is dependent on metabolic energy.

Little is currently known about the sites of action of ABA as it comes into contact with the plasma membrane of the guard cells. Hornberg and Weiler (1984) briefly reported binding proteins capable of binding ABA in the plasma membranes of guard cells, but there have been no further reports confirming this.

## CALCIUM IONS

Strong indications of mode of action of ABA in guard cells have been provided by studies of its interactions with calcium ions, and with known regulators of calcium activity in cells. The essential mechanism behind the modulation of responses by $Ca^{2+}$ is through induced changes in its intracellular concentration (McAinsh et al., 1990). In the cells, $Ca^{2+}$ is thought to bind to the small protein **calmodulin** (CaM) and then the $Ca^{2+}$-CaM complexes clicit responses of various kinds. Although it is known that the action of ABA is dependent on calcium which $Ca^{2+}$ inhibit stomatal opening in the absence of ABA and the response of stomata to other agents may also be dependent on calcium. $Ca^{2+}$ has the central role in the regulation of the turgor changes of guard cells that are involved in stomatal opening and closing. ABA is just one external agent that is capable of triggering changes in intracellular $Ca^{2+}$ but the importance of ABA in plant water relations is that it provides a means of cell to cell, or long distance signalling in the plant to indicate the need for change in stomatal aperture to regulate transpiration.

# CHAPTER 26

# Ethylene: Ripening Hormone

Ethylene is the simplest unsaturated hydrocarbon compound and is a major product of the petroleum industry. In Japan, about eight million tons of ethylene were produced in 1990 alone. It is of interest that this gas acts as a plant hormone. Today ethylene is the plant hormone, which is best understood with regard of biosynthesis, regulation of biosynthesis.

$$H_2C=CH_2$$

## HISTORY ETHYLENE RESEARCH

Progress in ethylene research has been slow and stepwise, because of the difficulty in making various physical and chemical determination of this compound. This is one of the reasons that workers have avoided ethylene research.

Harvey (1928) showed ethylene to be a ripener of fruits and vegetables. According to Zimmerman and Wilcoxon, (1935) auxin-induced ethylene production. Hanson (1939) and Kidd and West (1942) suggest that ethylene functions as a ripening hormone. The introduction of gas chromatography as an analytic technique in 1959 has enhanced the progress of research on ethylene. Burg and Thimann (1959) revived the study of the ethylene biology with the coming of new analytical techniques such as gas chromatography and radioisotope tracer. Little was known about the metabolism of $^{14}C$-ethylene prior to the publication of Burgs and Thimann's report (Hall et al., 1959).

K. Shimokawa initiated research on ethylene biosynthesis at the radiobiochemistry laboratory of Kyoto University, Japan. First attempt was made to develop a radioisotope tracer technique for $^{14}C_2H_4$ (Shimokawa and Kasai, 1965).

Morgan and Hall (1962) and Abeles and Rubinstein (1964) and demonstrated that auxin increased ethylene production. This was found 29 year later after the original discovery of Zimmerman and Wilcoxon, (1935). This report developed interest of many plant physiologist and horticulturists to know the biosynthetic mechanism and the role of ethylene in the effecting auxin.

Abeles et al. (1971) studied the abscission of cotton, bean and *Coleus* explants and developed an **aging-ethylene theory** on the basis of data on the explants of *Phaseolus vulgaris* L. An attempt was made to systematize the abscission as phytogerontology.

**TABLE 26.1:** Historical summary of ethylene research

| Age | Running title of the work |
|---|---|
| 1901 | Discovery of strange phenomenon induced by a gas (Neljubow, 1901) |
| 1934 | Identification of the gas as ethylene (Gane, 1934) |
| 1932-1935 | Investigation of ethylene physiology (e.g., Crocker et al., 1932) |

| Age | Running title of the work |
|---|---|
| 1940-1954 | Involvement of ethylene in fruit ripening (e.g., Kidd and West, 1945) |
| 1950 | Ethylene formation of infiltrated plants (Williamson, 1950) |
| 1959 | Metabolic fate of ethylene-$^{14}C$ (Hall et al., 1959) |
| 1959 | Gas chromatography of ethylene (Meigh, 1959) |
| 1959 | Beginning of research of ethylene biosynthesis (Burg and Thimann, 1959) |
| 1962 | Fruit ripening hormone (Burg and Burg, 1962) |
| 1964 | Auxin induced ethylene formation (Abeles and Rubinstein, 1964) |
| 1965 | Discovery of methionine as a precursor (Lieberman et al., 1965). |
| 1965 | Liquid scintillation counting method of ethylene-$^{14}C$ (Shimokawa and Kasai, 1965) |
| 1966 | Ethylene-induced abscission (Abeles, 1966) |
| 1967 | $CO_2$ -a competitive inhibitor of ethylene action (Burg ad Burg, 1967) |
| 1968 | Discovery of ethephon-an ethylene releasing compound (e.g., Cooke and Randall, 1968) |
| 1968 | Phytogerontology (a concept of abscission mechanism (Abeles, 1968) |
| 1970 | Preparation of an enzyme catalyzed ethylene formation from acrylate (Shimokawa and Kasai, 1970) |
| 1971 | Inhibition of ethylene formation with enol ether amino acid analogs (Owens et al., 1971) |
| | Autoinhibition of ethylene formation (Vendrell et al., 1971) |
| 1977 | Establishment of ethylene biosynthesis from methionine (Adams and Yang, 1977) |
| 1979 | Discovery of ACC as a precursor (Lurssen et al., 1979, Adams and Yang, 1979, Konze and Kende, 1979) |

The degreening of various citrus fruits with ethylene is widely done on a commercial scale through out the world. Shimokawa et al.(1978) in Japan and Cohen (1979) Israel developed a degreening technique of Citrus fruits. Plant physiologists in 80's investigated various aspects of ethylene-research such as ethylene-induced degreening (chlorophyll degradation) from morphological (and biochemical stand points, purification of ethylene induced chlorophyllase from ethylene-treated fruits of *Citrus unshiu*. Generally, the commercial application of ethylene is difficult since it is a gas. The discovery of ethephon an ethylene releasing compound has facilitated the practical use of ethylene. Now, many attempts are being made to find commercial applications for ethephon as a control for crop growth. A historical account of ethylene-research is presented in Table 26.1.

Lieberman, one of the leaders in ethylene research found that ethylene formation was inhibited by enol ether amino acid analogs (Lieberman et al., 1975). S. F. Yang and his coworkers (1977) postulated a biosynthetic pathway of ethylene via S-adenosyl methionine from methionine. Lurssen et al., (1979) and Adams et al., (1979) discovered independently that 1-aminocyclopropane- 1-carboxylic acid (ACC) is a precursor of ethylene. Their research is different in regard to the identification of the amino acid as a more direct precursor than methionine. Adams et al. (1979) identified ACC as a useful precursor of ethylene in apple tissues. On the other hand, Lurssen et al., (1979) found a high yield from ACC during a survey of ethylene forming activity to find ethylene-releasing compounds. They postulated a biosynthetic pathway from the non-protein amino acid.

## ETHYLENE AS A PLANT HORMONE

Ethylene is a plant hormone, however, about 15 years ago, it was unclear as to whether it was a plant hormone or a metabolite having unusual effects on growth.

In 1968, plant physiologists recognized it as a true growth regulator or plant hormone. Pratt and Goeschl (1969) reviewed ethylene biology in 1968 and applied the term plant hormone to ethylene. The rediscovery of auxin-induced ethylene formation was important for this recognition.

Indeed, ethylene is an organic compound consisting of carbons and hydrogens, and is produced within a plant. Ethylene promotes epinasty, prolification, root formation and abscission and inhibits growth and anthocyanin formation of morning glory seedlings. Ethylene freely moves about within a plant or sometimes among plants without definite direction in the manner of the polar transport of IAA. Also, one can not specify the place of origin. Response forward ethylene is everywhere evident in plant organs. But, exogenous ethylene exposed to stem or roots causes stem prolification in the upper parts of hypocotyl of morning glory seedlings. This suggest that exogeerous ethylene or its stimuli reach the responding part and modify plant organs quantitatively or qualitatively. Therefore, ethylene satisfied the definition of plant hormone. Presently ethylene has been firmly established as a plant hormone (Shimokawa, 1984).

## ETHYLENE BIOSYNTHESIS IN HIGHER PLANTS AND ITS CONTROL

### Biosynthetic Pathway

Ethylene formation in plants has been known since 1934 (Gane, 1934), but its study was initiated nearly three decades ago and several reviews, papers have been published (Yang, 1974; Sembdner *et al.*, 1980; and Peach *et al.*, 1992). In this discussion we shall cite only selected references and for details a reference may be made to the above works. Ethylene biosynthesis in higher plants differs from ethylene biogenesis in bacteria and fungi especially in respect of the pathway and other aspects (Primrose, 1976). Following factors contributed towards researches in ethylene biosynthesis; difficulties in detecting ethylene production, and use of non-enzymic model systems; in having simple chemical structure, ethylene has several potential precursors, and the biosynthetic route represents a very limited side-shoot of general metabolic pathways. Following substances are suggested as ethylene precursors - methionine, linolenic acid, propanol, ß-alanine, ß-hydroxypropionic acid, acrylic acid, propanol, ß-alanine, ß-hydroxypropionic acid, acrylic acid, ethanol, ethionine, glycerol, organic acids (acetic acid, fumaric acid), sucrose, and glucose. Of these methionine is a fully established physiological precursor in ethylene biosynthesis. The pathway of ethylene biosynthesis is from S-adenosylmethionine → 1-amino-cyclopropane-1-carboxylic acid (ACC) → ethylene (Pech *et al.*, 1993). The key enzymes involved are ACC synthase and ethylene forming enzyme (EFE or ACC oxidase), respectively. While ACC synthase has been successfully studied *in vitro* the EFE enzyme has been very difficult to characterize due to its instability or inactivity on homogenization. It is only recently that EFE has become amenable to *in vitro* biochemical study. Hamilton *et al.*, (1992) have demonstrated tomato genes encoding the ethylene forming enzyme by first inhibiting their expression with antisense genes and then by demonstrating their functional expression in a heterologous host. Their studies have shed much light on the nature of this enigmatic enzyme and also on its cellular location and mechanism of action.

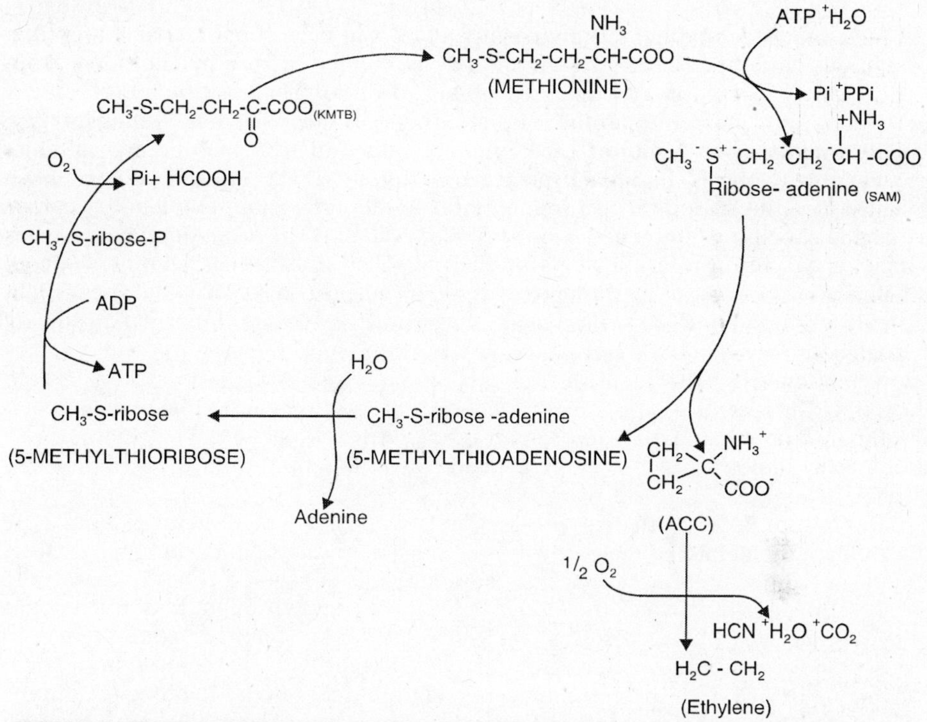

$NH_3$
$CH_3$-S-$CH_2$-$CH_2$-CH-COO
(METHIONINE)
ATP $^+H_2O$
Pi $^+$PPi
$+NH_3$
$CH_3$ $S^+$ $CH_2$ $CH_2$ CH -COO
Ribose- adenine (SAM)
$CH_3$-S-$CH_2$-$CH_2$-C-COO (KMTB)
O
$O_2$
Pi+ HCOOH
$CH_3$-S-ribose-P
ADP
ATP
$H_2O$
$CH_3$-S-ribose
(5-METHYLTHIORIBOSE)
$CH_3$-S-ribose -adenine
(5-METHYLTHIOADENOSINE)
Adenine
$CH_2$ $NH_3^+$
C
$CH_2$ $COO^-$
(ACC)
$^1/_2$ $O_2$
HCN $^+H_2O$ $^+CO_2$
$H_2C$ - $CH_2$
(Ethylene)

**FIG. 26.1:** Pathway of ethylene formation.

Recent studies have brought out that ACC oxidase is soluble enzyme (Rombaldi, 1992) though its exact subcellular location is still debated. These workers have also demonstrated that plant cells can generate ethylene from either apoplast or intracellular ACC, suggesting two potential sites for ethylene production. ACC synthase is encoded by a large gene family. In tomato, there are at least six ACC synthase genes some of which are specifically expressed during ripening.

## Ethylene formation in model systems

The $Cu^{2+}$-ascorbate rapidly converts methonine into ethylene and methionol seems to be intermediate, and ethylene is produced from carbons-3 and 4 of methionine. In this reaction $Cu^{2+}$ is reduced to $Cu^+$ by ascorbic acid and reoxidation of $Cu^+$ to $Cu^{2+}$ by $O_2$. Linolenic acid also forms ethylene with Cu-ascorbate system in some situations. The conversion of linolenic acid yields not only ethylene but also ethane.

A non-enzymatic conversion of methionine to ethylene is demonstrated by photo-chemical reaction based on FMN. However, under natural conditions, enzymes play the role of catalysts.

## Synthesis in vivo-physiological pathways

Slices of apple tissue are shown to synthesize ethylene under appropriate conditions. Apple tissues exposed to $^{14}$C-methionine produce ethylene. In banana methionine is shown to be precursor of ethylene, and also pea stem treated with

IAA including several other systems. Many additional facts support this contention: Compared with D-methionine, L-methionine is converted to ethylene; label from methionine is located in ethylene; the ability of different plant tissues to convert methionine into ethylene parallels the ability to produce ethylene endogenously, and Rhizobitoxine-an inhibitor of ethylene production from methionine, inhibits ethylene biosynthesis in higher plants but not in fungi. Further studies have revealed that the structural requirements for methionine as an ethylene precursor are highly specific. So ethylene is derived from C-3 and C-4 of methionine whereas C-2 yields $CO_2$ to form formic acid. Studies have also shown that 2, 4-dinitrophenol which is an uncoupler of methionine to ethylene suggests a requirement for ATP. S-adenosylmethionine (SAM) has been suggested to be an intermediate being degraded into ethylene. We have already stated that very recently the pathway has been completed by the identification of 1-aminocyclo-propane-1-carboxylic acid (ACC) as intermediate precursor of ethylene. Compounds like L-2-amino-4-precursor-(2-aminoethoxy)-trans-3-butanoic acid (aminoethoxyvinyl glycine) (AVG) inhibits conversion of ACC to ethylene, and conversion of ACC to ethylene requires oxygen.

The role of SAM in ethylene production is widely accepted. Figures 5.20-22, 35 36 elucidate the details of the pathway which leads to the production of ethylene. It may be noted that sulphur atom of the methionine is conserved by a cyclic, salvage process. But for this salvage, the amount of reduced sulphur might limit the amount of methionine and the rate of ethylene synthesis. We have already stated that ATP is essential for conversion of methionine to SAM and that $O_2$ is required in the final conversion of ACC to ethylene. It may be mentioned that under hypoxic conditions ethylene production nearly stops. There is evidence to show that four of the carbon atoms of the ribose unit of SAM are salvaged and reappear in methionine. Moreover an intermediate.$\alpha$-keto-$\gamma$-methyl-thiobutyric (KMTB) is essential in the salvage of these carbons. In addition to AVG, described previously, amino-oxyacetic (AOA) is also well known inhibitor of enzymes that require pyridoxal phosphate as a coenzyme. Both the inhibitors (AVG and AOA) block the conversion of SAM to ACC. Studies have shown that ACC synthase is a pyridoxal phosphate dependent enzyme.

We have already stated that ACC is finally converted to ethylene and this reaction is catalyzed by EFE. Studies involving isolated vacuoles of *Vicia faba* have shown that most of the ACC in cells is contained in them and that they form most of the ethylene. Suggestively, EFE is on or in the tonoplast. Bouyzayen *et al.*, (1990) have shown that both the plasma membrane and the tonoplast are concerned with the synthesis of ethylene. It may be added that ethylene formation is accompanied by formation of HCN and plants are reported to possess processes to metabolise the HCN which otherwise would inhibit mitochondrial respiration by poisoining cytochrome oxidase in mitochondria.

It may be added that linolenic acid as a possible source of ethylene is plants is not well accepted except that during ripening in the apple fruits lipoxidase activity enhances prior to the rise in ethylene evolution. In experiments where linolenic acid and lipoxidase were applied to apple discs there was stimulation of ethylene production. However, when labelled linolenic acid was used it was not converted to ethylene even though methionine served as an excellent precursor. This led to the general inference that linolate-propanol pathway did not operate in apples.

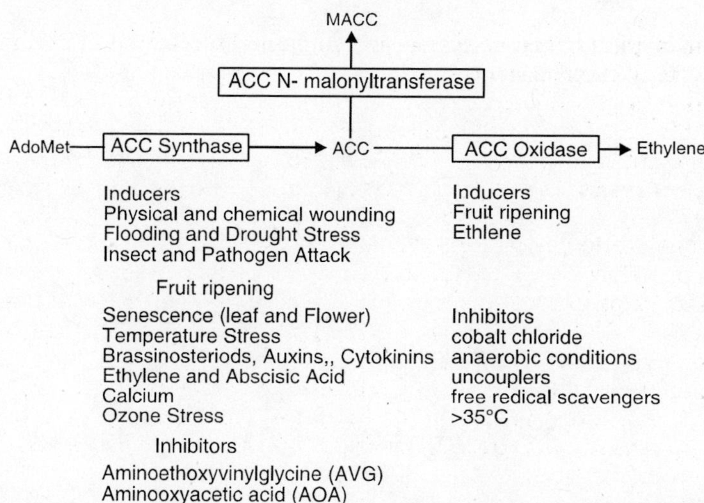


**FIG. 26.2:** Biosynthetic reactions from S-adenosylmethionine (AdoMet) to ethylene plus inducers and inhibitors.

In experiments where linolenic acid was added during homogenization of plant roots, both ethane production and ethylene liberation were stimulated. However, this pathway does not seem to operate in sugar beet leaf tissue or potato slices.

The role of ß-alanine-ß-hydroxy propionic acid-acrylic acid in ethylene production in plants has been suggested. However, confirmatory studies are desired to support this pathway of ethylene production in higher plants. We have already stated the role of several other substances as potential precursors of ethylene precursors of ethylene. However, none of them has been demonstrated to be direct percussor of ethylene *in vivo*.

## REGULATION OF ETHYLENE BIOSYNTHESIS

Rates of ethylene formation vary in different organs and tissues and depend upon growth and developmental stages of an organism. In general plenty of literature is available where ethylene production is correlated with several physiological processes like seed germination, vegetative growth, formation of flowers and fruits ripening, leaf senescence and abscission of leaf, flowers and fruit. Large number of studies also deal with the effects of different environmental factors (light, temperature, minerals) and stress conditions (water, temperature, wounding and chemical factors).

However, very little information exists on the factors which regulate ethylene biosynthesis. Presumably, ethylene production must be governed by production of substrate for ethylene forming enzyme system, its biosynthesis and the rate of degradation, cofactors and effectors. In ripening fruit tissues, the available evidence shows that methionine even though is a precursor of ethylene, its total content does not regulate the ethylene synthesis in ripening fruit tissues. Clearly ethylene production appears to be involved in regulation of ethylene biosynthesis during senescence of flower tissue.

Based on studies using inhibitors of nucleic acids and protein synthesis it has been suggested that biosynthesis and turnover of the ethylene forming system must be pivotal part of the regulatory system which could be affected in several ways. Here we shall discuss some selected instances;

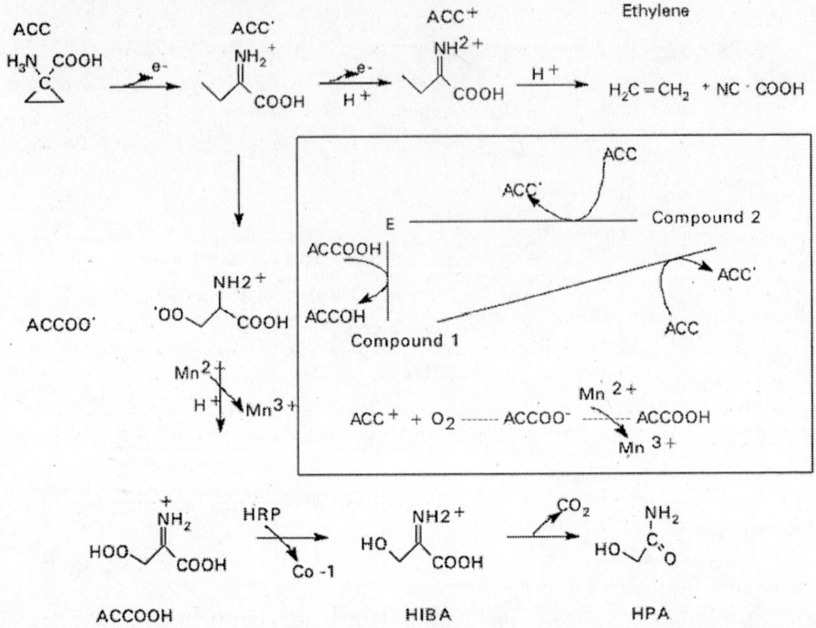

**FIG. 26.3:** Scheme I: Proposed mechanism for ethylene and HPA formation. **Inset**: Peroxidative cycle using ACC as substrate showing the formation of ACC-hydroperoxides. (Additional abbreviations used: HIBA: 4-hydroxy-2-imine-n-butyric acid).

## REGULATION OF ETHYLENE BIOSYNTHESIS AND EVOLUTION

The reaction catalyzed by ACC synthase, which converts SAM to ACC is the main site of control of ethylene biosynthesis. ACC synthase requires pyridoxal phosphate for maximum activity, and it strongly inhibited by AOA [(aminoxy) acetic acid] and AVG (L-e-amino-4-(2-aminoethoxy)-trans-3-butenoic acid), both inhibitors of pyridoxal phosphate- dependent enzymes. Application of ACC to various plant organs markedly increases ethylene production. The ethylene-forming enzyme (EFE) thus appears to be present in most tissues. In mung bean hypocotyls, ACC synthase is inactivated by its substrate, SAM, during its catalytic action, the reaction products, ACC and MTA (5-methyl thioadenosine) have no effect on ACC synthase activity.

Ethylene itself has an effect on its own synthesis. It promotes the last step in the pathway, the conversion of ACC to ethylene, apparently by inducing the synthesis of ethylene forming enzyme, EFE. This is the basis for the massive increase or **autocatalysis** of ethylene production that occurs with fruit ripening. Preclimacteric fruit is devoid of both ACC synthase and EFE. With the onset of ripening, activity of ACC synthase, ACC level as well as EFE activity increase. Both ethylene and ACC rapidly increase then decrease when the plant is under deficit water stress.

Ethylene biosynthesis is also controlled by the formation of polyamines from S-adenosyl methionine (SAM), a precursor of ACC and ethylene. On withdrawal of the ethylene precursor, the products of the reaction, polyamines, inhibit ethylene production. Conversely, ethylene inhibits polyamine production.

The oxidation of ethylene may take place in some plant and tissues under certain conditions to ethylene oxide or $CO_2$ There are suggestions that such ethylene oxidations increase during ripening, abscission and floral senescence and that ethylene may control the enzymes that oxidize it. Therefore, ethylene biosynthesis and evolution are controlled by many factors like substrates for the biosynthesis of ethylene, intermediates, end-products, side-paths and ethylene itself.

## Auxin-induced Ethylene Formation

Zimmerman and Wilcoxon (1935) first noted the stimulation of ethylene when tomato plants treated with IAA produced a gas that caused an epinastic response in plants. Later, Morgan and Hall (1962) presented evidence that this gas produced is ethylene. Abeles and Rubinstein (1964) concluded that ethylene evolution is closely allied with auxin levels and that the acceleration of abscission by NAA may be due ethylene produced following auxin applications. Burg and Burg (1966) also observed IAA-induced ethylene formation in etiolated pea stem sections. These studies confirmed that auxin induces ethylene formation. Auxin-induced ethylene production by etiolated bean hypocotyl section is inhibited by actinomycin-D and puromycin. These studies, therefore, suggest the involvement of RNA and protein synthesis in the induction of ethylene production.

A protein from the hypocotyls of etiolated mungbean seedlings has been purified which reversely inhibits auxin-induced ethylene synthesis. This proteinous inhibitor is found attaches to the cell surface of the plasma membrane.

A biosynthetic pathway of auxin induced ethylene production is similar to endogenous and wound ethylene production. Methionine is also an effective precursor in these productions. Yoshii and Imaseki (1981) postulated that ACC is an intermediate from methionine in auxin-induced ethylene production

## Cyanide and Ethylene Interrelationships

Exogenous application of cyanide can induce ethylene biosynthesis and it has been hypothesized that endogenous cyanide is a positive regulator of ethylene biosynthesis especially in climacteric tissue. It has been shown that with the onset of ripening in apples, increases occur in ACC, ethylene, cyanide and in beta-cyanoalanine synthase. There is evidence that beta-cyanoalanine synthase activity, which reduces levels of cyanide, plays a role in a plant's resistance to cyanide. Removal of cyanide could be expected to prevent cyanide-induced ethylene biosynthesis. The interrelationships between cyanide and ethylene remains, at present, a matter of conjecture.

## Ethylene Binding

Probably all plants produce ethylene, and all cells are exposed to ethylene at some stage. But all cells are not equally affected; and many cells that respond to ethylene do so only at certain stages of development. The affected cells at that particular time possess a particular molecular complement which recognizes and responds to ethylene. Ethylene is thought to bind to this receptor, enabling a primary reaction that then initiates other reactions, including modification of gene expression.

**What is the nature of ethylene binding?** No direct evidence is available, however, indirect evidence suggests that ethylene binds reversibly to a **metal containing receptor** site which exist in a variety of plant species. Whether some or all of them are requisite for the action of ethylene cannot be stated with certainty at present. In the cotyledons of *Phaseolus vulgaris.* It is shown to be hydrophobic integral membrane glycoprotein

Osborne et al. (1985) proposed that there are three general types of ethylene **target cells**, which have three distinct growth responses to ethylene and other hormones. Use of immunological techniques (including the use of monoclonal antibodies) helped us to identify the cellular targets of plant growth regulators.

Instead of supplying ethylene to plants, it is important to reduce the endogenous amount of ethylene present in plant cell. Silver ion also inhibits ethylene action in a great variety of plant responses, and is used commercially. **Norbornadiene**, a cyclic olefin, competitively inhibits ethylene action, and is a useful laboratory tool. Yang (1985) has summarized information on these inhibitors. There is also interest in further development of inhibitors of ethylene biosynthesis, particularly with a view to the reduction of flower and fruit abortion.

As we come to understand more about ethylene receptors, the mechanisms of action of ethylene, and biological structure-activity relationships, control of both ethylene biosynthesis and action for agricultural, forestry and biotechnological purposes may become more possible. As well, ethylene manipulation by external means may provide a quick, economical adjunct to advances that can be achieved with biotechnology and plant breeding.

The plant growth hormones (auxins, gibberellins, cytokinins, abscisic acid, ethylene and those yet to be identified) regulate many aspects of pant growth and development. With the techniques now at hand, we can explore the manipulation of natural regulatory mechanisms to produce improved crop varieties with desirable genetic traits. These advances will require continued investigations of the physiology and biochemistry of ethylene and other plant growth hormones, in order to effectively employ emerging biotechnological techniques.

Ethylene is capable of causing injury to plant roots. Crossett and Campbell (1975), using a solution culture, examined the shoot and root growth of barley plants after long-term exposure of the roots to ethylene. Both root and shoot dry weights were reduced slightly by ethylene treatment. Seminal root extension was greatly inhibited while lateral root growth was stimulated.

Although research in this field is slow because of the complexity of soil as an experimental system, it goes without saying that crop cultivation is of great practical importance.

## ANALYSIS OF ETHYLENE

Crocker and Zimmerman (1932) used biological responses for determination of ethylene concentration, but the reliability was questioned because of its specificity and sensitivity character. Young et al. (1952) used manometric methods but this method was not sufficient to analyze ethylene produced from plant sources. Gas chromatograph equipped with a hydrogen flame ionization were used to determine ethylene produced from plant sources. It can determine levels as low as 0.1 ml/liter. The determination of ethylene by gas chromatography is very simple, because no extraction is required prior to analysis.

## STRESS AND ETHYLENE PRODUCTION

Plants when subjected to stress produce ethylene. Such stresses may be disease, radiation, mechanical wounding and chemicals. This fact has occupied an important part of ethylene biology. The induction mechanisms of stress ethylene formation and the role of ethylene in these stresses involved many problems of practical nature.

### Disease

Ethylene is produced by leaves of *Physalis floridiana* infected with potato virus. Fruits treated with ethylene, in infected plant organs, the rate of respiration increased with that of ethylene production. This may be referred to as a **pseudo climacteric rise**- a sudden increase in respiration before death. A direct role of ethylene as phytotoxin has also been suggested. Pegg et al. (1976) concluded that ethylene production results from the **infection** but not a **cause of the infection**. Beijesbergen and Bergman, (1973) proposed that stress ethylene represses the formation of **tulipaline**- an antifungal compounds in plants infected by *Fusarium oxysporum* F. *tulipei*. According to Hall and Sexton (1972) that pathogen enhances cellulase and/or pectinase activity through the action of ethylene produced by pathogen, causing tissue disorganization in host plants.

Pathogen, however, is known to stimulate **peroxidase** activity but not cellulase activity. The stimulated peroxidase may participate in the reconstruction of disordered tissues, for instance through peroxidase-enhanced lignin biosynthesis Treatment of plant tissues with low concentrations of ethylene result in greater activity of peroxidases. This has been demonstrated with sweet potato infected with *Cerctocystis fimbriata*, a black-rot fungus, cow pea seedlings infected with cow pea mosaic virus and in many plants.

However, recent work shows that ACC is also a useful precursor of ethylene in wounded plant tissues and that ethylene formation is based on the rapid enhancement of the activity of ACC synthase,- a key enzyme of ethylene biosynthesis (Boller and Kende, 1980).

### Gene Expression and Disease Resistance

While resistance to disease is likely the result of a number of different processes, ethylene does appear to play a part in a number of instances. One of the earliest events in the plant pathogen interaction is a rapid increase in ACC synthetase. It has been suggested that the resulting increase in ethylene production may signal the plant to mobilize its defence systems.

Studies have been made of the effects of ethylene on two **plant defence response genes** in carrots (Ecker and Davis, 1987). The first is for PAL, the first enzyme in pathway to phenyl propanoid compounds, PAL activity results in the synthesis of photoalexins, and lignins and other phenolic compounds and correlates with increased resistance to pathogenic infection.

The second plant defence response gene is for hydroxy proline rich glycoproteins (HRGP), the major proteins components of the plant cell wall. The exact role of HRGP is not clear, but its accumulation is correlated with disease resistance. Ethylene treatment resulted in marked increase in the mRNAs for these two proteins.

### Physical Stress

Since the discovery of mechanically-stimulated changes in growth by Darwin, many investigators have observed this phenomenon. Jaffe (1973) observed a significant retardation in internode elongation by rubbing (a gentle mechanical stimulus), and termed this morphogenetic response as **thigmomorphogenesis** and the participation of ethylene in thigmomorphogenesis (for details see Jaffe, 1984. In: *Hormonal Regulation of Plant Growth and Development*, Vol. I, (ed.) S. S. Purohit).

Goeschl et al. (1966) observed an unexpected increase in ethylene evolution from epicotyl pressed with a container stopper. They investigated in detail this strange ethylene production and concluded that the production of ethylene by pea epicotyls increases by non wounding physical stress, and that ethylene acts as an endogenous growth regulator. The ethylene-induced thigmomorphogenesis is interesting form a practical standpoint because it makes possible growth regulation and developments without any requirement for growth retardants.

### Autocatalytic Ethylene Formation (Ethylene Stress)

As mentioned earlier in biosynthesis of ethylene that there is an **autocatalysis** of ethylene production that occurs with fruit ripening and its formation is autocatalytically regulated by ethylene itself. This characteristics has documented from time to time. First characterization of such formation was done by Vendrell and McGlasson (1971) using banana pulp slices and using propylene an ethylene analogue (McMurchie et al., 1972). Propylene suppressed ethylene formation and delayed the peak of ethylene production in intact green bananas.

**Autoinhibition** of ethylene production was more directly evidenced with the use of methionin-$^{14}$C. Such autoinhibition of ethylene production in *Citrus* flavedo discs results from a suppression of ACC formation through repression of the synthesis of ACC synthase and inhibition of its activity

### Ethylene Formation in Soils

Smith et al. (1971) reported that ethylene is produced in anaerobic soils by microbial activity, and can affect the growth of crop plants in waterlogged soils. They were of the opinion that ethylene acts as a soil fungistasis, and developed the concept of the $O_2$-$C_2H_4$ cycle (a self-regulating microbial cycle in soil). This cycle helps in regulation of organic matter turn over in soil, better availability of essential plant nutrients, and development of effective biological control of soil borne plant pathogens. The decomposition of ethylene by a group of aerobic microorganisms and the effects of ethylene on plant growth in waterlogged soils were also been examined (Cornforth, 1975). However, no information related to decomposed product was obtained.

### Abscission

The analysis of ethylene-enhanced abscission is possible using simple system constructed of explants prepared from plant seedlings. The action mechanism of plant hormones was studied in relation to the molecular biology involved. Studies on abscission resulting from the action of ethylene seem to indicate that the participation of an ethylene-enhanced enzyme activity is the action mechanism involved. Generally it is accepted that ethylene hastens abscission and auxin delays it. Also, abscisic acid (ABA) enhances the abscission of some plants as a third abscission regulator. In vivo, these plant hormones regulate abscission through complicated interrelationships with each other. In particular, the ethylene and auxin balance in the separation zone is very important. The ability of ethylene against abscission increases with a decrease in auxin concentration in the separation zone. This is a phytogerontological effect of ethylene, recently postulated by Abeles et al., (1971).

Abeles and his collaborates summarized their results as follows:

1. Ethylene-enhanced abscission depends on the synthesis of RNA and protein.
2. The addition of actinomycin-D results in the loss of this ability by inhibiting RNA synthesis.
3. Cycloheximide inhibits abscission through the inhibition of protein synthesis.

Horton and Osborne (1967) reported an ethylene-induced enzyme to be cellulase. The cellulase synthesis was inhibited by IAA, cytokinins, actinomycin-D, cycloheximide, and $CO_2$. These observations support the view that these substances inhibit abscission. Ethylene effects on the fine structure of abscission cells and initiate the development of a rough endoplasmic reticulum in sections of epidermal and cortical cells from the hook region of intact etiolated seedling of *Pisum sativum* L. This undoubtedly reflects an increase in protein synthesis and translocation.

## Ethylene and Duoyant Density of Plasma Membrane

Koechler and Lewis (1979) observed an increase in the buoyant density of a plasma membrane after ethylene treatment of bean abscission zones. These increase in density was observed by following the shift of cellulase and KCl-stimulated ATP activity. Such a change is consistent with the loss of lipids from the senescent cells of the abscission zone. Changes in density of cell organelles may possibly be a common phenomenon due to ethylene treatment.

## Chlorophyll Degradation

Degreening does not occur during exposure to ethylene in the peel of satsuma mandarin fruits, but does by exposing the ethylene-treated fruits to air. Chlorophyllase enzyme is responsible for degradation of chlorophyll in plants. Its activity is low during exposure to ethylene and during the first 6 h of the subsequent exposure to air. These observations indicate that a lag phase is a prerequisite for the activation of chlorophyllase and also that oxygen is required for the activation of chlorophyllase and also that oxygen is required for ethylene-enhanced chlorophyll degradation and chlorophyllase activity. The constructing effect of these two inhibitors of protein synthesis indicate that de novo synthesis of cytoplasmic chlorophyllase is essential for ethylene-enhanced chlorophyll degradation.

Ethylene degreening of *Citrus* fruits is a common commercial practice in USA, Japan, Israel, Australia to insure product quality in the market. Practical degreening methods are generally complete, but the role of ethylene is still not certain.

### *Ethylene and Chloroplast Structure*

Changes in chloroplast structure during degreening of *Citrus* fruits treated with ethylene were examined by Shimokawa (1978). The number of chloroplast decreased with ethylene treatment and tree-ripening. Reduction in chloroplast size was a characteristic feature of ethylene-treated fruits lost grana and lamellar systems. This disintegration of the membranes within the chloroplasts is expressed by the word **melt**. In the chloroplasts of ethylene-treated fruit, double-layered structures were degraded and the membrane layers become separated from each other. These observations are quite similar to the case of the ribosomes of the ethylene-treated separation zone (Freytag et al., 1977).

## ETHYLENE AS A RIPENING HORMONE

There were two schools of thought regarding the role of ethylene in fruit maturation:

1. Ethylene as a ripening hormone (Hansen, 1942; Kidd and West, 1945; Porritt, 1951).
2. Ethylene as a by-product of fruit ripening (Baile 1960) and it is a **by-product** of fruit maturation. According to by-product theory, in the absence or information correlating the internal ethylene content with the rate of ethylene production, one can advance the argument that small quantities sufficient to induce ripening are produced prior to the rise of respiration, but measurable amounts are detected only after the onset of the climacteric.

Burg and Burg (1968) have challenged his assertion on the basis of following arguments:

1. Internal ethylene content and the rate of ethylene production are determined for a variety of fruits including several previously been described as devoid of ethylene; e.g., the mango, pineapple, orange and lemon.
2. The application of ethylene hastens the ripening of mangoes even after the climacteric has began.

McGlasson et al. (1975) raised the question as to whether or not ethylene is a **true ripening hormone**. They examined the role of propylene in respiration and ethylene production of tomato fruits of rin-a non ripening mutant. Propylene stimulated respiration in the immature fruits of rin, but there was no change in endogenous ethylene production. It was concluded that the onset of ripening in normal tomato fruit is not controlled by endogenous ethylene, and increased ethylene production is probably an integral part of the ripening process. *This is the first work calling into question the fruit ripening hormone theory of ethylene.*

### Stimulation of Ripening

There are several aspects regarding the usefulness of stimulation of ripening. The most interesting consequence is the possibility of a once-over mechanical or manual harvest in crops that produce ripe fruits over a longer time period and therefore normally must be harvested several times. A second aspect is the possibility to speed up the time of cultivation of a crop, especially in greenhouses were follow up crops are planted immediately. A further advantage of stimulating ripening is the possibility to plan harvest and marketing times of fruits. In some crops it may also be of interest, to harvest and market fruits generally earlier in the year.

Ethephon is used for stimulation of tomato fruit ripening. The fruit quality is not negatively affected compared with normally ripened tomatoes. The application of the growth regulator about 8 days before harvest with 20% colored fruits in the field allows a once-over mechanical harvest without loosing a substantial amount of unripe fruits. Tomato fruit ripening cannot be enhanced by ACC sufficiently. However, a derivative of ACC, N-Formyl-ACC, is very effective in stimulation of tomato fruit ripening (Lurssen, 1982).

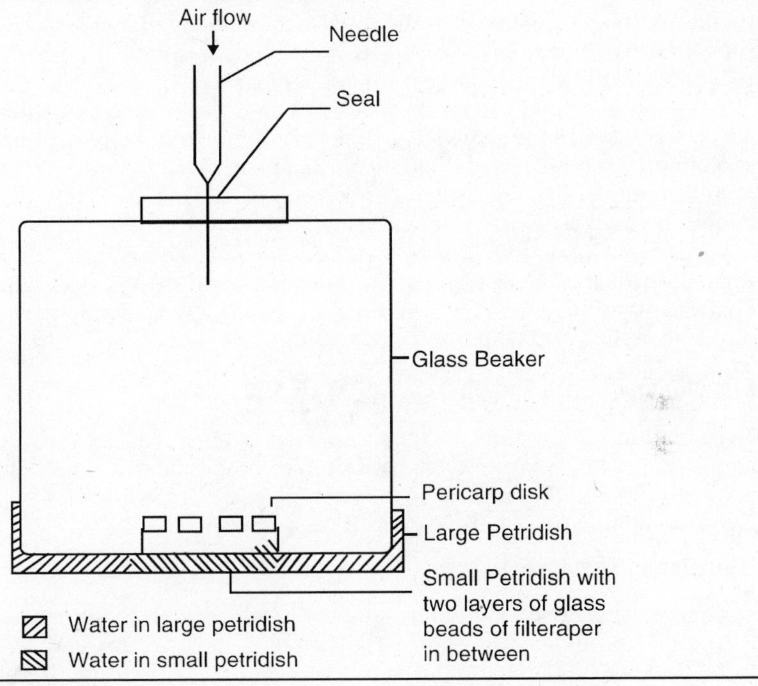


**FIG. 26.4:** System of ethylene induced ripening in tomato.

Ethephon enhances ripening also in other vegetables and fruits like peppers, fig, melon, peach, blueberry, papaya, apple, grape, etc. when applied shortly before harvest. In apple and grape an earlier time of application during fruit development also enhances the ripening process (Hale et al., 1970). However, the ripening of some fruits-possibly the non-climacteric ones-cannot be enhanced by ethylene or ethephon. In pineapple Ethephon is also used shortly before harvest to enhance ripening. The use of Ethephon to ripen coffee berries is complicated by the fruiting behaviour of coffee with fruits of different developmental stages being simultaneously on the tree. Ethephon causes ethylene-induced leaf drop. Ethephon is also used as a ball opening agent in cotton. The application allows a once-over harvest and increases total yield (Cothren, 1981).

Ethephon effect ripening of fruits (senescence) and enhance ripening of vegetative plant parts of sugarcane. The harvested material has a higher sugar content and higher juice purity after Ethaphon treatment. An enhanced leaf maturity is desirable in tobacco to reduce the harvest frequencies. With Ethephon it is possible to reduce the number of harvest from about 6 to 3. Furthermore the curing time after harvest is reduced.

## Gene Expression, Fruit Development and Ripening

The molecular basis of fruit development and ripening have been discussed with reference to gene expression. As would be anticipated, differential gene expression has been shown to be correlated with physiological changes in tomato fruit (Piechulla and Greussem, 1987). Both increases and decreases in individual (mature) RNA species occur during ripening (Morgan et al., 1987). Polygalacturonase, a cell wall softening enzyme associated with ripening (in tomatoes, one ripening specific cDNA is associated with a gene family). DNA sequence rearrangements appear to be associated with the time of expression of two of the cDNAs, one ripe specific and one green specific.

Ripening related genes have also been identified in *Avocado*. With *Avocado* slice, amino-ethoxyl vinyl glycine (AVG) completely suppressed ethylene production, the respiratory climacteric and ripening. There is at least a 50-fold increase in cellular message concentration during ripening and there are six cellulase genes. It is not known how the level of cellulase mRNA is regulated and whether ethylene which increases during the climacteric, is involved in the regulation of this gene expression. These genetic events resulting in synthesis of enzymes that lead to the softening of plant tissue do, however, fit well with ethylene ability to accelerate fruit ripening.

## Stimulation of Germination

Germination of a number of plant species is enhanced by ethylene. So there are some efforts to apply ethylene releasing compounds in combination or prior to application of herbicides to increase the effectiveness of the herbicide treatment. Even the application of ethylene as a gas seems promising.

## Induction of Female Flowers

Cucumber and melon normally produce male flowers earlier than female flowers. Ethylene stimulates the early production of female flowers. Often man flower formation is completely inhibited following ethylene treatment. Treatment of plants with Ethephon has the same effect. Female flower induction in Cucurbitacae is of commercial interest to achieve an earlier harvest and a yield increase.

## Male Sterility

In plant breeding of cereals the production of hybrid seed is of great interest to improve quality and to increase yields. Ethylene applied at a certain time of pollen development causes male sterility. Reports on the use of Ethephon to induce male sterility in wheat seem promising, however, the effect may not be sufficient for commercial application (Bennett and Hughes 1972; Hughes et al.,1974; Dotlacil and Apltauerova, 1978).

## Prevention of Lodging in Cereals

High level of cereal production is intensified by use of herbicides, fungicides and high amounts of fertilizer. Under such conditions cereals tend to lodge when rainfalls and storms occur later in the growing season. Mechanical harvest of lodging cereals is very time consuming. Furthermore, the quality of the crop is lower because of prolonged ripening. Under wet weather conditions the drying of the grains is severely affected, especially in rye germination of grains on the culm may occur.

In wheat the problem of lodging is solved by the use of chlorocholinchloride, a gibberellin antagonist. Apart from a growth inhibition a thickening of the culm is achieved. However, this compound does not prevent lodging in barley and is not very effective in other cereals. The ethylene-induced modification of cell size and shape offers the possibility to inhibit the growth of plants in sensitive developments phases. A stabilization of the culm by thicker cell walls according to the observed modification of cell walls may also occur. Under field conditions ethylene growth regulators cause a growth inhibition in cereals especially when they are applied at a late developmental stage after the appearance of the second node. In barley Ethephon alone or in combination with gibberellin antagonists can be applied successfully, however, the efficacy is not as good as that of chlorocholinchloride in wheat. In rye lodging prevention with Ethephon is also possible.

ACC inhibits the growth of cereals only at very high concentrations. Amazingly N-Formyl-ACC and similar derivatives of ACC are much more effective and prevent lodging at relatively low concentrations.

### Growth Inhibition

The growth inhibiting properties of ethylene in many plant species are moderate compared with those of plant growth inhibitors interfering with gibberellic acid biosynthesis. In water-plants even a stimulation of growth observed by ethylene. The use of ethylene growth regulators in cereal production to prevent lodging is based on growth inhibition. Some investigations show the usefulness of Ethephon in other crops for growth inhibition. In grape vigorous vegetative growth interferes with reproductive growth, berry size, coloration and ripening of the fruits. Therefore, special pruning systems have been developed for grape production. Ethephon may be useful in inhibiting vegetative growth of grape vines resulting in higher yields and better quality.

## ETHYLENE RELATED PLANT GROWTH REGULATORS

Ethylene as a gas can only be applied in more or less closed systems like glass houses or incubation chambers, limiting the usefulness of the hormone. Efforts of agrochemical industry to develop liquid or solid ethylene releasing compounds for easier handling in early 1960's is remarkable. These compounds degrade in presence of water evolving ethylene (Draber, 1977). The development of such compounds still is the only example of a rational design of plant growth regulators or pesticides with a new mode of action.

### Ethephon

2-chloroethylphosphonic acid (CEPA, Ethephon) is the best known compound of this group. It releases ethylene above pH 4.0 in aqueous solutions but is relatively stable below pH 4. The possibility of stabilisation under certain conditions is of course important for a commercial plant growth regulator, as such a compound is stored for some time before use. The degradation of 2-chloroethylphosphonic acid to ethylene is a base catalysed elimination reaction.

Several esters and other derivatives of 2-chloroethylphosphonic acid have also been described. Probably all these compounds are converted to 2-chloroethylphosphonic acid in the cell. Theoretically it is suggested that these derivatives penetrate better into the cell or by a different time course of ethylene evolution. However, these derivatives generally were not superior to ethephon itself.

$$Cl - CH_2 - CH_2 - \underset{OH}{\overset{O}{\overset{\|}{P}}} - OH + H_2O \rightarrow HCl + CH_2 = CH_2 + H_3PO_4$$

Degradation of 2-chloroethylphosphonic acid (Ethephon) to ethylene.

$$Cl - CH_2 - CH_2 - Si\ (O\text{-}CH_2\text{-}CH_2\text{-}O\text{-}CH_3)_3 + H_2O$$
$$HCl + CH_2{=}CH_2 + SiO_2 +_3 HO\text{-}CH_2\text{-}CH_2\text{-}O\text{-}CH_3$$

Degradation of 2-chloroethyl-tris-(2-methoxyethoxy) silan (Etacelasil) to ethylene.

$$Cl\text{-}CH2\text{-}CH2\text{-}SO\text{-}OH{+}H2O \rightarrow HCl + CH_2 = CH_2 + H_2SO_3$$

$$Cl\text{-}CH_2\text{-}CH_2SO_2\text{-}CH_2OH + H_2O \rightarrow HCl + CH_2 = CH_2 + H_2SO_3 + CH_2O$$

Degradation of 2-chloroethylsulfinic acid (upper formula) and (2-chloroethyl) sulfonyl-methanol (lower formula) to ethylene

$$HO\text{-}CH_2\text{-}CH_2\text{-}NH\text{-}NH_2$$

2-Hydroxyethylhydrazine is possibly converted to ethylene in the plant

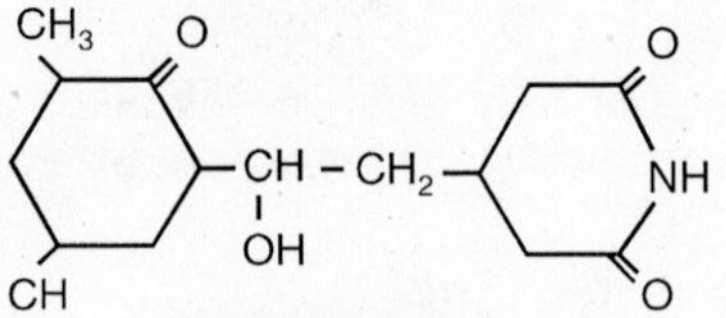

Cycloheximide

HO-N=CH-CH-N-OH

Glyoxaldioxim

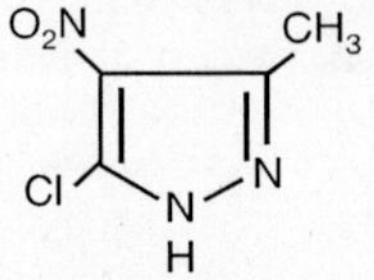

5-Chlor-3-methyl-4-nitro-1H-pyrazol

**FIG. 26.5:** Various kinds of ethylene-releasing compounds.

## Etacelasil

2-chloroethyl-tris-(2-methoxyethoxy)-silan (Etacelasil) belongs to the group of 2-chlorosilanes, also releasing ethylene in aqueous solutions. The chemistry of ethylene evolution is comparable to that of ethephon.

### Haloethylsulfinic acids

A third group of ethylene releasing compounds are the 2-haloethylsulfinic acids and derivatives. These compounds also have a similar mechanism of ethylene release as compared with ethephon.

### 2-Hydroxyethylhydrazine

The growth regulator 2-hydroxyethylhydrazine causes typical ethylene reactions in plants. However, the mode of action of this compound is not clear. The compound does not release ethylene spontaneously in aqueous solution (Block and Young, 1971). In more complex systems an ethylene generation can be measured and it could be assumed, that in plant tissue a conversion to ethylene can occur.

### Others Ethylene Releasing Compounds

Three chemically unrelated growth regulators should be mentioned; cycloheximide (trade name Act-Aid), glyoxal-dioxim (trade name Pik-off), and 5-chloro-3-methyl-4-nitro-1H-pyrazol (trade name Release). These compounds are useful only for a very limited field of application, namely citrus fruit abscission. They do not generate ethylene themselves but obviously slightly injure the fruit, causing wound ethylene production, which induces abscission layers.

1-aminocyclopropane- 1-carboxylic acid (ACC), known since 1957 as a constituent of ripe fruits is an intermediate of ethylene biosynthesis. ACC and its derivatives enhance ethylene biosynthesis and induce typical ethylene reactions in plants.

Chemical analogs of ethylene are far less active in biological systems. Propylene causes similar reactions, but 100 times the amount is needed compared with ethylene. Acetylene gives similar reactions only at about 3000 times the amount of ethylene

## BIOCHEMICAL ACTION OF ETHYLENE

The pigment metabolisms result from the ethylene-enhancement of related enzymes. The modification of endoplasmic reticulum indicates a change in protein synthesis or transportation of proteins. Eventually ethylene action results from DNA, RNA and protein synthesis of efficient utilization. A knowledge of molecular biology has facilitated the determination of the action mechanisms of plant hormones such as ethylene, auxin and gibberellins. However, the rapid response to the plant hormones as in the case of IAA, if it actually occurs, can not be explained by the molecular biological concept involving enzyme biosynthesis.

**TABLE 26.2.** Biochemical action of ethylene.

| RNA synthesis | Protein synthesis | Protein loss |
|---|---|---|
| Enzymes | Cellulase | Glucanase |
| Pectinase | Polygalacturonase | $\alpha$-amylase |
| Catalase | Chitinase | Phenylalanine ammonia lyase |
| Malic enzyme | Peroxidase | Leakage of peroxidase |
| Release of alfa-amylase | Stabilization of enzyme activity | Ribosomes |
| Modification of ribosomes | Mitochondria | ATPase stimulation |
| Swelling | Volume change | Inducement of alternative path of respiration |
| Chloroplast (Chromoplast) | Chlorophyll synthesis | Chlorophyll degradation |

| RNA synthesis | Protein synthesis | Protein loss |
|---|---|---|
| Degreening | Carotenoid synthesis | Pigments |
| Formation of anthocyanin | | |

## Molecular Requirement of Ethylene Action

Ethylene, simplest unsaturated hydrocarbon, regulates growth development and senescence to a considerable extent. Burg and Burg (1967) challenged the idea that this was due to the molecular structure of ethylene and discovered that biological activities requiring an unsaturated bond to be adjacent to a terminal carbon atom is related to molecular size, and that activity decreases with substitution. Furthermore, they observed that $CO_2$, a close structural analogue, is a competitive inhibitor of ethylene action. Although the application of a Linewever-Burk plot for enzyme kinetics is still questionable to some extent, it is recognized that the effects of $CO_2$ are the opposite of ethylene. The molecular requirements for abscission are in agreement with Burg's results (Abeles and Gahagan, 1968). It was not possible to adopt Lineweaver-Burk plots in the abscission of a bean petiole explant since abscission cannot be expressed as a rate phenomenon (Abeles and Gahagan, 1968). However, the idea that $CO_2$ is a competitive inhibitor in the abscission arises from the results of early research.

Dollwed and Seeman (1975) found that propylene competes with ethylene induced reduction in the epicotyl of an etiolated garden pea. They conclude that ethylene become loosely attached to a site.

## MECHANISM OF ETHYLENE ACTION

Ethylene biologists have also been interested in this action mechanism. But so far, there has not been developed a mechanism concept capable of explaining the wide scope of action such as ACC as a common precursor of ethylene biosynthesis under various conditions.

(i) DNA, RNA and protein biosynthesis, (ii) mitochondrial swelling, (iii) pigment metabolism, (iv) enzyme induction and (v) modification of endoplasmic reticulum.

## DNA, RNA and Protein Biosynthesis

There are evidences that several plant species have mechanisms for compartmentation of ethylene. In *Phaseolus* there is a high affinity binding site which has a dissociation constant of the order of $10^{-10}$ M. The relative physiological effectiveness of the structural analogues of ethylene was comparable to their capacity to compete with ethylene for the binding sites.

Dilworth (1986) ethylene influence the expression of genes at the transcriptional level (DNA to mRNA), the translation level (mRNA to protein) and the post translated level (modification of protein). The diversity of phenomena involving plant growth hormones has led to the suggestion that they do not act directly, but indirectly through a secondary messenger such as $Ca^{++}$ or certain oligosaccharides or certain polyamines.

Ethylene has been reported to mediate the auxin action since (like auxin) it enhanced DNA, RNA and protein synthesis when soybean hypocotyls were treated with ethylene. On the contrary, experiments done with chromatin isolated from soybean hypocotyls showed that ethylene is not a mediator of auxin action.

Ethylene has also been reported to affect the relative amounts of iso-accepting of leucyl-and lysyl-tRNA in tomato fruit, and rRNA and tRNA in ripe fig fruit.

The role of ethylene in inhibition of cell division and growth was shown to be due to a decrease in DNA synthesis in meristematic seedlings of pea. This decrease in DNA synthesis was associated with the decrease in DNA polymerase activity. No effect of ethylene on RNA synthesis was observed. A 60 per cent increase in poly A+ RNA was observed in ethylene-treated potato tubers. In vitro translation of this poly A+ RNA and SDS-PAGE of translational products indicated sharp differences between ethylene-treated and untreated tissues.

Several new proteins in the molecular weight range of 30-40 and 60-70 Kd were synthesized in addition to augmentation of some of the already existing protein bands. However, the nature and function of these proteins remains to be determined. The results indicate that ethylene increased the levels of translatable poly $A^+$ RNA in potato tuber by acting at transcriptional or post-transcriptional level.

Similar studies have showed that ethylene-induced changes in mRNA as well as in polyribosome size. It was proposed that ethylene could control gene regulation by selective expression of mRNAs or by translation control through proliferation of polyribosomes.

cDNA library prepared from poly $A^+$ RNA of carrots treated with ethylene was screened for picking up ethylene-specific mRNA complementary cDNA clones. These selective clones were employed to measure the kinetics of accumulation of ethylene-induced mRNA in carrot root tissues. The studies, however, did not provide any clue to the mechanism by which ethylene exerts control on gene expression. Similar results were observed in tomato leaves.

Recent studies indicate that ethylene can directly act as a signal for plants to induce genes involved in the biochemical defence against pathogens. Cloned carrot Phenylalanine Ammonia Lyase (PAL) and hydroxy proline rich glycoprotein (HRGP) genes were used as probes. These 2 genes were found to respond to defence mechanism of the carrot plant. Northern blot analysis of poly A+ containing RNA from 50-hr ethylene-treated roots revealed a 10-fold increase in the levels of PAL mRNA. A more complex pattern of HRGP mRNAs was observed. There was a dramatic 50 to 100-fold increase in the levels of 2 mRNAs (4.0 and 1.8 Kb) species, while the third mRNA (1.5 Kb) decreased. A gt 10 cDNA library prepared from ethylene treated root mRNA is being used to obtain probes specific to each of these ethylene regulated HRGP-transcripts.

## Ribosomes as a Target

Shimokawa et al. (1969) found that ethylene-$^{14}$C was fixed into high molecular immobile molecules in the leaves of morning glory plants. It was shown that ethylene has a high affinity for ribosome RNA and some affinity for ribosomal protein (Shimokawa and Kasai, 1968). They consider that the incorporation of ethylene into high molecules may induce a conformational change. This is the first evidence suggesting that a primary action site of ethylene may be ribosomes. In support of this, Freytag et al. (1977) published their research data. Ribosomes in ethylene-treated cotton and sugar beet radicles were larger than ribosomes in control and nitrogen treated cells. The ribosomes in ethylene-treated plants were swollen and less dense. The swollen ribosomes observed in electron micrographs of ethylene-treated materials underwent conformational changes as was indicated by significant differences in sedimentation patterns. The ribosomes from ethylene-treated material were less dense than those from the control material.

In experiments on the ethylene-enhanced chlorophyll degradation of satsuma mandarin, lamellar membranes of chloroplasts in ethylene-treated fruits were swollen and less dense (Shimokawa et al., 1976). These conformational changes in plant cell organelles are perhaps important to ethylene action.

Fruit ripening has been attributed to ethylene action. A cDNA library from ripe avacado fruit yielded a number of ethylene-related clones, one of which, pAV5, represented cellulase. Hybridization assays during the course of ripening indicated that different members of the putative cellulase gene family are expressed at different times.

### Effect of Ethylene on Gene Expression

The effects of ethylene on gene expression may be viewed in the light of following facts:

i) In potato tubers, ethylene enhances mRNA synthesis and results in the production of several new poly-peptides (Darnell, 1982).
ii) That ethylene treatment of carrot roots results in the appearance of new mRNA was demonstrated by Christoffersen and Laties (1982).

There was also evidence suggesting that ethylene increased the rate of translation of the message. In carrot roots ethylene induced transcription (DNA to RNA). A large respiratory response to ethylene occurs after the respiratory response to wounding subsides.

Plans are to create an ethylene-specific cDNA expression library and to acquire antibodies to proteins of interest (antibodies) that can be used to isolate the proteins in large amounts for study.

### Specific Hormones and their Action

Every hormone has its own affinity of binding, absorption, and action on cell at macro- or micro cellular levels. Therefore, it is essential to discuss every hormone's action individually.

## ETHYLENE RECEPTORS

A calculation of the amount of ethylene receptors in plant tissue by the Abeles (1973) revealed 10 mg. per metric ton or approximately 500 attachment sites per cell, assuming the receptor has a molecular weight of 100,000. Based on kinetic studies of ethylene binding and displacement in tobacco leaves, Sisler (1979) estimated 4000 binding sites per cell.

Although direct proof is lacking, some indirect evidence suggests that ethylene binds reversibly to a metal containing receptor site. Cell free extracts of unripe bean cotyledons exhibit ethylene binding activity with a high specificity and activity. The binding site seems to be membrane bound. It was possible to isolate the binding site without loss of activity by means of Triton x-100. It proved to be an asymmetrical protein with a molecular weight between 50,000 and 60,000. Binding sites for ethylene have been shown to exist in a variety of plant species (e.g. Smith and Hall, 1985). Whether some or all of them are requisite for the action of ethylene can not be stated with certainty at present. One of the best characterised, that in the cotyledons of *Phaseolus vulgaris* has been shown to be a hydrophobic integral membrane glycoprotein (Hall et al., 1987). Characterisation of sedimentation properties and identification of marker enzymes in the preparations suggest three different binding sites, possibly on protein body membranes, on smooth and rough endoplasmic reticulum (Evans et al., 1982a). These results were confirmed with high resolution autoradiography.

## PRACTICAL USE OF ETHYLENE-RELEASING COMPOUNDS

The diversity of present commercial uses of ethephon and other ethylene-releasing compounds indicates the substantial contribution that stimulation of ethylene production is making to agricultural productivity. For many years ethylene itself has been used commercially to elicit some of these effects, for example, fruit ripening. Compounds which initiate ethylene responses by inducing the plant to produce its own ethylene (a wound response) have been used in the field.

The most important Ethylene-releasing chemicals are shown in following table. The action of these materials, and their uses in agriculture, have been reviewed comprehensively by Kays and Beaudry.

**TABLE 26.3:** Common and chemical names of some important ethylene-releasing chemicals

| Common name | Chemical Name |
|---|---|
| Ethephon | (2-chloroethyl) phosphonic acid |
| Silaid | (2-chloroethyl) methylbis (Phenylmethoxy) silane |
| Alsol | (2-chloroethyl) tris (2-methoxyethoxy) silane |
| ACC | 1-aminocyclopropane-1-carboxylic acid |

## GROWTH AND DEVELOPMENT

### Prevention of Lodging

Neljubow observed prevention of lodging. It is the basis of one of the most important uses of ethephon. The prevention of lodging in small grains and other crops. The regulator treatment not only reduces length of the stems, but also stiffens the straw.

### Carbohydrate Content of the Straw

Knapp et al. examined the effects of ethephon treatment on carbohydrate content of the straw. Although straw from ethephon-treated plants contained more water soluble carbohydrates, they did not find the anticipated increase in structural carbohydrates. They concluded that the arrangement and interaction of the various structural carbohydrates and lignin in the culm cell walls. It is more important in lodging resistance than concentrations of these compounds. Researchers have now shown that similar benefits can be obtained from ethephon application to field corn where application greatly reduced lodging and accelerated maturation; a reduction in yield was also noted in some varieties.

### Reduction in Scape Length

The floral axis in bulb-type flowers is often too long for their use as potted flowers. Application of ethephon to hyacinth and narcissus plants when the leaves were 10 cm long reduced the length of the floral scape by 20-30%. The reduction in scape length was associated not only with a reduction in cell size, but, surprisingly, also with an increase in cell number, indicating that ethephon application had actually stimulated cell division.

### Reduced Yield and Mean Tuber Size

Hollow heart and brown centre of potatoes are disorders that accompany rapid plant and tuber growth. Although application of ethephon at tuber initiation reduced yield and mean tuber size, it also substantially reduced the incidence of these disorders, leading to increased yield of marketable tubers, relative to control fields which had high disorder incidence.

### Flowering Control

#### *Stimulation of Flowering in Pineapples*

For many years it has been known that ethylene stimulates flowering of Bromeliads, and treatment of pineapple plants with calcium carbide (which releases acetylene on hydration) was widely used commercially to stimulate flowering of pineapples. This response is now elicited in pineapples and flowering bromeliads by application of ethephon. A comparison of the effectiveness of ACC and ethephon in eliciting this response suggested that ACC was the preferred ethylene-releasing compound, but because of differences in application method and concentrations, this interpretation needs confirmation.

#### *Other Tropical Fruit Crops*

Flowering is also stimulated in many other fruit crops after ethylene treatment. Smoky fires are, for example, commonly lit in mango orchards to stimulate flowering. Application of ethephon has also been shown to stimulate flowering in *Litchi*, although the results vary widely depending on application time and cultivar.

Ethylene stimulate or accelerate flowering of many geophytes. In *Iris*, for example, a short treatment of harvested bulbs with ethylene will stimulate flowering in small-sized bulbs that would not normally flower. e.g., *Iris* and *Freesia*.

Ethephon application has been used to extend the flowering season of plantations of *Leucospermum*, a woody proteaceous ornamental.

### Leaf and Branch Abscission

The effects of ethylene in stimulating abscission are the basis of important commercial uses of ethylene-releasing compounds. Enhancement of fruit abscission in sour cherry, nut fall in macadamia, the removal of leaves from potato plants, cotton, and nursery stock, and the dehiscence of the shucks of walnuts and pecans are but a few examples of the potential uses of these materials in agriculture.

Defoliation of rutabaga required very high concentrations of ethephon; addition of 1% ammonium peroxydisulfate greatly increased the effectiveness of the ethephon treatment, perhaps due to accelerated oxidation of IAA in the abscising organ.

### Fruit Ripening

Ethylene gas is widely used commercially for ripening a variety of climacteric fruits and decoloring non-climacteric citrus fruits. However, the use of ethylene-releasing compounds to effect this response is confined to relatively new crops, and normally when they are used as a preharvest application. This is probably due to the ease with which ethylene gas can be applied to harvested fruit, and the dangers of applying aqueous solutions to these disease-susceptible organs.

**TABLE 26.4:** Uses approved for (2-chloroethyl)-phosphonic acid in one or more States of the USA

| Use | Approved crops and states (no postscript = all states) |
|---|---|
| Plant height control | Barley, daffodils, hyacinth, wheat |
| Stimulation of branching | Azaleas, geraniums |
| Flower induction | Pineapple and other bromeliads |
| Sex expression | Cucumber, squash |
| Flower bud development | Apple |
| Defoliation | Cotton, roses, buckthorn, apple |
| Fruit loosening | apples, blackberries (WA, OR), Cantaloupes, cherries (CA, AZ, TX) tangerines |
| Fruit removal | Apples, crabapples, carob, olive |
| Dehiscence | Walnuts |
| Postharvest fruit ripening | Bananas, tomatoes (FL) |
| Preharvest fruit ripening | Tomatoes, peppers |
| Maturity and/or colour development | Apples, cranberries (MA, JH, WI) figs (CA), filberts (OR), grapes, peppers, pineapple, tomatoes (CA, TX, etc.) |
| Degreening (preharvest) | Tangerines |
| Degreening (postharvest) | Lemons |
| Leaf curing | Tobacco |

### Fruit Abscission

Promotion of abscission as an aid in thinning or harvesting fruit crops is an important horticultural use of ethylene-releasing compounds. A recent study by Takeda and Peterson on mechanical harvesting of blackberries provides a good example of the benefits ethylene-releasing compounds provide in stimulating this response.

For many years, researchers have been investigating ways to promote abscission of olive fruits by using ethylene-releasing compounds. Although ethylene does loosen olive fruits, ethephon application causes undesirable leaf abscission. Trunk injection of ethephon accelerated shuck loosening, but caused unacceptable leaflet abscission, and reduced fruit set in the following year.

## COMBINATION WITH OTHER CHEMICALS

The rapid production of ethylene from the neutralized ethephon solution stimulated fruit abscission, but elevated ethylene concentrations were not present for long enough to cause leaf abscission. A mixture of 1% glycerin + ethephon is the most successful ethephon formulation for abscission uses. Combination with other PGRs has also proved to be useful in modifying plant responses to ethephon. Application to pecans of a spray combining 6 mM NAA with 3 mM ethephon resulted in accelerated shuck dehiscence without major leaflet abscission. Similarly, successful thinning of apples without undesirable leaf drop has been achieved by bloom-time sprays of ethephon combined with an auxin.

## CHAPTER 27

# Seed Physiology: Dormancy and Germination

### THE SEED

The seed stage of seed plants represents a unique developmental phase of the spermatophyte life-cycle, and as such involves structures, not characteristic of other stages of development. The essential structure of seed is defined as a *ripened ovule consisting of an embryo and its coats*. Anatomically a seed consists of some old or parental sporophyte tissue viz. the seed coats, which are derived from the integuments and nucleus: some endosperm, which may be either gametophytic tissue or fertilized tissue: and the embryo, the new young sporophyte. The normal seed contains materials which it utilizes during the process of its germination. These substances are frequently found in the endosperm. Thus endosperm may contain variety of stored materials such as starch, oils, proteins etc. In some plants, however, the reserve food material is present in cotyledons.

### IMPORTANCE OF SEEDS

The seed provides an expedient living unit for the study of wholeness that is a complex of biological factors which can be considered simultaneously. The seed occupies that sector of an organism's life-cycle from megasporogenesis (genetic) to the formation of seedling (ecological). However, a seed is not truly a reproductive structure, but rather an adaptive mechanism to facilitate suspending growth and interrupting the continuum of homeostasis in the life-cycle.

Seeds are the corner stone of agriculture because when seeds are planted in the soil and given water, nutrients, light and some protection from pests would reproduce plants and seeds identical to that planted; and also produce number of seeds which could be used for food or feed.

Therefore, it is apparent that, the knowledge seed structure, composition and its physiology is essential for increasing agricultural productivity.

For most of the annuals, seeds provide the only means of multiplication and their continued existence. These plants live only for a months and during they produce flowers, fruits and seeds and then die and disappear. Seeds are important means for establishing a species in new areas. They have also played very vital role in establishing human civilizations. Cereals and legumes constitute a very important part of human diet due to high nutritive value, the legendary *Dal-roti* of an average in sign.

### PHASES OF SEED MATURATION

After fertilization the embryo and endosperm begin a period of rapid growth and differentiation followed by seed maturation. The seed maturation and developmental phase involves different complex biochemical and physiological activity in both the embryo and food storage tissues and organs. It involves different sub phases like production of cells, their differentiation into specialized cells, tissue and organs, synthesis and deposition of storage nutrients in the seed, etc. The developmental seed phase also includes the deposition of reserve food in the endosperm (primarily of monocotyledons), the cotyledons (primarily in dicotyledons) and occasionally in the perisperm (nuclear tissue). At physiological maturity of the cell, the endosperm and perisperm consist of dead cells. These cells serve as *warehouse* for the storage of nutrients in the seed.

## SEED TYPES

### Dicotyledonous and Monocotyledonous seeds

In dicotyledonous seed (soybean) the endosperm is absent while in monocotyledonous seed (corn) the endosperm is present at maturity. The soybean, as in most other leguminous plants, is without endosperm at maturity, however, it consists of a seed coat surrounding the embryo. The seed coat is with a hilum or seed scar and a micropyle at one end. At the other end of the hilum, raphe is present. Each seed consists of two fleshy, large cotyledons. The hypocotyl is located below the cotyledons, and a short axis epicotyl, terminating in the shoot apex, is above the cotyledons.

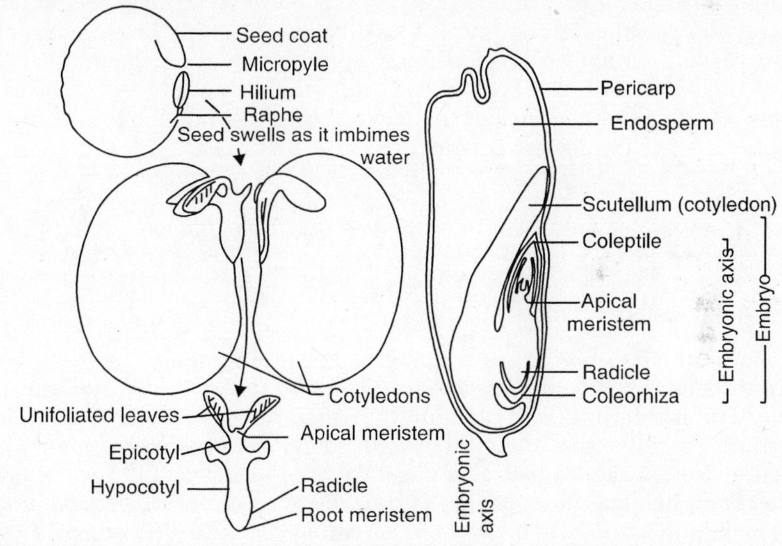


**FIG. 27.1:** (a) Dicotyledonous seed (soybean). (b) Monocotyledonous seed (corn).

Monocotyledonous seed or corn seed is a single-seeded fruit in which the pericarp (ovary wall) is tightly fused with the seed. Such seed is called a caryopsis and is common in grasses or cereal seeds. Embryo and endosperm are present, however, the true seed coat or testa is often absent. The endosperm is composed of an outer layer of cells, the aleurone layer, and the interior starchy endosperm. The cells of the aleurone layer contain proteins and fats, but little or no starch, whereas the cells of the starchy endosperm are filled with starch grains and protein bodies embedded in a proteinaceous matrix. The embryo is present just adjacent to the endosperm and consists of a scutellum or cotyledon and embryonic axis. The coleoptile is the shoot apex of the embryonic axis with several rudimentary leaves. The radicle or the rudimentary root is also surrounded by a sheath of cells called the coleorrhiza. The cotyledon, or scutellum, comprises a relatively large part of the embryo and lies in direct contact with the endosperm. A mature seed consists of the embryo, variable amounts of endosperm and the seed coat, or testa, derived from the integuments of the ovule.

## METABOLIC STATUS OF SEEDS

### Respiration in Developing Seeds

Seed utilize more oxygen and stored dry matter during respiration at the developmental phases. At early developmental phases the rate of respiration is high which gradually reduces as the seed matures and ultimately reaching a very low level at seed maturity. The rate of respiration remains relatively constant during the quiescent (dormant) period. The observed gradual fall in the respiratory quotient and oxygen consumption of a maturing seed occurs while the fresh weight of the seed remains constant.

In a dry and mature seeds the major cell organelles are often in a diffuse and ill-defined state. Mitochondria of dry seeds are considerably less active than those of germinating seeds. The tricarboxylic acid (TCA) cycle is not fully operative in the initial stages of germination because of this reason. The ATP level is almost nil in dry bean seeds but rapidly appears following imbibition and increases with the initiation of germination. The activity and structural changes of mitochondria have been shown to be related to the degree of swelling of the mitochondria. The activity of a number of enzymes like glycolytic pathway, pentose phosphate pathway, Krebs cycle and the cytochrome system, amino-acid-activating enzymes and the soluble enzymes of protein synthesis in the seed declines with advancing or final maturation of the seed. Cytochrome oxidase has been shown to be present in several types of seeds but is inactive in dry seeds.

### Changes in Sub-cellular Particles

During maturation various changes occur in the fine structure of the subcellular particles of cells. Usually at 60% moisture level, mitochondrial structure remains more or less intact until the end of the maturation period, after which the mitochondria rapidly lose their elongated shape and become rounded during dehydration. Structural changes also occur in chloroplasts as the seed approaches maturity. They become globular or bell-shaped with frequent invaginations; the internal membrane structure is lost and the grana disappear. In seeds of *Phaseolus lanatus*, at 60% or above moisture level the endosperm cells are vacuolated, and contain chloroplasts with well-developed grana structure, starch grains, and polysomes associated with the endoplasmic reticulum. On decrease of moisture content below 60%, protein bodies fill the vacuoles, the endoplasmic reticulum becomes less prominent, and polysomes, which are no longer associated with it, gradually disappear.

### Protein Synthesis

Proteins synthesized by maturing seeds are very different from those synthesized by germinating seeds. Maturing seeds contain *storage* proteins (prolamines, glutelins, and globulins) rather than metabolically active proteins. The change from synthesis of metabolically active to storage proteins during seed maturation requires at various stages during seed development and maturation a *switching-off* and a *switching-on* of the synthesis of, or translation of, some kinds of RNA and/or ribosomes and the accompanying complement of enzymes that produce these proteins. Quiescent seeds contain all the enzymes for amino acid incorporation into protein. The cell membrane also undergoes changes during development but how it influences the protein level and other metabolic activities needs research.

## Seed Composition

The reserve materials are accumulated in specialized cell compartments; lipids- in spherosomes (lipid bodies, oleosomes), proteins- in protein bodies, whereas starch forms granules originating from disintegrated plastids.

## Starch Biosynthesis

Photosynthesis by leaves is the source of glucose which is rapidly converted to sucrose and translocated to the developing seed for metabolic use or storage. Amylose and amylopectin-type starches are synthesized and stored in plastids.

The synthesis of these starches is two separate processes that are regulated by different enzyme systems. The starch (high-molecular-weight polymer of D-glucose) synthesized in the normal starch grain endosperm occurs via the starch synthetase pathway. Enzyme starch-synthetase uses a sugar nucleotide, ADP-glucose, as a D-glucose donor. Starch exists in a colloidal or water-insoluble form in the endosperm, permitting storage of large amounts in plant cells without influencing the osmotic pressure of the cell. The general chemical and physical characteristics of the two types of starch are as follows:

### *Amylose*

It is a linear polymer of glucose, consists of several hundred glucose units joined by α-D-1,4 linkages. Two hundred to 1000 glucosyl units are generally linked to form the molecule of amylose. It comprises approximately 10 to 30% of most starches. Its molecular weight ranges from 10,000 to 100,000. It is phosphate free and forms complexes with iodine and other polar agents; turns dark blue in presence of iodine. It is completely hydrolyzed to glucose with α-amylase and forms strong pliable films. It is insoluble in hot water (70 to 80°C), however, it is precipitable with *n*-butanol, with which it forms a crystalline complex.

### *Amylopectin*

It is a branched polymer of glucose. Its molecular size ranges from several hundred thousand to several million representing many thousands of D-glucose units per molecule. Glucose linkages in amylopectin consist of α-D-1,4 and α-D-1,6. It comprises from 70 to 90% of most starches, 100% in waxy type seeds. It contains phosphorous from 0.06 to 0.09% as glucose phosphate of terminal glucose units. It turns red in the presence of iodine and 60 per cent hydrolyzed to glucose with β-amylase. It forms brittle films and has amorphous X-ray pattern. It is soluble in hot water (70 to 80° C) and is not precipitated with *n*-butanol. Endosperm plastids also contain enzymes like starch phosphorylase and the Q enzyme (branching) that synthesize amylopectin.

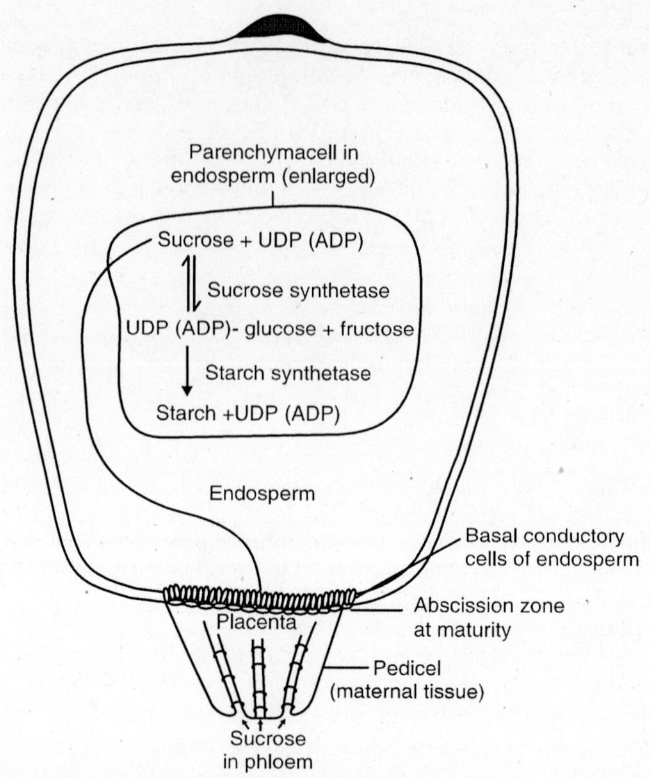


**FIG. 27.2:** Biosynthesis of starch in endosperm of corn seeds.

Genotype of the plant determines the shape, chemical composition, and molecular association of starch granules in seed. Genes regulate synthesis, quantity, quality and storage of starch in plants. For example, a range of 9 to 74 per cent in starch content of corn endosperm tissues has been observed in different genotypes.

### Protein Biosynthesis

A plant synthesizes the 18 amino acids and two amides that form the constituents common to nearly all protein molecules. However, about 100 other amides and amino acids are known to have a limited distribution in higher plants, existing either as amino acids or as low-molecular-weight peptides. Except in rare instances, these compounds are not incorporated into proteins and so are conveniently termed *non protein amino acids.*

## BIOMOLECULES PRESENT IN THE SEEDS

### Seed Storage Proteins: Classification

Seed contains stored proteins (often distinguished from metabolically active proteins by their disappearance during germination) in the form of definite bodies, protein bodies, protein granules or aleurone grains, or matrix proteins. The term *protein bodies* is used for storage protein that are located in cotyledons and endosperm, while the term *aleurone grains* is used for protein stored in the aleurone layer. The cell also contains other types of metabolic proteins, including membrane proteins, and proteins of cell organelles, nucleus, and cytoplasm. The sum of all these proteins (metabolic and storage) constitutes the total protein content of the seed. Protein bodies are present in the cotyledons and endosperm of both starch-bearing and oil-bearing seeds and have the following properties: They contain crystalline inclusions of phytic acid and are probably surrounded by a single membrane, remaining from the vacuole in which they are stored. They contain most of the cellular protein, but none of the oil. They swell, fuse and disappear early in germination. Enzymes are presumed to constitute the greater part of the albumin fraction, which is largely metabolically active proteins.

Usually the storage proteins of seeds contain large amounts of amide nitrogen that are formed by incorporation of asparagine and glutamine into the protein during synthesis and not as a result of amidation of free carboxylic groups of aspartic and glutamic acid bound in proteins.

The storage proteins in seeds of dicotyledonous plants are present in the form of globulin e.g., in pulses (edible seeds of legumes) with relatively small quantities of glutelin and prolamine but the quantity of each type of protein stored varies widely among crop species. In dicotyledonous plants (e.g., legumes), endosperm and the cotyledons usually serve as storage organs.

In the monocotyledonous plants (e.g., cereals) the primary storage organ is the endosperm, which is in close contact with the embryo proper. As a generalization, the seeds of cereals contain relatively high concentrations of prolamin and gluteins with much lower concentrations of globulin. The scutellum may also serve as a storage structure.

### *Zein*

In corn zein (a prolamin), is the major storage protein found in endosperm and contains a relatively large amount of alanine and leucine with little lysine and almost no tryptophan. Consequently, corn as the sole dietary source of protein is a poor quality protein for humans and other monogastrics.

### *Glutein*

It is the major storage protein of wheat and is comprised of many proteins with molecular weights ranging from 25,000 to millions. It has elastic property which accounts for the ability of wheat flour to *raise* and form a leavened loaf of bread. Rye also contains a relatively high concentration of glutein and is also widely used for bread.

### *Gliadin*

It is other major protein class in wheat which does not have the elastic properties like glutein. Thus, it is less valuable in wheat for bread-making purposes.

## Lipid Biosynthesis in Seeds

Lipids are insoluble in water and unable to diffuse from cell to cell. They are synthesized and stored in the cells in which they occur. Lipid synthesis in seeds occurs in cells of the embryo, primarily the cotyledons (especially the scutellum of cereals), and endosperm. The site of synthesis of triglycerides is located in the microsomes. The formation of a-glycerophosphate from free glycerol and ATP is essential for lipid synthesis. The glycerokinase that catalyzes this reaction occurs as a soluble protein in the cell. The thiokinases and the acylating enzymes appear to be associated with the microsomal particles. Glycerophosphate, synthesized by the phosphorylation of glycerol is acylated by acyl CoA to yield phosphatidic acid, which, in turn, is dephosphorylated to give diglycerides. The diglycerides are then further acylated in the presence of acyl to CoA to produce triglycerides.

The triglyceride is stored (10 to 50% of the tissue dry weight) in most of the crop seeds. Phospholipid and glycolipids normally represent less than 2% of the total seed lipid. Amount of lipids in crop seeds endosperm varies e.g., in wheat (1 to 2%), 8 to 15% of the germ, and about 6% of the bran, with an average of 2 to 4% for the whole kernel. Wheat germ oil consists primarily of triglyceride, whereas lipid extracts from endosperm and bran contain much higher proportions of phospholipids and glycolipids.

### Mineral Composition of Seeds

On the analysis of the total ash content of a plant one can find more than 60 elements in its ash. The amount of kinds of elements translocated to the grain are considerably less than those absorbed by the roots and translocated to the vegetative organs of the plant.

### Vitamin Composition of Seeds

The synthesis of vitamins in plants is dependent on the available supply of inorganic nutrient (particularly of macro elements) ions from soil. Deficiency of macro nutrient elements decreases the concentrations of riboflavin. How it decreases the amount and synthesis of vitamins is not understood. However, practical fertilization, to increase vitamin composition of grains, is still empirical. The level of vitamins B and E are high in cereal grains. Vitamin C is notably missing in significant amounts in these seeds.

## SEED GERMINATION

### What is Seed Germination?

Germination represents a dynamic period in the life cycle of plants as a seed makes the transition from a metabolically quiescent to an active and growing entity. Seed germination is one of the basic aspects of eco-physiological studies, because functionally a seed is device for reproduction, perpetuation, and multiplication of a particular species so as to ensure its continuation. Seed germination is the resumption of active growth on the part of the embryo resulting in the rupture of the seed coat (or peri-carp) with the subsequent emergence of the young sporophyte.

As such, seed germination represents the termination of a developmental phase and the commencement of a growth phase. The exact physiological and biochemical changes, which occur prior to, during and subsequent to the germination process, are complex and not too well understood and have been described differently by various investigators.

Germination is controlled by dormancy and various environmental factors. Given a viable, non dormant seed, the environmental conditions favorable for germination include adequate supply of water, suitable temperatures and oxygen, in some cases light, and the absence of external inhibitory such as inorganic salts or organic substances.

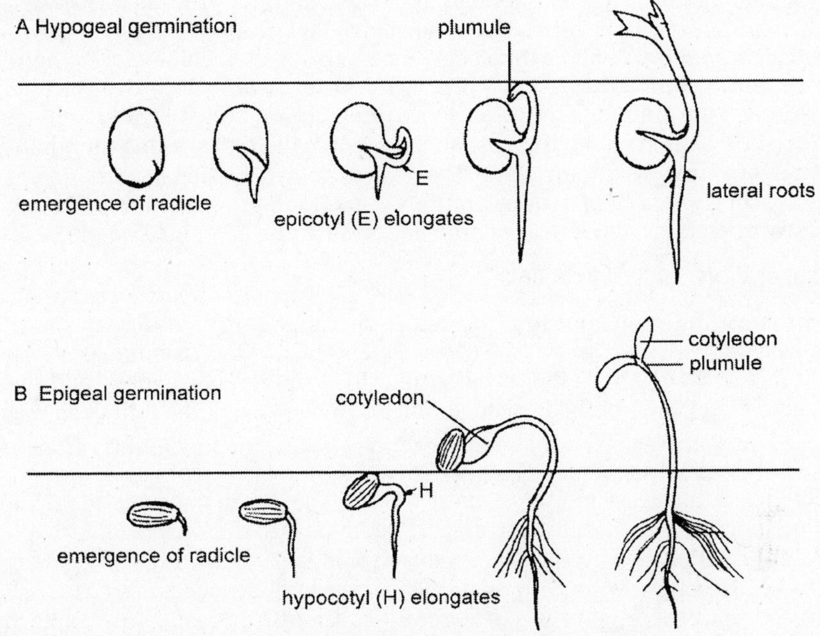


**FIG. 27.3:** Seed germination (A) Hypogeal germination, (B) Epigeal germination.

Summarizing, germination is a separate developmental phase in life-cycle of the higher plants. This phase is preceded by embryogenesis and seed development which is followed by seedling growth and development. Thus a seed has a specific programme for germination consisting differential gene activation, distinct from programmes for embryogenesis and for seedling development. In that programme the elimination of dormancy is included. Full understanding of the genetic programme of germination will enable to answer both questions: **what is germination? What is dormancy?**

## The Phasic Characters of Germination

There are three successive main phases which make up the process of germination. They are denominated in different manner. The distinction of phases is based upon recognition of processes dominating during a given period of germination. The distinction of the phase of germination is arbitrary. It is obvious that the processes that are typical for one phase can start during the preceding phase and can continue during the subsequent one. As a rule the phases of germination overlap one another. The sequence of germination in its simplest form is: water imbibition, enzyme activation, hydrolysis and catabolism of storage material, initiation of growth, anabolism and formation of new cell structures, rupture of seed coat, and emergence of the seedling.

## Water Imbibition or Catabolic Phases of Germination

Imbibition of seed is the first phase. The hydrophilic colloids present in seed coat became hydrated and their structure is rearranged. Several enzymes regain their activity. Hydration of cell colloids allows the diffusion of solutes inside of the seeds. The imbibation affects the non-colloidal structure of the seed. The structure of cell membranes undergoes rearrangement that assure their semipermeable properties. Imbibition enables the occurrence of processes belonging to the second phase of germination. During that phase, the true restoration of life in the seed takes place. The dominant processes are **hydrolysis** and **degradation** of food reserves. For this reason, this phase is often called the **catabolic phase of germination**. However, some syntheses, mainly of hydrolytic enzymes do occur in parallel to degradation of storage proteins, fats and carbohydrates. Respiratory processes and energy generation sharply increase at the beginning of this phase.

## Function of Water in germination

Imbibation is the **reactivation of metabolic systems** conserved, since seed maturation and storage at a moisture content of 10-14%, commences with the imbibition of water and the rehydration of proteins, enzymes, and cellular organelles. Water is essential in seed germination because of the following reasons:

1. Water acts as the solvent in which reactions occur, the medium for transport of enzymes, cofactors, and coenzymes.
2. It acts as a substrate in hydrolytic reactions, such as the enzymatic breakdown of the polymers of starches, proteins, and lipids into their basic units.
3. Water provides turgor pressure for increase in cell volume and accounts for the 30-40% increase in volume upon initial hydration of the seed.
4. Hydration of the seed to levels above 16-18% moisture results in a rapid rise in mitochondrial activity and activation of phytochrome.
5. Following initial imbibition rate of respiration increases and may be noted in as little as 10 min. The initial respiratory-rate-increase continues for 2 hours in crimson clover at 20°C, 6 hours in rye, 8 hours in corn, and 10-16 hours in dry bean at 25°C before a plateau or steady state is reached. The rate of respiration depends on substrates stored in the embryonic axis.
6. Imbibation or phase of induction of growth massive synthesis of cell wall material, as well as increased synthesis of protein and ribonucleic acids occur.
7. During this phase all essential capacities of a living plant are restored.

## Hydration and Enzyme Activation

On hydration of seed, various enzyme systems are activated, and cellular activity increases. The activation of enzymes following hydration may be from enzymes formed during maturation of the seed or newly synthesized enzymes. Enzymes formed during maturation are of two types:

1. Those requiring only hydration to become active, and this activity is reversed by drying. Triosephosphate isomerase, cytochrome reductase and adenylate cyclase are a few examples of these enzymes.
2. Those enzymes requiring the action of a hormone or other enzymes to gain activity within minutes to several hours.

Newly synthesized or *de novo* synthesized enzymes are also of two types:

1. Those enzymes produced by pre-existing mRNA utilizing a pool of pre-existing amino acids. Activity of these enzymes may occur in 2 to 4 hours.
2. Those enzymes produced from newly synthesized mRNA and utilize amino acids derived from degradation of storage proteins, which begins as early as 6 to 12 hours after hydration.

Activity of above mentioned enzymes varies from plant to plant because of the diversity found among plant species.

## ENZYMATIC DEGRADATION OF STORAGE MATERIALS

### Degradation of Carbohydrates

Seed contains simple sugars, free fatty acids and amino acids, starches, lipids, and proteins that are utilized during initial phases of germination. Starch catabolism is initiated by a hydrolyzing enzyme-a-amylase to glucose, which accounts for 90% of the starch degradation. The oxidative pentose phosphate pathway appears to function during the first germination phases of cereals but is quickly replaced by the hydrolytic pathway as synthesis of a- and b-amylase enzymes occurs, 24-48 hours after the imbibition of water. Sugars utilized by the growing embryo are principally glucose, sucrose, raffinose, and stachyose.

### Degradation of Proteins

Protein catabolism involves protease degradations of insoluble storage proteins. The soluble proteins are further hydrolyzed by peptidases to the component amino acids. The released amino acids are transported to the embryo where they are resynthesized into new cell proteins.

### Degradation of Fat Bodies

The scutellum of grasses and cotyledons or endosperm of dicots contain oil as fat bodies (glyoxysomes). They are degraded from triglycerides to glycerol and fatty acids by acid lipases. In the glyoxysomes monoglyceride is degraded by enzyme alkaline lipases. In castor bean such hydrolysis occurs early in germination. The fatty acids are oxidized via the b-oxidation pathway, which produces 2-carbon units of acetyl-CoA capable of entering the tricarboxylic acid (TCA) or the glyoxylate bypass, where fats are converted to sugars via reverse glycolysis. Glucose can be converted via the oxidative pentose phosphate pathway to provide the pentose (5-C) sugars needed for nucleic acid synthesis. Nucleic acid synthesis is a prerequisite for cell division associated with embryo growth.

### Synthesis and Growth of Embryo

Initial protein synthesis occurs within 30 min after imbibation begins. This protein is synthesized using RNA produced during seed maturation. New RNA synthesis begins 3 hours after initial imbibation, and DNA synthesis has been observed 15 hours after start of hydration. In onion first synthesized-RNA is mRNA.

### Endogenous Hormone Level of Seeds

Endogenous hormones regulate many facets of germination. The GA has dominant activator role following hydration of the seed through its effects on membrane permeability, ATP synthesis, and the interaction with cytokinin and ABA. Endogenous cytokinin is located in non embryonic regions of the seed which influences membrane permeability, overcome the effects of inhibitors such as ABA, function in the binding site of some tRNA or mRNA ribosome complexes, and interact with membrane-bound pigments such as phytochrome to mediate the germination process. Cytokinin, therefore, has a controlling role in the translation of genetic code into new proteins.

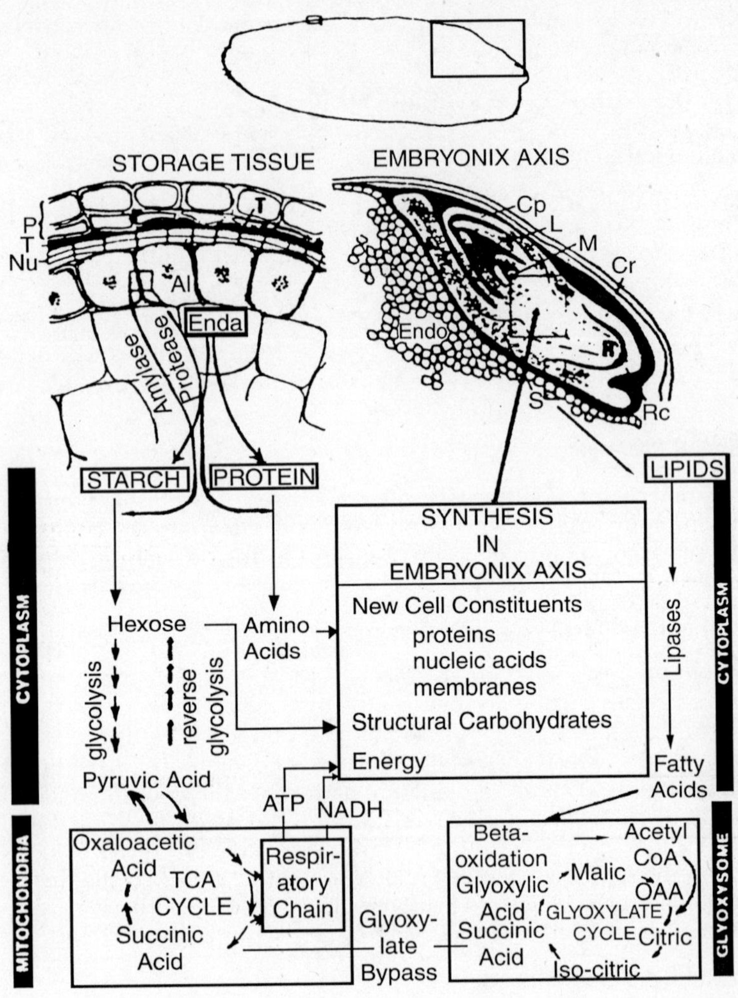


**FIG. 27.4:** Schematic representation of seed storage product degradation (light arrows), the participating organelles, the pathways of transformation, and synthesis (heavy arrows) in germinating seeds. (P-pericarp, T-testa, Nu-nucellar tissue, Al-aleurone, Endo-endosperm, Cp-coleoptile, L-embryo leaf, M-apical meristem, Cr-coleorhiza, R-radicle, Rc-rootcap, S-scutellum).

ABA is the inhibitor of germination. It blocks DNA synthesis and translation by mRNA for key enzymes like isocitratase and phosphorylcholine glyceride transferase.

## Role of GA in Germination

The role of GA in germination is well known. Various isomers of GA have been identified. The initial steps of GA synthesis GA occurs in the scutellum while after 3rd day of germination the embryonic axis produces GA. GA effects on germinating seeds can be divided into two following roles:

i) Germination (embryonic activity), and
ii) substrate mobilization

At primary level GA regulates of germination by changing membrane permeabilities and by influencing the level of initial ATP synthesis. Dormant seeds with low initial ATP levels and requiring stratification can be induced to germinate given a GA treatment, which changes membrane permeability and energy levels in the embryo. GA induces substrate mobilization. For example in barley GA increases the level of poly (A) RNA which is believed to contain the mRNA for the enzyme a-amylase synthesized by the aleurone cells. a-amylase hydrolyses starch to maltose. GA also stimulates ribonuclease and protease synthesis in aleurone cells. In addition, GA enhances the release of b-1, 3-glucanase, which dissolves the cell walls of the aleurone, permitting an increased supply of nutrients from the endosperm to the embryo.

## Do the Hormones Play any Role in the Control of Germination?

Plant hormones shift a tissue from one physiological state to another one. Thus, the entrance to a new development phase is supposed to be triggered by a hormonal factor. The major classes of hormones associated with seed physiology are gibberellins (GA) abscisic acid (ABA) and cytokinins (CK). Auxins (mainly IAA) and ethylene were also considered as involved in germination. The following evidences suggest the role of plant hormones in seed germination:

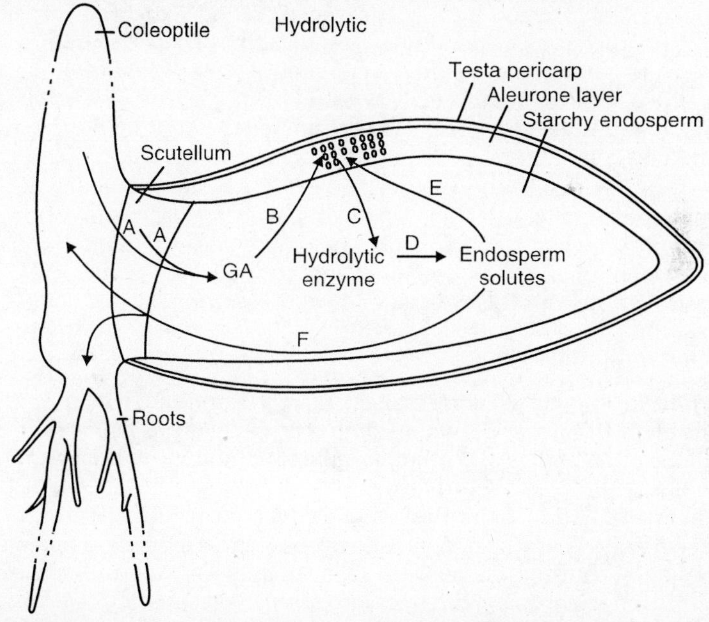


**FIG. 27.5:** Diagrammatic representation of the relationship between gibberellic acid (GA) production, $\alpha$-amylase production, and solute accumulation in germinating barley seeds.

## THE HORMONE HYPOTHESIS

1. Exogenous hormone has to provoke the specific effect when applied at concentrations similar these to present in the plant material under study.
2. Changes in the content of the endogenous hormone should correlate with the course of the studied process
3. Inhibitors of hormone biosynthesis, action or translocation have to inhibit the processes thought to be under control of the hormone.

The mechanisms of hormone action rely upon their control of differential gene activation (transcriptional or translational control) or in alteration of the properties of cell membranes. Taking into account the concept of specific receptors required for the action of hormones (Trewavas and Jones, 1981), one can assume that the primary event caused by a hormone is a conformational change of a specific macromolecule. Depending on character and location of that macromolecule (in membranes, cytosol, or in nucleus) further events may lead to different kinds of responses. Nevertheless, many authors arrived to similar conclusions using different approaches; so, the question: do the hormones play any role in the control of germination- ought to be answered positively.

### The Inhibitor Hypothesis

Various regulatory substances as coumarins, phenolics, sterols and cyanide play a role in germination of some seeds. It has been postulated that the dormancy of seeds is caused exclusively by the presence of germination inhibitors like coumarins or phenolics associated with seed coats. These inhibitors limit the oxygen access to the embryo, thus reducing cellular oxidation and energy generation needed for resumption of growth. Leaching of these substances out of the coat, (rain, imbibition conditions) enables immediate germination.

The germination inhibitors as principal agents responsible for seed dormancy was supported by the discovery of embryo-located substances of this kind. These were chromatographically isolated from different species and known as **B-inhibitor complex** in which several phenolics and abscisic acid were included. The embryo-located inhibitors are metabolized during removal of dormancy and it was postulated that their level is controlled by temperature, light, humidity, etc.

### ABA as Inhibitor of Germination

ABA or dormin is known as universal dormancy hormone which is commonly present in dormant tissues and remains active at concentrations of several orders lower than those of phenolics and other germination inhibitors. It has all features of a plant hormone.

**Distribution of ABA in plants:** ABA is located in different parts of seed; embryo axis, cotyledons and seed coats. Its presence in seeds is correlated with the depth of dormancy. During the removal of dormancy in *Prunus avium*, *Juglans regia*, *Fraxinus americana*, apple, *Acer saccarum*, *Acer pseudoplatanus* and *Acer platanoides* a step wise decrease of ABA content was observed down to amounts undetectable with bioassays. Its inhibitory action is supposed to be exerted upon different mechanisms and its decrease may also result from different causes such as catabolism, conjugation and leaching. Non-dormant, after ripened seeds of *Fraxinus excelsior* contain considerable amounts of ABA. Similar situation has been stated for seeds of *Acer saccharinum*. On the strength of above observations, however, ABA can not be considered as an exclusive factor responsible for physiological dormancy of seeds.

## IAA as Inhibitor of Germination?

In seeds of some woody species relatively high amounts of other germination-inhibiting hormone- indolylacetic acid (IAA), were found. It has been postulated by Nikolaeva (1967), that IAA, in parallel to ABA, plays a role in onset and maintenance of embryonal dormancy. She observed that the level of endogenous IAA gradually decreased during dormancy removal in *Acer tataricum*. Exogenous IAA not only inhibited germination of non-dormant embryos but, at lower concentrations, caused anomalies in seedling development similar to those observed during germination of embryos isolated from fully dormant seeds. It has, however, suggested that the role of IAA in dormancy may be restricted to some species, but it is not as common as that of ABA.

## Inhibitor-Promotor Equilibrium

Studies by various seed physiologists indicate that:

1. Seed germination is promoted by a number of chemicals like gibberellins and cytokinins, because of their pronounced germination-stimulatory effect and their common occurrence in seeds during dormancy removal and germination.
2. The seed can not germinate, i.e., it is dormant, not necessarily because of the presence of inhibitors(s), but also because of the absence of appropriate stimulators.

The involvement of hormonal stimulators in dormancy removal is well known. Gibberellins can replace light requirement in breaking the photodormancy in lettuce seeds, and cytokinins (benzyladenine, kinetin) are indispensable in elimination of thermodormancy in the same material. In cereal grains, which need several weeks of dry storage in order to achieve full germination ability, gibberellin treatment markedly shortens this period. The dark germination of isolated, dormant apple embryos (requiring a three months cold stratification) is stimulated by gibberellins, or by cytokinins.

Changes in endogenous gibberellin and cytokinin levels during the release of seed dormancy support promotor concept. e.g., GA level constantly increases in seeds of *Coryllus avellana* during stratification. In *Acer saccharum* and in apple seeds the contents of both GAs and CKs rise during stratification.

Similarly as found for ABA, the content of GA and CK may very in different parts of the seed. Thus, the inhibitor-promotor equilibrium may be different in each of these parts and the regulatory effect of that equilibrium may be of various character. This suggests that the mechanisms determining that equilibrium may work differently in various parts of the seed, however, their location is far from known. These are: selective leaching or translocation of hormones, hormone biosynthesis, reversible formation of inactive hormone conjugates, etc.

The concept of involvement of promotor in release of dormancy get supports from the increase in endogenous $GA_4$ level during the first 30 days of seed after-ripening and afterwards it gradually drops. On the other hand, ABA added to stratification medium delays the rise in gibberellin content in apple. *This seems to indicate that the inhibitor-promotor interactions may relay upon regulation of one hormone level by another one, rather than on an opposite action at the same regulator point.*

### Changes in the Inhibitor-Promotor Ratio

Changes in the inhibitor-promotor ratio during dormancy removal is obviously under more or less direct environmental control. In this respect an interesting hypothesis presented by Bulard's (Nice, France) group. This suggests that low temperature, necessary for deep dormancy removal in seeds of many species, differently affects the enzymes involved in formation and hydrolysis of inactive, bound forms of different hormones, so that it favours the formation of ABA conjugates and simultaneously promotes the release of GAs and CKs from their inactive forms. This hypothesis requires a full experimental confirmation.

### Ethylene as Germination Inhibitor

Involvement of ethylene in dormancy removal and germination of seeds was postulated by Toole et al., (1964), but the first data on the germination-stimulating activity of ethylene were reported by Haber as early as in 1926. Ethylene is probably another hormone that plays a role in the inhibitor-promotor equilibrium. Ethylene increases the germinability of seeds with different types of dormancy. The conditions that cause elimination of ethylene from atmosphere surrounding the seeds also decrease germination ability. Ethylene plays a supplementary role in the first two phases of germination as compared with the substantial role of ABA, GA or CK. Its involvement in the last period of germination, the growth initiation, appears to be more evident but still requires further studies. It is, however, doubtful weather ethylene is really involved in germination control or it affects the early growth of the seedling. Moreover, there are no data on the actual ethylene concentration in the seeds. The accumulation of $^{14}$C-ethylene in *Phaseolus* cotyledons was described (Hall et al., 1980; Evans et al., 1982); nevertheless, for the time being, these data do not help in understanding the involvement of ethylene in seed physiology.

### Cyanogenic Glucosides and Cyanolipids as Inhibitors

Seeds of some species contain considerable amounts of cyanogenic glucosides and cyanolipids, which release free hydrogen cyanide as a result of hydrolysis. The production of HCN takes place during the removal of dormancy in apple (Dziewanowska et al., 1979). Reports are also available that application of cyanide to various seeds markedly stimulates germination. There is, therefore, a serious probability that HCN may contribute to the complex regulatory mechanism involved in dormancy removal, at least in those seeds where its presence is proved.

## KHAN'S HYPOTHESIS ON THE SELECTIVE ROLES OF HORMONES

The inhibitor-promotor concept considered only two classes of factors- promotor and inhibitor. Moreover the term **promotor** was used for a large group of substances with various physiological activities and properties. On the other hand, in some seeds dormancy was broken without elimination of endogenous ABA. The effect of exogenous ABA or other inhibitors could be reversed by application of cytokinins, but not by gibberellins, cytokinin could synergize the effect of GA. From these observations Khan (1971, 1975) concluded that dormancy and germination could be determined in one of several ways, and not always by an excess of inhibitors or an excess of promotors. He attributed to gibberellins the primary role in the regulation of germination and the release of dormancy whereas the roles of inhibitors and cytokinins would be of secondary character: **preventive** and **permissive**, respectively.

Khan (1975) suggested that the selective action of hormones would be also valid for other physiological phenomena, but the particular roles in such cases would be played by different hormones. For example, in apical dominance, auxin would act as the primary factor, whereas cytokinins and gibberellins as preventive and permissive ones, respectively (Khan, 1975).

Khan's hypothesis assumes a very limited number of regulatory sites (points) affected by hormones that control dormancy. The competitive nature of ABA and CK interaction implies a common sites for these two hormones. In contrast, the non-competitive character of ABA and GAs effects suggests that they act at different regulatory points. The above conclusion from Khan's hypothesis, however does not fully comply with the current views on basic mechanism of hormone action.

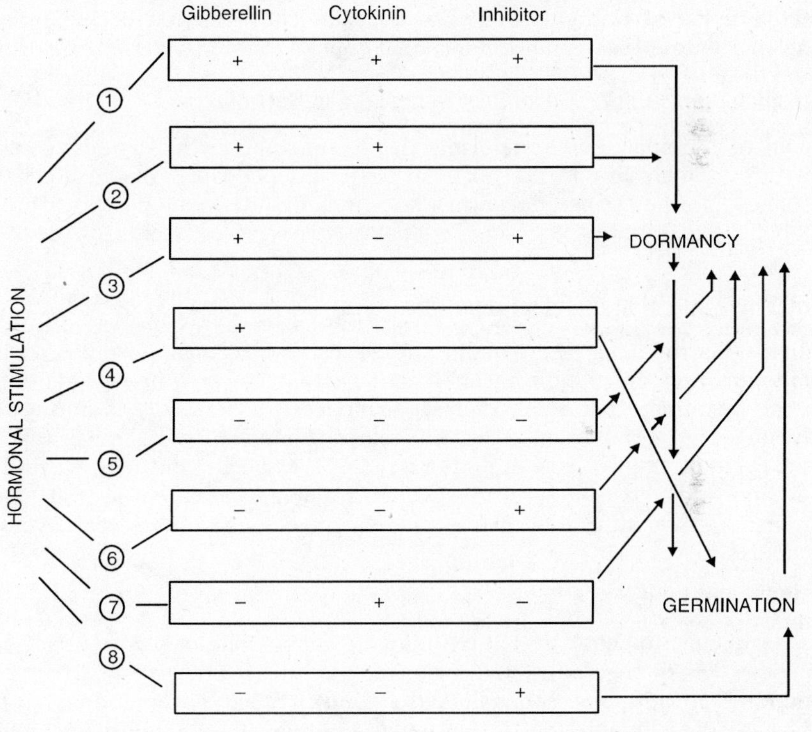


**FIG. 27.6:** Model explaining hormonal role in germination and dormancy (Khan's Hypothesis)

## Hormone Receptor Concept

The plant hormone, in order to act, has to be linked with a specific receptor (Trevawas and Jones, 1981). In other words: *the primary reaction of the hormone is its binding with a specific receptor.* The next events are mediated by the hormone-receptor complex. The variety of specific receptors can explain the variety of physiological response to a hormone. Changes in the number of the available receptor molecules in the tissue during ontogenesis reflect changes in sensitivity towards that hormone. There is a good evidence for the existence of such specific receptors, for hormones of all known classes (Kende, 1976). Nevertheless, evidence concerns, as yet, solely the occurrence, specificity and structure of receptors; there are no convincing reports on the physiological activity of a hormone-receptor complex.

The known early (not primary) responses to hormones in the seed may be different nature. There are well documented examples of induction or repression of different enzymes at the level of transcription as well as at that of translation, numerous enzymes are activated in result of hormone treatment. It is also known that hormones can affect the structure of cell membranes. It would be an unlikely assumption that in the control of all these phenomena only a few types of receptors are involved.

Since the only physical meaning of Khan's *regulatory site* of the hormone is a specific receptor molecule. Therefore, one has to consider the hypotheses of selective roles of hormones only as a useful simplification of the much more complex relations between hormones that control dormancy and germination.

## Metabolic Events during Dormancy Removal and Germination

A ready-to-germinate and non-dormant seed is metabolically very active while such activity extremely low in the dormant seed, independently of the mechanism of dormancy. During germination *sensu lato* (including dormancy removal) the entire metabolic apparatus of the seed becomes operative in a controlled, sequential fashion.

## Membrane- Site of Metabolic Processes

Membrane is the sites of metabolic processes. The most important properties of membranes are: the semipermeability, the activity of membrane-bound enzymes and the membrane potential. Indirect evidence is available that during the very early stages of germination structural rearrangement of membranes takes place. The leakage of amino acids and other solutes, that occurs during the first hours of imbibition indicate some injuries. This leakage decreases in the course of germination, thus indicating membrane rearrangement and the existence of repair mechanisms.

## GA and Membrane

There are some indications that GA may affect the state of plant membranes thus changing the permeability. GA also affects the synthesis of some essential membrane constituents (phosphatitylcholine). These observations suggest the involvement of membrane in germination specially in hormonal control of membrane repair mechanisms as well as the control of membrane development.

## ABA and Membrane

ABA also acts upon membrane potential and prevents the proton extrusion from cells. Although the role of cell membranes in germination seems to be unquestionable.

Following are the aspects related to the role of membrane in seed germination.

i) Which cell membrane are important at the early stage of germination?
ii) How for do the changes in membrane structure and function during

germination affect the metabolic events related to germination?

iii) Do some membranes play the role of sensors towards germination-triggering factors such as temperature and hormones?

## Nucleic acid and Protein Synthesis

Protein and nucleic acids synthesis is blocked in dormant seed and the genetic activity of such a seed is repressed. On the release of dormancy a gradual depression of genes takes place. It is still unclear that protein synthesis are identical for all kinds of seeds and all types of dormancy. Generalization concerning the sequence of appearance of different types of RNA during germination of cereal grains was described and proposed a **cascade model of genome activation** after the elimination of seed dormancy.

## Cascade Model of Genome Activation

The mRNA already present in a dry seed is activated on imbibation. This model describes that the depression of genes coding mRNA occurs first, thus allowing the onset of synthesis of some proteins. It is followed by the synthesis of ribosomal RNA and the last synthesized kind of RNA is the tRNA. The model further suggests that the initiation of DNA synthesis (replication) occurs just after activation of mRNA-coding genes. However, the convincing data coming from an elegant experimental series indicate that DNA replication takes place late during germination (Osborne, 1981). It does not occur until dormancy is completely broken.

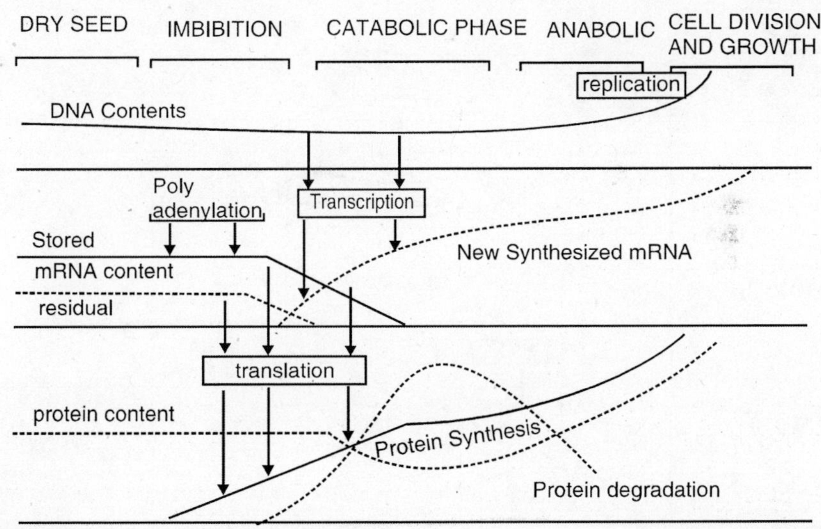


**FIG. 27.7:** Cascade model to show the sequence of appearance of new protein, mRNA and DNA and the changes in the rates of their syntheses.

## Enzyme Activation and Seed Germination

Various enzymes involved in metabolism of the germinating seed are synthesized de novo and many enzymatic activities appear early during germination without concomitant protein synthesis. Such enzymes are involved in RNA and protein synthesis (e.g., RNA polymerases) and enzymes engaged in mobilization of reserve (mainly hydrolyses). There is convincing evidence for activation of numerous enzymes during seed germination, coming mainly from experiments with the use of inhibitors of protein synthesis. The possible ways of activation are as follows:

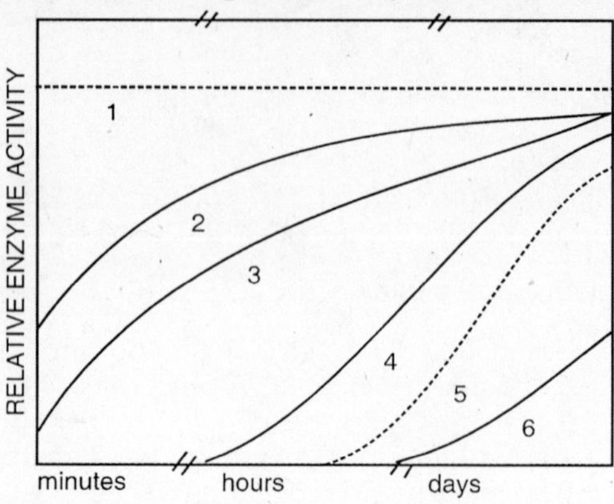


**FIG. 27.8:** The sequence of appearance of enzyme activities during germination. 1: enzymes continuously active. 2: activity rising during germination, but residual activity present in dormant seed. 3: as (2), but no residual activity. 4: enzymes activated in result of imbibition (dormancy breaking). 5: enzymes synthesized with the use of preformed mRNA. 6: enzymes synthesized with the use of newly formed mRNA.

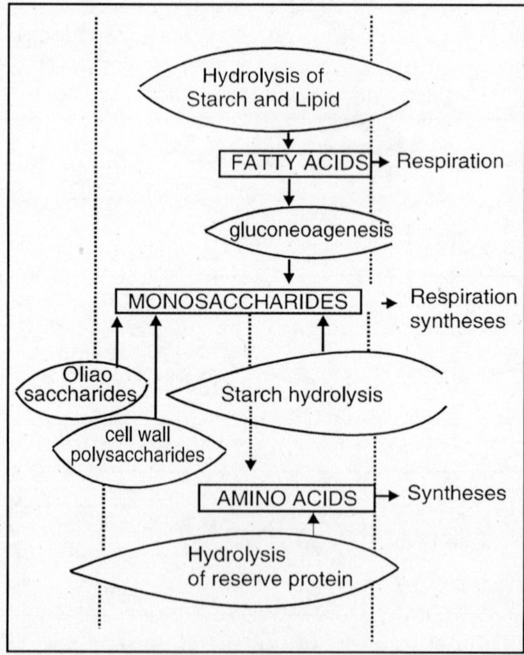


**FIG. 27.9:** Mobilization of reserve materials during the catabolic phase of germination.

(i) removal of an inhibitor, (ii) release from a subcellular compartment, (iii) release from a precursor by partial degradation, (iv) aggregation of enzyme subunits and (v) conformational changes of an inactive protein.

The process of enzyme activation during seed germination is at least as important as nucleic acid and protein synthesis. It seems, therefore, that it still deserves much more attention than it has been paid until now.

## MOBILIZATION OF STORAGE MATERIALS

In the sequence of metabolic events during germination, the first mass process is the mobilization of stored material; this consists in such a transformation of reserves that they can be used for energy generation and for building of the new compounds. Spherosomes and protein bodies are closed in a single membrane envelope. The lipid bodies disappear in result of hydrolysis of stored substances, whereas the protein bodies transform into vacuoles. Starch seems to be hydrolysed by cytoplasmic enzymes and the granules disappear when its hydrolysis is completed.

## ENERGY GENERATION DURING GERMINATION

The source of energy during germination of the seed is the oxidative breakdown of storage materials. In spite of limited reserves, the effective utilization of that energy (under the form of ATP) does not exceed 10%. Carbohydrates are the main direct substances for oxidative catabolism in seeds. The conventional pathway of oxidative glucose breakdown including glycolysis, tricarboxylic acid cycle, oxidative pentose phosphate pathway and cytochrome electron transport chain has been demonstrated in germinating seeds. In the beginning of germination glycolysis is the dominant whereas the contribution of pentose phosphate pathway increases later on. Roberts (1973) after using cyanide, azide, CO, $H_2S$, iodoacetate or malonate postulated the positive role of pentose phosphate pathway in release of seed dormancy. However, the role of oxidative catabolism in seed germination needs further studies.

## SEED DORMANCY

Dormancy has been an elusive phenomenon and difficult to define. Consequently, the term has been applied to a number of unrelated phenomena, resulting in considerable ambiguity. Amen (1963) restricted the term dormancy to an indigenously controlled but environmentally imposed temporary suspension of growth, accompanied by reduced metabolic activity. According to another definition, *dormancy is a state, in which viable seeds fail to germinate under conditions of moisture, temperature, and oxygen favourable for vegetative growth* (Evenari, 1956; Wareing, 1966). Sussman and Halvorson (1966) defined dormancy as *any rest period or reversible interruption of the phenotypic development of an organism*. Dormancy is a widespread phenomenon among non cultivated species, as it is a survival mechanism spreading germination over time and space. Dormancy of the seed is a very peculiar and persistent phenomenon. When all, or almost all, vital functions of a dormant organ are restored, the dormancy is broken. Which leads to growth of tissues; release of the seed from dormancy allows its germination.

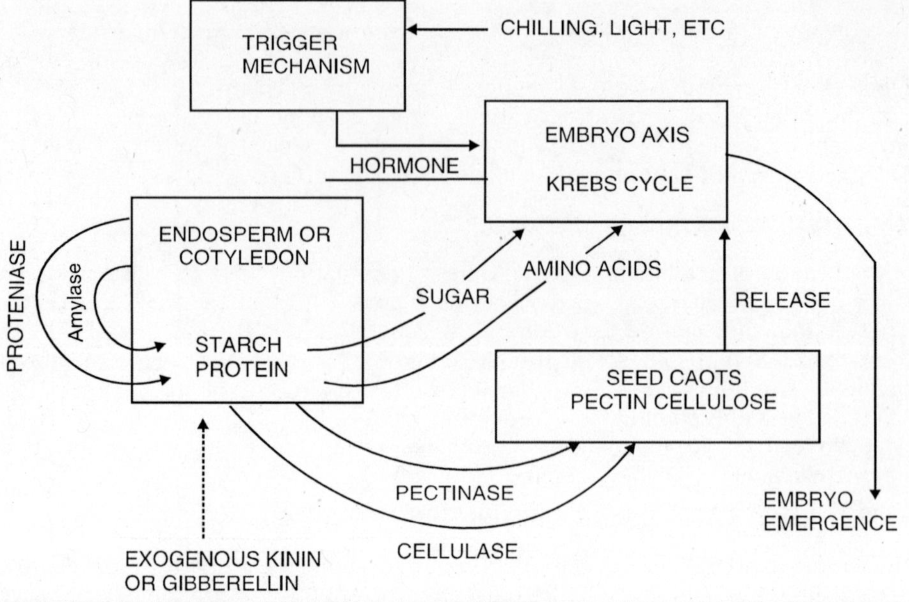


**FIG. 27.10:** Amen's model of germination after release of dormancy.

## CLASSIFICATION OF DORMANCY

There are several classifications of seed dormancy, proposed by different authors and all of them reflect the mechanisms involved in reduction of vital embryonal processes. The main types of dormancy are:

### The Imposed Dormancy (Quiescence)

Germination needs favourable environmental conditions like sufficient water, proper temperature, etc. In absence of such conditions seed does not germinate or becomes dormant which is sometimes considered as inhibition. This type of inhibition is known as imposed dormancy (quiescence). However, under favourable conditions this kind of dormancy disappears and the seed germinates immediately.

For seed of many species the limited water access is the only cause preventing germination, and hence the only cause of impose dormancy. One could pose the question; is imbibition the beginning of germination or is it the process of dormancy elimination?

### The Innate (Deep) Dormancy

Sometimes a seed unable to germinate due to structural, chemical or physiological properties of the seed itself, this phenomenon is known as Innate (deep) dormancy and is distinguished as physical, chemical, morphological and physiological types of dormancy Nikolaeva (1977). These terms describe the main mechanisms that are responsible for the inability of embryo to grow: physical barrier preventing water and/or oxygen access, presence of growth inhibitors, morphological or physiological immaturity of the embryo. These mechanisms can operate separately or in various combinations.

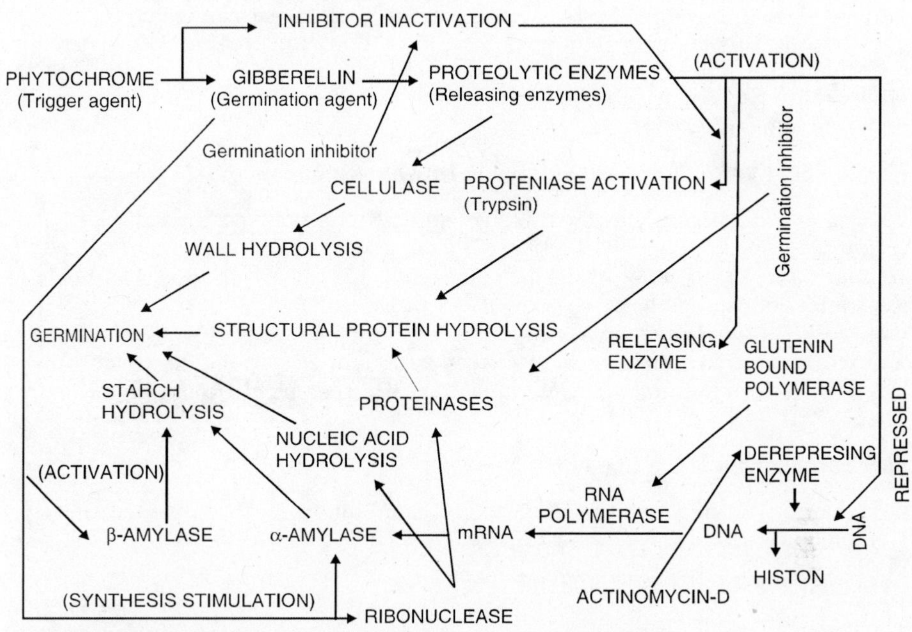


**FIG. 27.11:** Biochemical pathways in termination of seed dormancy

**Complex types of dormancy** are caused by several different mechanisms which makes germination difficult or hinder it altogether. These mechanisms act as a set of barriers which prevent germination. They can act upon whichever particular process, belonging to any phase of germination.

## CAUSES OF DORMANCY

Various important causes of seed dormancy, generally recognized, are rudimentary embryo, physiologically immature embryo (inactive enzyme system), mechanically resistant seed coat, impermeable seed coat, and presence of germination inhibitors (Bonner and Varner, 1965). As in most growth responses, dormancy and germination are regulated by a critical balance of inhibitor-promotor complexes. This idea is first attributed to Hemberg (1949). During seed maturation this balance may be shifted in favour of the inhibitor component, imposing dormancy. This may be accomplished by a curtailment in the synthesis of the promotor component, or by a building up of inhibitory intermediate metabolite, or by a direct antagonism.

Among cultivated seed crop species, genetic selection has occurred for uniform and rapid germination. Types of dormancy found in agronomic crops and weeds are as follows:

### Impermeability of Seed Coats to Water

Impermeability of seed coats to water is found in hard seeds and occur among members of the Leguminosae, Malvaceae, Chenopodiaceae, Liliaceae, and Solanaceae. This is an inherited character. It is influenced by environmental conditions during maturation. Conditions like low soil moisture, low relative humidity and high temperature during seed maturation increases the frequency of hard seeds, e.g., in alfalfa and soybean. Hard seed coat dormancy can be broken by chemical and mechanical methods. The most common treatment for legumes is scarification of the seed by rubbing the seeds against an abrasive surface. Cotton typically is scarified with sulfuric acid, as it is delineated, to improve germination and ease in planting.

### Mechanical Resistance of the Embryo to Growth

Several weed species have excessive strength of the seed coats like plantain (*Plantago* spp.), pigweed (*Amaranthus* spp.), and cocklebur (*Xanthium strumarium* L.). Seeds of such plants fail to germinate due to mechanical resistance of the embryo to growth. *Xanthium strumarium* seeds have two embryos. The dormant upper embryo is able to generate only 41 g of imbibitional and growth force while the testa requires a force of 56 g to pierce. By contrast, the lower, non dormant embryo generates 84 g of force and the testa requires 67 g to pierce.

### Low Permeability of the Seed Coat to Gases

Germination of most agricultural seeds is inhibited in the absence of oxygen. Notable exceptions are rice to wild rice.

### Dormancy Due to a Metabolic Block

The most notable examples among agricultural crops are the post harvest dormancy of some cereals, the stratification requirement of crops such as wild rice, and the requirement of light for the germination of some forage grass and weed species.

## FACTORS REGULATING EMERGENCE AND SEEDLING ESTABLISHMENT IN THE FIELD

Numerous factors interact to determine the number of healthy autotrophic plants obtained from planting a given sample of seed. Germination under laboratory conditions frequently is a poor simulation of the rigors encountered in the field. Factors influencing seedling establishment under field conditions include the physical, chemical, and biotic properties of the soil; method, data, depth and rate of seeding; and seed treatments. In addition, each of these factors interacts with the environmental factors of water, temperature, oxygen and light that regulate the rate of germination.

### Soil Moisture

Soil moisture conditions exert a dominant influence on stand establishment because of the modifying effects of moisture on soil properties. Soil structure, soil water potential, and seed-soil surface contact determine the rate of moisture uptake by the seed in soil. Increasing seed surface contact with liquid water decreases germination time and increases germination percentage. Any factor reducing soil hydraulic conductivity (determined by pore diam) or seed-soil-water contact (determined by aggregate size and matrix potential) reduces rate of water uptake and delays germination. Based on laboratory studies, the general guideline for preparing seedbeds in the row zone is to obtain a mean soil aggregate size of -1/5 the seed diam to insure rapid water uptake and germination.

Percent seedling emergence under field conditions varies with soil moisture content, soil type, and plant species. Work with many crops shows that total emergence is affected only slightly by moisture tension from 0.05-0.3 MPa. The number of emerging seedlings decreases rapidly at moisture tension greater than 0.7 MPa, and the time to maximum emergence increases as moisture stress increases. Many crop seeds are able to imbibe sufficient water at moisture near or slightly below the permanent wilting point to initiate germination but not elongate. Insufficient aeration presumably limits germination at very high soil moisture conditions, since diffusion of oxygen in air is 10000 times greater than in water at 20° C.

## Soil Strength

Soil strength is influenced primarily by bulk density (texture and soil particle aggregation) and moisture content, with some modification by organic matter and base saturation. As soils dry, soil strength increases, with soil strength restricting emergence before soil moisture limits seedling cell division and elongation. Seedlings growing horizontally under a crust reflect these phenomena.

Soil crusting is an increase in soil strength due to soil compaction and loss of soil structure. Crusting increases the resistance to shoot and root penetration and is most frequently a problem in stand establishment on fine-textured soil with poor soil aggregation and low organic matter. Seedling emergence through crusts is controlled by seedling size, number and spacing; soil cracking pattern; and crust strength around the emerging seedling.

## Seedling Emergence Strength

Emergence strength of individual seedlings varies from a low of 0.15 newton for alfalfa to a high of 2.9 newtons for corn. Multiple seedlings in a group are able to rupture higher-strength soil crusts. For example, studies show that the maximum thrust of one, two, and three cotton seedlings is 3.8., 5.8, and 8.5 newtons, respectively. Similarly, subterranean clover (*Trifolium subterraneum* L.), when seeded 2-cm deep in clumps of 1,2, and 5 seeds, has 28, 72, and 85% emergence.

Large seeds exhibit greater emergence thrust but also encounter more soil resistance due to the greater surface area of their emerging seedling structures. Emergence data for many species and cultivars generally favor the larger-seeded types when all other factors are equal. This is particularly true for the smaller-seeded grasses and legumes where emergence from deeper plantings is promoted by large seed size, even though emergence force per gram of seed weight is slightly greater for small seeds. Large-seeded dicots occasionally fail to establish themselves as well as smaller seeds of the same cultivar. However, this is usually due to differences in seed quality. Equal seed quality for comparison of seed size is difficult to obtain from machine-harvested seed lots because of mechanical injury problems. Large seeds frequently exhibit more evidence of injury than small seeds.

In cotton, hypocotyl elongation is more sensitive than the radicle to increasing soil impedance and decreasing soil moisture. Increased physical strength and decreased soil water reduce the depth from which seedlings can potentially emerge from over 7 cm at 0.03 MPa moisture and 0.02 MPa soil strength to less than 2 cm at 1 MPa moisture and 0.3 MPa soil strength.

Soybean exhibits genetic variation in hypocotyl length, with temperatures of 21-28°C inhibiting elongation of sensitive varieties. Thus, poor emergence of late plantings may be due to lack of hypocotyl elongation.

## Soil Salts and Fertilizer Effects on Emergence

High concentrations of fertilizer and soluble salts in soil solution restrict germination and field emergence due to osmotic and, in some cases, ion toxicity effects. Field crop establishment problems are greatest in arid and semiarid soils where soluble salts are more prevalent. A similar problem may occur in subhumid and humid regions when heavy fertilizer rates are concentrated in the seed zone.

Crops exhibit a wide salt tolerance range both among and within species. Saline soils contain more than 0.2% soluble salts and have conductivities greater than 4 dS $m^{-1}$. Conductivity greater than 6 dS $m^{-1}$ reduces emergence rate and final emergence percentage of sorghum, sugar beet, field bean, pigeon pea (*Cajanus indicus* Spreng), and cowpea. Salt tolerance is usually less during germination than for established plants. Sugar beet and alfalfa generally are classed as high-salt tolerant crops, while corn is intermediate. However, corn exhibits much higher salt tolerance than either sugar beet or alfalfa during laboratory germination and emergency.

Salt concentration in a saline soil increases as moisture content decreases. Thus, high moisture availability is required for satisfactory stand establishment. The interaction of soil moisture and osmotic (salt) potential on germinating seeds is expressed by the following equation:

$$OP = (SP/PW)0.36 \times EC_e \times 10^3$$

where OP = osmotic pressure, SP = % soil water at saturation, PW = % soil water at the tension studied, and $EC_e$ = electroconductivity of soil solution in dS $m^{-1}$.

Emergence is related to the combined stresses of moisture and osmotic potential. Emergence declines rapidly when the combined forces exceed 0.8 MPa. Approximately equal total stress and emergence is obtained at 0.4 MPa moisture stress and 1.4 dS $m^{-1}$ of electroconductivity or at 0.05 MPa of moisture stress and 6 dS $m^{-1}$ of electroconductivity.

Commercial fertilizers have an osmotic effect on germination. Urea and anhydrous ammonia are detrimental at concentrations of 0.0005 to 0.001 mole fraction, and potassium fertilizers cause injury at concentrations of 0.005 to 0.02 mole fraction. Phosphates are the least toxic, with concentrations of 0.05 to 0.1 mole fraction required for injury.

Applying anhydrous ammonia at high rates immediately before seedling may cause injury. Delaying time of planting after fertilizer application or increasing depth of fertilizer placement reduces injury. Concentrations of ($NH_3$ + $NH^{4+}$)-nitrogen greater than 0.0006 mole fraction inhibits corn radicle growth, while plumule growth inhibition occurs at 0.0009 mole fraction. Concentrations greater than 0.001 mole fraction reduced field stands. Applying 112 kg ha-1 immediately prior to planting can result in 0.0014 mole fraction ppm of ($NH_3$ + $NH^{4+}$)-nitrogen in the seed zone.

Seed and seedling damage from anhydrous ammonia, urea, and ammonium phosphates apparently is due to the mobile ammonia ions. In a fine sand at 3% moisture, ammonia moved 7.5 cm in 3 hours. In silt loam at 50% moisture saturation, ammonia from diammonium phosphate moved 2.5 cm in 24 hours and 7.5 cm in 27 hours. Ammonia moved at the rate of 2.5 cm per hour in a saturated soil.

Recommended seedling depths generally are greater on sandy than clay soils due to lower water-holding capacity and faster drying rate at a given depth. In regions where planting occurs during periods of low rainfall and low soil moisture availability, seedling depth may be double or more the depth commonly used in more moist regions. Increased soil compaction also increases emergence in drier regions. Wheat in the Palouse region of eastern Washington and Oregon may be planted more than 10 cm deep while seedling depth in the North Central States seldom exceeds 5-7 cm. Maximum depth for wheat planting is related to the coleoptiles ability to elongate and the mechanical resistance of the soil.

Soybean may emerge satisfactorily on sandy loam soils when planted as deep as 7.5 cm, while depths greater than 2.5 cm may reduce final emergence on clay soils. Also, soil temperatures near 25°C inhibit hypocotyl elongation of some soybean varieties, limiting emergence from depths greater than 7.5 cm.

Emergence of corn from 2.5-17.8 cm depends on moisture conditions. During dry years, placement up to 12.7 cm deep produces better stands than shallower placement, while no difference in final emergence is noted for 2.5-7.5 cm in moist years.

## SEED TREATMENTS

### Fungicides

Fungicides for seed treatment are of three common types:

1. Disinfectants that kill pathogens located under the seed coat, such as loose smut (*Ustilago nuda* [Jens.] Rostv.) of barley.
2. Disinfectants effective on seed surface-inhabiting organisms such as covered smuts
3. Protectants that protect the seed after placement in the soil from soil-inhabiting fungi causing seed rot and seedling blights, and from insects such as seed maggots and wire worms.

Seed treatments with fungicides are used most commonly with species exhibiting hypogeal emergence. The potential for invasion of seedling blight organisms is greater when the food storage structure remains below the soil surface in an environment more uniform in temperature and RH than the above-ground environment. Air-dried soils seldom have RH below 97-98%. Seed lots seeded into unfavorable environments or those of initial low quality, whether epigeal or hypogeal, show greater response to seed treatment than do vigorous, high quality seed lots planted at more favorable temperature and moisture conditions.

### Scarification

Scarification is the mechanical process of abrasively etching or scratching the seed coat to reduce the number of hard seeds. Hard seed problems occur frequently in alfalfa and other small-seeded legumes where viable seeds fail to germinate immediately when placed in moist soil due to impermeable seed coats. Generally, seed specialists and agronomists agree that more than 20-30% hard seeds are undesirable with the possible exception of self-reseeding annuals, where 5-60% hard seeds are considered acceptable.

## Preplant Hardening

Preplant hardening to hasten germination and precondition the seed to stress environments has been reported repeatedly in the Russian literature. The procedure consists of repeated inhibitions of the seed, short of radicle emergence, followed by redrying. The cycle may be repeated 3 to 5 times. The process reportedly activates latent physiological processes.

## Use of AMP

Adenosine 5' -monophosphate (AMP) has stimulated germination and emergence under low temperatures associated with early planting in cotton. Other germination-stimulating chemicals include potassium nitrate ($KNO_3$), hydrogen peroxide ($H_2O_2$) thiourea, gibberellins, auxins, cytokinin, and ethylene, but these are used primarily in seed laboratories.

## Electromagnetism

Electromagnetic seed treatments have been marketed for several years for *energizing* or *exciting* seeds of low vigor. Seeds either are placed in an electromagnetic field for a specified time, or the seed is passed through an electromagnetic force field via an auger. Results have been variable and yield increases usually non-significant.

## Herbicides

Herbicide protection by banding 165-330 kg $ha^{-1}$ of activated charcoal directly over the seed row at planting and prior to preemergence herbicide application has been effective in preventing movement of germination-inhibiting herbicides. The technique improves the selectivity of the chemical and/or permits the use of chemicals to control similar species, such as grass weeds in a grass-seed field.

Herbicide antidotes may be applied as a seed treatment prior to planting as with 1,8-naphthalic anhydride or as a herbicide spray tank mix, e.g., N, N-diallyl-2, 2-dichloroacetamide. Eradicane$^R$ and Sutan +$^R$ are herbicides with the latter antidote in a 12:1 ratio of herbicide to antidote. The antidotes permit the use of higher rates of herbicide for the control of difficult weeds with minimal damage to the germinating and growing crops. However, the antidotes may also alter the selectivity of herbicides.

## Inoculation

Inoculation of legume seeds with appropriate strains of *Rhizobium* bacteria is necessary for symbiotic nitrogen fixation when natural populations effective for the crop species planted to not inhabit the soil or are present in low numbers. Inoculation of seeds with *Rhizobium* bacteria has little or no effect on germination. However, problems of effective nodulation do occur when it is coupled with chemical seed treatment, because most fungicides are toxic to the rhizobia. Fortunately, epigeal-emerging legume species seldom show responses to fungicide application unless seed lots of low initial seed vigor are planted.

Pre inoculation of legume seeds by procedures such as pelleting, lyophilization, and vacuum infiltration are used in an attempt to avoid the time squeeze at planting. In general, the seed surface is a hostile environment for rhizobia bacteria, and some research indicates the need for over 5000 rhizobia per seed if inoculation occurs 2-3 months before planting. Pre inoculation is successful only where it is done during the cold periods preceding planting in areas, such as the Upper Midwest of the U.S. Coating with a peat-base inoculant has been most effective. Rhizobia will not survive prolonged elevated storage temperatures regardless of inoculation method.

# CHAPTER 28

# Physiology of Flowering

## WHAT IS A FLOWER?

From a botanist's view, a flower is defined as the *part of a plant with either male or female or both reproductive structures*. It is a group of reproductive organs of the flowering plants, which develops fruit and seeds and thus helps in forming new offspring of the plant. From morphological points of view flower is *a highly condensed and modified shoot*. Its function is reproduction.

In the course of evolution, plants have developed mechanisms to permit survival in harsh environments; one of these is the ability to flower at a time when environmental conditions will allow fruit and seed maturation.

The cryptogams (Gk. *kryptos*, hidden, plus gamos, marriage), including the ferns and other spore-producing plants, were predominant in the plant kingdom during the Paleozoic era, between 230 and 600 million years ago. During the subsequent Mesozoic and Cenozoic eras, the phanerogams (*phaneros*, visible), including both gymnosperms (*gymnos*, naked, plus *sperma*, seed) with sporophylls and angiosperms with true flowers, gained ascendance and today comprise the bulk of our economic plants. Flower formation is a transitional phase of the life cycle of a plant and it serves the functions of reproduction, seed production and in fruits formation. Flowering plants provide unlimited opportunity for genetic improvement.

It is the main part of a plant to produce blossoms. Although plants blossom seasonally for a short time, the flowers are considered as the most attractive part, when they are in full bloom. It is quite obvious that the garden enthusiast should know about the nature and structure of a flower. Nature has gifted the flowers with varied rich colours, beauty and elegance, exquisite delicacy, exotic shape and fragrance to attract the naturally pollinators such as flies, butterflies, moths, birds and the like. This creation has been developed to a great extent by man by producing elite flowers for his taste through hybridization. One can see many man-made creative, intergeneric, trigeneric and tetrageneric hybrids in orchids. Flowers, like several others living organisms, are delicate or fragile and must be handled with care to retain beauty and freshness.

## NATURE OF FLOWERING

Flowers likeaster, balsam, celosia, calandula, phlox are short-lived and bloom within two to three months of sowing seeds. They are herbaceous annuals. On the other hand, biennials such as *Althaea facifolia*, *Digitalis purpurea* and the like flower only in the second year of their life-cycle. However, most ornamental perennial shrubs and trees take more than two years to bloom after the seedlings are planted. Further, *kurinji* (*Strobilanthes kunthiyanus*), a shrubby plant, blossoms only once in 12 years. *Corypha umbrachifera* shows monocarpism as a 60-70 year-old giant palm suddenly changes its physiological behaviour to produce bamboo (Bambuseae) as some species take a longer time (20-120 years or so) to flower in their life span. *arundinaria metake* a common bamboo, native species from Japan known as *Metake*, blossoms once in 120 years. This shows bamboo has a long span of life. Perhaps no human being can witness flowering of this species during his lifetime planted by himself. Further *Fourcroya* sp. flower only once in their lifetime, i.e., nearly 400 years.

The flowers happen to be most conspicuous structure in the Angiosperms. The flowers appear on a particular species after a definite period of vegetative growth vegetative phase. The vegetative phase is specific for a particular species. On the basis of the length of vegetative period the plants have been classified into.

- **Annuals:** Plants like Mustard, Tobacco produce seed in one season or one year.
- **Biennials:** Plant like cabbage, beet, carrot require two seasons or year. First year vegetative growth takes place and next year flowering takes place.
- **Perennials**: Most trees are perennials. Their vegetative stage extends to some years and thereafter they flower year after year.

Some plants like species of *Palm* and Bamboo flower once in their life time. They are known as *Monocarpic plants.* They die after producing flowers.

## FLORAL STRUCTURE

Among the flowering plants *Magnolia grandiflora* is considered as the most primitive as the flower is simple, bisexual and complete with all the four component parts namely, sepals (calyx), petals (corolla), stamens and pistil. Further, members of the pea family Fabaceae (*Lathyrus odorotus*, *Pisum sativum*, etc.) and orchid family Orchidaceae (*Cattleya portia* and *Vanda coerulea*, etc.) have remarkable diversity in flower shape, size and colour but the flowers are still complete and more diversified.

There are large groups of plants with incomplete flowers, either petalless, sepalless or with a single sex (unisexual). Most of these diversified and incomplete flowers are usually cross-pollinated by wind, others by insects, birds, etc. Also, in a few cases, there are male (staminate) and female (pistilate) flowers either on the same plant members of Areacaceae such as palms.

In typical flower there are four types of structures called floral leaves. These are arranged in four whorls. Floral leaves are arranged on the swollen upper parts of flower stalk. The flower stalk is called **pedicel** and the swollen upper part is called Thalamus (= *torus*). In the thalamus are present nodes and highly reduced internodes. Floral leaves are borne on the nodes in whorls. The four floral whorls taken from periphery to centre are as follows:

### Calyx

It is the outermost whorl and floral leaves of this whorl are called **sepals**. Their function is that of protection of inner whorls and also that photosynthesis when green.

### Corolla

It is the second whorl of floral leaves which are called **petals**. These protect the inner two whorls and also function for attraction of insects for pollination.

Calyx and corolla are called **accessory whorls** of the flower. In some plants although two whorls are present they are not differentiated from one another. In such a case both whorls are called **perianth** and each part of the perianth is called **tepal**. If accessory whorls are absent in a flower it is called **achlamydous flower** (naked flower), if out of the two accessory whorls only one whorl is present it is called **monochlamydous flower** and if both whorls are present it is called **dichlamydous flower**.

### Androecium

This is third whorl of floral leaves and is the male reproductive organ of the flower. Each modified floral leaf is called **microsporophyll** or **stamen**.

### Gynoecium (Pistil)

This is the last (fourth) central whorl of floral leaves and is the female reproductive organ of the flower. Each modified floral leaf is called **megasporophyll** or **carpel**. Each gynoecium is made up of one or more megasporophyll (carpels).

The last two whorls are called **essential whorls** of the flower. In most flowers stamens and carpels are found in the same flower, such flowers are called **hermaphrodite** or **bisexual flowers** e.g., China rose, pea, cotton. In other flowers only one of the essential whorls is found. Such flowers are called **Unisexual flowers** e.g., Cucurbits, Mulberry etc. Flowers which have only stamens are called **staminate flowers** and those in which only carpels are present as **pistillate flowers.**

### Polygamous, Monoecious or Dioecious Plants

These terms are used for the plants. If a plant has bisexual as well as neutral flowers it is called **Polygamous plant** e.g., Mango. If a plant has either bisexual flowers e.g., Rose of although flowers are either staminate of pistillate but are borne on the same plant is called **Monoecious**. However, if plant bears only male flowers it is called male plant and if it bears only pistillate flowers is called female plant and such plants are called **Dioecious**. Thus a dioecious plant will always have unisexual flowers but a monoecious plant has either hermaphrodite or unisexual flowers on it.

In most flowers the thalamus is condensed but in some one or more internodes become long. If the internode between Calyx and Corolla becomes elongated it is called **anthophore** e.g., Dianthus. If the internode between corolla and androecium gets elongated it is called **androphore** or **gonophore** e.g., *Passiflora.* If the internode between androecium and gynoecium is elongated it is called **gynophore** e.g., *Capparis*. In *Gyandropsis pentaphylla* both androphore and gynophore are present and these taken together are called **gyanodrophore** or **androgynophore**. In *Coriandrum* and *Foeniculum* the thalamus in between to carpels get elongated and after bifurcation protrudes out of the two carpels. It is called **carpophore**. A flower is said to be **cyclic** when sepals, petals, stamens and carpels are arranged in circles or whorls and **acyclic** when these are sprially arranged.

## COLOUR RANGE IN FLOWERS

Nature has been lavish in distributing colourful pigments in blossoms. A number of pigments especially anthocynins, anthoxanthins, carotenoids and the like are responsible for the innumerable shades of wide colour range in flowers. For instance, anthocyanins are responsible for the wide colour range such as blue, purple, mustard, maroon, magenta and various shades or red and pink pigments are obtained from blossoms as well as from flower buds. Carotenoids, especially carotene, xanthophyll and its derivatives, give orange pigments. The green-pigments present in the foliage are chlorophyll-a and chlorophyll-b, in the ratio of 3:1 in most plants. Recently, with the help of genetic engineering scientists are trying to do what nature was unable to do i.e., growing blue roses as well as a creating rare colour blossoms among ornamentals, namely, chrysanthemums, carnations and gerberas.

## FRAGRANCE IN FLOWERS

Nature has also gifted some of the flowers with pervasive fragrance. Most roses are noted for their beauty, elegance and fragrance. Flowers of *Gardenia jasminoides* (cape jasmine), a tall growing shrub, give out a strong enticing smell in summer. Again, the fleshy, greenish yellow petalled lowers of *Artabotrys odoratissimus* (manoranjitham) have a strong pervading scent, while those of *Michelia champaca* (champak), *Conaga odorata* (ylang ylang or lavender champa) are mildly scented. A few climbers, *Jasminium grandiflorum* (jasmine) and *Hiptage modoblota* (madhavi), emit a sweet fragrance when they are in bloom. Further, the pervading fragrance of a few orchids lingers in air for weeks together, if they are kept inside the house.

## DEVELOPMENTAL STAGES OF FLOWERING

The transition from vegetative to reproductive development is clearly a very critical phase in the life cycle of higher plants, however, the actual mechanism by which the transition from vegetative to reproductive development occurs still remains unclear at the present time.

### First stage

The **first stage** is flower initiation which is an internal physiological change in the meristem which proceeds any morphological change. The first morphological change indicating enhanced cell division in the central zone immediately below the apical part of vegetative meristem. The divisions occurring here result in the differentiation of parachenchyma cells which surround the meristem giving rise to flower primordia.

### Second stage

The second stage is flower formation, the visible initiation of flower parts.

### Final Stage

The final stage is flower development which is differentiation of the flower structure including events from flower formation to anthesis (flowering). Each of these stages like any other physiological process is determined by the genotype. However, in many species the start of reproductive development is regulated by environmental factors such as day length (photoperiodism), temperature (vernilization) which vary on a fairlt regular basis throughout the year, and by specific plant growth substance(s) or intereaction between two or more.

## DEVELOPMENTAL STAGES

When the higher plant changes from vegetative to reproductive growth, a new set of structures, the flower, is formed and represents the expression of genes not previously expressed.

The formation of sepals, petals, stamens, and carpels occurs nowhere else in the plant and meiosis occurs only in the stamens and carpels. The transition to flowering, therefore, represents a change in determination of the shoot apical cells and their derivatives.

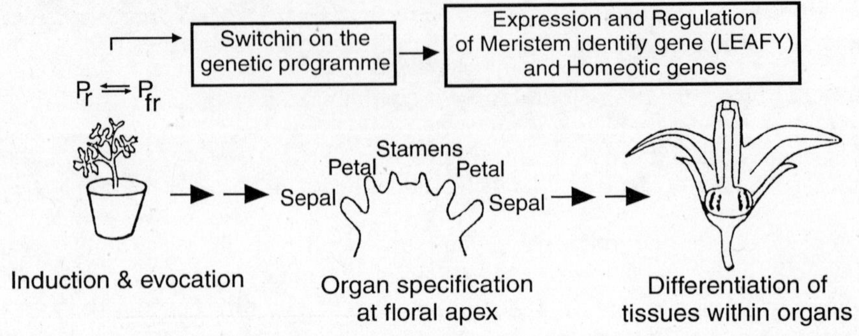


**FIG. 28.1:** An Examples of decision points in flowering. Discrete decisions must be made by tissues or individual cells during the formation of flowers. The first decision is whether or not to flower, represented in the left panel of the figure by the interconversion of the two forms of photochrome, a common mediator of floral induction. The cells of the apex are then directed to undergo evocation, resulting in the conversion of the vegetative meristem to a floral meristem. The meristem ma produce branches, but eventually the branches terminate in apexes for single flowers. As illustrated in the center panel, a series of organ primordia are produced at the floral apex. Early in the formation of the primordia, decisions must be made as to which floral organs the primordia will form (*center*). Genetic changes can result in fundamental alterations in this set of decisions. Finally, as the organs begin to take on their characteristic form each cell must be correctly directed to differentiate biochemically and anatomically so that all of the differentiated tissues in the flower form in proper spatial relationship to each other.

In a vegetative plant is competent to flower (ripe-to-flower), it only needs the change in determination. A lack of competence to flower is most clearly seen in those plants that show a juvenile phase, i.e. a phase in which it is not possible to cause flowering directly. Competence to flower may be defined as *the ability to develop along a particular pathway when given the appropriate signal, plants that flower in response to photoperiod, in which a signal is received from the leaves*. In such plants, the leaves are obviously competent to form the floral stimulus, but they may not be in some juvenile plants. Furthermore, it is often far from clear whether this signal from the leaves first has to make the apical meristem competent to flower or whether the meristem is already competent to flower but simply lacks the signal for determinations. We can see that competence may therefore reside in the leaves or in the apex, or neither, or both.

1. The leaves are competent to provide a signal but the apex is not competent to react. This is the case in *Bryophyllum*, in which the apex from a juvenile plant will form flowers if grafted to a mature plant.
2. The leaves are not competent to provide a signal, nor is the apex competent to react; this seems to be the case with plants requiring **vernalization** followed by a specific photoperiodic requirement. Vernalization (treatment at low

temperatures of about 5°C for several weeks), which acts on the apex and not on the leaves, seems to change the competence of the apex to react to signals produced by the leaves in the appropriate photoperiod.

3. Leaves and apex are both competent and the lants flower rapidly in response to an inductive signal.

Plants with a photoperiodic requirement to flower could fall into any one of these above four categories. Plants that flower in response to a signal photoperiod are likely to be those in which the apex is already competent to flower and requires a signal from the leaves for flower determination e.g., *Pharbitis*.

## Juvenile Phase

Plants are not capable of flowering upon germination but must pass through a juvenile lasting from a few days to many years depending upon species. The phenomenon of juvenility is important if problems are to be avoided in commercial production of certain crops. In areas with mild winter, cold crops such as cauliflower and cabbage (*Brassica oleracea* var. *capitata*) are sown in the fall for winter production. If sown at the proper time, cabbage will be in the (*bolting*) in adult plants and, therefore, produce heads, rather than seedstalks. If sown too early, bolting will occur. The juvenile phase of Kohlrabi is very short and will bolt whenever exposed to temperatures of less than 10° C for 1 week or longer.juvenile stage during exposure to temperatures low enough to induce flower stalk elongation

## Adult Phase

Following the juvenile period, the **adult phase** begins. It may last for as short a time as a summer in annuals or as long as several thousand years in sequoias (*Sequoiadendron giganteum* Buchholtz) and bristle-cone pines (*Pinus aristata* Engl.). **Senescence** then begins, the plant dies, and the cycle is repeated.

**FIG. 28.2:** Juvenile and adult forms of ivy (*Hedera helix*). The juvenile form (left) has lobed palate leaves arranged alternately, a climbing growth habit, and no flowers. The adult form (right) has entire ovate leaves arranged in spirals and an upright growth habit.

## Inflorescence Development

Light period may affect the rate of inflorescence development as well as induction per se. Long photoperiods following cold-temperature induction, hasten flowering in wheat and barley. In contrast, increase in daylength time increases heading in smooth-stalked meadow grass (*Poa pratensis*). Similarly, the photoperiod to which quantitative LDP's or SDP's are exposed prior to inflorescence initiation can affect subsequent development. Prolonging exposure to SD prior to induction by LD increases the number of flower primordia formed in *Lolium perenne* and wheat, both quantitative LDP's. In maize, a quantitative SDP, exposure to LD preceding initiation increases flower number in both male and female inflorescences.

### The Shoot Apex and Phase Changes

All multicellular organisms pass through a series of more or less defined developmental stages, each with its characteristic features. In humans, infancy, childhood, adolescence, and adulthood represent four general stages of development, and puberty is the dividing line between the nonreproductive and the reproductive phases. The development of higher plants likewise is characterized by developmental phases, but whereas in animals these changes take place throughout the entire organism, in higher plants they occur in a single, dynamic region, the *shoot apical meristem*.

## FLORAL MERISTEMS AND FLORAL ORGAN DEVELOPMENT

Floral meristems usually can be distinguished from vegetative meristems, even in the early stages of reproductive development, by their larger size. The transition from vegetative to reproductive development is marked by an increase in the frequency of cell divisions within the central zone of the shoot apical meristem. In the vegetative meristem, the cells of the central zone complete their division cycles slowly. As reproductive development commences, the increase in the size of the meristem is largely a result of the increased division rate of these central cells. Recently, genetic and molecular studies have identified a network of genes that control floral morphogenesis in *Arabidopsis*, snapdragon (*Antirrhinum*), and other species.

### Arabidopsis: An Example

Floral development in *Arabidopsis* (Fig. 28.3) has been studied extensively. The basic morphological changes that occur during the transition from the vegetative to the reproductive phase. The arrangement of the floral organs in four whorls on the meristem, and the types of genes that govern the normal pattern of floral development. According to the widely accepted ABC model, the specific locations of floral organs in the flower are regulated by the overlapping expression of three types of floral organ identity genes. Although floral development is genetically controlled, it is nevertheless sensitive to environmental and hormonal signals, although perhaps not to the same extent as vegetative development is. We end the chapter by discussing the roles of light and of gibberellin and other hormones in modifying the genetic program for floral morphogenesis.

#### *The Characteristics of Shoot Meristems in Arabidopsis Change with Development*

During the vegetative phase of growth in the first 25 days after germination, the wild-type *Arabidopsis* vegetative apical meristem produces about 12 phytomeres with very short internodes, resulting in a basal rosette of leaves.

As plants initiate reproductive development, the vegetative meristem is transformed into an indeterminate inflorescence meristem and no longer produces vegetative phytomeres. Instead, the primordial formed on the flanks of the inflorescence meristem develop into flowers. The leaf primordial that had been initiated before the floral transition of the meristem develop on the elongating flowering stalk and are called the cauline leaves (Hempel and Feldman 1994). Cauline leaves are smaller than rosette leaves and similar in size and structure to the bracks that subtend many angiosperm flowers. However, *Arabidopsis* flowers lack bracts. The lateral buds of the cauline leaves develop into secondary inflorescence meristems, and their activity repeats the pattern of the development of the primary inflorescence meristem.

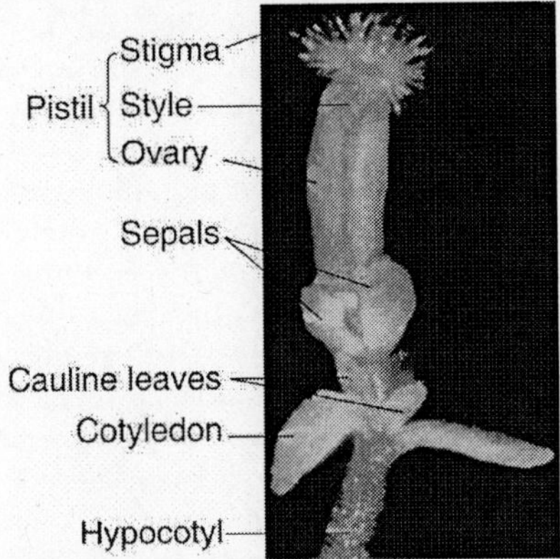


**FIG. 28.3:** The *emf* mutant of *Arabidopsis.* The striking phenotype of this mutant is that the seeding skips the vegetative phase of growth and forms an inflorescence directly. Perhaps because of their early appearance, the flowers are usually abnormal. The hypocotyls is shown at the base with two cotyledons. A short segment of inflorescence stem with a few reduced cauline leaves has formed immediately above the cotyledons. This example of an *emf* mutant flower lacks petals and stamens. A pistil with an enlarged stigma is the most prominent floral structure.

### *The Four Different Types of Floral Organs Are Initiated as Separate Whorls*

Floral meristems initiate four different types of floral organs: sepals, petals, stamens, and carpels (Coen and Carpenter 1993). Each set of organs is initiated as a **whorl** – that is, arranged in concentric rings around the flanks of the meristem. By the time the carpels are initiated, there are no more meristematic cells in the apical dome, and only the floral organ primordial are present. In the wild-type *Arabidopsis* flower, the first (outermost) whorl consists of four sepals, which are green at maturity. The second whorl is composed of four petals, which are white at maturity. The third whorl contains six stamens, two of which are shorter than the other four. Finally, what is considered to be the fourth whorl is a single complex organ, the gynoecium or pistil, which is composed of an ovary with two fused carpels, each containing numerous ovules, and a short style capped with a stigma.

## THE SHOOT APICAL MERISTEM PROGRESSES THROUGH THREE DEVELOPMENTAL PHASES

During postembryonic development, the shoot apical meristem passes through three more or less well-defined developmental stages in obligate sequence: the juvenile phase, the adult vegetative phase, and the adult reproductive phase (Poething 1990). The transition from one phase to another is called **phase change**.

The primary distinction between the juvenile and the adult vegetative phases is that the latter has the ability to form reproductive structures: flowers in angiosperms, cones in gymnosperms. As will be discussed later, the reproductive capacity of the shoot apical meristem is referred to as *competence*. However, expression of the reproductive competence of the adult phase (i.e., flowering) often depends on specific environmental and developmental signals. Thus, the absence of flowering itself is not a reliable indicator of juvenility.

The transition from juvenile to adult is frequently accompanied by changes in vegetative characteristics, such as leaf morphology, phyllotaxy (the arrangement of leaves on the stem), thorniness, rooting capacity, and leaf retention in deciduous plants. Such changes are most evident in woody perennials, but they are apparent in many herbaceous species as well. Unlike the abrupt transition from the adult vegetative phase to the reproductive phase, the transition from juvenile to vegetative adult is usually gradual, involving intermediate forms. A dramatic example is the progressive transformation of juvenile leaves of the leguminous tree, *Acacia heterophylla* into phyllodes, a phenomenon noted by the German poet and natural scientist Goethe (1749-1832). Whereas the juvenile leaves consist of petiole, blade, and leaflets, adult phyllodes are specialized structures consisting of flattened petioles in place of the blade (Figure 24.3). Intermediate structures form also during the transition from aquatic to aerial leaf types in heterophyllous (characterized by different leaf types) aquatic plantssuch as *Hippuris vulgaris* (common marestail). As in the case of *A. heterophylla*, these intermediate forms possess distinct regions with different developmental patterns. Whereas the apex of an intermediate leaf develops according to the earlier developmental program (juvenile or submerged), the base develops according to the new developmental program (adult or aerial) (Poethig 1990).

To account for intermediate forms during the transition from juvenile to adult in maize, Scott Poethig (1990) has proposed a **combinatorial model**. According to this model, shoot development can be described as a series of independently regulated, *overlapping* programs (juvenile, adult, and reproductive) that modulate the expression of a common set of developmental processes. In the transition from juvenile to adult leaves, the intermediate forms indicate that different regions of the same leaf can express different developmental programs. Thus, the cells at the tip of the leaf remain committed to the juvenile program, while the cells at the base of the leaf become committed to the adult program. The developmental fates of the two sets of cells in the same leaf are quite different.

## JUVENILE TISSUES

These tissues are produced first and are located at the base of the shoot. The temporal order of the three developmental phases results in a spatial gradient of juvenility along the shoot axis. Because growth in height is restricted to the apical meristem, the juvenile tissues and organs, which form first, are located at the base of the shoot. In rapidly flowering herbaceous species, the juvenile phase may last only a few days, and few juvenile structures are produced. In contrast, woody species have a more prolonged juvenile phase, in some cases lasting 30 to 40 years. In these cases, the juvenile structures can account for a significant portion of the mature plant. Once the meristem switches over to the adult phase, only adult vegetative structures are produced, culminating in floral evocation. The adult and reproductive phases are therefore located in the upper and peripheral regions of the shoot.

Attainment of a sufficiently large size appears to be more important than the plant's chronological age in determining the transition to the adult phase. Conditions that retard growth, such as mineral deficiencies, low light, water stress, defoliation, and low temperature tend to prolong the juvenile phase or even cause **rejuvenation** (reversion to juvenility) of adult shoots. In contrast, conditions that promote vigorous growth accelerate the transition to the adult phase (Poethig 1990). When growth is accelerated, exposure to the correct flower-inducing treatment can result in flowering. However, although size seems to be the most important factor, it is not always clear which specific component associated with size is critical. In some *Nicotiana* species, it appears that plants must attain a certain size to transmit a sufficient amount of the floral stimulus to the apex. In addition, the apex may need to undergo a phase transition that renders it more responsive to the floral stimulus (McDaniel et al. 1996).

**TABLE 28.1:** Length of juvenile period in some woody plant species

| Species | Length of juvenile period |
|---|---|
| Rose (*Rosa* [*hybrid* tea]) | 20-30 days |
| Grape (*Vitis* spp.) | 1 year |
| Apple (*Malus* spp.) | 4-8 years |
| *Citrus* spp. | 5-8 years |
| English ivy (*Hedera helix*) | 5-10 years |
| Redwood (*Sequoia sempervirens*) | 5-15 years |
| Sycamore maple (*Acer pseudoplatanus*) | 15-20 years |
| English oak (*Quercus robur*) | 25-30 years |
| European beech (*Fagus sylvatica*) | 30-40 years |

Once the adult phase is attained, it is relatively stable, and is maintained during vegetative propagation or grafting. For example, in mature plants of English ivy (*Hedera helix*), cuttings taken from the basal region develop into juvenile plants, while those from the tip develop into adult plants. Results for grafted plants in which the *scion* (upper graft partner) originated from different parts of an old flowering specimen of silver birch (*Betula verrucosa*) also indicate that the basal part retains the juvenile condition (Longman 1976). When scions were taken from the base of the flowering tree and grafted onto seedling rootstocks, there were no flowers on the grafts within the first 2 years. In contrast, the grafts flowered freely when scions were taken from the top of the flowering tree. In some species, the meristem appears to be competent to flower but does not receive sufficient floral stimulus until the plant becomes large enough. For example, in mango (*Mangifera indica*), juvenile seedlings can be induced to flower when grafted to a mature tree. In other species, however, grafting does not induce flowering.

## PHASE CHANGES

The change in phases are influenced by nutrients, hormones, and other chemical signals. The transition at the shoot apex from the juvenile to the adult phase may be affected by transmissible factors from the rest of the plant. In many plants, exposure to low-light conditions prolongs juvenility or causes reversion to juvenility. A major consequence of the low-light regime is a reduction in the supply of carbohydrates to the apex; thus, carbohydrate supply may play a role in the transition between juvenility and maturity. One obvious connection between nutritional status and the adult phase is the effect of nutrition on the size of the apex. For example, in the florist's chrysanthemum (*Chrysanthemum morifolium*), flower primordial are not initiated until a minimum apex size has been reached (Cockshull 1985).

The apex receives a variety of hormonal and other factors from the rest of the plant in addition to carbohydrates and other nutrients. Could any single factor promote the change between juvenility and maturity? Experimental evidence shows that the application of gibberellins causes reproductive structures to form in young, juvenile plants of several conifer families. For example, although most conifers require several years to attain maturity, male cones have been induced to form in Arizona cypress (*Cupressus arizonica*) plants that were only 2 months old by being sprayed with GAs. The involvement of *endogenous* GAs in the control of reproduction is also indicated by the fact that other treatments that accelerate flowering in pines (e.g., root removal, water stress, and nitrogen starvation) often also result in a buildup of GAs in the plant. On the other hand, while gibberellins promote the attainment of reproductive maturity in conifers and many herbaceous angiosperms, the application of $GA_3$ causes rejuvenation in *Hedera* and in several other woody angiosperms (Hackett and Srinivasan 1985). The role of gibberellins in the control of maturity is thus complex, varies among species and probably involves interactions with other factors.

## FLORAL EVOCATION

On the basis of extensive studies on tobacco, Carl McDaniel and his colleagues at the Rensselaer Institute have defined two general categories of developmental states associated with floral evocation: competent and determined (McDaniel et al. 1992). In general terms, a cell or group of cells is said to be competent if it can respond in the expected manner when given the appropriate developmental signal. For example, if a vegetative shoot (scion) is grafted onto a flowering stock, and the scion flowers immediately, it is obviously capable of responding to the level of floral stimulus present in the stock and is therefore competent. Failure of the scion to flower would indicate that the shoot apical meristem has not yet attained competence. This is another way of stating that the apical meristem is in the juvenile phase.

Flowering involves several processes that occur in different parts of the plant. In photoperiodic plants (discussed later in the chapter), the *inductive photoperiod* causes the leaves to export a floral stimulus to the rest of the plant that alters the developmental fate of the cells of the shoot apex. Once the meristem becomes committed to the new developmental program (flowering in this case), it is said to be florally determined. The meristem is **determined** if it follows the same developmental program even after being removed from its normal physical and environmental context.

For example, in an experiment to measure the floral determination of axillary buds of a flowering tobacco plant, decapitation of the donor promotes the outgrowth of the axillary bud (McDaniel 1996). After a certain number of leaves (in this case seven) have developed, flowers are produced. However, if the bud is either rooted or grafted onto stock, it produces many more leaves before it flowers. This result shows that the bud was not florally determined. In contrast, in another experiment, the axillary bud flowered after producing seven leaves in all three situations. This result demonstrates that the axillary bud is florally determined.

Using this approach with growing terminal buds, Singer and McDaniel (1986) showed that in a tobacco plant producing about 41 nodes, the terminal bud became florally determined after initiating 37 leaves. Extensive grafting of shoot tips among tobacco varieties established that the number of nodes a meristem produces before flowering is a function of the strength of the floral stimulus from the leaves and the competence of the meristem to respond to the signal (McDaniel et al. 1996). Thus, juvenility and phase change in tobacco are tightly correlated with, and perhaps explained by, two factors: the strength of the floral stimulus and meristem competence.

In some cases, the **expression** of flowering may be delayed or arrested even after the apex becomes determined, unless it receives a second developmental signal that stimulates expression. For example, intact *Lolium temulentum* (darnel ryegrass) plants become committed to flowering after a single exposure to a long day. If the shoot apical meristem is excised 28 hours after the beginning of the long day and grown in tissue culture, it will produce normal inflorescences in culture, but only if the hormone gibberellic acid (GA) is present in the medium. Since apices cultured from plants grown exclusively on short days never flower, even in the presence of GA, long days never flower, even in the presence of GA, long days are required for determination in *Lolium*, whereas GA is required for *expression* of the determined state (McDaniel and Hartnett 1996).

## CONTROL OF FLOWERING

Most people look forward to the spring season and the profusion of flowers it brings. Many vacationers carefully time their travels to coincide with specific blooming seasons: *Citrus* along Blossom Trail in southern California, tulips in Holland. In Washington, D.C., and throughout Japan, the cherry blossoms are received with spirited ceremonies. As spring progresses into summer, summer into fall, and fall into winter, wildflowers bloom at their appointed times. In the eastern United States, skunk cabbage flowers poke through the snow of forest floors in late February and early March. Violet and marsh marigold open their flower buds in April. Sometime in May one might see jack-in-the-pulpit, yellow clover, and painted trillium. In Jne, chicory, evening primrose, and Queen Anne's lace appear. July is the time for pickerelweed, pep-permint, and black-eyed Susan. By late July and early August, most of the spring blossoms are gone and in their places are monkey flower, spiked lobelia, and pearly everlasting. Toward the end of August, goldenrod and baneberry have their day in the sun. Finally, in September and October, aster, rose mallow, and gentian ring down the curtain on the blooming season.

Although the strong correlation between flowering and seasons is common knowledge, the phenomenon poses fundamental questions for plant physiologists. How do plants keep track of the seasons of the year and the time of day? Which environmental signals control flowering and how are those signals perceived? Can the environment be manipulated to alter the timing of flowering? How are environmental signals transduced to bring about the developmental changes associated with flowering? And what are the genes that control floral development? This chapter addresses these questions.

## The role of the root and shoot apical meristems in vegetative growth and development

The transition to flowering involves major changes in the pattern of morphogenesis and cell differentiation at the shoot apical meristem. Ultimately, this process leads to the production of the floral organs-sepals, petals, stamens, and carpels. Specialized cells in the anther undergo meiosis to produce four haploid microspores that develop into pollen grains. Similarly, a cell within the ovule divides meiotically to produce four haploid megaspores, one of which survivies and undergoes three mitotic divisions to produce the cells of the embryo sac. The embryo sac represents the mature female gametophyte; the pollen grain, with its germinating pollen tube, is the mature male gametophyte generation. The two gametophytic structures produce the gametes (egg and sperm cells), which fuse to form the diploid zygote, the first stage of the new sporophyte generation. Clearly, flowers represent a complex array of functionally specialized structures that differ completely from the vegetative plant body in form and cell types. The transition to flowering therefore entails radical changes in cell fate within the shoot apical meristem.

The events occurring in the shoot apex that specifically commit the apical meristem to produce *flowers* are collectively referred to as **floral evocation**. After discussing phase change, we will examine the types of signaling events that lead to floral evocation. The developmental signals that bring about floral evocation include endogenous factors, such as *circadian rhythms* and *hormones*, and external factors, such as *day length (photoperiod)* and *temperature*. In the case of photoperiodism, transmissible signals from the leaves, collectively referred to as the *floral stimulus*, are translocated to the shoot apical meristem. The interactions of these endogenous and external factors enable plants to synchronize their reproductive development with the environment.

Finally, at the end of the chapter we address the question of *floral development* itself. Recently, genes have been identified that play crucial roles in the formation of the floral organs. These studies have shed new light on the genetic control of reproductive development.

But before beginning our consideration of the changes that occur at the shoot apex during the transition to flowering, we need to distinguish between the two classes of developmental signals that regulate flowering: internal (autonomous) and external (environmental).

## AUTONOMOUS REGULATION VERSUS ENVIRONMENTAL CUES

A plant may flower within a few weeks after germinating, as in annual plants such as groundsel (*Senecio vulgaris*). In contrast, some perennial plants, especially many forest trees, may grow for 20 or more years before they begin to produce flowers. Thus, different species flower at widely different ages, indicating that age, or perhaps size of the plant, is an *internal* factor controlling the switch to reproductive development. When flowering occurs strictly in response to internal developmental factors and does not depend on any particular environmental conditions, it is referred to as *autonomous regulation*.

Some plants exhibit an absolute requirement for the proper environmental cues in order to flower. This condition is termed an *obligate* or *qualitative* response to an environmental cue. In other plant species, flowering is promoted by certain environmental cues but will eventually occur in the absence of such cues. This is called a *facultative* or *quantitative* response to an environmental cue. The flowering of this latter group of plants thus relies on both environmental and autonomous flowering systems. Photoperiodism and vernalization are two of the most important mechanisms underlying seasonal responses. *Photoperiodism* is a response to the length of day; *vernalization* is an effect on flowering brought about by exposure to cold. Other signals, such as total light radiation and water availability, can also be important external cues.

The evolution of internal (autonomous) and external (environment-sensing) control systems enables plants to carefully regulate the timing of flowering. It is critical for survival that plants flower at the optimal time for reproductive success. For example, in many populations of a particular species, flowering is synchronized. This synchrony favors crossbreeding and allows seeds to be produced in favorable environments, particularly with respect to water and temperature.

Plants unable to flower during the juvenile phase, regardless of environmental conditions like light, temperature, and to some extent moisture and nutrition, however, these factors can exert profound effects on flowering. However, in some plants, exogenous treatments of growth regulators and certain other chemicals may either promote or retard flowering.

A distinction should be made between qualitative, or absolute, *vs*. quantitative responses to photoperiod. Plants such as *Perilla* and *Lolium* exhibit qualitative responses; they do not flower if the dark period is too long or too short. The response is quantitative; if photoperiod is not optimum, the plants require longer to flower, or do not flower as profusely, but eventually they do flower. Light also affects flowering, such as those plants that require, for example, long days prior to short days (LSDP) or low temperature prior to long days.

Some plants flower following exposure to one long night or one long day but most require more than one cycle. Strawberry a SDP, requires six to 20 cycles depending upon the cultivar, while early flowering chrysanthemum [*Chrysanthemum morifolium* (Ramat) Hemsl.] requires 12 cycles. If chrysanthemum plants are transferred to long days too soon, the inflorescence stops developing and the axillary buds grow vegetatively. In at least one soybean cultivar (Biloxi), all plants flower after three cycles, but additional cycles up to seven increases the number of nodes bearing flowers.

**TABLE 28.2:** Photoperiod and temperature requirements for flowering of some economically important plants.* (Adapted from Salisbury, 1963)

| | Photoperiod requirement | | |
|---|---|---|---|
| **Temperature Requirement** | **Day Natural** | **Short Day** | **Long day** |
| **None** | Bean (*Phaseolus vulgaris*) | Coffee (*Coffea arabica*) | Carnation cvs.(*Dianthus caryophyllus*) |
| | Buckwheat (*Fagopyron esculentum* Moench.) | Pigweed (*Amaranthus retroflexus*) | Clover (*Trifolium* sp.) |
| | Corn (*Zea mays*) | Sweet potato (*Ipomoea batatas*) | Foxtail (*Alopecurus pratensis*) |
| | Cucumber (*Cucumis sativus*) | | Oat cvs. (*Avena sativa*) |

| Photoperiod requirement | | | |
|---|---|---|---|
| **Temperature Requirement** | **Day Natural** | **Short Day** | **Long day** |
| | + | Kentucky bluegrass | Ryegrass cvs. (*Lolium perenne*) |
| | (*Poa pratensis*) | | Spinach cvs. (*Spinacia oleracea*) |
| **High** | | | |
| Absolute | € | -- | -- |
| Quantitative | Fuchsia (*Fuchsia* spp.) | Cosmos (*Cosmos bipinnatus* Gav.Ann) | Phlox (*Phlox drummondii*) |
| | Summer rice (*Oryza sativa)* | Chrysanthemum cvs. | - |
| | - | [*Chrysanthemum morifolium* (Ramat.) Hemsl.] | - |
| | | Winter rice (*Oryza sativa*) | |
| **Low** | | | |
| Absolute | Cabbage (*Brassica oleracea* var. *capitata)* | Chrysanthemum cvs. (*Chrysanthemum mirifolium* (Ramat.) | (*Dianthus caryophyllus*) |
| | Celery (*Apium gravaeolens* var *dulce*) | | Ryegrass cvs. (*Lolium perenne*) |
| | Hydrangea [*Hydrangea macrophylla* (Thunb.) Ser.] | | Spinach cvs. (*Spinacia oleracea*) |
| | | | Sugar beet (*Beta vulgaris*) |
| Quantitative | Broad bean | Chrysanthemum cvs. | Carnation cvs.(Dianthus |
| | (Vicia faba) | [*Chrysanthemum morifolium* (Ramat.) Hems.] | Caryophyllus) |
| | Sweet pea (*Lathyrus odoratus*) | | Oat cvs. (*Avena sativa*) |
| | Vetch (*Vicia angustifolia*) | | Spinach cvs. (*Spinacia oleracea*) |
| | | | Winter barley (*Hordeum vulgare*) |
| | | | Winter wheat (*Triticum aestivum*) |

* Several species occur in more than one category because of cultivar differences in response. A somewhat different classification system is used by Vince-Prue (1975).

* No species are known which have these requirements.

**TABLE 28.3:** Number of inductive cycles required to saturate the flowering response in several species. Response varies with strain and cultivar

| Response type | Species | No. of cycles |
|---|---|---|
| **Short day plants** | Japanese morning glory | 1 |
| | (*Pharbitis nil* Chois.) | |
| | Cocklebur | 1 |
| | (*Xanthium strumarium*) | |

| Response type | Species | No. of cycles |
|---|---|---|
| | Soybean (*Glycine max*) | 3-7+ |
| | Strawberry | 6 to 20 |
| | [*Fragaria x ananassa* (Duch.)] | |
| | *Perilla ocymoides* | 8.5 |
| | Early flowering chrysanthemum | 12 |
| | [*Chrysanthemum morifolium* (Ramat.) Hemsl.] | |
| **Long day plants** | Spinach | 1 |
| | (*Spinacea oleracea*) | |
| | Darnel | 1 to 4 |
| | (*Lolium temulentum*) | |
| | Henbane | 4 to 5 |
| | (*Hyoscyamus niger*) | |

* Three cycles are sufficient to induce flowering, but the percentage of nodes flowering in creases with further cycles up to 7 or more.

Twelve cycles are sufficient for induction, but longer exposure to short photoperiods is necessary to prevent axillary buds from resuming vegetative growth on transfer to long photoperiods.

Martin et al. (1976) presented a more extensive listing of the photoperiodic behaviour of agronomic crops. The practical use of day length has its economic value and is evident in the florist industry. For example species of poinsettia (*Euphorbia pulcherimma*) and chrysanthemum (SDP) are covered with black cloth during part of the day to exclude light and thereby induce flowering in mid-summer. For flowering in carnation (*Danthus caryophyllus*) and coneflower (*Rudbeckia bicolor*) normal daylength is either extended with artificial lighting, or the dark period interrupted with a light break.

Commercial growers of field crops and vegetables must take light period into account for selection of cultivars.

## FLOWERING AND ENVIRONMENT

### Geographical Latitude and Flowering

Many ecological problems are linked with photoperiodism, since day length depends on geographic latitude. Towards the poles the days are longer during summer and shorter during winter. Near the equator days and nights are always of almost equal duration. During evolution many plant species have adapted their photoperiodic flowering to these different day lengths. Therefore near the equator we find almost exclusively SDP, while at latitudes 60° North or South, one finds predominantly LDP. Some soybean varieties depend on a specific day length to such an extent that they grow optimally only within a zone of about 80 km width. This effect also indicates that genetically determined, endogenous rhythms control the band width for measurement of exogenous factors, especially light. It is largely unknown which primary signal decides when the measurement is initiated.

### Flowering and Light Spectrum (Light quality)

Flowering is strongly influenced by light quality, which may be studied under laboratory conditions, although it is much less important under field conditions than is photoperiod. Red light (wavelength 660 nm) is as effective as white light in inhibiting flowering of SDP's when given in the middle of a long dark period, but far-red light (730 nm) has little or no effect. However, if a few minutes of far-red light are given immediately after the red light, the plants flower. The red/far red treatments can be repeated several times, flowering/no flowering depending upon the **last treatment** (red vs far-red). However, the ability of far-red to reverse the promotive effect of red declines with time, becoming nil after 1 to 90 min, depending upon species. Similar relationships hold for LDP's, except that flowering response is the opposite. For example, red light given in the middle of the long night promotes, rather than inhibits flowering.

These studies have laid the foundation for the discovery of **phytochrome**. The pigment that apparently plays a crucial role in photoperiodism. Many species are not as responsive as soybean to short interruptions of the dark period. In strawberry, the period of the light interruption should be more than 1 hour to inhibit flowering completely. Flowering is inhibited by as little as a few seconds to a few minutes of exposure to light during the long dark period in soybean and other sensitive species such as cocklebur (*Xanthium strumarium*) and Japanese morning glory (*Pharbitis nil*).

Very low levels of light exposure induces photoperiodic responses. It has been studied in cocklebur where flowering may be inhibited by as little as 0.02 to 0.02 m W $m^{-2}$ $nm^{-1}$ of red light during the inductive dark period. (For comparison, a full moon provides about 0.002 m W $m^{-2}$ $nm^{-1}$ at 660 nm). However, much higher irradiances- on the order of 10 m W $cm^{-2}$ are necessary during the light period for photosynthesis and optimum flowering.

Under laboratory conditions a few SDP's initiate flowering when grown in total darkness, provided they are supplied with an energy source, such as sucrose. Photoperiods between four and 14 to 15 hours initiate flowering in soybean, therefore, it is proposed that within limits, vigour is proportional to irradiance.

## Artificial Shading and Flowering

Artificial shading reduces flower production drastically. When light level was reduced to 37, 25, and 11% of full sunlight, flowering was reduced 40, 56, and 79%, respectively, in comparison with unshaded plants of wheat, carnation, apricot (*Prunus armeniaca*), blueberyy (*Vaccinium corymbosum*), peach (*Prunus persica* Batsch.), pear (*Pyrus communis*), grape (*Vitis vinifera*), and pine (*Pinus* spp.). However, just opposite to these observations short-term shading can reduce flowering, mature plants of cacao (*Theobroma cacao*), vanilla (*Vanilla planifolia* Andr), and coffee (*Coffea arabica*), are intolerant of high light levels. Day light exposure (full sun) causes leaf abscission, and death of twigs results into reduced yields. Banana (*Musa acuminata* Colla) or plantain (*Musa paradisiaca*) trees are often used for shading coffee plantations in the tropics.

## Flowering and Endogenous Rhythms

Flowering in SDP's is based on the hypotheses: (a) light initiates the photoperiodic reaction, (b) the length of time in darkness is measured, and (c) the message is relayed from the leaf to the meristem where flowering actually occurs. The explanation most often advanced for the first mechanism is based on the observation that regular oscillations or **endogenous rhythms** occur in plants and other organisms. For example, movement of leaves of many plants following an approximate 24-hour (*circadian*, from the Latin *circa* = approximate, and *dies* =

day) cycle, and time of illumination can alter this pattern. Students of the flowering process, therefore, have proposed that the effect of prior illumination on the plant's response to a critical dark period varies, depending on the timing of the exposure relative to an endogenous rhythm. Furthermore, both the **light on** or **dawn** signal and the **light off** or **dusk** signal after response, and the two must be properly timed with respect to one another. Although this explanation may seem fanciful, further analysis of the process may prove it to be correct.

## Availability of Water and Flowering

Influence of availability of water affects yields of grains, although effects on flowering per se are not as well established. Non-availability of moisture or deficits moisture prolong the time of flower primordia formation in barley and grain sorghum. If the deficit is prolonged, spikelet numbers can be reduced. Sorghum is relatively tolerant of moisture stress, and development resumes on rewatering, although anthesis may be delayed for 10 days or more. In rice, deficit moisture reduces both number of spikelets per panicle and spikelet fertility; greatest effect on fertility is observed when treatment is applied during the 2 week period prior to heading. Sugarcane is grown for its stems, rather than its fruit. However, flowering inhibits growth and reduces sugar content, but can be prevented by imposing a mild moisture stress in August-September in Hawaii.

## Carbon: Nitrogen Ratio

Nutrition is considered as one of the controlling factors of flowering. High ratio of carbohydrate to inorganic nutrients, particularly nitrogen (C/N ratio), favoured flowering. Heavy applications of nitrogen (= low C/N ratio in the plant) delayed flowering in young fruit trees, while old, weak trees (having a high C/N ratio) often flower profusely. However, there appears to be no correlation between critical C/N ratio, similar to a critical photoperiod, for flowering. The effect of nitrogen status varies with species in photoperiodically induced plants, nitrogen deficiency often promoting flowering in LDP's, while delaying it in SDP's.

Nitrogen may affect flower development following induction. The application of nitrogen stimulates production of **female flowers** (strobili) in pine and tung (*Aleurites fordii*), but has little or no effect on production of male strobili. In Application of Nitrogen in Douglas-fir (*Pseudotsuga menziesii*) increases the production of both male and female strobili. Nitrogen deficiency reduces the numbers of spikelets and fertile florets in wheat and rice. Nitrogen also increases the numbers of spikelets per panicle in rice when applied 3 to 8 weeks prior to heading.

## Temperature

Low temperatures (0 to 10°C) exposure for several days to several weeks promote flowering in many species and with respect to photoperiodism, there are qualitative and quantitative responses depending upon species and strain. Cold treatment (0 to 10°C) promote flowering and the phenomenon is called **vernalization** (literally, *springization*). Cold temperature treatment is a requirement for a large group of herbaceous plants such as biennials and including carrots (*Daucus carota*). These plants have an absolute requirement for vernalization. They normally do not flower until the second year of growth, being induced to flower by the low temperatures of water between first and second growing seasons.

The effect of low temperatures on biennial plants often creates problems for the farmer. If celery, grown for its petioles, or beets, grown for their roots, flower prematurely, the crop is often worthless. Entire celery fields have been known to flower when temperatures of 5 to 10°C have prevailed for 10 or more days in the early part of the season.

The requirements for vernalization often differs within a cultivars of species. Winter cultivars of wheat and barley tiller freely without vernalization, but will not form heads unless imbedded seeds or young plants are subjected to temperatures of 3 to 10° C for 2 to 10 weeks. Plants older than 3 months do not respond. Spring cultivars do not require vernalization. Some bulb crops require chilling, not for flower induction, but for stimulating subsequent elongation of the inflorescence. Flowering of many temperate zone trees is likewise dependent on adequate chilling to break bud dormancy.

### Temperature and Daylength

Interactions between temperature and photoperiod has been established. The temperature effects upon the plant's response to daylength. In Petkus winter rye flowering does not occur under SD, regardless of germination temperature. When grown under LD, plants developing from seeds germinated at 18°C require 20 weeks to flower, but those from seeds germinated at 1°C require only 9 weeks. The requirement for cold treatment is quantitative in this case, rather than qualitative, in that lack of cold only delays rather than prevents flowering under LD. Short photoperiods can replace vernalization in some wheat cultivars, provided plants are subsequently transferred to long days.

One can observe interesting interaction between temperature and photoperiod in a **single** species of morning glory (*Ipomoea purpurea*)- which is a LDP at low temperature (13°C), a SDP at high temperature (21 to 24°C) and day-neutral at intermediate temperature (17 to 18.5°C).

### High Temperature and Flowering

In some species, high temperatures are required for flower induction although the requirement is generally quantitative one. Here also temperature interacts with photoperiod. High temperatures also hasten flowering in the short-day species such as in winter rice and cosmos (*Cosmos bipinnatus* Cav. Ann.), and in phlox (*Phlox drummondii* Hook)- a LPD. On the other hand, camellias (*Camellia japonica*) exhibit an absolute requirement for high temperature and a quantitative requirement for long days.

High temperatures initiate and also determine the time of flowering. **Heat units** are required from sowing to heading in wheat and rye. Head units are calculated by determining the mean temperature for each day, subtracting a base temperature (usually 40°F or 5°C), and accumulating the remainders (**degree-days**) for each day from sowing (spring wheat) or 1 March (winter wheat) to heading. In the USA as a whole, spring wheat requires an average of 1142 units from sowing to heading, whereas winter wheat requires 978 from 1 March to heading.

## PHOTOPERIOD AND FLOWERING

As we have seen, the circadian clock Time of day (photoperiodism) enables organisms to determine the particular molecular of biochemical event in a flower. The ability of an organism to detect day length, makes it possible for an event to occur at a particular *time of year*, thus allowing for a *seasonal* response. Both photoperiodism and circadian rhythms have the common property of responding to cycles of light and darkness.

Photoperiodic phenomena are found in both animals and plants. In the animal kingdom, day length controls such seasonal activities as hibernation, development of summer or winter coats, and reproductive activity. Plant responses controlled by day length are numerous, including the initiation of flowering, asexual reproduction, the formation of storage organs, and the onset of dormancy (Vince-Prue 1975). Perhaps all plant photoperiodic responses utilize the same photoreceptors, with subsequent specific signal transduction pathways regulating different responses. Since it is clear that monitoring the passage of time is essential to all photoperiodic responses, a timekeeping mechanism must underlie both the time-of-year and the time-of-day responses. The circadian oscillator is thought to provide an autonomous time-measuring mechanism that serves as a reference point for the response to incoming light (or dark) signals from the environment.

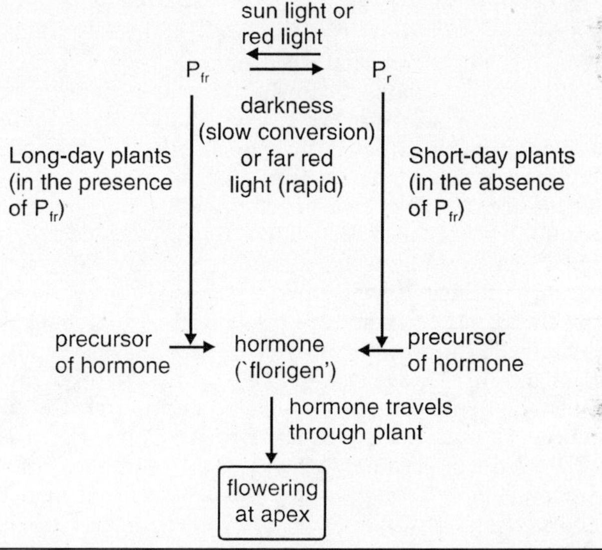


**FIG. 28.4:** Hypothetical scheme summarizing the photoperiodic control of flowering.

Although we now take for granted the ability of plants to detect seasonal change by measuring day length, this was a revolutionary concept when it was first proposed in the 1920s. In this section we will briefly describe the discovery of photoperiodism and the early physiological studies, which showed the following aspects:

1. That night length is more important than day length for the response.
2. That a circadian rhythm is part of the photoperiodic mechanism.
3. That phytochrome is the photoreceptor for many photoperiodic phenomena.
4. That the blue light receptor, cryptochrome, also appears to be involved.

## CLASSIFICATION OF PHOTOPERIODIC RESPONSES

Many plant species flower during the long days of summer, and for many years plant physiologists believed that the correlation between long days and flowering was a consequence of the accumulation of photosynthetic products synthesized during long days. This hypothesis was shown to be incorrect by the work of Wightman Garner and Henry Allard, conducted in the 1920s at the U.S. Department of Agriculture laboratories in Beltsville, Maryland. They found that a mutant variety of tobacco, Maryland Mammoth, grew profusely to about 5 m in height, but it failed to flower in the prevailing conditions of summer. However, the plants flowered in the greenhouse during the winter.

These results ultimately led Garner and Allard to test the effect of artificially providing short days by covering plants grown during the long days of summer with a light-tight tent late in the afternoon. These artificial short days also caused the plants to flower. This requirement for short days was difficult to reconcile with the idea that longer periods of radiation and the resulting increase in photosynthesis promote flowering. Garner and Allard concluded that the length of the day was the determining factor in flowering and were able to confirm this hypothesis in many different species and conditions. This work laid the foundations for the extensive subsequent research on photoperiodic responses.

The classification of plants according to their photoperiodic responses is usually made on the basis of flowering, even though many other aspects of their development may also be affected by day length. The two main photoperiodic response categories are **short-day plants (SDPs)**, in which flowering occurs only in short days (*qualitative* SDP) or is accelerated by short days (*quantitative* SDP), and **long-day plants (LDPs)** in which flowering occurs only in long days (*qualitative* LDP). A few plants have more specialized day length requirements. **Intermediate-day plants** flower only between narrow day length limits (for example between 12 and 14 hours in one variety of sugarcane). Plants in another special category, called **ambiphotoperiodic** species, flower in long days or short days but not at intermediate day lengths. The flowering of many species is not regulated by day length; these plants are called **day-neutral plants**.

The essential distinction between long-day and short-day plants is that flowering in LDPs is promoted only when the day length *exceeds* a certain duration in every 24-hour cycle, whereas promotion of flowering in SDPs requires a day length that is *less than* a critical value. The value of the critical day length varies widely between species and only when flowering is examined for a range of day lengths can the correct photoperiodic classification be established.

Long-day plants can effectively measure the length-ening days of spring or early summer and delay flowering until the critical day length is reached. Many varieties of wheat (*Triticum aestivum*) behave in this way. SDPs often flower in fall, when the days shorten below the critical day length, as in many varieties of *Chrysanthemum morifolium*. However, day length alone is an ambiguous signal because it cannot distinguish between spring and fall. Plants exhibit several strategies to avoid this ambiguity. One is the coupling of a temperature requirement to a photoperiodic response. Certain plant species, such as winter wheat, do not respond to photoperiod until after vernalization (overwintering) has occurred. (We will discuss vernalization a little later in the chapter.) Another example is found in some short-day varieties of strawberry (*Fragaria x ananassa*), in which flowers are initiated when the critical day length is reached in fall but do not emerge until spring. A different strategy for avoiding seasonal ambiguity would be a mechanism for distinguishing between *lengthening* and *shortening* days. Some animals appear to have this capability, but such a mechanism is not known to operate in plants.

## DAY LENGTH BY MEASURING THE LENGTH OF THE NIGHT

Under natural conditions, day and night lengths configure a 24-hour cycle of light and darkness. A plant thus could perceive a critical day length by measuring the duration of either light or darkness. Much experimental work in the early studies of photoperiodism was devoted to establishing which part of the light-dark cycle is the controlling factor in flowering. Results showed that flowering of SDPs is determined primarily by the duration of darkness. Thus, it was possible to induce flowering in SDPs with light periods longer than the critical value, provided that these were followed by sufficiently long nights. Similarly, SDPs did not flower when short days were followed by short nights. More detailed experiments demonstrated that photoperiodic timekeeping in SDPs is a matter of measuring the duration of darkness (Hamner and Bonner 1938; Hamner 1940). For example, the plants flowered only when the dark period exceeded 8.5 hours in cocklebur (*Xanthium strumarium*) or 10 hours in soybean (*Glycine max*). The duration of darkness was also shown to be important in LDPs.

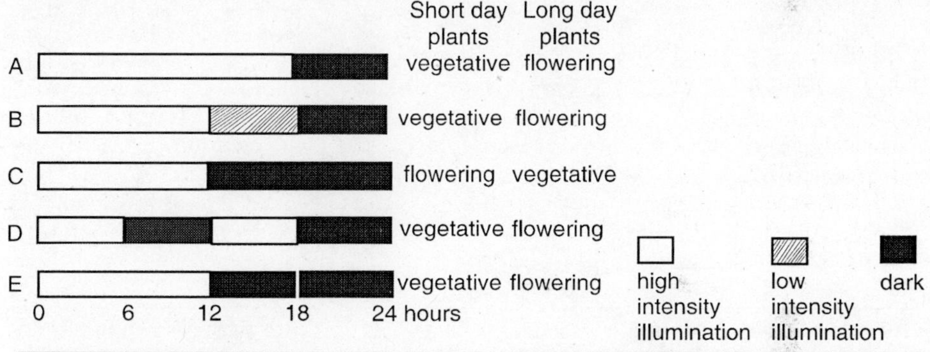


**FIG. 28.4A:** Effect of breaks of light during the dark period on flowering in LDP and SDP. Critical dark period is also marked out.

These plants were found to flower in short days, provided that the accompanying night length was also short; however, long days followed by long nights were ineffective.

A feature that underlines the importance of the dark period is that it can be made ineffective by interruption with a short exposure to light, called a **night break**. In contrast, interrupting a long day with a brief dark period does not cancel the effect of the long day. Night break treatments of only a few minutes are effective in *preventing* flowering in many SDPs, including *Xanthium* and *Pharbitis*, but much longer exposures are often required to *promote* flowering in LDPs. In addition, the effect of a night break varies greatly according to the time when it is given. For both LDPs and SDPs, a night break was found to be most effective when given near the middle of a dark period of 16 hours. However, when considerably longer dark periods were given, it was established that the time of maximum sensitivity to a night break is related to the time from the beginning of darkness. In *Xanthium*, for example, the maximum effect of a night break still occurred about 8 hours after the beginning of a 24-hour dark period, even though the subsequent 16-hour period of unbroken darkness considerably exceeds the critical night length of 8.5 hours.

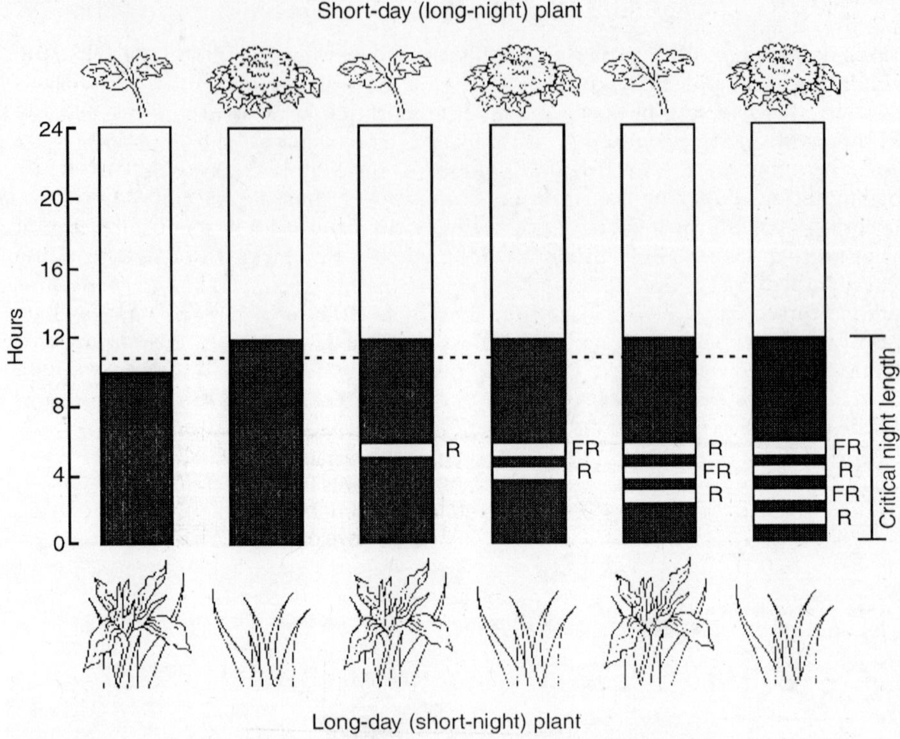


**FIG. 28.5:** Control of flowering by red (R) and far-red (FR) light. Red light is the most effective wavelength in night breaks. A flash of red light during the dark period induces flowering in an LDP, and the effect is reversed by a flash of far-red light. This response indicates the involvement of phytochrome. In SDPs, a flash of red light prevents flowering, and the effect is reversed by a flash of far-red light.

The discovery of the night break length of 8.5 hours. Dependence, had several important consequences. It established the central role of the dark period and provided a valuable probe for studying photoperiodic timekeeping. Because only small amounts of light are needed, it became possible to study the action and identity of the photoreceptor without the interfering effects of photosynthesis and other nonphotoperiodic phenomena. This discovery has also led to the development of commercial methods for regulating the time of flowering in crop plants such as chrysanthemum and poinsettia (*Euphorbia pulcherrima*).

## ENDOGENOUS OSCILLATOR AND PHOTOPERIODIC TIMEKEEPING

The decisive effect of night length on flowering indicates that measuring the passage of time in darkness is central to photoperiodic timekeeping. Two different hypotheses have been proposed to explain how plants measure night length. According to the **hourglass hypothesis**, time is measured by a unidirectional series of biochemical reactions that start at the beginning of the dark period. If not interrupted by light, this series of reactions would lead to induction of flowering in SDPs or to inhibition of flowering in LDPs. This type of mechanism can measure only one interval of time and, by analogy with an hourglass device, would be reset each day by the photoperiod.

Some of the characteristics of flowering induction are consistent with the accumulation or decay of biochemical products, as predicted by the hourglass hypothesis (Thomas and Vince-Prue 1984). However, most of the available evidence favors a different mechanism, called the **clock hypothesis** (Bunning 1960). This hypothesis proposes that photoperiodic timekeeping depends on the endogenous circadian oscillator of the type involved in the daily rhythms described earlier in the chapter. In this view, the role of the photoperiod is to set the phase of the rhythm in a way leading to the measurement of time in darkness and of the critical night length.

Measurements of the effect of the night of the night break on flowering may be used to investigate circadian rhythms in flowering and many studies have shown that those rhythms exist. For example, when soybean plants are kept under a long dark cycle after 8 hours of light, and the dark cycle is interrupted at different times by a 4-hour night break, the flowering response to the night break shows a circadian rhythm. Many other rhythmic responses of both LDPs and SDPs have been reported (Thomas and Vince-Prue 1984). Further evidence for the presence of a circadian oscillator is the observation that the photoperiodic response can be phase-shifted by a light treatment.

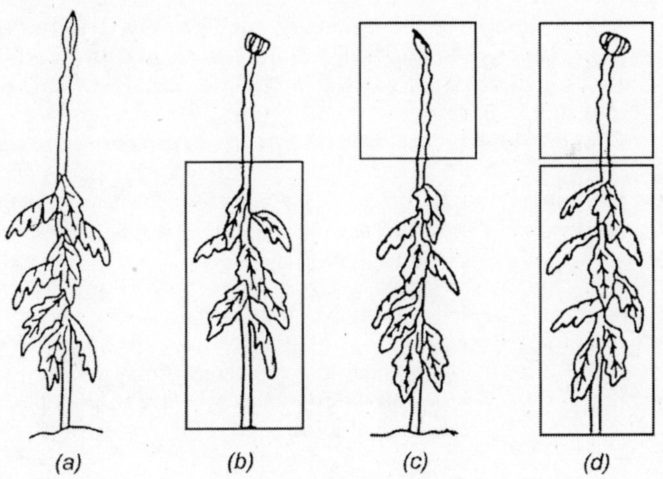


**FIG. 28.6:** Experiment demonstrating that the flowering of chrysanthemum (a short day plant) is induced by exposure of the leaves to short days (b and d), irrespective of the day length under which the apex is maintained.

The involvement of a circadian oscillator in photoperiodism poses an important question. How does an oscillation with a 24-hour periodicity measure a critical duration of darkness of, say, 8 to 9 hours, as in the SDP *Xanthium*? Erwin Bunning, working in Tubingen, Germany, proposed in 1936 that the control of flowering by photoperiodism is achieved by an oscillation of phases with different sensitivities to light. Bunning's hpothesis invokes two alternating phases, one requiring 12 hours of darkness, the other 12 hours of light (Bunning 1960).

According to Bunning's hypothesis, light has two distinct roles in the flowering process. Light signals at dawn and dusk, acting as zeitgebers, set the phase of the photoperiodic rhythm. In addition, the rhythm has a light-sensitive phase called the **inducible phase**. When a light signal is received during the inducible phase of the rhythm, the effect is to promote or prevent floral induction. Thus, in nature the phase the phase of the rhythm is initially set by the light-on signal at dawn. The rhythm then continues to run until it reaches a specific phase point at which it is suspended while the plant remains in the light. At the transition to darkness at dusk, the rhythm is released and runs freely. This entrainment at dusk ensures that the rhythm is released at a specific phase point and that the inducible phase always occurs at a constant real time after transfer to darkness. In SDPs, exposure to light prevents flowering, so flowering would be induced only when dawn occurs after the completion of the inducible phase of the rhythm. In this way, the photoperiod establishes the time of sensitivity to light in the subsequent dark period and sets the conditions for the measurement of the critical night length.

## PERCEPTION OF PHOTOPERIODIC STIMULUS

Leaf is the site of photoperiodic stimulus. One well-established fact about the photoperiodic stimulus is that it is perceived by leaves. Treatment of a single leaf of the SDP *Xanthium* with short photoperiods is sufficient to cause the formation of macroscopically visible flowers, even when the rest of the plant is exposed to long days (Hamner and Bonner 1938). For both LDPs and SDPs, many experiments have confirmed that the photoperiod experienced by the leaves determines the response at the apex. Thus, in response to photoperiod the leaf transmits an unknown signal that regulates the transition to flowering at the shoot apex. The photoperiod-regulated processes that occur in the leaves resulting in the transmission of a floral stimulus to the shoot apex are referred to collectively as **photoperiodic induction**.

Photoperiodic induction can also take place in a leaf that has been separated from the plant. In the SDP *Perilla crispa* (red-leaved perilla), Jan Zeevaart at the Department of Energy Plant Research Lab at Michigan State University has shown that an excised leaf exposed to short days can cause flowering when subsequently grafted to a noninduced plant maintained in long days (Zeevaart and Boyer 1987). Thus, photoperiodic induction appears to depend on events that take place in the leaf. When several photoperiodic cycles are required, these must be given to the same leaf. The cumulative effect of repeated cycles appears to take place in the leaf and not to be a consequence of the accumulation of a metabolite at the apex.

Leaves have also been shown to play an important role in plants in which flowering is under autonomous control. For example, grafting a single leaf of Agate, a day-neutral variety of soybean, caused flowering of the short-day variety Biloxi when the latter was maintained in noninductive long days (Heinze et al. 1942). Thus flowering in day-neutral plants may be regulated by the same transmissible signals that control flowering in photoperiod-responsive plants.

The flowering stimulus appears to be transported via the phloem and to be chemical, rather than physical, in nature. For example, grafted leaves cause flowering only when a functional phloem connection is established between the donor and the receptor plants. Treatments that restrict phloem transport, such as girdling or localized heating, prevent movement of the floral signal.

Grafting studies have also demonstrated that in certain situations transmissible *inhibitors* of flowering are produced. For example, when a day-neutral variety of tobacco is grafted to an LDP tobacco species, *Nicotiana sylvestris*, and the graft is maintained in short days, flowering of the day-neutral partner is strongly inhibited. Conversely, when this graft is maintained in long days, flowering of the day-neutral partner is promoted. Thus *N. sylvestris* is capable of transmitting both flowering promoters and flowering inhibitors via a graft junction (Lang et al. 1977). Similar studies in peas have led to the identification of several genetic loci that regulate steps in the biosynthetic pathways of both floral activators and floral inhibitors.

It is possible to measure rates of transport of the flowering stimulus by removing a leaf at different times after induction, and comparing the time which it takes for the signal to reach two buds located at different distances from the induced leaf. The rationale for this type of measurement is that a threshold amount of the signaling compound has reached the bud when flowering takes place despite the removal of the leaf. Studies using this method (Evans and Wardlaw 1966; King et al. 1968) have shown that the rate of transport of the flowering signal is comparable to, or somewhat slower than, the rate of translocation of sugars in the phloem. These rates of transport are consistent with a chemical message and make an electric signal unlikely.

## FLOWERING AND HOUR-GLASS HYPOTHESIS

One of the original hypotheses to explain time measurement, or how the plant senses when the dark period has exceeded a critical length, was that this period represented the time required for conversion of $P_{fr}$ to $P_r$ (the **hour-glass hypothesis**). Thus flower induction (or inhibition of induction in the case of LDP's) could proceed only after the level of $P_{fr}$ fell below a critical level. But indirect evidence indicates that this conversion is complete a short time after the dark period begins. If this is true, some other mechanism must measure time. For example, the ratio of $P_{fr}$ to $P_r$ could be the determining factor in **starting the timer** in an oscillating system dependent on endogenous rhythms. Provided at least 6 hours have elapsed in darkness irradiation with red light will set the clock back to zero, while subsequent irradiation with far red restores the clock to its position prior to irradiation.

The processes in LDP's, while similar to those occurring in SDP's, are not as well-defined.

## CIRCADIAN RHYTHMS

It is the clock within the plant. The topic of measurement of time period in living organism is known as **chronobiology**, or the study of **biological clocks**. The best understood biological clock is the circadian rhythm.

Organisms are normally subjected to daily cycles of light and darkness, and both plants and animals often exhibit rhythmic behavior in association with these changes (Brady 1982; Sweeney 1987). Examples of such rhythms include leaf and petal movements (day and night positions), stomatal opening and closing, growth and sporulation patterns in fungi (e.g., *Pilobolus* and *Neurospora*), time of day of pupal emergence (the fruit fly *Drosophila*), and activity cycles in rodents, as well as metabolic processes such as photosynthetic capacity and respiration rate.

When organisms are transferred from daily light-dark cycles to continuous darkness (or continuous dim light), many of these rhythms continue to be expressed, at least for several days. Under such uniform conditions the periodicity of the rhythm is then close to 24 hours, and consequently the term **circadian** (from the Latin for "about one day") **rhythm** is applied. Because they continue in a constant light or dark environment, these circadian rhythms cannot be direct responses to the presence or absence of light but must be based on an endogenous pacemaker that is self-sustaining. This pacemaker is often called an **endogenous oscillator**. The endogenous oscillator is coupled to a process, such as leaf movement or photosynthesis, and maintains the rhythm. For this reason the endogenous oscillator is considered to represent the clock itself, while the physiological functions that are being regulated, such as leaf movements or photosynthesis, can be thought of as the *hands* of the clock.

## Circadian Rhythms: Characteristic Features

There are several characteristic features circadian rhythms in plants. Rhythms derive from cyclic phenomena, and the pattern repeats over and over. The time between comparable points in the repeating cycle is known as the **period**. The term **phase** is used for any point in the cycle recognizable by its relationship to the rest of the cycle. The most obvious phase points are the maximum (peak) and minimum (trough) positions. The **amplitude** of a biological rhythm is usually considered to be the distance between peak and trough, and it can often vary while the period remains unchanged.

In constant light or darkness, rhythms depart from an exact 24-hour periodicity. The rhythms then drift in relation to solar time, either gaining or losing time depending on whether the period is shorter or longer than 24 hours. Under natural conditions, the endogenous oscillator is entrained (synchronized) to a true 24-hour periodicity by environmental signals, the most important of which are the light-to-dark transition at dusk and the dark-to-light transition at dawn. Such signals are termed **zeitegebers** (from the German for *time giver*). When the entraining signals are removed-for example, by transfer to continuous darkness-the rhythm is said to be **free-running** and reverts to the circadian period that is characteristic of the particular organism.

Although the rhythms are innate, they normally require an environmental signal, such as exposure to light or a change in temperature, to initiate their expression. In addition, many rhythms damp out (i.e., the amplitude decreases) when the organism is in a constant environment for some time and then require an environmental zeitgeber, such as a transfer from light to dark or a change in temperature, to restart them. It is important to note that the clock itself does not damp out; only the coupling between the clock (endogenous oscillator) and the hands of the clock (physiological function) is affected. Temperature has little or no effect on the periodicity of the free-running rhythm. This insensitivity to temperature is an essential characteristic; the clock would be of no value if it could not keep accurate time under the fluctuating temperatures experienced in natural conditions.

## Phase Shifting and Day-Night Cycles

Phase Shifting sdjusts circadian rhythms to various Day-Night Cycles in plants. In circadian rhythms, the operation of the endogenous oscillator sets a response to occur at a particular time of day. For example, the unicellular alga *Gonyaulax polyedra* generates its maximum "glowing" bioluminescence toward the end of a 12-hour night, and its maximum "flashing" bioluminescence near the middle. How do such responses remain on time when the daily durations of light and darkness are changing? The answer to this question lies in the fact that the phase of the rhythm can be changed if the whole cycle is moved forward or backward in time without its period being altered.

As an example, we can take a phase point that occurs 1 hour after the onset of darkness in cycles of 12 hours light and 12 hours dark. If daylight is extended for 1 hour, this phase point must have drifted forward relative to the dusk zeitgeber, and the appropriate response would be a correcting backward shift (or phase *delay*) such that this phase point still occurs 1 hour after the end of the light period and the rhythm stays on local time. Similarly, the appropriate answer to light perceived at a phase point that would normally occur 1 hour before dawn would be a phase *advance*. Adaptively, therefore, different *directions* of phase shifting are to be expected in response to light signals given at different phase points in the circadian cycle, and experimentally that is what is observed.

Investigators test the response of the endogenous oscillator usually by placing the organism in continuous darkness and examining the response to a short pulse of light (usually less than 1 hour) given at different phase points in the free-running rhythm. When an organisms is entrained to a cycle of 12 hours light and 12 hours dark and then allowed to free-run in darkness, the phase of the rhythm that coincides with the light period of the previous entraining cycle is called the **subjective day** and the phase that coincides with the dark period is called the **subjective night**. If a light pulse is given during the first few hours of the subjective night, the rhythm is delayed; in contrast, a light pulse given toward the end of the subjective night advances the phase of the rhythm. As already pointed out, this is precisely the pattern of response that would be expected if the rhythm is to stay on local time. Therefore, these phase-shifitng responses enable the rhythm to be entrained to cycles with different durations of light and darkness, and they demonstrated that the rhythm will run differently under different natural conditions of day length. It has been demonstrated that different photoperiodic rhythms show a very similar pattern of phase shifting when tested in the same way.

The physiological mechanism whereby a light signal causes phase shifting is not yet known. However, the low levels and specific wavelengths of light that have an effect indicate that the light response must be mediated by a photoreceptor and not by rates of photosynthesis. The role of phytochrome in phase shifting has been discussed in next chapter. Light reception by phytochrome in the rhythmic leaf movements in *Samanea*, a semitropical leguminous tree, is well characterized (Satter and Galston 1981). Phytochrome is also involved in the phase shifting of photoperiodic rhythms. For example, the circadian rhythm in the flowering response of seedlings of Japanese morning glory (*Pharbitis nil*) can be phase-shifted by exposure to a few seconds of red light (Lumsden et al. 1986). On the other hand, phytochrome does not appear to be the universal photoreceptor for phase shifting, since it is not involved in the circadian rhythm of bioluminescence in *Gonyaulax*.

## ISOLATION OF CIRCADIAN CLOCK MUTANTS IN ARABIDOPSIS

The isolation of clock mutants has been an important tool for the identification of clock genes in other organisms, such as *Neurospora* and *Drosophila*. Isolating clock mutants in plants requires a convenient assay that allows monitoring of the circadian rhythms of many thousands of individual plants to detect the rare aberrant phenotype. This requirement has proven to be difficult because plants that exhibit externally visible circadian rhythms, such as nyctinastic leaf movements, are not suitable for molecular genetic analysis. In *Arabidopsis*, which is well suited for molecular genetic approaches, nyctinastic leaf movements can be measured, but they are more subtle and are difficult to monitor on a mass scale. Circadian rhythms in the expression of certain genes, such as the gene that encodes the chlorophyll *a/b*-binding protein (*CAB* or *LHCB*), have been characterized in *Arabidopsis* (Millar and Kay 1991).

To allow screening for clock mutants in *Arabidopsis*, the promoter region of the *LHCB* gene was fused to the gene that encodes luciferase, an enzyme that emits light in the presence of its substrate, luciferin. This reporter gene construct was then used to transform *Arabidopsis* via the Ti plasmid of *Agrobacterium* as a vector. Using a video system, investigators were able to monitor the temporal and spatial regulation of bioluminescence in individual seedlings in real time (Millar et al. 1992). Using bioluminescence as the circadian phenotype, Millar and colleagues (1995) were able to isolate 21 independent **toc** (timing of CAB expression) mutants, including both short-period and long-period lines. The *toc* mutants hold great promise for helping us identify the genes that regulate circadian rhythms in plants. As we will discuss later in the chapter, the luciferase imaging technique has also been useful for studying the role of blue light in the control of flowering.

## PHYTOCHROME AND FLOWERING

Phytochrome is the pigment that receives the photoperiodic signal and, therefore, plays a key role in flowering, as well as many other processes. Its structure and properties change with the wavelength of incident radiation. The isolation of phytochrome (Gr. *Phytos*, plant, and *chroma*, a colour) by Sterling B. Hendricks and Harry Borthwick was preceded by extensive investigations into the effects of different wavelengths of light on physiological processes that were light-dependent, including flowering of both SDP's and LDP's and germination of lettuce and other seed. The studies were conducted by using a large spectrograph for separating light into its various components by means of a prism, and partially defoliated plants were exposed to the various wavelengths. Flowering response was plotted as a function of incident energy at each wavelength. It was observed that the wavelengths most active in promoting the germination of lettuce seeds at 660 nm (red light) were also maximally effective in inhibiting the flowering of SDP's, and promoting the flowering of LDP's, when given in the middle of a long night. It was interesting to note that far-red light (730 nm) given immediately after the red nullified its effects; that is, lettuce seeds did not germinate, SDP's flowered, and LDP's did not flower.

These results indicated the presence of a **pigment** in plants which can be held responsible for all of these phenomena, and was assumed to exist in two forms, one ($P_r$) predominantly absorbing red light, the other ($P_{fr}$) predominantly absorbing far-red light. The $P_r$ was converted to $P_{fr}$ on absorption of red light; $P_{fr}$ was converted back to $P_r$ on absorption of far-red light. In many plants $P_{fr}$ converts to $P_r$ also during long periods of darkness. This conversion of $P_{fr}$ to $P_r$ was considered to be one of the time-measuring reactions in photoperiodism, with LDP's requiring certain relatively high levels of $P_{fr}$ for flowering, while SDP's required relatively low levels. This concept was an oversimplification, however, the present status is:

a) $P_{fr}$ may be destroyed, as well as being converted to $P_r$,
b) the $P_r/P_{fr}$ ratio may be more important than are absolute levels of their, and
c) other time-measuring reactions are far more important than is conversion of $P_{fr}$ to $P_r$.

**TABLE 28.4:** Inhibition of flowering in cocklebur by a 2 min exposure to red light (R) in the middle of an inductive dark period, and the effect of subsequent 2 min exposures to far-red (FR), or alternating R and FR, light in counteracting the inhibition. Note that the effect of far-red declines with time. (From Downs, 1956).

| Exposure | Floral stage+ |
|---|---|
| Control - no night-break | 6.0 |
| R | 0.0 |
| R-FR | 5.6 |
| R-FR-R | 0.0 |
| R-FR-R-FR | 4.2 |
| R-FR-R-FR-R | 0.0 |
| R-FR-R-FR-R-FR | 2.4 |
| R-FR-R-FR-R-FR-R | 0.0 |
| R-FR-R-FR-R-FR-R-FR | 0.6 |

+0 = Vegetative; 8 = maximum flowering response.

**TABLE 28.5:** Effect of length of interval between exposure to red and far-red light during inductive dark period on flowering of cocklebur. (From Down, 1956.)

| Treatment | Floral stage+ |
|---|---|
| Control-no interruption | 7.0 |
| Control-red without far-red | 0.0 |
| Red, followed by far-red after (min) | |
| 0 | 6.5 |
| 20 | 3.8 |
| 30 | 1.8 |
| 40 | 0.5 |
| 60 | 0.0 |

+0 = Vegetative; 8 = maximum flowering response.

The process as a whole in cocklebur may be separated into 3 more sequences.

i) The light period preceding the critical dark period.
ii) The critical dark period itself.
iii) The subsequent light period.

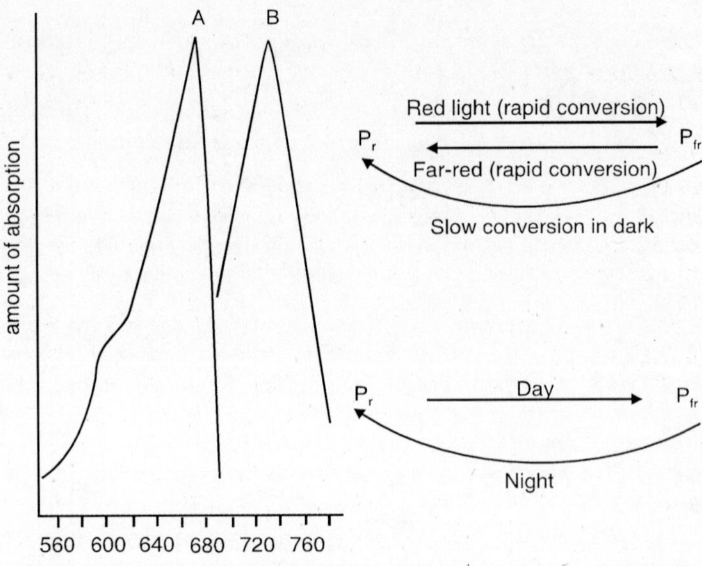


**FIG. 28.7:** Absorption of light by [hytochrome. (A) Red absorbing form. (B) Far-red absorbing form. **A** has its maximum absorption at 665 nm wavelength, where **B** has its maximum absorption at 725 nm. (C) Interconversion of the forms of phytochrome ($P_r$ and $P_{fr}$) in response to light of different wavelengths and in dark.

## PHYTOCHROME AND PHOTOPERIODISM

Phytochrome is the main photoreceptor in photoperiodism. Night break experiments are well suited for studying the nature of the photoreceptors involved in the reception of light signals during the photoperiodic response. The inhibition of flowering in SDPs by night breaks was one of the first physiological processes shown to be under the control of phytochrome (Borthwick et al., 1952). In many SDPs, a night break becomes effective only when the supplied dose of light is sufficient to saturate the photoconversion of Pr (phytochrome that absorbs red light). A subsequent exposure to far-red light, which photoconverts the pigment back to the physiologically inactive Pr form, restores the flowering response (Downs 1956). Red and far-red reversibility has also been demonstrated in some LDPs, in which a night break of red light promoted flowering and a subsequent exposure to far-red light prevented this response.

Action spectra for the inhibition and restoration of the flowering response in SDPs reveals that a peak at 660 nm, the absorption maximum of Pr, is obtained when dark-grown *Pharbitis* seedlings are used to avoid interference from chlorophyll. In contrast, the spectra for *Xanthium* provide an example of the response in green plants, in which the presence of chlorophyll can cause some discrepancy between the action spectrum and the absorption spectrum of Pr (Vince-Prue and Lumsden 1987). These action spectra and the reversibility between red light and far-red light confirm the role of phytochrome as the photoreceptor that is involved in photoperiod measurement in SDPS. In LDPs the role of phytochrome is more complex, and a blue-light photoreceptor (which we will discuss shortly) also plays a role in controlling flowering.

## FAR RED LIGHT AND MODIFICATION OF FLOWERING IN SOME LDPS

Circadian rhythms have also been found in LDPs. A circadian periodicity in the promotion of flowering by far red light has been observed in barley (*Hordeum vulgare*) (Deitzer et al. 1979) and in darnel ryegrass (*Lolium temulentum*). In both cases, when the plant is exposed to far-red light for 4 to 6 hours, flowering is promoted compared with plants maintained under continuous white or red light. The rhythm continues to run in the light. In SDPs, on the other hand, an essential feature of the circadian timing mechanism seems to be that the rhythm is suspended after a few hours in continuous light and released, or restarted, upon transfer to darkness. The response to far-red light is not the only rhythmic feature in LDPs. Although relatively insensitive to a night break of only a few minutes, many LDPs can be induced to flower with a longer night break, usually of at least 1 hour. A circadian oscillation in the flowering response to such a long night break has been observed in LDPs, showing that a rhythm of responsiveness to light continues to run in darkness.

Thus, circadian rhythms that modify the flowering response in LDPs have been shown to run both in the light (promotion by far-red light) and in the dark (promotion by red or white light). However, we do not yet know how these rhythms affect the photoperiodic response.

## PHYTOCHROMES A AND B

Phytochromes A and B have contrasting effects on flowering. Phytochrome is encoded by members of a multigene family, and it is possible that the effects of red and far-ed light described in the previous section are mediated by different types of phytochrome (A and B). In LDPs such as *Arabidopsis*, pea (*Pisum sativum*), and sorghum (*Sorghum bicolor*), phytochrome B (phyB) appears to be an *inhibitor* of flowering, since mutations in the gene *PHYB* that eliminate or reduce the amount of the phyB protein cause more rapid flowering (Childs et al. 1992; Reed et al. 1994).

The promotion of flowering by far-red light may result from a decrease in the amount of the Pfr form of phyB, which would have the same effect on the *phyB* Pfr levels as the mutations that reduce the total amount of *phyB*. As already discussed, red-light interruptions of a dark period prevent flowering in SDPs; therefore, the Pfr form of *phyB* may be an *inhibitor* of flowering in SDPs as well. In the LDPs *Arabidopsis* and pea, mutations in the *PHYA* gene interfere with the promotion of flowering by long days (Reed et al. 1994; Weller et al. 1997); thus the Pfr form of phyA may promote flowering in LDPs. The effect of the *phyA* mutations is quite strong in pea (the mutant plants are essentially unable to respond to inductive photoperiods) whereas the *phyA* mutation in *Arabidopsis* has only a slight effect on the ability to respond to inductive photoperiods. The species difference in the effect of *phyA* mutations is probably due to the fact that pea relies entirely on phyA to sense inductive photoperiods, whereas *Arabidopsis* also uses a blue light photoreceptor for this purpose.

## BLUE-LIGHT PHOTORECEPTOR AND PHOTOPERIODISM

In some LDPs, blue light can promote flowering, suggesting the possible participation of a blue-light photoreceptor in the control of flowering. The role of blue light in flowering and its relationship to circadian rhythms has been investigated by use of the luciferase reporter gene construct described earlier. In continuous white light, the cyclic luminescence has a period of 24.7 hours, but in constant darkness the period lengthens to 30 to 36 hours. Both red and blue light, given individually, shorten the period to 25 hours.

To distinguish between the effects of phytochrome and a blue-light photoreceptor, researchers transformed phytochrome-deficient *hy1* mutants, which are defective in chromophore synthesis and are therefore deficient in *all* phytochromes, with the luciferase construct to determine the effect of the mutation on the period length (Millar et al. 1995). Under continuous white light, the *hy1* plants had a period similar to that of the wild type, indicating that little or no phytochrome is required for white light to affect the period. Furthermore, under continuous red light, which would be perceived only by phyB, the period of *hy1* was significantly lengthened (i.e., became more like constant darkness), whereas the period was not lengthened by continuous blue light. These results indicate that both phytochrome and a blue-light photoreceptor are involved in period control.

The role of blue light in regulating both circadian rhythmicity and flowering is also supported by studies with an *Arabidopsis* flowering-time mutant: **elf3** (early flowering 3) (Hicks et al. 1996; Zagotta et al. 1996). The *elf3* mutant flowers earlier than the wild type under a variety of growth conditions and is insensitive to photoperiod; that is, it is day-neutral. Transformation of *elf3* with the luciferase reporter construct revealed a defect in circadian rhythmicity. The mutant displayed no circadian rhythmicity when grown in continuous light. Significantly, the *elf3* mutant also has a long hypocotyls when grown in constant white light, similar to the *hy* mutants, and is less responsive than the wild type to all wavelengths of light, but especially to blue light. This finding suggests that the circadian clock is intact in the mutant, but that the transduction of light signals to the clock is impaired (Hicks et al. 1996; Zagotta et al. 1996). The fact that the *elf3* mutant is more impaired in its response to blue light than in its response to red light further suggests that a blue-light photoreceptor plays a role in the photoperiodic control of flowering.

Confirmation that a blue light photoreceptor is involved in sensing inductive photoperiods in *Arabidopsis* was recently provided by experiments demonstrating that mutations in one of the cryptochrome genes, *CRY2*, caused a delay in flowering and an inability to perceive inductive photoperiods (Guo et al. 1998). *CRY1* encodes a possible blue light photoreceptor controlling seedling growth in *Arabidopsis*. Thus, various *CRY* family members have, through evolution, become specialized for different functions in the plant.

## VERNALIZATION

Vernalization promotes flowering with cold temperature. Vernalization is the process whereby flowering is promoted by a cold treatment given to an imbibed (fully hydrated) seed or to a growing plant. Dry seeds do not respond to the cold treatment. Without the cold treatment, vernalization-requiring plants show delayed flowering or remain vegetative. In many cases these plants grow as rosettes with no elongation of the stem.

Another environmental factor which affects several aspects of the plant lifecycle is *winter chilling*. The role of winter chilling in the control of flowering was first discovered from studies on rye which, in common with other cereals such as wheat and barley, exists in two forms:

a) winter rye, which is normally sown in the late autumn, and which then flowers and matures in the following summer; and

b) spring rye, which is normally sown in the spring and flowers in the same year.

If the sowing of winter rye is delayed until the spring it frequently fails to flower in the summer although it produces abundant leaf growth. The question therefore arises as to why winter rye needs to be shown in the autumn to obtain flowering in the following summer.

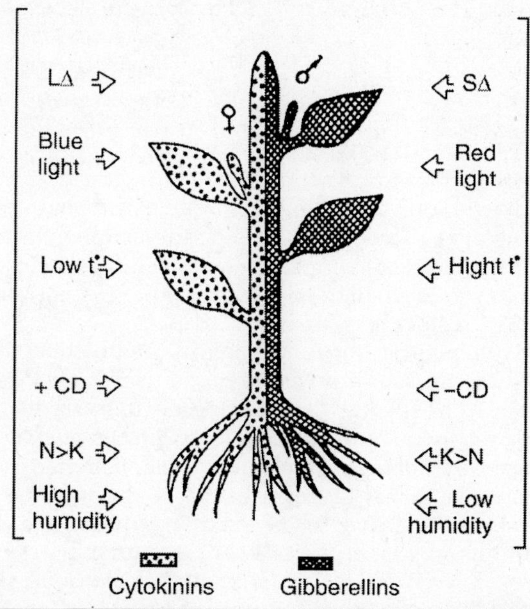


**FIG. 28.8:** The role of environmental factors and phytohormones in sex expression of plants.

Gassner investigated this problem by germinating winter rye seeds at a range of different temperatures (1-2, 5-6, 12 and 24°C) on successive dates from 10 January to 3 July. After germination the seedlings were planted out of doors. He found that seedlings germinated at 1-2°C always flowered in the following summer whereas those germinated above 1-2°C only flowered if they were planted out of doors before early April, that is, if they were exposed to chilling temperatures after planting out. On the basis of these results Gassner concluded that winter rye required to be exposed to a period of winter chilling for flower initiation to occur. It should be noted, however, that in winter rye flower initiation does not occur during the actual chilling period but several weeks later, after the plant has formed a number of leaves. It was later shown by Lysenko that the chilling period can be given at the seed stage by keeping the imbibed grains at 0-5°C for several weeks, after which the would flower when planted in the field. This technique he referred to as *vernalisation*, a term which has since been extended to the general phenomenon of flower induction by winter chilling.

Following these early experiments with winter rye it has become apparent that a considerable number of spring-flowering species require a period of chilling for flower initiation, and these include both biennial and perennial species. Indeed it is now clear that biennials are plants which remain vegetative during their first year of growth because of a chilling requirement which is met during the winter, so that they flower in the following spring. Perennial species which show a chilling requirement include some varieties of garden chrysanthemum, Michaelmas daisies (*Aster nova-belgae*), rye grass (*Lolium perenne*), and spring-flowering species such as primrose (*Primula vulgaris*) and violets (*Viola* spp.) and many other species.

**TABLE 28.6:** Species with a vernalisation requirement

| Winter rye (*Secale cereale*) | Primrose (*Primula vulgaris*) |
|---|---|
| Winter wheat (*Triticum aestivum*) | Violets (*Viola* spp.) |
| Beet (*Beta vulgaris*) | Sweet *William (Dianthus*) |
| Cerery (*Apium graveolens*) | Honesty (*Lunaria biennis*) |
| Cabbage and related forms (*Brassica oleracea*) | Foxglove (*Digitalis purpurea*) |
| Mustard (*Sinapis alba*) | Chrysanthemum (some varieties) |

Some species have not only a chilling requirement but also a specific daylength requirement for flowering. Thus, the garden chrysanthemum requires chilling followed by short days, whereas rye grass requires chilling followed by long days. We shall see later that winter chilling has also a marked effect upon the dormancy of certain seeds and the resting buds of woody plants.

The most effective temperatures for vernalisation are about 1-2°C and freezing is not required. Vernalisation is a slow process and usually required several weeks. Thus, in rye, the time from sowing to flowering decreases with the length of the vernalisation period up to about six weeks. Several pieces of evidence indicate that the vernalising processes occur in the shoot apical meristem. Thus it is still possible to vernalise rye embryos if they are dissected out of the grain and placed upon a sterile culture medium. It is even possible to vernalise the excised shoot apices of embryos in this way. The apices later regenerate roots and can be grown into mature plants which flower in response to the previous chilling treatment. Similarly, celery plants can be induced to flower if only the shoot apex is chilling by a cooling coil placed around it.

The vernalisation process is apparently dependent upon active metabolism since it does not occur when rye grains are kept under anaerobic conditions. Moreover isolated embryos can only be vernalised if they are supplied with a source of sugar.

The vernalised condition is apparently transmitted `undiluted' through cell division, since once a rye embryo has been fully vernalised all tissues subsequently formed by it appear to be irreversibly vernalised. However, vernalised embryos can be *de-vernalised* by exposing to high temperatures (40° C) immediately after the chilling treatment, and such de-vernalised embryos can then be revernalised. Perennial species, such as chrysanthemum and perennial rye grass require fresh vernalisation each winter.

By contrast with the photoperiodic induction of flowering, the vernalised condition cannot be transmitted by grafting an unvernalised shoot on to a vernalised stock. An apparent exception is presented by the observation that flowering can be induced in unvernalised biennial henbane (which normally requires vernalisation) by grafting into it a leaf from the annual form (which requires long days but not vernalisation). Even leaves from a quite different genus, such as tobacco, will induce flowering in the biennial henbane. However, it appears that it is not the vernalised state which is transmitted by grafting in this way but the flowering stimulus which is produced by vernalisation of the biennial form or by long days in the annual form.

As in the case of the flowering stimulus of photoperiodic plants, the nature of the flowering stimulus involved in vernalisation is unknown: however, it is of considerable interest that flowering may be induced in unvernalised biennial henbane by prolonged treatment with rather high amounts of the hormone gibberellic acid. This treatment is also effective in other species, such as carrot, in which flowering is normally stimulated by winter chilling. Similarly, a number of long day plants can be stimulated to flower under short days by treating them with gibberellic acid. However, there are several reasons for thinking that gibberellic acid is not the natural flowering stimulus.

a) The stimulus appears to be identical in long day and short day plants, and yet gibberellic acid is ineffective in stimulating flowering in most short day plants.
b) Gibberellic acid will not stimulate flowering in all long day plants.

Some of the plant physiologists hold the view that a specific flowering stimulus does not exist and that flowering is controlled by a specific balance of general metabolites such as growth hormones. However, against this view is the fact that flowering cannot be induced in many short day plants by any combination of the known hormones, and the problem remains unresolved at present.

## Cold Induction

Vernalization implies the conversion of winter varieties to the summer varieties by cold treatment. The low temperature treatment shortens the vegetative period and promotes flowering. Russian geneticist T. D. Lysenko observed that artificial cold treatment of seeds of winter wheat permitted them to behave like spring wheat planted in spring. This process is called vernalization.

In the annual plants, growth is started in the spring, flowers are developed in the summer and fruits and seeds are produced in the fall of the same year. The flowering is primarily under the control of photoperiod and the influence of temperature is secondary. Biennials, on the other hand, need a cold winter before flowering in its second growing season. Without subjecting to a cold treatment, the majority of these plants maintain their vegetative stage. Hence, they have a qualitative or an absolute requirement for cold e.g. *Hyoscyamus niger*. However, in *Secale cereale* the requirement is quantitative or facultative. For instance, under continuous light unvernalized winter rye will flower in 15 weeks, and if vernalized flowering is observed in 7.5 weeks. Obviously in this variety, vernalization shortens the time to flower.

The plants that require vernalization are *Secale cereale*, *Triticum aestivum*, *Lolium perenne*, *Brassica oleracea*, *Beta vulgaris*, *Apium graveolens*, *Sinapsis alba*, etc.

For most biennials an **artificial** cold treatment followed by the correct photoperiod and temperature treatment will cause flowering during the first growing season. Thus in a biennial species, the flowering may be induced in the same time period as required for flowering of annuals. The stage of vernalization also varies. For example, *Secale cereale* is vernalized in the seed stage while *Hyoscyamus* has to be vernalized when the plant is 10 days old and in the rosette stage. This is done by soaking and sprouting the seed in winter and keeping them in a vegetative by a low temperature or freezing until it can be sown in spring. Flowering of man biennials is promoted by LD treatment following vernalization as in *Hyoscyamus niger*. Other biennials are day neutral following low temperature treatment. In perennials grasses flowering is promoted by SD treatment following the low temperatures. Thus many different kinds of plants are stimulated to flower by cold periods which may have qualitative or quantitative effect. The flowering of some species also is promoted by a suitable day length.

## Site of Vernalization

Whenever vernalization occurs in the mature plants, the receptor of stimulus is the stem apex. In other words, the meristematic cells in the bud respond to cold treatment. Thus, if a hormone-like-substance is produced following vernalization, this occurs in the same cells in which it acts. This fact is borne by the following experiment. If a plant is vernalized and its meristem is grafted on to an unvernalized plant, the latter will flower, indicating as if it had been vernalized. Conversely, if a meristem of an unvernalized plant is grafted on to a vernalized plant where the vernalized meristem was removed, the transplanted, (unvernalized) meristem remains vegetative. Obviously, vernalization is restricted to the meristematic tissues themselves.

Some workers have reported that the excised leaves on isolated roots of *Lunaris biennis* could also be vernalized. On close examination, it was observed that new buds were formed on the leaf petioles. Further, these meristematic tissues had to appear before the vernalization was effective. It was proposed that vernalization requires not only the presence of a meristem but also the actively dividing cells. Indeed, replicating DNA was necessary for the perception of vernalization, no matter what their location in the plant was.

It is also suggested that the vernalization process requires energy. Thus, low temperature was effective only in the presence of oxygen, energy substrate (sucrose) and water contents at 40%. The percentage of flowering depends on the duration of cold treatment, temperature used and the age of plant which vary with species. Gregory and Purvis have shown that embryo by itself can also be vernalized. In short receptor or perceptive mechanism of vernalization seems to be present in different parts of different plants.

## Shoot Apex as the Site of Perception

Plants differ considerably in the age at which they become sensitive to vernalization. Winter annuals, such as the winter forms of cereals (which are sown in the fall and flower in the following summer), respond to low temperature very early in their life cycle. They can be vernalized before germination if the seeds have imbibed water and become metabolically active. Other plants, including most biennials (which grow as rosettes during the first season after sowing and flower in the following summer), must reach a minimal size before they become sensitive to low temperature for vernalization.

The effective temperature range for vernalization is from just below freezing to about 10°C, with a broad optimum usually between about 1 and 7°C (Lang 1965). The effect of cold increases with the duration of the cold treatment until the response is saturated. The response usually requires several weeks of exposure to low temperature, but the precise duration varies widely with species and variety. Vernalization can be lost as a result of exposure to devernalizing conditions, such as high temperature, but the longer the exposure to low temperature, the more permanent the exposure to low temperature, the more permanent the vernalization effect.

Vernalization appears to take place primarily in the meristematic zones of the shoot apex (Thomas and Vince-Prue 1984). Localized cooling causes flowering when only the stem apex is chilled, and this effect appears to be largely independent of the temperature experienced by the rest of the plant. Excised shoot tips have been successfully vernalized, and where seed vernalization is possible, fragments of embryos consisting essentially of the shoot tip are sensitive to low temperature.

## Interaction of Photoperiodism and Vernalization

A vernalization requirement is often linked with a requirement for a particular photoperiod (Napp-Zinn 1984). The most common combination is a requirement for cold treatment *followed* by a requirement for long days-a combination that leads to flowering in early summer at high latitudes. Examples of plants that exhibit this behavior are winter wheat and the biennial form of *Hyoscyamus niger* (henbane). In biennial *H. niger*, prior vernalization renders plnts competent to respond to long-day inductive photoperiods for up to 300 days after the vernalization treatment is complete. In this case, vernalization may be necessary for the meristem to become competent to respond to flowering signals produced in long days, because shoot tips of vernalized plants will flower when grafted onto nonvernalized stock plants (Lang 1985).

A different type of behavior is exhibited in *Campanula medium* (Canterbury bell): Short days acting in the leaf can completely substitute for vernalization at the apex. In this case vernalization and photoperiod appear to act as alternative pathways to flowering. Many plant species can substitute short days for vernalization, but the relationship between the two processes is not known. Vernalization of the meristem may cause that meristem to produce leaves capable of making a flowering signal without exposure to inductive photoperiods. Alternatively, vernalization may cause flowers to be initiated at the apex without the product(s) of photoperiodic induction in the leaves.

## Vernalization and DNA Demethylation

Much remains to be learned about the metabolic processes associated with vernalization. Both sugars and oxygen are required, so it appears unlikely that low temperatures simply arrest some metabolic reactions that inhibit flowering. The sugar and oxygen requirements suggest activation of an aerobic metabolic reaction that is essential for flowering. On the other hand, the rates of most metabolic reactions *decrease* with temperature, so the induction of flowering by low temperatures is likely to be more complex than the simple activation of a metabolic pathway.

The vernalized state of the shoot apex is stable through many rounds of cell division. As discussed in the previous section, *Hyoscyamus niger* remains vernalized for at least 300 days after the end of a cold treatment (Lang 1985). (Investigators determined the stability of the vernalized state by maintaining vernalized plants in short days for various lengths of time before transferring them to inductive long days.) The susceptibility of dividing cells to vernalization and the stability of the vernalized state suggest that cold treatment causes a permanent change in the pattern of gene expression.

It has been proposed that changes in the pattern of DNA methylation can account for altered gene expression induced by vernalization. For example, late-flowering ecotypes of *Arabidopsis* flower early if vernalized. If nonvernalized plants are treated with the DNA-demethylating agent 5-azacytidine, they flower significantly earlier than untreated controls (Burn et al. 1993). *Arabidopsis* lines that do not require vernalization were found not to respond to 5-azacytidine. Moreover, both 5-azacytidine and vernalization caused DNA demethylation in the late-flowering ecotype (Burn et al. 1993). Thus, genes that are needed for early flowering could be blocked by DNA methylation in the late-flowering *Arabidopsis* ecotypes.

Since gibberellins can substitute for the vernalization requirement perhaps a gene that encodes an enzyme in the gibberellin biosynthetic pathway is demethylated during vernalization. Although the demethylation model of vernalization is considered controversial, like all good hypotheses it provides a framework for thinking about the problem and suggests additional experiments in which it can be tested.

## DEVERNALIZATION

Once a plant is vernalized, it may be devernalized by immediate exposure to high temperature (e.g. 30°C or above) and in some cases, revernalized with another exposure to low temperature. Devernalization does not occur if the plant is allowed a period of 4 to 5 days at 20°C. During this time vernalization is stabilized.

## FLORIGEN CONCEPT

In the 1930s, Mikhail Chailakhyan, working in Russia, postulated the existence of a universal flowering hormone on the basis of grafting experiments with chrysanthemum and other plants and called it **florigen** (hormone). The many attempts to isolate and characterize this *hypothetical hormone* have been largely unsuccessful. If a non-induced plant were grafted onto an induced plant, the former would flower, even if it were never exposed to inductive daylengths. Such effect can be **transmitted** from plant to plant through several **graft generations**. Thus grafting a lateral shoot from an induced plant on a non-induced plant induces flowering of the latter. If another shoot is then removed from the second plant and grafted on a third, it too flowers, and soon. In several genera, notably *Nicotiana*, LD species or cultivars will flower under SD when grafted to induced plants of SD species, and vice versa. These observations suggest that a florigen is produced during induction and which move through a graft union to induce flowering in the graft partner. Florigen was found to be at least physiologically identical in LDP's, SDP's, and DNP's. It is non-specific with reference to genus and species.

The availability of florigen in plants is supported by the studies of the effect of removing the photoinduced leaf at various intervals after induction. If one leaf of a cocklebur plant is darkened to provide a **short day** while the remainder of the plant is illuminated, flowering will result provided all leaves between the darkened leaf and the apex are removed. If the induced leaf is removed soon after the beginning of the inductive dark period, no flowering occurs. Removal between 20 and 36 hours results in progressively greater responses. Thus, the hormone appears to be translocated from the induced leaf to the apical meristem, which responds by changing from the vegetative to the reproductive state.

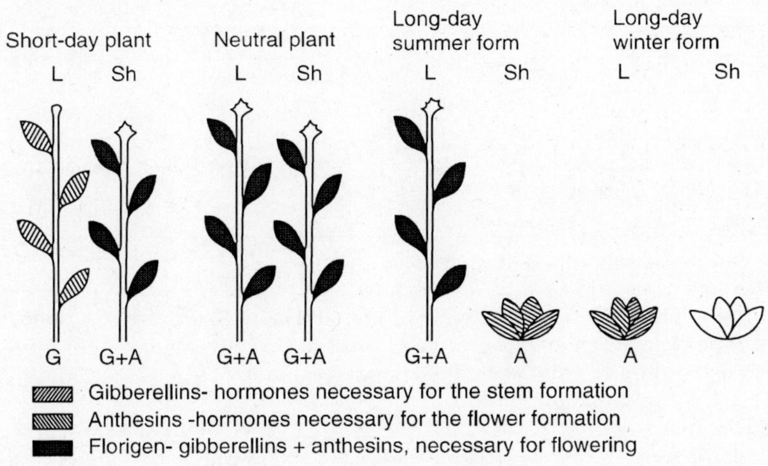


**FIG. 28.9:** The concept of flowering Stimulus (after Chailakhyan).

## Isolation of Florigen?

Attempts have been made to isolate florigen, however, no reports are confirmed with unequivocal evidence for its chemical nature. Therefore, alternative hypotheses have been proposed. One being that flowering is controlled by a **balance** between promoters produced by photoinduced leaves and inhibitors produced by leaves exposed to non-inductive conditions. This concepts is supported by the fact that the presence of non-induced leaves reduces the flowering response when only one portion of the plant is exposed to inductive photoperiods. Their removal early in the inductive cycle permits maximum response. This evidence for inhibitors has been discounted by some investigators on the grounds that leaves between the induced leaf and the apex act as competing **sinks** for assimilates, thus reducing the quantity of florigen reaching the apex. The direct evidence for such inhibitors was provided by Anton Lang (1977) and Soviet co-workers. According to them grafting scions of *Nicotiana sylvestris* (a LDP) on a day-neutral cultivar of tobacco inhibited flowering under SD. Inhibitors produced under SD, therefore, must have been translocated from the LDP.

## Biochemical Signaling Involved in Flowering

The morphological changes associated with floral evocation occur at the apical meristems of the shoots, the events that result in the morphological changes are triggered by biochemical signals arriving at the apex from other parts of the plant, especially from the leaves. Mutants have been isolated (e.g., *gigas* in pea) that are deficient in the floral stimulus.

The biochemical signals arrive from the leaves and other parts of the plant in response to photoperiodic stimuli. Such signals may serve either as activators or as inhibitors of flowering. After years of investigation, no single substance has been identified as the universal floral stimulus, although certain hormones, such as gibberellins and ethylene, can induce flowering in some species. Hence, current models of the floral stimulus are based on multiple factors.

**TABLE 28.7:** Examples of the successful transfer of flowering signals by grafting in cold-requiring plants

| Flowering donor plant | Receptor plants induced to flower |
|---|---|
| *Hyoscyamus niger*, cold-requiring, vernalized<br>*Hyoscyamus niger*, non-cold-requiring annual form<br>*Petunia hybrida*, non-cold-requiring annual | *H. niger*, cold-requiring, nonvernalized |
| *Beta vulgaris*, cold-requiring, vernalized<br>*Beta vulgaris*, non-cold-requiring annual form<br>*Beta procumbens*, non-cold-requiring annual | *B. vulgaris*, cold-requiring, nonvernalized |

## Identification of the Hypothetical Florigen Remains Elusive

Production by leaf cells of a biochemical signal that acts on a distant target tissue (the shoot apex) satisfies an important criterion for a hormonal effect.

The most common approach has been to make extracts from induced leaf tissue and test for their ability to elicit flowering in noninduced plants. In other experiments, investigators have extracted and analyzed phloem sap from induced plants. In some studies, extracts from one of these sources have induced flowering in test plants, but these results have not been consistently reproduced (Zeevaart and Boyer 1987). Attempts to isolate a specific, graft-transmissible inhibitor of flowering have also been unsuccessful. Thus, despite unequivocal data showing that transmissible factors regulate flowering (Lang et al. 1977; Zeevaart 1984), the substances involved are yet to be characterized.

**TABLE 28.8:** Examples of flowering induction by gibberellins in plants with different environmental requirements for flowering

| Flowering requirement | Plant | Effect of gibberellins |
|---|---|---|
| **Long-day plants** | Arabidopsis | Promotes in SD |
| | Lolium | Promotes in SD |
| | *Fuchsia* | Inhibits in LD |
| | Anagallis | No effect |
| **Short-day plants** | *Zinnia* | Promotes in LD |
| | *Fragaria* | Inhibits in SD |
| | *Xanthium* | No effect |
| **Dual-day length plants** | *Bryophyllum (LSDP)* | Promotes in SD |
| | *Coreopsis (SLDP)* | Promotes in SD |
| | *Cestrum (LSDP)* | Inhibits |
| **Day-neutral plants** | *Many conifers* | Promotes |
| | *Many woody angiosperms* | Inhibits |
| **Plants requiring vernalization** | *Daucus* | Promotes |
| | *Oenothera* | No effect |

*Note:* LSDP, long-short-day-plants; SLDP, short-long-day platns; LD, long days; SD, short days.

Grafting experiments have also been done to investigate possible signals from vernalized plants. Cold-requiring, nonvernalized plants have been induced to flower when grafted with a vernalized donor, pointing to a transmissible stimulus. Whether the transmitted message is a specific compound involved in vernalization or is the same substance that induces flowering in plants that do not require vernalization remains to be determined (Thomas and Vince-Prue 1984).

## MULTIPLE-HORMONE SYSTEM AND FLOWERING

Chailakhyan (1968) postulated another multiple-hormone system flowering. He suggests that high levels of two hormones- **gibberellins** and **anthesins** (Gk. *anthein*, to bloom)- must be present in a plant before flowering. Short-day plants kept on long photoperiods contain sufficient gibberellins, but little if any anthesins. Transfer to short days stimulates the production of anthesins and the plants flower. Conversely, long day plants kept on SD produce anthesins in abundance, but are deficient in gibberellins. During long day length gibberellins are synthesized, and flowering ensues.

## THE FUTURE TASK

Despite the difficulties in isolating florigen or other flowering promoters or inhibitors, the concept of such hormones is still a challenge for plant physiologists. However, there remains following unanswered questions which need further research:

i) How does the plant measure the length of sequential light and dark periods?
ii) How is the signal that is received by phytochrome translated into chemicals messenger?
iii) What is florigen?
iv) What is the chemical nature of inhibitors of flowering?
v) How can systems which have the same receptor (phytochrome) and messengers (hormones which promote/inhibit flowering) have opposite effects in different plants (i.e., LDP's vs. SDP's)?

## MOLECULAR ASPECTS OF FLOWERING

S.P. Sen (1978) has brought out some molecular aspects of flowering. Besides other factors, he tasted several aspects which demand careful analysis. For instance, in *Bombax*, *Ceiba* plants only flower when all the leaves are shed. As if some flower-inhibiting substance is present in the leaves. In seeking answers to the role of gibberellins the same author has proposed the metabolism of gibberellin may be linked with bound and inactive state. Recent studies have shown that membranous structures are also sites of gibberellin action and phytochrome is supposedly concerned with membrane control of permeability. Several of the steroids are also reported to promote the flowering of many species where inhibitor of steroids biosynthesis inhibits the flowering of both photoperiod sensitive and day neutral plants. Some workers suggest that protein component of phytochrome undergoes some conformational change during photo transformation. There are also some evidences that such changes were associated with pH shifts. The effect of such localized pH drifts either on leaf cells which perceive the photoperiodic stimulus in the shoot apex and buds which manifest the photoperiodic effect. Phytochrome is also shown to affect the activity and synthesis of several enzymes. Even isoenzyme changes are phytochrome mediated. Phytochrome mediated responses also include anthocyanin biosynthesis.

The specific need of nucleic acid and protein synthesis in plant flowering is well accepted. Recent studies have indicated that the two processes may be causing effects indirectly. Photoperiods influence precursors incorporation which point towards their direct effect. Incidentally inhibitors of steroid synthesis also inhibit RNA synthesis. Based on DNA-RNA hybridization studies Sen (1978) has demonstrated that a small fraction of RNA was synthesized during the induction period and it was different from that produced under non-inductive conditions. There is a specific need to investigate the specific effects of photoperiod and phytochrome on the initiation and termination of polynucleotide, polypeptide chain synthesis. Sen (1978) has reported that the RNA molecules synthesized in presence of different levels of metabolites cause qualitative changes in RNA synthesis.

J. Brulfert and O. Queiroz (1978) have made a comparative study of the photoperiodic induction of morphogenetic and metabolic processes. The comparative study of CAM system and of flowering suggests similar types of mechanism which governs induction by photoperiodism at morphogenetic and enzymic levels. Phytochrome mediates the endogenous rhythm of sensitivity to the information, balance between promotors and inhibitors is controlled by photoperiodism. These authors have advanced the concept of threshold which seemingly depends on the level of biological organisation which the response is made out. Further, a distinction is necessary between threshold effects from one developing programme into another. The study of CAM gives information on possible regulatory metabolic mechanisms intervening between the inductive process and the terminal response. Photoperiod exerts a two fold control of CAM operation. Thus, under LD; PEPc is low and production of PEPc inhibitors is high. Under SD treatments, there is high PEPc and low production of its inhibitors. Enuments possibly modify the concentration of these inhibitors.

**FIG. 28.10:** Genes (gi and elf3) that control flowering time: (A) Long days (B) Short days

## Genes Regulating Flowering

Genetic studies with Arabidopsis and pea have identified a large number of genes that regulate flowering. In *Arabidopsis*, at least two flowering pathways are thought to exist, a photoperiod-sensitive pathway and an autonomous pathway. GA may function as part of the autonomous pathway or could define a third pathway. Each of the flowering pathways includes both activator and repressor genes, and epistasis tests and physiological experiments have led to the idea that these pathways function in parallel. In Arabidopsis several genes that serve as promoters of flowering have been recently cloned, and the sequences of the proteins encoded as well as their expression patterns have suggested possible functions for these genes. For example the genes *CO* and *LD* both appear to encode transcription factors, whereas *FCA* possibly affects RNA metabolism suggesting that a cascade of gene activation events likely controls flowering.

Genes that function to repress the transition to adult and/or reproductive phases have been identified in Arabidopsis. elf3 (*early-flowering 3*) and lhy (*long hypocotyls*) are photoperiod-insensitive mutants that show alterations in circadian clock function and appear to be involved in the repression of flowering in the LD pathway. The HST (*HASTY*) gene is thought to promote a juvenile pattern of development, and loss-of-function mutations in this gene result in early transition to the adult phase an early flowering. Arabidopsis mutants defective in phytochorme synthesis or phytochorme function flower early, indicating that this light receptor is involved in repression of floral initiation. At least some of the effects of phytochrome may be mediated by GA, because recent studies show that the hypocotyls tissue of *phyB* (*phytochrome* B) mutant seedlings is more responsive to exogenous GA than wild-type seedlings. The tfl (Terminal Flower) mutant flowers early and produces a determinate inflorescence, often generating only one or a few terminal flowers.

A new gene, *eafl* (*early flowering* 1), that appears to function in the autonomous pathway to repress the transition from juvenile to adult development was identified. *eafl* mutant plants exhibit truncated juvenile and adult phases, resulting in early flowering. *eafl* mutant plants exhibit phenotypes similar to the GA response mutants *spy* and *rga* and *eafl* seed show increased resistance to paclobutrazol.

## Genes That Control Flowering Time

As We Have Seen, the Maryland Mammoth mutation converts a day-neutral variety of tobacco into an SDP. Analysis of this **flowering-time mutant** by W. Garner and H. Allard led to the discovery that plants can measure photoperiod as an environmental cue for floral induction, and it identified the Maryland Mammoth genes as one of the genes that regulates this process.

The most extensive studies of **flowering-time genes** have been performed in pea (*Pisum sativum*) and *Arabidopsis thaliana* (Weigel 1995; Reid et al. 1996). In these species a range of mutations can alter various aspects of the photoperiodic control of flowering. For example, mutations in certain genes convert these LDPs into day-neutral plants that can flower rapidly. Mutations in other genes result in day-neutral plants that exhibit a long delay in flowering. Because these mutations eliminate the response to photoperiod, the normal (wild-type) products of these genes are thought to be involved in photoperiodism. Other mutations can delay or promote flowering without affecting the photoperiodic response. These mutants are though to affect genes involved in the *autonomous* regulation of flowering.

Since most biological regulatory systems are under both positive and negative control, it is not surprising to find mutations that promote or delay flowering. Inasmuch as most mutations result in a loss or reduction of gene activity, the wild-type role of genes that delay flowering when mutated is to promote flowering. Similarly, mutations that promote flowering are likely to occur in genes that normally inhibit the process. These genes are promoters and inhibitors of flowering in genetic terms, but the mutant phenotype alone does not usually reveal the flowering mechanism that is affected. For example, a mutation that eliminates the response to photoperiod could either block the production of a flowering hormone (floral stimulus) or, alternatively, interfere with the ability of the target cells in the meristem to receive the hormonal signal (meristem competence).

*Arabidopsis* plants containing mutations that promote or delay flowering have been isolated. Part A of the figure shows the wild type and a *gigantean* (*gi*) mutant that were grown in inductive photoperiods (long days). The plants are shown at the stage when flower buds have become visible. Flowering in the *gigantean* mutant is delayed; thus this plant has been growing approximately three times as long as the wild type and has formed many more leaves from the primary apical meristem. Because a mutation in the *GIGANTEA* gene delays flowering, the normal role of this gene is to promote flowering. Part B of the figure shows the wild type and an *elf3* mutant plant that were grown in noninductive photoperiods (short days). The mutant phenotype indicates that the normal role of the *ELF3* gene is to repress flowering.

One way to explore the role of these genes is through grafting studies. For example, if the apical bud from a mutant plant with a delayed-flowering phenotype flowers normally when grafted onto a wild-type shoot, the mutant shoot apex is clearly competent to receive and respond to flowering signals. One might then suspect that the mutation affects the production of flowering signals by leaves. If this were the case, a wild-type shoot apex region should exhibit delayed flowering when grafted onto a mutant shoot. Grafting studies of this sort have identified two types of genes in pea: those that control the production of transmissible promoters and inhibitors of flowering, and those that control the sensitivity of the shoot apical meristem to these signals (Reid et al. 1996).

For example, the *GIGAS* gene is thought to be involved in the production of a transmissible flowering promoter. The wild-type pea exhibits a quantitative LDP phenotype. Plants homozygous for the recessive *gigas* mutation exhibit severely delayed flowering. For example, when grown under a 16-hour photoperiod, the wild type flowers after producing 16 leaves, whereas the *gigas* mutant flowers after producing 43 leaves. Delayed flowering in *gigas* plants could be caused either by the presence of inhibitors or by the absence of a floral stimulus.

To test for inhibitors, researchers grafted the shoot of an early flowering day-neutral line onto *gigas* stock. The time of flowering of the scion was unaffected by grafting to the mutant, indicating that the mutant does not produce flowering inhibitors. However, when a *gigas* shoot was grafted onto either the early-flowering or the wild-type stock, flowering was greatly accelerated. Instead of flowering after producing 40 leaves or more, the plants flowered after producing only 14 or 23 leaves, respectively (Beveridge and Murfet 1996). Because the *gigas* mutant shoot apex flowers readily after grafting, its ability to respond to the flowering stimulus from the leaves is not affected by the mutation. The mutation is therefore thought to affect the production of a floral stimulus.

Another approach to elucidating the floral induction process is to isolate the genes identified by mutations and to study their properties. Once the gene is cloned, the encoded protein can be overexpressed in *E. coli* and purified in sufficient quantities to study its biochemical properties and to produce specific antibodies. The antibodies can be used to determine the tissues in which the protein is produced, as well as its location within the cell. The localization may provide clues to the function of the protein. For example, a protein in the shoot apex may be involved in the reception of flowering signals, whereas a protein present in the leaves may be involved in the production of flowering signals.

The *constans* mutants of *Arabidopsis* are insensitive to day length and flower later than the wild type under long-day conditions. The *CONSTANS* gene of *Arabidopsis* has been cloned and shown to ncode a protein that shows similarities to zinc finger transcription factors, suggesting that it acts as a transcriptional regulator. The wild-type *CONSTANS* gene is expressed in leaves, and the level of *CONSTANS* mRNA increases during inductive photoperiods (Putterill et al. 1995). This pattern of expression, along with the photoperiod-insensitive mutant phenotype, implicate this gene in the photoperiodic control of flowering.

## Three Types of Genes Regulate Floral Development

Mutations have identified three classes of genes that regulate floral development (Coen 1991). One group of genes directly controls **floral organ identity.** The proteins encoded by these genes are transcription factors that likely control the expression of other genes whose products are involved in the formation and/or function of these organs. Genes of the second group act as spatial regulators of the floral organ identity genes by setting boundaries for their expression. These have been called **cadastral genes** (from the word "cadastre," which is a map or survey showing property boundaries for taxation purposes). The third group of genes is necessary for the initial induction of the organ identity genes. These genes, which are the positive regulators of floral organ identity, are called **meristem identity genes**.

## Meristem Identity Genes in Antirrhinum and Arabidopsis

Meristem identity genes must be active for floral meristems to be formed by inflorescence or vegetative meristems. For example, the *floricaula* mutant of *Antirrhinum* (snapdragon) develops an inflorescence but does not produce flowers. Instead of floral meristems in the axils of the bracts, it develops additional inflorescence meristems. The wild-type *FLORICAULA* gene controls the determination step in which floral meristem identity is established.

*APETALA1* (*AP1*) and *LEAFY* (*LFY*) are the most important genes involved in establishing floral meristem identity in *Arabidopsis*. These genes are expressed in floral meristems and appear to be part of a genetic pathway that must be activated to establish floral meristem identity. Floral induction triggers the expression first of *LFY* and then of *AP1* (Simon et al. 1996). Recent studies suggest that gibberellin promotes flowering in *Arabidopsis* by activating the expression of the *LEAFY* gene (Blazquez et al. 1998). This suggests that GAs stimulate flowering in some species through a pathway that regulates LFY transcription.

In addition to being controlled by positive regulators of floral meristem identity, floral development is subject to negative regulation. An *Arabidopsis* gene designated *EMBRYONIC FLOWER* (*EMF*) has been identified with a mutant phenotype in which plants flower almost immediately after germination (Sung et al. 1992). The shoot apical meristem initiates no vegetative structures in the *emf* mutant, but immediately initates floral organs as germination proceeds. While the *EMF* gene could regulate the establishment of vegetative meristem identity, it also is possible that it is a negative regulator of the floral meristem identity genes. The flowers also exhibit various abnormalities, suggesting that the *EMF* gene is required for normal inflorescence development (Chen et al. 1997).

## Homeotic Mutations Help in Identification of Floral Organ Identity Genes

The genes that determine floral organ identity were discovered as floral **homeotic mutants**. Recall that in the fruit fly *Drosophila*, homeotic mutations led to the identification of a set of homeotic genes encoding transcription factors that determine the locations at which specific structures develop. Such genes act as major developmental switches that activate the entire genetic program for a particular structure. The expression of homeotic genes thus gives organs their identity.

The dicot flowers consist of successive whorls of organs that form as a result of the activity of floral meristems: sepals, petals, stamens, and carpels. These organs are produced when and where they are because of the orderly, patterned expression and interactions of a small group of homeotic genes that specify floral organ identity. The floral organ identity genes were identified through homeotic mutations that altered floral organ identity so that some of the floral organs appeared in the wrong place. For example, *Arabidopsis* plants with mutations in the *APETALA2* (*AP2*) gene produce flowers with carpels where sepals should be and stamens where petals normally appear.

The homeotic genes that have been cloned so far encode transcription factors-proteins that control the expression of other genes. Most plant homeotic genes belong to a class of related genes known as **MADS box genes.** In animals, however, most homeotic genes, which control segment identity, belong to a class of transcription factor-encoding genes known as *homeobox genes*. Plants also have homeobox genes, but at this point none of these genes are known to have homeotic function. For example, both *KN1* and *STM* genes are homeobox genes. Although they are important in plant development, most plant biologists do not consider them to be homeotic genes.

Many of the genes that determine floral organ identity are MADS box genes, including the *DEFICIENS* gene of snapdragon and the *AGAMOUS*, *APETALA1*, and *APETALA3* genes of *Arabidopsis*. Both homeobox and MADS box genes encode proteins that function as transcription factors, but they differ in structure. The MADS box genes share a characteristic, conserved nucleotide sequence known as a MADS box, which encodes a protein structure known as the MADS domain. The MADS domain enables these transcription factors to bind to DNA that has a specific nucleotide sequence.

## Three Types of Homeotic Genes Control Floral Organ Identity

Five different genes are known to specify floral organ identity in *Arabidopsis: APETALA1 (AP1)*, *APETALA2 (AP2)*, *APETALA3 (AP3)*, *PISTILLATA (PI)*, and *AGAMOUS (AG)* (Bowman et al. 1989; Weigel and Meyer owitz 1994). The organ identity genes initially were identified through mutations that alter the identity of the floral organs produced in two adjacent whorls. Plants bearing *ap3* mutations, for example, produce sepals instead of petals in the second whorl, and carpels instead of stamens in the third whorl. Since mutations in these genes change floral organ identity without affecting the initiation of flowers, they are homoetic genes. These homeotic genes fall into three classes-types A, B, and C-defining three different kinds of activities.

Type A activity controls organ identity in the first and second whorls. Loss of type A activity, encoded by *AP1* and *AP2*, results in the formation of carpels in the first whorl instead of sepals, and of stamens in the second whorl instead of petals. Type B activity, encoded by *AP3* and *PI*, controls organ determination in the second and third whorls. Loss of type B activity results in the formation of sepals in the second whorl instead of petals, and of carpels instead of stamens in the third whorl. The third and fourth whorls are controlled by type C activity, encoded by AG. Loss of type C activity results in the formation of petals instead of stamens in the third whorl, and replacement of the fourth whorl by a new flower so that the fourth whorl of the *ag mutant* flower is occupied by sepals.

The role of the organ identity genes in floral development is dramatically illustrated by an experiment in which all three activities are eliminated by loss-of-function mutations. After floral induction, these triple-mutant plants produce floral meristems that develop as pseudoflowers; all the floral organs are replaced with green leaflike structures, although these organs are produced with a whorled phyllotaxy. Evolutionary biologists, even the eighteenth-century German philosopher Goethe, have speculated that floral organs are highly modified leaves, but this experiment gives direct support to these ideas.

Although floral homeotic genes have been identified in several other species, the work of Enrico Coen and Zsusana Schwarz-Sommer leading to the identification of the floral organ identity genes in *Antirrhinum* has been especially important for our general understanding of flowering (Schwarz-Sommer et al. 1990; Coen et al. 1991). Four of the five *Arabidopsis* floral identity genes have close homologs in *Antirrhinum*, in terms of both gene structure and activity.

### The ABC Model Explains the Determination of Floral Organ Identity

In 1991, Elliot Meyerowitz and Enrico Coen proposed a model called the **ABC model**, to explain how homeotic genes control organ identity (Coen and Meyerowitz 1991). The ABC model proposes that organ identity in each whorl is determined by a unique combination of the three organ identity gene activities. Activity of type A alone specifies sepals; formation of petals requires both A and B activities. Stamens are formed by a combination of activities B and C, and activity C alone specifies carpels. The model further proposes that activities A and C mutually repress each other (Figure 24.29B and C); that is, both A and C type genes have cadastral function in addition to their function in determining organ identity.

The pattern of organ formation in the wild type and most of the mutant phenotypes are predicted and explained by this model. The model is supported not only by genetic evidence, but also by *in situ* hybridization studies that demonstrate that most of the organ identity genes are expressed in the areas predicted by their mutant phenotype. The challenge now is to understand not only how the expression of these organ identity genes, which encode transcription factors, alters the pattern of genes expressed in the developing organ, but also how this altered pattern of gene expression results in the development of a specific floral organ.

### The TEOPOD (TP) Genes Regulate Juvenility in Maize

As noted previously, regulation of the transition from juvenile to adult (i.e., the attainment of competence) is influenced by a wide variety of conditions that affect the growth rate, and by transmissible factors such as carbohydrates and gibberellins. Hence, it has been difficult to determine exactly what regulates the phase change during normal development. In recent years, plant developmental biologists have begun to take a genetic approach to the problem. The analysis of mutations that affect the timing of phase changes, called **heterochronic mutations**, provides the opportunity to identify specific genes that regulate juvenility. The best-studied example is in maize.

**TABLE 28.9:** Characteristics of juvenile and adult phases of English ivy (*Hedera helix*) and maize (*Zea mays*)

| Traits | Juvenile | Adult |
|---|---|---|
| **Hedera helix** | | |
| Leaf shape | Entire | Lobed |

| Traits | Juvenile | Adult |
|---|---|---|
| Leaf thickness | 230 μm | 330μm |
| Phyllotaxy[a] | Alternate | Spiral |
| Plastochron[b] | 1 week | 2 weeks |
| Growth habit | Horizontal | Vertical |
| Anthocyanin | Present | Absent |
| Aerial roots | Present | Absent |
| Rooting ability | Good | Poor |
| Flowers | Absent | Present |
| **Zea mays** | | |
| Cuticle thickness | 1 μm | 3μm |
| Epidermal cell shape | Circular | Rectangular |
| Epicuticular wax | Present | Absent |
| Aerial roots | Present | Absent |
| Bulliform cells[c] | Absent | Present |
| Lateral buds | Tillerlike[d] | Ears or absent |

*Source*: Poethig 1990. [a] Arrangement of leaves around the stem. [b] The time period between the successive initiation of leaves. [c] Large cells on the upper epidermis whose turgor changes control leaf unrolling in members of the grass family. [d] Branches forming at the base of the stem.

The juvenile and adult phases of corn plants are characterized by differences in vegetative morphology. Scott Poethig and his colleagues at the University of Pennsylvania have been characterizing the heterochronic *teopod* (*tp*) mutants of maize: *tp1*, *tp2* and *tp3* (Poethig 1990; Dudley and Poethig 1991, 1993). These semidominant mutations cause cells and tissues that would normally develop as adult structures to become more juvenile. In *tp* plants, leaves in the normally adult part of the shoot are morphologically intermediate between juvenile and adult leaves. Whereas in the wild type, tassels bear male flowers and ears bear female flowers, in *tp* mutants both tassels and ears possess leaves in addition to flowers. However, *tp* mutations affect neither the rate nor the duration of growth of the plant, nor do they interfere with the timing of tassel formation. The *teopod* mutations thus specifically affect the transition from juvenile to vegetative adult phase without affecting the transition to the reproductive phase.

The possible interaction between the *TEOPOD* genes and the gibberellins has recently been explored (Evans and Poethig 1995). In GA-deficient dwarf mutants of maize, the transitions from juvenile to adult vegetative development and from adult vegetative to reproductive development occur later than in wild-type plants. Thus, endogenous GAs are required for the normal timing of phase changes.

In general, once plants attain the adult phase, they exhibit an increasing tendency to flower with age. For example, in plants controlled by day length, the number of short-day or long-day cycles necessary to achieve flowering is often fewer in older plants. These observations suggest that the effect of day length could be to accelerate a flowering process that is already occurring at a slow pace.

## PLANT GROWTH REGULATORS AND FLOWERING

### Phytohormones and Flower Initiation

Phytohormones change the flowering behaviour of several plant species and their effects varies with the plant species, its age, concentration of growth hormone employed as well as temperature, etc. The mechanism of phytohormone action on control of flower initiation is still not clear. However, like many morphogenetic processes, flower initiation involves interactions between phytohormones leading to the synthesis of some **flowering stimulating compounds**.

Use of growth regulators for flower induction, under environmental conditions not conducive to flowering, has considerable interest in agriculture. Cultivars can save many hours of labour if poinsettias and other SDP's could be sprayed once or twice with a chemical, rather than having to cover them with black cloth each evening and to remove the cloth every morning.

It is proposed on experimental grounds that photoinductive conditions for flowering are related with the regulation of levels of the endogenous hormones which in specific concentration cause flower initiation. In the present text the role of different phytohormones in flowering is presented:

## Auxin and Flowering

Research in the 1940s and early 1950s stressed the role of auxins in the photoperiodic induction of flowering, and many studies were made on the relative levels of these compounds under long vs. short days. Although auxin levels increase temporarily when plants are moved from short to long days, no conclusive evidence has been obtained for their role in flowering. Auxin promote flowering in *Citrus*, *Litchi* and *Ananas*. Exogenous treatment of auxin inhibits flower initiation in shortday plants like *Kalanchoe*, *Xanthium* and *Chenopodium* and the effect is pronounced when it is applied during the dark period.

If auxin is applied after the photoinductive cycle, it increases floral differentiation. *Calendula* and *Namesia* are long day plants (LD) and exogenous application of auxin on these plants inhibits flowering.In LD plants like *Silene*, when the auxin treatment was given under-low-light intensity flowering was promoted. It is also reported that auxin-induced effect could be replaced by increasing the number of photoinductive cycles.

Exogenous application of antiauxin also counteracted the auxin effect. It is now well accepted that auxin(s) interferes with the photoperiodic reaction which occurs in the leaves and hinders the synthesis of florigen. Further it is indicated the auxin play important role during later stages of flower induction in shoot apex. Therefore, it may be proposed that auxin is metabolized in the new biosynthetic pathways at the apex and these biosynthetic pathways are triggered by photoperiodism. Similarly when the seeds of different cereals are soaked in auxin flower initiation is induced. As discussed under vernalization, auxin treatment also increases vernalizing effects in some plant species.

Use of naphthaleneacetic acid (NAA) in pineapple (*Ananas comosus* Merr.) industry successfully induced more concentrated flowering and fruiting, thereby, facilitating, harvesting. Initially it was thought that this success was due to the role of natural auxins in flowering, but now it is known that the NAA had indirect effect through the stimulation of ethylene production. Therefore, presently ethylene-generating compounds (ethephon) are currently used for this purpose. Many other species in the same family (Bromeliads) can be induced to flower by treating with ethylene, but the response does not occur in most plants.

## GA and Flowering

Gibberellic acid could induce flowering in certain long-day plants (Lang, 1956). This concept gave great impetus to research on the role of the native gibberellins in flowering. As with auxins, comparisons of the gibberellin content of plants kept under long vs. short photoperiods has not until recently revealed differences sufficient to account for flowering responses. It has also been demonstrated that gibberellins increase dramatically in spinach plants when they are transferred from SD to LD. Spinach grows as a rosette under SD and bolts under LD, thus changes in GA content may be associated with the elongation of the stem, rather than with flowering, per se. Exogenous supply of GA causes flower initiation in LD plants like *Beta*, *Brassica*, etc. The effect of GA is not similar on all the plants, it respond differently to different species. For example in some LD plants the increased flowering may be associated with the elongation of internodes caused by GA and in L-S-D plants gibberellins can substitute for LD. On the contrary SD plants fail to respond to GA. $GA_3$ has been shown to substitute for cold treatment in some *Chrysanthemum* species. Although high levels of abscisic acid (ABA) is known to occur in plants grown under short days, subsequent investigations with more refined techniques have not confirmed this observations.

*Impatiens* is an obligate SD plant. Studies conducted by Prof. K.K. Nanda et al. revealed that $GA_3$ could completely substitute for the dark period. *Kalanchoe* shows similarly flowering pattern after GA-treatment. From studies conducted on several species, it has been inferred that GA affects the expression of flower stimulus at the shoot apex rather than leaves. GA increases the mitotic activity in the subapical meristem and thus becomes more responsive to photoinductive conditions.

GA promotes flowering both in many LDP's and in plants that require vernalization and certain synthetic growth retardants, which promote flowering in woody plants. External application of gibberellin $A_3$ ($GA_3$) induced flowering of the LDP's lettuce and spinach under short days, and of the cold-requiring plants beet, carrot, cabbage, and parsley (*Petroselinum crispum* Nym.) in the absence of chilling temperatures. The commercial application of GA's has been limited.

## Cytokinins

Application of benzyladenine (BA) and $GA_5$ accelerate flowering in *Chrysanthemum* sp. BA is also able to overcome the inhibitory effect of auxin in flower initiation. Cytokinins have been shown to promote flowering in *Wolfia*. There is a view that cytokinins redirect the flow of assimilates essential for the formation of flowers to the shoot species.

## Ethylene

Exogenous application of ethylene inhibits flowering when applied in inductive short periods. The effect was pronounced when given to the cotyledons. The effect of ethylene is confined to the leaves. Flowering in several cereals is induced by external application of ethylene. Ethylene also stimulate flowering in pineapple and some mango varieties.

## Other Plant Growth Substances and Flowering

The exogenous application of other major growth substances like morphactins, MH and TIBA have been shown to influence the development of sex organs in monoecious, dioecious and hermaphrodite species. TIBA (2,3,5 triiodobenzoic acid) interferes with auxin transport, promotes flowering in apple, tomato, and several other species. Daminozide [butanedioic acid mono-(2,2-dimethyl-hydrazide)] hastens flowering in young fruit trees when applied early in the growing season. Growth retardants, which may inhibit the synthesis of endogenous gibberellins, promote flowering in some herbaceous species like tomato. They also promote flowering in woody plants such as azalea (*Rhododendron* spp.), apple and stone fruits. However, the compound slows fruit growth and cannot be used at this time on bearing trees if fruit size is a problem.

### Both Photoperiod and Hormones can Influence Floral Development

As we saw in Chapter 20, gibberellins can affect sex determination in flowers. Gibberellin has also been shown to interact with photoperiod in the regulation of meristem identity in *Arabidopsis*. For example, when *agamous* mutants or plants heterozygous for the *leafy* mutation are grown under short days, they undergo a process called floral reversion. In **floral reversion** the floral meristem reverts to an inflorescence meristem, causing an inflorescence shoot to grow directly out of the developing flower (Figure 24.30). Jack Okamuro and his colleagues at the University of California, Santa Cruz, have shown that phytochrome is the photoreceptor involved in the response and that exogenous gibberellin prevents floral reversion under short-day conditions. Floral reversion is also blocked by the *spindly* mutation, which activates the GA signal transduction pathway in the absence of the hormone (Okamuro et al. 1996, 1997) (see Chapter 20). These results indicate that both the GA and the phytochrome signal transducation pathways participate in regulating meristem identity in *Arabidopsis*.

### Gibberellins Can Induce Flowering in Some Plants

Among the naturally occurring growth hormones, gibberellins (GAs) can have a strong influence on flowering. Exogenous gibberellin can substitute for photoperiodic induction when applied to long-day plants that grow as rosettes in short days (Lang 1965; Vince-Prue 1985). In these plants, the flowering response (either to GAs or to long days) is accompanied by elongation of the flowering stem. However, it is important to note that flower formation and stem elongation are independent processes (Zeevaart 1984). In addition, application of GAs can evoke flowering in a few short-day plants in noninductive conditions and can substitute partly or completely for a low-temperature signal in several cold-requiring plants. As previously discussed, cone formation can also be promoted in juvenile plants of several gymnosperm families by addition of GAs (Pharis and King 1985).

Thus, exogenous GAs can sometimes substitute for the endogenous trigger of age in autonomous flowering and for the primary environmental signals of day length and low temperature. A mutation in GA biosynthesis renders the quantitative LDP *Arabidopsis thaliana* unable to flower in noninductive short days but has little effect on flowering in long days, demonstrating that endogenous GA is required for flowering in specific situations (Wilson et al. 1992).

Plants contain many GA-like compounds. Many of these compounds may be precursors to, or inactive metabolites of, the active forms of GA. In some situations different GAs have markedly different effects on flowering and stem elongation, such as in the long-day plant *Lolium temulentum* (darnel ryegrass). These observations suggest that the regulation of flowering may be associated with specific GAs.

Considerable attention has been given to the effects of day length on GA metabolism in the plant. For example, in the long-day plant spinach (*Spinacia oleracea*), the levels of gibberellins are relatively low in short days, and the plants maintain a rosette form. After the plants are transferred to long days, the levels of all the gibberellins of the 13-hydroxylated pathway ($GA_{53} \rightarrow GA_{44} \rightarrow GA_{19} \rightarrow GA_{20} \rightarrow GA_1$) increase. However, the fivefold increase in the physiologically active gibberellin, $GA_1$, is what causes the marked stem elongation that accompanies flowering. Whether other GAs specifically affect flower formation has not yet been determined.

In addition to GAs, other growth hormones can either inhibit or promote flowering (Vince-Prue 1985). One commercially important example is the striking promotion of flowering in pineapple (*Ananas comosus*) by ethylene and ethylene-releasing substances, a response that appears to be restricted to members of the pineapple family (Bromeliaceae).

## FLORAL STIMULUS MAY HAVE SEVERAL COMPONENTS

Ultimately, floral primordia form and develop at the shoot apex as a result of biochemical and cellular changes leading to the production of the floral organs. The transition to flowering may involve a complex system of interacting factors that include, among others, carbohydrates, gibberellins, and cytokinins (Bernier 1988). Sometimes changes normally associated with flowering can be caused by treatments that do not themselves bring about floral initiation. For example, one of the earliest events observed at the shoot apex following photoperiodic induction is a transitory increase in the number of cells undergoing mitosis. When cytokinins are applied to the stem apex of the long-day plant white mustard (*Sinapis alba*), they cause an increase in mitotic activity similar to that caused by exposure to a single long day. However, whereas the long day induces flowering, the cytokinin treatment does not. Cytokinin may, therefore, be a component of the floral stimulus, although it is insufficient by itself to bring about flowering.

Other potential components include the polyamines. For example, photoperiodic induction of the LDP *Sinapis alba* was correlated with a large increase in the concentration of putrescine, the major polyamine, in the phloem exudates from the leaves. Spraying the leaves with an inhibitor of putrescine biosynthesis lowered the level of putrescine in the phloem exudates and also inhibited flowering. This result suggests that putrescine is a component of the floral stimulus in *S. alba* (Havelange et al. 1996).

In addition to receiving signals from the leaves, the shoot apex may accumulate flowering signals from the roots. Classic experiments with cuttings have shown that the initiation and growth of adventitious roots inhibit flowering in the shoot. In many species, root removal promotes flowering. Thus, at least in certain species, the root appears to produce one or more flowering inhibitors. In other cases, the roots may contribute floral activators. For example, both GA and cytokinin are synthesized in the root and transported to the shoot via the xylem. During photoperiodic induction in *Sinapis alba*, the amount of cytokinin present in the xylem exudates from the root increases considerably (Lejeune et al. 1994). This finding indicates that signals emanating from the leaf are transported both to the shoot apex and to the root during photoperiodic induction. In response to the signal from the leaves, the roots produce and export additional chemical signals that modulate the flowering response.

If the floral stimulus does have multiple components, the limiting component may vary among species. Moreover, exposure to a high level of one component, such as a particular gibberellin, may compensate for the lack of a limiting component and induce flowering in its absence. Such ability to compensate could explain the wide range of substances and conditions that can cause flowering. However, many physiological experiments show that the stimulus exported from an induced leaf has some specific properties. Establishing the biochemical basis of flower induction remains a major challenge.

## SEX EXPRESSION

On the basis of sex organs the flowers are classified into two categories.

a) **Bisexual** (dioecious) flowers which bear stamens as well as pistils.
b) **Unisexual** flowers which bear either stamens or pistils. Stamen bearing flower is known as *andromonoecious* or male flower whereas pistil bearing flower is known as *gynoecious* or female flower.

The sex of the flowers and distribution of andromonoecious and gynoecious flowers is a genetic character which is fixed for a species. In most of the dioecious flowers, for example in *Coccinia*, the sex is determined by distribution of X and Y chromosome, similar to that in man. XY combinations of chromosomes determine male and XX combinations female. According to Janick and Stevenson (1955) monoecious habit of some plants is an expression of a modified gene on the X chromosome.

Chemical treatment alters the ratio of male to female flowers in cucurbits. The primordia contain both pistil and stamen initials, but one or the other is suppressed in the formation of pistillate or staminate flowers. Auxin-treatment favours the production of the former; however, gibberellin-treatment favours the latter. Gibberellins $A_4$ and $A_7$ are particularly effective and are used by breeders to induce gynoecious (female) mutants to produce pollen. Responses to both auxins and gibberellins are probably mediated through their effects on ethylene biosynthesis.

64 Cytokinin also affect sex expression and causes the development of perfect flowers in some male (androecious) grape clones.

**TABLE 28.10:** Effects of ethephon, an ethylene-releasing chemical, gibberellin on sex expression in cucumber flowers.

| Node of appearance of first male or female flower | | |
|---|---|---|
| **Treatment** | **Male** | **Female** |
| None | 2.0 | 8.0 |
| Ethephon, 250 mg $L^{-1}$ | 14.3 | 2.0 |
| $GA_3$ 2000 mg $L^{-1}$ | 1.3 | 17.0 |

### Growth Regulators and Sex Expression

Investigations of the problem of sex formation and transformation in plants are of great importance for selection and raising of crop yields of many plants, especially of those which separate male and female flowers. Most authors have investigated the effect of spraying and pre-sowing treatment of seeds with phytohormone solutions on shifts in sex expression. It was shown in 1949 by Laibach and Kribben that treatment of cucumber plants with auxins greatly increases the ratio of female to male flowers. Late it was discovered that auxin in this case acts indirectly, by increasing the production of ethylene by the plants, and that ethylene the actual sex-modifying agent.

This work with auxin was the first clear report on effects of a phytohormone on sex expression in a monoecious plant. Somewhat later, it was shown that auxin treatment of hemp, a dioecious plant, increased the number of female individuals. Auxin effects on sex expression were also found in maize. Later it was shown in cucumber and hemp that spraying with ethrel (2,9-chloroethylphosphonic acid), a compound releasing ethylene changes sex expression in the female direction, whereas in contrast treatment of hemp with gibberellic acid (GA) caused a change in the male direction.

Furthermore, this latter effect was not obtained when the plants were treated with abscisic acid (ABA) at the same time. Results with other physiologically active substances have been less clear or even contradictory One reason for such different results probably is that different investigators used plants at different stages of development.

Hormonal regulation of flowering, male sterility, self Incompatibility, sex Expression and other related aspects have been studied in various economically important plants like, Apple, Arachis, Brassica, Brinjal, Budwood, *Calendula officinalis*, *Cannabis sativa*, *Centurea*, China aster (*Callestephus chinensis*), *Chrysanthemum*, *Citrus*, Conifers, *Coriandrum sativum*, *Cosmos sulphurens*, Cotton, Cucumber, Cuscuta, *Dahlia variabilis*, *Datura innoxia*, *Euphorbia pulcherrima*, Gladiolus, Grape, Guava, *Helianthus annuus*, *Impatiens balsamina*, *Ipomoea carica*, Jasmine, *Kalanchoe*, *Leman pancicostata*, *Lilium*, Litchi, *Macrosperma lentils*, Mango, Marigold, *Mocardonia procumberns*, *Morus alba*, Moss, *Nicotiana*, Okra, Papaya, Pea, Peach, Pearl millet, Pineapple, Plume (*Prunus salicina*), *Polianthes tuberosa*, *Ranunculus muricatus*, Rice, *Ricinus communis*, Rose, and *Scirpus affinis*. These aspects have also been studied in short day plants like, *Solanum khasianum*, *Sorghum*, Soybean, *Spirodela polyrrhiza*, *Tagetes erecta*, *Thevetia peruviana*, Tomato, *Trichosanthes dioica*, *Vicia faba*, *Vigna radiata* and Wheat.

# CHAPTER 29

# Photomorphogenesis

## LIGHT AS A BIOLOGICAL AGENT

Light is the ultimate source of free energy for virtually all life on our plant. Solar energy is captured by the chlorophyll of the green plant and transformed through photosynthesis into potential chemical energy in the form of complex organic substances which directly or indirectly form the food sources for heterotrophic organisms. Through this process, the green plants constitutes a fundamental, inescapable link between the biosphere and its energy supplies. The importance of light was apparent even to early mankind as is evident from the ancient of religious of sunworship, in which the superstitious show a primitive realization of a life giving force. In the present the role of green plant as the transducer of this vital energy is fully recognized, although the molecular mechanisms involved may still appear mystical to many.

Light, as we understand it in everyday life, comprises a small region of the continuous electromagnetic spectrum of radiant energy emitted by sun. Radiant energy emitted by sun are of dual native. During propagation it behaves as a waveform while on interaction with matter it behaves as a stream of discrete packets of energy known as light Quanta or photon. All biological process initiated by light depend, on the absorption of photon by specific photoreceptive substances which become activated of energized and chemical reactivity is thereby changed. By virtue of its changed chemical reactivity the photoreceptor then initiates a sentence of metabolic process which lead ultimately to the observed developmental changes. Three major categories of light responses have been recognized. Each of these has its own vitally important part to play in the developmental adoption to the environment.

### *Phototropism*

Phototropism is defined as a directional growth response to a directional light stimulus. In the natural environment phototropism is of great importance in orienting the growth of the plant so that it is able to optimize its interception of the available solar energy for photosynthesis. Such a mechanism provides a plant with a great adaptive advantage over non-phototropic plants.

### *Photoperiodism*

It is defined as a non-directional response to a non-directional periodic light stimulus. The most intensively investigated photoperiodic response is flowering. Several other developmental processes also come under photoperiodic control in certain species including seed germination, bud dormancy and the rate and pattern of vegetative development.

### *Light Regulated Gene Expression*

The effect of light on gene expression was first observed in experiments showy that different polypeptides were produced by in vitro translation of mRNA extracted from light or dark-green plants. Photoregulation of gene expression was confirmed by run off transcription and Northern blot analysis indirectly that the steady-state mRNA and transcriptional levels of many genes are directly affected by light. Both chloroplast and nuclear gene expression including photosynthesis and the non-photosynthesis associated genes are regulated by light. The regulation by light of planted gene expression is rather complex and light can influence transcription initiation, RNA processing, RNA stability and translation of existing mRNAs. In contrast, the effect of light on nuclear gene expression in at the transcriptional level. It has been proposed that is the absence of light a posttranscriptional regulatory mechanism is active which prevents protein, accumulation probably at the level of protein turnover or translation and this negative regulation of protein accumulation is overcome by light.

### *Photomorphogenesis*

It is defined as a non-directional developmental response to a non-directional, non-periodic light stimulus. It Mohr (1964) has defined photomorphogenesis as *the control which may be exerted by light over the growth development and differentiation of a plant independently of photosynthesis*. Photomorphogenetic responses occur throughout the whole life of the plant and may be responsible for controlling seed germination, stem elongation, leaf expansion phototropic and geotropic sensitivity, the formation of stomata and the hairs, the synthesis of chlorophyll, the development of chloroplast structure and the synthesis of a wide range of secondary products including pigments such s the anthocyanins.

The control of morphogenesis by light is called **photomorphogenesis** that is a *non-directional developmental response to a non-directional, non-periodic light stimulus*. This also includes all those ways in which light regulates development, not covered by the more precisely defined phenomena of phototropism and photoperiodism. This definition is not universal because some plant physiologists have tended to use a much broader definition including photoperiodism and phototropism within its frame of reference. Prof. Hans Möhr (University of Freiburg) defined photomorphogenesis as *the control that may be exerted by light over the growth, development, and differentiation of a plant, independently of photosynthesis*.

Photomorphogenic responses occur throughout the whole life of the plant and may be responsible for controlling seed germination, stem elongation, leaf expansion, phototropic and geotropic sensitivity, the formation of stomata and leaf hairs, the synthesis of chlorophyll, the development of chloroplast structure, and the synthesis of a wide range of secondary products including pigments such as the anthocyanins. Clearly, not all of these responses are regulated by light at the same time in any one species.

## Nature of Photomorphogenic Reactions

Biological responses are evoked by light energy and the basic sequences of the events must occur. The light is absorbed by a specific photoreceptive molecule whose chemical reactivity is changed and the photoreceptor then initiates a sequence of metabolic processes that lead ultimately to the observed developmental changes in plants. It is essential to know the following informative aspects of photomorphogenesis:

I. nature of the required light stimulus;

II. the chemical identity of the photoreceptor and the photochemical changes which occur upon light absorption,
III. the immediate mechanism through which the photoreceptor triggers the developmental responses,
IV. the nature of the ultimate developmental changes, and
V. the molecular and biochemical nature of the partial processes that intervene between the photoreceptor.

Above phenomena stimulate morphogenic responses.

## *Types of Photoreceptors which Affect Photomorphogenesis in Plants*

### Phytochrome

It absorbs most strongly red and far-red light; and also absorb blue light. At least two major kinds of phytochrome are known. Phytochrome seems to be the most important photoreceptor in vascular plants.

### Cryptochrome

It is a group of similar, unidentified pigments that absorb blue light and long-wave ultraviolet wavelengths (the UV-A region, about 320 to 400 nm). It has special importance in cryptogams (non flowering plants).

### UV-B Photoreceptor

These are unidentified compounds (technically not pigments) and absorb ultraviolet radiation between about 280 and 320 nm.

### Protochlorophyllidea

It is a pigment, absorbs red and blue light and becomes reduced to chlorophyll a.

Phytochrome and other photoreceptors control morphogenic processes beginning with seed germination and seedling development and ending with formation of new flowers and seeds.

## *Photomorphogenesis at which Stage of Life Cycle?*

Photomorphogenesis is controlled at many stages in a plants life cycle, individual processes are highly specific for a particular plant part at a particular developmental stage. Light itself carries no morphogenic information, however, kind of photoreceptor is a specific information carrier. Hans Möhr, emphasized that photomorphogenesis has two following important stages:

### Pattern specification

In these stage cells and tissues develop and become competent to react to light, and pattern realization, during which time the light dependent process occurs.

### Need for amplification system

For photomorphogenesis, there is the need for an amplification system. The numbers of molecules in a plant that become altered by light may be several thousand or several million times the number of photons that cause the response. In many (but far from all) cases, gene activation represents part of the amplification process. This hypothesis is under revision and role of proteins is gaining more importance as a step between gene activation and response.

Some photomorphogenic effects of light can easily be noted by comparing seedlings grown in light and dark because large seeds with abundant food reserves eliminate the need for photosynthesis for several days. Dark-grown seedlings are etiolated. Following differences caused by light are:

1. Chlorophyll production is promoted by light.
2. Leaf expansion is promoted by light, but less so in the monocot (maize) than the dicot (bean).
3. Stem elongation is inhibited by light in comparison to etiolation caused under dark.
4. Root development is promoted by light.

Above differences seem advantageous for a seedling if it is to extend its stem through the soil and if its leaves are to reach the light. More of the food reserves in the endosperm (maize) or cotyledons (bean) are used to extend the stem upward in darkness than in light, and less food is used to develop leaves and roots and to form chlorophyll, all of which is less important for a dark-grown plant.

## Informational Photoreceptors

Plants are able to detect the quality, quantity and direction of light and use it as an external segnal to optimize their growth and develop-throughout the entire life cycle of plants are influenced by light. Plants possess at least two types of photoreceptors whose main physiological function is the acquisition of information. These photoreceptors include, the phytochrome which absorbs maximally in the red (R) and far-red (FR) regions of the spectrum and the blue (B) or UV-A (315-400 nm) photoreceptors. The blue/UV-A photoreceptors sense the blue and UV-A part of the spectrum and some of them have recently been cloned (Ahmad and Cashmore, 1993). Nothing is known yet about the molecular nature of the UV-B photoreceptors which regulate the formation of UV-Shielding pigments (Christie and Jenkins, 1996).

### *Cryptochrome, the Blue/UV-A Photoreceptor*

Julius von Sachs (1864) discovered the effects of blue light on plants and observed that phototropism is caused only by blue (and violet) wavelengths. Since then, many effects of blue light in many kinds of plants and fungi have been discovered. The near UV range consists of wavelengths just shorter than our eyes can detect; it is commonly called **UV-A radiation** and extends from 400 nm down to 320 nm.

Cryptochrome is likely a **flavoprotein** (a protein with riboflavin attached). It probably exists associated with a cytochrome protein in or tightly bound to the plasma membrane. Cryptochrome absorbs UV-A radiation, the largest peak in action spectra is usually in the blue-violet region near 450 nm. Most photo responses caused by cryptochrome probably result from absorbance of blue and violet wavelengths, hereafter simply called blue.

Blue light affects several process such as growth of the hypocotyls, stem, cotyledons and the leaves, stomata opening, phototropism, flowering and gene expression. Most of these processes are also activated by phytochrome, but several lines of evidence demonstrate the involvement of specific blue/UV-A photoreceptors. Action spectroscopy indicated the either flavins pterins or caroteroids could act as chromophores of blue light receptors up to now three blue light receptors or likely blue light receptor cardidates have been cloned from Arabidopsis and mutants have been identified for each.

The Arabidopsis CRY1 and CRY2 genes encode homologous proteins that been similarity to the two chromophorc microbial protolyases. The photoreceptors, cryptochromes 1 and 2 (Cry 1 and Cry 2) also have two chromophores (a flawn and either a pterin or a deazaflavin) but have no protolyase activity. Both Cry 1 and Cry

2 are involved in the compost of hypocotyls elongation responses to Blue light in Arabidopsis (Ahmed and Cashmore, 1993; Lin et al., 1998). Plant phototropic responses to blue light are controlled by a different kind of blue light photoreceptor which has been recently characterized as an anto-phosphorylatry flavoprotein in Arabidpsis (Photokinase NPH1; Christie et al., 1998).

## Phytochrome

### *Discovery*

Harry A. Borthwick (1972), Briggs (1976), and Furuya (1987) described the history of phytochrome discovery and made important contribution in the field of phytochrome research.

W.W. Garner and H. A. Allard in about 1920 (Beltsville) found that the relative duration's of light and dark periods control flowering in certain plants. Thereafter, (1938) it was discovered by others in cocklebur (it requires nights that are longer than some critical minimum length to flower) that flowering is prevented by a brief interruption of its dark period with light. Red light proved much more effective than other wavelengths, not only for interrupting the long nights that otherwise induce cockleburs and Biloxi soybeans to flower, but also for promoting expansion of pea leaves. Red light interrupting a dark period was also the most effective colour of light in promoting flowering of Wintex barley and other long day plants that require nights that are shorter and days that are longer than some critical length.

Borthwick and Hendricks obtained an action spectrum with a peak in the red wavelengths for promotion of germination of Grand Rapids lettuce seeds, only 5 to 20 per cent of which will usually sprout in darkness.

It was noted that when seeds were exposed to far-red light just after a promotive treatment with red light, promotion was nullified, but if red light was given after far red light, germination was enhanced. During the course of investigation it was realized by the plant physiologists that a blue pigment was present that absorbed red light (it was named $\mathbf{P_r}$); but its concentration was too low to give colour to etiolated maize seedlings, in which it was first detected by absorption changes with a spectrophotometer. The pigment could be converted by red light to a different form (called $\mathbf{P_{fr}}$) that absorbed far-red light (a form that eventually proved to be olive green in colour) and that the blue pigment could be regenerated with far-red light. The olive form produced by red light was deduced to be the active form, whereas the blue form seemed to be inactive.

H. W. Siegelman and other protein chemists in early 1960s purified phytochrome from homogenates of cereal-grain seedlings by using column chromatography and other techniques routinely used to purify proteins. They demonstrated that isolated phytochrome changes colour reversebly upon exposure to either red or far-red light. Essentially all the early deductions based only upon physiological experiments with whole plants had been verified. Even the absorption spectrum of each form of phytochrome was measured. These investigators' conclusions are shown by the photo reversible scheme below:

$$P_r \underset{\text{far red light}}{\overset{\text{red light}}{\longleftrightarrow}} P_{fr}$$

### *Physical and Chemical Properties of Phytochrome*

Physical and chemical properties of phytochrome were studied after obtaining pure phytochrome pigment from etiolated seedlings. Seedlings were selected as a source material because when they were grown in total darkness, they contain 10 to 100 times as much total phytochrome as seedlings grown in light; they have no chlorophyll that absorbs blue and red light and interferes with spectrophotometric studies of phytochrome. Only recently, phytochrome in green plants has been studied much.

### *Distribution of Phytochrome in Plants*

Presence of phytochrome has been known in gymnosperms, liverworts, mosses, ferns, and some green algae. Recently it has been reported to exist in certain red and brown algae and probably absent in bacteria and fungi. Little is known about chemical properties of phytochrome in species other than angiosperms.

Distribution of phytochrome in grass shoots is variable, but oat, rye, rice, and barley seedlings all have high concentrations in the apical regions of the coleoptile, near the shoot apex, and in the growing leaf bases. Phytochrome is present in most organs of all plants investigated, including roots. In all plants phytochrome is present and synthesized entirely as $P_r$; apparently no $P_{fr}$ can be synthesized in darkness.

When seeds germinate and seedlings develop in darkness, the amount of type 1 phytochrome increases by roughly 100 times; thus, dark-grown seedlings contain far more total phytochrome than do light-grown seedlings, as mentioned earlier. The abundance of type 1 phytochrome probably allows these seedlings to intercept even very weak light and develop into normal green plants. When seedlings receive light, one of their responses is to lose most of their type 1 phytochrome because of the following reasons:

1. They stop making more mRNA needed to synthesize that phytochrome.
2. Type-1-phytochrome mRNA appears to be an unstable mRNA (is rapidly hydrolyzed).
3. Most of the type-1-phytochrome protein is rapidly degraded.

**Destruction and Dark Reversion Processes:** When etiolated plants are exposed with a brief red-light treatment, some of the $P_{fr}$ that is formed disappears gradually, even in darkness. Two processes account for this property

First process is **destruction**, because after an interval of time in darkness it is no longer possible to regenerate as much Pr in tissues by an exposure to far-red light; thus the total amount of detectable phytochrome is less than before.

The second process is **dark reversion** to $P_r$, which usually requires a few hours. (It should be emphasized that destruction and reversion also occur in the light but are not directly caused by light.

#### **Ubiquitin**

Degradation of type 1 $P_{fr}$ is now known to occur by a highly selective process in which $P_{fr}$ is first attached to a small protein called ubiquitin. It is found in all eukaryotes examined and absent in prokaryotes. It is a small protein of only 76 amino acids, and its amino-acid sequence is highly conserved in various organisms. Attachment of ubiquitin to $P_{fr}$ or to other proteins targets them for degradation, and to carry out degradation's ATP and three enzymes are required. Proteases then recognize and hydrolyze the targeted proteins and release free ubiquitin.

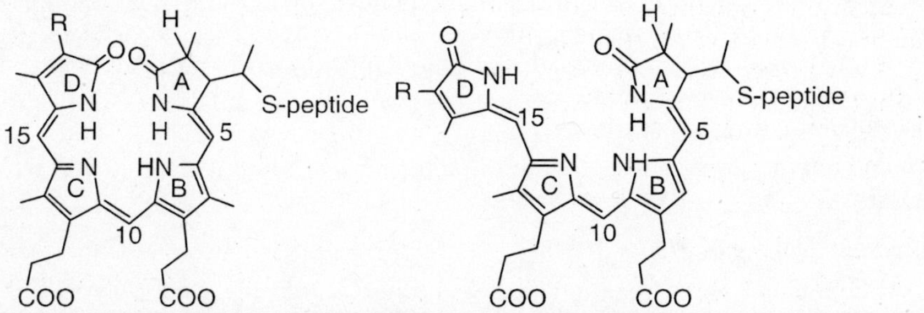

**FIG. 29.1:** Chemical configuration of phytochrome chromophore (a) $P_r$ (B) $P_{fr}$.

## Intracellular Distribution

Phytochrome has been found in all parts of plant including stems, leaf blades, petioles, cotyledons, coleoptiles, hypocotyle, vegetative buds, floral receptacles, inflorescences and developing fruits. When the intracellular localization in etiolated seedlings was studied, the Pr form was found to be evenly distributed is the cytoplasm. On conversion into Pfr from it is rapidly bound to discrete as yet unidentified sites in the plasmamembrane. In green lower plants, phytochrome is permanently localized in a dichroic conformation in or near the plasmamambrane.

### *Absorption spectra of Phytochrome*

The photoreceptor molecule is reversibly inter-convertible between its inactive (Pr) and active (Pfr) forms by sesnatial absorption of red and far-red photon of light. This was subsequently confirmed by isolation of photoreceptor. Only the Pfr form is believed to be responses induced with red light can be reversed with far-red light. However, this dogma has been challenged by studies that have implicated a role for Pr in regulatory seed germination (Reed et al., 1994) and gravitropism (Mancinelli, 1994). Thus, with the exception of seed germination and gravitropism under certain light conditions, Pfr is considered to be the active form and Pr the inactive form of the photoreceptor (Chory et al., 1996). It is suggested that the Pyr form of phytochrome is less stable than the Pr form. The basic observation is that the Pyr form of phytochrome in solution undergoes dark reversion to the Pr from. The dark reversion is pH-dependent. A variety of reducing agents greatly accelerates the dark reversion.

The absorption spectra of highly purified phytochrome molecules from etiolated angiosperms have maxima in red wavelengths at about 666 nm for the red absorbing bright blue form ($P_r$) and at about 730 nm for the far-red-absorbing olive form ($P_{fr}$).

### *Phytochrome is a Pigment System: Some Evidences*

Following are the evidences which reveal that phytochrome is a pigment system:

1. There are similarities in the absorption spectra of phytochrome and the action spectra of plant responses is one important evidence suggesting that phytochrome is the pigment causing such responses.
2. The responses caused by red light are, always nullified by an immediate subsequent exposure to far red light.
3. Only low irradiance levels of either red or far-red light that are capable of inter converting phytochrome from one form to another cause these responses.
4. In leaves and stems, light absorbed by chlorophyll alters both the action

spectra for phytochrome responses caused by $P_{fr}$ and the amount of light (absorbed by $P_r$) required for the response.

5. Both $P_r$ and $P_{fr}$ absorb violet and blue light, but low irradiance levels of these wavelengths are much less effective than red or far-red light for the physiological processes.

In general, green safelights and green filters are highly useful in phytochrome studies.

### *Chemical Nature of Phytochrome*

1. Chemically, phytochrome is a homodimer of two identical polypeptides, each with a molecular weight of about 120 kDa.
2. Each polypeptide has a prosthetic group called a **chromophore** that is attached via a sulfur atom in a cysteine residue of the polypeptide. This chromophore is an open-chain tetrapyrrole similar to the photosynthetic phycobilin pigment of red algae and cyanobacteria.
3. The chromophore, not the protein, that absorbs the light and causes phytochrome responses.
4. When $P_r$ is converted to $P_{fr}$ by red light, there is apparently a *cis-trans* isomerization in the chromophore. Alteration of the chromophore in phytochrome then causes several unidentified subtle changes in the structure of the phytochrome protein.
5. The changes in protein structure are somehow responsible for the physiological activity of $P_{fr}$ and the inactivity of $P_r$.

### *Two Types of Phytochrome Systems*

In plant system, there exist two major types of phytochrome:

I. **Type 1** predominates in green plants and seeds (at least oat seeds).

II. **Type 2** from green oat seedlings is slightly smaller (118 kDa) than that from etiolated oats (124 kDa), but it too exists as a dimer in vivo and has spectral properties similar but not identical to those of etiolated type 1.

#### Spectral Difference

The important spectral difference in types 1 and 2 of oats is that the $P_r$ form of type 2 has an absorption maximum near 654 nm, as opposed to the 666 nm peak of type 1 $P_r$. If type 2 $P_r$ absorbed maximally at 654 nm in all species, that would help explain the differences in action spectra of dark-grown and green plants; however, type 2$P_r$ from green pea seedlings has an absorption peak at 667 nm, the same as type 1 from dark-grown seedlings. Type 2 $P_{fr}$ formed from both green pea and oat seedlings has an absorption maximum in vitro at 724 nm, as opposed to 730 nm for type 1 phytochrome from various plants.

#### Immunological Reactions

Based on immunological reactions, the proteins of phytochrome types 1 and 2 are quite different, even though some important similarities exist. However, even within a single green plant more than one type 2 form occurs. The difference between the type 2 phytochromes is as great as that between either of them and type 1. The structures of these genes and the phytochromes they code for are now being studied intensively.

### Photomorphogenesis and Dose-Response Relationships

A **short exposure** of light promotes formation of the purple pigment (anthocyanin) in white mustard (*Sinapis alba*) seedlings which exhibits a peak near 660 nm, where $P_r$ absorbs most effectively. However, when exposures of **several hours**, there is a large peak near 725 nm in the far-red region and a smaller peak in the blue region. Also, when etiolated lettuce seedlings are given **extended periods** of light, hypocotyl elongation is inhibited and again the action spectrum has a distinct peak in the far-red region near 720 nm. Other peaks occur in the blue and UV-A region, but here (and frequently in other examples) there is no action in the red region where $P_r$ absorbs. Plant responses that require relatively high irradiance levels and that have action spectra that are a typical of commonly observed phytochrome responses came to be known as **high irradiance reactions** or **responses**; **HIR**.

### Synthesis of Phytochrome

Phytochrome is synthesized in the Pr form. In a dark-green seeding only Pr (the inactive form) is present. The physiologically active Pyr originates under the influence of red light and a large part of Pr is transformed into Pyr. Both forms of Phytochromes have different spectral absorption ranges but there is considerable overlap between the two forms. Due to this overlape the photoequilibium of Pfr/Pr depends on the wavelength and is about 80% in red (660 nm), 3% in far-red (730 nm) and about 40% in blue (450 nm) light. Due to the activation of phytochrome by blue light, a photoresponse induced by blue can be caused by phytochrome or specific blue light receptors. The photoconversions of Pr to Pfr and of Pyr to Pr involve a series of short-lived intermediates.

### Biochemistry of Phytochromes

Phytochromes are cytosolically located dimes composed of two ~ 125 KD Polypeptides, each carrying a covalently linked tetrapyrrole chromophore in the $NH_2$-terminal domain and dimerization determinates in the COOH-terminal domain. The N-terminal portion confers the spectral property characteristic to phytochrome. In the C-terminal domain is critical for transferring the light signal to downstream components. A small N-terminal extension of variable length has also been implicated in signaling (Fankhauser and Corey, 1997).

Phytochromes are biliproteins similar to whole of phycobiliproteins in cyanobacteries and red algae. The results indicate that cyanobacteria harbour useful information about the phytochrome system. Action spectroscopy reveled photo-reversible effects analogous to those of plant phytochrome but in plants the responses are maximally induced by red light and reverted by green light. Thus, the phytochrome appeared before eukaryotes, evolving over vast tracts of time and under changing selection pressures to glean and then transmit pertinent information about the light environment to allow the organism to respond appropriately (Hughes and Lampartner, 1999).

### Phyto-Conversion of Pr form to Pyr form

Rudiger and Thummler (1991) proposed a model reaction which involves Z.E. isomerization of tetrapyrrotes. The position of isomerization in the Pfr peptide was traced to the C15-C16 dmble bond by 1H NMR spectroscopy hence, the Pr chromophore is assigned as 5Z, 10Z, 15Z configuration 20 and the Pyr chromophore as 5Z, 10Z, 15E configuration 21.

Resonance Raman spectroscopy and surface Enhanced Resonance Raman Scattering Spectroscopy studies also confirm a 15Z configuration for the chromophore in the Pr form and a 15E configuration for the Pyr form. In addition to thus, the phytochrome polypoptide undergoes conformational changes throughout its length. It has been detected by protease susceptibility, phosphorylation by exogenous kinases and immunoreactivity. Taken together these data indicate that the conformational differences better Pr and Pyr involve the N-terminal portion of the phytochrome molecule and the N-terminal 6KDa region has important effect on the properties of the photoreceptor.

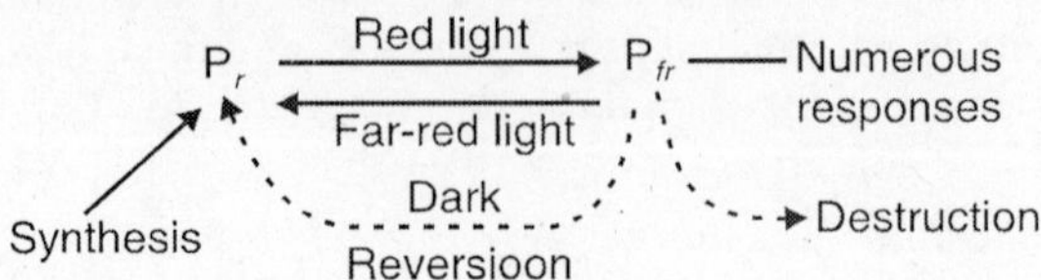


**FIG. 29.2:** Effect of red and far red light on phytochrome.

Protcolytie removal of the N-terminal 6 to 10 KDa results in the following:

1. a shift in the Pyr absorption maximum from 730 to 722,
2. a decrease in the Quantum yield for the photoconversion of Pr to Pyr,
3. a greater than 10-fold increase in the rate of dark reversion from Pyr to Pr,
4. increased reactivity of the chromophore with tetranidromethane, and
5. changes in the Pyr circular dichroism spectrum.

## Biosynthesis of Phytochrome

Elzch and Lagarias (1989) provided evidence that the phytochromobilin chromophore in synthesized in vivo from home with biliverdin IXλ as an intermediate. In seed plants the enzyme for bilir synthesis are located in the plastid but are nuclear encoded. The pathway begins with the opening of the home tetrapyrrole range between Tyrroles A and D by home oxygenase to form the linear tetrapyrrole biliverdin IXλ. Genes for this enzyme known from Synechocytis Pcc 6803 from the plastome of red algae and from the genomes of trabidopsis and several animals. In trabidopsis home oxygenize is encoded by the nuclear gene HY1 (Muramoto et al., 1990). The subsequent steps are not fully understood biochemically nor have any further associated genes have been described.

## Cyno Cyanobacterial Phytochromes

Phytochrome operates in green algae, mosses, Pteridophytes and spermophytes. It is lacking is fungi. Recently, genome sequencing of the cyanobacterium *Synechocystis* sp. PCC 6803 at the Kazusa Institute is Japan has provided wealth of valuable new data. Among the 3,168 open leading frames identified, two phytochrome-like open reading framers were recognized. The one that resembles highs plant phytochrome genes most closely is called Cph1 (cyanobacterial phytrchrome). The Cph1 sequence encodes on putative protein trits an N-terminal Chromophone domain resembling that of higher plant phytochromes. The C-terminal domain resembles a histidine-kinase domain of the bacterial two-component family. It is shorts than the corresponding domain in higher plant phytochromes, which have a cryptic duplication of the C-terminal region of second open reading frame, called rep1 (response regulator for cyanobacterial phytochrome), ten nucleotide away from Cph1 ercodes a putative response regulator. Yeh et al., (1997) expressed rcp1 as a fusion protein in E. coli, pure-fired it and included it in the phosphorylation reactions with cph1. They found that cph1 could can transfer a phosphate group to rcp1. Intriguingly, the Pr form of chp1 carry out thus phosphotransfer reaction but prephosphorylated cph1 that was converted to Pfr can not. Thus the Pr from of cph1 has both a to phosphorylatoin and phosphotransfer activities whereas the Pyr-form has little or none of either activity.

### Bacteriophylochromes

The reach detection of phytochrome-like proteins in lower plants, algae, cyanobacteria and purple bacteria suggests that all photosynthetic organized contain phytochrome (Schneider-Poetsch et al., 1998; Jiang et al., 1999; Hughes et al., 1997). The distribution of these chromoproteins has been expanded beyond photoantotrophs with the discovery of phytochrome-like proteins in the non-photosynthetic eubacteria, *Deinococcus radiodurans* and *Psendomonas aeruginosa* (Davis et al., 1999). The genes, designated BphP for bacteriophytochrame photoreceptor, were discovered by scanning genomic databases for coding regions similar to whole of phytochrome. The in coded D. radiodurans BphP (Dr BphP) and *P. aernginosa* BphP (Pa BhpP) proteins are 755 and 728 amino acids respectively with an overall amino acid sequence identity of 37% and a similarity of 48% each other. Their $NH_2$ -terminal ~ 500 amino acids are similar to the chromophore-binding region of Cph1, RcaE (Fremyelia diplosiphon), Ppr (Purple bacterium *Rhodospirillum continuum*) and plant phytochromes. Like photosynthetic bacteria sequences, the COOH-terminal ~ 250 amino acids of Dr BphP and PaBphP are related to histidene kinase domains present in environmental sensors that function as two-component regulators (Hoch and Scilhavy, 1995). This region contains the four conserved motifs that compose the catalytic center including a positionally conserved hestidine that served as the phosphorylation lite.

Like plant phytochromes, the D. radio-durians receptor covalently bonds linear tetrapyrroles autocrtalytically to generate a photochromic holoprotein. Mutational analysis suggest that D. radio durans bactariophytochrome functions as a light-regulated histidine kinase which helps protect the bacterium from visible high. The D. radio-durans BphP pathway would provide a useful paradigm for studying the mechanism (s) of phytochrome action because in offers for the fresh time simplified phytochrome-like response in an organism that is naturally devoid of photosynthesis (Davis et al., 1999).

### Molecular Species of Phytochrome

Physiological, spectroscopic and immunological studies have revealed that plants contain two operationally defined type of phytochrome. The predominant phytochrome species isolated from green oat seedlings shoos an absorption maximum in the red region of the spectrum at 652 nm raths than at 660 nm. The absorption maximum of green phytochrome in the far-red is same (730 nm) as that for etiolated phytochrome.

Etiolated and green phytochrome are immunologically district the results show that immunologically distinguishable pool of phytochrome is also present in green oat seedlings shoot. An absorption maximum in the red region of the spectrum at 652 nm rather than at 660 nm. The absorption maximum of green phytochrome in the far-red is same (730 nm) as that for etiolated phytochrome. Etiolated and green phytochrome are immunologically distinct.

The immunologically distinguishable pool of phytochrome is also present in green pea seedlings. Green and etiolated oat phytochrome molecule can also be distinguished by peptide mapping. The spectrophotometrically detectable phytochrome present in oat seeds is relatively stable and is immunologically district from etiolated oat seedling phytochrome. The phytochrome present in oat seed is similar to the phytochrome molecule isolated from green oat seedlings. The phytochrome existing predominantly in etiolated tissues is designated as Type 1 add is rapidly degraded as Pyr in the cell. Type II is most abundant in light-green tissues and is stable as Pyr.

### *Phytochrome Genes*

Conclusive evidence for the presence of distinct types of phytochrome in a single plant species was provided by the cloning and sequencing of genes from the *Arabidopsis* genome (Sharroek and Quail, 1989; Clack et al., 1994). In a further step, cDNA and genomic DNA of phytochrome genes was cloned and sequenced from various seed plants, like ferns, mosses and algae. In *Arabidopsis* there are five genes and are known as PHY A PHYB, PHY C, PHY D and PHY E. The PHY A gene, encoding the type 1 phytochrome species, is highly expressed in etiolated plants and is repressed by Pyr (Somers and Quail, 1995).

Furthermore the Phy 4 protein is rapidly degraded in the light was ubiquities-dependent pathway. PHy B encodes a type II phytochrome species. A comparison of the expression of phytochrome B gene promoter-B-glucuronidase (PHYB : GUS) gene fusion in transgene *Arabidopsis* demonstrated that PHY B expression in etiolated seedlings is suppressed two fold by while light. However, phytochrome B is stable in the Pyr form and as a result of Phy B is the most abundant species in light-growth plants. The remaining phytochrome species (phytochrome B to phytochrome E) constitute the type II phytochrome which are the predominant pool of phytochrome is green blanks.

Phylogenetic analysis of the phytochrome genes of higher plants suggests that duplication of an ancestral gene at about the time of origin of the seed plants led to the PhyA and PhyC horologes and the others giving rice to PhyB PhyD and PhyE horologes. Subsequent duplications are proposed to have occurred near the time of the origin of flowering blank (Mathers and Sharrock, 1997).

In *Arabidopsis* the PhyB and PhyD proteins Share approximately 80% amino acid sequence identity and are thought to result from a gene duplication in a recent progenitor of the cruciferous. The PHY B and PHY D proteins are more closely related to PHY E (with approximately 55% identity) than they are either the PHY A or PHY C proteins (with approximately 17% identity).

#### **Structure of the Phytochrome Genes**

Phytochrome cDNA and genomic clones have been isolated and sequenced from various seed plant and the primary structure of genes has been determined. The phytochrome genes consists of several exons interspersed by Shorter introns. For example, the oat PHY A gene is 5.94 KB long and consists of Six exons and five introns. The PHY A genes also show long intron sequences in the 5'- and 3'-flanking, untranslated regions of the gene. In rice, the 5'-terminal inform is 2585 bare pains in length and represents one of the longest introns in a plant gene. In contrast to the phytochrome genes founding monocot plants, the pea phytochrome gene consists of five exons and form short. The organization of the PHY C gene (from moss cerrtodon) differs from the pattern of the PHY A gene. The PHY C gene compries just two large coding exons on either on either side of a short, control exon.

The PHY A genes in oat maize, rice, and pea contain sequence motifs in which promotion upon play an important role in regulation of their expression. These include a TATA motif, aCCAAT-motif and a GC element, which are found in the promoter region of many eukaryotic genes. These genes also show AATAAA or related mosifs in the 3' untranslated region which are responsible for phyadenylation of the corresponding transcripis.

The PHY A promoters of oat and rice have been functionally characterized. A combination of deletion analysis and linker-scan mutagenesis identify three cis-regulating elements in oat. These are designated as PE 1 (Positive element-1), PEB (Positive element-3) and RE 1 (repressor element-1), PE 1 and PE 3 gel synergistically to support maximal expression under derepressed conditions (100 Pyr levels). Mutation of either elements decrease expression to based levels. In contrast, mutation of the RE 1 element results in maximal transcription under PHY A transcription through this negatively acting element. The nice PHY A promoter contains a triple of GT-element which are functionally equivalent to the oat PE 1 elements. These elements are boiled by a transcription factor known as GT-2.

### *Phytochrome Function*

The classical criterion for a phytochrome response is reversibility. If the effect of red light is called by a subsequent pulse of far-red light, the effect can be classified as a low-flense response (LFR). These are this phytochrome responses which are induced by such 100 intensities of red light that the Pfr/Ptot ratio is wither the same range as when as saturating far-red pulse alone is given. Such a response can not be reversed by far-red treatment. The response to such 100 light intensities is called VSY 100 fluence response (VLFR). VLFR can be induced by red light in the range between $10^{-10}$ and $10^{-7}$ mol $m^{-2}$. In addition, there are phytochrome responses which depend on prolonged exposure to relatively high light intensities. It is called high irradiance response (HIR). Like the VLER, the HIR is not photoreversible. The classical HIR described by Hartmann (1966) operators in continuous far-red light but these is also an HIR is Red and other light qualities since HIRs depend on prolonged light exposure, the reciprocity law is not valid for these response involves.

Establishing the voles of the individual phytochrome species has been the subject of extensive research and much has been revealed from the identification of mutants deficient in defined phytochromes and over expression of different phtochromes in wild type and mutant backgrounds. The most extensive range of phytochrome mutants has been isolated in *Arabidopsis*. Phytochrome A-deficient (PHY A) *Arabidopsis* do not germinate in high intensity far-red and poorly in very low intensity light over the whole range from UV-A to far-red (Shinomura et al., 1996).

*Arabidopsis* PHY A and PHY B double mutants show phytochrome responses. For example, seed germination in the PHY A/PHY B double mutant is still red-far-red reversible indicating that the involved phytochorme belongs to the LFR type but shows a fluence rate depending similar to PHY A (Poprpe and Schafer, 1977). PHY C function has been studied by over expression. Transgenic tobacco and *Arabidpsis* lines over expressing PHY C both show on red light dependent increase in leaf expansion Qin et al., (1997).

### *Heterodimer-Homodimer Hypothesis for Phytochrome A-and-B Specific Response*

Biochemical and biophysical experiments have demonstrated that the purified samples of both PHY A and PHY B form dimers. A phytochrome monomer dimerizes with a another identical phytochrome monomer such as PHY A with PHY A or PHY B with PHY B (Furuys and Schafer, 1996). Mutants studies show that a photoirreversible VLFR results from PHY A and a photoreversible LFR results from PHY B. Furuya and Schafer (1996) proposed a hiterodimer-homodimer model for phytochrome activity. According to this model PHY A combines as a heterodimer (PHY $A_r$ PHY $A_{yr}$) to induce the response, while PHY B is snitched on in the homodimeric PHY $By_r$ PHY $By_r$ form and off in the PHY Br PHY Br form. In this model, the heterodimer of PHY B would be unable to snitched on or off in the terms of PHY B signaling. The heterodimens of PHY A is functional upon very low fluence irradiation and approximately the same amounts of PHY A heterodimene are present at the VLFR threshold as PHY B homodimere at the LFR threshold. This implies that PHY A and PHY B transdince the light-induced.

### *Domains Involved in Phytochrome Function*

As outlined above, different phytochromes have different response modes: PHY A mediates VLFR and FR-HIR; PHY B mediates the LFR. The different response modes of phytochromes may suggest that different phytochromes use different signaling pathways but Domain Swap experiments performed by Quail and Co-workers (1996) argued against it. It indicates that the N-terminal domain determines the photosensory specificity whereas the C-terminal domain is reciprocally interchangeable between PHY A and PHY B suggesting a common molecular function (Wagnor et al., 1996).

### *Degradation of Phytochrome A*

Phytochrome A is a short-lived protein and is rapidly degraded in the light. It is highly expressed in young seedlings and is rapidly defined once converted to Pyr. The short half-life of phytochrome A Pyr helps seedlings adapt to continuously fluctuating light conditions as they grown through the soil by removing previous light signals stored as Pyr. Shanklin et al. (1987) investigated the possible involvement of the urbanite pathway in PHY A degradation in oat and found that soon after Pyr formation, the chromoprotein in vivo. Although a direct link between ubiquitination and involvement. Selective ubiquitination of Phy A could involve specific structural differences and/or differential aggregation between the Pr and Pfr form (Vierstra, 1994). The extreme C-terminus of Phy A, appears to be liquefied as its removal stabilizes the chromoprotein as Pyr.

### *Regulation of Gene Expression by Phytochrome*

Phytochrome plays a crucial role in regulating plant development in response to light. Considerable evidence has been obtained linking the change in expression of specific light-responsive genes to the photoconversion of phytochrome from the Pr to the Pyr form. Prominent light-induced genes controlled by phytochrome include those encoding the small subunit of ribulrse-1,5 bisphosphate carboxylase and the chlorophyll a/b binding (CAB) protein of the light harvesting chlorophyll-protein complex.

Phytochrome negatively regulates the expression of its non gene(s) as well as gene(s) for protochlorophyllide reductase in some plants. In addition to phytochrome, one or more blue/UV photoreceptor play an important role in plant gene expression as has been demonstrated for chaconne synthase (Kuhlemeier et al., 1987; Watson, 1989; Smith and Bhitelam, 1990). On the other hand, more than 50 different enzymes are under photocontrol. The photocontrol is exerted at the level of selective gene expression that has been examined by inhibitor treatments, in vivo labeling immunological and molecular biological techniques.

The participation of gene activation and repression in phytochrome-mediated photomorphogenesis was suggested as early as 1964 by Mohr and collaborator's (Hock and Mohr, 1964; Mohr, 1972). Mohr Suggests that the genome of each cell, at any pent is development, consists of four closes of genes as characterized by their state of activity and their potential for reacting with Pfr. The four classes are 1. inactive genes 2. active genes 3. inactive genes which are activated by Pyr and 4. active genes which are repressed by Pyr. Gene in category (1) are not required for the particular cell either in the presence or absence of Pyr (Eg. genes related to root development being inactive in a shoot cell). Category (2) genes are required for the cell irrespective of Pyr availability (Eg. genes related to synthesis of the enzyme of basic metabolism). The other two categories of genes are regulated by Pyr either directly or indirectly.

### *The CAB Gene: A Model for Phytochrome Regulation of Transcription*

Expression of the plant CAB gene family encoding the chlorophyll a and b binding proteins of the light-harvesting complex of phtosystem II, is one of the most extensively studied plants gene families whose expression is phytochrome regulation at the level of transcription (Karlin-Neumann et al., 1988). In plants grown from seed complete darkness (etiolated seedlings), little or no CAB mRNA can be detected upon transfer of etiolated seedlings to the light, CAB transcription is induced to high level trits concomitant in CAB protein level s. A brief pulse of red light is sufficient for the induction of CAB expression in etiolated seedlings. The red light induction of CAB expression is attenuated by subsequent far-red illumination, demonstrating that phytochrome mediates this response (Quail, 1991): Similarly, a single pulse of blue light acting through a nonphytochrome phtoreceptor system induces CAB expression in etiolated pea seedlings (Marrs and Kaufman 1991).

By using a type-specific Phy A null mutant and a Phy B null mutant of *Arabidopsis*, Reed et al. (1994) demonstrated phytochrome induction of CAB expression in etiolated seedlings of the mutants after an inductive red light pulse that was photoreverseble by far-red light. CAB mRNA accumulated in these mutants to a level similar to that observed in wild-type seedlings. These results suggest that either Phy A and Phy B have overlapping roles in the acetic phytochrome induction of CAB expression by red light in etiolated tissue. Sun and Tobin (1990) observed equivalent levels of CAB mRNA in etiolated seedlings of both wild type and an *Arabidopsis* Phy B mutant after an inductive red light pulse. Reed et al (1994) observed less CAB mRNA accumulation in etiolated seedlings of the Phy A-Phy B double mutant after a red light pulse. It suggests that Phy A and Phy B may have redundant roles in mediating CAB expression and a species other than Phy A or Phy B (eg. Phy C, Phy D Phy E) may mediate CAB expression.

The promoters of CAB genes are well characterized. *Arabidopsis* contains three well-characterized copies of CAB encoding identical mature proteins as well as several other CAB gene family members.

### *Interaction of Phytochrome with Transcription System*

Transcriptional activators or repressors generally consists of several functional modules with DNA-binding, activation and repression domains. On the basis of their usually highly conserved DNA-binding domains most eukaryotic transcriptional factors can be assigned to one of several classes, these include basic region – leucine zipper (bZIP) proteins, MYB-live proteins MADS – domain proteins, helix-loop – helix proteins zinc-finger proteins and homeo-box proteins (Meshi and Iwabuchi, 1995). Within these classes, DNA-binding specificity is brought about by suitable changes in the amino acid sequence of the DNA-binding domain.

Transcriptional factor activity is also affected by interaction with the other DNA-binding proteins rization increases the number of transcription factor complexes that can be formed from a smaller number of homeomenie components and can have a positive or negative influence on their DNA-binding characteristics activation potential or nuclear realization. The dimerization process is regulated by the availability of their respective partners, their affinity for each other and the affinity of the dimmer for other components (Such as DNA) – these sensitizes responses to miner fluctuations in effector concentration. The dZIP and the MADS-domain class transcription factors are well characterized dimerizing proteins in plants.

The G-box is a ubiquitous, cis-acting DNA regulatory element found in plant genomes. The G-box plays a role in the responsiveness of plant promoters to light anaerobiosis, P-coumaric acid and hormones such as abscisic acid examples, deletion or mutation of the G-box dramatically reduces the overall promoter activity and/or renders the promoter unresponsive to a given stimulus.

Recently, Peter Quail's group (Ni et al., 1998) demonstrated that the physical association between interesting factor 3 (PIF 3) is strictly light-dependent in vitro. PIF3 belongs to the basic helix-loop-helix (bHLH) class of transcriptional regulators which specifically bends to various light-regulated promoters containing a G-box DNA sequence motif. PIF3 contains a single PAS-like domain in its N-terminal region and specifically interacts trits the C-terminated domains of both Phy A and B. To corroborate these findings, Pets, Quail et al., (1998) concluded the following:

1. Transgenic plants overexppressing the PFI3 gene in antisense orientation showed reduced responsiveness to red and far-red treatments.
2. Interaction of PIF3 trits the mutant derivatives of the C-terminal domain of Phy A and Phy B, which has been found previously to affect Phy A and Phy B specific signaling in vitro, was also severely compromised in vitro.

### *Phytochrome Signaling Molecules*

For optimal growth and development, plants require the ability to complex developmental processes and at the same time sense and respond to endogenous physiological factors and environmental cues. The response can be divided into three main exacts: *stannous perception, generation and transmission* of a signal and the subsequent changes in downstream biochemical process. For this process to be successfully executed specialized receptor molecules, which perceive the endogenous incoming cue, are required to generate chemical and/or protein effectors molecules to transmit the signal in order to trigger subsequent downstream responses. Due to the complexity of signaling cascades, transduction pathways require sensitivity and specificity that are coordinated and integrated by positively and negatively acting signaling components. The combination of genetic and biochemical approaches to gain insight into plant signaling networks, s revealing the inherent complexity of how plants respond to developmental and environmental cues.

Light is an extremely important environmental cue and because of their photosynthetic and non-motile nature, plants need to be particularly adaptive to their light environment. Light provides both energy for photosynthesis and informational signals to optimize growth and development. The perception and transduction of these light signals is accomplished through the use of specialized photoreceptors of the three major classes of plant photoreceptors, the phytochromes are the best characterized and are important in controlling morphological responses that are underbinned by changes in gene expression (Kendrick and Kronenberg, 1994).

Studies based on biochemical and genetic approaches have dramatically advanced our understanding of phytochrome signal transduction. Even through the immediate signaling partners of phytochrome have remained obscure, recent progress has provided a wealth of potential players.

1. **First**, the phytochrome mutants and over expression of plants suggest that individual subtypes of the phytochrome family play a distance but overlapping roles, with evidence for cross talk between the pathways triggered by Phy A and Phy B.
2. **Second**, genetic screens have uncovered mutants in over two dozer genes that delimit both positive and negative steps involved in phytochrome legalizing.
3. **Third**, cellular biological and pharmacological.

These studies have implicated specific biochemical intermediates is phytochrome action.

### *Involvement of Second Messengers in Phytochrome Signaling*

In terms of Phy A signaling, Phy A mutants have been used extensively in defining primary signaling intermediates, mostly small molecules between Phy A and gene expression using microinjection approach it has been demonstrated that putative heterotrimenic G proteins acted as an upstream component of the Phy A signal transduction pathway (Bowler et al., 1994, Neuhaus et al., 1993). Treatment of etiolated oat seedlings with cholera toxin, an activator of G-prolicins induced CAB gene expression and inhibited Phy A expression (Romero et al., 1991).

In elegant series of experiments involving the microinjection of various putative signaling molecular as well as agonists and antagonists of these molecular into single hopocotyl cells of the aura mutant of tomato, which leaves functional phytochrome A (Sharms et al., 1994), has been used to further identify signaling molecular functioning in the phytochrome-mediated pathway controlling gene expression.

Results from several studies demonstrate that both a $Cr^{2+}$/calmodulin-dependent step (Neuhaus et al., 1993) and CGMP (Bohler et al., 1994) function in the phytochrome signal transduction pathway. Injection of either $Ca^{2+}$ or $Ca^{+}$-activated calmodulin is sufficient to induce the expression of a CAB: GUS reporter gene construct and partial chloroplast development including the accumulation of components of PS II, LHC I and II, ATP syntheses and Rubisco.

On the basis of the preceding results a model for phytochrome signal transduction pathways has been proposed in which the Pfr form of Phy A activates one or more G-proteins, which in turn activate downstream signaling pathways regulating subsets of phytochrome-mediated responses (Bowler et al., 1994). In pathway, G-protein mediated increases in CGMP levels results in the activation of genes encoding proteins involves in anthocyanins production. In the second pathway, G-protein activation result sin increased cellular $Cr^{2+}$ levels, producing $Cr^{2+}$- activated calmodulin, which subsequently induces the transcription of PSII, ATPase structural genes CAB and RBCS and Partial chloroplast development. Signals from both the $Cr^{2+}$/calmodulin and cGMP pathways must converge at some porch and are required for the transcription of genes encoding components of PSI and cytochrome b6f complexes and for complete chloroplast development.

The downstream target of these positively acting pathways are divided along functional lines, with the cGMP pathway controlling the production of photoprotective anthocyanins and the $Cr^{2+}$ anthocyanins and the $Cr^{2+}$- and $Cr^{2+}$/cGMP – dependent pathways controlling the expression of genes encoding chloroplast-localized proteins and chloroplast development.

Differential regulation of the respective pathways has been proposed as a mechanism for regulating the relative levels of photosynthetic complexes and photoprotective anthocyanins (Bowler et al., 1994). Accordingly, manipulation of signal flow through these respective pathways has been used to demonstrate that high cGMP levels negatively regulate the $Cr^{2+}$- and $Cr^{2+}$/cGMP-dependent pathway and, conversely, that high $Cr^{2+}$/calmodulin levels negatively the cGMP pathway. Termed reciprocal control (Bowler et al., 1994), the negative interactions between these pathways may be both ecologically and physiologically relevant. For example, the negative regulation of the $Cr^{2+}$ - and $Cr^{2+}$/cGMP-dependent pathways by high cGMP may prevent the development of photosynthetic competency prior to the production of sufficient photo-protectants, and the $Cr^{2+}$/calmodulin pathway may suppress the cGMP pathway once high levels of anthocyanins are no longs required. Moreover, the $Cr^{2+}$/cGMP-dependent has a 10-fold lower requirement for cGMP than does the cGMP-dependent pathway, which would allow for chloroplast development in the absence of the anthocyanin biosynthesis (Bowler et al., 1994).

Phosphorylation events downstream of cGMP and $Cr^{2+}$/calmodilin are likely to play a critical role in transducing the phytochrome signal. In animal signaling pathways, cGMP and $Cr^{2+}$/calmodulin function by the activation of regulation molecules such as kinases and phosphorylases and changes in phosphorylation state have been well documented as a means to alter the activity of transcriptional regulators (Huater, 1995).

### *Phytochrome as a Light-Regulated Enzyme*

In mid 1980s, serine kinase activity was detected in highly purified phytochrome preparation from plants and speculated to be intrinsic to the photoreceptor (McMichael and Lagarias, 1990) but this controversial proposal remains unsolved because phytochrome locks the consensus sequences that define eukaryotic protein kinases.

### *Cyanobacterial Phytochrome as a Model for Phytochroml Action*

Protein phosphorylation is key mechanism for intracellular signal tansduction in bots eukaryotic and prokaryotic cells. This process is catalyzed by protein kinases and these are classified into there major groups based on their substrte specificities: sauce/threonine kinases, tyrosine kinases and histidine kinases.

Histidine kinases in bacteria play key roles in sensing and transuding extracellular signals including chemotactic factors, changes in osmolarity and nutrient deficiency. This signal transduction on system is mediated by phosphotransfer between two types of signal transducers and is therefore referred to as the two component system.

The two component system has been well defined in bacteria and in 1993, the existence of a bacterial-type histidine kinase was reported in yeast and in Aradidopsis. Since then, many genes encoding two-component regulated have been identified in plants, yeast and Dictyostclium (Chang and Stewart, 1998).

## Interaction between Phytochrome and Plant Hormones

Plant hormones can induce germination bolting, flowering, gene expression and other responses identified to those initiated by phytochrome. The overlapping role of light and plant hormones in development raises the interesting Question of whether light and phytochrome act independently or whether plant hormones are involved in the sequence of events initiated by photoactive phytochrome. Considerable evidence is amassing that phytochrome and hormone metabolism or signal transduction are intimately entered. For instance, in several plants alterations in gibberllin metabolism or response can cause phenotypes that elongation and flowering phenotypes of Phy B mutants (Devlin et al., 1992). The induction of flowering by long days in rosette plants have been correlated trits increased gibberellin levels and gibberellin deficient *Arabidopsis* mutants fail to flower in short days.

Transgenic tobacco and tomato plants over expressing Phy A have a dwarf phenotype (Boylan and Quail, 1989). The tobacco lines contain lower levels of several gibberellins than the trits type imphying that Phy A can inhibit gibberellin biosynthesis (Jordar et al., 1995).

Phytochrome may also act through auxins to control stem elongation. It is indicated by the study of hy 5 mutation which results in seedlings trits long hypocotyl under light. The HY5 gene encode bZIP transcription factor that is involved in auxin signaling. Recently, Shy 2 was cloned from *Arabidopsis* and shown to encode IAA3, a member of the auxin-indicate gene family (Tian and Reed, 1999). Since Shy 2 was identified as a suppressor of the long hypocotyl phenotype in hy2, it is feasible that SHY 2 plays a key role in auxin regulatory processes and represses a communications link between light and auxin signaling FIN 219 gene plays a role in Phy A-mediated inactivation of COP 1 and in promoting photomorphogenesis. The sequence of FIN 219 gene is highly homologous to a family of proteins defined by the soybean early auxin-indicable gene $GH_3$. Similar to $GH_3$, the expression of FIN 219 is rapidly inducible by auxin. It clearly indicates that a cross-talk between auxin response and light.

A possible role of brassinosteroids in these interactions created a lot of excitement but the lack of genetic tools has made studies difficult. However, a number of dwarf mutants such as def 2, dwf1 and cpd (light signaling mutants) are defective in brassinosteroid biosynthesis. It has opened up new insight into interaction between light and brassinosteroid pathways (Ecker, 1997). Similarly the identification of bri1, a brassinosteroid insensitive mutant and subsequent cloning of BRI1 (Li and Chory, 1997) revealed interesting clones concerning possible interactions between these pathways. det2 a pleotropic mutant shows a number of light-insensitive traits such as a show and thick hypocotyl, anthocyanin accumulation open expanded cotyledons and primary leaf binds in dark.

# CHAPTER 30

# Fruit Ripening

## INTRODUCTION

A fruit is the product of the gynoecium (entirety of the carpels) with accessory tissues involved in fruit formation. In the course of its life cycle the fruit passes through various development phases. One of these is fruit ripening. It commences when the growth of the fruit is complete. The ripening of a fruit is associated with a series of perceptible changes involving the colour, the aroma, the consistency and the flavour. These contribute to making the fruit suitable for eating. Fruit ripening constitutes a natural irreversible process of aging. Correctly, it should therefore be referred to as **fruit again** and not fruit ripening. It many fruits, the aging process begins immediately after they are picked from the plant. For some fruits, picking is indeed an essential precondition for ripening (e.g., in the avocado pear). Others (e.g., apples) exhibit a marked acceleration of ripening compared to those left on the plant. Recent research shows that numerous biochemical processes run parallel in time during fruit ripening. These processes are not necessarily causally linked with each other.

### Climacteric

The perceptible changes in ripening which have been mentioned are correlated in time with a rapid increase in respiration. This phenomenon, termed the climacteric, was already known towards the end of the last century. It is characteristic for many fleshy fruits. The typical climacteric fruits include apples, apricots, avocados, bananas and pears. Besides these there are fruits (or compound fruits) lacking a climacteric (for example, cherries, figs and strawberries).

However, the non-climacteric group by no means always shows a lower intensity of respiration: strawberries, figs and cherries respire just as actively or indeed more actively than many climacteric fruits.

There is a (probably fortuitous) relation between the occurrence of a climacteric or non-climacteric and the chemical nature of the pigments: climacteric fruits contain carotenoids and non-climacteric fruits contain anthocyanins.

Elucidation of the physiological basis of the sudden rise in respiration was long regarded as the key to understanding the entire ripening process. One of the first phenomena at the beginning of the climacteric in apples is the disappearance of free fructose from the cytoplasm. This is attributable to a phosphorylation.

A simultaneous change in tonoplast permeability is thought to permit further fructose to reach the cytoplasm from the vacuole. This stream of fructose is viewed as the cause of the increased respiration. A further attempt to explain the respiratory climacteric assumes that the rate of respiration is controlled by the concentration of phosphate receptors, mainly ADP. The rate of respiration is low if the ATP/ADP ratio is very great. This situation obtains if the energy requirement of the system is low. The climacteric rise in respiration would be interpreted as resulting from a raised energy requirement at the beginning of ripening. Splitting of

ATP occurs and the resulting elevation of the ADP level accelerates respiration.

Uncoupling agents such as 2,4-dinitrophenol can prevent ripening, e.g., in tomatoes. The occurrence of natural uncouplers of oxidative phosphorylation has also been discussed as the factor inducing the rise in respiration. Extracts from climacteric fruits do in fact act as uncouplers on mitochondrial preparations of other plants. However, this concept of an uncoupled respiration only partly coincides with the true facts. It was shown in avocado pears that the climacteric respiration is accompanied by an increase in oxidative phosphorylation.

Today there is a tendency to the view that the respiratory climacteric is to be seen quite generally in the context of the raised energy needs in ripening.

## RNA Metabolism

In many fruits the maturation process is accompanied by an increased RNA synthesis (the banana is an exception). In apples which have been picked, the RNA content at the beginning of the climacteric increases by about 50%. This raised RNA synthesis is restricted to a relatively short time period at the beginning of the climacteric; it has died down again at the height of the climacteric.

Paradoxically, there is a raised activity of RNA-splitting in parallel with the rise in RNA content. It was concluded from this finding that the RNA and RNase are located in different compartments. In the fruits concerned, new synthesis of RNA appears to be a necessary precondition of the ripening process. Pears can be inhibited in ripening by actinomycin D (an inhibitor of transcription). In this respect, there is a distinct difference between aging leaves and ripening fruits. As a rule, a degradation of RNA and of proteins in the course of aging is found in leaves, in which the fall in RNA concentration is correlated with a marked rise in RNase activity. Interestingly, the fall in the RNA content in leaves of *Nicotiana tabacum* and *Vicia faba* first concerns the cytoplasmic RNA, while the RNA in the chloroplasts is largely preserved. RNA can still be measured in the organelles of these two species even when the leaf has already become yellow.

The retention of chloroplast RNA is thought to be one of the factors enabling leaves which have already become yellow to green up again under suitable conditions (e.g., through the action of kinetin). In fruits there is a similar process, at least phenotypically. Valencia oranges grown in southern California are intense orange in colour during the winter months and can become green again in spring and summer. This colour change involves a conversion of chromoplasts into chloroplasts.

Fruits which show a rise in the RNA concentration at the beginning of the ripening process also exhibit an elevation of the protein content. Results of studies using labelled precursors make it probable that this extra content of protein is attributable to new synthesis. This new synthesis is also necessary for the ripening of many fruits. Ripening can be hindered by cycloheximide (an inhibitor of translation). Cycloheximide is less effective the later in the climacteric it is applied. This can be evaluated as an indication that the proteins necessary for ripening are synthesis at a very early stage.

## Enzymes

In recent years there has been highly intensive research on the changes in enzyme activities and the enzyme pattern during fruit ripening. Quite generally, there is an increase in the activity of many hydrolytic enzymes. Among others, polygalacturonase, pectin methyl esterase and cellulase show a marked increase. Softening of the fruit is associated with the rise in activity of these enzymes. Alterations in taste are also attributable to the activities of enzymes. Many fruits become sweet through degradation of starch into soluble sugar and a fresh synthesis of sucrose.

An accumulation of free fatty acids in many fruits is caused by an elevated lipase activity. However, the rise in the activities of many enzymes is not yet quite clear in physiological terms, e.g. the significance of the lipoxidase activity in ripening apples and that of peroxidase. It is thought that among other things these enzymes have a role in the biosynthesis of ethylene. It could be shown with the help of polyacrylamide gel electrophoresis that enzymes can also undergo qualitative changes during the process of ripening. In ripening bananas, for example, differences appear in the specificity of peroxidase isoenzymes. This altered protein pattern has been termed globally **ripening enzyme**. A coordinated synthesis of specific enzymes with a simultaneous degradation of proteins which are not required is essential for the initial stages of the ripening process. It is obviously essential for the initial stages of the maturation processes that there is a coordinated synthesis of specific enzymes with a simultaneous degradation of proteins which are not required.

## Pigment Formation

The most striking change which a ripening fruit can undergo concerns its colour. A pigment which is already present is not always demasked in all cases by the disappearance of chlorophyll. An impressive example is the tomato. In the course of ripening, tomatoes exhibit the following sequence of colours: green - white - yellow - orange or red-orange - red. The change from green to white is caused by breakdown of chlorophyll. Various carotenoids occur in the coloured intermediate stages; the red colour of ripened tomatoes is due to lycopene.

Carotenoid formation in ripening fruits is associated with the conversion of chloroplasts into chromoplasts. Electron microscopic investigations as well as biochemical analyses show that changes in the enzyme complement occur during formation of chromoplasts. Chromoplast formation commences with a disruption of the granathylakoids and the disappearance of the chlorophylls. A rise in the chlorophyllase activity can be measured in this phase. In many chloroplasts compartments are formed in place of the grana in which the carotenoids are synthesized. In fully differentiated chromoplasts of the tomato, long elongated carotenoid crystals filling a large part of the plastid lumen appear. The double membrane of the original chloroplasts is preserved.

In many fruits, the chlorophyll breakdown is displaced in time in relation to carotenoid biosynthesis. In the red pepper fruit, the major part of the isoprenoid plastid pigments is only formed after degradation of the chlorophylls.

The colour change of fruits during ripening is not caused in all cases by the formation of carotenoids. Red apples, berries, cherries, and strawberries are examples of fruits which synthesize anthocyanins during ripening. Anthocyanins are water-soluble pigments which arise outside the plastids and are accumulated in the vacuole.

Synthesis of both carotenoids and anthocyanins in ripening fruits is controlled by the phytochrome system. In both cases, the phytochrome does not act directly on the pigment formation, but probably influences in the first instance phytohormones which in turn regulate the pigment synthesis.

## Changes in Permeability

It has been known for a long time that tissue slices from ripening fruits release increased amounts of dissolved substances into the medium. This is attributable to raised permeability and is quite generally an early sign of incipient senescence, and is thus not restricted to ripening fruits. In some cases, the elevation of the permeability in the fruit tissue begins long before the start of the climacteric and is not causally linked with it. By ethylene treatment, the respiratory climacteric can be induced without a simultaneous change in tissue permeability. The reasons for the alteration of membrane properties during ripening of fruits and during the senescence of other plant organs are largely obscure. One of the causes might be sought in the decrease in particle-bound unsaturated fatty acids found in ripening cherries and pears.

## Hormonal Regulation of Fruit Ripening

Five categories of plant hormones are known today which regulate fruit ripening: **auxins**, **cytokinins**, **gibberellins**, **abscisic acid** and **ethylene**.

The auxins isolated from fruits and determined in the chemical structure are indole derivatives. Citrus and a few other fruits are thought to contain auxins without an indole moiety. The occurrence of indoleacetic acid (IAA) in fruits was disputed for a long time, but has meanwhile been proved beyond doubt. In young fruits, the seeds are the main site of auxin synthesis. In later phases of fruit development, the fruit flesh (pericarp and associated accessory tissue) can also synthesize auxin to a certain extent.

It has been known for a long time that auxin delays the aging of leaves. The knowledge is more recent that auxins can also slow down fruit ripening, though this statement cannot be generalized: auxins can also accelerate the ripening process. However, this auxin effect may well constitute the exception and only apply under special conditions. As a rule, auxin acts as an inhibitor of ripening. It has been known for some time that auxin prevents ethylene formation in fruits. Ethylene is today discussed as the ripening-inducing hormone. Auxins must be broken down to make ripening possible. It is considered today that this process is catalysed by peroxidases. The peroixdases, primarily IAA oxidase, degrade the endogenous auxins and thus control fruit ripening. Only when the auxins have largely disappeared is the tissue sensitive to ethylene. In many climacteric and non climacteric fruits, ripening is accompanied by a rise in auxin-degrading enzymes.

Cytokinins also slow down ageing processes in plants. For example in leaves they contribute to keeping the protein and chlorophyll content largely constant. Protein degradation is stopped by cytokinins rather than protein synthesis being stimulated. Analogous effects are also ascribed to cytokinins in fruits. This information was gained by addition of exogenous cytokinins. On the other hand, little is known on the endogenous cytokinin content and the metabolism of cytokinins in fruits.

Giberellins act in a similar way to auxins and cytokinins. The first information on a relationship between giberellins and the process of fruit ripening derives from studies on oranges. It was found that giberellins inhibit the degradation of chlorophyll. It was later shown that giberellins also delay the accumulation of carotenoids. Giberellins appear quite generally to prevent the colouring of fruits. In bananas, a treatment with giberellins inhibits yellowing, while all other maturation processes are not affected.

Before the first visible phenomena of ripening, there is in many fruits an accumulation of *abscisic acid* (ABA). A series of findings underlines the importance of this phytohormone as regulator of fruit ripening. In harvested apples, the ABA content increases tenfold within five days. If the apples are left on the tree, an increase in ABA cannot be measured. At the beginning of the ripening process, higher ABA concentrations are found in the inner part of the fruit flesh of green tomatoes than in the outer part. Tomatoes ripen in a centrifugal direction. With progressive ripening, the relationship is reversed. In the inner parts which are already ripe, the ABA concentration falls again. Ripening of tomatoes is thus associated with an ABA wave across the sectional area. In ripening tomatoes a relationship was also discerned between the phytochrome system and the endogenous ABA concentration. The proportion of lycopene constitutes a measure of the degree of ripeness. It can be seen from the diagram that when the fruits are illuminated with red, the ABA content first rises, reaches a maximum after five days and then falls. Lycopene synthesis parallels the rise in ABA.

ABA thus appears to be a trigger of lycopene synthesis. In contrast to this, illumination with for red leads to a fall of the endogenous ABA level. The change in the ABA in the tomatoes must be attributed to a phytochrome effect, since a rise in ABA induced by red can be reversed by an irradiation with far red. From the presently available data it is concluded that Pfr (the morphogenetically active form of phytochrome) allows the ABA concentration to rise, resulting in induction of the enzyme of carotenoid synthesis.

Numerous studies have shown that *ethylene* is a hormone of ripening. No fruit ripening occurs in the absence of ethylene. Ethylene probably brings about the climacteric, since in many fruits the rise in respiration is directly preceded by an elevation in the ethylene concentration. This concept is supported by the observation that non-climacteric fruits also show an increase in respiration after treatment with ethylene; this increase is comparable with that of climacteric fruits. The differentiation between climacteric and non-climacteric fruits would thus be based on the different ethylene production during ripening.

In pears and apples, the time of exposure to a given ethylene concentration which is necessary to initiate the ripening process becomes shorter during the course of the development of the fruits. The tissue becomes increasingly sensitive to ethylene. The formation of the ethylene and its influence on the ripening process depends on relatively high oxygen concentrations. The auxins, cytokinins or giberellins probably have a senescence-delaying action by blocking the point of attack by ethylene. Ripening can be induced by ethylene only when auxin has largely disappeared through the action of degradative enzymes.

It has recently been reported that ethylene alters the proportion of individual transfer-RNA species. This effect of ethylene may influence the translation of mRNA and thus initiate ripening. There is a correlation between ABA and ethylene production in ripening fruits. Ethylene formation in tomatoes can be accelerated by treatment with oxygenous ABA. However, it is not clear whether ABA induces ethylene synthesis under natural conditions. Light can also trigger ethylene formation. The light quality is important here. Illumination with red accelerates ethylene formation, while far red slows it. However, ethylene formation in plants is not exclusively induced by light. Ethylene is formed for example in injuries, diseases, physical or chemical stress and by the action of metal ions (e.g., $Cu^{2+}$ and

$Ca^{2+}$).

Studies in recent years have shown that the phenomenon of fruit ripening is based on extraordinarily complex physiological events. In very few cases is there a reasonably well-defined conceptions as to the cause and effect of individual processes of ripening. Future research will show whether a generally applicable picture of fruit ripening can be developed.

## CHAPTER 31

# Senescence: A Suicidal Impulse

One of the most puzzling events in a living system is its death. Plants and their parts develop continuously from germination to death. The latter part of the developmental process that leads from maturity to the ultimate complete loss of organization and function is termed senescence. In living organisms there occur programmed cell death (PCD). It refers to a process by which cell promotes their own death through the activation of self-destruction system.

PCD is a physiological cell death process involved in the selective elimination of unwanted cells. In animals these unwanted cells include those that have served temporary functions, cells that are overproduced and cells that are unwanted or present in inappropriate positions. PCD in animals depends on the induction and action of specific genes that bring about the controlled disassembly of a cell. It involves the condensation, shrinkage and fragmentation of the cytoplasm and the fragmentation of the nuclear DNA into 50 kb and in some cases 0.14 kb pieces that represent oligonucleosomes. These fragments can be detected by terminal deoxy-nucleotidyl transferase – mediated dUTP nick end labelling (TUNEL) of DNA 3' – OH groups in sections of cells. Many of the proteins responsible for PCD have now been identified. Proteases of the interleukin - 1$\beta$ - converting enzymes (ICE) family bring out cell condensation and shrinkage and endonucleases causes the fragmentation of nuclear DNA. The basic leucine Zipper transcription factor, ces-2 (cell death specification-2) and oncoproteins in the Bcl-2 family regulate the onset of PCD in many animal cells. Growth factor deprivation and reactive oxygen species can trigger PCD. $Ca^{2+}$ also participates in PCD.

It is now generally accepted that many plant developmental processes and stress responses are achieved through the operation of PCD. These include senescence, xylogenesis, embroygenesis, sex determination in monoecious plants such as maize, abscission zone formation and the hypersensitive reaction to pathogen infection. In plans selective cell death is necessary for growth and survival and can occur on local or large scale. Morphological observations of dead and dying cells show that the cytoplasm and nuclei condense and shrink. TUNEL experiments reveal oligonucleosome sized DNA fragments that are characteristic of PCD. It appears that the mechanism controlling PCD in plants and animals may be similar. However, that is no molecular evidence for this.

During the process of cell death, various proteins that are normally sequestered in the intermembrane space of mitochondria including Cytochrome C, apoptosis-inducing factor and certain procapspase proteins are released into cytosol. Depending upon the stimulus, this event can lead to necrosis through irreversible mitochondrial damage and energetic catâstrophe C-dependent Caspase activation. In the cytosol, cytochrome C activates the caspase, a family of killer protease through formation of a complex with Apaf-1 (apoptotic-protease activating factor-1), procaspare-g and ATP or dATP (Martinou et al., 2000). Release of cytochrome C from mitochondria during apoptosis is controlled by proteins of the Bcl-2 family: those that inhibit cell death and prevent the release of cytochrome C. little is known about the mechanism by which cytochrome C escapes from mitochondria during cell death. It is reported that the release of cytochrome C is a rapid, complete and kinetically invariant event.

Animal and plant senescence differ in a variety of ways although both are apparently the manifestations of the same basic phenomenon. In animals senescence is largely due to the cumulative deleterious effects of the environment upon the organism. In contrast, in plants it is largely hereditary. This is because the animals, which due to their heterotrophic mode of nutrition are rather a delicate system, depend upon other sources for food as against plant which being autotrophs are rather hardy and relatively less liable to the harmful inflictions of the environment. In most instances, plants senesce not because of adverse climatic conditions but due to and after the ripening of fruits, since they start senescing even at a time when climatic conditions are still favourable for their normal growth.

Senescence is a genetically programmed sequence of biochemical and physiological changes. Its initiation is subject to regulation by many environmental and autonomous (internal) factors. The environmental include stresses such as extremes of temperature, drought, ozone, nutrient deficiency, pathogen infection, wounding and shading where as the antonomous factors include age, reproductive development and phytohormone levels (Fig. 31.1).

Senescence is an essential aspect of plant development as after the senescence leaves die and withdraw their photosynthate to reproductive part. As new leaves are initiated from the shoot apical meristem, older leaves often are shaded and lose the ability to function efficiently in photosynthesis. Since the plant invested valuable resources in leaf formation, senescence recovers a portion of those resources. Hydrolytic enzymes, produced as part of the senescence program, will break down many cellular proteins, carbohydrates, and nucleic acids. The component sugars, nucleosides, and amino acids are then transported back into the plant via the phloem, where they will be reused for synthetic processes. Many minerals also are transported out of senescing organs, back into the main body of the plant. Senescence is not just the running down of tissues or cells but a programmed part of development. Senescence is often of great advantage to plants, which would be serious difficulties if did not occur. Loss of leaves in deciduous trees is an essential part of winter cold avoidance. Older leaves in herbaceous plants often senesce and die, and their nutrient contents are withdrawn for the nutrition of the growing parts of the plant. Senescence and death are essential to the functioning of xylem and sclerenchyma cells.

Plants have evolved mechanisms by which leaf senescence can be induced by many stresses to reallocate nutrients to reproductive organs and to eliminate water consumption by older, less productive leaves. This regulation of leaf senescence has an obvious adaptive value, allow in the plant to complete its life cycle even under stressful condition.

The early stage of senescence becomes harmful because at this stage leaves have minimum photosynthates, which results in loss of yield. It means senescence after complete photosynthesis may helpful in improving the yield.

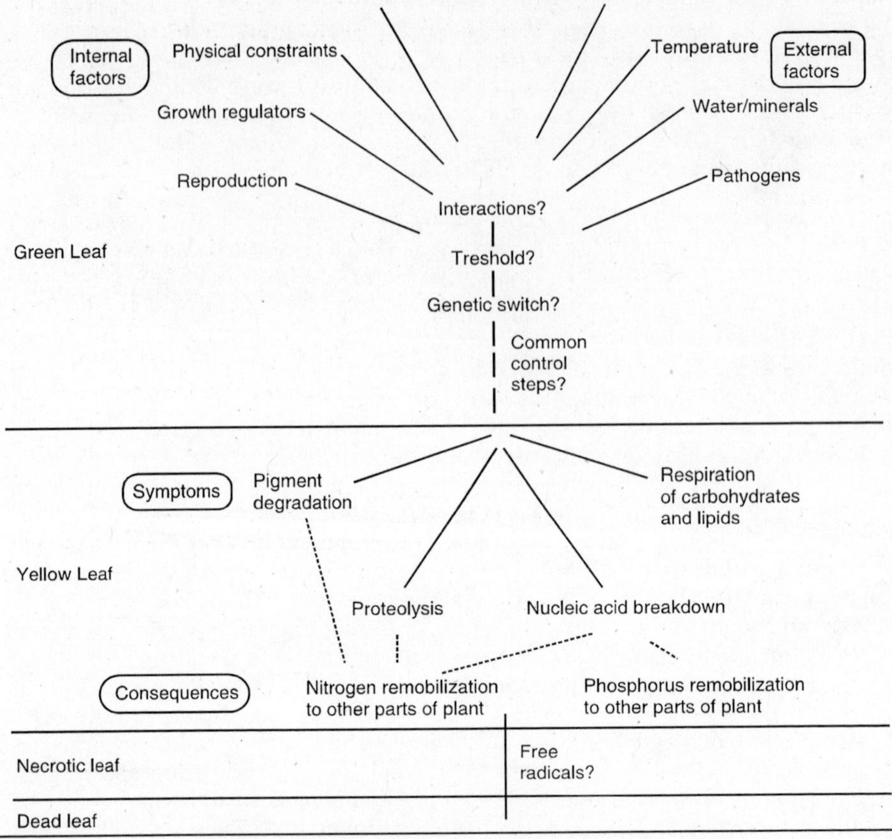


**FIG. 31.1:** Factors implicated in the initiation of leaf senescence (After Smart, 1994).

Thus, senescence is a correlative event and is the cumulative effect of the formation of flowers, fruits, apices and roots. Senescence is a mechanism, which works in favour of maintaining an ecological balance, both for the internal milieu and the external environment of a species.

## Is Senescence is a Universal Phenomenon?

Processes that are similar to programmed cell death have been described in prokaryotes and lower eukaryotes. Some strains of *E. coli* harbor plasmid encoded genes that trigger proteolysis and cell death after infection by bacteriophage (Ameisen, 1996) and activation of the AMP pathway in non-differentiating cells of Dictyostelium discoides induces cell death process closely resembling programmed cell death. Culture of several kinds of unicells at low density also induces a process resembling programmed cell death (Welburn et al. 1996). Programmed cell death is also thought to occur in the green alga Volvox and the fern Marsilea. Degreening in Chlorella is accompanied by the loss of proteins and is associated with ultrastructural changes of the chloroplasts such as the loss of thylakoids and production of granular bodies. These changes resemble the changes occurring during the transition of chloroplasts into gerontoplasts in leaf senescence.

Its universality has been questioned primarily because *Chlorella*,

*Chlamydomonas* and other unicellular organisms including prokaryotes like bacteria can be kept indefinitely in media or chemostatic cultures without any exhibition of the symptoms of senescence. If so, then why do multicellular forms, obviously derived from these so-called immortal unicellular forms, undergo senescence?

## Aging *vs.* Senescence

According to Medawar (1957) senescence may be defined as the *deteriorative processes that are natural causes of death.* Aging, by contrast, refers to processes of occurring maturity with the passage of time. By these definitions, then aging includes a much wider span of physiological changes, some of which may lead to the weakening of the organism while others may be quite neutral with respect to capability of the biological organism to survive. Senescence, on the other hand, refers to those changes that provide for the endogenous regulation of death. As examples, the physiological changes in a plant, which cause its conversion from a seedling to juvenile plant, from a juvenile plant to a mature plant, or the gradual decline in vigor with increasing age, would be examples of aging. On the other hand, physiological changes involved in the autumnal colouring (and subsequent death) of leaves, or in the dieback of tulips and other spring bulbs in early summer, or in the yellowing and death of annual and biennial species after the completion of fruiting or after a brief period of drought would be examples of senescence.

Relationship between aging and senescence is also worth considering. Aging is mostly taken to be a continuous process from embryogenesis to death, whereas senescence always starts after reaching the reproductive maturity. However, some people still take aging and senescence as synonyms.

Longevity is a natural correlate of aging and senescence. The ranges of longevity over the plant kingdom are enormous, ranging from a few weeks (in the case of desert annuals) to perhaps 4000 years (in case of Bristlecone Pines). Generally, it might be said that the shortest-lived plant species tend to have longevities limited by programmed senescence, whereas, the perennials tend to have much less precise limits of longevity and tend to die through a gradual attribution associated with aging.

## Various levels of senescence

Senescence can be expressed at various levels.

### *Organism senescence*

It is a kind of senescence terminalized by the death of the plant. The life span varies to a very great extent from about a few days to 4,500 years. Annuals and biennials have a very short life span. They lack elaborate modifications to escape detrimental effects of the forthcoming unfavorable season. Their flowering is determined by the photoperiodic stimulus and it finally results in the death of individual. Their growth curve is typically sigmoid, consisting of a lag phase with little or no growth, a log phase with steep increase in the growth rate, a period of zero growth or growth plateau, followed by a negative curve or the phase of senescence. Trees are unique in avoiding the detrimental effects of the season by shedding their leaves. The life span varies a great deal among trees. *Betula* starts senescing at the age of 50 years, Redwood (Sequoia) lives for 3000 years, and one specimen of Bristlecone Pine has been proved to be 4,500 years old. Trees show a staircase like growth curve: log phase represented by the spring season growth a growth plateau by the autumn growth. Trees in general show a continuity of life span for thousands of years, but it is led by a rapid death of cells followed by continuous cell replacement by meristems.

### *Clonal senescence*

Clonal senescence is decline in vigour through successive generations in vegetatively propagated species. Its occurrence still rests in question. *Cryptomeria*, *Musa* and so many other fruit trees have been grown for centuries by stem cuttings, still no aging symptoms are reported. However, there is a reasonable evidence for aging in *Citrus* clones.

### *Organ senescence*

Shedding of cotyledons from a seedling in epigeal germination, periodic shedding of leaves in deciduous trees, shedding of floral parts (calyx, corolla, stamens etc.) and fruit ripening come under this category.

### *Tissue senescence*

It is an important part of tissue differentiation. Cork formation by suberization, sclerenchyma formation in general, and xylogenesis, all are outcome of this process.

### *Cell senescence*

Cell is the basic unit of life and cells are involved in all kinds of senescence, but the senescence of a cell in particular is also met with. It may be a part of differentiation, for example sclereid differentiation, cork cell formation or may be due to some irrepairable defect arising in a cell. In the latter case the cell simply undergoes autophagic digestion without causing any harmful effects to neighbouring cell. Cells with longest life span so far reported are the pith cells of *Carnegia gigntea*, which live for about 100 years. In trees, the pith and ray parenchyma cells are the oldest living cells.

### *Organelle senescence*

All living cells constantly show organelle senescence and organelle renewal, even in the so-called immortal unicellular forms. As an organelle falls in, due to some irrepairable changes or defects, it undergoes senescence and is finally digested by autophagy. One of the primary and easily seen phenotypic symptoms of senescence is yellowing of leaves. Cotyledons and fruits also show yellowing with the arrival of senescent phase. Another common symptoms are the loss of rigidity of stems, coupled with yellowing in dicotyledonous herbs and some monocots. Rigidity of leaves and fruits are also considerably decreased due to hydrolysis of organizational framework of the plant.

## Monocarpy *vs.* Polycarpy

In considering the senescence of the whole plants, most species can be classified into two groups.

**Polycarpic plants.** Plants that normally have more than one reproductive phase before death.

**Monocarpic plants.** Plants that have one reproductive phase followed by death.

According to Woolhouse (1982) monocarpy has polyphylatically evolved from polycarpy through natural selection and hybridization. Indeed, crop plants (mostly monocarpic) are usually developed from their monocarpic progenitors through artificial selection in breeding programme where in most cases achievement of increase in the ratio of reproductive: vegetative portion at harvest is an obvious goal. However, the above ratio is found to be lower in a wider range of polycarpic plants in a particular season (personal observation). Polycarpic plants are subjected to annual organ as well as cellular formation of secondary growth leading to the formation of dead cork cells, tracheae etc.). Senescence is lacking in herbaceous monocarpic individuals. It may be speculated that efficient processes of recycling of nutrients along with periodic detoxification through abscission of leaves and other useless organs (which might act as a potential portal for infection) in the polycarpic plants provide means for greater survival compared to the monocarpic plants. Perhaps stem of the polycarpic deciduous plants serves as an efficient intermediary storage organ where nutrients are temporarily stored after recovery from the senescing leaves for further use by the newly developed foliar organs.

Monocarpy occurs in a wide range of plant taxa and evidently has evolved independently numerous times. Secondly, while many annuals and biennials are monocarpic, some perennials (e.g., bamboo century plant) are also monocarpic. Thirdly, monocarpy is very common in cultivated plants, especially field crops, where the abbreviated life span and complete desiccation prior to harvest and/or synchronous production have apparently been selected (Schwanitz, 1966). Monocarpic field crops in fact put on the world's most dramatic displays of senescence, for vast fields of individuals undergo senescence and die essentially simultaneously. It is convenient to call this degeneration following the reproductive phase, monocarpic senescence. In monocarpic (those that flower only once and then die) senescence (Noodén and Lindoo, 1978) the entire plant senesces at the time of reproductive development. Leaf senescence precedes the fall of dead leaves in deciduous trees. The maturation of xylem cells derived from vascular cambium may be considered as a senescence process, which is under hormonal control. However, it is difficult to state definitely what processes naturally terminates the life of an organ or a plant and what processes are simply reflection of the changes associated to the passage of time.

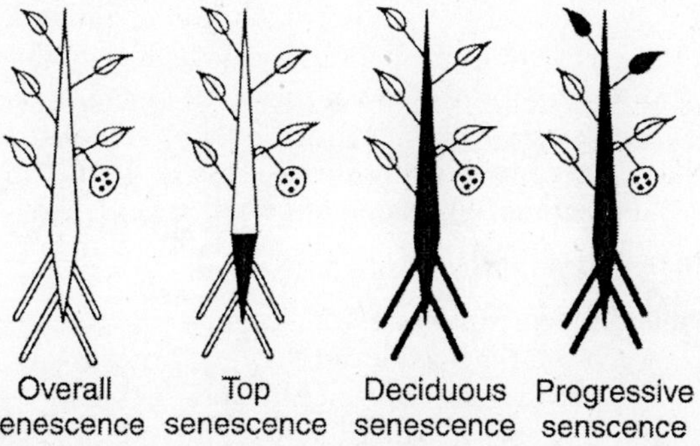


**FIG. 31.2:** Representation of various patterns of senescence in plants.

Plants do not grow in pieces. They age, as they grow and develop, in orderly fashion. In some plants it occurs as the gradual enrichment of deterioration process and in other plants there may be a fairly abrupt deterioration leading to death. The study of plant senescence is concerned with deteriorative changes which occur during plant development and which result in death of either entire plant or portion of the plant. All cells and tissues in the plant ultimately undergo senescence.

According to Leopold (1961) there are four morphological patterns of senescence in plants, viz., overall senescence-where the individual dies in an abrupt way (most of the monocarpic plants); top senescence-where senescence of the aerial parts of the plants occurs; deciduous senescence-where it includes the annual senescence of the leaves as in woody perennials; and progressive senescence-where it concerns the sequential senescence of the leaves in an acropetal order.

The entire plant development can be viewed as one series of events. The successive stages of development, starting with juvenility, vigorous vegetative growth at its peak, maturity of plant when mobilization of nutrients out of the vegetative parts into the flowers and fruits occurs and all this followed by senescence, are attended by numerous advantages. During senescence, this process is imposed in regular patterns due to widespread occurrence of cellular and organ senescence, which slowly but steadily spreads through the entire plant in a natural fashion.

## Plant Growth and Senescence

During the course of their lives, all multicellular organisms and their organs and tissues reach a peak in terms of their physiological function, and then they decline until they die. According to Molisch (1938), the death of the whole plant commonly occurs (annual and biennial species) upon completion of fruiting. The reproductive activity is lethal to the plant and removal of the flower and fruit of several species of plants prevents senescence.

Senescence is almost certainly a complex of many sequential and parallel steps. Cessation of vegetative growth, particularly stem elongation or senescence of apical meristem, and the transition from juvenile to adult phase though interesting phenomena, may not be measures of whole plant senescence. Among the different types of senescence (organelle, cellular, organ and whole plant), the whole plant senescence in monocarpic annuals seems to be unique in that further development of the plant ceases in an abrupt but coordinated way, seemingly for the formation of reproductive propagules, while in cellular along with their organelles (during the formation of sieve tube, tracheae, lysigenous cavity, etc.) and organ senescence (abscission) further developmental processes are tagged, whole plant senescence terminates the life process providing means for adaptation (e.g. in ephemerals) and evolution (e.g. in annuals).

### *Effect of Senescence on Vegetative Growth*

Whole plant senescence is accompanied or preceded by a decrease or cessation of both shoots and root growth. The reduction in vegetative growth that occurs during the reproductive phase has been recognized for a long time in both monocarpic and polycarpic plants, and is generally attributed to a shift in distribution of growth limiting nutrients within the plant. Thus, it is of importance to determine when the reduction in root or shoot growth occurs relative to flowering, fruit development, and senescence. Where there shoot apex is converted to an inflorescence (e.g., corn and cocklebur), vegetative node production ceases coincident with the appearance of flower buds, though stem elongation may continue. In other species (e.g., pea and cotton) where the apical meristem of the main shoot does not become and inflorescence, it may stop growing and senescence during their reproductive phase. In soybean, (Borst and Thatcher, 1931; Lindoo and Noodén, 1977; Noodén, 1977) cotton (Crowther, 1935) and pea (Lockhart, 1961), stem elongation decreased but continues after flowering has started and stops while the fruit are growing

Flowering and vegetative growth are not always antagonistic, for under long days (16 hr), the shoot apex of morning glory, *Ipomoea caerulea*, continues vegetative growth and flowers are produced from lateral buds, but short days cause the terminal bud to become a flower (Ashby, 1950). Similarly, Marx (1958) has found a variety of pea (G2) that continuous to grow and produce fruit under short days, whereas these plants terminate growth and die under long days. It is important to know if the rate of fruit development and nutrient demand (e.g., dry weight or nitrogen accumulation in seeds) is the same in senescing and non-senescing plants. In soybean root growth has been reported to terminate at about the same time as shoot growth (Mitchell et al., 1971) or a bit sooner. (Borst, 1931) Cotton root growth also declines during the reproductive phase, apparently ahead of shoot growth. The roots of radish cease growth at flowering. The roots of common perilla (*Perilla frutescens*), continues to grow after node production ceased, but start to slow about the time when the first flowers open (Beever and Woolhouse, 1975). In rice and wheat (Asana and Singh, 1967; Brenehley and Jackson, 1921) the roots may stop growing later, during heading, and degenerate even lateral during grain development. Likewise, corn roots grow little after seed growth has started (Foth, 1962). Plentiful supply of water and nutrients may however, sustain wheat roots. (Carpenter et al., 1952) Pea nodules reach a maximum mass at flowering, (Pate, 1958) and substantial losses may occur at fruiting. (Pate, 1958; Salter and Drew, 1965) Soybean nodule also found to senesce during fruit development. (Hardy et al., 1973) Thus, at least in some cases, root growth may be affected early in reproductive development, apparently too early to due to a nutrient deficiency resulting from massive diversion of the nutrient supplies to developing fruit.

The prime candidates for control centers for growth cessation are the reproductive structures. Indeed, the nutrient diversion explanation of the cessation of growth carries with it the prediction that removal of the reproductive sinks should restore vegetative growth, but this is only partially true.

### *Effect of Senescence on Reproductive Growth*

Many plants produce far more flowers than fruit and initiate more fruit than they carry to maturation. Considerable evidence from a wide range of monocarpic and polycarpic plants indicates that the reproductive structures themselves exert an inhibitory effect on the development of other reproductive structures and may even induce their abortion, deflowering or depodding can offset these effects. Normally, this inhibition of excess flower and fruit development does not cause whole plant senescence. Nonetheless, it seems important to determine whether or not the flowers or fruit produce these effects via the same mechanism by which they induce senescence or inhibit vegetative growth. In the case of soybeans, at least, there seem to be some differences between these inhibitory effects.

### *Senescence Related to Stress*

A new fad is abroad in the biological world- stress physiology. There are those who would put disease-induced senescence among the stress induced phenomena. It involves mechanical damage, chemical damage, and the metabolic load associated with the presence of pathogen, etc. None of these is a well defined phenomenon at the biochemical level.

However, these factors do play a role in most diseased tissue and they are important from the point of view of the plant pathologist and physiologist because they lead to striking changes in metabolism; they can easily obscure the more subtle, specific changes and are often related to symptom develop. They are responsible for the fact that some symptoms, e.g., senescence, are widespread in disease plants.

## Models Proposed for Mechanism of Senescence

It is difficult to propose any single mechanism of senescence in chloroplasts for two main reasons. First, it is difficult to define a senescence mechanism because of the contradictions in the accumulated data resulting from system specificity and a lack of uniformity in experimental conditions. Second, our knowledge of chloroplast senescence is still inadequate and the available data are not sufficient to develop concepts, generalize them, and propose a definite explanation of organelle degradation. Three possible mechanisms are discussed.

The loss of the efficiency of the Calvin cycle as the first major step during chloroplast senescence. The signals for the loss of Rubisco, the crucial enzyme of the Calvin cycle, comes from the to oxygen to generate oxygen free radicals such as superoxide that cause peroxidation of thylakoid lipids including lipids on the oxygen-evolving complex.

Damage to the oxygen-evolving system results in the generation of singlet oxygen and other free radicals. Free radical-induced lipid peroxidation may activate membrane-specific of membrane lipids. Alternatively, a loss of lipids may expose intrinsic membrane proteins and make them accessible to already existing active proteases. Lipid peroxidation may inactivate membrane-bound enzymes and electron-transport carriers by oxidation and cross-linking of proteins. This may ultimately lead to complete breakdown of the chloroplast. The difficulty in accepting the Calvin cycle as the first major step of chloroplast degradation is the lack of data on the activity of Rubisco *in vivo* during organelle senescence.

Most of the results regarding the rapid loss of the enzyme activity is measured *in vitro*, which may not simulate the conditions occurring in leaves. In addition, there are data available from some plant systems that suggest that the loss of chlorophyll and the electron-transport system occurs before any significant loss of Rubisco activity which would mean that a loss in the efficiency of the Calvin cycle may not be a limiting factor for chloroplast senescence. It is possible that a senescence-specific loss of thylakoid lipids triggered by a lipase is the first step of chloroplast degradation. The oxygen-evolving system has a very delicate structural organization and is highly susceptible to even mild stress. The role of thylakoid lipids in oxygen evolution has been established (Akabori et al., 1984). The loss of lipids may cause the release of extrinsic proteins and/or $Mn^{2+}$ from the oxygen-evolving complex. The release of extrinsic proteins under stress of $Mn^{2+}$ during senescence have been reported. This obviously leads to a decline in electron-transport efficiency and a change in the light-induced concentration of $Mg^{2+}$ and pH of the stroma, which in turn affect the efficiency of carbon dioxide fixation in the Calvin cycle.

There is a loss of some of the components of the electron-transport chain linking the oxygen-evolving complex to NADP reduction. The loss of the efficiency of the electron-transport system is attributed to the quantitative loss of quinones, reaction centers, plastoc-anin, Fd-NADP reductase, or the coupling factor for the ATP synthesis. This may lead to loss of the efficiency of the water-splitting system or Calvin cycle resulting in the breakdown of the organelle.

Of the three above proposed schemes, the first one seems to be more logical than the other two. The rapid and early loss of Rubisco has physiological significance. Rubisco protein constitutes about 50% of the total protein of green leaves, the major source of nitrogen for other growing parts of the plant system. During chloroplast senescence, the degradation products of this protein move to the growing parts. However, transport of these products needs a continuous supply of ATP. It is, therefore, reasonable to assume relative stability of the electron-transport system over the Calvin cycle. When the economy of the system is considered, the second target of the degenerative process obviously should be the oxygen-evolving complex, which should degrade earlier than the intrinsic proteins of the electron-transport chain. As long as the oxygen-evolving system remains intact, it logically requires a totally efficient electron-transport chain to link NADP reduction and ATP synthesis. Therefore, in the sequence, the loss of intrinsic electron transport components should follow the senescence-induced damage to the water-splitting system.

## Senescence Induced Structural and Ultrastructural Modifications

Senescence is the last phase of development of a whole organism, organ, cell, or organelle. It is basically a degenerative process that leads to the death of a living system. Although it's precise mechanism is not known, recent experiments suggest that the process is genetically programmed (Woolhouse, 1984). Senescence, although deteriorative in nature, is a requirement to facilitate turnover processes and is an essential process for growing parts of living organisms. Senescence of green leaves in plants has been extensively investigated in the last two decades (Thomas and Stoddart, 1980; Woolhouse and Jenkins, 1983; U.C. Biswal and Biswal, 1984). Diffierent tissues and cells of the leaves have their own pattern and timing of senescence. The initiation of senescence in mesophyll cells may not necessarily be synchronized with the process in vascular or epidermal tissues. Even different organelles of a single leaf cell, namely chloroplasts, mitochondria, endoplasmic reticulum, ribosomes, and the nucleus do not show synchrony in the induction and progress of senescence (Biswal et al., 1983b). Different organelles of a leaf cell degrade sequentially; the sequence may change depending on many factors (Woolhouse, 1984a). Chloroplasts are the first organelles to show symptoms of disorganization when all other organelles are normal (Dodge, 1970). This is followed by a change in the structure of the endoplasmic reticulum and loss of ribosomes. The next organelle that is affected is the mitochondrion, although at this time the nucleus, plasma membrane, and plastid envelope are shown to be intact. Mitochondria show symptoms of degradation that are reflected in the alteration and loss of the cristae and the reduction in the size of the organelle. Nuclear contents show shrinkage and the membrane becomes structurally irregular (Butler and Simon, 1971). Finally, senescence results in the rupture of the plastid envelope followed by the breakdown of the plasma membrane (Woolhouse, 1984a).

### *Senescence Induced Structural Modifications*

The deterioration processes in the leaf that leads to senescence begin as soon as the leaf attains its full size. Several types of structural changes are associated with leaf senescence. Before discussing the changes occurring due to senescence, we should deal with chloroplast structure.

**Structure of chloroplast**

The structure and organization of chloroplasts are described in detail by Kaplan and Arntzen (1982). The lamellar structure of chloroplats varies between photosynthetic organisms. The structure is relatively simple in less evolved plants such as blue-green algae and photosynthetic bacteria and more complex in higher angiosperms. Electron microscopic studies reveal the following basic features of a typical higher plant chloroplast. The organelle is lens shaped and the lamellar system is encircled by a double membrane envelope. The envelope acts as a selective barrier to the transport of various metabolites into or out of chloroplasts. The lamellar system of the organelle is basically composed of lipoprotein disks that are stacked at regular intervals to form structures known as grana. The unstacked loose lamellae, which connect stacked grana zones, are intergranal or stroma lamellae. The entire lamellar system is suspended in a granular stroma that is mainly composed of plastid genetic materials and ribosomes responsible for chloroplast replication and self-regulation. In addition, starch grains also constitute a part of the granular structure. The proteins present in the stroma are mostly the enzymes of the Calvin cycle. The other components of the stroma are plastoglobuli, which in electron micrographs appear as darkly stained structures consisting of lipids.

The grana and stroma lamellae are the structural units for primary photochemical reactions and contain the photosynthetic pigments in addition to lipids and proteins. The lipids of the thylakoid membrane consist of a complex mixture. The major lipids of the thylakoid membrane are galactolipids, phospholipids, and sulpholipids.

Chlorophylls, carotenoids, and phycobilins are the major photosynthetic pigments and are involved in the process of light absorption by the organelle. Chlorophylls a and b, which are chemically different from each other, are abundantly present in the higher plant chloroplasts. According to light absorption and fluorescence characteristics, chlorophyll a *in vivo* can have a number of optical species as well as different chemical species. The formation of different optical species of chlorophyll a is attributed to the complex interaction of the pigment with the membrane environment. A change in membrane environment could lead to an alteration in the optical behavior of the pigments. Therefore, light absorption and fluorescence spectroscopy are used as sensitive tools to probe the microenvironment of pigments in the membrane. Carotenoids and phycobilins found in lower plants are accessory pigments that absorb light energy and ultimately transfer it to chlorophyll for photochemical reactions. In addition, carotenoids protect chlorophyll molecules from photo-destruction. Chlorophyll and carotenoids are bound to thylakoid proteins to form different types of pigment-protein complexes.

Proteins associated with the thylakoid membrane are coupling factor proteins, light-harvesting chlorophyll-protein (LHCP) complex, proteins of the electron transport carrier, proteins associated with the reaction center and the polypeptides of the oxygen-evolving complex. The proteins of the coupling factor and some of the polypeptides of the water-splitting system are extrinsic, but most of the thylakoid proteins are intrinsic in nature.

## *Loss of Pigments, Lipids, and Macromolecules*

Leaf senescence causes a loss of all types of photosynthetic pigments. However, depending on the plant system and environmental conditions, differential rates of loss of chlorophyll *a*, chlorophyll *b*, and carotenoids have been demonstrated. Carotenoids are relatively stable in comparison to chlorophylls. In contrast, equal rates of degradation have been demonstrated for chlorophyll a, chlorophyll b, and carotenoids of the chloroplasts of fern leaves. Leaf senescence is characterized by a massive loss of green pigments, a process referred to as degreening. The degreening process of mature leaves that occurs either in nature or induced by darkness in laboratory conditions is accompanied by the disorganization of various cellular and subcellular structures.

### Chlorophyll breakdown in senescent leaves

The biosynthetic pathway of Chl probably the most pigment in the biosphere, has been studied in considerable detail and is mentioned in plant physiology textbooks. Oddly enough, however, the breakdown of Chl is barely mentioned. Yet, as pointed out by Hendry et al. (1987), an estimated one billion tons of Chl are broken down every year, seemingly without leaving a trace. The only enzyme that is taught is chlorophyllase, discovered by Arthur Stoll in 1912, which cleaves Chl into phytol and Chlide the Mg-porphyrin moiety of Chl. For most of the subsequent 80 years, chlorophyllase has been the only known part of the catabolic system. Until quite recently all the subsequent steps of Chlide breakdown remained undiscovered because no one was able to identify the break-down products.

Most of the biochemical steps of the catabolic pathway of Chlorophyll are now known. The catabolism of chlorophyll has been studied in senescing leaves and ripening fruits of a large number of plant species. The pathway of chlorophyll degradation as depicted in Fig.(31.3) summarizes current knowledge. The catabolic machinery comprises at least six reactions, whereby the first four are responsible for the conversion of chlorophyll to a primary colourless catabolite (pFCC).

Chlorophyll is first dephytylated to chlide by the action of chlorophyllase and subsequently the central Mg atom is removed by Mg dechelatase. The product of the reaction, Pheide, is the last coloured (green) intermediary catabolite of the pathway. Its porphyrin macrocycle is oxygenolytically opened in a two-step reaction with a red catabolite (RCC) as an intermediate, the product, pFCC, is a fluorescent linear tetrapyrrole. This key step in chlorophyll breakdown is catalyzed by the joint action of Pheide-a-oxygenase (PaO) and RCC reductase and it formally occurs by the addition to Pheide a of two atoms of oxygen and four atoms of hydrogen. The structures of the final products of chlorophyll breakdown, NCC, from different plant species suggest that two further common reactions take place, namely hydroxylation and non-enzymic tautomerization of the fluorescent to the non-fluorescent catabolites.

Breakdown of chlorophyll in higher plants is initiated in the thylakoid membrane and it ends with NCCs located in the central vacuole of senescent mesophyll cells. Hence, catabolism takes place in several subcellular compartments and requires transport processes across membranes. The topographical model shown in Fig. 31.3.

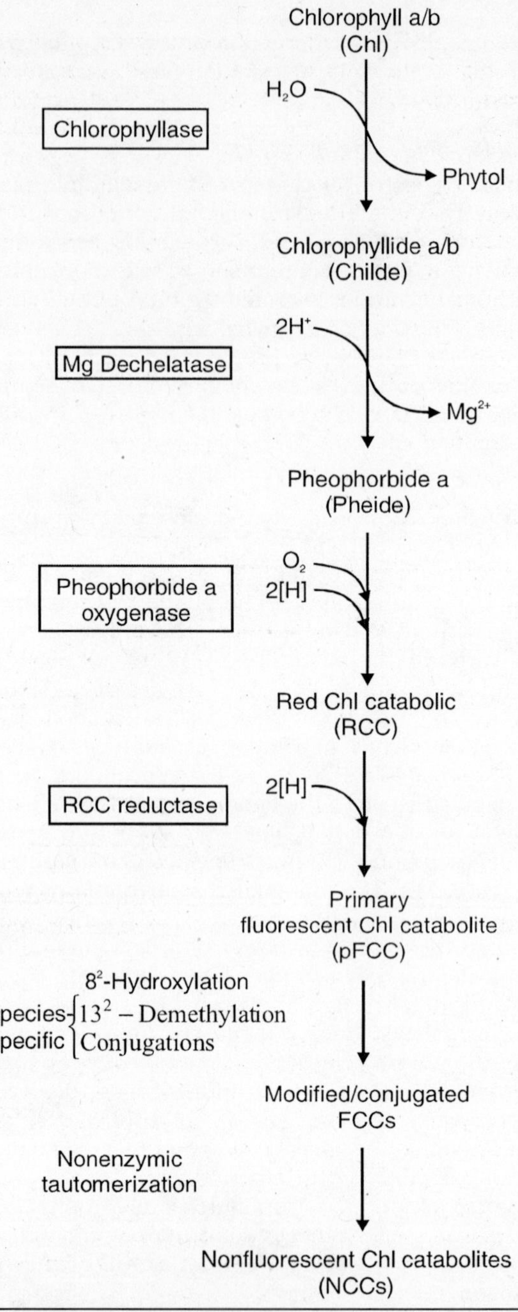


**FIG. 31.3:** The pathway of Chl breakdown in higher plants. Enzymes identified and characterized to date are boxed (after Hortensteiner, 1999).

It summarizes the current knowledge of the catabolic pathway of chlorophyll as organized within a degreening mesophyll cell.

The enzymes chlorophyllase and PaO (probably Mg dechelatase) are located in the inner envelope membrane, where as RCC reductase is a soluble protein of the stroma. It has been speculated that a specific transport protein for chlorophyll or plastoglobules plays a role in shuttling pigment molecules between thylakoids and the chloroplast envelope (Thomas, 1997). The factor "X" appears to act upstream of chlorophyllase and is nowly synthesized as chlorophyll breakdown is induced. The nature of "X" is unknown. Recently, a water – soluble apoprotein of chlorophyll has been cloned and characterized from cauliflower. The purified protein was able to remove chlorophyll from pigment protein complexes (Satoh et al., 1998). The protein is located in plastids and is a candidate of a putative shuttle for chlorophyll from thylakoids to the site of chlorophyllase.

The final catabolites of chlorophyll breakdown are localized in the vacuole of senescent cells (Hinder et al., 1996). A specific carrier of the tonoplast is responsible for transport across the tonoplast have been identified as members of the family of ATP binding cassette (ABC) transporters. The entire metabolism of catabolites including modifications and conjugations takes place at the level of FCCs prior to transport into vacuoles. The disposal of NCCs in the vacuole is reminiscent of the stepwise detoxification in plant cells of xenobiotics and herbicides. It is reasonable to state that chlorophyll is detoxified rather than metabolized during leaf senescence.

Some questions on different aspects of chlorophyll breakdown remain unanswered. The factor (s) initiating and the mechanisms regulating chlorophyll degradation are elusive. The isolation of the genes encoding the catabolic enzymes and the characterization of their respective promoters will be necessary for addressing these questions.

### Significance of Chl Breakdown

If it is justifiable to extrapolate from barley, rape, and a few other species to higher plants in general, then it can be stated that Chl is not catabolized beyond the level of linear tetrapyrroles and that these catabolites are stored in the mesophyll cells down to the end of the senescence period. In other words, the four N atoms contained in each molecule of Chl are unlikely to be recycled from senescent leaves to other parts of the plant. They may contribute to the diet of soil microorganisms when leaves are finally shed. Ant yet, the question of why plants catabolize Chl at all and in such a complicated and energy-consuming fashion remains to be answered. A look at stay-green mutants that are characterized by high retention of Chl in senescent leaves or in ripening fruits (Thomas and Smart, 1993) is enlightening. Chl stability in such plants is associated with the presistence of Chl-binding proteins in the thylaoids (Thomas and Hilditch, 1987; Guiamet et al., 1991; Cheung et al., 1993; Bachmann et al., 1994).

Apoproteins seem to be protected from proteolysis as long as they are properly complexed with pigments. Since the proteins of pigment complexes in the thylakoids account for over 30% of the total protein of chloroplasts, their recycling contributes substantially to the overall N budget. When *F. pratensis* plants are cultivated under nutritional conditions that promote internal N recycling, the stay-green line pays a significant penalty compared with the wild type in terms of rates of growth and development, largely because thylakoid protein N is immobile in senescent leaves. It would be expected therefore, that stay-green genotypes might be less competitive under natural conditions of N limitation than are wild types with a more efficient internal N economy.

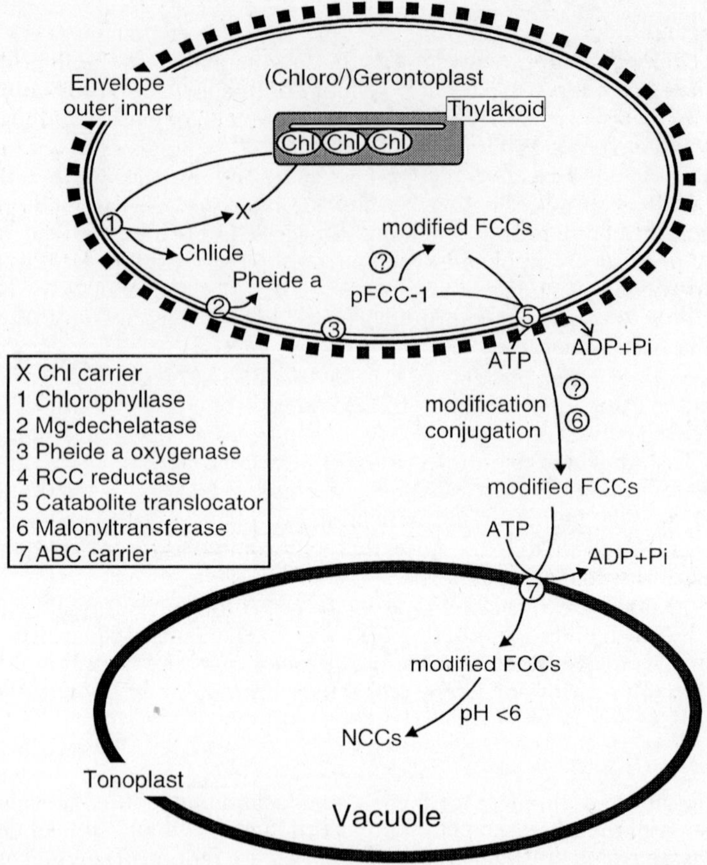


**FIG. 31.4:** Topographical model of the pathway of Chl breakdown involving several cellular components. Characterized enzymic reactions and transporters are numbered. Question marks indicate proposed but as yet unknown reactions. An unidentified factor (X) is responsible for the removal of Chl from the thylakoids (after Hortensteiner, 1999).

It is no wonder then that stay-greens are found above all in cultivated plant species, particularly in legumes such as soybean, French bean, and pea (Thomas and Smart, 1993).

Another important reason why the cellular machinery for breakdown in senescent leaves needs to be so elaborate concerns the photodynamic nature of Chl. Disassembly of pigment-protein complexes is associated with the separation of Chl from the various mechanisms that, in the intact thylakoid, prevent photooxidative damage by activated oxygen. Therefore, it is crucial for the viability of senescent mesophyll cells that dismantling of the photosynthetic membranes remains tightly coupled with the photodynamic inactivation of Chl. In this sense, Chl breakdown may be regarded as a process of detoxification. It is significant that the metabolic sequence that transports Chl catabolites to the vacuole and sequesters them as conjugates is also reminiscent of the fate of xenobiotics and toxic secondary compounds in plant cells.

**Loss of Carotenoid**

All photosynthetic pigments, namely Chla, Chlb, carotenes and xanthophylls are found to decline during leaf senescence [Britton et al., 1989; Biswal et al., 1994; Sanger et al., 1971; Biswal et al., 1994). Studies relating to leaf senescence in higher plants shows a greater loss of Chls than carotenoids, which results in yellow colouration of senescing leaves. Relative stability of carotenoids is demonstrated during leaf senescence in gymnosperms (Toyama, 1965; Matile, 1994) and angiospermic plants (Biswal et al., 1983, Biswal et al., 1976, Joshi et al., 1991).

Chloroplasts produced during leaf senescence (known as gerontoplasts) are characteristically different from chloroplasts of green mature leaves, the transformation is expected to bring changes in the composition and distribution of carotenoids.

Parallel to a quantitative loss of carotenoids, some of the remaining carotenoids in gerontoplasts turn out to be qualitatively different. Leaf senescence causes formation of xanthophylls acyl esters.

Senescence-induced qualitative alteration of xanthophylls may also occur, leading to the appearance of other carotenoid species. The possibility of senescence-induced formation of *B*-carotene epoxide has been discussed by Britton and Young (1989). The epoxide is known to be a product of photo-oxidative events associated with oxidation of β-carotene (Young et al., 1989). Although its presence could not be traced in dark-induced senescence of detached barley leaves (Britton and Young, 1989). Biswal et al. (1994) have demonstrated its appearance during an advanced stage of senescence in the leaves of *Festuca pratensis.*

Thylakoid breakdown is accompanied by the formation of which increase in number and size with the progress of which are found to contain thylakoid breakdown products including carotenoid esters and free carotenoids such as β-carotene and xanthophylls during senescence (Biswal et al., 1988).

Tevini and steinmuller (1985) have examined the formation of plastoglobuli in relation to deposition of thylakoid breakdown products, including free carotenoids and their esters, in them in detail during the course of senescence of beech leaves in autumn. In non senescing leaves, plastoglobuli are demonstrated to contain small amounts of carotenoids but the pigment exhibits an increase in its content during senescence. With the progress of senescence, there is an increasing tendency towards the appearance of carotenoid esters which constitute the major component of total carotenoids when the leaves turn completely yellow.

In addition to the production of heat, carotenoids play a crucial role in residual photosynthesis in gerontoplasts to generate ATP molecules for ordered degradation of cellular structure during leaf senescence (Biswal et al., 1994)

## *Mechanism of carotenoid degradation*

Definite mechanism of carotenoid degradation is not known. Although the enzymes of carotenoid biosynthetic pathways are well characterized, and their genes are even cloned (Hundle et al., 1991; Hundle et al., 1991).

**Enzymatic degradation**

Senescence is an integral part of plant development. The following experimental findings support enzyme participation for its degradation

1. Dark-induced senescence of attached and detached leaves results in loss of both Chl and carotenoids (Biswal et al., 1976; Joshi et al., 1991. Biswal et al., 1983; Mishra et al., 1981). In continuous darkness the participation of free

radicals in destroying carotenoid molecules is minimized and its degradation may, therefore, be attributed to the involvement of enzymes. This is further supported by the findings of Tetley and Thimann (1974), who have demonstrated degradation of carotenoids in etiolated seedlings grown in the dark even with the same rate of degradation of Chl in green leaves. Since phytohormones are known to act at the level of enzymes, further evidence supporting the involvement of enzymes in the breakdown of carotenoids comes from their results on kinetin –induced retardation of the pigment loss. A similar retarding effect of the hormone on carotenoid loss has been reported by Biswal and Mohanty during senescence of barley leaves (1976).

2. The data on carotenoid metabolism during leaf senescence of non- yellowing mutants of *Festuca pratensis* also favour enzymatic degradation of the pigment. Thomas and Matile (1988) have demonstrated almost full retention of the photosynthetic pigments in the mutant despite complete loss of the scavenging system and protective mechanism against photo-oxidative destruction, which would mean involvement of a process other than photo-oxidation for degradation of the pigment. Since the mutation of a *unclear gene* in *Festuca* results in high retention of carotenoids, the control of carotenoid degradation by genes through enzyme system (s) is suggested.
3. The proposition that carotenoids degrade independent of action of oxygen free radicals in supported by the observations that light, which is known to generate these radicals, rather retards the pigment loss during leaf senescence of higher plants (Biswal et al, 1976; Joshi et al., 1991) and ferns (Whitified, 1974).

While searching for the location and nature of enzymes (Friend, 1960) and Friend and Acton (1966) have done extensive experimentation with chloroplasts of sugar beet leaves. In their experiments using water soluble carotenoid glycoside crocin as a model carotenoid, they have demonstrated degradation of crocin by isolated chloroplasts. While examining the response of different extracts of chloroplasts to various additives including inhibitors, metal salts and metal chelating agents on crocin oxidation, they suggested the involvement of a haematin type of copper protein enzyme that could possible be responsible for pigment catabolism. The possibility of a peroxidase-catalyzing carotenoid degradation has been suggested by Matile and Martinoia (1982). They have found peroxidase activity in the thylakoid preparation which could efficiently bleach lutein, a major xanthophylls in higher plants. These in vitro experiments with thylakoid preparation indicate two things: that enzymes participate in carotenoid breakdown: and that these enzymes are located in chloroplasts.

However, in spite of these findings the exact nature of the enzyme (s) largely remains unclear. Review of the literature indicates that is most likely that both enzymatic and photoxidative events are operating during senescence, each contributing to the total loss of carotenoids. It is also possible that both the photo-oxidative process and the enzymatic participation are the two components of a single mechanism operating for degradation of the pigment.

### *Carotenoid degradation: a coordinated event*

Chl, carotenoids and leaf proteins exhibit simultaneous loss during induction of senescence. Although the loss of proteins and Chl is relatively rapid compared to the loss of carotenoids the general kinetic patterns of loss in all these cases remain almost the same. The pattern of retardation of loss of both these pigments by different senescence modulators including phytohormones also remains more or less the same (Biswal et al., 1976; Joshi et al., 1991; Mishra et al., 1981). Retention of Chl in non-yellowing mutants of *Festuca pratensis* is also accompanied by a similar retention of carotenoids. These findings may suggest a coordinated breakdown of photosynthetic pigments during senescence.

Since carotenoids are part of different Chl-protein complexes, the stability of these complexes may contribute to relative stability of the pigments. An increase and decrease in Chl a/b suggests the relative stability of RC II complexes and LHCP respectively.

Whatever the location of carotenoids, since these pigments exist in the form of complexes with proteins, Chl and presumeably in association with specific lipids, breakdown of any individual component may lead to the collapse of the complex. Disassembly of the complex is in fact the prerequisite for enzymatic degradation of individual components including carotenoids. However, what is the component that degrades first during senescence, is still to be answered.

## Loss of Lipids and Membranes

The outstanding change, which occurs in the cells of senescing plant, is the deterioration of membranes. Changes in cell's membrane composition and membrane permeability occur during senescence. Many studies have been conducted on changes in the ultra structure of leaves and cotyledons during senescence. In photosynthetic tissue, the chloroplast thylakoids are typically the first membrane to show signs of deterioration. As deterioration of thylakoid continues, the ER, mitochondria, and tonoplast show evidence of deterioration. Such changes also occur in non-photosynthetic tissues like cotyledons, roots, xylem, and phloem during senescence. In photosynthetic tissue, lipids associated with the chloroplast are the first lipids to decline, while the non-chloroplast lipids decline later. In plasma, microsomal, and chloroplast membranes, it has been reported that there are 400% increase in the sterol: phospholipid ratio as senescence progresses. Changes in fatty acid content have also been reported in many tissues undergoing senescence. Degradation of lipid and the subsequent action of lipoxygenase on fatty acids play an important role in destruction of the various membranes of the cell. The activity of superoxide dismutase decrease during senescence. Leshem (1984) proposed a model for the degradation of membrane based upon acyl hydrolase and lipoxygenase activities.

Many studies reveal that membranes in mature tissue are in liquid crystalline state, while in senescent tissue the membranes, or portion thereof, are in crystalline gel phase at physiological temperature. Therefore, during senescence the membrane lipid progressively becomes crystalline. Senescence process increases the permeability of the membrane and so contributes to loss of compartmentation. It is very likely that the increase in permeability is at least partial due to the occurrence of the crystalline gel phase in membrane lipids. It also helps in translocation of materials from cotyledons to other tissue.

## Loss of Macromolcules

### Nucleic acids

The characteristic decrease in protein content in senescing leaves is accompanied by a parallel decline in RNA content. This decline in RNA content is associated with a reduced capacity of senescing leaves to incorporate radioactive precursor into nucleic acids. In contrast, leaves treated with growth regulators which retard senescence the capacity to utilize precursor is sustained. This evidence suggests that decline in RNA content during senescence is due to the decomposing capacity for RNA synthesis and it has been suggested that senescence is caused by failure of DNA to provide an effective templates for RNA synthesis.

If senescence is associated with decomposition in DNA template, availability might be expected that there would be a reduction in the synthesis of RNA components in order to detect changes in mRNA. This method for determining relative levels at polysomes to monosomes in ribosomal preparation has been used. Ribosomal preparation has been obtained from senescing leaves which, showed that though the polysomes level declines during the growth of the leaf, the mature leaf retain cytoplasmic polysomes but chloroplast polysomes are absent only in younger leaves. The absence of cytoplasmic polysomes at later stages of senescence may indicate that mRNA is long lived or alternative mRNA synthesis may be sustained even when the leaves are yellowing and senescing. The mRNA synthesized now could be loading for the production of hydrolytic enzymes during senescence although polysomes may still be absent.

The total ribosome content of leaves declines during senescence. In initial stages, the properties of chloroplast ribosome declines, later on free cytoplasmic ribosomes are depleted whereas the membrane bound ribosomes are sustained during senescence there is an increase in the rate of soluble RNA to rRNA suggesting a greater liability or reduced synthesis of rRNA. However, the evidence supporting this concept is extremely conflicting.

**Proteins/Amino acid**

The protein molecules in the plants are in a constant state of turnover and at steady state condition when protein content remains constant, the rate of synthesis is equal to the rate of degradation. In leaves of annual plants, the protein content does not remain constant for a long period. Instead the leaf protein content reaches high at the time when the leaf becomes fully expanded and thereafter it begins to decline. The initial decline in leaf protein occurs before there are any visible indications of chlorophyll depletion or other outward signs of the onset of senescence. The decline in protein content and onset of senescence are accelerated by leaf attachment. It was observed that the capacity to incorporate exogenously supplied labeled amino acids decline as detached leaves are immersed in water. However, when these leaves are treated with growth regulators, which retard senescence, the ability to incorporate amino acids into protein a retained. This led scientists to conclude that senescence caused by a failure or a declining capacity is protein synthesis.

During senescence amino acids are accumulated in detached leaves and with time there is an increase utilization of these amino acids as respiratory substrates, while the carbon skeletons are used in respiratory metabolism. The nitrogen components appear to be accumulated in the amino residues of glutamine and aspartate. In early senescence glutamine formation predominates while in the later stages of senescence aspartate biosynthesis predominates.

Kao (1981) observed a striking increase in proline content in the senescing leaves of rice. He noticed that first there is a slight but significant increase in the proline content during first day of dark incubation, but subsequently proline accumulation becomes relatively rapid and after 4 days of excision, 30 to 50 folds increase the proline level in the leaves. This proline accumulation during senescence was observed in many other plants besides rice leaves.

Protein degradation determines, or at least is parallel, with a diminution of the photosynthetic capacity of the leaf and with ability of the leaf to export carbohydrates. Carbohydrates are also required as energy source and as carbon precursors for biosynthesis in reproductive structures. One can think that an adjustment of the nitrogen and carbohydrate demands of developing reproductive structures may control leaf senescence in monocarpic plants, but the problem may be difficult for an experimental examination. In fact, frequently reproductive structures can carry considerable photosynthesis.

### *Senescence-induced Ultrastructural Changes*

Senescence-induced ultrastructural modifications, including changes in the molecular composition of thylakoids, have been probed by electron microscopy, X-ray diffraction, immunotechniques, absorption and fluorescence spectroscopy, and phase properties of membranes, of all these probes, electron microscopy has been used extensively by many authors to examine the changes in size, shape and ultrastructural organization of chloroplasts during senescence. Chloroplasts of completely yellow leaves appear oval and exhibit minimal lengths and widths indicating senescence-induced shrinkage of the organelle.

During senescence, the organelle shows disorganization at the level of both the envelope and lamellar structures. Lamellar breakdown, as revealed by electron microscopy, exhibits unstacking of grana thylakoids. This is followed by the formation of loose, elongated lamellae scattered in the organelle. In the next phase, these lamellae undergo further degradation with the concomitant appearance of plastoglobuli. The size of the globuli increases as senescence advances. Although these ultrastructural modifcations have been qualitatively studied, attempts have been made by Naito et al. (1981) to quantify some of the structural parameters. A decrease in the number of grana per chloroplast section and in granum length with an increase in the number, length, and width of plastoglobuli during senescence.

The sequence of ultrastructural changes during chloroplast senescence varies depending on the environmental conditions and plant species.

The pattern of ultrastructural changes of chloroplasts differs with plant species, even though the experimental conditions remain same. For example, rupture of the plastid envelope is followed by the degradation of the thylakoid system during senescence of wheat leaves (Hurkman, 1979), however in the chicory leaf, the thylakoid system degrades first and is followed by the rupture of the envelope, although in both cases, detached leaves were used for the experiments.

Ultrastructural degradation consists of three major events: thylakoid breakdown, formation of plastoglobuli, and rupture of the envelope. The precise mechanism of senescence-induced thylakoid breakdown is not known.

Cytokinin treatment prevents an increase in the number of plastoglobuli and simultaneously prevents thylakoid disintegration. This suggests a correlation between the level of plastoglobuli and the structural status of thylakoid membranes. On the other hand, some reports suggest that plastoglobuli may only in part be the product of lamellar breakdown and that the rest of the rest of their structure may represent a nonspecific storage of insoluble lipids. This view is supported by the observation that the globuli from mature chloroplasts of *Vicia faba* and spinach contain a mixture of lipids and other substances that are not specific to the lamellar membrane system.

## Physiological and Biochemical Changes Associated with Leaf Senescence

Senescence in general denotes the onset of a complex series of detrimental changes ultimately leading to the death of a living organism. Leaf senescence is an integral part of the leaf development. It is like other developmental processes- a well-organized and regulated process rather than a passive decay of the leaf. It has been studied and worked upon extensively by a large number of scientists from time to time as these organs form a useful experimental system in which senescence can be studied properly.

Senescence is regarded as the series of event concerned with cellular disassembly in the leaf and the mobilization of materials released during this process. This may also be defined as the resultant of a number of deteriorative processes set in motion according to hereditary instruction, which are aimed at terminating the functional life of an organ or organism under natural condition. Considerable quantities of fatty acids may be liberated during the breakdown of thylakoid lipids, which can be further metabolized in the peroxisomes and finally fuel the tricarboxylic acid cycle in the mitochondria. These suggest that the sequential senescence is the sequential loss of photosynthesis and respiration, a well organized process and not a decay of leaf.

Various biochemical and photochemical changes occur during senescence like photoelectron transport system, oxygen evolving system, photophosphorylation, photosynthetic rate, respiration rate, nutritional status, hormone level, mineral uptake, enzymes, stomatal conductance, seed filling period and yield

Most of the biochemical work deals with detached organs rather than whole plants or attached organs, which may represent a more natural situation. In leaves, and presumably other organs, many components (protein, nucleic acid, chlorophyll, etc.) or process (photosynthesis) may reach a maximum about the time maximum size (growth) is attained; this peak may be followed by an early, slow decline and then a later, rapid decline. The exact significance of the early, slow phase is still uncertain. The early, slow phase could be a metabolic step down following cessation of growth, but there is at present no reason to believe that it is not part of the continuum of degeneration leading to death.

### *Photoelectron Transport System*

The pigments associated with thylakoid membranes are characteristically assembled into two photosystems, namely photosystem I (PS I) and photosystem II (PS II), that are linked by a series of electron-transport carriers (Cramer et al., 1985). The two photosystems have specific orientations across the thylakoid membrane (Haehnel, 1984). It is proposed that PS I is located in stroma lamellae and is primarily responsible for cyclic electron flow. PS I in grana, on the otherhand, is linked with PS II and is largely involved in the reduction of $NADP^+$. Normally, higher wavelength-absorbing pigments are present in PS I that do not provide any significant contribution to the fluorescence emitted by chloroplasts. The reaction center of PS I is known as P700, which is not directly associated with the oxygen-evolving complex. The major electron-transport carriers associated with PS I are plastocyanin on the oxidizing side, ferredoxin, P430, and NADP reductase on its reducing side. The pigments associated with PS II are the short wavelength form of chlorophyll *a* and other accessory pigments that contribute significantly to the total fluorescence of chloroplasts. PS II is responsible for oxygen evolution, although the oxygen-evolving complex has been identified as a separate structural unit. P680 is the reaction center of PS II. The major electron-transport carriers of this photosystem are cytochromoe $b_{559}$, pheophytin, and quinines.

Various steps of photochemical reactions, namely light absorption by antenna pigments, migration of light quanta to the reaction center, separation of charge, and movement of electrons, are best explained by **Z scheme**. The scheme explains the participation of PS I and PS I in the synthesis of ATP and formation of NADPH. Water is the ultimate natural electron donor that donates electrons to Q. The process if photosensitized by the oxidized to a strong oxidant capable of splitting water to liberate oxygen through the oxygen-evolving enzyme system. Similarly, P700, the reaction center of PS I on excitation loses electrons to P430 and is oxidized. The reduced Q and the oxidized P700 are reoxidized and reduced, respectively, due to the continuous flow of electrons from water of $NADP^+$ in the presence of NADP reductase. In addition to water, a number of exogenous electron donors have been identified and are known to donate electrons at specific sites of the electron-transport chain. Exogenous electron donors, acceptors, and inhibitors are used extensively to characterize the specific sites of the electron-transport chain of the thylakoid membrane.

Photoelectron transport in thylakoid membranes is coupled to ATP synthesis through coupling factor ($CF_1$), a well-characterized extrinsic protein complex of the thylakoid membrane (Strotmann and Bickel-Sandkotter, 1984). Thus, light-driven electron transport in thylakoid membranes generates NADPH and ATP, the two essential molecules required for fixation of carbon dioxide to sugar through a series of enzymes in the Calvin cycle. Although most of the enzymes of the Calvin cycle are well studied, rubulose 1,5-bisphosphate carboxylase/oxygenase (Rubisco) has drawn the attention of many workers because its complex structure, regulation of its synthesis, and susceptibility to different environmental stress. The enzyme is also a major protein of green leaves.

### *Disassembly of Ps I and Ps II and Loss of Electron-Transport Efficiency*

Senescence causes an extensive disruption of the thylakoid system including the unstacking of grana and disappearance of photosynthetic pigments. This kind of structural degradation could logically lead to a loss in the capacity of chloroplasts for photoinduced electron transport. A decline in the rates of photochemical reactions in chloroplasts undergoing senescence has been demonstrated by many authors using different plant systems. There are, however, different views on the nature of the loss of photochemical reactions. For example, Camp et al. (1982) have shown relative stability in whole chain electron-transport efficiency when it is measured per chloroplast as compared to a leaf area basis at the initial stage of senescence.

The above findings may suggest that a loss in photochemical potential during senescence could be a due to loss of whole chloroplasts rather than the loss of thylakoid organization. On the other hand, the loss in the capacity for photochemical reactions has also been attributed to senescence-induced temporal disorganization of thylakoid ultrastructure, loss of chlorophyll, and physical changes in thylakoid membranes. A gradual decay of membrane structure and consequently a decline in the photochemical potential of chloroplasts are, however, compatible with sequential degradation of thylakoid structure as revealed by electron microscopic studies.

Senescence induced-loss of the photochemical reactions associated with PS I and PS II has been demonstrated to be sequential. Senescence of detached barley leaves was shown to cause a relative instability of PS II of thylakoid membranes. A preferential loss of chlorophyll *b*, a major pigment of PS II, is demonstrated during senescence of mustard cotyledons treated with continuous darkness suggesting a loss of LHCP and disassembly of PS II of thylakoid membranes. This could be correlated with the loss of PS II photochemical reactions during senescence of leaves, algae, and isolated chloroplasts. Severe damage to the oxygen-evolving complex might also be a factor in the loss of the photochemical potential of PS II of thylakoid membranes. During senescence of chloroplasts in detached wheat leaves, a relatively rapid loss of chlorophyll *a* occurs indicating the disintegration of the PS I chlorophyll-protein complex.

### Damage to the Oxygen-Evolving Complex

A decline in oxygen evolution and a restoration of the senescence-induced loss of PS II-mediated DCIP photoreduction with exogenous electron donors are indicative of the limitations imposed by a damaged oxygen-evolving system on the efficiency of the electron-transport system of thylakoid membranes during senescence. Although the precise geometry and the mode of function of the oxygen-evolving complex are not yet clearly known, recent findings propose the participation of a number of polypeptides that constitute the $O_2$-evolving complex. Some of the polypeptiedes are known to be intrinsic and are embedded in the lipophilic environment of the membrane, and some are extrinsic, obviously vulnerable to externally applied reagents.

### Loss of Photophosphorylation

A few reports are available on the senescence-induced loss of the ability of plastid membranes to synthesize ATP. There are several possible explanations for the mechanism of senescence-induced loss of photophosphorylation. A decline in the efficiency of the electron-transport system during senescence as reviewed earlier may limit the ATP-synthesizing ability of the membrane. This is also evident from the observation of the senescence-induced loss of activity of the thylakoid-bound $Mg^{2+}$-ATPase. The activity of this enzyme reflects the energization status of thylakoid membranes induced by photoelectron transport.

An inactivation or quantitative loss of coupling factor proteins could also lead to a decline in photophosphorylation. A change in the aggregation pattern and conformation of the coupling factor may result in a decline in its functional efficiency.

### Loss of Rubisco Activity

The modification of enzymes of the stroma has been examined during chloroplast senescence. The activities of NADP-dependent glyceraldehydes 3-phosphate dehydrogenase, phosphoglycerate kinase, ribose 5-phosphate isomerase, phosphoribulokinase, and ribulose 1,5-biphosphate carboxylase reportedly decrease during senescence. Ribulose 1,5-biphosphate carboxylase has been shown to be most susceptible to senescence (Lauriere, 1983). There is, however, controversy concerning the relative susceptibility of Rubisco compared to that of chlorophyll and photochemical reactions.

### *Oxygen evolving Complex*

**Photophosphorylation**

**Photosynthetic Changes:** Decrease in rate of photosynthesis with the age of a leaf to decline in dry weight of leaves and their eventual death are the symptoms of leaf senescence. The rate of photosynthesis starts declining as soon as the leaf reaches its full size or becomes mature (senescence leads to profound changes in the pigments composition of the cell, thereby affecting photosynthesis rate). There is gradual loss of chlorophyll and carotenoids, which, results in yellowing of leaves and/or cotyledons particularly during the later stages of senescence.

**Respiration:** When a plant tissue has reached maturity, its respiration rate will either remain roughly constant or decrease slowly as the tissue ages and ultimately senesces. One might except that if senescence is viewed as a general decline in function, the respiration rate would steadily decline. In fact such a decrease does generally occur, but only quite late in the senescence process, and at first there is a sharp increase. Low RQ values were detected in senescing leaves. This makes clear that lipids rather than carbohydrates were primary source of respiration substrates.

**Nutritional Status:** The general simultaneity between monocarpic senescence and reproductive processes, led to Molisch, 1938 to postulate that senescence of laged organs is produced by a competence between these and the reproductive structures for nutrient uptake. At the end the competence would produce a withdraw of nutrients from vegetative to reproductive organs. In relation with the nutrient redistribution associated with plant senescence, the degradation of proteins and the export of nitrogen by senescing leaves is, probably, the most studied process.

**Hormone Levels:** According to Osborne and Thompson (1992) the response of any cell to a change in the level of any of the regulatory substance is dictated not by the chemical alone, but first and foremost by the mode of the genetic expression of the cell. However, the latter is controlled by the environment, hormones and developmental stages of the plant; which also determines the way in which, and the extent to which, the signal is perceived and transduced (Bressan and Handa, 1992). The balance among the different growth regulators controls senescence. Change in hormone level influenced by the development stage of leaves and by other structures of the plant. Hormone level changes may be a primary cause or a secondary effect of senescence in leaves.

It is generally agreed that all plant hormone groups influence senescence (Kelley and Davies, 1988; Noodén and Leopold, 1988). In most plants cytokinins are antisenescence compounds that can postpone plant death (Noodén et al., 1989), gibberellins also have similar effects, but in fewer plant species. Auxins influence senescence and abcission in very complex manners, apparently depending on the auxin source, or depending on the age and receptivity of the tissue. When applied to the leaf, auxin often retards senescence, but paradoxically when auxin enters the leaf from the stem, it sometimes promotes senescence.

Two plant hormones have been suggested as candidates for death hormones: ethylene and abscisic acid. Ethylene is intimately involved in many senescence phenomena, especially flower wilting and fruit ripening. However, there is no clear correlation between ethylene and monocarpic plant death. Also the mobility of ethylene as a gas does not fit that of a death hormone. Abscisic acid was isolated on the basis of its promotion of abcission and dormancy.

**Abscisic acid (ABA)**

Abscisic acid was originally isolated as an abscission-causing factor. It is clearly involved in leaf senescence, and through its promotion of senescence it might indirectly increase ethylene formation and stimulate abscission.

The endogenous concentrations of ABA in developing fruits of barley, wheat, soybean, and cherry tomato change in a similar pattern; ABA rises during the rapid growth stage of the fruit or seed, peaks at the point of maximum fresh and dry weight of fruit, and then declines rapidly with senescence of the plant and with ripening or drying of the fruit. Developing cotton and pea fruit show a slightly different pattern, with low levels of ABA during rapid fruit growth and a rise during fruit maturation. Senescent bean leaves contain less free ABA than young leaves; however, the senescent leaves contain more bound ABA, thereby equalizing the total ABA in the young and senescent leaves.

Water stress-induces ABA in different subcellulár compartment (cytoplasm?) from the nonstress ABA. The stress-induced increases in ABA may be large enough to mask the senescence-related changes, or if variable, they could obscure them. Clearly, stress-induced senescence and ABA need to be controlled if other patterns, especially monocarpic senescence, are to be studied.

The exogenous supply of ABA promotes chlorophyll loss in leaf discs collected outdoors from seventeen tree species plus bracken fern. ABA also accelerates yellowing of detached leaves in radish, barley, nasturtium and *Rumex pulcher*.

Translocation, inactivation, poor uptake, and competition from endogenous hormones are all possible reasons for this difference between attached and detached leaves. In fact, attached leaves sprayed once (or even several times) with ABA solutions may still receive less of the hormone than leaves floating on the solutions. Nevertheless, ABA sprays can accelerate senescence of attached leaves including tree of heaven, Brussels sprouts, and soybean.

The function of ABA as a promoter of flower tissue senescence, including such parameters as wilting, color fading of `blueing', and protein loss, has been established. Here, ABA may act through hastening of autocatalytic ethylene production. In some cases, ABA actually exhibits a senescence-retarding effect.

ABA sprays cause persistence of grape inflorescences, which mimics cytokinin sprays and have been termed the cytokinin sparing effects (Mullins and Osborne, 1970). At $10^{-4}$ M, ABA retards senesces in radish leaf discs, but this retardation may be due to the disruption of the normal senescence process (Colquhoun and Hillmann, 1972).

In addition, ABA may cause short-term inhibition of senescence of cut flowers by closing the stomata, thereby water loss.

Acceleration of RNA and protein loss in senescing tissues by ABA has been reported several times.

ABA elevates protease and nuclease activity through increased synthesis (or the activation of stored enzyme), but not all hydrolases are increased ABA has been found to block the rise of β-1,3-glucan hydrolase in tobacco leaf disks, while this enzyme increases during the senescence of the water controls. Overall, ABA decreases protein synthesis and increases protein degradation, but the relative contributions of these processes and their responses to exogenous ABA are not clear. Not all of the studies on protein synthesis effects are corrected for changes in the specific activity of the precursor pool or decreased precursor uptake over time. Many pieces of evidences suggest that ABA may play an important role as an inducer of senescence, including monocarpic senescence; the picture is far from complete.

Zacarias and Reid (1990) compared the effects of ABA and ethylene on the senescence of leaf discs excised from wild-type *Arabidopsis* and an ethylene-insensitive mutant. Chlorophyll loss from untreated leaf discs of the ethylene-insensitive mutant was much slower than that from the wild type, in the light and in the dark.

**Ethylene**

Unfortunately, little work has been done to examine the involvement of ethylene in whole plant senescence; perhaps that is because it appears not to play a very important part. The senescence in flower petals (fading, rolling, etc.) is accompanied by the increase of production of ethylene. Production of ethylene in leaves generally rises slowly until the leaves become senescent and abscise. Flowers also synthesize ethylene, especially just before they fate and wither, and in most species this gas causes their senescence and abcission.

Although ethylene has profound role in fruit ripening, excised organ senescence and flower initiation in some plants, its precise role in whole plant senescence is still elusive. Recently, Morgan et al (1992) have demonstrated that intact leaves of cotton experiences a climatric like storage in ethylene after visible symptom of senescence appears, which consequently signals for abscission. L-serine promotes the senescence of detached oat leaves possibly through stimulation of the formation of two proteases, which normally increase throughout the senescence of the leaf. L-serine has only a slight effect on the senescence of isolated chloroplasts, the protease synthesis is assumed to occur on cytoplasmic polysomes. L-cysteine, L-glycine, and L-alanine also promote senescence and protease activity, but they act only at higher concentrations than serine. The role of endogenous serine in control of senescence remains to be clarified.

Ethylene has well characterized effects on the abscission process, stimulating the synthesis and secretion of cellulases and polygalacturonases. Ethylene (7 ppm) has no effect on leucine incorporation in *Rhoeo* leaf disks, but there may be a slight inhibition of uridine incorporation into RNA.

Etherel (which breaks down into ethylene) sprays have no effect on foliar senescence in wheat seedlings. Exogenous ethylene (applied as Etherel or the gas) increases endogenous ethylene and ABA production and overcomes the senescence retarding effects of exogenous cytokinins in flowers.

Recently Morgan et al. (1992) have demonstrated a climacteric like rise of ethylene in intact leaves of cotton after the appearance of visible symptom of senescence, which consequently signals for abscission. L-serine, L-cysteine, L-glycine, L-alanine are also known as senescence promotors.

**Cytokinin**

Cytokinin activity in the leaves decreases during senescence and may parallel chlorophyll loss. Cultivars of rice and soybean with early senescence show correspondingly earlier decreases in foliar cytokinin. While free zeatin declines in rice and soybean leaves, the zeatin derivatives dropped less rapidly, or even rose. Similarly, senescing *Streptocarpus* leaves show a decline in butanol-extractable cytokinins (free zeatin and nucleosides), while the residual water-soluble cytokinins (Zeatin nucleotides) increase. Treating isolated leaves of many species with cytokinins will delay their senescence. Although applied cytokinins do not prevent senescence completely, their effects can be dramatic, particularly when the cytokinin is sprayed directly on the intact plant.

Probably, root apices are major sites of cytokinin synthesis and from root they are transported to aerial parts of plant. As the formation of roots halts senescence in detached leaves starts. It was suggested (Chibnall, 1954) that roots produce some senescing-retarding factors. Later, Kulaeva (1962) demonstrated that root exudates of substances retard senescence.

Cytokinin influence the movement of nutrients into leaves from other parts of the plant, a phenomenon known as cytokinin-induced nutrient mobilization. This process is revealed when nutrients (sugar, amino acids etc.) radiolabeled with $^{14}C$ or $^{3}H$ are fed to plants after one leaf or part of a leaf is treated with a cytokinin. Later the whole plant is subjected to autoradiography to reveal the pattern of movement and the sites at which the labeled nutrients accumulate. Experiments of this nature have demonstrated that nutrients are preferentially transported to, and accumulate in, the cytokinin- treated tissues. It has been postulated that the hormone causes nutrient mobilization by creating a new source-sink relationship.

Leshem (1984) established a causative relationship among cytokinin, lipoxygenation, anti-oxidation and lipid-associated free radical scavenging. During both normal and induced senescence there is a rise in lipoxygenase which catalyzes the oxidation of polyunsaturated fatty acids. These include linoleic and linolenic acids which are of common occurrence in plants and inter alia may be situated in membrane phospholipids. Lipoxygention causes the production of free redicals such as the superoxide, fatty acid and peroxy species. Cytokinin lowers lipoxygenase and superoxide dismutase activities significantly in senescing foliage and is mimicked by the endogenous lipid antioxidant, tocopherol, in its chlorophyll-retaining and lipoxygenase – lowering effects. He indicated that cytokinin interaction with free radicals may occur in following two ways:

1. Cytokinin may act as a direct free radical scavenger.
2. Cytokinin may serve as an incipient preventative of free radical formation by inhibiting oxidation of plant purine compounds which at certain stages of breakdown release superoxide and hydroxyl free redicals.

These observations collectively indicate that prevention of free radicals formation and (or) their direct scavenging should be included in the multifactorial antisenescence mode of action of cytokinin.

Foliar applications of cytokinin have now been shown to retard senescence in attached leaves of several species including common bean, oat, soybean, and wheat. Spraying with the auxin NAA delays senescence of attached soybean leaves. The roots are the main sources of cytokinins for the leaves, a decrease in the level of cytokinin during leaf senescence must be due:

1. To an accelerated cytokinins metabolism in the leaves before their senescence.
2. To an accelerated transport from leaves to other structures.
3. To a decrease in the supply of cytokinins by the roots.

The cytokinins content, in turn, determinates if a structure will be a sink or a source. However, the fate of root-cytokinin during the reproductive phase is not clear. At early fruit development, its cytokinins content is high, and then decreases. In soybean fruit development no correlation was found between fruit cytokinin decrease and leaf senescence. Root-cytokinin reduction is not the primary causal factor in monocarpic senescence in soybean. In soybean seedling, leaf senescence is caused by a competition among leaves for root cytokinin. While endogenous level of cytokinin is reduced in intact older leaves prior to senescence. Exogenous application of the hormone is also effective in the deferment of the senescence of foliar organs.

### Auxin

Auxin level decrease with age in leaves of several species. Auxin activity drops to low levels in the senescing first-trifoliate leaves of the common bean but an increase is reported in senescing primary leaves of the same species. Measurements of auxin by modern physical methods show that the IAA content in leaves of cotton and common bean decreased with age, thereby supporting the former on bean leaves.

α-Naphthalene-acetic acid retard senescence of leaf discs and bean pods. 2,4-D butyl ester (a synthetic auxin largely used as a herbicide) retards the proteolysis of detached leaves of *Prunus serrulata* (one of the flowering cherries). Auxins have not, however, been generally found very potent in retarding leaf senescence.

In oat leaf, the auxin (IAA was used) retarded senescence only at concentrations around 500 times that of kinetin. In wheat seedlings naphthalene-acetic acid has a small effect, but only at 100 mg/L, i.e., over 100 times the effective concentration of BAP, and one which would soon be toxic. A complication is that auxin, at levels only slightly above the physiological, stimulates the production of ethylene, which in some tissues promotes senescence; however, its influence in leaves is not strong, and often indeed it seems to show no effect at all. Thus the influence of auxin, though slight, is probably to be ascribed to the auxin it self.

Auxins retard senescence of detached leaves from some species, but it was clear right from the start that auxin treatments do not work with all species. Therefore, it can be said that although auxin is not strong as cytokinin but have a retarding effects on senescence in some assay system.

### Chloroindole auxins

4-chloroindole acetic acid (ClIAA) derivatives may be death hormone in *Pisum*, *Vicia*, *Lens* and *Lathyrus*, because of:

1. ClIAA induces very strong, almost irreversible ethylene production (Ahmad et al. 1987)
2. ClIAA induces death of the apical meristem in pea cuttings (Ahmad et al. 1987)
3. Auxin in lanolin placed in deseeded bean pods induce leaf senescence, and ClIAA is the strongest natural auxin known.
4. Five different chloroindole auxin derivatives are present in large amounts in maturing pead seeds, and reach concentrations of several mg $kg^{-1}$ for methyl 4-chloroindole acetate.

The five naturally occurring chloroindoles known are 4-chloroindole acid, its methyl ester, the methyl and ethyl esters of malonyl-4-chlorotryptophan, and monomethyl ester of 4-chloroindoleacetyl aspartic acid. There are of course problems with chloroindoles as death hormones or endogenous herbicides. One problem is the postulation of two distinct auxin groups with opposite physiological properties: growth inducers and endogenous herbicides. Another problem is that chloroindole auxins have so far only been found in pea and related species, but not in soybean or kidney bean. Despite some attempts, none of these have been identified yet; and they may be death hormones.

Chlorine containing compound has both auxin and death hormones like characteristics in the slime mold *Dictystelium discoideum*. The chlorinated phenolic compound was first named DIF-1 (differentiation inducing factor); together with cyclic AMP it induces stalk cell morphogenesis, including cellulose wall formation, cell elongation, vacuolization, and eventually the stalk cell dies. At present, it is not clear if the compound induces cell death as such or just launches the cell on and irreversible path.

**Gibberllins**

Parallel decrease in content of gibberellins in senescence has been reported in *Tropaeolum* leaves. A similar decrease occurs in lettuce leaves under water stress. Endogenous gibberellin-like components decline during the senescence of detached *Nastursium* and *Dandelion* leaves. The amino acid arginine reverses the senescence promoting effect of serine in oat leaves, but the effective concentration is probably too high for an *in vivo* significance of the experiments. The retention of chlorophyll in barley leaf discs is enhanced by the polyamines - putrescine, spermidine and spermine.

In leaves of *Taraxacum* and in fronds of *Lemna trisulca* gibberellic acid powerfully retards senescence, but kinetin has no or only a low effect on senescence. Gibberellic acid was demonstrated to delay development of autumn coloration when applied to several species of deciduous trees. Subsequently, gibberellins were shown to delay senescence of detached leaves and leaf segments, but not all species respond to gibberellins.

Similar to cytokinins, gibberellins also act as retardents of senescence in leaves. The role of auxin is not so clear probably auxin can sometimes interacts with other retardants in retarding senescence.

Some recent results also support the above hormone receptor theory (Nissen, 1988; Jones, 1990; Mishra, 1992). Saunders (1990) has presented a four-component model of phytohormone action. These are as follows:

1. Perception of stimulus by a receptor.
2. Activation of an intercellular second messenger.
3. Cascade of amplification of the message.
4. Physiological response.

Nevertheless, the progress of hormone research on the above lines is far less rewarding, as yet. In addition to the isolation of some specific hormone binding protein (Hall et al 1984; Sisler, 1984; Goodwin and Mercer, 1986; Romanov et al., 1990) some second messengers in hormone action have also been isolated. The role of Ca-calmodulin as second messenger for the regulation of senescence has been attracting interest of some plant physiologists. Cytosolic calcium hastens the process of senescence through calmodulin: Phospholipase-$A_2$ complex. Intrusion of Ca into cytosol appears to play a pivotal regulatory mechanism of plant catabolism because calmodulin increases free fatty acid and ethylene and both are involved in senescence development. Interestingly, ABA also plays role in intrusion of Ca into the cytosol. Some recent reports indicate that transporting of Ca into the cell also activates ABA biosynthesis through signal-transduction chain. This might be attributed to an autocatalytic effect of ABA.

**Jasmonic acid**

Jasmonic acid derivatives may be death hormones have following effects:

1. Jasmonic acid methyl ester is a senescence indurcer in some assays it is

stronger that ABA (Sembdner and Klose, 1985).

2. Jasmonic acid widely distributed in plants (Sembdner and Klose, 1985).
3. Soybeans contain high concentration in vascular bundles of the pericarp (Lopez et al. 1987).
4. Jasmonic acid effects can often be alleviated or reversed by cytokinins. This is in agreement with effects of cytokinins on monocarpic senescence (Sembdner and Klose, 1985).
5. Jasmonic acid methyl ester induces ethylene formation (Saniewski et al. 1987)

**Brassinosteroids**

In 1979 a novel plant growth-promoting steroidal lactone termed brassinolide was found in the pollen of rape (*Brassica napus*). Since then chemists and biologists have isolated brassinolide and indentified its related compounds in various higher and lower plants. As a result of extensive investigations, brassinosteroids were found to show characteristic physiological actions on the growth of plants in micro-quantities. Therefore, brassinosteroids can be regarded as a new class of plant hormones in addition to auxin, GA, cytokinin, ABA, ethylene and jasmonic acid with an independent role in the growth regulation of plants. Recently, it has been reported that Arabidopsis mutants deficient in brassinosteroid biosynthesis exhibit delayed leaf senescence (Clouse, 1996).

## *Free Radicals and Ions*

Some of the inorganic ions reportedly regulate chloroplast senescence. The senescence-dalaying action of $Mg^{2+}$ is explained in terms of its stabilizing effect on the stacking of thylakoids, which are known to unstuck during senescence cations such $Mg^{2+}$ and $Mn^{2+}$ have been shown to regulate the level of thylakoid lipid peroxidation and modify the process of chloroplast senescence. $Ca^{2+}$ and $Mg^{2+}$ may interact with the surface of plasma membranes, rigidify them, and consequently delay the process of cellular senescence. The action of $Ca^{2+}$ however, is not comparable with that of $Mg^{2+}$ because of the specific affinity of the former to calmodulin. Calcium ion is also known to stimulate the process of senescence when it enters inside the cell.

## *Cytosolic $Ca^{2+}$ levels during senescence*

Senescence in a deteriorative process that is acociated with a wide range of biochemical changes. Most notable in green leaves is the decrease of Chl content, who is accompanied by a parellel degradation of polar ion from the matrix of the thylakoid membrane (Thomas as Stoddart, 1980; Thomas, 1986; Meir and Philosoph-Haes, 1995) and by proteolysis (Philosoph-Haes et al., 1994). One of the factors that may participate in regulation of leaf senescence is [Ca2+]cyt (Leshem, 1987), the regulation of which is considered essential in all eukaryotes (Bush, 1995).

Cellular $Ca^{2+}$ homeostasis is normally maintained at submicromolar levels by membrane associated $Ca^{2+}$ transport pumps, secondary transporters, and ion channels (Bush, 1995). Stimulus-induced changes in $[Ca^{2+}]$cyt may be transient, sustained, or oscillatory, and the time required for full response varies from a few seconds to several hours (Bush, 1995). A transient increase in $[Ca^{2+}]$cyt may result either, from a decrease in plasma membrane $Ca^{2+}$, ATPase activity or form the opening of $Ca^{2+}$ channels in the cell membranes.

### *Role of $Ca^{2+}$ and calmodulin mediated processes in senescence*

Studies on the role of Calcium in mediating diverse physiological and development processes of plants are now appearing frequently. Many new roles are assigned to this element.

The possible role of Calcium in plant senescence may be inferred from the fact that calcium is widely known to play a major role in membrane structure and function. Its effects on the maintenance of RNA and protein have been well demonstrated and these components are considered as the indices of senescence. Each of the parameters of senescence, such as loss of chlorophyll, decrease in protein and increase in apparent free space and hydraulic conductivity is affected by the calcium status in the tissue.

Calcium effects on various parameters of foliar senescence, such as loss of chlorophyll and protein, increased membrane leakage, dissolution of middle lamella, associated changes well established. As such the present study is aimed at understanding the $Ca^{2+}$ and calmodulin related processes associated with senescence of primary leaves of cowpea.

The results obtained in the present study are briefly summarized below:

1. Calmoduline concentration for unit leaf area is estimated in all the reference stages along with levels of $Ca^{2+}$ Calmodulin is identified, quantified and characterised, Mol.wt. is determined through electrophrosis and identification is carried out through phosphodiesterase activity.
2. To characterise the specific nature of Calmodulin, amino acid analysis is carried out. An attempt is made to correlate the activities of SOD, catalase; lipoxygenase, phospholipids tp $Ca^{2+}$ and calmodulin related events during natural foliar senescence.
3. It is observed that $Ca^{2+}$ and CaM levels per unit leaf area of the primary leaves showed unaltered concentration throughout the period of senescence.
4. Lipid peroxidation, Lipoxygenase activity, Phospholipase activity showed an increase with the progress of senescence.
5. Total protein content increased from $7^{th}$ to $12^{th}$ day and gradually decreased but calmodulin level remained constant.
6. The ratio of calmodulin to total protein content per unit leaf area showed a decrease upto 12th day and followed by an increase till 32nd day. $Ca^{2+}$ level also showed a similar trend.
7. The unaltered levels of $Ca^{2+}$ and calmodulin indicate their active role in the progress of senescence.

### *Mineral Uptake*

Redistribution of minerals probably occurs to some extent in all multicellular plants and during many stages of development, but it is particularly prominent during fruit growth. Mineral distribution in soybean has been studied most extensively. N, P, K, and Mg, but not Fe, Ca, S, and Mn seem to be redistributed from the leaves to the developing fruit of soybean. Mg, however, is not readily translocated.

Senescence is characterized by the transition of the metabolism from nitrogen assimilation to nitrogen remobilization (Feller and Keist, 1986; Peoples and Dalling, 1988). Nitrogen is redistributed in maturing cereals from senescing leaves to the maturing grains, allowing an efficient use of this frequently limiting macronutrient (Feller, 1990). The importance of nitrogen remobilization from senescing leaves for the nutrient economy is obvious.

### *Enzymes*

There is loss of Rubisco activity in dark induced senescence of mustard cotyledons. The activity of protease increases during senescence, which increase degradation of protein. There is an increase in peroxidase during senescence, which takes part in degradation of lipid and bleaching or carotenoids. Decrease in the level of mRNA responsible for the synthesis of the small subunit of Rubisco during dark-induced senescence of mustard cotyledons. A temporal correlation between enzyme protein breakdown and senescence-induced activity of proteases may suggest proteolytic degradation of the enzyme. A senescence-induced protease capable of hydrolyzing soluble proteins, including Rubisco, has been demonstrated in *Festuca* chloroplasts. The protease is characterized by its ability to hydrolyze different substrates, including Rubisco protein, at different pH during senescence.

As far as enzymes involved in nitrate reduction and amino acid metabolism, the enzymes nitrate reductase and nitrite reductase are only needed for the assimilation of inorganic nitrogen. Nitrate reductase is located in the cytosol and requires NADH or NADPH for its function, whie nitrite reductase is a chloroplast enzyme and depends on reduced ferredoxin.

### *Stomatal Conductance*

In soybean De Souza (1997) found that water stress condition declines in stomatal conductance. De-fruiting caused stomatal resistance to increase yet depodding in soybean inhibited rather than promoted senescence, along with the inhibition of transpirational flux and partitioning of nutrients.

### *Seed filling period and Yield*

In soybean, De Souza (1997) found that water stress during seed filling period causes shortening the seed filling period. Water stress during seed filling could reduce soybean (*Glycine max* L.) yield by accelerating leaf senescence and shortening the seed filling period. Water stress during seed filling reduces yield by reducing seed size. These reductions in seed size have been associated with earlier maturity and a shorter seed filling period. It is generally assumed that the shorter seed filling period results from accelerated leaf senescence, but direct measures of leaf senescence supporting this assumption are lacking.

Leaf senescence in soybean. Characterized by declines in leaf N, chlorophyll, and photosynthesis, begins early in seed filling and is usually complete by physiological maturity. During leaf senescence, C and N are redistributed from the leaf and other vegetative plant parts to the developing seed. This redistributed N can be a significant source of N for the seed, contributing from 20 to 100% of the mature seed N in cultivars of differing maturity. The amount of redistributed N in the seed and the N available for redistribution at the beginning of seed fill are directly related to the vegetative mass at beginning seed fill, which is largest in late-maturing cultivars. Similar results have been obtained when the variation in vegetative mass and redistributable N at beginning seed fill was created by water stress. Water stress during seed filling could reduce yield by accelerating the decline in leaf photosynthetic activity and increasing the remobilization of C and N to the seed.

# CHAPTER 32

# Movements in Plants

## INTRODUCTION

When plants, although rooted to the soil, exhibit a range of fascinating movements that often- but not always- can be interpreted as adaptations to their environments. The concept of movement in plants was developed by Darwin and later on by Julius von Sachs (1832-1897) who invented the clinostat and studied phototropism and gravitropism in plants. Movements are classified into two following natural categories:

1. The **tropisms** in which the direction of the environmental stimulus determines the direction of movement. Stems growing away from gravity or toward a light source are examples of tropisms.
2. The **nastic** movements, which are triggered by an external stimulus (sometimes interacting with an internal timing mechanism) but in which the stimulus direction does not determine the direction of movement.

Both these movements are often the result of **differential** (irreversible) growth, but they can also be caused by differential and reversible uptake of water into special cells, called motor cells, that collectively form a **pulvinus**. Migration of slime molds, gliding of blue-green algae, cytoplasmic streaming, orientation of chloroplasts and nuclear migration, and action of cilia and flagella are also included in movement, however for the purpose of this book we shall not discuss them. Other movements are examples of taxis, in which an organism (usually single or few celled; rarely an organ such as the pollen tube) moves toward or away from some stimulus e.g., chemical gradient, light, warmth or cold.

Plants are influenced by the energy and mass parameters of their environments which induce either tropistic or nastic responses, or both. An environmental change that induces a plant movement (or other response) is a **stimulus**. Often a stimulus **induces** in the plant a process that continues after the stimulus no longer exists in its initial form, e.g., daily leaf movements. In this case leaf position is changed in response to the rising or setting of the sun; similarly, a stem continues to bend toward the light or away from gravity for some time after the light or gravitational stimulus is removed.

Stimuli are received by plant receptor. Once the stimulus has been perceived, it is changed (**transduced**) into some other form, known as **signal**, which then leads to some **motor response** (growth or pulvinar action). This is the actual cause of the plant movement. Plant movements can be explained in three following steps:

### *Perception*

When a plant or plant receive any environmental stimulus that causes the response. It is very difficult to answers following questions related to plant response because such plant organs as leaves, stems, and roots, unlike eyes and ears, are not specialized to respond **only to one stimulus**.

1. What pigment absorbs the light that causes phototropism?
2. What in the cells or tissues responds to gravity?
3. Where in the plant is the perception mechanism located?
4. What is the mechanism?

### *Transduction*

Similarly with reference to transduction there are following unanswerable facts:

1. How does the perception mechanism transduce its perceived stimulus to cells in the organ where movement occurs?
2. What **signal** does it send- that is, what biochemical or biophysical changes occur in response to the environmental stimulus?

Plant physiologists are actively working to find out the answers specially with reference to electrical signals and for chemical messengers (hormones). Any hypothesis put forth to explain the mechanisms of perception and transduction must account for the observed motor responses.

### *Response*

Response is the final expression of movement, but it is difficult to answer what actually happens during movement? It has been established that similar mechanisms within a plant often cause different responses e.g., changes in cell volume caused by $K^+$ and then water movement *in and out* of cells accounts for such diverse responses as stomatal opening and closing and thigmonastic leaf folding in the sensitive plant; $K^+$ could also be important in some tropisms. Different mechanisms can produce similar responses in different or even the same organisms e.g., different pigment systems are apparently responsible for phototropic bending in different organisms (with certain exception).

## Nastic Movements

Nastic movements can be observed in leaves or leaflets of compound leaves. It may be of two types:

1. **Hyponasty:** An upward bending of an organ.
2. **Epinasty:** Downward bending of an organ.

Such leaf movements are caused by pulvini at the base of the petiole, blade, or leaflet. In absence of pulvini such movement can also occur. Epinasty occurs, for example when cells on the top of the petiole or blade, especially in the main veins, grow (elongate irreversibly) more than those on the bottom. Generally speaking, nastic movements are reversible, whether they are controlled by pulvini or by changing relative growth rates on top and bottom of an organ.

When epinastic leaf bending is caused by more rapid growth of cells on top of the petiole and blade than on the bottom (when cells on the lower side begin to grow faster) such bending changes to hyponastic movement.

### Nyctinasty

When leaf (e.g., doubly compound leaves) movements occur from nearly horizontal during the day to nearly vertical at night called exhibit striking **sleep movements**. It occurs in silk tree (*Albizzia julibrission*), the sensitive plant (*Mimosa pudica*), and the raintree (*Samanea saman*). At night, the tips of opposite leaflets of *Albizzia julibrissio* fold together (close), rise upward, and point toward the distal end of the rachilla. *Samanea* leaflets fold down instead of up.

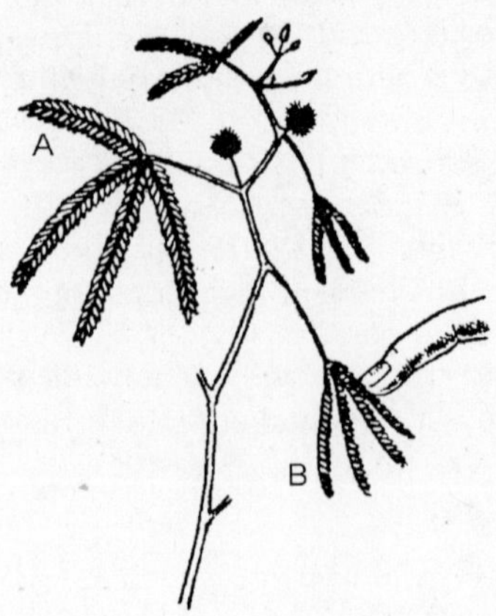


**FIG. 32.1:** (A) *Mimosa pudica* - showing seismonasty (A) Normal leaf (B) Dropping leaf.

In either case, cells in the pulvinus that swell during opening are called **extensors**, whereas those that shrink are called **flexors**. The pulvinus of *Samanea* is a straight cylinder during the day when the leaflets are open and a curved cylinder or hook at night when leaflets are closed.

Hideo Toriyama (1955) observed in pulvini of *Mimosa pudica* that $K^+$ moves out of the cells that lose water. Satter and coworkers observed that the $K^+$ concentration of *Albizzia* pulvini is unusually high (almost 0.5 M) and that leaflet closing is accompanied by loss of $K^+$ from ventral cells and absorption of $K^+$ by dorsal cells; and changes in $Cl^-$ parallel to those of $K^+$. Similar changes have been observed in *Samanea*. The above results further show that in *Mimosa*, *Albizzia*, and *Samanea*, water movement occurs in response to the osmotic driving force of ion transport, just as in stomatal opening and closing.

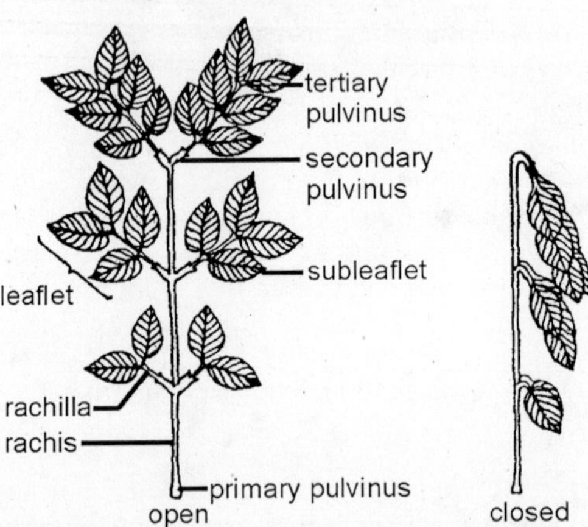


**FIG. 32.2:** (A) *Albizzia julibrissin* leaves in the normal daytime position. (B) in typical night (sleep) position.

Satter and coworkers (1988) tried to describe the mechanism of ion transport into and out of extensor and flexor cells. The observations suggest the presence in membranes of depolarization activated, $K^+$ selective channels that provide outward pathways for $K^+$ diffusing from shrinking cells. $Cl^-$ appears to be co-transported into motor cells via $H^+$/ anion symport. $H^+$ transport appears to be the primary mechanism. In some manner, changes from light to dark or dark to light, acting via phytochrome and a blue-absorbing pigment, activate these mechanisms, possibly via effects on the phosphatidylinositol cycle.

Light acts in opposite ways in the extensor and flexor cells. For example, the transition from white light to darkness activates the $H^+$ pump in flexor cells and inactivates the $H^+$ pump in extensor cells. Because leaves open and close on an approximately 24-h schedule under constant conditions, the biological clock must also activate these mechanisms. Light acts not only directly on the ion absorption mechanisms but also indirectly by resetting the clock.

### Hydronasty (or Hygronasty)

**Hydronasty** is the folding or rolling up of leaves in response to water stress which helps to reduce the exposure of the leaf surface to dry air, supplementing stomatal closure to reduce transpiration rate. This also minimizes the danger of photoinhibition. Hydronasty (folding and rolling movements) is caused by loss of turgor in thin-walled motor cells called **bulliform cells**. The bulliform cells have little or no cuticle. These cells have higher transpiration rate than other epidermal cells. As their turgor pressure decreases, constant turgor in cells on the abaxial of the leaf causes the folding. This is one of the mechanisms by which plants resist drought.

### Thigmonasty

Thigmonasty is a type of nastic movement which results from touch- (Greek, *thigma*, touch). These movements are common in plants especially in certain members of the subfamily Mimosoideae in the family Fabaceae (Leguminosae). The most notable example is *Mimosa*, the sensitive plant. Upon being touched, shaken, heated, rapidly cooled, or treated with an electrical stimulus, the leaflets and leaves rapidly fold together. When only one leaflet is stimulated, a stimulus moves throughout the plant, collapsing the other leaflets. The folding is caused by water transport out of motor cells in pulvini, an occurrence associated with an efflux of $K^+$. The exact utility of this movement (response) in plant is uncertain, however, the folding leaflets startle away insects before they begin to eat the foliage.

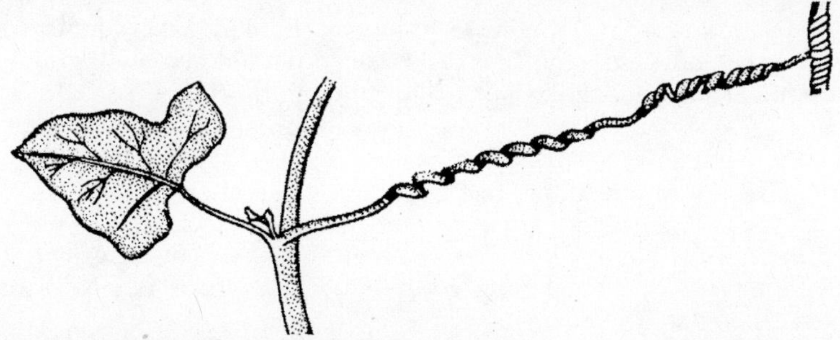

**FIG. 32.3:** *Cucurbita* tendril showing thigmotropic movements.

Several studies have been made to find out the mechanism of signal transmission in *Mimosa*. Two distinct mechanisms have been identified - (i) electrical and (ii) chemical mechanisms.

**Electrical Response Mechanism:** Jagadish Chander Bose first observed (1907 and 1914) electrical response and then by A. L. Houwink (1935) in Holland. An **action potential** is a change in voltage that forms a characteristic peak when plotted as a function of time. Action potentials in *Mimosa* are similar to those occurring in animal nerve cells but are much slower. In both plant and animal cells, they are caused by fluxes of specific ions across cell membranes. In *Mimosa*, action potential travels through parenchyma cells (connected by plasmodesmata) of the xylem and phloem at velocities up to about 2 cm $s^{-1.}$

**Chemical Response Mechanism:** The action potential is linked with chemical response, because action potential does not pass through a pulvinus from one leaflet to another unless the chemical response is also elicited, in which case several leaflets may fold up. The chemical response, is caused by a substance that moves through the xylem vessels along with the transpiration stream. The active substance, formerly called **Ricca's factor** but now identified as a **turgorin** has been extracted from wounded cells. When applied to a cut branch, its folding effects can be measured. Such **wound signals** are rapidly transported. They have been observed in pea tissues.

**Turgorins:** Schildknecht suggested that the chemicals which act on the turgor of pulvinus cells should be named as **turgorins**. They are active at low concentrations ($10^{-5}$ to $10^{-7}$ M) (similar with other phytohormones), and meet the criterion of translocation (with certain exception). These hormones control nastic movements in plants and is also known as Ricca's factor. An extensive work has been done in isolating and identifying this compound which activate pulvini in the leaves of *Mimosa* and *Acacia karroo*.

## Thigmo- and Seismomorphogenesis

M. Jaffe in the early 1970s, began to investigate developmental effects of mechanical stimulation, especially rubbing, on plants. Most vascular plants studied responded with slower stem elongation and increased stem diameter, which produced short, stocky plants, sometimes only 40 to 60 per cent as tall as controls. Jaffe called these and similar developmental responses to mechanical stress **thigmomorphogenesis**. Clearly, they are related to thigmonasty, but rapid plant movements are not involved.

Flexing stems also causes these responses, and in nature the bending effects of wind influence plant development in this way. Similar experiments were carried out to study the effects, some scientists placed plants on a shaker table such that they are mechanically shaken but not physically contacted. The same basic responses appear, and it was termed **seismomorphogenesis** (Greek, *seismos*, from *seiein*, to shake).

Rubbing by farm machinery and by animals almost certainly contributes to the hardening. Spraying tomato plants with water for 10 seconds once each day reduced their growth in a greenhouse to about 60 percent of the growth of controls. The stockier plants were desirable in greenhouse culture, but unfortunately fruit yields were significantly reduced.

Such decrease in stem elongation and increased stem thickening suggest that ethylene production plays a role, and indeed increased ethylene has been observed following mechanical stimulation. Auxin transport is also apparently inhibited and calcium may be involved. How a change in growth-regulator balance might come about is unknown.

### A Model of the Involvement of Phytohormones

Jaffe, constructed a first approximation of the possible way in which phytohormones may help to mediate thigmomorphogenesis in bean internodes. It should be stressed that this is only a working model, and that some of it is still speculative. However, it is useful, since it helps to place this aspect of the syndrome in the perspective necessary for further analysis. According to the model, proposed by Jaffe, mechanical perturbation causes membrane changes which induce a burst of ethylene production. The ethylene in turn may block basipetal auxin transport and the endogenous auxin level may increase in the stimulated internodes to such a high level that it may induce further ethylene production via a kind of feedback loop. In the case of the latter, an increased ABA titer could be, at least in part, responsible for the observed retardation of elongation. At the same time, gibberellin drops. The integrated action of the hormones may in turn be responsible for profound changes in the balance of the lipids and proteins of the endomembranes. These changes are almost surely related causally to the anatomical and morphological changes which constitute thigmomorphogenesis.

Recently, Braam and Davis (1990) observed that mRNA in *Arabidopsis* plants increased 10 to 100 per cent 30 minutes after the plants were mechanically stressed by spraying with water or by other means. Typically such stressed plants became less than 50 per cent as tall as unstressed controls. Mechanical stress activated at least 4-5 genes which produce mRNA and one of those genes coded for calmodulin (calmodulin may activate the gene for its own production) and two others coded for two closely related proteins. Calcium levels increased in stressed cells, so it is possible that calcium-calmodulin is a **secondary messenger** in the gene activation.

### Thigmonasty in Tendrils

Tendrils are long thin organs, often modified from leaf primordia which respond thigmonastically when their ventral is rubbed. This response enables them to wrap around a support which they encounter during circumnutation. In peas, as well as in many other species, the ventral side contracts following MP (mechanical pertubation), and as the tendril bends around the support, the dorsal side then begins to grow faster than the ventral side. Thus, the tendril grows round the support. Jaffe (1970) applied ethephon asymmetrically by painting it onto the ventral or dorsal surfaces of pea tendrils. The ventral ethylene treatment mimiced the reaction to ventral MP. In the same study, Jaffe also showed that rubbed tendrils produced three times as much ethylene as did resting tendrils. This suggested that ethylene may mediate contact coiling. Support for this hypothesis came from the work of Bangerth (1974). Using cucumber tendrils, he found that auxin-stimulated ethylene production in the ventral side was several times greater than in the dorsal side. Since auxin can move basipetally in the tendril, it may be that MP, sensitizes the ventral side so that translocating auxin induces ethylene production; and that the localized ethylene production is, in turn, responsible for the curvature.

The role of auxin in tendril coiling is, to some extent, clear. Exogenous auxin both enhances MP-induced coiling and is capable of inducing coiling. An interesting observation has been made concerning the tendril tip which is sensitive to auxin. Hertel et al. (1972) have shown that auxin can be specifically bound to a putative receptor which can be recovered in a membrane-rich fraction from various plants. Auxin may act as an immediate trigger at the distal end of the tendril, responding to

MP.

## Tropisms

Tropism is the plant response to the direction of an environmental stimulus in which unequal (differential) growth (usually cell elongation) occurs on different sides of an organ. The responses may be of different types:

1. **Gravitropism:** Roots grow downward and stems upward in response to gravity.
2. **Phototropism:** Stems and leaves frequently orient themselves with respect to light rays, but roots seldom exhibit phototropism.
3. **Thigmotropism:** It is a response to contact with a solid object that is exhibited by climbing plants that grow around a pole or the stem of another plant.

## Phototropism

### *Coleoptiles and Stems*

The classic experiments of Darwin (1880) on plant tropisms are well known. The experiment was based on the fact that canary grass (*Phalaris canariensis*) coleoptiles would not bend toward dim light if the tips were cut off or shaded. They also studied dicots and some other monocots. It was their experiments (plus those of others) localizing phototropic sensitivity in the tips of coleoptiles. These studies eventually led to the experiments of Frits Went and thus to the discovery of the first known auxin.

The Darwins observed phototropic sensitivity in *Avena* coleoptiles bending toward light when their tips were covered; that is, some occurs below the tips, but the **tip response** is about a thousand times more sensitive than the **base response**. Thus, if dim light is used, most of the response is localized in the tip, and this is evident because curvature toward the light begins at the tip and moves gradually down the coleoptile as the stimulus is transmitted from the tip to the tissues below. If higher light levels are used, however, bending begins simultaneously along the entire length of the coleoptile. **Hypocotyls** (seedling stem below the cotyledons) of such dicot seedlings as sunflower show no apical-bud response.

### *Perception: Dose-response relations*

There are several aspects of phototropic response. Of primary importance aspect is the **reciprocity**: This is based on the fact that the response occurs in plant part is proportional to the duration of the exposure, that is dependent on light **energy** or **photon flux** (energy or photons per unit area per unit time = **irradiance** or **fluence rate**; often loosely called intensity), or to the total dose; that is, the product of exposure duration times flux (called **fluence**). If the response is to total dose, then flux and duration bear a reciprocal relationship to each other. Therefore, the above action follows the **law of reciprocity**.

#### Blue Light Exposure

Zimmermann and Briggs (1963) conducted experiments on *Avena* coleoptiles using three different photon fluxes of blue light for varied exposure times. They observed that the first, ascending part of the curve is the same for all three photon fluxes (irradiances), so reciprocity holds over this low-fluence range. This part of the curve is called **first positive curvature**. At higher irradiance (shorter exposure times) first-positive curvature is followed by decreasing curvature with increasing fluence, producing a bell-shaped curve. In *Avena* coleoptiles, this trend continues until the organs bend away from the light, this response is called **first-negative curvature**. With further increase in fluence, a **second-positive curvature** and even a **third positive curvature** become evident.

The above experiments suggest that light really has two following effects in phototropism.

1. Light acts as a trigger for the bending response, which is what we have been emphasizing.
2. Light decreases the sensitivity of the organ to subsequent light. This effect is non directional and is referred to as a **tonic effect**.

### Red Light Exposure

When *Avena* coleoptiles are exposed with red light just prior to blue light, it shifts the region of first-positive and first negative curvatures to tenfold higher irradiances. The second positive curvature was shifted to threefold lower irradiances by pre treatment with red light. Thus, red light changes the sensitivity of the tissue to the blue light that actually causes the bending. Zimmermann and Briggs, using brief exposures of red light, were able to reverse the red effect (maximum at 660 nm) with subsequent exposures to far-red light (about 730 nm). This shows the involvement of phytochrome pigment system in determining sensitivity of coleoptiles to the blue light that causes bending.

### *Perception: Action Spectra and the Photoreceptor Pigment*

Plant pigments are responsible for any photochemical process. It can be examined by comparing the action spectrum for the process with the absorption spectra of pigments suspected to be involved. Blaauw (1909) found that blue light was most effective in causing phototropic curvature. It is now established that one or both of two common yellow pigments, carotenoids and flavins, after light absorption, exhibit phototropic movement. The action spectra for the monocot and the dicot are identical. Therefore, these results suggest that the same blue-absorbing pigment functions in phototropism in both plant groups. The pigment is known as **cryptochrome**.

The results of series of experiments further suggest that a flavin pigment is the primary photoreceptor in phototropism. *Phycomyces* is phototropically active fungi which have action spectra for various responses similar to that of higher plants. The response is dependent on pigment **flavoprotein** (flavin attached to a protein). This pigment after light absorption becomes oxidized by reducing a b-type cytochrome in the plasmalemma.

### Transduction in Phototropism

Light acts directly on stem or coleoptile and inhibits growth on the irradiated side. Later, Cholodny (1926) proposed a theory and Went successfully demonstrated that auxin apparently migrates from the irradiated to the shaded side of a unilaterally irradiated coleoptile tip. Briggs further demonstrated that the mount of auxin transported into agar blocks at the base of a coleoptile tip is indicated by the degrees of curvature elicited when the agar blocks were placed on decapitated coleoptiles (as performed in the standard *Avena* curvature test for auxin).

When the coleoptile tips are exposed to light they can export as much auxin as those kept in the dark. This experiment suggests that light does not affect the destruction of auxin, as suggested by Blaauw in his model.

When the agar and part of the coleoptile were divided with a thin piece of mica and the tip exposed to light from one side, the amount of curvature was twice as great for the block under the shaded side as for that under the irradiated side. Division of the entire tip with mica prevented lateral auxin transport in the tip, so auxin amounts collected from both sides were the same.

### *Cholodny-Went Model (Hypothesis)*

Cholodny and -Went proposed a model which suggests that light from one side in some way causes a transport of auxin toward the shaded side. This is the cause for the basic transduction mechanism in phototropism. The model was verified by several plant physiologists till 1980. Fern and Digby (1980) proposed following four criteria for further verification of Cholodny-Went hypothesis.

1. In a phototropically curving organ, acceleration of growth on the shaded side should accompany retardation of growth on the irradiated side.
2. The development of a lateral auxin gradient must accompany or precede the appearance of differential growth.
3. It must be shown that auxin is indeed a factor limiting growth in the responsive organ.
4. It must be shown that the auxin differential established is sufficient to account for the growth differential that is observed.

Using $^{14}C$ radioactive tracer technique on transport of $^{14}C$-IAA suggested that the direction and rate of transport of IAA in coleoptiles could, indeed, be influenced by light. These results are still controversial. It has been observed in maize coleoptile that when curvature began in the apical regions, IAA moves downward and the rate of movement is consistent with known rates of auxin transport. When auxin is applied to coleoptile tips, auxin becomes a limiting factor for coleoptile growth. It shows that there is an approximately linear relation between auxin concentration and growth rate over a range that spanned the rates occurring in phototropic bending.

### *Phototropism in Dicots*

There is a different mechanism which operates in at least some dicots. Franssen and Bruinsma (1981) were unable to record auxin gradient across a phototropically stimulated sunflower hypocotyl, but they did demonstrate a gradient of an inhibitor, xanthoxin, across the hypocotyl, with 60 to 70 per cent of the inhibitor on the lighted side. Hasegawa et al. (1989) critically tested the Cholodny-Went model by repeating Went's experiment and obtained the same results when the agar blocks were placed on properly prepared coleoptiles: The block from the dark side caused the most bending. The blocks were analyzed and it was observed that equal amounts of IAA in both blocks were present, and there was nearly 2.5 to 7 times as much as was indicated by the curvature test. These evidences are in support of Cholodny-Went hypothesis, however, these results strongly support an inhibitor model for grass coleoptile as suggested by Blaauw.

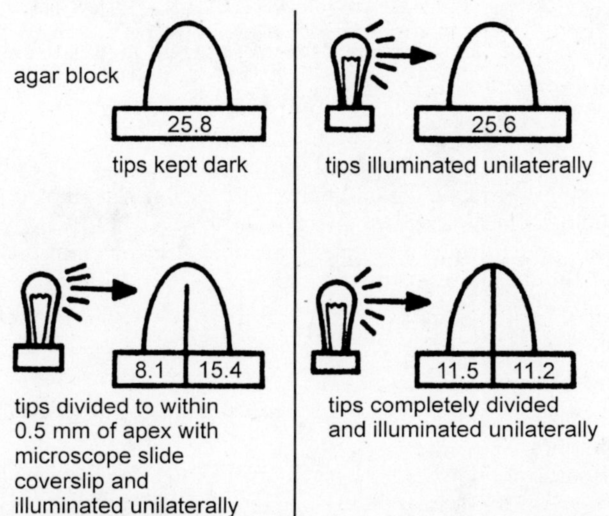


**FIG. 32.4:** Experiments showing that unileteral illumination of maize-coleoptile tips leads to a transport of auxin from the illuminated to the shaded side of the tips and apparently does not cause a destruction of auxin. Numbers on the agar block indicate the degrees of curvature caused by application of the blocks unilaterally to decapitated oat-coleoptile stumps.

## *The Unequal Growth Response*

What actually happens during phototropic bending of a coleoptile or hypocotyl?. This may be due to following reasons:

1. The growth on the light and shaded sides of the bending organ.
2. Much growth reduction on the light and shaded sides of the bending organ.
3. Much growth reduction on the light side as promotion on the shaded side.

This is in support with Blaauw hypothesis which suggests the inhibition on the lighted side with little change (or some inhibition) on the shaded side. Unequal growth can be measured manually or by using photographic techniques.

### Phototropic Movement and Overlapping of Leaves (Leaf Mosaics)

Leaves also respond phototropically to light. When half of a cocklebur leaf blade in the light is covered with aluminium foil (simulating natural shading), two following responses occur

1. An elongation of the petiole on the covered side.
2. An upward bending (hyponasty) of the shaded side of the leaf.

These responses in a number of leaves would **presumably move** them from the shade into the light whenever this is physically possible. For example the leaves of barley often overlap (perhaps one part of a leaf blade is shaded by another) this overlapping structure is called **leaf mosaics**. Such mosaics appear in leaf canopies can be observed in of many trees and climbers. When one part of a leaf blade is shaded by another, the shaded part transports more auxin to its side of the petiole than the brightly illuminated side transports to its corresponding petiole part, or probably the irradiated side transmits an inhibitor.

### *Solar Tracking*

When the flat blade of the leaf remains nearly at right angles to the sun throughout the day and maximizing the light harvested by the leaf the phenomenon is known as **solar tracking**. Wainwright (1977) studied a desert lupine (*Lupinus arizonicus*) and Ehleringer and Forseth (1980) observed solar tracking in desert plants and in other natural ecosystems. Solar tracking has been observed in leaves of such species of cotton, soybean, beans, alfalfa, and various wild members of the Malvaceae, such as *Malva neglecta* and *Lavetera cretica*.

It is a true phototropic movement because the orientation of the leaves is determined by the direction of the sun's rays, but it is neither positive nor negative and the organ orients at right angles to the light rays (**diaphototropism** or **diaheliotropism**).

#### Mechanism of Solar Tracking

Leaf orientation is controlled by motor cells in a pulvinus where the blade joins the petiole. Movement of water in and out of these motor cells is completely reversible and is almost certainly controlled by osmotic solutes, including $K^+$. Solar tracking, like other examples of phototropism, is most sensitive to blue light. Negative solar tracking has also been observed in leaves of some desert plants. The leaves remain parallel to the sun's rays during dry conditions. This helps to minimize leaves exposure to solar radiation, gain less heat, and transpire less.

### Gravitropism

Gravitropism is the growth movements toward or away from the earth's gravitational pull are the examples of positive and negative gravitropism. Roots usually grow away from light (positively gravitropic). The primary roots grow more vertically than secondary roots while tertiary roots and roots of higher orders are less to not gravitropic at all, because they grow in random directions. This helps the root system to explores the soil more thoroughly.

Stems and flower stalks are negatively gravitropic and their response is highly variable. The main stem or tree trunk usually grows 180° away from the earth's center of gravity, but branches, petioles, rhizomes, and stolons are usually more horizontal. It may be of different types:

#### Orthogravitropism

Orth-and ortho- are combining forms, meaning straight, upright, vertical. Hence, vertical growth (e.g., stem or root).

#### Diagravitropism

*Dia* = through or across meaning horizontal growth.

#### Plagiotropism or Plagiogravitropism

When growth of an organ occurs at any determined angle ($\alpha$) to the vertical ($0< \alpha<180°$) is called plagiotropism (or plagiogravitropism; Greek, *plagio* = placed sideways, oblique). Organs with no response to gravity are known as **agravitropic**.

### *Nature of the Gravitropic Response in Higher Plants and Its Adaptive Significance*

One of the best, and agriculturally important, manifestations of the gravitropic curvature is seen in shoots of cereal grasses which have become prostrated by the action of wind or rain. This phenomenon is called lodging. Once lodged, and provided that the shoots are not too heavy with grain, they begin to manifest an upward bending response. In cereal grasses, this curvature response takes place at swollen localized regions of the leaf sheath bases and, in some grasses (mainly C-4) grasses), also at the bases of the internodes. Such regions are referred to, respectively, as leaf sheath and internodal pulvini. It requires almost 60 hours for the shoots of lodged cereal grasses to attain an upright position (90° curvature response).

## *Primary Components of the Gravitropic Response Mechanism in Higher Plants*

As with phototropism, we can conveniently subdivide the gravitropic response mechanism into three sequential components: gravity perception, gravity transduction, and asymmetric growth response. For convenience, we shall discuss the gravitropic responses in roots and shoots of higher plants in terms of these three components.

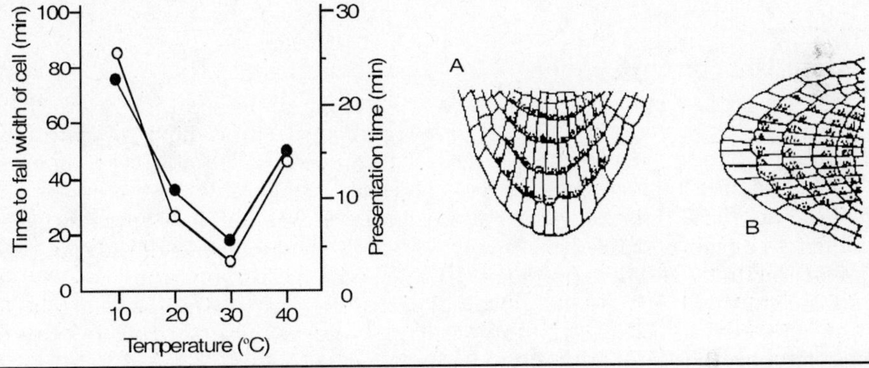


**FIG. 32.5:** Gravitropism (A) Positions of the starch grains (amyloplasts) in the root cap cells of a root orientated vertically and horizontally. (B) The graph represents the efect of temperature on the time taken for the starch statoliths to fall to the bottom of the cells, and the presentation time required for the gravity induced bending response, in the stem of a pea plant. Statolith movement, open circles; presentation time, solid dots.

## *Root Gravitropism*

The role of hormones in regulating root gravitropism has been controversial for the past decade IAA has been primary hormone cited as regulating the process. Many experiments have shown that the root cap produces a growth inhibitor and that this inhibitor causes a reduction of growth on the lower side of the root. The Cholodny-Went hypothesis proposes that this inhibitor is auxin, which moves from the root cap to the lower side of the growing zone of a horizontal root. A primary source of auxin for the root is the shoot, but apparently auxin synthesis can also take place in the root tip itself. In addition, IAA applied to the root tip can migrate back to the growing zone (Davies, et al., 1976), and, if placed laterally on the tip, it can induce a bending toward that side. When IAA is applied laterally to the growing zone of a horizontal root, three times more IAA moves downward to the lower side of the root than in the reverse direction (Klann, et al., 1992). This polarity of lateral auxin movement is abolished by the removal of the root cap. If there is a higher level of endogenous IAA in the lower side of a horizontal root (a point not yet conclusively settled), then it would probably inhibit growth via the synthesis of ethylene, because if ethylene synthesis is blocked, than IAA no longer inhibits root growth, but only causes a stimulation (Mulkey, et al., 1982).

According to a model proposed by (Hasenstein and Evans, 1986), IAA synthesized in the shoot is transported to the root cap through vascular system. In vertically held roots, the statoliths (starch-filled amyloplasts) in root cap cells are at the basal ends of the cells. Under this situation, IAA and calcium ions are symmetrically distributed in the root cap tissue. Then, IAA and Ca2+ are transported acropetally to the elongating zone of the root cortex.

When roots are placed horizontally, the statoliths settle down to the lateral side of the plasma membranes which face the source of gravity. Somehow, this event initiates the polar transport of Ca2+ and IAA in the root cap results in the accumulation of IAA in the lower side of the root elongation zone. Such a supraoptimal concentration of IAA in the lower side of the elongation zone inhibits cell elongation in this region either by enhanced synthesis of ethylene or by other unknown mechanisms. The IAA role is thus still a viable one. In fact, today, IAA receives most support as being the primary effector hormone that regulates the root gravitropic curvature response.

### Perception: The Statolith Concept

The site of gravity perception is the root cap. When the root cap of maize or other species is removed by careful microsurgical techniques. The removal that do not inhibit growth of the root and there is no gravitropic response until the same or another cap has been replaced or a new cap has been regenerated.

Haberlandt (1902) Bohumil (1901) proposed that it was the **amyloplasts** (each containing two or more starch grains) which settle in the large cells of the root cap in response to gravity and thus provide the basic perception mechanism. Evidences reported during the 1970s support amyloplasts as the **statoliths** (Greek, *lithos*, stone) in **statocytes** (cells with statoliths) of many tissues. The following evidences support the statolith theory:

1. There is a close correlation between the presence of sedimentable amyloplasts in organs and the organs ability to respond gravitropically.
2. The observable gravitropic bending after the root is returned to the vertical, and amyloplasts are known to settle a perceptible distance in this brief time.
3. If roots (or coleoptiles) are treated with gibberellin and kinetin at elevated temperatures, all starch in the amyloplasts disappears, as does the response to gravity.
4. **Gravitropic sensitivity** reappears at the same time that starch grains reappear either in the amyloplasts after destarching or in the newly regenerated root cap after removal of the cap. Sensitivity reappears as soon as the new starch grains are able to settle; they usually remain suspended in the cytoplasm for a few hours after they first appear.

Could other cell organelles settle and cause the gravitropic response? There is evidence that some organelles besides amyloplasts or starchless plastids do settle, but the rates do not correlate with presentation times.

### Transduction: The Role of An Inhibitor (ABA or IAA?)

Root cap sends an inhibitor to the lower side of the root which either inhibit or slow down the growth. This initiates root bending downward. The inhibitor of root caps was identified as ABA? ABA inhibits root growth at higher concentrations (ABA is not present at higher concentration under natural condition in root) When roots are grown in the presence of a compound that inhibits ABA synthesis, they continue to respond to gravity. Roots will even curve down when they are immersed in a high concentration of ABA.

Evidences have also been accumulated which indicate IAA is the inhibitor. IAA is present in roots at possibly at supra-optimal concentration it inhibits root growth (100 to 1,000 times lower concentration than those of ABA). When root cap is removed and a small block of agar containing IAA is added to one side, the root bends toward that side. This also supports the role of IAA as an effective inhibitor in root gravitropism. Acid growth hypothesis may also be implicated in root gravitropism, which is suggestive of a promotive auxin action.

### *Root Gravitropism: Hormonal Regulation of the Process*

The role of hormones in regulating root gravitropism has been controversial for the past decade IAA has been primary hormone cited as regulating the process. Many experiments have shown that the root cap produces a growth inhibitor and that this inhibitor causes a reduction of growth on the lower side of the root. The Cholodny-Went hypothesis proposes that this inhibitor is auxin, which moves from the root cap to the lower side of the growing zone of a horizontal root. A primary source of auxin for the root is the shoot, but apparently auxin synthesis can also take place in the root tip itself. In addition, IAA applied to the root tip can migrate back to the growing zone, and, if placed laterally on the tip, it can induce a bending toward that side. When IAA is applied laterally to the growing zone of a horizontal root, three times more IAA moves downward to the lower side of the root than in the reverse direction. This polarity of lateral auxin movement is abolished by the removal of the root cap. If there is a higher level of endogenous IAA in the lower side of a horizontal root (a point not yet conclusively settled), then it would probably inhibit growth via the synthesis of ethylene, because if ethylene synthesis is blocked, than IAA no longer inhibits root growth, but only causes a stimulation.

According to a model proposed by (Hasenstein and Evans, 1986), IAA synthesized in the shoot is transported to the root cap through vascular system. In vertically held roots, the statoliths (starch-filled amyloplasts) in root cap cells are at the basal ends of the cells. Under this situation, IAA and calcium ions are symmetrically distributed in the root cap tissue. Then, IAA and Ca2+ are transported acropetally to the elongating zone of the root cortex.

When roots are placed horizontally, the statoliths settle down to the lateral side of the plasma membranes which face the source of gravity. Somehow, this event initiates the polar transport of $Ca^{2+}$ and IAA in the root cap results in the accumulation of IAA in the lower side of the root elongation zone. Such a supraoptimal concentration of IAA in the lower side of the elongation zone inhibits cell elongation in this region either by enhanced synthesis of ethylene or by other unknown mechanisms. The IAA role is thus still a viable one. In fact, today, IAA receives most support as being the primary effector hormone that regulates the root gravitropic curvature response.

### *Root Gravitropism: Non-Hormonal Regulation of the Process*

Root gravitropism without mentioning an important non-hormonal regulator of the process, namely calcium. Exogenously applied $^{45}Ca$ typically moves to the lower side of tips of gravistimulated roots. When $Ca^{2+}$ is placed on the lower side of the tip, the root will curve downward. EDTA or EGTA, calcium chelators, will block the effect of $Ca^{2+}$ in inhibiting root curvature. Thus, its role in regulating the curvature response is undisputed. One of its primary physiological roles is to activate the enzyme modulator, calmodulin; the calmodulin activates plasma membrane bound Ca-ATPases in plant cells. $Ca^{2+}$ is also essential for the basipetal transport of IAA in both roots and shoots, and it may have important regulatory effects on the cell elongation process itself. It may also be important in gravity perception because the starch statoliths themselves have significant amounts of $Ca^{2+}$ associated with them. The descent of the starch grains in the root cap in gravistimulated roots may provide a mechanism for getting $Ca^{2+}$ to the lower side of the root where it exerts its regulatory effect on cell elongation back of the root apex. The means by which a $Ca^{2+}$ redistribution causes downward bending may be via IAA since the presence of $Ca^{2+}$ on the lower side of the root enhances the downward movement of IAA applied to the upper side.

In addition $Ca^{2+}$ has been found to be necessary for the growth-inhibitory action of IAA in roots. $Ca^{2+}$ redistribution in the root cap may cause auxin to move upward from the root cap. Alternatively, the $Ca^{2+}$ might possibly move to the growing zone and there influence the distribution and inhibitory action of auxin. This auxin could have arrived in the growing zone either by downward movement from the stem, or by upward movement from the root cap. As auxin is only inhibitory in the presence of $Ca^{2+}$, the gravitropic response could, in fact, be caused by evenly distributed auxin in the presence of an asymmetric distribution of $Ca^{2+}$. The exact mechanism awaits elucidation.

### *Proton Efflux*

When roots are gravistimulated, one of the initial events which takes place is a rapid proton efflux on the upper side of the elongation zone of the root. This is easily seen with the proton efflux indicator dye, bromocresol purple, which changes from red at pH 6.5 to yellow at more acid pHs; the shift from red to yellow indicated proton efflux. No such color change occurs on the lower side of the downward bending root. Such a change on the upper side of the root would make the pH in the vicinity of the cell walls lower, and this would favor the enhancement in the activities of one or more cell wall-loosening enzymes. This has not yet been demonstrated experimentally, but it is a likely event. It could explain the rapid increase in cell elongation that occurs on the upper side, in the growing zone, of a gravistimulated root.

## Shoot Gravitropism

In shoots, several hormones are implicated in the regulation of negative gravitropic curvature (Kaufmann et al., 1995). These include IAA and its inositol and amide-linked conjugates, GAs and their glucosyl esters and glucoside conjugates, and ethylene. The hormone (s) involved apparently vary with the experimental system (plant) being considered, and as expected, multihormonal control occurs in most of them.

The upward bending response in maize seedlings occurs very rapidly, usually within one hour at 30° C (Bandurski, 1984). It starts within three minutes in the coleoptile and five minutes in the mesocotyl (first internode). The onset of auxin asymmetry is also very rapid. Analyses for free IAA in gravistimulated seedlings indicates that top/half bottom half asymmetry first appears within three minutes in the mesocotyls. This is two minutes before the mesocotyls start to bend upward;

thus IAA could be an important effector hormone in triggering the upward bending response. Full IAA asymmetry is established within 15 minutes after seedlings are oriented horizontally. Fifty-seven percent of the free IAA occurs in the lower halves of the mesocotyls at this time and remains at this level at 30, 60 and 90 minutes after seedlings are first gravistimulated. We can only speculate on how the asymmetry in IAA is established. Further experiments are necessary to determine whether or not basipetal transport is fast enough. It is also possible that the IAA asymmetry arises as a result of release of IAA from its conjugates to a greater extent on the lower side, and/or that more IAA synthesis begins to occur here soon after seedlings are gravistimulated.

### *The Role of Ethylene in Shoot Gravitropism*

Ethylene has been implicated as being another one of the regulatory hormones in shoot gravitropism. Following gravistimulation of cereal grass shoots, one finds very large increases in ethylene evolution from the tissues where curvature takes place, with more ethylene emanating from the lower halves than from the upper halves. In *Avena* shoots that have been gravistimulated, the burst in ethylene production brought about by this treatment occurs some six hours after shoots are first placed horizontally; this is almost five and one-half hours after initiation of upward bending. Thus, one can conclude that ethylene in this system does not play a primary role in the initiation of curvature. Its enhanced production and the greater amount produced on the lower sides of gravistimulated shoots may be brought about, respectively, by the greatly enhanced levels of IAA synthesized in pulvini of gravistimulated shoots and the greater amounts of IAA found in the lower halves. Perhaps ethylene plays a role in the differential expansion of cells that occurs later during the course of upward bending.

The ethylene evolution precedes the initiation of shoot up turning, so it could be important in causing this hyponastic response. Additional evidence in support of this idea comes from exogenous ethylene applications to the shoots; here, such applications cause upward bending. Furthermore, removal of endogenous ethylene from the plants with an external mercuric perchlorate trap causes the branches to grow downwards. Whether ethylene or GAs or IAA or a combination of these hormones initiates the upward bending response in these plants is not yet clear. Studies are needed on the kinetics for ethylene evolution and on changes in levels of endogenous IAA and GAs during the course of hyponastic growth of these lateral branches.

# CHAPTER 33

# Circadian Rhythms in Plants

## INTRODUCTION

Nature does not reveal all her secretes at once. Man has curiously observing the multitude of activities of the living objects of unbound and mysterious nature since developed aesthetic sense. The periodicity in the physical environment, specially involving periodic fluctuations of light and temperature has favoured most of the organisms who have adopted physiologically to this temporal order. The photosynthesis of green plants and the activities of animals dependent on visual information for orientation are necessarily restricted to the day time. Therefore, organisms show a 24 h rhythms in their activities. Similar rhythms can be found in bioluminescence of *Gonyaulax*, the growth patterns and spore discharge of *Neurospora*, leaf movements of *Phaseolus multiflorus* (syn. *P. coccineus*), *Gossypium hirsutum*, petal movements of *Kalanchoe*, odour output in flowers, carbon dioxide output of *Bryophyllum*, enzyme activities in *Chenopodium*, photosynthesis, transpiration, dark respiration and carbon dioxide compensation point of peanut leaves, etc. Rhythmically, therefore, is an essential feature of physiology, development and behaviour of all living systems, and characteristics of life which expresses itself at all levels of organization from unicellular systems to man (Bunning, 1973).

### Historical Aspects

The grandeur of the opening of the blossom, concomitant with release of perfume and nectar to attract bees and butterflies, the unfolding of the horizontal position of the foliage in response to beam of the sunlight and reverse activities in the night has been speculated to be metaphysical.

The retrospection of the glorious past unveils that the horizontal and vertical position of the bean leaf in relation with time was known to Androsthenes, a historian in the regime of Alexander, the great in 400 BC.

Later on French astronomer, Demairan (1729) discerned that the leaf movements might occur in response to some internal time measurement, even when conditions remained constant. Consequently this puzzle of biological clock was tackled with the modern tools including biophysics and biochemistry. Rhythmicity, of living organisms persist, sometimes for indefinite period, when organisms are exposed to constant temperature and continuous light conditions. For example although nyctinastic (sleep movement) leaves usually open in the light and fold together when darkened. An alternating periods of light and darkness were not found necessary for such leaf movements. Though this phenomenon was well known to 19th Century. The daily rhythm lacked function and mostly reflected *memory* of previously experienced environmental cycles or driven by some mysterious *Factor X*. This factor was ascribed variously to electrical charges in the air, cosmic rays and radioactive radiations. Brown (1960) and Enright (1965) presented evidences against the control of rhythms by known external factors after careful analyses of several published experiments and concluded that there is no role of subtle geophysical factors in the synchronization of rhythms as advocated by the earlier. Cumming and Wagner, (1968) and Bunning (1973) proposed that the rhythmicity exhibited by organisms in their activity is truly an endogenous one. In 1960 a Cold Spring Harbour Symposium *on Biological Clock* the term *circadian* (from Latin *circu* = about and *dies* = day) was accepted.

## Definitions

A circadian rhythm (circadian clock) is defined as an oscillation in a biochemical, physiological or behavioural function which under conditions in nature has a period of exactly 24 hours, in phase with the environmental light and darkness, but which continues to oscillate under constant but permissive conditions of light and temperature with a period of approximately but usually not exactly 24 hours (Sweeney, 1976).

## The Occurrence of Rhythms in Plants

For a survey of literature dealing with physiological rhythms in plants reveal that this phenomenon occurs in plants with period lengths ranging from a second to hourly, daily and even annual cycles and has been observed in members of each major plant group with the exception of Bryophytes (Wilkins, 1973). Circadian rhythms are claimed to be confined to the eukaryotes. In bacteria and blue green algae circadian rhythms are unknown with one possible exception of *Escherichia coli* culture, however, it needs confirmation.

**TABLE 33.1:** Rhythmic phenomena in plants

| Phenomenon | Period length |
|---|---|
| Bioluminescence photosynthetic capacity and cell Division | circadian |
| Phototaxis | circadian |
| Sporulation | circadian |
| Growth pattern and spore discharge | circadian |
| **Bioelectricity** | |
| Action potentials | |
| Separation interval | 0.1-10 S |
| Refractory period | 12 mm |
| Change in propagation direction | circadian |
| Current, transcellular | circadian |
| **Leaf movements** | |
| Rotational | 53 min. |
| Up and down | circadian |
| Petal movements | circadian |
| Stomatal opening | min. to circadian |
| Transpiration | min. to circadian |
| Water potential | min. to circadian |
| Water uptake | 30 min. |
| Respiration | circadian |
| Photosynthesis | circadian |
| $CO_2$ fixation in darkness | circadian |
| Plastid shape | circadian |

| Phenomenon | Period length |
|---|---|
| Enzyme activities | 4-6 h; semi circadian |
| Energy charge | circadian |
| Redox potential | circadian |
| Nuclear size | circadian |
| Growth | circadian |
| Flowering stimulus | circadian |
| Flower induction | circadian |
| Seed germination | circadian |

**TABLE 33.2:** Examples for circadian rhythms in plants partially at different temperatures. (After Bunning, Wagner modified).

| Group of organisms | Organism | Reaction | Rhythm |
|---|---|---|---|
| Photosynthetic Flagellates | *Gonyaulax polyandra* | luminescence | 23-25 h |
| | | photosynthesis rate, growth | 22.5 h |
| | | luminescence (15.9° C) | 26.8 h |
| Algae | *Oedogonium cardiacum* | spore formation | 22 h |
| Fungi | *Pilobolus sphaerosporus* | sporangial discharge | 26.4 h |
| Ferns | *Selaginella serpens* | plastid morphology | ca. 24 |
| Higher plants | *Phaseolus multiflorus* | leaf movements | 27 h |
| | *Kalanchoe blossfeldiana* | petal movement | 26 h |
| | *Avena sativa* | coleoptile growth | 23.3 h |
| | *Chenopodium* | flower induction | 30 h |
| | | dark respiration | 21-24 h |
| | | betacyanine accumulation | 24-30 h |
| | | adenylate kinase activity | 30 h |
| | | chlorophyll accumulation | 15 h |
| | | *net photosynthesis* | *15 h* |

*Acetabularia* cells show persistence of circadian rhythmicity with out the presence of nuclei. With an array of current researches, it has been experimentally proved that rhythmicity is an inbuilt quality of almost all the living systems which is expressed at all levels of organization and such circadian rhythms prepare the organisms for predictable environmental changes.

## Exogenous Factors Influencing Rhythms

The movements observed in bean leaves are either nyctinastically or photonastically induced. **Nyctinastic movements** are circadian reactions in which the petioles lower in the primary pulvini at dawn while the leaf blades raise in the secondary pulvini. During dusk the movements are reversed. These are rhythmic, antagonistic changes in the turgor of the upper and lower halves of the pulvinus. The movements induced by light are known as photonastic movements. Green leaves also show photonastic **growth movement** responses.

Light causes the lowering of the leaf blade in the secondary pulvinus by decreasing the turgor pressure in both pulvinus halves. Even the decrease in the fluence rate caused by a cloud can induce a photonastic reaction. How these photonastic growth movements occur is still a matter of research, however, but the involvement of the phytochrome system has been found in many cases, such as the leaflet movements of *Mimosa*.

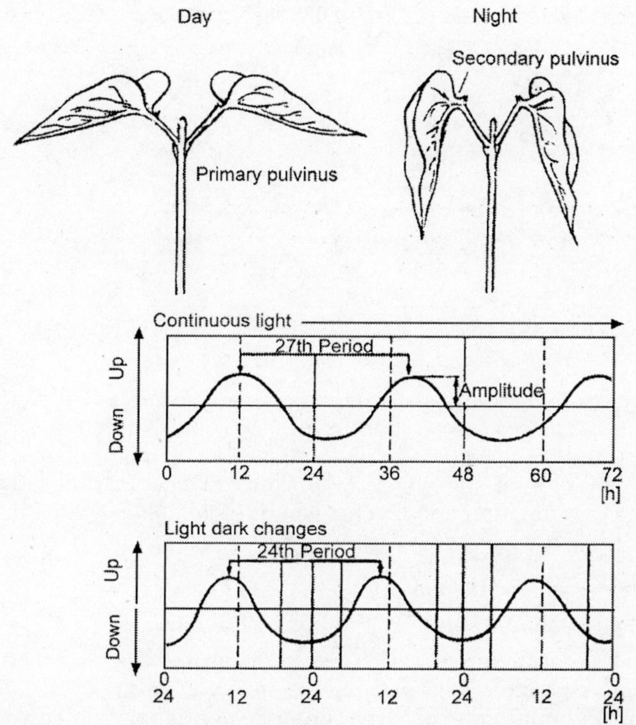


**FIG. 33.1:** Rhythmic movement in *Phaseolus vulgaris.* (A) Day and night position. (B) Circadian, free running rhythm with a period of 27 h in continuous light; (C) rhythm synchronized by light-dark regime with a period of 24 h.

Stomatal movement has also been shown to be controlled by light and they also show a very sensitive photonastic reaction. The mechanics also depends on turgor changes. The petals movement of *Nuphar*, *Gentiana* sp. and cacti (plant species) flower is dependent on the decrease in the fluence rate.

In many cases both photic and thermic stimuli are effective. Young leaves of *Impatiens* species respond to a sudden shading with an increase of the growth rate on the leaf surface. Older leaves show only turgor movements rather than growth changes.

## Endogenous Rhythms

Well-studied examples of endogenous rhythmic (ER) leaf movements especially among the Fabales: at constant light and temperature beans show a free running ER with a period of 27-29 h. In the natural day-and-night cycle the free-running biological clock is synchronized to 24 h. Endogenous rhythms may occur in a very short periods e.g., in the second or millisecond range also. Since, in nature, one can not measure such short period with a timer (**Zeitgeber**). Endogenous rhythms with a circadian period is very common in almost all eukaryotes but not in prokaryotes.

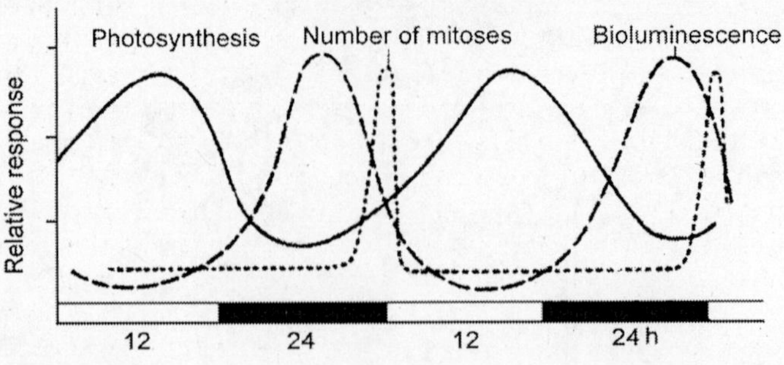


**FIG. 33.2:** Circadian rhythms of photosynthesis, meiosis frequency and bioluminescence of the marine flagellate *Gonyaulax polyedra*.

In plants circadian rhythms occur simultaneously and independent of each other, like mitosis frequency, photosynthesis and bioluminescence in the dinoflagellate *Gonyaulax polyedra*. In higher plants and animals, functions of either single organs or of the whole organism can be controlled by a circadian periodicity. The coordination and mechanism of individual rhythms in an organism is not very clear.

### *Endogenous Nature of the Rhythm*

Endogenous rhythm should satisfy the following conditions:

1. The rhythm should continue in an environment in which as many permissive environmental parameters as possible are held constant.
2. In constant conditions the free-running period of the rhythm is only approximately 24 h.
3. The phase of the rhythm can be altered by a brief disturbance of the constant regimen (phase shifting).
4. In contrast to some biorhythms with other frequencies, circadian rhythms are temperature compensated and almost unsusceptible to chemical manipulation.

## Molecular Mechanisms of Circadian Rhythms

The basic molecular phenomena of induction and continuation of the circadian rhythms are largely unknown, however, we known about the visible or measurable reactions. It is proposed that not one **single substance** controlled this phenomenon whose concentration oscillates between two extremes in a daily rhythm, acting as a **zeitgeber**. These are called biological oscillators which are probably complicated sequences of biological reactions. **Belousov-Zhabotinski reaction** is an example of this process. This is a chemical model system for oscillating reactions. Malonate reacts with bromide and bromate, producing bromine malonate. When a high concentration is reached bromine malonate disintegrates to $CO_2$, acetic acid and bromide, which again reacts with malonate. Ferroin is used as an indicator for this reaction. Depending on the concentration of the initial substances (especially malonic acid, which is consumed) the oscillations can continue for more than an hour. Similar mechanisms on an enzymatic or biochemical basis are considered for the explanation of biological oscillations, some of which are discussed below. Occurrence of several enzymes in plant and animal tissues whose activity is controlled by rhythmic changes which can be synchronized by external factors. The activity of the enzyme is inhibited by light. In chicken *N*-transferase enzyme activity followed by circadian rhythm and proposed that such **enzymatic rhythms** are the molecular basis for circadian rhythms.

Protein synthesis in biological oscillations also control several circadian rhythms with translation blockers. For example involvement of the nucleus in circadian rhythms has also been proposed. Since a rhythmic DNA and RNA accumulation was found in liver cells. In contrast to it *Acetabularia* shows continued circadian rhythms for several weeks under constant conditions after removal of the nucleus. After implantation of a nucleus from another plant the cell follows the rhythm of the **new nucleus**. For the generation of rhythms interactions between the nucleus, organelles and/or membranes are the basic requirements for induction of rhythm. This hypothesis is supported by the fact that circadian rhythms have not been demonstrated in prokaryotes.

### *Membrane Participation*

The participation of membrane processes has been shown in almost all models. These models are based either on changes in the membrane itself or on diffusion processes. Photoreceptors are located in membranes which play their role in rhythmic movements. Experimentally induced modifications in the fatty acid composition of the membrane in *Neurospora* (for example by temperature changes or by application of various membrane-active substances such as some steroids) induce a significant alteration in the period length of the circadian rhythm.

### *Ionic Concentration*

The concentration of various ions in a cell also control endogenous rhythm in plants. Presence of calcium ions has been studied well in this respect because they play the dominant role in circadian rhythm. Lithium ions- in competition with calcium and magnesium ions- inactivate ATPases, membrane receptors and several enzymes, and thus change the period length.

Presence of potassium has also been shown to involved in several membrane processes oscillates with a circadian rhythm. Pulses of increased potassium concentrations change the period length in several organisms. The same effect is found after application of **valinomycin**, an ionophore which increases the flux rates through the membrane.

### *Burgoyne Model*

The involvement of the membrane in the rhythmic transfer of ions is proposed by Burgoyne. The Burgoyne model is based on the following observations:

1. Monovalent ions, such as potassium, are transported through membrane proteins (Na-K-ATPase) into the cell.
2. High potassium concentrations blocks the synthesis of membrane proteins.
3. Since the membrane proteins are continuously degraded the *potassium influx into the cell* slows down.
4. This causes decreased potassium level which induces a renewed synthesis of membrane proteins.
5. This induces an increased potassium influx and the process continues.

The lifetime of the membrane proteins, and the time needed for synthesis and transport, could explain a rhythm with circadian periods. Plant physiologists are also of the opinion that similar processes at the chloroplast or mitochondria membrane occur but the *de novo* synthesis is replaced by a specific inhibition of membrane proteins.

### $D_2O$ (Heavy Water)

It always induces a lengthening of the period and it is not known whether $D_2O$ effects the rhythm by changing the ion transport through the membrane or by an exchange of H-atoms with D-atoms which could stabilize the protein structure.

### Photoperiodism

Plants and animals able to adapt their inner clock to the changing environmental conditions like changing length of the daylight period. The regulation of growth and environmental processes by a certain day length, the photoperiod, is called **photoperiodism**. The best-known example is the photoperiodic regulation of flowering. In addition to reproduction, senescence and dormancy phases as well as germination are controlled by the photoperiod.

# CHAPTER 34

# Stress Physiology

## CONCEPT

Every plant has some requirements of some physical factors, viz. light, temperature, air, water, etc. and for some chemical factors, for example, inorganic nutrients, organic molecules etc. in optimum amount and in available form at a right time as its nutritional requirements for its proper growth and development.

Any deviation from this optimal condition of any factor essential for its growth will lead to aberrant change in physiological processes and due to this plant body will experience tension and this state can be referred as plant under **stress**. When this condition prevails for some time stress causes injury with or without producing visible symptoms which can be referred as **strain**. Injury occurs as a result of aberrant metabolism and may be expressed as reduction in growth, yield, or value, or death of the plant or plant part.

Thus, we can say that each and every species of plants has some specific environmental and chemical (nutritional) requirements in a particular range in available form (optimum condition) at a specific times of its life cycle for its normal and proper growth and development. If conditions are not so that is if these factors are in deficient or excess amount than its requirement plant will experience bodily tension (stress) which has potential to produce injury with or without visible symptoms (strain) which cause a definite loss in growth, yield, or value of the product, or death of the plant or plant part due to aberrant changes in physiological process brought about by one or a combination of environmental and biological factors. This, we can define stress, zero stress and strain as follows.

**Zero stress**: In concept that level of exposure to an environmental factor that leads neither to injury nor to reduction in growth, yield or value.

**Stress**: A physical or chemical factor that causes bodily tension and may be a factor in disease causation, or

1. a state or condition caused by factors that tend to alter an equilibrium, or
2. the state or condition of strain, or
3. expressed quantitatively in units of force per unit.

**Strain**: An injury of a body part or organ resulting from excessive tension, force, influence, or factors causing excessive physical tension.

**TABLE 34.1:** Sources of Environmental Stress for Plants

| ABIOTIC | | BIOTIC |
|---|---|---|
| **Physical** | **Chemical** | |
| 1. Drought | 1. Air pollution | 1. Competition |
| 2. Temperature | 2. Allelochemicals (organic) | 2. Allelopathy |
| 3. Radiation | 3. Nutrients (inorganic) | 3. Lack of symbiosis |
| 4. Flooding | 4. Pesticides | 4. Human activities |
| 5. Mechanical | 5. Toxins | 5. Disease |

| ABIOTIC | | BIOTIC |
|---|---|---|
| Physical | Chemical | |
| 6. Electrical | 6. Salts | 6. Insects |
| 7. Magnetic | 7. pH of soil solution | |

## Wind

Whole plants can be resistant and survive stress injury, or some parts of a single plant may be resistant (seeds, buds, dormant cells) while other parts (meristems, succulent organs, seedlings) are susceptible. Through the process of evolution of plant species can become fit or adapted to an environment in which it thrives. Natural selection causes those paths of evolution that are successful to survive and those that are not to perish. Survivors have a tolerance of injury from environmental factors that enables them to overcome partially or completely any adverse effects.

Levitt (1980) has classified plants into three categories according to their strategy for resistance to stress: the three basic modes are explained.

### *Escaper*

Because of the timing of development in relation to the occurrence of a stress, some plants escape injury.

### *Plants Resistance to Stress*

#### Stress Avoidance

If a plant does not come to thermodynamic equilibrium with the stress or can exclude the stress by means of a physical or metabolic barrier, the plant is called stress avoider.

#### Plants tolerant to stress

Stress tolerance occurs if the plant comes to thermodynamic equilibrium with the stress but no injury occurs or injury that does occur is repaired. Thus, the tolerance to a stress is the capacity of a plant to survive and grow ever though subjected to an unfavourable environment; the plant can sustain the effects of stress without dying or suffering irreparable injury.

In the process of evolution, the selection has been toward avoidance mechanisms that are more efficient than tolerance mechanisms in causing resistance to stresses.

Examples of resistance to individual stresses are:

1. **Temperature:** Plants, with few exceptions, attain the temperature of the ambient environment. They are poikilotherms. Because of this, they must have some form of tolerance to temperature stress.
2. **Drought:** Terrestrial plants are normally turgid and thus, resistance results from avoiding loss of turgor. Drought resistance should probably be divided into dehydration avoidance or postponement and dehydration tolerance (Kramer, 1980).
3. **Irradiation:** Because of the penetrating nature of radiation, plants cannot escape it. Tolerance of irradiation stress depends upon the intensity and duration of the radiation and the amount of energy absorbed.
4. **Salts:** Plants that grow with root systems in soil of high salt content have low osmotic potentials as a result of an increased concentration of solutes and are

salt tolerant. Some plants are resistant because they have mechanisms by which they exclude salt or by which the salt is concentrated in vacuoles.

5. **Nutrient Deficiency:** Tolerance of deficiency of a nutrient may depend on the ability of the roots to exude metabolites enabling them to obtain more of the nutrient from the soil, as with iron, or to use substitute ions such as sodium for potassium.
6. **Nutrient Toxicity:** Much of the tolerance of toxicity is the result of the processes of exclusion at the root or by concentration in vacuoles of leaf cells. Either process prevents the toxic concentration from interfering with metabolism.

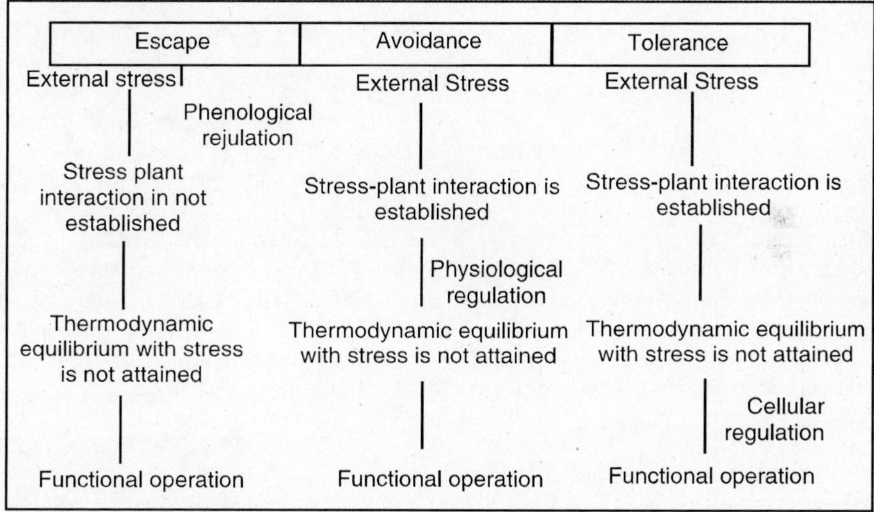


**FIG. 34.1:** Types of regulation in stress escape, avoidance and tolerance.
***Note:*** In escape and avoidance mechanisms, thermodynamic equilibrium with stress is not attained, and these plants do not experience stress *per se*, but in tolerance, a thermodynamic equilibrium with stress is achieved, but the functional optimization is magnified despite the equilibrium with external stress and experiencing the strain.

The reaction of plants to stress conditions may be pathological. It has been estimated that more than 50% of all plant diseases caused by improper environment, nutritional, or physical conditions.

Such diseases are distinguished from those caused by infectious organisms or viruses and are referred to by term such as non infectious, abiotic, or physiogenic. Organisms that cause diseases in plants are called pathogens. Perhaps the term physiogens should be used for those conditions of the physical environment that cause disease and the process of development of such diseases is physio genesis as distinguished from pathogenesis.

The study of the physiology of plants under stress thus, becomes the study of physiogenesis.

**Water Stress:** Two conditions of water availability are termed as follows:

1. **Drought:** It is a meteorological term that means a lack of precipitation over a prolonged period of time. According to an agronomist drought refers to a situation where any area receives annual rainfall less than average rainfall it

receives generally. Here, an amount of rainfall may create a situation of drought in an area but in other area it may be sufficient. While a physiologist refers drought to a situation where transpiration rate exceeds absorption rate so plant experience a stress. Thus, drought has different meaning for different disciplines.

2. **Water logging:** It refers to a condition where water is present in excess amount than its optimum requirement. Its excess creates an anaerobic situation in rhizosphere due to which a plant experience stress.

## Drought Stress

### *Basic Water Relations Concepts and Drought Stress*

Water comprises 85 to 90 of the fresh weight of most living herbaceous plants. In higher plants, water is absorbed by roots from soil and is translocated to the shoots as a result of pressure gradients developed either from root pressure or transpiration. Whenever the rates of water loss by transpiration exceed the rates of water absorption by roots, water in the conducting tissues is subject to a tension (negative pressure), that is, its potential is lowered and competition for water among the various tissues and organs of the plant takes place because of the equilibria among the separate water potentials have been disturbed.

The water potential ($\Psi$) in plants is the sum of turgor potential ($\Psi p$), osmotic potential ($\Psi s$), and matric potential ($\Psi m$). The relationship is given by the equation: $<\Psi w = \Psi s + <\Psi p + \Psi m$. Osmotic potential is created by dissolved particles, either molecules or ions, and is lowered in proportion to the number of particles in solution. Matric potential is the result of water adhering to surfaces and interfaces where the molecules of water become more ordered in arrangement and give up some of their kinetic energy. Activity of the water molecules is lost and the kinetic energy may be dissipated as thermal energy.

Turgor potential is created by water molecules bombarding the surfaces of membranes and cell walls retaining water in a closed system such as a vacuole. Turgor pressures are usually positive and are opposed by the membranes, the cell walls, or hydrostatic pressure caused by gravity in columns of water in the rufem tissues. Maintenance of turgor is necessary for growth and, if decreases, wilting of parts of the plant may be visible. Turgor potential is the first component of water potential to be affected by water or drought stress.

A brief recap of the status of water in the soil and how water moves through the soil may be useful in understanding how water reaches the surface of the root where it may be absorbed. Soil is composed mainly of particles and aggregates of various sizes interspersed with space filled with gases and water vapour or, under conditions of saturation, with liquid water. As water is removed from the soil by drainage, evaporation, or absorption by roots the continuity of the liquid water is interrupted. Some water remains on the surface of soil particles and some changes into vapour in the soil pore spaces. As more and more water is removed from the soil, that which remains is more lightly held on the soil particle surfaces. One can say that the water potential becomes lower and lower until it reaches a point at which a plant root can no longer remove enough water to overcome the deficit imposed by the loss in transpiration or other water using processes. The plant may wilt and not recover even if transpiration is stopped by placing the plant in an atmosphere of high humidity. The plant has permanently wilted and the amount of water in the soil is at the permanent wilting point (PWP). The water potential in the soil at this point is somewhere between -1.0 and -2.0 MPa (megapascals, 1 MPa = 10 bars). Frequently crop plant exist in a soil moisture range between -0.03 MPa and the permanent wilting point (-1.5MPa).

**FIG. 34.2:** Orientation of leaflets of field-grown soybean (*Glycine max*) plants in the normal, unstressed, position (A); during mild water stress (B); and during severe water stress (C). The large leaf movements induced by mild stress are quite different from wilting, which occurs during severe stress. Note that during mild stress (B), the terminal leaflet has been raised, whereas the two lateral leaflets have been lowered; each is almost vertical.

At the permanent wilting point, the water potential of the leaves remains below the water potential of the soil. At this point, water can no longer move to the roots rapidly enough to overcome the deficit in the plant even though the stomatas are closed. One can say that the permanent wilting point is controlled by the osmotic potential of the leaves and death by dessication results as this severe stress continues.

### *Effects of Drought on Growth and Yield*

Drought has profound effects on growth, yield, and plant quality. The first effect of the stress may well be a loss of turgor that affects the rate of cell expansion and ultimate cell size. Loss of turgor is probably the process most sensitive to water (drought) stress. The result is a decrease of growth rate, of stem elongation, of leaf expansion, and of stomatal aperture.

According to Hsiao (1973) the mechanisms underlying the responses of plants to water stress may be divided into five categories:

1. Reduction of water potential or activity of cellular water.
2. Decrease of cell turgor pressure.
3. Concentration of small molecules and macromolecules as cell volume decreases with reduced turgor.
4. Alteration of spatial relations in the plasmalemma, tenoplast, and organelle membranes by volume changes.
5. Changes in structure or configuration of macromolecules by removal of water of hydration or through modification of structure of adjacent water.

**TABLE 34.2:** Classification of drought stress according to Hsiao (1973)

| Type of Stress | Ψw (MPa) | Reduction in RWC |
|---|---|---|
| Mild Stress | 0.1 MPa | 8-10% |
| Moderate Stress | (-1.2)-(-1.5 MPa) | >10 <20% |
| Severe Stress | <(-1.5 MPa) | >20% |

If the RWC is lowered more than 50% that is, if plants lose more than half of their tissue water, one can say they are dessicated. Of course, such designations are purely arbitrary, but they do help in discussing the status of the stress at a point in time.

Moderate water stress affects translocation indirectly by altering the source to sink relationships for assimilates. For example, because cell expansion is reduced, the source (leaves) is smaller and less photosynthate is available for translocation to fruits. On the other hand, the sink (fruits) size may be reduced, and if drought stress occurs after leaf expansion, the result is that the competition between leaves and fruits is lessened. Effects on yield depend upon which part of the plant is harvested.

As stress develops a competition for water occurs within the plant. What factors control the distribution of water? One factor is the stage of growth and development of various plant parts. Meristems may compete favourably because synthesis of cellular material, that is, protein, creates a matric potential that contributes to lowering the water potential and the establishment of a strong Ψw gradient toward the meristematic tissue. Vacuolation also require water. Those parts of the plant in which solute content is increasing, for example, photosynthesizing leaves, are able to compete favourably for water when plants are under mild stress. Organs in which insoluble organic material is converted to soluble organic material may also compete favourably for water.

In plants that have sizable canopies, the location of the plant part in relation to its exposure to incident radiant energy and to the atmosphere plays a role in internal water competition. Direct exposure to sunlight, as opposed to shade, affects temperature of the plant part, which in turn affects the water gradients. Water loss from shaded leaves may be less because of the increased relative humidity in the atmosphere brought about by reduced temperature.

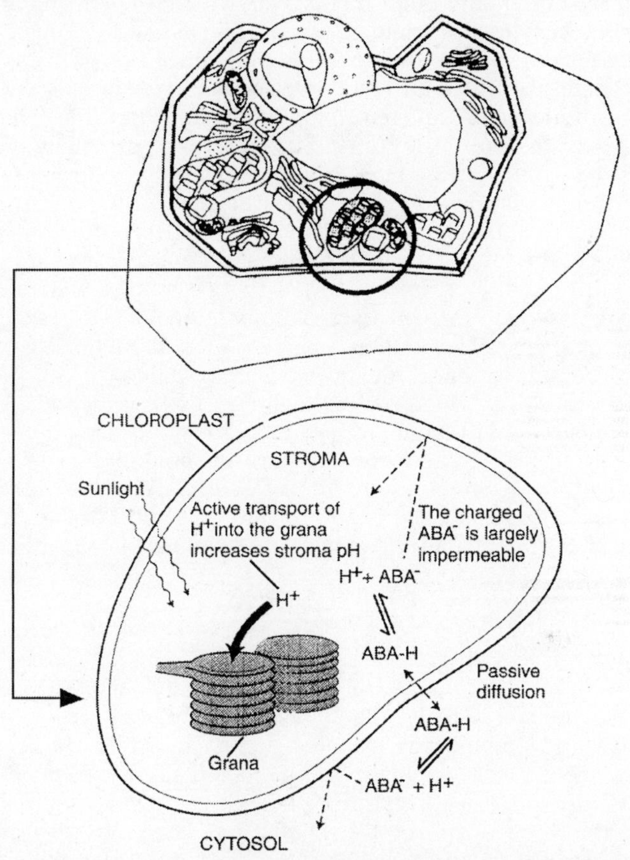


**FIG. 34.3:** Accumulation of ABA by chloroplasts in the light. Dashed lines indicate passive movement of substances by diffusion; the wide arrow indicates movement by active processes. Light stimulates proton uptake into the grana, making the stroma more alkaline. The increased alkalinity causes the weak acid ABA•H to dissociate into $H^+$ and the ABA- anion. The concentration of ABA•H in the stroma is lowered below the concentration in the cytosol, and the concentration difference drives the passive diffusion of ABA•H across the chloroplast membrane. At the same time, the concentration of $ABA^-$ in the stroma increases, but the chloroplast membrane is almost impermeable to the anion, which thus remains trapped. This process continues until the ABA•H concentrations in the stroma and the cytosol are equal. But as long as the stroma remains more alkaline, the total ABA concentration (ABA•H + $ABA^-$) in the stroma greatly exceeds the concentration in the cytosol.

The timing of drought stress conditions in relation to the stage of plant development is also important in terms of internal competition is also important in terms of internal competition for water. Water stress is especially critical during reproductive development. Fruits and grains may not enlarge because rapidly transpiring leaves create lowered water potentials in the xylem, which may result in water actually leaving the fruits. Of critical significance is the availability of water at the time of grain filling in cereals. What happens in upland rice plants is a good example. The last 55 to 60% of the growth is characterized by the four developmental phases consisting of panicle initiation, production of gametes, anthesis and fertilization, and grain filling. Water deficit occurring at any of these crucial stages causes loss of yield. Perhaps the gamete production stage and anthesis are the most sensitive. Reduction in number of flower primordia occurs if water deficits occur at panicle initiation and results in lower spikelet numbers per panicle. Stress at anthesis or gamete production means infertility can occur with an increase in percentage of unfilled grains. Stress during grain filling produces lower weight of individual grains. Drought stress of relatively short duration, as much as 7 days to 2 weeks, may not have a drastic effect because of the difference in time of development of tillers in the field and because of a redistribution of water stored in the plants (RWC) at the time of drought occurs.

As a result of reduction or loss of turgor there may be a reduction in cell size accompanied by a reduction of leaf area. Reduction in leaf area as a result of drought stress has secondary effects because of a reduction of the irradiated surface area. There is an accompanying overall reduction of the amount of photosynthates produced, which in turn contributes to the reduction of growth. Smaller leaves, in addition to reducing carbon assimilated, also reduced water loss.

The reduction of leaf expansion occurs at threshold leaf water potentials that vary with species. For example, in sunflower, leaf water potentials of -0.2 to -0.4 MPa cause cessation of leaf expansion (Boyer, 1970), in corn the threshold leaf water potential is -0.7 MPa (Acevedo et al., 1979), and in soybean it is -1.2 MPa.

### *Effects Water Stress on Ultrastructure of a Cell*

Under mild to severe drought stress there is a disruption of compartmentation with the concomitant release of hydrolases that act on substrates that are normally protected by isolation in compartments. The acid and alkaline lipases that are released destroy membranes in plants susceptible to stress, while in plants resistant to stress the membrane systems remain intact and a recovery system allows repair of what little damage is done. There is, however, a **point of no return** in both susceptible and resistant plants that if exceeded, results in death. After reaching the point of no return, rehydration aggravates the damage. Under mild stress in corn the tonoplast may lose structure, releasing the vacuolar contents. Chloroplasts burst when they come into contact with the vacuolar sap and do not recover activity upon rehydration.

Several lines of evidence indicate that there are changes in the photosynthetic mechanism in the chloroplasts and that a change in the ultrastructure may occur under water stress. When wheat plants were subjected to water stress the reduction of photosynthetic rate was not directly proportional to the stress imposed (Wardlaw, 1967). The reduction in stomatal aperture accounted for only about one half of the reduction in photosynthesis and even when $CO_2$ was forced into the leaf, photosynthesis did not increase (Redshaw and Meidner, 1972). When the technique became available to isolate individual chloroplasts from leaf tissue it was demonstrated that free chloroplasts showed reduced photosynthesis at $\Psi w$ of -0.8 to -1.2 MPa (Boyer and Bowen, 1970).

Some chloroplasts, such as those from sunflower, are irreversibly altered at -2.5 MPa, probably because of changes in the chloroplast membranes. In contrast, the chloroplasts from sorghum were still able to function after they were stressed to -3.7 MPa and then removed from the stress.

### *Effects of Water Stress on Photosynthesis*

The effects of water stress on photosynthesis are not well understood. Under water stressed conditions that cause closure of stomatas, radiant energy continues to be intercepted and absorbed by leaves. Reducing substances are produced that ordinarily would be used in reducing the carbon of $CO_2$ but with stomatas closed. $CO_2$ entrance into the leaf is limited. What then, happens to the reductant?

With limited $CO_2$ available, the reducing compounds must either be stored and accumulate or be used. The subsequent use of carbohydrate reserves results in a deficiency of carbon compounds for both growth and maintenance processes.

### *$C_3$, $C_4$ and CAM Plants under Water Stress*

All three types of plants differ from each other not only in anatomy of leaf and timings of opening of stomatas but also differ in metabolic pathways of $CO_2$ fixation.

Illuminated leaves of $C_3$ plants evolve $CO_2$ in photorespiration at rates above 20% of the net photosynthesis rate and are, as a result, less efficient than $C_4$ plants. Dark respiration rates of mature leaves do not change markedly during mild stress, so the difference in rate of photorespiration between $C_3$ and $C_4$ plants become an important factor in survival during drought stress.

Under drought stress, the $CO_2$ exchange of the illuminated leaves is altered by changes in stomatal aperture and by differences in resistance to diffusion of $CO_2$ in the mesophyll cells. Since photorespiration in $C_3$ plants does not decrease as rapidly as gross photosynthesis, under severe stress photorespiration rates remain high. The leaf under these conditions uses carbon sources within the leaf instead of synthesizing them. With the net result of death from starvation if the severe stress endures. Although $C_4$ plants are able to survive longer because of lower photorespiration rates, they too eventually succumb to severe drought stress from causes other than starvation.

Whereas $C_3$ and $C_4$ plants open their stomatas during day (sunlight) the third group of plants, the CAM plants, mostly succulents show a unique feature of opening of stomatas in dark at night and closed during the day which enable them to survive under arid regions. Malic acid is synthesized and stored until light energy is available to generate photosynthetic reducing power that fixes the $CO_2$ released from malate decarboxylation. If the stomatas were open during the day, photosynthesis would proceed by the $C_3$ pathway. At least one plant, Mesembryanthemum crystallinum, shifts from $C_3$ to CAM metabolism as the growing season advances and the conditions change from mesoic to xeric in the normal cycle.

It has been estimated that $CO_2$ fixation is about 40% less efficient in $C_3$ plants than in $C_4$ plants but there may be some advantages of $C_3$ metabolism for survival (Laing et al., 1974). The recycling of $CO_2$ generated from photorespiration in $C_3$ plants uses the reductant from the Hill reaction to refix the carbon. Such a cycle may have limited utility in stressed $C_4$ plants. Alternative pathways for the energy dissipation are fluorescence or reactions with oxygen. In either event, large reserves of carbohydrates are consumed.

In $C_3$ plants the photosynthetic carbon reducing cycle is linked to the photorespiration carbon oxidation cycle by the activity of ribulose diphosphate carboxylase- oxygenase. The proportion of carbon flowing is determined by the relative concentrations of $CO_2$ and $O_2$. In $C_4$ plants the mesophyll cells lack the ribulose diphosphate carboxylase- oxygenase enzyme and many of the enzymes of the photosynthetic carbon reduction and photorespiratory carbon oxidation pathways. All the enzymes are contained in the bundle sheath cells, however, where low $O_2$ level may prevent photorespiration. What processes act as sinks for the photochemical energy from the Hill reaction when the stomatas are closed under water stressed condition? It is not clear at this time whether or not metabolites from the bundle sheath cells function as a sink.

There appears to be a $CO_2$ homeostasis, which is an important feature of plants that adapt to a slowly developing water stress. Processes come into play that help maintain a constant intercellular concentration of $CO_2$ when leaves are illuminated under water stressed conditions. The slowly adapting features of both stomatal and nonstomatal components of the photosynthetic system are involved. Under rapidly developing stress, which seldom occurs in the field, the system does not adapt and photo-inhibition occurs in the chloroplasts (Osmond et al., 1980).

### *Nitrogen Metabolism under Water Stress*

The distribution of carbon to metabolites in drought stressed leaves has been studied using $^{14}CO_2$. Although less of the carbon enters stressed leaves because of reduced stomatal aperture, it has been possible to ascertain that more $^{14}CO_2$ goes to certain amino acids than to organic acids and sugars. In unstressed corn, for example, more of the fixed carbons is in organic acids than in amino acids. Glycine and serine increase. Under severe stress the competition for carbon skeletons between alanine synthesis and the glycolate pathway increases with more of the carbon going to glycine synthesis, which causes an increase in concentration.

Nitrate reductase activity declines in water stressed leaves. The decrease may be related to a lowered translocation of nitrate in the xylem (Shaner and Boyer, 1976 a,b). Still unclear is whether the change in nitrate reductase activity is the result of loss of enzyme activity, decreased rate of enzyme synthesis or an increased rate of enzyme degradation.

Free proline may accumulate in plants under water stress. Boggess et al. (1976) proved that the increase in proline is the result of synthesis de novo and not protein degradation. It is doubtful that proline accumulation contributes to the survival of plants even though it does contribute some solute to the osmoticum (Hansen et al., 1977).

Betain accumulation has a similar pattern to proline in relation to stress. According to Hanson and Nelson (1978) the betain arises as a result of de novo synthesis from one and two carbon precursors during water stress. Under drought stress barley leaves can accumulate betain at the rate of 200 nM per 10 cm leaf per day upon wilting. The betain is synthesized from serine.

Protein synthesis is also affected by water stress. Even under mild stress there is a shift from polyribosome content to monoribosomes that are inactive in protein synthesis (Hsiao, 1973).

### *Water Logging and Anaerobiosis (Anoxia)*

Creation of an anaerobic environment in the vicinity of roofs is the first and foremost effect of waterlogging, from which plants and especially plant roots, cannot obtain oxygen. The diffusion rate of oxygen in water is approximately $10^{-4}$ that of air and consequently the rate of replenishment to plant surfaces that are submerged is extremely low. Soils at field capacity have 10 to 30% of the volume composed of air-filled spaces but the percentage decreases as water content increases. So the oxygen supply is greatly reduced, and even sometime it is reduced so greatly as to cause death of the plants, e.g. In marshes and swamps and along the shores of ocean, lakes, rivers and ponds, the pore spaces become saturated with water and oxygen is reduced to such an extent that causes death of terrestrial plants. Although some anaerobic respiration may occurs in roots, the decrease in energy released from respiratory substrates is sufficient to maintain vital processes. The rates of water and nutrient absorption are curtailed and death of the roots occurs if flooded conditions continue.

In tomato plants after 24 h of flooding stomatal conductance and translocation were reduced by 30-40%. Although leaf water potential remained unchanged, root conductance decreased. The production of ethylene caused petiole epinasty, which reduced the amount of light intercepted by the leaves. Three hypotheses were proposed to explain the effect of flooding on stomatal conductance: the water logged roots could export some factor that caused closure of stomatas; the roots could fail to export sufficient quantities of some factor normally present in xylem sap that prevents stomatal opening; or there could be a reduction in phloem transport to anaerobic roots, which could lead to a build-up of assimilates or growth regulators in the leaves.

Other symptoms associated with flooding injury are:

1. Pattern of yellowing leaves from base to the top of plant.
2. Drooping of the petioles while plant is still turgid.
3. Leaf epinasty.
4. Hypertrophy (a swelling of cells).
5. New root formation from the stems.
6. Wilting under severe conditions of flooding.

Ethylene production has been implicated in the production of some of the symptoms of flooding injury. e.g., Drew et al. (1979) linked ethylene production to adventitious roots and aerenchyma formation in corn plants subjected to reduced aeration.

## Temperature Stress

Plants are poikilotherms; that is, they assume the temperature of their environment. Any change in temperature affect plants greatly, so stress imposed by temperature have important implications for agriculture. For example, it has been conjectured that a 1°C decrease in the world mean temperature would result in a 40% reduction in rice production. Similarly, a 2°C increase in hardiness of wheat might extend production to areas now in spring wheat. Other examples could be cited to show that temperature is a major uncontrollable climatic factor and point up the need for research on inheritable characteristics for temperature stress resistance and on the manipulations of crop plant physiology to increase tolerance of temperature extremes.

Current research effort can be divided into the following topics, in each of which there is a need for more knowledge:

1. Mechanisms of freezing avoidance and tolerance.

2. Effective cryoprotectants.
3. Biochemical changes in nucleic acids, polyribosomes, amino acids, sugars, and ultrastructural comparisons.
4. Changes that occur in membranes that cause them to rupture, to alter structure leading to protoplast lysis, to bring about the disruption of active transport mechanisms, and to change lipid phases that affect membrane fluidity.
5. Acclimation processes involving changes in proteins and the effects of antimetabolites, surfactants, regulators and other solutes on acclimation processes.
6. Supercooling events occurring primarily in woody species. What kinds of ice nucleators occur and what are the genetic differences in degree of super cooling?
7. Survival traits and breeding for the control of hardiness.

Temperature stress on plants can be divided into the effects of temperatures that cause

(i) Chilling Injury, (ii) Freezing Injury and (iii) high Temperature Injury.

Also involved are the mechanisms bringing about hardiness or tolerance of temperature stresses.

## *Chilling Injury*

Some plants, such as those of tropical origin, are injured when the temperature drops to some point above freezing but low enough to cause damage to tissues, cells, or organs of the plant. This injury which causes due to low but above zero degree centigrade temperature is called chilling injury.

### Direct effects of injury

(i) necrosis, (ii) discolouration, (iii) tissue breakdown & browing, (iv) reduced growth, or (v) failure to germinate in the case of seeds.

More indirect and subtle events may also occur as following:

(i) Reduced grain set, (ii) delayed harvest, (iii) reduced photosynthesis, (iv) reduced water absorption.

### Effect of low temperature on ultrastructure

When tomato hypocotyl cells were exposed to 5°C for 3 days, severe damage occurred at all levels of cellular organisation. The membrane systems are disorganised and 1 cell in 10 dies. The effects were found to be progressive with time.

### Effects on Membranes

In both chilling and freezing there appears to be a critical temperature below which injury occurs. The freeze-induced dehydration cannot be the cause of injury during chilling. In both chilling and freezing injury the membranes of cells and tissues are of pivotal importance. The membrane hypothesis to explain chilling injury implies that cellular membranes in sensitive plants undergo a physical phase transition from a normal flexible liquid crystalline to a solid gel structure at the temperature critical for chilling injury. Apparently there is some molecular reordering in the membranes and a change in fluidity occurs.

### The Freezing Process

At 0°C there is a phase transition in water from solid to liquid or from liquid to solid. Dilute solutions, such as are found in plant cells, under go melting or freezing at different temperatures. As the phase separation occurs, both liquid and solid coexist. The melting points of the solutions are lowered to below 0°C. The solutions usually super cool before actual freezing occurs and the magnitude of the super cooling varies with the influence of a number of factors.

Freezing may be triggered by ice nucleators, which are materials that cause crystallization of the water. Ice crystals themselves can be nucleators and the process of crystallization, once started, may proceed rapidly until a supercooled solution is completely frozen. Usually ice nucleation of water occurs between 0 and -10°C but, if water occurs in small droplets of less than 10 μm diameter, the water may supercool to -38.1°C, which is the homogenous nucleation temperature. Th. if the melting point depression temperature is Tm, then Th = (-2 Tm + 38.1). For most plants Tm is -41 to -47°C. Whole plants usually supercool to the limit. Since the solutions in the cells are more concentrated than the water or the dilute solutions outside the cells or plasmalemma, a phase separation occurs during exposure to freezing temperature.

**What are the events that occur as plants or parts of plants freeze?**

If one starts with temperature decreasing below the freezing point of water then:

1. Water in the cells and in the intercellular spaces super cools.
2. Ice forms extracellularly because the solute concentration is lower and there are more effective ice nucleators present.
3. Intracellular nucleation is prevented by the plasmalemma.
4. Since there is no contact of ice with intracellular water, a vapour pressure gradient is formed from inside to outside.
5. An equilibrium is established either by the water evaporating out of the cell to the extracellular ice or by the formation of intracellular ice.
6. The way in which the equilibrium is established depends on the rate of cooling in relation to the permeability of the plasmalemma and the surface area to cell volume ratio.
7. The amount of dehydration of the protoplasm that occurs depends on the intracellular osmolarity. Whether the maximum amount of water is removed or not depends upon the permeability of the plasmalemma and the surface area available for efflux. If the flux is adequate the equilibrium is established by intracellular ice formation, which is usually lethal to the cell.
8. Some liquid water or solution is not frozen.
9. Injury becomes apparent only upon thawing and some of the injury may be the result of the thawing process.

At equilibrium the difference in Gibbs free energy across the phase separation boundaries must be zero.

## Freezing Injury to Membranes

In freezing injury to membranes research involves the change in the cellular membranes when cell volume contracts because of dehydration caused by the freezing process. If a membrane can be envisioned as a mono or bilayer of lipid, when the surface is reduced there is a tangential pressure induced within the interface. The potential energy stored in this fashion eventually exceeds the hydrophobic forces holding the layer together and lipid is lost to the environment. This suggests that the point at which tangential pressure causes irreversible loss of membrane material coincides with the limit of reversible plasmolysis. In tolerant plants the lipid is stored in the cytoplasm and then restored to the membrane upon rehydration of the cells when they are thawed.

**Effects of Freeze-Thaw Cycles on Plasmamembranes**

Since the contraction and expansion are caused by dehydration and rehydration, the effects of freezing thaw cycles can be predicted by osmotic manipulations. From studies of this type it has been learned that a lytic lesion occurs that is a function of surface area of the plasma membrane during expansion. The freeze thaw injury is a result of two strains: (i) a contraction induced membrane alteration that decreases the maximum critical surface area, and (ii) an expansion induced dissolution of the plasma membrane that occurs when the maximum critical surface area is exceeded.

There is still little knowledge of what happens to the molecular structure of the membranes during the contraction and expansion processes that could be used to explain why the protoplast does not resume the same volume after expansion that it had before contraction. What is the nature of lesions? What happens when tissues in the intact plant freeze?

Changes in protein molecules of plasma membrane during freezing of protoplast also occur and is somehow related to freezing tolerance. For example, the membrane protein is implicated as aggregation of the protein particles occurs in membranes of the frost resistant Solanum acaule but not in the membranes of the frost resistant Solanum tuberosum.

Tolerance to chilling has been linked to the degree of unsaturation of membrane lipids, other lipid factors, and soluble proteins or enzymes. Because there is no resynthesis of lipids in older tissues, any change in the degree of unsaturation occurs through molecular reorientation.

Palta and Li (1978) have formulated a hypothesis for a sequence of events that leads to death of cells during freezing and thawing cycles. They linked the breakdown of membranes to the inactivation of potassium and sugar pump mechanisms, which leads to a change in cellular metabolism and death of the cell. Death may also be brought on by the lowered oxygen concentration as a result of infiltration of tissue by water when membrane integrity is destroyed.

Hypothesized sequence of events in injury and death during slow freezing followed by slow thawing.

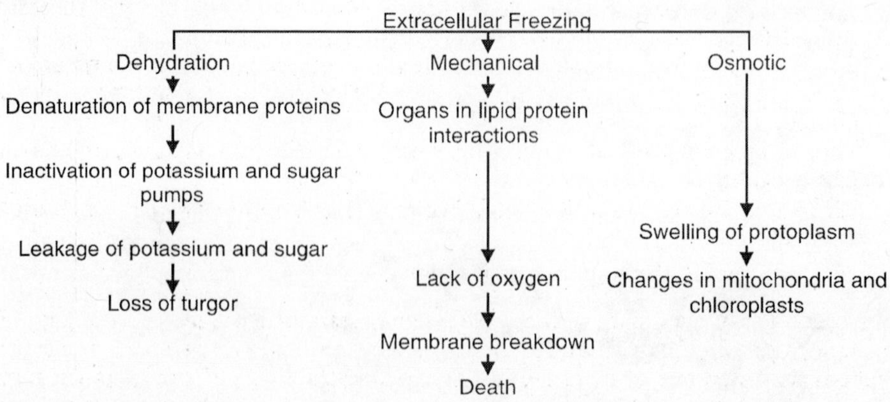


**FIG. 34.4:** Hypothetical series of events that lead to injury and death during freeze-thaw cycles.

**Tolerance of Freezing Stress:** Plants with little or no freezing tolerance, such as corn, cucurbits, beans and others, are subject to frost damage. The foliage becomes flaccid and water soaked. Examination of the tissue usually reveals that compartmentation of cells is destroyed and the membranes are leaky. External temperatures need be no lower than -1 to -3°C. There may be slight supercooling, but when nucleation happens intracellular freezing of susceptible tissues of these types of plants occurs.

Plants with limited tolerance (hardiness) withstand some ice formation. This group of plants includes spring wheat, peas, potatoes, and cabbage. Plants with the growing points below the surface of the soil or snow cover are somewhat protected. Plants in this group seldom supercool more than -1 to -2°C unless the water content is low. Intracellular ice formation is rare. Death results from dehydration as water moves to the intercellular ice.

Woody plants are capable of deep supercooling. Most deciduous forest species and fruit free cultivars have some tissues that may supercool to about -40°C. Let us use apple tree as an example. During summer growth, the tissues are sensitive and behave like susceptible herbaceous species. During the winter, changes have occurred and the bark and the bud tissues may freeze extracellularly. Water from the cambium and phloem migrates to the cortex where it is intercellularly frozen. The xylem ray cells undergo deep supercooling. The buds in the dormant state also deep supercool.

Some woody plants do not supercool and must have other mechanisms for tolerating subzero temperatures. Tolerance in these species, many of which occur in the arctic and subarctic climates, must be on a protoplasmic level. There may similarities between low temperature tolerance and the protoplasmic dessication tolerance.

### Frost Resistance and Cold Hardiness

Frost resistant may not be the correct term to use because of the poikilothermic state of plants. Rather, frost tolerance should be the term used. The tolerance is inducible and may be brought about by a number of possible mechanisms. All of the following have been implicated in tolerance.

1. There may be a decreased concentration of potentially toxic compounds so that, when the solutes become concentrated, they are non-toxic.
2. The ratio of non-toxic compounds to toxic compounds may be such that those that might be toxic are rendered non effective by dilution.
3. Special "protective" compounds may shield membranes to toxic compounds.
4. There may be a decrease in the sensitivity of the membranes to toxic compounds.
5. The colligative protection of solutes such as sugars and amino acids may prevent injury.
6. There may be synthesis of soluble proteins that in some way protect the cells from injury.

## *Effects of High Temperature*

In tropics the plants may reach a high temperature with resulting injury. For example, the emerging seedlings are subjected to high temperature of soil that exceeds the tolerance of emerging plant part and the injury that occurs frequently mistaken for damping off fleshy fruits or vegetative parts of the plants may have a build-up of temperature inside because the dissipation of energy is not efficiently carried out. When stress, the evaporative cooling that would normally reduce the temperature does not occur and leaf temperatures may be several degrees above the ambient temperature.

Symptoms of high-temperature injury are the appearance of necrotic lesions, particularly on stems and hypocotyls, a chlorotic molting of leaves, fruits, and needles (of conifers), and death. The cause of these symptoms are varied. Because of the elevated temperature, respiration rate increases and photosynthetic rate decreases; the tissue may deteriorate because of the lack of life supporting energy. There are also alteration in the proteins or enzymes and membranes. Enzymes have specific ranges of temperature at which they are active. The denaturation and aggregation of proteins may also occur. With the disruption of membranes there is a destruction of the compartmentation of cells, enzymes came into contact with substrates that they ordinarily would not, and reactions occur that alter ion, water, and organic solute movement.

The differential temperature lability of proteins (enzymes) causes changes in the basic metabolism of cells at high temperatures.

Lipid behaviour, particularly that in membranes, changes with a rise in temperature at low temperatures there is solidification of fatty acids with increasing chain length or decreasing number of saturated bonds. A rise in temperature therefore causes a change in the viscosity of the lipid components of the cells to a more fluid state. There is some evidence that high temperature tolerance can be increased by the application of chemicals that increase the ratio of saturated to unsaturated fatty acids.

High-temperature tolerance is usually expressed as the temperature required to produce a defined level of injury or a defined level of dead plants (usually 50%). Other criteria used are plasmolysis of the protoplasm, cytoplasmic streaming, and photosynthetic rate. Tolerance of high temperatures stems from the ability to maintain high rates of photosynthesis, the stability of proteins that must resist conformational changes at high temperatures, the ability to repair or synthesize structural protein, and the possession of substances that protect the proteins from change.

There is evidence that the lower the hydration of protoplasm the greater the hydrogen bonding and the greater the tolerance of high temperatures.

Thus, dormant tissues, such as in seeds and buds, are more tolerant than rapidly growing tissues. A reduction in nitrogen fertilization, therefore, produces more tolerance. Cross linkages by divalent cations may also cause more tolerance.

Some plants avoid high-temperature injury because of the angle and arrangement of the leaves toward the incident radiation. Colouration and reflactance properties of the leaves are important. Desert species may reflect as much as 70% of the indigent infrared radiation. The presence of a thick cuticle and leaf hairs alters the reflectance properties of the surface and improves conductive coolings. Factors that improve the evaporative cooling of transpiration when water is available is also improve the high-temperature tolerance.

## Nutrient Stress

Approximately 10% of the plant dry weight is composed of minerals or about 1.5% of the fresh weight. Currently 15 to 17 elements are considered to be essential for plant growth and reproduction. Each of these essential elements has a direct and very pivotal (indispensable) role in the plants normal life cycle. But the range in which these are required (optimum amount) is very narrow so the less or more than this range for any of these elements produce stress of that mineral and cause injury. Approximately one fourth of the earth's soils are considered to produce some kind of mineral stress (Dudal, 1976).

If plant takes less than optimum quantity of any essential mineral than the situation is referred as deficiency of that mineral or if availability and absorption is more than optimum than situation is referred as toxicity of that mineral. Arbitarily determined ranges are severe deficiency, moderate deficiency, luxury consumption, and toxic. The concept of critical concentration range has been described as that range of concentrations that occurs between deficiency and luxury consumption. Such information is useful in measuring the degree of stress a plant may be experiencing and whether it is in the deficient or toxic range of concentrations.

### Conditions Causing Nutrient Stress

Deficiency of an element results from a number of conditions among which are the amounts and concentrations present in the soil, the form in which they exist, the processes by which they become available to the plant, the content of the soil solution, and the soil pH. Nutrient concentrations in plants vary with age, plant part, species and the rooting medium or soil type. The amounts of nutrients in soil vary widely and depend upon the parent material from which the soil is derived and the processes in cycling the elements in natural or agricultural ecosystems. Absorption of minerals from soils by plants is also depend upon the efficiency and volume of roots. Growth of roots through the soil also brings them into contact with fresh areas of soil and ions become available by this process of interception. Ability of plant roots to exude into the rhizosphere materials that affect the solubility and form of nutrients is also well known.

Interference with any of the processes involved in changing the availability of ions or of their absorption by the plant roots may result in stress. Besides this, mobility of ions within the plant body may also cause deficiency or toxicity of that element in any part of the plant irrespective of the absorption of that. Movement in the xylem shows little or no difference in rate of translocation of the various ions and is controlled by transpiration rates and the pressure gradients created by differences in water potential. In the phloem there are distinct rate differences for translocation of the various ions and such differences may produce deficiencies in some parts of the plant.

Elements such as calcium, boron, and iron often become deficient in the younger parts of the plant because of little or no retranslocation and poor phloem mobility. Similarly, if such ions are applied foliarly, there is little basipetal movement. Toxic concentrations of ions often appear in older tissues of the plants as a result of accretion with time.

### Deficiency Causes and Symptoms

A symptom is any perceptible change in known structure, appearance or function. Such things as leaf yellowing (chlorosis), death (necrosis), lesions, malformations, malfunctions, or reduced growth and yield are all considered to be symptoms and occur as a result of nutrient deficiency.

### Nutrient and Metal Toxicity

The range of concentrations of the essential nutrients between deficiency and toxicity is often small and depends on a variety of factors in the environment that affect availability and growth of the plant. The transition metals (micrometabolic nutrients) are most often implicated in toxicity reactions. Toxicity implies a sensitivity of an "essential to life" process or processes. Toxic levels of Zinc, Copper, and Nickel occur frequently in soil. Toxic concentrations of lead, cobalt, beryllium, and cadmium can occur but only under unusual conditions. Lead and cadmium are of particular interest because plant accumulation introduces them into the food chain. Aluminium and manganese toxicities are the most studied and the information about them can serve as a basis for some concepts and principles that may apply to others.

**TABLE 34.3:** Plant nutrient deficiency symptoms in tobacco grown in sand culture: A key

| Symptoms | Deficiency Elements |
|---|---|
| A. Non parasitic troubles; never infectious; caused by element deficiencies. | |
| B. Older or lower leaves of plant mostly affected; effects localised or generalised. | |
| C. Effects mostly generated over whole plant; more or less drying or firing of lower leaves; plant light or dark green. | |
| D. Plant light green; lower leaves yellow; drying to light brown colour; stalks short and slender if element is deficient in later stages of growth. | Nitrogen |
| D. Plant dark green, of ten developing red and purple colours; lower leaves sometimes yellow, drying to greenish brown or black, stalks short and slender if element is deficient in later stages of growth.... | Phosphorus |
| C. Effects mostly localised; molting or chlorosis with or without spots of dead tissue on lower leaves; little or no drying up of lower leaves. | |
| D. Lower leaves mottled or chlorotic, with or without dead spots; leaf margins sometimes tucked or cupped upward or downward. | |
| E. Mottled or chlorotic leaves typically, may redden, as with cotton; sometimes with death spots; tips and margins turned or cupped upward; stalks slender.... | Magnesium |
| E. Mottled or chlorotic leaves with large or small spots of dead tissue. | |
| F. Spots of dead tissue small, usually the tips and between veins, more marked at margins of leaves, stalk slender......... | Potassium |
| F. Spots generalized; rapidly enlarging, generally involving areas between veins and eventually involving secondary and even primary veins; leaves thick; stalks with shortened internodes. | Zinc |
| B. Newer or bud leaves affected; symptoms localized | |
| C. Terminal bud dies, following appearance of distortions at tips or bases of young leaves. | |
| D. Young leaves of terminal bud at first typical hooked, finally dying back at tips and margins, so that later growth is characterized by a cut out appearance at these points; stalk finally dies at terminal bud. | Calcium |
| D. Young leaves of terminal bud becoming light green at bases, with final breakdown here; in later growth, leaves become twisted; stalk finally dies back at terminal bud. | Boron |
| C. Terminal bud commonly remains alive; wilting or chlorosis of younger or bud leaves with or without spots of dead tissues; veins light or dark green. | |
| D. Young leaves permanently wilted (wither-tip effect) without spotting or marked chlorosis; twig or stalk just below tip and seedhead often unable to stand erect in later stages when shortage is acute. | Copper |
| D. Young leaves not wilted; Chlorosis present with or without spots of dead | |

| Symptoms | Deficiency Elements |
|---|---|
| tissue scattered over the leaf; | |
| E. Spots of dead tissue scattered over the leaf; Smallest veins tend to remain green, producing a checkered or reticulate effect. | Manganese |
| E. Dead spots not commonly present; chlorosis may or may not involve veins, making them light or dark green in colour. | |
| F. Young leaves with veins and tissue between veins light green in colour. | Sulphur |
| F. Young leaves chlorotic, principal veins typically green; stalks short and slender. | Iron |

### *Aluminium Toxicity*

Stunting, small dark green leaves, and late maturity with a purpling of the leaves and stems. Often there is a yellowing and death of the leaf tips. On other plants there is an induced calcium deficiency that results in curling or rolling of young leaves, and a collapse of growing points or petioles. Roots are injured and become brittle and stubby with a spatulate shape. Root tips become thickened and turn brown and there are many stubby lateral roots with no fine branching which gives rise to a coralloid appearance. Such roots are predisposed to fungal infection and the absorption processes are inefficient.

Little information is available about the effects of aluminium toxicity on the physiological processes. Interference with cell division should be suspected because of the effect on root tips. The cells become binucleate and if aluminium penetrates to the nuclei, it is precipitated there by the phosphorus containing compounds. There is an interference with the enzymes governing the deposition of cell wall polysaccharides and, because of the cross linking of aluminium in the pectins, the walls become rigid. Cross linkages within proteins also occur. Probably because of the change in pattern of root growth, there is interference with the uptake and transport of other essential elements such as calcium, phosphorus and potassium.

### *Manganese Toxicity*

The element accumulates in the shoots of plants and the injury is in proportion to the amount accumulated. Symptoms include a marginal chlorosis and necrosis of the leaves leaf puckering, and formation of necrotic spots. The symptoms may be produced without a reduction of the vegetative growth. Severe injury to the shoot may later result in the roots turning brown. The physiological effects vary from species to species and have been described using a variety of species. Manganese toxicity has been associated with the destruction of auxin in cotton (Morgan et al., 1966; 1976) and with an amino acid imbalance in potato (Rabinson and Hodgson, 1961).

### *Copper Toxicity*

Copper toxicity generally results in chlorosis and stunting. The chlorosis is attributed to iron deficiency caused by the high concentrations of copper in the tissues. Toxicity is experienced first in the root tips followed by inhibition of the development of lateral roots. Reduction of the root system has implications for the absorption of other nutrients and the subsequent growth inhibition. The effect of copper on roots has not been explained.

### *Chelation as a Mechanism of Tolerance*

Chelation is the coordination between a metal and a ligand resulting in the formation of a ring structure. Usually two or more ligands are necessary for the ring structure. Ligands are compounds containing functional groups of atoms that share electrons with metals. They are basis with unshared electron pairs. The metals act as acids and accept electron pairs. The neutralization reaction between these Lewis bases and Lewis acids is called coordination.

The relationship of chelates to mineral element stress tolerance lies in the structure and stability of the chelate ring, which is a function of the rings and the composition of the ligands. Because chelation has an effect on the reactivity of the metal, plants may be protected from excessive concentrations of free ions and may have increased stabilization of structural elements.

Chelation results in the formation of metalloenzymes, which perform vital functions in plant metabolism. Changes in metabolism may come about because of the changes in enzyme composition as a result of stress and the effects on available ligands, metal ion concentrations, and the proteins associated with enzymes. Some metalloenzymes are located in membranes where deficient or toxic concentrations of metal ions may affect permeability. Calcium, for example, reacts with acid phosphatase or acid proteins in the membranes.

### *Genetics of Mineral Element Stress Tolerance*

Considerable research has been done on the genetics of metal toxicities. In barley and wheat, for example tolerance to aluminium has been found to be controlled by a single major, dominant gene in some cultivars but a more complex situation existed in other cultivars (Clark, 1982). Apparently, there is sufficient evidence that for cereals and other plants genetic improvement for aluminium and manganese tolerance can be achieved.

The technology of tissue culture has been used to produce lines that contain chromatin material from other lines brought about by a transfer manipulation that cannot happen naturally. For example, the copper efficiency factor from rye was transferred to triticale and then to wheat by modern technology.

The genetics of iron efficiency have been investigated extensively. Plants that exude hydrogen ions and reducing compounds from their roots or are capable of reducing iron at the root surfaces can survive on soils with low iron availability, but plants that do not contain these traits may not survive. Iron efficient plants under iron deficiency conditions develop epidermal transfer cells in the subapical region of the root. The transfer cells are involved in the extrusion of protons into the soil, which results in a lower pH. They may also be involved in the production of reducing substances. An abundance of root hairs develops that may contain transfer cells as well. A rapid increase in iron uptake occurs as a result of these changes in root morphology and anatomy (Romheld et al., 1983).

Another factor in tolerance is that plants that are tolerant or iron deficiency conditions frequently contain higher concentrations of elements such as phosphorus, calcium, zinc, and manganese than the less tolerant plants.

Because of the importance of overcoming nutrient stress in soils of countries where there is promise of increase in yield, there is an international cooperation in research on ways of bringing about adaptation of important crop plants to the stress.

## Salt Stress

Salinity has been an important factor in the history of mankind and in the agricultural systems upon which mankind has relied. Civilization have been destroyed by the encroachment of salinity on the soils as a result of poor water management. If there is limited rainfall, salt is not leached out of the soil volume in which crop plants take root and yields are reduced as the salt concentration increases. Irrigation water contains dissolved salts that are concentrated as the water evaporates and build up in the soil over time. Vast areas of the land surfaces are rendered unfit for agriculture because of the present low tolerance of crop plants to salt and the lack of knowledge about the tolerance mechanisms available for plant breeders in the selection processes. In some areas with agricultural potential, two approaches may be taken. One approach is to try to find inexpensive and cost-efficient ways of desalting seawater to be used for irrigation purposes. Another approach is to select crop plants that can grow and survive under highly saline conditions.

Salt stress affects plants adversely in two following ways:

i) High solute content in rooting medium create water stress by decrease in osmotic potential.
ii) Direct toxic effect of high concentrations of ions.

Some plants have evolved mechanisms for dealing with these stresses and others can become adapted to them. But, the majority of crop plants are susceptible and will not survive under conditions of high salinity or will survive but with decreased yields.

Plants can be divided according to their tolerance for salt stress

i) **Halophytes**: Grow in saline soils, in high concentrations of salts, and are either facultative or obligate halophytes. They can again be divided according to extreme of tolerance into
   a) **Euhalophytes:** Tolerate extreme salinity.
   b) **Oligohalophytes:** Tolerate moderate salinity.
ii) **Glycophytes:** Cannot grow in the presence of high concentrations of salts or possess some mechanism by which the protoplasm is not exposed to high salt concentrations.

Levitt (1980) separated salt stress from ion stress.

I. Salt Stress: If the salt concentration is high enough to lower the water potential by 0.05 to 0.1 MPa then the plant is under water stress.
II. If the salt concentration is not this high, the stress is ion stress and may be caused by one particular species of ion.

### *Mechanisms of Tolerance*

Some of the characteristics of halophytes that may give clues as to the physiological mechanisms involved in tolerance are:

1. Succulence that may lead to dilution of intracellular salt such as occurs in *Salicornia sp.*
2. The presence of salt excreting glands that reduce the concentration of salt in the plant, such as occur, for example in species of *Frankenia*, *Cressa*, *Spartina*, and *Glaux*.
3. Development of small leaves, water storage hairs, and aerenchyma.
4. A water potential range from -2 to -4 MPa, which is equivalent to an ion concentration of 400 to 700 mM. (In species *Salicornia rubra* the sodium and

chloride ions make up 75 to 93% of the total osmotic potential).

5. Species that rely mainly on a higher ion uptake and internal concentration of ions to maintain turgor.
6. Sometimes there is synthesis and accumulation of organic solutes that aids in the maintenance of turgor.

These are some adaptations by which plants manage to grow in places with high salt concentration. But the glycophytes do not enjoy of these features so they are sensitive to salt stress. Glycophytes manage to grow in saline conditions by adopting some mechanisms as follows:

1. Accumulation of sugars in leaves to maintain turgor under conditions of salt stress.
2. Membranes of root hairs of plants selectively permeable in nature so in some plants absorption across membranes in root cells and leaf cells, the entrance of sodium becomes limited.
3. Some plants tolerate to salt stress have comparrtmentilization so that toxic ions can be checked from reaching to the photosynthetic parts of the plants where they can disturb this. For, example there is sometimes high concentrations of salt in roots and sometimes in the stems of the plants, but low concentrations in the leaves (Jacoby, 1965).

### *Irradiation Stress*

All life on earth is supported by the radiant energy from the sun, which plants convert into chemical energy by the process of photosynthesis. Light is one of the most important and variable components of the plant environment. The autotrophic plant is directly influenced by the intensity of light, which drives photosynthesis, and thereby, provides the chemical energy and carbon needed for plant growth and development. Thus, an alteration in light intensity, whether a deficit or excess, will result in a disruption of plant metabolic processes. Besides affecting photosynthetic rate, solar radiation also affects plant temperature and photomorphogenic responses in ways that can produce stress.

### *Effects of Light Deficit (Shade)*

Since plants cannot move about they must adapt to new light intensities as they occur. Light intensities below the compensation point lead to starvation with carbohydrates used as substrate for respiration first and then other source of energy. Adaptation to shade comes in two ways:

i) Increase of leaf area in a way that minimizes the use of metabolites (for example, by increasing leaf area at the expense of photosynthate allocation to root growth), and
ii) A decrease in the amount transmitted and reflected light. Shade leaves are thinner but larger in surface area than sun leaves. Increased light absorption is brought about by increased number of chloroplasts per unit leaf area and by increased chlorophyll concentration of the chloroplasts which is accompanied by a lowered concentration of other pigments that interfere with the light absorption process. In addition, there is sometimes a change in the orientation of the chloroplasts so that the broad dimension is oriented toward the light.

Resistance to chlorophyll degradation in shade plants serves to increase chlorophyll concentration. In some plants there is also a reduction in the respiratory rate, which conserves photosynthate.

## Effect of Bright Light

The photosynthetic performance of plants in high intensity light is not dependent on the chlorophyll content or on absorption and trapping of light quanta as in shade plants but rather on the relationship between photoinhibition of photosynthesis and the degree of chlorophyll destruction. The extent of photoinhibition is dependent on the intensity and duration of exposure to high light intensities. Photoinhibition is caused by a transfer of the excess quanta produced during the light reaction to the reaction centres instead of being disposed of as heat of fluorescence. Creation of singlet oxygen or free radicals destroys the reaction centres unless a receptor for the increased reducing power is present. Usually the receptor is the carbon dioxide fixation processes involving RuBP case (Ribulose Biphosphate Carboxylase) that use the quanta. Shade plants have low levels of RuBP case a while sun plants, especially $C_3$ plants, have higher levels of enzyme. Sun plants with $C_4$ metabolism are able to use photosynthetic energy more efficiently without an increase in RuBP case.

Obligate shade plants have a lower capacity for electron transport brought about by lower content of the carriers cytochrome $f_1$ cytochrome $b_{559}$ and $b_{563}$. Plastoquinone, ferredoxin, and ATP synthetase. Injury to such plants by high light intensities is dependent upon the concentration of carbon dioxide and oxygen present as receptors of the increased reducing power generated. Low temperatures cause a reduction of carbon dioxide fixation again decreasing the availability of receptors of reducing power.

Leaf growth and development is changed by high light intensities in such a way that there is an increase in the elongation of the pollisade cells and an increase in the number of cells across the leaf section and in the average cell diameter.

### *Resistance to High Light Intensity Injury*

Shade plants have a limited capacity for adjustment to high light intensities and are usually injured or killed. Sun plants, on the other hand, gradually increase the capacity for light saturated photosynthesis by (1) increasing the level of carbon metabolism enzymes, (2) increasing photosynthetic electron transport chain carriers, or (3) increasing the capacity for carbon dioxide transport.

Resistance to high light stress also occurs in a number of other ways. Increased suberin, wax, cuticle, and cell wall materials of the leaves can cause an increase in the amount of reflected light. The photodestruction of chlorophyll can be reduced by an increase in the content of ß-carotene, xanthophyll, and phospholipids of the leaves. An increase in the stacking of grana accompanied by a stabilization of the chloroplast membranes aids in preserving chlorophyll. And, the heliotropism of leaves can cut down on the amount of light intercepted.

### *Ultraviolet Radiation*

The electromagnetic radiation from the sun contains about 7% ultraviolet at sea level. Ultraviolet radiation is absorbed by plant chromophores, which include nucleic acids, proteins, indoleacetic acid, abscissic acid, and flavoproteins. Absorption by nucleic acids may after protein synthesis but proteins are also absorbers of radiation and may react at a primary level. Alteration of protein and lipid membrane components by ultraviolet may alter membrane permeability and ionic balance. Such events also cause an inhibition of photosynthesis and respiration.

The most common inactivation of nucleic acids by ultraviolet radiation is through photochemical lesions involving polymers of pyrimidine base in the deoxyribo nucleic acid (DNA). The result is the production of pyrimidine dimers and loss of DNA biological activity. It is evident that ultraviolet radiation absorption cause an acceleration in the mutation rates and aberrations of chromosomes.

Reduced plant growth with exposure to ultraviolet radiation can occur as a result of inhibition of photosynthesis as a reduction in leaf expansion. Since both indoleacetic acid and abscisic acid are receptors for ultraviolet radiation, alterations of concentrations of these growth regulators may result in irregular growth. Such irregularities may take the form of depressed flower development, loss of apical dominance, abscission of leaves, and alteration of mineral nutrient concentrations in tissues.

### *Resistance to Ultraviolet Radiation*

Resistance to ultraviolet radiation damage consists of the ability of a plant to repair the damage to the DNA, or to synthesize screening compounds, such as flavonoids and flavones, in the epidermis. Cuticular waxes may also be effective in absorbing harmful ultraviolet radiation. Changing the angle of leaves to the incidence of radiation or increasing the reflecting properties of leaf surfaces are other mechanisms of resistance.

### *Ionizing Radiation as Stress*

Ionizing radiations are radiations that produce ions in the media they traverse. Energy is transferred from the moving particles and causes the ejection of an electron or of a proton or capture of the particle by the nucleus of the atom that gets in the way. Charged particles directly produce many ionizations per particle. Photons in X rays, _ rays, or neutrons produce few energetic charged particles but each in turn produce many ionizations. In biological materials the production of electrons causes all of the ionizations. Excitation of matter is considered relatively unimportant in causing injury in biological material. It is the free radicals formed that are injurious.

Free radicals contain unpaired electrons and are extremely reactive in chemical reactions within the system in which they are produced. The resultant injury to plant may arise from alteration of essential components of protoplasm such as the genetic material in which mutations can be created- Alterations, deletions, and additions to the DNA acid bases can occur. Such changes in the genetic potential of cells has been used in some breeding programmes to increase the rates of mutations.

The symptoms of radiation damage in plants take many forms. A change in ploidy with the formation of polyploids can lead to loss of reproductivity, mitotic delay, lengthened cell cycle time, and change in cell morphology. The processes underlying differentiation may be changed, resulting in gall formation or even death. At the whole plant level, the symptoms take the form of early senescence, somatic mutations, increased photosensitivity of seed germination, change in resistance to stress, prevention of geotropic responses, production of ethylene, and many others depending upon the condition of the plant at the time of exposure and the amount or mass of dead tissue present as in the woody species. Some of the physiological factors that determine the extent of radiation damage are water content, ambient temperature, oxygen concentration, and the state of nutrition.

Seeds are more susceptible to radiation when they have a high water content. Presence of oxygen increase the free radical production. P and Ca deficiencies enhance radiosensitivity low temperature slow or prevent repair while relatively high temperatures enhance the recovery processes.

## Drought Resistance

Different kinds of plants manage to tolerate drought in different ways. According to their way by which they tolerate or avoid in drought they were classified by many workers into following categories:

**TABLE 34.4:** Terms used in drought stress resistance

| Drought Resistance | Dehydration Avoidance | Ludlow (1980) |
|---|---|---|
| | Dehydration Tolerance | |
| Drought Tolerance | Dehydration Postponement | Kramer (1980) |
| | Dehydration Tolerance | |
| Drought Resistance | Drought Avoidance | Levitt (1980) |
| | Drought Tolerance | |
| Drought Escape | | |

So, on the basis of Levitt's classification plants can be classified into following categories:

**Drought Escape**: Plants can escape drought by various mechanisms that enable them to remain dormant during the drought or to complete their life cycle during favourable moisture conditions. Desert ephemerals, for example, grow and reproduce during periods of adequate rainfall and produce seeds that remain dormant during periods of adequate rainfall and produce seeds that remain dormant during periods of drought. Plants with a rapid rate of development and an indeterminant growth habit are best adapted for escaping drought.

**Drought Resistant**: Levitt divides resistance to stress into avoidance and tolerance.

**Drought Avoidance**: Drought Avoidance occurs if a plant does not come to thermodynamic equilibrium with the stress or can exclude the stress by means of a physical or metabolic barrier. In this plants maintain high tissue water potential that results from either a reduction of water loss or a maintenance of water uptake. On this strategy drought avoiders can be classified into two categories.

1. **Water Savers**: Reduction of water loss follows from an increase in stomatal and cuticular resistance, a reduction in absorbed radiation or a reduction in leaf area.
   Reduced radiation absorption ariser through movement of leaves so that the angle of incidence to incoming radiation provides less area for absorption, the development of pubescence, which insulates the leaf surface, or the development of increased waxiness and reflecting qualities of the leaf surface. Reduction in leaf area comes about as the result of a reduction in number of leaves under the influence of drought stress
   A reduction in leaf area also reduces water loss. Drought stress causes an increase in rate of senescence of older leaves and premature abscission, which still further reduces transpiration. In addition, these older leaves may not contributing much carbohydrate to developing grains or fruits. These kinds of adaptations are largely irreversible.

2. **Water Spenders**: Drought avoiders may also maintain high water potential by spending more water. The main strategy to avoid drought by spending more water is the efficient root system of these plants. Deeply rooted plants can continue to absorb water until the drought becomes severe and reaches deep into the soil. The rate of growth of roots affect the root shoot ratio. Increase rate of root growth may result in less shoot growth and increase in length and density of roots. Low resistance to water movement through the roots and the conducting system to the shoots by an increase in vessels or vessel diameter also aids in maintaining water uptake.
3. **Drought Tolerance:** Some scientists hold the view that the true tolerance to drought stress occurs when plants are subjected to drought and the water potential is reduced. Plants having capacity to tolerate drought at low water potential can be grouped under two categories:
4. **Dehydration (Desiccation) Avoidance:** Plants coming in this category tolerate drought by maintaining high turgor. They manage to maintain high turgor by following means: (i) a lowering of the osmotic potential, (ii) a capacity to actively accumulate solutes, (iii) highly elastic cells or tissues, and (iv) small cells.

   Low water potential is the result of a low osmotic potential probably produced through osmotic adjustment and the addition of soluted to cell sap. As stress develops there is a gradual lowering of the osmotic potential as the solute concentration increases because of the loss of solvent water.

   Factors that affect osmotic adjustment are:

1. The rate of development of stress: A water potential decrease of -0.1 to -0.5 MPa per day allows for adjustment. A decrease of -1.0 to -1.2 MPa per day is too rapid.
2. Degree of stress: Early in the stress cycle full turgor may be maintained but continuation of stress results in less capacity to adjust fully.
3. The environmental conditions: Those factors that affect the rate of drying such as temperature and light intensity have direct effects. The day and night temperatures and other factors that affect the rate of photosynthesis and production of solutes have an indirect effect.
4. Difference in cultivars and in tissue or organ exposed.
5. The age of the plants.
6. **Dehydration (Desiccation) Tolerance:** This category of plants tolerate desiccation even at negative turgor. This type of plants includes lichens, mosses, and resurrection plants which almost completely dehydrated and yet recover when they are rehydrated. Such tolerance must occur at the protoplasmic level and has been termed dessication tolerance. Some parts of plants may posses dessication tolerance. For example, seeds are very resistant to drying as are some rhizomes and leafless stems.

# CHAPTER 35

# Plant Physiology in Agriculture and Industry

## PHYSIOLOGICAL ADAPTATIONS TO WATER STRESS AND BREEDING FOR DROUGHT TOLERANCE

Levitt's classification of drought resistance (1980) linked adaptive strategies to plant water relation parameters. Drought avoidance and tolerance were defined as the ability to survive or thrive at high and low water potential. Osmotic adjustment constituted the major strategy of dehydration avoidance and stomatal behaviour was associated with water saving and water spending strategies. Thus, the ability to quantify plant or crop water deficit became fundamental to research on the response of plant communities to water stress. Great progress towards this end has been made in the 1960's or 1970's particularly because of the development of psychrometers (Manohar, 1971), pressure chamber and steady-state-porometer for measurement of leaf water potential and stomatal characteristics (Slavik, 1974).

Number of important metabolic processes namely, photosynthesis, respiration and the conversion of sucrose to starch in developing grains of wheat have been found related to drought tolerance, however, a stable cell membrane system that remains functional during water stress appears central to adaptation to high temperature. Wallner et al. (1982) detected genetic differences in heat tolerance in turf grass using this procedure. Thermostability test can be used for screening heat and drought tolerant genotypes (Kumar et al., 1994).

The employment of plant and canopy temperature measurements in detecting crop water stress, assessment of irrigation requirement (Hatifield et al., 1983), crop performance under optimum and stressed environment is constituting IRT technology.

The Crop Water Stress Index (CWSI) computed on the basis of canopy temperature is based on the relationship between foliage-air temperature gradient (TF-TA) and air vapour pressure deficits (VPD) for the plant in question when it is well watered and transpiring at the potential rate. The non-water stressed base line, which is a linear regression between TF-TA and VPD for non-stressed plants, is very important for the computation for CWSI. Such a relationship has been worked out for a large number of plant species and almost universal. It has been reported that with increasing VPD the TF-TA decrease, apparently on account of increase in transpiration and stomatal opening.

In contrast, the non-stressed plants, an upper line response has also been determined which theoretically means that water stressed plants kept their stomata close maintaining transpiration at low level with constantly high TF-TA. The CWSI at a given point of time or VPD is computed on the basis of a ratio based on the differential of TF-TA with non-stress base line and stressed upper limit response. Alternatively the CWSI can also be computed with the help of an Idso's diagram (Idso et al., 1981).

Breeding genotypes for the water limiting environments by empirical approach based on grain yield is difficult because of annually varying soil moisture and consequently the variance in drought tolerance attributes of crop plants. It is believed that better adapted and high yielding genotypes could be breed more efficiently and effectively, if analytical approach based on physiological and biological criteria is adopted. However, there are few examples where this approach, has been used and even fewer, where it proved successful (Richards, 1982). Nevertheless, the search for new physiological traits continued, e.g., on the basis of epicuticular wax (Rajgopal, 1995).

### Glaucousness and Reflectance

Glaucousness is the waxy covering which imparts a dull white or bluish-green cast commonly referred to as bloom. Glaucousness is one characteristic that has been cited as a plant adaptation to drought. Isogenic lines of sorghum with or without glaucousness or waxy bloom has been developed (Blum, 1975). Glaucous lines had greater adaxial leaf reflectance in the visible and near infra red wavelengths than non-glaucous lines. Higher reflectance of radiant energy will result in lower leaf temperature, thus reducing thermal injury. To identify genetic and environmental sources of variation in residual transpiration of wheat, the effects of glaucousness, ECW, leaf age, plant height and growth environment were determined using excised leaves. Glaucousness reduced residual transpiration by an average of 10%. Differing quantities of epicuticular wax (ECW) and glaucousness in an isogenic pair of genotypes affected residual transpiration, the rate being 33% greater in low wax non-glaucous line than in the high wax glaucous line.

It is possible that the difference in yield between bloom and bloomless sorghum under semi-arid conditions, as reported by Ross (1972), is a result of the difference in gas exchange ratio. The importance of glaucousness in wheat was suggested by Fischer and Wood (1979), according to whom the best prediction of yield under drought from traits measured in the absence of drought was given by a linear model containing an index for degree of leaf waxiness or glaucousness. Near isogenic lines that differ in the presence or absence of glaucousness were developed in durum wheat (Triticum turgidum) and common wheat (T. aestivum) grown in plots in three field environments, varying in three soil water availability.

In view of the role of ECW in controlling cuticular transpiration (CT), several authors have used the determination of ECW as an effective method of screening varieties for drought resistance/tolerance in different crops. Six varieties of oat were screened for CT and ECW and were arranged according to decreasing order of drought resistance.

### Physiological Adaptation to Salt Stress and Breeding for Salt Tolerance

Plant responses to salinity are controlled by genes but with environmental stresses such as salinity or drought, a genetic approach that is divorced from an understanding of plant physiology is unlikely to succeed (Munns et al., 1990). There are two main reasons for this. First, resistance to salinity or drought is not confined to cellular level but it is expressed at tissue and whole plant level. Secondly, there is paucity of rapid methods of screening for salt or drought resistance at seedling stage. The existing methods involving measurements of reduction in biomass or yield on account of stress are time consuming and empirical. In order to formulate a comprehensive breeding and selection strategy for salt affected areas the protocol is to:

a) identify salt tolerant germplasm,
b) compare germplasm responses to identify a morpho-physiological trait,
c) compute tolerance indices,
d) study the heritability of physiological trait, and
e) use the characteristics in breeding for salt tolerance.

It is well documented that salt tolerance is a holistic response and the behaviour under field and laboratory testing may or may not match. But still, establishment of relationship between laboratory responses of young seedlings with salt tolerance behaviour was found useful. Ionic Regulation Indices (IRINa) was found positively related with simpler indices (Kumar, 1994). Selection pressure has been employed at gametophyte stage assuming that some genes which are expressed in sporophyte growth are also expressed during growth of the gametophyte. Thus use of gametophyte, particularly the pollen, have been demonstrated in the screening for salt tolerance.

Among other physiological adaptations specialized proteins are promising. In the last decade substantial progress has been made toward understanding the heat shock (HS) response, a ubiquitous response of organisms to abrupt increases in temperature above the normal growing temperature. In general terms the HS response is characterized by a decrease in normal protein synthesis, and synthesis of new HSmRNAs and consequently the accumulation of new HS proteins (hsp). This leads to acquisition of thermotolerance to otherwise lethal temperatures. Gradually normal protein synthesis is restored and hsp synthesis declines during prolonged heat treatment. Salt adaptation proteins in cultured carrot cells in citrus and tomato cells and soybean cells have been reported. Among the newly synthesized proteins are one major protein of low molecular weight of 17 KDa in soybean, 26 KDa in citrus and tomato cells and 66 KDa in carrot cells. The 26 KDa protein was first markedly suppressed by a salt shock and the time required for its recovery was dependent upon degree of salt tolerance. Biological adaptation resulted in an increase of this proteins. Many salt induced changes in proteins are species specific and do not resemble with each other.

Various mechanisms for salt tolerance have been proposed (Muuns, 1994). The one most likely to explain genotypic differences in salt tolerance of crop plants is better regulation of salt transport preventing excessive accumulation of salt in mature leaves. Despite the fact that salinity responses are mediated through water imbalance but water relation parameters do not correlate consistently well with salt tolerance. Contrasting genotypes of pearl millet and wheat differ in morpho-physiological characters but genotypic differences in salt tolerance correlate fairly well with low salt concentrations in leaves, because the roots of the more tolerant genotype are able to discriminate between the various salts in the soil solution and allow water to enter the conducting vessels. Negative correlation between leaf sodium content and grain yield was found in five bread wheat cultivars.

An ion exclusion mechanism has been postulated in tolerant cultivars of rice. A comparision with iso-osmotic effect of PEG 6000 and NaCl suggested the role of ionic toxicity in the salt mediated anomalies. Wheat and barley appear to have diverse mechanisms of salt tolerance. Essentially wheat employs compartmentation but possibly ions are freely transported upto flag leaf in barley.

One more approach which has been adopted in wheat is to screen the current collections, select and back cross the tolerant lines. Using Kharchia 65 as source of salinity tolerance, efforts have been made to eliminate undesirable features of Kharchia 65 like susceptibility to rust, red colour of grains and lodging.

Cyto-physiological studies pertaning to mapping of gene(s) responsible for Na/K discrimination have been attempted in wheat. The salt tolerance of hexaploid (AA BB DD) wheat has been attributed to a large extent to genes in the D genome and this, therefore, widens the gene pool to *Aegilops squarrosa* (*Triticum tauschii*), the donar of D genome, and other D genome carrying *Aegilops* spp. Forster et al. (1988) succeeded in obtaining a hybrid between *Agropyron junceum* and wheat. Subsequent work has involved the production of wheat- *A. junceum* disomic addition lines. It was concluded that genes on the group 2 chromosomes conferred susceptibility to salt and chromosome 5J of *A. junceum* carries major gene(s) for tolerance to salt.

Transfer of genomes from halophytic allies of wheat like *Agropyron elongatum* and from wild progenitors like *Triticum tauschii* (originally called *Aegilops squarrosa*) have been shown to result in salt tolerant amphidiploids.

The genes for production of osmoprotective molecules (*osm* genes) are present in plants but these have been studied in detail in bacteria. In nut shell, it has been shown that:

a) It is possible to convert osmo-sensitive bacteria into osmotolerant one,
b) *osm* genes conferring stress tolerance are carried as a segment of DNA comprising about 10,000 base pairs,
c) the segment codes for enzymes catalysing first two steps in proline pathway, and
d) enzyme synthesised through *osm* gene loses its sensitivity to feed back inhibition by proline, which is the requisit property important for overproduction of metabolites and imparting salt tolerance.

Novel symbiotic vectors like gall forming *Agrobacterium tumefaciens* (Ti-plasmids) and root nodule bacteria (*Rhizobia*) have delivered *osm* genes to leguminous plants. *Agrobacterium* transfers foreign genes to host cells; where as the *Rhizobium* does not transfer DNA to the host itself but functions as a chemical factory converting raw material into useful products that are passed to the host plants. In addition, a segment of DNA has also been identified that permit high level synthesis of foreign genes in root nodule bacterioids. These foreign genes are expected to enhance energy efficiency of nitrogen fixation, delivery of plant growth regulators and perhaps even pest resistance. It is indeed queer to notice that role of proline in imparting salt tolerance in higher plants is not unequivocal.

## Biosalinity-Industry Interface

The biosaline concept has proved its relevance in the modern context of biotechnology based industry. The concept was promulgated as *poor soils, high solar insolation and saline water, which prevail in lands should be viewed as useful resources rather than as disadvantages*, and that these can be used for non-traditional production of food, fuels and chemicals. Since then much information has been generated on Bio-saline concept (Pasternak and San Pietro, 1985). In addition to the application of genetics and molecular biology in developing salt tolerant glycophytes and halophytes. Reed et al. (1985) have described that filamentous blue green alga *Spirulina* spp. can be cultured on saline water as a major source of single cell protein for animal nutrition. At NFTRI, Mysore it has been demonstrated that the alga is good for human consumption also.

The presence of the organic storage compound poly-β-hydroxy-butyrate in blue green algae grown in saline media may find further commercial applications in plastic industry. Along other potentially useful chemicals which can be obtained through algal culture on saline water are sulfated polysaccharides from unicellular red alga- *Porphyridium alginates* and carrageenan from seaweeds and glycerol from *Dunaliella* which inhabit saline waters upto 5M NaCl concentration.

## From Hydroponics to Industrial Biotechnology

In 1860, it was discovered in Germany that many plants could be grown, in aqueous solutions of four salts: calcium nitrate, potassium dihydrogen phosphate, magnesium sulfate, and a trace of iron sulfate. Given with oxygen, carbon dioxide, and sunlight most plants grew well. Later it became evident that plants require eight other nutrients in trace amounts, in addition to the six macro nutrients (calcium, magnesium, potassium, sulfate, nitrate, and phosphate). These discoveries are the basis of all research in plant mineral nutrition and the formulation of inorganic fertilizers for agriculture. Thus, hydroponics, became the biotechnological bye product of this research.

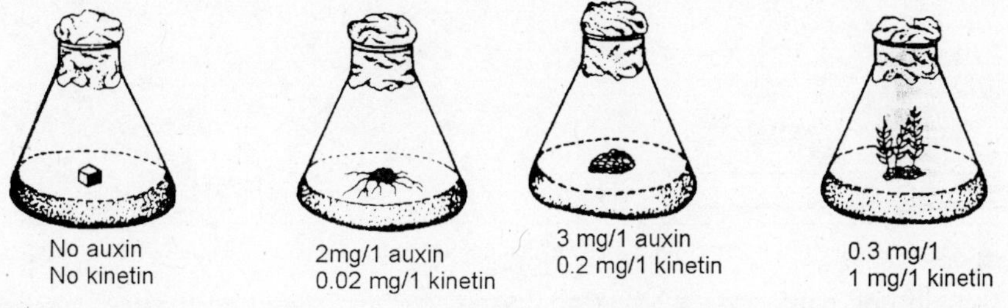


**FIG. 35.1:** Stages of callus growth in culture medium. Note the ratio of auxin and kinetin in culture media.

The nutrient film technique, an advanced form of hydroponics, is being used for commercial cultivation of vegetables like tomato, pepper and lettuce. In U.S.A. this is becoming a major industry.

The possibility of growing plant in laboratory in correctly formulated solutions opened new vistas. Among which micro-propagation through meristem culture is gaining prominence. A correct ratio of auxin and CK leads to formation of multiple shoots or clonal propagation. Clone is a group of genetically identical individual. The term cloning can be applied to organisms, cells and molecules (DNA). The cloning of plants in tissue culture, as just described, forms the basis of a whole new worldwide micropropagation industry that multiplies plants by this clonal method. Within a year, a single plant can produce millions of cloned plants that are essentially identical.

Sometimes, in meristem culture an undifferentiated mass of tissue (callus) is developed. At the surface of callus embryo like structures are formed (somatic embryo) which make shoots, roots and plants. More than a thousand different plant species have been propagated in tissue culture. The method has been used in hundreds of small and large nurseries and laboratories throughout the world. Micropropagation has been especially successful with ornamental plants, and many ornamentals sold in supermarkets are thus propagated.

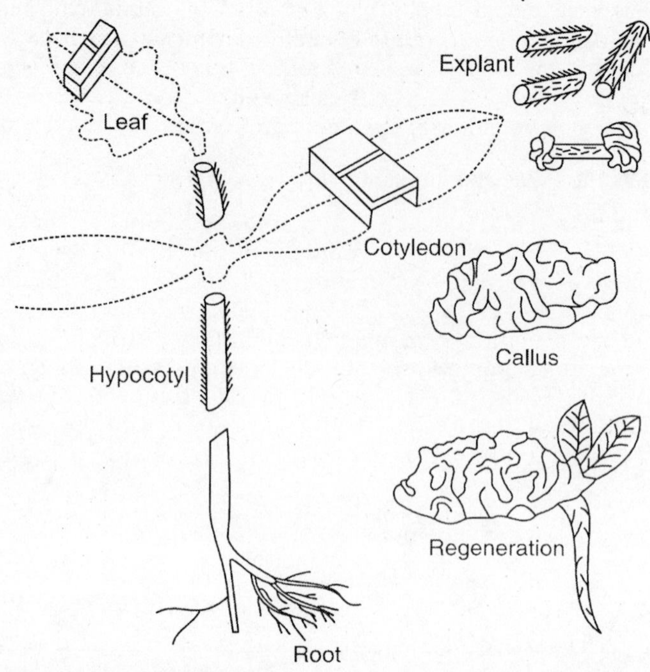


**FIG. 35.2:** Schematic representation of static callus culture of plant tissue and sources for plant parts.

The method has been extended to agricultural plantation crops, such as potato, strawberry, oil palm, and banana, as well as to medicinal and aromatic plants and trees. One obvious advantage of micropropagation is that elite specimens- especially in the case of trees- can quickly be multiplied without fear that the genetic makeup will change. When plants reproduce sexually, change in genetic makeup usually occurs. Improving tree crops via traditional plant-breeding approaches is very slow and uncertain, and micropropagation has considerably reduced the time needed. The oil palm was one of the first crops improved by applying tissue culture research.

Viruses live within plant cells and are transmitted from one generation to the next in plants, such as potatoes, strawberries, or sugarcane that are vegetatively propagated (without seeds). As viruses cause serious crop losses by weakening the plants, it is important to start with plantlets that are virus free. Procuring disease and virus free plants is another major advantage of micropropagation. For reasons that are not clear, there are generally no viruses in plant meristems, and starting from a meristem ensures that all plants derived from it also lack viruses. Commercial strawberry growers and potato producers start every year a new with virus-free planting material produced via micropropagation.

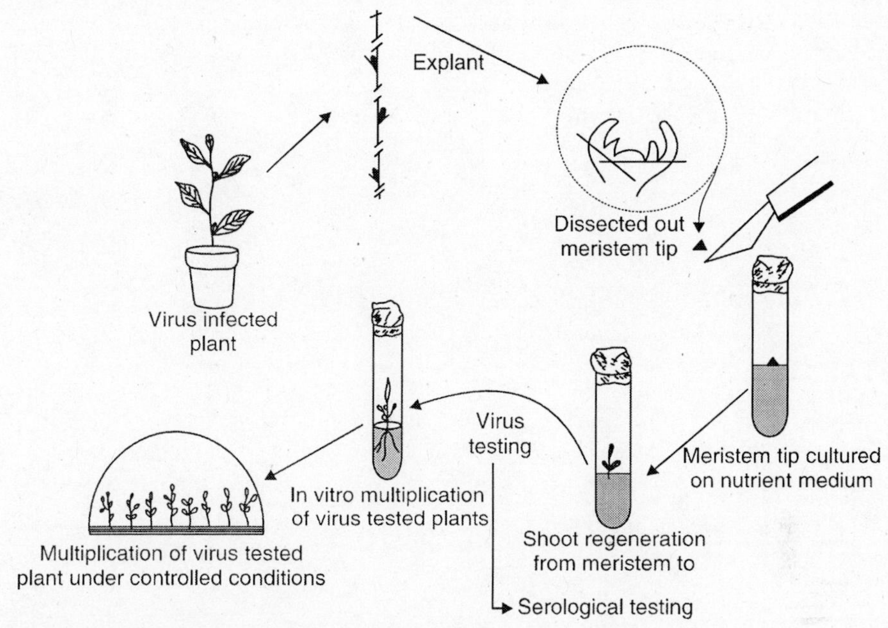


**FIG. 35.3:** Diagrammatic representation of meristem tip culture.

## Synthetic Seeds and Plant Propagules

In order to combine the benefits of clonal propagation with handling somatic embryos are converted to artificial seeds. During normal plant development, a fertilized egg cell in the female reproductive part of the flower develops into an embryo and then into a seed. Seeds are usually dehydrated and quiescent, have a strong protective coating that contains antibiotics, and contain the food reserves that are needed for initial growth of the embryo.

The technology of choice at the moment is to encapsulate each embryo in a protective hydrated gel. The kind of packaging needed will depend on the plant species, the specific agricultural application, and the physiological state of the embryo-whether it is dehydrated or not, and quiescent or not. The capsule must provide physical protection and can carry nutrients, growth regulators, antibiotics, and fungicides that all help establish the seedling in the field. Technologists envision both hydrated and dry encapsulation methods, but widespread use of such synthetic seeds with either method has not yet occurred. The technical problems that still need to be solved are mainly in the area of seedling establishment in the field.

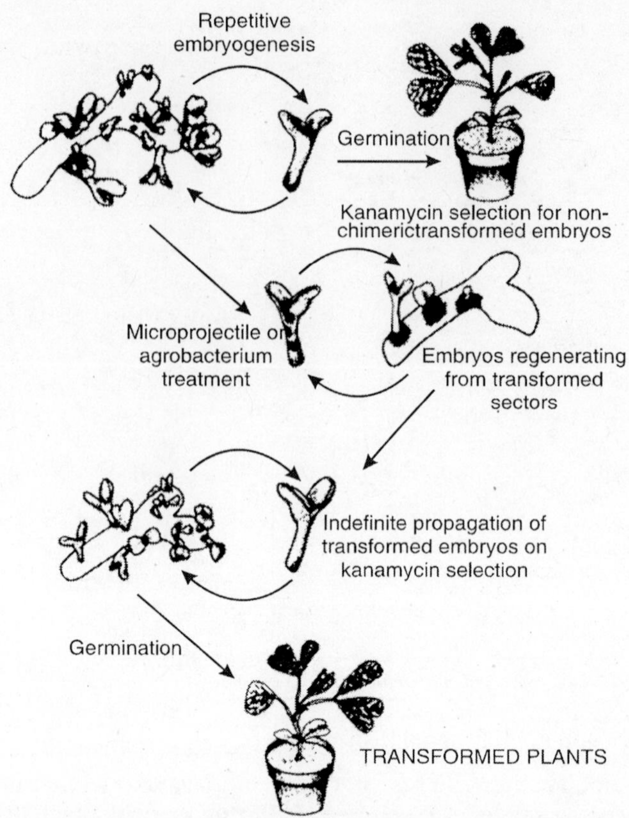


**FIG. 35.4.** Recurrent embryogenesis and its use in genetic transformation and mass propagation. Although new embryos can form from older embryos at any stage of development, this example depicts recurrent embryogenesis occurring from cotyledonary stage embryos. Developing embryos can be exposed to *Agrobacterium* or bombarded with microprojectiles, transforming individual epidermal cells. As the embryo continues to grow and develop, transformed cells give rise to patches of transformed tissue from which transformed embryos develop on selection medium. In this example, kanamycin is used as the selection agent, but several other agents could be used, depending on the vector used in transformation. As long as the cycles of recurrent embryogenesis are maintained, transformed or nontransformed embryos can be propagated indefinitely.

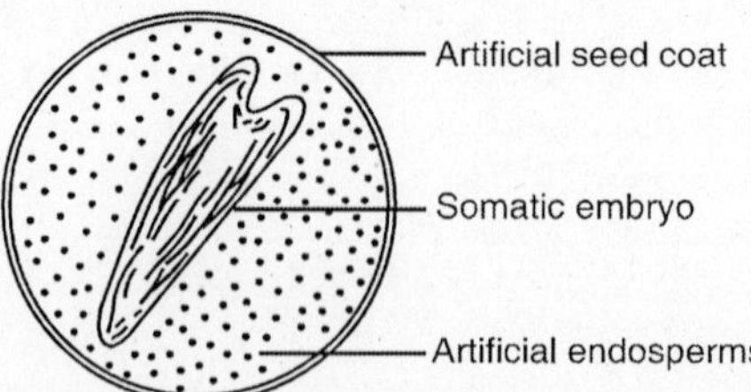


**FIG. 35.5:** The concept of artificial seed. Capsule gel with hydrophobic membrane, the (A) artificial seed coat, (B) somatic embryo, (C) artificial endosperms.

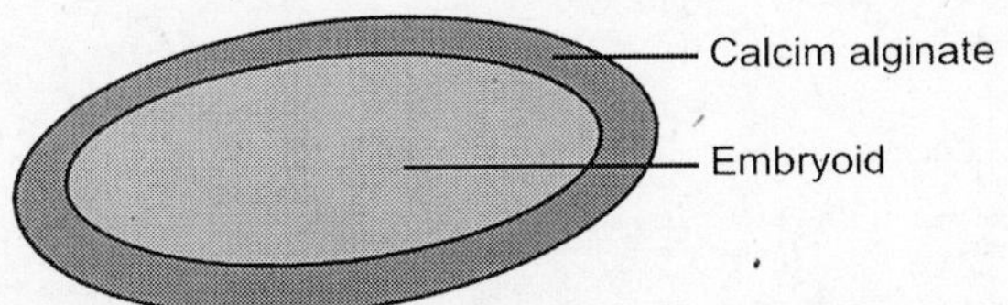


**FIG. 35.6.** Cross section of synthetic seed.

### *Steps of Commercial Artificial Seed Production*

The following steps are needed for commercial synthetic seed production:

1. Production of embryogenic tissue from transformed cells or tissue.
2. Large-scale production of synchronous somatic embryos.
3. Maturation of somatic embryos.
4. Non-toxic encapsulation/coating process.
5. Artificial endosperm/megagametophyte, depending on species.
6. Storage capability of artificial seeds.
7. High frequency, direct green house/nursery field conversion, depending on production requirements.
8. Low genetic and epigenetic variation.
9. Appropriate expression of engineered trait.
10. Artificial Seed Propagation
11. Artificial seed propagation could potentially reduce the time needed to insert a desirable gene into a productive forest, as compared to using seed as the propagation method. A considerable advantage would be to eliminate or minimise the requirements for seed production using the following process:
12. Production of large-scale embryogenic tissue from genetically engineered cells;
13. Concurrent plant regeneration, confirmation of transformation, and *progeny* testing;
14. Cryogenic storage of potential superior lines;
15. Scale-up production and maturation of somatic embryos;
16. Encapsulation of somatic embryos as artificial seeds;
17. Either greehouse/nursery establishment, growth, and transplanting into the field or direct seedling.
18. The final stage will be the evaluation of production plantations for increased yield/ performance due to engineered trait.

Unlike herbaceous species, the growth and maturity of trees far exceed the time required for tissue culture manipulations. If the engineered trait is expressed only in the mature tree and is not stable during meiosis, then mass clonal propagation would be accomplished only through tissue culture methods (or, by traditional relatively low-volume ramet production).

Cryopreservation of desirable genotypes is one of the key components for artificial seed propagation of tree species. It retains the genetic gains from genetically engineered tree species without having to establish clonal orchards. Once the progeny tests are completed, then embryogenic tissue corresponding to the superior clones can be thawed for rapid scale-up production via somatic embryogenesis.

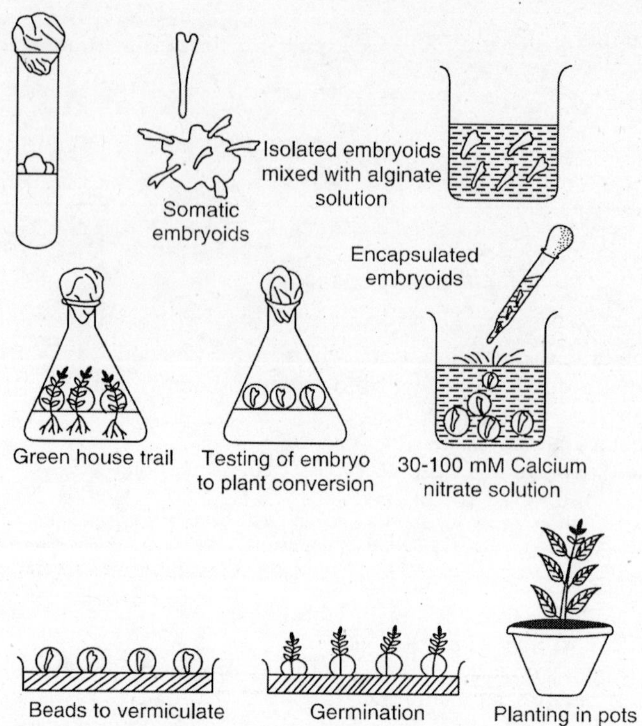


**FIG. 35. 7:** Flow diagram presenting the procedure of synthetic seed production.

Although the use of artificial seed technology should be extremely valuable for the rapid introduction of genetically engineered material into production of forests.

## Synthetic Seed and Forest Trees

The use of biotechnological approaches in forestry may be greatly enhanced and considerable time could be saved by using artificial seed technology. Genetic engineering in forestry will be similar to that for field and horticultural crops. Desirable genes should be identified, cloned and inserted into the tissue (protoplasts, cells-pollen, zygotic embryos, needle tissue, etc.). The putatively transformed tissue will be regenerated to plants and tested for expression of the genes. With annuals, the transformed individual plants can then be backcrossed with the original population for large-scale production of transformed seed within one to few years. However, for most tree species, after adequate gene expression is confirmed, scale-up production of transformed seeds deviates significantly at this point from that of annual and biennial crops because of the very long generation time for trees, particularly conifers.

Plain steps in genetic engineering of tree species focus on transformation, followed by traditional forest seed production and tree evaluation are as under:

1. Regeneration of genetically engineered tissue;
2. Confirming presence and expression of engineered gene;
3. Bulking up propagules through vegetative propagation or seed production;
4. Greenhouse/nursery establishment and growth;

5. Seed orchard establishment with concurrent progeny testing, including evaluation of the engineered trait;
6. Seed orchard roguing;
7. Seed production in seed orchards and;
8. Either nursery production with transplanting into the field or direct seedling.
9. The final stage will be the evaluation of plantations for increased performance due to the engineered trait.

## Tissues and Organ Culture as Chemical Factory

Vigorously growing calluses can also be cut into small pieces that can be shaken in a liquid nutrient medium (usually the same medium, but without the agar) in a rotary shaker. Single cells or small clumps of cells will separate from the surfaces of the callus pieces and swirl around in the nutrient medium. These cells continue to divide and enlarge, and the small clumps continuously break apart.

This process establishes a cell suspension culture. Such cultures can be handled in the laboratory or the factory in the same way as bacterial or fungal cultures can be. Bacterial and fungal cultures are widely used in the biotechnology industry for producing industrial feed stocks (chemicals used in manufacturing), pharmaceuticals (penicillin), and enzymes. Over the years, micro-organisms have been modified (domesticated) to maximize their productivity in culture. Similar developments are now under way in the plant cell suspension culture field.

Many species of plants, especially tropical plants, produce unusual chemicals that have found uses as drugs, cosmetics, flavouring compounds, or agrichemicals. Until now, these chemicals have always been extracted from the plants grown in open fields, often in tropical or semitropical regions. However, it is also possible to establish cell suspension cultures from the organs that produce these chemicals, and the plant biotechnology industry aims to grow the cells that produce these chemicals in cell suspension cultures, maximize the production of the chemicals, and eventually replace farms with factories. This process imitates the advances in microbial biotechnology that have taken place over the past 50 years.

**TABLE 35.1:** Natural products from plants and their associated industrial uses

| Plant Species | Plant Product | Industrial Uses | Industry |
|---|---|---|---|
| *Papaver somniferum* | Codeine (alkaloid) | Analgesic | Pharmaceuticals |
| *Dioscorea deltoidea* | Diosgenin (steroid) | Antifertility agents | |
| *Cinchona ledgeriana* | Quinine (alkaloid) | Antimalarial | |
| *Digitalis lanata* | Digoxin (cardiac glycoside) | Cardiatonic | |
| *Datura stramonium* | Scopolamine (alkaloid) | Antihypertensive | |
| *Catharanthus roseus* | Vincristine (alkaloid) | Antileukemic | |
| *Chrysanthemum cinerariaefolium* | Pyrethrin | Insecticide | Agrochemicals |
| *Chinchona ledgeriana* | Quinine (alkaloid) | Bittering agent | Food and drink |
| *Thaumatococcus danielli* | Thaumatin (protein) | Nonnutritive sweetener | Sweetener |
| *Lithosperum erythrorhizon* | Shikonin | Dye | Cosmetics |

## Molecular Farming

Modern plant physiologists see plants as small, cheap factories that need only water, sunlight, minerals and the right combination of additional genes to produce exactly what industry wants. Given the right genes, plants to change some of the amino acids in this variable region without affecting the overall structure of the protein. A series of such modifications, either to produce biological peptides or to change the amino acid composition of the seed has now been made. For example, a storage protein gene was synthesized in which the information encoding a stretch of seven amino acids was replaced by new information encoding a stretch of seven amino acids was replaced by new information encoding the amino acids lysine-tyrosine-glycine-glycine-phenylalanine-leucine-lysine. This modified seed storage protein gene, which already contains sequences that direct seed-specific expression, was used to transform canola, via *Agrobacterium tumefaciens*. The researchers obtained seeds from the transformed plants, purified the storage proteins, and treated them with trypsin, an economical digestive enzyme that cuts proteins next to lysine. This treatment produces new sequence, which was subsequently treated with a second enzyme to remove the terminal lysine to give authentic Leu-enkephalin.

## Somaclonal Variations in Crop Improvement

### *Somaclonal and Gametoclonal Variant Selection*

In plant cell, genetic variability develops spontaneously during tissue cultures. Larkin and Scowcroft (1981) introduced the term somaclonal variation. Plant variant obtained from tissue culture is called **somaclones**. Evans et al (1984) coined the term **gametoclonal** variation for variant clones specifically raised from gametic or gametophytic cells. Occurrence of somaclonal variation in regenerated plants is considered a rule rather than exception.

Historically, plant cell culture has been viewed by most to be a method for rapid cloning. In essence, it was seen as a method of sophisticated asexual propagation, rather than a technique to add new variability to the existing population. For example, it was believed that all plants arising from such tissue culture were exact clones of the parent, such that terms like *calliclone*' *mericlone*, and *protoclone* were used to describe the regenerants from callus, meristems, and protoplasts, respectively. Although phenotypic variants were observed among these regenerants, often they were considered as artifacts of tissue culture. Such variation was though to be due to 'epigenetic' factors such as exposure to plant growth regulators (PGRs) and prolonged culture time.

Studies concerning different aspects of somaclonal variation are important for several reasons.

1. **First**, it is hailed as a novel source of genetic variation. However, successful utilization of somaclonal variation heavily depends upon its systematic evaluation and judicious utilization in breeding programmes. This necessitates appropriate experimentation.
2. **Second**, somaclonal variation is of interest as a basic genetic process, since it contradicts the concept of **clonal uniformity**. The cells and tissues which are expected to produce true to type plants through the processes of de-differentiation, division and re-differentiation, possibly perceive the whole process as stress, as a result of which the genome, known for its plasticity, restructures itself to modulate the expression of genes as demanded by the *in vitro* conditions.
3. **Third**, somaclonal variation is unwanted when the objective is micropropagation of elite genotypes or genetic transformation that partly

involve tissue culture. Under such circumstances, prevention or at least minimisation of variation is of utmost importance. To achieve this, the frequency, nature and magnitude of somaclonal variation in relation to manipulation of media components, explant source, culture conditions etc. should essentially be understood.

Majorities of studies undertaken on somaclonal variation are confined to early generations of somaclones. Therefore, information on the nature, inheritance pattern and stability of morphological and molecular changes expressed in the advanced generations of somaclones is lacking.

The different aspects of somaclonal variation investigated so far are as follows:

1. Generation of variation
2. Characterization of variants for morphological traits.
3. Analysis of biochemical and chromosomal basis of variation.
4. Relating the variations to alterations in DNA.

### *Genetic Variations and Crop Improvement*

Genetic variation appears during or after culture in vitro. It may occur in undifferentiated cells, isolated protoplasts, calli, tissues and morphological traits of regenerated plants. Variant selected in tissue cultures have been referred as **calliclones** (from callus culture), or **protoclones** (from protoplast culture) The change occurs because of variation in chromosome number and structure. Cytological heterogeneity in cultures develops due to the following reasons:

1. The expression of chromosomal mosaicism or genetic disorders in cells of the initial explant. The chromosomal mosaicism may be pre-existing mosaicism (polysomaty - the phenomenon of polyploidisation of body cells) in explants used for culture initiation, or by nuclear fragmentation associated with first cell division of callus initiation or endoreduplication or endomitosis occurring during culture initiation and/or abnormalities of the mitotic process leading to aneuploidy.
2. New irregularities brought about by culture conditions.

Somaclonal variation is causes problem for plant propagationists, since their objective is to reserve a specific plant genotype. While for plant geneticist's somaclonal variation is a valuable source of new genetic information? It leads to crop plant improvement and a deeper understanding of the biochemical and molecular basis of inheritance (Chaleff, 1983).

Somaclonal variation is one of the aspects of tissue culture technology and is widely recommended for crop improvement especially of desired traits for the salt, drought, temperature and disease tolerance. The method refers to heritable changes that accumulate in the callus from a somatic explant and express in the progeny of in vitro regenerants obtained from callus. Even though relative data from deliberately designed experiments is not available, indications of possible advantages of somaclonal *vis-a-vis* induced mutagenesis are:

1. The frequency of variation seems to be far greater than the yield of induced mutations.
2. The changes are very subtle and may not involve drastic alteration in the genetic background.
3. Somaclonal variation occurs for trait of both nuclear and cytoplasmic origin. The variation of cytoplasmic genes obtained by this method is a distinct

advantage.

In wide crosses somaclonal variations provide a mechanism of gene introgression. Immature embryos of the wide cross can be callused and plants with the introgressed desired gene (or gene complex) are selected among the regenerants of their progenies.

### *Why Variations Occur?*

Variants may also arise as a result of more suitable changes due to single gene mutations in cultures that have cells apparently showing no karyological changes. Every possible factor that could result in a genetic change has been accounted for as a cause for somaclonal variation. Recessive mutations are not detected in plants regenerated in vitro from any cell or tissue, but expressed in progeny. This shows that variants are the mutants. Single gene mutation responsible for somaclonal variation relates to transposable elements. Transposon-induced changes have been observed in maize, tobacco and wheat. Somaclonal variation may also be due to changes caused by mitotic crossing over in regenerated plants. Such changes may also occur due to changes in organelle DNA, isoenzyme and protein profiles e.g., in wheat, potato, maize, barley and flax, etc. Changes in cytoplasmic genome have also been observed in somaclones.

Factors that contribute to somaclonal variation are of three major categories, i.e., physiological, genetic, and biochemical.

### *Physiological Causes of Variation*

Variation induced by physiological factors were identified quite earlier. Such variation are those induced by habituation to PGRs in culture and culture conditions and are *epigenetic*. They may not be inherited in a Mendelian fashion. Prolonged exposure of explanted tissue to powerful auxin such as phenoxyacetic acids (e.g., 2,4-D or 2,4,5-T) often results in variation among the regenerants. In oil palm (*Elaeis guineensis* Jacq.), plants generated from long-term callus cultures in the presence of 2,4-D show significant amounts of variability in the field. In grapevine (*Vitis vinifera* L.), embryogenic cells that have been maintained in culture for several years gradually lose their ability to differentiate and regenerate into plants over time.

### *Genetic Causes of Variation*

Tissue culture regenerants certain variations which are the result of alterations at the chromosomal level. Although the explanted tissue may be phenotypically similar, plants often have tissue comprised of diverse *cell types* of cells that is there are cytological variations among the cell types within an explanted tissue. Such a preexisting condition often results in plant, regenerated from the tissue that are dissimilar. These species are referred as *polysomatic* species. Species such as barley (*Hordeum vulgare* L.) and tobacco (*Nicotiana tabacum* L.) have been documented to possess such polysomatic tissues.

Chromosomal mutations (deletion, duplication, and somatic recombination) are the sources of genetic variations which are expressed by somaclones. Lee and Phillips (1988) have described the possible mechanisms these chromosomal changes. They pointed out that late replicating heterochromatin is the primary cause of somaclonal variation in maize (*Zea mays* L.) and broad beans (*Vicia faba* L.).

*Transposable elements* are activated during culture in explanted tissue and results in altered genotypes among the regenerated plants. The well-known transposable element complex of *Ac-Ds* in maize which activates in vitro culture.

### Biochemical Causes of Variation

Biochemical deviations are the most predominant type of variation in tissue culture. They have been noticed in barely (unless a specific test is performed). In tissue culture, several biochemical variations have been identified in various crop plants and some of these variants show Mendelian inheritance (many may be *epigenetic* and may be lost in the plant regenerated). Biochemical variations also include alterations in carbon metabolism leading to lack of photosynthetic ability (albinos in cereals such as rice), starch biosynthesis, carotenoid pathway, nitrogen metabolism, and antibiotic resistance.

Genomic DNA exhibits normal methylation patterns. Methylation is a process where a particular nucleotide – usually adenine (A) or cytosine (C) – has a methyl group ($CH_3$) attached to it. Prolonged exposure of plant tissues to in vitro culture has resulted in the alteration of normal methylation patterns (e.g., maize, potato and grapevine). However, at present we do not know why this process happens.

### Induced (or Directed) Causes of Variation

Plant improvement techniques involve screening of a large number of plants in the greenhouse or field for selection of a particular trait of interest. If one has to develop a salt-tolerant line of a particular species, a large number of individual plant must be grown and subjected to various doses of salt to eventually identify the plants that can withstand the screening process. For this purpose limited material, space and time will be available. In addition, environmental factors will also interfere with the selection process. Cell culture systems provide the breeder with the ability to select from a very large amount of genetically uniform material and to conduct the screening quickly in a few Petri dishes or flasks. This provides much greater control over the selection process.

### In Vitro Mutation

Certain crop plants such as bananas and plantains (*Musa* spp.) do not have a large genetic base and have been propagated by asexual means for thousands of years. Genetic improvement in these species is very difficult because they seldom produce fertile seeds. Therefore, one has to look either for natural somatic mutations (do occur at an extremely low frequency), or induce mutations. In such vegetatively propagated crops, even inducing mutations is rather difficult as their propagules are large as in the case of banana suckers. Attempts to avoid such large vegetative propagules result either mosaics or fatalities. Development of a cell regeneration system, such as somatic embryogenesis, provides an opportunity to expose a large number of regenerative cells to either gamma irradiation or chemical mutagen such as ethylmethyl sulfonate etc., in a very controlled manner, and thus widen the existing germplasm base (Novak, 1992).

### Applications of Tissue Culture-derived Variation

Cell culture systems offer plant breeders a well-defined environment where selection pressures can be imposed on thousands of genetically uniform single cells, each capable of growing into a whole plant. The effect of environment variation is minimized, so that escapes or adaptations that can revert back to original genetic background are also reduced. This controlled growth atmosphere in a minimal space provides the plant breeder new options for introducing variation. In addition, a scientist can study a tropical species in a temperate region or vice versa because specialized environmental conditions can be provided anywhere. Some of the important applications for induced variation are discussed below with one or two classical form the literature.

**TABLE 35.2:** In Vitro selection for disease resistance against phytotoxins

| Crop | Pathogen | Phytotoxin | Results |
|---|---|---|---|
| Alfalfa *Medicago sativa* | *Colletotrichum gloeosporioides* | Culture filtrate | Enhanced disease resistance |
| Asparagus *Asparagus offcinalis* L. | *Fusarium oxysporium f.* sp. *asparagi* | Pathogen inoculation | Increased disease resistance |
| Banana *Musa* spp. | *Fusarium oxysporum f.* sp. *cubense race 4* | Fusaric acid | Resistance clones to the disease |
| Citrus/lemon *Citrus limon* L. | *Phoma tracheiphila* | 'Mal secco' toxin | Resistant embryogenic cultures |
| Coffee *Coffea arabica* L. | *Colletotrichum kahawae* | Partially purified culture filtrate | Plants with increased resistance |
| Maize *Zea mays* | *Helminthosporium maydis* | T-toxin | Phytotoxin-resistant cells and plants |
| Mango *Mangifera indica* | *Colletotrichum gloeosporioides* | Colletotrichin and culture filtrate | Resistant embryogenic cultures |
| Oat *Avena sativa* L. | *Helminthosporium victoriae* | Victorin | Inheritable disease resistance |
| Peach *Prunus persica* L. | *Xanthomonas campestris pv. Pruni* | Culture filtrate | Resistant clones to the disease |
| Potato *Solanum tuberosum* | *Phytophthora infestans* | Culture filtrate | Phytotoxin-resistant plants |
| Rice *Oryza sativa* | *Xanthomonas oryzae* | Culture filtrate | Filtrate-resistant plants |
| Strawberry *Fragaria* sp. | *Fusarium oxysporum f.* sp. *fragariae* | Fusaric acid | Resistant shoots |
| Sugarcance *Saccharum officianarum* L. | *Helminthosporium sacchari* | Culture filtrate | Disease-resistant clones |
| Tobacco *Ncotiana tabacum* | *Psedomonas tabaci* | Methionine sulfoximine | Phytotoxin-resistant plants |
| Tomato *Lycopersicon esculentum* Mill. | *Fusarium oxysporum* | Fusaric acid | Plants with elevated resistance |

**Development of disease-resistant plants from tissue culture**

Hammerschlar (1992) pointed out the effectiveness in vitro selection for disease resistance can proceed. An effective selection agent, that can be produced and utilized in an in vitro system, must be identified. The identified selection agent should act at the cellular level and should be an important factor in the disease process. There must be a reliable protocol for regenerating whole plant from single cells for the species in question. The protocol must allow the cells to withstand several cycles of selection in a stringent environment and still be able to regenerate whole plants.In addition to these important factors, effective tools to determine if selected cells are truly resistant to the pathogen at the cellular level and whole plant level are necessary.

Phototoxins (in late 1970s and early 1980s) were employed as selection agents to impart disease resistance. How in vitro-derived resistance occur is not well known. One possible reason is that these phytotoxins are produced by pathogen in a very timely and specific manner and in very low quantities during the disease process. When plant cells are subjected to higher doses of these toxins, they not only affect the ability of the cells to resist the phytotoxin, but also cause some unwarranted genetic damage to the cells. Another problem is that sometimes regenerants resistant to the phytotoxin were not resistant to the pathogen. These results exposed problems of using phytotoxins to select for disease resistance.

One of the main findings is that there are compounds other than phytotoxins produced by the pathogen that are involved in the disease process. For instance, 'harpin' proteins produced by the pathogen have been shown to elevate plant resistance against a diverse group of pathogens. Using these compounds as a whole unit (as in a crude culture filtrate) in suspension cultures could be a better approach to bring out the true genetic resistance of the plant.

**TABLE 35.3:** In vitro selection for abiotic stress

| Crop | Selection factor | Result |
|---|---|---|
| Almond (*Prunus amygdalus* Batsch.) | Iron deficiency | Resistant shoots |
| Citrus/sweet orange *Citrus sinensis* (L.) Osbeck | Sodium chloride | Resistant plants |
| Flax (*Linum usitatissimum* L.) | Sodium sulfate | Resistant plants; inheritable |
| Grape (*Vitis rupestris* Scheele) | Sodium chloride | Resistant callus |
| Potato (*Solanum tuberosum)* | Sodium chloride; seawater | Resistant callus |
| Rice (*Oryza sativa)* | Seawater; sodium chloride | Resistant plants; not inheritable |

### *Induction of salt heavy metal tolerance through tissue culture*

Crop development for saline regions (salts such as sodium chloride or heavy metals like aluminum) is still a high priority for agriculturist. The availability of arable land is continuously shrinking. Tolerance to sodium chloride using in vitro selection has been achieved in several crop species, such as rice, potato, sugarcane, and tomato. However, in most cases the resistance was epigenetic. A few reports are available for heavy metal tolerant somaclonal variation. The identification and cloning of genes that could elevate resistance to salt and heavy metals is a major breakthrough for plant geneticists.

## *The Mechanism of Somaclonal Variation*

According to Bhaskaran (1985) variations in somaclons occur due to the following reasons:

1. The pre-existing genetic variation in the explant tissue,
2. The spontaneous mutation that can accumulate during the many division cycles that cells of the explant go through before differentiating into an in vitro plant. The recessive mutations will naturally require a method by which they can express even diploid cells. Somatic crossing over followed by segregation is a likely mechanism, for the homozygosity and thus phenotypic expression of the recessive (Chopra and Sharma, 1988).
3. Numerical and structural changes in chromosomes during in vitro growth.
4. Intracellular mutagenic agents produced during in vitro growth.
5. Activation of transposable elements or **jumping genes**, are genetic entities

which have the locus at which they get integrated is matured.

Agriculturists are very hopeful about practical advantages of somaclonal variations and they are waiting when this technique is fully integrated with the conventional plant breeding procedures.

### Source Material and Culture Conditions

Plant cell, tissues and somatic embryos developed from various explants sources for generating somaclonal variation. Explants are generally taken from any tissue, namely leaves, internodes, ovaries, roots and inflorescence. The source of explant has often been considered a critical variable for somaclonal variation. Initiation of callus and suspension cultures as described in previous chapter.

### Determination of Cell Number

Take an aliquot of suspension and filter off the cultures through a wire mesh (300 μm). Note the volume of the filtrate (F) containing single cells and small clumps and place the drop of this suspension to haemocytometer to determine the number of cells by the equation:

$$N = \frac{P \times 100 \times F}{0.1 \text{ mm}}$$

Where, N = total number of cells and clumps, P = number of cells in the squares of the haemocytometer, F = volume of the filtrate.

### Forms of Somaclonal Variation

Many different forms of somaclonal variation arise. The most common forms include point mutations, chromosomal aberrations, and increases or decreases in the number of nuclear chromosomes. It is important to realize that not all forms of variability that arise in vitro are heritable. Some morphological and biochemical variants are due to physiological effects and are not exhibited in subsequent generations.

### Detection and Isolation of Somaclonal Variants

There are several different approaches to detecting and isolating somaclonal variants from cultured plant cell populations.

1. Morphologically distinct cells such as nonphotosynthetic (nongreen) cells or cells that accumulate anthocyanins and other plant pigments are detected visually.
2. To isolate herbicide- and antibiotic-resistant variants, plant cells are simply grown on media containing of the wild-type cells in a culture.
3. The surviving cells are then subcultured and retested for growth on herbicide or antibiotic supplemented medium.
4. Through this method, one can eliminate any remaining wild-type cells that may have inadvertently survived the first round of selection.

**TABLE 35.4:** Somaclonal variation induced in morphological traits in some crop plants.

| Crop | Characters |
|---|---|
| Sugarcane | Cane diameter, stalk length and weight Cane yield, sugar yield, stalk number, length diameter, volume, density the weight |
| Potato | Growths habit, maturity period, tuber uniformity and skin color. Photoperiod requirement, fruit production. |

| Crop | Characters |
|---|---|
| Tobacco | Yield, Days to flowering, plant height, stem diameter, leaf number, leaf length, leaf width and yield Leaf Shape, leaf number, plant height, type of inflorescence and yield. |
| Rice | Number of tillers per plant, number of fertile fillers per plant, averages panicle length, frequency of fetile seed, plant stature and flag leaf length. Flowering period, plant height, seed fertility and heading date. |
| Oats | Plant height, heading date, twin culms, yellow leaf stripe, awn morphology and fertility. |
| Maize | Twin stalks from a single node, reduced pollen fertility and male sterility. |
| Brassica | Altered leaf way, multiple branching of the stem, precocious flowering from the apex, stem or leaf, abnormal leaves, reduced lamina in leaves, spontaneously absorbing vegetative buds, slow growth, failure to flower, large pollen grains. Delayed flowering altered growth habit and gross morphology. |
| Pelargonium | Leaf shape, size and form, flower morphology, plant height, fasciation, pubescence and anthocyanin pigmentation. |
| Pineapple | Leaf colour, foliage density, leaf width and leaf spine formation. |
| Tomato | Male sterility, joint-less pedicel, tangerine virescent leaf, flower and fruit colour, lethal chlorophyll deficiency mottled leaf appearance, fruit ripening and growth habit. |

We can also use this direct selection technique for isolation of temperature-resistant variants because those cells which survive in an extended incubation periods at abnormally high or low temperatures are temperature-resistant.

Those somaclonal variants that cannot be detected visually or selected directly are isolated by indirect means.

Those auxotrophic plant cells which unable to survive in absence of specific nutrient supplements not required by sensitive cells, which will not survive at temperatures above or below a certain threshold. This threshold will not affect normal wild-type cells. In absence of the necessary nutrient supplement or when grown at excessive temperatures, these cells become very weak and at this weakened state, these cells assimilate exogenously supplied cytotoxic compounds (arsenate, bromodeoxyuracil, or fluorodeoxyuridine) at lower rate than that of wild-type cells. Thus, treating a cell culture under restrictive growth conditions with one of these substances will tend to favor short-term survival of the weaker auxotrophic or temperature-sensitive cells.

### *Mutagenesis and Somaclonal Variation*

Mutagenic agents (chemical and physical) are used to produce particular heritable forms of somaclonal variation e.g., ultraviolet (UV) radiation, ethyl-methane sulfonate (EMS), nitrosuguanidine, and sodium azide, etc.

**FIGURE 35.8:** The biochemical reduction of nitrate to ammonia.

UV radiation induces dimer formation between adjacent thymine residues in DNA. This produces lesions that cause frameshift mutations and base pair substitutions during DNA replication. EMS and nitrosoguanidine are alkylating agents that cross-link and sever DNA molecules, often resulting in gross DNA alterations. In contrast, sodium azide induces single base pair substitutions or deletions, resulting mostly in small point mutations.

### Somatic Genetics of Nitrogen Metabolism

Plant cells in culture medium can metabolize ammonium, nitrate, and nitrite sources of inorganic nitrogen. The cells use ammonium directly while the nitrate is first reduced to nitrite (enzyme nitrate reductase, NR) and then nitrite is reduced to ammonium (enzyme nitrite reductase NiR). NR needs a molybdenum-containing cofactor (MoCo) for proper enzyme activityas follows:

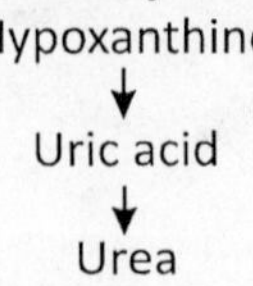


**FIGURE 35.9:** The biochemical reduction of hypoxanthine to urea.

For isolation of variants, which unable to use nitrate, plant cells are grown on chlorate-supplemented media. Chlorate is a close analog of nitrate and will be reduced to chlorite; a patent cytotoxin that accumulates internally and eventually kills the cell. Certain metabolic mutants can survive chlorate treatment. Some of these mutant cells carry mutations affecting NR expression or assimilation. Other mutations may affect MoCo expression or chlorate assimilation.

### Method for Isolation of Desired Variant Cells for NaCl-Tolerant from Callus/ Suspension Cultures

1. Semi-solid liquid media is prepared and NaCl (500 mg/l) is added to the medium, pH is adjusted at 5.6. Distribute the medium in 30-ml aliquots into 150-ml flasks or 50-ml aliquots into 250-ml flasks and grow the callus on NaCl containing medium. Select the granular friable callus for further subculture.
2. 1 g of callus is inoculated to each 250 ml flask containing 50 ml of liquid medium then incubate the cultures on a gyratory shaker (100 rpm) for 21 days at 25 ± 2°C.
3. Pass the cell suspension through a stainless steel wire mesh (300 µm) and collect filtrate in 15-ml centrifuge tubes. Examine the filtrate for single cells and small aggregates (5-20 cells). Centrifuge the filtrate at 100 g for 5 min and discard the supernatant, add 5 ml of the sterilized medium to the pellet to obtain a cell suspension, and maintain the cell density of the suspension about 0.5-2.5 × $10^5$ cell/ml. Add to the flasks containing with or without NaCl.
4. Incubate the cultures at 100 rpm at 25 ± 2°C and centrifuge the cultures grown in control and NaCl (500 mg/l) at 100 rpm for 5 min.
5. Transfer the pellets to fresh medium of the same composition. Incubate at 25 ± 2°C for 21 days and repeat these steps for 2-3 times to get a good growth rate of the NaCl tolerant cell line
6. Grow cultures on 1000 mg/l NaCl and follow the same steps, as carried out at 500 mg/l NaCl until a salt concentration beyond which the cells cease to grow. It is suggested that the salt concentration should be increased gradually by repeating steps
7. Regenerate variant cells into whole plants. Transfer 0.5 ml of cell suspension onto the surface of the regeneration medium gelled with agar by a wide mouth graduated sterile pipette.
8. In case of callus, inoculate 500 mg piece onto the surface of the organ/shoot

forming medium having 500 mg/l NaCl

9. In most of the plant species, the differentiation of shoots and roots occurs on different media. Hence once the shoots are obtained, they are transferred to the rooting medium. Transfer the selected NaCl tolerant plantlets to soil.
10. Use NaCl solution to water the salt tolerant plants. The concentration of NaCl should be the same as used in the regeneration medium. This procedure can also be used to develop tolerant plants of alkali, high temperature, drought, disease; herbicide can be developed by using in vitro selection system.

## *Applications in Plant Breeding*

Somaclonal variation and gametoclonal variation are the important sources of introducing genetic variations that could be of value to plant breeders. Single gene mutation in the nuclear or organelle genome usually provides the best available variety *in vitro* which has a specific improved character. Somaclonal variations are used to uncover new variants retaining all the favourable characters along with an additional useful trait, e.g., resistance to diseases or a herbicide. These variants can then be field-tested to ascertain their genetic stability. Gametoclonal variation is induced by meiotic recombination during the sexual cycle of the $F_1$ hybrid results in transgressive segregation to uncover unique gene combinations.

Various cell lines selected *in vitro* and plant regenerated through it prove potentially applicable to agriculture and industry specially resistance to herbicide, pathotoxin, salt or aluminium, useful in the synthesis of secondary metabolites on a commercial scale, etc.

**TABLE 35.5:** Examples of gametoclonal variation among some plants raised from gametophytic tissue cultures

| Species | Altered characteristics observed | Genetic basis | Androgenetic (A) or Callus Mediated (CM) |
|---|---|---|---|
| *Brassica napus* | Time to flowe, glucinolate content, leaf shape and colour, flower type, pod size and shape | Mutation | A |
| *Hordeum vulgare* | Plant height, yield, fertility, neck length, days to maturity | Not examined | CM |
| | Grain yield, culm length, heading date | Transgressive segregation, gametophytic selection | CM |
| *Nicotiana sylvestris* | Leaf colour, leaf shape, growth rate, DNA content, repeated sequences of DNA | Crumpled leaf is nuclear, germ amplification | A |
| *N. sylvestris (several cycles)* | Flower length, plant height, leaf shape, leaf colour, exerted stigma, capsule weight, foliar outgrowths | Nucleus genes | A |
| *N. tabacum* | Yield, grade, time to flower, plant height, number of leaves, leaf shape, total alkaloids, % reducing sugars | Not examined | A |
| | Yield, plant height, number of leaves, % nornicotine | Not residual heterozygosity | A |
| | TMV resistance, root knot resistance | Variation in expected segregation ratios | A |

| Species | Altered characteristics observed | Genetic basis | Androgenetic (A) or Callus Mediated (CM) |
|---|---|---|---|
| | Days to flower, total alkaloids, yield per plant, leaves per plant | Residual heterozygosity; no cytoplasmic effects | A |
| | Yield reduction, plant height, days to flower, total alkaloids | Nuclear mutations except plant height is maternal; not residual heterozygosity | A |
| | Yield reduction, alkaloid content, leaf size, time of flowering, leaf quality | Induced changes, not residual heterozygosity | A |
| | Heterochromatin content, nuclear DNA content | Gene amplification | A |
| | Time of flowering, plant size, leaf shape, alkaloid content, number of leaves | Residual heterozygosity | A |
| | Altered temperature sensitivity | Cytoplasmic mutant | A |
| *Oryza sativa* | Chloroplast content, time to flower, plant height, fertility | Some single genes induced in culture | CM |
| | Plant height, seed and leaf size, seed weight, tillering, flowering, % open hulls | Transmitted to progeny | CM |
| | Seed size, seed protein level, plant height, level of tillering | Transmitted to progeny | CM |
| | Waxy mutant | Single gene change | CM |
| *Saintpaulia ionantha* | RuBCase activity | Not examined | A |
| *Secale cereale x S. vavilotrii* | Fertility | Not analyzed | CM |

**TABLE 35.6:** Aspects of improvement desired in some crop plants through somaclonal variation

| Crop plants | Improvement desired |
|---|---|
| **Cereals** | |
| Rice | Tungro virus and leaf hopper resistance |
| Wheat | Quality, high temperature tolerance |
| Sorghum | Borer and shoot fly resistance |
| Maize | Delinking of phytohormone effect from reproductive development |
| **Legumes** | |
| Pigeon-pea | Resistance to pod borer, fusarium wilt, high protein content |
| Bengal gram | |
| **Oilseeds** | |
| Groundnut | Resistance to *Cercospora*, *Fusarium, Aspergillus flavus* |
| Rapeseed | Male sterility, low pungency combined with insect resistance |
| Sunflower | Self compatibility, resistance to *Alternaria* leaf blight, *Rhizoctonia*, *Fusarium* complex |
| Cotton | Insect resistance to gossypol free lines |
| **Vegetables** | |
| Potato | Virus and *Phytophthora* resistance |
| Tomato | Resistance to disease |

| Crop plants | Improvement desired |
|---|---|
| **Fruits** | |
| Grapes | Resistance to disease, short duration |
| Citrus | Higher solid content, increased shelf life |
| Watermelons | |
| **Other (s)** | |
| Sugarcane | Resistance to disease, short duration |
| Castor | Mutants with high fatty acids; oleic acid, ricinoleic acid |

The techniques used for development of somaclonal and gametoclonal variation are relatively easier than recombinant DNA technology and is the appropriate technology for genetic manipulation of above cited crops.

**TABLE 35.7:** Somaclonal variation induced in morphological traits in some crop plants

| Crop | Characters |
|---|---|
| Sugarcane | Cane diameter, stalk length and weight Cane yield, sugar yield, stalk number, length diameter, volume, density the weight |
| Potato | Growth habit, maturity period, tuber uniformity and skin color. Photoperiod requirement, fruit production. |
| Tobacco | Days to flowering, plant height, stem diameter, leaf number, leaf length, leaf width and yield Leaf Shape, leaf number, plant height, type of inflorescence and yield. |
| Rice | Number of tillers per plant, number of fertile fillers per plant, average panicle length, frequency of fertile seed, plant stature and flag leaf length. Flowering period, plant height, seed fertility and heading date. |
| Oats | Plant height, heading date, twin culms, yellow leaf stripe, awn morphology and fertility. |
| Maize | Twin stalks from a single node, reduced pollen fertility and male sterility. |
| Brassica | Altered leaf way, multiple branching of the stem, precocious flowering from the apex, stem or leaf, abnormal leaves, reduced lamina in leaves, spontaneously absorbing vegetative buds, slow growth, failure to flower, large pollen grains. Delayed flowering altered growth habit and gross morphology. |
| Pelargonium | Leaf shape, size and form, flower morphology, plant height, fasciation, pubescence and anthocyanin pigmentation. |
| Pineapple | Leaf colour, foliage density, leaf width and leaf spine formation. |
| Tomato | Male sterility, jointless pedicel, tangerine virescent leaf, flower and fruit colour, lethal chlorophyll deficiency mottled leaf appearance, fruit ripening and growth habit. |

### *The Future*

Induced variation still is the best route in perennial crop improvement, although one can argue in favor of the currently untapped potential of genetic transformation. However, it must be noted that tissue culture-induced variation does not have the socio-ethical hurdles like GM crops (see Chapter two). In addition, there are not have the any significant technology ownership issues as has become problematic with genetic engineering. Further, gene transfer techniques, though successful in herbaceous species, still have not been commercialized in perennial and woody species. Molecular techniques have greatly aided in understanding the plant cell response to biotic and abiotic stresses at the subcellular level. The future lies in utilizing these techniques to induce the species' own resources, such as disease resistance, to our benefit.

## Transgenic Plants

Words and terms are frequently the root of problems associated with GE. *Transgenic* sounds very much like *eugenic* and could, therefore, be condemned by association, and *engineered* sounds more sinister than *domesticated* though they are virtual synonyms.

Transgenic plants refer to plants in which functional foreign genes have been inserted. They help understanding the mechanisms of gene expression. This could be possible only by using tissue culture and recombinant DNA techniques (efficient gene cloning methods, improved vector systems, organ specific promoters, marker genes, cloned DNA etc.). The techniques for transferring foreign genes into plant cell regenerating the transformed cells in fertile transgenic plants were established by Bevan et al, (1983) and Fraley et al, (1983). Thus, the year 1983 was a watershed for plant biotechnology. The technical break through fueled not only activities in basic research on the mechanism of plant gene regulation, but also intensive exploitation of its biotechnological application for crop improvement (Chua, 1994, 1996).

During the last 17 years, dramatic progress has been made in the application of transgenic technology for crop improvement. Because the transfer of one or two genes can be done reproducibly and routinely with this technology it is not surprising that the first generation experiments in plant biotechnology entailed the transfer and modification of single gene traits. These experiments were directed towards two commercial target areas: plant protection and quality enhancement.

### *Advantages of Transgenic Plants*

1. Agronomic traits can be improved by producing plants with increased resistance to pest, disease, stress along with increased vigor, yield.
2. Enhanced efficiency of physiological process like photosynthesis and improved nutritional qualities.
3. Transgenic plants are also used as an analytical tool to explore aspects of gene regulation.
4. Major advantage of molecular breeding is that when a particular gene has been isolated and reconstructed it can be first in model plants and later it can be used in a variety of cultivars of different crops.

### *Other Uses*

#### Environment

Transgenic plants may be used as a means of bioremediation i.e. decontamination of polluted land. Transgenic Liriodendrom tulipifera (yellow poplar) expressed bacterial mercuric reductase gene (Rugh et al., 1988). These plants survive in high Hg levels and release elemental Hg.

#### Industry

**Biomass/fuel:** Biomass producing crops include rapidly growing tress used for fuel or products (ethanol). Transgenic plants for this aim either overexpress cellulose synthase gene (Arioli et al., 1998) or cellulases that convert cellulose to ethanol (Lebel et al., 1998).

**Lognin/paper:** Cellulose is raw material for paper. Lignin is separated from cellulose fibers during pulping. Transgenic plants with modified lignin (Lapierre et al., 1999) show ease of lignin solubilization and fragmentation. The only difference is the red brown coloured wood.

**Plastics:** High value polymeric compounds like carbonyl include non calorific sweetness like fructan produced in transgenic sugar beet (Seveneir et al., 1998). Biodegradable plastics include polyesters of 3-hydroxy acids i.e. polyhydroxy alkanoates (PHA).

**Terpenoids:** Aromatic qualities of essential oils is a feature of monoterpenes and sesquiterpenes belonging to bionging to bio prenoid group of compounds. Construct of c-DNA library from oil gland secretory cells of peppermint has allowed menthol. Another way increasing essential oil synthesis is by increasing density of mint glandular trichomes.

**House hold- (clothes and ornamentals)**

Transgenic approach on cotton include introduction of pigment compounds and synthesis of plastics like PHB in same cells. Coloured fibres avoid dying process whereas plasticity lowers thermal conductance. Genetic manipulation of ornamentals can produce coloured flowers that don not occur naturally. Vase life is increase by generating the anti sense down regulation of an ethylene biosynthesis gene.

**Food and Drink**

**Protein, starch, oil and vitamins:** Proteins can be modified either by changing the amino acids in a protein or by adding entirely new protein. Another application is expression of human milk proteins in edible plants. For example, gene for beta casein protein can be introduced in potato under the control of an amino auxin inducible promoter. Similarly lactoferrin (Salmon et al., 1998) proteins have been expressed in transgenic tobacco. It has antibacterial, anti fungal and antiviral properties.

Degree of branching in carbohydrates can be altered (Heyer et al., 1999) and degree of saturation and chain length of oils can be altered (Murphy, 1999) to produce transgenic plants useful for human and industry.

To overcome vitamin A deficiency psy gene encoding enzyme phytoene synthase can be introduced in plants deficient in vitamin A, production of carotenoid (vit A) is enhanced. Similarly, content of micronutrients can be increased. Endosperm of transgenic rice had high levels of Fe when gene encoding soyabean ferritin (an iron storage protein) was introduced and expressed under control of rice seed storage protein promoter (Goto et al., 1999).

**Allergens/Toxins:** Allergens can be successfully reduced in transgenic plants. Example, level of allergenic proteins (eg. AraHl) in peanut can be reduced. Similarly, caffiene level can be reduced by antisense down regulation.

**Beer:** Transgenic barely is generated involving the modification of the glucanase enzyme as means of improving the efficiency of brewing process. (Jensen et al., 1996). Introduction of leuco anthocyanidin reductase gene improves malt quality.

**Medical and veterinary applications**

**Vaccines/antigenes/antibiotics:** This area include production of plant based vaccines and other medically related compounds. Transgenic potato have been successful in expressing human insulin (auto antigen) either alone or as a protein in which insulin was linked to c-terminus of cholera toxin B sub-unit (CTB).

**Enzymes:** Production of active enzymes is another area of interest. "World's most expensive drug" is the human lysosomal enzyme glucocerebrosidase, the only treatment for patients suffering from Gaucher's disease. It is purified from human placentae. Thus, it's production in plants is an attractive option.

### *Agriculture development and transgenic crops*

The 2nd Biotechnology India-2000 conference concluded in Delhi with the hope that at least a number of transgenic varieties in cotton, rice, brassica, pigeonpea, moongbean, improved varieties of wheat, tissue culture regeneration for important species culture like mango, saffron, citrus, neem would soon be produced in the country in the coming years for major commercial activity.

It is estimated that in the 21st century, biotechnology based industrial based industrial revolution will surpass will all other areas and may contribute between 40 to 50 per cent of the overall industrial production. The micro-propogation technology would provide high quality planting materials to the farming community. New biocontrol agents and bio-fertiliser packages would be made available to the farmers.

India would be in a position to fully utilize on a sustainable basis, the bio-resource particularly the medicinal and aromatic plants. Plants as bio-reactors would be able to produce large number of proteins of therapeutic value and many other important items. The recent discovery of the gene for recalcitrant species is a landmark event. The molecular biology route for development of new diagnostic kits and vaccines for major diseases, emerging and re-emerging infections would make the healthcare system much more efficient and low-cost.

Viewed in the global context, this millennium would usher in an era to conquer cancer, grow new blood vessels in the heart, block the growth of blood vessels in tumour, create new cells from stem cells and have designer crops with desired traits to give plants of nutritional value and higher productivity plants as bio-reactors. New vaccines and diagnostics would be produced through recombinant DNA technology.

With the introduction of new innovations in biotechnology in the form of novel transgenics of all types, bacterial factories to use the plant genes as source of bio-chemicals, pharmaceuticals, enrichment of genetic diversity, oral vaccines, other recombinant vaccines, tissue culture as a major commercial venture, application of DNA fingerprinting techniques especially for endangered species and many others would be available to mankind. It is expected that the innovations in biotechnology would lead to the development of easily biodegradable plastics, pre-cursors from plants substituting the petroleum as the current source of plastic. Bio-processing techniques would have a major role in the production of fuels from renewable sources such as plant materials animal waste.

India has developed core competence in handling sterile fermentation processes in handling microbes and isolation methods, in extraction and isolation animal products, in plant and animal breeding and infrastructure in fabricating bio-reactors and processing equipment. (Financial Express 04-12-2000).

### *Transgenic are not terminator crops*

The most debated topic in agriculture and biotechnology is genetically modified (GM) crops, that is, transgenics. Genetic engineering (GE) techniques allow the transfer of a gene of interest from one species to another. It is nothing but the extension of conventional plant breeding. This technology helps the breeders and molecular biologists to introduce the only gene of interest with more selective modification and represents a significant advance.

**Terminator technology** is a technique of genetic engineering by which the plant is altered or modified in such a way that it produces sterile seeds, which will not germinate in the next generation. Normally, seeds have the ability to germinate generation after generation. In terminator seed, a toxic gene becomes active and is forced to die after one generation of survival. In terminator technology, three gene constructs are introduced through GE techniques to express the desired character, sterility.

The three gene constructs or stretches of DNA are referred as gene X (lethal gene), Y and Z. This entire system is under the control of an external inducer, that is, tetracycline. If only the lethal gene is inserted, then the first generation itself will not germinate. Initially, the lethal gene is prevented from being expressed for seed multiplication and thereafter, gene Y helps to excise out the 'blocking sequence' to make the gene X (ledder the control of a different gene, Z. The lethal gene becomes functional only at a later stage of seed development and allows the gene to produce toxin that kills the embryo. Ultimately, the seed harvested by the farmer lacks germination ability but according to paten claim, the seed or fruit can be used as food.

Terminator technology is developed to serve following two main purposes:

1. To force the farmers to buy seed each year for raising the next crop.
2. To prevent the piracy of the introduce special genes (that is biotic and abiotic resistance or nutritional characteristics) by researchers and multinational companies.

## Legacy of Green Revolution

Green Revolution- the introduction of high-yielding varieties of wheat and rice- had about run its course, not only in Third World countries, but also in the developed world, where semidwarfs greatly boosted wheat production. Countries such as India, which once faced chronic food shortages, today have substantial grain reserves. Indonesia, once the leading rice importer, became a rice exporter. Yet the benefits of the Green Revolution were not felt everywhere. The high-yielding varieties were adopted by farmers who had access to markets, and who could afford fertilizers, pesticides, irrigation machinery, and small tractors. These inputs are required to reap the most benefits from the Green Revolution strains. According to a report by the Worldwatch Institute of Washington, DC, high-yielding varieties were being grown on 36% of the grain-producing lands in Asia and the Middle East, 22% of Latin America's grain lands, and only 1% of Africa's grain lands. Whereas per capita food production has increased in Asia and Latin America, it has declined in Africa. Africa gained the least from the Green Revolution, in part because it has large areas with sufficient rainfall but poor soils, and other areas with reasonable soil but insufficient rainfall. In addition, its population cannot afford the necessary inputs. The Green Revolution emphasized wheat and rice, but the staples of Africa are maize, sorghum, millet, cassava, sweet potatoes, and yams- all crops that do well on marginal lands. On a worldwide scale, as many as 1.5 billion people live on such marginal lands, where they practice a rainfed subsistence agriculture, in which food crops are raised for local consumption and not for export.

In nut shell, the Green Revolution is the breeding of high-yielding varieties of wheat and rice and their use in Asia and Latin America to raise food production. These new varieties have performed well with the agricultural inputs (fertilizers, pesticides, and mechanical energy) that are characteristic agriculture in developed countries and quickly replaced traditional crop varieties. The breeding of these high-yielding varieties was a planned international effort supported by U.S. foundations and governments of developing countries. Quick adoption of these varieties led to dramatic increases in grain production. Grain importers became grain exports. The export of grain was possible, even necessary, because the local people did not always have the system. Although long-term effects on humans have not been studied, the known environmental side effects of DDT led to a severe restriction of its use in the United States in 1972 and to its replacement by the nonpersistent organophosphates. However, although the organophosphates are not persistent, they are extremely toxic to humans and this is their principal drawback. Following are the important residual effects of Green Revolution.

### Human Health Risks

These are associated with the use of these potent synthetic insecticides. Although many consumers are concerned about pesticide residues on the produce they buy in the store- at least in developed countries- that is probably not the major health problem associated with pesticide use. Every year nearly 1,000 cases of pesticide poisoning are reported in the United States alone, and two-thirds of these are caused by organophosphates. The actual incidence is undoubtedly much higher. The resurgence (bounding back) of the pest population, after the crop has been sprayed, occurs naturally because the food supply-the crop-is still present. Resurgence is often fast because the pesticide killed the predators of the target pest along with the target pest itself. For example, when potatoes are sprayed with guthion or methoxychlor to kill the green peach aphid, the pesticides also kill lady bird beetles and lacewing larvae, two organisms that prey on green peach aphids.

### Secondary Pests

These are created when the populations of insects that did no damage start to increase because their natural enemies have been destroyed. For example, spraying of pesticides on cotton fields to kill the destructive boll weevil also killed the natural enemies of the cotton budworm and the cotton bollworm, which now became major pests of cotton. Nearly all the species of insects now considered serious pests in California were not pests before the days of chemical pest control. They became pests as a result of chemical pest control mesures. Sometimes insects are kept in check by fungal diseases, and fungicide spraying can cause insect pest numbers to rise, as happened with velvet bean caterpillars and cabbage loopers in soybean fields.

### The Emergence of Resistant Pest Strains

This is another major consequence of chemical pest control, although resistant pest strains also arise regularly when genetic pest control (new resistant crop strains) is used. Any regimen that causes extensive mortality and acts on a large population of organisms is a strong selective agent. Already more than 600 species of insects, weeds, plant pathogens and nematodes have evolved resistance to chemical pesticides. In many cases, the pests have evolved resistance to more than one pesticide. Development of resistance is unavoidable and as natural as evolution.

### Land Degradation

The term *land degradation* is used to describe the physical and chemical changes that reduce long-term soil productivity. Such changes are sometimes very obvious, but are often difficult to measure because they occur over long periods of time. They are often compensated for by other improvements, such as more fertilizer or new genetic strains, and as a result the effect of land degradation does not always show up as decreased yield- at least, not in the short run.

Physical degradation takes the form of erosion, the carrying off of soil particles by wind and water. In an annual cropping system, the soil is alternatively covered with plants and almost totally bare. When the soil is bare, it is exposed to higher wind velocities and to the force of raindrops, which destroys the structure of the uppermost layer of topsoil. The net result is muddy runoff that carries away soil particles and/or dust storms.

Chemical degradation can take several forms, including (1) acidification from acid rain, (2) alkalinization and salinization (building of salts) as a result of irrigation, (3) exhaustion of the mineral nutrients when insufficient fertilizers are used to replace the minerals that are removed with the crop, and (4) leaching of excess mineral fertilizers in the streams, lakes, and groundwater.

### *Erosion of Traditional Wisdom*

#### In Energy Conservation

In old days, farmers used their own seeds or other planting material for vegetatively propagated crops such as potatoes and employed animal power and human labor. They bought very few things, and the value of the purchased inputs was only 5-10% of the value of the outputs (agricultural products). In high-input farming, this value is now 50% or higher. Farmers buy seeds, fertilizers, pesticides, herbicides, tractors and other machinery, and irrigation equipment, and often must provide and maintain housing for seasonal workers. All these purchased inputs depend heavily on the use and therefore on the price of energy. Most of this energy comes ultimately from coal, natural gas, or oil, our primary energy sources.

#### In Germplasm Conservation

A main feature of our high-input system of agriculture is genetically uniform crops, because new strains, more resistant to diseases are more responsive to fertilizers. Those farmers who start planting seeds furnished by seed companies or state agencies, stop using and usually discard their own varieties which were well adapted to the microenvironment on their farm. Local varieties contain valuable genes that are thus lost to the plant breeders. Second, genetic uniformity means crops are more susceptible to sudden outbreaks of disease. When a pathogen mutates and overcomes the plants resistance, the entire crop over a wide geographical area can be ravaged.

### *Dependence on Subsidies*

A final area of concern is government policies that actively promote high-input agriculture and prevent sustainable practices and technologies, is the growing dependence of farming community on government aids, subsidies and insurance.

## Exploring Sustainable Alternatives and Role of Plant Physiologists

The high-input agricultural system now used in most developed and developing countries like India has many undesirable side effects, and the high price of purchased inputs makes it ever more difficult for farmers to earn a fair return on their labor, in spite of dramatic increases in subsidies. To protect their resource base- the soil- and to decrease costs, a number of farmers in the United States and elsewhere have begun to adopt alternative practices.

In Indian states namely, Gujarat, Tamil Nadu, Maharashtra and Rajasthan people are getting inclined toward eco-friendly agriculture, use of indigenous herbal plant protectants, preservation of traditional farming wisdom and organic farming. The more widely used term *organic farming* describes two major aspects of alternative agriculture:

1. the substitution of manures and other organic matter for inorganic fertilizers, and
2. the use of biological and herbal pest control instead of chemical pest control.

The objective of sustainable agriculture is to sustain and enhance, rather than reduce and simplify the biological interactions on which production agriculture depends. Alternative agriculture is not a single system of farm practices, but encompasses many farming systems also called *organic*, *biological*, *low input*, *regenerative*, or *sustainable* agriculture.

Many farming techniques are useful in sustainable agriculture. Important among these are crop rotation, green manuring, animal manuring, conservation tillage and integrated pest management.

### *Sustainability Concern and Plant Physiologists Role*

Although skeptically, a room for plant genetic engineering in sustainable agriculture is being envisaged. Plant growth regulators, largely because of their use in low concentrations, re a group of agrochemicals which are benign and non traumatic. Thus, their use does not contradict priorities of sustainability.

Among other ideas likely to contribute in sustainability are-

1. Eco-friendly agriculture
2. Conservation of indigenous plant biodiversity and its employment as herbal plant protestants.
3. Resurrection of traditional farming practices and their on farm and laboratory verification.
4. Promotion of individual and collective plant physiologists (plant physiology) can participate in these activities both as a scientists and a human being concerned about environment and sustainability.

## Practical Uses of Plant Growth Regulators

### *PGRs of Microbial Origin*

Microorganisms are sources of biologically active, secondary metabolites that are diverse in structure. These are fermentation products that may be used as pesticides or which may be used as novel templates for the synthetic production of agro-chemicals. There are many biologically active natural products to be discovered and the impetus for accessing new compounds will depend upon two factors; the protection of the environment and the necessity for new agrochemical templates.

With the exception of the brassinosteriods, all of the major plant growth regulators have ben isolated from microorganisms: IAA from *Rhizopus suinis* and *Neurospora crassa*; $GA_3$ from *Gibberella fujikuori* (*Fusarium moniliformae*) and other GA's from *Sphaceloma* and *Elsinoe* species; ABA from *Cercospora cruenta*, *Botrytis cinerea*, *Rhizoctonia solani*, *Fusarium oxysporum* myriad fungi; trans-zeatin; trans-zeatin riboside, and 6-(4-hydroxy-1,3-dimethyl-but-trans-2-enylamino)-

9-ß-D-ribofuranosylpurine from *Pseudomonas syringae* pv. *savastanoi*. However, there exist many microbial metabolites that have either plant growth regulatory or phytotoxic activity, e.g., the cytocholasins are a potent group of compounds. Cytochalasin H inhibits flowering in tobacco while its congener, pyrichalasin H, inhibits root and shoot growth in rice.

More complex molecules, specifically AM toxins I, II, and III are selectively phytotoxic at 0.1 ppb against apple and represent a class of cyclic peptides that are under investigation. Tentoxin, while it induces chlorosis in certain plants, is a cyclic peptide whose synthetic fragments induce auxin-like activity in plants. Moniliformin, a phytotoxic metabolite of *Fusarium moniliforme* has been derivatized and its analogs patented as herbicides. These examples, taken from an ever growing list of biologically active products, suggest that parent molecules and their synthetic, or biotransformed products may be useful in agriculture and plant physiology.

### The Herbicidins

Bacteria also offer unique sources of potential herbicides and some of the most potent and target specific are the herbicidins from *Streptomyces saganonensis* which are purine analogs. While only five have been elucidated to date, and these are A, B, E, F, and G, the implications are that at least two more exist (C and D). Their structures have been fully reported but revisions of the structures have been made. Unfortunately, only the effects of herbicidins A and B have been reported but these have been extensively tested against both monocotyledonous and dicotyledonous weeds and economic crops. Barnyard grass, goose grass, manna grass, and green panicum were controlled with treatments of herbicidin A and B at concentrations from 37.5 to 300 ppm. Both compounds controlled common purslane, *Achyranthes*, white goose foot, smart weed, wild amaranth, Asiatic day flower, tomato and radish when they were treated with metabolites is that they were not toxic to rice plants and they controlled a major rice pathogen, *Xanthomonas oryzae*.

ABA has been disappointing as an agrochemical for field application. However, the recent discovery of ABA and its metabolites in fungi has led to synthetic modification of the molecule primarily for establishing the exact conformation of these novel congeners. This, in turn, has led to derivatives possessing biological activity. Of these, the (2Z) deoxy-ABA and (2E) deoxy-ABA have been tested on lettuce and rice seed. The (2Z)-isomer weakly inhibited rice seedling growth at $1 \times 10^{-3}$M while germination of lettuce seed was greatly inhibited at $1 \times 10^{-4}$M (~100%). The (2E)-isomer inhibited rice ~30% at $10^{-3}$ M and inhibited lettuce ~100% at $5 \times 10^{-4}$M. Because of the increased activity in the synthetic modification of ABA and its analogs the possibility of finding a suitably active product for practical application is possible.

### The Evolution of Single Herbicide Resistances

Herbicide resistances have evolved in a large variety of weeds to two types of herbicides having single modes of action:

1. Herbicides inhibiting electron transport in PSII (mainly s-triazines, uracils and some but not all phenylureas),
2. Herbicides inhibiting microtubule formation by complexing with tubulin (dinitroanilines).

One type of cropping system coupled with one particular property of the herbicides seem to be responsible for the rapid evolution of resistance. Rapid is a relative term; it took 8 to 10 generations of selection with both DDT on flies, and atrazine on weeds to evolve resistant populations; in the first case this happened in a year, in the second 8-10 years. All cases of resistance appeared in either monoculture where a single herbicide was used, or in rotation where only herbicides with the same site of action were used. In all cases the herbicide had high persistence and thus a high season long selection pressure that did not allow later germination of weeds.

Resistance has yet to evolve where crop-rotations were used; where herbicides of different sites of action were used as mixtures or sequentially, or where herbicides with low persistence were used not more than few times per season. Resistance to atrazine had been predicted to evolve in rotational situations in maize, based on the paucity of information available a few years ago. It is now clear that the genetic unfitness of triazine-resistant weeds causes a decimation of selected plants during the rotational periods. This results in a much lower rate of evolution than had previously been considered possible, and accentuates the value of rotating both crops and herbicides to delay resistance.

### *Biochemical Mimicry in the Weeds*

As the grass weeds evolved resistance to all the wheat selective herbicides, irrespective of their site of action, the most likely hypothesis is that they evolved the same biochemical detoxification mechanism as wheat, i.e. evolved a biochemical *mimicry*. This is supported by evidence that compounds that suppress the herbicide degradation in wheat also suppress their degradation. The weeds evolved spectra of cross resistances that are broader than the resistance in wheat. The debate is on whether we should employ PGRs or broaden genetic base to nullify effects of evolved cross resistance.

### *PGRs as Grass Mowers*

PGRs are reported to suppress grass growth any where from 4 to 10 weeks after application, occasionally with a period of post-retardation enhancement. Seedhead suppression has ranged from none to complete season long control. The various types of foliar injury reported include chemical phytotoxicity, discolouration, droughty appearance, off colour and tip dieback. However, injury often does not occur, but has not been predictable. Suppression of new leaves, tillers, stolons, rhizomes and roots have been observed. Because biological responses have been erratic, commercial acceptance of PGRs has been limited.

These PGRs suppress both growth and development including most of the stages. The chemicals in this group act as inhibitors of cell division or cell cycle. Examples include amidochlor, EPTC, mefluidide, chlorflurenol, and maleic hydrazide. These regulators suppress leaf growth and initiation of new turfgrass leaves for about six weeks. High doses do not kill the meristems of the plant, but some result in direct and immediate injury to the leaves.

These PGRs suppress growth only. The developmental sequence of the plant continues, however, new plant organs develop in miniature size. Examples of this type include paclobutrazol, fluprimidol, uniconazole, and tetcyclasis. These compounds have primary action as GA biosynthesis inhibitors and are far more effective in suppressing internode elongation than leaf size.

Fungicide Type II growth regulators are those that are primarily used as a fungicide but have been observed to modify the host. Darker green colour and growth reduction has been observed to modify the host. Darker green colour and growth reduction has been observed with triadimefon and fenarimol`. The latter is recommended for use on putting greens for selective suppression of annual bluegrass.

These PGRs will inhibit development only. There are no examples to date, but the possibilities warrant a continued interest in their discovery. A Type III PGR would be extremely valuable in turf culture as proper timing would result in elimination of the stemmy phase of mowed turfs and the unsightly seedheads of unmowed turfs.

### *Practical Utility of Ethylene Releasing Compounds*

The stunting of ethylene-treated plants noted by Neljubow is now the basis of one of the most important uses of ethephon, the prevention of lodging in small grains and other crops. The regulator treatment not only reduces length of the stems, but also stiffens the straw. Knapp et al. examined the effects of ethephon treatment on carbohydrate content of the straw. Although straw from ethephon-treated plants contained more water soluble carbohydrates, they did not find the anticipated increase in structural carbohydrates. They concluded that the arrangement and interaction of the various structural carbohydrates and lignin in the culm cell walls may be more important in lodging resistance than concentrations of these compounds. Researchers have now shown that similar benefits can be obtained from ethephon application to field corn where application greatly reduced lodging and accelerated maturation; a reduction in yield was also noted in some varieties. The floral axis in bulb-type flowers is often too long for their use as potted flowers. Application of ethephon to hyacinth and narcissus plants when the leaves were 10 cm long reduced the length of the floral scape by 20-30%. The reduction in scape length was associated not only with a reduction in cell size, but, surprisingly, also with an increase in cell number, indicating that ethephon application had actually stimulated cell division.

For many years it has been known that ethylene stimulates flowering of Bromeliads, and treatment of pineapple plants with calcium carbide (which releases acetylene on hydration) was widely used commercially to stimulate flowering of pineapples. This response is now elicited in pineapples and flowering bromeliads by application of ethephon. A comparison of the effectiveness of ACC and ethephon in eliciting this response suggested that ACC was the preferred ethylene-releasing compound, but because of differences in application method and concentrations, this interpretation needs confirmation.

**TABLE 35.8:** Uses approved for (2-chloroethyl)-phosphonic acid

| Use | Approved crops |
|---|---|
| Plant height control | Barley, daffodils, hyacinth, wheat |
| Stimulation of branching | Azaleas, geraniums |
| Flower induction | Pineapple and other bromeliads |
| Sex expression | Cucumber, squash |
| Flower bud development | Apple |
| Defoliation | Cotton, roses, buckthorn, apple |
| Fruit loosening | apples, blackberries, Cantaloupes, cherries |
| Fruit removal | Apples, crabapples, carob, olive |

| | |
|---|---|
| Dehiscence | Walnuts |
| Postharvest fruit ripening | Bananas, tomatoes (FL) |
| Preharvest fruit ripening | Tomatoes, peppers |
| Maturity and/or colour development | Apples, figs, grapes, peppers, pineapple, tomatoes |
| Degreening (postharvest) | Lemons |
| Leaf curing | Tobacco |

Other tropical fruit crops are also stimulated to flower by treatment with ethylene. Smoky fires are, for example, commonly lit in mango orchards to stimulate flowering. Application of ethephon has also been shown to stimulate flowering in Litchi, although the results vary widely depending on application time and cultivar.

The effects of ethylene in stimulating abscission are the basis of important commercial uses of ethylene-releasing compounds. Enhancement of fruit abscission in sour cherry, nut fall in macadamia, the removal of leaves from potato plants, cotton, and nursery stock, and the dehiscence of the shucks of walnuts and pecans are but a few examples of the potential uses of these materials in agriculture. Defoliation of rutabaga required very high concentrations of ethephon; addition of 1% ammonium peroxydisulfate greatly increased the effectiveness of the ethephon treatment, perhaps due to accelerated oxidation of IAA in the abscising organ.

Ethylene gas is widely used commercially for ripening a variety of climacteric fruits and decoloring non-climacteric citrus fruits. However, the use of ethylene-releasing compounds to effect this response is confined to relatively new crops and normally when they are used as a preharvest application. This is probably due to the ease with which ethylene gas can be applied to harvested fruit, and the dangers of applying aqueous solutions to these disease-susceptible organs.

The major commercial application of ethephon for ripening fruits is in processing tomatoes. Ethephon treatment of tomato plants was shown to have adverse effects on the germinability of older seeds, indicating that there was no benefit of ethephon application to tomatoes destined solely for seed production.

Promotion of abscission as an aid in thinning or harvesting fruit crops is an important horticultural use of ethylene-releasing compounds. A recent study by Takeda and Peterson on mechanical harvesting of blackberries provides a good example of the benefits ethylene-releasing compounds provide in stimulating this response. These researchers found that the separation pull force of thornless blackberries (Rubus spp.) decreased too slowly to allow differentiation between black ripe and red fruit by mechanical shaking. When they applied sufficient force to remove 80% of the ripe fruit, 31% of the harvested fruit were red and green. When the vines were shaken four days after application of 500 ppm ethepon, only 7% of the harvested fruit were red and green. The ethephon treatment doubled the number of ripe berries on the plants and this increase, as well as a concomitant reduction in attachment force, presumably explains the satisfactory results of the treatment.

Scarification of seed (mechanically, and/or with H2SO4) is a common method of promoting germination of seeds with hard coats. Sometimes this may release a mechanical inhibition of expansion, but in some cases it overcomes a physiological dormancy. Such is the case of Pelargonium seed, where germination is improved simply by pricking the seed coat with a needle. Typical germination percentages for unscarified seed of inbred lines of Pelargonium hortorum range from 30% to 50%. Soaking imbibed seed for 12 h in 50 ppm ethephon increased germination to 94%. Unimbibed seed responded similarly, but required higher concentrations (> 250 ppm).

The stimulation of seed germination by ethephon also suggests the possibility of using this material as part of an herbicidal program, using it to ensure pre-plant germination of dormant weed seeds. Researchers have therefore been studying ethephon effects in the germination of weed seeds.

The fact that ethylene-resistant plants grow perfectly normally raises the questions of the role of ethylene as a PGR. It has been suggested that ethylene's gaseous nature makes it an ideal regulator for enabling plants to respond to environmental stresses and attacks from pathogens and insect pests. Several recent studies have addressed the possibility that ethylene-releasing compounds might stimulate defense reactions and thus prevent pathogen attack. The stimulation of phenolic biosynthesis by ethephon applications to grapes and apples may represent such a reaction. Treatment of mung bean seedlings with ethephon decreased their susceptibility to infection by Rhizoctonia solani, concomitantly with increased activities of peroxidase and polyphenoloxidase. In contrast, ethylene stimulates germination of teliospores of Urocystis agropyri. Low-temperature stress is an important limitation in the production of many deciduous fruit trees. Ethephon application to peach trees during the fall delayed bloom in the following spring (thereby reducing the potential risk of frost damage), and also increased the intrinsic winter hardiness of the buds.

A wound response also seems implicated in the stimulation by ethephon of latex flow during tapping of rubber trees, a response which has been shown to be associated with a change in the trans-tonoplast membrane potential. The increased production of secondary metabolites such as latex is probably a common response to stress in plants, and ethephon may prove to be useful in increasing their concentration in important economic plants. Cho et al., found that addition of ethephon to cell suspension cultures of Coffea arabica and Thalictrum rugosum stimulated alkaloid production in both species, even though the biochemical pathways were widely different. They suggest that ethylene-releasing compounds may be valuable in increasing secondary metabolite levels in a variety of cell-suspension culture systems.

### Commercial Applications of GA and CKs

Since the discovery of GAs in the 1920s and cytokinins (CKs) in the 1950s, a considerable body of information has emerged regarding the roles these hormones play in plant growth and development. As these compounds and analogs became available, either by fermentation or chemical synthesis, a number of potential practical applications were identified. However, despite extensive research with GAs and CKs, the number of commercial uses has fallen short of their predicted impact and growth in agriculture. Unique opportunities may still result from new research and development with GAs and CKs to meet the future demands for increased food quality and production.

### Growth Retardants in Agriculture and Horticulture

Plant growth retardants reduce the shoot length of plants without being phytotoxic or causing malformations. This is primarily achieved by reduced cell elongation but also be a lowered rate of cell division. Some growth retardants are of economic importance in agriculture and horticulture, since they can be employed to affect the morphology of cultivated plants in a desired way. Reviews on plant growth retardants have been presented by Dicks and Nickel.

**TABLE 35.9:** Physiological response and current commercial uses of gibberellin an cytokinin products registered

| Physiological response | Crop | PGR Used* |
|---|---|---|
| Breakage of dormancy | Seed potatoes | $GA_3$ |
| | Rhubard | $GA_3$ |
| Acceleration of seed germination/hydrolytic enzyme production | Barley malting | $GA_3$ |
| Stimulation of bud break and increased lateral branching | Tart cherries | $GA_3$ |
| | Apples | $GA_{4+7}$/BA |
| | Conifers | BA |
| | Carnations/Roses | PBA |
| Stimulation of fruit set | Tangerine hybrids | $GA_3$ |
| | Blueberries | $GA_3$ |
| Reduction of fruit set | Grapes | $GA_3$ |
| | Tart and sweet cherries | $GA_3$ |
| Increased fruit size and quality | Grapes | $GA_3$ |
| | Apples | $GA_{4+7}$/BA |
| | Sweet cherries | $GA_3$ |
| | Apples | $GA_{4+7}$ |
| Increased vegetative growth and yield | Sugar cane | $GA_3$ |
| | Spinach | $GA_3$ |
| | Hops | $GA_3$ |
| Delayed senescence/ fruit ripening | Banana | $GA_{4+7}$ |
| | Sweet cherries | $GA_3$ |
| | Citrus | $GA_3$ |
| | Bermuda grass | $GA_3$ |
| Stimulation of abscission/ Senescence | Cotton | TDZ |

*Abrreviations = $GA_3$: gibberellic acid; $GA_{4+7}$: gibberellin $A_4$ and $A_7$; BA: benzyladenine; PBA: tetrahydropyranylbenzyladenine; TDZ: thidiazuron.

At present the most relevant practical applications of growth retardants are:

1. Control of lodging in small grains such as wheat, barley and rye with CCC, mepiquat chloride and ethephon,
2. Reduction of vegetative growth and enhanced uniformity of ripening in cotton with mepiquat chloride,
3. Regulation of the ratio between vegetative growth and fruit production and improvement of fruit quality in orchard trees with daminozide,
4. Control of excessive vegetative growth of peanuts with daminozide,
5. Reduction of stalk length or shoot height in ornamentals with compounds such as CCC, daminozide and ancymidol.

Several additional indications are, however, conceivable which cannot be covered by the established compounds. Therefore, continuous efforts are being made to find and develop new and better suited retardants, also as possible substitutes for existing products.

Several new types of plant growth retardants have been detected and developed during recent years. Norbornanodiazetines (e.g. tetcyclacis), triazoles (e.g. paclobutrazol, uniconazole, triapenhenol, BAS 111 W), isonicotinanilides (e.g. inabenfide) have to be mentioned here as well as flurprimidol (a pyrimideine derivative) which is closely related to ancymidol. In general, the growth retarding activity of these compounds is relatively high in a far broader spectrum of plant species as compared, for instance, with CCC, mepiquat chloride and daminozide. As their basic mode of action, the norbornanodiazetines, triazoles, isonicotinanilides and pyrimidines inhibit the endogenous formation of GAs by blocking cytochrome P-450-type oxygenases that catalyze the reactions between ent-kaurene and ent-kaurenoic acid. Clear evidence is available that other cytochrome P-450 oxygenases are also affected by these compounds, although most of them at a far lower degree of activity. In particular, the inhibition of sterol formation by the blocking of 14-demethylation and the inhibition of the oxidative inactivation of ABA must be mentioned. It has also been reported that the oxidative metabolism of certain xenobiotics can be inhibited and one may forecast that further such effects will be detected in the future. The situation is even more complex since these side activities can be fairly different from compound to compound and from species to species.

### *Potentials for Practical Uses*

In most cases the growth-retarding effect is the only point of interest. This would primarily exploit the potential of the new growth retardants to lower the endogenous levels of vegetative-growth active GAs. In addition to this, other physiological processes also affected by the new types of compounds may cause beneficial effects as well. The usefulness of a given compound in a given crop may therefore be determined by the combination of such side effects with morpho regulation. In certain cases growth retardation may even be of secondary importance.

## Methanol: A New Fertilizer

Recently a Japanese-American farmer of Arizona desert, Nonomura, has discovered that spraying plants with a dilute (20%) solution of methanol, stimulates their growth. Since methanol has long been known to be toxic to plants, its positive effects might appear to be unexpected. Nonomura, noticed that his cotton plants wilted during the midday heat. Such wilting is caused by the more rapid loss of water from the leaves than its replacement by water taken up by the roots from the soil. Wilting causes the stomata to close and reduces photosynthesis because no new $CO_2$ can enter the leaves. At the same time, the wasteful process called photorespiration becomes more pronounced because of the lower $CO_2$ level inside the leaves.

Nonomura applied some dilute methanol to his wilted cotton. Within hours, the sprayed plants were once again turgid, their stomates were open, photosynthesis resumed at high levels, and photorespiration went down. Not surprisingly, the growth rate increased significantly, and the plants produced cotton bolls two weeks sooneer than usual, thereby reducing their overall need for water. The next candidates for methanol spraying were cabbage plants, resulting in heads twice as big as those on untreated plants grown for the same length of time then watermelon (yields up by 36%). Methanol also stimulates the growth rates of wheat and barley, and this should result in increased yields.

Not all plants are positively affected by the methanol spray. For example, corn and other C4 plants do not exhibit the midday reduction in photosynthesis and do not have photorespiration, so these crops would not be expected to be stimulated by methanol- and they are not. Nor are plants growing in the shade, an indication that whatever methanol is doing to the plant is mediated by processes involving light. The discovery of this effect of methanol is so recent that almost no research has been done so far to understand the underlying mechanism.

# Collateral Readings


Abadia, A. Poc, A. and Abadia, J. 1991. Could iron nutrition status be evaluated through photosynthetic pigment changes? *J. Plant. Nutri.* 14: 987-999.

Abeles, F. B. Morgan, P. W. and Saltveit, M. E. Jr. 1992. *Ethylene in Plant Biology*, 2nd ed. Academic Press, San Diego.

Addicott, F. T. Ed. 1983. Abnscisic Acid. Praeger, New York.

Addicott, F. T. 1965. Physiology of abscission. Pages 1094-1126 in W. Ruhland ed. Encyclopedia of Plant Physiology, Vol. 15, Part 2. Springer-Verlag, Berlin.

Agren, G. I. and T. Ingestad. 1987. Root: shoot ratio as a balance between nitrogen productivity and photosynthesis. Plant, Cell and Environment 10: 579-586.

Al-Ani, A. F. Bruzau, P. Raymond, V. Saint-Ges, J. M. Leblanc, and A. Pradet. 1985. Germination, respiration, and adenylate energy charge of seeds at various oxygen partial pressures. Plant Physiology 79: 885-890.

Aloni, R. 1987. The induction of vascular tissues by auxin. Pages 363-374 in P. J. Davies Ed. Plant Hormones and Their Role in Plant Growth and Development. Martinus Nijhoff Publishers, Boston.

Altman A. 1982. Retardation of radish leaf senescence by polyamines. *Physiol. Plant.* 54, 189-193.

Ameisen, J. C. 1996. The origin of programmed cell death. *Science.* 272: 1278-1279.

Amir-Shapira D, Goldschmidt EE, Altman A 1986. Autolysis of chlorophyll in aqueous and detergent suspensions of chloroplast fragments. *Plant Sci.* 43: 201-206.

Amthor, J. S. 1989. Respiration and Crop Productivity. Springer-Verlag, New York.

Anderson, J. M. and B. Andersson. 1988. The dynamic photosynthetic membrane and regulation of solar energy conversion. Trends in Biochemical Sciences 13:351-355.

Anderson, J. M. Spilatro, S. R. Klauer, S. F. and Franceschi, V. R. 1989. Jasmonic acid-dependent increase in the level of vegetative storage proteins in soybean. *Plant Sci.* 62: 45-52.

Andreasson, L. -E. and T. Vanngard. 1988. Electron transport in photosystems I and II. Annual Review of Plant Physiology and Plant Molecular Biology 39: 379-411.

Andrews, M. 1986. The partitioning of nitrate assimilation between root and shoot of higher plants. Plant, Cell and Environment 9: 511-519.

Andrews, T. J. and G. H. Lorimer. 1987. Rubisco: Structure, mechanisms, and prospects for improvement. Pages 131-218 in M. D. Hatch and N. K. Boardman eds. Photosynthesis. The Biochemistry of Plants, Vol. 10. Academic Press, New York.

Apelbaum A, Icekson I, Burgoon A C and M. Lieberman 1982. Inhibition by polyamines of macromolecular synthesis and its implication for ethylene production and senescence processes. *Plant Physiol.* 70, 1221-1223.

Arnon, D. I. 1984. The discovery of photosynthetic phosphorylation. Trends in Biochemical Sciences 9: 258-262.

Arnon, D. I. 1988. The discovery of ferredoxin: The photosynthetic path. Trends in Biochemical Sciences 13: 30-33.

Atherton, J. G. ed. 1987. *Manipulation of Flowering.* Butterworths, London.

Audus, L. J. 1979. Plant geosensors. Journal of Experimental Botany 30: 1051-1073.

Avers, Charlotte J. 1981. Cell Biology, Second Edition. Van Nostrand Reinhold, New York.

Avigad, G. 1982. Sucrose and other disaccharides. Pages 217-347 in F. A. Loewus and W. Tanner eds. Encyclopedia of Plant Physiology, New Series, Vol. 13A, Plant Carbohydrates I. Intra-cellular Carbohydrates. Springer-Verlag, Berlin.

Ayub, R. Guis, M. Amor, B. Gillot, L. Roustan, J. P. Latche, A. Bouzayen, M. and Pech, J. C. 1996. Expression of ACC oxidase antisense gene inhibits ripening of Cantaloupe melon fruits. *Nature-biotech.* 14: 862-866.

Azumi, Y. and Watanabe, A. 1991. Eviodence for a senescence-associated gene induced by darkness. *Plant Physiol.* 95: 577-583.

Bachmann A, Fernandez-Lopez J, Ginsburg S, Thomas H, Bouwkamp JC, Solomos T, Matile P 1994. Stay-green genotypes of *Phaseolus vulgaris* L: chloroplast proteins and chlorophyll catabolites during foliar senescence. *New Phytol.* 126: 593-600.

Baker, D. A. and J. L. Hall. Eds. 1987. Solute Transport in Plant Cells and Tissues. Longmans, Harlow, England.

Bandurski, R. S. 1984. Metabolism of indole-3-acetic acid. Pages 183-200 in A. Crozier and J. R. Hillman eds. The Biosynthesis and Metabolism of Plant Hormones. Cambridge University Press, Cambridge.

Bandurski, Robert S. A. Schulze, P. Dayanandan, and P. B. Kaufman. 1984. Response to gravity by *Zea mays* seedlings. I. Time course of the response. Plant Physiology 74: 284-288.

Bariol PA, Howard CJ, Taylor CB, Verburg MT, Jaglan VD, Green PJ. 1994. The limitation, Arabidopsis ribonuclease gene RNSI is tightly controlled in response to phosphate. *The Plant Journal.* 6, 673-85.

Barnabas, A. D. V. Butler, and T. D. Steinke. 1986. Phloem structure and transport pathways in the leaves of a seagrass. Pages 177-180 in James Cronshaw, William J. Lucas, and Robert T. Giaquinta eds. Phloem Transport. Alan R. Liss, New York.

Bassham, J. A. 1965. Photosynthesis: The path of carbon. Pages 875-902 in J. Bonner and J. E. Varner eds. Plant Biochemistry, Second Edition. Academic Press, New York.

Bate NJ, Rothstein SJ, Thompson JE. 1991. Expression of nuclear and chloroplast photosynthesis-specific genes during leaf senescence. *Journal of Experimental Botany.* 42, 801-11.

Battle, R. W. D. L. Laidman, and J. K. Gaunt. 1977. The relationship between floral induction and gamma-tocopherol concentations in leaves of *Xanthium strumarium* L. Biochemical Society of London Transactions 5: 322-324.

Bazzaz, MB, Rebeiz, CA 1978. Chloroplast culture: the chlorophyll repair potential of mature chloroplasts incubated in a simple medium. *Biochem Biophys Acta.* 504: 310-323.

Beale, S. I. 1990. Biosynthesis of the tetrapyrrole pigment precursor delta-aminolevulinic acid, from glutamate. Plant Physiology 93: 1273-1279.

Becker W, Apel K. 1993. Defferences in gene-expression between natural and artifically induced leaf senescence. *Planta.* 189, 74-9.

Beevers, L. 1976. Nitrogen Metabolism in Plants. American Elsevier, New York.

Behrens, H. M. M. H. Weisenseel, and A. Sievers. 1982. Rapid changes in the pattern of electric current around the root tip of *Lepidium sativum* L. following gravistimulation. Plant Physiology 70: 1079-1083.

Belknap, W. and Garbarino, J. 1996. The role of ubiquitin in plant senescence and stress responses. *Trends in Plant Science.* 1: 331-335.

Bernhard W, Matile P. 1994. Differential expression of glutamine synthetase genes during the senescence of *Arabidopsis thaliana* rosette leaves. *Plant Science.* 98, 7-14.

Bernier, G. 1988. The control of floral evocation and morphogenesis. *Annu. Rev. Plant Physiol. Plant Mol. Biol.* 39: 175-219.

Bernier, G. Havelange, A. Houssa, C. Petitjeab, A. and Lejeune, P. 1993. Physiological signals that induce flowering. *Plant Cell* 5: 1147-1155.

Berns, M. W. 1984. Cells, Second Edition. Modern Biology. Saunders, New York.

Berry, J. A. G. H. Lorimer, J. Pierce, J. R. Seemann, J. Meek, and S. Freas. 1987. Isolation, identification, and synthesis of 2-carboxyarabinitol-1-phosphate, a diurnal regulator of ribulose-bisphosphate carboxylase activity. Proceedings of the National Academy of Sciences USA 84: 734-738.

Beveridge, C. A. and Murfet, I. C. 1996. The *gigas* mutant in pea is deficient in the floral stimulus. *Physiol. Plantarum* 96: 637-645.

Bewley, J. Derek and M. Black. 1978. Physiology and Biochemistry of Seeds in Relation to Germination. Vol. 1, Development, Germination, and Growth Springer-Verlag, Berlin, New York.

Bewley, J. D. and A. Marcus. 1990. Gene expresion in seed development and germination.

Progress in Nucleic Acid Research and Molecular Biology 38: 165-193.

Beyer, E. M. Jr. 1976. A potent inhibitor of ethylene action in plans. *Plant Physiol.* 58: 268-271.

Bhandal, I. S. and C. P. Malik. 1988. Potassium estimation, uptake, and its role in the physiology and metabolism of flowering plants. International Review of Cytology 110: 205-254.

Bienfait, F. and U. Luttge. 1988. On the function of two systems that can transfer electrons across the plasma membrane. Plant Physiology and Biochemistry 26: 665-671.

Biswal U C and Sharma R, 1976. Phytochrome regulation of senescence in detached barley leaves. *Z. pflanzenphysiol.* 80: 71-73.

Biswal, B. Choudhury, N K, Sahu P and Biswal U C 1993. Senescence of detached fern leaves. *Plant Cell Physiol.* 24. 1203-1208.

Biswal, U C, Kasemir H and Mohr H, 1982. Phytochrome control of degreening of attached cotyledons and primary leaves of mustard *Sinapis alba* L. seedlings. *Photochem. Photobiol.* 35: 237-241.

Black, C. C. W. H. Campbell, T. M. Chen, and P. Dettrich. 1973. The monocotyledons: Their evolution and comparative biology. III. Pathways of carbon metabolism related to net carbon dioxide assimilation by monocotyledons. Quarterly Review of Biology 48: 299-313.

Black, M. and P. F. Wareing. 1955. Photoperiodism in the light inhibited seed of Nemophila insiginis. J. Exp. Bot. 11: 28-39.

Blakemore, Richard P. and Richard B. Frankel. 1981. Magnetic navigation in bacteria. Scientific American 2456: 58-65.

Blazquez, M. A. Green, R. Nilsson, O. Sussman, M. R. and Weigel, D. 1998. Gibberellins promote flowering of *Arabidopsis* by activating the LEAFY promoter. *Plant Cell* 10: 791-800.

Bloom, A. J. 1988. Ammonium and nitrate as nitrogen sources for plant growth. ISI Atlas of Science: Animal and Plant Sciences 55-59.

Blowers, D. P. and A. J. Trewavas. 1989. Second messengers: Their existence and relationship to protein kinase. Pages 1-28 in W. F. Boss and D. J. Morre eds. Second Messengers in Plant Growth and Development. Alan R. Liss, New York.

Blumwald, E. 1987. Tonoplast vesicles as a tool in the study of ion transport at the plant vacuole. Physiologia Plantarum 69: 731-734.

Bolitho, K. M. Lay-yee, M. Knighton, M. L. and Ross, G. S. 1997. Antisense apple ACC oxidase RNA reduces ethylene production in transgenic tomato fruit. *Plant Sci.* 122: 91-99.

Boller, T. 1988. Ethylene and the regulation of antifungal hydrolases in plants. Oxford Survey of Plant Molecular and Cell Biology 5: 145-174.

Boller, T. 1989. Primary signals and second messengers in the reaction of plants to pathogens. Pages 227-255 in D. J. Morre and W. Boss eds. Second Messengers in Plant Growth and Development. Alan R. Liss, New York.

Borochov A, Woodson WR. 1989. Physiology and biochemistry of flower petal senescence. *Horticultural Review.* 11, 15-43.

Borthwick, H. A. S. B. Hendricks, E. H. Toole, and V. K. Toole. 1954. Actuion of light on lettuce seed germination. Botanical Gazette 115: 205-225.

Bortlik K 1990. Chlorophyllabbau: Charakterisierung von Kataboliten in seneszenten Gerstenblattern. Ph. D. Thesis, University of Zurich, Switzerland.

Bortlik K, Peisker C, Matile P 1990. A novel type of chlorophyll catabolite in senescent barley leaves. *J. Plant Physiol.* 136: 161-165.

Bould, C. E. J. Hewitt, and P. Needham. 1984. Diagnosis of Mineral Disorders in Plants, Vol. 1, Principles. Chemical Publishing Co. New York.

Bowen ID. 1993. Apoptosis or programmed cell death? *Cell Biology International.* 17, 365-80.

Bowsher, C. G. D. P. Hucklesby, and M. J. Emes. 1989. Nitrite reduction and carbohydrate metabolism in plastids purified from roots of Pisum sativum L. Planta 177: 359-366.

Brady, C. J. 1987. Fruit ripening. Annual Review of Plant Physiology 38: 155-178.

Brenner, M. L. 1981. Modern methods for plant growth substance analysis. Annual Review of Plant Physiology 32: 511-538.

Briggs, W. R. 1976. H. A. Borthwick and S. B. Hendricks-pioneers of photomorphogenesis.

Pages 1-6 in H. Smith ed. Light and Plant Development. Butterworths, London.

Brown, P. H. R. M. Welch, and E. E. Cary. 1987. Nickel: A micronutrient essential for higher plants. Plant Physiology 85: 801-803.

Buchanan-Wollaston V, Ainsworth C. 1997. Leaf senescence in Brassica napus: cloning of senescence-related genes by subtractive hybrization. *Plant Molecular Biology* in press.

Buchanan-Wollaston V. 1994. Isolation of cDNA clones for genes that are expressed during leaf senescence in Brassica napus. Plant Physiology 105, 839-46.

Buchanan-Wollaston, V. 1997. The molecular biology of leaf senescence. *J. Exp. Bot.* 48: 181-199.

Budde, R. J. A. and D. D. Randall. 1990. Protein kinase in higher plants. Pages 351-367 in D. J. Moore, W. F. Boss, and F. A. Loewus eds. Inositol Metabolism in Plants, Wiley-Liss, New York.

Budde, R. J. A. and R. Chollet. 1988. Regulation of enzyme activity in oplants by reversible phosphorylation. Physiologia Plantarum 72: 435-439.

Bunning, Erwin. 1937. The endogenous daily rhythm as the basis of the photoperiodic reaction. Berichte der Deutschen Botanischen Gesellschaft 54: 590-607.

Burg, S. P. and Dijkman, M. J. 1967. Ethylene and auxin participation in pollen induced fading of vanda orchid blossoms. *Plant Physiol.* 61: 812-815.

Burgess, Jeremy. 1985. An Introduction to Plant Cell Development. Cambridge University Press, Cambridge, New York, London.

Burn, J. E. Bagnall, D. J. Metzger, J. D. Dennis, E. S. and Peacock, W. J. 1993. DNA methylation, vernalization, and the initiation of flowering. *Proc. Natl. Acad. Sci. USA* 90: 287-291.

Butt VS, Lamb CJ. 1981. Oxygenases and the metabolism of plant products. In: Conn EE, ed. The biochemistry of plants, Vol. 7. New York: Academic Press, 626-65.

Callis J, Viestra RD, 1989. Ubiquitin and ubiquitin genes in higher plants. *Oxford Surveys of Plant Molecular Biology.* 6, 1-30.

Carfts-Brandner SJ. 1991. Non-structural carbohydrate metabolism during leaf ageing in tobacco *Nicotiana tabacum. Physiologia Plantarum.* 82, 299-305.

Carpita, N. C. 1985. Tensile strength of cell walls of living cells. Plant Physiology 79: 485-488.

Cech, T. R. and B. L. Bass. 1986. Biological catalysis by RNA. Annual Review of Biochemistry 5: 599-629.

Chapin, F. S. III. 1987. Adaptations and physiological responses of wild plants to nutrient stress. Pages 15-25 in H. W. Gabelman and B. C. Loughman eds. Genetic Aspects of Plant Mineral Nutrition. Martinus Nijhoff, Boston.

Chapin, F. S. III. 1988. Ecological aspects of plant mineral nutrition. Advances in Plant Nutrition 3: 161-191.

Chatterton, N. J. P. A. Harrison, J. H. Bennett, and K. H. Asay. 1989. Carbohydrate partitioning in 185 accessions of gramineae grown under warm and cool temepratures. Journal of Plant Physiology 134: 169-179.

Chen QG, Bleecker AB. 1995. Analysis of ethylene signal-transduction kinetice associated with seedling-growth response and chitinase induction in wild-type and mutant Arabidopsis. *Plant Physiology.* 108, 597-607.

Chen, L. Cheng, J. C. Castle, L. and Sung, Z. R. 1997. *EMF* genes regulate *Arabidopsis* inflorescence development. *Plant Cell* 9: 2011-2024.

Chen, S. J. and Kao, C. H. 1998. Methyl jasmonate, ammonium and leaf senescence in rice. *J. Plant Physiol.* 152: 353-357.

Cheung AY, McNellis T, Piekos B 1993. Maintenance of chloroplast components during chromoplast differentiation in the tomato mutant green flesh. *Plant Physiol.* 101: 1223-1229.

Childs, K. L. Cordonnier-Pratt, M. M. Pratt, L. H. and Morgan, P. W. 1992. Genetic regulation of development in *Sorghum bicolor* VII. ma3 flowering mutant lacks a phytochrome that predominates in green

Chitnis, P. R. and J. P. Thornber. 1988. The major light-harvesting complex of photosystem II: Aspects of its molecular and cell biology. Photosynthesis Research 16: 41-63.

Chubatsu L, Meneghini R. 1993. Metallotheionein protects DNA form oxidative damage. *Biochemical Journal.* 291, 193-8.

Ciechanover, A. 1994. The ubiquiten – proteasome proteolytic pathway. *Cell.* 79: 13-21.
Clarkson, D. T. 1985. Factors affecting mineral nutrient acquisition by plants. Annual Review of Plant Physiology 36: 77-116.
Clarkson, D. T. and J. B. Hanson. 1980. The mineral nutrition of higher plants. Annual Review of Plant Physiology 31: 239-298.
Cleland, R. 1971. Cell wall extension. Annual Review of Plant Physiology 22: 197-222.
Cleland, R. E. 1987. The mechanism of wall loosening and wall extension. Pages 18-27 in D. J. cosgrove and D. P. Knievel eds. Physiology of Cell Expansion during Plant Growth. American Society of Plant Physiologists, Rockville, Md.
Clouse, S. D. 1996. Molecular genetic studies confirm the role of brasinosteroids in plant growth and development. *Plant. J.* 10: 1-8.
Coen, E. S. 1991. The role of homeotic genes in flower development and evolution. *Annu. Rev. Plant Physiol. Plant Mol. Biol.* 42: 241-279.
Coen, E. S. and Carpenter, R. 1993. The metamorphosis of flowers. *Plant Cell* 5: 1175-1181.
Coen, E. S. and Meyerowitz, E. M. 1991. The war of the whorls: Genetic interactions controlling flower development. *Nature* 353: 31-37.
Coen, E. S. Doyle, S. Romero, J. M. Elliott, R. Magrath, R. and Carpenter, R. 1991. Homeotic genes controlling flower development in *Antirrhinum. Development* 112 Suppl. 1: 149-155.
Comai L, Dietrich RA, Maslyar DJ, Baden FS, Harada JJ. 1989. Coordinate expression of transcriptionally regulated isocitrate lyase and malate synthease genes in Brassica napus L. *The Plant Cell.* 1. 293-300.
Cooke, R. and I. D. Kuntz. 1974. The properties of water in biological systems. Annual Review of Biophysics and Bioengineering 3:95-126.
Cosgrove, D. J. and D. P. Knievel. Eds. 1987. Physiology of Cell Expansion during Plant Growth. American Society of Plant Physiologists, Rockville, Md.
Cotton, J. L. S. C. W. Ross, D. H. Byrne, and J. T. Colbert. 1990. Down-regulation of phytochrome mRNA abundance by red light and benzyl-adenine in etiolated cucumber cotyledons. Plant Molecular Biology 14: 707-714.
Coupe SA, Taylor JF. Roberts JA. 1995. Characterization of an mRNA encoding a metallothionein-like protein that accumulates during ethylene-promoted abscission of Sambucus nigra L. leaflefts. *Planta.* 197, 442-7.
Crafts, A. S. 1961. Translocation in Plants. Holt, Rinehart & Winston, New York.
Cram, W. J. 1988. Transport of nutrient ions across cell membranes in vivo. Pages 1-53 in B. Tinker and A. Lauchli eds. Advances in Plant Nutrition, Vol. 3. Praeger, New York.
Creelman, R. A. 1989. Abscisic acid physiology and biosynthesis in higher plants. Physiologia Plantarum 75: 131-136.
Cronshaw, J. 1981. Phloem structure and function. Ann. Rev. Plant Physiol. 32: 465-484.
Csupor L 1971. Das Phytol in vergilbten Blattern. *Planta Med* 19: 37 – 41.
Cumming, Bruce G. 1969. *Chenopodium rubrum* L. and related species. Pages 156-185 in L. T. Evans ed. The Induction of Flowering, Some Case Histories. Macmillan of Australia, South Melbourne, Australia.
Curry, G. M. 1969. Phototropism. Pages 241-273 in M. B. Wilkins ed. The Physiology of Plant Growth and Development. McGraw-Hill, New York.
Curty C, Engel N 1996. Detection, isolation and structure elucidation of a chlorophyll *a* catabolite from autumnal senescent leaves of *Cercidiphyllum japonicum. Phytochemistry.* 42: 1531-1536.
Curty C, Engel N, Gossauer A 1995. Evidence for a monooxygenase-catalyzed primary process in the catabolism of chlorophyll. FEBS Lett 364: 41-44.
Czeisler, Charles A. James S. Allan, Steven H. Strogatz, Joseph M. Ronda, Ramiro Sanchez, C. David Rios, Walter O. Freitag, Gary S. Richardson, and Richard E. Kronauer. 1986. Bright light resets the human circadian pacemaker independent of the timing of the sleep-wake cycle. Science 233: 667-671.
Dai, Y. R. and Galston A W 1981. Simultaneous phytochrome controlled promotion and inhibition of arginine decarboxylase activity in buds and epicotyls of etiolated peas. *Plant Physol.* 67: 266-269.
Davies KM, Grierson D. 1989. Identification of cDNA clones for tomato Lycopersicon

esculentum Mill. mRNAs that accumulate during fruit ripening and leaf senescence in response to ethylene. *Planta*. 179, 73-80.

Davies, D. D. Ed. 1987. Physiology of Metabolism. The Biochemistry of Plants, Vol. 12. Academic Press, New York.

Davies, P. J. Ed. 1987. Plant Hormones and Their Role in Plant Growth and Development. Martinus Nijhoff Publishers, Boston.

de Duve, Christian. 1983. Microbodies in the living cell. Scientific American 248 5: 74-84.

De Greef, J A, Butler W L, Roth T. F and Frederieq H, 1971. Control of senescence in *Marchantia* by phytochrome. *Plant Physiol.* 48: 407-412.

Delmer, Deborah P. 1987. Cellulose biosynthesis. Ann. Rev. Plant Physiol. 38: 259-290.

Delrot, Serge. 1987. Phloem loading: Apoplastic or symplastic? Plant Physiol. Biochem. 255: 667-676.

Dixon, R. O. D. and C. T. Wheeler. 1986. Nitrogen Fixation in Plants. Chapman and Hall, New York.

Douce, R. and D. A. Day. Eds. 1985. Higher Plant Cell Respiration. Encyclopedia of Plant Physiology, Vol. 18, New Series. Springer-Verlag, Berlin.

Drew, M. C. 1988. Effects of flooding and oxygen deficiency on plant mineral nutrition. Pages 115-159 in B. Tinker and A. Lauchli eds. Advances in Plant Nutrition, Vol. 3. Praeger, New York.

Dudley, M. and Poethig, R. S. 1991. The effect of a heterochronic mutation, *Teopod2*, on the cell lineage of the maize shoot. *Development* 111: 733-740.

Dudley, M. and Poethig, R. S. 1993. The heterochronic *Teopod1* and *Teopod2* mutations of maize are expressed non-cell-autonomously. *Genetics* 133: 389-399.

Duggelin T, Bortlik K, gut H, Matile P, Thomas H 1988a. Leaf senescence in *Festuca pratensis*: accumulation lipofuscin-like compounds. *Physiol Plant.* 74: 131 –134.

Duggelin T, Schllenberg M, Bortlik K, Matile P 1988b. Vacuolar location of lipofuscin – and proline-like compounds in senescent barley leaves. *J Plant Physiol.* 133: 492-497.

Dugger, W. M. 1983. Boron in plant metabolism. Pages 626-650 in A. Lauchli andR. L. Bieleski eds. Encyclopedia of Plant Physiology, New Series, Vol. 15B, Inorganic Plant Nutrition. Springer-Verlag, Berlin.

Ecker JR. 1995. The ethylene signal transduction pathway in plants. *Science*. 268, 667-75.

Edwards, G. E. and M. S. B. Ku. 1987. Biochemistry of $C_3$-$C_4$ intermediates. Pages 275-325 in M. D. Hatch and N. K. Boardman eds. Photosynthesis. The Biochemistry of Plants, Vol. 10. Academic Press, New York.

Ellis, R. J. and S. M. Hemmingsen. 1989. Molecular chaperones: Proteins essential for the biogenesis of some macromolecular structures. Trends in Biochemical Sciences 14: 339-342.

Engel N, Jenny TA, Mooser V, Gosauer A 1991. Chlorophyll catabolism in *Chlorella protothecoids*: isolation and structure elucidation of a red bilin derivative. FEBS Lett 293: 131-133.

Engel, P. C. 1977. Enzyme Kinetics. Wiley, New York.

Engvild KC. 1989. The death hormone hypothesis. *Physiologia Plantarum*. 77, 282-5.

Erickson, Ralph O. 1976. Modeling of plant growth. Ann. Rev. Plant Physiol. 27: 407-434.

Eschrich, Walter. 1986. Mechanisms of phloem unloading. Pages 225-230 in James Cronshaw, William J. Lucas, and Robert T. Giaquinta eds. Phloem Transport. Alan R. Liss, New York.

Evan-chen, Z, Mattoo A K and Goren R, 1982. Inhibition of ethylene biosynthesis by aminoethoxyvinyl-glycine and bypolyamines shunts label from 3,4-[$^{14}$C] methionine into spermidine in aged orange peel disc. *Plant Physiol.* 69: 385-388.

Evans, M. M. S. and Poethig, R. S. 1995. Gibberellins promote vegetative phase change and reproductive maturity in maize. *Plant Physiol.* 108: 475-487.

Feldman, L. J. 1988. The habits of roots. What's up down under? BioScience 38: 612-618.

Feller U, Fischer A. 1994. Nitrogen metabolism in senescing leaves. *Critical Reviews in Plant Sciences.* 133. 241-73.

Feller U, Keist M. 1986. Senescence and nitrogen metabolism in annual plants. In: Lambers H, Neeterson J. J. Stulen I, eds. Fundámental, ecological and agricultural aspects of nitrogen metabolism in higher plants. Dordrecht: Martinus Nijhoff Publishers, 219-34.

Ferguson, D. L. A-Khatib, K. Guikema, J. A. and Ulsen, G. M. 1993. Degradation of proteins from the thylakoid membranes in senescing wheat leaves at high temperature. *Plant, Cell and Environ.* 16: 421-428.

Fincher, G. B. 1989. Molecular and cellular biology associated with endosperm mobilization in germinating cereal grains. Annual Review of Plant Physiology and Plant Molecular Biology 40: 305-346.

Flowers, T. J. 1988. Chloride as a nutrient and as an osmoticum. Advances in Plant Nutrition 3: 55-57.

Forrester, M. L. G. Krotkov, and C. D. Nelson. 1966. Effect of oxygen on photosynthesis, photorespiration, and respiration in detached leaves. I. Soybean. Plant Physiology 41: 422-427.

Foyer CH, Descourvieres P, Kunert KJ. 1994. Nucleotide-sequence and expression of a ripening and water stress-related cDNA from tomato with homology to the MIP class of membrane channel proteins. *Plant Molecular Biology.* 24, 539-43.

Foyer, Christine H. 1987. The basis for source-sink interaction in leaves. Plant Physiol. Biochem. 255: 649-657.

Fukuchi-Mizutani M, Savin K, Cornish E, Tanaka Y, Ashikari T. Kusumi T, Murata N. 1995. Senescence-induced expression of a homologue of delth 9 desaturase in rose petals. Plant Molecular Biology 29, 627-35.

Galston, A W, 1983. Polyamines as modulators of plant development. *Bioscience*, 33: 382-388.

Gan S, Amasino RM. 1995. Inhibition of leaf senescence by autoregulated production of cytokinin. Seience 270, 1966-7.

Gan, S. Adnm Amasino, R. M. 1997. Making sense of senescence. *Plant Physiol.* 113: 313-319.

Garbarino J. E. Oosumi T, Belknap WR. 1995. Isolation of a polyubiquitin promoter and its expression in transgenic potato plants. *Plant Physiology.* 109, 1371-8.

Garcia A. L. Galindo L 1991. Chlorophyllase in *Citrus* leaves: localization and partial purification of the enzyme. *Photosynthetica.* 25: 105-111.

Garner, W. W. and H. A. Allard. 1920. Effect of the relative length of day and night and other factors of the environment on growth and reproduction in plants. Journal of Agricultural Research 18: 553-606.

Gasser, C. S. and Robinson-Beers, K. 1993. Pistil development. *Plant Cell* 5: 1231-1239.

Geiger, Donald R. and Bernadette R. Fondy. 1980. Phloem loading and unloading: Pathways and mechanisms. What's New in Plant Physiology 117: 25-28.

Genschik P, Durr A, Fleck J. 1996. Differential expression of several E2-type ubiquitin carrier protein genes at different developmental stages in *Arabidopsis thaliana* and *Nicotiana sylvestris. Molecular and General Genetics.* 244, 548-56.

Genschik, P. et al. 1994. Structure and promoter activity of a stress and developmentally regulated polyubiquitin encoding gene of *Nicotiar tabacum. Gene.* 148: 195-202.

Gepstein S and Thimann K V 1980. Changes in the abscisic acid content of oat leaves during senescence. *Proc. Natl. Acad. Sci. USA* 77: 2050-2053.

Gepstein S. 1982. Light induced $H^+$ secretion and the relation to senescence of oat leaves. *Plant Physiol.* 70: 1120-1124.

Gepstein, S. and Thimann K. V, 1981. The role of ethylene in the senescence of oat leaves. *Plant Physiol.* 68: 349-354.

Giaquinta, R. T. 1983. Phloem loading of sucrose. Ann. Rev. of Plant Physiol. 34: 347-387.

Gillissen, L. J. W. and Hoekstra, F. A. 1984. Pollination induced wilting in Petunia hybrida rapid transfer through the style of a wilting inducing substance. *Plant Physiol.* 74: 496-498.

Ginsburg S. Matile P. 1993. Identification of catabolites of chlorophyll-porphyrin in senescent rapre cotyledons. *Plant Physiol.* 102: 521 – 527.

Ginsburg S. Schellenberg M. Maitle P. 1994. Cleavage of chlorophyll-porphyrin. Requirement for reduced ferredoxin and oxygen. *Plant Physiol.* 105: 545-554.

Glass, A. D. M. 1989. Plant Nutrition. An Introduction to Current Concepts. Jones and Bartlett, Boston.

Glazer, A. N. and A. Melis. 1987. Photochemical reaction centers: Structure, organization, and function. Annual Review of Plant Physiology 38: 11-45.

Goldthwaite J. J. and Laetsch W. M. 1967. Regulation of senescence in bean leaf discs by light and chemical growth regulators. *Plant Physiol.* 42: 1757-1762.

Goren R, Palavan N, Flores H, and Galston A W, 1982. Changes in polyamine titer in etiolated pea seedlings following red light treatment. *Plant Cell Physiol.* 23: 19-26.

Gosauer A 1994. Catabolism of tetrapyrroles. *Chimia.* 48: 352-361.

Gough, D. and Hemming, G. 1974. The stereochemistry of butalaprenoids biosynthesis. *Biochem.* J. 117: 309-318.

Graf, P. and E. W. Weiler. 1989. ATP-driven $Ca^{2+}$ transport in sealed plasma membrane vesicles prepared by aqueous two-phase partitioning from leaves of Commelina communis. Physiologia Plantarum 75: 469-478.

Graham IA, Denby KJ, Leaver CJ. 1994. Carbon catabolite repression regulates glyoxylate cycle gene expression in cucumber. *The Plant Cell.* 6, 761-72.

Graham IA, Leaver CJ, Smith SM. 1992. Induction of malate synthase gene expression in senescent and detached organs of cucumber. The Plant Cell 4, 349-57.

Granell A, Harris N, Pisabarro AG, Carbonell J. 1992. Temporal and spatial expression of thiolprotease gene during pea ovary senescence and its regulation by gibberellin. *The Plant Journal.* 26. 907-15.

Grbic V, Bleecker AB. 1995. Ethylene regulates the timing of leaf senescence in *Arabidopsis. The Plant Journal.* 84. 595-602.

Green P. J. 1994. The ribonucleases of higher plants. *Annual Review of Plant Physiology and Plant Molecular Biology.* 45, 421-45.

Greene, E. M. 1980. Cytokinin production by microorganisms. The Botanical Review 46: 25-74.

Grierson D. Kader A. 1986. Fruit ripening and quality. In: Atherton J. Rudich J. eds. The tomato crop: a scientific basis for improvement. London: Chapman and Hall, 241-80.

Grossmann, K. 1990. Plant growth retardants as tools in physiological research. Physiology 38: 419-466.

Guiamet J. J. Schwatz E. Pichersky E. Nooden L. D. 1991. Characterization of cytoplasmic and nuclear mutations affecting chlorophyll and chlorophyll-binding proteins during senescence in soybean. *Plant Physiol.* 96: 227-231.

Guo, H. Yang, H. Mockler, T. C. and Lin, C. 1998. Regulation of flowering time by *Arabidopsis* photoreceptors. *Science* 279: 1360-1363.

Gut H. Martile P. 1988. Apparent induction of key enzymes of the glyoxylic acid cycle in senescent barley leaves. *Planta.* 176, 548-50.

Haber, A.H., Thompson P J, Waline P L and Triplett L L 1969. Non-photosynthetic retardation of chloroplast senescence by light. *Plant Physiol.* 44: 1619-1628.

Hagen, G. 1987. The control of gene expression by auxin. Pages 149-163 in P. J. Davies ed. Plant Hormones and Their Role in Plant Growth and Development, Martinus Nijhoff Publishers, Boston.

Hall, J. L. D. A. Brummell, and J. Gillespie. 1985. Does auxin stimulate the elongation of intact plant stems? New Phytologist 100: 341-345.

Hamilton AJ, Lycett GW, Grierson D. 1990. Antisense gene that inhibits synthesis of the hormone ethylene in transgenic plants. *Nature.* 346, 284-7.

Hamilton, A. J. Lycett, G. W. and Grierson, D. 1990. Antisense gene that inhibits synthesis of the hormone ethylene in transgenic plants. *Nature.* 346: 284-287.

Hamner, K. C. and J. Bonner. 1938. Photoperiodism in relation to hormones as factors in floral initiation and development. Botanical Gazette 100: 388.

Hamner, K. C. and Bonner, J. 1938. Photoperiodism in relation to hormones as factors in floral initiation and development, *Bot. Gaz.* 100: 388-431.

Hanfrey C, Fife M, Buchanan-Wollaston V. 1996. Leaf senescence in *Brassica napus*: expression of genes encoding pathogenesis-related proteins. *Plant Molecular Biology.* 30: 597-609.

Hanson, J. B. 1984. The function of calcium in plant nutrition. Advances in Plant Nutrition 1: 149-208.

Harris, N. 1986. Organization of the endomembrane system. Ann. Rev. Plant Physiol. 37: 73-92.

Hart, J. W. 1990. Plant Tropisms and Other Growth Movements. Unwin Hyman, London.

Hatch, M. D. and N. K. Boardman Eds. 1987, Photosynthesis. The Biochemistry of Plants, Vol.

10. Academic Press, New York.

Havelange, A. Lejeune, P. Bernier, G. Kaur-Sawhney, R. and Galston, A. W. 1996. Putrescine export from leaves in relation to floral transition in *Sinapis alba. Physiol. Plantarum* 96: 59-65.

Hedrich, R. and J. I. Schroeder. 1989. The physiology of ion channels and electrogenic pumps in higher plants. Annual Review of Plant Physiology and Plant Molecular Biology 40: 539-569.

Hemming, F. W. 1969. Polyprenols. *Biochem. J.* 113: 23-25.

Hemming, F. W. 1981. Biosynthesis of dolichol and related compounds. In: *Biosynthesis of isoprenoid compounds*, Porter, J. W. and Spurgein, S. L. eds. John Wiley and Sons, New York. 305-354.

Hempel, F. D. and Feldman, L. J. 1994. Bidirectional development in *Arabidopsis thaliana*: Acropetal initiation of flowers and basipetal initiation of paraclades. *Planta* 192: 276-286.

Hendry G. A. F. Houghton J. D. Brown S. B. 1987. The degradation of chlorophyll: a biological enigma. *New Phytol.* 107: 255-302.

Hensel LL, Grbic V, Baumgarten DA, Bleecker AB. 1993. Developmental and age-related processes that influence the longevity and senescence of photosynthetic tissues in Arabidoposis. *The Plant Cell.* 5, 553-64.

Hewitt, E. J. and T. A. Smith. 1975. Plant Mineral Nutrition. Wiley, New York.

Hicks, K. A. Millar, A. J. Carre, I. A. Somers, D. E. Straume, M. Meeks-Wagner, D. R. and Kay, S. A. 1996. Conditional circadian dysfunction of the *Arabidopsis* early-flowering 3 mutant. *Science* 274: 790-792.

Hillman, J. R. 1984. Apical dominance. Pages 127-148 in M. B. Wilkins ed. Advanced Plant Physiology. Pitman, London.

Hinder B, Schellenberg M, Rodonis, Ginsburg S, Vogt E, Martinoia E, Matile P, Hortensteinder S 1996. How plants dispose of chlorophyll catabolites: transport of linear tetrapyrroles into isolated vacuoles. J. Biol. Chem. in press.

Hinder, B. Schellenberg, M. Rodoni, S. Ginsburg, S. Vogt, E. Martinoia, E. et al. 1996. How plants dispose of chlorophyll catabolites. Directly energized uptake of tetrapyrrolic breakdown products into isolated vacuoles. *J. Biol. Chem.* 271, 27233-27236.

Ho, Lim C. 1988. Metabolism and compartmentation of imported sugars in sink organs in relation to sink strength. Annual Review of Plant Physiology and Plant Molecular Biology 39: 355-378.

Hoober, J. K. 1987. The molecular basis of chloroplast development. Pages 1-74 in M. D. Hatch and N. K. Boardman eds. The Biochemistry of Plants, Vol. 10, Photosynthesis. Academic Press, New York.

Hope, A. B. and D. B. Matthews. 1988. Electron and proton transfers around the *b/f* complex in chloroplasts: Modelling the constraints on Q-cycle activity. Australian Journal of Plant Physiology 15: 567-583.

Hortensteiner S, Vicentini F, Matile P 1995. Chlorophyll break down in senescent leaves: enzymic cleavage of phaeophorbide an *in vitro. New Phytol.* 129: 237-246.

Hortensteiner, S. 1999. Chlorophyll breakdown in higher plants and algae. Cell Mol. Life Sci. 56: 330-347.

Hsieh H. M. Liu W. K. Huang P. C. 1995. A novel stress-inducible metallothionein-like gene from rice. *Plant Molecular Biology.* 28, 381-9.

Huber, S. C. J. A. Huber, and K. R. Hanson. 1990. Regulation of the partitioning of products of photosynthesis. Pages 85-101 in I. Zelitch ed. Perspectives in Biochemical and Genetic Regulation of Photosynthesis. Alan R. Liss, New York.

Huff, A. K. and C. W. Ross. 1975. Promotion of radish cotyledon enlargement and reducing sugar content by zeatin and red light. Plant Physiology 56: 429-433.

Huffaker R. C. 1990. Proteolytic activity during senescence of plants. *New Phytologist.* 116, 199-231.

Ingestad, T. and G. I. Agren. 1988. Nutrient uptake and allocation at steady-state nutrition. Physiologia Plantarum 72: 450-459.

Iordanov, I. T. and Merakchiiska-Niikolova M. G. 1983. Influence of darkening on the stability of plastid pigments and photosynthetic rate of primary bean leaves of different age of physiological state. *Photosynthetica.* 17: 176-181.

Ito H, Tanaka A. 1996. Determination of the activity of chlorophyll *b* to chlorophyll *a* conversion during greening of etiolated cucumber cotyledons by using pyrochlorophyllide *b*. *Plant Physiol Biochem.* 34: 35-40.

Ito H. Takaichi S. Tsuji H. Tanaka A. 1994. Properties of synthesis of chlorophyll *a* from chlorophyll *b* in cucumber etioplasts. *J Biol Chem.* 269: 22034-22038.

Ito H. Tanaka Y. Tsuji H. Tanaka A. 1993 Conversion of chlorophyll *b* to chlorophyll *a* by isolated cucumber etioplasts. *Arch Biochem Biophys.* 306: 148-151.

Iturraspe J. Engel N. Gossauer A. 1994. Chlorophyll catabolism. Isolation and structure elucidation of chlorophyll *b* catabolites in *Chlorella protothecoides. Phytochemistry.* 35: 1387-1390.

Iturraspe J. Moyano N. 1995. A new 5-formylbilinoen as the major chlorophyll *a* catabolite in tree senescent leaves. *J Org Chem.* 60: 6664-6665.

Jacobs, M. and P. H. Rubery. 1988. Naturally occurring auxin transport regulators. Science 241: 346-349.

Jacobs, W. P. 1979. Plant Hormones and Plant Development. Cambridge University Press, Cambridge.

Janistyn, Boris. 1982. Gas chromatographic-mass spectroscopic identification of prostaglandin $F_{2a}$ in flowering *Kalanchoe blossfeldiana* v. Poelln. Planta 154: 485-487.

Jensen, M. H. and W. L. Collins. 1985. Hydroponic vegetable production. Horticultural Reviews 7: 483-558.

John I. Drake R. Farrell A. Cooper W. Lee P. Horton P. Grierson D. 1995. Delayed leaf senescence in ethylene-deficient ACC-oxidase antisense tomato plants-molecular and physiological analysis. *The Plant Journal.* 73. 483-90.

Jones M. L. Larsen P. B. Woodson W. R. 1995. Ethylene-regulated expression of a carnation cysteine proteinase during flower petal senescence. *Plant Molecular Biology.* 283: 505-12.

Jones, J. B. Jr. 1982. Hydroponics: Its history and use in plant nutrition studies. Journal of Plant Nutrition 5: 1003-1030.

Kamachi K. Yamaya T, Hayakawa T, Mae T, Ojima K. 1992. Changes in cytosolic glutamine synthetase polyeptide and its mRNA in a leaf blade or rice plants during natural senescence. *Plant Physiology.* 98: 1323-9.

Kamachi K. Yamaya T. Mae T, Ojjima K. 1991. A role for glutamine synthetase in the remobilization of leaf nitrogen during natural senescence in rice leaves. *Plant Physiology.* 96: 411-17.

Kasemir H and Mohr H, 1982. Coaction of three factors controlling chlorophyll and anthocyanin synthesis. *Planta.* 156: 282-288.

Kasemir H. 1983. Light control of chlorophyll accumulation in higher plants. In *Encyclopedia of Plant Physiology* Edited by W. Shropshire, Jr and H. Mohr. Vol. 16. Pp. 662-686. Springer-Verlag, Berlin.

Kasemir, H. 1983. Light control of chlorophyll accumulation in higher plants. Pages 662-686 in W. Shropshire, Jr. and H. Mohr eds. Encyclopedia of Plant Physiology, New Series, Vol. 16B, Photomorphogenesis. Springer-Verlag, Berlin.

Kaur-sawhney R. Shin L. M. Flores H. E. and Galston A. W. 1982. Relation of polyamine synthesis and titer to aging and senescence of oat leaves. *Plant Physiol.* 69: 405-410.

Kawakami N. Watanabe A. 1988. Senescence-specific increase in cytosolic glutamine synthetase and its mRNA in radish cotyledons. *Plant Physiology.* 88: 1430-4.

Kegi J. 1991. Overview of metallothioneins. In: Riordan J. Vallee B, eds. *Methods in enzymology.* 613-26.

Kelly, G. J. J. A. M. Holtum, and E. Latzko. 1989. Photosynthesis. Carbon metabolism: New regulators of $CO_2$ fixation, the new importance of pyrophosphate, and the old problem of oxygen involvement revisited. Progress in Botany 50: 74-101.

Kelly, M. O. and P. J. Davies. 1988. The control of whole plant senescence. CRC Critical Reviews in Plant Sciences 7: 139-173.

Kendrick, R. E. and G. H. M. Kronenberg. Eds. 1986. Photomorphogenesis in Plants. Martinus Nijhoff, Boston.

Kim D. J. Smith S. M. 1994a. Molecular cloning of cucumber phosphoenol pyruvate carboxykinase and developmental regulation of gene expression. *Plant Molecular*

*Biology*. 26, 423-34.

Kim D. J. Smith S. M. 1994b. Expression of a singal gene encoding microbody NAD-malate dehydrogenase during glyoxysome and peroxisome development in cucumber. *Plant Molecular Biology*. 26: 1833-41.

Kinet, Jean-Marie, Roy M. Sachs, and Georges Bernier. 1985. The Physiology of Flowering. III. The Development of Flowers. CRC Press, Boca Raton, Fla.

King G. A. Davies K. M. Stewart R. J. Borst W. M. 1995. Similarities in gene-expression during the post-harvest-induced senescence of spears and natural foliar senescence of asparagus. *Plant Physiology*. 108, 125-8.

Klebs, G. 1910. Alterations in the development and forms of plants as a result of environment. Proceedings of the Royal Society of London 82: 547-548.

Klee, H. J. Hayford, M. B. Kretzmer, K. A. Barry, G. F. and Kishore, G. M. 1991. Control. Of ethylene synthesis by expression of a bacterial enzyme in transgenic tomato plants. *Plant Cell*. 3: 1187-1193.

Knaff, D. B. 1990. The cytochrome $bc_1$ complex of photosynthetic bacteria. Trends in Biochemical Sciences 15: 289-291.

Koiwai A. Matsuzaki T. Suzaki F. Kawashima N. 1981. Changes in total polar lipids and their fatty acid composition in tobacco leaves during growth and senescence. *Plant and Cell Physiology*. 22, 19059-65.

Koizumi M. Yamaguchi-Shinozaki K. Tsuji H. Shinozaki K. 1993. Structure and expression of two genes that encode distinct drought-inducible cysteine proteinases in *Arabidopsis thaliana*. Gene. 129: 175-82.

Kornberg H. L. Beevers H. 1957. The glyoxylate cycle as a stage in the conversion of fat to carbohydrate in castor beans. *Biochimica et Biophysica Acta*. 26. 531-7.

Koukkari, W. L. and S. B. Warde. 1985. Rhythms and their relations to hormones. Pages 37-77 in R. P. Pharis and D. M. Reid eds. Encyclopedia of Plant Physiology, New Series, Vol. 11. Hormonal Regulation of Development III. Role of Environmental Factors. Springer-Verlag, Berlin, Heidelberg.

Kramer, Paul J. 1983. Water Relations of Plants Academic Press, Santa Clara, Calif.

Kramer, Paul J. 1983. Water Relations of Plants. Academic Press, Santa Clara, Calif.

Krautler B. Jaun B. Amrein W. Bortlik K. Schellengerg M. Matile P. 1992. Breakdown of chlorophyll: constitution of a secoporphinoid catabolite isolated from senescent barley leaves. *Plant Physiol Biochem*. 30: 333-346.

Krautler B. Jaun B. Bortlik K. Schellenberg M. Maitle P. 1991. On the enigma of chlorophyll degradation: the constitution of a secoporphinoid catabolite. *Angew Chem Int Ed Engl.* 30: 1315-1318.

Kutschera, U. and P. Schopfer. 1985. Evidence against the acid-growth of maize coleoptiles. Planta 163: 483-493.

Lagendorf, A. T. 1967. The chemiosmotic hypothesis of photophosphorylation. Pages 69-78 in A. San Pietro, F. A. Greer, and T. J. Army eds. Harvesting the Sun, Photosynthesis in Plant Life. Academic Press, New York.

Lanahan M. B. Yen H. C. Giovannoni J. J. Klee H. J. 1994. The Never Ripe mutation blocks ethylene perception in tomato. *The Plant Cell*. 6: 521-30.

Langmeier M. Ginsburg S. Matile P. 1993. Chlorophyll break-down in senescent leaves: demonstration of Mg-dechelatase activity. *Physiol Plant*. 89: 347-353.

Law, C. N. and Rachel Scarth. 1984. Genetics and its potential for understanding the action of light in flowering. Pages 193-209 in D. Vince-Prue, B. Thomas, and K. E. Cockshull eds. Light and the Flowering Process. Academic Press, London.

Lejeune, P. Bernier, G. Requier, M-C. and Kinet, J-M. 1994. Cytokinins in phloem and xylem saps of *Sinapis alba* during floral induction. *Physiol. Plantarum* 90: 522-528.

Leshem, Y. 1984. Interaction of cytokinins with lipid-associated oxy free radicals during senescence: a prospective mode of cytokinin action. *Can. J. Bot*. 62: 2443-2449.

Letham, D. S. 1971. Regulators of cell division in plant tissues. XII. A cytokinin bioassay using excised radish cotyledons. Physiologia Planarum 25: 391-396.

Lichtenthaler H. K. Burgstahler R. Buschmann C. Meir D. Prenzek U. and Schonthal A. 1982. Effect of high light and high light stress on composition, function and structure of the photosynthetic apparatus. In *Stress Effects on Photosynthesis*. Edited by R. Marcelle.

Pp. 353-370. Dr. W. Junk Pubishers, The Hague.

Lincoln J. E. Fischer R. L. 1988. Regulation of gene expression by ethylene in wild-type and rin tomato fruit. *Plant Physiology.* 88, 370-4.

Liu, X. Shiomi, S. Nakatsuka, A. Kubo, Y, Nakamura, R. and Inaba, A. 1999. Characterization of ethylene biosynthesis associated with ripening in banana fruit. *Plant Physiol.* 121: 1257-1265.

Lloyd, Clive W. 1987. The plant cytoskeleton: The impact of fluorescence microscopy. Ann. Rev. Plant Physiol. 38: 119-139.

Lloyd, E. J. 1980. The effects of leaf age and senescence on the distribution of carbon in *Lolium temulentum. J. Exp Bot.* 31: 1067-1079.

Lohman K. N. Gan S. John M. C. Amasino R. M. 1994. Molecular analysis of natural leaf senescence in *Arabidopsis thaliana. Physiologia Plantarum* 92: 322-8.

Lovell, P. H. Lovell, P. J. and Nichols, R. 1987. The importance of the stigma in flower senescence in petunia *Petunia hybrida. Ann. Bot.* 60: 41-47.

Lucas, W. J. and M. A. Madore. 1988. Recent advances in sugar transport. Pages 35-84 in Jack Preiss ed. The Biochemistry of Plants, Vol. 14. Academic Press, New York.

Luttge, U. and D. T. Clarkson. 1989. Mineral nutrition: Potassium. Progress in Botany 50: 51-73.

MacMillan, J. and B. O. Phinney. 1987. Biochemical genetics and the regulation of stem elongation by gibberellins. Pages 156-171 in D. J. Cosgrove and D. P. Knievel eds. Physiology of Cell Expansion During Plant Growth. American Society of Plant Physiologists, Rockville, Md.

Malik N. S. A. and Thimann K. V. 1980. Metabolism of oat leaves during senescence VI. Changes in ATP levels. *Plant Physiol.* 65: 855-858.

Malik N. S. A. 1987. Senescence in oat leaves: changes in translatable mRNAs. *Physiologia Plantarum.* 70: 438-46.

Malkin, R. 1988. Structure-function studies of photosynthetic cytochrome b-$c_1$ and $b_6$-f complexes. ISI Atlas of Science: Biochemistry 10: 57-64.

Marchner, H. 1986. Mineral Nutrition of Higher Plants. Academic Press, London.

Markides S. C. Goldthwaite J. 1981. Biochemical changes during bean leaf growth, maturity and senescence. Contents of DNA, polyribosomes, ribosomal RNA, protein and chlorophyll. *Journal of Experimental Botany.* 32: 725-35.

Marme, D. 1989. The role of calcium and calmodulin in signal transduction. Pages 57-80 in W. F. Boss and D. J. Morre eds. Second Messengers in Plant Growth and Development. Alan R. Liss, New York.

Marre, E. 1979. Fusicoccin: A tool in plant physiology. Annual Review of Plant Physiology 30: 273-312.

Matile P. 1992. Chloroplast senescence. In: Baker NR, Thomas H, eds. *Crop Photosynthesis: Spatial and Temporal Determinants.* Amsterdam: Elsevier, 413-40.

Matile P. Duggelin T. Schllenberg M. Rentsch D. Bortlik K. Peisker C. Thomas H. 1989. How and why is chlorophyll broken down in senescen tleaves? *Plant Physiol Biochem.* 27: 595-604.

Matile P. Ginsburg S. Schellenberg M. Thomas H. 1987. Catabolites of chlorophyll in senescent leaves. *J. Plant Physiol.* 129: 219-228.

Matile P. Ginsburg S. Schellenberg M. Thomas H. 1988. Catabolites of chlorophyll in senescing barley leaves are localized in the vacuoles of mesophyll cells. *Proc. Natl. Acad. Sci.* USA. 85: 9529-9532.

Matile P. Schellenberg M. 1996. The cleavage of phaeophorbide *a* is located in the envelope of barley gerontoplasts. *Plant Physiol Biochem.* 34: 55-59.

Matile P. Schellenberg M. Vicentini F. 1996. Localization of chlorophyllase in the chloroplast envelope. *Planta.* In press.

Mattoo, A. K. J. B. Marder, and M. Edelman. 1989. Dynamics of the photosystem II reaction center. Cell 56: 241-246.

Maunders M. J. and Brown S. B. 1983. The effect of light on chlorophyll loss in senescing leaves of sycamore *Acer Pseudoplatanus L. Plant.* 158: 309-311.

Maxie, E. C. Farnham, D. S. Mitchell, F. G. Sommer, N. F. Pearson, R. A. Snyder, R. G. and Rae, H. L. 1973. Temperature and ethylene effects on cut flowers of carnation *Dianthus caryophyllus* L. *J. Am Soc. Hort. Sci.* 98: 568-572.

McClendon, John H. 1982. Water relations curves for plant cells: Toward a realistic Hofler diagram for textbooks. What's New in Plant Physiology 135: 17-20.

McDaniel, C. N. 1996. Developmental physiology of floral initiation in *Nicotiana tabacum L. J. Exp. Bot.* 47: 465-475.

McDaniel, C. N. and Hartnett, L. K. 1996. Flowering as metamorphosis: Two sequential signals regulate floral initiation in *Lolium temulentum. Development* 122: 3661-3668.

McDaniel, C. N. Hartnett, L. K. and Sangrey, K. A. 1996. Regulation of node number in day-neutral *Nicotiana tabacum*: A factor in plant size. *Plant J.* 9: 56-61.

McDaniel, C. N. Singer, S. R. and Smith, S. M. E. 1992. Developmental states associated with the floral transition. *Dev. Biol.* 153: 59-69.

McLaughlin J. C. Smith S. M. 1994. Metabolic regulation of glyoxylate-cycle enzyme synthesis in detached cucumber cotyledons and protoplasts. *Planta.* 195: 22-8.

McLaughlin J. C. Smith S. M. 1995. Glyoxylate cycle enzyme synthesis during the irreversible phase of senescence of cucumber cotyledons. *Journal of Plant Physiology.* 146, 133-8.

Meier D. and Lichtenthaler H. K. 1983. Ultrastructure of chloroplasts from wheat and maize plants grown at high and low qunata fluence rates. In *Photosynthesis and Plant Productivity.* Edited by H. Metzner. Pp. 2356-238. Wissenschftl. Verlags-gesellschaft. Stuttgart.

Merzlyak M. N. Hendry G. A. F. 1994. Free radical metabolism, pigment degradation and lipid peroxidation in leaves during senescence. *Proceedlings of the Royal Society of Endinburgh.* 102B: 459-71.

Metraux, J-P. 1987. Gibberellins and plant cell elongation. Pages 296-317 in P. J. Davies ed. Plant Hormones and Their Role in Plant Growth and Development. Martinus Nijhoff Publishers, Boston.

Meyerowitz, E. M. 1994. The genetics of flower development. *Sci. Am.* 271: 56-65.

Milburn, John A. 1979. Water Flow in Plants, Longman Group Limited, London. Published in the U. S. by Longman, Inc. New York.

Millar, A. J. and Kay, S. A. 1991. Circadian control of cab gene transcription and messenger RNA accumulation in *Arabidopsis. Plant Cell* 3: 541-550.

Millar, A. J. and Kay, S. A. 1997. The genetics of phototransduction and circadian rhythms in *Arabidopsis. Bioessays* 19: 209-214.

Millar, A. J. Carre, I. A. Strayer, C. A. Chua, N-H. and Kay, S. A. 1995. Circadian clock mutants in *Arabidopsis* identified by luciferase imaging. *Science* 267: 1161-1163.

Millar, A. J. Short, S. R. Chua, N-H. and Kay, S. A. 1992. A novel circadian phenotype based on firefly luciferase expression in transgenic plants. *Plant Cell* 4: 1075-1087.

Mistuhashi W. Feller U. 1992. Effects of high and external solutes on the catabolism of nuclear-encoded stromal proteins in intact chloroplasts isolated from pea leaves. *Plant Physiology.* 100: 2100-5.

Mitchell, P. 1966. Chemical coupling in oxidative and photosynthetic phosphorylation. Biological Reviews 41: 445-502.

Mittler R. Lam E. 1995. Identification, characterization, and purification of a tobacco endonuclease activity induced upon hypersensitive response cell death. *The Plant Cell.* 7, 1951-62.

Mohr H. and Shropshire W. Jr, 1983. An introduction to phtomorphogenesis for the general reader. In *Encyclopedia of Plant Physiology* Edited by W. Shropshire, Jr. and H. Mohr. Vol. 16. Pp. 24-38. Springer-Verlag. Berlin.

Monson, R. K. and B. D. Moore. 1989. On the significance of $C_3$-$C_4$ intermediate photosynthesis to the evolution of $C_4$ photosynthesis. Plant, Cell and Environment 12: 689-699.

Morison, J. I. L. and Jarvis P. G. 1983. Direct and indirect effects of light on stomata. 1. In scotpines and sitka spruce. *Plant Cell Environ.* 6: 95-101.

Mueller, B. 1952. *Goethe's Botanical Writtings.* University of Hawaii Press, Honolulu.

Muhitch, M. J. Edwards L. A. and Fletcher J S 1983. Infuence of diamines of polyamines on the senescence of plant suspension cultures. *Plant Cell. Rep.* 2: 82-84.

Muhlecker W. Hortensteiner S. Ongania K. H. Matile P. Krautler B. 199. A fluorescing chlorophyll catabolite from higher plans. *Chimia.* 50: 346.

Muhlecker W. Krautler B. Ginsburg S. Matile P. 1993. Break-down of chlorophyll: a tetrapyrrolic chlorophyll catabolite from senescent rape leaves. *Helv Chim Acta.* 76: 2976-2980.

Muller-Uri, F. Parthier, B. and Nover, L. 1988. Jasmonate-induced alteration of gene expression in barely leaf segment analyzed by in-vivo and in-vitro protein synthesis. *Planta.* 176: 241-247.

Nagatani, A. J. B. Reid, J. J. Ross, A. Dunnewijk, and M. Furuya. 1990. Internode length in *Pisum.* Type I and II levels in *lv* plants. Journal of Plant Physiology 135: 667-674.

Nakamura H. Kishi Y. Shimomura O. Morse D. Hastings J. W. 1989. Structure of dinoflagellate luciferin and its enzymatic air-oxidation products. *J Am Chem Soc.* 111: 7607-7611.

Nakamura H. Musicki B. Kishi Y. 1988. Structure of the light emitter in krill *Euphausia pacifica.* bioluminescence. *J. Am. Chem. Soc.* 110: 2683-2685.

Napier, R. M. and M. A. Venis. 1990. Receptors for plant growth regulators: Recent advances. Journal of Plant Growth Regulation 9:113-126.

Napp-Zinn, K. 1984. Light and vernalization. In *Light and the Flowering Process*, D. Vince-Prue, B. Thomas, and K. E. Cockshull eds. Academic Press, London, pp. 75-88.

Nejidat A. Itai C. and Bejerano N. R. 1983. Stomatal response to ATP mediated by phytochrome. *Physiol. Plant.* 57: 367-370.

Neves, M. C. P. and M. Hungria. 1987. The physiology of nitrogen fixation in tropical grain legumes. CRC Critical Reviews in Plant Sciences 6: 267-321.

Newton, K. J. 1988. Plant mitochondrial genomes: Organization, expression, and variation. Annual Review of Plant Physiology and Plant Molecular Biology 39: 503-532.

Nichols, R. Buffer, G. Mor, Y. Fujino, D. W. and Reid, M. S. 1983. Changes in ethylene and 1-amino-cylopropane-1-carboxylic acid content of pollinated canation flowers. J. Plant Growth Regul, 21: 8.

Nissen, P. 1991. Multiphasic uptake mechanisms in plants. International Review of Cytology 126: 89-134.

Nonami, Hiroshi, John S. Boyer, and Ernst Steudle. 1987. Pressure probe and isopiestic psychrometer measure similar turgor. Plant Physiology 83: 592-595.

Noodén L. D. 1988. The phenomena of senescence and aging. In: Nooden LD, Leopold AC, eds. *Senescence and Aging in Plants*. San Diego: Academic Press.

Nooden LD, Singh. S. Letham DS, 1990. Correlation of xylem sap levels with monocarpic senescence in soybean. Plant Physiology 93, 33-9.

Nooden, L. D. and A. C. Leopold. Eds. 1988. Senescence and Aging in Plants. Academic Press, New York.

Nooden, L. D. and J. J. Guiamet. 1989. Regulation of assimilation and senescence by the fruit in monocarpic plants. Physiologia Plantarum 77: 267-274.

Nurnberger, T. Abel, S. Jost, W. and Glund, K. 1990. Induction of an extracellular ribonuclease in cultured tobacco cells upon phosphate starvation. *Plant Physiol.* 93: 33-39.

Oeller, P. W. Li, M. W. Taylor, L. P. Pike, D. A. and Theologis, A. 1991. Reversible inhibition of tomato fruit senescence by antisense RNA. *Science* 254: 437-439.

Ogawa, Yukiyoshi. 1981. Stimulation of the flowering of *Pharbitis nil* Chois. By gibberellin $A_3$: Time dependent action at the apex. Plant & Cell Physiol. 224: 675-681.

Oh, S. A. Lee, S. Y. Chung, I. K. Lee, C.H. and Nam, H. G. 1996. A senescence-associated gene of Arabidopsis thaliana is distinctively regulated during natural and artificially induced leaf senescence. *Plant Mol. Biol.* 30: 739-754.

Okamuro, J. K. den Boer, B. G. W. and Jofuku, K. D. 1993. Regulation of *Arabidopsis* flower development. *Plant Cell* 5: 1183-1193.

Okamuro, J. K. den Boer, B. G. W. Lotys-Prass, C. Szeto, W. and Jofuku, K. D. 1996. Flowers into shoots: Photo and hormonal control of a meristem identity switch in *Arabidopsis. Proc. Natl. Acad. Sci. USA* 93: 13831-13836.

Okamuro, J. K. Szeto, W. Lotys-Prass, C. and Jofuku, K. D. 1997. Photo and hormonal control of meristem identity in the *Arabidopsis* flower mutants *apetala2* and *apetala1. Plant Cell* 9: 37-47.

Osborne, B. A. and R. J. Geider. 1987. The minimum photon requirements for photosynthesis. New Phytologist 106: 631-634.

Osborne, D. J. M. T. McManus, and J. Webb. 1985. Target cells for ethylene action. Pages 197-

212 in J. A. Roberts and G. A. Tucker eds. Ethylene and Plant Development. Butterworths, London.

Oshio Y. Hase E. 1969. Studies on red pigments excreted by cells of *Chlorella protothecoides* during the process of bleaching induced by glucose or acetate. I. Chemical properties of the red pigments. *Plant Cell Physiol.* 10: 51-59.

Owen, J. H. 1988. Role of abscisic acid in a $Ca^{2+}$ second messenger system. Physiologia Plantarum 72: 637-641.

Panigrahi P. K. and Biswal U. C. 1983. Aging of chloroplasts *in vitro*-III. Comparison of the effects of aging *in vitro* and senescence on the red absorption band and primary photochemical reaction of sunflower chloroplast. *Photosynthesis Res.* 4: 183-189.

Papenfuss, H. D. and Salisbury, F. B. 1967. Aspects of clock resetting in flowering of *Xanthium. Plant Physiol.* 42: 1562-1568.

Park K. Y. Drory A. Woodson W. R. 1992. Molecular cloning of a 1-aminocyclopropane-1-carboxylate synthase from senescing carnation flower petals. *Plant Molecular Biology.* 18: 377-86.

Park Y. Morris M. M. Mackinney G. 1973. ON chlorophyll break down in senescent leaves. *J. Agric. Food. Chem.* 21: 279-281.

Parthier, B. 1990. Jasmonates: Hormonal regulators or stress factors in leaf senescence? Journal of Plant Growth Regulation 9: 57-63.

Passioura, J. B. 1982. Water in the soil-plant-atmosphere continuum. Pages 5-33 in O. L. Lange, P. S. Nobel, C. B. Osmond, and H. Ziegler eds. Encyclopedia of Plant Physiology, New Series, Vol. 12B, Physiological Plant Ecology II. Springer-Verlag, New York, Berlin, Heidelberg.

Passioura, J. B. 1988. Response to Dr. P. J. Kramer's article, "Changing concepts regarding plant water relations," Volume 11, Number 7, pp. 565-568.

Pastori G. M. del Rio L. A. 1994. An activated-oxygen-mediated role for peroxisomes in the mechanism of senescence of *Pisum sativam* L. leaves. *Planta.* 193, 385-91.

Pastori, G. M. and Del Rio, L. A. 1997. Natural senescence of pea leaves: An activated-oxygen mediated function for peroxisomes. *Plant Physiol.* 113: 411-418.

Patrick, J. W. 1990. Sieve element unloading: Cellular pathway, mechanism and control. Physiologia Plantarum 78: 298-308.

Patterson, B. D. and D. Graham. 1987. Temperature and metabolism. Pages 153-199 in D. D. Davies Ed. The Biochemistry of Plants, Vol. 12, Physiology of Metabolism. Academic Press, New York.

Peisker C. 1991. *Chlorophyllabbau in vergilbenden Gerstenblattern: Entwicklung und Chlorophyll.* Ph. D. Thesis, University of Zurich, Switzerland.

Peisker C. Thomas H. Matile P. 1990. Radiolabelling of chlorophyll for studies on catabolism. *J. Plant Physiol.* 136: 544-549.

Peisker, C. Duggelin T. Rentsch D. Matile P. 1989. Phytol and the breakdown of chlorophyll in senescent leaves. *J. Plant. Physiol.* 135: 428-432.

Pennazio, S. 1981. Changes in the soluble protein constitution of Gomphrena globosa leaves showing spontaneous local lesions. *Rivista di Patologia Vegetale.* 17: 127-135.

Permadasa, M. A. 1981. Photocontrol of stomatal movements. Biology Review 56: 551-588.

Pfeiffer H. and Kleudgen H. K. 1980. Investigations on the phytochrome control of senescence in the photosynthetic apparatus of *Hordeum Vulgare* L *Z. Pflanzenphysiol.* 100: 437-445.

Pharis, R. P. and King, R. W. 1985. Gibberellins and reproductive development in seed plants. *Annu. Rev. Plant Physiol.* 36: 517-568.

Pharis, R. P. Evans, L. T. King, R. W. and Mander, L. N. 1987. Gibberellins, endogenous and applied, in relation to flower induction in the long-day plant *Lolium temulentum. Plant Physiol.* 84: 1132-1138.

Pharis, Richard P. Lloyd T. Evans, Roderick W. King, and Lewis N. Mander. 1989. Gibberellins and flowering in higher plants-differing structures yield highly specific effects. Pages 29-41 in Elizabeth Lord and Georges Bernier eds. Plant Reproduction: From Floral Induction to Pollination. American Society of Plant Physiologists Symposium Series Vol. 1, Amer. Soc. of Plant Physiol. 15501-A Monona Drive, Rockville, Md. 20855.

Picton S. Barton S. L. Bouzayen M. Hamilton A. J. Grierson D. 1993. Altered fruit ripening and leaf senescence in tomatoes expressing an antisense ethylene-forming enzyme transgene. *Plant Journal.* 3: 469-81.

Pilbeam, C. J. 1992. Effect of nitrogen supply on the growth and senescence of leaves of *Lilium perenne* with contrasting rates of leaf respiration. *Ann. Bot.* 70: 365-370.

Pilbeam, D. J. and E. A. Kirkby. 1983. The physiological role of boron in plants. Journal of plant Nutrition 6: 563-582.

Pjon C. J. 1981. Effects of cyclohexamide and light on leaf senescence in maize and hydrangea. *Plant Cell Physiol.* 22: 847-854.

Plumb, R. C. and W. B. Bridgman. 1972. Ascent of sap in trees. Science 176: 1129-1131.

Poethig, R. S. 1990. Phase change and the regulation of shoot morphogenesis in plants. *Science* 250: 923-930.

Pogson B. J. Downs C. G. Davies K. M. 1995a. Differential expression of two 1-aminocyclopropane-1-carboxylic acid oxidase genes in broccoli after harvest. *Plant Physiology.* 108, 651-7.

Pogson B. J. Downs C. G. Davies K. M. Morris S. C. Buchanan-Wollaston V. 1995b. Nucleotide sequence of a cDNA clone from broccoli with high identity with the PSST aubunit of NADH: ubiquinone oxidoreductase. *Plant Physiology.* 108, 859-60.

Portis, A. R. Jr. 1990. Rubisco activase. Biochimica et Biophysica Acta 1015: 15-28.

Preiss, J. 1984. Starch, sucrose biosynthesis and partition of carbon in plants are regulated by ortho-phosphate and triose-phosphates. Trends in Biochemical Sciences 9: 24-27.

Preston, R. D. 1952. Movement of water in higher plants. Pages 257-321 in A. Frey-Wyssling ed. Deformation and Flow in Biological Systems. Elsevier North-Holland, Amsterdam.

Prince, R. C. 1988. The proton pump of cytochrome oxidase. Trends in Biochemical Sciences 13: 159-160.

Purohit, S. S. ed. 1984-1993. Hormonal Regulation of Plant Growth and Development. Agro Botanical Publishers, Bikaner.

Purvis, O. N. and Gregory, F. G. 1952. Studies in vernalization of cerals. XII. The reversibility by high temperature of the vernalized condition in Petkus winter rye. *Ann. Bot.* 1: 569-592.

Putterill, J. Robson, F. Lee, K. Simon, R. and Coupland, G. 1995. The *CONSTANS* gene of *Arabidopsis* promotes flowering and encodes a protein showing similarities to zinc finger transcription factors. *Cell* 80: 847-857.

Quatrano, R. S. 1987. The role of hormones during seed development. Pages 494-514 in P. J. Davies ed. Plant Hormones and Their Role in Plant Growth and Development. Martinus Nijhoff, Boston.

Ranjan, R. Chojnacki, T. and Swiezewska, E. 2001. On the occurrence of polyprenols in plants. *Polish J. Bot.* accepted for publication.

Ranjan, R. Marczewski, A. Chojnacki, T. Hertal, J. and Swiezewska, E. 2001. Polyprenols and the phenomenon of the senescence in plants. *Acta Biochim. Polon.* Accepted for publication.

Rao G. N. 1981. Delaying of petal senescence in excised flowers of Dombeya by certain plant growth regulators. *Ind J Expp Biol.* 21: 47-48.

Raschke, K. 1975. Stomatal action. Ann. Rev. of Plant Physiol. 26: 309-340.

Raschke, K. 1976. How stomata resolve the dilemma of opposing priorities. Phil. Trans. R. Soc. Lond. B. 273: 551-560.

Raschke, K. 1987. Action of abscisic acid on guard cells. Pages 253-270 in E. Zeiger, G. D. Farquhar, and I. R. Cowan eds. Stomatal Function. Stanford University Press, Stanford Function. Stanford University Press, Stanford, Calif.

Raval M. K. Mishra A N and Biswal U. C. 1982. Changes in the membranes of wheat chloroplasts aging *in vitro*. *Photobiochem Photobiophys.* 4: 293-297.

Raven, J. A. 1988. Requisition of nitrogen by the shoots of land plants: Its occurrence and implications for acid-base regulation. New Phytologist 109: 1-20.

Raven, J. A. L. L. Handley, J. J. MacFarlane, S. McInroy, L. McKenzie, J. H. Richard, and G. Samuelsson. 1988. Tansley Review No. 13: The role of $CO_2$ uptake by roots and CAM in acquisition of inorganic C by plants of the isoetid life-form: A review, with new data on Eriocaulon decangulare L. New Phytologist 108: 125-248.

Rechsteiner M. 1991. Natural substartes of the ubiquitin proteolytic pathway *Cell.* 66, 615-18.

Reed, J. W. Nagatani, A. Elich, T. D. Fagan, M. and Chory, J. 1994. Phytochrome A and

Phytochrome B have overlapping but distinct functions in *Arabidopsis* development. *Plant Physiol.* 104: 1139-1149.

Reid, J. B. 1987. The genetic control of growth via hormones. Pages 318-340 in P. J. Davies ed. Plant Hormones and Their Role in Plant Growth and Development. Martinus Nijhoff, Boston.

Reid, J. B. Murfet, I. C. Singer, S. R. Weller, J. L. and Taylor, S. A. 1996. Physiological genetics of flowering in *Pisum. Sem. Cell Dev. Biol.* 7: 455-463.

Reid, M. S. and Wu, M-J. 1992. Ethylene and flower senescence. *Plant Growth Regul.* 11: 37-43.

Reilly, P. and N. Nelson. 1988. Photosystem I complex. Photosynthesis Research 19: 73-84.

Reinberg, A. 1971. La chronobiologie, Recherche 2: 242-250.

Reinbothe S. Mollenhauer B. Reinbothe C. 1994. JIPs and RIPs: the regulation of plant gene expression by jasmonates in response to environmental cues and pathogens. *The Plant Cell.* 6: 1197-1209.

Rhie G. Beale S. I. 1992. Biosynthesis of phycobilins. Ferrdoxin-supported NADPH-independent oxygenase and phycobilin-forming activities from *Cyanidium caldarium. J. Biol. Chem.* 267: 16088-16093.

Ricard, J. 1980. Enzyme flexibility as a molecular basis for metabolic control. Pages 31-80 in D. D. Davies Ed. The Biochemistry of Plants, Vol. 2, Metabolism and Respiration. Academic Press, New York.

Ricard, J. 1987. Enzyme regulation. Pages 69-105 in D. D. Davies Ed. The Biochemistry of Plants, Vol. 11, Biochemistry of Metabolism. Academic Press, New York.

Richards, F. C. 1969. The quantitative analysis of growth. Pages 2-76 in F. C. Stewart ed. Plant Physiology, Vol. 5A, Analysis of Growth: Behaviour of Plants and Their Organs. Academic Press, New York.

Robin, G. 1982. Movements and bioelectrical events induced by photostimulation in the primary pulvinus of *Mimosa pudica*. Zeitschrift fur Pflanzenphysiologie 106: 299-303.

Rohmer, M. Seeman, M. Horbach, S. Bringer-Meyer, S. and Sahm, H. 1996. Glyceraldehyde 3-phosphate and pyruvate as precursors of isoprenic units in an alternative non-mevalonate pathway for terpenoid biosynthesis. *J. Am. Chem. Soc.* 118: 2564-2566.

Rohwer, F and Scheirle J j1982. Efect of light on ethylene production red light enhancement of 1-amino cyclopropane-1-carboxylic acid concentration in etiolated pea shoots. *Z. Pflanzenphysiol.* 107: 295-300.

Rolfe, B. G. and P. M. Gresshoff. 1988. Genetic analysis of legume nodule initiation. Annual Review of Plant Physiology and Plant Molecular Biology 39: 297-319.

Ross, Stephen D. Mark P. Bollmann, Richard P. Pharis, and Geoffrey B. Sweet. 1984. Gibberellin $A_{4/7}$ and the promotion of flowering in *Pinus radiata*. Plant Physiol. 76: 326-330.

Rousseaux, M. C. Hall, A. J. and Sanchez, R. A. 1996. Farred enrichment and photosynthetically active radiation level influence leaf senescence in field growth sunflower. *Physiol Plant.* 96: 217-224.

Ryu S. B. Wang X. M. 1995. Expression of phospholipase-D during castor bean senescence. *Plant Physiology.* 108, 713-19.

Sabater, M. and P. H. Rubery. 1987. Auxin carriers in Cucurbita vesicles. Planta 171: 507-513.

Salisbury, Frank B. 1963. Biological timing and hormone synthesis in flowering in Xanthium. Planta 49: 518-534.

Salisbury, Frank B. 1981. Response to photoperiod. Pages 135-167 in O. L. Lange, P. S. Nobel, C. B. Osmond, and H. Ziegler eds. Physiological Plant Ecology I. Vol. 12A. In A. Pirson, M. H. Zimmermann eds. Encyclopedia of Plant Physiology, New Series. Springer-Verlag, Berlin, Heidelberg.

Salisbury, Frank B. George G. Spomer, Martha Sobral, and Richard T. Ward. 1968. Analysis of an alpine environment. Botanical Gazette 1291: 16-32.

Sanchez, R. A. L. De Miguel, and O. Mercuri. 1986. Phytochrome control of cellulase activity in *Datura ferox* L. seeds and its relationship with germination. Journal of Experimental Botany 37: 1574-1580.

Satler S. O. and Thimann K. V. 1983. Metabolism of oat leaves during senescence VII. The interaction of carbon dioxide and other atmospheric gases with light in controlling chlorophyll loss and senescence. *Plant physiol.* 71: 67-70.

Satoh, H. Nakayama, K. and Okada, M. 1998. Molecular cloning and functional expression of a water soluble chlorophyll molecules in cauliflower. *J. Biol. Chem.* 273: 30568-30575.

Scaife, A. and M. Turner. 1984. Diagnosis of Mineral Disorders in Plants, Vol. 2, Vegetables. Chemical Publishing Co. New York.

Schafer, E. T-U. Lassig, and P. Schopfer. 1982. Phytochrome-controlled extension growth of *Avena sativa* L. seedlings. II. Fluence rate response relationships and action spectra of mesocotyl and coleoptile responses. Planta 154: 231-240.

Scheffner, M. Nuber, U. and Huibregtse, J. M. 1995. Protein ubiquitination involving and E1-E2-E3 enzyme ubiquitin thioester cascade. *Nature.* 373: 81-83.

Schellenberg M. Matile P. Thomas H. 1993. Production of apresumptive chlorophyll catabolite in vitro: requirement for reduced ferredoxin. *Planta.* 191: 417-420.

Schellenberg M. Matile P. Thoms H. 1990. Breakdown of chlorophyll in chloroplasts of senescent barley leaves depends on ATP. *J. Plant Physiol.* 136: 564-568.

Schmidt A, Jager K. 1992. Open questions about sulfur metabolism in plants. Annual Review of Plant Physiology and Plant Molecular Biology 43, 325-49.

Schmutz D, Wyss H R and Brunold C 9183. Activity of sulfate-assimilating enzymes in primary leaves of *Phaseolus vulgaris L.* during dark induced senescence *Z. Pflanzenphysiol.* 110: 211-220.

Schrempf, M. Ruth L. Satter, and Arthur W. Galston. 1976. Potassium-linked chloride fluxes during rhythmic leaf movement of *Albizzia julibrissin.* Plant Physiology 58: 190-192.

Schulz, A. F. Wengenmayer, and H. M. Goodman. 1990. Genetic engineering of herbicide resistance in higher plants. CRC Critical Reviews in Plant Sciences 9:1-15.

Schulze, E-D. E. Steudle, T. Gollan, and U. Schurr. 1988. Response to Dr. P. J. Kramer's article, "Changing concepts regarding plant water relations," Volume 11, Number 7, pp. 565-568. Plant, Cella nd Environment 11: 573-576.

Schulze, E-D. J. Cermak, R. Matyssek, M. Penka, R. Zimmermann, E. Vasicek, W. Gries, and J. Jucera. 1985. Canopy transpirationa nd water fluxes in the xylem of the trunk of Larix and Picea trees- a comparison of xylem flow, porometer and cuvette measurements. Oecologia Berlin. 66: 475-483.

Schwarz-Sommer, Z. Huijser, P. Nacken, W. Saedler, H. and Sommer, H. 1990. Genetic control of flower development: Homeotic genes in *Antirrhinum majus. Science* 250: 931-936.

Shackel, Kenneth A. 1987. Direct measurement of turgor and osmotic potential in individual epidermal cells. Plant Physiology 83: 719-722.

Shankin J. DeWitt N. D. Flanagan J. M. 1995. The stroma of higher plant plastids contain Clpp and ClpC, functional homologs of *Escherichia coli* Clpp and ClpA: an archetypal two-component ATP-dependent protease. *The Plant Cell.* 7, 1713-22.

Shepard, James F. 1982. The regeneration of potato plants from leaf-cell protoplasts. Scientific American 2465: 154-166.

Shih, L. M. Sawhney R. K. Fuhrer J. Samanta S. and Glaston A. W. 1982. Effects of exogenous 1,3-Diamino propane and spermidine on senescence of oat leaves I. Inhibition of protease activity, ethlene production and chlorophyll loss as related to polyaminc content. *Plant Physiol.* 70: 1592-1596.

Shioi Y. Sasa T. 1986. Purification of solubilized chlorophyllase from *Chlorella protothecoides. Methods Enzymol.* 123: 421-427.

Shioi Y. Tatsumi Y. Shimokawa K. 1991. Enzymatic degradation of chlorophyll in *Chenopodium album. Plant Cell Physiol.* 32: 87-93.

Shioi Y. Tomita N. Tsuchiya T. Takamiya K. 1996. Conversion of chlorophyllide to pheeophorbide by Mg-dechelating substance in extracts of *Chenopodium album. Plant Physiol. Biochem.* 34: 41-47.

Sibaoka, Takao. 1969. Physiology of rapid movements in higher plants. Annual Review of Plant Physiology 20: 165-184.

Simon, R, Igeno, M. I. and Coupland, G. 1996. Activation of floral meristem identity genes in *Arabidopsis. Nature* 384: 59-62.

Sinclair, T. R. C. B. Tanner, and J. M. Bennett. 1984. Water-use efficiency in crop production. Bio-Science 341: 36-40.

Singh, S. Letham, D. S. and Palni, L. M. S. 1992. Effect of growth substances and mineral nutrients on cytokinin levels and senescence of tobacco *Nicotiana rustica*. Leaves. In: Karssen, C. M. van loon, L. C. Vreugdenhil, D. eds. *Progress in Plant Growth Regulation*. Dordrecht: Kluwer Academic Publishers, 628-634.

Sisler, E. C. 1990. Ethylene-binding receptors- Is there more than one? Pages 193-200 in S. Rood and R. P. Pharis eds. Plant Growth Regulators 1988. Springer-Verlag, Berlin.

Skoog, F. and D. J. Armstrong. 1970. Cytokinins. Annural Review of Plant Physiology 21: 359-384.

Slatyer, Ralph O. 1967. Plant-Water Relationships, Academic Press, London, New York.

Smart C. M. 1994. Gene experssion during leaf senescence. *New Phytol*. 126: 419-448.

Smart C. M. 1994. Gene expression during leaf senescence. *New Phytology*. 126, 419-48.

Smart C. M. Hosken S. E. Thomas H. Greaves J. A. Blair B. C. Schuch W. 1995. The timing of maize leaf senescence and characterization of senescence-related cDNAs. *Physiol Plant*. 93: 673-682.

Smart C. M. Hosken S. E. Thomas H. Greaves J. A. Blair B. G. Suhuch W. 1995. The timing of maize leaf senescence and characterization of senescence-related cDNAs. *Physiologia Plantarum*. 93, 673-82.

Smart C. M. Scofield S. R. Bevan M. W. Dyer T. A. 1991. Delayed leaf senescence in tobacco plants transformed with tmr, a gene for cytokinin production in Agrobacterium. *The Plant Cell*. 3, 647-56.

Smart, C. M. 1994. Gene expression during leaf senescence. *New Phytol*. 126: 419-448.

Solomonson, L. P. and M. J. Barber. 1990. Assimilatory nitrate reductase: Functional properties and regulation. Annual Review of Plant Physiology and Plant Molecular Biology 41: 225-253.

Spalding, E. P. and D. J. Cosgrove. 1988. Large plasma-membrane depolarization precedes rapid blue-light-induced growth inhibition in cucumber. Planta 178: 407-410.

Sponsel, V. M. 1987. Gibberellin biosynthesis and metabolism. Pages 43-75 in P. J. Davies ed. Plant Hormones and Their Role in Plant Growth and Development. Martinus Nijhoff Publishers, Boston.

Stein, W. D. 1986. Transport and Diffusion across Cell Membranes. Academic Press, New York.

Steinback, K. E. S. Bonitz, C. J. Arntzen, and L. Bogorad. Eds. 1985. Molecular Biology of the Photosynthetic Apparatus. Cold Spring Harbor Laboratory, Cold Spring, New York.

Steintz B. Cohen A. and Leshem B. 1980. Factors controlling the retardation of Chlorophyll degradation during senescence of detached statice *Limonium sinuatium*. Flower stalks. *Z. Pflanzenphysiol*. 100: 343-349.

Stitt, M. 1990. Fructose-2,6-bisphosphate as a regulatory molecule in plants. Annual Review of Plant Physiology and Plant Molecular Biology 41: 153-185.

Stitt, M. and W. P. Quick. 1989. Photosynthetic carbon partitioning: Its regulation and possibilities for manipulation. Physiologia Plantarum 77: 633-641.

Stoddart J. L. and Thomas H. 1982. Leaf senescence. In *Encyclopedia of Plant Physiology* Edited D. Boutler and B. Parthier. Vol. 14A. pp. 592-636. Springer-Verlag. Berlin.

Stoddart L. Thomas H. 1982. Leaf senescence. In: *Encyclopedia of Plant Physiology*, New series, Vol. 14A. Springer Verlag, 592-636.

Stoddart, J. L. and H. Thomas. 1982. Leaf senescence. Pages 592-636 in D. Boulter and B. Parthier eds. Encyclopedia of Plant Physiology, New Series, Vol. 14A, Nucleic Acids and Proteins in Plants I. Springer-Verlag, Berlin.

Sung, Z. R. Belachew, A. Shunong, B. and Bertrand-Garcia, R. 1992. EMF, an *Arabidopsis* gene required for vegetative shoot development. *Science* 258: 1645-1647.

Sweeney, Beatrice M. 1987. Rhythmic Phenomena in Plants, Second Edition. Academic Press, San Diego.

Swiezewska, E. Dallner, G. Andersson, B. and Ernster, L. 1993. Biosynthesis of ubiquinone and Plastoquinone in the endoplasmic reticulum-Golgi membranes of spinach leaves. *J. Biol Chem*. 268: 1494-1499.

Szamosi I. Shaner D. L. Singh B. K. 1993. Identification and characterization of a biodegradative form of threonine dehydratase in senescing tomato Lycopersicon esculentun leaf. *Plant Physiology*. 101, 999-1004.

Takahashi, N. B. O. Phinney, and J. MacMillan. Eds. 1990. Gibberellins. Springer-Verlag,

Berlin.

Taylor C. B. Bariola P. A. Delcardayre S. B. Raines R. T. Green P. J. 1993. RNS2-a senescence-associated RNAse of Arabidopsis that diverged from the sRNAses before speciation. *Proceedings of the National Academy of Sciences* USA. 90: 5118-22.

Taylor, S. A. and R. O. Slatyer. 1962. Proposals for a unified terminology in studies of plant-soil-water relations. Arid Zone Res. 16: 339-349.

Taylor, S. A. and Murfet, I. C. 1996. Flowering in *Pisum*: Identification of a new *ppd* allele and its physiological action as revealed by grafting. *Physiol. Plantarum* 97: 719-723.

Teramoto H. Toyama T. Takeba G, Tsuji H. 1995. Changes in expression of cytokinin-repressed genes, CR9 and CR20, in relation to aging, greening and wounding in cucumber. *Planta.* 196, 387-95.

Theil E. C. 1987. Ferritin: structure, gene regulation, and cellular function in animals, plants and microoragnisms. *Ann. Rev. Bioch.* 56: 289-315.

Theologis A. 1993. One rotten apple spoile the whole bushel: the role of ethylene in fruit ripening. *Cell.* 70, 181-4.

Thimann K. V. and Satler S. O. 1979. Relation between leaf senescence and stomatal closure: senescence in light. *Proc. Natl. Acad Sci USA.* 76: 2295-2298.

Thimann K. V. 1980. The senescence of leaves. In: Thimann KV, ed. *Senescence in Plants.* Boca Raton: CRC Press, 85-115.

Thimann, Kenneth V. and Z.Y. Tan. 1988. The dependence of stomatal closure on protein synthesis. Plant Physiology 86: 341-343.

Thomas H, Hilditch P 1987. Metabolism of thylakoid membrane proteins during foliar senescence. *In* WW Thompson, EA Nothnagel, RC Huffaker eds. *Plant Senescence: Its Biochemistry and Physiology. The American Society of Plant Physiologists*, Rock-ville, MD, pp. 114-122.

Thomas H, Smart CM. 1993. Crops that stay green. *Ann. App. Biol.* 123: 193-219.

Thomas H. Bortlik K. Rentsch D. Schellenberg M. Matile P. 1989. Catabolism of chlorophyll in vivo: significance of polar chlorophyll catabolites in a non-yellowing senescence mutant of *Festuca pratensis* Huds. New Phytol 111: 3-8.

Thomas H. Schellenberg M. Vicentini F. Matile P. 1996. Gregor Mendel's green and yellow pea seeds. *Bot Acta.* 190: 3-4.

Thomas H. Smat C. M. 1993. Crops that stay green. *Ann Appl Biol.* 123: 193-219.

Thomas, H. 1997. Chlorophyll: a symptom and a regulator of plastid development. *New Phytol.* 136: 163-181.

Thomas, H. Ougham H. J. Davies T. G. E. 1992. Leaf senescence in a non-yellowing mutant of Festuca partensis-transcripts and translation products. *Journal of Plant Physiology.* 139, 403-12.

Thomas, H. Stoddart, J. L. and Potter, J. F. 1980. Temperature responses of membrane associated activities from spring and winter oats. *Plant, Cell and Environ.* 3: 271-277.

Thomas, T. H. 1989. Gibbrellin involvement in dormancy-break and germination of seeds of celery *Apium graveolens* L. Journal of Plant Growth Regulation 8: 225-261.

Thompson JE, Paliyath G. Brown JH, Duxbury CL. 1987. The involvement of activated oxygen in membrane deterioration during senescence. In: Thomson WW. Nothnagel EA. Huffaker RC. eds. Plant senescence: its biochemistry and physiology Rockville, MD: The American Society of Plant Physiologists, 146-55.

Tibbitts, T. W. and W. M. Hertzberg. 1978. Growth and epinasty of marigold plants maintained from emergence on horizontal clinostats. Plant Physiology 61: 199-203.

Tinker, B. and A. Lauchli. Eds. 1988. Advances in Plant Nutrition, Vol. 3. Praeger, New York.

Titus, J. S. 1989. Nitrogen recycling in the apple *Malus domestica* Borkh. *Ann. Sci. Fores.* 46 Suppl.: 654-659.

Tomsett, A. B. and D. A. Thurman. 1988. Molecular biology of metal tolerances of plants. Plant, Cell and Environment 11: 383-394.

Trebitsh T. Goldschmidt E. E. Riov J. 1993. Ethylene induces de novo synthesis of a chlorophyll degrading enzyme in *Citrus* fruit peel. *Proc. Natl Acad Sci.* USA 90: 9441-9445.

Trewavas, A. J. Ed. 1986. Molecular and Cellular Aspects of Calcium in Plant Development. Plenum, New York.

Trewavas, A. J. 1982. Growth substance sensitivity: The limiting factor in plant development. Physiologia Plantarum 55: 60-72.

Trewavas, A. J. 1987. Sensitivity and sensory adaptation in growth substance responses. Pages 19-38 in g. V. Hoad, M. B. Jackson, J. R. Lenton, and R. K. Atkin eds. Hormone Action in Plant Development. Butterworths, London.

Trewavas, A. J. and R. E. Cleland. 1983. Is plant development regulated by changes in the concentration of growth substances or by changes in the sensitivity to growth substances? Trends in Biochemical Sciences 8: 354-357.

Tucker D. J. 1981. Phytochrome regulation of leaf senescence in cucumber and tomato. *Plant Sci Lett.* 23: 103-108.

Turgeon, Robert. 1989. The sink-source transition in leaves. Ann. Rev. Plant Physiol. Plant Mol. Biol. 40: 119-138.

Tyree, M. T. and P. G. Jarvis. 1982. Water in tissues and cells. Pages 35-77 in O. L. Lange, P. S. Nobel, C. B. Osmond, and H. Ziegler eds. Encyclopedia of Plant Physiology, New Series, Vol. 12B, Physiological Plant Ecology. Springer-Verlag, New York, Berlin, Heidelberg.

Tyree, M. T. and P. G. Jarvis. 1982. Water in tissues and cells. Pages 35-78 in Lange, O. L. P. S. Nobel, C. B. Osmond and H. Ziegler eds. Encyclopedia of Plant Physiology, New Series, Vol. 12B, Physiological Plant Ecology II. Springer-Verlag, Berlin, Heidelberg, New York.

Tyree, M. T. and J. S. Sperry. 1989. Vulnerability of xylem to cavitationa nd embolism. Annual Review of Plant Physiology 40: 19-38.

Tyree, Melvin T. and Michael A. Dixon. 1986. Water stress induced cavitationa dn embolism in some woody plants. Physiologia Plantarum 66: 397-405.

Valpuesta C, Lange NE, Guerrero C, Reid MS. 1995. Regulation of a cysteine protease accompanies the ethylene-insensitive senescence of day lily Hemerocallis. Flowers Plant Molecular Biology 28, 575-82.

Valtersson, C. van Duijn, G. Verkleij, A. Chojnacki, T. de Kruijf, B. and Dallner, G. 1985. The influence of dolichol, dolichol esters and dolichyl phosphate on phospholipid polymorphism and fluidity in model membrane. J. Bio. Chem. 260: 2742-2751.

van Staden J. Cook E. L. Nooden L. D. 1988. Cytokinins and senescence. In: Nooden LD, Leopold AC, eds. *Senescence and Aging in Plants*. San Diego: Academic Press, 281-328.

Van Staden, J. E. L. Cook, and L. D. Nooden, 1988. Pages 281-328 in L. D. Nooden and A. C. Leopold eds. Senescence and Aging in Plants. Academic Press, New York.

Vanderhoef, L. N. C. Stahl, N. Siegel, and R. Zeigler. 1973. The inhibition by cytokinin of auxin-promoted elongation in excised soybean hypocotyl. Physiologia Plantarum 29: 22-27.

Vein, H. and van de Geijn, S. C. 1978. Mobility and ionic form of silver as related to longevity of cut carnations. *Planta.* 140: 93-96.

Vicentini F. Hortensteiner S. Schellenberg M. Thomas H. Matile P. 1995. Chlorophyll breakdown in senescent leaves: identification of the biochemical lesion in a stay-green genotype of *Festuca pratensis* Huds. *New Physiologist* 129, 247-52.

Vicentini F. Hortensteiner S. Schellenberg M. Thomas H. Matile P. 1995a. Chlorophyll breakdown in senescent leaves: identification c of the lesion in a stay-green genotype of *Festuca pratensis. New Phytol.* 129: 247-252.

Vicentini F. Iten F. Matile P. 1995b. Development of an assay for Mg-dechelatase of oilseed rape cotyledons, using chlorophyllin as the substrate. *Physiol Plant.* 94: 57-63.

Vicentini F. Matile P. 1993. Gerontosomes, a multi-functional type of peroxisome in senescent leaves. *Journal of Plant Physiology.* 142: 50-6.

Vierstra R. D. 1993. Protein degradation in plants. *Annual Review of Plant Physiology and Molecular Biology.* 44: 385-410.

Vierstra, R. D. and P. H. Quail. 1986. The protein. Pages 35-60 in R. E. Kendrick and G. H. M. Kronenberg eds. Photomorphogenesis in Plants. Martinus Nijhoff, Boston.

Vince-Prue, Daphne and K. E. Cockshull. 1981. Photoperiodism and crop production. Pages 175-197 in C. B. Johnson ed. Physiological Process Limiting Plant Productivity. Butterworths, London.

Vogelmann, T. C. 1989. Yearly review: Penetration of light into plants. Photochemistry and Photobiology 50: 895-902.

Wagner E. 1976. The nature of photoperiodic time measurement: energy transduction and phytochrome action in seedlings of *Chenopodium rubrum.* In *Light and Plant Development* Edited by H. Smith. pp. 419-443. Butterworths, London.

Walsh, K. B. M. E. McCuly, and M. J. Canny. 1989. Vascular transport and soybean nodule function: Nodule xylem is a blind alley, not a throughway. Plant, Cell and Environment 12: 395-405.

Wang, J. W. and Kao, C. H. 1999. Endogenous abscisic acid levels are not linked to methyl jasmonate promoted senescence of detached rice leaves. Plant Growth Regul. 28: 17-20.

Wanner L. Keller F, Matile P. 1991. Metabolism of radiolabelled galactolipids in senescent barley leaves. *Plant Science.* 78, 199-296.

Wardle K. and Short K. C. 1983. Stomatal responses and the senescence of leaves. *Ann. Bot.* 52: 411-412.

Wareing, P. F. and Thompson A. G. 1976. Rapid effects of red light on hormone levels. In *Light and Plant Development* Edited by H. Smith. pp. 285-294. Butterworths, London.

Waring, R. H. and B. D. Cleary. 1967. Plant moisture stress: Evaluation by pressure bomb. Science 155: 1248-1254.

Watanabe H. Abe K. Emori Y. Hosoyama H. Arai S. 1991. Molecular cloning and gibberellin-induced expression of multiple cysteine proteinases of rice seeds oryzains. *Journal of Biologial Chemistry.* 266, 16897-902.

Watanabe, S. and T. Sibaoka. 1983. Light- and auxin-induced leaflet opening indetached pinae of *Mimosa pudica.* Plant and Cell Physiology 24: 641-647.

Watson D. J. 1952. The physiological basis for variation in yield. *Advances in Agronomy.* 4, 101-45.

Weidhase, R. A. Kramell, H. M. Lehmann, J. Liebisch, H. W. Lerbs, W. and Parthier, B. 1987. Methyl jasmonate induced changes in the polypeptide pattern of senescing barley leaf segments. *Plant. Sci.* 51: 177-186.

Weigel, D. 1995. The genetics of flower development: From floral induction to ovule morphogenesis. *Annu. Rev. Genet.* 29: 19-39.

Weigel, D. and Meyerowitz, E. M. 1994. The ABCs of floral homeotic genes. *Cell* 78: 203-209.

Welburn, S. C. Dale, C. Ellis, D. Beecroft, R. and Pearson, T. W. 1996. Apoptosis in procyclic *Trypanosoma brucei rhodesiense* in vitro. *Cell Death Differ.* 2: 229-236.

Weller, J. L. Murfet, I. C. and Reid, J. B. 1997. Pea mutants with reduced sensitivity to far-red light define an important role for phytochrome A in day-length detection. *Plant Physiol.* 114: 1225-1236.

Whitehead, C. S. Fujino, D. W. and Reid, M. S. 1983. Identification of the ethylene precursor, ACC in pollen. *Sci. Hort.* 21: 291-297.

Wiesman, Z. J. Riov, and E. Epstein. 1989. Characterizationa and rooting ability of indole-3-butyric acid conjugates formed during rooting of mung bean cuttings. Plant Physiology 91: 1080-1084.

Wilkins, M. B. 1984. Gravitropism. Pages 163-185 in M. B. Wilkins ed. Advanced Plant Physiology. Pitman, London and Marshfield, Mass.

Wilkins, Malcolm B. 1975. The role of the root cap in root geotropism. Current Advances in Plant Science 8: 317-328.

Willekens H. Langebartels C. Tire C. Van Montagu M. Inze D. Van Camp W. 1994. Differential expression of catalase genes in *Nicotiana plumbaginifoila* L. *Proc. Nat. Acad. Sci.* US A. 91, 10450-4.

Williams, J. H. H. and J. F. Farrar. 1990. Control of barley root respiration. Physiologia Plantarum 79: 259-266.

Wilson, R. A. Heckman, J. W. and Sommerville, C. R. 1992. Gibberellin is required for flowering in *Arabidopsis thaliana* under short days. *Plant Physiol.* 100: 403-408.

Wilson, S. B. and D. T. D. Nicholas. 1967. A cobalt requirement for nonnodulated legumes and for wheat. Phytochemistry 6:1057-1066.

Wingate V. P. M. Lawton M. A. Lamb C. J. 1988. Glutathione elicits a massive and selective induction of the plant defense genes. *Plant Physiol.* 87, 206-10.

Woltering, E. J. and Harren, F. 1989. Role of rostellum desiccation in emasculated – induced phenomenon in orchid flowers. *J. Exp. Bot.* 40: 907-914.

Woltering, E. J. and van Doorn, W. G. 1988. Role of ethylene in senescence of petals: Morphological and taxonomical relationships. *J. Exp. Bot.* 208: 1605-1616.

Woltering, E. J. Harren, F. and Boerrigter, A. M. 1988. Use of a laser-driven photoacoustic detection system for the measurement of ethylene production in *Cymbidium* flowers. *Plant Physiol.* 88: 506-510.

Woodson, W. R. Lawton, K. A. and Goldsbough, P. B. 1889. Ethylene-regulated gene expression during carnation petal senescence. Acta Hort. 261: 137-144.

Woolhouse H. W. and Jenkins G. I. 1983. Physiological response, metabolic change and regulation during leaf senescence. In *Growth and Functioning of Leaves* Edited by J. E. Dale and F. L. Milthorpe. pp. 449-487. Cambridge University Press.

Woolhouse, H. W. 1983. Toxicity and tolerance in the responses of plants to metals. Pages 245-300 in O. L. Lange, P. S. Nobel, C. B. Osmond, and H. Ziegler eds. Encyclopedia of Plant Physiology, New Series, Vol. 12C, Physiological Plant Ecology III, Responses to the Chemical and Biological Environment. Springer-Verlag, Berlin.

Woolhouse, H. W. 1984. Senescence in plant cells. In: Davis, I. and Sigee, D. C. eds. *Cell Ageing and Cell Death.* Cambridge: Cambridge University Press, 123-153.

Xing, T. Higgine, V. J. and Blumwald, E. 1997. Race-specific elicitos of Cladosporium fulvum peromote translocation of cytosolic components of NADPH oxidase to the plasmamembrane of tomato cells. *Plant Cell.* 9: 249-259.

Yang HS, Huang HC, Collmer A. 1993. *Pseudomonas syringae* pv. *syringae* Harpin: a protein that is secreted via the Hrp pathway and elicits the hypersensitive response in plant. *Cell.* 73: 1255-66.

Yopp, J. H. L. H. Aung, and G. L. Steffens. Eds. 1986. Bioassays and Other Special Techniques for Plant Hormones and Plant Growth Regulators. Plant Growth Regulator Society of America, Beltsville, Md.

Zagotta, M. T. Hicks, K. A. Jacobs, C. I. Young, J. C. Hangarter, R. P. and Meeks-Wagner, D. R. 1996. The *Arabidopsis* ELF3 gene regulates vegetative photomorphogenesis and the photoperiodic induction of flowering. *Plant J.* 10: 691-702.

Zarembinski T. I. Theologis A. 1994. Ethylene biosynthesis and action: a case of conservation. *Plant Molecular Biology.* 26: 1579-097.

Zeevaart, J. A. D. 1984. Photoperiodic induction, the floral stimulus and flower-promoting substances. In *Light and the Flowering Process*, D. Vince-Prue, B. Thomas, and K. E. Cockshull, eds. Academic Press, London, pp. 137-142.

Zeevaart, J. A. D. and Boyer, G. L. 1987. Photoperiodic induction and the floral stimulus in *Perilla.* In *Manipulation of Flowering*, J. G. Atherton, ed. Butterworths, London, pp. 269-277.

Zeevaart, J. A. D. T. G. Heath, and D. A. Gage. 1989. Evidence for a universal pathway of abscisic acid biosynthesis in higher plants from $^{18}O$ incorporation patterns. Plant Physiology 91: 1594-1601.

Zeevaart, Jan A. D. 1976. Phytoghormones and flower formation. In D. S. Letham et al. eds. Plant Hormones and Related Compounds. ASP Biological and Medical, Amsterdam.

Zeevaart, Jan A. D. 1982. Transmission of the floral stimulus from a long-short-day plant, *Bryophyllum daigremontianum*, to the short-long-day plant *Echeveria harmsii.* Ann. Bot. 49: 549-552.

Zhang, J. and W. J. Davies. 1990. Changes in the concentration of ABA in xylem sap as a function of changing soil water status can account for changes in leaf conductance and growth. Plant, Cell and Environment 13: 277-285.

Zhang, Y. 1989. The Influence of $GA_3$ on Fructosyl Carbohydrates in Wheat Plants. M. S. Thesis, Colorado State University.

Zhang, Y. C. W. Ross, and G. L. Orr. 1990. Effects of $P_{fr}$ on NAD kinase and nicotinamide coenzymes in lettuce seeds. Plant Physiology supplement. 93: 30.

Ziegler R. Blaheta A. Guha N. Schonegge B. 1988. Enzymic formation of pheophoribide and pyropheophorbide during chlorophyll degradation in a mutant of *Chlorella fusca* Shihira et Kraus. *J. Plant Physiol.* 132: 327-333.

Ziegler, H. 1975. Nature of transported substances. Volume 1, pages 59-100 in M. H. Zimmermann and J. A. Milburn volumes eds. Transport in Plants. I. Phloem Transport. In A. Pirson and M. H. Zimmermann series eds. Encyclopedia of Plant Physiology, New Series. Springer-Verlag, Berlin.

Zimmer, R. 1962. Phasenverschiebung and andere Storlichtwirkungen auf die endogen tagesperiodischen Blutenblattbewegungen von *Kalanchoe blossfeldiana* Phase shift and other light interruption effects on endogenous diurnal petal movements. *Planta* 58: 283-300.

Zimmerman, R. H. Hackett, H. P. and Pharis, R. P. 1985. Hormonal aspects of phase change and precocious flowering. In *Encyclopedia of Plant Physiology*, New Series, Vol. II III. R. P. Pharis and D. M. Reid, eds. Springer, Berlin, pp. 79-115.

Zimmermann, B. D. and W. R. Briggs. 1963. Phototropic dosage-response curves for oat coleoptiles. Plant Physiology 38: 248-253.

Zimmermann, M. H. 1961. Movement of organic substances in trees. Science 133: 73-79.

Zimmermann, M. H. 1983. Xylem Structure and the Ascent of Sap. Springer-Verlag, Berlin, Heidelberg, New York.

# Glossary

**α-Amylase.** An enzyme with endo--(1, 4)-glucanase activity. Bacterial a-amylase is produced by submerged fermentation techniques using specially selected strains such as *Bacillus subtilis*. Fungal enzyme is produced by strains of *Aspergillus oryzae* or *A. niger*. This enzyme is used commercially in food preparations, thinning of starch for alcohol fermentations.

**2,4,5,-T.** 2,4,5-trichlorophenoxacetic acid, a translocated hormone weedkiller used to control scrub and woody vegetation.

**2,4-D.** 2,4-dichlorophenoxyacetic acid; a synthetic translocated hormone weed killer, used to control broad leaved herbaceous plants, and as a defoliant.

**Abietic acid.** A diterpene carboxylic acid, and the isomeric neoabietic acid, can easily be interconverted. There two resin acids are the main components of resin (up to 90%), from which they could be obtained by treatment with heat or acids, possibly as products of the rearrangement of other diterpene carboxylic acids. Amber contain derivatives of abietic acid.

**Abscisic acid (ABA).** A plant growth substance associated with leaf senescence, fruit drop, seed and bud dormancy, some aspects of apical dominance and inhibition of flowering in long-day plants under short-day conditions.

**Abscission.** The shedding of leaves or fruit as a result of cambial activity of cells in abscission zone (abscission layer) that extends across the base of the petiole; premature formation of abscission layers (proleptic abscission) is a hyperplastic symptom of some plants.

**Absorbance.** A measure of the extent to which light is attenuated during passage through a coloured liquid or solid.

**Absorptiometer.** A device used to measure the absorbance of a sample.

**Absorption spectrum.** The image recorded when electromagenetic radiation from a source emitting a continuous spectrum is passed through a substance. If the material is in the gaseous phase, bands will appear in the same position as the lines that occur in the characteristic emission spectrum of that substance. In solids or liquids, these bands are broadened and can be used to identify or quantify the material.broadened and can be used to identify or quantify the material.

**Acetic Acid, Ethanoic Acid:** $CH_3$ COOH. A very common monocarbosylic acid which occurs in the free forms as the end product of fermentation and oxidation reactions in some organisms. Acetate is formed metabolically by dehydration of acetaldehyde, catalysed either by aldehyde oxidase (EC 1.2.3.1) or a NAD(P) - dependent aldehyde dehydrogenase (EC 1.2.1.3).

**Acetone.** An important industrial solvent made by some bacteria, notably *Clostridium acetobutylicum*, as a result of fermentation. Provide a means of producing cosolvents for addition to lead-free petroleum blends.

**Acetone/butanol fermentation**. An industrial process, based on the fermentation of sugars (e.g., molasses), using *Clostridium acetobutylicum* to produce a mixture of acetone and butanol. This was the first fermentation to be carried out on a large scale using pure culture and aseptic methods. At one time, fermentation was major source of these solvents. However, almost all plants were closed with the advent of the petrochemical industry. Interest in this process is now increasing as it may

**Acetyl coenzyme A** (*acetyl-CoA*). A complex of an acetyl group with coenzyme A produced during the oxidative decarboxylation of pyruvate and during oxidation of fatty acids. It functions as the link compound between glycolysis and the TCA cycle in respiration, is an intermediate of the glyoxylate cycle, and is involved in the synthesis of a wide variety of primary and secondary products including fatty acids, some amino acids and terpenoids.

**Acetyl-DL-amino acid.** The acetylated form on an amino acid produced by treating hydrolyzed protein with acetic anhydride.

**Acid Medium**. A culture medium of pH betwen 1 and 5.

**Acid protease**. A protein-hydrolyzing enzyme characterized by low levels of basic amino acids and have a low isoelectric point. They are insensitive to sulphydryl reagents, heavy metals and metal chelators, but are inactivated at pH values above 6.0. They are widely used in the food and beverage industries. Acid proteases of commercial importance are prepared rom fungal sources and are of two types: pepsin-like and rennin-like enzymes. Pepsin like proteases are generally prepared from *Aspergillus* species, whereas rennin-like proteases are produced from *Mucor* species.

**Acidification.** A process in which the pH falls or acid is added to decrease the pH. In anaerobic digestion, problems may arise due to acidification following shock loading. Acid-producing bacteria are capable of faster rates of growth and metabolism than are methanogenic bacteria, which are also less tolerant of acid conditions. Hence, if the rate of acid production increases for any reason, the imbalance is enhanced as the acid producers grow faster and the acid-utilizing bacteria are inhibited. This can lead to the collapse of the biogas-producing system.

**Action spectrum.** A plot of the intensity or magnitude of a biological reaction as a function of wavelength of electromagnetic radiation. Action spectra are usually plotted over the visible and/or ultraviolet regions. The magnitude of the response will, for a photomediated or light-dependent reaction, correspond to the absorption spectrum of the compound involved. Hence tha action spectrum can be used to infer a casual relationship between a given pigment or photoreceptor and a biological reaction.

**Activating enzymes.** The enzymes which activate amino acids in the presence of ATP before they are covalently linked to tRNAs are called activating enzymes. They are also known as aminoacyl tRNA synthetases.

**Active site.** The region of the enzyme with which the substrate binds during catalytic conversion to a product. The basic theory of enzyme activity assumes formation of an enzyme-substrate complex through binding at the active site. Electrostatic attraction, repulsion and other intermolecular physical forces contribute to a lowering of the activation energy, thus facilitating intermolecular conversions.

**Active transport.** A biological mechanism that results in the transport of molecules across a membrane against a concentration gradient. Active transport requires metabolic energy (usually ATP or an electrochemical gradient) and may be associated with a specific carrier protein or lipoprotein molecule.

**Adaptive enzyme.** An enzyme that is formed in response to an outside stmulus during adaptation. This term is now generally obsolete and has been replaced by inducible enzyme.

**Adenosine monophosphate (AMP).** A nucleotide comprising adenine and D-ribose and one phosphate group. Addition of one or two further phosphate groups leads to the formation of ADP or ATP, respectively. See cyclic adenosine monophosphate.

**Adenosine.** A nucleoside comprising adenine linked to D-ribose through a ß-glycoside bond. Phosphorylated forms of adenosine (AMP, ADP and ATP) are the major compounds involved in energy transfer in biological systems.

**Adsorption**. (1) In physical chemistry, the adhesion of molecules to solids. This process forms the basis of adsorption chromatography, which is used to purify proteins and other macromolecules, as well as some method of cell or enzyme immobilization. (2) In microbiology, the process whereby bacterophage attaches to specific receptors on the host cell prior to injection of their nucleic acid upon infection.

**Aerobe.** An organism that requires oxygen for respiration and hence growth.

**Aerobic respiration.** The overall process in which carbohydrates are completely oxidized to carbon dioxide and water using molecular oxygen. In the initial stage (glycolysis), which is common to both aerobic and anaerobic respiration, glucose phosphate is degraded to pyruvate. Pyruvate is then converted to acetyl-CoA which enters the TCA cycle by combining with oxaloacetic acid to form citric acid. Citric acid is then metabolized to reform the acceptor molecule oxaloacetic acid in a cyclic process. This results in the release of two molecules of carbon dioxide and the passage of hydrogen atoms and their electron equivalents to an intermediate electron transport chain which ends with the formation of water from molecular oxygen. During passage through the series of redox couples that constitute the intermediate electron transport chain energy is conserved in the form of ATP.

**Agarose gel.** An inert matrix used in the electrophoretic separation of nucleic acid molecules on the basis of their size or conformation. Gels may be cast as tuubes or slabs although the latter now predominate. Molecules are visualised in the gel by the ultraviolet fluorescence of ethidium bromide which is either included in the running buffer or used to stain the gel after electrophoresis (see comb, LGT agarose, power pack, Tris-acetate buffer, Trisborate buffer).

**Ageing.** The biochemical and genetical changes that lead to the gradual senescence and death of a cell or organism.

**Alcoholic Fermentation.** An aerobic (occurring in the absence of oxygen) formation of ethanol and carbon dioxide from glucose. The most important fermenting organisms are yeasts and other microorganisms. For lack of pyruvate decarboxylase, there is no alcoholic fermentation in animals. The process produces energy under anaerobic conditions. The fermentation of 2 mol glucose yields 1 mols ATP. The starting point for alcoholic fermentation has been glucose 6-phosphate, which is converted by the glycolysis reactions to pyruvate. Pyruvate is the branching point for the last step of carbohydrate degradation and gets decarboxylated by pyruvate decarboxylase (EC 4.1.1.1) to acetaldehyde, which is then reduced to ethanol by alcohol dehydrrogenase (EC 1,1,1,1 balance.:

**Alginic acid.** A polysaccharide consisting of various mixture of b-1, 4-linked d-mannuronic acid and 1, 4-linked 1-guluronic acid units fouund in the cell walls of brown algae, Fucus, Laminaria, Macrocystis etc., used as an emuulsifying agents in preparation of polishes and paints; can stop bleeding effectively an is employed as a highly efficient gauze in internal operations; derivatives fo alginic acid used in preparation of soups creams and also antibioic capsules.

**Alkaline phosphatase.** An enzyme which removes the 5' terminal phosphate groups from linear DNA molecules. It is used to prevent the religation of plasmid vector molecules following cleavage with a restriction endonuuclease. This increases the chance that intract circular molecules generated by the ligase reaction are recombinant.

**Allosteric effector.** A comopound that induces a conformational change in a protein which changes its catalytic activity, by binding at a specific recognition site. The effector molecule may be a normal substrate, the position of the binding site that causes the conformation changes differs from the position of the substrate binding site.

**Allosteric effectors.** Small molecules, usually metabolites that binds at sites other than the active site.

**Allosteric enzyme.** An enzyme that shows unusual kinetic properties.

**Amino acid.** Any one of a class of organic compounds containing the amino ($NH_2$) groups and the carboxyl (-COOH) group. Amino acids are building blocks of proteins. Alanine, proline, threonine, histidine, lysine, glutamine, phenylalanine, tryptophan, valine, arginine, tyrosine, and leucine are among the common amino acids. Several hundered different amino acids are known, but only 20 are normally found in proteins. These are classified on the basis of similarity in structure of route of biosynthesis as shown in the list. Other biologically important amino acids include ornithine and citruline which are intermediates of the urea cycle, T-aminobutyric acid which functions as a neuurotransmitter, ß-alanine which is a precursor of pantothenic acid, and D-glutamate which is found in bacterial cell walls.

**Amino purine.** A base analogue that may be incorporated into the DNA of an organism in place of adenine, resulting in a mutation due to the incorrect base pairing in cell generations some time after the incrorporation of the base analogue.

**Amino sugar.** Monosaccharide in which a hydroxyl group has been replaced by an amino group (-$NH_2$). The amino group is often acetylated. Examples are D-galactosamine, D-glucosamine, D-mannosamine, neuraminic acid and muramic acid.

**Aminoacyl synthetase.** Any one of at least 20 different enzymes that catalyze (1) the reaction of a specific amino acid with ATP to form aminoacyl-AMP (activated amino acids) and pyrophosphate and (2) the transfer of the activated amino acid to tRNA forming aminoacyl-tRNA and free AMP.

**Ammonia** ($NH_3$). A gas freely soluble in water that is readily absorbed by plants, usually as the ammonium ion. The major route incorporation of nitrogen into all living organsims is by assimilation of ammonium ions either through the GS/GOGAT pathway or in the reaction catalyzed by glutamate dehydrogenase. In both cases, the nitrogen forms an amino group in glutamic acid from where it is transferred to other compounds by transmination. A few types of bacteria and the cyanobacteria (nitrogen fixers) are capable of reducing atmospheric dinitrogen to ammonium. However, most bacteria and plants use nitrate, which they reduce to ammonia prior to assimilation into amino acids.

**Ammonia Detoxification.** The detoxification of the toxic, non-dissociated compound by formation of ammonium salts and nitrogen-excretion compounds. The ammonia produced by catabolism in animals is either excreted directly, or it is converted into other nitrogenous compounds for excreted directly, or it is converted into other nitrogenous compounds for exvretion; it is not ressimilated.

**AMO-1618.** (2 isopropyl-5 methyl-4-trimethyl ammonium chloride)-phenyl-1-piperidonocarboxylate, a synthetic growth retardant of plants. It is an antagonist of gibberellin and inhibits the biosynthesis of gibberellin $A_3$.

**Anabolic.** Descriptive of a biological reaction that results in net synthesis of more complex molecules.

**Analyzer.** An instrument used to determine the nature and/or quantity of an element or compound.

**Anthranilic acid.** Amino benioic acid.

**Antiauxin.** These are chemicals which are strong inhibitors of polar transport of auxin e.g., 2, 3, 5-triodobenzoic acid (TIBA) and naphtylphylphthalamic acid (NPA).

**Antimetabolite.** A substance which compeitively inhibites the utilization, by an organism, of an exogenous or endogenous metabolite. Sulphanilamide (for example) acts as an antimetabolite in respect of PABA.

**Apical dominance.** The inhibition of the development of lateral buds by the growth of the apical meristem.

**Apoenzyme.** The enzyme or protein part of an enzyme cofactor complex that has lost the cofactor and is thus inactive.

**Apoplast.** Part of the plant tissue matrix outside the membranes enclosing the cell contents, i.e., outside the plasmalemma.

**Arid crop.** A crop plant that is adapted to conditions of low water availability. Such crops include the euphorbias, which have been suggested as a potential source of hydrocarbon fuels, and guayule, a plant that may be used for the production of rubber.

**Aril.** An extra seed envelope, often coloured and fleshy found in some angiosperms. The aril is produced from the tissue of the funicle or base of the ovule; **arillate.**

**Armed.** Having thorns or spines.

**Aromatic amino acids.** Amino acids whose side chains include a derivative of phenyl group. The aromatic amino acids in protein are phenylalanine, tyrosine and tryptophan.

**Aromatic.** A drug which is fragrant, spicy and mildly stimulant.

**Aromatic.** Descriptive of chemical compound with a cyclic arrangement of six carbon atoms as found in benzene.

**Artefact.** An apparent structure caused by the method of preparation, and not native to the organ or organism.

**Arthrospore.** A jointed spore; a spore resulting from the fragmentation of a hypha, as in some actinomycetes.

**Articulated.** Jointed.

**Artificial classification.** A classification structured for convenience, using easily observed phenotypic characters and not necessarily indicating phylogenetic relationships; key classification; *cf.* natural classification.

**Artificial key.** A way of identifying plants by steps. Each step involves a choice between at least two different characters, each of which leads to another choice of two characters, finally leading to the right identification.

**Artificial selection.** The process by which man selects plant varieties or individuals with useful characters or traits e.g., wheat plants with large seeds, in order to breed them for his own purposes. Because of this, many cultivated plant appear to be very different from their wild ancestors.

**Asaccharolytic.** Used of microorganisms that are unable to break down carbohydrates, alcohols or polysaccharides.

**Ascigerouus.** Ascus-bearing (perfect) stage of development of an Ascomycete.

**Ascocarp.** An organ of a fungus which produces asci and ascospores.

**Ascogenous hyphae.** Hyphae producing asci and ascospores and arising from an ascogonium.

**Ascogenous.** Producing, or giving rise to, asci.

**Ascogoniuum.** Spirally coiled filament inside young ascocarps, later producing asci.

**Ascomycetes.** The sac fungi; a class of fungi in which the meiospores, called ascospores, are produced in asci.

**Ascorbic acid** (Vitamin C). A water-soluble compound structurally related to hexose. Plants and most animals are able to synthesize ascorbic acid.

**Ascorbic acid.** Vitamin C, the water-soluble antiscorbutic vitamin.

**Ascospore.** A spore produced by free-cell formation following meiosis in an ascus.

**Ascostroma.** A stromatic ascocarp bearing asci directly in locules within the stroma.

**Ascus.** A sac-like cell of the Ascomycota in which karyogamy is followed immediately by meiosis and in which ascospores of definite number arise by free-cell formation.

**Asepsis.** The absence of microorganisms that produce disease or contamination of other culture, or process characterized by the used of feedstocks and other materials that are free from such organisms. Asepsis may be achieved by sterilization of media and equipment.

**Aseptate,** Lacking transverse walls, as in some fungi, e.g., *Mucor*, or some algae, e.g., *Vaucheria*, or laticiferous tissue in some flowering plants.

**Aseptic technique.** Produucers used to prevent the introduction of fungi, bacteria, viruses, mycoplasma or other microorganisms inot cell, tissue and organ cultures. Althouth these procedures are used to prevent microbial contaimation of cultures, they also prevent cross contamination of cell culture as well. These procedures may or may not exclude the introduction of infectious molecules.

**Aseptic.** Of techniques designed to inhibit actively putrefactive micro-organisms; by extension, of techniques designed to inhibit or exclude all micro-organisms.

**Asexual.** Of structures and reproductive processes lacking gametes for conjugation or fertilisation.

**Asexuual reproduction.** Development of new organisms without the fusion of gametes. This may occur in plants by either spore formation or vegetative reproduction.

**Asparagine** (Asn). It is ne of the 20 common amino acids found in proteins.

**Asparate.** The salt of aspartic acid.

**Aspartame.** L--Aspartyl-Phenylalanine methyl ester; a colourless crystalline material that is 160 times sweeter than sucrose.

**Aspartic acid Production.** L. Aspartic acid is produced from ammonium fumarate in a reaction catalyzed by aspartase derived from bacteria such as *E. Coli.* and *Pseudomonas fluorescens. Alaligenes* can produce aspartate from ammonium maleate.

**Aspergillic acid.** An antibiotic, $M_r$224.3, produced by *Aspergillus flavous*.

**Aspergillosis.** One of a group of diseases of animals and human beings caused by various species of *Aspergillus.* (usually *A. fumigatus*).

**Asporogenic.** Not producing, or developing from spores; asporogenous.

**Asporogenous yeast.** A yeast for which the life cycle is not known, or one that does not have a sexual stage.

**Assimilate.** In the wider sense, a product of assimilation. In the narrower sense, a stabilized end product of photosynthesis.

**Assimilation Number.** It is the amount of $CO_2$ absorbed in grams per hour for each gram of chlorophyll during phtosynthesis.

**Assimilation.** A general term for constructive metabolism; usually of carbon-assimilation (photosynthesis). The conversion of simple moleucles into the complex constituents of the living body.

**Assimilator.** Leaf-like upright part of some Chlorophycean members.

**Aster.** A cytoplasmic radiation found at the poles of certain cells during mitosis. In the cytoplasm of dividing cells of certain algae, fungi, and animals, asters are the sources of radiating, fiberlike processes. They appear to serve as the centers to which chromosomes move during anaphase.

**Asyngamy.** Reproductive isolation by virtue of different breeding or flowering seasons.

**Atactostele.** A scattered arrangement of xylem and phloem groups (vascular bundles).

**ATEE.** N-Acetyl-L-tyrosine ethyl ester.

**Athelete"s foot.** An infection of the skin between the toes, caused by a ringworm type of fungus. In severe cases the disease may spread to the whole foot or the other parts of the body.

**Atmobios.** Organisms living in the air; aerial organisms.

Atmophyte. An epiphyte that obtains water by aerial assimilation over its entire surface.

Atokous. Without offspring; non-reproductive; vegetative;

**Atom.** One of the smallest parts of an element that can exist.

**Atomic absorption spectroscopy.** A method used for the quantitative determination of metals. The sample is burnt in a flame through which ultraviolet light, generated from a lamp specific for the element under investigation, is passed. Attenuation of the beam is detectyed using a photomultiplier and converted into an electrical signal which is concentration or amount using a microprocessor.

**ATP.** Adenosine triphosphate, an energy-rich compound participating in energy-storing and energy-using reactions in the cell.

**ATPase.** Also called myosin, an enzyme catalysing TTP, CTP, CTP and ATP to yield orthophosphate and the corresponding diphosphates.

**Atrazine.** A triazine used as an herbicide that inhibits photosynthetic electron transport.

**Atrophy.** Dwindling or abortion of tissues or organs.

**Atropine.** An alkaloid which can be extracted from the plant *Atropa belladona*, also called deadly nightshade plant. It is found in a number of other plants of the family Solanaceae and is also available as a synthetic product. Altropine causes dilation of pupil and is used as an aid in the examination of the retina. It is also used in the treatment of inflammation in the cornea. It also reduces the secretion of tears, sweat, saliva and digestive juices. As atropine inhibits the secretion of the bronchial glands, it is sometimes used in the treatment of asthma.

**Attenuation.** A regulatory mechanism employed by bacterial cell. Whereas enzyme repression allows the cell to respond to extreme concentrations of metabolites, it probably represents a means of "fine tuning" to relatively mild fluctuations in the concentrations of metabolities.

**Attenuent.** An agent that dilutes fluids.

**Aulophyte.** A non parasitic plant living within a hollow cavity in another plant.

**Aureomycin.** An antibiotic obtained from one of the actinomycetes.

**Auricle.** A small outgrowth at the side of the base of the leaf in some grasses.

**Aurone,** 2-benzylidene-3-cumarones. The yellow flavonoid plant pigments. These are very common in the petals of Compositate, Leguminosae and Scrophulariaceae.

**Autecology.** The study of an organism in relation to its environment.

**Authority.** The name of the author who was the first to give a species or other taxon its name. In the case of species, the authority is given after the binomial.

**Autoallogamy.** The condition of a species in which some individuals are adapted to cross fertilization and others to self-fertilization; homodichogamy; allautogamy; autallogamia, autallogamy.

**Autoanalyzer.** Registered name applied to fully automate analytical system marketed by Technicon. The autoanalyzer is based on automatic sampling and continuous flow through a variable manifold (which can be adopted as required for colorimetric, enzymic or other reactions) to a detector (colorimeter, absorptiometer, fluorometer, etc.). The samples are kept separate by bubbles of gas.

**Autochorus.** Having motile spores or propagules disseminated by the action of the parent plant; autochore, autochory.

**Autoclave.** A strong metallic vessel, gastight when closed, using steam under pressure for sterilization.

**Autoclave.** An appliance, essentially a pressure cooker, used to obtain sterile media or to destroy microbially contaminated material such as used petri culture plates. The material is placed in a container which is sealed and heated, having been vented of air. Any living organism is killed by the superheated steam generated under pressure.

**Autocolony.** A miniature daughter colony formed by coenobic algae.

**Autoecious.** Of parasites living their whole life-history on one host.

**Autogamy.** Self fertilisation; found in some flowers, and in some gametes derived from the same gametangium.

**Autograft.** A graft derived from the tissue of the recipient.

**Autolysis.** The breakdown of animal or plant products as a result of the action of enzymes contained within the cells or tissue affected.

**Autolyzate.** A product of cell decomposition produced by autolysis.

**Automated enzyme analysis.** A technique in which an instrument is used to carry out a number of assays based on an enzyme-dependent procedure. It is used to determine levels of enzymes using specific substrates. Analyzers continuously record the reaction or record its extent after a fixed time. In the latter case, the maximum extent of the reaction should be set within the substrate/enzyme concentrations such that not more than 15 per cent of the substrate is used.

**Autoneried.** An autotrophic water plant.

**Autonomic Movement.** (Spontaneous Movement). Movement due to internal causes or influences. Movements independent of changes in the external environment; movements that are the result of internal growth changes.

**Autoorthotropism.** The tendency of a plant to grow in a straight line; autoorthotrophic.

**Autoparasite.** A parasite living on another parasite; hyperparasite; superparasite.

**Autoparthenogenesis.** The development of an unfertilized egg that has been activated by a chemical or physical stimulus.

**Autophagy.** The interacellular digestion of endogenous material of the cell involving lysosomes.

**Autophytic.** Capable of making its own food, as a green plant.

**Autopolyploid.** Of nuuclei and of plants having multiples of two completely identical sets of chromosomes.

**Autoradiograph** (*radioautography*). The picture produced by the exposuure of a film emulsion to radioactive particles. When a photographic emuulsion is placed in contact with radioactive material (e.g., thin section of a cell), the radiation exposes the film, revealing details of the location and geometry of the radioactive components.

**Autoradiography.** A technique for detecting the location of a radioisotope in a tissue, cell or molecule. The sample is placed in contact with a photographic emulsion, usually an X-ray film. The emission of particles from the sample activates the silver halide grains in the emulsion and allows them to be reduced to metallic silver when the film is developed. In genetic engineering, autoradiography is most commonly used to detect the hybridisation of a radioactive probe molecue to denatured DNA in either the Southern transfer or colony hybridisation procedures. In general, both macromolecules and other chemicals are separated by any of a number of a electrophoretic or chromatographic techniques. The autoradiograph then reveals the position of the compound on the carrier matrix. In molecular biology, autoradiography is used to reveal the position of hybridized nucleic acids in blot tests.

**Autosome.** Those chromosomes of the dividing nucleus carrying genes for any factor except that of sex.

**Autosporangium.** A sporangium containing autospores.

**Autospore.** A nonmotile, asexual daughter cell which lacks the ontogenetic potentiality of motility.

**Autotroph.** An organism that utilizes carbon dioxide as its major source of carbon and obtain energy from the sun through photosynthesis (photoautotroph) or from organic or inorganic reduced chemicals (chemoautotroph). An organism which can generate its own food supply from simple organic and inorganic elements. (e.g., green plants).

**Autotrophic cells.** Cells that can synthesize macromolecules from simple nutrient molecule, such as carbon dioxide, ammonia and water.

**Autotrophic plant.** A self-nourishing plant; one capable of making its own food.

**Autotrophic.** Capable of synthesizing protoplasm from entirely inorganic substances.

**Autotrophy.** Refers to the ability to synthesize all organic components from simple inorganic compounds like carbon dioxide, ammonia, nitrate and sulfate.

**Auxin antagonists.** Inhibitors of auxins whose effects can be at least partly reversed by auxins.

**Auxin.** A plant growth substance. The natural auxins are indoleacetic acid (IAA) and indoleacetonitrile (IAN). Synthetic auxins, which are used for weed control or as herbicides, include naphthaleneacetic acid, indolebutyric acid and 2, 4-dichlorophenoxyacetic acid. Natural auxins are formed in plant apices, from where they diffuse through the tissue stimulating cell elongation in growth and tropic movements.

**Auxoautotrophic.** Used of an autotrophic organism requiring certain specific additional nutrients for growth; auxoautotroph.

**Auxotrophy.** Refers to the condition of requiring nutritional supplements for growth. It can arise through mutation.

**Available electrons.** Those electrons for pyrimidine and purine. Thosee electrons present in any compound or element, which is used as a substrate that are not involved in orbitals with oxygen and are thus available weither for transfer of oxygen or synthetic reductive reactions. The number of available electrons per mole of substrate is equal to four times the total oxygen demand, since four electrons are required to reduce one molecule of oxygen.

**Available water.** That part of the soil water that can be readily taken up by plants; cf. holding capacity.

**Bacterial photosynthesis**. A primitive form of photosynthesis using $H_2S$, thiosulphate, fatty acids or other organic reducing agents rather than water as the source of hydrogen for carbon fixation.

**Bacteriochlorophyll.** $C_{52}H_{70}O_6N_4Mg$. A tetrahydroporphyrin chlorophyll compound occurring in the forms *a* and *b* in photosynthetic bacteria.

**Basic amino acids.** Amino acids having a net positive charge at neutral pH.

**Beta carotene $C_{40}H_{56}$.** A carotenoid hydrocarbon pigment found widely in nature, always associated with chlorophylls; converted to vitamin A in the liver of many animals.

**Beta counter.** A device for quantifying the radioactive decay from a b-emitting radioisotope. This can be either a Geiger-Muller counter or a liquid scintillation counter.

**Binding proteins.** Soluble proteins which specifically and reversibly bind a veriety of substances (including sugars, amino acids, inorganic ions and vitamins) but which appear to have no enzymic activity. Binding proteins are believed to occur in the periplasmic space of many gram-negative bacteria, and have also been found in certain yeasts. They are believed to play some role in the transport of the substances bound, but the precise nature of this role remains uncertain; it is thought that they may act as sites of recognition for the substance to be transported. Osmotic shock has been used to release the binding proteins from gram-negative bacteria.

**Binding site.** A specific arrangement of atoms or molecules that is recognized by and forms the point of attachment for, an ion compound, antibody, virus, cell or organism on another structure or organism.

**Bioassay** (biological assay). It is the use of an organism (or other living system) for assay purposes. Thus, the pathogenicity of an entomopathogenic microorganism can be assayed by the use of susceptible insects, and microorganisms may be used for the measurement of low concentrations of certain vitamins, amino acids, etc. For the assay of Pantothenic acid use is made of an organism (e.g., *Tetrahymena pyriformis*) for whom this vitamin is an essential growth requirement. The organism is inoculated into a mediuum containing adequate concentrations of all growth requirements **except** pantothenic acid (which is present at a known, sub-optimal concentration) and incubated under standard conditions (care must be taken to avoid carrying over any pantothenic acid in the inoculum). After incubation the amount of growth is determined -

the amount of growth being proportional to the concentration of pantothenic acid in the medium.

**Bioautography.** A method for detecting small amounts of one or more substances (e.g., vitamins) in a complex mixture, such substance(s) being essential growth requirements for the test organism(s) used. The components of the mixture are initially separated by chromatography. The presence of a particular component (in a given chromatographic band) is then detected by extracting the band, with an appropriate solvent, and adding the extract to a medium which lacks the relevent growth factor. The medium is inoculated with an appropriate test organism, and incubated. Growth indicates that the particular component was present in the original mixture.

**Biocatalyst.** A catalyst that consists of, or is derived from, a living organism or tissue, or cell culture, Biocatalysts may be categorized as follows: 1. ***Cells*** *(a) Growing.* The use of growing cells in suspension as biocatalysts is generally termed fermentation. Such processes may be distinguished as batch, fed batch or continuous. (b) *Non-growing.* Non growing systems include the use of suspensions, cells retained within a semipermeable membrane, or cells immobilized in or on a solid support. Such immobilized cells may be used in a fixed bed, expanded bed, fluidized bed or fully mixed system. 2. ***Enzymes*** *(a) soluble enzymes systems.* Soluble enzymes of varying purity may be used in solution, generally in a batch system, and are lost at the end of the reaction. (b) *Immobilized enzymes.* Enzymes may be entrapped, absorbed on to surfaces or cross-linked to support material and then used in fixed bed, expanded bed or fully mixed system.

**Biochemistry.** The chemistry of living matter.

**Biodegradable.** Descriptive of substances that can be decomposed by the activity of microorganisms.

**Biodegradation.** A process in which material is broken down to its chemical constitutions by the action of living organisms. The term is often restricted to the breakdown of wastes release into the environment. In general biodegradation is regarded as a desirable process in contrast to biodeterioration.

**Bioenergetics.** Thermodynamics as applied to living cells or whole organisms.

**Biological clock.** A hypothetical mechanism usde to account for the occurrence of circadian and biorhythms. It is the mechanism which allows expression of certain biological structures (genes) at periodic intervals.

**Biological screen.** A test or series of test using biological material to determine the effects of newly synthesised chemicals on growth.

**Biological spectrum.** A list of plant species in a given region or community together with percentage abundances of individual species.

**Biomembrane.** A structure having lipids, glycolipid, proteins and glycoproteins, bounding the cell (cell membrane) or subdividing it into compartments. It is a sheet-like structure about 60 to 100AÉ thick. The lipids have hydrophilic "head" groups, and hydrophobic "tail" regions. The currently accepted model of the structure of biomembrane is based on the fluid mosaic concept. Physically, the surface of a biomembrane is an interface between a polar solvent (water) and a nonpolar solvent (the hydrophobic tails of the lipid molecules). Hydrophobic molecules may cross biomembrane because they can dissolve in the hydrophobic milieu of the membrane interior.

**Biomethanation.** The anaerobic fermentation of organic material to produce biogas (a mixture of methane and carbon dioxide).

**Biosynthesis.** The synthesis of organic molecules by living organisms using linked reactions, in which the energy and reducing power are derived from ATP and reduced pyridine nucleotides generated by fermetnation, respiration, chemosynthesis or photosynthesis.

**Biotechnology.** It can be defined as the use of living organisms in systems or processes for the manufacture of useful products; it may involve algae, bacteria, fungi, yeasts, cells of higher plants or animals of subsystems of any of these isolated components from such living matter. The term biotechnoilogy is quite old. A Bulltein of the Bureau of biotechnology was published in July, 1920 from the office of the same name in Leeds in Yorkshire. The articles in this bulletin described the varied roles of microbes in leather industry to pest control. Biotechnology means the application of microbes and biochemical technology to applied biology, i.e., to the exploitation of biological systems and processes for our own use. In recent years, with the greater understanding of biological processes and the ability to manipulate the genes, it has become possible to conceive processes and systems not hitherto contemplated. The extent of biotechnology-application is wide ranging, from the development of improcved pharmaceuticals (antibiotics, vaccines and hormones) to the replacement of oil and gas hydrocarbons for fuel and chemical feedstock from renewable source. Unfortunately, biotechnology, being still in the process of early development, does not possess a sharp and easily defined form. Only in last five years progress has been made by microbiologists and genetic engineers, and we are hopeful to solve energy crisis.

**Biotin.** A vitamin of the B complex that functions as a coenzyme in some reactyions associated with the fixation of carbon dioxide.

**Biotype.** Distinct physiological race or strain within morphological species. A population of individuals with identical genetic constitution. A biotype may be made up of homozygotes, of which only the formed would be expected to breed true.

**Boand.** Into polysachharides or the synthesis of sugars from $CO_2$ and $H_2O$. The force of attraction binding two atoms in a molecule; a force uniting atoms when they are sharing electrons is called Bond-covalent, and a force uniting atoms that bear opposite electrical charge is called Bond-ionic.

**Brassicasterol,** Ergosta-5, 22-dien-3ß-ol. A plant sterol (Sterols). M.p. 148ÉC. It was first isolated from rapeseed (*Brassica campestris*) oil.

**Buffer.** A solution containing a mixture of a weak acid and a base which resists changes in pH and is therefore able to provide a favourable environment for enzymic reactions. Traditionally buffers based on inorganic salts (phosphate, carbonate) and organic acid salts (acetate, citrate, succinate, glycine, maleate, barbiturates etc.) were used in biological experiments. In many cases, these have been replaced by specific zwitterionic buffers.

**Butyric acid.** An organic acid produced in some bacterial fermentations.

**C/N ratio.** The ratio of carbon to nitrogen in a feedstock, substrate or culture medium. Various systems require a C/N ratio that falls within certain specific limits in order to optimize the process and ensure its reliability.

$C_3$ **Plants.** The majority of oxygen-producing photoautotrophs. In $C_3$ plants, carbon dioxide assimilated during photosynthesis is fixed directly by the enzyme ribulose bisphosphate carboxylase and further metabolized through the photosynthetic carbon reduction cycle. The term $C_3$ arises from the fact that the first stable product of carbon metabolism recognized in rediolabelling experiments in the three-carbon compound 3-phosphoglyceric acid. Most aquatic plants, trees, shrubs and temperate herbaceous plants are $C_3$. Photosynthesis in $C_3$ plants is inhibited by oxygen. They are characterized by low levels of the enzyme phosphoenol pyruvate (PEP) carboxylase in green leaves. Also they display significant rates of photorespiration and have compensation points of around 40-100 ppm carbon dioxide. They show a discrimination against the heavy isotope of carbon ($^{13}$**C**) of around -30 percent. Short-term photosynthetic yields are about 20-30 grams per square metre per day and annual productivities about 10-30 tonnes per hectare per year.

$C_4$ **plants.** Higher plants in which the normal photosynthetic carbon reduction cycle is supplemented by another pathway of carbon assimilation. The term $C_4$ is derived from the fact that the first stable products obtained in the light - identified using $^{14}C$ labelling - are the four-carbon organic acid malic acid and the four-carbon amino acid aspartic acid. Most $C_4$ plants are tropical or subtropical grasses, although some temperatate grasses and some dicotyledons with $C_4$ photosynthesis are known.

**Calmodulin.** A regulatory protein found in all eukaryotic cells that binds with absolute speciticity to a number of enzyme groups involved in cyclic nucleotide transport, calcium ion transport, contractile processes, neurotransmission and some phosphatases. Such binding is directly dependent on the presence of calcium ions. This feature enables calmodulin to be used as an affinity product to or from which selective binding or elution of the above type of proteins may be effected with a high degree of specificity.

**Calorie.** A unit of energy equal to 4.1868 Joules; equivalent to the amount of heat required to raise the temperature of one gram of water one degree centigrade. The calorie or kilocalorie, equal to 100 calories, is often used to expressthe energy content of foods. The amount of heat required to raise the temperature of one gram of water from 14.5<F128>É<F255> to 15.5x C.

**Carbohydrate.** A compound consisting of carbon, hydrogen and oxygen with the general formula $C_x(H_2O)_y$. Carbohydrates are described as monosaccharides, disaccharides, oligosaccharides or polysaccharides, depending on the degree of polymerization of the basic sugar units. They are the early products of photosynthesis and function as transport and storage materials in animals, plants and microorganisms. They are also components of food, animal feed and bacterial growth media, providing calories for growth and movement. Phosphorylated derivatives of monosaccharides are important intermediates in energy metabolism.

**Carbon balance.** An analysis carried out on a fermentation or other bioreaction in which the distribution of carbon, supplied in the substrate, is determined in relation to the distribution into cell biomass, carbon dioxide and other metabolites on either a theoretical or an experimental basis as a means of investigating process efficiency.

**Carbon dioxide concentration.** The amount of carbon dioxide, expressed in terms of equivalent volume under standard conditions, moles or weight per unit volume of a gas or liquid phase.

**Carbon source.** The major substrate present in a culture medium which provides the bulk of the carbon structure of the growing culture of organism. For photosynthetic and autotrophic organisms, the carbon source is generally carbon dioxide. For many heterotrophic organisms the carbon source is usually glucose (derived from starch or cellulose) or sucrose.

**Carboxylase.** An enzyme that catalyzes a reaction in which carbon dioxide is incorporated into another molecule. In plants, the major carboxylases are ribulose bisphosphate carboxylase/oxygenase (rubisco), which catalyzes the primary assimilation of carbon dioxide in photosynthesis, and phosphoenol pyruvate carboxylase, which is active in the ancillary C4 metabolism associated with C4 photosynthesis and CAM plants.

**Carboxymethyl cellulose.** A cellulose derivative bearing substituted carboxymethyl groups which, depending on the molecular weight and the degree of substitution, may comprise a fully soluble polymer or an insolube polymer or an insoluble weak ion.

**Carotenoids.** Plant pigments based on a tetraterpene structure. A number of different carotenoids are associated with photosynthesis as accessory pigments or protective agents, preventing photooxidation of chlorophyll. The carotenoids may be divided into the carotenes (hydrocarbon tetraterpenes) and the xanthophylls (those containing oxygen functions, such as hydroxyl groups).

**Catabolic.** Descriptive of an enzymic reaction that leads to the breakdown of a complex biological molecule into less complex components. Such reactions may either yield energy in the form of ATP or generate metabolic intermediate used in subsequent anabolic reactions.

**Catabolism.** The overall processes involved in the breakdown of complex organic compounds to provide energy in the from of ATP. Catabolism involves such process as glycolysis, glycogenolysis, proteolysis and oxidation of fatty acids.

**Catalase.** An enzyme of wide distribution in plants and animals that accelerates the decomposition of hydrogen peroxide with the liberation of water and free oxygen.

**Cation.** A positively charged ion which is attracted to the cathode in electrophoresis, or any positively changed ion, radical or molecule.

**cDNA.** Complementary DNA; a single-stranded DNA that is formed from a mRNA template by the enzyme reverse transcriptase. Using radioactive nucleotides (generally $^{32}P$), cDNAs can be obtained in a highly radioactive form and used as probes in hybridization assays to determine the number and location of genes in a chromosome, to construct physical maps of different genomes, to study the organization of eukaryotic genes, and to locate coding sequences in a portion of DNA. It is the DNA complement of an RNA sequence. It is synthesised by the enzyme RNA-primed DNA polymerase or reverse transcriptase. The sigle-stranded DNA product of this enzyme (the reverse transcript) may be converted into the double-stranded form by DNA-primed DNA polymerase (DNA polymerase) and inserted into a suitable vector to make a cDNA clone. cDNA cloning is commonly used to achieve the expression of mammalian genes in bacteria or yeast. Since the cDNA clone is a copy of the mature mRNA molecule it contains no introns to act as a barrier to expression.

**Cell affinity chromatography.** A method of obtaining defined, functionally homogenous cell populations from mixed cultures using affinity chromatography. The affinity absorbents are prepared by immobilizing proteins (antibodies, lectins, etc.) on to macrobeads. The sample is applied to a column packed with this material, impurities are washed out, and the cells selectively desorbed by washing in a buffer with excess competitive agent (e.g., free monosaccharides for elution from immobilized lectins).

**Cell cycle**. The growth cycle of an individual cell. Starting from a cell that has just undergone mitosis, and thus has only one chromatid and half the normal DNA content, the first phase is termed G 1. This is followed by the S phase in which each chromatid is duplicated, and the G2 phase in which duplication is complete but mitosis has not been initiated again. Mitosis then occurs in the M phase.

**Cellulase.** A complex enzyme system that hydrolyzes cellulose to sugars of lower molecular weight, including cellobiose and glucose.

**Cellulose ion exchanger.** A substituted cellulose derivative that is used in Bead form a weak ion exchangeer. The most frequently used products are DEAE-cellulose (diethylaminoethyl) and CM-cellulose (carobxymethyl), which are weak base (anion) and weak acid (cation) exchangers, respectively. They are prepared from cellulose by substituution of some primary and secondary hydroxyl groups of the anhydroglucose units with diethylaminoethyl or carboxymethyl groups, respectively, joined through ether linkages.

**Chemiosmotic theory.** A biochemical mechanism suggested as the route of phosphorylation (formation of ATP) linked to electron transport in either respiration or photosynthesis. The theory suggest the hydrogen ions are pumped across the inner mitochondrial membrane, or across the chloroplast thylakoid membrane, as a result of the passage of electrons through the electron transport chain. This requires that the proteins and other intermediates of the electron transport chain are arranged so hydrogen ions are only taken up on one side of the membrane and discharged on the other side. The electrochemical gradient so formed (the proton motive force) is the driving force for phosphorylation. The proton motive force is composed of two elements: (1) a pH difference or gradient across the membrane of about 1.5 pH units; (2) an electrical potential difference of about 0.15 V. The formation of ATP, which requires the removal of the elements of water between the two reactants, is catalyzed by an ATPase built into the membrane. The ATPase is a complex molecule consisting of a stalk and a head piece, which contains the ATPase activity. When ADP and orthophosphate bind to the ATPase, the $H^+$ and $OH^-$, removed in the formation of ATP are pulled in opposite directions under the influence of the electrochemical gradient and react with the opposite ion favoured by the stalk of the ATPase which passes through the membrane providing a channel for passage of the $OH^-$ ions.

**Chemoautotroph.** An organism that obtains energy for growth by oxidation of reduced compounds of sulphur, hydrogen, ammonia or nitrite.

**Chemostat**. A fermentaion vessel or bioreactor in which steadystate growth is maintained by providing the cells with a constant input of nutrients. The rate of growth can be regulated by the level of a specific nutrient and the dilution rate (rate of removal of organisms from the vessel) matched to the cell growth rate.

**Chlorophyll.** A green pigment composed of a tetrapyrrole (porphyrin) ring containing magnesium complexed with a molecule of the long-chain alcohol phytol. Chlorophylls are the primary photosynthetic pigments found in all autotrophic organisms. Several distinct structural forms occur; chlorophyll a (found in all oxygen producing phototoautotrophs); chlorophyll b (found in higher plants and green algae); chlorophylls c and d (found in various algal groups); bacteriochlorophyll (found in the green and purple bacteria). Various forms of chlorophyll a and b may be distinguished by their absorption spectra. These differ in vivo from those of isolated chlorophyll in organic solvents. Differnces are the result of the association of chlorophyll molecules, or chlorophyll dimers, with specific proteins and include the so-called reaction centre complexes P700 and P680, which form the energy traps of photosystem I and photosystem II, respectively.

**Chloroplast genes.** Genetic information carried on the DNA contained within the chloroplast. Since the chloroplast are cytoplasmic and during fertilization cytoplasm comes only from the ovum characteristics coded for in the chloroplasts do not show Mendelian inheritance; characteristics are only passed on through the female line.

**Chloroplast genome.** The genetic information of the chloroplast, which occurs as multiple copies of a circular double-stranded DNA molecule. This DNA codes for rRNA, tRNA and a number of proteins, including the large subunit of rubisco and some membrane proteins (e.g., photosystem I chlorophyll protein). The mechanism of replication of chloroplast DNA is semiconservative and involves rolling circle intermediates.

**Chloroplast mRNA.** An mRNA that is only translated on 70S ribosomes.

**Chloroplast proteins.** The soluble stroma proteins and the insoluble lamellar proteins. These are encoded by nuclear genes and by chloroplast genes. The enzyme rubisco is the best studied of the stroma proteins.

**Chloroplast ribosome.** A 70S ribosome found in eukaryotic chloroplasts. Protein synthesis of 70S ribosomes is inhibited by chloramphenicol, but not by cycloheximide; the reverse is true for 80S cytoplasmic ribosomes. Chloroplast ribosmes dissociate into 50S and 30S subunits.

**Chloroplast.** The membrane-bound organelle of plants that contains chlorophyll. It is present in the cytoplasm of eucaryotic photosynthetic cells; site of photosynthesis. The outer envelope encloses a complex lamellar structure embedded in an amorphous protein stroma. The internal volume can thus be divided into three distinct regions with different functions:(1) the stroma containing soluble enzymes associated with the dark reactions of carbon assimilation and further metabolism; (2) the membrane system, made up of thylakoids, bearing the photosynthetic pigments, the intermediates of the electron transport chain and many of the components associated with the production of oxygen, the formation of ATP and the reduction of NADP - collectively known as the light reactions; (3) the lumen of the thylakoids, an aqueous region containing proteins, magnesium ions and other charged inorganic molecules that result in the formation of a potential gradient across the thylakoid membrane. Movement of ions across the membrane produces changes in potential and changes in pH and magnesium concentration which control the activities of some of the enzymes associated with the dark reactions of carbon assimilation. Chloroplasts also contain DNA and ribosomes that code for and synthesize some of the chloroplast proteins; others are encoded by nuclear genes and formed by cytoplasmic ribosomes. In C4 plants, the activities of photosynthesis are divided between the chloroplasts of the mesophyll cells and the bundle sheath. Such chloroplasts differ in structure and/or function.

**Chramatographic column.** A tubular container used to hold and support the solid stationary phase in chromatographic separations. Columns are made of glass, plastic or metal, and vary in size and length according to the technique used.

**Chromatographic detector.** It is an instrument used to monitor and quantify the products eluted from a chromatography column. Techniques used are based on the physical, chemial or biological properties of the molecules.

**Chromatography.** Chromatography has been classically defined as a separation methods in which mixtures are resolved by differential migration of their constitutents during passage through a column. It is essentially a physical methods in which the components to be separated get distributed between two phases: a stationary phase bed and a mobile phase which percolates through this bed. The chromatographic process is a result of repeated sorption/desorption acts during the movement of the sample components along the stationary phase and the separation is due to the differences in the distribution constants of the individual components. Both the stationary and mobile phases are highly variable and the differences in them result in the different types of chromatography. The stationary phase can be a solid (Column chromatography. Thin layer chromatography, Ion exchange chromatography etc.) or a liquid (Paper chromatography, Gas chromatograghy etc.). The mobile phase generally is a liquid except for gas chromatography. The distribution of the components in a chromatogram may be due to partition (PC, HPLC, GLC), adsorption (Column TLC), ion exchange (ion-exchange chromatography).

**Coenzyme A (CoA).** A coenzyme consisting of pantothenic acid linked to a mononucleotide based on adenine. It is an intermediate involved in the transfer of acetyl groups (with which it combines to form acetyl-CoA) and other fatty acyl groups which are bound to a thiol group on the molecule. CoA plays an important role in fatty acid synthesis and oxidation, as well as in the first step of the TCA cycle.

**Coenzyme Q.** A substituted benzoquinone (ubiquinone) that plays a role in the respiratory electron transport chain, where it receives hydrogen or electrons from a flavoprotein and is reduced to a quinol, which then reduces the next number of the chain (a cytochrome).

**Coenzyme.** A small organic molecule assoicated with the protein portion of a holoenzyme, weakly bound at the active site of the enzyme and required for enzyme activity. Cooenzymes may be regarded as dissociable prosthetic groups. Their protein partners are termed apoenzymes. Neither the apoenzyme not the coenzyme is, in itself, a complete catalyst. In the course of the catalyzed reaction, the coenzyme may be changed chemically in the same way as a substrate and dissociate from the apoenzyme. However, all coenzymes can be regenerated in assoicated reactions and can complex again with the apoenzyme. Some coenzymes function with many different apoenzymes, and hence play an important role in coupling different biochemical reactions. Coenzymes of importance in central metabolism include the pyridine nucleotides (NAD and NADP), coenzyme A, flavin adenine dinucleotide (FAD) and pyridoxal phosphate.

**Cofactor.** It is a non-protein compound which is essential for the catalytic activity. An enzyme-cofactor complex is termed as holoenzyme. The protein on its own is termed as apoenzyme.

**Compensation point.** Light intensity at which respiration equals photosynthesis over a 24-hour period.

**Conjugated protein.** A protein that contains or is linked to other non-protein moieties; for example, metalloproteins (contain metals) lipoprotines (contain lipids) and nuceloproteins (contain nucleic acids).

**Coordinated enzyme synthesis.** Enzyme whose rates of production are observed to vary together. (For example, in *E.coli* cells growing in the absence of ß-galactosides, the addition of lactose to the medium causes the coordinated induction of ß-glactosidease and ß-galactoside permease).

**Coupled reactions.** Two chemical reactions that have a common intermediate through which energy can be transferred from one reaction to the other. It is a thermodynamically unfavourable reaction which by association with a thermodynamically favourable reaction is driven in the direction of product formation.

**Coupling factor.** ($F_1$factor) The headpiece of the mitochondrial inner membrane subunit that has ATPase activity, associated with phosphorylation in both respiration and photosynthesis. Coupling factor is fouund on the membranes of mitochondria and chloroplasts.

**Crabtree effect.** A form of substrate inhibition see in yeasts such as **Saccharomyces cervisiae** in which media with a glucose content of over 5 percent totally inhibit the formation of respiratory enzymic, even in the presence of air.

**Crassulacean acid metabolism (CAM).** Higher plants in which the photosynthetic carbon reduction cycle is supplemented by a second carbon assimilation pathway based on an initial fixation of carbon dioxide by the enzyme phosphenol pyruvate carboxylase. Many CAM plants live in arid environments, and CAM metabolism may be associated with a need to conserve water. In the classic CAM plant, the stomata are only open in the dark (at night) when carbon dioxide is incorporated into malate. In the light (day time) the stomata closes, and malate is decarboxylated releasing carbon dioxide internally. This carbon dioxide is then refixed through the normal photosynthetic carbon reduction pathway.

**CTP.** Cytidine triphosphate like ATP it is an energy-rich compound. It is also a precursor of

RNA.

**Cytochrome oxidase.** A copper-containing complex associated with cytochromes $a_1$ and $a_2$ in the terminal electron transport chain of oxidative respiration. This complex reduces molecular oxygen. It is inhibited by carbon monoxide, cyanide and hydrogen.

**Cytochrome $P_{450}$.** An intermediate in electron transport chains, found in microsomes for example. It is associated with desaturation of fatty acids and hydroxylation of steriods. ATP is not produced duuring electron transfer through these electron transport chains.

**Cytochrome.** Electron-transport intermediates containing heme of related prosthetic groups that undergo valency changes or related prosthetic groups that undergo valency changes of the iron atom.

**Cytokinin.** A naturally occurring plant growth substance; a kinin.

**Cytoplasm.** A term used to include all the living parts of a cell bounded by the plasma membrane, but excluding the nucleus.

**Cytosol.** Th part of the cytoplasm that does not contain any membrane or particulate structures.

**Dark reaction.** The carbon assimilation through the photysynthetic carbon reduction cycle (PCR) and associated pathways, including the $C_4$ cycle and $C_2$ photorespiratory cycle. The dark reactions are catalyzed by soluble enzymes found in the stroma of the chloroplasts, and use ATP and NADH generated in the light reactions of phtosynthesis as a source of energy and reducing power.

**Dark respiration.** Respiration in photosynthetic plants occurring during the night.

**Death.** The end of life from an organic complex of otherwise known structure.

**Dehydration.** The drying or removal of water. It is used as a means of preserving biological material or preparing specimens for examination, under a microscope. Dehydration using careful freeze-drying techniques, is used to preserve bacteria, yeasts and fungi in culture collections.

**Dehydrogenation.** Removal of hydrogen from a reduced substrate, which thereby becomes oxidized. Enzymatic dehydrogenation is catalysed by dehydrogenases or oxidases.

**Denitrifying bacteria.** The soil bacteria which break nitrate and nitrite with the evolution of free nitrogen in an oxygenless atmosphere.

**Densitometer.** A device used to such chromatographic plates or electrophoretic gels to evaluate the amount of various compounds present on the plate, or gels. The essential parts consist of a light source, a sample holder and a detector (photomultiplier). The sample is passed continually through the light (visible or ultraviolet) beam and the changes in density, reflected light or fluorescence (depending on the mode of operation) are recorded on a chart or using a data logger. The samples being scanned are often stained using a suitable dye prior to analysis.

**Determinant growth.** Plant growth, in which growth of season ends in a bud.

**Development.** The process of growth and differentiation of individual cells into tissues, organs and organisms.

**Devernalization.** A process where stimulus to flowering received by treatment with low temperature (i.e., vernalization) can be reversed.

**Differentially permeable membrane.** A membrane whereby the passage of diverse material is regulated into and out of cell, organelles and vacuole at different rates depending on the relative affinity of the material for lipid and protein constituents of the membrane.

**Digital Glycosides.** The cardiac glycosides of the cardenolide group found in the leaves of foxgloves, **Digitalis purpurea** and Digitalis lanata. They are indispensible cardiotonic agents, used for longterm treatment of chronic heart weakness are replacing the leaf powders or extracts formerly used.

**Dissolved carbon dioxide.** The amount of carbon dioxide, expresed in terms of equivalent volume under standard conditions, moles or weight, in true solution in unit volume of culture medium or other liquid.

**Distribution coefficient** ($K_{sv}$). In gelfiltration chromatography, the fraction of the total gel volume that is available to the solute. $K_{av}$ =($V_e$-$V_o$)/$V_t$-$V_o$) Where $V_e$ = the evolution volume, $V_o$= the void volume and $V_t$= the bed volume. There is a linear relationship between $E_{av}$ and log molecular weight of many substances, including globular proteins.

Gel filtration can be used in a column calibrated with standard proteins to detrmine the approximate molecular weight of an unknown protein sample.

**Diurnal acid rhythm.** The fluctuation in the content of organic acids in the chlorophyll-containing parts of succulent plants, such as cacti. At night, $CO_2$ is fixed by carboxylation of phosphoenolpyruvate to oxaloacetate, which is then reduced to malate, leading to a considerable acidification of the tissue. By day, the malate is oxidatively decarboxylated to puruvate, and the $CO_2$ is fixed photosynthetically. The mechanism allows the plants to absorb $CO_2$ from the air at night, when they can open their stomata without suffering the high water loss which would occur during the day. It is thus an adaptation to a hot, arid climate.

**Eadie plot.** A method of treating data from investigations of enzyme kinetics in order to obtain straight line plots from which the various kinetic constants can be calculated. In such plots, the value of obtained by dividing the substrate concentration (S) by the observed velocity (v) at the substrate concentration is plotted as a function of substrate concentration. This gives a slope with a value equal to the reciprocal of the maximum velocity and an intercept on the x-axis equal to the negative of the $K_m$.

**Elaioplast.** A plastid that stores fats or oils.

**Electrolysis.** The decomposition of a compound by an electrical current or the destruction of tumours, hair roots, etc. in a similar manner.

**Electrolyte.** (1) A conducting medium in which the flow of electrons is accompained by the flow of matter. (2) A substance which when dissolved in water dissociates into ions forming a conductor of electricity.

**Electron carriers.** Intermediates such as flavoproteins and cytochromes that reversibly gain or lose electrons.

**Electron Transfer Flavins,** *abb.* ETF. The flavoproteins which mediate electron transport from reduced **FADH** to the cytochrome system. Flavoproteins can oxidize those substrates of the Respiratory chain whose oxidation does not involve pyridine nucleotides.

**Elongation factors, Transport factors.** The proteins catalysing the elongation of peptide chains in protein biosynthesis.

**Emerson effect.** This is a kind of photosynthetic enhancement discovered by Emerson et al. where the efficiency of phtosynthesis at wavelengths exceeding 680 nm can be restored by a simultaneous application of a shorter wavelength.

**Emulsifier**. A compound that assists in the formation and stabilization of an emulsion.

**Endogenic reaction.** A chemical reaction with a positive standard free energy change; an energy-consuming reaction.

**Endogenous.** Arising from within the generating structure e.g., endoconidia of **Thielaviopasis basicola**. *cf.* exogenous.

**Endoglucanase.** An enzyme that cleaves a polysaccharide such as cellulose at a point within the macromolecular substrate, rather than at the ends.

**Endomembrane system.** A system of cytoplasmic membranes found in eukaryotic cells. It participates in the synthesis, transport and storage of chemical compounds within the cell and facilitates export from cells. It includes the plasma membrane and tonoplast of vacuoles.

**Endoplasmic reticulum.** A system of membranes found in most eukaryote cells that is continuous with the outer membrane of the nucleus. It is termed rough endoplasmic reticulum when ribosomes are present and smooth endoplasmic reticulum when ribosomes are absent.

**Energy balance.** An analysis carried out on a process or reaction system, in which all inputs and outputs of energy are determined experimentally or by calculation. Such balance may be used to determine the overall efficiency of the system or to identify areas for process improvement.

**Energy crop.** Selected plant species grown and harvested specifically as biomass to be used as a source of energy and converted into a fuel or bulk chemical. The major energy crops include fast-growing or coppiced trees (willoows, poplars and eucalyptus species), sugar crops (sugar cane and sugar beet), starch crops (maize or cassava), inulin crops (chicory or *Jerusalem artichoke*), aquatic plants including algae and water weeds (water hyacinth and emergent species such as rushes and reeds).

**Enthalpy.** A thermodynamic expression for heat content dfined by the equation.

**Enzyme amplification.** A technique used to visualize or quantify an immunoreaction in an assay procedure in which the enzyme label in the immnoassay is used to provide the trigger substance for a secondary ststem that can produce a large amount of a coloured product.

**Enzyme unit.** The amount of an enzyme that will catalyz the transoformation of one micromole of substrate n a given time (e.g., one ninute) under defined conditions of temperature, pH and substrate concentration.

**Enzyme.** Any substance, protein in whole or in part that regulates the rate of a specific biochemical reaction in living organisms. It is capable of catalyzing a reaction in which substrate(s) are converted to product(s) through the formation of an intermediate enzyme - substrate complex. As with other catalysts, enzymes are responsible for accelerating the rate of a chemical reaction.

**Epigeal.** A condition in plants where the cotyledons are raised above the ground during seed germination.

**Ethylene**. An unsaturated hydrocarbon gas that acts as a plant growth substance. It is associated with the ripening of fruit and the climacteric (rise in respiratory activity), plays a role in the breaking of seed dormancy and can inhibit growth of some plant organs.

**Exoenzyme.** An enzyme localized on the outer side of the cell membrane. In bacteria, for instance, exoenzymes may be released into the outer medium or lie in the sape between the cell wall and the cell membrane.

**Exothermic process.** A process in which heat is evolved.

**Exothermic reaction.** A chemical change in which heat is given off; a reaction with a negative enthalpy change.

**Exponential growth.** A type of growth in which the increase in population is proportional to the number of individuals present.

**Facultative aerobe.** An anaerobic microorganism which can grown under aerobic conditions.

**Facultative anaerobe.** An organism, generally a bacterium or fungus that can adapt its metabolism to survive and grow either in the presence or the absence of oxygen.

**Facultative halophytes.** Plants capable of growing in salty as well as in non-salty soils, e.g., *Atriplex allenorflfea*, grow best where salt levels in the soil are high, as in deserts or in soils saturated with bracish waters on the sea coasts. They also grow in non-salty soils.

**Facultative photoheterotroph.** Any bacterium that usually grows anaerobically in light but can also grow aerobically in the dark.

**Facultative saprophyte.** An organism capable of growing on dead organic matter, or of infecting another living organism according to circumstances.

**Facultative thermophile.** An organism requiring a temp-erature between 50 and 65°**C** for optimum growth but capable of growth at lower temperatures; cf. obligate thermophile.

**Farnesol**. A colourless liquid extracted from oils of plant such as citronella, neroli, cyclamen, and tuberose. It has delicate floral odor.

**Fat biosynthesis (Triacylglyceride synthesis).** Refers to biosynthesis of neutral fats from fatty acids and glycerol. Fatty acids and glycerol are synthesized by separate pathways and then combined in a series of reactions in which the acyl moieties of fatty acyl-CoA molecules get transferred to glycerol phosphate.

**Fat degradation.** It refers to the hydrolysis of the ester bonds in netural fats to form fatty acids and diacylgycerol, which is further degraded to fatty acids and glycerol. The enzymes responsible are lipase.

**Fatty acid bio synthesis.** Long-chain fatty acid synthesis is divided into three distinct phases: (1) acetyl-CoA is carboxylated to yield malonyl-CoA; (2) the two-carbon units derived from decarboxylation of malonyl-CoA are condensed to form a fatty acid of intermediate chain length, eventually palmitic acid ($C_{16}$) (this is a multi-step process catalyzed by a multienzyme complex termed fatty acid synthetase); (3) the wide range of longer-chain, saturated and unsaturated, fatty acids is derived from plamitic acid by the concerted action of fatty acid elongation and desaturation systems.

**Fatty acid oxidation.** The catabolic pathways for fatty acids. The most common to ß.oxidation; minor pathways are and oxidation. 1. Beta-Oxidation, the stepwise degradation of fatty acids from the carboxyl end. In each turn of the cycle, two carbon atoms are removed as acetyl-**CoA,** the beta-carbon atom 2. alfa-Oxidation of fatty acids occurs in germinating plant seeds. A fatty acid peroxidase (EC 1.11.1.3) catalyses the decarboxylation and simuultaneous formation of an aldehyde, in which $H_2O_2$ acts as hydrogen acceptor. The aldehyde can either be oxidized to a fatty acid or reduced to a fatty acid or reduced to a fatty alcohol. 3. The terminal methyl group of a fatty acid is oxidized by enzymes located in the microsomal fraction of animal and microbial cells. The substrate is usually a $C_8$ to $C_{12}$ fatty acid in two steps. The first is the hydroxylation to a $^{w}$- hydroxy fatty acid. Which requires oxygen and NADPH. The second step is catalysed by a soluble, non-microsomal enzyme, which is usually $NAD^+$- dependent.

**Fatty acid.** A long-chain (hydrocarbon) carboxylic acid. Fatty acids are components of many lipids including glyceride. Long-chain fatty acids are insoluble in water. However, their sodium and potassium salts are soluble as soaps. The most common naturally occurring fatty acids are monocarboxylic acids which have an even number of carbon atoms (16 or 18) and which may be saturated or unsaturated (possess cis double bond(s) between the carbon atoms). In polyunsaturated acids (i.e. those with more than one double bond) the double bonds are arranged in a methylene-interrupted system ($-CH=CH-CH_2-CH=CH-$).

**Fermentation.** (1) A metabolism, the degradation of organic substances by organisms or cells to provide chemical energy as ATP in reaction that do not require molecular oxygen. In anaerobic cells, it is the only energy-producing process. (2) Descriptive of processes in which cells or other microorganisms are cultured in a container (bioreactor or fermenter) in liquid or solid medium for experimental or commercial processes. The term was applied only to anaerobic cultures, but now is often used to describe all cultures of microorganisms or cell in fabricated vessels.

**Ferredoxin.** An iron-sulphur protein. Ferredoxins have a very negative redox potential and are constituents of electron transport chains. A ferredoxin is associated with the reduced side of photo- system I in photosynthesis, donating electrons to NADP in a reaction mediated by ferredoxin-NADP oxidoreductase.

**Flavanol.** (i) Any class of flavonoid compounds which are hydroxy derivatives of flavone. (ii) $C_{16}H_{10}O_2$, a colour less crystalline compound from which many yellow plant pigments are derived.

**Flavone.** (1) Any of a number of ketones composing a class of flavonoid compounds. (2) $C_{15}H_{10}O_2$ A colourless crystalline compound occurring as dust on the surface of many primrose plants.

**Flavonoids.** Compounds based on two benzene rings joined by a three-carbon chain. Flavonoid glycosides are important plant pigments and include the anthocyanins, benzalcoumarins, chalcones, flavones and flavanols, as well as the colourless leucoanthocyanidins and isoflavones.

**Florigin.** A hypothetical plant growth substance supposedly associated with the onset or regulation of flowering.

**Flow.** (1) To move along in stream or to circulare. (2) In fluid mechanics, a deformation of which at least part is nonrecoverable.

**Flowering hormone (Anthesin, Florigen, Vernaline).** A principle which has been demonstrated physiologically but never isolated or chemically characterized.

**FMN.** Flavin mononucleotide (oxidized form); a derivative of riboflavin that functions as a prosthetic group in flavoproteins and acts as an acceptor in oxidation - reduction reactions. $FMNH_2$ is the reduced form.

**Free energy.** A Thermodynamic function representing the energy evolved or absorbed during a reversible reaction. It is defined as: $G = H - TS$ It is a component of the total energy of a system that can do work under conditions of constant temperature and pressure.

**Fumaric acid.** A chemical compound that occurs naturally in most cells. It is an intermediate in the TCA cycle.

**Labelling.** The tagging of compound with an isotopic tracer so that its fate within an organism can be followed.

**Lactic acid.** The end product of fermentation in vertebrate that binds (chelates) iron.

**Lag phase.** The period between the initial in oculation of a culture medium with an organism and the resumption of normal growth.

**Latent period.** This may be regarded as the penultimate stage of the replication cycle of the bacteriophages. This period begins with the formation of the first new phase particle in the host cell and ends with the lysis of the same (q.v.).

**Leucoplast.** A starch-storing organelle in the cytoplasm. It is a colourless plastid made up of concentric layers of starch around a central body.

**Leucosin or chrysolaminarin.** Food storage polysaccharide of golden-brown algae composed mostly of β-1,3 linked glucose residues.

**Life cycle.** The series of events from the production of gametes in one generation to the same stage in the subsequent generation. In some plants the life cycle only involves the production of one type of individual.

**Light compass reaction.** A reaction of many invertebrates in which the angle between the direction of movement and the direction of the sun is kept constant.

**Light reaction.** The light-dependent sequence of photosynthetic reactions that, taking place on the thylakoid membrances of chloroplast, produce the ATP and NADPH used in the carbon dioxide fixation of the subsequent dark reactions.

**Light.** Light from the sun provides the energy to fuel the photosynthetic fixation of carbon dioxide by green plants is thus the basis of life on earth.

**Lignin.** A complex carbohydrate polymer making up about 25% of the wood of trees and also found in the cell walls of schlerenchyma tissues and vessels, fibres, and tracheids at maturity.

**Lipase.** Any enzyme involved in the breakdown of storage fats.

**Lipopolysaccharide (LPS).** The lipid and sugar complex that makes up part of the outer membrane of the gram-negative cell wall.

**Lipoprotein.** An association of lipids and proteins in a system that combined aspects of the properties of both types of molecule.

**Liposome.** Synthetic phospholipid sacs (vesicles) having properties of biological membranes.

**Lithotrophic or autotrophic.** Needing only light and/or inorganic compounds for growth.

**Log phase of growth.** Growtg phase characterized by rapidly dividing cells.

**Log phase.** The period of most rapid reproduction in the growth phases of a culture; the generation time is constant when plotted on a graph, and the log of the number of organisms appears as a straight line.

**Logistic growth.** Growth, especially in the size of an organism or in the number of organisms that constitute a population, which slows steadily as the entity apporaches its maximum size (contrast with exponential growth).

**Lysosome.** Single-membrane-bounded cytoplasmic particle containing destructive enzymes.

**Lysozyme.** Enzyme which hydrolyzes the bond between N-acetylglucosamine and N-acetylmuramic acid of the peptidoglycan of the bacterial cell wall; this causes the cell wall to break, and lysis of the cell usually follows.

**Lytic.** Condition in which a bacterium lyses shortly after infection by a virus; the viral genome does not become stabilized within the bacterial cell. (Contrast with lysogenic).

**Macrophyllous.** Having leaves with branching veins, the traces of which leaves are associated with gaps in the stem stele.

**Macrophyte.** A large macroscopic plant, used especially of an aquatic forms such as kelp; macrophytic.

**Macrophytophagous.** Feeding on higher plant material only; macrophytophage, macrophytophagy.

**Magnesium, Mg.** A cation widely distributed in biological systems, with many different biological functions. Mg has great biological significance as a constituent of the prophyrin system of chlorophyll. Mg is an essential nutrient, and it plays an important part is metabolism by acting as a cofactor of many different enzyme system.

**Magnetotaxis.** The reaction of certain aquatic bacteria in response to magnetic field lines. Such bacteria are influenced by the Earth's magnetic field. Magnetotactic bacteria (discovered in 1975) are able to synthesise tiny crystals of magnetite, $Fe_3O_4$. They are either north-seeking or south seeking. Magnetotactic bacteria are found in both freshwater and sea water. They include cocci, bacilli, and spirals.

**Maintenance energy.** Energy required by an organism or a cell culture in order to survive in a viable condition, but without further increase in size or cell number. This may be expressed in terms of the weight of substrate, or energy in joules, required per dry weight of cells or body mass.

**Major element.** An element which is essential for life and is required for the formation of the major structural parts of living organisms. They include carbon, nitrogen, oxygen, hydrogen, sulphur, phosphorus, calcium and, in some instances, silicon.

**Malate.** A salt of malic acid.

**Maleic Hydrazide, MH.** 1, 2-dihydro-3, 6-pyridazinedione. A synthetic plant growth retardant which is used as an herbicide against grasses. It causes a unique depression of growth, inhibition being marked but temporary. It inhibits germination of seeds, suppresses growth of roots and terminal shoots and retard flower and bud development. It prevents the formation of suckers in tobacco and tomatoes, and prevents the sprouting of onions and potatoes.

**Malic acid fermentation.** A metabolic process that results in the degradation of malic acid to ethanol. There are two possible pathways based on malic enzyme or malate dehydrogenase. In the former, malic acid is decarboxylated to pyruvic acid, and then to acetaldehyde and ethanol. In the latter, the malic acid is oxidized to oxaloacetic acid, decarboxylated to pyruvate and then metabolized to ethanol as above. These pathways are characteristic of yeasts such as *Schizosaccharomyces pombe.*

**Malic Acid.** Monohydroxysuccinic acid,-HOOC-COOH-$CH_2$-COH. A dicarboxylic acid found in many plants juices. The malate ion is formed in the Tricarboxylic acid cycle and the Gloxylate cycle. Malate plays an important role in the Diurnal acid rhythm of the Crassulaceae (stone-crop family).

**Malic enzyme(s), L-malate-NADP Oxidoreductase, Decarboxy-lating** (EC 1.1.1.40). An important enzyme found in most organisms, which catalyses the decarboxylation of L-malate to pyruvate and $CO_2$ with concomitant reduction of $NADP^+$ to NADPH (or the synthesis of malate by the reverse reaction.

**Manganese, Mn.** A bioelement present in all living cells, which usually conain less than 1 ppm on a dry weight basis, or less than 0 01 mM in fresh tissue. Bone contains 3.5 ppm. Mn (II) is necessary for sporulation in *Bacillus subtillis*, and these bacteria can maintain an intracellular Mn concentration of 0.2 mM against an external concentration of 1 uM. Mn is an essential nutrient for animals and plants. In plants, Mn deficiency results in chlorosis and mottling. Mn may also be important in the regulation of enzyme activity e.g., nonadenylylated glutamine synthetase requires Mg (II), but the adenylated form binds Mn (II). Also the specificity of nucleases and DNA polymerases is changed when Mg (II) is replaced by Mn (II). The physiological significance of these differences is not clear.

**Mannitol.** A sugar, fermentation of which is carried out by a set of genes produced by *E. coli.*

**Mannose.** An aldohexose sugar, an epimer of glucose, occurring frequently in polysaccharides.

**Mass spectrometry.** A process which sorts streams of electrically charged particles according to their different masses by using deflecting fields. The device which accomplishes this, a mass spectrometer, consists generally of a long tube which generates a magnetic field. Particles of varying masses are passed through the tube and the degree of deflection by the magnetic field is recorded on a photographic film at the other end.

**Mass transfer.** It is the process which results in the movement of molecules from one place to another. In biological systems, mass transfer results from diffusion, facilitated diffusion or active transport.

**Matrix.** (*pl.* matrices). (1) A ground substance consisting of non-living intercellular material in which living cells are embedded. (2) The liquid inside a mitochondrion, (3) The substratum in or on which a fungus or lichen grows.

**Matter.** That which has weight and occupies space.

**Maturation region.** The area within a young root (several centimeters from the tip) where the cells reach their full size and become completely specialized.

**Maximum enzyme velocity ($V_{max}$).** It is the theroretical maximum rate at which a given enzyme can catalyzed a specific reaction when supplied with substrate at an infinite concentration. $V_{max}$ is determined experimetally using for example, Lineweaver - Burk kinetics.

**Maximum oxygen transfer rate.** It is a measure of the fastest possible rate of uptake of oxygen by a growing culture expressed in terms of the volumetric mass transfer coefficient times the oxygen solubility.

**Maximum specific growth rate**. The theoretical maximum rate of growth of a culture under optimal conditions of nutrient supply. <108> max is determined experimentally from a double reciprocal derivation of the Monod equation analogous to the Lineweaver -Burk plot for enzyme kinetics.

**MCPA.** 2-Methyl-4-chloro-phenoxyacetic acid; a synthetic auxin used as a weed killer.

**Mechanism.** The view of living organisms which holds that life is wholly explicable in terms of physical and chemical laws.

**Medium.** A composite formulation of substrates, minerals growth factors and vitamins used in the culture of microorganisms (bacteria, yeast, fungi), animal cell lines or plant tissue cultures. Media are liquid or solidified with polymers such as agar or gelatin.

**Membrane enzymes.** The enzymes present on the surface of the many different types of biological membranes. There are, e.g., glucose-6-phosphatase of the endoplasmic reticulum, galactosyl transferase of the Golgi apparatus, oligomycinsensitive ATPase of inner mitochondrial membrane, monoamine oxidase and rotenone-insensitive NADP-cytochrome c-reductase of the outer mitochondrial membrane, and the sodium-independent ATPase and the 5-nucleotidase of the cell membrane. A very large number of reactions occur on membranes, and the membrane enzymes play a central part in cell metabolism.

**Membrane filter bioreactor.** A bioreactor in which the fermenting broth is circulated continuously through a membrane filter. The product diffuses from the reaction zone through the membrane and is withdrawn continuously. This method is also applied to enzyme and resting cell reactors. The hollow fibre filter fermenter is one variation of this type of bioreactor.

**Membrane.** In most plants and animals, a bilayer of lipids (e.g., phospholipids, glycolipids and sterols) in which globular proteins are embedded.

**Mesophyll.** The chloroplast-containing tissue comprising the bulk of the leaf between the upper and lower epidermis.

**Metabolic heat.** Energy lost as heat during the growth of an organism. In aerobic biocatalysis and fermentation processes, knowledge of the amount of heat produced is important in bioreactor design since provisions have to be made to remove the heat. Heat production may be calculated from an enthalpy balance for a growing culture. However, for aerobic systems the amount of heat released can be derived on the basis of oxygen consumption. In general, 460 KJ of heat are released per mole of oxygen consumed. In fermentation, to produce ethanol, around 70 KJ are produced per mol of glucose converted to biomass or to ethanol.

**Metabolic pathway.** A set of consecutive interacellular enzymatic reactions that converts one molecule to another

**Metabolic poison.** Any substance e.g., cyanide, that prevents the production of ATP is cells. Without ATP as a source of energy for metabolism, cells and organisms quickly die.

**Metabolic rate.** A measure of the rate of chemical reactions in an organism. For instance, in an aerobic organism, it may be expressed as the rate of oxygen consumption.

**Metabolic shunt, Metabolic Bypass.** A metabolic or amphibolic pathway which bypass some reactions and exploits others of a primary metabolic pathway.

**Metabolism.** Sum total of all chemical processes in living cells by which energy is provided and used.

**Metabolite.** A substance produced by metabolism.

**Methane ($CH_4$).** Natural gas; a major component of biogas produced by methanogenic bacteria from carbon dioxide, hydrogen and acetate.

**Methane digester.** A synonym for anerobic digester.

**Methanogen.** A strictly anaerobic bacterium that is capable of producing methane. Methanogens include the genera. *Methanobacterium, Methanobacillus, Methanococcus* and *Methanosarcina.* Methane bacteria are found in anaerobic soils (marshes and bogs), sewage, sludge, landfill sites and anaerobic digestion systems. They also play an important role in the metabolism of ruminants.

**Methanogenesis.** The final step in the production of methane during which acetate, hydrogen and bicarbonate are converted to methane due to the activity of methanogens.

**Michaelis constant.** A constant $K_m$ such that the initial rate of reaction V, produced by an enzyme when the substrate concentration is high enough to saturate the enzyme, is related to the rate of reaction $v$ at a lower substrate concentration $c$ by the formula $V=v(1 + K_{m/c})$.

**Micro nutrients.** Trace nutrients.

**Microaerophilic.** Descriptive of a microorganism that grows best in the presence of only low concentrations of molecular oxygen.

**Microbial leaching.** A method of recovering such metals as copper and uranium from ores of low content. This is carried out using bacteria of the genus **Thiobacillus,** such as *T.thioxidans.* These organisms grow chemoautotrophically on energy released by oxidation of sulphur compounds, generating sulphuric acid and copper ions.

**Microenvironment.** The specific environmental factors in a microhabitat.

**Millon's test.** A qualitative test for protein. The sample is heated with a solution of mercuric nitrate and nitrous acid (Millon's reagent). In the presence of protein, a red precipitate of coagulated protein is formed.

**Minimal medium.** A medium containing only those elements absolutely essential for the growth of a particular organism, and which the organism cannot synthesize itself. Generally, minimal media contain a carbohydrate source, various inorganic salts, and sometimes a growth factor such as biotin.

**Minimum lethal dose (MLD).** The minimum dose of an agent sufficient to cause 100 per cent mortality of the test population.

**Mitochodrial replication.** The formation of new mitochondria by division of existing ones. Division is accomplished by invagination of the inner membrane.

**Mitochondria, Chondriosomes** (obsolete). Organelles present in all eukaryotic cells.

**Mitochondrial ATPase.** An enzyme found on the inner mitochondrial membrane that catalyzes the formation of ATP from ADP and othophosphate.

**Mitochondrial enzymes.** The mitochondrial matrix contains all the enzymes of the TCA cycle, with the exception of the succinate dehydrogenase which is built into the inner surface of the inner membrane. Mitochondria also contain enzymes for fatty acid metabolism and interconversion of some amino acids.

**Mitochondrial membranes.** The two membranes of a mitochondrion differ in structure and composition. The outer membrane has a higher lipid content and a distinct enzyme complement.

**Mitochondrial mRNA.** mRNA transcribed from mitochondrial DNA. **Mitogen.** A compound which stimulates the activity of lymphoid cells leading to the production of soluble factors (e.g., lympokines, interferon, etc.) changes in fuctional state and blast cell formation.

**Mitochondrial respiratory chain.** A sequence of enzymes and proteins present in the inner mitochondrial membranes of eukaryotic cells and in the cell membranes of prokaryotes. The chain receives electrons or hydrogen atoms derived from reactionsof the TCA cycle.

**Mitochondrial ribosome.** A ribosome associated with the mitochondrion having a sedimentation constant of 70S.

**Mitochondrion.** A double-membrane-bounded cytoplasmic organelle, site of energy release in cellular respiration.

**Mitosis.** The process whereby a cell nucleus divides into two daughter nuclei, each having the same genetic complement as the parent cell.

**Molality.** The expression of the number of moles of solute per litre of solution.

**Molar (M).** A solution which contains one mole of a compound to give one litre of final solution.

**Molar extinction coefficient.** The absorbance (extinction) of a 1 M solution using a 1-cm light path.

**Molybdenum, Mo.** An important trace element for plants and bacteria, and for the symboiotic fixation of nitrogen. As a constituent of various molybdoenzymes, Mo plays an important part in the nitrogen metabolism of plants and bacteria. In biological systems there is a partial antagonism beteem Mo and Cu.

**Mucoprotein.** Any of a group of glycoproteins containing a sugar, usually chondroitinsulphuric or mucoitinsulphurgic acid, combined with amino acids or polypeptides.

**Multisubstrate enzymes.** Enzymes that require two or more substrates in order to catalyse a particular reaction. Accordingly, the enzymes forms a ternary (two substrates), quaternary (three substrates) etc. complex.

**NAD.** Nicotinamide adenine dinucleotide, oxidized form; a coenzyme that functions as a hydrogen carrier in oxidation-reduction reactions associated with fermentation and aerobic respiration. NAD accepts hydrogen atoms from suitable substrates in reactions catalyzed by dehydrogenases to produce the reduced form NADH. The reducing equivalent is used in intermediate metabolism to reduce other compounds or is oxidized to water via the terminal electron transport chain.

**NADH.** Nicotinamide adenine dinculeotide, reduced form.

**NADP**. Nicotinamide adenine dinucleotide phosphate, oxidized form; a coenzyme that performs a similar function to that of NAD. It is particularly important in the oxidation-reduction reactions of photosynthesis.

**NADPH.** Nicotamide adenine dinucleotide phophate, reduced form.

**Naphthaleneacetic acid (NAA).** A synthetic plant growth substance with auxin-like activity.

**Negative feedback.** A mechanism of self-regulation whereby a change in a system in one direction is converted into a command for a change in the opposite direction. A means of helping maintain a biological system in equilibrium.

**Neutral fats**. Glycerides; fatty acid esters of glycerol; a major storage form of fats.

**Nick translation**. A technique for radioactively labelling a DNA molecule to high specific activity in order to produce a probe which may be used in some hybridisation technique such as Southern blotting. Nicks or gaps are introduced into the DNA molecule using a limited DNase digest. Each nick is then extended by the 5' to 3' exonuclease activity of DNA polymerase I. At the same time as 5' nucleotides are being removed by the exonucleolytic activity of the enzyme, they are also replaced by its polymerising activity. The supply of lablled $^{32}P$ deoxynucleotide triphosphates to the reaction ensures that the newly synthesised DNA will be highly radioactive.

**Nick**. A break or cut in one of the two strands of a DNA duplex.

**Nick-closing enzyme**. A form of DNA polymerase that can restore base sequences. A nick is introduced adjacent to an incorrect base. The base is then removed and the correct base inserted.

**Nickel, Ni**. Ni occurs only in traces in living system. In particular, it seems to be associated with RNA. A nickel metalloprotein named 'nickeloplasmin' has been isolated from human and rabbit serum, but its function is not known.

**Nif genes**. A set of genes that encodes the information needed for the formation of the enzymes and other proteins associated with the fixation of atmospheric nitrogen.

**Nif**. The genetic notation for the genes involved in nitrogen fixation. The nif operon is a complex array of seventeen genes. The proteins encoded by the nif genes will fix atmospheric nitrogen ($N_2$) into ammonia ($NH_4^+$) and nitrate ($NO_3^-$). Many soil bacteria will fix nitrogen and there is much interest in manipulation of nif genes from bacteria to allow plant to fix nitrogen.

**Nigeran**. A linear glucan which cotains alternating -(1-3) and -(1-4) linkage. It is synthesized by ***Aspergillus niger***.

**Ninhydrin reaction**. A chemical test or spray used in chromatography to detect amino acids. A chromatogram is sprayed with a 0.2 per cent solution in acetone or an acidified 0.5 per cent solution in butanol. When the chromatogram is heated at 105<144> C, all compounds containing a primary or secondary amino group attached to an apliphatic carbon given a purple colour. However, amino acids will react more specifically in about three hours at room temperatuure to give similar colours.

**Ninhydrin.** An organic chemical used in detecting amino acids, with which it gives blue colour on heating. It is used in several types of chromatography.

**Nitrate assimilation**. The conversion of nitrate to ammonia. For most plants in their natural environment nitrate is the usual source of nitrogen.

**Nitrate bacteria.** Bacteria, e.g., *Nitrobacter*, which convert nitrite, $NO_2^-$. to nitrate, $NO_3^-$; they are found in soil.

**Nitrate reductase**. An enzyme that catalyzes the conversion of nitrate to nitrite. Eukaryotic nitrate reductase in an enzyme complex containing flavin (FAD), heam (cytochrome $b_{557}$) and molybdenum. Electrons are provided by NAD (P)H derived from photosynthesis or carbodhydrate oxidation. Prokaryotic nitrate reductase is a smaller protein containing molybdenum, but no haem or cytochrome. The enzyme has a fast rate of turnover and is present at high concentrations when nitrate is available, but is repressed by the presence of ammonium ions.

**Nitrate reduction**. In the narrow sense, nitrate reduction is the reduction of the nitrate ion ($NO_3$ -) to nitrite ($NO_2$), catalysed by Nitrate reductase. In the broad sense, it is the reduction of nitrate to gaseous products (nitrous oxide or nitrogen), or to ammonia.

**Nitrate respiration**. (assimilatory nitrate reduction). Under anaerobic or microaerophilic conditions, some bacteria can use nitrate as a terminal electron acceptor for respiratory metabolism.

**Nitrification**. The conversion of ammonia, liberated from decaying organic matter in the soil, to nitrate. Two genera of bacteria are involved; *Nitrosomonas* converts ammonia to nitrate and *Nitrobacter* converts nitrite to nitrate.

**Nitrifying bacteria.** Bacteria in the soil which oxidize ammonia ($NH_3$) to nitrate ($NO_3^-$). Examples include, *Nitrococcus, Nitrosomonas*. This is one of the important stages in the nitrogen cycle, which makes nitrate available for plants; nitrifiers.

**Nitrite bacteria.** Bacteria, e.g., *Nitrococcus* and *Nitrosomonas*, which convert ammonium, $NH_4^+$, or ammonia, $NH_3$, to nitrite $NO_2^-$; they are found in soil.

**Nitrite reductase**. An enzyme that converts nitrite to ammonia. Plant and photosynthetic prokaryotes possess a ferredoxin-linked nitrite reductase which accepts electrons from reduced ferredoxin, but not from NAD (P)H. The enzyme contains a tetranuclear iron sulphur centre and a special haem (siro) similar to that found in sulphite reductase. Nitrite is reduced directly to ammonia without the formation of free intermediates.

**Nitrites** (as food additives and preservatives). are used e.g., in bacon and canned meats; they inhibit bacterial growth- particularly under acid conditions-and maintain an attractive coloration in various types of processed meat. The permitted concentrations of nitrites are limited since they are physiologically active substances.

**Nitrogen balance.** Difference between the total nitrogen intake of an organism and its total nitrogen loss. Young growing animals are in positive nitrogen balance i.e., they retain more nitrogen (as protein added during growth) than they excrete. Mature, healthy adults show a zero nitrogen balance i.e., nitrogen intake is exactly balanced by nitrogen excretion. A negative nitrogen balance results from a deficiency of an essential amino acid; a decrease in the concentration of any proteogenic amino acid in the body's amino acid pool impairs total protein synthesis, so that the concentration of all other free amino acids in the pool is increased; this leads to an exaggeration of degradative pathways and an increase in urea formation.

**Nitrogen catabolite repression.** Repression by ammonia of the synthesis of various enzymes of nitrogen metabolism.

**Nitrogen cycle**. The overall circulation of nitrogen through the biosphere. The cycle involves the nitrogen-fixing organisms (bacteria and cyanobacteria), organisms that produce ammonia from decaying matter, the nitrifying bacteria which oxidize ammonia to nitrite or nitrite and plants which assimilate nitrogen as ammonia or nitrate and form amino acids, as well as the denitrifying bacteria which convert nitrate back to nitrogen gas through nitrite.

**Nitrogen fixation.** The assimilation of gaseous (atmospheric) nitrogen by reduction to ammonia and incorporation into amino acids. Only some groups of organisms (soil bacteria and some cyanobacteria) are capable of nitrogen fixation. The bacteria include mainly those of the genus *Rhizobium*, which develop a symbiotic relationship with higher plants forming nodules on their roots.

**Nitrogen.** A chemical element present in amino acids and proteins, and hence important in the diet of animals and man. It forms about 80 per cent of the atomosphere.

**Nitrogenase**. An enzyme complex that catalyzes the reduction of atmospheric nitrogen (dinitrogen) to ammonia in nitrogen-fixing organisms. The enzyme has two major components, each of which has two or four subunits. The large component contains molydenum, non-heme iron and sulphur, and the smaller contains iron and sulphur. The enzyme, which is strictly anaerobic and is easily inactivated by molecular oxygen, also acts as a hydrogenase. Hydrogen production is a wasteful reaction, and those species with a low level of hydrogen production are more efficient nitrogen fixers.

**Nitrogen-fixing bacteria.** Bacteria capable of drawing free nitrogen from the atmosphere and converting it into soluble nitrates. These nitrates can then be used by plants.

**Nodulation and fixation genes.** The genes involved in the **Rhizobium** - legume symbiosis, which is one of the most effective nitrogen fixing systems. Identification of some *Rhizobiuum* nif genes has been achieved by hybridization with the nif YKDF fragment of *Klebsiella pneumoniae*, with cloning and analysis of special *Rhizobium* nif fragments in *E.coli*. Fixation and nodulation genes can also be identified by transposon mutagenesis. By this technique it is possible to analyze the whole symbiotic process, and hence to identify the genes involved by cloning the genes in **E.coli.** The next step is to alter the fixation and nodulation genes by genetic engineering techniques in order to construct *Rhizobium* strains that are more effective in nitrogen fixation.

**Nodule.** An enlargement on the roots of certain plants in which nitrogen-fixing bacteria live. Swelling on the roots of certain leguminous plants where nitrogen-fixing bacteria reside in a symbiotic relationship with the plant.

**Nonheme iron proteins, NHI-proteins.** Proteins containing iron that is not bound in a heme system. In these proteins, the iron is bound by the sulfur of cystein residues, and it is often also associated with inorganic sulfur. They are also called iron-sulfur proteins.

**Normal curve.** A smooth, symmetrical, bell-shaped curve of distribution.

**Normal distribution curve.** A bell-shaped curve, more or less symmetrical, with the greatest value in the center (mean) with values decreasing equally on both sides. Also called a normal curve.

**Nutrient.** An element or compound that is needed by an organism to provide it with energy and a building material for growth. Nutrients include the major elements, trace elements and a wide range of carbon sources, as well as (to varying degrees) essential metabolities including vitamins, amino acids, fatty acids and other growth factors.

**Nutritional requirement.** The minimal nutritional need for the proper growth and maintenance of an organism or culture. It includes both macronutrients and micronutrients (trace elements, vitamins, essential amino acids, etc.).

**Oat Coleoptile Test, Avena Test.** A biotest for the quantitative determination of auxins. In this the tip of the coleoptile is cut off, and the cotyledon within it is removed by putting on its base. An agar block with the test substance is set on one side of the coleoptile stump. The auxin diffuses into the side of the coleoptile covered by the agar and, due to the one-sided stimulation of growth, causes it to bend. The angle of the bend is a function of the auxin concentration.

**Obligate anerobe.** An organism whose growth is inhibited or that is killed in the presence of oxygen.

**Obligate gamete.** A gamete that is unable to develop parthenogenetically.

**One gene-one enzyme hypothesis.** A supposition that the gene is equivalent to the amount of genetic material required to encode the sequence for one specific enzyme. However, this is now known not to be correct, since some genes do not encode protein structures and many enzymes are composed of several subunits which may be coded for by separate genes. A more correct interpretation would be one gene-one polypeptide chain.

**Open water.** Lake water that is free from emergent vegetation, artificial obstructions, or tangled masses of underwater vegetationat very shallow depths.

**Optical activity.** The rotation of the plane of vibration of a beam of polarized light when passed through a solution of an optically active compound. Compounds that contain one or more asymmetric carbon atoms can exist in two different configuration known as enantiomers, which are mirror images of one another. One enantiomer rotates the beam of polarized light to the life (1-form or leavorotatory) and the other rotates the beam to the right (d-form or dextrorotatory).

**Optical isomerism.** A form of isomerism in which the isomers differ in their optical activity.

**Osmosis.** The phenomenon which results in a difference in rate of movement (or passage through the membrane) of solvent molecules from opposite sides of selectively permeable membrane separating two solutions of different concentration or separating a solution from its pure solvent.

**Osmotic potential.** A measure of the pressure that needs to be applied to a solution in order to make its water potential equal to that of pure water. When this pressure is applied to a solution, pure water will not pass into it across a semi-permeable membrane. When there is no matric potential or pressure potential, i.e., under experimental conditions, osmotic potential is equal to water potential.

**Osmotic pressure.** Of a solution, the maximum hydrostatic pressure that this solution could develop when separated from its pure solvent by a membrane permeable only to the solvent. A molar solution of any nonelectrolyte has an osmotic pressure of 22.4 atm.

**Osmotic.** In the manner of osmosis.

**Osmotrophic.** Term describing a heterotrophic organism that absorbs organic molecules in a soluble form.

**Oxidation.** A type of chemical reaction involving the removal of electrons. Frequently, but by no means necessarily, the element oxygen is involved.

**Oxidative phosphorylation.** The production of ATP from ADP and orthophosphate, using energy released in the oxidation of organic compounds in the electron transfer chain. Oxidative phosphorylation takes place in the mitochondria and is the main source of ATP in heterotrophic organisms. In the final reaction of the process, molecular oxygen ($O_2$) is reduced to water.

**Oxygen respiration.** Respiration requiring the presence of free oxygen.

**Oxyphilous.** Thriving in acidic habitats; acidophilic; oxyphile, oxyphilt; *cf.* oxyphobous.

**Oxyphobous.** Thriving in alkaline habitats; intolerant of acidic conditions; basophilous; oxyphobe, oxyphoby; *cf.* oxyphilous.

**P/O ratio.** In oxidative phosphorylation (acerobic respiration) the ratio of phosphate incroporated into ATP to the oxygen consumed.

**Palisade mesophyll.** The main region of photosynthetically active cells found in the leaves of dicotyledons. The palisade layer is characterized by long closely fitting columnar cells containing many chloroplasts.

**Palisade tissue.** Tabular chlorenchyma comprising the upper mesophyll of most leaves.

**Palisade.** A term used to describe the arrangement of cells seen in the stained smears of certain bacteria; two three or more bacilli in contact, with their long axes parallel.

**Pantothenic acid.** Vitamin of B group, forming a co-enzyme; required by a cariety of organisms, e.g., some yeasts, some bacteria, insects, vertebrates.

**Paper chrmatography.** Paper chromatography is one type of partition chromatography in which the compounds spotted on paper, a porous medium, get separated when a solvent/solvent system migrates over the spots to the ends of the apaper by capillary actions. The compounds, soluble in both water and organic solvents, are separated based on their partition coefficients in the respective solvents. Partition usually takes place between two immiscible solvents. A compound, soluble in such solvents for e.g., in water and butanol, if shaken with a mixture of these solvents, gets distributed unequally between the two phase. The partition coefficient - is the ratio of the concentration of the compound is the solvents at equilibrium. Compounds having well separated partition coefficients get separated easily. The chromatographic paper is almost pure (99%) cellulose. Cellulsoe is referred to as an inert support, but it is found that the solvents get bound to the free hydroxyl groups of cellulose through secondary valence forces and/or through hydrogen bonds. The amorphous regions of cellulose micelle absorb and retain more water than the crystalline regions. Partion takes place between the solvent (usually water) held in the amorphous regions and interstitial spacees (which act as a stationary phase) and the mobile phae, (usually a water saturated-mostly lipophilic-organic-solvent). When the latter flows past the stationary phase, a compound more soluble in water is retained while lesser solubles are carried forward. At the end of the run, various compounds are deposited at different positions in the chromatograms.

**Pectic material.** Chemical substances derived from pectin, a jellylike substance found in fruits and other parts of plants; the chief cementing material of plant cells.

**Pectic substances.** Polysaccharides extracted from plant cells walls by prolonged boiling in water. They are formed largely from galacturonic acid, arabinose, galactose and methanol covalently linked to oner another and to other polysaccharides and cell wall components. The galacturonan of higher plants is of commercial calue in the food industry, being used under the name of pectin as a setting agent in jam making.

**Pectinase.** An enzyme that catalyzes the hydrolytic breakdown of pectins to pectic acids.

**Pectinesterase.** An enzyme that catalyzes the hydrolytic breakdown of pectins to pectic acids.

**Pectins.** A group of complex polysaccharides which consist basically or -(1-4)-linked D-galacutonic acid residues (**pectic acid**) in which a high proportion of the (C-6) carboxyl groups are methylated. Residues of L-rhamnose may occur at intervals within the chain, and sides chains of neutral sugars (e.g., arabinose, galactose) may be present. Structural details vary with source. Pectins form gels under certain conditions of e.g., ionic concentration.

**Pectolytic enzymes.** Enzymes capable of degrading pectic substances. Pectolytic enzymens, which occur in a wide variety of bacteria and fungi.

**Peroxisome.** A small organelle surrounded by a membrane, containing the enzyme catalase, which catalyses the breakdown of hydrogen peroxide ($H_2O_4$) to water and oxygen. Catalase prevents the build-up of $H_2O_2$ in cells; $H_2O_2$ is toxic, and is known to be a product of some metabolic reactions. Peroxisomes also contain enzymes involved in the oxidation of glycolic acid ($COOHCH_2OH$).

**pH electrode.** The primary standard used for measurement of hydrogen ion concentrations. It is a platinum electrode which is immersed in the solution to be measured with gaseous hydrogen bubbled over it surface at a pressure of 1 atmospher. The electromotive force of this electrode in 1.0 M $H_3\ O^+$ ions is zero. The potential difference between the electode and a standard electrode (e.g., a calomel electrode) is measured and used to calculate the $H_3\ O^+$ ion concentration. This type of electrode is not suitable for everyday use and is generally replaced by a device consisting of a glass electrode surrounded by a reference electrode with a potassium chloride solution forming a salt bridge. This design gives a practical, compact and simple probe which is electrostatically screened over most of its length.

**pH index.** The pH index is a number used to express the hydrogen ion concentration of a solution. The conception of pH is due to Sorensen 8 who in 1909 suggested that numerical calues should be used as the basis for an acidity scale. He used the symbol $pH^+$ for his scale but this was subsequently abbreviated to pH. The pH index is defined as the common iogarithm, with sign reversed, of the hydrogen ion concentration expressed in gramme ion per litre of solution and it lies between 0 and 14. Alkalis have a pH value of less than 7, acids have an index greater than 7, and 7 represents the index for a neutral solution[9]. Values for pH can be determined electrically or by chemical indicators, the latter giving suufficient accuracy for normal laboratory work. Other scales have been proposed fro expressing the acidity of a solution. These include the pR unit of Giribaldo (1938) the Nn unit of Gestle (1938), sometimes called the hydron, the rA number of Catani, the paH unit (1924), the $p^wH$ unit (1930) and the ptH unit (1938). The last three units are of the same order of magnitude as the pH indes, but in the pR and $N_n$ units alkalis have valudes between 0 and 14, whereas acids lie in the range 0 to -14 and zero represents a neutral solution[44].

**pH meter.** A device that combines a pH electrode with electronic circuits in order to determine the pH of a solution.

**pH.** A symbol denoting the relative concentration of hydrogen ions in a solution; pH values extend from 0 to 14; the lower the value, the higher the acidity or the more hydrogen ions the solution contains. pH is actually the negative logarithm of the hydrogen-ion concentration. Water at 25°C has a concentration of H ion of $10^{-7}$; the pH therefore is 7.

**Phosphoglyceric acid (PGA).** A compound with three atoms of carbon, which is the first product of the reaction between $CO_2$ and ribulose-diphosphate in the reductive pentose pathway of photosynthesis.

**Phospholipid.** A polar lipid fouund in plant and animal membranes phospholipids consist of two groups; phosphglycerides and sphingomyelins.

**Phosphorylase.** An enzyme that catalyzes the formation of glucose-1-hposphate from glycogen and inorganic phosphate; it is widely distributed in animals, plants, and microorganisms.

**Phosphorylation potential.** The relationship between the levels of ADP and ATP, which control the balance of anabolic and catabolic processes in living organisms, tissue, cells or organelles.

**Phosphorylation.** Bringing about the combination of an organic acceptor molecule with phosphate by means of enzyme action. In phosphorylation, sugars, for instance, are commonly combined with phosphate before being broken down; and phosphorylation temporarily alters the activity of many proteins. The combination with phosphat, which is usually done by ATP (q.v.), produces compounds that are highly reactive in water with other organic molecules in the presence of appropriate enzymes; when their phosphate is in turn transferred, energy becomes available for biochemical and other work. Such processes from the fundamental energy-transfer system in metabolism. Most important high-energy compound in ATP (q.v.), universally present in living organisms.

**Photoautotroph.** An organism that uses light energy in photosynthesis to form nutrients from inorganic substrates (carbon dioxide, water, nitrate, sulphate, etc.).

**Photobiological hydrogen production.** A system that uses light and a biocatalyst to form hydrogen and oxygen.

**Photoheterotroph.** A photosynthetic organism that uses organic compounds as electron donors. Photoheterophs include the green, purple and brown algae that do not form oxygen.

**Photolysis of water.** The part of the light reaction of photosynthesis in which water molecules are split to give hydrogen and oxygen.

**Photomorphogenesis.** The control exerted by light over growth, development, and differentiation of plants that is independent of photosynthesis.

**Photonasties.** Nasties caused by changes in light intensity.

**Photoperiod.** The number of hours of daylight needed by a plant before it will begin to flower.

**Photoperiodism.** The response of an organism to the length of a light period. The most notable photoperiodic responses occur in plants where flowering, bud dormancy, leaf fall, tuber formatio, etc. are controlled by day length. Plants are distinguished as long-day, short-day or day-neutral, depending on their response to variations in the period of darkness to which they are subjected. Responses in plants are the result of phytochromemediated phenomena; a reaction to red light may be reversed by a flash of far red light, and vice versa. Many animals also show photoperiodic responses.

**Photophillic.** Thriving in full light.

**Photophobic.** (1) Avoiding light. (2) Exhibiting negative phototropism.

**Photophosphorylase.** An enzyme that is associated with the surface of a thylakoid membrane and is involved in the final stages of adenosinetriphosphate production by photosynthetic phosphorylation.

**Photophosphorylation.** Process of formation of ATP from ADP and inorganic phosphate in the light reactions of photosynthesis; occurs by a cyclic or noncyclic pathway involving phtosynthems I and II.

**Photoreactive chlorophyll.** Chlorophyll molecules which receive light quanta from antenna chlorophyll and constitute a photo reaction center where light energy conversion occurs.

**Photoreceptor.** The part of the cell that receives the stimulus in phototaxis, usually a dense area in a flagellar swelling.

**Photorespiration.** A ligh-dependent process that occurs in the green leaves of higher plants. It is characterized by the uptake of oxygen and the release of carbon dioxide, as well as the formation of ATP.

**Photorespiration.** Uptake of oxygen and release of carbon dioxide by photosynthetic cells or whole plants in the light.

**Photosynthate.** Organic product of photosynthesis.

**Photosynthesis.** The enzymatic conversion of light enrgy into chemical energy by forming carbohydrates and oxygen from $CO_2$ and $H_2O$ in green plant cells. It is one of the most fundamental biological functions. In this process, the chlorophyll traps the energy of sunlight emitted as photons (i.e., excitons) as transform it into chemical energy. This energy is stored in the chemical bonds that are produced during the synthesis of various food stuffs. Photosynthesis consists of two interconnected sets of reactions (i) Phase I reactions, and (ii) Phase II reactions. Phase I reactions take place in the membranes inside the chloroplast and phase II reactions take place in the open space (stroma) inside the chloroplast.

**Photosynthetic efficiency.** A measure of the extent to which light energy intercepted by a leaf, leaf canopy or photosynthetic cell suspension is converted into organic matter. Efficiency is expressed in terms of total incidenmt light or photosynthetic active radiation (PAR).

**Photosynthetic phosphorylation.** The enzymatic formation of ATP from ADP in green plants coupled to light-dependent transport of electrons from excited chlorophyll

**Photosynthetic pigments.** Chlorophylls, carotenoids and phycobilins. All of these classes of pigments occur as chromoproteins (pigment-protein complexes). All oxygen-producing plants possess chlorophyll a. Higher plants also contain chlorophyll b. and other forms (chlorophylls c and d) are found in algae. The carotenoids and phycobilins are accessory pigments.

**Photosynthetic productivity.** It is an of the yield, in terms of harvested dry weight, produced by a given crop or within a tonnes dry weight produced per hectare per year.

**Photosynthetic rate.** It is the measure of the capacity of a leaf or plant to carry out photosynthesis. It is expressed as the amount of carbon dioxide fixed or oxygen evelved in relation to leaf area, leaf weight, chorophyll content, protein content, area of land occupied by the plant, etc.

**Photosystem I (PS I).** A photchemical reaction system in photosynthesis; coupled with photosystem II.

**Photosystem II (PS II).** A photochemical reaction system in photosynthesis; coupled to photosystem I.

**Phototaxis.** Movement of a whole organism toward (positive) or away from (negative) light.

**Phototroph.** An organism that uses light as an energy source. Phototrophs are divided into two groups; photoheterotrophs and photoautotrophs.

**Phototrophic bacteria.** Primarily aquatic bacteria comprising two principal groups; purple bacteria and green sulfur bacteria; all contain bacteriochlorophylla.

**Phototropism** (Heliotropism). Tropism (q.v.) in which stimulus is light. E.g. bending of stems of indoor plants towards a window. Brought about by increased elongation of cells in growth region at tip of stem on shaded side. Generally accepted that effect is due to greater concentration of auxin (q.v.) on that side than on side facing light but some recent evidence suggests as the explanation asymmetric distribution of some cofactor that influences auxin effect.

**Phycobilins (Biliproteins).** Red (phycoertythrin) and blue (phycocyanin) pigments present with chlorophyll in red and blue-green algae.

**Phycobiliprotein or phycobilin.** Water-soluuble blue-green or pink pigment in the Cyanophyta, Rhodophyta, and Cryptophyta.

**Phycobilisome.** An aggregation of phycobiliproteins on the surface of a thylakoid.

**Phycocyanin.** A secondary blue pigment often present in the Cyanophyceae, the blue-green algae.

**Phycocyanobilin.** $C_{31}H_{38}O_2N_4$. Phycobilin with an ethylidene side chain (=$CH$-$CH_3$) and only one asymmetric carbon atome ($C_1$).

**Phycoerythrin.** A secondary red or purple pigment often present in the Rhodophyceae, the red algae.

**Phycology.** The study of algae.

**Phycomycetes.** A group of fungi formerly considered to be a class but now divided into six classes as follows; Chytridiomycetes, Hyphochytridiomycetes; Oomycetes, Plasmodiophoromycetes, Zygomycetes, Trichomycetes.

**Phycophaein.** A secondary brown or yellow pigment often present in the Phaeophyceae, the brown algae.

**Phycoplast.** Type of cell division in which the mitotic spindle disperses after nuclear division with the two daughter nuclei coming close together; another set of microtubules arising perpendicular to the former position of the microtubules of the mitotic spindle, and the new cell wall forming along these microtubules.

**Phyllobiology.** The study of leaves.

**Phylloclade.** A flattened stem, usually of one or two internodes, presenting a superficial appearance like a leaf.

**Physiologic rate.** A biotype, or group of closely related biotypes, that differ from other biotypes of that species in the ability to infect particular varieties of the susceptible plant species; *cf.* forma specialis.

**Physiologic specialization.** The existence of a number of races or form of one species of pathogen.

**Physiological anisogamy.** A type of isogamy in which the fusing isogametes differ physiologically.

**Physiological environment.** All the elements surrounding an organism, excluding other living organisms.

**Physiological.** Pertaining to the normal behavior of cells, tissues, organs, or organisms.

**Physiology.** The study of the functions of organism.

**Phytochrome.** It is a chromoprotein found in plants. The prosthetic group of phytochrome is an open tetrapyrrole related to phycocyanobilin. This molecule mediates various physiological responses and red light (730 nm) in the same condition.

**Pigment.** Any coloured substance present in the tissues of an organism. Pigments absorb energy from light. Some of them, like chlorophyll, are important in photosynthesis. Others, e.g., phytochromes, help to control growth.

**Plant fermentation.** A form of plant metabolism in which carbohydrates are partially degraded without the consumption of molecular oxygen.

**Plant physiology.** The branch of botany concerned with the processes which occur in plants.

**Plant/nif gene transfer.** A procedure used to transfer *Klebsiella* nif genes directly to plants in order to establich them in the plant nuclear genome. this involves the following steps: (1) transferring the *Klebsiella* nif region to ***E.coli***; (2) cloning the whole nig gene cluster in a multicopy plasmid cloning vehicle; (3) introducing the **Klebsiella** nif gene cluster into the nucleus of plant protoplasts; (4) regenerating the whole plant thus leading to the expression of the nif genes. Such transfer of the nif gene cluster to a plant nucleus is accomplished by using the Ti plasmid of *Agrobacterium tumerfaciens*.

**Plasma membrane.** A lipoprotein of a definite structuure which surrounds and contains the living matter within a cell. The membrane has three layers two outer protein surfaces and an inner core of lipid. The polar part of each lipid molecule is associated with the protein on the surface, while the nonpolar portion points into the middle of the structure.

**Plasmalemma.** The outer membrane that surrounds the protoplasm of a cell.

**Plastid.** A cytoplasmic body found in the cells of plants and some protozoans. Chloroplastids produce chlorophyll which is involved in photosynthesis. It is a class of membrane-bound organelles found within the cells of higher plants and algae (though not in blue-green algae or other procaryotes); plastids contain pigments and/or certain products of the cell (e.g., starch, lipid). Chloroplasts are familar examples.

**Polyamine.** Polyamines are low molecular weight aliphatic nitrogenous polycations. They occur unbiquitously in living organisms in free or conjugated form. The diamine putresine and the triamine spermidine may be present in all living cells. In addition, the eukaryotes contain the tetramine, spermine. Cadaverine, another diamine, is found is some plants, whicle numerious other polyamines have also been reported from algae and higher plants. In addition, conjugates of polyamines with hydroxycinnamic acids occur widely in plants, especially in flowers.

**Polymerase.** An enzyme which will form oligomeric molecules from monomers. A DNA polymerase will synthesise DNA from deoxynucleoside triphosphates using a complementry DNA strand.

**Polysaccharide.** A high molecular weight polymer formed from sugar units (monosaccharides) joined through glycosidic linkages.

**Polysome:** It is a complex of ribosomes bound together by a single messenger ribonucleic acid molecule; polyribosomes.

**Polysomy.** The occurrence in a nucleus of one or more individual chromosomes in a number higher than that of the remainder.

**Potential energy.** Energy capable of doing work.

**Primary protein structure.** The number of polypeptide chains in a protein, the sequence of amino acids, within htem, and the location of interchain and intrachain disulfide birdges.

**Proplastids.** Proplastids, the progenitors of other plastids are found in root and shoot meristems, embryos, endosperms and in young developing leaves. The progenitor function of proplastids should be regarded as secondary to their simple structure which is a primary distinguishing feature. They are not structurally complex. They consist of an envelope, a nucleiod with DNA fibrils, a small number of ribosome, thylakoids, some time small starch molecule. Proplastids are varied in shape. They are usually only slightly larger than mitrochondria. In light grown plants the proplastids in stems and leaves usually differentiate into chloroplasts. Proplastids may also have important biochemical functions such as the synthesis and storage of starch and other metabolites

**Prosthetic group.** An organic compound of conjugated proteins. Conjugated proteins are a combination of simple protein and another substance called the prosthetic group. Example: Nucleoproteins have as prosthetic groups nuclei acids.

**Protamines.** Proteins rich in basic amino acids and are soluble in water. Found in the nucleus and may have an association with nucleic acids. Tyrosine, tryptophan and sulfur are not present.

**Protease.** An enzyme that catalyzed the hydrolysis of proteins to amino acids.

**Protein efficiency ratio (PER).** A term used expressing the growth promoting ability of a protein i.e. gain in weight (in gms) per gram of protein consumed.

**Protein synthesis.** An elaborate process of protein formation in plant cells consisting of (i) activation of amino acids, (ii) attachment of amino acids activated to S-RNA and (iii) formation of polypeptides on ribosome. Transcription and translation are the processes involved at molecular level in the passage of message from DNA.

**Protein.** A complex organic molecule composed of amino acids joined in specific sequence by peptide bonds. Proteins serve both structural and enzymatic functions.

**Proteins.** Nitrogen containing molecules of huge size and complexity made of repeating amino acid units linked together by peptide bonds.

**Proteomics.** It refers to chloroplast genetics. It is the large scale study or protein, usually by biochemical methods. The word protromics has been associated traditionally with displaying a large number of proteins from a given cell line or organism on two-dimentional polyacrylamide gels. In 1990s, biological mass spectrometry; matrix assisted laser desorption/ionization-time of fright (MALDI-TOF, MS) and Electrospray ionization mass spectrometry (ESI-MS/MS), emerged as a powerful analytical method that remove most of the limitations of protein analysis. To day the term covers much of the functional analysis of gene products of functional genomics, including large scale identification or localization studies of proteins and interaction studies. Therefore, proteomics is the study of all the expressed proteins of an organism. The information that protections studies can provide cellular localizations, protein-protein interactions and protein-nucleic acid interactions, Recently, Peltier (2000) studied proteimics of the chloroplast, the soluble and peripheral proteins in the thyladkoids of pea.

**Proton.** The nucleus of an atom, bearing a charge of positive electricity.

**Protopectins.** Substances having higher molecular weight than pectic acid and pectins and are insoluble in water. They are intermediate between pectic acid and pectins in the degree of methylation of carboxyl groups. They are found in apples and oranges.

**Protoplasm.** The living substance of the plant; it is largely protein, but contains assemblages of food material, enzymes, water, and in its nucleus the material and mechanism for nuclear division and transmission of heritable characters.

**Prototroph.** (i) Plants with primitive nutritive mechanisms, e.g., nitrogen-fixing bacteria, (ii) in genetics, plants lacking the necessary genes for particular biochemical syntheses.

**Prototrophic.** (1) Autotrophic; protrophic. (2) Obtaining nourishment from a single source only; prototroph, prototrophism. (3) Used of microorganisms capable of growth on simple sugars and salts without specific additional nutrient requirements.

**Pyruvic acid.** The final product of glycolysis.

**Rad.** The unit of absorbed radiation dose; which results in the absorption of 100 ergs per gram of specified material.

**Radical.** The embryonic root, continuous with the hypocotyl, that occurs in germinating angiosperm and gymnosperm seedings.

**Repressible enzyme.** Enzyme synthesized in the absence of its substrate which then represses further synthesis of the enzyme.

**Repression.** The machanism that prtecents the expression of a gene (and hence the synthesis of a particular enzyme). See reprosor.

**Reprossor.** The substance which represses the functioning of an operon. It is the product of the regulatory gene and has been identified as a protein. When it combines with the operator the operon is in the repressed state. Respression is relieved when the repressor combines with the inducer.

**Respiration.** The overall oxidative metabolism of food materials to produce metabolic energy. In general, the term is applied to aerobic (oxygen-requiring processes, although the term anaerobic respiration may also be used. Various partial process may also be used. Various partial process in animals associated with respiratory gas exchange are also referred to as respiration: for example, iinhalation and exhalation (breathing) or the overall processes by which carbohydrate and oxygen interact to produce carbon dioxide and water with the generation of ATP (i.e.l, the combined action of glycolysis and the respiratory electron transport chain).

**Respiratory pigment.** A compound that increases the oxygen carrying capacity of an animals's circulatory system. The pigment occcurs in solution in the plasma or is confined to cell (corpuscles). Respiratory pigments incluude haemoglobin, haemocyanin and heamerythrin.

**Respiratory quotient.** The ratio of the volume of carbon dioxide produced by an organism during respiration to the volume of oxygen taken up. The respiratory quotient varies according to the type of food reserves being oxidized. The theoretical values for various classes of substrate are as follows; carbohydrate, 1.0; protein, 0.8; fats, 0.7 In practice, however, these values are modified due to the interconversion of metabolites and the creation and discharge of an oxygen debt.

**Respirometer.** An instrument used to measure the rate of respiration or gas exchange of a culture of tissue. Most respirometers use manometric principles or are the based on oxygen electrodes.

**Response.** The reaction to a stimulus.

**Salt.** A compound which upon dissociation yields cations of a metal and anions of an acid radical; the product of a reaction of an acid and a base.

**Saturation deficit.** A measure of humidity, derived by subtracting the actual water vapour pressure from the maximum possible vapour pressure at a given temperature, expressed as a percentage of total saturation in mm of mercury; water vapour deficity; *cf.*relative humidity.

**Secondary cycle.** A cycle initiated by inoculus from a primary from another secondary cycle without an interposed resting or dormant period for the pathogen; *cf.* primary cycle.

**Seed coat.** The envelope which encloses the seed except for a tiny pore, the micropyle.

**Seed dormancy.** The growth of a seed is completely arrested after it is fully developed. Sometimes the growth is suspended because of the lack of favourable conditions in the environment. Such seeds will germinate if they are supplied with water and suitable temperature. In many cases, however, the seeds would not germinate, even if they are provided with water etc. In such cases, the completely dry ripe seed is physiologically inactive and is said to be in resting stage. The seed is called dormant and the phenomenon is termed dormancy. There are several factors which control dormancy like. Hard seed coat (seed coat impermeable to water, seed coats impermeable to oxygen, mechanically resistant seed coats), immature embryo, period ofter ripening, germination inhibitors, etc.

**Seed leaf.** A cotyledon.

**Seed plants.** The spermatophyta; plants that form true seeds. These are divided into two groups; the Gymnospermae and the Angiospermae.

**Seed test.** A series of tests on purity, moisture content and germination of a seed sample to determine its quality.

**Seed.** (1) In general, the propagative part of a plant, including true seeds, bulbs, corms, stem cuttings, etc. (2) Specifically, the structure that develops from an ovule after fertilization and contains the embryo plant plus food reserves. (3) A microbial culture held back as a starter for subsequent cultures.

**Seedling.** The young plant developing from a germinated seed.

**Semipermeable.** Of membranes which let some substances pass through them, but not others. The membranes in plant cells are mostly permeable to small molecules, e.g., water ($H_2O$), monosaccharides and amino acids, but not to large molecules, e.g., polypeptides.

**Senescence.** The terminal phase in the life cycle of an organism, during which degradation gradually takes place. In some organisms, senescence of discrete areas of the organism is part of the normal pattern of growth. Senescence may be a genetically programmed event and under hormonal control or may result from the accumulation of errors in genetic expression as the organism ages.

**Separation layer.** A structurally distinct layer of the abscission zone of a plant containing abundant starch and dense cytoplasm

**Separation.** Disk dead cells, also called concave cells that delimit the hormogonia in the filament of *Oscillatona*.

**Sephadex.** Sephadex is one of the dextran gels used in affinity chromatography. A free aminogroup is attached to the gel by reacting it with azide ($H_2$ N-$NH_2$) to produce carboxymethyl sephadex azide to which affininants with basic aminogroups are further coupled. The protein gets bound to the amino group. Their low porosity partly restricts the use of dextran gels. They are widely used, without any modification, as specific sorbents for the isolation of a series of lectins. Agarose and its derivatives fulfill all the requirements of an ideal carrier and are the most commonly used gel in affinity chromatography. Sepharose is an agarose gel with spherical particles often available in the swollen state suspended in water containing 0.02% sodium azide as the bacteriostatic reagent. Sephadex is a registered trademark of *Pharmacia Biotechnology* AB.

**Shikimic acid. $C_7H_{10}O_5$.** A crystalline acid that is a plant constituent, and an intermediate in the biochemical pathway from phosphoenolpyruvic acid to tyrosine.

**Short-day plant (SDP).** A plant which has a flowering period initiated and accelerated by exposure to a short photoperiod of less than 12 h duration. A short day plant may have an obligate requirement of a favourable day length for flowering and may remain permanently vegetative otherwise (absolute SDP) e.g., *Xanthium pennsylyanicum*; other species may show hastened flowering under SD (quantitative SDP) e.g., *Cannabls sativa*.

**Sieve element.** A cell in the sieve-tube of the phloem. Sieve elements are long, thin, living cells with thin cell walls and sieve plates at their ends. Translocation of substances takes place in the sieve elements.

**Sieve membrane.** Wall structure covering a loculus in the Bacillariophyta.

**Sieve plate.** The perforated wall area of a sieve-tube member. Perforated parts of transverse and sometimes lateral walls, of sieve tubes.

**Sieve tubes.** Elongated cells of the phloem concerned with transport of synthesised food; they have cellulose walls and no protoplasm, and are associated with narrow companion cells, which possess protoplasm.

**Single substrat enzymes.** Enzymes which catalyse reactions involving only one substrate. They are usually isomerases or hydrolytic enzymes; in the latter case, the water involved in the reaction is regarded as a constant, and there is frequently no special enzyme-water complex formed.

**Sodium pump.** An active transport mechanism associated with cell membranes. The pump moves sodium ions out of the cell against a concentration gradient producing the resting potential across the membrane. The sodium pump is important in neurones.

**Solar propagation.** A method of rooting plant cuttings involving the use of a modified hotbed; bottom heat is provided by radiation of stored solar heat from bricks of stones in the bottom of the hotbed frame.

**Specific enzyme activity.** The activity of an enzyme expressed in term of the amount or number of moles of substrate converted per unit time per unit weight of protein.

**Specific growth rate constant** (u). A proportionality constant for the growth of unicellular microogranisms undergoing binary fussion. In a well mixed batch culture system, growth is proportional to the biomass concentration.

**Specificity Constant, Physiological Effectivity.** A measure of the turnover of a substrate. it is the ratio of the catalytic constant and the Michaelis constant: $K_{cat}/K_m$. It is equal to the rate constant of a reaction for the rate equation:

**Spectrophotometer.** An analytical device consisting of a light source, diffraction grating, sample holder and detection system that is used to measure absorption spectra or changes in absorbance at fixed wavelength.

**Spectrum.** An image formed when radiant energy is dispersed so that its rays are arranged in the order of their wavelengths, e.g., the visible spectrum, a rainbow, the electromagnetic spectrum.

**Splicing enzyme.** A ligase associated with joining of length of DNA or RNA in vivo or in vitro.

**Starch sheath.** A layer of endodermal cells that surrounds the stele in some leaves and stems of dicotyledons.

**Starch.** A nutrient polysaccharide $(C_6H_{10}O_5)_n$, found in plant cells, prootists and certain bacteria; occurs in cells in the form of granules several ums in diameter; contains a mmixture of two different polysaccharides, amylose, a linear polymer of glucose and amylopectin, a glucose polymer with interlinked chains. The relative amouns of these two polysacccharides vary according to the souurce of the starch. Iodine solution turns blue black on contact with starch; this reaction is used as a test for either starch or iodine.

**Steroids.** A large group of terpenoids, including many important biological compounds, e.g., Sterols, Steroid hormones, Bile acids, Cardiac glycosides, Steroid alkaloids and steroid saponins Synthetic S.

**Sterols.** A group of naturally occurring steroids, possessing a 3-hydroxyl group and a 17 - aliphatic side chain. These are structural derivatives of the parent hydrocarbons, cholestane, ergostane and stigmastane. They occur in animal and plant cells as free sterols, and as glycosides and ester. According to their origin, they are classified as zoosterols (animals), phytosterols (plants), mycosterols (fungi) and marine. Sterols (marine animals and plants).

**Stomatal movement.** Stomatal movement causes changes in the pore sizee of the stomata in the epidermis of higher plants. These changes control both the rates of phtosynthetic gas exchange and the rates of water use (loss) through transpiration. In general stomata close in response to water deficit and open in response to low concentrations of carbon dioxide. Stomata also respond to changes in light intensity and temperature, as well as showing both blue-light and phytochrome-type responses. Closure is affected by the plant growth regulation abscisic acid.

**Stomatal resistance.** The property of the stomata in restricting the free exchange of carbon dioxide ($CO_2$) by a plant leaf; the major constraint on $CO_2$ uptake into the plant leaf governed largely by the diameter of the stomatal pores.

**Stomatal transpiration.** The loss of water vapour through the open stomata of leaves; *cf.* cuticular transpiration.

**Stroma matrix.** Colorless, homogenous (by light microscopy) ground substance of cell organelles, like Chloroplasts and Mitochondria.

**Stroma.** The loosely packed membrane regions within the chloroplast where the dark reaction occurs.

**Succinic Acid 2, 2-dimethylamide, Alar 85, B9**: A synthesis growth regulator. It is used to stimulate blossoming in apples and to acclerate the development of fruit color and fruit loosening in cherries.

**Sucrose gradient.** A density gradient which is preformed by mixing sucrose solutions of different concentrations from the top to the bottom of the tube. Sucrose gradients are commonly used to separate molecules or particles on the basis of their sedimentation coefficients, which depend on their size and shape rather than their density.

**Sucrose.** A white crystalline disaccharide that occurs as the major transport carbohydrate in most plants and as a storage carbohydrate in sugar cane and sugar beet from which it is isolated commercially.

**Sun plant.** A plant tolerant of exposure to much bright light or needling more light.

**Temperature coefficient $Q_{10}$.** A measure of the increase in rate of a reaction or process produced by an increase in temperature of 10°C.

**Temperature efficiency index.** For a give location, a muasuure of the long rage effectiveness of temperature (thermal efficiency) in promoting plant growth. Abbreviated T-E index; thermal efficiency index.

**Temperature efficiency rato.** For a given location and month, a measure of thermal efficiency; it is equal to the departure, in degrees Fahernheit, of the normal monthly temperature above 32ÉF (0ÉC) divided by 4: (t-32)/4. Abbreviated T-E ratio, thermal-efficiency ratio.

**Temperature quotient ($Q_{10}$).** The ratio of the rate of reaction at a given temperature to that of the same reaction carried out at a temperature that is 10 degrees centrigrade lower.

**Thermonasty.** A nastic response to temperature.

**Thermophil.** A microorganism with a temperature otpimum for growth of between 40 and 70ÉC or higher. Thermophilis are generally associated with hot springs.

**Thin layer Chromatography, TLC.** A form of chromatography in which the solid carrier material is spread in a thin layer on a glass or plastic plate. The advantages of the method are the short distances required for good separation and the correspondingly short development times, high sensitivity, separation of very small amounts of substances, and, if inorganic carrier materials are used, the possibility of using caustic detection reagents. TLC can be used preparatively as an open column. Depending on the size of the plate, upto 100 g of material can be separated by preparative TLC while analytical TLC can be used for amounts between 10 mg and 10 mg.

**Thin layer gel filtration.** A technique in which molecules are separated on a flat plate (typically 20 x 20 cm) on which a layer of xerogel 0.5-1.0 mm thick is spread.

**Thylakoid.** Originally denoted the basic subunit of the chloroplast lamellae, which in cross-section has the apperance of a vesicle surrounded by a single lipoprotein membrane. Although it is now known that the internal membranes of most chloroplasts are derived from only one or two highly folded or convoluted sheets, the terms is still used for convenience. The appearance of thylakoids in cross-section varies in the different groups of plants. In red algae, the thylakoids are single and lie parallel to one another. In other algae groups, the thylakoids are paired or occur in threes. In green algae and higher plants, the thylakoids become appressed in stacks known as grana. An early concept of the chloroplast of higher plants suggested that the thylakoids were of two types; larger intergranal thylakoids, which ran from one end to the other of the chloroplast; small granal thylakoids, which formed the stacks. This concept suggested that each thylakoid contained a discrete lumen. This is now known that the folder chloroplast lamellae system in fact divides the stroma from a single complex cavity. The thylakoid membranes contain or have assoicated with them the intermediates of light-harvesting systems, the reaction centres and other intermediates of photosystem I and II, the intermediates associated with water splitting and oxygen production, components of the intermediate electron transport chain, the iron-sulphur proteins and enzymes associated with the reduction of NADPH and an ATPase known as coupling gactor assoicated with the process of photophyophorylatrion.

**Tissue culture.** The growth and maintanance of cells from higher organisms *in vitro*, outside the tissue of which they are normally a part.

**Tolerance.** (1) An ability of a plant to withstand adverse environment conditions. (2) An ability of a plant to withstand the development of a parasite within it, without showing serious symptoms of disease.

**Tonoplast.** The membrane that limits vacuoles in plant cells.

**Total carbon.** The total amouunt of carbon in a chemical compound.

**Total organic carbon (TOC).** A measure of the carbon content of a sample. It is used as an indication of the amount of organic pollution in a water sample.

**Total oxygen demand.** The amount of oxygen consuumed in the combustion of all the organic carbon in a sample at about 900ÉC.

**Trace element.** An element that is required in low concentrations to support growth of all living organisms. Trace elements participate in metabolism as cofactors in enzyme reactions, electron transport and membrane transport phenomenon.

**Transpiration stream.** The flow of water from the soil through the xylem of the roots and stems to the leaves. The movement of water is the result of loss from the leaves dur to transpiration.

**Transpirational flux.** It is the driving force for the movement of water vapour out of a leaf and the total resistance to the movement in the pathway. The driving force is the difference in concentration of water vapour between a leaf and the external atmosphere. The total resistance is the sum of stomatal resistance and the boundary layer resistance.

**Transport** (across bacterial membranes). Only small, non-ionic molecules (especially lipophilic ones) can diffuse passively across the hydrophobic barrier of the cell membrane; such molecules include oxygen, carbon dioxide, enthanol and certain amino acids. Most substances, therefore, enter or leave the cell as a result of specific transport mecalınisms which may or many not require metabolic energy, and which are though to involve specific protein `carriers' located within the cell membrane.

**Tricarboxylic acid cycle.** A respiratory pathway located in the mitochondria that oxidizes acetyl-CoA derived from pyruvate (formed by glycolysis) completely to carbon dioxide.

**Triose phosphate.** D-glyceraldehyde 3-phosphate ($PO_3H_2$-$OCH_2$-CHOH-CHO) and dihydroxyacetone phosphate ($PO_3H_2$-$OCH_2$-CO-$CH_2$ OH). Important intermediates in glycolysis and alcoholic fermentation. The two triose phosphates are interconvertible via the ene-diol form, by the action of triose phosphate isomerase; the equilibrium mixture contains 96 per cent ketotriose phosphate and 4 per cent aldotriose phosphat.

**Tropism.** It is the directional growth movement of plants that occurs in response to a directional stimulus. Growth towards the stimulus is termed positive and growth away from the stimulus negative. Stimuli causing tropic growth (and the responses) include light **(phototropism),** chemicals (**chemotropism**), gravity **(geotropism),** water **(hydrotropism)** and touch **(haptotropism).**

**Tryptophan (Trp).** One of the 20 common amino acids which occur in proteins

**Turgid.** Of the state of a cell which can absorb no more water by osmosis, because the cell wall prevents an increse in size. Turgid cells are the main means of support in herbaceous plants.

**Turgor movement.** A reversible change in the position of plant parts due to a change in turgor pressure in certain specialized cells; movement of *Minosa* leaves when touched is an example.

**Turgor pressure (TP).** The pressure caused by water in the vacuole which pusheds against the cytoplasm and the cell wall. The turgor pressure is balanced by the osmotic pressure when the cell is well supplied with water when the cell loses water, the turgor pressure decreases and more water can enter the cell by osmosis.

**Turgor.** Stat eof cell in which the cell-wall is rigid, stretched by increase in volume of vacuole and protoplasm during absorption of water. The cell is described as turgid. Essential feature of mechanical support of plant tissues, loss of turgor (when water loss exceeds water absorption) being followed by wilting; and aids expansion of young cells in growth. Within a fully turgid plant cell osmotic pressure tending to force water into cell is balanced by hydrostatic pressure (turgor pressure) within the cell against its wall. When water is lost from cell, turgor pressur is reduced and more water enters by osmosis untill turgidity is restored. Other illustrations of effects of alteration of turgidity include opening and closing of guard cells of stomata and sudden seismonastic (*q.v.*) droping of leaves in the sensitive plant.

**Warberg's atmungsferment.** Cytochrome oxidase.

**Warburg's respiratory enzyme.** Cytochrome oxidase.

**Water balance.** The difference between the rate of water intake (absorption) by a plant and water loss (transpiration).

**Water capacity.** The amount of water that a soil can retain against gravity; usually expressed as a percentage per volume of soil.

**Water cultures.** Aqueous solutions containing different concentrations of various mineral salts. They are used in experiments to show the relative importance of individual elements to plant growth.

**Water potential.** In osmosis, the tendency for a system (a cell or solution) to take up water from pure water, through a differentially permeable membrane. Water flows toward the system with a more negative water potential. (contrast with osmotic potential, pressure potential).

**Water-use efficiency of photosynthesis.** The ratio of photosynthesis to transpiration; calculated as $W_f$= photosynthesis/ Amount of water transpired.

**Water-use efficiency of productivity.** The ratio of dry matter produced to amount of water consumed; calculated as $W_D$ = Dry matter production/Water consumption.

**Wavelength.** The length of a wave of light. Different wavelengths have different colours and different levels of energy.

**Wood sugar.** Xylose.

**Wound parasite.** A parasite which can only gain entry into the body of a plant through a wound.

**Wound tissue.** A pad of parenchymatous cells formed by a cambium after wounding. It may give rise to groups of meristematic cells from roots and buds form.

**X-ray crystallography.** The use of diffraction patterns produced by x-ray scattering from crystals to determine the 3-D structure of molecules. It is used to determine the three-demensional structure of molecules, especially proteins and nucleic acids.

**Xylose.** (Wood sugar). It is aldopentose sugar widely distributed in plants, particularly in woody tissues.

**Yield stress.** The value of shear stress below which there is no flow.

**Yield.** The net recoverable amount of the desired product from a reaction or process. This may be expressed in terms of the percentage yield or the ratio of final product to starting material.

**Zinc, Zn.** An essential biolement for the growth and development of plants, animals and microorganisms. It has a high affinity for nitrogen and sulfur ligands, and occurs in the cell in association with many different compounds, e.g., proteins (insulin), amino acids, nucleic acids.

**Zymase.** A mixture of numerous enzymes (at least 14). Zymase was extracted from yeast in 1903 and was then believed to be a single enzyme. Zymase catalyses the breakdown of glucose to ethanol and $CO_2$ in fermentation, requiring the presence of a number of coenzymes and certain metal ions in order to act completely.